Steel Heat Treatment Handbook

Steel Heat Treatment Handbook

edited by

George E. Totten

Union Carbide Corporation
Tarrytown, New York

Maurice A. H. Howes

IIT Research Institute
Chicago, Illinois

MARCEL DEKKER, INC.   NEW YORK · BASEL

First Indian Reprint 2005

Library of Congress Cataloging-in-Publication Data

Steel heat treatment handbook / [edited by] George E. Totten, Maurice
   A. H. Howes.
      p. cm.
   Includes index.
   ISBN 0-8247-9750-7
   1. Steel—Heat treatment. I. Totten, George E. II. Howes,
Maurice A. H.
TN751.S69   1997
672.3'6—dc21

96-52020
CIP

MARCEL DEKKER, INC.
270 Madison Avenue, New York, New York 10016

Printed and bound by Replika Press Pvt. Ltd., India.

FOR SALE IN THE INDIAN SUBCONTINENT ONLY.

PREFACE

Heat treatment is one of the oldest manufacturing processes, having first been practiced approximately 5000 years ago. It continues to be one of the most important fundamental processes in a modern industrial economy. Thermal processing is involved in almost every market sector of the economy, including ground transportation, agriculture, aerospace, the military, oil drilling, fastener production, tools and machinery, manufacturing, and many other industries.

In view of its global importance, research and development is conducted in nearly every advanced and developing economy to identify better materials and more efficient and improved thermal technologies, and to integrate them into modern production practices. Therefore, there is a continuing need for instructional and reference materials for both students and practitioners that not only present classically important information but also integrate continuing developments in this important technology.

In this handbook we have endeavored to provide the reader with well-referenced information of sufficient breadth and depth to serve as an excellent teaching resource for advanced undergraduate and graduate metallurgical, mechanical, and materials engineers. For skilled practitioners, this text includes a thorough update of classically important heat treatment technologies such as steel alloy chemistry, hardenability, carburizing, and quenching. To enhance clarity and provide long-term reference value, each chapter contains numerous tables, charts, and figures.

There are 16 chapters.

Chapter 1. Fundamental Concepts in Steel Heat Treatment.   This chapter provides a fundamental overview of steel heat treatment metallurgy.

Chapter 2. The Effects of Alloying Elements on the Heat Treatment of Steel.   This is one of the most thorough treatments of the effects of steel alloy chemistry that we are aware of.

Chapter 3. Hardenability.   Not only are the classical approaches to steel hardenability thoroughly reviewed, this subject has been updated to reflect current developments such as computer modeling.

Chapter 4. Quenching and Quenching Technology.   This is the only modern and thorough one-chapter treatment of this ancient technology to be published in recent years.

Chapter 5. Distortion of Heat-Treated Components. This subject is one of the most important and least understood technologies in heat treatment. This chapter provides an overview of the causes of distortion, empirical prediction methods, and methods of measurement of residual stresses and distortion.

Chapter 6. Heat Treatment Equipment. Selected technologies related to heat treatment equipment and thermal processing are discussed in detail. These include furnace classification and selection, furnace heating, heat transfer, thermocouples, furnace atmospheres and atmosphere generation, atmosphere sensors, refractories, fans, fixtures, parts washing, quenching systems, furnace safety, salt baths, and fluidized-bed furnaces.

Chapter 7. Vacuum Heat Treatment. This chapter presents an overview of vacuum furnace technology. Some of the subjects covered are volatilization, dissociation and degassing, furnace equipment, hardening, atmosphere heat treatment, plasma nitriding, vacuum brazing, sintering, tempering, and stress relieving.

Chapter 8. Steel Heat Treatment. Subjects discussed in detail in this chapter include heat transfer; role of steel lattice defects; TTT and CCT diagrams—their generation and use; oxidation; decarburization; and thermal, transformational, and residual stresses.

Chapter 9. Heat Treatment with Gaseous Atmospheres. This is the most comprehensive, single-chapter treatment of this important subject that has been published.

Chapter 10. Nitriding Techniques and Methods. This chapter provides an excellent reference to a subject that has been the focus of increased interest in recent years.

Chapter 11A. Induction Heat Treatment. Basic principles, computation, coil construction and design considerations are discussed in this chapter.

Chapter 11B. Induction Heat Treatment. Modern power supplies, load matching, process control and monitoring are discussed in detail. This is a continuation of induction heat treatment in Chapter 11A.

Chapter 12. Heat Treatment of Powder Metallurgy Steel Components. Increasingly, heat treaters are encountering powdered metal parts, particularly in the automotive and fastener industries. In view of this increased interest, this chapter is included to provide the reader with a basic overview of this rapidly growing technology.

Chapter 13. Metallurgical Property Testing. The objective of this chapter is provide a thorough treatment of microstructural testing, including testing methods and how they are conducted, and the physical principles involved.

Chapter 14. Mechanical Property Testing Methods. This chapter provides a comprehensive overview of various mechanical properties such as resistance to tensile stress and creep, modulus, fracture toughness, resistance to fatigue, and hardness.

Chapter 15. Steel Nomenclature. The various systems used to classify steels are discussed, and extensive tabular data relating to international steel designations are provided.

Chapter 16. Environmental and Safety Regulations Affecting Heat Treaters. A thorough treatment of this subject is provided.

The preparation of a text of this scope was a tremendous task. We are indebted to the vital assistance of the various international experts whose contributions are included here.

Special thanks go to Alice Totten and Susan Meeker, who typed and assisted in the editing of much of this material; to Glenn Webster and Yinghua Sun, who assisted with many of the figures; and to Union Carbide Corporation and the Manufacturing Department of IIT Research Institute for their continued support.

George E. Totten
Maurice A. H. Howes

CONTENTS

CONTRIBUTORS

Micah R. Black   Inductor Design Supervisor, INDUCTOHEAT, Inc., Madison Heights, Michigan

Jan W. Bouwman   Research and Development Department, Ipsen Industries International GmbH, Kleve, Germany

Joseph M. Capus, B.Sc., Ph.D., C.Eng., F.I.M.   Consultant in Powder Metallurgy, Beaconsfield, Quebec, Canada

Raymond L. Cook, P.E., B.S., M.B.A.   Vice President of Engineering, INDUCTOHEAT, Inc., Madison Heights, Michigan

Bernd Edenhofer   Director of Research and Development, Research and Development Department, Ipsen Industries International GmbH, Kleve, Germany

Howard Ferguson, A.B., B.S.   Director/Technology, Technical Resource Center, Metal Powder Products Company, Coldwater, Michigan

Gary R. Garsombke, M.T., B.I.M., M.T.L.R.   Quality Control Manager, Assembly Components Division, ITW Shakeproof, Milwaukee, Wisconsin

Johann Grosch, Dr.-Ing.   Professor, Institut für Wekstofftechnik, Technische Universität Berlin, Berlin, Germany

Maurice A. H. Howes, Ph.D., M.Sc., C.Eng.   Chief Scientist, Manufacturing, IIT Research Institute, Chicago, Illinois

D. Randy Junkins, P.E., M.S.C.E., D.E.E., C.H.M.M.   President, Junkins Engineering, Inc., Morgantown, Pennsylvania

Božidar Liščić, Ph.D., Mech.Eng. Full Professor, Faculty of Mechanical Engineering and Naval Architecture, University of Zagreb, Zagreb, Croatia

Don L. Loveless, B.S.E.E. Group Vice President of Research and Development, INDUCTOHEAT, Inc., Madison Heights, Michigan

D. Scott MacKenzie, B.S., M.S. Consultant, Villa Ridge, Missouri

Arnold R. Ness, Ph.D. Associate Professor, Industrial and Manufacturing Engineering and Technology, Bradley University, Peoria, Illinois

David Pye Pye Metallurgical Consulting Inc., Meadville, Pennsylvania

Ray W. Reynoldson, C.P.Eng., F.I.E., F.R.M.I.T., A.F.A.I.M. Managing Director, Quality Heat Treatment Ltd., North Bayswater, Victoria, Austrialia

Valery I. Rudnev, Ph.D. Chief Scientist, INDUCTOHEAT, Inc., Madison Heights, Michigan

Anil Kumar Sinha, Ph.D., M.Tech., B.Sc. Met.Eng. Thompson Steel Company, Inc., Baltimore, Maryland

Anton Stich, Ph.D. Technical University of Munich, Munich, Germany

Alexey V. Sverdlin, Ph.D. Professor, Industrial and Manufacturing Engineering and Technology, Bradley University, Peoria, Illinois

Hans M. Tensi, Dr.-Ing. Technical University of Munich, Munich, Germany

George E. Totten, Ph.D. Senior Research Scientist, Union Carbide Corporation, Tarrytown, New York

Xiwen Xie Professor, Department of Materials Science and Engineering, Beijing University of Aeronautics and Astronautics, Beijing, People's Republic of China

1

FUNDAMENTAL CONCEPTS IN STEEL HEAT TREATMENT

Alexey V. Sverdlin and Arnold R. Ness
Bradley University, Peoria, Illinois

I. INTRODUCTION

The purpose of heat treatment is to cause desired changes in the metallurgical structure and thus in the properties of metal parts. Heat treatment can affect the properties of most metals and alloys, but ferrous alloys, principally steels, undergo the most dramatic increases in properties, and therefore structural changes in iron-carbon alloys are considered in this chapter. In general, the most stable steel structures are produced when a steel is heated to the high-temperature austenitic state (to be defined later) and slowly cooled under near-equilibrium conditions. This type of treatment, often referred to as annealing or normalizing, produces a structure that has a low level of residual stresses locked within the part, and the structures can be predicted from an equilibrium diagram. However, the properties that interest heat treaters the most are those exhibiting high strength and hardness, usually accompanied by high levels of residual stresses. These are metastable structures produced by nonequilibrium cooling or quenching from the austenitic state.

Most of this chapter discusses equilibrium and nonequilibrium structures, their properties, and the tools that we have at our disposal to predict different types of phase formations and their properties. It is essential that heat treaters have a clear understanding of the structures that can be produced in steel under different treatment conditions that they can apply in their equipment.

II. CRYSTAL STRUCTURE AND PHASES

A. Crystal Structure of Pure Iron

Iron in the solid state is known in two allotropic states. (Allotropy is the phenomenon of an element having different crystal lattices depending on the particular temperature and pressure.) Starting from low temperatures and up to 910°C (1670°F), iron possesses a body-centered cubic (bcc) lattice and is called α-iron (α-Fe). At 910°C α-iron crystals turn into γ-iron crystals possessing a face-centered cubic (fcc) lattice. The γ crystals retain stability up to a temperature of 1400°C (2550°F). Above this temperature they again acquire a bcc lattice and are usually called δ crystals. The δ crystals differ from α crystals only in the temperature region of their existence. Iron has the following lattice constants: 0.286 nm for bcc lattices (α-Fe, δ-Fe) and 0.364 nm for fcc lattices (γ-Fe).

At low temperatures, α-Fe exhibits a strongly ferromagnetic character. When it is heated to about 770°C (1418°F), ferromagnetism vanishes. In accordance with the latest findings, this is due to the fact that the lattice loses its ferromagnetic spin ordering. The state of iron above 770°C is called β-Fe. The lattice of paramagnetic β crystals is identical to the lattice of α crystals.

The points at which one allotropic form of iron transforms to another are conventionally symbolized by the letter A with subscripts indicating the ordinal number of the transformation. The subscripts 0 and 1 signify transformations that are absent in pure iron but are observed in carbon alloys of iron. The subscript 2 denotes a magnetic transformation of the α phase, while the subscripts 3 and 4 stand for transformation of α to γ and γ to δ.

In going from one form to another, iron is capable of undercooling. This causes a difference in the position of transformation points on heating and cooling. The difference depends on the cooling rate (Figure 1) and is termed hysteresis. The letters "c" and "r" indicate whether the transformation is due to heating or cooling.

A change in the density of α-Fe as it transforms to γ-Fe results in an abrupt change in the volume of the material. Sometimes this gives rise to stresses that exceed the elastic limit and lead to failure. The density of γ-Fe is about 4% higher than that of α-Fe.

B. The Iron–Carbon Equilibrium Diagram

The structure of iron-carbon alloys can contain either pure carbon (graphite) or a chemical compound (cementite) as the carbon-enriched component. Cementite is present even in relatively slowly cooled alloys: a long holding at elevated temperatures is required to decompose cementite to iron and graphite. For this reason the iron–carbon diagram is usually treated as the iron–iron carbide diagram. The former is stable, whereas the latter is metastable.

The iron–carbon diagram is shown in Figure 2. Dashed lines stand for the stable Fe–C diagram, and solid lines denote the metastable Fe–Fe₃C diagram.

1. The Metastable Fe–Fe₃C Equilibrium Diagram

As shown in Figure 2, the lattices of allotropic forms of iron (δ, γ, and α) serve as sites of formation of δ, γ, and α solid solutions of carbon in iron (the same symbols are adopted for the designation of solid solutions).

When carbon-depleted alloys crystallize, crystals of the δ solid solution precipitate at the liquidus AB and solidus AH. The δ solid solution has a bcc lattice. At the maximum carbon temperature of 1490°C (2714°F), the δ solution contains 0.1% C (point H). At 1490°C a

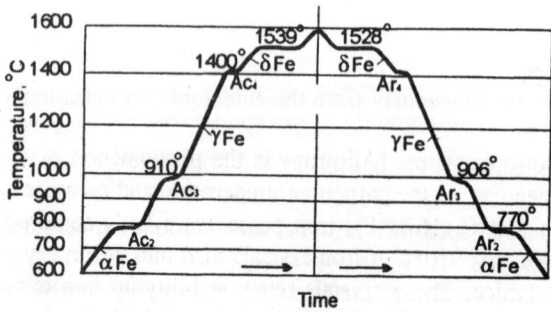

Figure 1   Heating and cooling curves for pure iron.

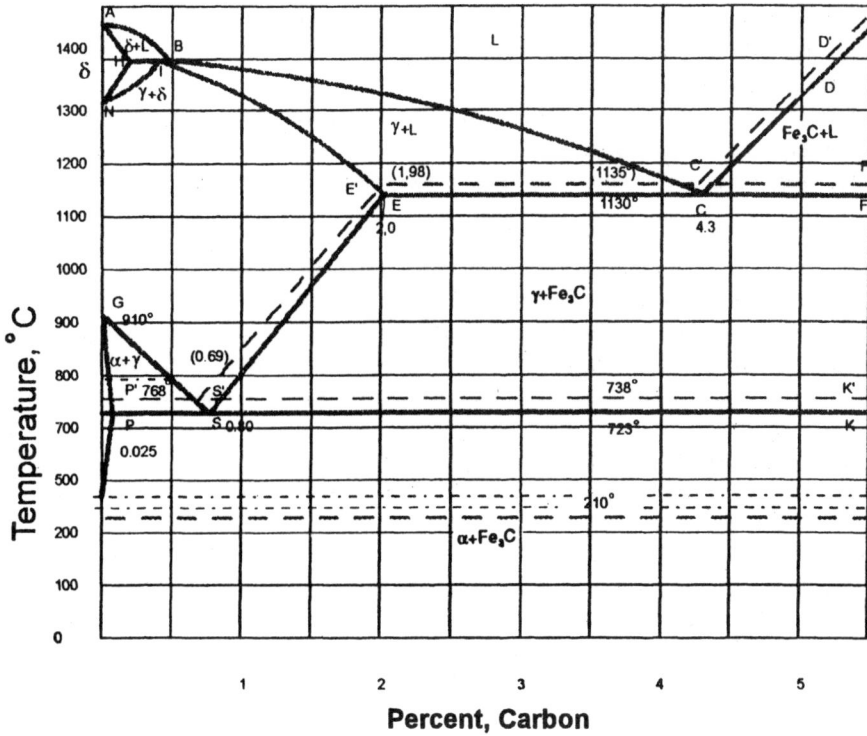

Figure 2 Iron–carbon diagram.

peritectic reaction takes place between the saturated δ solution and the liquid containing 0.5% C (point B). As a result, the γ solid solution of carbon in γ-Fe is formed. It contains 0.18% C (point I).

If the carbon content is higher than 0.5%, the γ solid solution crystallizes directly from the liquid (at the liquidus BC and solidus IE). At 1130°C (2066°F) the limiting solubility of carbon in γ-Fe is close to 2.0% C (point E). Decreasing the temperature from 1130°C (2066°F) leads to lowering the carbon solubility in γ-iron at the line ES. At 723°C (1333°F) the solubility is 0.8% C (point S). The line ES corresponds to precipitation of iron carbide from the γ solution.

As the carbon content is raised, the temperature at which the γ lattice transforms to the α lattice lowers, and the transformation takes place over the temperature interval corresponding to the curves GS and GP.

The α-phase precipitation curve GS intersects the iron carbide precipitation curve ES. The point S is a eutectoid point with the coordinates 723°C (1333°F) and 0.80% C. At this point a saturated α solution and Fe₃C precipitate simultaneously from the eutectoid concentration γ solution.

The lattice of the α solid solution is identical to the lattice of the δ solid solution. At the eutectoid temperature of 723°C (1333°F) the α solid solution contains 0.02% C (point P). Further cooling leads to lowering of the carbon solubility in α-Fe, and at room temperature it equals a small fraction of a percent (point D).

When the carbon content is 2.0–4.3%, crystallization starts with precipitation of the γ solution at the line BC. An increase in the carbon content to over 4.3% causes precipitation of iron carbide at the line CD.

Precipitation of the surplus primary phase in all alloys containing over 2.0% C is followed by a eutectic crystallization of the γ solution and iron carbide at point C, whose coordinates are 1130°C (2066°F) and 4.3% C.

The line MO is associated with a magnetic transformation, that is, a transition from the ferromagnetic to the paramagnetic state.

Table 1 describes structural components of the iron–carbon system.

2. Stable Fe–C Equilibrium Diagram

Given very low rates of cooling, carbon (graphite) can crystallize directly from the liquid. In this case a eutectic mixture of austenite and graphite is formed instead of the eutectic of austenite and cementite. As is seen in Figure 2, the dashed lines symbolizing the iron–graphite system are at higher temperatures than the lines of the iron–cementite system. This testifies to the greater stability and closeness to a full equilibrium of the iron–graphite system. The conclusion is also supported by the fact that heating of high-carbon alloys with a large amount of cementite leads to its decomposition: $Fe_3C \rightarrow 3Fe + C$.

At intermediate rates of cooling, part of the alloy can crystallize according to the graphite system and the other part according to the cementite system.

Phase equilibrium lines in the diagrams of both systems can be displaced depending on particular cooling rates. A most pronounced displacement can be observed for the lines of precipitation of the carbon solid solution in γ-Fe (austenite). For this reason the diagram holds completely true only with respect to the alloys that are subjected to a relatively slow cooling rate.

C. Effect of Carbon

A maximum solubility of carbon in α-Fe is observed at 721°C (1330°F) and is equal to 0.018% C. Subject to quenching, carbon can remain in the α solid solution, but soon precipitation of phases commences, by an aging mechanism.

In the α solid solution, carbon can form either (1) a homogeneous solution, a statically uniform interstitial distribution (a rare case), or (2) an inhomogeneous solution, with the formation of clusters at places where the crystal lattice structure is disturbed (grain boundaries, dislocations). The latter is the most probable state of the solid solution. The clusters thus formed represent an obstacle to movement of dislocations during plastic deformation and are responsible for an inhomogeneous development of the deformation at the onset of plastic flow.

To analyze the influence of the carbon content on iron-carbon alloys, every structural component should be characterized. Slowly cooled alloys comprise ferrite and cementite or ferrite and graphite. Ferrite is plastic. In the annealed state ferrite has large elongation (about 40%), is soft (Brinell hardness is 65–130 depending on the crystal dimension), and is strongly ferromagnetic up to 770° C (1418°F). At 723°C (133°F), 0.02% C dissolves in ferrite. But at room temperature only thousandths of a percent of carbon is left in the solution.

Table 1 Components of the Iron–Carbon System

Phase or mixture of phases	Name
Solid solution of carbon in α (δ)iron	Ferrite
Solid solution of carbon in γ-iron	Austenite
Iron carbide (Fe_3C)	Cementite
Eutectoid mixture of carbon solid solution in γ-iron with iron carbide	Ledeburite
Eutectoid mixture of carbon solid solution in α-iron with iron carbide	Pearlite

Cementite is brittle and exhibits great hardness (the Brinell hardness is about 800), it is weakly magnetic up to 210°C (410°F), it is a poor conductor of electricity and heat, and it has a complicated rhombic lattice. Usually a distinction is made between primary cementite, which crystallizes from the liquid at the line CD; secondary cementite, which precipitates from the γ solution at the line ES; and tertiary cementite, which precipitates from the α solution at the line PQ.

Graphite is soft. It is a poor conductor of current but transfers heat well. Graphite does not melt even at temperatures of 3000–3500°C (5430–6330°F). It possesses a hexagonal lattice with the axis relation

$$\frac{c}{a} > 2$$

Austenite is soft (but is harder than ferrite) and ductile. Elongation of austenite is 40–50%. It has lower conductivity of heat and electricity than ferrite and is paramagnetic. Austenite possesses an fcc lattice.

The structure of the steel containing 0–0.02% C comprises ferrite and tertiary cementite (Figure 3). A further increase in the carbon content leads to the appearance of a new structural component—a eutectoid of ferrite and cementite (pearlite). Pearlite appears first as separate inclusions between ferrite grains and then, at 0.8% C, occupies the entire volume. Pearlite represents a two-phase mixture, which usually has a lamellar structure (Figure 4).

As the carbon content of steel is raised to over 0.8%, secondary cementite is formed along with pearlite. The secondary cementite is shaped as needles (Figure 5). The amount of cementite increases as the carbon content is increased. At 2% carbon it occupies 18% of the field of vision of the microscope. A eutectic mixture appears when the carbon content exceeds 2%.

If rapidly cooled steels, not all the surplus phase (ferrite or cementite) has time to precipitate before a eutectoid is formed.

Alloys with 3.6% C contain ledeburite (a eutectic mixture of carbon solid solution in γ-Fe and iron carbide). An electron microscopic image of the carbides is shown in Figure 6. These alloys would be more properly classified with hypoeutectic white cast irons.

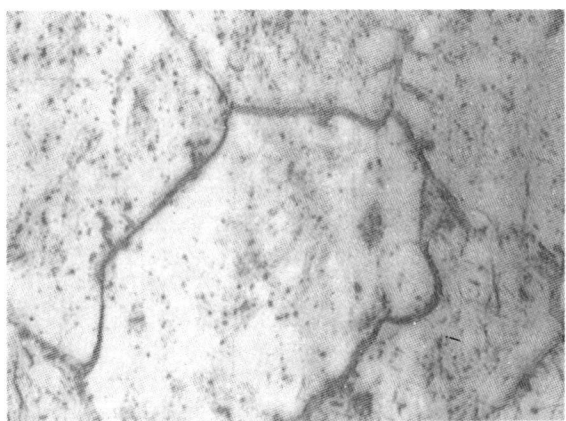

Figure 3   Steel microstructure of ferrite and tertiary cementite at grain boundaries, 500×.

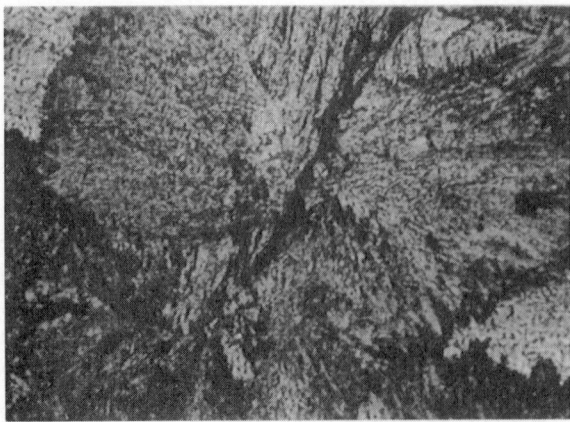

Figure 4   Steel microstructure of pearlite, 500×.

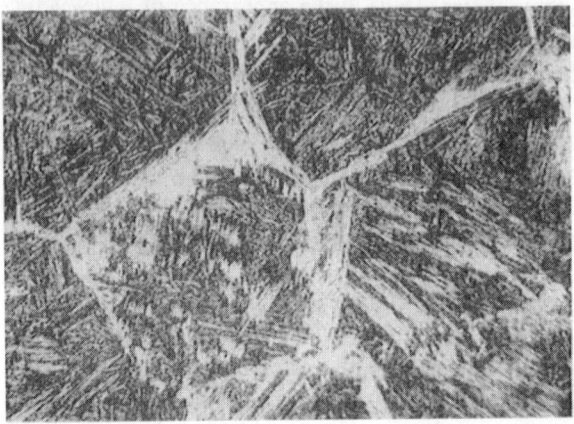

Figure 5   Steel microstructure of secondary cementite (needles) and pearlite, 500×.

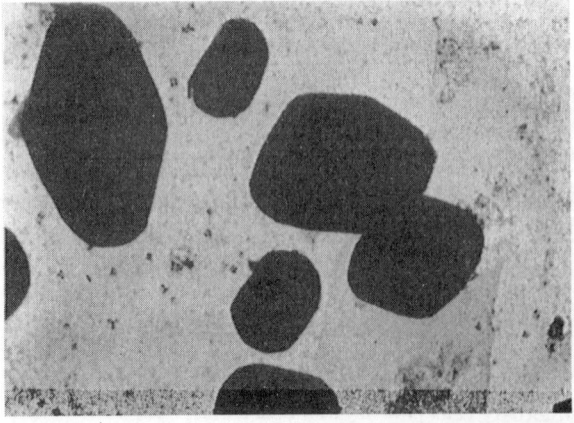

Figure 6   Steel microstructure of electron microscopic image of iron carbides, 3000×.

D. Critical (Transformation) Temperatures

Carbon has a pronounced effect on transformations of iron in the solid state. The positions of the lines GS and NI in the iron–carbon equilibrium diagram show that an increase in the carbon content leads to lowering of the point A_3 and raising of the point A_4 with respect to their counterparts depicted in Figure 2 for pure carbon. So carbon extends the temperature range of the δ phase.

When a eutectoid (pearlite) is formed, heating and cooling curves exhibit a stop, which is designated as the point A_1 (A_{c1} on heating and A_{r1} on cooling). This phenomenon takes place at 0.9% C (point S in the Fe–C diagram). Precipitation of ferrite in hypoeutectoid steels (on crossing the line GOS) shows up in heating and cooling curves as an inflection symbolized as the point A_3. The point corresponds to the $\gamma \rightarrow \alpha$ transformation in pure iron. Precipitation of cementite (crossing of the line ES), which precedes the eutectoid precipitation, is seen in the cooling curve as a weak inflection designated as the point A_{cm} ($A_{c,cm}$ on heating and $A_{r,cm}$ on cooling). Addition of carbon has little influence on the magnetic transformation temperature (point A_2). Therefore the line MO corresponds to the magnetic transformation in alloys with a small carbon content. In alloys containing greater amounts of carbon, this transformation occurs at the line GOS, which corresponds to the onset of ferrite precipitation.

If the carbon content is higher than the one corresponding to point S, then the magnetic transformation coincides with the temperature A_1.

Cementite undergoes a magnetic transformation. Whatever the carbon content, the transformation takes place at a temperature of 210–220°C (410–430°F). It occurs without a marked hysteresis, as does the magnetic transformation of pure iron at point A_2.

III. STRUCTURAL TRANSFORMATIONS IN STEEL

When a steel part is hardened, it is heated to a high temperature in order to convert the entire structure to the austenite phase. As discussed earlier, austenite is a single-phase structure of iron and carbon stable at high temperatures. If the steel were cooled slowly, the austenite would transform to pearlite, which is the equilibrium phase at room temperature. A pearlitic structure is an annealed structure and is relatively soft with low physical properties. If the steel is cooled very rapidly, a very hard and strong structure called martensite forms that is a metastable phase of carbon dissolved in iron. It may be tempered to produce lower hardness structures that are less brittle. Intermediate cooling rates will produce other structures referred to as bainites, although this type of structure is only produced in quantity in an alloy steel. Eutectoid carbon steels produce predominantly martensite or pearlite, depending on the cooling rate.

A. The Austenite–Pearlite Transformation

Transformation of the face-centered lattice of austenite to the body-centered lattice of ferrite is hampered due to the presence of dissolved carbon in austenite. The austenite lattice has enough space to accommodate carbon atoms at the centers of unit cells. The body-centered lattice of ferrite has no such space. For this reason the solubility of carbon is lowered considerably on transition from austenite to ferrite. During the $\beta \rightarrow \alpha$ transformation, almost all carbon precipitates from the austenite lattice. In accordance with the metastable Fe–Fe_3C diagram it precipitates as iron carbide (cementite). This transformation can be described by three interconnected processes:

1. Transformation of the γ-Fe lattice to the α-Fe lattice
2. Precipitation of carbon as the carbide Fe_3C (cementite)
3. Coagulation of the carbides

At the temperature of point A_1 processes 1 and 2 proceed almost simultaneously, with the formation of a lamellar mixture of ferrite and cementite.

Atoms of dissolved carbon are distributed randomly in the lattice. For this reason cementite nucleates in carbon-rich regions and ferrite in carbon-depleted regions that have little if any carbon. Such a redistribution of carbon is realized though diffusion and depends on temperature and time.

When hypoeutectoid steels containing less than 0.8% carbon are subjected to slow cooling, the transformation starts with the formation of ferrite at grain boundaries. The boundaries act as ferrite crystallization centers. Carbon is forced inside the crystallite. As ferrite precipitates, a concentration necessary for the ferrite formation is achieved in central volumes.

When hypereutectoid steels (carbon content less than 0.8%) are subjected to slow cooling, on crossing the line ES cementite starts precipitating at grain boundaries. Here the grain boundaries also serve as crystallization sites.

The carbon diffusion rate in the lattices of γ- and α-iron decreases rapidly as the temperature is lowered, since the diffusion coefficient depends on temperature as

$$D = D_0^{-Q/RT}$$

Presetting an appropriate cooling rate, undercooling can be enhanced to such an extent that formation of pearlite becomes impossible.

In the range of low temperatures, the transformation mechanism and the character of the formed structure depend solely on the temperature at which the transformation takes place. Considering the degree of undercooling, three transformation temperature ranges are distinguished: (1) the pearlite range, (2) the intermediate range, and (3) the martensite range. A continuous transition from one transformation mechanism to another can take place over these temperature ranges. The processes strongly depend on the content of alloying elements, especially of carbon, in steel. They can commence by a more rapid mechanism and end by a slower one.

In the pearlite range the transformation is characterized by the simultaneous formation of a mixture of ferrite and carbide. Free ferrite or carbides can precipitate at the austenite grain boundaries. Here the formation and growth of both phases are controlled by diffusion processes (diffusion crystallization). Diffusion of iron and other alloying elements also plays a significant part. The structure fineness is enhanced as the temperature is lowered, until a longer time is required for diffusion crystallization of ferrite and carbides.

B. The Structure of Pearlite

A mechanical mixture of ferrite and carbide plates is formed on transformation in the pearlite range. The rate at which nuclei of pearlite crystallization are formed depends on supersaturation of austenite with carbide, which increases as the temperature is lowered. The rate also depends on the diffusion rate, which decreases with temperature. The growth of pearlite islets depends in the main on the diffusion rate of carbon and iron atoms. The other decisive factors are the degree of supersaturation and the free energy advantage during the ferrite formation. Pearlite islets grow not only through the formation of new plates but also by way of further growth of old plates in all directions. Carbide plates grow faster than ferrite ones. The process can start, however, with the formation of ferrite nuclei. Multiple alternations of nucleation of ferrite and cementite plates and branching of the plates of both phases lead to the formation of plane-parallel and fan-shaped pearlite plates.

Pearlite nuclei appear predominantly in the lattice regions with crystal structure defects: grain boundaries, insoluble carbides, or nonmetal inclusions such as sulfides. A very significant characteristic of pearlite is the plate-to-plate spacing. Strength properties of steel improve with a decrease in that spacing.

The formation rate of cementite and ferrite crystallization centers in the pearlite range accelerates as the temperature is lowered. The plate-to-plate spacing decreases, and the fineness of the structure increases.

In the eutectoid steel the pearlite transformation takes place on cooling to 600–700°C (1100–1300°F). In this case the plate-to-plate spacing equals 0.5–1 μm.

Precipitation of austenite over the temperature interval of 650–600°C (1200–1100°F) provides the plate-to-plate distance of 0.4–0.2 μm. In this case the eutectoid is finer pearlite.

When austenite precipitates over the temperature interval of 600–500°C (100–930°F), an extremely fine eutectoid mixture is formed, where the plate-to-plate spacing equals ~0.1 μm.

An important characteristic that influences the properties of steel is the dimension of the pearlite colony. A decrease in the colony dimension is accompanied by a growth of the impact strength and decrease of brittleness. The critical brittleness temperature depends on the pearlite morphology as
where d is the pearlite colony dimension. Thus a relatively high strength pearlite is formed in the case of the breaking of ferrite and cementite plates, forming a high density of dislocations inside the ferrite.

A better fracture strength of pearlite is achieved through spheroidization of cementite particles. The spheroidization can be facilitated by deformation of pearlite, subsequent heating, and holding at a temperature near A_{c1}. Another method providing relatively high strength and ductility of pearlite consists in deformation during pearlite transformation. This leads to the formation of a polygonal structure and spheroidization of cementite.

The yield stress of the ferrite–pearlite mixture depends on the properties of ferrite and pearlite in an additive manner:

$$\sigma_0^2 = f_\alpha \sigma_\alpha + (1 - f_\alpha)\sigma_p$$

where f_α is the volume fraction of ferrite, σ_α is the yield stress of pearlite and σ_p is the yield stress of pearlite.

C. The Transformation of Austenite in Hypo- and Hypereutectoid Steels

The transformation of austenite in eutectoid composition steels was considered above. In hypo- and hypereutectoid steels the pearlite transformation should be preceded by precipitation of excess phases—ferrite and secondary cementite (see the Fe–C equilibrium diagram in Figure 2).

The relative amount of the structurally free excess phase depends on the degree of austenite undercooling. The amount of excess ferrite or cementite decreases with an increase in the cooling rate. Given a sufficient degree of undercooling, the formation of an excess phase as an independent structural component can be avoided.

When a hypoeutectoid steel containing a small amount of eutectoid austenite is subjected to slow cooling, eutectoid ferrite grows on the grains of excess ferrite and eutectoid cementite is left as structurally free interlayers at grain boundaries.

In a hypereutectoid steel the eutectoid can also be subject to structural degeneration. Cementite, which is formed as a result of the eutectoid precipitation under a very low cooling below the point A_1 (above ~700°C or 1300°F), is deposited on secondary cementite. Areas of structurally free ferrite are found alongside.

This eutectoid transformation, which is accompanied by separation of the phases, is referred to as abnormal. In normal eutectoid transformation, ferrite and cementite grow cooperatively in the form of colonies with a regular alternation of the two phases. In the case of abnormal transformation, a coarse mixture of ferrite and cementite does not have a characteristic eutectoid structure. During a eutectoid transformation the mechanism can change from abnormal to normal. Therefore, with a rapid cooling and a correspondingly great undercooling of austenite, the abnormal transformation can be suppressed altogether.

Consider the forms and structure of excess ferrite in hypoeutectoid steels. The ferrite is found in two forms: compact equiaxial grains and oriented Widmannstatten plates (Figure 7). Compact precipitates of hypoeutectoid ferrite appear predominantly at austenite grain boundaries, whereas Widmannstatten plates are formed inside grains. The Widmannstatten ferrite is observed only in steels with less than 0.4% C and rather coarse grains of austenite. As the dimensions of austenite grains decrease, the share of ferrite in the form equiaxial grains grows. The Widmannstatten ferrite is formed over the temperature interval from A_3 (50°C or 90°F) to 600–550°C (1112–1022°F). With an increase in the carbon content of steel, the share of the Widmannstatten ferrite in the structure lowers.

It is assumed that the Widmannstatten ferrite is formed owing to a shear $\gamma \rightarrow \alpha$ rearrangement of the lattice, which is accompanied by an ordered interrelated movement of atoms. Equiaxial grains of ferrite grow by means of a normal diffusive rearrangement of the lattice with a disordered transition of atoms across the γ/α boundary.

One of the methods used to strengthen steels consists in providing a structure with hypoeutectoid ferrite containing dispersed carbide precipitates. To produce such a structure, the steel should be heated until special carbides dissolve in austenite and then cooled rapidly so as to preclude the usual precipitation of carbide directly from austenite before hypoeutectoid ferrite starts forming.

D. Martensite Transformation

The martensite transformation takes place on quick cooling of the high-temperature phase, a process that is referred to as quenching. The most characteristic features of the martensite transformation in carbon steels are as follows.

1. The martensite transformation is realized on rapid cooling of steel from a temperature above A_1 in, e.g., water. In this case diffusive precipitation of austenite to a mixture of two phases (ferrite and carbide) is suppressed. The concentration of carbon in martensite corresponds to that in austenite. The main difference between the

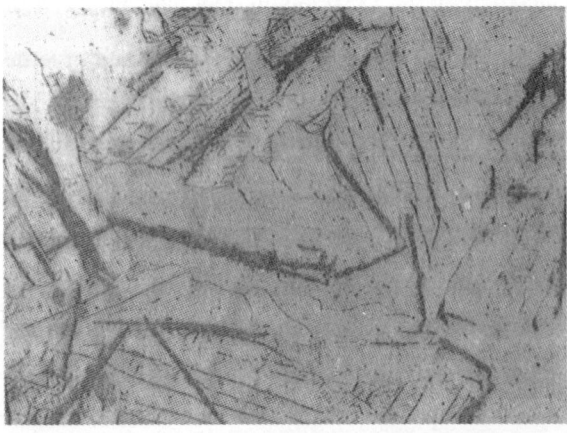

Figure 7 Structure of excess ferrite in hypereutectoid steel, 500×.

martensite transformation and the pearlite transformation is that the former is diffusionless.

2. Transformation of austenite to martensite starts from the martensite start temperature (M_s). Whereas the pearlite start temperature lowers with an increase in the cooling rate, the martensite start temperature depends little if at all on the cooling rate. Martensite is formed over a certain temperature interval. The particular temperature is determined by the carbon content of the steel (Figure 8).

3. Termination of cooling over the temperature interval M_s–M_f suspends formation of martensite. This feature distinguishes the martensite transformation from the pearlite transformation. In the latter case transformation continues to the end at a constant temperature below the point A_1, and the final result is a complete disappearance of austenite given a sufficient isothermal holding time. With the martensite transformation, a certain amount of retained austenite is left.

4. As distinct from the pearlite transformation, the martensite transformation has no incubation period. A certain amount of martensite is formed instantaneously below the temperature M_s.

5. On cooling below M_s, the amount of martensite increases rapidly owing to the quick formation of new plates. The initially formed plates do not grow with time. This feature also distinguishes the martensite transformation from its pearlite counterpart; in the latter case new colonies nucleate and old colonies continue growing.

6. The martensite lattice is regularly oriented relative to the austenite lattice. A certain orientation relationship exists between the lattices. With the pearlite transformation, lattices of the phases comprising the eutectoid mixture exhibit a random orientation with respect to the starting austenite grain.

The temperature M_s characterizes an alloy of a certain composition that has been subjected to a particular pretreatment. In a given steel the martensite transformation starts at the same temperature whatever the cooling rate. That temperature depends on the alloy composition and decreases greatly as the carbon content of the steel is raised (see Figure 8). Part of the carbon enters carbides, which coexist with austenite. The carbides dissolve in austenite if the quenching temperature is elevated. Consequently, the carbon concentration of austenite increases and the M_s point lowers.

The martensite formation is characterized by a shear mechanism of the austenite lattice rearrangement. The martensitic (shear) mechanism of phase transformation is distinguished by an ordered interrelated movement of atoms to distances shorter than the interatomic spacing, and the atoms do not exchange places. An atom in the initial phase preserves its neighbors in the martensite phase. This is the main feature specific to a shear rearrangement of the lattice.

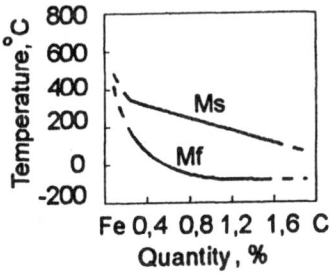

Figure 8   Martensite start M_s and finish M_f temperatures vs. carbon content.

This character of the lattice rearrangement provides coherence of the boundary between the old and new phases. Coherence, or an elastic conjugation of lattices at the boundary between martensite and the initial phase, ensures a very fast movement of the boundary toward the matrix even at low temperatures. The atoms move cooperatively to distances shorter than the interatomic spacing. Hence the growth of the martensite crystal.

As the martensite crystal grows, an elastic strain accumulates at the coherence boundary. On reaching the yield stress, coherence is disturbed. Atoms become disordered at the boundary between the martensite crystal and the starting matrix. "Slipping" movement of the boundary is rendered impossible. Hence, growth of the crystal by the martensitic mechanism is terminated, and subsequently the crystal can grow by diffusion only. But the martensite transformation takes place at low temperatures, where the diffusion rate is very small. Therefore, after coherence is broken, little if any growth of the martensite crystal is observed.

The polymorphous transformation of solid solutions by the martensitic mechanism is characterized by the absence of diffusive redistribution of the components.

In what follows we consider the conditions necessary for the martensitic mechanism by which the high-temperature phase transforms to the low-temperature phase. The martensite transformation is impossible at a small undercooling. This is explained by the fact that in the case of a disordered rearrangement of the lattice, elastic deformation is determined by changes in the volume only, whereas with the martensite transformation it additionally depends on coherence of the lattices of the initial and martensite crystals. As the degree of undercooling is increased, the disordered rearrangement rate of the lattice increases, achieves a maximum, and then drops. When γ-Fe is undercooled to 911–750 °C (1670–1380°F), the normal $\gamma \to \alpha$ transformation is realized, while below 750°C (1380°F) the martensite $\gamma \to \alpha$ transformation takes place. To realize the martensitic mechanism of polymorphous transformation in iron, samples should be strongly overheated in the γ range and then cooled very quickly to suppress development of the normal transformation.

E. The Morphology of Ferrous Martensites

Consider the crystallo geometry of the rearrangement of the face-centered lattice of austenite to the body-centered tetragonal lattice of martensite, which is similar to the bcc lattice of α-iron.

The austenite lattice transforms into the martensite lattice through the Bain deformation. The deformation consists in compression of the tetragonal cell of austenite along the c axis and a simultaneous increase in dimensions along the a axis. The degree of the tetragonal distortion of the martensite lattice, c/a, grows directly as the carbon concentration of martensite. The martensite lattice retains tetragonality at room temperature.

The orientation relationship of the initial and martensite phases has been established. Three basic orientation relationships are known for austenite and martensite lattices in iron alloys: those due to Kurdyumov and Zacks, Nishiyama, and Treninger and Trojano.

The Kurdyumov–Zacks relationship: (111)A||(101)M;[110]A||[111]M.

The Nishiyama relationship: (111)A||(101)M;[121]A||[101]M.

The Treninger–Trojano relationship is intermediate between the first two relationships.

Several hypotheses are available as to the character of martensite nucleation. Most of them suggest a heterogeneous nucleation at special defect sites in the starting matrix. It was shown experimentally that the sites do not include grain and subgrain boundaries, as these are not places of preferable nucleation of martensite. They might be stacking faults arising in the γ

phase during splitting of dislocations. According to other hypotheses, the sites include special configuration dislocation pile-ups or separate dislocations, which are the sources of fields of internal stresses. This decreases the work on critical nucleus formation.

By morphology, martensite can be divided into two basic types: plate and massive martensite. They are different in shape, mutual arrangement of crystals, substructure, and habit plane.

Plate (needle) martensite is found most frequently in high-carbon steels and carbon-free iron alloys. Martensite crystals are shaped as thin lenticular plates (Figure 9). Neighboring plates are not parallel to one another.

Plates that appear first pass throughout the unit, dividing it into separate parts. But they cannot cross the matrix grain boundary. Therefore the plate dimension is limited by the dimension of the austenite grain. New martensite plates are formed in austenite sections. Here the plate dimension is limited to the dimension of the section (see Figure 9). If the austenite grain is small, martensite plates are so fine that the needle structure of marensite cannot be seen in it microsection specimens. Such martensite is called "structureless" martensite, and its is most desirable.

Massive (lath) martensite can be observed in low- and medium-carbon steels. Crystals of this type of martensite are shaped as interconnected plates having approximately the same orientation. The habit plane of laths is close to the {111}A plane. Plates of massive martensite are separated with low-angle boundaries. An electron microscopic image of massive martensite is given in Figure 10. As is seen, a package of plates is the main structural component. Several martensite packages can be formed in an austenite grain.

F.   The Bainite Transformation

The bainite transformation is intermediate between pearlite and martensite transformations. The kinetics of this transformation and the structures being formed exhibit features of both diffusive pearlite transformation and diffusionless martensite transformation.

A mixture of the α-phase (ferrite) and carbide is formed as a result of the bainite transformation. The mixture is called bainite.

The bainite transformation mechanism involves $\gamma \to \alpha$ rearrangement of the lattice, redistribution of carbon, and precipitation of carbide.

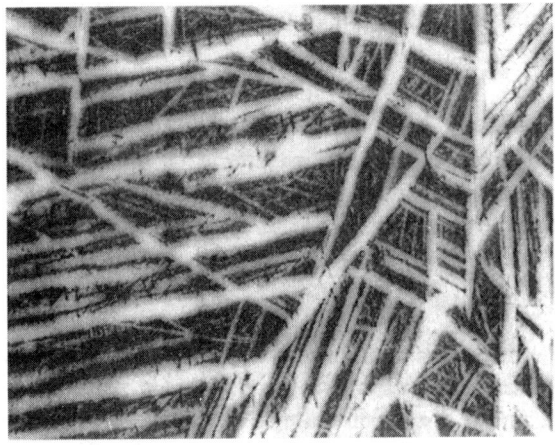

Figure 9   Martensite plates and retained austenite (dark) in quenched steel, 1000×.

Figure 10 Electron microscopic image of lath martensite, 20,000×.

Most researchers are of the opinion that ferrite precipitates from austenite by the martensitic mechanism. This is attested to by the presence of retained austenite in alloyed steels, a similarity in the structure of lower bainite and martensite, and the resemblance of upper bainite to low-carbon martensite.

Closeness of the bainite transformation to its pearlite and martensite counterparts can be explained as follows. The diffusive movement of atoms of the basic component—iron—is almost completely suppressed over the bainite transformation range. Then the $\gamma \rightarrow \alpha$ formation of ferrite is difficult because pearlite precipitation is suppressed. However, carbon diffusion is rather active and causes precipitation of carbides.

Over the intermediate range the γ-phase crystals are formed through coherent growth similarly to martensite plates. But the α-phase plates are formed slowly rather than "instantaneously." This is due to the fact that over the intermediate temperature range the α phase can precipitate only from the carbon-depleted γ phase. Thus the growth rate of the α-phase crystals depends on the carbon diffusive removal rate. In this case the martensite start point M_s in austenite rises and the martensite $\gamma \rightarrow \alpha$ transformation takes place at temperatures above the temperature M_s typical of the steel with a given composition.

At the instant of martensite transformation the carbon concentration remains unchanged. Only the crystal lattice is altered and a supersaturated α solution is formed. Carbide precipitates after $\gamma \rightarrow \alpha$ transformation.

G. Morphology of the Bainite Transformation

A distinction is drawn between upper and lower bainite, which are formed in the upper and lower parts of the intermediate temperature range. The conventional boundary between the bainites is close to 350°C (660°F). Upper bainite has a feathery structure, whereas lower bainite exhibits an acicular morphology, which is close to that of marensite. The difference in the structures of upper and lower bainites is attributed to a different mobility of carbon in the upper and lower parts of the bainite temperature range.

An electron microscopic analysis showed that the α-phase substructure of upper bainite resembles the substructure of massive martensite in low-carbon steels, while the α-phase structure of lower bainite approximates the structure of martensite in high-carbon steels. In upper bainite, carbide particles can precipitate both at lath boundaries and inside laths. This fact suggests that here carbides precipitate directly from austenite. In lower bainite, carbide is found inside the α-phase. This means that carbide is formed during precipitation of a supersaturated solid solution of carbon in the α-phase. Both upper and lower bainites exhibit a high density of dislocations inside the α-phase.

Cementite is the carbide phase in upper bainite, and ε-carbide in lower bainite. As the holding time is increased, ε-carbide turns into cementite. The austenite grain dimensions have no effect on the martensite transformation kinetics.

H. Tempering

The main processes that take place during tempering are precipitation and recrystallization of martensite. Quenched steel has a metastable structure. If subjected to heating, the structure becomes closer to equilibrium. The character of the processes that occur during tempering is determined by three major features of quenched steel: strong supersaturation of the martensite solid solution, high density of crystal lattice defects (dislocations, low- and large-angle boundaries, twin interlayers), and the presence of retained austenite.

The main process taking place during tempering of steels is precipitation of martensite accompanied by formation of carbides. Depending on the temperature and duration of tempering, the martensite precipitation may involve three stages: pre-precipitation, precipitation of intermediate metastable carbides, and precipitation and coagulation of cementite. Retained austenite can precipitate simultaneously.

Owing to a high density of dislocations in martensite, its substructure is similar to the substructure of a work-hardened (deformed) metal. Hence, polygonization and recrystallization can develop during tempering.

When carbon steels are tempered, supersaturation of the γ' solution in austenite increases with an increase in the carbon content of steel. This leads to lowering of the temperature M_s and transition from massive martensite to plate martensite. The amount of retained austenite also increases.

Carbon segregation represents the first structural changes that take place during tempering of carbon steels. The segregated carbon can nucleate heterogeneously at lattice defects or homogeneously in the matrix. The heterogeneous nucleation of the segregated carbon occurs either during quenching or immediately after it.

Flat homogeneous clusters of carbon atoms that are not connected with lattice defects are formed at tempering temperatures below 100°C (212°F). Their formation is due to considerable displacements of iron atoms and the appearance of elastic distortions. As the tempering temperature is increased, the clusters become larger and their composition is close to Fe_4C. This process depends on carbon diffusion. Metastable ε-carbide (Fe_2C) is formed above 100°C (212°F). It possesses a hexagonal lattice and appears directly from carbon clusters when the carbon concentration is increased. Metastable ε-carbide can also precipitate directly from the α solution. At low temperatures ε-carbide precipitates as very fine (10–100 nm) plates or rods (Figure 11). With an increase in tempering temperature or time, ε particles become coarser. This carbide precipitates in steels containing a minimum of 0.2% C.

In steels having a high M_s temperature, i.e., in all structural steels, partial precipitation of martensite accompanied by deposition of excess carbide is accomplished during quench cooling in the martensite range. Then self-tempering of these steels occurs during their quenching.

Cementite, Fe_3C, is formed at a temperature above 250°C (482°F). Two mechanisms of cementite nucleation have been known. First, it precipitates directly from a supersaturated α solid solution. Cementite particles grow at the expense of the dissolution of less stable carbides. Second, cementite appears as a result of transformation of the intermediate carbide lattice to the Fe_3C lattice.

Figure 11   Electron microscopic image of the ε-carbide, 50,000×.

The final stage of the carbide formation during tempering is coagulation and spheroidization of carbide. These processes develop intensively starting from 350–400°C (660–750°F). Above 600°C (1112°F), all cementite particles have a spherical shape and undergo coagulation only.

A considerable part of the tempering process is devoted to the precipitation of retained austenite accompanied by deposition of carbides. Precipitation occurs over the temperature interval 200–300°C (400–570°F). During tempering, retained austenite transforms into lower bainite.

A decrease in the carbon concentration of the α phase during carbide formation causes changes in the phase structure. Martensite precipitation can conventionally be divided into two stages. The first stage of precipitation is realized below 150°C (300°F). At these temperatures, the mobility of carbon atoms is sufficient for the formation of carbide plates. However, it is insufficient for the carbide plates to grow by diffusion of carbon from the areas of unprecipitated martensite with a high carbon concentration. This results in a nonuniform content of carbon in different areas of the martensite and consequently inhomogeneity of martensite with respect to its tetragonality. In areas with precipitated carbide, tetragonality is lower than in unprecipitated areas. Two solid solutions with different carbon concentrations coexist. For this reason the precipitation is referred to as a "two-phase precipitation." The two-phase precipitation of martensite results from the deposition of new carbide particles in areas containing martensite with the initial carbon concentration. Carbide particles do not grow at this stage.

At the second stage of martensite precipitation (150–300°C; 300–570°F) the α solution is depleted of carbon owing to diffusive growth of carbide particles. But the process proceeds very slowly. Therefore the precipitation kinetics are characterized by a rapid depletion of the α solution in carbon (the time span decreases as the annealing temperature is increased). Subsequently, depletion of the solid solution in carbon stops. At 300°C (570°F) about 0.1% C is left in the α solution. Above this temperature no difference between the lattice of the α solution and that of the α-Fe is detected. Below 300°C the degree of tetragonality ($c/a > 1$) is still measurable. Above 400°C (750°F) the α solution becomes completely free of excess carbon and transformation of martensite to ferrite is finished.

As mentioned earlier, plates (needles) of quenched martensite have a high density of dislocations, which is comparable to the density of the deformed material. However, recrystallization centers and their development to recrystallized grains have not been observed. This is due to the fact that carbide particles pin dislocations and large-angle boundaries. It is only above 600°C (1112°F), when the density of the particles decreases owing to their coagulation, that recrystallization growth of grains takes place at the expense of migration of large-angle boundaries. Therewith the morphological features of lath martensite vanish. These processes are hampered in high-carbon steels in comparison with low-carbon alloys, because the density of carbides is greater in high-carbon steels. The acicular structure is retained up to the tempering temperature of about 650°C.

The structural changes that occur during tempering cause alteration of steel properties. They depend on the tempering temperature and time. Hardness lessens as the tempering temperature is raised.

IV. KINETICS OF AUSTENITE TRANSFORMATION

A.   Isothermal Transformation (IT) Diagrams

To understand the kinetics of transformations to austenite, it is important to follow the process at a constant temperature. To this end, a diagram was constructed that characterizes the isothermal process of austenite precipitation. In this diagram the transformation time is the abscissa on the logarithmic scale and the temperature is plotted on the ordinate (see Figure 12). From this diagram, the incubation period (left-hand curve) can be determined and also the time required for completion of the process (right-hand curve). The instant an alloy passes the points A_3 and A_1 during quenching is usually taken as the zero time reference. The time required to achieve the temperature of the quenching medium is often neglected. The start and finish of the transformation are difficult to determine from the transformation curve behavior at the initial and final sections of the curve. Therefore the lines of the isothermal transforma-

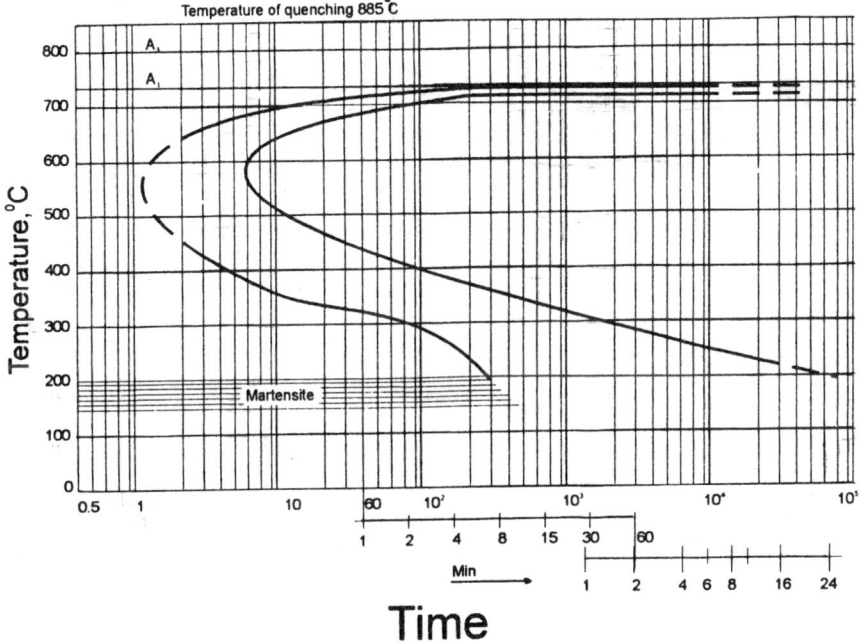

Figure 12   Isothermal transformation diagram.

tion diagram usually correspond to a certain final volume that underwent transformation, e.g., 3% and 97% for the transformation start and finish, respectively. The volume value is usually not shown in the diagram.

In addition to the above-mentioned curves, the diagram often contains intermediate curves that correspond to certain values of the transformed volume, e.g., 10, 50, or 90%.

A decrease in the transformation rate causes displacement of the transformation start and finish curves to the right, i.e., toward greater duration. This phenomenon can be observed if the quenching heating temperature rises as a result of a decrease in the number of foreign inclusions, enlargement of austenite grains, etc.

An increase in the transformation rate leads to displacement of the IT curves to the left. This phenomenon can be accounted for by a decrease in the quenching heating temperature, the presence of carbides or foreign inclusions, and refinement of the austenite grain.

For a given steel the temperature that corresponds to a maximum transformation rate (the so-called "nose" of the sigmoid curve) does not, as a rule, change significantly.

B. Continuous-Cooling Transformation (CCT) Diagrams

Continuous-cooling transformation (CCT) diagrams consider the transformation kinetics of a eutectoid steel. The major transformation that takes place during annealing cooling of steel is a eutectoid precipitation of austenite into a mixture of ferrite and carbide. The eutectoid transformation kinetics are given by IT diagrams of austenite (Figure 13) at a temperature of 727°C (1340°F). The structure obtained after tempering below 300°C (572°F) is called tempered martensite. An acicular structure is observed after tempering at 300–450°C (572–842°F). Tempering over the temperature interval of 450–600°C (842–1112°F) exhibits a pronounced

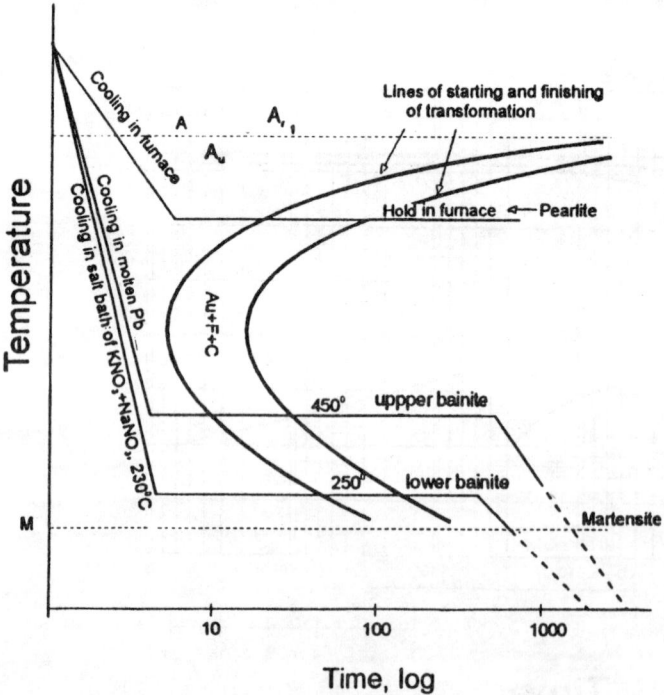

Figure 13 Isothermal transformation diagram for a eutectoid composition steel. A, stable austenite; A_u, undercooled austenite; F, ferrite; C, carbide.

dot structure. The structure obtained after tempering below 300°C (572°F) is called tempered martensite. Austenite is in a thermodynamically stable equilibrium with the ferrite–cementite mixture. Stability of undercooled austenite is defined by a period of time during which the appearance of precipitation products in the diagram cannot be registered by conventional methods (Figure 13). It is equal to the distance from the Y axis to the left-hand curve. The degree of austenite undercooling is the main factor that determines the steel microstructure. The necessary degree of undercooling is provided by either continuous cooling or isothermal treatment. The diagram in Figure 13 shows the entire range of structures formed in a eutectoid steel depending on particular undercooling conditions.

1. *Transformations That Take Place Under Continuous Cooling of Eutectoid Steels*

As mentioned above, in hypoeutectoid steels the formation of pearlite is preceded by precipitation of hypoeutectoid ferrite. With a decrease in the transformation temperature and a rise in the degree of undercooling, precipitation of hypoeutectoid ferrite is suppressed. The amount of pearlite increases and the carbon content becomes less than that in pearlite of the eutectoid steel. In the region of the maximum transformation rate, the two curves merge. Thus, a purely pearlitic structure is formed in steel with 0.4% C (Figure 14). In steels containing greater amounts of carbon, the precipitation of ferrite cannot be suppressed even if the carbon content decreases. Ferrite precipitation precedes the formation of pearlite even at a maximum transformation rate, but the amount of ferrite will be less than was formed at small undercooling.

These propositions are valid for the precipitation of cementite in hypereutectoid steels, but it can be suppressed even at relatively small undercooling. In this case the carbon content of pearlite becomes higher than that in the eutectoid steel.

As a result of suppression of the hypoeutectoid ferrite precipitation under continuous cooling from the region of the γ solid solution, the point A_{r3} lowers much faster than the point A_{r1} as the cooling rate is increased. Given a certain cooling rate, both points merge into one point A_2' (Figure 14), which corresponds to the formation of a fine plate structure of the pearlite type free of ferrite.

Under continuous cooling the transformation process can also be pictured as diagrams in temperature–time coordinates (Figure 15). Hence the behavior of cooling curves should be analyzed to obtain characteristics of the transformation processes. In this diagram the ferrite and pearlite start lines are shifted toward longer periods of time compared to the IT diagram of Figure 13. This is due to an increase in the temperature interval necessary for preparing

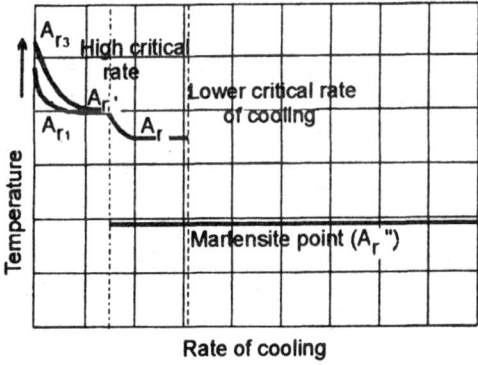

Figure 14 Schematic diagram showing changes in location of the critical points depending on the particular cooling rate.

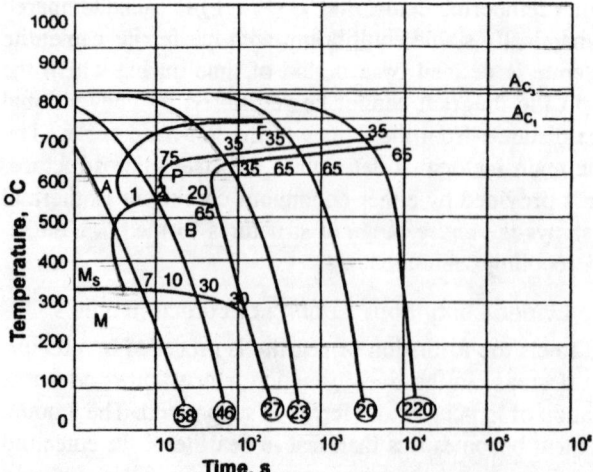

Figure 15 Continuous cooling transformation (CCT) diagram. A, Austenite range; F, ferrite range; P, pearlite range; B, bainite range; M, martensite range. Shown in circles is hardness HV or RC. Numerals at the curves denote the relative amounts of structural components.

the transformation processes in the austenite lattice. As a result, only part of the incubation period, which is required for the isothermal transformation to start, is effective. In this case the incubation period is the mean of the effective lengths of time corresponding to different periods of time in the given range. This proposition can be used to calculate the behavior of the transformation start line in the pearlite range from the IT diagram. The reverse calculation is also possible.

Similar to the pearlite range, in the bainite temperature range, the precipitation of under-cooled austenite starts after a certain incubation period. Resemblance of the bainite and pearlite transformation kinetics consists not only in the presence of an incubation period but also in the character of the volume increase during isothermal soaking: the fraction of the transformed volume of austenite increases first with acceleration and then with deceleration. At the same time, as in the case of the martensite transformation, retained austenite does not disappear completely during the bainite transformation. Every point in the bainite finish curve corresponds to a certain amount of retained austenite. Similar to the pearlite transformation, the bainite transformation can take place both during isothermal soaking and under continuous cooling (see Figure 15). Austenite that has not been transformed over the bainite range turns partially into martensite when the steel is cooled to room temperature. Owing to the fact that after the bainite transformation austenite is inhomogeneous with respect to the carbon content, martensite is formed predominantly in carbon-enriched regions.

In the case of high-alloy steels, isothermal curves can be separated by a temperature interval in which undercooled austenite is highly stable. In this interval pearlite precipitation does not take place for many hours, while undercooling is insufficiently great for the bainite transformation (Figure 16). In carbon steels the bainite transformation proceeds concurrently with the pearlite transformation. Products of the pearlite transformation dominate at higher temperatures, and those of the bainite transformation at lower temperatures.

2. Transformations of Austenite on Cooling in the Martensite Range

The martensite component in the steel structure appears when the cooling rate achieves a certain value. The minimum cooling rate at which the marensite component is formed is called

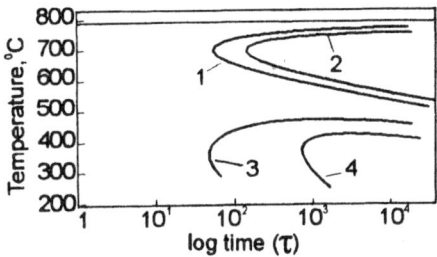

Figure 16 Diagram of isothermal precipitation of austenite in steel with 0.43% C and 3% C. Curve 1, pearlite formation start; curve 2, pearlite formation finish; curve 3, bainite formation start; curve 4, bainite formation finish.

the lower critical rate of cooling. The rate at which transformations by the pearlite and bainite mechanisms are suppressed completely is referred to as the upper critical rate of cooling (quenching). If the conditions of austenite formation (austenitization temperature and the holding time at this temperature) and the cooling conditions (cooling rate should exceed the upper critical rate) are constant, the location of the martensite point M_s depends only on the content of carbon and alloying elements in the steel.

If the cooling rate is high, the formation rate of separate needles of martensite is also high, and transformation of austenite to martensite commences on reaching M_s. It continues on subsequent cooling to lower temperatures. As the temperature of the quenching medium is lowered, the amount of formed martensite rises first rapidly and then slowly. With an increase in the quenching heating temperature (austenitization temperature), the transformation also shifts toward lower temperatures (Figure 17) as more of the alloying elements are taken into solution. A certain amount of martensite may be formed during isothermal holding, but it is not high in carbon steels. Retained austenite is stabilized during isothermal holding. As a result, more martensite is formed during subsequent cooling. Formation of martensite stops at the point M_f.

Figure 18 shows a relationship between some factors that influence the stabilization of martensite. As is seen, if continuous cooling is stopped at the temperature T_{h1} and a holding time is allowed at this temperature, the formation of martensite starts after passing through a certain temperature interval rather than immediately when cooling is resumed. Subject to cooling below the point M'_s, further formation of martensite takes place. If holding is realized at a lower temperature, T_{h2}, the effect of stabilization is enhanced, because further formation of martensite commences at the temperature M_{s2} after passing through a greater temperature in-

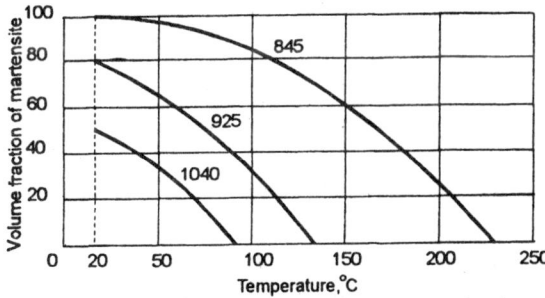

Figure 17 Curves showing variation of the relative amount of martensite as a function of the transformation in steel with 1.1% C and 2.8% Cr for different homogenization temperatures.

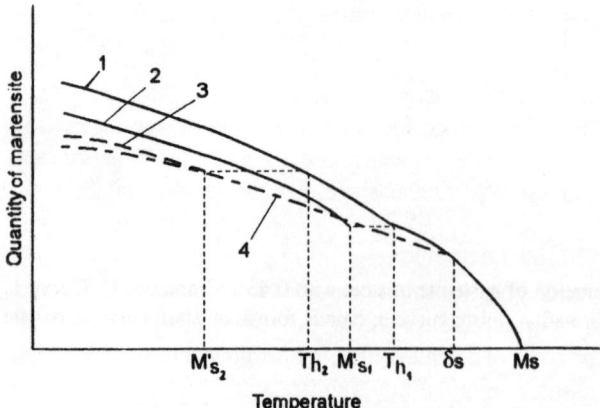

Figure 18 Curves showing the relative amount of martensite as a function of the austenite stabilization temperature. 1, Under continuous cooling; 2, after isothermal holding at T_{h1}; 3, after isothermal holding at T_h; 4, M'_s curve, τ_s is the limiting stabilization temperature.

terval (curve 3, Figure 18). The effect of stabilization increases with the amount of martensite in the structure or, the amount of martensite being equal, with temperature. Joining of the points M'_s determined after holding at different temperatures yields a curve that intersects the curve that corresponds to the relative amount of martensite formed under continuous cooling. The point of intersection of the curves, σ_s means that stabilization of austenite is impossible at a higher temperature.

C. Derivation of the CCT Diagram from the IT Diagram

When solving practical problems involved in thermal treatment of steel, it is often necessary to know how the continuous cooling rate affects the structure formed as a result of austenite transformation. To this end, attempts were made to establish the relationship between the transformation kinetics of austenite under isothermal conditions (IT diagram) and under continuous cooling (CCT diagram). The attempts stared from the concept of additivity of the transformation processes at different temperatures. It was assumed that holding of undercooled austenite at a preset temperature is part of the incubation period. It was found, however, that calculated and experimental data coincide satisfactorily only if the pearlite transformation is continuous.

  If the pearlite transformation is preceded by precipitation of eutectoid pearlite or the pearlite and bainite transformations occur concurrently, calculated data are at a discrepancy with the experimental. It was found that the discrepancy is due to the following factors:

1. Holding of austenite during the time accounting for fractions of the incubation period causes acceleration of the subsequent intermediate transformation at the expense of preparatory processes.
2. Precipitation of hypoeutectoid ferrite alters the austenite composition. This delays the subsequent intermediate transformation.
3. Partial transformation of austenite over the intermediate range decreases the rate of the said transformation at lower temperatures and facilitates an increase in retained austenite. This is due to a redistribution of carbon and enrichment of the non-transformed part of austenite in carbon.

4. A change in the cooling rate over the martensite range affects stabilization of austenite in different ways.

For this reason, special methods of constructing thermokinetic transformation diagrams of
austenite subject to continuous cooling were elaborated for noneutectoid steels (see Figure 15).
From these diagrams it is possible to determine the critical rate of quenching cooling or continuous cooling that is necessary to complete a particular stage of austenite precipitation.

D. CCT Diagram as a Function of the Bar Diameter

When steel is subjected to martensitic hardening, it should be cooled form the quenching temperature so that on undercooling to a temperature below the M_s point austenite has no time to
precipitate and form a ferrite–carbide mixture. To achieve this, the cooling rate should be less
than the critical value. The critical cooling rate is the minimum rate at which austenite does
not precipitate to a ferrite–carbide mixture. In actual fact, of course, the cooling rate of steel
products is nonuniform over their cross section. It can be higher than the critical rate on the
surface and lower than the critical rate at the center.

   The critical cooling rate at different points of a product can be directly determined from
an IT diagram (Figure 19). In the first approximation it is given by the slope of the tangent
to the C curve that denotes the austenite precipitation onset. This method gives a value that is
about 1.5 times the true critical rate. The cooling rate can be determined more accurately if
one uses thermokinetic diagrams (Figure 20). Intercepts of the cooling curves with the lines
of the thermokinetic diagrams show the start and finish temperatures of the corresponding
transformation.

   From the transformation diagram it is possible to determine, for example, the rate that
will provide 50% martensite in the structure or the rates at which the entire transformation
occurs in the pearlite range, i.e., hardening is excluded altogether. Since data on the critical
hardening rate depend on a certain cooling time law and should be associated with a particular temperature (while direct measurements of the hardening rate are practically impossible),
it is appropriate to specify the cooling time for a specific temperature interval, for example,
from the point A_3 to 500°C (932°F). Point A_3 in the diagram is the time reference. Then it is
possible to straightforwardly determine the critical cooling time K: K_m for fully martensitic

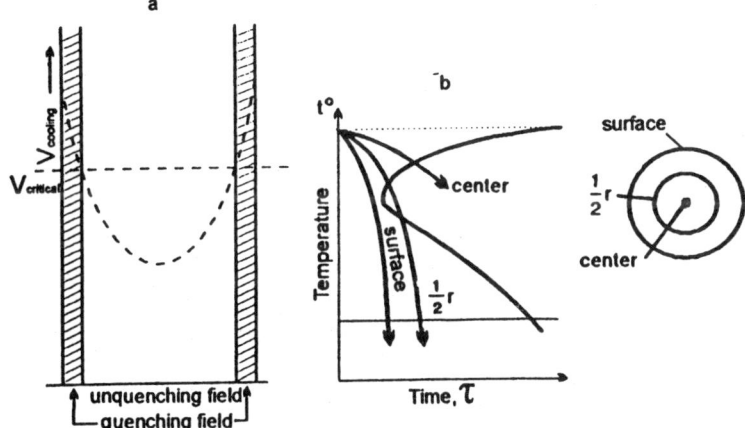

Figure 19 Diagram showing distribution of the cooling rate over the cross section of a sample and the
corresponding isothermal transformation (IT) curve.

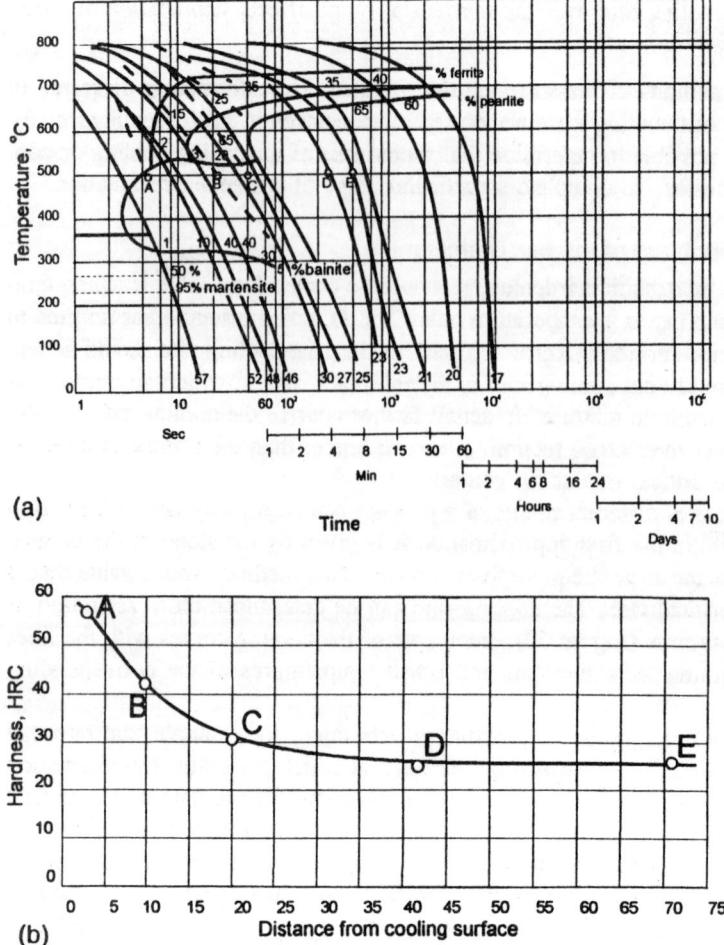

(a)

(b)

Figure 20　(a) Time–transformation temperature diagram for continuous cooling of steel containing 0.38% C compared with (b) the process of the sample's cooling during face quenching. Numerals at the bottom of the curves denote hardness (RC) after cooling to room temperature.

hardening; K_f for initial appearance of ferrite; K_p for full transformation in the pearlite range. Since the cooling time (Figure 20) and the progress of the subsequent cooling of the sample during end-face hardening are known, the outcome of hardening can be determined from the transformation diagram. It should be remembered that a transformation diagram is valid only for particular conditions of melting and homogenization. Deviations in the composition or grain dimensions cause changes in the trend of thermodynamic curves. This is explained by the fact that an increase in the homogenization temperature and time and, consequently, enlargement of the grains enhance the stability of austenite. Conversely, refinement of grains lowers the critical cooling rate, because stability of austenite decreases with an increase in the extent of grain boundaries.

E.　Definition of Hardenability

The depth of the hardened zone is termed hardenability. This is one of the most important characteristics of steel. Since the cooling rate is nonuniform along the cross section of a sample

(Figure 19), austenite can pass into martensite in surface layers only, while at the center of the sample austenite undergoes the pearlite transformation. In the first place, hardenability depends on the critical cooling rate. An examination of the temperature curves (Figure 20) plotted for different areas of the sample shows that the cooling rate of the core of a large diameter product is lower than the critical value and therefore the core is not martensitically hardened. Martensite is present in the surface layer only.

After hardening treatment, a bulky part with a large cross section may exhibit the entire range of structures: a smooth transition from martensite near the surface through troostite-martensite and troostite to pearlite at the center.

The geometry of samples can influence the character of the cooling curves. However, given the same surface-to-volume ratio, the curves coincide in the main. The greatest changes in the cooling rate are incurred by the diameter of samples.

Considering what has been said above, to achieve a through hardening of bulky products or full martensitic hardening to the core of a product, one has to provide the critical hardening rate along the entire cross section of the product. IT and CCT diagrams can be used to determine this rate. The diagrams were plotted for different grades of steel taking into account the progress of cooling in different sections and in different hardening media.

Note that hardenability depends on the steel composition, specifically on the carbon content. Hardenability of each grade of steel is presented as a "hardenability band" (Figure 21). These diagrams have been plotted for almost all existing grades of steel. They show how to achieve through hardening of a product made of a particular steel.

Hardenability of steel is also characterized by transformation time–temperature curves (IT curves). The more the curve is shifted to the right along the abscissa axis, the greater is the hardenability of the steel. This is explained by the fact that the rightward shift of the IT curve is due to better stability of austenite.

An improvement in the stability of undercooled austenite and hence an increase in the critical hardening rate lead to a greater depth of hardening. Then hardenability depends on all the factors that improve the stability of undercooled austenite. For example, the stability of austenite can be raised by alloying steel with chromium and tungsten. These elements lower the austenite precipitation rate and can make a steel an air-hardening one. Steel with a usual (commercial) content of impurities is hardened to a strength 10 times that of a pure iron-carbon alloy.

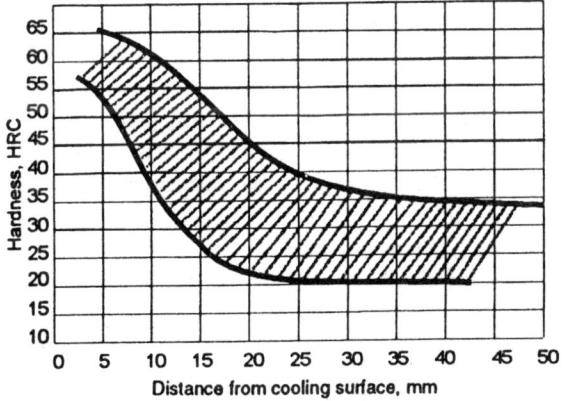

Figure 21   Steel hardenability band.

Elevation of the hardening temperature favors an increase in the hardening depth thanks to the homogenization of austenite and enlargement of austenite grains. Refinement of grains impairs hardenability as grain boundaries affect the stability of austenite.

The hardening depth also depends on the hardening medium used. The greater the intensity of cooling, the greater the depth of hardening. Besides, the hardening depth depends on the cross-sectional diameter of the products. The critical diameter is that of the greatest cross section that lends itself to through hardening in a given hardening medium. The critical diameter is different for different hardening media and characterizes the hardenability provided by a particular method only.

Hardenability has an effect on the mechanical properties of steel. In the case of through hardening the properties do not differ along the cross section of a product. Otherwise they decrease from the surface to the center. Let us analyze the influence of hardenability on the properties of steels that were tempered after hardening. A high-temperature tempering favors equalization of hardness along the cross section. However, the structure of weakly hardenable steels remains inhomogeneous; a grain structure will appear on the surface, where martensite is formed during quenching, while a lamellar structure will remain at the center. A grain structure will be present along the entire cross section of a through-hardening steel. This determines the character of changes in the properties of steels with different hardenability. The properties that are independent of the cementite form (yield stress, specific elongation, impact strength) will differ. A decrease in σ_s and a_k is observed at the center of non-through-hardening steels, while in a through-hardening steel these quantities remain unchanged along the cross section.

The properties of tempered steels (fracture stress, yield stress, impact strength, reduction of area) are impaired if ferrite precipitates during quenching.

A product's mechanical properties depends on its cross-sectional area. To obtain the best mechanical properties in the tempered state, a grain structure should be provided along the entire cross section; i.e., through hardenability should be ensured in the quenched state.

V.   GRAIN SIZE

A.   Structure of Grain Boundaries

When analyzing any processes or properties associated with grain boundaries, it is necessary to know the structure of the material. The overwhelming majority of structural materials are polycrystalline. They comprise a set of grains separated by boundaries. The grain boundary is one of the basic structural elements in polycrystalline materials.

The grain boundary represents an interface between two differently oriented crystals. This is the region of crystal imperfection. It is capable of moving and adsorbing impurities. The boundary has a high diffusive permeability.

In polycrystalline materials the boundaries determine the kinetics of many processes. For example, movement of grain boundaries controls the process of recrystallization. A high diffusive permeability of grain boundaries determines the kinetics of diffusion-dependent processes at moderate temperatures. Grain boundaries adsorb impurities. Embrittlement of metal materials is connected with enrichment of grain boundaries in impurities.

Grain boundaries may conventionally be divided into two large groups: low-angle and large-angle boundaries. Low-angle boundaries (or subgrain boundaries with an angle of less than 10°) represent networks or walls of dislocations. The structure of large-angle boundaries is much more complicated. Figure 22 shows both types of grain boundaries.

The progress in understanding the structure of grain boundaries (GB) is connected with elaboration of the models describing the observed microscopic properties of the boundaries.

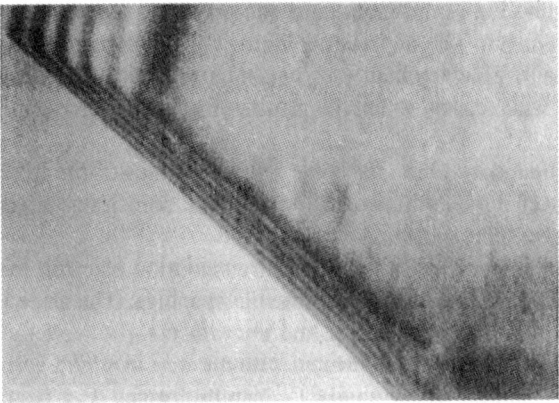

(a)

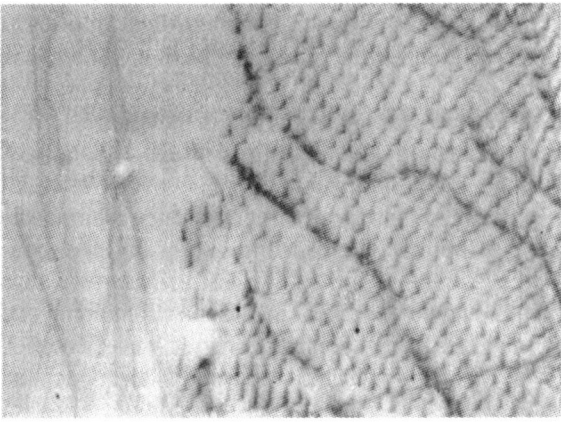

(b)

Figure 22 Electron microscopic image of (a) low-angle boundary and (b) large-angle boundary. 50,000×.

1. *Structural Models*

The pioneering structural model is the model of an amorphous boundary. It allows an explanation of the value of the surface tension Γ_3 and the grain boundary slip. In terms of this model it was assumed that the usual boundary with a large angle has random regions of incontingency similar to a liquid. The width of the regions does not exceed three atomic diameters. In later models, amorphous portions of the boundaries were added with crystalline portions. According to Mott [1,2], Γ_3 represents portions of good and poor contingency. In the opinion of Smoluchowski [3,4], even when the boundary angle exceeds 15°, dislocations combine themselves into groups and form incontingency regions separated by undistorted areas. If the misorientation angle is greater than 35°, then Γ_3 is a solid region of incontingency. Geisler and Hill [5] and Hargreaves [6] described the grain boundary in terms of the model of a transition lattice. According to this model, a certain system in the arrangement of atoms exists in Γ_3. The arrangement corresponds to a minimum energy possible under given conditions.

At certain mutual orientations of neighboring grains (special orientations), a "superlattice," which is common for both grains, may appear. The superlattice sites will be atoms that are common to the crystal lattices of both grains. The boundaries lying in close-packed planes of such superlattices will be most favorable with respect to energy. If the misorientation angle is small, the coincidence is upset.

Coinciding atoms are present in the boundary plane for some discrete values of the grain misorientation angles. Boundaries that meet the conditions required for the coinciding atoms to appear are called partial contingency (or special) boundaries.

Direct experimental studies of Γ_3 are scarce. Microscopy and transmission electron microscopy have shown that the transition zone occupies 2–3 interatomic spacings. The zone is saturated with defects like grain boundary dislocations, steps, and microfacets.

Particular grain boundary characteristics are closely connected with the way in which grain boundaries are formed. A grain structure, and correspondingly Γ_3, can be formed as a result of crystallization from the liquid state, phase transformations in the solid state, or recrystallization annealing of a deformed material during deformation.

Only conjectures can be made as to the formation of grain boundaries during crystallization. Under real conditions of crystallization the growth of crystals often exhibits an oriented rather than chaotic character. Correspondingly, the spectrum of boundaries in a cast material should differ from the random distribution.

In the case of recrystallization in the solid state, i.e., polymorphous transformations of metals and alloys, the new phase has certain orientation relationships with the initial phase. Obviously, when transformations within a single grain of the matrix phase are completed, the formed boundaries should have strictly defined and crystallographically determined misorientations rather than random orientations. Many boundaries that appear during a polymorphous transformation are close to special boundaries of coincident sites. Experimental studies into misorientations of the crystals formed during phase transformations in chromium-nickel steel and titanium alloy showed that misorientations at real boundaries agree with theoretical ones (what is meant here is the calculation of crystallographically determined misorientations for the fcc-hcp and hcp-fcc, hcp-bcc and bcc-hcp, and fcc-bcc transformations). Then the crystallo geometry of the boundaries resulting from polymorphous transformations is controlled by orientation relationships of the phases being formed.

In the case of recrystallization processes the grain structure depends on the stage of recrystallization at which annealing was stopped. During primary recrystallization the formation of the structure starts with the appearance of nuclei, that is, dislocation-free portions of the matrix. They are surrounded by large-angle Γ_3. The proposed models of nucleation assume that nuclei of new grains are formed near the initial Γ_3 owing to a rearrangement of intergrain lattice dislocations. However, it has been established recently that new grains with large-angle boundaries can be formed without participation of intergrain dislocations but rather during splitting of initial boundaries. This process can be accounted for in the following way. After plastic deformation the grain boundaries are in a nonequilibrium state owing to the trapping of lattice dislocations. During annealing the grain boundary structure regains the equilibrium state at the expense of splitting of the boundaries. Splitting of the boundaries during recrystallization is caused by lowering of the total energy of the grain system because high-energy boundaries are replaced by low-energy ones. Here mutual misorientations depend on misorientation of the nuclei in the deformed matrix.

At subsequent states of recrystallization the grains become coarser owing to migration of the boundaries. One would expect that the average statistical trend of the process should be toward formation of low-energy special boundaries. However, the available experimental data are contradictory. This fact suggests that in addition to the tendency to a thermodynamic equilibrium, kinetic factors (different mobility of the boundaries, their pinning by impurities and precipitates) play an important role in the process of structure formation during annealing. It was found, for example, that at the stage of collecting recrystallization, random boundaries dominate in iron and molybdenum alloys. However, in ultrapure aluminum the fraction of these boundaries decreases with an increase in the recrystallization temperature and time. In contrast, in commercially pure aluminum the fraction of special boundaries decreases.

The state of grain boundaries in a material depends not only on their misorientation but also on the content of lattice defects. For this reason the boundaries of recrystallization nuclei are not in equilibrium; they are formed in the regions of the deformed matrix with an excess density of dislocations of like sign. Rearrangement of the dislocations within the nucleus boundaries is not complete. A nonequilibrium state of the boundaries is also preserved during their migration through the deformed matrix as the matrix absorbs lattice dislocations. Then the boundaries are nonequilibrium in ultrafine grain materials formed at the early stage of recrystallization.

The degree of boundary nonequilibrium decreases at later stages of recrystallization during collecting growth of the grains.

Grain boundaries of the deformation origin can be divided into two groups: grain boundaries formed at a low-temperature ($< 0.3-0.4 T_{melt}$) deformation and those formed under deformation at high temperatures.

At low temperatures new boundaries are formed at relatively large degrees of deformation. First they represent broken boundaries, which appear nonuniformly in separate grains of a polycrystal. A continuous network is formed in the areas adjacent to the initial boundaries. It is only under large deformations that a network of these boundaries covers the whole volume. On the average, two out of three boundaries are large-angle ones. In this case the fraction of special boundaries is small.

When subject to deformation at high temperatures, the formation of grain boundaries is due to development of recrystallization processes directly during deformation. This phenomenon is called a dynamic recrystallization. The grains formed during a dynamic recrystallization are large-angle ones. Data on crystallogeometrical parameters of these boundaries are very scarce.

From the above discussion it appears that, depending on their origin, grain boundaries have different structures and therefore possess different properties. The properties of polycrystalline materials are largely determined by the extent of these structural components, which is controlled by the grain size.

B. Determination of Grain Size

The size of the grain that is formed under a given treatment is determined from microsections after their etching. For carbon and alloyed steels the following reagent is used: 1–5 mL HNO_3 + 100 mL ethyl or methyl alcohol. Austenitic steel is etched in a copper sulfate chloride solution containing 10 g copper sulfate, 50 mL hydrochloric acid, and 50 mL water. When carbon low-alloy steels are etched, the reagents turn pearlite dark and make visible the ferrite grain boundaries, the martensite structure, and tempering products. The etching rate rises with the amount of nitric acid. The etching time is from several seconds to a minute.

Etching of austenitic steel reveals the austenite structure and the austenite grain boundaries.

Carburization is also used to establish the austenite grain boundaries. In this case samples are heated to 930°C (1700°F) in a carburizing medium (for example, a mixture of 40% $BaCO_3$ and 60% charcoal), cooled, and etched.

In addition, an oxidation method is used according to which microsections are heated in vacuum to a temperature 20–30°C (35–55°F) higher than the quenching temperature and are soaked for 3 h. Subsequently air is fed to the furnace for 30–60 s, and the samples are cooled in water. Prior to quenching it is recommended to heat samples in a borax melt at 930–950°C (1700–1750°F) for 30–40 sec. and then cool them in water. After these treatments micro-sections are polished and etched in a 15% solution of hydrochloric acid in ethyl alcohol. Grain boundaries are seen as the oxide network.

Apart from this, use is made of the method of etching austenite grain boundaries, the method of the network of ferrite (for steels with a carbon content of up to 0.6%) or cement-ite (for hypereutectoid steels), and the method of the pearlite network for steels that are closer in composition to eutectoid steels.

The grain size is determined by comparing the observed microstructure at a $100\times$ mag-nification with standard scales (the scales are elaborated so that at a magnification of $100\times$ the grain number N corresponds to the formula $n = 8 \times 2^n$, with n being the number of grains per 1 mm^2 of the microsection area) or by counting the number of grains per unit area of the microsection, or by calculating the mean nominal diameter of the grains or their number per cubic millimeter.

The number of grains (at least 50) is counted on the focusing screen of the microscope or from a photomicrograph within the area bounded by a circle 79.8 mm in diameter. At $100\times$ magnification this value corresponds to a microsection area of 0.5 mm^2. The total number of grains is calculated from the formula $m_{100} = m + 0.5m_1$, where m is the number of grains inside the circle and m_1 is the number of grains intersected by the circle. The number of grains per square millimeter of the microsection is $M - 2m_{100}$. If a magnification other than 100 power is used, $M = 2(g/100)^2 m_g$ (g being the magnification power used and m_g the number of grains counted at this magnification power). The mean number of grains (M_{mean}) is calculated using three characteristic areas.

The mean area (S_{mean}) and diameter (d_{mean}) of the grains are calculated using the formu-las

$$S_{mean} = 1/M_{mean} \qquad \text{and} \qquad d_{mean} = 1(M_{mean})^{1/2}$$

The values of equiaxial grains are characterized by the mean nominal diameter, which is determined on the focusing screen of the microscope or from a photomicrograph. For this purpose several arbitrary straight lines are drawn so that every line intersects at least 10 grains. Then the number of intersections on the length of all the lines is counted. Finally, the mean diameter of the grains is calculated.

Statistical methods are used, and bar charts are plotted to obtain quantitative characteris-tics of the structure, particularly grain dimensions. The mean diameter of grains is calculated using the distribution curve.

It is possible to calculate the mean area of grains (S_{mean}) from the formula used to determine d_{mean} if one assumes that the grain is spherical in shape. ($x = \pi D^2/4$): $S_{mean} = k^2 \varepsilon\,(\pi D^2/4)m/\varepsilon m$. Then the number of grains (N) per square millimeter is found from the formula $N = 1/S_{mean}$.

C. Austenite Grain Size Effect and Grain Size Control

The austenite grain boundary structure that is produced on heating above the critical points is important because the austenite transformation products formed during cooling (martensite, pearlite, etc.) appear inside austenite crystals. A coarse austenite grain determines a coarse plate structure of martensite during quenching or a coarse cellular network of ferrite (cementite) precipitates at the boundary of the initial austenite grains during annealing or normalization. The pearlite structure is also the coarser, the larger the pearlite grain.

As is known, a coarse grain structure of steel (ferrite-pearlite, martensite, etc.) is characterized by lower mechanical properties. For this reason a fine grain structure of steel is preferable in practice. Then the primary task is to produce fine-grain austenite. Since austenite appears during heating of a ferrite–carbide mixture, growth centers of the austenite phase are very numerous, and initially austenite grains are extremely small, on the order of 10–20 μm. But with an increase in the heating temperature or holding time in the austenite range, the grains begin to grow intensively.

Two types of steels exist: hereditarily coarse-grained steel and hereditarily fine-grained steels. This difference is due to the grain growth kinetics with an increase in temperature. In hereditarily coarse-grained steels a grain gradually and rather uniformly becomes larger as the temperature is raised above A_{c3}. In hereditarily fine-grained steels, fine grains are preserved up to about 950°C (1750°F). On transition through the "coarsening" temperature, separate grains start growing intensively and variations in grain size arise. Near 1100–1200°C (2000–2200°F), grains of hereditarily fine-grained steels may be even larger than those of hereditarily coarse-grained steels.

Such differences in the growth of grains in steels is explained by the differences in number and state of disperse nonmetal inclusions such as, above else, aluminum nitrides, certain carbides, and oxides. These particles retard movement of grain boundaries until temperatures are reached at which the particles dissolve in austenite. The barrier effect of the particles diminishes nonuniformly, which leads to variations in grain size.

A standard test can be used to distinguish between the steel classes. If a noticeable growth of austenite grains is not observed for 8 h after carburization at 925°C (1700°F), the steel is assumed to be a hereditarily fine-grained one. Extrapure steels, those produced with a minimum amount of foreign impurities, nitrogen and oxygen, are distinguished by a rapid growth of grains above the critical point A_{c3}.

In the case of the usual commercial steels, a grain 20–50 μm in size corresponds to standard heating for quenching, normalization, or annealing. As the temperature is elevated to 1200–1250°C (2200–2250°F), the grain size reaches 0.1 mm, and in large forgings and welds, grains several millimeters in size occur. In ingots and castings, grains can be as large as several centimeters.

If a steel is heavily alloyed with elements that stabilize austenite, the austenite structure is fixed during cooling to or below room temperature and the steel grain is equal to the initial austenite grain. If austenite passes to pearlite, then, for example, for a hypereutectoid steel one should take into account the size of the pearlite colony, which is characterized by the same crystallographic orientation of ferrite and cementite plates. A pearlite colony usually differs in size from an austenite grain. Several pearlite colonies are formed in every grain. So an austenite grain is broken into several grains. This is also true of the ferrite-pearlite structure of a hypoeutectoid steel. But in the latter case a network of excess ferrite is formed at grain boundaries. This suggests a connection between a grain of a thermally treated steel and the initial austenite grain.

When a steel undergoes quenching, a large number of martensite crystals appear in every austenite grain. They are connected with the initial austenite grain by certain orientation relationships. For this reason a correlation is easily seen between the initial austenite grain and grains of the quenched steel. Refinement of the initial grain under heating above A_{c3} results in refinement of grains in the quenched steel. Then it is possible to correct a coarse-grained structure by heating to the austenite state.

However, correction of a coarse-grained austenite or bainite structure may be complicated by a structural inheritance. When crystallographically ordered structures of bainite or martensite are heated, austenite can also be formed, under certain conditions, in a crystallographically ordered way. Therefore under heating above A_{c3} the austenite grain is equal in size to a coarse grain of steel. In this case refinement of the crystal structure as a result of phase recrystallization during the $\alpha \rightarrow \gamma$ transformation does not take place. After the $\alpha \rightarrow \gamma$ transformation the structure of the initial austenite is restored. Both the grain size and its crystallographic orientation are reestablished. However, the restored austenite is structurally unstable. If the temperature or holding time is increased, the austenite structure changes. But the grain is refined rather than becoming larger as is normally the case. The degree of grain refinement is different, but the structure changes completely above a certain temperature in the range of the stable γ phase. The austenite structure is altered within the temperature interval where phase transformations do not occur. Therefore this phenomenon is attributed to a spontaneous recrystallization of austenite. It is caused by the $\gamma \rightarrow \alpha \rightarrow \gamma$ transformation hardening.

The primary recrystallization of austenite, which is due to the transformation hardening, is followed, with a further increase in temperature, by collective recrystallization. The grain becomes still coarser. Note once again the unusual character of this two-stage process, which includes first a reestablishment of the initial austenite grain and then a refinement of the grain with temperature.

Plastic deformation inhibits the structural inheritance. This is due to the appearance of globular austenite in deformed steel that is subjected to either rapid or slow heating. Besides, deformation intensifies the austenite recrystallization in the α state.

If a hypoeutectoid steel undergoes sufficiently slow heating, austenite is often formed as same-oriented sections. As the temperature is raised, excess ferrite dissolves in these sections. When ferrite dissolves completely, newly formed grains of austenite fully duplicate the initial austenite grains. With an increase in the heating rate, sections of austenite with a different orientation appear. If isothermally heated these sections grow larger and absorb the restored austenite and excess ferrite. The greater the number of such sections formed at a faster rate of heating, the finer the austenite grain.

Formation of these sections cannot be explained in terms of the austenite recrystallization because their growth stops as soon as the excess ferrite dissolves completely. They appear in a somewhat overheated and therefore nonequilibrium ferrite-austenite structure. Note an anomalous dependence of the point A_{c3} on the heating rate. Increasing the heating rate of the steel allows completion of austenitization at lower temperatures.

D.   Grain Size Refinement

Tchernov [7] was the first in 1868 to show that it was possible to refine a coarse-grained structure. Since that time the procedure has been widely used for treatment of steel products. The grain refinement, which takes place upon heating steels above the temperature A_{c3}, is related to a transition to the austenite state through nucleation of numerous centers of the austenite phase. Development of these centers leads to formation of a relatively fine-grained structure. Above A_{c3} the cross-sectional size of the grain is 10–30 μm. Initially the grain size is independent of the grain of the starting structure, it can be very fine irrespective of whether the starting structure of the steel was fine or coarse. A fine-grained structure of the restored austenite provides a fine-grained structure of cooled steel whatever structural components are formed—pearlite, bainite, or martensite. This is due to the fact that all the transformation products nucleate within each separate grain of austenite.

Excess phases (ferrite in hypoeutectoid steels and cementite in hypereutectoid steels) precipitate at boundaries of small austenite grains, and the pearlite transformation is accompanied by the appearance of smaller pearlite colonies. Fine austenite grains determine the formation of fine-needle martensite. This underlies the grain refinement effect that is associated with heating above A_{c3}. Heating the steel above A_{c3} during full annealing, normalization, or quenching is followed by recrystallization. Given an initially coarse-grained structure, recrystallization results in refinement of grains at a heating temperature of about A_{c3}. If the heating temperature is much higher than A_{c3}, the grain is enlarged again, and the expected correction of the structure during the $\gamma \rightarrow \alpha$ transformation does not take place. Refinement of crystallites is especially pronounced when transformation to the austenite state starts in many centers inside the initial structure. The formed centers should have a random orientation, which is not connected with the orientation of the α phase in the initial structure. Normally such centers are sufficiently great in number that the grain size does not exceed 15–30 μm.

Breaking of an austenite grain into pearlite colonies, each of which can be considered an independent grain, also represents refinement of steel during pearlite precipitation of austenite.

VI. STRENGTHENING MECHANISM IN STEEL

A. Solid Solution Strengthening

Solid solution strengthening is a phenomenon that occurs when the number of impurity atoms in the lattice of the basic element is so small that they are incapable of forming both stable and metastable precipitation phases under any thermal treatment conditions. Nevertheless the impurity atoms favor improvement of mechanical properties. This can be accounted for by the following. The presence of impurity atoms in the matrix lattice leads to distortion of the lattice because of the difference in size between the atomic radii of the impurity and the basic component. This in turn leads to the appearance of elastic deformation fields, which retard movement of dislocations in slip planes under the action of applied stresses. In addition, the impurity atoms can inhibit movement of dislocations by forming impurity atmospheres around them.

Both of the above factors play a leading role in solid solution strengthening.

Consider the influence of carbon, which is statistically uniformly distributed in the lattice of the α-iron, on the structure and properties of α-iron. Solubility of carbon in α-iron is much lower than in the γ-iron. It forms interstitial solid solutions with both irons. However, whereas the γ-iron lattice has sufficiently large pores for implantation of carbon atoms, the cubic lattice of the α-iron suffers, upon introduction of carbon atoms, a tetragonal distortion similar to the one of the martensite lattice, except that in the former case the distortion is much smaller. In addition, implantation of carbon atoms causes the entire lattice of the α-iron to expand somewhat. For example, at a carbon content of 0.015% the lattice constant increases at room temperature by $0.025c$.

As follows from the foregoing, carbon should affect properties of the α-phase. Indeed, a dependence of the yield stress on the carbon concentration in the solid α solution was detected. The yield stress rises most dramatically with an increase in the carbon concentration from 10^{-7} to 10^{-4}–10^{-3}%. The influence that carbon exerts on plastic deformation resistance of the α phase is due to both its strong interaction with dislocations and pinning of the dislocations and elastic deformations arising as a result of the tetragonal distortion of the α-phase lattice after implantation of carbon atoms.

What is more, the presence of carbon in lattices of different structural components formed during thermal treatment of steel also leads to changes in their mechanical characteristics. For example, the location of implanted carbon atoms predominantly in one of the sublattices of interstitial sites during the martensite formation brings about additional tetragonal distortions of the martensite crystal lattice. This enhances plastic deformation resistance owing to the interaction between the stress fields around carbon atoms and those at dislocations.

The flow stress grows linearly or in proportion to the square root of the percent carbon with an increase in the carbon content. This is accompanied by impairment of the plastic characteristics of the steel and lowering of the fracture stress. For example, if the carbon content is raised from 0.25% to 0.4% in a steel with 5% Cr, after quenching and low tempering the tensile strength increases from 1600 MPa to 2000 MPa, the fracture stress of samples with a purpose produced crack decreases from 1300 MPa to 1000 MPa, and the impact strength drops from 0.3 to 0.04 $J \cdot m^2$.

The influence of carbon dissolved in the α phase on the mechanical properties of steel is also observed in the case of the ferrite–pearlite transformation. The factors responsible for this phenomenon were analyzed above. Both in the homogeneous α phase and the ferrite–pearlite mixture, the yield stress rises most sharply when the carbon concentration of ferrite is raised from 10^{-7} to 10^{-4}–10^{-3}%.

A direct examination of the crystal structure of the α phase formed over the temperature interval of 250–300°C (482–572°F) during the intermediate (bainite) transformation also revealed a tetragonal structure with the c/a axis ratio equal to 1.006 and 1.008 at carbon contents of 1% and 1.2%, respectively. This attests to dissolution of part of the carbon in the α phase and suggests that the solid solution strengthening of the phase is one of the factors providing the high strength properties of intermediate transformation products.

B. Grain Size Refinement

In Section V.D the possibility of refining steel grains by means of phase recrystallization under heating to a temperature above A_{c3} was considered. Although austenite passes to other phases during cooling, its grain size represents an important characteristic of steel. This is due to the fact that all structural components are formed within each separate crystal. The smaller the austenite grains, the finer the network of excess ferrite at their boundaries and the smaller the pearlite colonies and martensite crystals. Therefore a fine grain corresponds to a fine-crystal fracture of steel and vice versa at the temperatures where austenite has already precipitated. Impact strength is especially sensitive to the austenite grain size, it decreases with grain enlargement. A decrease in the dimensions of pearlite colonies inside the initial austenite grain favors a rise in impact strength also.

Although the grain size has a considerable effect on impact strength, its influence is small if any on the statistical characteristics of mechanical properties such as hardness, fracture stress, yield stress, and specific elongation. Only the actual grain size affects steel properties, the inherited size has no effect. However, the technological process of heat treatment is determined by the inherited grain. For example, a hereditarily fine-grained steel may be deformed at a higher temperature with the assurance that the coarse-grained structure will not occur.

C. Dispersion Strengthening

In the majority of metal alloys, precipitation of supersaturated solid solutions formed during quenching is followed by precipitation of disperse particles enriched in atoms of the alloying components. It was found that the strength (hardness) of the alloys increases with the precipitation of these particles. The increment in the value of these characteristics increases as the dispersion and volume fraction of the particles increase. This phenomenon has been referred to as "dispersion strengthening."

Precipitation of supersaturated solid solutions occurs during heating (aging) of quenched alloys. The study of precipitation processes is ultimately aimed at elaboration of the most efficient methods for strengthening aging materials. In a general case, strengthening results from an increase in resistance to the movement of dislocations in a crystal when obstacles (barriers) of any type are formed. In aging alloys, dislocations meet regions [ordered atomic clusters (ГП zones) or different-structure precipitate particles] that retard their movement. The character of interaction between moving dislocations and precipitates of the second phase can be different depending on the phase morphology and structure.

The total effect of aging on the strength properties of alloys is determined by (1) the strength of the precipitates formed, (2) the volume fraction of precipitates, (3) the degree of precipitate dispersion, (4) morphology, structure, and type of binding with the matrix, and (4) test temperature.

When a solid solution of carbon in α-iron is cooled below point A_1, carbon should precipitate as cementite with lowering of the carbon solubility and a decrease in temperature. This process is realized under sufficiently slow cooling, which is accompanied by diffusion processes leading to the formation of cementite.

In the case of abrupt cooling, e.g., water quenching, carbon has no time to precipitate. A supersaturated α solid solution appears. At room temperature the retained amount of carbon can correspond to its maximum solubility of 0.018%. During subsequent storage at room temperature (natural aging) carbon tends to precipitate from the solid solution. Carbon-enriched regions appear predominantly in defective sections of the matrix. Precipitation of carbon from a supersaturated solid solution during natural aging results in improvement of its strength characteristics and hardness. However, plastic characteristics—reduction of area, specific elongation, and impact strength—are impaired (see Figure 23). A clearly pronounced yield stress appears after a long natural aging. Hardness may increase by 50% over that of the as-quenched state. The phenomenon of dispersion strengthening is observed.

As the heating temperature is increased (artificial aging), dispersion strengthening accelerates. At 50°C (122°F) the precipitation rate of carbon from the α-iron increases to such an extent that in several hours of aging it reaches the value obtained after several days of natural aging (Figure 24). This is due to intensification of diffusion processes with an increase in tem-

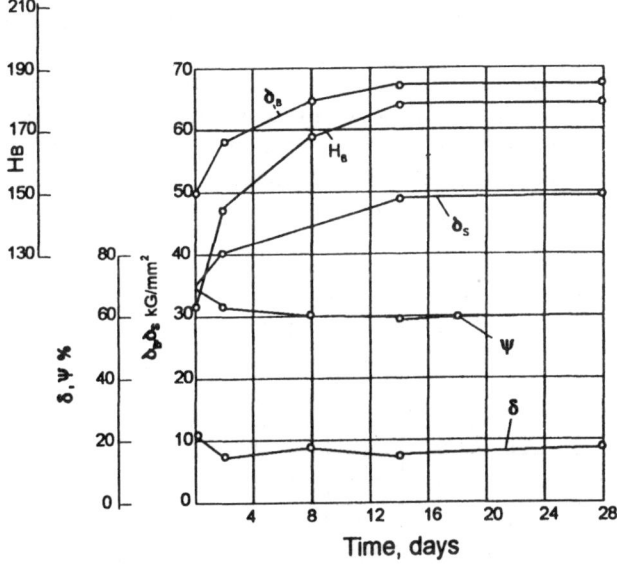

Figure 23   Curve showing strengthening of fcc crystals.

perature. As the temperature is elevated further, precipitation of the supersaturated α solid solution proceeds still faster.

The total process of carbon precipitation from the supersaturated solid solution in α-iron comprises several consecutive processes. Mechanical characteristics and hardness are most sensitive to structural changes that take place during the aging of alloys. Sharp changes in properties indicate alterations in the structural state of the steel.

A maximum change in mechanical properties during precipitation is achieved only if excess crystals in a highly-disperse state precipitate. Subsequent coagulation of the crystals leads to degradation of the properties (Figures 24 and 25).

As the temperature is raised above 100°C (212°F), carbides begin to homogeneously precipitate directly from the solid solution. The precipitating phase has the carbide lattice below 200°C and the cementite lattice above 300°C (572°F). A transition from one phase to the other is realized over the temperature interval of 200–300°C (392–572°F). The onset of transition from atomic clusters near dislocations to precipitation of the ε-carbide remains to be ascertained. The temperature and time during which the ε-carbide crystals precipitate from the inhomogeneous solid solution depend on the degree of the solution supersaturation and concentration of vacancies.

Coagulation of the ε-carbide crystals lowers the increment in hardness, fracture stress, and yield stress as the effect of braking the slip dislocations diminishes. Above 200°C (392°F), where precipitation of the particles is detected even by metallographic methods, hardness stops increasing.

If a naturally aged sample of steel is heated at a temperature of 100–200°C (212–392°F), a decrease in hardness can be observed. This is due to the phenomenon of recovery where the phase nuclei that were formed at room temperature dissolve on heating to higher temperatures.

The influence of different solubility of carbon in α-iron on the properties of the alloy (dispersion strengthening) during low-temperature aging is pronounced in steels with a very low content of carbon. In steels containing over 0.4% C, the effects considered above are obscured by the influence of cementite particles formed during the pearlite transformation. Besides, nucleation of the precipitating phase can be inhibited owing to migration of carbon to the cementite/ferrite interfaces. As a result, the amount of carbon concentrated at lattice defects decreases.

Cold plastic deformation greatly accelerates precipitation of a supersaturated solid solution. This is due both to an increase in the density of dislocations, which are preferable sites of heterogeneous nucleation of precipitates, and to an increase in the concentration of vacancies, which facilitates the diffusion of carbon to clusters. The phenomenon has also been ob-

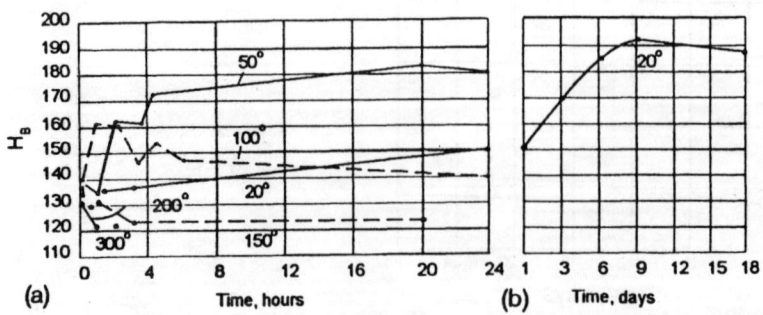

Figure 24   The aging temperature dependence of hardness of carbon steel.

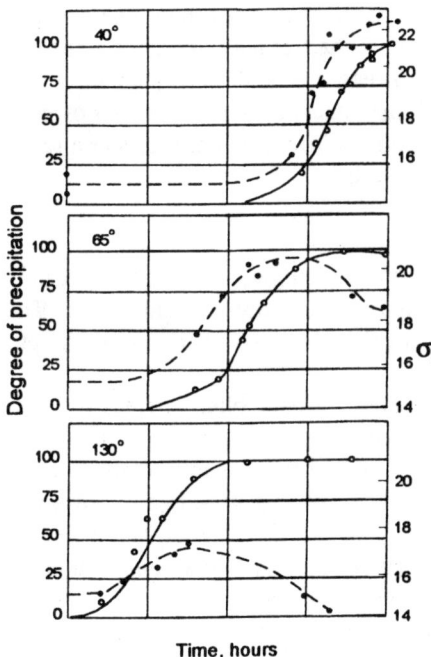

Time, hours

Figure 25 Variation of the yield stress as a result of carbon precipitation from the α-iron lattice at different temperatures.

served in other aging alloys. Mechanical properties change during aging after cold deformation in the same way as after quenching. That is, the yield stress, the fracture stress, and hardness are altered. With an increase in aging time, specific elongation and reduction of area decrease and the tendency to brittle fracture is enhanced. The rate of change is greater than in a quenched alloy. What is more, the character of the changes is different. Whereas in the case of aging after quenching, hardness reaches a maximum and then drops, after cold deformation hardness does not decrease with the aging time (Figure 26). As the aging temperature is raised, the maximum hardness of a quenched alloy lowers, while after cold deformation hardness is independent of the aging temperature. This is explained by the fact that a considerable amount of carbon is concentrated near dislocations. Few if any clusters nucleate in the matrix homogeneously. Consequently, clusters cannot grow at the expense of other clusters, i.e., they cannot coagulate.

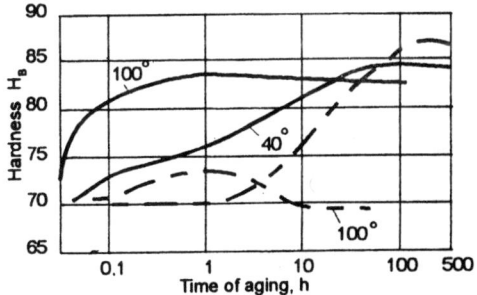

Time of aging, h

Figure 26 Aging after quenching from 720°C (1328°F) (----) and after 10% cold deformation (——) of cast steel.

Since the solubility of carbon in γ-iron is also susceptible to changes, here one can also expect the effect of dispersion strengthening. However, the γ phase is not fixed during quenching but undergoes the martensite transformation. For this reason an additional amount of carbon transferred to the solid solution at the line ES will just enhance the precipitation of martensite and retained austenite during tempering of steel. Still, an increase in hardness as a result of carbide precipitation is observed in purely austenitic steels.

D. Work Hardening (Dislocation Strengthening)

An important method used to strengthen steels is deformation strengthening. Strengthening achieved with crystal deformation can be judged from the shape of stress–strain curves. The actual shape of these curves largely depends on the crystal lattice type of the metal, its purity, and thermal treatment.

In the case of cubic lattice metals, strengthening curves are parabolic, whereas for hexagonal lattice metals a nearly linear dependence is observed between the stress and the strain. This fact suggests that plastic deformation strengthening is determined mainly by the interaction of dislocations and is associated with the structural changes that impede the movement of dislocations. Metals with a hexagonal lattice are less prone to deformation strengthening than cubic lattice metals because the hexagonal lattice has fewer easy slip systems. In cubic lattice metals the slip proceeds in several intersecting planes and directions.

Examinations of fcc crystals showed that their strengthening curve is complicated (Figure 27). Three stages can be distinguished in this curve.

Stage I is characterized by easy slip. It depends on the orientation of the crystal relative to external forces and on the presence of impurities. This stage is characterized by a linear dependence of strain stresses on the strain at a small work-hardening rate. Dislocations slip in primary systems.

The work-hardening rate is much greater at stage II than at stage I. Dislocations move in intersecting slip planes and, on colliding, form additional obstacles to their movement. This state is most extensive in the stress–strain curve. The ratio between the work-hardening rate and the shear modulus (or any other elastic constant) is almost independent of the applied stress and temperature. It depends little on the crystal orientation and presence of impurities. For most fcc metals the ratio between the work-hardening rate and the shear modulus is about 4×10^3.

A cellular structure is formed at stage II. Cells 1–3 mm in diameter are practically free of dislocations (Figure 28). Groups of like dislocations represent subboundaries of the cells. During their movement, dislocations overcome stress fields of different groups.

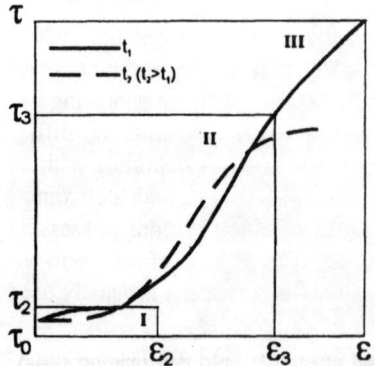

Figure 27 Curve showing strength of fcc crystals.

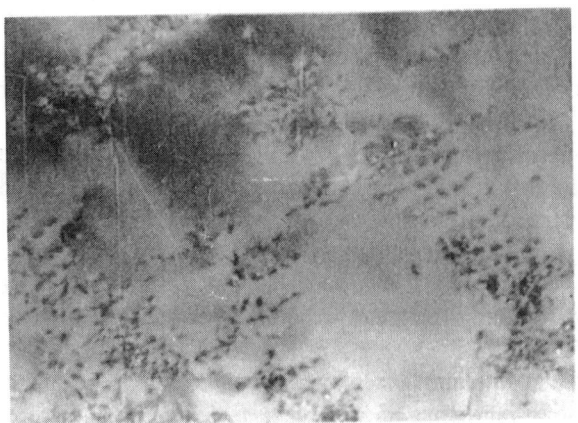

Figure 28   Electron microscopic image of a cellular structure, 50,000×.

The formation of obstacles that inhibit propagation of the shear in slip planes and cause a high degree of strengthening at stage II leads to a nonuniform distribution of the strain over the crystal volume.

At stage III, changes are possible in the distribution of dislocations. They can either get around obstacles that retard their movement at stage II or interact with dislocations. As a result, the work-hardening rate lowers compared to that observed at stage II. At this stage a partial relaxation of stresses may occur owing to the appearance of the secondary slip other system. The diminishment of distortion may have the result that deformation continues in the primary system, which "gets rid" of a certain number of dislocations passing to the system. A characteristic feature of deformation at stage III is the development of a cross-slip representing the main mechanism by which dislocations bypass the obstacles formed at stage II.

E.   Thermal Treatment of Steels

There are three basic types of thermal treatment of steels: annealing, quenching, and tempering.

1.   Annealing

Annealing has the following forms: (a) diffusion annealing; (b) softening; (c) phase-recrystallization annealing or full annealing (normalization, high-temperature or coarse-grain annealing, pearlitization); (d) stress relief annealing and recrystallization annealing.

Diffusion Annealing   The goal of diffusion annealing is to eliminate, insofar as possible, inhomogeneities of the chemical composition, in particular liquation inhomogeneities, which appear during crystallization of alloys. This annealing is usually carried out in the range of the γ solid solution at a temperature of 1100–1300°C (2012–2372°F). Diffusion annealing can be used primarily to smooth out a difference in the content of alloying elements, the difference being due to the intercrystal liquation. This shows up as smearing of dendrites with an increase in temperature and heating time. Differences in microhardness are eliminated simultaneously. The overall hardness of the alloy decreases because liquation regions possessing high hardness are removed. Some average hardness is obtained. The success of diffusion annealing largely depends on the steel purity and liquation. This type of annealing is usually used to improve properties of medium-purity steels.

Softening   Softening is used to produce the structure of globular pearlite. This structure is very soft and readily lends itself to deformation during drawing, cold rolling, etc. Steels with a low carbon content become too soft after this annealing treatment. The globular pearlite structure is favorable in steels with a carbon concentration of more than 0.5%. Another goal of softening is to produce a uniform fine structure with finely dispersed carbon after quenching.

The simplest method of softening consists in holding for many hours at a temperature slightly above A_{c1}. In this case martensite that is left from the previous treatment is removed and the work hardening caused by deformation (e.g., forging) is eliminated.

Carbide plates of pearlite fully coagulate only after a long annealing time. Since fine-plate pearlite transforms more easily to globular pearlite, it is recommended that normalization (see next subsection) be carried out prior to softening treatment. Cooling after softening can be done in air starting from 600°C (1112°F).

Refinement of the structure subjected to softening is achieved only above the point A_1. In practical applications this type of annealing represents an intermediate treatment, and therefore no strict requirements are imposed on the mechanical properties of annealed materials.

Phase Recrystallization Annealing (Normalization, High-Temperature Annealing or Coarse-Grain Annealing, Pearlitization) A twofold $\gamma \rightarrow \alpha$ transformation, which takes place during phase recrystallization annealing, leads to the appearance of a fine-grained uniform structure differing completely from the initial structure. Refinement of the grain during normalization results in the disappearance of the Widmannstatten and coarse-grained cast structures, which have poor mechanical properties. Inhomogeneity of the structure in the deformed state is eliminated.

The closer the annealing temperature is to A_{c3} and the shorter the holding time at this temperature, the finer the grain. The mechanisms of this phenomenon are analyzed in detail in Section V.C. Refinement of the grain structure is also facilitated if the heating rate to the annealing temperature and the cooling rate from this temperature are increased.

In the case of normalization, cooling is done in air. Here it is important to allow for different rates of cooling along the cross section of large-diameter products. The arising thermal stresses are removed by stress relief annealing or high-temperature tempering.

To obtain a fine-grained structure, rapid cooling is realized only over the transformation temperature interval.

The normalization heating temperature should not be much higher than the transformation point, since otherwise the grain may be too coarse (overheating). An excessively long holding time will have the same result.

The optimal heating temperature is determined by the carbon content. For steels with a carbon concentration of up to 0.9%, it is 20–30°C higher than A_{c3}. For eutectoid steels, heating between A_{c1} and A_{cm} suffices. In the case of low- and medium-carbon steels, the best results are obtained if ferrite and plate pearlite are formed during subsequent heat treatment.

Stress Relief Annealing and Recrystallization Annealing Dislocation pile-ups and crystal lattice distortions arising in cold-deformed metals may result in the appearance of macroscopic stresses (stresses of the first kind). Usually these stresses are very high.

Changes in properties that occur under cold deformation can be rectified during subsequent heating. The greater the degree of deformation, the lower the heating temperature.

Depending on the temperature and time of annealing, various structural changes take place in a cold-deformed material. The changes are divided into recovery and recrystallization processes.

Recovery is a totality of any spontaneous processes of variation in the density and distribution of defects before the onset of recrystallization. If recovery proceeds without the formation and migration of subgrain boundaries inside the recrystallized grains, it is called restoring. If subgrain boundaries are formed and migrate inside the crystallites, recovery is referred to as polygonization.

Restoring does not include an incubation period. Properties start changing right at the beginning of annealing. Restoring is accompanied by a redistribution of point defects whose concentration decreases subsequently from excess concentration to the equilibrium concentration. Simultaneously, dislocations are redistributed and unlike-sign dislocations are annihilated. The total density of dislocations decreases during restoring. Restoring is realized at a temperature below $0.3T_{melt}$.

The main process that takes place during polygonization is the redistribution of dislocations accompanied by formation of walls. A dislocation wall does not have long-range stress fields, and therefore the wall formation process is energetically favorable. A wall composed of like-sign dislocations represents a low-angle boundary separating neighboring subgrains with a small misorientation of the lattices. As the annealing time and temperature are increased, the subgrains tend to become coarser. They may be as large as ~ 10 μm. However, the subgrains grow within the old deformed grains.

In iron, polygonization starts at 200°C (392°F) (block boundaries appear in the structure). Clearly delineated boundaries of the blocks are retained up to 850°C (1562°F) and persist even after long holding at this temperature.

Starting from a certain annealing temperature, the structure changes drastically. New rather equilibrium grains are observed along with extended deformed grains. They differ from the grains of the deformed matrix by having a more perfect internal structure. While the density of dislocations in a strongly deformed matrix is $10^{11}-10^{12}$ cm^{-2}, after recrystallization it lowers to 10^6-10^8 cm^{-2}. As distinct from the polygonized structure, recrystallized grains are separated from the matrix with large-angle boundaries.

The formation and growth of grains with a more perfect structure that are surrounded by large-angle boundaries, at the expense of initially deformed grains of the same phase, is called primary recrystallization. Recrystallization begins with an incubation period. The recrystallization rate first increases from zero to a maximum and then decreases owing to an ever rising number of new grains in contact with one another.

Inclusions of insoluble impurities (carbides, nitrides) lower the tendency to growth of recrystallized grains. This is especially important in the case of ferritic steels, which are prone to grain growth. Another phase may precipitate during recrystallization in alloys that were subjected to a strong cold deformation.

Sometimes the intensive growth of individual crystals can be observed after a strong deformation and long holding (for several days) at temperatures close to the melting point. This phenomenon is called secondary or collective recrystallization.

The carbon content of steel affects the polygonization and recrystallization kinetics. With an increase in the carbon content, polygonization slows down or shifts toward higher temperatures. Given a large initial grain size, recrystallization commences the earlier, the greater the degree of deformation. At a preset degree of deformation the recrystallization temperature is the higher, the coarser the initial grain. After recrystallization an initially coarse-grained structure gives a larger grain than a fine-grained structure does. In iron-carbon alloys coarse grains are formed until the appearance of new grains associated with a polymorphous transformation.

Under critical conditions of recrystallization the grain size decreases with an increase in the carbon content. This is due to lowering of the point A_3 and narrowing of the recrystallization temperature range. Besides, the number of γ-phase crystals formed between A_{c1} and A_{c3} increases. They impede the growth of the α-phase grains at temperatures above A_{c1}. Carbides also retard growth of the grains.

As recrystallization proceeds, strengthening lowers. A fine-grained material possesses an improved long-time strength at lower temperatures, while a coarse-grained material exhibits this property at higher temperatures. A required size can be obtained by a proper choice of the deformation and recrystallization conditions. In the case of steels where no transformations take place (pure ferritic or austenitic steels) this combination of technological operations is the only opportunity to influence the grain size.

2. Quenching (Strengthening Treatment)

"Quenching" refers to cooling from the temperature range of the solid solution at such a rate that transformations in the primary and bainite ranges are suppressed and martensite is formed. In this state, steels are characterized by the greatest hardness. A distinction is made between (a) normal quenching, which is used mainly for treatment of medium- and high-carbon steels, and (b) quenching after a thermochemical treatment (carburization, high-temperature cyaniding), which is used for low-carbon steels.

Normal Quenching   To provide a required cooling rate during quenching, various cooling media and methods are employed. Water, oil, or air can serve as the cooling medium. Many alloyed steels, which are characterized by a high stability of austenite, are subjected to step quenching. With this method of quenching the temperature drop is less than in the case of direct cooling to room temperature and consequently quenching stresses are less.

A certain amount of austenite is retained during quenching even in steels with a relatively small content of carbon. For this reason it is impossible to impart the maximum hardness to a product. Since austenite is stable at room temperature and passes to martensite at lower temperatures, steels undergo a subzero treatment. Under this treatment quenching is continued and steels with a high content of retained austenite are immersed in liquid air or quenching mixtures whose temperature is below room temperature.

For surface quenching (if it is necessary to harden only the surface layer to a preset depth), special quenching heating regimes are used: The surface of the product is fully heated, while the core is cold and remains unquenched on subsequent rapid cooling. The selection of steel for surface quenching must be governed by the sensitivity of the metal to quick heating and cooling. For this reason the carbon concentration is limited to 0.7%. Otherwise cracks are formed.

Among the main quenching defects are excessive holding and overheating. They show up as enlargement of martensite needles and coarse-grain fracture. This leads to a high brittleness of quenched products and the formation of cracks. Cracks often form at the boundaries of initial austenite grains. A low quenching temperature or too short a holding time at the given temperature causes incomplete quenching. In this case a quenched metal is insufficiently hard.

Carburization is associated with surface saturation of steel with carbon and nitrogen. These elements quickly dissolve in iron by the interstitial method and are capable of rapid diffusion to a considerable depth. Products made of low-carbon (up to 0.25%) steels are subject to carburization. Carburization is carried out at 900–950°C (1650–1750°F) and sometimes 1000–1050°C (1800–1900°F). Gas carburization is used mostly, under which steel is heated in the atmosphere generated from natural gas (containing predominantly CH_4) or from liquid hydrocarbons (kerosene, gasoline, etc.). Carburization is aimed at enrichment of the surface layer in carbon. The required strengthening of the surface layer is achieved by means of quenching, which is performed after carburization. The specific volume of the quenched carburized layer is greater than the specific volume of the core, and therefore considerable compression stresses arise in the layer. This enhances the fatigue strength of products.

Cyaniding is the saturation of the surface of products with carbon and nitrogen in a cyanide-containing salt bath. The carbon/nitrogen ratio in the diffusion layer is controlled by changing the medium's composition and the processing temperature. Advantages of cyaniding over carburization consist in a shorter process time and improved wear and corrosion resistance (owing to the presence of nitrogen in the surface layer).

3. Tempering

The main purpose of tempering is to provide a disperse structure at a preset degree of cooling. In the case of low-carbon steels, quenching serves as tempering; even if not subjected to high-temperature tempering, the steel has a high viscosity and a relatively great strength.

When certain steels are quenched in oil, a structure is formed even during transformation in the bainite range that is more disperse than the one formed after cooling in air. But the most disperse distribution of carbides and the most favorable properties are obtained after martensite tempering. The structure dispersion has the greatest effect on the yield stress. An improvement of the fracture stress and yield stress and an increase in the fracture stress/yield stress ratio may be taken as a measure of the tempering efficacy. The tempering efficiency depends on the cross-sectional area and on the content of carbon and alloying elements in the steels.

Although to achieve a through quenching the critical quenching rate has to be exceeded over the entire cross section, full tempering does not require this procedure. For example, in a quenched steel that has martensite in the surface zone and pearlite in the core, the hardness of the core sometimes may be higher than that of the surface zone after tempering. This is especially the case during a short tempering when precipitation of carbides from martensite proceeds faster than the coagulation of pearlite plates.

Tempering of hypoeutectoid steels, which do not contain free ferrite, provides a uniform improved structure. In the presence of ferrite precipitates, the fracture stress/yield stress ratio decreases and the impact strength is smaller than in the surface zone. Therefore in selecting the content of carbon and alloying elements and particular conditions of austenitization and cooling, the size of the product to be tempered must be considered. For tempering to yield adequate properties, it often suffices to suppress the formation of ferrite during continuous cooling. Only when a very high fracture stress is required is an abrupt cooling used for tempering. In this case susceptibility to full tempering can be improved by raising the quenching temperature and thus enlarging the austenitic grain size.

REFERENCES

1. N.F. Mott, *Imperfections in Nearly Perfect Crystals,* Wiley, New York, 1952, p. 173.
2. N.F. Mott, and F.R.N. Nabarro, *Proc. Phys. Soc.* 52:8 (1940).
3. R. Smoluchowski, *Physica* 15:179 (1949).
4. R. Smoluchowski, *Phase Transformations in Solids*, Wiley, New York, 1952, p. 173.
5. A.H. Geisler and J.K. Hill, *Acta Cryst.* 11:238 (1948).
6. M.E. Hargreaves, *Acta Cryst.* 4:301 (1951).
7. D.K. Tchernov, *Metals Science*, Metallurizdat, Moscow, 1950 (in Russian).

FOR FURTHER READING

Belous, M.V., V. T. Cherepin, and M.A. Vasiliev, *Transformations During Tempering of Steel*, Metallurgiya, Moscow, 1973 (in Russian).
Bernshtein, M.L., and A.G.M. Richshtadt (Eds.), *Physical Metallurgy and Thermal Treatment of Steels—Handbook*, 3rd ed., Vols. 1–3, Metallurgiya, Moscow, 1983 (in Russian).
Blanter, M.E., *Phase Transformations During Thermal Treatment of Steel*, Metallurgiya, Moscow, 1962 (in Russian).
Blanter, M.E., *Physical Metallurgy and Thermal Treatment*, Mashinostroyeniye, Moscow, 1963 (in Russian).
Delle, V.A., *Structural Alloy Steel*, Metallurgiya, Moscow, 1959 (in Russian).

Goldshtein, M.I., S.V. Grachev, and Yu.G. Veksler, *Special Steels*, Metallurgiya, Moscow, 1985 (in Russian).

Gudreman, E., *Special Steels*, Vols. 1 and 2, Metallurgiya, Moscow, 1959 (in Russian).

Gulyaev, A.P., *Physical Metallurgy*, Metallurgiya, Moscow, 1976 (in Russian).

Gulyaev, A.P., *Pure Steel*, Metallurgiya, Moscow, 1975 (in Russian).

Hardy, H.K., and T.J. Heal, *Prog. Met. Phys.* 5:143 (1954).

Kaschenko, G.A., *Fundamentals of Physical Metallurgy*, Metallurgiya, Moscow, 1964 (in Russian).

Kurdyumov, G.V., L.M. Utevski, and R.I. Entin, *Transformations in Iron and Steel*, Nauka, Moscow, 1977 (in Russian).

Meskin, V.S., *Fundamentals of Steel Alloying*, Metallurgiya, Moscow, 1964 (in Russian).

Mott, N.F., *Proc. Phys. Soc. B* 64:729 (1951).

Mott, N.F., *Phil. Mag.* 8(1):568 (1956).

Novikov, I.I., *Theory of Thermal Treatment of Metals*, Metallurgiya, Moscow, 1986 (in Russian).

Popov, A.A., *Phase Transformations in Metal Alloys*, Metallurgiya, Moscow, 1963 (in Russian).

Vinograd, M.I., and G.P. Gromova, *Inclusions in Alloy Steels and Alloys*, Metallurgiya, Moscow, 1972 (in Russian).

Zimmerman, R., and K. Gunter, *Metallurgy and Materials Science—Handbook*, Metallurgiya, Moscow, 1982 (in Russian).

2

THE EFFECTS OF ALLOYING ELEMENTS ON THE HEAT TREATMENT OF STEEL

Alexey V. Sverdlin and Arnold R. Ness
Bradley University, Peoria, Illinois

I. EFFECTS OF ALLOYING ELEMENTS ON HEAT TREATMENT PROCESSING OF IRON–CARBON ALLOYS

A steel that contains, in addition to iron and up to 2% carbon, specially introduced chemical elements not found in a usual carbon steel is called an alloy steel. Chemical elements purposely added into steel are termed alloying elements. Steels may contain various numbers of alloying elements, and accordingly they are classified as ternary steels, which have, along with Fe and C, one specially introduced alloying element; quaternary steels, which contain two additional alloying elements, and so on. Alloying elements cause a steel to have a wide variety of microstructures after heat treatment that cause a wide range of properties to be obtainable.

The following elements, arranged in descending order of their application in practice, are usually used for alloying of steel: Cr, Ni, Mn, Si, W, Mo, V, Co, Ti, Al, Cu, Nb, Zr, B, N, and Be.

The alloying elements interact with iron, carbon, and other elements in the steel, resulting in changes in the mechanical, chemical, and physical properties of the steel. Improvement of the properties of steel in accordance with its designated purpose is the main goal of alloying. The level to which the properties of steel are changed by alloying depends on the amount of alloying elements introduced and the character of their interaction with the main elements of the steel, that is, Fe and C. That is why an analysis of the influence of alloying elements on the properties of steel should begin with consideration of the relationship between particular alloying elements and Fe and C. What should be considered is the effect of alloying elements on the critical points of iron and steel and also the distribution of the alloying elements in the steel.

A. The γ-and α- Phase Regions

The position of the critical points A_3 and A_4 and the location of the eutectoid temperature A_1 are of great significance because they determine the lowest temperature to which a steel should be heated for quenching, annealing, or normalization as well as the temperatures of the maxima in the curve showing the precipitation rate of undercooled austenite. The processes that take place at the critical temperatures in steels are associated with the Fe$_\gamma$ ⇌ Fe$_\alpha$ transformations and dissociation of carbides.

Different alloying elements have different effects on the position of the critical points A_3 and A_4. The alloying elements are accordingly divided into two large groups, each in turn being broken down into two subgroups.

Addition of the elements from the first group is followed by lowering of the critical point A_3 and a simultaneous rise of the point A_4. This effect is shown schematically in Figure 1 and is most vividly pictured in Fe–Mn and Fe–Ni equilibrium diagrams. It is seen that with an increase in the content of the alloying element, the γ-phase region broadens considerably, and starting from a certain concentration the alloys are found in the state of the γ solid solution until they melt. This shift of the critical points is brought about by such elements as Ni, Co, Mn, Pt, Pd, Rh, and Ir (Ni group).

The other subgroup of the first group includes elements that in general have a limited solubility in iron. Given a certain concentration of these elements in iron alloys, chemical compounds are formed and eutectic or eutectoid transformations are observed. In other words, heterogeneous regions appear in diagrams of the iron–alloying element system. The heterogeneous regions limit the γ-phase occurrence range. This type of phase equilibrium diagram of alloys is exemplified in Figure 2. As is seen, with an increase in the concentration of the alloying element in the alloy, the critical point A_3 lowers and A_4 rises. As a result, the range of γ solid solutions widens. But then, owing to the formation of heterogeneous regions, the γ phase narrows and, finally, vanishes. Equilibrium diagrams of this type (exhibiting first a wide range of the γ phase and then a narrowing of the phase caused by the appearance of heterogeneous regions) are found for N, C, Cu, Zn, Au, Re, etc.

As distinct from the elements of the first group, elements entering the second group elevate the point A_3 and lower the point A_4 as their content in the alloy is raised. This leads initially to narrowing and then to a complete closing of the region of the γ solid solution as shown schematically in Figure 3. This shift of the critical points of alloys is induced by such elements as Cr, Mo, W, Si, Ti, Al, and Be (Cr group). These elements can be placed in the first subgroup of the second group of alloying elements. The second subgroup includes elements whose introduction causes the appearance of other phases in the equilibrium diagrams before the γ-phase range is closed. It follows from Figure 4 that in this case the narrow range

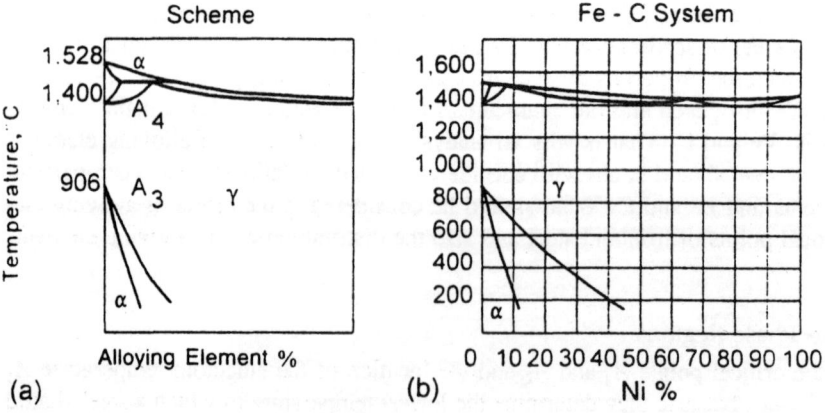

Figure 1   Scheme (a) and equilibrium diagram (b) for Fe and alloying elements with extended γ-phase range and unlimited solubility.

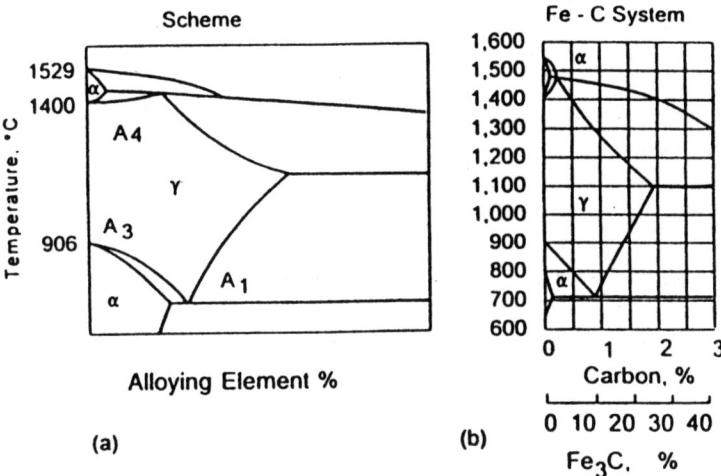

Figure 2   Scheme (a) and equilibrium diagram (b) for Fe and alloying elements with extended γ-phase range and limited solubility.

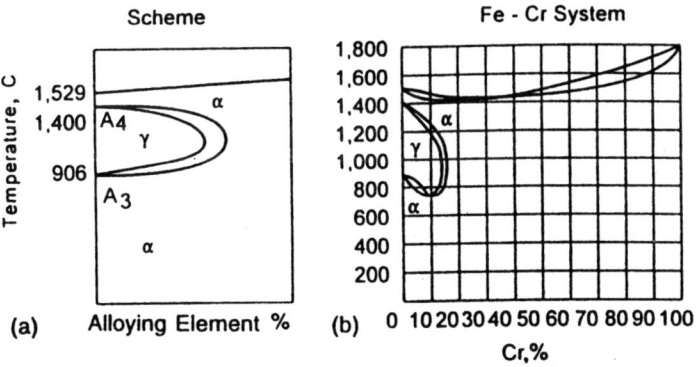

Figure 3   Scheme (a) and equilibrium diagram (b) for Fe and alloying elements with closed γ-phase range.

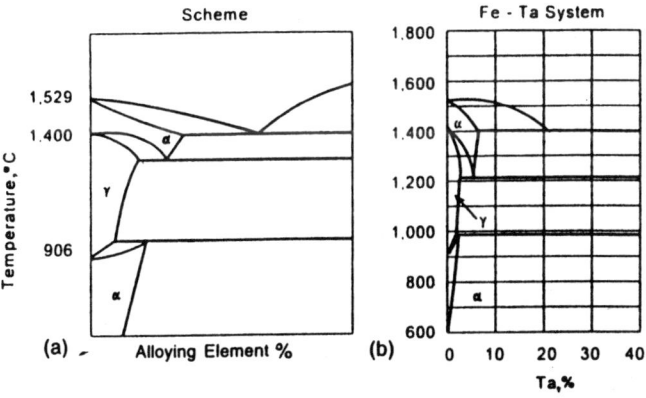

Figure 4   Scheme (a) and equilibrium diagram (b) for Fe and alloying elements with narrow γ-phase range limited by adjacent heterogeneous region.

of the γ phase is limited by adjacent heterogeneous regions. Equilibrium diagrams of this type (with a narrow range of the γ phase and its limitation by an adjacent heterogeneous region) are due to Zr, Ta, Nb, Ce, etc.

The above-described division of alloying elements into two large groups can be applied to ternary and more complex systems. The first basic ternary diagram is obtained when iron is alloyed with two elements, each leading to broadening of the γ-phase range in binary iron alloys. Such alloys can be exemplified by Fe-Co-Ni, Fe-Co-Mn, and Fe-Ni-Mn.

The second basic diagram covers iron alloys with two elements, which close the γ-phase range. An example of these alloys is the Fe-Cr-Mo system, but it includes, along with regions of solid solutions, intermetallic compounds that are formed at high concentrations.

The third basic type of equilibrium diagram applies to a ternary system where one of the elements widens the γ-phase range and the other element closes it. An example is the Fe-Cr-Ni system, which is important in technical terms. Thus even ternary systems may include purely ferritic (α-phase) and purely austenitic (γ-phase) alloys as well as alloys possessing a multiphase structure.

B. Eutectoid Composition and Temperature

The aforementioned division of alloying elements into groups according to their influence on allotropic transformations in alloys of the iron–alloying element system allows one to predict to some extent the effect of the elements on the critical points of carbon steel. For example, considering the direction of the diagram lines that correspond to the transition of Fe from one allotropic form to another, it can be expected that the elements extending the γ-phase range (Ni group) will lower the $\alpha \to \gamma$ iron transition point A_{c3}, while the elements narrowing the γ-phase range (Cr group) will elevate that point.

A similar effect of the elements is observed, to a certain extent, in the pearlite transformation A_{c1} as in this case, too, an allotropic change of iron takes place: Fe_α transforms to Fe_γ. Figure 5 illustrates the influence of the most important alloying elements on the position of the critical point A_{c1}. As is seen, the elements narrowing the γ-phase range do raise the critical point A_{c1}, while the elements broadening the γ-phase range lower it.

It should be noted that in the case of Cr group elements one observes a known relationship between the limiting concentration necessary to close the γ-phase range in iron–alloying

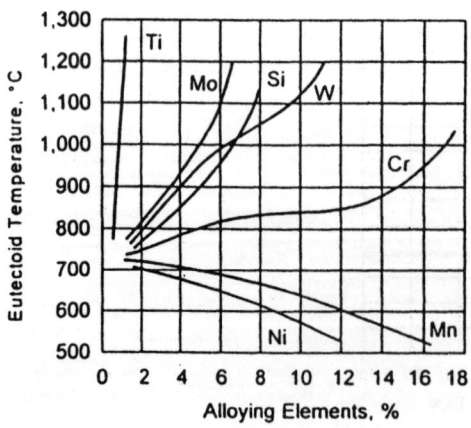

Figure 5   Effect of alloying elements on the eutectoid transformation temperature A_{c1}.

element alloys and the degree of elevation of the point A_{c1}. The lower the concentration of the element at which the γ-phase range is closed, the more abrupt the rise of the critical point A_{c1}.

If a steel simultaneously contains two or more alloying elements that influence its critical points in the same direction, the critical points usually lower or elevate to a greater extent than would be the case if only one of the elements exerted its influence. But here the result cannot be presented by a simple dependence. If a carbon steel contains alloying elements that have an opposite effect on the position of the critical points during heating, the influence of the elements shows up differently depending on their quantitative ratio. Table 1 gives values of the critical points during heating and cooling for some multialloy steels. As is seen, rather high A_{c1} and A_{c3} points are characteristic of chromium-silicon and chromium-silicon-molybdenum steels of the heat-resistant type, high-chromium steels of the stainless type, chromium-molybdenum-vanadium steels, and others. The data of Table 1 are also interesting in that they show a simultaneous effect of the most significant alloying elements on the critical points A_{r3} and A_{r1} under constant-rate cooling. In particular, a very sharp lowering of these points in multialloy steels is caused by molybdenum. Molybdenum is responsible for a drastic drop of the critical points under cooling in steels that containing chromium and nickel at the same time. This last fact is especially important for structural steel.

The effect of alloying elements manifests itself in a shift of the critical points with respect not only to temperature but also concentration. Figure 6 illustrates how the content of alloying elements in steel affects the carbon concentration at the eutectoid point. As can be seen from the figure, all the alloying elements shift the eutectoid point S to the left, i.e., toward lowering of the C concentration, and consequently decrease the carbon content of alloy pearlite. In analogy to the shift of the eutectoid point to the left, the addition of most alloying elements in steel is followed by a leftward displacement of the point E in the Fe–C equilibrium diagram, which determines the solubility limit of carbon in austenite. The point E is shifted most by Cr, Si, W, Mo, V, and Ti, which are arranged here in ascending order of their influence. All these elements narrow the γ-phase range in alloys of the iron–alloying element system.

If a carbon steel contains a certain amount of an alloying element, point E is displaced to the left to such an extent that even at a carbon concentration of several tenths of a percent the steel structure may have ledeburite, which is present in pure iron-carbon steels only when the carbon content is over 1.7%.

It is of interest to note that the more strongly an element shifts the points E and S, the lower the element concentration at which it closes the γ-phase range in the iron–alloying element diagram. Therefore a leftward shift of the points E and S can be considered as the tendency of a specific alloying element to narrow the γ-phase (austenite) range.

Therefore the introduction of alloying elements into a carbon steel is accompanied by a shift of the equilibrium critical points with respect to both temperature and carbon concentration. The shift is greater, the larger the amount of the elements introduced.

C. Distribution of Alloying Elements

In commercial alloy steels, which are multicomponent systems, alloying elements can be found (1) in the free state; (2) as intermetallic compounds with iron or with each other; (3) as oxides, sulfides, and other nonmetal inclusions; (4) in the carbide phase as a solution in cementite or in the form of independent compounds with carbon (special carbides); or (5) as a solution in iron.

Table 1   Position of Critical Points During Heating and Cooling of Some Multialloy Steels

No.	Chemical composition[a] (%)								Critical points[b] (°C)			
									Heating		Cooling	
	C	Mn	Si	Cr	Ni	Mo	V	W	A_{c1}	A_{c3}	A_{r3}	A_{r1}
1	0.20	—	2.25	1.50	—	—	—	—	820	860	800	715
2	1.10	—	1.55	4.83	—	0.51	—	—	845	880	865	760
3	0.06	—	—	5.33	—	0.57	—	—	830	880	820	750
4	0.36	—	—	0.96	—	0.20	—	—	770	810	745	680
5	0.45	—	—	1.00	—	0.35	0.26	—	775	790	700	650
6	0.30	—	—	4.82	—	2.20	0.42	—	885	920	745	710
7	0.34	—	—	4.72	—	1.65	—	0.93	780	820	310	225
8	0.25	1.26	—	—	—	0.41	—	—	745	840	750	690
9	0.22	1.15	—	—	—	1.18	—	—	750	840	540	430
10	0.35	—	—	—	—	0.39	—	0.19	710	765	495	400
11	0.36	—	—	—	2.73	0.39	—	—	715	760	515	400
12	0.29	—	—	0.64	2.87	—	—	—	715	770	600	525
13	0.60	—	—	0.75	3.44	0.22	—	—	730	765	540	630
14	0.28	—	—	0.86	1.40	0.35	—	—	740	785	490	390
15	0.15	—	—	11.27	1.97	—	—	—	810	860	770	715
16	0.12	—	—	12.19	0.13	0.58	—	—	830	870	700	—
17[c]	0.13	—	—	13.15	—	—	—	—	890	900	810	785

[a]The content of Mn and Si is standard if not specified otherwise. Residual Cr and Ni are also present in all steels.
[b]The critical points were determined by the dilatometric method; the cooling rate ~2°C/s. (~4°F/s).
[c]No. 17 also has 0.25% Al.

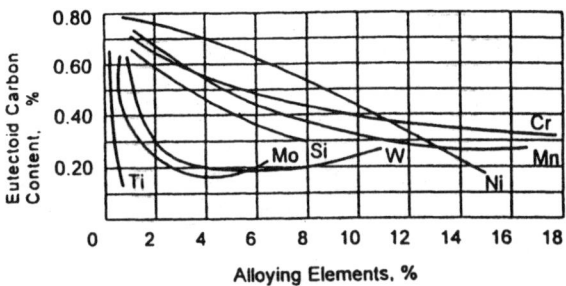

Figure 6   Effect of alloying elements on the concentration of carbon in eutectoid.

As to the character of their distribution in steel, alloying elements may be divided into two groups:

1.   Elements that do not form carbides in steel, such as Ni, Si, Co, Al, Cu, and N.
2.   Elements that form stable carbides in steel, such as Cr, Mn, Mo, W, V, Ti, Zr, and Nb.

The law determining the manner in which elements of the first group are distributed in steel is very simple. These elements do not form chemical compounds with iron and carbon, and consequently the only possible form in which they can be present in steel is in solid solutions with iron. The only exceptions are Cu and N. Copper dissolves in α-iron at normal temperatures in amounts of up to 1.0%. If the Cu content exceeds 7%, iron will contain copper in the free state as metal inclusions. Similar behavior is typical of the alloying elements that do not dissolve in solid iron at all (e.g., Pb or Ag). Nitrogen also has a limited solubility in ferrite. When the N content is higher than 0.015%, nitrogen is found in steel in the form of chemical compounds with iron or some alloying elements (V, Al, Ti, Cr). These chemical compounds are called nitrides.

Most alloying elements can form intermetallic compounds with iron and with each other. But these compounds are formed only at concentrations of the alloying elements, that are not used in the usual commercial steels. Therefore it can be assumed that the common quantity-produced steels do not have intermetallic compounds of alloying elements with iron or with each other. Intermetallic compounds are formed in high-alloy steels, a fact that is of great significance for these steels.

Alloying elements whose affinity for oxygen is greater than that of iron are capable of forming oxides and other nonmetal compounds. When added at the very end of the steel melting process, such elements (e.g., Al, Si, V, Ti) deoxidize steel by taking oxygen from iron. The deoxidizing reaction yields Al_2O_3, TiO_2, V_2O_5, and other oxides. Owing to the fact that alloying elements that are deoxidizers are introduced at the final stages of the steel melting process, the majority of oxides have no time to coagulate or to pass to slag, and as a result they are retained in the solid steel as fine nonmetal inclusions. In addition to a great affinity for oxygen, some alloying elements have a greater affinity for sulfur than iron does, and upon being introduced into steel, they form sulfides.

Compared to non-carbide-forming elements, alloying elements that form stable carbides in steel exhibit a much more complicated distribution. They can be found in the form of chemical compounds with carbon and iron or be present in the solid solution. The distribution of these elements depends on the carbon content of steel and the concurrent presence of other carbide-forming elements. If a steel contains a relatively small amount of carbon and a great quantity of an alloying element, then, obviously, carbon will be bound to carbides before the carbide-forming elements are completely used. For this reason excess carbide-forming elements will be found in the solid solution. If a steel has a large amount of carbon and little of the alloying elements, the latter will be present in steel mainly as carbides. Carbide formation is treated in detail in the next section.

Note in conclusion that most alloying elements, except C, N, O, B, and metalloids standing far from iron in the periodic table, dissolve in great amounts in iron. The elements standing to the left of iron in the periodic table are distributed between iron (base) and carbides; those to the right of iron (Co, Ni, Cu, etc.) form solutions with iron only and do not enter into carbides. Thus one can state that alloying elements dissolve predominantly into basic phases (ferrite, austenite, cementite) of iron-carbon alloys or form special carbides.

D. Alloy Carbides

Carbides are formed in steels only by iron and metals that stand to the left of iron in the periodic table: Mn, Cr, W, V, Zr, Nb, Ti. Here the elements are arranged in accordance with their affinity for carbon. The elements at the left end of the row form relatively unstable carbides that dissociate readily on heating. In contrast, the elements at the end of the row form extremely stable carbides that dissociate at temperatures much higher than the critical points of steel.

Similar to iron, the above-mentioned carbide-forming elements refer to the elements of transition groups but possess a less perfect d electron band. The further left a carbide-forming element stands in the periodic table, the less perfect is its d band.

There is reason to believe that during carbide formation carbon donates its valence electrons to fill the d electron band of the metal atom, while valence electrons of the metal form a metal bond, which determines the metallic properties of carbides. At the same time numerous experiments show that the more to the left an element stands in the periodic table, the less perfect is its d electron band and the more stable is the carbide. Proceeding from these facts, it is possible to formulate general principles of carbide formation in steels: Only metals whose d electron band is filled less than that of iron are carbide-forming elements. Their activity as carbide-forming elements is the greater and the stability of the carbide phases being formed is the higher, the less perfect is the d band of the metal atom. This principle allows specifying conditions of carbide formation in steels in the presence of several carbide-forming elements, the sequence of dissolution of various carbides in austenite, and other factors that are important for the theory of alloying, manufacturing practice, and application of alloy steels. The formation activity and stability of carbides in alloy steels will increase in going from Mn and Cr to Mo, W, V, Ti, and other elements whose d bands are less perfect than those of Mn and Cr. This means that if a steel contains, e.g., both chromium and vanadium, one should expect vanadium carbides to form first (under equilibrium conditions).

If the atomic radius of carbon is taken equal to 0.079 nm, it is easy to calculate that for all carbide-forming elements except Fe, Mn, and Cr, the ratio of atomic radii of carbon and metal is less than 0.59. It is known that if the ratio of atomic radii of a transition group metal and a metalloid with a small atomic radius (C, N, H) is less than 0.59, special types of compounds called "interstitial phases" can be formed.

The carbon/metal ratio of most carbide-forming alloying elements is lower than 0.59, and therefore the elements and carbon are capable of forming interstitial phases. It was found that the following carbide compounds may be formed in steels:

Carbides of group I	Carbides of group II
Fe_3C	Mo_2C
Mn_3C	W_2C, WC
$Cr_{23}C_6$, Cr_7C_3	VC
Fe_3Mo_3C	TiC
Fe_3W_3C	NbC
	TaC, Ta_2C
	ZrC

However, the above carbides are not found in steels in pure form. Carbides of all alloying elements contain iron in solution, and if other alloying elements are present, they include these elements too. Thus, in a chromium-manganese steel the carbide $(Cr, Mn, Fe)_{23}C_6$, which contains iron and manganese in the solution, is formed instead of the pure chromium carbide $Cr_{23}C_6$.

Owing to the fact that carbides with the same chemical formula mutually dissolve, in the presence of titanium and niobium, for example, rather than two kinds of carbides forming, a single carbide will be formed that includes titanium and niobium on "equal terms." For this reason possible variants of carbide formation are fewer, and actually we have only six kinds of carbides in steels:

Carbides of group I	Carbides of group II
M_3C	MC
$M_{23}C_6$	M_2C
M_7C_3	
M_6C	

where M denotes a sum of carbide-forming (metal) elements.

The carbides placed in group I possess a complicated crystal structure; an example is cementite. A specific structural feature of the carbides of group II as interstitial phases is a simple crystal lattice. They usually crystallize with a great carbon deficiency.

It is worth noting that interstitial phases dissolve poorly in austenite and may not pass into solid solution even at high temperatures. This distinguishes interstitial phases from the carbides of group I, which readily dissolve in austenite on heating. All carbide phases have a high melting temperature and high hardness. Interstitial phases surpass the carbides of group I in this respect.

II. EFFECTS OF ALLOYING ELEMENTS ON AUSTENITE TRANSFORMATIONS

The overwhelming majority of alloy steels are used after heating to the austenite stage, quenching, and subsequent annealing. During quenching and annealing, austenite transforms, with three types of transformations being possible: pearlite transformation (often called diffusive transformation, precipitation to the ferrite–carbide mixture, or stage I transformation), intermediate transformation (bainite or stage II transformation), and martensite transformation (stage III transformation). The precipitation stability of undercooled austenite is characterized by the diagrams of isothermal and thermokinetic austenite transformation.

The isothermal diagrams characterize the precipitation kinetics of austenite at constant temperature of undercooling. Such diagrams are useful for comparative evaluation of different steels and also for clarifying the influence of alloying and other factors (heating temperature, grain size, plastic deformation, and so on) on the precipitation kinetics of undercooled austenite.

Thermokinetic diagrams characterize the precipitation kinetics of austenite under continuous cooling. These diagrams are less illustrative but have great practical importance, because, when subjected to thermal treatment, austenite precipitates under continuous temperature variation rather than under isothermal conditions. Under continuous cooling, transformations occur at a lower temperature and take a longer time than in the isothermal case. Alloying elements have considerable influence on the kinetics and mechanism of all three types of transformation of undercooled austenite: pearlite, bainite, and martensite transformations. However, these elements influence austenite precipitation in different ways.

Alloying elements that dissolve only in ferrite and cementite without the formation of special carbides exert just a quantitative effect on the transformation processes (Figure 7). Cobalt speeds up a transformation but the majority of elements, including Ni, Si, Cu, Al, etc., slow it down

Carbide-forming elements produce both quantitative and qualitative changes in the kinetics of isothermal transformations. Thus, the alloying elements forming group I carbides influence the austenite precipitation rate differently at different temperatures:

At 1300–930°F (700–500°C) (pearlite formation), they slow the transformation.

At 930–750°F (500–400°C), they dramatically slow the transformation.

At 750–570°F (400–300°C) (bainite formation), they speed up the transformation.

Thus, steels alloyed with carbide-forming elements (Cr, Mo, Mn, W, V, etc.) have two maxima of the austenite isothermal precipitation rate separated by a region of relative stability of undercooled austenite (Figure 7). The isothermal precipitation of austenite has two clearly defined intervals of transformation, (1) to a lamellar structure (pearlite transformation) and (2) to a needle-like structure (bainite transformation). At temperatures lower than those indicated

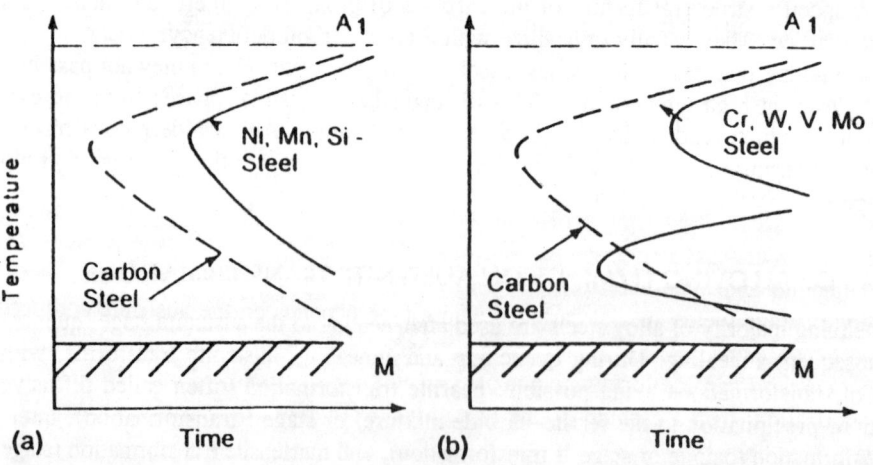

Figure 7 Diagrams of isothermal austenite precipitation. (a) Carbon steel and steel alloyed with non-carbide-forming elements; (b) carbon steel and steel alloyed with carbide-forming elements.

above and given greater degrees of undercooling as in the $\gamma \to \alpha$ transformation start temperature, the martensite transformation develops in alloy steels. As a result, a supersaturated α-iron-based solid solution is formed.

The remainder of this section considers in more detail the influence of alloying elements on the mechanism and kinetics of austenite precipitation for all three types of transformations.

A. The Influence of Alloying on Ferrite and Pearlite Interaction

The most important practical feature of alloying elements is their capacity to decrease the austenite precipitation rate in the region of the pearlite transformation, which shows up as a rightward shift of the line in the isothermal austenite precipitation diagram. This favors a deeper hardening and undercooling of austenite up to the range of martensite transformation under slow cooling such as air cooling.

In alloy steels the pearlite transformation consists of a polymorphous $\gamma \to \alpha$ transformation and diffusion redistribution of carbon and alloying elements. As a result, special carbides and a ferrite–cement mixture (pearlite) are formed. Particular alloying elements and their amounts in the initial γ solid solution determine the rates of the individual steps of pearlite transformation and consequently its kinetics as a whole.

The polymorphous $\gamma \to \alpha$ transformation in iron under small undercooling of austenite (near the temperature of stage I) proceeds by means of disordered displacement of atoms, as distinct from the martensite transformation (under greater undercooling), which proceeds through ordered shear. As mentioned above, all alloying elements dissolved in austenite, except Co, slow the pearlite transformation and shift the top section of the isothermal austenite precipitation curve to the right.

The nature of the increase in stability of undercooled austenite under the influence of alloying elements is rather complicated. Whereas in carbon steels the pearlite transformation is associated with the $\gamma \to \alpha$ rearrangement of the lattice and diffusion redistribution of carbon, in alloy steels these processes can be supplemented with the formation of special carbides and diffusion redistribution of alloying elements dissolved differently in ferrite and carbide.

Not only do austenite-dissolving elements have small diffusion coefficients of their own which are sometimes several orders of magnitude smaller than that of carbon, but some of them (e.g., Mo, W) slow the diffusion of carbon in the γ lattice. Besides, some of the elements (e.g., Cr, Ni) retard the $\gamma \to \alpha$ rearrangement, which is part of the pearlite transformation.

Depending on the steel composition and degree of undercooling, the decisive role may belong to one of the above-mentioned factors.

The formation of carbides during pearlite transformation in steel results from a redistribution of carbon and alloying elements between the phases being formed: ferrite and carbides. In the presence of dissolved strong carbide-forming elements (Nb, V, Cr, etc.) special carbides are formed in undercooled austenite before the $\gamma \to \alpha$ transformation begins, in excess ferrite (in eutectoid and hypereutectoid steels this stage is absent); and in eutectoid ferrite (pearlite). Every stage yields special carbides whose type depends on the austenite composition.

If a steel contains carbide-forming elements (V, Nb, Ti, Zr) that pass into solid solution during austenitization, then a carbide of one type, MeC (VC, NbC, TiC, ZrC), is formed at all the stages.

The scheme of austenite precipitation and the carbide formation process during pearlite transformation in steels with V, Nb, Ti, and Zr is as follows:

$$\gamma \rightarrow MeC_A + \gamma'$$

$$\alpha_{se} + \gamma_e$$

$$MeC_f + \alpha_{ee} \ Fe_3C + \alpha_{se.} \rightarrow MeC_e + a_{ee}$$

Here γ is the initial undercooled austenite; MeC_A is the carbide precipitated in austenite; γ' is austenite after carbide precipitation; α_{se} is excess ferrite supersaturated with a carbide-forming element and carbon; α_{ee} is equilibrium excess ferrite; MeC_f is the carbide precipitated in excess ferrite; γ_e is austenite of eutectoid composition; Fe_3C is the eutectoid cementite (pearlite); α_{se} is the supersaturated eutectoid ferrite (pearlite); a_{ee} is the equilibrium eutectoid ferrite; MeC_e is the carbide precipitated in the eutectoid ferrite.

The formation of carbide (MeC_A) in undercooled austenite before the $\gamma \rightarrow \alpha$ transformation starts is due to the fact that solubility of the carbide-forming element and carbon in austenite decreases with decreasing temperature. As is seen from the scheme, after the polymorphous $\gamma \rightarrow \alpha$ transformation the ferrite (both excess and eutectoid) is first supersaturated with the carbide-forming element and carbon, then a carbide is formed from ferrite, and subsequently the state of ferrite approximates the equilibrium condition. This process lasts a few seconds.

In steels containing other carbide-forming alloying elements (Cr, Mo, W), the carbide formation process is much more complicated. Depending on their content in austenite, these elements can form several types of carbides: alloy cementite $(Fe, Cr)_3C$ and special carbides $(Fe, Cr)_7C_3$ and $(Fe, Cr)_{23}C_6$ in Cr steels and carbides $(Fe, Mo)_{23}C_6$, MoC, Mo_2C, and $(Fe, Mo)_6C$ in steels with Mo (W forms analogous carbides).

Non-carbide-forming elements (Ni, Co, Si, etc.) do not participate directly in carbide formation. As a rule their amount in cementite equals their average concentration in steel. These elements can indirectly influence the thermodynamic activity of other elements, i.e., the process of their redistribution during carbide formation. As mentioned above, the process of carbide formation is limited by the mobility of the carbide-forming elements. With a decrease in temperature their diffusion mobility diminishes and special carbides are not formed below 400–500°C (750–930°F). At lower temperatures the intermediate (bainite) transformation takes place, and at higher undercooling rates martensite (diffusionless) transformation occurs.

B. Effect on Martensite Transformation

As in carbon steels, the martensite transformation in alloy steels takes place under rapid cooling from temperatures higher than the equilibrium temperature of the $\gamma \rightarrow \alpha$ transformation (A_1). At the martensite transformation temperature both the diffusion movement of metal atoms of iron and alloying elements and that of metalloid atoms of carbon and nitrogen are suppressed. For this reason the martensite transformation in steels proceeds by a diffusionless mechanism.

The martensite transformation can take place in carbon-containing alloy steels, non-carbon-containing alloy steels, and binary iron–alloying element alloys. The martensite transformation usually leads to formation of a supersaturated α-iron-based solid solution. In carbon-containing steels the solid solution is supersaturated mainly with carbon, and in non-carbon-containing alloy steels, with alloying elements. The content of carbon and alloying elements in martensite is the same as that in the initial austenite.

The transformation of austenite into martensite during cooling starts at a certain temperature called M_s. This temperature is independent of the cooling rate over a very wide range of cooling rates.

The martensite transformation kinetics of most carbon and structural and tool alloy steels is athermal in character. The athermal martensite transformation is characterized by a smooth increase in the amount of martensite as the temperature is lowered continuously in the martensite interval M_s–M_f, where M_f is the martensite finish temperature. As a rule, this transformation takes place in steels with the martensite point M_s higher than room temperature.

A version of athermal martensite transformation is explosive martensite transformation, where a certain quantity of martensite is formed instantly at or a little below the temperature M_s. This transformation is observed in alloys with the martensite point below room temperature.

The position of the martensite point also determines the microstructure and substructure of the martensitic quenched steel. At temperatures M_s below room temperature, lamellar (plate) martensite is formed in quenched iron-carbon and alloy steels. Crystals of this martensite are shaped as fine lenticular plates. In steels with the martensite point M_s higher than room temperature, lath martensite is formed during quenching. Crystals of this martensite have the form of approximately equally oriented thin plates, which are combined into more or less equiaxial packets. The substructures of needle and lath martensite are qualitatively different.

From what has been said above it might be assumed that the martensite transformation kinetics, the morphological type of martensite, the substructure of martensitic quenched steels, and other phenomena are connected to a great extent with the martensite start temperature M_s. Thus the influence of the elements on martensite transformation is determined primarily by their influence on the position of the martensite point M_s. Of practical importance is also the martensite finish temperature M_f.

Experiments concerned with the influence of alloying elements on the position of the martensite point show that Co and Al elevate the martensite start temperature, Si has little if any effect, and all the other elements decrease M_s (Figure 8).

The quantitative influence of alloying elements is approximately as follows (per 1 w % of the alloying element):

Element	Mn	Cr	Ni	V	Si	Mo	Cu	Co	Al	
Shift of point M_s	−45	−35	−26	−30	0	−25	−7	+12	+18	(°C)
	−81	−63	−47	−54	0	−45	−13	+22	+32	(°C)

These data are given for carbon steels containing 0.9–1.0% C. For a wider range of C content, the quantitative influence of the elements can be different. In particular, it was established that the smaller the C content, the weaker is the influence of the alloying elements on the position of point M_s. The martensite start temperature of medium-carbon alloy steels can be estimated using empirical formula

$$M_H(°C) = 520 - 320(\%C) - 50(\%Mn) - 30(\%Cr) - 20[\%(Ni + Mo)] - 5[\%(Cu + Si)]$$

where %C, %Mn, etc. are the contents of the corresponding elements in weight percent.

The results of calculations by this formula for steels containing 0.2–0.8% C are in good agreement with the experimental data. However, for multialloy steels this formula does not always yield reliable data because if a steel contains several alloying elements it is impossible to determine their combined effect on the martensite point by simple summation. Thus, for example, Mn lowers the point M_s to a greater extent than Ni, but in a steel with a high Cr content its effect is weaker than that of Ni.

The martensite point M_s is affected most by C dissolved in austenite (Figure 8). The transformation finish temperature M_f intensively decreases, too, as the C content is increased up to 1% [to 100°C (212°F)] and remains constant at higher amounts of carbon.

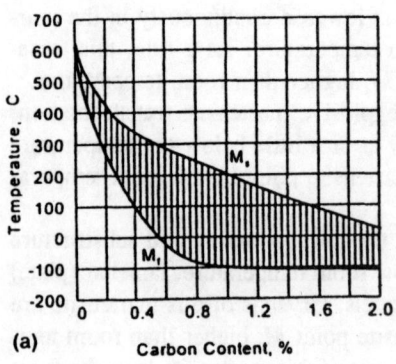

(a)

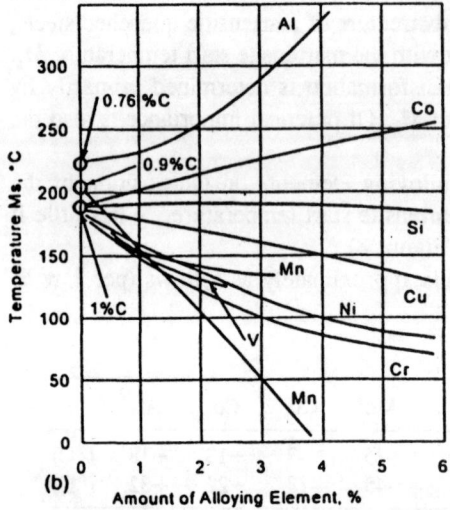

(b)

Figure 8 The influence of the content of (a) carbon and (b) alloying elements at 1% C on the marten-site point position.

The reason carbon and alloying elements influence the position of the martensite point is mainly a change in the relative thermodynamic stability of γ and α phases of iron, because the martensite transformation itself is the $\gamma \to \alpha$ transformation considering the conditions it takes place in.

C. Retained Austenite

A characteristic feature of the martensite transformation in steels, whatever its character (athermal, explosive, or fully isothermal), is that transformation of austenite to martensite is never complete. Figure 9 shows the amount of martensite formed when the temperature is decreased continuously in the martensite range M_s–M_f (martensite curve) for the athermal type of martensite transformation. The transformation starts at the point M_s, and the amount of martensite increases with decrease in temperature. The end of the transformation corresponds to the temperature M_f. At this temperature a certain amount of austenite is still left (retained

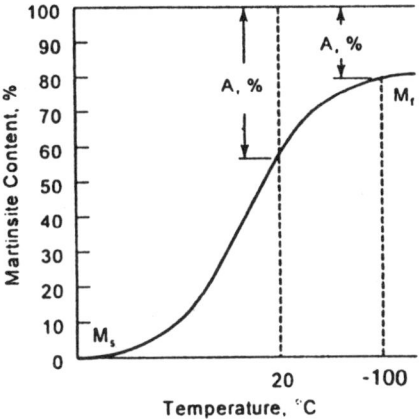

Figure 9   Martensite curve.

austenite, A, %). Cooling below M_s does not lead to further transformation or lower the amount of retained austenite.

Investigations show that martensite curves of different steels, both carbon steels and steels alloyed with different elements and in different amounts, exhibit approximately the same behavior. Then, if the martensite finish temperature is below room temperature, the amount of retained austenite should, in a general case, be higher, the lower the martensite point M_s. Strictly speaking, the amount of retained austenite depends on the martensite temperature range, i.e., on the M_s–M_f temperature difference; it increases as the range narrows. But the martensite range itself depends on the position of the martensite point M_s: the range narrows as the point M_s lowers.

Thus the influence of alloying elements on the amount of retained austenite formed during the quenching of steels should qualitatively and, to a great extent, quantitatively correspond to their influence on the position of the martensite start point M_s. Available experimental data show that alloying elements that lower and raise the martensite start point increase and decrease the amount of retained austenite, respectively (Figure 10). Besides, a certain sequence in the arrangement of the elements is observed from the point of view of their quantitative influence. In particular, the largest amount of retained austenite in accordance with their influence on the position of the martensite point is due to Mn, Cr, Ni, etc. As illustrated in Figure 11, the influence of these elements on the martensite range follows the same sequence.

In alloy steels the martensite point M_s lowers most and the martensite range narrows under the influence of carbon. Therefore the influence of carbon on the amount of retained austenite is much stronger than that of alloying elements. An increase in the C content of chromium-nickel steel from 0.4 to 0.6% increases the amount of retained austenite to ~8.5% after quenching; an increase in the Ni content of the same steel from 1 to 4% brings the amount of retained austenite to ~6% only. The fact that carbon promotes the greatest retention of austenite during quenching is especially unfavorable for low-alloy tool steels.

In multiple alloy steels a given element favors the formation of a greater amount of retained austenite than the law of summation suggests. However, in multiple alloy steels, too, the relationship between lowering of the martensite point under the influence of a given element and an increase in the amount of retained austenite caused by the same element persists in the main.

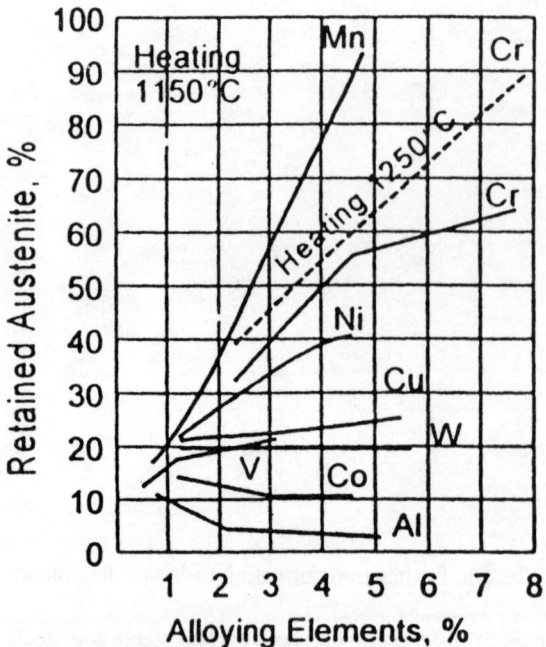

Figure 10 The influence of alloying elements on the amount of retained austenite in quenched steel (1% C).

In addition to the content of carbon and alloying elements, other factors can influence the amount of retained austenite formed during quenching of steel. The most important of these is the rate of cooling below the martensite point M_s and the quenching temperature.

The steel cooling rate has no influence on the position of the martensite point, but it affects the martensite transformation process in a certain way. A little below the point M_s, slower cooling enhances the transformation of austenite to martensite. The ability of austenite to isothermally formate martensite at temperatures a little lower than the point M_s is realized here.

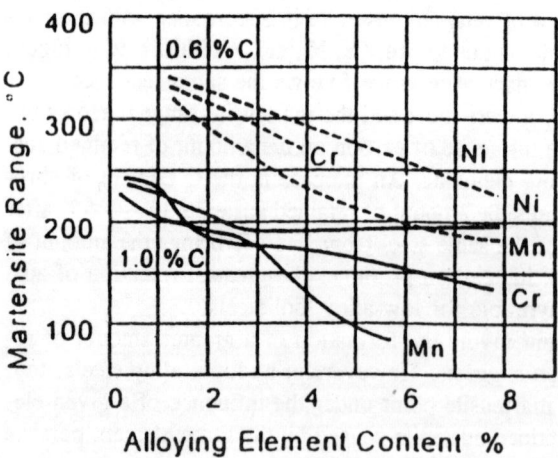

Figure 11 The influence of alloying elements on the martensite range.

At temperatures close to the martensite finish temperature M_f but within the interval M_s–M_f, when a rather significant amount of martensite has been formed already, acceleration of cooling favors a more complete transformation. Here a phenomenon called the stabilization of austenite comes into play. Holding in the region of the martensite finish temperature makes retained austenite less prone to subsequent transformation. With slow cooling the austenite stabilizing processes have time to near completion and the transformation proceeds more slowly. Austenite stabilization is associated with stress relaxation. The longer the holding time, the greater the stress relaxation and the greater the degree of the metal cooling needed to accumulate stresses required for the martensite transformation to continue.

The quenching temperature can influence largely, either directly or indirectly, the amount of retained austenite. Its direct effect can be connected with thermal stresses facilitating the transformation of austenite. An indirect effect of the quenching temperature is associated with enrichment of intercrystallite boundaries of austenite in carbon and alloying elements and, primarily, with the transfer of carbides, ferrite, and other phases to the solution. If a steel is heated to a temperature falling within the interval between the critical point A_{c1} and the temperature of full dissolution of ferrite or carbides, the heating temperature will determine the content of carbon and alloying elements in austenite. If carbides dissolve above A_{c1}, then the amount of retained austenite will increase with quenching temperature. If the quenching temperature is elevated above A_{c1} and excess ferrite dissolves (with resulting decrease in the austenite concentration), then the martensite point will occupy the lowest position when a steel is quenched from temperatures slightly higher than the point A_{c1}. Correspondingly, the amount of retained austenite must be the largest at these temperatures and must subsequently decrease until the temperature of full ferrite dissolution is reached.

D. Effect on Bainite Transformation

The bainite transformation (stage II transformation) takes place in carbon steels under the precipitation curve of undercooled austenite (C curve) in the interval of approximately 500–250°C (930–480°F). This is called the intermediate transformation: It occurs in between the pearlite and martensite transformations. The kinetics of this transformation and the structures produced are similar to those observed during the diffusion pearlite or diffusionless martensite transformation. A mixture of the α phase (ferrite) and carbide is formed as a result of the bainite transformation, the mixture being referred to as bainite.

The kinetics of the intermediate transformation are characterized by a number of peculiarities, such as an incubation period; in the bainite temperature range, precipitation of undercooled austenite begins with a certain time delay. The temperature of the maximum transformation rate (minimum incubation period) depends mainly on the chemical composition of the steel.

For alloy steels, C curves of pearlite and bainite transformations can be separated by a temperature interval of a highly stable undercooled austenite where pearlite does not precipitate for many hours, while undercooling is insufficient for the bainite transformation (Figure 12).

Alloying elements affect the kinetics of the intermediate transformation, although to a lesser degree than in the case of the pearlite transformation. In some alloy steels the isothermal transformation is retarded over the entire range of the intermediate transformation, whereas in other steels it is inhibited only at temperatures in the upper part of that range. In steels alloyed with 2% Si or Cr, the transformation of austenite stops even at the lowest temperatures of the intermediate transformation. When a steel is alloyed with Ni or Mn, the transformation is retarded only at high temperatures of the intermediate transformation, whereas at lower temperatures austenite transforms almost completely.

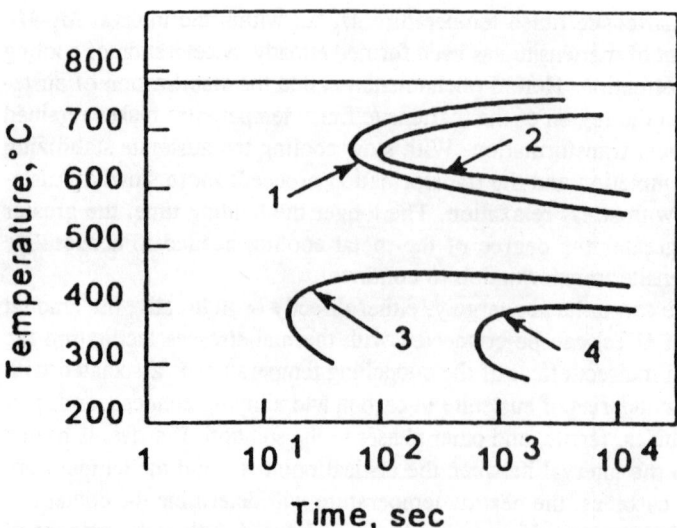

Figure 12   The austenite precipitation diagram of alloy steels with separate C curves of pearlite and bainite transformations. 1, Start of pearlite formation; 2, finish of pearlite formation; 3, start of bainite formation; 4, finish of bainite formation.

Many alloying elements produce a marked effect on the duration of the incubation period, the temperature of minimum stability of austenite, and the maximum transformation rate in the intermediate range. Figure 13 shows the influence of some alloying elements on these parameters for high-carbon steels with 1.0% C. As is seen, Mn and Cr strongly influence the kinetics of the intermediate transformation, increasing the duration of the incubation period and lowering the temperature of minimum stability of austenite and the maximum transformation rate. At the same time, alloying with Mo and W, which markedly delays the pearlite transformation, does not have a pronounced effect on the kinetics of the intermediate transformation.

The intermediate transformation in alloy steels consists of a diffusion redistribution of carbon in austenite, diffusionless $\gamma \rightarrow \alpha$ transformation, and formation of carbides, namely ε-carbide (a type of Fe carbide) and cementite. Owing to the low diffusion mobility of metallic alloying elements, which are substitutional impurities, special carbides are not formed during the intermediate transformation. The content of alloying elements in the ε-carbide and cementite of bainite is the same as in the initial austenite. Alloying elements do not undergo redistribution during the bainite transformation.

E.   Transformation Diagrams for Alloy Steels

The kinetics of austenite transformation, i.e., the form of the precipitation diagram, depends on a variety of factors, primarily on the chemical composition of austenite.

Depending on the alloying of a steel, it is possible to distinguish six basic versions of the diagram of isothermal precipitation of austenite (see Figure 14). In carbons steels and some low-alloy steels containing basically non-carbide-forming elements such as Ni, Si, and Cu, the isothermal precipitation is characterized by C-shaped curves with one maximum (Figure 14a). The pearlite and intermediate stages are not separated. When these steels are subjected to continuous cooling, three types of structures—martensite, martensite and a ferrite–carbide mixture, and only a ferrite–carbide mixture—can be formed depending on the cooling rate.

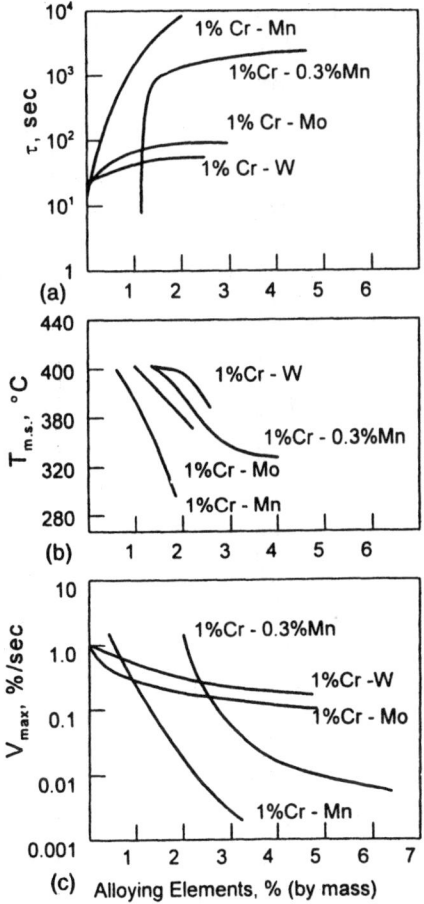

Figure 13 The influence of Cr, Mo, W, and Mn (a) on the bainite period τ at a minimum stability of austenite; (b) on the temperature of minimum stability t_{ms}, and (c) on the maximum transformation rate v_{max} in the intermediate range.

In the case of alloy steels containing carbide-forming elements such as Cr, Mo, W, and V (Figures 14b and d), the precipitation diagrams have two clearly separated ranges of pearlite and intermediate transformations. Each of the ranges is characterized by its own C-shaped curves. When the carbon content of structural steels is up to 0.4–0.5%, the stage I transformation is shifted to the right relative to the stage II transformation (Figure 14b); if the carbon content is higher, stage I is found to the left of stage II (Figure 14d).

Chromium-nickel-molybdenum and chromium-nickel-tungsten steels containing 0.15–0.25% carbon (Figure 14c) are characterized by a rather high stability of undercooled austenite in the pearlite range and a low stability of undercooled austenite in the bainite range. As a consequence, stage I is absent from the austenite precipitation diagram.

In high-alloy chromium steels the intermediate transformation may be strongly inhibited and shifted to the martensite temperature range. For this reason the austenite precipitation diagrams have only pearlite transformation and no intermediate transformation (Figure 14e).

In steels of the austenitic class (high-alloy steels), the martensite start temperature is below room temperature and stage I and II precipitation practically does not take place owing

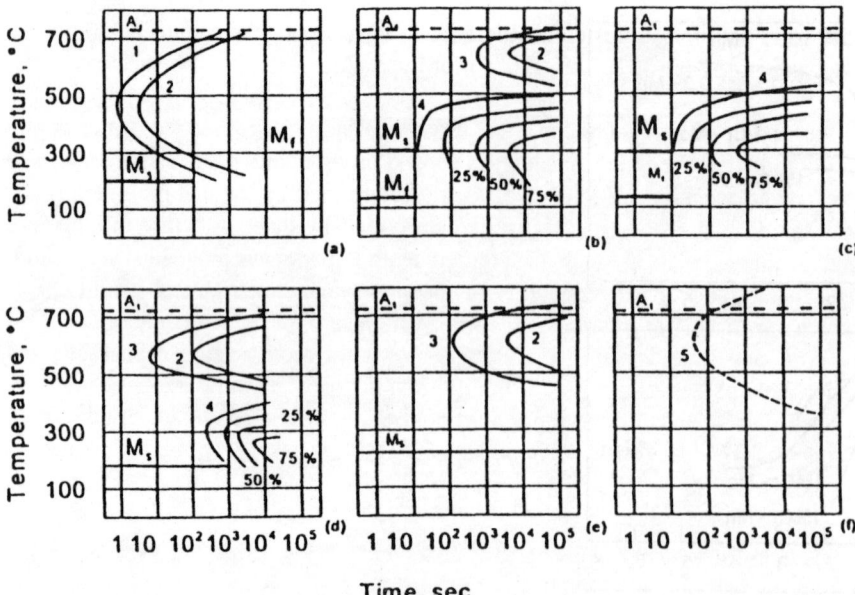

Figure 14 Basic versions of precipitation diagrams of undercooled austenite. (a) Carbon and low-alloy steels containing no carbide-forming elements; (b) alloy steels (up to 0.4–0.35% C) containing carbide-forming elements; (c) steel alloyed with Cr, Ni, Mo, and W and having a low content of carbon (up to 0.2–0.25% C); (d) alloy steels containing carbide-forming elements (over 0.4–0.5% C); (e) high-alloy steels with a high content of Cr; (f) high-alloy austenitic steels. 1, transformation start; 2, transformation finish; 3, start of formation of a ferrite–carbon mixture; 4, start of formation of the intermediate transformation products; 5, start of carbide precipitation.

to a high content of Cr, Ni, Mn, and C (Figure 14f). Thanks to the high content of carbon in the austenite of these steels, excess special carbides may be formed on undercooling.

It is worth noting that the aforementioned distinction of the diagrams is conventional to a certain measure as they do not cover a great variety of isothermal and thermokinetic precipitation diagrams of supersaturated austenite.

III. HARDENING CAPACITY AND HARDENABILITY OF ALLOY STEEL

As noted in Section II, at great rates of cooling when the cooling curves do not touch the region of isothermal transformation even at inflection points where austenite is least stable, the latter is undercooled to the martensite range (below the point M_s) and steel is fully martensitically hardened. Martensitic transformation of austenite results in a supersaturated solid solution of carbon in α-Fe; the higher the carbon content of the austenite, the more supersaturated the solution. Compared with other austenitic transformation products (pearlite and upper and lower bainite), martensite possesses the greatest hardness and gives very hard steels.

The ability of a steel to increase in hardness during quenching is called its hardenability or hardening capacity. The hardening capacity is characterized by the maximum hardness that can be obtained on the surface of a given steel product by quenching. To achieve maximum hardness it is necessary to observe basic conditions: the rate of cooling should be equal to or higher than the critical rate at which quenching gives martensite alone (inevitably with some retained austenite, of course, but without bainite); all carbon at the quenching temperature should be in the solid solution in austenite (the quenching temperature should be above the critical points A_{c1} and A_{c3} by 30–50°C (80–120°F) for hypereutectoid and hypoeutectoid steels, respectively).

Alongside the notion of hardening capacity, broad use is made in practice of the notion of "hardenability," though these two characteristics depend on different factors and are achieved in different ways. The hardening capacity of a steel is determined by the factors affecting the hardness of martensite, while its hardenability is determined by those affecting the quantity of the martensite obtained and the hardness penetration depth. Upon quenching, steel can feature high hardening capacity and low hardenability at the same time. Such a steel would correspond to the schematic curve 1 in Figure 15. If for a workpiece of the same diameter D cooled under the same conditions the distribution of hardness over the cross section is characterized by curve 2, such a steel possesses medium or poor hardening capacity but good hardenability. Finally, a steel that corresponds to curve 3 would possess high hardening capacity and high hardenability.

A. Hardness and Carbon Content

The hardening capacity of a steel whose general characteristic could be maximum hardness depends mainly on the carbon content and, to a lesser extent, on the amount of alloying elements and austenite grain size. Increasing the carbon content of martensite increases its hardness (Figure 7). Note that the hardness of a quenched steel and the hardness of martensite crystals are not the same thing because quenched steel contains retained austenite. The hardness of a steel quenched from austenite temperatures passes its maximum at a carbon concentration of 0.8–0.9% C and then decreases due to an increase in the volume fraction of soft retained austenite (Figure 16). For the above carbon content of steel, the martensite point M_s drops significantly, which leads to an increase in the proportion of retained austenite in quenched steel. Steel with 1.9% C quenched from a temperature higher than A_{st} has the same hardness as quenched steel with 0.1% C. If hypereutectoid steels are quenched from a temperature of A_{c1} + (20–30)°C (70–90°F), as is common practice, all hypereutectoid steels would have practically the same austenite composition at the same quenching temperature and level of hardness (Figure 16, curve b).

Another important feature of the dependence of steel hardness on carbon content is that an increase in the carbon content to ~0.6% results in a most dramatic rise in the maximum hardness; then the curve becomes less steep. This is probably associated with the very nature of high martensite hardness in steel.

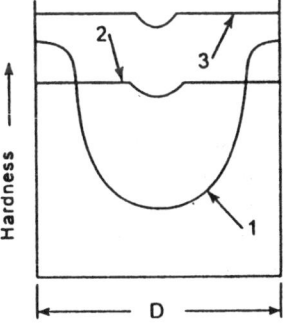

Figure 15 Distribution of hardness over the cross section of a workpiece for three steels differing in hardenability and hardening capacity.

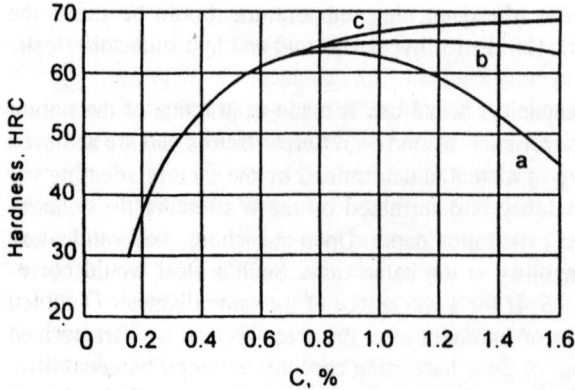

Figure 16 Hardness of carbon and alloy steels depending on the carbon content and quenching temperature. (a) Quenching above A_{c3}; (b) quenching above A_{c1} (770°C); (c) microhardness of martensite.

The martensite transformation of austenite results in a supersaturated solid solution of carbon in α-Fe. An increase in the carbon content of martensite weakens, rather than strengthens, the interatomic bonds. This is due to an increase in the distance between iron atoms brought about by implanted carbon atoms. Carbon nevertheless increases the hardness of martensite, which is explained primarily by the fact that carbon atoms implanted into the α-Fe lattice impede the slip of dislocations in martensite (the so-called solid solution strengthening mechanism).

During quenching or during the aging of quenched steel, carbon atoms in martensite crystals surround dislocations (atmospheres around dislocations), thus pinning them. This leads to a general increase in plastic deformation resistance despite the fact that carbon weakens interatomic bonding in the martensite lattice. In steels with a high martensite start point M_s such as carbon steels containing less than 0.5% C [$M_s > 300°C$ (570°F)], quench cooling over the martensite range is characterized by the most favorable conditions for partial precipitation of martensite with the release of disperse carbide particles. Moreover, in all steels hardened at normal rates, carbon has time to segregate as the steel cools above the point M_s. The carbon segregates of austenite are inherited by martensite, and since the latter is already supersaturated with carbon, these segregates become nucleation sites of carbide particles. This is in agreement with the fact that at very high cooling rates the hardness of martensite crystals is two-thirds of that obtained at normal cooling rates. High hardness of martensite may also be due to the fact that carbon makes a noticeable contribution to covalent bonding whose main property is high plastic deformation resistance.

Owing to the above strengthening mechanisms, carbon has such a strong strengthening effect on martensite that the hardness of a quenched steel does not depend on the concentration of alloying elements dissolved in the martensite by the substitutional mechanism but is determined solely by the concentration of carbon.

To conclude, extreme strengthening of steels during martensitic hardening is due to the formation of a carbon-supersaturated α solution, an increase in the density of dislocations during the martensite transformation, the formation of carbon atom "atmospheres" around dislocations, and precipitation of disperse carbide particles from the α solution.

B. Microstructure Criterion for Hardening Capacity

Studies of the structure of hardened carbon steels and carbon-free iron-based alloys revealed two main morphological types of martensite: plate and lath. These two types of martensite differ in the shape and arrangement of crystals, substructure, and habit plane.

Plate martensite (which is also called needle-type, low-temperature, and twinned) is a well-known "classical" type of martensite that is most pronounced in quenched high-carbon iron alloys with a high concentration of the second element—for instance, Fe-Ni alloys with a Ni content higher than 28%. Martensite crystals are shaped as thin lenticular plates. Such a shape corresponds to the minimum energy of elastic distortions when martensite is formed in the austenite matrix; it is similar to the shape of mechanical twins (Figure 17).

Neighboring plates of martensite are commonly not parallel to each other and form frame-like ensembles. Plates that are formed first (near the point M_s) extend through the entire length of the austenite grain, dividing it into sections. A martensite plate cannot, however, cross the boundary of the matrix phase; therefore the maximum size of the martensite plate is limited by the size of the austenite grains. As the temperature is lowered, new martensite plates are formed in the austenite sections, the size of the plates being limited by the size of the matrix sections. In the course of transformation the austenite grain splits into still smaller sections, in which smaller and smaller martensite plates are formed. In the case of a small austenite grain caused by, for instance, a small overheating of steel above A_{c3}, the martensite plates are so small that the needle-type pattern cannot be observed in the microsection and the martensite is usually called structureless. It is this type of martensite that is most desirable.

After quenching, martensite retains some austenite between its plates at room temperature (Figure 18).

Lath martensite (which is also called massive, high-temperature, or nontwinned) is a widespread morphological type that can be observed in quenched low-carbon and medium-carbon steels, the majority of structural alloy steels, and comparatively low alloy non-carbon-containing iron alloys—for instance, Fe-Ni alloys with a Ni concentration of less than 28%. Crystals of lath martensite are shaped like thin plates of about the same orientation, adjacent to each other and forming more or less equiaxial laths.

The plate width within the lath is about the same everywhere, ranging from several micrometers to fractions of a micrometer (commonly 0.1–0.2 μm); i.e., it can reach or even exceed the resolution limit of a light microscope. Inside the martensite laths there are interlayers of retained austenite 20–50 nm thick. One austenite grain can contain several martensite laths.

The formation of lath martensite has all the main features specific to the martensite transformation, including the formation of a relief on a polished surface.

Transmission electron microscopy reveals a rather complicated fine structure of martensite crystals with a lot of dislocations and twins in many iron alloys.

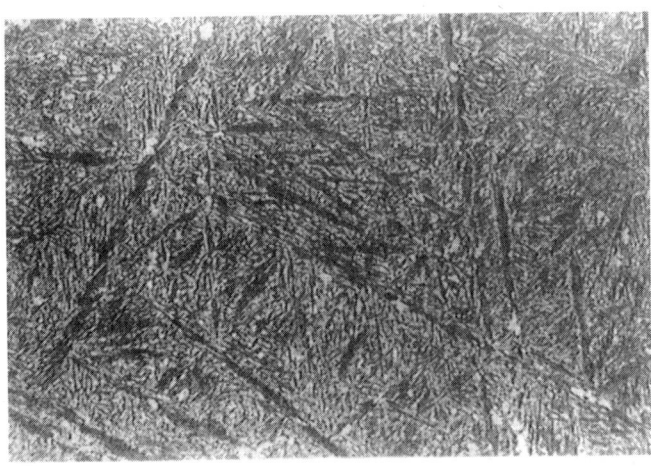

Figure 17   Microstructure of plate martensite. Light shading = retained austenite 50,000×.

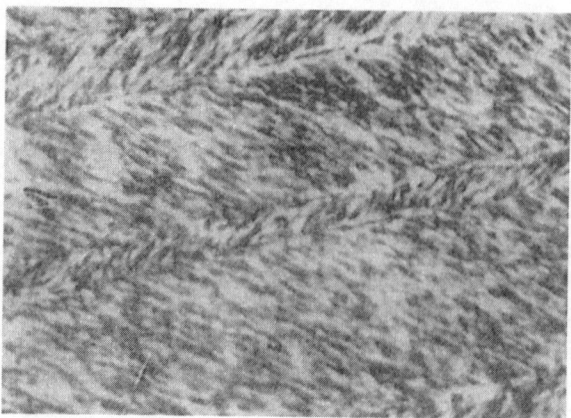

Figure 18   Microstructure of lath martensite. 50,000×.

The substructure of the plate martensite shows an average zone of elevated etchability, also called a midrib, even under a light microscope. Electron microscopy has shown that the midrib is an area with a dense arrangement of parallel fine twin interlayers. The twinning plane is martensite of iron-based alloys is commonly $\{112\}_M$. Depending on the composition of the alloy and martensite formation conditions, the thickness of the twinned interlayers may form several tenths of a nanometer to several tens of nanometers. On both sides of the central twinned zone there are peripheral areas of martensite plates that contain dislocations of relatively low density (10^9–10^{10} cm^{-2}).

The substructure of lath martensite is qualitatively different from the substructure of plate martensite in that there is no zone of fine twin interlayers. It is a complex dislocation structure characterized by high-density dislocation pileups with densities on the order of 10^{11}–10^{12} cm^{-2}, i.e., the same as in a metal subjected to strong cold deformation. The laths of lath martensite often consist of elongated slightly misoriented subgrains. Twin interlayers can occur in lath martensite, but their density is much lower than in the midrib in plate martensite, while many of the laths do not contain twins at all.

The substructure of retained austenite differs from that of the initial austenite by a higher density of imperfections occurring under local plastic deformation due to martensite crystals. Flat dislocation pileups, dislocation tangles, and stacking faults may be observed in austenite around martensite crystals.

At present it is believed that the decisive role in the formation of plate martensite belongs to accommodating (complementary) twinning deformation, whereas for the lath type this role is played by slip. As temperature is decreased, resistance to slip increases at a higher rate than that of resistance to twinning; therefore, the martensite transformation at low and high temperatures results in twinned and lath martensite, respectively. In alloys a decrease in temperature causes the morphology of martensite to change from the plate to lath type. The composition of iron-based alloys has a substantial effect on the martensite morphology. Shown below is the effect of some alloy compositions on the formation of plate and lath martensite; the numerals show the second component content in percent.

System	Fe-C	Fe-N	Fe-Ni	Fe-C$_2$
Lath type, M	≤0.6	≤0.7	≤29	≤10
Plate, M	0.6–2.0	0.7–2.5	29–34	—

Chromium-nickel, manganese, chromium-manganese, and other steel alloys with low-energy stacking faults contain hexagonal ε-martensite with plates in parallel to planes $\{111\}\gamma \| \{011\}\varepsilon$ (Figure 19). Some alloy steels have a mixture of ε- and α-martensites.

In conclusion it should be noted that the structure of a metal or alloy that has undergone martensite transformation features many more imperfections than after a disordered rearrangement of its crystal lattice: more developed grain boundaries and subboundaries, greater density of dislocations and twin interlayers.

C. Effect of Grain Size and Chemical Composition

As noted in Section III.A, the hardening capacity of a steel, i.e., its ability to undergo martensitic hardening, depends mainly on its carbon content and to a lesser extent on its content of alloying elements and the size of austenite grains. At the same time these two factors—grain size and chemical composition of the steel (or austenite, to be more exact)—can produce a substantial effect on hardenability, i.e., the depth to which the martensite zone can penetrate. It is reasonable, therefore, to consider the effects of grain size and chemical composition of austenite on hardening capacity and hardenability separately.

In the case of austenite–martensite transformation, martensite plates develop inside the austenite grain, extending from one side to the other. If the steel is considerably overheated above the critical point A_{c3}, then coarse grains of austenite are formed; this results in larger martensite plates than normally. In quenched steel the volume fraction of retained austenite is higher if its structure is dominated by large martensite plates. Although the hardness of martensite is practically independent of the plate size, an increase in the total soft austenite content of quenched steel leads to a decrease in its maximum hardness, i.e., impairs its hardening capacity. Moreover, the plastic properties of steel, particularly its toughness, also deteriorate with the coarsening of the structure. It is therefore advisable to obtain a fine-needle structure after quenching if a set of good mechanical properties is needed; this may be achieved with a fine-grain austenite structure, which is produced at small overheating of steel at temperatures higher than A_{c3}.

Alloying elements can have direct or indirect effects on the hardening capacity of steel. Indirect effects are associated primarily with hardenability. Since the majority of the elements tend to shift the isothermal austenite precipitation curves to the right and hence decrease the critical rate of cooling, it is easier to obtain maximum hardness for an alloy steel than for an ordinary carbon steel. Specifically, the presence of alloying elements facilitates the achieve-

Figure 19 Microstructure of ε-martensite. 50,000×.

ment of maximum hardness by cooling the steel in more lenient media (in oil, for instance) when the mass of the workpiece to be hardened is comparatively large.

A second indirect effect is related to the carbide-forming elements. If hardening is to be applied to an alloy steel containing elements that form stable carbides by heating below the carbide dissolution point, then the carbon content of the major martensite mass will be lower than the total carbon content of the steel. As a result, the maximum achievable hardness for this carbon content will not be attained because the hardening capacity of the steel will deteriorate.

It should be noted that high-carbon steel, such as tool grade steel, features high hardness even if it is quenched from a temperature somewhat lower than the carbide dissolution point, despite the fact that part of the carbon stays outside the solution. The martensite hardness decreases also. this decrease is small and is due mainly to the fact that hardness vs. carbon curves for carbon steel and alloy steels are rather smooth at carbon concentrations higher than 0.6% (Figure 16); it is compensated for by a considerably lower amount of retained austenite and the high hardness of the carbides themselves.

The situation is different for structural steels containing less than 0.4% C. The maximum hardness curve is so steep (Figure 16) that even a small decrease in the concentration of carbon in martensite in an alloy steel due to incomplete dissolution of special carbides would lead to a considerable reduction in the hardness of the martensite. The steepness of the curve over the range of 0.1–0.4% C shows that from the viewpoint of hardening capacity it is essential to heed even the smallest fluctuations in the carbon content, specifically fluctuations within the quality limits, up to individual ingots. Therefore the carbon content limits for specific structural steels should be as narrow as possible, although this may encounter some technological difficulties.

The indirect effect of alloying components on hardenability is not great. It is therefore possible to construct a general curve for the dependence of maximum hardness on the carbon content for carbon and alloy steels (Figure 16). This is understandable because martensite of an alloy steel is a "combined" solid solution in which atoms of the alloying elements replace iron atoms in the lattice, while carbon atoms are implanted into this lattice. Carbon atoms introduced into the α-Fe lattice impede the slip of dislocations in martensite and thereby increase its hardness.

A certain increase in the hardness of martensite due to alloying elements can be expected only because of the strengthening of α-Fe during quenching. The possibility of quench strengthening an alloyed ferrite has been studied in detail. From Figure 20 one can judge roughly how different elements increase the hardness of ferrite with $\sim 0.01\%$ C upon quenching from 1200°C in water. Alloying elements causing substantial strengthening of ferrite should also increase the hardness of martensite in a quenched steel, though this increase in hardness should be comparatively low.

During the quenching of steel products, the rate of cooling is greatest for the surface, decreasing steadily toward the center of the section. Evidently, the depth of the hardened zone (hardenability) will be determined by the critical rate of quenching; thus, hardenability will increase with a decrease in the critical rate of quenching. This rate, in turn, depends on the resistance of austenite to precipitation at temperatures higher than the martensite point M_s. The farther to the right the lines in the isothermal austenite precipitation diagram, the lower the critical rate of quenching and the higher the hardenability of the steel products.

Thus, the factors that affect the stability of undercooled austenite will affect the hardenability as well.

The main factors that produce a decisive effect on the hardenability of steel are (1) the chemical composition of the steel (composition of austenite, to be more exact); (2) austenite

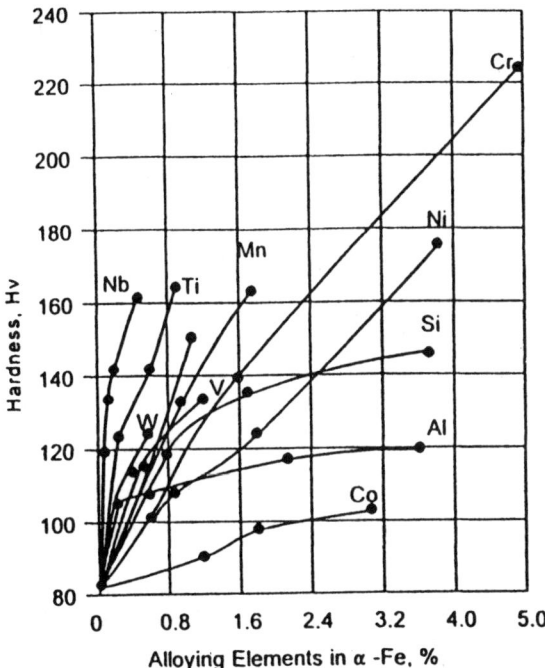

Figure 20 Effect of dissolved alloying elements on hardness of ferrite after quenching from 1200°C in water.

grain size; and (3) the homogeneity of austenite. Under otherwise equal conditions, coarse austenite grains improve the hardening capacity of steel. This circumstance is connected with the extent of grain boundaries: the extent is less, the coarser the grain. Since nucleation centers are formed primarily along the austenite grain boundaries during austenite precipitation above the point M_s, it is always easier to undercool austenite with coarse grains, thereby increasing hardenability.

To estimate hardenability, in practice use is made of the quantity called the critical diameter. The critical diameter (D_{cr}) is the maximum diameter of a bar permitting through hardening for a given cooling medium. To avoid putting hardenability in dependence on the method of cooling and type of coolant, use is made of still another notion, the ideal critical diameter (D_∞), which is the diameter of a maximum section allowing through hardening in an "ideal" cooling liquid that is absorbing heat at an infinitely great rate. It was established that the grain size and quantity D_∞^2 have an approximately linear inverse dependence on one another.

The chemical composition of a steel has the strongest impact on its hardenability. This is due primarily to the fact that carbon and alloying elements affect the critical rate of quenching.

Figure 21 shows the effect of carbon on the critical rate of quenching for carbon steel. It can be seen that the minimum rate of quenching is observed in steels that are close to eutectoid with respect to carbon content (curve I). A decrease in the carbon content of steel below 0.4% leads to a sharp increase in the critical rate to the extent that at a certain minimum carbon content martensitic hardening becomes virtually impossible. An increase in the critical rate of hypereutectoid steel with an increase in the carbon content is explained by the presence of cementite nuclei facilitating the austenite precipitation. Hence, the trend of curve I is related to the incomplete hardening of hypereutectoid steels. If completely hardened (sufficient hold-

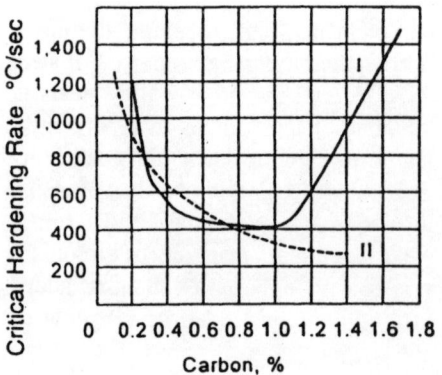

Figure 21 Curves showing the effect of carbon on the critical rates of quenching. I, low heating temperatures; II, high heating temperatures.

ing at a temperature higher than A_{st}), an increase in the carbon content leads to a continuous decrease in the critical rate of quenching (curve II), with a resulting rise in hardenability.

The effect of alloying elements on hardenability can be estimated by the degree of increase or decrease in stability of undercooled austenite in the pearlite and intermediate ranges. With the exception of cobalt, all alloying elements dissolved in austenite impede its precipitation, decrease the critical rate of quenching, and improve hardenability. To this end, broad use is made of such additives such as Mn, Ni, Cr, and Mo. Particularly effective is complex alloying whereby a combination of elements enhances their individual useful effects on hardenability. Figure 22 shows the effect of third-element alloying on the hardenability of an iron-nickel steel. It can be seen that Mn, Cr, and Mo additives improve hardenability to a considerable extent.

The improving effect of alloying on hardenability is used in two ways. First, alloying ensures through hardening across sections inaccessible for carbon steels. Second, in the case of small-section products, replacing carbon steel with an alloy steel permits less radical cooling regimes. Small-diameter carbon steel products can be hardened by quenching in water. This, however, may result in impermissible residual stresses, deformations, and cracks, particularly in products of a complicated shape. If an alloy steel is used, quenching in water can be replaced with softer hardening in emulsion, oil, or even air.

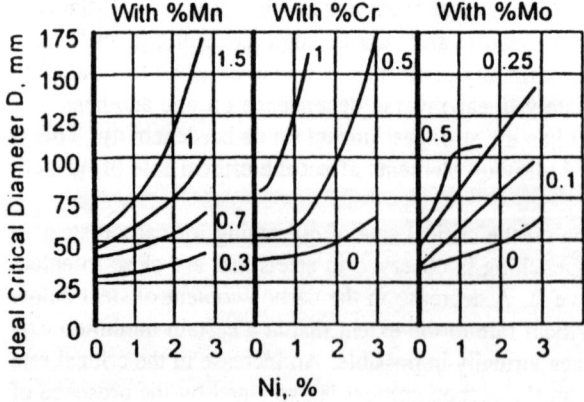

Figure 22 The effect of manganese, chromium, and molybdenum on the hardenability of steels with different nickel contents.

D. Boron Hardening Mechanism

It has long been observed that small additions of certain elements, e.g., titanium, aluminum, vanadium, zirconium, and boron, can considerably improve the hardening properties of steel. The most effective in this respect is boron. Hardenability of carbon and low-alloy steels increases considerably upon introduction of boron in amounts of thousandths of a percent. A further increase in the boron content does not produce any further improvement in hardenability. The improving effect of boron is noticeable only where steel has been preliminarily well deoxidized and denitrified, because boron has good affinity for oxygen and nitrogen. Therefore, before introducing boron into steel it is necessary to add aluminum, titanium, or zirconium. Figure 23 shows hardenability curves for a low-carbon steel without boron, with boron added, with boron and vanadium, and with boron, vanadium, and titanium. It can be seen that the extent of the martensite range for all the steels with boron added is greater than that in steels without boron; however, it is practically the same for steels with several additives. At the same time, the extent of the half-martensite structure range is much greater if the steel contains other additives along with boron.

The effect of boron on hardenability decreases with an increase in the carbon content. If the carbon content exceeds 0.9%, boron does not have any measurable effect on hardenability. Boron alloying for the purpose of improving hardenability is therefore useful only for low-carbon steels of various applications. It does not have any effect on the hardenability of tool steels or on high-carbon carburized layers. The above relationship between the carbon content and the effect of boron on the hardenability of steel is due to the fact that the two elements have the same effect on austenite precipitation. Boron increases the length of the austenite precipitation incubation period, thereby decreasing the critical temperature of quenching. Like carbon, boron facilitates the enlargement of austenite grains under heating. These two factors have a positive effect on the hardenability of steel. Therefore, in high-carbon steels the effect of small doses of boron is practically negligible.

It should be noted also that the positive effect of boron on hardenability is full only if the quenching temperature is sufficiently high [850–900°C (1560–1650°F)].

Various suggestions have been put forward with respect to the mechanism by which boron affects the hardenability of steel. Quite probably, this is a very special mechanism because boron can products its effect at very low concentrations above which no further effect on hardenability is observed. Some researchers believe that boron increases hardenability just because it facilitates the increase in the size of austenite grains under heating. Although bo-

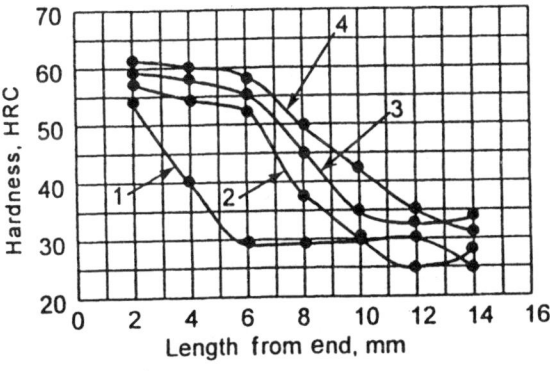

Figure 23 Curves of hardenability for steels with 0.44–0.43% C and various small amounts of additives. 1, Without boron; 2, with an addition of boron; 3, with boron and vanadium added; 4, with boron, vanadium, and titanium added.

ron does tend to increase the grain size, it produces the same hardenability-improving effect in steels with small grains. Others explain the fact that boron increases the stability of austenite and consequently improves hardenability by the fact that it increases the coefficient of surface tension at the "austenite/new phase nucleus" interface. Therefore more energy is required for the formation of a nucleus of a critical size capable of growth. Because of this stability of austenite increases. An increase in hardenability following the addition of small doses of boron is more frequently explained by the fact that this element is surface-active in austenite. There is experimental evidence that boron is segregated at the boundaries of austenite grains and dissolves in insignificant amounts in this layer. Since boron forms an interstitial solid solution with iron, interaction between its atoms and iron atoms must, evidently, be the same as in carbon, which leads to a decrease in the difference in free energy between the γ and α phases. This impedes the formation of new phase nuclei of critical size, which are formed primarily at the austenite grain boundaries. The resistance of austenite to precipitation improves, thereby increasing the hardenability of the steel. The cessation of the boron effect on hardenability with increasing boron content is due to its low limiting solubility in γ-iron at a given temperature [about 0.003% at 1000°C (1832°F)]. As soon as the boron solubility limit at the grain interface is achieved, any further increase in its total content leads to the formation of iron-boron compounds such as Fe_2B within austenite grain boundaries and to the distribution of boron over the bulk of the grain. Being centers of crystallization, chemical compound particles cause an earlier onset of austenite precipitation, which results in lower hardenability. At the same time, an earlier onset of austenite precipitation at grain boundaries is compensated for by a delay in the formation of critical new-phase nuclei in the bulk of the grain caused by boron due to an increase in its content up to the solubility limit for austenite. Therefore, alloying with boron in amounts exceeding thousandths of a percent does not have any effect on hardenability and even may impair it.

Based on this viewpoint one can explain some other specific effects of boron on the hardenability of steel that were noted above. Thus, boron increases the duration of the austenite precipitation incubation period only, the duration being determined by the formation of critical nuclei at grain boundaries but not affecting the length of austenite precipitation. From this viewpoint it is clear also that boron, like carbon, facilitates enlargement of austenite grains under heating. An increase in the quenching temperature first improves hardenability owing to an increase in the concentration of boron at the austenite grain boundaries to its solubility limit at these boundaries. A further increase in the quenching temperature and enrichment of the grain boundaries in boron can lead to the formation of boron-iron compounds. The hardenability of the steel will not increase further and may even decrease. Finally, the reduction of the boron effect by carbon can be explained by the fact that boron and carbon have virtually the same effect on hardenability. In high-carbon steels, therefore, the effect of boron becomes practically negligible owing to its poor solubility in iron.

E.   Austenitizing Conditions Affecting Hardenability

The austenite condition prior to quenching (chemical composition, grain size, homogeneity of austenite) has the decisive effect on the hardening capacity and, especially, hardenability of steel. Other factors are secondary or derivative from the basic three. These factors, in turn, are determined by the carbon content, type and amount of alloying elements at the time of quenching, the quenching temperature (austenitizing temperature), and the holding time at a given temperature.

Austenitizing of a heated alloy steel consists in polymorphous α–γ transformation, the dissolution of cementite, special carbides, nitrides, and intermetallics in austenite, and recrystallization of the austenite grains.

To improve the hardening capacity and hardenability of steel, the austenitizing conditions should be such as to ensure that a maximum amount of carbon passes from the ferrite–carbide mixture to the solution and, at the same time, no marked growth of grains occurs as a result of overheating, as this would lead to a high brittleness and the formation of quenching cracks. The quenching temperature should be maintained as constant as possible, and the holding time should be just enough to ensure uniform heating of the workpiece and dissolution of carbides. For their complete dissolution in austenite, coarse-plate and coarse-grain carbides need more time than thin-plate and fine-grain ones. Steels alloyed with elements forming special carbides should be heated to a temperature considerably exceeding A_{c3}. Small carbides available in the structure impede enlargement of grains and, being nuclei of the new phase, facilitate transformation of austenite in the pearlite range and increase the critical rate of quenching, thus decreasing the hardenability of the steel. As the quenching temperature and time are increased, the critical rate of quenching decreases and, accordingly, hardenability rises, because carbides and other inclusions playing the role of new-phase nuclei dissolve most.

The degree of austenite homogeneity and dispersion of local carbon pileups (which can act as nuclei during transformation in the pearlite range) can have a strong effect on hardenability and hardening capacity. When the quenching temperature is increased, carbides dissolve together with other minute, sometimes hardly measurable, quantities of inclusions such as nitrides and sulfides, which can also serve as nuclei during transformation.

Finally, when the quenching temperature and holding time are increased, enlargement of austenite grains has its effect on the process of transformation. Since the pearlite transformation begins at grain boundaries, an increase in the austenite grain size causes a decrease in the critical rate of quenching and hardenability improves.

Nearly all of the alloying elements impede the growth of austenite grains. The exception is manganese, which adds to the growth of grains. The strongest growth retardants are V, Ti, Al, Zr, W, Mo, and Cr; Ni and Si produce a weaker retarding effect. The main cause of this retarding effect is believed to be the formation of low-soluble carbides, nitrides, and other phases, which may serve as barriers for the growth of austenite grains. Such active carbide-forming elements as Ti, Zr, and V impede growth more strongly than Cr, W, and Mo do, because the carbides of the former elements are more stable and less soluble in austenite. Experimental studies on the solubility of V, Nb, Ti, and Al carbides and nitrides in austenite show that the carbides of these elements (Ti in particular) are more soluble than nitrides. Titanium nitrides are virtually insoluble in austenite no matter what the temperature is. Niobium and aluminum nitrides also have poor solubility in austenite. Carbon has a great effect on the solubility of carbides. Figure 24 shows relevant data on V, Nb, Ti, and Zr carbides. An increase in the temperature of carbide solubility in austenite with a rise in the carbon content is due to the greater activity of carbon at its higher concentrations in a solid solution and higher thermodynamic activity. It should be noted that C, N, and Al not bound into carbides or nitrides but found in the solid solution of austenite facilitate the growth of austenite grains. The elements B, Mn, and Si also favor the growth of grains. Therefore, addition of these elements into steel improves its hardenability.

Different heats of steels of the same quality may considerably differ in their tendency toward the growth of austenite grains because they contain different amounts of low-soluble disperse particles of carbides, nitrides, and other phases, which are barriers to the growth of austenite grains. The distribution and size of these particles depend both on steelmaking conditions and preliminary heat treatment. Thus, the tendency of steel to grain size growth under heating depends on, in addition to its composition, the metallurgical quality and process, i.e., its history preceding the thermal treatment.

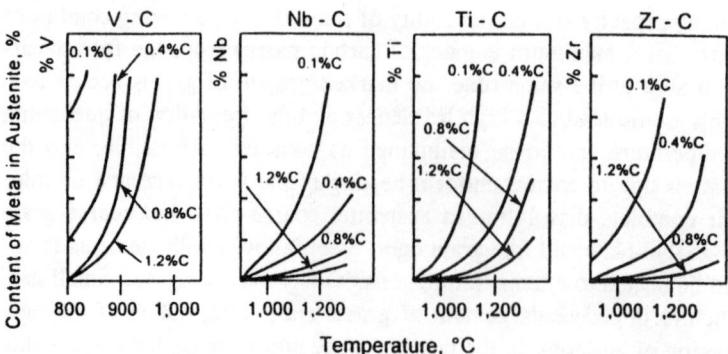

Figure 24 Solubility of carbides in austenite at various temperatures depending on the carbon content (shown as % C on the curves) for (a) vanadium; (b) niobium; (c) titanium; (d) zirconium.

Liquation also has a considerable effect on hardenability. In order to obtain homogeneous austenite in steel exhibiting liquation, it is necessary to keep the steel at the quenching temperature for a sufficiently long time. This refers to cast steel where liquation is the highest and also to forged and rolled steels. Longer quenching times increase hardenability owing to the elimination of residual liquation and fluctuation inhomogeneity of austenite.

Note in conclusion that hardening capacity and hardenability are not important by themselves in practical applications. They are important if they can improve overall properties of steels in accordance with practical needs.

IV. TEMPERING OF ALLOY STEELS

A. Structural Changes on Tempering

Structural changes on tempering were considered in detail in Chapter 1. Therefore, this section only briefly considers the most characteristic effects.

Tempering is a thermal martensitic treatment of quenched steels. The basic process that takes place during tempering is martensite precipitation. The first structural change during tempering is carbon segregation at dislocations. The second stage of tempering is precipitation of intermediate ε-carbide with a hexagonal lattice, which forms under heating above 100°C (212°F). During the third stage, cementite precipitates above ~250°C (~480°F). At the final stage of tempering above 350°C (660°F), cementite particles coagulate and spheroidize.

Consider the changes that takes place in martensite and austenite structures at different stages of tempering. During the first stage, beginning at 80°C (175°F) and up to 170°C (330°F), the c parameter of the martensite lattice decreases. The ratio c/a becomes close to unity. Tetragonal martensite transforms to a cubic form called tempered martensite. The decrease of tetragonality is connected with precipitation of carbon from the solution.

Heating of steels over 200°C (390°F) and up to 300°C (570°F) activates transformation of retained austenite to a heterogeneous mixture composed of a supersaturated α-solution and the Fe_3C carbide. This means that retained austenite transforms to tempered martensite,

$$Fe_\gamma(C) \rightarrow Fe_\alpha(C) + Fe_3C$$

By the end of transformation (~300°C; 570°F) the retained austenite contains about 0.15–0.20% C.

Heating above 300°C (570°F) leads to a further precipitation of carbon and the relaxation of internal stresses arising from previous transformations. Complete precipitation of carbon was found to occur at 400°C (750°F). A further increase in temperature leads only to coagulation of ferrite and cementite particles.

During tempering, cementiite acquires a globular form when a ferrite–cementite mixutre is formed from martensite. The different form of cementite in the ferrite–cementite mixture determines the difference in properties.

B. Effect of Alloying Elements

The influence of alloying elements on transformations during tempering depends on whether they dissolve in ferrite and cementite or form special carbides.

The diffusion mobility of atoms of alloying elements dissolved in α-Fe by the substitutional method is many orders of magnitude lower than the diffusion mobility of carbon atoms dissolved by the interstitial method. So at a temperature below 400°C (750°F) no diffusion redistribution of alloying elements in the matrix takes place. First the ε-carbide and then cementite precipitate from the α-solid solution. The concentration of alloying elements in them is the same as in martensite. Atoms of alloying elements in the ε-carbide and cementite lattice formed below 400°C (750°F) partly replace iron atoms. Complex carbides such as (Fe, Cr)$_3$C and (Fe, V)$_3$C are formed.

The first stage of transformations in martensite (formation of tempered martensite) at a temperature below 150°C (300°F) is affected little by alloying elements. At this stage of tempering, nucleation of carbide particles depends basically on supersaturation of the α-solution with carbon.

The second stage of martensite precipitation is strongly influenced by a number of alloying elements. They retard the growth of carbide particles, and consequently supersaturation of the α-solution with carbon is preserved. Thus the state of tempered martensite is retained up to temperatures of 450–500°C (840–930°F). Additions of Cr, W, Mo, V, Co, and Si bring about this effect.

A delay in martensite precipitation can be explained by two factors. First, one of the alloying elements lowers the rate of carbon diffusion in the α-solution. Second, the other elements can increase the strength of interatomic bonds in the α-solution lattice. This will prevent the atoms from crossing the α-solution carbide interface. Both factors impede precipitation of martensite.

Alloying elements affect carbide transformation under tempering under 450°C (840°F) when their diffusion movement becomes possible. In this case special carbides are formed. With an increase in the tempering temperature, intermediate metastable carbides stabilize. For example, when molybdenum and tungsten steels are tempered, Me_2C(Mo_2C and W_2C) is formed first, then $Me_{23}C_6$ appears, and finally Me_6C emerges. The sequence of their formation can be written as

$$Fe_3C \rightarrow Me_2C + Me_{23}C_6 \rightarrow Me_6C$$

Alloying elements affect the coagulation rate of carbide particles. Nickel accelerates the coagulation rate while chromium, molybdenum, vanadium, and other elements slow it down. Owing to the low diffusion rate of alloying elements, the coagulation of special carbides proceeds slowly. Even alloyed cementite (Fe, Cr)C$_3$ coagulates much more slowly than Fe$_3$C in a carbon steel.

Additions of alloying elements slow down recrystallization and polygonization. Atoms of these elements form impurity atmospheres near dislocations and prevent their movement during polygonization. Disperse particles of special carbides retard movement of large-angle boundaries during polygonization.

C. Transformations of Retained Austenite (Secondary Tempering)

Alloying elements have the greatest influence on the martensite transformation temperature. This affects the amount of retained austenite in alloy steel. Some elements (e.g., cobalt) raise the point M_s, thus decreasing the amount of retained austenite. Others (e.g., silicon) have no influence on M_s. However, the majority of elements decrease the martensite point and increase the amount of retained austenite in quenched steel. Up to 60% of retained austenite is left in high-carbon steels during quenching and and 10–15% in a large number of structural alloy steels.

During tempering of carbon and low-alloy steels, retained austenite transforms over the temperature interval of 230–280°C (440–540°F) or at lower temperatures if the holding time is extended. Alloying elements, especially Cr and Si, inhibit that transformation, shifting it to higher temperatures and longer tempering time. The transformation kinetics of retained austenite during tempering are similar to those of undercooled austenite. Steels with two clearly distinguished transformation ranges (pearlite and bainite) also exhibit two regions of fast transformation of retained austenite during tempering that are separated by a zone of high stability of retained austenite.

When alloy steels are tempered at 500–600°C (930–1110°F), in many cases the transformation of retained austenite is not complete. The retained austenite that did not precipitate at these tempering temperatures transforms during cooling from those temperatures (secondary quenching). The phenomenon is most pronounced in high-speed and high-chromium steels. The secondary martensite transformation during cooling after tempering is caused by the depletion of austenite in carbon and alloy elements in the course of tempering. As a result, the temperature of retained austenite M_s during cooling is increased.

Secondary quenching (double tempering) is also observed in structural steels. It takes place if the primary quenching is accompanied by a partial intermediate transformation leading to an increase in the carbon content of austenite. During tempering at 500–550°C (930–1020°F), retained austenite with a high content of carbon yields carbides intensively, and the martensite transformation temperature M_s increases. As a result, the secondary martensite transformation takes place during cooling after tempering.

In high-alloy steels, for example high-speed steels, even a very long tempering at high temperatures does not completely eliminate the retained austenite. To obtain a full transformation, it is necessary to perform double tempering. Double tempering favors additional precipitation of special carbides and decreases the degree of austenite alloying. This causes another increase in the transformation temperature M_s. Sometimes multiple tempering is required to realize the most complete transformation of retained austenite.

D. Time–Temperature Relationships in Tempering

The kinetics of structural transformations during tempering are described by temperature–time curves similar to the curves shown in Figure 25. After quenching from 900°C (1650°F), steels with 0.7% C, 1% Cr, and 3% Ni contain 30% of retained austenite. When plotting the curve and diagram, the nontransformed austenite (30%) was taken to be 100%. It is found that at 600°C (1110°F) 5% of primary austenite transforms in 7 min and 5% of retained austenite transforms in 30 s. However, after 10–15% of the retained austenite is transformed, the transformation rate of retained austenite becomes smaller than that of the initial austenite. The transformation of retained austenite is not complete. It is inhibited on reaching 45% at 500°C (930°F) and 60% at 550°C (1020°F). The retained austenite that does not precipitate immediately at these tempering temperatures transforms during cooling after tempering. The transformation rate of retained austenite in the intermediate range is much higher than that of the

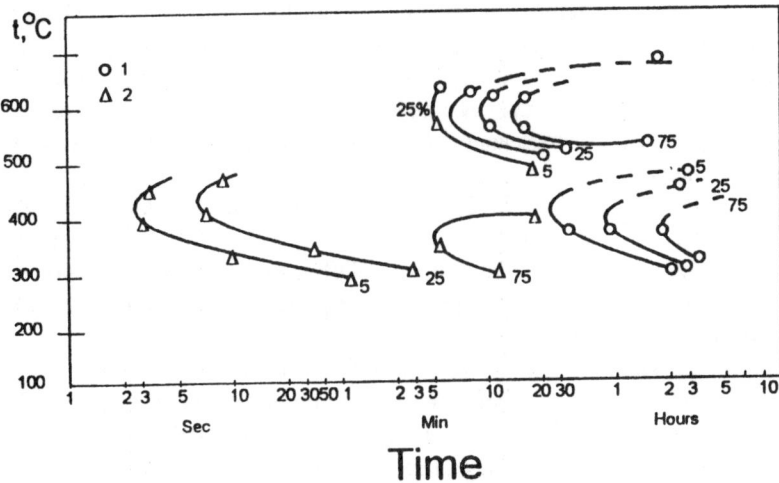

Figure 25 Time–temperature transformation curves.

initial austenite: 25% of initial austenite transforms at 300°C (570°F) in 75 min and the same amount of retained austenite in 15 s; 75% of the initial austenite transforms in 220 min and 75% of the retained austenite in 9.5 min. In a typical structural steel (0.37% C, 1% Cr, 1% Mn, 1% Si), 5% of the initial austenite transforms in 19 min at 600°C (1110°F) or in 5 min at 400°C (750°F); 5% of retained austenite at the same temperatures transforms in a few seconds.

Another specific feature of retained austenite transformation of this steel in the intermediate range is the lowering of the transformation limit. For example, at 350°C (660°F), 70% of the initial austenite and only 40% of retained austenite transform. This difference decreases with increasing temperatures.

E. Estimation of Hardness After Tempering

Hardness decreases noticeably when alloy steels and addition-free steels are subjected to tempering at 500–600°C (930–1110°F). This decrease is due to the precipitation of martensite and coagulation of cementite. However, when the tempering temperature is higher, the hardness of steels with additions of titanium, molybdenum, vanadium, or tungsten increases. This phenomenon is called secondary hardening.

Secondary hardening is caused by the formation of clusters of atoms of alloying elements and carbon (a maximum hardness often corresponds to the clusters) and the replacement of relatively coarse particles of cementite by much more disperse precipitates of special carbides (TiC, VC, Mo_2C, W_2C). When these particles coagulate, hardness decreases. Particles of Me_6C are rather coarse and do not add to strengthening.

The chromium additive causes a small secondary hardening. This is connected with a rapid coagulation of the Cr_7C_3 carbide at 550°C (1020°F) as opposed to Mo_2C and especially W_2C. During secondary hardening an increase in the yield stress is accompanied by an increase in toughness owing to dissolution of coarse cementite particles.

F. Effect of Tempering on Mechanical Properties

The manner in which structural changes that take place during tempering affect the properties of steels depends on the particular tempering conditions. The general tendency of changes in mechanical properties of carbon steels during tempering is that as the tempering temperature is elevated, the strength parameters σ_B and $\sigma_{0.2}$ (fracture stress and yield stress) decrease, while the elasticity parameters δ and ψ (percent elongation and percent reduction of area) are improved. However, these properties change nonmonotonically, and the variation depends on the tempering temperature intervals.

Low-temperature tempering (120–250°C; 250–480°F) is used for treatment of high-strength structural and tool steels. Medium-temperature tempering (350–450°C; 660–840°F) is applied mainly to spring steels to achieve high elasticity. High-temperature tempering (450–650°C; 840–1200°F) is widely used for products made of structural steels combining a relatively high strength with resistance to dynamic loads.

Alloying of high-strength steels preserves high strength characteristics up to 400°C (750°F). In steels containing additions of chromium, nickel, tungsten, and aluminum it is possible to obtain a very favorable combination of strength (σ_B, $\sigma_{0.2}$), ductility (δ, ψ), and impact strength under low-temperature tempering (160–200°C; 320–390°F). In low-carbon martensitic steels containing chromium, manganese, nickel, and molybdenum, the tensile strength remains unchanged up to 400–500°C (750–930°F). In steels with secondary hardening (e.g., in steels with 0.26% C, 5% Cr, 1% Mo, 1.2% V, and 1.4% Si), strength and impact strength increase under high-temperature tempering.

These data suggest that mechanical properties of every type of steel exhibit certain specific features that vary with the tempering temperature. These features are determined by the influence of alloying elements on the kinetics of phase transformations: change of the martensite point M_s, stabilization of retained austenite, and carbide formation.

The influence of structural evolution on properties during tempering can be most fully understood through the example of a maraging (martensite aging) alloy containing 0.02–0.03% C and also Co, Mo, Ti, and Al. Alloying with cobalt increases the temperature M_s and provides 100% martensite after cooling. Tensile strength reaches 1000–1100 MPa. The subsequent tempering (aging) at temperatures of 450–500°C (840–930°F) results in considerable strengthening. Thus, σ_B can reach 1900–2100 MPa, $\sigma_{0.2}$ = 1800–2000 MPa, and δ = 8–10%. Such high strength properties are due to the segregation of impurity atoms during aging (initial stages) and then to Ni_3Ti, Ni_3Mo, Fe_2Mo, etc. phases coherently bound with the matrix. The size of particles is approximately 100 nm. Coagulation of the precipitates with an increase in temperature leads to lowering of the strength characteristics and increasing of the ductility.

G. Embrittlement During Tempering

When carbon and alloy steels are tempered over the temperature interval of 250–400°C (480–750°F), a dramatic drop in impact strength is observed. If the steel is subjected to a higher temperature tempering and then tempering is repeated at 250–400°C, the brittle state is not recovered. Therefore this phenomenon has been called irreversible tempering brittleness. Such tempering brittleness is typical of almost all carbon steels and alloys. High-temperature mechanical treatment and refinement of grains weaken this type of brittleness. In high-purity steels it does not occur at all. The embrittlement may be caused by nonuniform precipitation of martensite at the second stage of tempering. This structure has a lower resistance to dynamic loads. This effect is enhanced when the initial grain boundaries of austenite get saturated with impurities under quenching heating. The alloying elements, which retard the second stage of martensite precipitation, shift the interval of irreversible brittleness toward higher temperatures.

Another drop in impact strength is found at tempering temperatures of 450–600°C (840–1110°F). A very significant feature of embrittlement is its reversibility under high-temperature tempering. If a steel that has undergone tempering embrittlement is heated to a temperature above 600°C (1110°F) and then cooled rapidly, its impact strength is restored. Therefore such brittleness is termed reversible.

In the state of reversible tempering embrittlement, steel possesses a structure that consists of ferrite and carbide. When subjected to impact tests, fracture occurs mainly along the boundaries of the initial austenite grains.

Embrittlement over a certain temperature interval is typical not only of martensitically hardened steels. It also shows up, although to a lesser degree, in steels with the bainite structure and is least pronounced in steels with the pearlite structure. Additions of chromium, nickel, and manganese facilitate tempering embrittlement. Small additions of molybdenum (not more than 0.2–0.3%) weaken tempering embrittlement.

The presence of Sb, P, Sn, and As in industrial steels makes these steels most susceptible to tempering embrittlement.

V. HEAT TREATMENT OF SPECIAL CATEGORY STEELS

A. High-Strength Steels

Low-alloy steels are most often used as construction materials. The combination of high strength and ductility with high resistance to destruction is of particular importance for steels. The mechanical properties of such steels can be improved after hot rolling or normalization and after quenching with tempering. Alloying makes it possible to perfect the properties of steels without using quenching with tempering because

1. The properties of ferrite are changed when alloying elements are dissolved in it (solid solution strengthening).
2. Disperse strengthening phases precipitate in the process of cooling after hot rolling or normalization.
3. The steel grains and microstructure components become finer, and changes occur in the morphology and location of structural components.

The overall content of alloying elements in low-alloy steels does not exceed 2.5%. In accordance with carbon content and principles of strengthening they can be divided into three groups:

1. Low-carbon steels (0.11–0.22% C) used in the hot-rolled or normalized states. Thermal treatment of such steels (quenching and tempering) only slightly improves their strength characteristics.
2. Low-carbon steels (0.05–0.18% C) strengthened by disperse precipitation of carbides and carbonitrides are used in the normalized or hot-rolled states.
3. Medium-carbon steels (0.25–0.50% C). The required level of properties is achieved in such steels by quenching and high tempering.

The main alloying element in these steels is manganese. Additional alloying of manganese steels with Mo, Nb, and/or V results in formation of "needle" ferrite. Owing to their "needle structure," these steels combine high strength, high viscosity, and cold resistance. Therefore, the steels with 0.05% C, 2% Mn, 0.4% Mo, 0.01–0.02% Nb, or 0.07% V exhibit the following strength characteristics: σ_B = 650–750 MPa, $\sigma_{0.2}$ = 520 MPa, δ = 33%.

These steels are more often used in the normalized state and seldom in the hot-rolled state.

Low-alloy steels with improved strength characteristics include steels containing mainly manganese and silicon. Tensile strength σ_B in these steels is more than 600 MPa and can be as high as 1800 MPa. Their ductility and viscosity depend on their carbon content and on the types of treatment. High-strength low-alloy steels with $\sigma_B = 1800$ MPa are used in the hot-rolled or cold-worked state. However, in this case they are characterized by low impact strength. In the normalized state their strength decreases to 800 MPa, and after quenching and tempering to 600 MPa, the impact strength increasing simultaneously.

B. Boron Steels

Alloying of austenitic steels with rather high amounts of boron results in disperse hardening. The maximum hardness of such steels is attained upon quenching from 1230°C (2250°F) and tempering at 800°C (1470°F). The strength and yield limits increase simultaneously. At the same time the viscosity of such steels decreases more than in the usual austenite steels. The precipitation of borides, because of the high temperature of disperse hardening (800°C; 1470°F), has a beneficial effect on the properties of refractory alloys (chromium, chromium-nickel, chromium-nickel-cobalt alloys).

At test temperatures up to 700°C (1290°F) but still below the temperature of boride precipitation, the refractory characteristics of steels appreciably improve even at a very low boron content.

In low-carbon steels with 16–30% Cr and 6.5–30% Ni, the effect of disperse hardening associated with the presence of boron was not observed. But boron binds the elements stabilizing austenite, thus favoring the formation of martensite.

Carbon steels with 0.2–0.3% C, on the other hand, are hardened considerably owing to the formation of boride-carbide precipitates.

Addition of boron to cemented steels improves their hardenability and increases the strength of the core. Boron somewhat accelerates carburizing, but its influence on the case lessens with increasing carbon content. The greatest effect of boron was observed in steels with 0.7–0.8% C.

The influence of boron is enhanced as the quenching temperature is raised. However, the sensitivity of steel to overheating increases also. Therefore, boron steels usually contain small quantities of titanium and vanadium, which have a favorable effect on the structure of steels when they are heated to high temperatures.

C. Ultrahigh-Strength Steels

Low-carbon steels with a martensite structure have been developed recently that, upon cooling in air, undergo subsequent dispersion hardening at 400–500°C (750–930°F). The tensile strength of such steels is in the range 2200–2500 MPa. As a rule, they are alloyed with 12–18% Ni, up to 10% Cr, 3–5% Mo, and 0.6–1.0% Ti. These martensite-aging steels are distinguished by having a low temperature of brittle fracture, very low sensitivity to cracks, and high strength characteristics.

The strengthening of martensite-aging alloys is a result of three processes: strengthening of substitutional solid solution in the course of alloying, strengthening brought about by the martensite $\gamma \rightarrow \alpha$ transformation, and strengthening connected with different stages of the solid solution precipitation accompanied by formation of segregates and disperse particles of metastable and stable phases, the main contribution being made by the second and third processes. As a consequence of the martensite $\gamma \rightarrow \alpha$ transformation (during cooling in air), a fine substructure with a high density of dislocations is formed. The particles of intermetallics 50–100

Å in size are found at the stage of maximum strengthening. These particles are coherently connected with the matrix.

The high resistance of martensite-aging alloys to brittle fracture is determined by the high viscosity of the matrix—the low-carbon martensite alloyed with Ni and Co, which enhance the mobility of dislocations. In addition, a high density of dislocations in martensite is responsible for high dispersion and homogeneous distribution of phases precipitated during aging.

Alloying with Mo suppresses the precipitation of particles of strengthening phases at the grain boundaries and prevents intergrain brittle fracture. Alloying with Co increases the martensite start temperature M_s and ensures a 100% martensite structure. At the same time, alloying with cobalt reduces the solubility of molybdenum and fosters dispersion hardening.

At low carbon content and moderate cooling rates, the martensite structure in martensite-aging alloys is obtained through relatively high degrees of alloying. At 10–18% Ni, the point M_s lowers so significantly that the γ transformation can be realized only according to the martensite mechanism.

Compared with the high-strength manganese steels considered in Section V.A, the martensite-aging steels are distinguished by a greater degree of alloying of the γ-solid solution. This promotes almost complete transformation of austenite to martensite. A wide range of alloying elements ensures a stronger solid solution strengthening and increases the volume fraction of disperse particles precipitated during phase aging. The above-mentioned three factors are responsible for considerable enhancement of strength characteristics of low-alloy steels.

The high-strength state of alloys can be obtained by using various external means of affecting their structure. The most advantageous of these is low-temperature thermomechanical treatment (LTMT), which consists of deformation of the undercooled austenite in the region of high stability and subsequent quenching. Undercooling of austenite is used to achieve deformation below the temperature of its recrystallization. Such treatment allows the attainment of advanced mechanical properties. The results gained at LTMT can be achieved by such factors as the composition of the steel, the temperature of austenitization, the rate of cooling to the deformation temperature, the temperature of deformation and holding time at this temperature, the degree and rate of deformation, the rate of cooling to room temperature, and final tempering conditions. The most important are the composition of steel, the temperature, and the degree of steel deformation.

Deformation of the undercooled austenite should be completed prior to the beginning of the bainite transformation. In conformity with this, the steels undergoing LTMT should contain austenite-stabilizing elements. LTMT strengthening is usually employed for high-alloy steels with 1–7% Cr, 1–5% Ni, ≤0.5% V, ≤2.5% Mo, and ≤2% Si and sometimes with other additions as well.

The strengthening of steels in LTMT depends on their carbon content. The strengthening effect of carbon is more pronounced in LTMT than in conventional quenching.

With an increasing amount of deformation the yield stress of steel increases continuously. When the thickness of a billet decreases by 1% in LTMT, by rolling, the tensile strength is increased by 7 ± 2 MPa.

A decrease in deformation temperature results in a more intensive strengthening of steel in LTMT. The strength characteristics of steel after a small (up to 30%) deformation are lower and less sensitive to changes in the deformation temperature than at high degrees of compression.

At deformations of up to 20–30%, LTMT leads to a sharp drop in ductility; with a rise in the degree of compression above this value, the ductility begins to increase. There is a critical degree of deformation in LTMT above which the ductility of steel is sufficient. As the temperature of deformation increases, the ductility also increases.

In LTMT, the steel should be tempered as after the usual quenching. It is the opinion of the majority of researchers that the strengthening effect of LTMT is retained up to 350–400°C (660–750°F). If steel is alloyed with Mo, V, or W, the strengthening effect of LTMT persists up to 500°C (930°F). The tensile strength of steel alloyed with tungsten is 2600 MPa after quenching at 350°C (660°F) and 2450 MPa after quenching at 500°C (930°F).

The rate of cooling after deformation affects the properties of steel undergoing LTMT only if nonmartensitic structures are formed at insufficiently strong cooling.

The study of the fine structure of alloys subjected to low-temperature thermal treatment has allowed us to explain the appearance of superhigh strength properties at rather satisfactory degrees of ductility by two structural factors: considerable reduction of size of martensite crystals and changes in their morphology. This can be attributed to the emergence of a cellular structure during deformation of undercooled austenite. The sites of dislocation pileups in austenite remain the sites where dislocations accumulate in martensite after the transformation. Upon LTMT deformation, the fragmentation of austenite crystals results in the fragmentation of the martensite structure. Individual fragments measuring fractions of a micrometer mutually disoriented through 10–15° are joined with each other by dense dislocation pileups. These fragments, in turn, consists of 100–200 Å fragments disoriented relative to each other through angles greater than 1°.

Thus, one of the possible mechanisms of strengthening in LTMT is connected with the creation of a high density of structural imperfections in austenite as a result of deformation and the inheritance by martensite of the dislocation structure of the work-hardened austenite. This mechanism provides the most comprehensive explanation for the high strength of martensite obtained with LTMT.

D. Martensitic Stainless Steels

Pure iron and low-alloy steels are not resistant to corrosion in the atmosphere, water, or many other media. The resistance of steel to corrosion can be enhanced by alloying it with various elements. High strength of such steels is achieved primarily by quenching to obtain the martensite structure and through its subsequent aging.

In martensitic stainless steels, the amount of martensite necessary for strengthening is formed after high-temperature heating and subsequent cooling to room temperature at a relatively small content of alloying components. The majority of alloying additions improve the resistance of martensite by lowering the point M_s. The possibilities for anticorrosion alloying of martensitic steels are limited.

In austenitic-martensitic steels (transition class), quenching does not lead to the complete transformation of austenite to martensite because of the low position of the point M_s. Consequently, no considerable increase in strength occurs. The degree of the $\gamma \rightarrow \alpha$ transformation in these steels can be increased by means of (1) deep freezing treatment to temperatures below M_s; (2) plastic deformation below M_s; and (3) heating in the region of the most intensive precipitation of alloyed carbides from austenite (700–750°C; 1290–1380°F); when the matrix is depleted in alloying elements, the resistance of austenite decreases.

The austenitic-martensitic steels admit a high degree of alloying and therefore afford more possibilities for achieving total corrosion resistance and high strength. Such alloying elements as copper, tungsten, nickel, molybdenum, silicon, and chromium lower the martensite point at direct $\gamma \rightarrow \alpha$ transformation. The intensity of the influence of one or another element depends on their combination.

Cold plastic deformation initiates the martensite transformation. The less stable the austenite and the lower the deformation temperature, the quicker the transformation. The strength of austenitic-martensitic and martensitic stainless steels increases with decreasing deformation temperature.

If after treatment these steels contain 70–90% martensite, their yield stress can amount to 700–1000 MPa and their tensile strength to 1100–1400 MPa. Further improvement of strength is achieved by aging the martensite. This enhancement of strength is attributed to segregation of the Guinier–Preston regions type.

The effect of martensitic aging is observed when steels are alloyed with titanium, beryllium, aluminum, manganese, zirconium, niobium, copper, or certain other elements. Depending on the alloying elements, intermetallic phases of the types A_3B (Ni$_3$Ti), Ni$_3$Al, Ni$_3$Mn, Ni$_3$Be), A_2B [Fe$_2$Mo(Fe, Ni, Co)$_2$], or AB (NiTi, NiAl, NiMn) precipitate during aging.

Greater strength values can be achieved if the deformation of steel upon quenching proceeds below the temperature M_{c1} under rather high compression. On the one hand, this accelerates the martensite transformation, and on the other hand, aging takes place in the martensite strengthened by deformation (sometimes also in the presence of the deformation-strengthened austenite). After complete thermal treatment, steels have the following characteristics: $\sigma_{0.2} =$ 830–1200 MPa, $\sigma_B =$ 1200–1300 MPa.

The conditions of quenching are set with allowance for complete dissolution of carbides subject to the absence of excessive grain growth. Deep freezing treatment after quenching ensures a more complete transformation of austenite to martensite. The amount of martensite can be as high as 70–90%.

The conditions of aging should provide the required set of mechanical properties and corrosion resistance. The maximum strength values are attained as a result of aging in the temperature range 450–500°C (840–930°F). At the same time, the best corrosion resistance is attained at the lower aging temperature range of 350–280°C (660–540°F) (high total corrosion resistance is obtained at the stage preceding precipitation of strengthening phases).

E. Precipitation-Hardening Steels

As is known, steels are classified into structural, spring, tool, and heat-resistant alloy steels in conformity with their application. This section considers the behavior of precipitation-hardening alloys in each of these groups.

1. Structural Steels

Low-carbon manganese steels (0.1–0.2% C) containing 1.3–1.7% Mn, 0.10–0.20% V, about 0.1% Ti, and ~0.05% Al can be classified as steels strengthened with disperse precipitates. Such compositions favor the formation of disperse precipitates of vanadium and titanium carbonitrides or aluminum nitrides. These disperse precipitates can improve not only the strength of the steel but also, owing to grain refinement, its viscosity and cold resistance. A number of industrial alloys with carbonitride strengthening have been developed. Usually, these steels are used in the normalized state. Their properties are determined by the degree of dissolution of strengthening phases in the process of heating.

Another group of low-alloy precipitation-hardening steels includes "low-pearlite" steels, which contain up to 0.1% C and up to 2% Mn as well as vanadium (~0.1%), niobium (~0.06%), and sometimes molybdenum (~0.15–0.3%). Aluminum (up to 0.05%) can also be present in these steels. The properties of the steels under consideration are formed in the process of rolling during precipitation of disperse particles of the strengthening phase and grain refinement. The conditions of rolling should ensure maximum dissolution of components that subsequently cause the formation of disperse particles. These particles strengthen ferrite, which leads to grain refinement. The rolling temperature of low-pearlite steels depends on their composition and the strength and viscosity requirements. A high heating temperature ($\sim$1200°C; 2190°F) ensures more complete dissolution of vanadium and niobium. This contributes to the strengthening effect during precipitation of phases containing these elements. However, owing to the grain growth, the strength of these alloys is lower than in the case of heating at 1050–1100°C (1920–2010°F).

The temperature at the end of rolling of low-pearlite steels is usually reduced to 700–800°C (1300–1475°F). This is due to (1) a decrease in the austenite grain size; (2) an increase in the degree of dispersion of the strengthening phase and, hence, enhancement of the hardening effect; and (3) displacement of the $\gamma \rightarrow \alpha$ transformation to the region of lower temperatures, which results in a finer ferrite grain.

2. Spring Steels

Alloys based on Fe-Ni, Fe-Ni-Cr, Co-Ni-Cr, Ni-Cr, and other systems, predominantly with titanium and aluminum or niobium additions, are used for spring steels strengthened by precipitation hardening. The particles of strengthening phases in these alloys precipitate during aging (tempering). Additional improvement of the strength properties of these alloys can be achieved through plastic deformation between quenching and aging. In this case the precipitation of the supersaturated solid solution may proceed according to a discontinuous mechanism. If the discontinuous precipitation cells completely occupy each grain (which is possible for a very fine grain structure), a very strong strengthening of alloys takes place. In the process of aging, additional refinement of the initial grain occurs during discontinuous precipitation.

Strengthening is observed in alloyed martensite-aging steels under developing of disperse particles of precipitating phases. A large number of steels differing in composition and properties are used in industry. In addition to 0.4–0.8% C, they contain at least two of such alloying elements as Si, Cr, V, Mo, Mn, more rarely Ni and W. Isothermal quenching with subsequent tempering is advantageous for these steels, especially for those containing silicon. The maximum elastic limit in alloyed steels is attained with tempering at 300–350°C (570–660°F). These tempering conditions correspond to the conditions of a sufficiently complete precipitation of austenite accompanied by preservation of a high density of dislocations, since disperse particles of carbides hammer the redistribution and annihilation of dislocations. In addition, the carbide particles increase the resistance to low plastic deformation.

For carbon steels, the amount of carbides can be increased and the martensite point can be lowered owing to the higher carbon content. This brings about a significant improvement of the strength characteristics of such steels, the highest properties being achieved when strong carbide-forming elements (e.g., vanadium) enter into their composition.

Martensite-aging alloys containing nickel and titanium possess the best set of properties. Such alloys are quenched from 870–1150°C (1600–2100°F) depending on their titanium content; the greater the titanium content, the higher the quenching temperature. The finer the grain, the higher the properties of these steels. Fine grain can be attained either by multiple $\gamma \rightarrow \alpha$ transformations or through deformation and recrystallization processes. To reduce the quantity of austenite retained upon quenching, a deep freezing treatment (–70°C; –94°F) is employed. Then comes aging at 450°C (840°F) for 6 h, during which NiTi or Ni_3Ti phases precipitate.

Austenitic steels are also strengthened as a result of precipitation hardening. They may contain chromium, nickel, titanium, or molybdenum. Upon quenching, these alloys have the structure of the γ solid solution with chromium, titanium, and titanium carbonitride inclusions. The properties of aged alloys depend on the quenching temperature, which determines the degree of supersaturation of the solid solution, and on the cooling rate, which should be as high as possible. Aging may proceed by the discontinuous or continuous mechanism.

3. Tool Steels

Tool steels strengthened by precipitation-hardening tempering on the basis of the initial martensite structure are used to manufacture dies for cold deformation of steels. As a result of tempering, the hardness and strength characteristics of steels are enhanced when strengthening phases (carbides) precipitate from martensite. The retained austenite, a phase with low hardness, transforms to martensite. These processes increase the yield stress under compression but reduce viscosity.

Precipitation-hardening strengthening is also characteristic of heat-resistant steels. The structure of these steels represents a martensite matrix with particles of the strengthening phases—carbides or intermetallics—precipitating during tempering.

The main principles of heat treatment of precipitation-hardening tool steels are now considered in greater detail. The basic operations of heat treatment are annealing, quenching, and tempering.

The annealing heating temperature is chosen a little higher than A_1. It is 760–780°C (1400–1435°F) for carbon steels, 780–810°C (1435–1490°F) for alloy steels, and 830–870°C (1525–1600°F) for high-alloy chromium steels, the holding time being 2–3 h.

Quenching of tool steels is aimed at obtaining martensite with a high concentration of carbon and alloying components with retained fine-grain structure. That is why quenching is carried out at temperatures corresponding to complete dissolution of the basic carbides in austenite. These temperatures, however, should not be conducive to austenite grain growth. Usually, the quenching temperature corresponds to the temperature of heating: it is a little higher than A_1 for steels in which the main carbide phase is cementite; up to 1000–1060°C (1832–1940°F) for steels with a chromium-based carbide phase of types Me_7C_3 and $Me_{23}C_6$, and 1080–1100°C (1975–2010°F) for steels with greater carbide content of the type $Me_{23}C_6$.

At tempering, the assigned level of properties is achieved by changing the structure of the quenched steel. Heating a quenched steel during tempering to 150–200°C (300–390°F) causes the precipitation of small ε-carbide plates from martensite and reduces the carbon concentration. Such tempering only slightly impairs the steel's hardness but significantly improves its strength and viscosity. Heating to 250–280°C (480–535°F) during tempering noticeably decreases the carbon concentration in martensite and enhances the strength and viscosity characteristics of the steel. This tempering permits almost complete removal of the retained austenite.

An appreciable increase in the hardness of steels results from the precipitation of a large number of small carbide particles (intermetallics of alloying elements of the types Me_2C, $Me_{23}C_6$, MeC, and Me_7Me_6) from martensite.

4. Heat-Resistant Alloys

A vast group of heat-resistant alloys consists of austenitic steels strengthened with carbides and intermetallics. To increase heat resistance, elements that strengthen the solid solution and induce precipitation hardening are introduced into iron-nickel-based alloys. These elements include Cr, Mo, W, Nb, V, Ti, and Al. To acquire high heat resistance, the alloys undergo double quenching. The purpose of the first quenching is to obtain grains of a certain size and to transform the excess γ' phase [intermetallic phases γ'-$Ni_3(Al,Ti,B)$] to a solid solution. Quenching is followed by cooling in air. The γ' phase partially precipitates in this process. The aim of the second quenching (1050°C; 1920°F) with subsequent aging is to obtain disperse precipitates of the γ' phase 200–500 Å in diameter. As a result of these quenching procedures, the strengthened alloy contains a certain number of larger precipitates along with fine inclusions. This structure ensures high strength and the necessary margin of ductility.

F. Transformation-Induced Plasticity (TRIP) Steels

There are numerous examples of improving the plasticity of load-bearing samples under the influence of phase transformations of the diffusion and shear types. The term "transformation-induced plasticity" was proposed to denote an improvement of plasticity under martensite transformation. High plasticity of steels below the critical point was called by A. P. Gulyaev "subcritical superplasticity."

The TRIP effect appears under the action of high stresses that exceed the yield stress of austenite. In the segment where localization of flow sets in, martensite deformation occurs. This segment is stronger than austenite, and because of this the flow extends to the neighboring segments of the sample. Thus, the quasiequilibrium flow in TRIP steels is due to high deformation strengthening. The index of the flow stress rate sensitivity remains low.

Transformation-induced plasticity is observed at fixed test temperatures. In the case of isothermal transformations, the volume that undergoes such a transformation reaches a certain level and does not increase further. Therefore, the greatest overall effects of plasticity improvement are observed at cyclic temperature changes that lead to multiple occurrences of the phase transformation.

Transformation-induced plasticity is found when the temperature of the sample is changed (within limits exceeding the temperature range of the phase transformation), the load being constant. The load applied is usually lower than the yield stress of any of the phases involved in the transformation.

In each temperature-changing cycle, the value of deformation is in tenths of a percent. With a large number of cycles it may amount to several hundred percent. The deformation value in one cycle is directly proportional to the applied stress. An increase in the applied stress above a certain limit disrupts the linear dependence of deformation (per cycle) on stress. This is caused by transition from the plastic deformation to the usual deformation at comparatively high stresses. As the volumetric effect of phase transformation increases, the deformation value per cycle increases.

Transformation-induced plasticity can be observed in metals with any grain size, among them coarse-grain metals, and at any temperature, including low temperatures. For example, after 150 temperature-changing cycles in the range of 204–648°C (400–1200°F), a sample of Fe-15.4%Ni alloy with an initial grain size of 150 μm became 160% longer, with no neck formed, owing to the reversible martensite transformation under load.

The deformation mechanisms typical of transformation-induced plasticity have not been clearly established because of difficulties in using direct structural methods during phase transformation when the structure of a sample changes constantly. Among the proposed hypotheses the following may be quoted:

1. Accelerated transfer of dislocations owing to an excess of vacancies formed during volumetric changes
2. Weakening of bonding forces between atoms at the interface at the moment of transformation
3. Changes in the form associated with realization of particular orientations of the formed martensite
4. Summation over the phase and applied stress, which determines the plastic deformation of the weaker phase (ferrite in the case of $\alpha \rightleftharpoons \gamma$ transformation)
5. Formation of ultrafine grain in the course of phase transformation

G. Tool Steels

Tool steels can be classified into four groups according to their application: (1) steels for cutting tools used in mild conditions, (2) steels for cutting tools used in severe conditions, (3) measuring tools; and (4) die steels.

Steels for cutting tools must have high hardness exceeding 60 R_c. Therefore, such tool steels contain a minimum of 0.6% C. The main requirement imposed on steels used in severe conditions (high-speed steels) is stable hardness under long heating. All tool steels fall into four categories: (1) carbon tool steels, (2) alloy tool steels, (3) die steels, and (4) high-speed steels.

1. Carbon Tool Steels

Carbon tool steels contain 0.60–0.74% C, 0.25–0.35% Mn, and 0.30% Si. The quenching temperature of these steels is chosen in conformity with the Fe–C equilibrium diagram. The tetrahedral structure of martensite and internal stresses in quenched steels bring about considerable brittleness. That is why tempering after quenching is an obligatory operation. The tempering temperature is determined by the required working hardness of the tools. Usually it ranges between 180–240°C (350–465°F).

Of great importance in terms of machinability is the structure of annealed steels. Steels with the structure of lamellar pearlite are difficult to machine. Therefore, with the help of annealing at a temperature slightly above A_{c1}, easily worked steels with the structure of globular pearlite are obtained. As a rule, carbon steels are quenched in water. Because of this, tools made of such steels have a soft unannealed core and are less brittle than tools made of through-hardened steels.

2. Alloy Tool Steels

Compared with carbon steels, alloy tool steels possess greater hardenability and wear resistance. This is achieved by the introduction of small quantities of alloying elements, predominantly chromium. For chromium steels, it is imperative that quenching be accompanied by subsequent tempering. If it is necessary to preserve hardness at the level of the quenched state, the tempering temperature should not exceed 150–170°C (300–340°F).

In all cases where quenching should be accompanied by minimum deformation during the pearlite → martensite transformation (pearlite is the initial structure in this process), low-deformation tool steels are used. Such steels can be obtained by alloying with elements that increase the amount of retained austenite in the quenched state, namely, chromium and manganese. These steels contain about 12% Cr and ~1.5% C. The formation of a large amount of carbides $(Cr,Fe)_7C_3$ significantly improves their wear resistance.

These high-chromium steels belong to the ledeburitic class. In the cast state, the initial carbides form the eutectic ledeburite. In forging, the eutectic breaks down and the structure of the steel consists of sorbite-forming pearlite with inclusions of excess carbides. When heated for quenching, the carbides dissolve in austenite. The highest hardness of the steel is achieved upon quenching at ~1050°C (~1920°F).

To obtain high hardness, the steel is quenched in oil. The retained austenite precipitates in the process of cold treatment and tempering. Owing to the greater stability of martensite compared to other steels, the tempering temperature is increased to 200–220°C (390–430°F).

3. Die Steels

Dies operating in the cold state need high-hardness steels. The steel to be used in hot pressing should have low sensitivity to local heating.

Different grades of steels—from carbon to complex alloy steels—are used in the production of dies. Carbon steel is used for dies operating under mild conditions and alloy steel for dies operating under severe conditions. Carbon steel contains 0.6–1% C. Alloy steel includes 0.3–0.7% C, Cr, Si, and sometimes Ni.

Dies made of alloy steels are usually quenched from ~850°C (1560°F) in oil with subsequent tempering at 500–550°C (930–1020°F). Hardness of the steel amounts to 350–400 H_B.

Dies for cold pressing are quenched in the temperature range of 860–1050°C (1580–1920°F) (depending on the steel grade) in oil with subsequent tempering at 200–300°C (390–570°F). Depending on the tempering temperature and the steel grade, the steel hardness is within $R_c = 56$–62.

4. High-Speed Steels

High-speed steels must not only possess high hardness in the hot state, they must also be able to retain it during long heating (red hardness). To preserve hardness during heating, it is necessary to hamper the process of carbide coagulation. For this purpose, special carbides should be formed. Such carbides can be produced if the steel is alloyed with 3% Cr. The special carbide Cr_7C_3 coagulates at high temperatures to a lesser degree than cementite. Noticeable precipitation and coagulation of special Cr, Mo, W, and V carbides occur at temperatures over 500°C (930°F).

All high-speed steels are rated in the ledeburite class and in the cast state have the structure of white hypoeutectic cast iron. As a result of forging, the structure of the high-speed steel changes and the eutectic is broken down into individual carbides. In the annealed state, three types of carbides are observed: coarse primary carbides, smaller secondary carbides, and fine grain carbides entering into the composition of upper bainite. Ferrite, which is found in upper bainite, also contains some alloying impurities.

Heating of the high-speed steel to the point A_{c1} (800–850°C; 1470–1560°F) is not accompanied by structural changes. Above this point, the eutectoid transforms to austenite, the secondary carbides dissolve in the austenite, and it is saturated with carbon and alloying elements.

Solubility of carbides depends on how long the steel is held at the quenching temperature. With an increase in the holding time, there is more complete dissolution of carbides in austenite.

Carbon and alloying elements contained in austenite lower the martensite point and increase the content of retained austenite. At quenching temperatures above 1000°C (1832°F) the martensite point decreases to 0°C (32°F) or lower. This peculiarity is taken advantage of in the heat treatment of tools made of high-speed steels.

In the process of steel tempering, the following structural changes take place. Heating to 100–200°C (212–390°F) causes a small compression, since the tetragonal martensite transforms to the cubic modification. At 300–400°C (570–750°F), hardness deteriorates owing to a decrease in the work hardening of retained austenite. At 500–600°C (930–1110°F), fine disperse carbides precipitate from austenite. Cooling the steel from these temperatures brings about the secondary formation of martensite: depleted austenite transforms to martensite in larger quantities.

The higher the tempering temperature or the longer the tempering time, the greater the amount of retained austenite transformed to martensite. Complete transformation of austenite can be attained by multiple tempering. The microstructure of the quenched and tempered steel should consist of finely dispersed martensite and carbides.

The quenching temperature of high-speed steel should be as high as possible, but at the same time it should not allow intensive grain growth (1260–1280°C; 2300–2340°F). During quenching the steel may be cooled comparatively slowly owing to a low critical rate of quenching (in air or oil). Tempering is an obligatory operation and is usually realized at 560–580°C

(1040–1075°F) for 3 h. To obtain still better properties, two- or three-fold tempering is used, with the holding time at each stage being at least 1 h.

BIBLIOGRAPHY

Belous, M. V., V. T., Cherepin, and M. A. Vasiliev, *Transformations During Tempering of Steel*, Metallurgiya, Moscow, 1973.

Bernshtein, M. L., and A. G. Richshtadt, (Eds.), *Physical Metallurgy and Thermal Treatment of Steels, Handbook*, Vols. I, II, and III, 3rd issue, Metallurgiya, Moscow, 1983.

Blanter, M. E., *Phase Transformations During Thermal Treatment of Steel*, Metallurgiya, Moscow, 1962.

Blanter, M. E., *Physical Metallurgy and Thermal Treatment*, Mashinostroyeniye, Moscow, 1963.

Delle, V. A., *Structural Alloy Steel*, Metallurgiya, Moscow, 1959.

Goldshtein, M. I., S. V. Grachev, and Yu. G. Veksler, *Special Steels*, Metallurgiya, Moscow, 1985.

Gudreman, E., *Special Steels*, Vols. I and II, Metallurgiya, Moscow, 1959.

Gulyaev, A. P., *Physical Metallurgy*, Metallurgiya, Moscow, 1976.

Gulyaev, A. P., *Pure Steel*, Metallurgiya, Moscow, 1975.

Kaschenko, G. A., *Fundamentals of Physical Metallurgy*, Metallurgiya, Moscow, 1964.

Kurdyumov, G. V., L. M. Utevski, and R. I. Entin, *Transformations in Iron and Steel*, Nauka, Moscow, 1977.

Meskin, V. S., *Fundamentals of Steel Alloying*, Metallurgiya, Moscow, 1964.

Novikov, I. I., *Theory of Thermal Treatment of Metals*, Metallurgiya, Moscow, 1986.

Popov, A. V., *Phase Transformations of Metal Alloys*, Metallurgiya, Moscow, 1963.

Vinograd, M. I., and G. P. Gromova, *Inclusions in Alloy Steels and Alloys*, Metallurgiya, Moscow, 1972.

Zimmerman, R., and K. Gunter, *Metallurgy and Materials Science Handbook*, Metallurgiya, Moscow, 1982.

3

HARDENABILITY

Božidar Liščić
*Faculty of Mechanical Engineering and Naval Architecture,
University of Zagreb, Zagreb, Croatia*

I. DEFINITION OF HARDENABILITY

Hardenability, in general, is defined as the ability of a ferrous material to acquire hardness after being austenitized and quenched. This general definition comprises two subdefinitions: the ability to reach a certain hardness level (German: Aufhärtbarkeit) and the hardness distribution within a cross section (German: Einhärtbarkeit).

The ability to reach a certain hardness level is associated with the highest attainable hardness. It depends first of all on the carbon content of the material and more specifically on the amount of carbon dissolved in the austenite after the austenitizing treatment, because only this amount of carbon takes part in the austenite-to-martensite transformation and has relevant influence on the hardness of martensite. Figure 1 shows the approximate relationship between the hardness of the structure and its carbon content for different percentages of martensite [1].

The hardness distribution within a cross section is associated with the change of hardness from the surface of a specified cross section toward the core after quenching under specified conditions. It depends on carbon content and the amount of alloying elements dissolved in the austenite during the austenitizing treatment. It may also be influenced by the austenite grain size. Figure 2 shows the hardness distributions within the cross sections of bars of 100 mm diameter after quenching three different kinds of steel [2].

In spite of quenching the W1 steel in water (i.e., the more severe quenching) and the other two grades in oil, the W1 steel has the lowest hardenability because it does not contain alloying elements. The highest hardenability in this case is that of the D2 steel, which has the greatest amount of alloying elements.

When a steel has high hardenability it achieves a high hardness throughout the entire heavy section (as D2 in Figure 2) even when it is quenched in a milder quenchant (oil). When a steel has low hardenability its hardness decreases rapidly below the surface (as W1 in Figure 2), even when it is quenched in the more severe quenchant (water).

According to their ability to reach a certain hardness level, shallow-hardening high-carbon steels may reach higher maximum hardness than alloyed steels of high hardenability while at the same time achieving much lower hardness values across a cross section. This can be best compared by using Jominy hardenability curves (see Section III.B). Hardenability is an

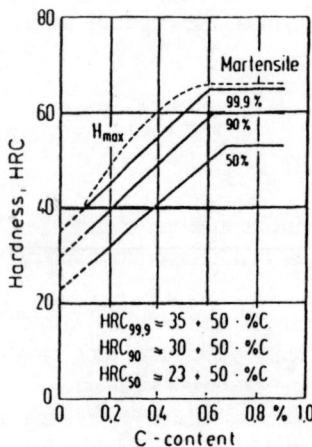

Figure 1 Approximate relationship between hardness in HRC and carbon content for different percentages of martensite. (From Ref. 1.)

inherent property of the material itself, whereas hardness distribution after quenching (depth or hardening) is a *state* that depends on other factors as well.

II. FACTORS INFLUENCING DEPTH OF HARDENING

Depth of hardening is usually defined as the distance below the surface at which a certain hardness level (e.g., 50 HRC) has been attained after quenching. Sometimes it is defined as the distance below the surface within which the martensite content has reached a certain minimum percentage.

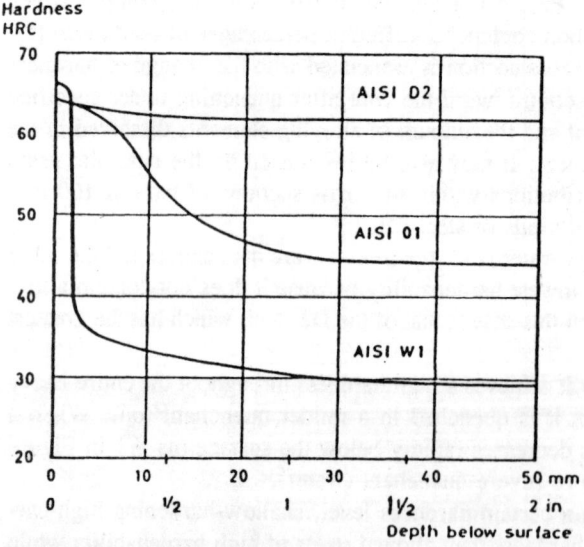

Figure 2 Hardness distributions within cross sections of bars of 100 mm diameter for three different kinds of steel, after quenching. Steel W1 was water-quenched; the rest were oil-quenched. (From Ref. 2.)

Being a consequence of the austenite-to-martensite transformation, the depth of hardening depends on the following factors:

1. Shape and size of the cross section
2. Hardenability of the material
3. Quenching conditions

Quenching conditions include not only the specific quenchant with its inherent chemical and physical properties, but also important process parameters such as bath temperature and agitation rate.

The cross-sectional shape has a remarkable influence on heat extraction during quenching and consequently on the resulting hardening depth. Bars of rectangular cross sections always achieve less depth of hardening than round bars of the same cross-sectional size. Figure 3 is a diagram that can be used to convert square and rectangular cross sections to equivalent circular cross sections. For example, a 38 mm square and a 25 × 100 mm rectangular cross section are each equivalent to a 40 mm diameter circular cross section; a 60 × 100 mm rectangular cross section is equivalent to an 80 mm diameter circle [2].

The influence of cross-sectional size when quenching the same grade of steel under the same quenching conditions is shown in Figure 4A. Steeper hardness decrease from surface to core and substantially lower core hardness values result from quenching a larger cross section.

Figure 4B shows the influence of hardenability and quenching conditions by comparing an unalloyed (shallow-hardening) steel to an alloyed steel of high hardenability when each is quenched in (a) water or (b) oil. The critical cooling rate (v_{crit}) of the unalloyed steel is higher

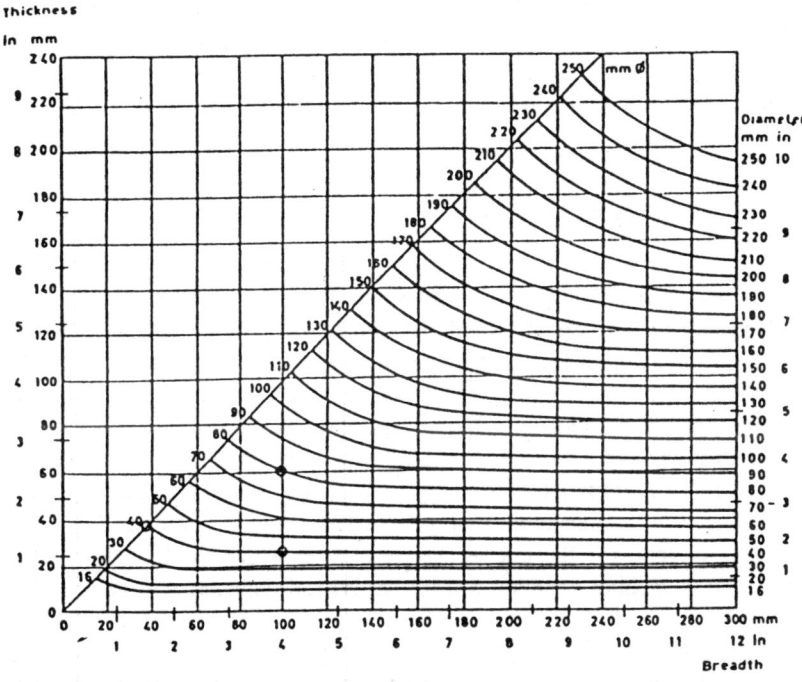

Figure 3 Correlation between rectangular cross sections and their equivalent round sections, according to ISO. (From Ref. 2.)

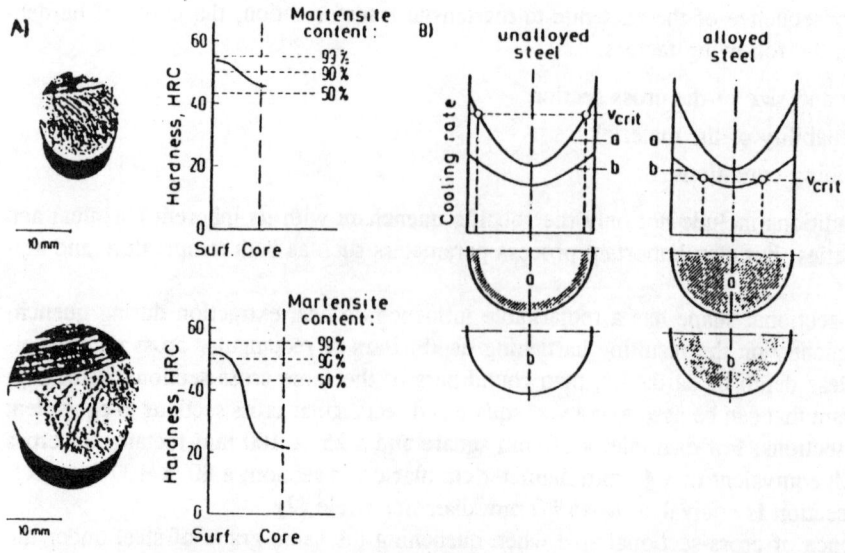

Figure 4 Influence of (A) cross-sectional size and (B) hardenability and quenching conditions on the depth of hardening. (a) Water quenching; (b) oil quenching. v_{crit}, critical cooling rate. (From Ref. 1.)

than the critical cooling rate of the alloyed steel. Only those points on the cross section that have been cooled at a higher cooling rate than v_{crit} could transform to martensite and attain high hardness. With unalloyed steel this can be achieved up to some depth only by quenching in water (curve a); oil quenching (curve b) provides essentially no hardness increase. With alloyed steel, quenching in water (because of the high cooling rate of water) produces a cooling rate greater than v_{crit} even in the core, resulting in through-hardening. Oil quenching (curve b) provides, in this case, cooling rates higher than v_{crit} within quite a large depth of hardening. Only the core region remains unchanged.

III. DETERMINATION OF HARDENABILITY

A. Grossmann's Hardenability Concept

Grossmann's method of testing hardenability [3] uses a number of cylindrical steel bars of different diameters hardened in a given quenching medium. After sectioning each bar at midlength and examining it metallographically, the bar that has 50% martensite at its center is selected, and the diameter of this bar is designated as the *critical diameter* (D_{crit}). The hardness value corresponding to 50% martensite will be determined exactly at the center of the bar of D_{crit}. Other bars with diameters smaller than D_{crit} have more than 50% martensite in the center of the cross section and correspondingly higher hardness, while bars having diameters larger than D_{crit} attain 50% martensite only up to a certain depth as shown in Figure 5. The critical diameter D_{crit} is valid for the quenching medium in which the bars have been quenched. If one varies the quenching medium, a different critical diameter will be obtained for the same steel.

To identify a quenching medium and its condition, Grossmann introduced the quenching intensity (severity) factor H. The H values for oil, water, and brine under various rates of agitation are given in Table 1. From this table, the large influence of the agitation rate on the quenching intensity is evident.

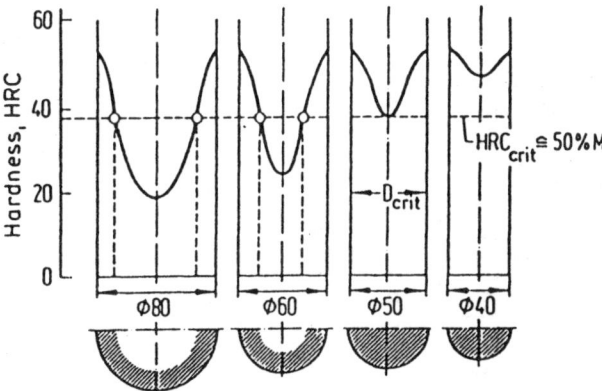

Figure 5   Determination of the critical diameter D_{crit} according to Grossmann. (From Ref. 1.)

To determine the hardenability of a steel independently of the quenching medium, Grossmann introduced the *ideal critical diameter* D_I, which is defined as the diameter of a given steel that would produce 50% martensite at the center when quenched in a bath of quenching intensity $H = \infty$. Here, $H = \infty$ indicates a hypothetical quenching intensity that reduces the surface temperature of the heated steel to the bath temperature in zero time. Grossmann and his coworkers also constructed a chart, shown in Figure 6, that allows the conversion of any value of critical diameter D_{crit} for a given H value to the corresponding value for the ideal critical diameter (D_I) of the steel in question [5].

For example, after quenching in still water $(H = 1.0)$, a round bar constructed of steel A has a critical diameter (D_{crit}) of 28 mm according to Figure 6. This corresponds to an ideal critical diameter (D_I) of 48 mm. Another round bar, constructed of steel B, after quenching in oil $(H = 0.4)$, has a critical diameter (D_{crit}) of 20 mm. Converting this value, using Figure 6, provides an ideal critical diameter (D_I) of 52 mm. Thus, steel B has a higher hardenability than steel A. This indicates that D_I is a measure of steel hardenability that is independent of the quenching medium.

If D_I is known for a particular steel, Figure 6 will provide the critical diameter of that steel for various quenching media. For low- and medium-alloy steels, hardenability as determined by D_I may be calculated from the chemical composition after accounting for austenite grain size. First, the "basic" hardenability of the steel as a function of carbon content and austenite grain size is calculated from Figure 7. The influence of each alloying element is then

Table 1   Grossmann Quenching Intensity Factor H

Method of quenching	H value (in.$^{-1}$)		
	Oil	Water	Brine
No agitation	0.25–0.30	1.0	2.0
Mild agitation	0.30–0.35	1.0–1.1	2.0–2.2
Moderate agitation	0.35–0.40	1.2–1.3	
Good agitation	0.40–0.50	1.4–1.5	
Strong agitation	0.50–0.80	1.6–2.0	
Violent agitation	0.80–1.10	4.0	5.0

Source: Ref. 4.

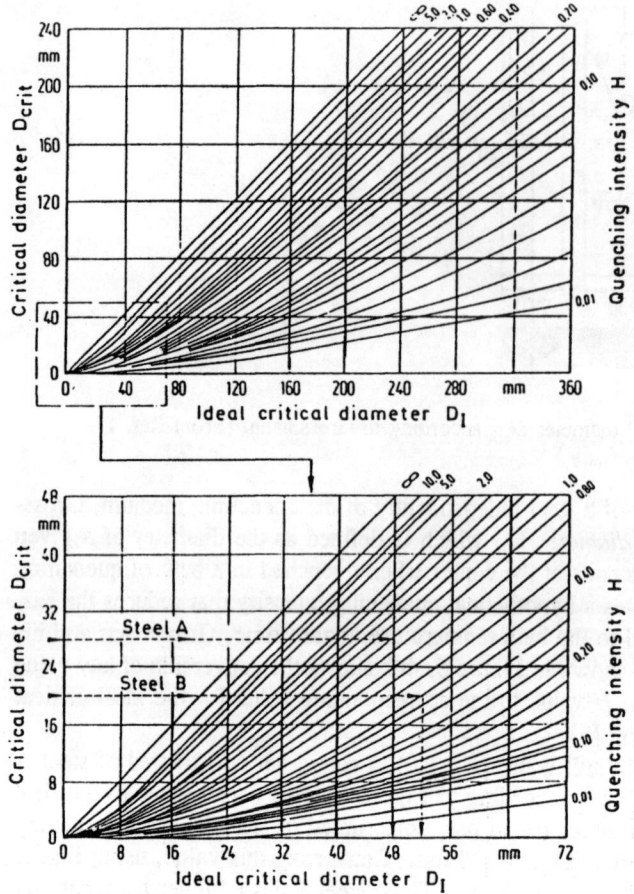

Figure 6 The chart for converting the values of the critical diameter D_{crit} into the ideal critical diameter D_I, or vice versa, for any given quenching intensity H, according to Grossmann and coworkers. (From Ref. 2.)

included by multiplying the "basic" value from Figure 7 by the corresponding multiplying factor, which is determined from Figure 8 according to the weight percent of each element present. For example: If a steel has an austenite grain size of ASTM 7 and the chemical composition C 0.25%, Si 0.3%, Mn 0.7%, Cr 1.1%, Mo 0.2%, then the basic value of hardenability from Figure 7 (in inches) is $D_I = 0.17$. The total hardenability of this steel is

$$D_I = 0.17 \times 1.2 \times 3.3 \times 3.4 \times 1.6 = 3.7 \text{ in.} \tag{1}$$

For these calculations, it is presumed that the total amount of each element is in solution at the austenitizing temperature. Therefore the diagram in Figure 8 is applicable for carbon contents above 0.8% C only if all of the carbides are in solution during austenitizing. This is not the case, because conventional hardening temperatures for these steels are below the temperatures necessary for complete dissolution of the carbides. Therefore, decreases in the "basic" hardenability are to be expected for steels containing more than 0.8% C, compared to values in the diagram. Later investigations by other authors produced similar diagrams that account for this decrease in the basic hardenability that is to be expected for steels with more than 0.8% C, compared to the values shown in Figure 8 [6]. Although values of D_I calcu-

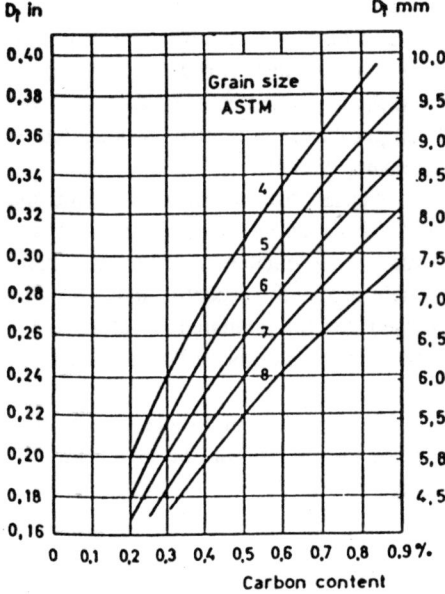

Figure 7 The ideal critical diameter (D_1) as a function of the carbon content and austenite grain size for plain carbon steels, according to Grossmann. (From Ref. 2.)

lated as above are only approximate, they are useful for comparing the hardenability of two different grades of steel.

The most serious objection to Grossmann's hardenability concept is the belief that the actual quenching intensity during the entire quenching process can be described by a single H value. It is well known that the heat transfer coefficient at the interface between the metal surface and the surrounding quenchant changes dramatically during different stages of the quenching process for a vaporizable fluid.

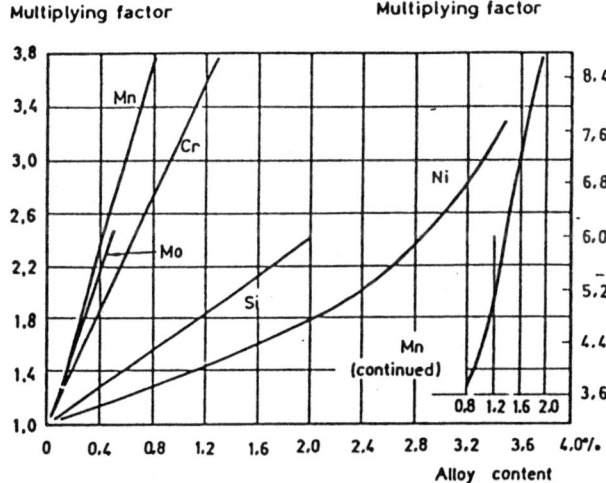

Figure 8 Multiplying factors for different alloying elements when calculating hardenability as D_I value, according to AISI. (From Ref. 2.)

Another difficulty is the determination of the H value for a cross-sectional size other than the one experimentally measured. In fact, H values depend on cross-sectional size [7]. Figure 9 shows the influence of steel temperature and diameter on H values for an 18Cr8Ni round bar quenched in water from 845°C [7]. It is evident that the H value determined in this way passed through a maximum with respect to terminal temperatures. It is also evident that H values at the centers of round bars decreased with increasing diameter.

Values of the quenching intensity factor H do not account for specific quenchant and quenching characteristics such as composition, oil viscosity, or the temperature of the quenching bath. Tables of H values do not specify the agitation rate of the quenchant either uniformly or precisely; that is, the uniformity throughout the quench tank with respect to mass flow or fluid turbulence is unknown. Therefore, it may be assumed that the tabulated H values available in the literature are determined under the same quenching conditions. This assumption, unfortunately, is rarely justified.

In view of these objections, Siebert et al. [8] state: "It is evident that there cannot be a single H-value for a given quenching bath, and the size of the part should be taken into account when assigning an H-value to any given quenching bath."

1. Hardenability in High-Carbon Steels

The hardenability effect of carbon and alloying elements in high-carbon steels and the case regions of carburized steels differ from those in low- and medium-carbon steels and are influenced significantly by the austenitizing temperature and prior microstructure (normalized or spheroidize-annealed). Using Grossman's method for characterizing hardenability in terms of the ideal critical diameter D_I, multiplying factors for the hardenability effects of Mn, Si,

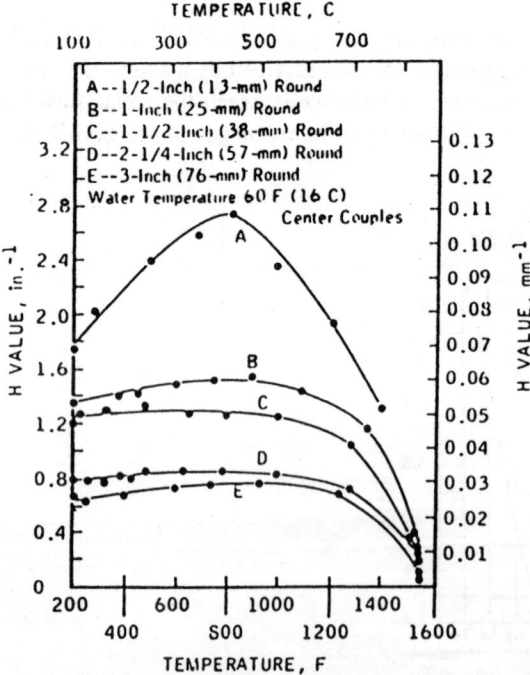

Figure 9   Change of the H value with temperature and size of the round bar. Calculated from cooling curves measured at the center of bars made of 18Cr8Ni steel quenched in water from 845°C, according to Carney and Janulionis. (From Ref. 7.)

Cr, Ni, Mo, and Al were successfully derived [9] for carbon levels ranging from 0.75 to 1.10% C in single-alloy and multiple-alloy steels quenched at different austenitizing temperatures from 800 to 930°C. These austenitizing temperatures encompass the hardening temperatures of hypereutectoid tool steels, 1.10% C bearing steels, and the case regions of carburized steels. All of these steels, when quenched, normally contain an excess of undissolved carbides, which means that the quantity of carbon and alloying elements in solution could vary with the prior microstructure and the austenitizing conditions. The hardenability of these steels is influenced by the carbide size, shape, and distribution in the prior microstructure and by austenitizing temperature and time. Grain size exhibits a lesser effect because hardenability does not vary greatly from ASTM 6-9 when excess carbides are present.

As a rule, homogeneous high-carbon alloy steels are usually spheroidize - annealed for machining prior to hardening. Carburizing steel grades are either normalized, i.e., air-cooled, or quenched in oil directly from the carburizing temperature before reheating for hardening. So different case microstructures (from martensite to lamellar pearlite) may be present, all of which transform to austenite rather easily during reheating for hardening. During quenching, however, the undissolved carbides will nucleate pearlite prematurely and act to reduce hardenability.

In spheroidize-annealed steel, the carbides are present as large spheroids, which are much more difficult to dissolve when the steel is heated for hardening. Therefore the amount of alloy and carbon dissolved is less when one starts with a spheroidized rather than a normalized or quenched microstructure. Nevertheless, it has been demonstrated that a spheroidized prior microstructure actually yields higher hardenability than a prior normalized microstructure, at least for austenitizing temperatures up to approximately 855°C. This effect occurs because larger carbides are not as efficient nuclei for early pearlite formation upon cooling as fine and lamellar carbides and the nuclei are present in lower numbers. With either prior microstructure, if strict control is maintained over austenitizing temperature and time, the solution of carbon and alloy can be reproduced with sufficient consistency to permit the derivation of multiplying factors.

For all calculations, it was important to establish whether pearlite or bainite would limit hardenability because the effects of some elements on these reactions and on hardenability differ widely.

The multiplying factors were calculated according to a structure criterion of D_I to 90% martensite plus retained austenite (or 10% of nonmartensitic transformation) and with reference to a base composition containing 1.0% C and 0.25% of each of the elements Mn, Si, Cr, and Ni, with 0% Mo to ensure that the first transformation product would not be bainite. The 50% martensite hardenability criterion (usually used when calculating D_I) was selected by Grossmann because this structure in medium-carbon steels corresponds to an inflection in the hardness distribution curve. The 50% martensite structure also results in marked contrast in etching between the hardened and unhardened areas and in the fracture appearance of these areas in a simple fracture test. For many applications, however, it may be necessary to through-harden to a higher level of martensite to obtain optimum properties of tempered martensite in the core.

In these instances, D_I values based on 90, 95, or 99.9% martensite must be used in determining the hardenability requirements. These D_I values can be either experimentally determined or estimated from the calculated 50% martensite values using the relationships shown in Figure 10, which were developed for medium-carbon low-alloy steels [10]. A curve for converting the D_I value for the normalized structure to the D_I value of the spheroidize-annealed structure as shown in Figure 11 is also available. New multiplying factors for D_I values were

102   LIŠČIĆ

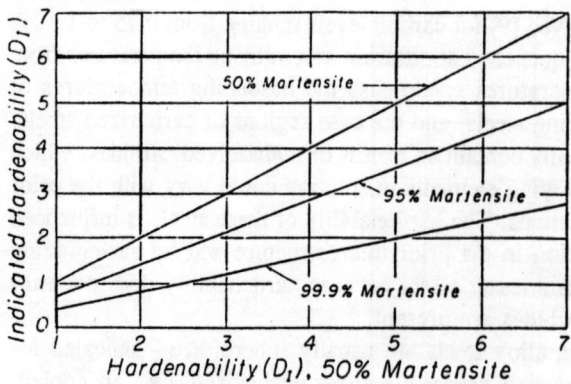

Figure 10 Average relationships among hardenability values (expressed as D_1) in terms of 50%, 95%, and 99.9% martensite microstructures. (From Ref. 10.)

obtained from the measured Jominy curves using the conversion curve modified by Carney shown in Figure 12.

The measured D_1 values were plotted against the percent content of various elements in the steel. These curves were then used to adjust the D_1 value of the steels whose residual content did not conform to the base composition. Once the D_1 value of each analysis was adjusted for residuals, the final step was to derive the multiplying factors for each element from the quotient of the steel's D_1^* and that of the base as follows:

$$f_{Mn} = \frac{D_1^* \text{ at } x\% \text{ Mn}}{D_1} \tag{2}$$

where D_1 is the initial reference value.

Excellent agreement was obtained between the case hardenability results of carburized steels assessed at 1.0% carbon level and the basic hardenability of the 1.0% C steels when quenched from the normalized prior structure. It was thus confirmed that all multiplying factors obtained

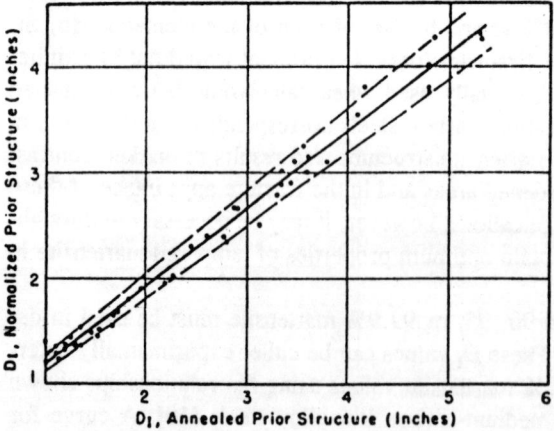

Figure 11 Correlation between hardenability based on normalized and spheroidize-annealed prior structures in alloyed 1.0% C steels. (From Ref. 9.)

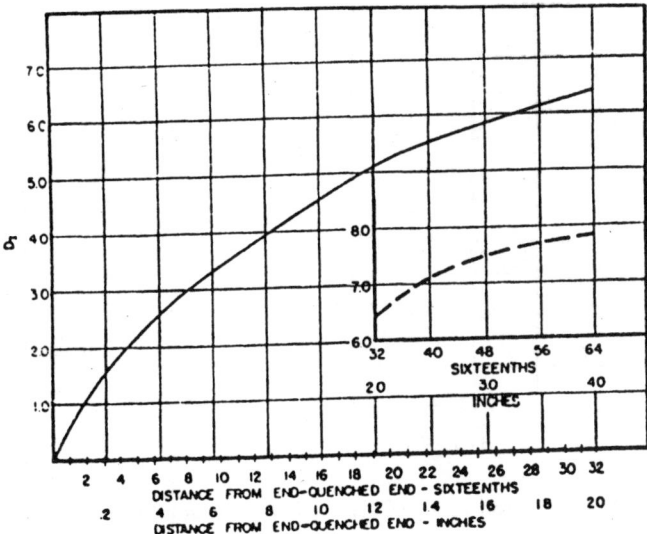

Figure 12 Relationship between Jominy distance and D_I. (From Ref. 9.)

with prior normalized 1.0% C steels could be used to calculate the hardenability of all carburizing grades that are reheated for hardening following carburizing.

Jatczak and Girardi [11] determined the difference in multiplying factors for prior normalized and prior spheroidize-annealed structures as shown in Figures 13 and 14. The new
or revised multiplying factors (which all begin with a value of 1.0 at 0% of the elements) for
C, Mn, Si, Cr, Ni, Mo, and Al are shown on Figures 15–18. The influence of austenitizing
temperature on the specific hardenability effect is evident. The multiplying factors shown on
Figures 15–18 were principally determined in compositions where only single-alloy additions
were made and that were generally pearlitic in initial transformation behavior. Consequently,
these multiplying factors may be applied to the calculation of hardenability of all singly al-

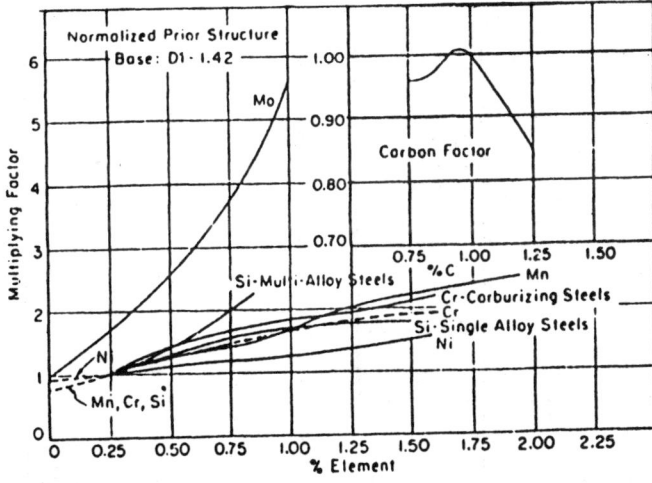

Figure 13 Multiplying factors for calculation hardenability of high-carbon steels of prior normalized
structure. (From Ref. 11.)

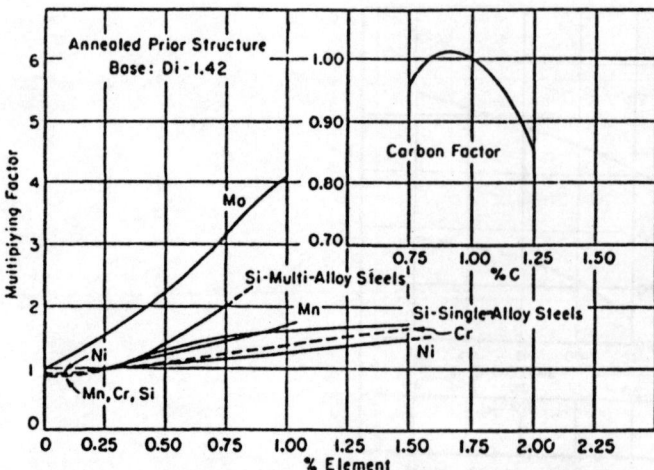

Figure 14 Multiplying factors for calculation of hardenability of high-carbon steels of prior spheroidize-annealed structure. (From Ref. 11.)

loyed high-carbon compositions and to those multialloyed compositions that remain pearlitic when quenched from these austenitizing conditions. This involves all analyses containing less than 0.15% Mo and/or less than 2% total of Ni plus Mn and also less than 2% Mn, Cr, or Ni when they are present individually. Of course, all of the factors given in Figures 15–18 also apply to the calculation of case hardenability of similar carburizing steels that are rehardened from these temperatures following air cooling or integral quenching.

For steels containing more Mo, Ni, Mn, or Cr than the above percentages, the measured hardenability will always be higher than calculated with the single-alloy multiplying factors

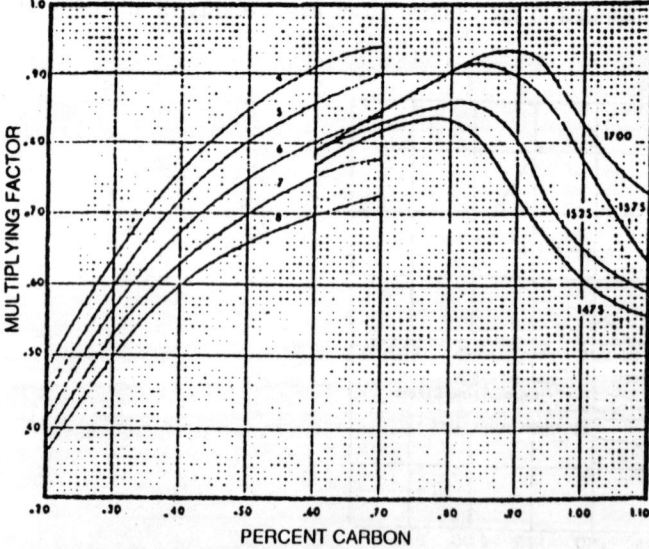

Figure 15 Multiplying factors for carbon at each austenitizing condition. Data plotted on the left-hand side are data from Kramer for medium-carbon steels with grain size variation from ASTM 4 to ASTM 8. (From Ref. 9.)

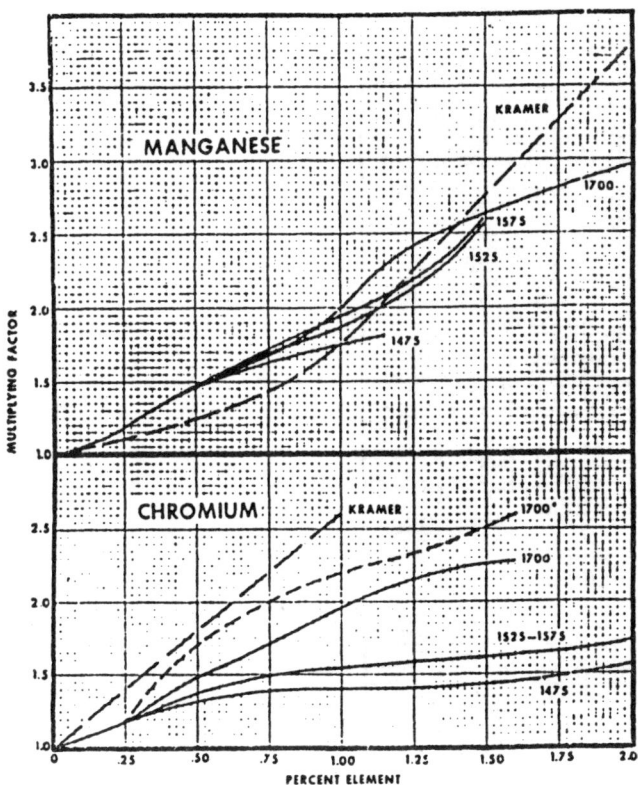

Figure 16 Effect of austenitizing temperature on multiplying factors for Mn and Cr at high carbon levels (Kramer data for medium-carbon steels). (From Ref. 9.)

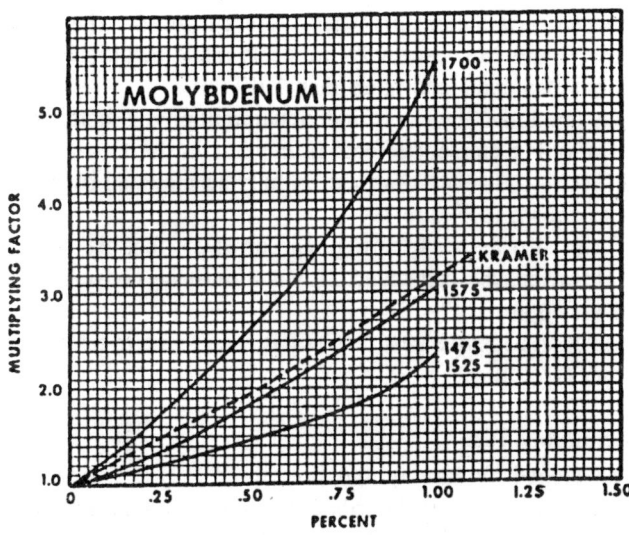

Figure 17 Effect of austenitizing temperature on multiplying factors for Mo at high carbon levels. (From Ref. 9.)

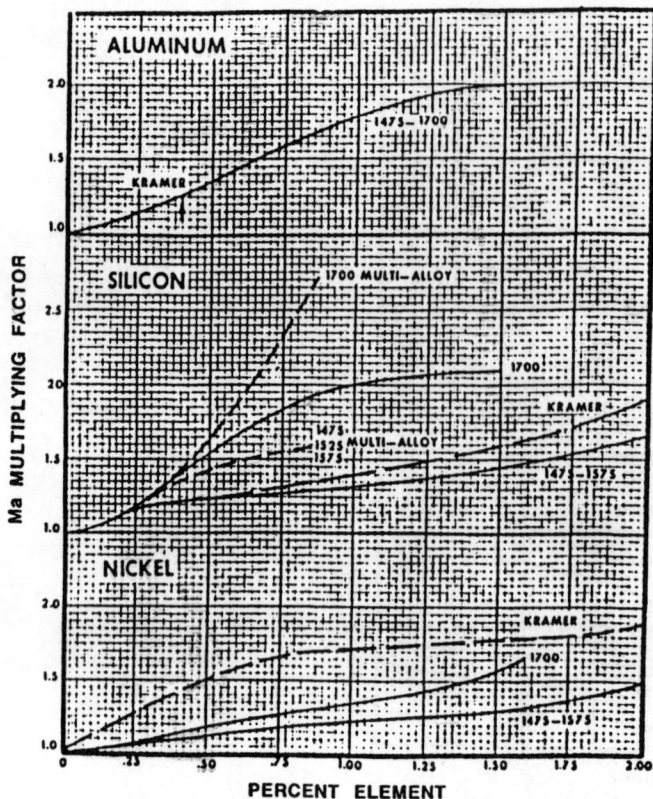

Figure 18 Effect of austenitizing temperature on multiplying factors for Si, Ni, and Al at high carbon levels. (Arrow on Al curve denotes maximum percentage studies by Kramer.) (From Ref. 9.)

because these steels are bainitic rather than pearlitic and also because synergistic hardenability effects have been found to occur between certain elements when present together. The latter effect was specifically noted between Ni and Mn, especially in steels made bainitic by the addition of 0.15% or more Mo and that also contained more than 1.0% Ni.

The presence of synergistic effects precluded the use of individual multiplying factors for Mn and Ni, as the independence of alloying element effects is implicit in the Grossmann multiplying factor approach. This difficulty, however, was successfully surmounted by computing combined Ni and Mn factors as shown in Figure 19.

The factors from Figures 15–18 can also be used for high-carbon steels that are spheroidize-annealed prior to hardening. However, the calculated D_I value must be converted to the annealed D_I value at the abscissa on Figure 10.

The accuracy of hardenability prediction using the new factors has been found to be within $\pm 10\%$ at D_I values as high as 660 mm (26.0 in.).

B. The Jominy End-Quench Hardenability Test

The end-quench hardenability test developed by Jominy and Boegehold [12] is commonly referred to as the Jominy test. It is used worldwide, is described in many national standards, and is available as an international standard [13]. This test has the following significant advantages:

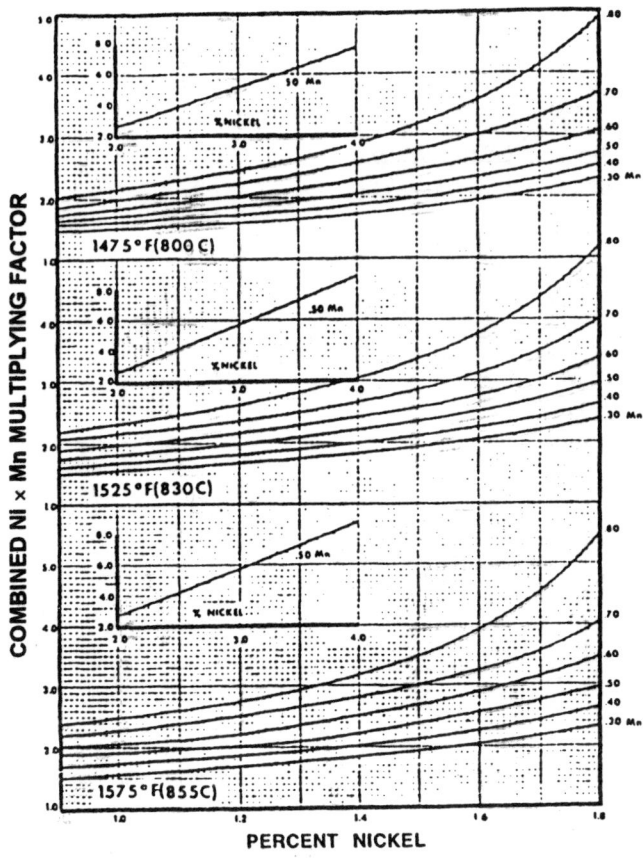

Figure 19 Combined multiplying factor for Ni and Mn in bainitic high-carbon steels quenched from 800 to 855°C, to be used in place of individual factors when composition contains more than 1.0% Ni and 0.15% Mo. (From Ref. 9.)

1. It characterizes the hardenability of a steel from a single specimen, allowing a wide range of cooling rates during a single test.
2. It is reasonably reproducible.

The steel test specimen (25 mm diameter × 100 mm) is heated to the appropriate austenitizing temperature and soaked for 30 min. It is then quickly transferred into the supporting fixture (Jominy apparatus) and quenched from the lower end by spraying with a jet of water under specified conditions as illustrated in Figure 20. The cooling rate is the highest at the end where the water jet impinges on the specimen and decreases from the quenched end, producing a variety of microstructures and hardnesses as a function of distance from the quenched end. After quenching, two parallel flats, approximately 0.45 mm below surface, are ground on opposite sides of the specimen and hardness values (usually HRC) are measured at 1/16 in. intervals from the quenched end and plotted as the Jominy hardenability curve (see Figure 21). When the distance is measured in millimeters, the hardness values are taken at every 2 mm from the quenched end for at least a total distance of 20 or 40 mm, depending on the steepness of the hardenability curve, and then every 10 mm. On the upper margin of the Jominy

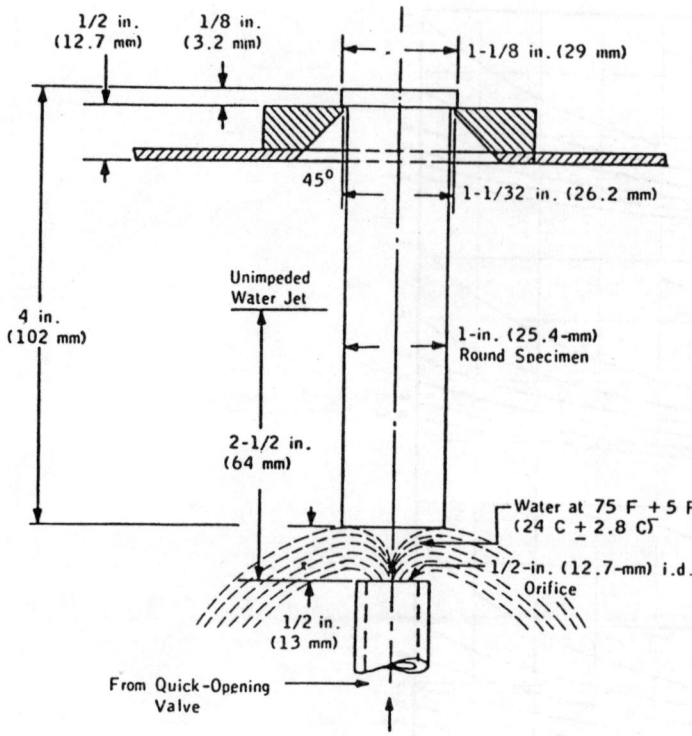

Figure 20 Jominy specimen and its quenching conditions for end-quench hardenability test.

hardenability diagram, approximate cooling rates at 700°C may be plotted at several distances from the quenched end.

Figure 22 shows Jominy hardenability curves for different unalloyed and low-alloyed grades of steel. This figure illustrates the influence of carbon content on the ability to reach a certain hardness level and the influence of alloying elements on the hardness distribution expressed as hardness values along the length of the Jominy specimen. For example, DIN Ck45, an unalloyed steel, has a carbon content of 0.45% C and exhibits a higher maximum hardness (see the value at 0 distance from the quenched end) than DIN 30CrMoV9 steel, which has only 0.30% C. However, the latter steel is alloyed with Cr, Mo, and V and shows a higher hardenability by exhibiting higher hardness values along the length of the specimen.

The Jominy end-quench test is used mostly for low-alloy steels for carburizing (core hardenability) and for structural steels, which are typically through-hardened in oils and tempered. The Jominy end-quench test is suitable for all steels except those of very low or very high hardenability, i.e., $D_I < 1.0$ in. or $D_I > 6.0$ in. [8]. The standard Jominy end-quench test cannot be used for highly alloyed air-hardened steels. These steels harden not only by heat extraction through the quenched end but also by heat extraction by the surrounding air. This effect increases with increasing distance from the quenched end.

The reproducibility of the standard Jominy end-quench test was extensively investigated, and deviations from the standard procedure were determined. Figure 23 shows the results of an end-quench hardenability test performed by nine laboratories on a single heat of SAE 4068 steel [8]. Generally, quite good reproducibility was achieved, although the maximum differ-

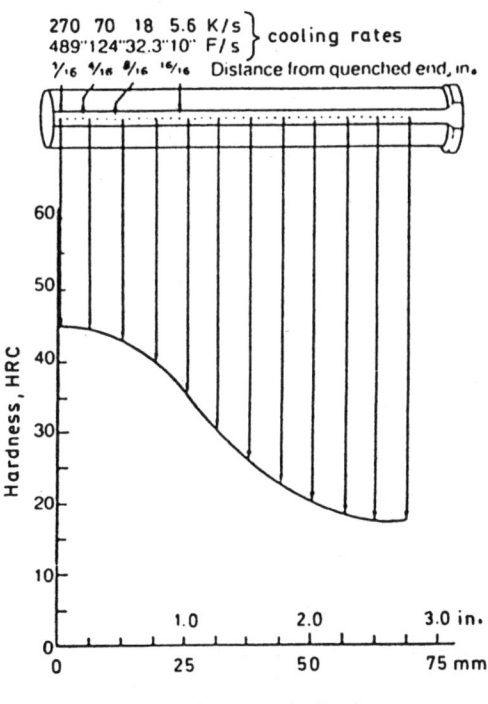

Figure 21 Measuring hardness on the Jominy specimen and plotting the Jominy hardenability curve. (From Ref. 20.)

ence may be 8–12 HRC up to a distance of 10 mm from the quenched end depending on the slope of the curve. Several authors who have investigated the effect of deviations from the standard test procedure have concluded that the most important factors to be closely controlled are austenitization temperature and time, grinding of the flats of the test bar, prevention of grinding burns, and accuracy of the measured distance from the quenched end. Other variables such as water temperature, orifice diameter, free water-jet height, and transfer time from the furnace to the quenching fixture are not as critical.

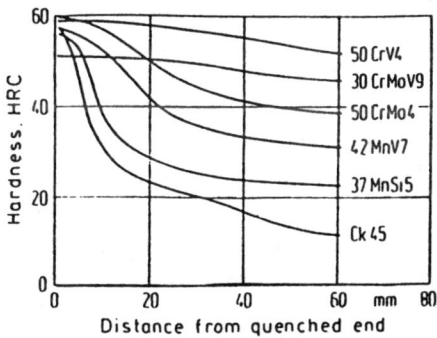

Figure 22 Jominy hardenability curves (average values) for selected grades of steel (designations according to German DIN standard). (From Ref. 1.)

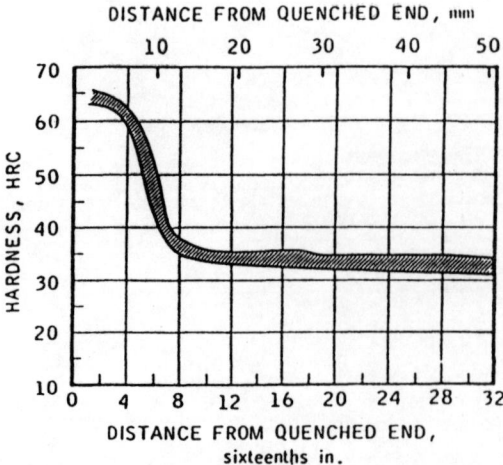

Figure 23 Reproducibility of the end-quench hardenability test. Hardenability range (hatched area be-tween curves) based on tests by nine laboratories on a single heat of SAE 4068 steel. (From Ref. 8.)

1. Hardenability Test Methods for Shallow-Hardening Steels

If the hardenability of shallow-hardening steels is measured by the Jominy end-quench test, the critical part of the Jominy curve is from the quenched end to a distance of about 1/2 in. Because of the high critical cooling rates required for shallow-hardening steels, the hardness decreases rapidly for every incremental increase in Jominy distance. Therefore the standard Jominy specimen with hardness readings taken at every 1/16 in. (1.59 mm) cannot describe precisely the hardness trend (or hardenability). To overcome this difficulty, it may be helpful to (1) modify the hardness survey when using standard Jominy specimens or (2) use special "L" specimens.

Hardness Survey Modification for Shallow-Hardening Steels.   The essential elements of this procedure, described in ASTM A255, are as follows.

1. The procedure in preparing the specimen prior to making hardness measurements is the same as for standard Jominy specimens.
2. An anvil that provides a means of very accurately measuring the distance from the quenched end is essential.
3. Hardness values are obtained from 1/16 to 1/2 in. (1.59–12.7 mm) from the quenched end at intervals of 1/32 in. (0.79 mm). Beyond 1/2 in., hardness values are obtained at 5/8, 3/4, 7/8, and 1 in. (15.88, 19.05, 22.23, and 25.4 mm) from the quenched end. For readings within the first 1/2 in. from the quenched end, two hardness traverses are made, both with readings 1/16 in. apart: one starting at 1/16 in. and being completed at 1/2 in. from the quenched end, and the other starting at 3/32 in. (2.38 mm) and being completed at 15/32 in. (11.91 mm) from the quenched end.
4. Only two flats 180° apart need be ground if the mechanical fixture has a grooved bed that will accommodate the indentations of the flat surveyed first. The second hardness traverse is made after turning the bar over. If the fixture does not have such a grooved bed, two pairs of flats should be ground, the flats of each pair being 180° apart. The two hardness surveys are made on adjacent flats.
5. For plotting test results, the Standard Form for Plotting Hardenability Curves should be used.

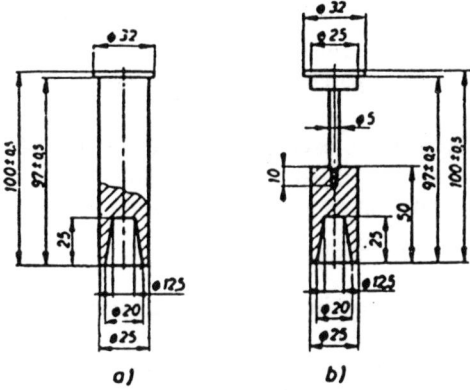

Figure 24 L specimens for Jominy hardenability testing of shallow-hardening steels. All dimensions in millimeters. (From Ref. 14.)

The Use of Special "L" Specimens.   To increase the cooling rate within the critical region when testing shallow-hardening steels, an "L" specimen, as shown in Figure 24, may be used. The test procedure is standard except that the stream of water rises to a free height of 100 ± 5 mm (instead of the 63.55 mm with a standard specimen) above the orifice, without the specimen in position.

The SAC Hardenability Test.   The SAC hardenability test is another hardenability test for shallow-hardening steels, other than carbon tool steels, that will not through-harden in sizes larger than 25 mm (1 in.) in diameter. The acronym SAC denotes Surface Area Center and is illustrated in Figure 25. The specimen is 25.4 mm (1 in.) in diameter and 140 mm (5.5 in.) long. After normalizing at the specified temperature of 1 h and cooling in air, it is austentized by being held at temperature for 30 min and quenched in water at 24 ± 5°C, where it is allowed to remain until the temperature is uniform throughout the specimen.

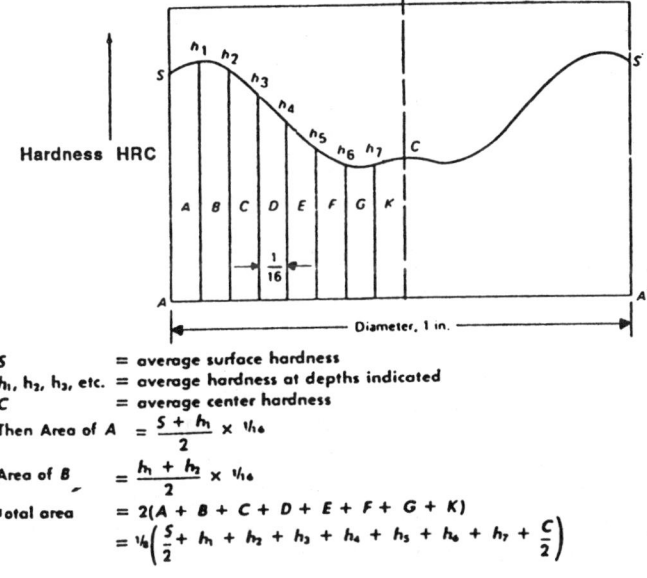

S = average surface hardness
h_1, h_2, h_3, etc. = average hardness at depths indicated
C = average center hardness

Then Area of A = $\dfrac{S + h_1}{2} \times \frac{1}{16}$

Area of B = $\dfrac{h_1 + h_2}{2} \times \frac{1}{16}$

Total area = $2(A + B + C + D + E + F + G + K)$

$= \frac{1}{8}\left(\dfrac{S}{2} + h_1 + h_2 + h_3 + h_4 + h_5 + h_6 + h_7 + \dfrac{C}{2} \right)$

Figure 25   Estimation of area according to SAC method. (From Ref. 15.)

After the specimen has been quenched, a cylinder 25.4 mm (1 in.) in length is cut from its middle. The cut faces of the cylinder are carefully ground parallel to remove any burning or tempering that might result from cutting and to ensure parallel flat surfaces for hardness measuring.

First HRC hardness is measured at four points at 90° to each other on the surface. The average of these readings then becomes the surface reading. Next, a series of HRC readings are taken on the cross section in steps of 1/16 in. (1.59 mm) from the surface to the center of the specimen. From these readings, a quantitative value can be computed and designated by a code known as the SAC number.

The SAC code consists of a set of three two-digit numbers indicating (1) the surface hardness, (2) the total Rockwell (HRC)-inch area, and (3) the center hardness.

For instance, SAC 60-54-43 indicates a surface hardness of 60 HRC, a total Rockwell-inch area of 54, and a center hardness of 43 HRC. The computation of the total Rockwell-inch area is shown in Figure 25.

Hot Brine Hardenability Test. For steels of very low hardenability, another test has been developed [16] that involves quenching several specimens 2.5 mm (0.1 in.) thick and 25 mm (1.0 in.) square in hot brine at controlled temperatures (and controlled quench severity) and determining the hardness and percent martensite of each specimen. The brine temperature for 90% martensite structure expressed as an "equivalent diameter" of a water-quenched cylinder is used as the hardenability criterion. Although somewhat complex, this is a precise and reproducible method for experimentally determining the hardenability of shallow-hardening steels. By testing several steels using this method, a linear regression equation has been derived for estimating hardenability from chemical composition and grain size that expresses the relative contribution of carbon and alloying elements by additive terms instead of multiplicative factors.

2. Hardenability Test Methods for Air-Hardening Steels

When a standard Jominy specimen is used, the cooling rate at a distance of 80 mm from the quenched end (essentially the opposite end of the specimen) is approximately 0.7 K/s. The hardenability of all steel grades with a critical cooling rate greater than 0.7 K/s can be determined by the standard Jominy end-quench hardenability test as a sufficient decrease in hardness will be obtained from increasing amounts of nonmartensite transformation products (bainite, pearlite, ferrite). However, for steels with a critical cooling rate lower 0.7 K/s there will be no substantial change in the hardness curve because martensite will be obtained at every distance along the Jominy specimen. This is the case with air-hardening steels. To cope with this situation and enable the use of the Jominy test for air-hardening steels, the mass of the upper part of the Jominy specimen should be increased [17] by using a stainless steel cap as shown in Figure 26. In this way, cooling rates of the upper part of the specimen are decreased below the critical cooling rate of the steel itself.

The complete device consists of the conical cap with a hole through which the specimen can be fixed with the cap. When austenitizing, a "leg" is installed on the lower end of the specimen as shown in Figure 26 to equalize heating so that the same austenitizing conditions exist along the entire test specimen. The total heating time is 40 min plus 20 min holding time at the austenitizing temperature. Before quenching the specimen according to the standard Jominy test procedure (together with the cap), the leg should be removed, Figure 27 illustrates cooling rates when quenching a standard Jominy specimen and a modified specimen with added cap. This diagram illustrates the relationship between the cooling times from the austenitizing temperature to 500°C and the distance from the quenched end of the specimen for different austenitizing temperatures.

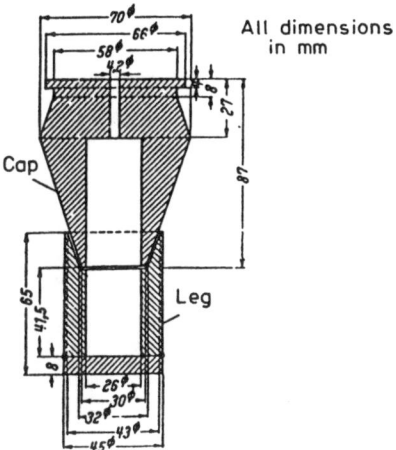

Figure 26 Modification of the standard Jominy test by the addition of a cap to the specimen for testing the hardenability of air-hardening steels. (From Ref. 17.)

Figure 27 shows that at an austenitizing temperature of 800°C up to a distance of 20 mm from the quenched end, the cooling time curves for the standard specimen and the modified specimen have the same path and thus the same cooling rate. At distances beyond approximately 20 mm, the cooling time curve for the modified specimen exhibits increasingly slower cooling rates relative to the standard specimen. By adding the cap, the cooling time is nearly doubled, or the cooling rate is approximately half that exhibited by the unmodified test piece.

Figure 28 shows two Jominy hardenability curves, one obtained with the standard specimen and the other with the modified specimen, for the hot-working tool steel DIN 45CrMoV67 (0.43% C, 1.3% Cr, 0.7% Mo, 0.23% V). Up to 20 mm from the quenched end, both curves are nearly equivalent. At greater distances, the retarded cooling exhibited by the modified specimen causes the decrease in hardness to start at 23 mm from the quenched end, while the decrease in hardness for the standard specimen begins at approximately 45 mm.

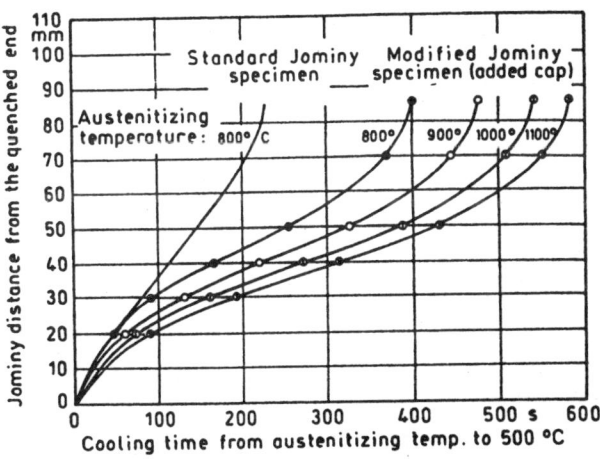

Figure 27 Cooling times between austenitizing temperature and 500°C for the standard Jominy specimen and for a specimen modified by adding a cap. (From Ref. 17.)

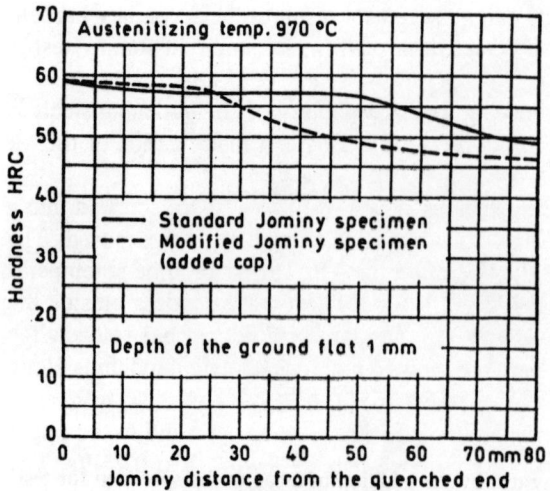

Figure 28 Jominy hardenability curves of grade DIN 45CrMoV67 steel for a standard specimen and for a specimen modified by adding a cap. (From Ref. 17.)

The full advantage of the test with modified specimens for an air-hardening steel can be seen only if a quenched Jominy specimen is tempered at a temperature that will result in a secondary hardening effect. Figure 29 illustrates this for the tool steel DIN 45CrVMoW58 (0.39% C, 1.5% Cr, 0.5% Mo, 0.7% V, 0.55% W). After tempering at 300°C, the hardness near the quenched end decreases. Within this region martensitic structure is predominant. At about 25 mm from the quenched end the hardness curve after tempering becomes equal to the hardness curve after quenching. After tempering to 550°C, however, the hardness is even more decreased up to a distance of 17 mm from the quenched end, and for greater distances

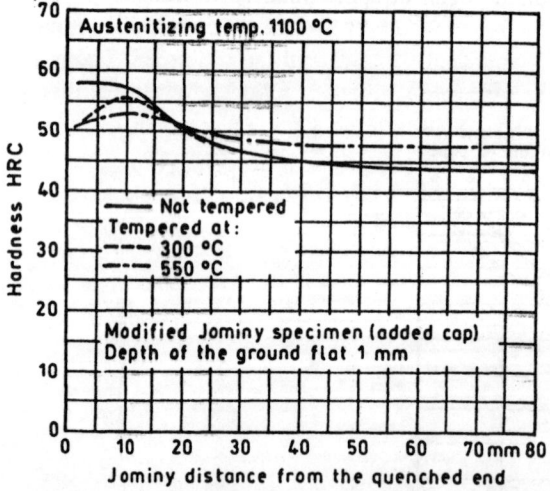

Figure 29 Jominy hardenability curves of grade DIN 45CrVMoW58 steel after quenching (solid curve) and after quenching and tempering (dashed curves) for a specimen modified by adding a cap. (From Ref. 17.)

a hardness increase up to about 4 HRC units can be seen as a result of the secondary hardening effect. This increase in hardness can be detected only when the modified Jominy test is
conducted.

Another approach to measuring and recording the hardenability of air-hardening steels is
the Timken Bearing Company Air Hardenability Test [18]. This is a modification of the air
hardenability testing procedure devised by Post et al. [19].

Two partially threaded test bars of the dimensions shown in Figure 30 are screwed into a
cylindrical bar 6 in. in diameter by 15 in. long, leaving 4 in. of each test bar exposed. The
total setup is heated to the desired hardening temperature for 4 h. The actual time at temperature is 45 min for the embedded bar sections and 3 h for the sections extending outside the
large cylinder. The test bar is then cooled in still air. The large cylindrical bar restricts the
cooling of the exposed section of each test bar, producing numerous cooling conditions along
the bar length.

The various positions along the air hardenability bar, from the exposed end to the opposite end (each test bar is 10 in. long), cover cooling rates ranging from 1.2 to 0.2°F/s. The
hardenability curves for six high-temperature structural and hot-work die steels are shown in
Figure 31. The actual cooling rates corresponding to each bar position are shown. Each bar
position is equated in this figure to other section sizes and shapes producing equivalent cooling rates and hardnesses at the section centers when quenched in air. To prevent confusion,
equivalent cooling rates produced in other media such as oil are not plotted in this chart.
However, position 20 on the air-hardenability bar corresponds to the center of a 13 in. diameter bar cooled in still oil and even larger cylindrical bars cooled in water.

C. Hardenability Bands

Because of differences in chemical composition between different heats of the same grade of
steel, so-called hardenability bands have been developed using the Jominy end-quench test.
According to American designation, the hardenability band for each steel grade is marked by
the letter H following the composition code. Figure 32 shows such a hardenability band for
1340H steel. The upper curve of the band represents the maximum hardness values, corresponding to the upper composition limits of the main elements, and the lower curve represents the minimum hardness values, corresponding to the lower limit of the composition ranges.

Hardenability bands are useful for both the steel supplier and the customer. Today the
majority of steels are purchased according to hardenability bands. Suppliers guarantee that 93

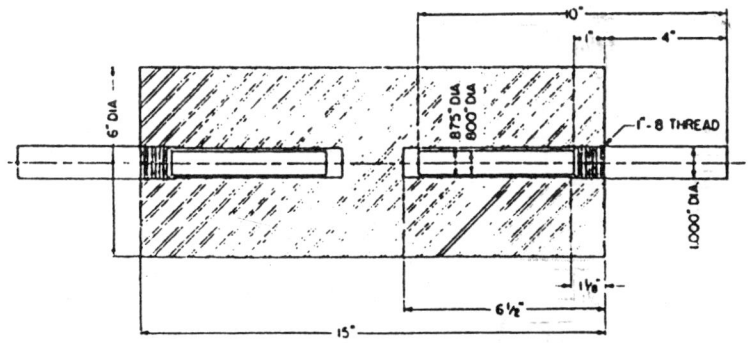

Figure 30 Timken Roller Bearing Co. air hardenability test setup. Two test specimens with short
threaded sections as illustrated are fixed in a larger cylindrical bar. (From Ref. 18.)

Type	Heat no.	Code	C	Mn	Co	W	Si	Cr	Ni	Mo	V	Norm Temp °F	Quench Temp °F
1722 AS	10420	●●●	.29	.61	-	-	.67	1.30	.18	.47	.26	Ann	1750
"·Co	A120	▫▫▫	.31	.54	1.06	-	.53	1.26	-	.52	.27	"	1750
H-11	12887	▲▲▲	.38	.40	-	-	.85	4.87	.11	1.34	.60	"	1850
Halmo	A115	△△△	.39	.52	-	-	.86	5.12	-	5.10	.68	"	1850
Lapelloy	18287	○○○	.31	1.07	-	-	.27	11.35	.43	2.85	.24	"	1900
HTS-1100	A117	■■■	.44	.42	-	1.70	.51	1.39	-	1.48	1.01	"	1900

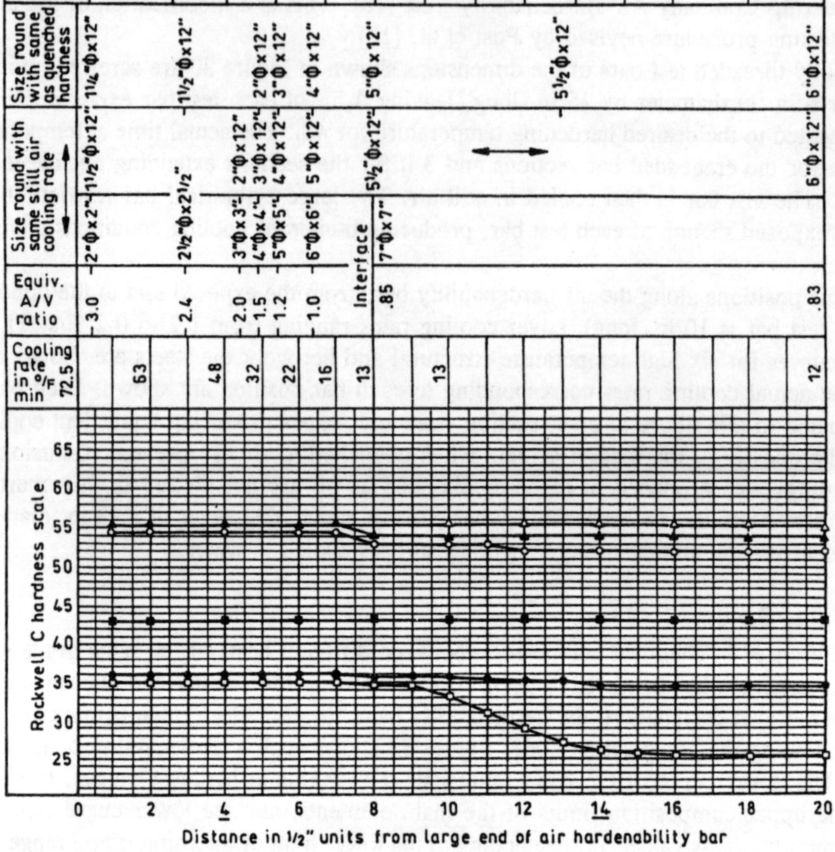

Figure 31 Chemistry and air-hardenability test results for various Cr-Mo-V steels. (From Ref. 18.)

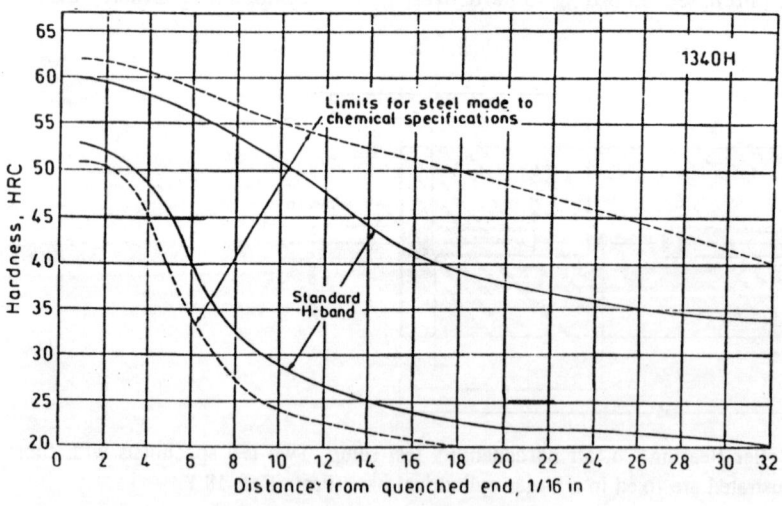

Figure 32 Hardenability band for SAE 1340H steel.

or 95% of all mill heats made to chemical specification will also be within the hardenability band. The H bands were derived from end-quench data from a large number of heats of a specified composition range by excluding the upper and lower 3.5% of the data points. Steels may be purchased either to specified composition ranges or to hardenability limits defined by H bands. In the latter case, the suffix H is added to the conventional grade designation, for example 4140H, and a wider composition range is allowed. The difference in hardenability between an H steel and the same steel made to chemical specifications is illustrated in Figure 32. These differences are not the same for all grades.

High-volume production of hardened critical parts should have close tolerance of the depth of hardening. The customer may require, at additional cost, only those heats of a steel grade that satisfy, for example, the upper third of the hardenability band. As shown in Figure 33, the SAE recommended specifications are

A minimum and a maximum hardness value at any desired Jominy distance. For example,

$$J_{30-56} = 10/16 \text{ in.} \quad \text{(A-A, Figure 33)} \tag{3}$$

If thin sections are to be hardened and high hardness values are expected, the selected Jominy distance should be closer to the quenched end. For thick sections, greater Jominy distances are important.

The minimum and maximum distance from the quenched end where a desired hardness value occurs. For example,

$$J_{45} = 7/16\text{-}14/16 \text{ in.} \tag{4}$$

Two maximum hardness values at two desired Jominy distances. For example,

$$J_{52} = 12/16 \text{ in. (max);} \quad J_{38} = 16/16 \text{ in. (max)} \tag{5}$$

Two minimum hardness values at two desired Jominy distances. For example,

$$J_{52} = 6/16 \text{ in. (min);} \quad J_{28} = 12/16 \text{ in. (min)} \tag{6}$$

Minimum hardenability is significant for thick sections to be hardened; maximum hardenability is usually related to thin sections because of their tendency to distort or crack, especially when made from higher carbon steels.

If a structure–volume fraction diagram (see Figure 34) for the same steel is available, the effective depth of hardening, which is defined by a given martensite content, may be deter-

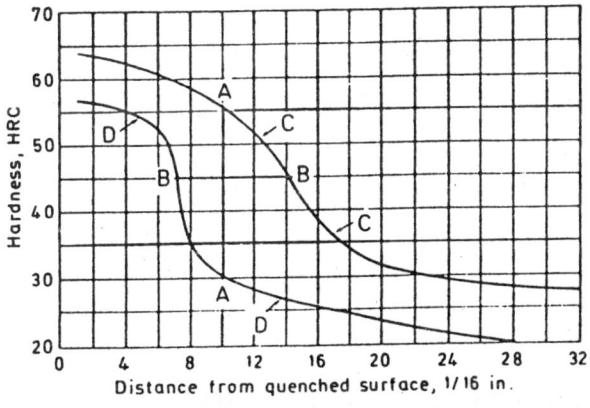

Figure 33 Different ways of specifying hardenability limits according to SAE.

mined from the maximum and minimum hardenability curves of the band. The structure–volume fraction diagram can also be used for the preparation of the transformation diagram when limits of the hardenability of a steel are determined. If the structure–volume fraction diagram is not available, the limit values of hardness or the effective depth of hardening can be estimated from the hardenability band using the diagram shown in Figure 35. Hardness depends on the carbon content of steel and the percentage of martensite after quenching. Figure 36 shows the hardenability band of the steel DIN 37MnSi5; the carbon content may vary from a minimum of 0.31% to a maximum of 0.39%.

The tolerance in the depth of hardening up to 50% martensite between a heat having maximum hardenability and a heat with minimum hardenability can be determined from the following examples. For C_{min} = 0.31% and 50% martensite, a hardness of 38 HRC can be determined from Figure 35. This hardness corresponds to the lower curve of the hardenability band and is found at a distance of 4 mm from the quenched end. For C_{max} = 0.39% and 50% martensite, a hardness of 42 HRC can be determined from Figure 35. This hardness corresponds to the upper curve of the hardenability band and is found at 20 mm from the quenched end.

In this example, the Jominy hardenability (measured up to 50% martensite) for this steel varies between 4 and 20 mm. Using conversion charts, differences in the depth of hardening for any given diameter of round bars quenched under the same conditions can be determined.

Effective depth of hardening is not the only information that can be derived from the hardenability band. Characteristic features of every hardenability band provide information on the material-dependent spread of hardenability designated "the maximum hardness difference"

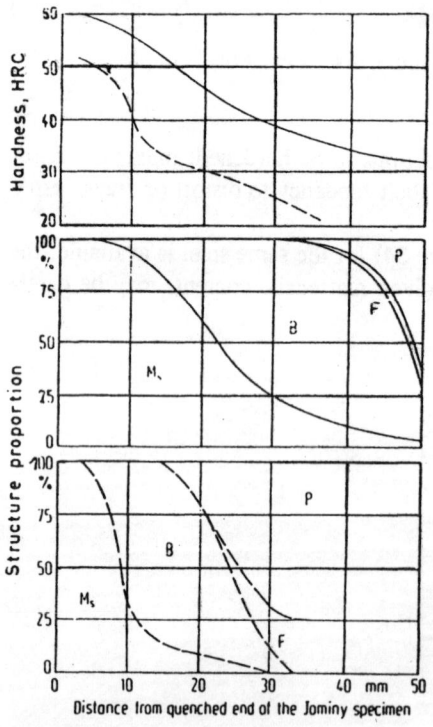

Figure 34 Hardenability band and structure–volume fraction diagram of SAE 5140 steel. F = ferrite, P = pearlite, B = bainite, M_s = martensite. (From Ref. 21.)

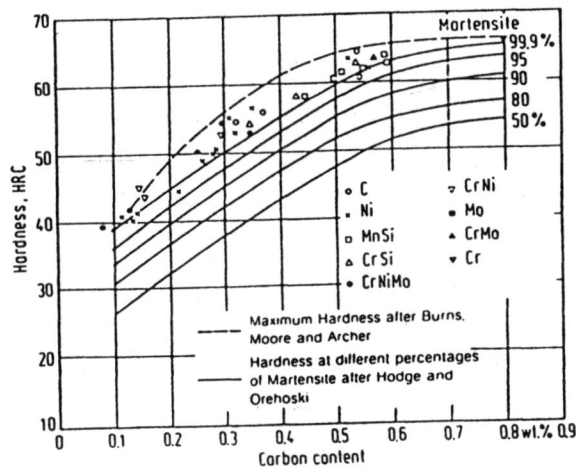

Figure 35 Achievable hardness depending on the carbon content and percentage of martensite in the structure. (From Ref. 21.)

shown in Figure 36. The hardness difference at the same distance from the quenched end, i.e., at the same cooling rate, can be taken as a measure of material-dependent deviations. Another important technological point that can be derived from the hardenability band is the "hardness gradient." In Figure 36, this is illustrated by the minimum hardenability curve for the steel in question where there is a high gradient of hardness (22 HRC for only 5 mm difference in the Jominy distance). High hardness gradients indicate high sensitivity to cooling rate variation.

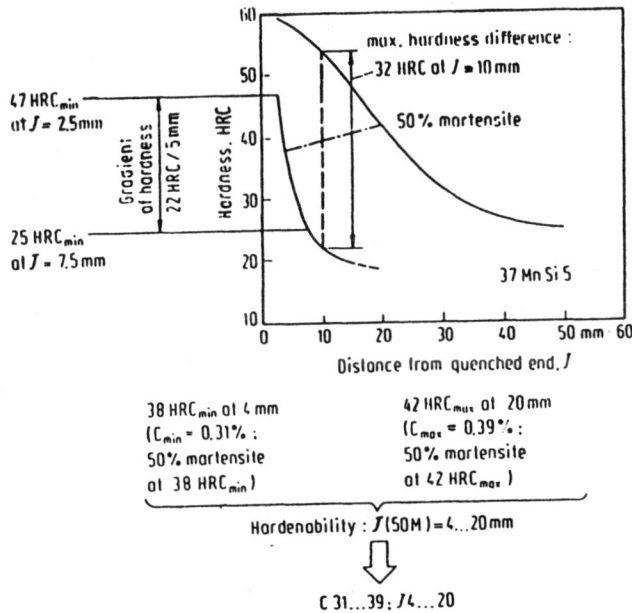

Figure 36 Hardenability band of DIN 37MnSi5 steel and the way technologically important information can be obtained. (From Ref. 21.)

IV. CALCULATION OF JOMINY CURVES FROM CHEMICAL COMPOSITION

The first calculations of Jominy curves based on the chemical composition of steels were performed in the United States in 1943 [22,23]. Later, Just [24], using regression analysis of fictitious Jominy curves from SAE hardenability bands and Jominy curves of actual heats from the USS *Atlas* (USA) and *MPI-Atlas* (Germany), derived expressions for calculating the hardness at different distances (E) from the quenched end of the Jominy specimen. It was found that the influence of carbon depends on other alloying elements and also on the cooling rate, i.e., with distance from the quenched end (Jominy distance).

Carbon starts at a Jominy distance of 0 with a multiplying factor of 50, while other alloying elements have the factor 0 at this distance. This implies that the hardness at a Jominy distance of 0 is governed solely by the carbon content. The influence of other alloying elements generally increases from 0 to values of their respective factors up to a Jominy distance of about 10 mm. Beyond this distance, their influence is essentially constant. Near the quenched end the influence of carbon prevails, while the influence of other alloying elements remains essentially constant beyond a Jominy distance of about 10 mm. This led Just to propose a single expression for the whole test specimen, except for distances shorter than 6 mm:

$$J_{6...80} = 95\sqrt{C} - 0.00276E^2\sqrt{C} + 20\text{ Cr} + 38\text{ Mo} + 14\text{ Mn} + 5.5\text{ Ni} + 6.1\text{ Si}$$
$$+ 39\text{ V} + 96\text{ P} - 0.81K - 12.28\sqrt{E} + 0.898\text{ }E - 13\text{ HRC} \tag{7}$$

where J = Jominy hardness, HRC; E = Jominy distance, mm; K = ASTM grain size; and the element symbols represent weight percentage of each.

In Equation (7), all alloying elements are adjusted to weight percent, and it is valid within the following limits of alloying elements: $C < 0.6\%$; $Cr < 2\%$; $Mn < 2\%$; $Ni < 4\%$; $Mo < 0.5\%$; $V < 0.2\%$. Calculation of hardness at the quenched end (Jominy distance 0), using the equation for the maximum attainable hardness with 100% martensite, is

$$H_{max} = 60\sqrt{C} + 20\text{ HRC}, \qquad C < 0.6\% \tag{8}$$

Although Equation (7) was derived for use up to a distance of 80 mm from the quenched end of the Jominy specimen, other authors argue that beyond a Jominy distance of 65 mm the continuous decrease in cooling rate at the Jominy test cannot be ensured even for low-alloy steels because of the cooling effect of surrounding air. Therefore, newer calculation methods rarely go beyond a Jominy distance of 40 mm.

Just [24] found that a better fit for existing mutual correlations can be achieved by formulas that are valid for groups of similar steels. He also found that multiplying hardenability factors for Cr, Mn, and Ni have lower values for case-hardening steels than for structural steels for hardening and tempering. Therefore, separate formulas for case-hardening steels were derived:

$$J_{6...40(\text{case-hardening steels})} = 74\sqrt{C} + 14\text{ Cr} + 5.4\text{ Ni} + 29\text{ Mo} + 16\text{ Mn} - 16.8\sqrt{E}$$
$$+ 1.386E + 7\text{ HRC} \tag{9}$$

and for steels for hardening and tempering,

$$J_{6...40(\text{steels for hardening and tempering})} = 102\sqrt{C} + 22\text{ Cr} + 21\text{ Mn} + 7\text{ Ni} + 33\text{ Mo}$$
$$- 15.47\sqrt{E} + 1.102E - 16\text{ HRC} \tag{10}$$

In Europe, five German steel producers in a VDEh working group jointly developed formulas that adequately define the hardenability from different production heats [25]. The goal was to replace various existing formulas that were used individually.

Data for some case-hardening steels and some low-alloy structural steels for hardening and tempering have been compiled, and guidelines for the calculation and evaluation of formulas for additional families of steel have been established. This work accounts for influential factors from the steel melting process and for possible deviations in the Jominy test itself. Multiple linear regression methods using measured hardness values from Jominy tests and actual chemical compositions were also included in the analyses. The number of Jominy curves of a family of steel grades necessary to establish usable formulas should be at least equal to the square of the total number of chemical elements used for the calculation. Approximately 200 curves were suggested. To obtain usable equations, all Jominy curves for steel grades that had similar transformation characteristics (i.e., similar CCT diagram) when hardened were used. Therefore, precise equations for the calculation of Jominy hardness values were derived only for steel grades of similar composition [25].

The regression coefficients for a set of equations to calculate the hardness values at different Jominy distances from 1.5 to 30 mm from the quenched end are provided in Table 2 [25]. The chemistry of the steels used for this study is summarized in Table 3. The regression coefficients in Table 2 do not have the same meaning as the hardenability factors in Equations (7), (9), and (10); therefore, there is no restriction on the calculation of Jominy hardness values at less than 6 mm from the quenched end. Because the regression coefficients used in this method of calculation are not hardenability factors, care should be taken when deriving structural properties from them.

The precision of the calculation was determined by comparing the measured and calculated hardness values and establishing the residual scatter, which is shown in Figure 37. The upper curve for a heat of DIN 41Cr4 steel, having a residual scatter of $s = 2.94$ HRC, shows an adequate consistency, while the lower curve for another heat of the same steel, with a residual scatter of $s = 7.45$ HRC, shows inadequate consistency. Such checks were repeated for every Jominy distance and for every heat of the respective steel family. During this process it was found that the residual scatter depends on the distance from the quenched end and that calculated Jominy curves do not show the same precision (compared to measured curves) at all Jominy distances. For different steel grades with different transformation characteristics, the residual scatter varies with Jominy distance, as shown in Figure 38. In spite of the residual scatter of the calculated results, it was concluded "that properly calibrated predictors offer a strong advantage over testing in routine applications" [26].

When judging the precision of a calculation of Jominy hardness values, hardenability predictors are expected to accurately predict (± 1 HRC) observed hardness values from the chemical composition. However, experimental reproducibility of a hardness value at a fixed Jominy distance near the inflection point of the curve can be 8–12 HRC (see Figure 23 for J_{10mm}). Therefore it was concluded "that a properly calibrated hardenability formula will always anticipate the results of a purchaser's check test at every hardness point better than an actual Jominy test" [26].

A. Hyperbolic Secant Method for Predicting Jominy Hardenability

Another method for predicting Jominy end-quench hardenability from composition is based on the four-parameter hyperbolic secant curve-fitting technique [27]. In this method, it is assumed that the Jominy curve shape can be characterized by a four-parameter hyperbolic secant (sech) function (SECH).

Table 2 Regression Coefficients for the Calculation of Jominy Hardness Values for Structural Steels for Hardening and Tempering Alloyed with About 1% Cr

Jominy distance (mm)	Regression coefficients										
	Constant	C	Si	Mn	S	Cr	Mo	Ni	Al	Cu	N
1.5	29.96	57.91	2.29	3.77						-2.65	83.33
3	26.75	58.66	3.76	2.16		2.86				-2.59	59.87
5	15.24	64.04	10.86		-41.85	12.29					-115.50
7	-7.82	81.10	19.27	4.87	-73.79	21.02				4.56	-176.82
9	-27.29	94.70	22.01	10.24	-37.76	24.82				8.58	-144.07
11	-39.34	100.78	21.25	14.70		25.39		6.66	38.31	7.97	
13	-42.61	95.85	20.54	16.06		26.46	30.41		52.63	9.0	
15	-42.49	88.69	20.82	17.75		25.33	38.97		54.91	8.89	
20	-41.72	78.34	17.57	20.18		23.85	26.95	7.51	47.16	9.96	
25	-41.94	72.29	18.62	20.73	-65.81	24.08	35.99	7.69		9.64	
30	-44.63	72.74	19.12	21.42	-81.41	24.39	27.57	10.75		9.71	

Source: Ref. 25.

Table 3 Limiting Values of Chemical Composition of Structural Steels for Hardening and Tempering Alloyed with About 1% Cr[a]

| | \multicolumn{11}{c}{Content (%)} |
	C	Si	Mn	P	S	Cr	Mo	Ni	Al	Cu	N
Min.	0.22	0.02	0.59	0.005	0.003	0.80	0.01	0.01	0.012	0.02	0.006
Max.	0.47	0.36	0.97	0.037	0.038	1.24	0.09	0.28	0.062	0.32	0.015
Mean	0.35	0.22	0.76	0.013	0.023	1.04	0.04	0.13	0.031	0.16	0.009
s	0.06	0.07	0.07	0.005	0.008	0.10	0.02	0.05	0.007	0.05	0.002

[a]Used in calculations with regression coefficients of Table 2.
Source: Ref. 25.

The SECH curve-fitting technique utilizes the equation

$$DH_x = A + B \{\text{sech}[C(x - 1)^P]-1\} \tag{11}$$

or alternatively

$$DH_x = (A - B) + B \text{ sech}[C(x - 1)^P] \tag{12}$$

where the hyperbolic secant function for any y value is

$$\text{sech}_y = \frac{2}{e^y + e^{-y}} \tag{13}$$

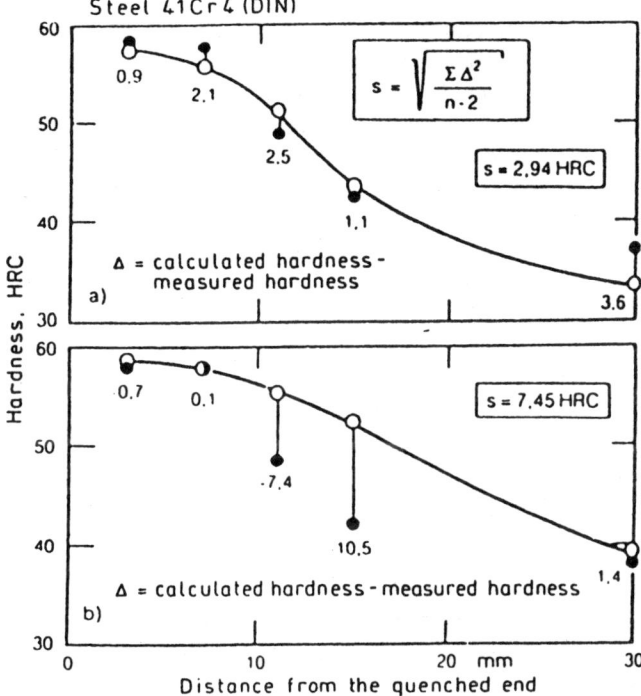

Figure 37 Comparison between measured (O) and calculated (●) hardness values for a melt with adequate consistency (top) and with inadequate consistency (bottom). (From Ref. 25).

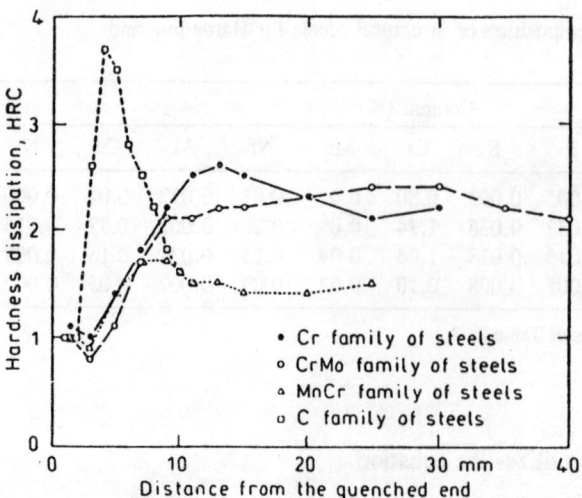

Figure 38 Residual scatter between measured and calculated hardness values vs. distance to the quenched end, for different steel grade families. (From Ref. 25.)

where x is the Jominy distance from the quenched end, in sixteenths of an inch; DH_x is the hardness at the Jominy distance x; and A, B, C, D are the four parameters, which can be set such that DH_x conforms closely to an experimental end-quench hardenability curve.

The relationship between parameters A and B and a hypothetical Jominy curve is illustrated in Figure 39.

The parameter A denotes the upper asymptotic or initial hardness (IH) at the quenched end. The parameter B corresponds to the difference between the upper and lower asymptotic hardness values, respectively (DH_∞). This means that for a constant value of A, increasing the value of B will decrease the lower asymptotic hardness, as shown in Figure 40a.

The parameters C and D control the position of, and the slope at, the inflection point in the calculated Jominy curve. If the $A, B,$ and C parameters are constant, lowering the value

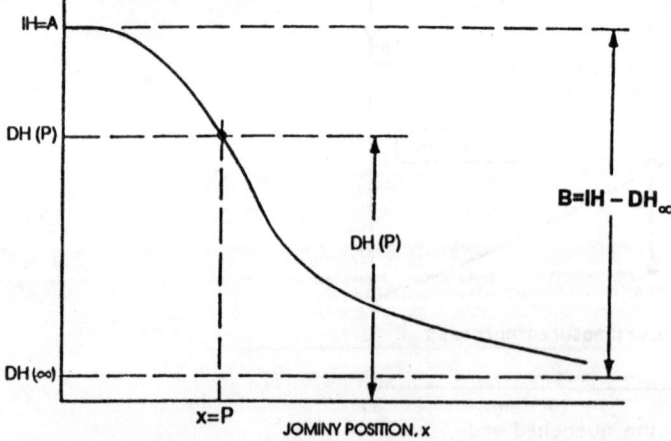

Figure 39 Schematic showing the relationships between the hyperbolic secant coefficients A and B and Jominy curve characteristics. (From Ref. 12.)

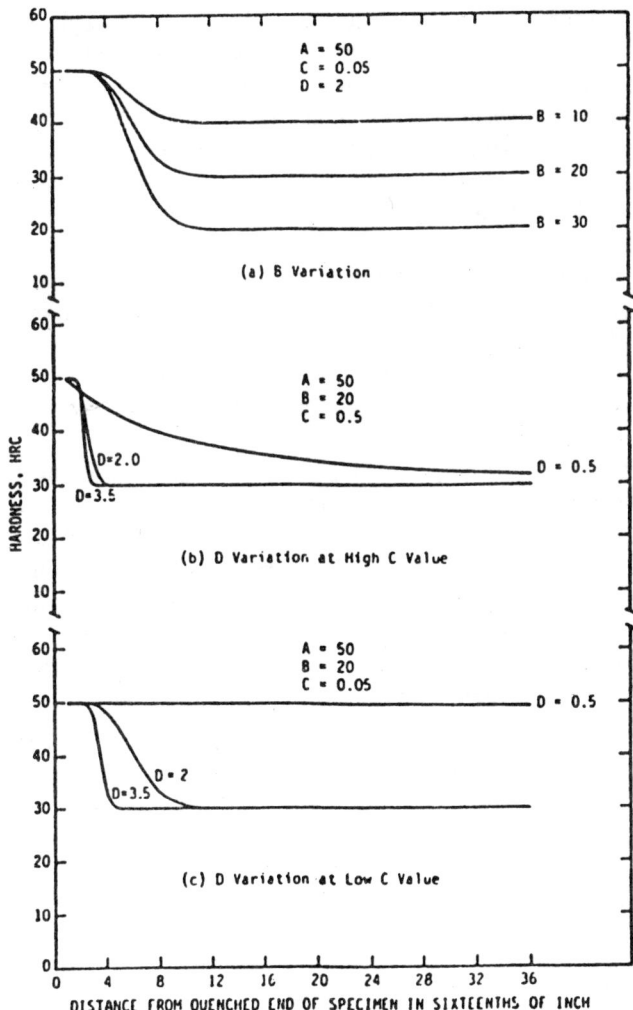

Figure 40 Effect of SECH parameter variation on Jominy curve shape. (From Ref. 12.)

of parameter D will cause the inflection point to occur at greater Jominy distances, as shown in Figures 40b and 40c. A similar result will be obtained if parameters A, B, and D are kept constant, and parameter C is shown by comparing Figures 40b and 40c. In fact, it may be appropriate to set C or D to a constant value characteristic of a grade of steels and describe the effects of compositional variations within the grade by establishing correlations with the other three parameters.

It should be noted that some Jominy curves cannot be well described by a general expression such as Equation (11) or (12). For example, if a significant amount of carbide precipitation were to occur in the bainite or pearlite cooling regime, a "hump" in the Jominy curve might be observed that could not be calculated.

To calculate the values of the four SECH parameters for each experimentally obtained Jominy curve, the minimum requirement is a data set from which the predictive equations will be developed. This data set should contain compositions of each steel grade (or heat), with associated values of Jominy hardness at different end-quench distances, as determined by the experiment. Other metallurgical or processing variables such as grain size or austenitizing temperature can also be included. The data set must be carefully selected; the best predictions will be obtained when the regression data set is both very large and homogeneously distributed over the range of factors for which hardenability predictions will be desired.

A linear/nonlinear regression analysis program using least squares was used to calculate separate values of the four parameters for each experimental Jominy curve in the regression data set by minimizing the differences between the empirical and analytical hardness curves, i.e., obtaining the best fit.

Figure 41 provides experimental end-quench hardenability data and best fit hyperbolic secant function for one steel in a data set that contained 40 carburizing steel compositions. Excellent fits were obtained for all 40 cases in the regression data set. Once the four parameters A, B, C, and D have been obtained for each heat as described above, four separate equations with these parameters as dependent variables are constructed using multiple regression analysis by means of a statistical analysis computer package.

Table 4 provides multiple regression coefficients obtained with the "backward elimination" regression analysis of the above-mentioned 40 cases. In this elimination process, 31 variables were arbitrarily defined for possible selection as independent variables in the multiple regression analysis. The list of these variables consisted of all seven single-element and grain size terms, the seven squares and seven cubes of the single-element and grain size terms, and all 10 possible two-way element interaction terms that did not include carbon or grain size.

Based on the multiple regression coefficients from Table 4, the following four equations for SECH parameters were developed for the regression data set of 40 carburizing steels:

$$A = 481\ C^3 + 41.4 \tag{14}$$

$$B = -28.7\ CrMo - 61.6\ MnSi - 1.72\ GS - 1.35\ Ni^3 + 60.2 \tag{15}$$

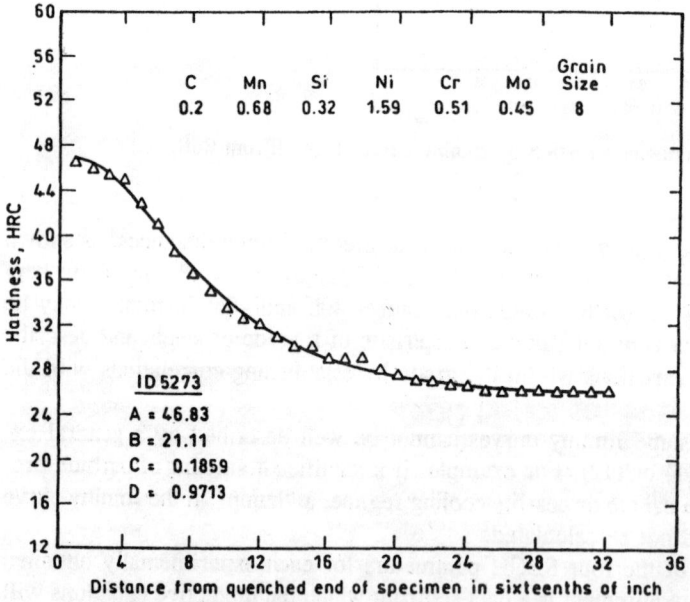

Figure 41 Experimental end-quench hardenability data and best-fit hyperbolic secant function. (From Ref. 12.)

Table 4 Multiple Regression Coefficients for Backward-Elimination Regression Analysis

				Dependent variable (SECH parameter)				
A		B		C		D		
Ind. var.	Coeff.	Ind. var	Coeff.	Ind. var.	Coeff.	Ind. var.	Coeff.	
C*C*C	481.27031	Cr*Mo	-28.71764	Cr*Mo	-0.79950	Cr*Mo	1.19695	
(Constant)	41.44362	Mn*Si	-61.55499	Mn*Si	-1.04208	Mn*Si	1.97624	
		GS	-1.71674	Ni*Ni*Ni	-0.04871	Ni*Ni*Ni	0.09267	
		Ni*Ni*Ni	-1.35352	C*C*C	-14.85249	C*C*C	33.57479	
		(Constant)	60.23736	33.57479	0.92535	(Constant)	-0.26580	

$$C = -0.8 \text{ CrMo} - 1.04 \text{ MnSi} - 0.05 \text{ Ni}^3 - 14.9 \text{ C}^3 + 0.93 \tag{16}$$

$$D = 1.2 \text{ CrMo} + 1.98 \text{ MnSi} + 0.09 \text{ Ni}^3 + 33.6 \text{ C}^3 - 0.27 \tag{17}$$

where an element name denotes percentage of that element in the steel and GS denotes grain size. Equations (14)–(17) are valid for steel compositions in the range of 0.15–0.25% C, 0.45–1.1% Mn, 0.22–0.35% Si, 0–1.86% Ni, 0–1.03% Cr, and 0–0.76% Mo, with ASTM grain sizes (GS) between 5 and 9.

After the four parameters are calculated, they are substituted into Equation (11) or (12) to calculate distance hardness (DH) at each Jominy distance x of interest.

To validate this method, the Jominy curves were predicted for an independently determined data set of 24 heats, and this prediction was compared with those obtained by two other methods (AMAX [28] and Just [29] prediction methods). The SECH predictions were not as accurate as distance hardness predictions based on the two methods developed earlier because of the limited size and sparsely populated sections (not homogeneously distributed) of the initial data set.

B. Computer Calculation of Jominy Hardenability

The application of computer technology has greatly enhanced the precision of these calculations. Commercial software is available for the calculation of Jominy hardness. For example, the Minitech Predictor [26] is based on the initial generation of an inflection point on the Jominy curve. Figure 42 shows a typical output of the Minitech Predictor operating in the data processing mode. Input values are Jominy hardness values, chemical composition, and estimated grain size.

The Minitech program generates a predicted Jominy curve (J_n) and a predicted inflection point distance from quenched end x' and displays a comparison of the predicted and experimentally obtained curves as shown in Figure 42a. A weighting pattern J_n is accessed that specifies a weight of 1.5 for all J'_n distances from $n = 1$ to $n = 2x'$ and a weight of 0.75 for $n > 2x'$ to $n=32$ mm (or any limit of the data). Using an effective carbon content and grain size as adjustable parameters, the theoretical curve is then iterated about J'_n and x' to minimize the weighted root mean square deviation of the calculated curve from the experimental curve. The final best-fit calculated curve is plotted along with the main processed data as shown in Figure 42b.

Calculated Jominy hardness curves are used to replace Jominy testing by equivalent predictions for those steel grades (e.g., very shallow hardening steels) that it is difficult or im-

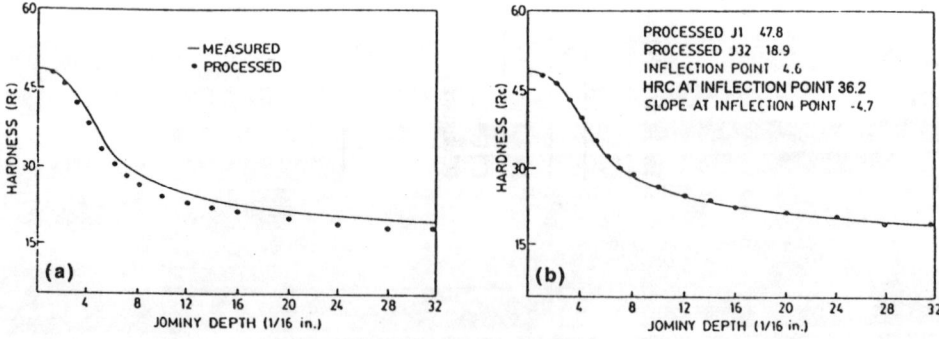

Figure 42 Outputs from Minitech Predictor data processing program for best fit to measured Jominy data. (a) Initial trial; (b) final trial. (From Ref. 26.)

possible to test. Although the accurate prediction of hardenability is important, it is more important for the steel manufacturer to be able to refine the calculations during the steelmaking process. For example, the steel user indicates the desired Jominy curve by specifying three points within H-band for SAE 8620H:

Jominy distance	Hardness
1.5 mm	460 HV
5.0 mm	370 HV
9.0 mm	270 HV

HV = Vickers pyramid hardness

as shown in Figure 43 [30].

Using these data, the steel mill will first compare the customer's specification against two main criteria:

1. That the hardenability desired is within limits for the steel grade in question.
2. That the specified points fall on a Jominy curve permissible within the analysis range for SAE 8620H, i.e., the specified points must provide a physically possible Jominy curve.

When the actual heat of steel is ready for production, the computer program will automatically select the values for alloy additions and initiate the required control procedures. The samples taken during melting and refining are used to compute the necessary chemical adjustments. The computer program is linked directly to the ferroalloy selection and dispensing system. By successive adjustments, the heat is refined to a chemical composition that meets the required hardenability specification within the chemical composition limits for the steel grade in question.

The use of calculated Jominy curves for steel manufacturing process control is illustrated in the following example. Quality control analysis found that the steel heat should have a manganese value of 0.85%. During subsequent alloying, the analysis found 0.88% Mn. This

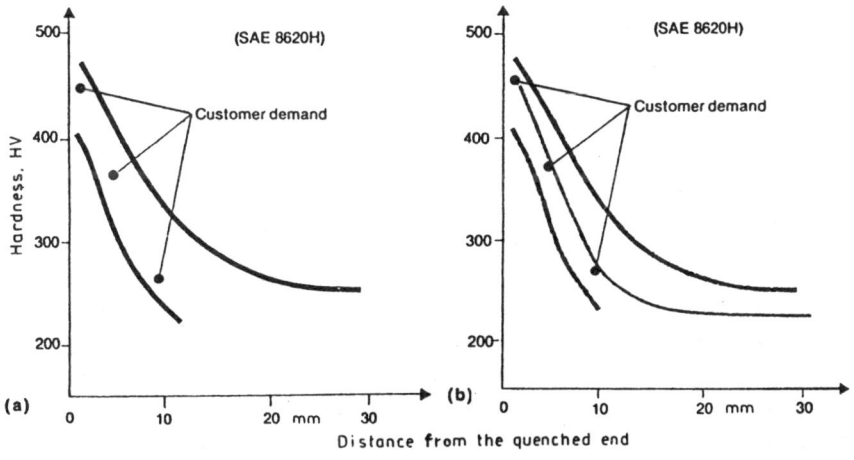

Figure 43 (a) Customer's specification of hardenability within an H band for SAE 8620H. (b) Jominy curve for finished heat. (From Ref. 30.)

overrun in Mn was automatically compensated for by the computer program, which adjusted hardenability by decreasing the final chromium content slightly. The resulting heat had the measured Jominy curve shown in Figure 43b. In this case, the produced steel does not deviate from the required specification by more than ± 10 HV at any Jominy distance below 19 mm.

V. APPLICATION OF HARDENABILITY CONCEPT FOR PREDICTION OF HARDNESS AFTER QUENCHING

Jominy curves are the preferred method for the characterization of steel. They are used to compare the hardenability of different heats of the same steel grade as a quality control method in steel production and to compare the hardenability of different steel grades when selecting steel for a certain application. In the latter case, Jominy curves are used to predict the depth of hardening, i.e., to predict the expected hardness distribution obtained after hardening parts of different cross-sectional dimensions after quenching under various conditions. Such predictions are generally based on the assumption that the rates of cooling prevailing at different distances from the quenched end of the Jominy specimen may be compared with the cooling rates prevailing at different locations on the cross sections of bars of different diameters. If the cooling rates are equal, it is assumed that equivalent microstructure and hardness can be expected after quenching.

The diagrams shown in Figure 44 have been developed for this purpose. These diagrams provide a correlation of equivalent cooling rates at different distances from the quenched end of the Jominy specimen and at different locations (center, half-radius, three-quarter radius, surface) on the cross section of round bars of different diameters. They are valid for the specified quenching conditions only. Figure 44a is valid only for quenching in water at an agitation rate of 1 m/s, and the diagram in Figure 44b is valid only for quenching in oil at an agitation rate of 1 m/s.

Another diagram showing the relation between cooling rates at different Jominy distances and cooling rates at different distances below the surface of round bars of different diameters, taken from the ASTM standard, is shown in Figure 45. From this diagram, the same cooling rate found at a Jominy distance of 14 mm prevails at a point 2 mm below the surface of a 75 mm diameter bar, at 10 mm below the surface of 50 mm diameter bar, and at the center of a 38 mm diameter bar when all the bars are quenched in "moderately" agitated oil. Using this diagram, it is possible to construct the hardness distribution curve across the section after hardening. This type of diagram is also valid for only the specified quenching conditions.

To correlate the hardness at different Jominy distances and the hardness at the center of round bars of different diameters that are quenched in various quenchants under different quenching conditions, the critical diameter (D_{crit}), the ideal critical diameter (D_I), and Grossmann's quenching intensity factor H must be used. The theoretical background of this approach is provided by Grossmann et al. [5], who calculated the "half-temperature time" (the time necessary to cool to the temperature halfway between the austenitizing temperature and the temperature of the quenchant). To correlate D_{crit} and H, Asimow et al. [32], in collaboration with Jominy, defined the half-temperature time characteristics for the Jominy specimen also. These half-temperature times were used to establish the relationship between the Jominy distance and ideal critical diameter D_I, as shown in Figure 46. If the microstructure of this steel is determined as a function of Jominy distance, the ideal critical diameter can be determined directly from the curve at that distance where 50% martensite is observed as shown in Figure 46. The same principle holds for D_{crit} when different quenching conditions characterized by the quenching intensity factor H are involved. Figure 47 shows the relationship between the

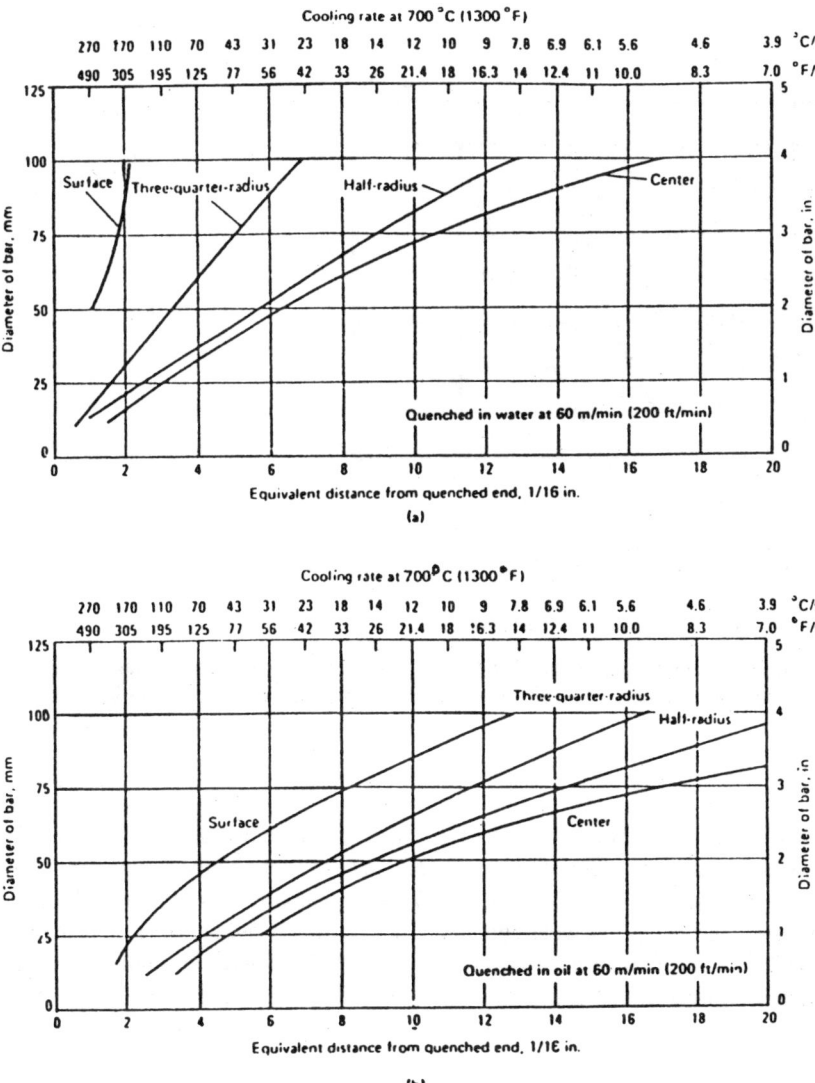

Figure 44 Correlation of equivalent cooling rates at different distances from the quenched end of the Jominy specimen and at different locations on the cross section of round bars of different diameters, quenched in (a) water agitated at 1 m/s and in (b) oil agitated at 1 m/s. (From Ref. 31.)

diameter of round bars (D_{crit} and D_I) and the distance from the quenched end of the Jominy specimen for the same hardness (of 50% martensite) at the center of the cross section after quenching under various conditions [32].

The application of the Figure 47 diagram can be explained for the two steel grades shown in Figure 48. The hardness at 50% martensite for the unalloyed steel grade Ck45 (0.45% C) is 45 HRC, while for the low-alloy grade 50CrMo4 steel (0.5% C) the hardness is 48 HRC. The lower part of the diagram depicts two H curves taken from the diagram in Figure 47. One is for vigorously agitated brine ($H = 5.0$), and the other for moderately agitated oil ($H = 0.4$). From both diagrams in Figure 48, it is seen that quenching the grade 50CrMo4 steel

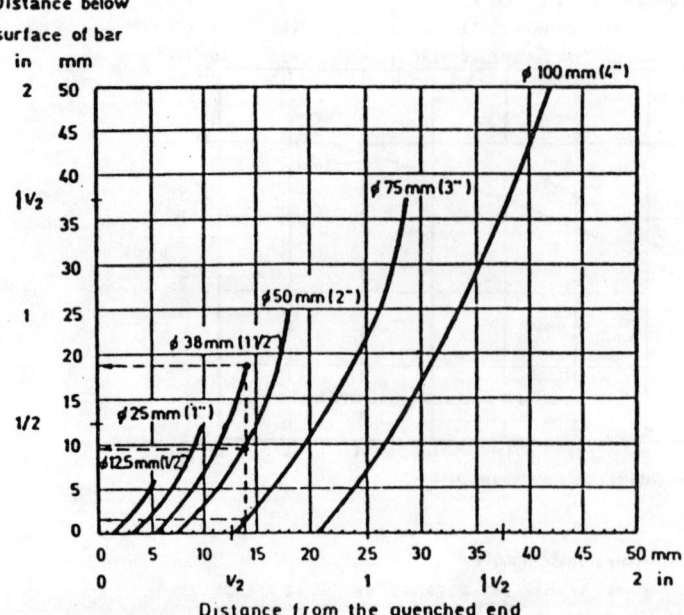

Figure 45 Relationship between cooling rates at different Jominy distances and cooling rates at different points below the surface of round bars of different diameters quenched in moderately agitated oil. (From Ref. 2.)

in vigorously agitated brine provides a hardness of 48 HRC in the center of the cross section of a round bar of 110 mm diameter. Quenching the same steel in moderately agitated oil provides this hardness at the core of round bars of only 70 mm diameter. The unalloyed grade Ck45 steel, having lower hardenability when quenched in vigorously agitated brine, provides a hardness of 45 HRC in the center of a 30 mm diameter bar. Quenching in moderately agitated oil provides this hardness in the center of a round bar of only 10 mm diameter.

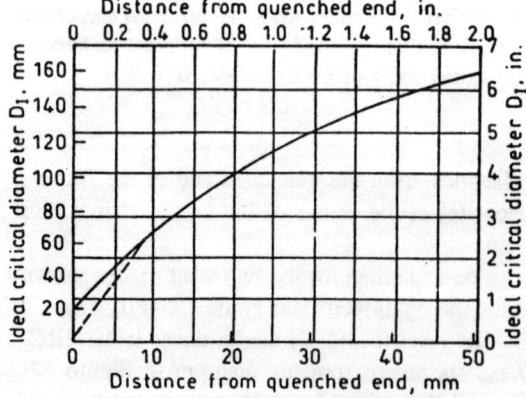

Figure 46 Relationship between the distance from the quenched end of the Jominy specimen and the ideal critical diameter. (From Ref. 32.)

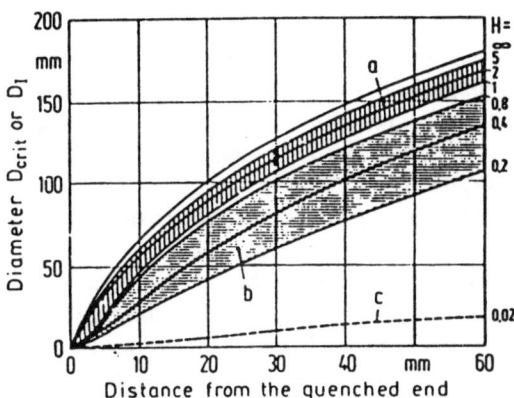

Figure 47 Relationship between the round bar diameter and the distance from the quenched end of the Jominy specimen, giving the hardness in the center of the cross section after quenching under different quenching conditions, a, water; b, oil; c, air. (From Ref. 32.)

A. The Lamont Method

The diagram shown in Figure 47 permits the prediction of hardness only at the center of round bars. Lamont [33] developed diagrams relating the cooling rate at a given Jominy distance to that at a given fractional depth in a bar of given radius that has been subjected to a given Grossman quenching intensity (H) factor. Analytical expressions have been developed for the Lamont transformation of the data to the appropriate Jominy distance J:

$$J = J(D, r/R, H) \tag{18}$$

where D is the diameter of the bar; r/R is the fractional position in the bar ($r/R = 0$ at the center; $r/R = 1$ at the surface), and H is the Grossmann quenching intensity factor. These expressions [34] are valid for any value of H from 0.2 to 10 and for bar diameters up to 200 mm (8 in.).

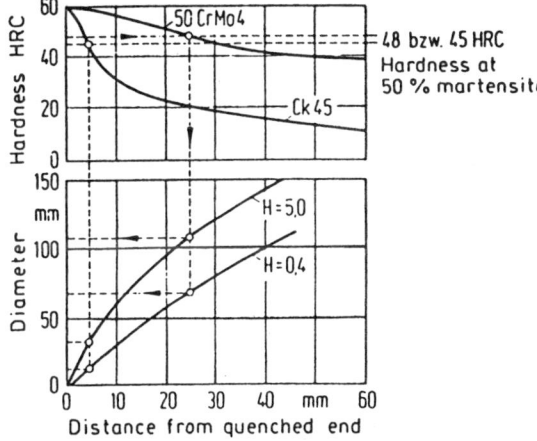

Figure 48 Determining the critical diameter of round bars (i.e., the hardness of 50% martensite at the center) from the Jominy hardenability curves of two steel grades quenched in vigorously agitated brine ($H = 5.0$) and in moderately agitated oil ($H = 0.4$). (Steel grade designation according to DIN.) (From Ref. 1.)

Lamont developed diagrams for the following points and fractional depths on the cross section of round bars: $r/R = 0$ (center), $r/R = 0.1$, $r/R = 0.2, \ldots, r/R = 0.5$ (half-radius), $r/R = 0.6, \ldots, r/R = 1.0$ (surface). Each of these diagrams is always used in connection with Jominy hardenability curve for the relevant steel. Figures 49–51 show the Lamont diagram for $r/R = 0$ (center of the cross section), $r/R = 0.5$, and $r/R = 0.8$, respectively.

The Lamont method can be used for four purposes.

1. To determine the maximum diameter of the bar that will achieve a particular hardness at a specified location on the cross section when quenched under specified conditions. For example, if the Jominy hardenability curve of the relevant steel grade shows a hardness of 55 HRC at a Jominy distance of 10 mm, then the maximum diameter of the bar that will achieve this hardness at half-radius when quenched in oil with $H = 0.35$ will be 28 mm. This result is obtained by using the diagram in Figure 50 for $r/R = 0.5$ and taking the vertical line at a Jominy distance of 10 mm to the intersection with the curve for $H = 0.35$, giving the value of 28 mm on the ordinate.

2. To determine the hardness at a specified location when the diameter of the bar, the quenching intensity H, and the steel grade are known. For example, if a 120 mm diameter bar is quenched is still water ($H = 1.0$), the hardness at the center ($r/R = 0$) will be determined at a distance of 37.5 mm from the quenched end on the Jominy curve of the relevant steel grade (see Figure 49).

3. To select adequate quenching conditions when the steel grade, the bar diameter, and the location on the cross section where a particular hardness should be attained are known. For example, a hardness of 50 HRC, which corresponds to the distance of 15 mm from the quenched end on the Jominy curve of the relevant steel grade, should be attained at the center of a 50 mm diameter bar. The appropriate H factor can be found by using Figure 49. In this case, the horizontal line for a 50 mm diameter and the vertical line for a 15 mm Jominy distance intersect at the point that corresponds to $H = 0.5$. This indicates that the quenching should be done in oil with "good" agitation.

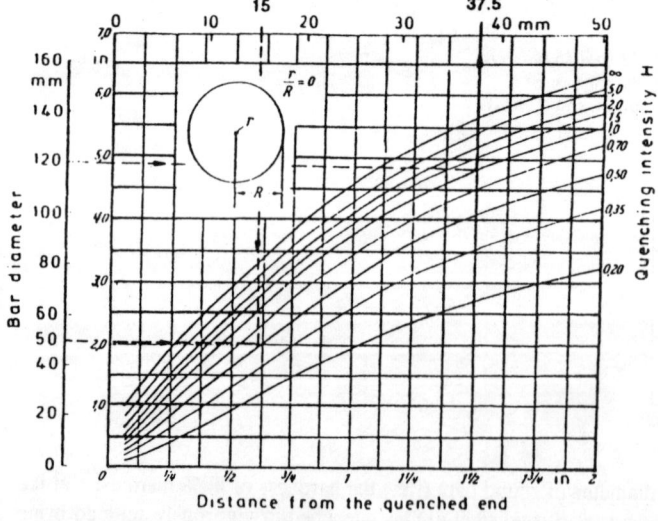

Figure 49 Relation between distance from the quenched end of Jominy specimen and bar diameter for the ratio $r/R = 0$, i.e., the center of the cross section, for different quenching intensities. (From Ref. 33.)

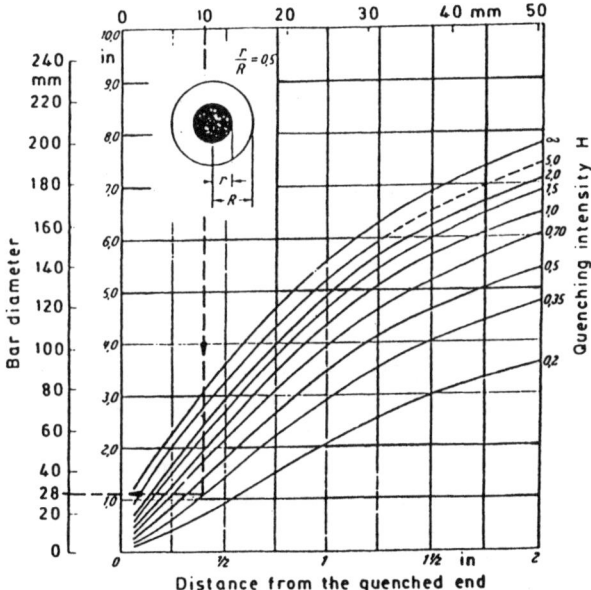

Distance from the quenched end

Figure 50   Relation between distance from the quenched end of Jominy specimen and bar diameter for the ratio $r/R = 0.5$, i.e., 50% from the center, for different quenching intensities. (From Ref. 33.)

If the required hardness should be attained only up to a certain depth below the surface, the fractional depth on the cross section must first be established to select the appropriate transformation diagram. For example, if 50 HRC hardness, which corresponds to a 15 mm distance from the quenched end on the Jominy curve of the

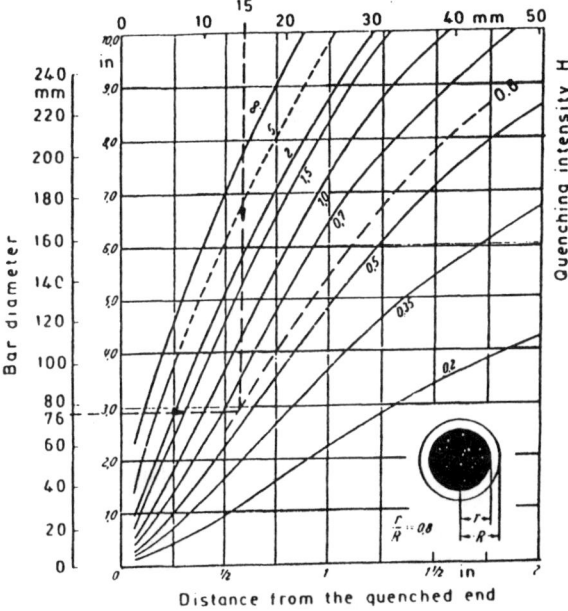

Distance from the quenched end

Figure 51   Relation between distance from the quenched end of Jominy specimen and bar diameter for the ratio $r/R = 0.8$, i.e., 80% from the center, for different quenching intensities. (From Ref. 33.)

relevant steel grade, should be attained at 7.6 mm below the surface of a 76 mm diameter bar, then

$$\frac{r}{R} = \frac{38 - 7.6}{38} = 0.8 \tag{19}$$

This calculation indicates that the diagram for $r/R = 0.8$ (Figure 51) should be used. In this case, the horizontal line for 76 mm diameter intersects the vertical line for 15 mm Jominy distance on the interpolated curve $H = 0.6$. This indicates that quenching should be performed in oil with "strong" agitation (see Table 1).

4. To predict the hardness along the radius of round bars of different diameters when the bar diameter and steel grade and its Jominy curve and quenching intensity H are known. For this calculation, diagrams for every ratio r/R from the center to the surface should be used. The following procedure should be repeated with every diagram. At the point where the horizontal line (indicating the bar diameter in question) intersects the relevant H curve, the vertical line gives the corresponding distance from the quenched end on the Jominy curve from which the corresponding hardness can be read and plotted at the corresponding fractional depth. Because some simplifying assumptions are made when using Lamont diagrams, hardness predictions are approximate. Experience has shown that for small cross sections and for the surface of large-diameter bars, the actual hardness is usually higher than predicted.

B. Steel Selection Based on Hardenability

The selection of a steel grade (and heat) for a part to be heat-treated depends on the hardenability that will yield the required hardness at the specified point of the cross section after quenching under known conditions. Because Jominy hardenability curves and hardenability bands are used as the basis of the selection, the method described here is confined to those steel grades with known hardenability bands or Jominy curves. This is true first of all for structural steels for hardening and tempering and also for steels for case hardening (to determine core hardenability).

If the diameter of a shaft and the bending fatigue stresses it must be able to undergo are known, engineering analysis will yield the minimum hardness at a particular point on the cross section that must be achieved by hardening and tempering. Engineering analysis may show that distortion minimization requires a less severe quenchant, e.g., oil. Adequate toughness after tempering (because the part may also be subject to impact loading) may require a tempering temperature of, e.g., 500°C.

The steps in the steel selection process are as follows

Step 1. Determine the necessary minimum hardness after quenching that will satisfy the required hardness after tempering. This is done by using a diagram such as the one shown in Figure 52. For example, if a hardness of 35 HRC is required after hardening and then tempering at 500°C at the critical cross-sectional diameter, the minimum hardness after quenching must be 45 HRC.

Alternatively, if the carbon content of the steel and the percentage of as-quenched martensite at the critical point of the cross section is known, then by using a diagram that correlates hardness with percent carbon content and as-quenched martensite content (see Figure 53), the as-quenched hardness may then be determined. If 80% martensite is desired at a critical position of the cross section and the steel has 0.37% C, a hardness of 45 HRC can be expected. Figure 53 can also be used to determine the necessary carbon content of the steel when a particular percentage of martensite and a particular hardness after quenching are required.

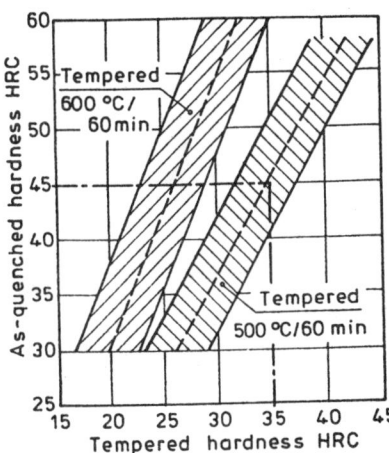

Figure 52 Correlation between the hardness after tempering and the hardness after quenching for structural steels (according to DIN 17200).

Step 2. Determine whether a certain steel grade (or heat) will provide the required as-quenched hardness at a critical point of the cross section. For example, assume that a shaft is 45 mm in diameter and that the critical point on the cross section (which was determined from engineering analysis of resultant stresses) is three-fourths of the radius. To determine if a particular steel grade, e.g., AISI 4140H, will satisfy the requirement of 45 HRC at (3/4) R after oil quenching, the diagram shown in Figure 54a should be used. This diagram correlates cooling rates along the Jominy end-quench specimen and at four characteristic locations (critical points) on the cross section of a round bar when quenched in oil at 1 m/s agitation rate (see the introduction to Section V and Figure 44). Figure 54a shows that at (3/4) R the shaft having a diameter of 45 mm will exhibit the hardness that corresponds to the hardness at a distance of 6.5/16 in. (13/32 in.) from the quenched end of the Jominy specimen.

Step 3. Determine whether the steel grade represented by its hardenability band (or a certain heat represented by its Jominy hardenability curve) at the specified distance from the quenched end exhibits the required hardness. As indicated in Figure 54b, the minimum hardenability curve for AISI 4140H will give a hardness of 49 HRC. This means that AISI 4140H has, in every case, enough hardenability for use in the shaft example above.

This graphical method for steel selection based on hardenability, published in 1952 by Weinmann and coworkers can be used as an approximation. Its limitation is that the diagram

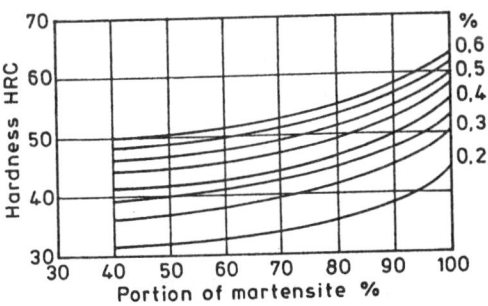

Figure 53 Correlation between as-quenched hardness, carbon content, and percent martensite (according to Hodge and Orehovski). (From Ref. 15, p. 481.)

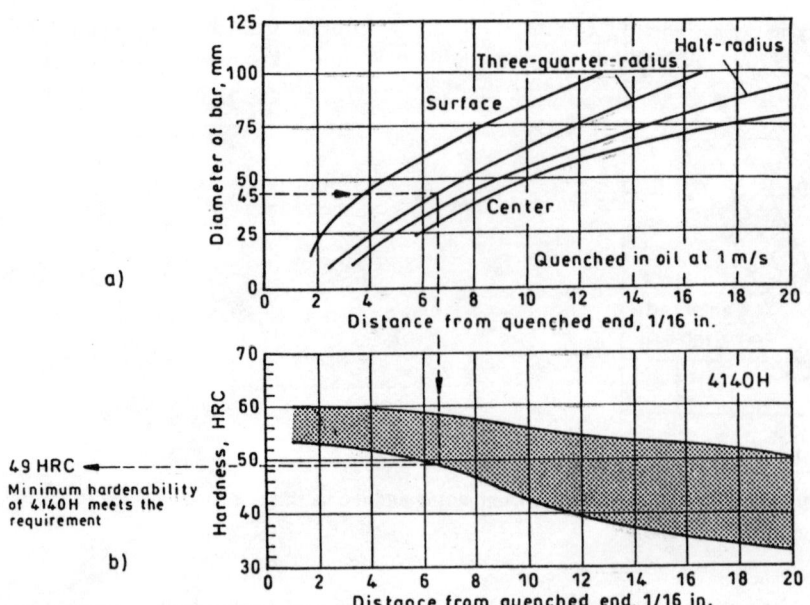

Figure 54 Selecting a steel of adequate hardenability. (a) Equivalent cooling rates (and hardness after quenching) for characteristic points on a round bar's cross section and along the Jominy end-quench specimen. (b) Hardenability band of AISI 4140H steel. (From Ref. 15, p. 493.)

shown in Figure 54a provides no information on the quality of the quenching oil and its temperature. Such diagrams should actually be prepared experimentally for the exact conditions that will be encountered in the quenching bath in the workshop; the approximation will be valid only for that bath.

C. Computer-Aided Steel Selection Based on Hardenability

As in other fields, computer technology has made it possible to improve the steel selection process, making it quicker, more intuitive, and even more precise. One example, using a software package developed at the University of Zagreb [36], is based on a computer file of experimentally determined hardenability bands of steels used in the heat-treating shop. The method is valid for round bars of 20–90 mm diameters. The formulas used for calculation of equidistant locations on the Jominy curve, described in Ref. 24, were established through regression analysis for this range of diameters.

The essential feature of this method is the calculation of points on the *optimum Jominy hardenability curve* for the steel being calculated. Calculations are based on the required as-quenched hardness on the surface of the bar and at one of the critical points of its cross section [(3/4)R, (1/2)R, (1/4)R, or center]. The input data for the computer-aided selection process are the following:

Diameter of the bar (D mm)

Surface hardness HRC

Hardness at a critical point HRC

Quenching intensity factor I (I equals the Grossmann quenching intensity factor H as given in Table 1)

Minimum percentage of martensite required at the critical point

The first step is to calculate the equidistant locations from the quenched end on the Jominy curve (or Jominy hardenability band). These equidistant locations are the points on the Jominy curve that yield the required as-quenched hardness. The calculations are performed as follows [24]:

On the surface:

$$E_s = \frac{D^{0.718}}{5.11 \ I^{1.28}} \tag{20}$$

At $(3/4)R$:

$$E_{3/4R} = \frac{D^{1.05}}{8.62 \ I^{0.668}} \tag{21}$$

At $(1/2)R$:

$$E_{1/2R} = \frac{D^{1.16}}{9.45 \ I^{0.51}} \tag{22}$$

At $(1/4)R$:

$$E_{1/4R} = \frac{D^{1.14}}{7.7 \ I^{0.44}} \tag{23}$$

At the center:

$$E_c = \frac{D^{1.18}}{8.29 \ I^{0.44}} \tag{24}$$

[*Note*: The calculated E values are in millimeters.]

After calculating the equidistant locations for the surface of the bar (E_s) and for one of the critical points (E_{crit}), using the hardenability band of the relevant steel, the hardness values achievable with the Jominy curve of the lowest hardenability (H_{low}) and the hardness values achievable with the Jominy curve of the highest hardenability (H_{high}) for both E_s and E_{crit} locations are then determined as shown in Figure 55.

The *degree of hardening* S is defined as the ratio of the measured hardness after quenching (at a specified point of the cross section) to the maximum hardness that can be achieved with the steel in question:

$$S = \frac{H}{H_{max}} \tag{25}$$

It can be easily calculated for the equidistant location E_{crit} on the upper and lower curves of the hardenability band, taking the value for H_{max} from the relevant Jominy curve at distance 0 from the quenched end ($J = 0$). In this way, two distinct values of the degree of hardening, S_{upper} and S_{lower}, are calculated. Each corresponds to a certain percentage of martensite in the as-quenched structure as shown in Table 5.

It is also possible to determine whether the required percentage of martensite can be achieved by either Jominy curve of the hardenability band. Instead of providing the percentage of martensite in the as-quenched structure as input data, the value of S (degree of hard-

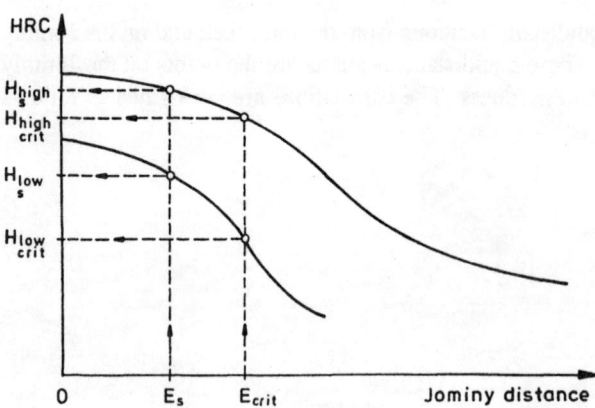

Figure 55 Determination of minimum and maximum hardness for equidistant locations E_s and E_{crit} from a relevant hardenability band. (After Ref. 35.)

ening) may be given. For statically stressed parts, $S < 0.7$; for less dynamically stressed parts, $0.7 < S < 0.86$; and for highly dynamically stressed parts, $0.86 < S < 1.0$. In this way, a direct comparison of the required S value with values calculated for both Jominy curves at the E_{crit} location can be performed. There are three possibilities in this comparison:

1. The value of S required is even lower than the S value calculated for the lower curve of the hardenability band (S_{lower}). In this case all heats of this steel will satisfy the requirement. The steel actually has higher hardenability than required.

2. The value of S required is even higher than the S value calculated for the upper curve of the hardenability band (S_{upper}). In this case, none of the heats of this steel can satisfy the requirement. This steel must not be selected because its hardenability is too low for the case in question.

3. The value of required degree of hardening (S) is somewhere between the values for the degree of hardening achievable with the upper and lower curves of the hardenability band (S_{upper} and S_{lower}, respectively).

Table 5 Correlation Between Degree of Hardening S and Percentage of Martensite in As-Quenched Structure

Percent martensite	Degree of hardening S
50–60	0.70–0.74
60–70	0.74–0.76
70–80	0.76–0.78
80–85	0.78–0.81
85–90	0.81–0.86
90–95	0.86–0.91
95–97	0.91–0.95
97–100	0.95–1.00

Source: Ref. 36.

In the third case, the position of the S required, designated as X, is calculated according to the formula

$$X = \frac{S - S_{lower}}{S_{upper} - S_{lower}}$$

where X is the distance from the lower curve of the hardenability band on the ordinate E_{crit} to the actual position of S required, which should be on the optimum Jominy curve. This calculation divides the hardenability band into three zones:

The lower third, $X \leq 0.33$

The middle third, $0.33 < X \leq 0.66$

The upper third, $0.66 < X$

All heats of a steel grade where the Jominy curves pass through the zone in which the required S point is situated can be selected as heats of adequate hardenability. This zone is indicated in a graphical presentation of the method. Once the distance X is known, the *optimum Jominy hardenability curve* can be drawn. The only requirement is that for every distance from the quenched end the same calculated ratio (X) that indicates the same position of the Jominy curve relative to the lower and upper hardenability curves of the hardenability band is maintained.

The following example illustrates the use of this method in selecting a steel grade for hardening and tempering.

A 40 mm diameter shaft after hardening and tempering should exhibit a surface hardness of $H_s = 28$ HRC and a core hardness of $H_c = 26$ HRC. The part is exposed to high dynamic stresses. Quenching should be performed in agitated oil.

The first step is to enter the input data and select the critical point on the cross section (in this case the core) as shown in Figure 56. Next, the required percentage of martensite at the critical point after quenching (in this case 95%, because of high dynamic stresses) and the quenching intensity I (in this case 0.5, corresponding to the Grossmann value H) are se-

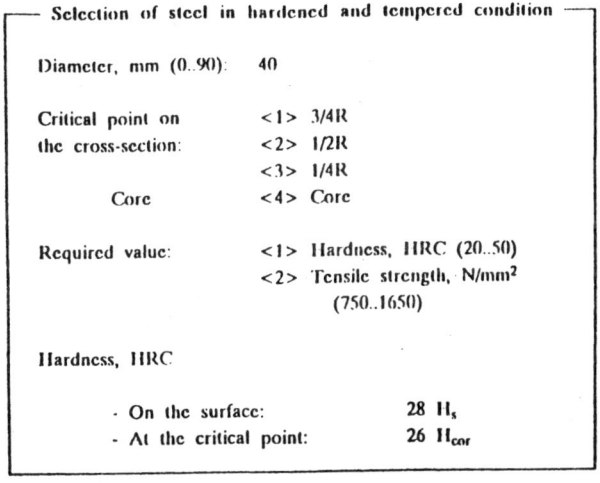

Figure 56   Input data for computer program. (From Ref. 36.)

```
─────────────── Results of steel selection ───────────────

   JUS          AISI
   Č4181                   Not suitable
   Č4730        4130       Not suitable
   Č4731        E4132      Suitable heats from upper third of band
   Č4781                   Suitable heats from upper third of band
   Č4732        4140       Suitable heats from middle third of band
   Č4782                   Suitable heats from middle third of band
   Č4733        4150       Suitable heats from middle third of band
   Č4738                   Too high hardenability
   Č4734                   Too high hardenability
```

Figure 57   List of computer results. (From Ref. 36.)

lected. The computer program repeats the above described calculations for every steel grade for which the hardenability band is stored in the file and presents the results on the screen as shown in Fig. 57. This is a list of all stored steel grades regarding suitability for the application being calculated. Acceptable steel grades, suitable from the upper, middle, or lower third of the hardenability band, and unacceptable steel grades with excessively high hardenability are determined.

For each suitable steel grade, a graphical presentation as shown in Figure 58 can be obtained. This gives the optimum Jominy hardenability curve for the case required and indicates the desired zone of the hardenability band.

In addition, the necessary tempering temperature can be calculated according to the formula

$$T_{temp} = 917 \sqrt[6]{\frac{\ln \dfrac{H_{crit}^{-8}}{H_{temp}^{-8}}}{S}} \tag{26}$$

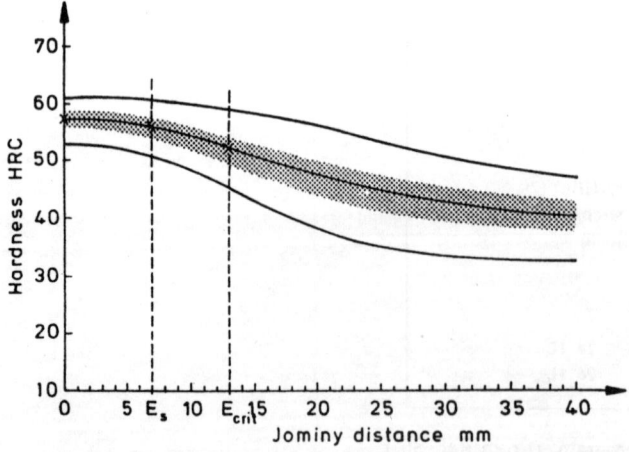

Figure 58   Graphical presentation of the optimum Jominy hardenability curve. (From Ref. 36.)

where

T_{temp} = absolute tempering temperature (K) (valid for $400 < T_{temp} < 660°C$)

H_{crit} = hardness after quenching at the critical point HRC (taken from the optimum Jominy curve at the distance for the critical point)

H_{temp} = required hardness after tempering at the critical point HRC

S = degree of hardening (ratio between hardness on the optimum Jominy curve at the distance E_{crit} and at the distance $E = 0$)

Tensile strength (R_m, N/mm²) is also calculated at the relevant points using the formula

$$R_m = 0.426 \, H^2 + 586.5 \quad [N/mm^2] \tag{27}$$

where H is the corresponding hardness value in HRC. Knowing the tensile strength (R_m), other mechanical properties are calculated according to the formulas

Yield strength:

$$R_{p0.2} = (0.8 + 0.1S)R_m + 170S - 200 \quad [N/mm^2] \tag{28}$$

Elongation

$$A_5 = 0.46 - (0.0004 - 0.00012S)R_m \quad [\%] \tag{29}$$

Contraction:

$$Z = 0.96 - (0.00062 - 0.00029S)R_m \quad [\%] \tag{30}$$

Bending fatigue strength:

$$R_d = (0.25 + 0.45Z)R_m \quad [N/mm^2] \tag{31}$$

Impact energy (toughness):

$$KU = [460 - (0.59 - 0.29S)R_m](0.7) \quad [J] \tag{32}$$

For every steel grade (and required zone of the hardenability band) that has been found suitable, the mechanical properties for the surface and for the critical cross section point can be calculated. The computer output is shown in Figure 59.

(AISI-4140)

Heats from the middle third of the band

Calculated tempering temperature: 643 °C

Mechanical properties:	Surface	Critical point
Yield strength: $R_{p0.2}$, N/mm²	793	735
Tensile strength: R_m, N/mm²	920	874
Bending fatigue strength: R_d, N/mm²	499	474
Elongation: A5, %	20	21
Contraction: Z, %	65	65
Impact energy: KU, J	125	123

Figure 59 Computer display of calculated mechanical properties. (From Ref. 36.)

Compared to the previous steel selection processes, these computer-aided calculations have the following advantages:

1. Whereas the previously described graphical method is valid for only one specified quenching condition for which the relevant diagram has been plotted, the computer-aided method allows great flexibility in choosing concrete quenching conditions.
2. The selection of the optimum hardenability to satisfy the requirements is much more precise.
3. Calculations of the exact tempering temperature and all mechanical properties after tempering at the critical point, that give much more information and facilitate the steel selection, are possible.

VI. HARDENABILITY IN HEAT TREATMENT PRACTICE

A. Hardenability of Carburized Steels

Carburized parts are primarily used in applications where there are high surface stresses. Failures generally originate in the surface layers where the service stresses are most severe. Therefore, high case strength and high endurance limits are critical factors. High case hardness improves the fatigue durability. Historically, it was thought that core hardenability was required for the selection of carburizing steels and heat treatment of carburized parts and that core hardenability would ensure adequate case hardenability. Equal additions of carbon, however, do not have the same effect on the hardenability of all steel compositions; therefore the historical view of core hardenability may not be correct. In fact, hardenability of both case and core is essential for proper selection of the optimum steel grade and the heat treatment of carburized parts.

It is now also known that the method of quenching after carburizing, i.e., "direct quenching" or "reheat and quench," influences case hardenability. The case hardenability of carburized steel is determined by using the Jominy end-quench test.

Standard Jominy specimens are carburized in a carburizing medium with a high C potential for sufficient time to obtain a carburized layer of the desired depth. In addition to the Jominy specimens, two bars of the same steel and heat, the same surface finish, and the same dimensions (25 mm diameter) are also carburized under identical conditions. These bars are used to plot the carbon gradient curve shown in Figure 60a, which is produced by chemical analysis of chips obtained from machining of the carburized layer at different layer thicknesses. In this way, as shown in Figure 60a, the following carbon contents were found as a function of case depth:

1.0% C at 0.2 mm depth (distance from the surface of the bar)—d_1
0.9% C at 0.32 mm depth—d_2
0.8% C at 0.45 mm depth—d_3
0.7% C at 0.57 mm depth—d_4

One of the carburized Jominy specimens should be end-quenched in the standard way using the Jominy apparatus directly from the carburizing temperature (direct quenching), and the other should first be cooled to room temperature and then reheated and quenched from a temperature that is usually much lower than the carburizing temperature (reheat and quench).

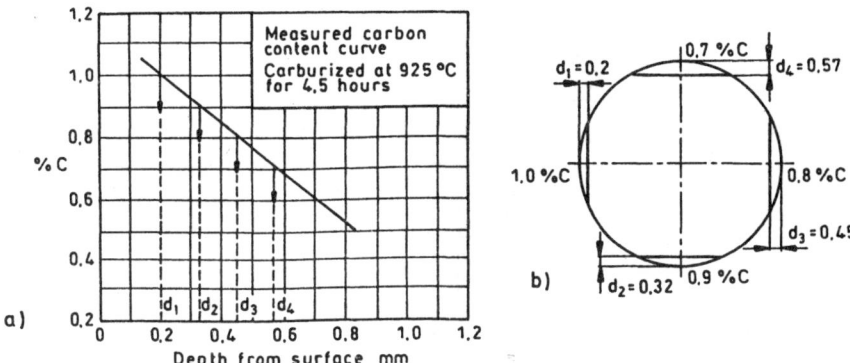

Figure 60 (a) Measured carbon gradient curve after gas carburizing at 925°C for 4.5 h. (b) Grinding of the carburized Jominy specimen. (From Ref. 37.)

After quenching, all Jominy specimens should be ground on four sides of the perimeter to the depths d_1, d_2, d_3, and d_4, as shown in Figure 60b. Hardness is measured in the standard way on each of ground surfaces, and the corresponding Jominy curves are plotted. Figure 61a provides an example of Jominy hardenability curves for the carburizing steel grade DIN 16MnCr5 (0.17% C, 0.25% Si, 1.04% Mn, 1.39% Cr). The carbon contents in the case were 1.0, 0.9, 0.8, and 0.7% C, and the core carbon content was 0.17% C after direct quenching from the carburizing temperature, 925°C. Figure 61b provides Jominy curves for the same carburized case after indirect quenching (reheated to 820°C). From both diagrams of Figure 61 the following conclusions can be drawn:

1. The hardenability of the core is substantially different from the hardenability within the carburized case.

2. The best hardenability of the carburized case is found for this steel grade at 0.9% C with direct quenching and at 0.8% C with indirect quenching (reheat and quench).

Consequently, the carburizing process should be controlled so that after carburizing a surface carbon content of 0.9% is obtained for direct quenching and one of 0.8% for indirect quenching.

B. Hardenability of Surface Layers When Short-Time Heating Methods Are Used

When short-time ("zero" time) heating processes for surface hardening are used, e.g., flame hardening, induction hardening, or laser hardening, the same metallurgical reactions occur as in conventional hardening except that the heating process cycle must be much shorter than that of conventional hardening. Heating times for these processes vary by one to three orders of magnitude; approximately 100 s for flame hardening, 10 s or less for induction hardening, and 1 s or less for laser hardening. This means that the heating rates are very high. Problems associated with these high heating rates are twofold.

1. The transformation from the bcc lattice of the α-iron to the fcc lattice of the γ-iron does not occur between normal temperatures Ac_1 and Ac_3 as in conventional hardening because the high heating rate produces nonequilibrium systems. The Ac_1 and Ac_3 temperatures are displaced to higher temperatures as shown in Figure 62. Although an austenitizing temperature may be sufficiently high to form austenite under slow

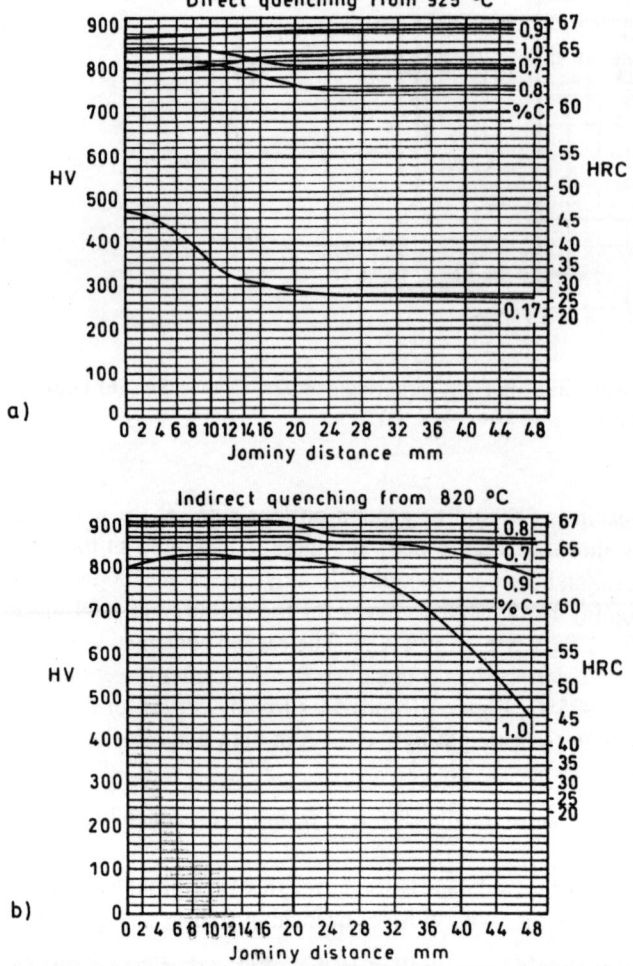

Figure 61 Jominy case hardenability curves of carburized DIN 16MnCr5 steel (a) after direct quenching from 925°C and (b) after reheating followed by quenching from 820°C. (From Ref. 37.)

heating conditions (conventional hardening), the same temperature level may not be sufficient to even initiate austenization under high heating rates [38]. Therefore, substantially higher austenitizing temperatures are used with flame, induction, and laser hardening (especially the latter) than for conventional hardening of the same steel.

2. For quench hardening, the austenitization must dissolve and uniformly distribute the carbon of the carbides in the steel. This is a time-dependent diffusion process (sometimes called homogenization), even at the high temperatures used in short-time heating methods. At very high heating rates, there is insufficient time for diffusion of carbon atoms from positions of higher concentrations near carbides to the positions of lower concentrations (areas that originated from practically carbon-free ferrite). This diffusion depends on the path length of carbon atoms and therefore is dependent on the distribution of carbon in the starting structure. Coarse pearlitic structures, spheroidized structures, and (particularly) nodular cast iron with a high content of free ferrite are undesirable in this regard. Tempered martensite, having small and

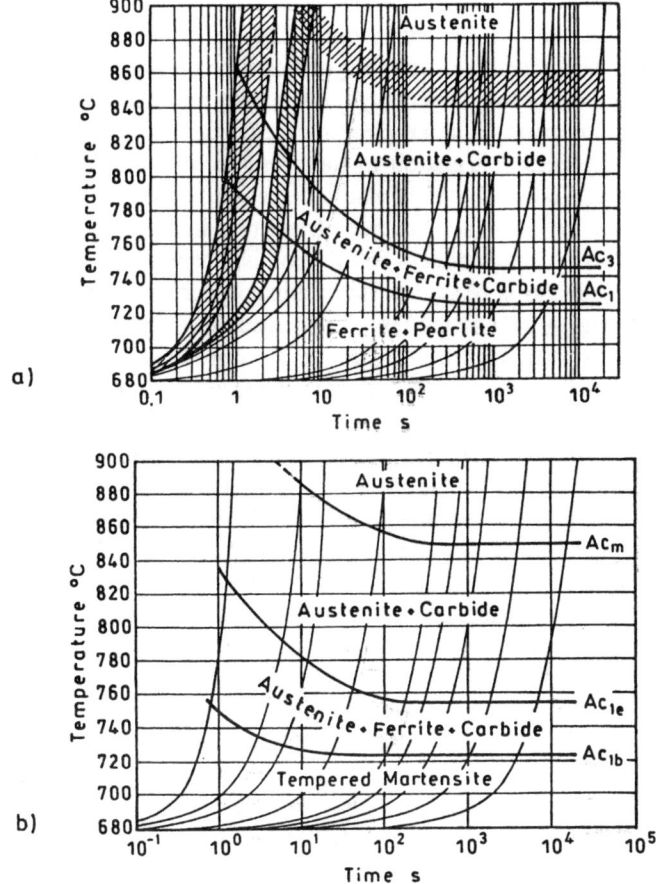

Figure 62 Time–temperature–transformation diagram for continuing heating with different heating rates, when austenitizing an unalloyed steel with 0.7% C. (a) Starting structure, ferrite and lamellar pearlite; (b) starting structure, tempered martensite. (From Ref. 39.)

finely dispersed carbides, provides the shortest paths for carbon diffusion and is therefore most desirable.

Figure 62a illustrates a time–temperature–transformation diagram for continuous heating at different heating rates when austenitizing an unalloyed steel with 0.7% C with a starting structure of ferrite and lamellar pearlite. Figure 62b shows a similar diagram for a starting structure of tempered martensite. A comparison of the two diagrams illustrates the influence of starting structure on the austenitizing process. Whereas for the ferrite–pearlite starting structure at maximum heating rate the upper transformation temperature Ac_3 is 865°C, for the starting microstructure of tempered martensite, the Ac_3 temperature is 835°C. This means that the austenite from a starting structure of tempered martensite has a better hardenability than the austenite of a pearlite–ferrite starting structure. The practical consequence of this is that prior to surface hardening by any short-time heating process, if the steel is in the hardened and tempered condition, maximum hardened case depths are possible. If the annealed material has a coarse lamellar structure, or even worse, globular carbides, minimum hardening depths are to be expected.

C. Effect of Delayed Quenching on the Hardness Distribution

Delayed quenching processes have been known for a long time. *Delayed quenching* means that austenitized parts are first cooled slowly and then after a specified time they are quenched at a much faster cooling rate. Delayed quenching is actually a quenching process in which a discontinuous change in cooling rate occurs. In some circumstances, depending on steel hardenability and section size, the hardness distribution in the cross section after delayed quenching does not have a normal trend (normally the hardness decreases continuously from the surface toward the core) but instead exhibits an "inverse" trend (the hardness increases from the surface toward the core). This inverse hardness distribution is a consequence of the discontinuous change in the cooling rate and is related to the incubation period (at different points in the cross section) before changing the cooling rate. This process has been explained theoretically by Shimizu and Tamura [41,42]. Figure 63 shows the measured hardness distribution for the cross section of 50 mm diameter × 200 mm bars made of the same heat of AISI 4140 that had been quenched according to Table 6.

In every experiment, the delay in quenching was measured as the time from immersion to the moment when maximum heat flux density on the surface (t_{qmax}) occurred. As shown in Figure 63 and Table 6 for AISI 4140 steel with a section 50 mm in diameter, when the delay in quenching (due to high concentration of the PAG polymer solution and corresponding thick film around the heated parts) was more than 15 s ($t_{qmax} > 15$ s), a completely inverse or inverse to normal hardness distribution was obtained. In experiments where t_{qmax} was less than 15 s, a normal hardness distribution resulted.

Besides the inherent hardenability of a steel, delayed quenching may substantially increase the depth of hardening and may compensate for lower hardenability of the steel [40]. Interestingly, none of the available software programs for predicting as-quenched hardness simulates the inverse hardness distribution because they do not account for the length of the incubation period before the discontinuous change in cooling rate at different points in the cross section.

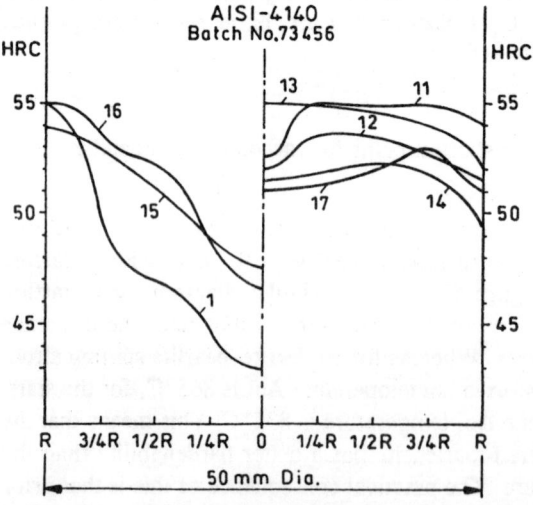

Figure 63 Measured hardness distribution on the cross section of 50 mm diameter × 200 mm bars made of AISI 4140 steel quenched according to conditions given in Table 6. (From Ref. 40.)

Table 6 Time from Immersion (t_{qmax}) Until Maximum Heat Flux Density Under Various Quenching Conditions for AISI 4140 Bars (50 mm Diameter × 200 mm)[a]

Figure 63 curve No.	Quenching conditions	t_{qmax} (s)
1	Mineral oil at 20°C, without agitation	14
11	Polymer solution (PAG) 5%; 40°C; 0.8 m/s	16
12	Polymer solution (PAG) 15%; 40°C; 0.8 m/s	33
13	Polymer solution (PAG) 25%; 40°C; 0.8 m/s	70
14	Polymer solution (PAG) 20%; 35°C; 1 m/s	30
15	Polymer solution (PAG) 10%; 35°C; 1 m/s	12
16	Polymer solution (PAG) 5%; 35°C; 1 m/s	13
17	Polymer solution (PAG) 20%; 35°C; 1 m/s	47

[a]See Figure 63.
Source: Ref. 40.

D. A Computer-Aided Method to Predict the Hardness Distribution After Quenching Based on Jominy Hardenability Curves

The objective here is to describe one method of computer-aided calculation of hardness distribution. This method, developed at the University of Zagreb [45], is based on the Jominy hardenability curves. Jominy hardenability data for steel grades of interest are stored in a databank. In this method, calculations are valid for cylindrical bars 20–90 mm in diameter. Figure 64 shows the flow diagram of the program, and Figure 65 is a schematic of the step-by-step procedure:

Step 1. Specify the steel grade and quenching conditions.

Step 2. Harden a test specimen (50 mm diameter × 200 mm) of the same steel grade by quenching it under specific conditions.

Step 3. Measure the hardness (HRC) on the specimen's cross section in the middle of the length.

Step 4. Store in the file the hardness values for five characteristic points on the specimen's cross section [surface, (3/4)R, (1/2)R, (1/4)R, and center]. If the databank already contains the hardness values for steel and quenching conditions obtained by previous measurements, then eliminate steps 2 and 3 and retrieve these values from the file.

Step 5. From the stored Jominy hardenability data, determine the equidistant points on the Jominy curve (E_s, $E_{3/4R}$, $E_{1/2R}$, $E_{1/4R}$, E_c) that have the same hardness values as those measured at the characteristic points on the specimen's cross section.

Step 6. Calculate the hypothetical quenching intensity I at each of the mentioned characteristic points by the following regression equations, based on the specimen's diameter D_{spec} and on known E values.

$$I_s = \left[\frac{D_{spec}^{0.718}}{5.11 \, E_s} \right]^{0.78} \tag{33}$$

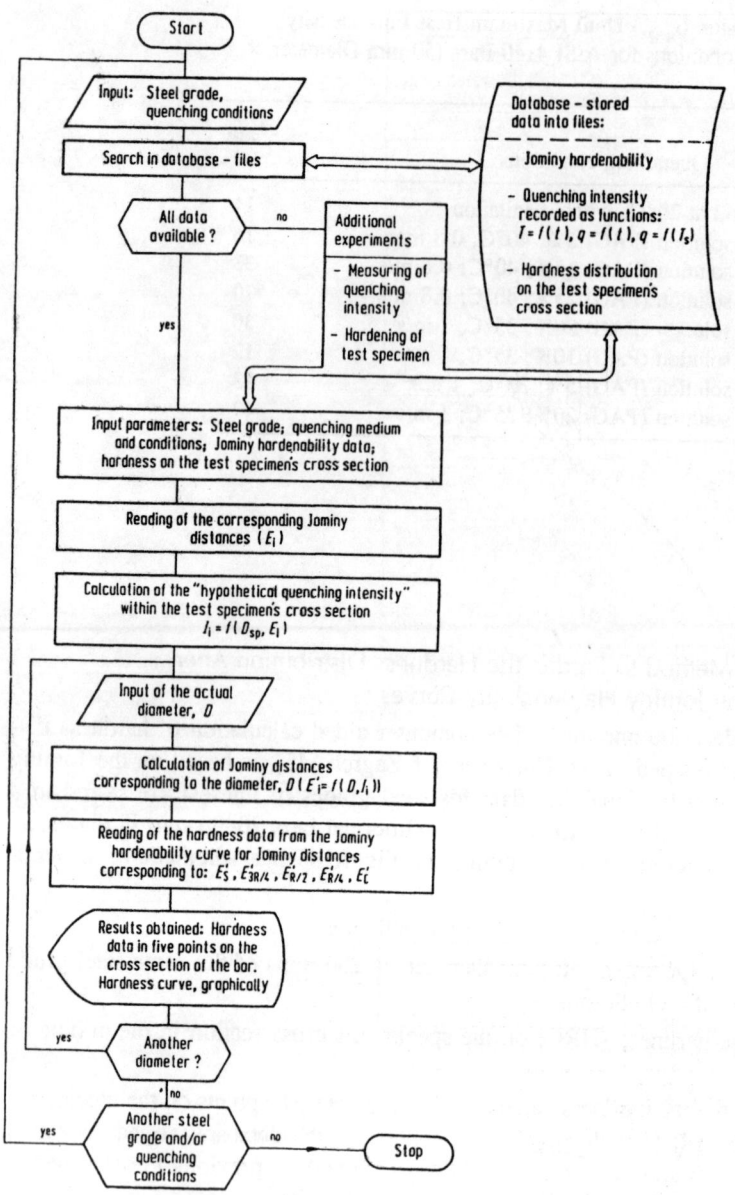

Figure 64 Flowchart of computer-aided prediction of hardness distribution on cross section of quenched round bars. (From Ref. 21.)

$$I_{3/4R} = \left[\frac{D_{spec}^{1.05}}{8.62 \ E_{3/4R}}\right]^{1.495}$$

(34)

$$I_{1/2R} = \left[\frac{D_{spec}^{1.16}}{9.45 \ E_{1/2R}}\right]^{1.495}$$

(35)

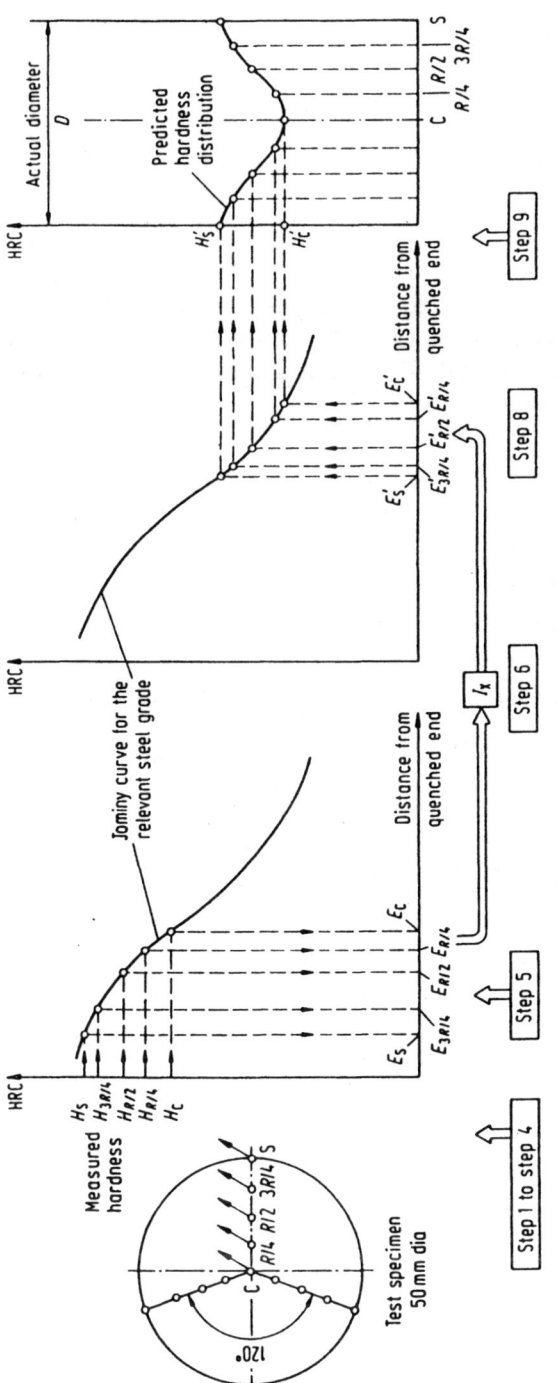

Figure 65 Stepwise scheme of the process of prediction of hardness distribution after quenching. (From Ref. 21.)

$$I_{1/4R} = \left[\frac{D^{1.14}_{spec}}{7.7 \ E_{1/4R}} \right]^{2.27} \tag{36}$$

$$I_c = \left[\frac{D^{1.18}_{spec}}{8.29 \ E_c} \right]^{2.27} \tag{37}$$

Equations (33)–(37) combine the equidistant points on the Jominy curve, the specimen's diameter, and the quenching intensity and were derived from the regression analysis of a series of Crafts–Lamont diagrams [23]. This analysis is based on Just's relationships [43] for the surface and the center of a cylinder:

$$E_i = A \frac{D^{B_1}}{I^{B_2}} \tag{38}$$

where

E_i = corresponding equidistant point on the Jominy curve
A, B_1, B_2 = regression coefficients
D = bar diameter
I = quenching intensity (H according to Grossmann)

Step 7. Enter the actual bar diameter D for which the predicted hardness distribution is desired.

Step 8. Calculate the equidistant Jominy distances $E'_s, E'_{3/4R}, E'_{1/2R}, E'_{1/4R}, E'_c$ that correspond to the actual bar diameter D and the previously calculated hypothetical quenching intensities I_s–I_c using the formulas

$$E'_s = \frac{D^{0.718}}{5.11 \ I^{1.28}} \tag{39}$$

$$E'_{3/4R} = \frac{D^{1.05}}{8.62 \ I^{0.668}} \tag{40}$$

$$E'_{1/2R} = \frac{D^{1.16}}{9.45 \ I^{0.51}} \tag{41}$$

$$E'_{1/4R} = \frac{D^{1.14}}{7.7 \ I^{0.44}} \tag{42}$$

$$E'_c = \frac{D^{1.18}}{8.29 \ I^{0.44}} \tag{43}$$

Step 9. Read the hardness values $H'_s, H'_{3/4R}, H'_{1/2R}, H'_{1/4R}$, and H'_c from the relevant Jominy curve associated with the calculated Jominy distances and plot the hardness distribution curve over the cross section of the chosen actual diameter D.

Figure 66 provides an example of computer-aided prediction of hardness distribution for 30 mm and 70 mm diameter bars made of AISI 4140 steel quenched in a mineral oil at 20°C without agitation. Experimental validation using three different steel grades, four different bar diameters, and four different quenching conditions was performed, and a comparison to predicted results is shown in Figure 67. In some cases, the precision of the hardness distribution prediction was determined using the Gerber–Wyss method [44]. From examples 2, 3, 5, and 6 of Figure 67 it can be seen that the computer-aided prediction provides a better fit to the experimentally obtained results than the Gerber–Wyss method.

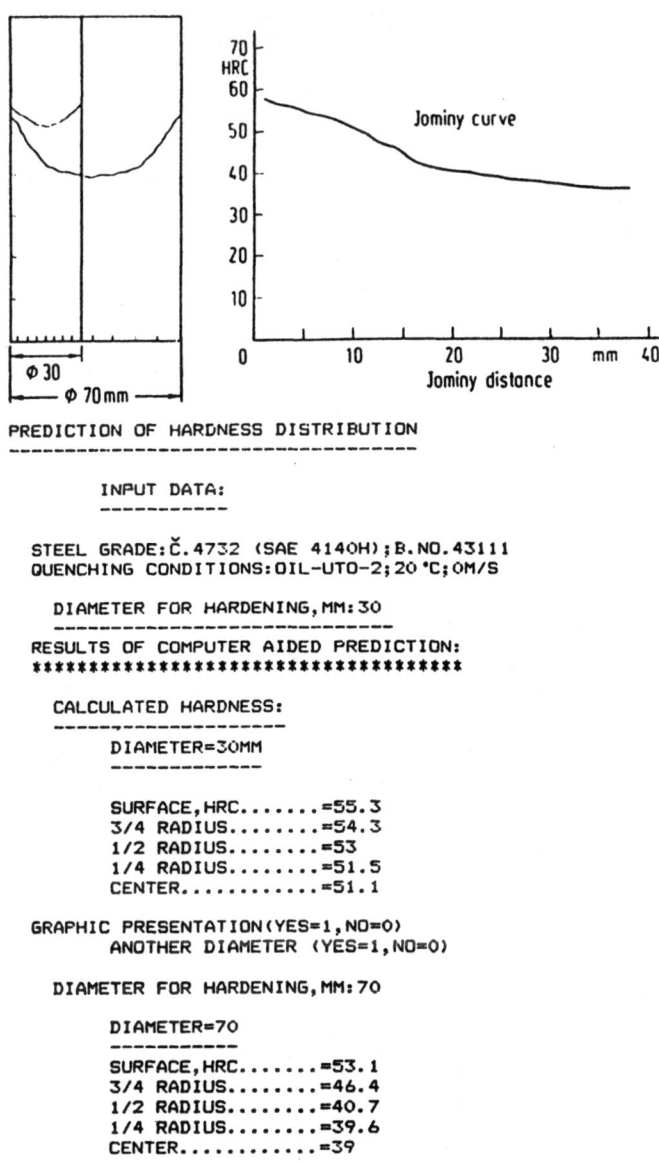

```
PREDICTION OF HARDNESS DISTRIBUTION
------------------------------------

        INPUT DATA:
        -----------

   STEEL GRADE:Č.4732 (SAE 4140H);B.NO.43111
   QUENCHING CONDITIONS:OIL-UTO-2;20 °C;OM/S

     DIAMETER FOR HARDENING,MM:30
     -------------------------------
   RESULTS OF COMPUTER AIDED PREDICTION:
   *************************************

     CALCULATED HARDNESS:
     ----------------------
          DIAMETER=30MM
          -------------

          SURFACE,HRC.......=55.3
          3/4 RADIUS........=54.3
          1/2 RADIUS........=53
          1/4 RADIUS........=51.5
          CENTER............=51.1

   GRAPHIC PRESENTATION(YES=1,NO=0)
          ANOTHER DIAMETER (YES=1,NO=0)

     DIAMETER FOR HARDENING,MM:70

          DIAMETER=70
          -----------
          SURFACE,HRC.......=53.1
          3/4 RADIUS........=46.4
          1/2 RADIUS........=40.7
          1/4 RADIUS........=39.6
          CENTER............=39

   GRAPHIC PRESENTATION(YES=1,NO=0)
          ANOTHER DIAMETER (YES=1,NO=0)
```

Figure 66 An example of computer-aided prediction of hardness distribution for quenched round bars of 30 and 70 mm diameter, steel grade SAE 4140H. (From Ref. 21.)

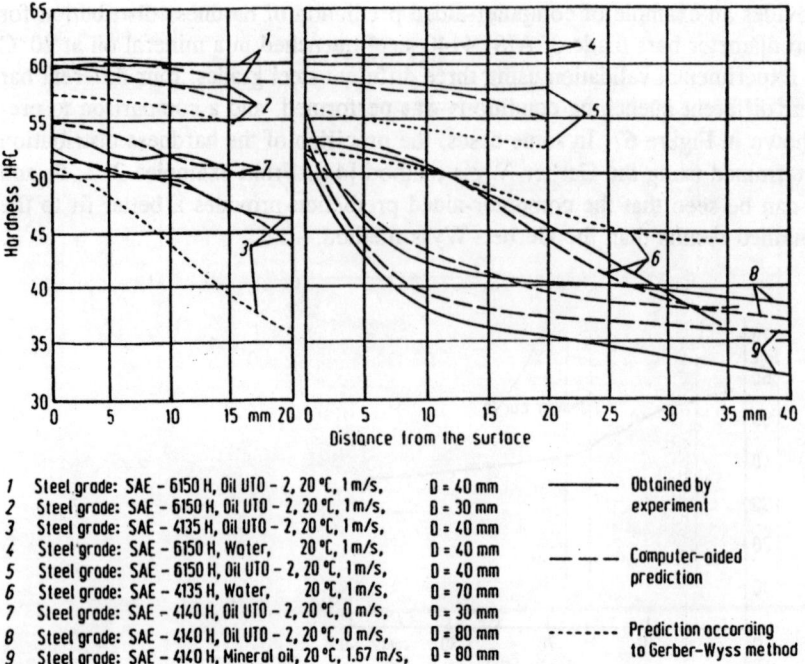

#	Steel grade:		
1	Steel grade: SAE – 6150 H, Oil UTO – 2, 20 °C, 1 m/s,	D = 40 mm	———— Obtained by
2	Steel grade: SAE – 6150 H, Oil UTO – 2, 20 °C, 1 m/s,	D = 30 mm	experiment
3	Steel grade: SAE – 4135 H, Oil UTO – 2, 20 °C, 1 m/s,	D = 40 mm	
4	Steel grade: SAE – 6150 H, Water, 20 °C, 1 m/s,	D = 40 mm	— — — Computer–aided
5	Steel grade: SAE – 6150 H, Oil UTO – 2, 20 °C, 1 m/s,	D = 40 mm	prediction
6	Steel grade: SAE – 4135 H, Water, 20 °C, 1 m/s,	D = 70 mm	
7	Steel grade: SAE – 4140 H, Oil UTO – 2, 20 °C, 0 m/s,	D = 30 mm	
8	Steel grade: SAE – 4140 H, Oil UTO – 2, 20 °C, 0 m/s,	D = 80 mm	········ Prediction according
9	Steel grade: SAE – 4140 H, Mineral oil, 20 °C, 1.67 m/s,	D = 80 mm	to Gerber–Wyss method

Figure 67 Comparison of the hardness distribution on round bar cross sections of different diameters and different steel grades, measured after experiments and obtained by computer-aided prediction as well as by prediction according to the Gerber–Wyss method. (From Ref. 21.)

1.   Selection of Optimum Quenching Conditions

The use the above relationship and stored data permits the selection of optimum quenching condition when a certain hardness value is required at a specified point on a bar cross section of known diameter and steel grade. Figure 68 illustrates an example where an as-quenched hardness of 51 HRC (H_q) at $(3/4)R$ of a 40 mm diameter bar made of SAE 4140H steel is required. Using the stored hardenability curve for this steel, the equivalent Jominy distance $E_{3/4R}$ yielding the same hardness can be found. Using $E_{3/4R}$ and the actual diameter D, the hypothetical quenching intensity factor $I_{3/4R}$ can be calculated according to Equation (34). That equation also applies to the test specimen of 50 mm diameter and can be written as

$$I_{3/4R} = \left[\frac{7.05}{E_{3/4R}} \right]^{1.495} \tag{44}$$

By substituting the calculated value of $I_{3/4R}$ and $D = 50$ mm, the equivalent Jominy distance $E'_{3/4R}$ corresponding to $(3/4)R$ of the specimen's cross section, can be calculated:

$$E'_{3/4R} = \frac{10.85}{0.668} I_{3/4R} \tag{45}$$

For calculated $E'_{3/4R}$, the hardness of 48 HRC can be read off from the Jominy curve as shown in Figure 68. This means that the same quenching condition needed to produce a hard-

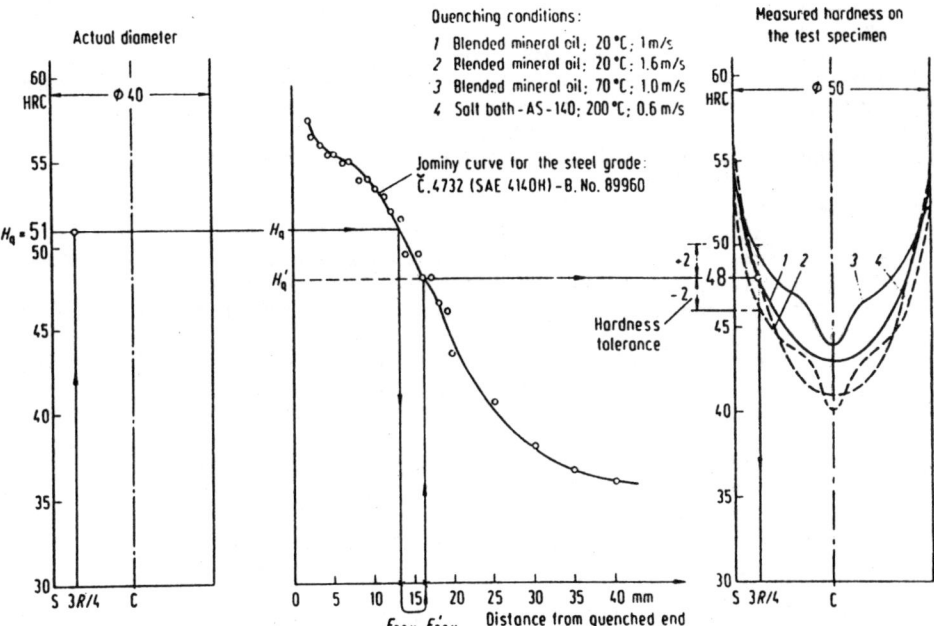

Figure 68 An example of computer-aided selection of quenching conditions (From Ref. 21.)

ness value H_q = 51 HRC at (3/4)R of a 40 mm diameter bar will yield a hardness H'_q of 48 HRC at (3/4)R of the 50 mm diameter standard specimen.

The next step is to search all stored hardness distribution curves of test specimens made of the same steel grade for the specific quenching condition by which the nearest hardness H'_q has been obtained (tolerance is ± 2 HRC). As shown in Figure 68, the required hardness may be obtained by quenching in four different conditions, but the best suited are conditions 1 and 2.

The special advantage of computer-aided calculations, particularly the specific method described, is that users can establish their own databanks dealing with steel grades of interest and take into account (by using hardened test specimens) the actual quenching conditions that prevail in a batch of parts using their own quenching facilities.

REFERENCES

1. G. Spur (Ed.), *Handbuch der Fertigungstechnik*, Band 4/2, *Wärmebehandeln*, Carl Hanser, Munich, 1987, p. 1012.
2. K. E. Thelning, *Steel and Its Heat Treatment*, 2nd ed., Butterworths, London, 1984, p. 145.
3. M. A. Grossmann, M. Asimov, and S. F. Urban, *Hardenability of Alloy Steels*, ASM, Cleveland, OH, 1939.
4. *Metals Handbook*, 8th ed., Vol. 2, American Society for Metals, Cleveland, OH, 1964, p. 18.
5. M. A. Grossmann, M. Asimov, and S. F. Urban, Hardenability its Relation to Quenching and Some Quantitative Data, *Hardenability of Alloy Steels*, ASM, Cleveland, OH, 1939.
6. A. Moser and A. Legat, Determining hardenability from composition, *Härterei Tech. Mitt.* 24(2):100–105 (1969) (in German).
7. D. J. Carney and A. D. Janulionis, An examination of the quenching constant *H*, *Trans. ASM* 43:480–496 (1951).

8. C. A. Siebert, D. V. Doane, and D. H. Breen, *The Hardenability of Steels,* ASM, Cleveland, OH, 1977.
9. C. F. Jatczak, Hardenability in high carbon steel, *Metall. Trans. 4*: 2267–2277 (October 1973).
10. *Metals Handbook,* ASM, Cleveland, OH, 1948, p. 499.
11. C. F. Jatczak and D. J. Girardi, *Multiplying Factors for the Calculation of Hardenability of Hypereutectoid Steels Hardened from* 1700°*F,* Climax Molybdenum Company, 1958.
12. W. E. Jominy and A. L. Boegehold, *Trans. ASM 26:* 574 (1938).
13. These procedures are described in ASTM A 255, SAE J 406, and ISO/R 642 (1967).
14. JUS-Standard C.A2.051/1959.
15. *Metals Handbook,* 9th ed., Vol. 1, ASM, Metals Park, OH, 1978, pp. 473–474.
16. R. A. Grange, Estimating the hardenability of carbon steels, *Metall. Trans. 4:* 2231 (1973).
17. A. Rose and L. Rademacher, Weiterentwicklung des Stirnabschreckversuches zur Prüfung der Härtbarkeit von tiefer einhartenden Stählen, *Stahl Eisen 76*(23):1570–1573 (1956) (in German).
18. C. F. Jatczak, Effect of microstructure and cooling rate on secondary hardening of Cr-Mo-V steels, *Trans. ASM 58:*195–209 (1965).
19. C. B. Post, M. C. Fetzer, and W. H. Fenstermacher, Air hardenability of steel, *Trans. ASM 35:*85 (1945).
20. G. Krauss, *Steels Heat Treatment and Processing Principles,* ASM International, 1990.
21. B. Liscic, H. M. Tensi, and W. Luty, *Theory and Technology of Quenching,* Springer-Verlag, Berlin, 1992.
22. J. Field, Calculation of Jominy end-quench curve from analysis, *Met. Prog. 1943:*402.
23. W. Crafts and J. I. Lamont, *Hardenability and Steel Selection,* Pitman, London, 1949.
24. E. Just, Formel der Härtbarkeit, *Härterei Tech. Mitt. 23*(2):85–100 (1968).
25. R. Caspari, H. Gulden, K. Krieger, D. Lepper, A. Lübben, H. Rohloff, P. Schüler, V. Schüler, and H. J. Wieland, Errechnung der Härtbarkeit im Stirnabschreckversuch bei Einsatz und Vergütungsstählen, *Härterei Tech. Mitt. 47*(3):183–188 (1992).
26. J. S. Kirkaldy and S. E. Feldman, Optimization of steel hardenability control, *J. Heat. Treat. 7:*57–64 (1989).
27. J. M. Tartaglia. G. T. Eldis, and J. J. Geissler, Hyperbolic Secant Method for Predicting Jominy Hardenability; an Example Using 0.2 C-Ni-Cr-Mo Steels, *J. Heat. Treat. 4*(4):352–364 (1986).
28. J. M. Tartgalia and G. T. Eldis, *Met. Trans. 15A*(6):1173–1183 (1984).
29. E. Just, *Met. Prog. 96*(5):87–88 (1969).
30. T. Lund, *Carburizing Steels: Hardenability Prediction and Hardenability Control in Steel-Making,* SKF Steel, Tech. Rep. 3, 1984.
31. *Metals Handbook,* 9th ed., Vol. 1, ASM Int., 1978, p. 492.
32. M. Asimov, W. F. Craig, and M. A. Grossmann, Correlation between Jominy test and quenched round bars, *SAE Trans. 49*(1):283–292 (1941).
33. J. L. Lamont, How to estimate hardening depth in bars, *Iron Age 152:*64–70 (1943).
34. D. V. Doane and J. S. Kirkaldy (Eds.), *Hardenability Concepts with Applications to Steel,* Proceedings of a Symposium, Chicago, Oct. 24–26, 1977, The Metallurgical Society of AIME, 1978.
35. T. Filetin, A method of selecting hardenable steels based on hardenability, *Strojarstvo 24*(2): 75–81 (1982). (in Croatian).
36. T. Filetin and J. Galinec, Software programme for steel selection based on hardenability, Faculty Mech. Eng., University of Zagreb, 1994.
37. T. Filetin and B. Liscic, Determining hardenability of carburizing steels, *Strojarstvo 18*(4):197–200 (1976). (in Croatian).
38. *ASM Handbook,* 9th ed., Vol. 4, *Heat Treating,* ASM Int., 1991, p. 287.
39. A. Rose, The austenitizing process when rapid heating methods are involved, Der Peddinghaus Erfahrungsaustausch, Gevelsberg, 1957, pp. 13–19 (in German).
40. B. Liscic, S. Svaic, and T. Filetin, Workshop designed system for quenching intensity evaluation and calculation of heat transfer data, *ASM Quenching and Distortion Control,* Proc. First Int. Conf. on Quenching and Control of Distortion, Chicago, IL, 22–25 Sept. 1992, pp. 17–26.
41. N. Shimizu and I. Tamura, Effect of discontinuous change in cooling rate during continuous cooling on pearlitic transformation behavior of steel, *Trans. ISIJ 17:*469–476 (1977).
42. N. Shimizu and I. Tamura, An examination of the relation between quench-hardening behavior of steel and cooling curve in oil, *Trans. ISIJ 18:*445–450 (1978).
43. E. Just, Hardening and tempering—influencing steel by hardening, *VDI Ber. 256:*125–140 (1976) (in German).
44. W. Gerber and U. Wyss, Hardenability and ability for hardening and tempering of steels, *Von Roll Mitt. 7*(2/3):13–49 (1948) (in German).
45. B. Liščić and T. Filetin, Computer-aided evaluation of quenching intensity and prediction of hardness distribution, *J. Heat. Treat. 5*(2):115–124 (1988).

4

QUENCHING AND QUENCHING TECHNOLOGY

Hans M. Tensi and Anton Stich
Technical University of Munich, Munich, Germany

George E. Totten
Union Carbide Corporation, Tarrytown, New York

I. INTRODUCTION

The outstanding importance of steels in engineering is based on their ability to change in mechanical properties over a wide range when subjected to controlled heat treatment. For unalloyed carbon steels, for example, the hardness can be increased by up to 500% just by changing the cooling rate from the austenitizing temperature from extremely slow to extremely fast. But quenching (i.e., cooling at a rate faster than in still air) does not only set the desired mechanical properties. An important side effect of quenching is the formation of thermal and transformational stresses that lead to changes in size and shape and thus may result in quenching cracks'that damage the workpiece.

Figure 1 schematically represents the coupling effects between the three different characteristics of quenching—cooling rate, metallic structure, and internal stresses. The cooling rate influences the phase transformation of the metallic structure, whereas the latent heat due to structural changes affects the cooling rate. All phase transformations of austenite during quenching are accompanied by volume expansion. In addition, steels contract with decreasing temperature. As a consequence, locally and temporally different changes of structure and temperature cause nonuniform volumetric changes in the quenched part that can result in transformational and thermal stresses. These stresses accelerate or hinder the phase transformation and influence the volume expansion. While the phase transformation brings out a defined metallic structure, the volumetric dilatation and the thermal and transformational stresses result in deformations and residual stresses. At room temperature both characteristics influence the material properties, which is the only aim of the quenching performance.

The coupling effects between temperature distribution during quenching, metallic structure, and stresses require correct handling of the cooling rate during heat treatment. This includes sufficient reproducibility and predictability of the quenching performance as well as the ability to exactly control the quenching intensity by varying the type of quenchant and its physical state. The main objective of the quenching process is to achieve the desired microstructure, hardness, and strength while minimizing residual stresses and distortion.

157

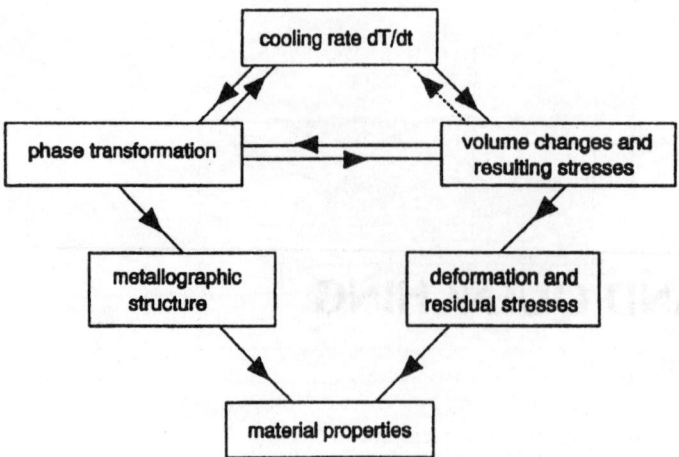

Figure 1 Coupling effects between cooling rate, phase transformation and stresses, and their influence on the material properties.

The most common quenchants in hardening practice are liquids including water, water that contains salt, aqueous polymer solutions, and hardening oils. Inert gases, salt and metal melts, and fluidized beds are also used.

The quenching techniques used for liquid media are immersion quenching and spray quenching. Immersion quenching, where the part is submerged in a nonmoving or nonagitated quenchant, is the most widely used practice. The part may be quenched directly from the austenitizing temperature down to room temperature (direct quenching) or to a temperature above the M_s temperature, where it is held for a specified period of time, followed by cooling in a second medium at a slower cooling rate (time quenching or interrupted quenching; see Section III). The quenching intensity can be changed by varying the type of quenchant, its concentration and temperature, and the rate of agitation. "Spray quenching" refers to spraying the liquid through nozzles onto those areas of the hot workpiece where higher cooling rates are desired. The heat transfer is mainly determined by the impingement density and its local distribution.

II. METALLURGICAL TRANSFORMATION BEHAVIOR DURING QUENCHING

A. Influence of Cooling Rate

The transformation behavior during very slow cooling of an unalloyed steel with 0.45 wt % carbon (1040 steel) from the austenitizing temperature [about 850°C (1560°F)] is exactly described by the metastable iron–cementite equilibrium diagram [1–3] (Figure 2a). The stable iron–carbon system is not of interest for steels. At 850°C (1560°F) the face-centered cubic γ-iron, called austenite after Sir Robert Austin, is the stable phase. The transformation starts at 785°C (1445°F) (A_3 temperature) with a precipitation of the body-centered cubic α-iron, called ferrite. The transformation into proeutectoid ferrite is finished when the eutectoid temperature of 723°C (1333°F) (A_1 temperature) is reached. The concentration of carbon in the austenite grains increases from the initial concentration of 0.45% at 785°C to 0.8% at 723°C (1333°F), the concentration of the eutectoid metastable equilibrium. At 723°C (1333°F) the austenite transforms into a lamellar-like structure of ferrite and cementite called pearlite[1] (see

---

[1]After the pearly appearance of the metallographically prepared surface.

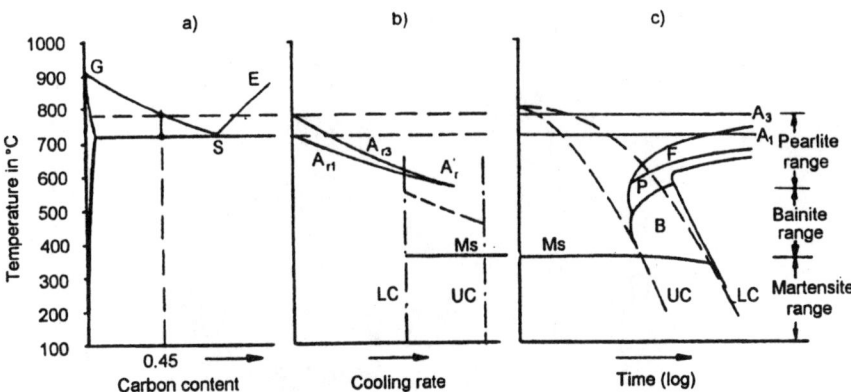

Figure 2 Influence of cooling rate on the transformation temperatures of austenite. (a) Metastable iron–cementite equilibrium diagram; (b) change of transformation temperatures of a 1040 steel with increasing cooling rate; (c) CCT diagram of a 1040 steel.

Figure 3c). At room temperature there is a metastable equilibrium between ferrite and pearlite (Figure 3b), which, strictly speaking, does not correspond to the exact thermodynamic equilibrium. Tempering this metallographic structure over a long time (2 or 3 days) at a temperature just below the eutectoid temperature A_1, a temperature where no phase transformations occur according to the iron–cementite equilibrium diagram, produces a stable structure of globular cementite in a ferritic matrix by minimizing the surface energy between ferrite and carbide (Figure 3a).

Nearly all technical kinds of quench treatments, especially the quenching from austenitizing temperature, produce structures that are not in accordance with the thermodynamic equilibrium due to a specified cooling rate. The effect of an increasing cooling rate on the transformation behavior of a 1040 steel from austenitizing temperature is described in Figure 2b. The temperature of proeutectoid ferrite precipitation (A_{r3}) and the eutectoid temperature (A_{r1}) fall with increasing cooling rate; the A_{r3} point descends faster than the A_{r1} point, so the distance between them decreases until they join in a single curve at what is called the A_r' point. Consequently, the volume fraction of proeutectoid ferrite permanently decreases with increasing cooling rate while the volume fraction of pearlite increases [1]. At the A_r' point, ferrite precipitation is suppressed and the structure consists only of pearlite with an average carbon content of 0.45 wt % (deviating from the carbon concentration of the eutectoid equilibrium of 0.8 wt %). This quenching process is called pearlitizing. The mechanical behavior of the single pearlite structure (Figure 3c) strongly differs from that of the ferritic/pearlitic one (Figure 3b). Ferrite is nearly pure iron with low tensile and yield strengths R_m and R_p and high elongation A. Pearlite, which is a mixture of ferrite and cementite, has high strength values and low elongation. Therefore, the strength distribution of ferritic pearlitic structures is inhomogeneous.

If the cooling rate is further increased and reaches a limit, which is called the lower critical (LC) cooling rate, the diffusion-controlled transformation in the pearlite range is first partly and then completely suppressed. The solid solution of austenite is maintained down to lower temperatures and transforms into bainite and martensite (Figures 3d and e). The microstructures of the bainite range grow partly diffusion-controlled; the microstructures of the martensite range are mainly diffusionless [4]. Similar to the A_{r3} and A_{r1} points, the beginning of bainite transformation, A_b, shifts to lower temperatures with increasing cooling rate. The formation of martensite starts at a temperature called M_s, which does not depend on the cooling rate.

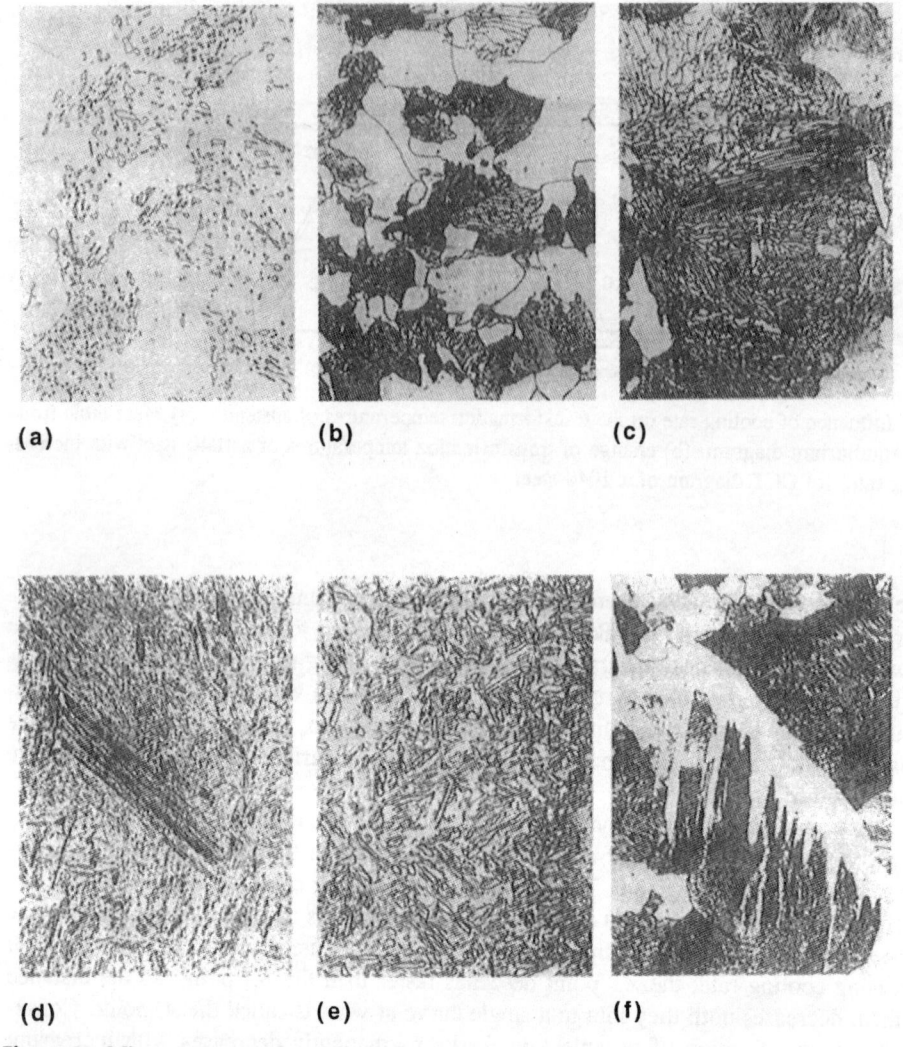

(a)   (b)   (c)

(d)   (e)   (f)

Figure 3   Microstructures of a 1040 steel obtained with increasing cooling rate from austenitizing temperature (850°C). Magnification 1000×. (a) Globular cementite (gray) in a ferritic matrix (white), after tempering at 680°C over 60 h starting from a ferrite-pearlite microstructure. (b) Proeutectoid ferrite (white) and pearlite (dark) in a normalized structure. Heat treatment: 850°C/still air. (c) Pearlite-ferrite (white) and carbide (dark) in an arrangement of parallel plates, and small amounts of proeutectoid ferrite (white). (d) Bainite. Heat treatment: 850°C/oil. (e) Martensite. Heat treatment: 850°C/water. (f) Widmannstätten structure. Heat treatment: 1100°C/still air.

The upper critical (UC) cooling rate is the upper limit of the cooling rate at which plain martensite is formed. Martensitic structures are the aim of most hardening processes, because martensite has the highest tensile and yield strength a steel of a given chemical composition can achieve.

If the same 1040 steel is quenched at a definite cooling rate from an overheated austenitizing temperature, ferrite may be built inside large grains at crystallographically preferred glide plains in spite of the under-eutectoid carbon concentration (Figure 3f). After precipitation of ferrite, the retained austenite transforms into pearlite. This structure, which is practically useless for technical applications, is called the Widmannstätten arrangement.[2]

The correlation between cooling rate and microstructure for a given chemical composition and definite austenitizing conditions is described in the continuous cooling transformation (CCT) diagrams, which can also contain data on hardness [5]. An example for a 1040 steel is shown in Figure 2c. For an infinitely slow cooling rate the transformations described by the CCT diagram are identical with that of the equilibrium diagram. The S-shaped curves describe the beginning (B) and the completion (C) of the transformation of austenite into the indicated microstructure. The microstructures that are formed with increasing cooling rate from the austenitizing temperature of a steel are ferrite and pearlite, pearlite, bainite, and martensite with retained austenite. For certain cooling rates these microstructures can occur in combinations. The prediction of the microstructures and microstructural combinations formed during cooling is restricted by the fact that the shape of the cooling curves has an essential influence on the course of phase transformation [6]. Therefore the transformation diagrams are valid only for the specified temperature/time cycles. In addition, deviations from the specified chemical composition and the metallographic structure before austenitizing as well as differences in the austenitizing conditions (temperatures and time) have a very strong influence on the transformation behavior.

An uncomplicated experimental procedure to describe the dependence of phase transformation on cooling rate is given by hardenability curves derived from end-quench (Jominy) tests (see Section III of Chapter 3). In the end-quench test the cooling rate continuously decreases the increasing distance from the water-quenched front end. The change of hardness as a function of the distance from the quenched end face describes the hardenability of a steel. In addition to the hardness values, the structural changes along the mantle line can be documented. For a 1040 steel, for example, the hardness reaches from about 850 Vickers hardness (HV) (HRC 66) at the water-quenched end face (martensite microstructure) to 200 HV (HRA 56) at the upper end of the sample (ferrite and pearlite microstructure). Deviations in the chemical composition can be taken into account in the form of hardenability bands that describe the upper and lower limits of measured hardness values.

B. Influence of Carbon Concentration

The hardness of steels rises with the concentration of carbon dissolved in austenite before quenching, as shown in Figure 4a. The increase of hardness is caused by the rising dislocation density and the distortion of the body-centered martensite lattice due to the inserted carbon atoms. If the carbon concentration exceeds approximately 0.5 wt %, the hardness rises only slightly [7] because the extremely high distortion of the formed martensite hinders the further transformation of austenite; the temperature for completing the martensite formation M_f falls below room temperature (Figure 4b), and growing amounts of retained austenite remain in the martensite structure with increasing content of dissolved carbon (Figure 4c) [8]. Compared to martensite, the retained austenite has a very low yield strength and hardness and is so unstable that it immediately transforms to martensite or bainite if energy is induced in the structure during the technical application. The volume expansions that accompany this transformation process cause changes of shape and size and can result in cracks that damage the workpieces. The volume expansion of roll barrels in bearings during the rolling motion, for example, can burst the ball races. Moreover, with rising carbon content the toughness of

---

[2]Dr. Widmannstätten investigated the structure of parts of a meteor fell in Czechien in 1808. The friction of air in the atmosphere caused extreme overheating of the alloy; the quenching was done by the marshy ground.

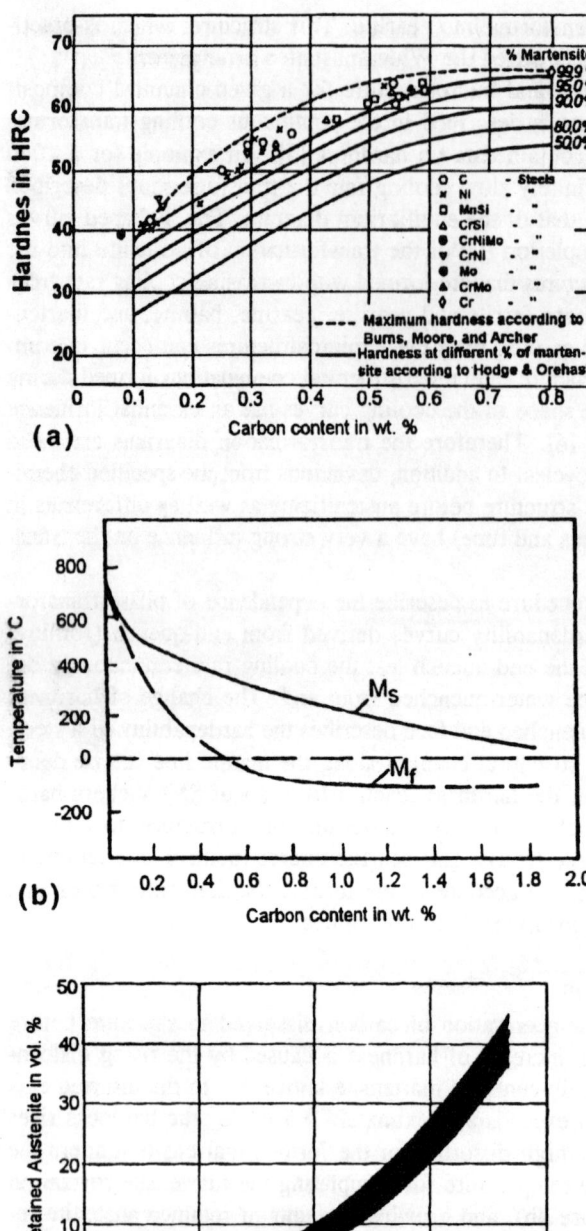

Figure 4 Effect of carbon concentration on (a) hardness for structures with different martensite content, (b) temperature for starting and completing the martensite formation M_s and M_f, and (c) retained austenite.

martensite decreases with high local stresses develop because of the carbon supersaturation. The consequence is that unalloyed steels with carbon concentrations higher than 0.5 wt % are used only after tempering. Tempering reduces the volume fraction of retained austenite and releases the tension of martensite by reducing the dislocation density and carbon supersaturation by precipitation. In addition, tempering leads to a reduction of quenching stresses.

C. Influence of Alloying Elements

As described in the previous section, the concentration of carbon dissolved in austenite before quenching has a great effect on hardness and strength values. While carbon is placed at interstitial locations in the iron lattice, alloying elements are dissolved at the original lattice sites. The additional warping of the iron lattice due to the alloying elements causes only a slight increase in hardness but strongly affects the movability of carbon, which results in a drastically reduced transformation rate. This influence can be used to produce steels whose transformation characteristics are adapted to the desired microstructure and to the geometry of the parts to be quenched.

The influence of alloying elements on the beginning and end of austenite transformation with regard to transformation time and temperature is graphically described in Figure 5 for the alloying element chromium. Figures 5a–5c represent CCT diagrams of three different steels with similar carbon content and increasing chromium content [5]. For the unalloyed 1040 steel, the incubation period of beginning austenite transformation is very short. Fully martensitic structures can be achieved only for very fast quenching rates with suppressed transformation in the pearlite and bainite range. The incubation period of the subcooled austenite drastically rises with increasing chromium concentration, as illustrated by Figures 5b and 5c. In the unalloyed 1040 steel, a low cooling rate, according to a cooling curve that crosses the temperature of 500°C (932°F) at a time of about 300 s, produces a structure consisting only of ferrite and pearlite (Figure 5a). After quenching at the same cooling rate, the austenite of the low-alloy 5140 steel transforms to mainly bainite and martensite and small amounts of ferrite and pearlite (Figure 5b). After the same cooling, bainite and martensite are formed in the high-alloy DIN 45CrMoV67 steel (Figure 5c).

To illustrate the effect of a reduced transformation rate on the hardenability of steels, Figure 6 shows the measured hardness distributions over the cylindrical cross section of a 1040 and a 5140 steel for two sample diameters (15 and 40 mm) after identical quenching in water. With the smaller diameter of 15 mm and the unalloyed 1040 steel, the hardness falls from about 850 HV (HRC 66) at the surface (martensite microstructure) down to 500 HV (HRC 49) in the core (ferrite, pearlite, bainite, and martensite microstructure). In contrast, the hardness of the 5140 steel is constant over the cylindrical cross section because of the only martensitic transformation in the sample volume due to the longer incubation period of the subcooled austenite. Since the cooling rate dramatically decreases with increased section thickness, the cooling rate in the core of the larger diameter is limited by the rate of heat conduction from the interior to the surface. The consequence is that for the 1040 steel and the 40 mm diameter, the austenite transforms only to ferrite and pearlite in the inner regions because of the reduced cooling rate. Compared to the 15 mm diameter, this results in a drop of hardness in the core; the hardness values of about 220 HV (HRA 57) nearly correspond to the initial hardness of the steel. Martensitic hardening occurs only in a thin outer shell of the cylindrical sample (Figure 6b). The higher carbon concentration of about 1 wt % in the 5140 steel clearly delays the transformation in the pearlite and bainite range and results in only a slight decrease in hardness with increasing distance from the cylinder surface.

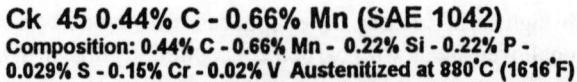

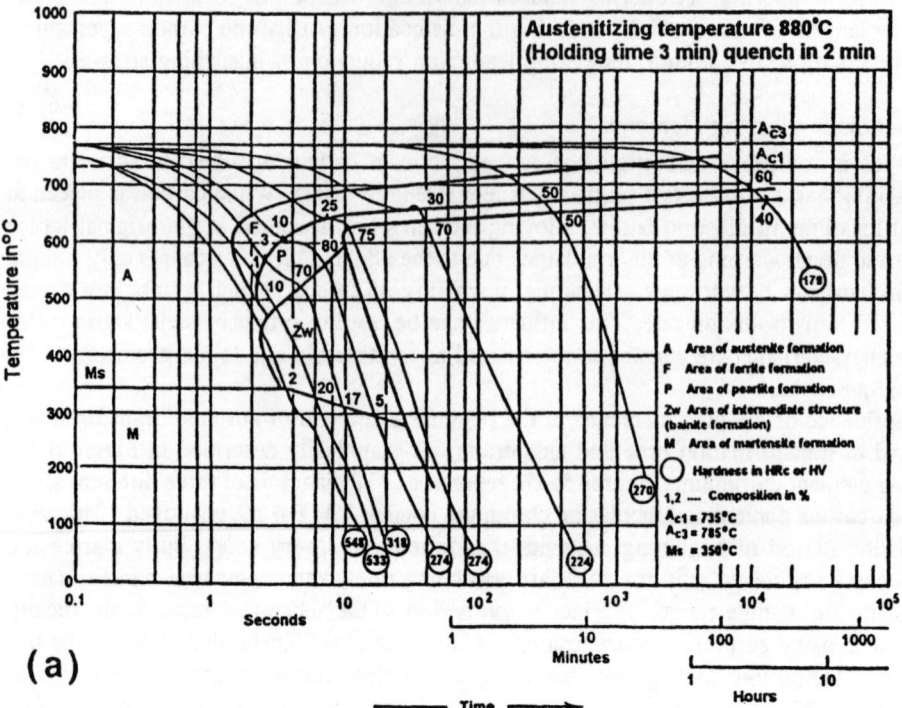

(a)

Figure 5   Influence of alloying elements, here chromium, on the transformation of subcooled austenite described according to CCT diagrams of (a) a 1040 steel with about 0.15 wt % Cr (German grade Ck 45); (b) a 5140 steel with about 1 wt % Cr (German grade 41 Cr 4), and (c) a yyyy steel with about 1.5 wt % Cr and 0.7 wt % Mo (German grade 45CrMoV67).

D.   Influence of Stresses

The internal stresses formed during quenching have a decisive influence on the transformation behavior of austenite as already described in Figure 1. Depending on the chemical composition and the cooling rate, the austenite can transform in the pearlite, bainite, or martensite range. All these phase transformations are accompanied by volume expansions. The change of length during slow and rapid cooling of unstressed and tensile stressed austenite is shown in Figure 7. After cooling austenite at a very slow cooling rate, close to the equilibrium ferrite and pearlite are formed in the temperature range between A_{r3} and A_{r1} temperatures (continuous line). With fast cooling rates (broken lines), the diffusion-controlled transformations are suppressed, and below the martensite start temperature M_s the unstressed austenite transforms into martensite with an increase in specific volume. If tensile stresses occur within the sample, the martensite start temperature rises from M_s to M_s' and larger changes of length occur [9]. The opposite is true for compressive stresses. This phenomenon is called transformation plasticity.

The volume changes of the ferritic–pearlitic transformation as well as those of the martensitic transformation are due to the transformation of the face-centered cubic austenite crys-

41 Cr 4 (SAE 5140)
Composition: 0.44% C - 0.80% Mn - 0.22% Si - 0.030% P -
0.023% S - 1.04% Cr - 0.17% Cu - 0.04% Mo - 0.26% Ni -
<0.01% V Austenitized at 840°C (1544°F)

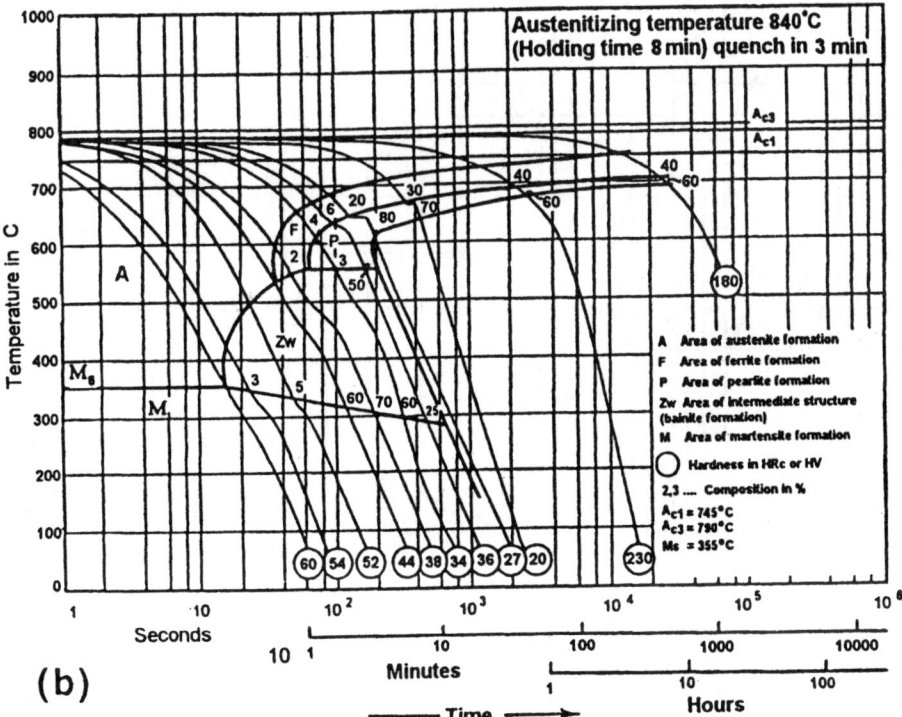

Figure 5 Continued

tal lattice into the body-centered cubic ferrite lattice or the tetragonal deformed martensite lattice. The ferritic and martensitic crystal lattices have a higher specific volume.

According to Kobasko [10], with increasing cooling rate within the martensitic transformation range, the probability of quench crack formation rises to a maximum value and then decreases to zero. This phenomenon can be explained by transformation plasticity and the change of specific volume during phase transformation. The higher the cooling rate, and thus the temperature gradient, the greater the surface layer expansion at the moment of superplasticity. During very fast cooling, the surface layer compresses and tensile stresses occur because of the heated and expanded core. At room temperature high compressive stresses arise at the surface of the part. In this way martensitic hardening of big parts made of unalloyed steels, e.g., a lorry axis, is possible.

III. QUENCHING PROCESSES

Previously it was shown that the cooling rate and the shape of the cooling curve influence the course of phase transformations and residual stresses and distortion. In quench hardening, fast

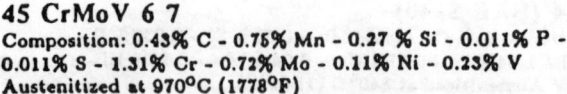

45 CrMoV 6 7
Composition: 0.43% C - 0.75% Mn - 0.27 % Si - 0.011% P -
0.011% S - 1.31% Cr - 0.72% Mo - 0.11% Ni - 0.23% V
Austenitized at 970°C (1778°F)

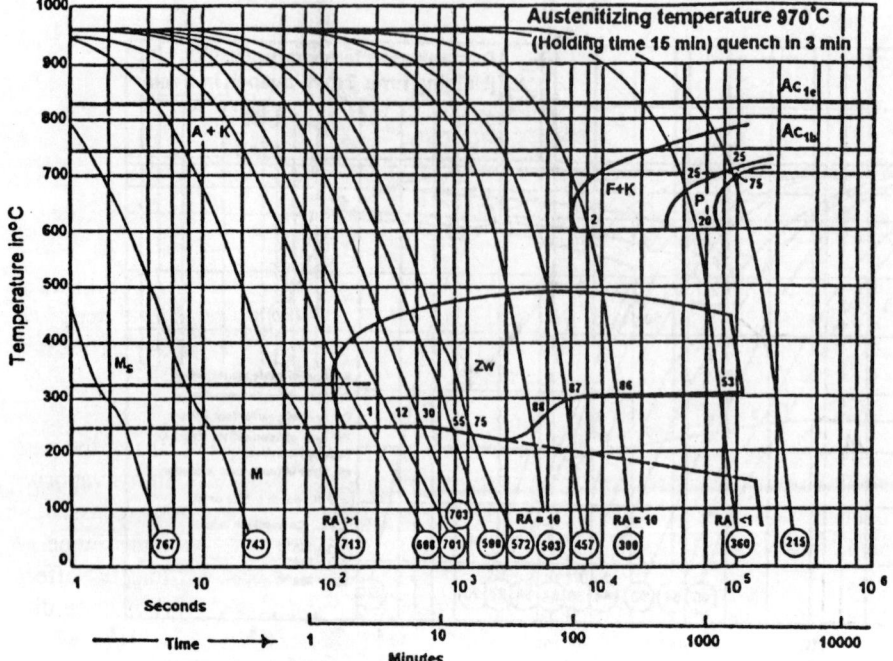

A + K	Area for austenite and carbides	F	Area for ferrite formation
K	Area for carbide formation	F + K	Area of nonlamellar eutectoids
P	Area for pearlite formation	Z	Cementite
O	Hardness in HV	10Z	10% Cementite
Zw	Area for intermediate structure	1 ; 2	(refers to numbers on curves)
M	(bainite formation)		proportion of structure formed, in percent
M	Area for martensite formation		
RA	Residual austenite		

(c)

$Ac_{1b} = 745°C$
$Ac_{1e} = 830°C$
$M_s = 325°C$

Figure 5   Continued

cooling rates, depending on the chemical composition of the steel and its section size, are frequently applied to prevent diffusion-controlled transformations in the pearlite range and to obtain
a structure consisting mainly of martensite and/or bainite. However, the reduction of undesirable thermal and transformational stresses due to volume changes usually requires slower
cooling rates. Quenching processes therefore require the selection of cooling rates that are fast
enough to permit the desired microstructure to form but slow enough to minimize residual
stresses and distortion. These considerations have resulted in different quenching methods such
as direct quenching, interrupted quenching, spray quenching, and gas and fog quenching. The
time–temperature cycles that can be obtained with different quenching techniques are shown
in Figure 8 for the center and surface of the quenched part together with the time–temperature–transformation diagram [5].

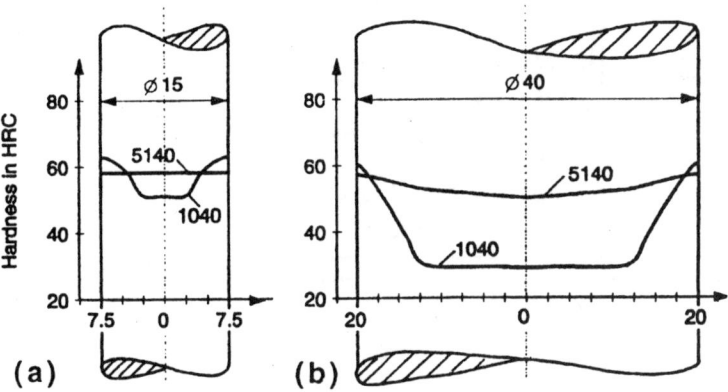

Figure 6   Influence of alloying elements on hardness distribution over the cross section of cylindrical samples with a diameter of (a) 15 mm and (b) 40 mm, described for a 1040 steel and a 5140 steel. The samples were quenched in water at 30°C and an agitation rate of 0.3 m/s.

Direct quenching, the most common quenching technique, refers the quenching of the part from the austenitizing temperature directly to room temperature by immersion into a vaporizable liquid quenchant (Figure 8a). Petroleum solutions are usually used for this process.

Interrupted quenching consists of rapidly quenching steel from the austenitizing temperature to a temperature above the M_s temperature, where it is held for a time sufficient to affect the desired transformation and then cooled in air. Interrupted quenching comprises three different quenching techniques—marquenching, austempering, and isothermal annealing—which differ in the temperature at which quenching is interrupted and the time for which the steel is held at this temperature. The quenchants usually used for interrupted quenching are molten salt baths and specialty oils.

Marquenching consists in rapidly quenching the steel to a temperature just above the M_s temperature, holding it at this temperature to equalize the temperature throughout the workpiece, and then removing it from the bath before transformation into bainite begins (Figure 8b). The martensite structure formed during marquenching is the same as after direct quenching;

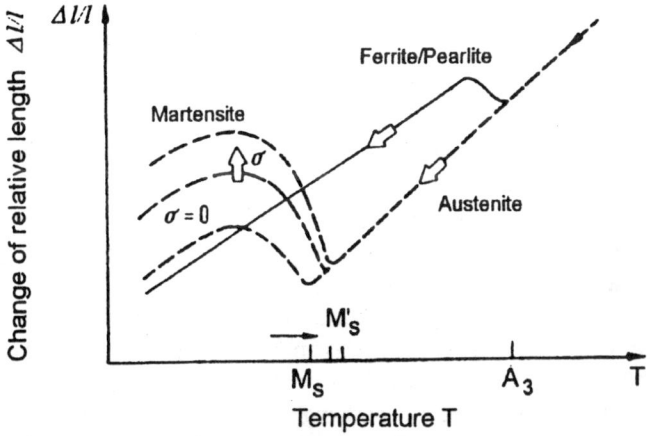

Figure 7   Effect of tensile stresses on the change of relative length during very fast cooling compared to a slow cooling rate, shown schematically.

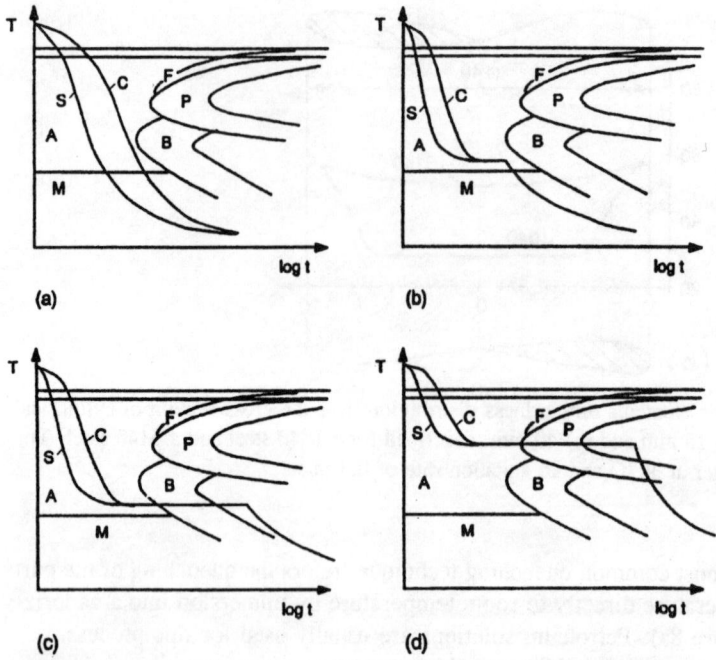

Figure 8 Cooling curves for the center and the surface of quenched parts for different quenching methods correlated to a time–temperature–transformation schematic diagram. (a) Direct quenching; (b) marquenching; (c) austempering; (d) isothermal annealing or pearlitizing.

however, residual stresses are minimized owing to the more homogeneous temperature distribution and the slower cooling rate during martensite formation.

Austempering is similar to marquenching in that the steel is rapidly quenched from the austenitizing temperature to a temperature above M_s but differs in that the workpiece is held at temperature for sufficient time to allow an isothermal transformation into bainite (Figure 8c). Relative to untempered martensite bainite has adequate toughness and strength than would be formed by marquenching. Martensite formed during direct quenching and marquenching is often tempered, because tempered martensite has more homogeneous elemental distribution and improved toughness and strength.

Isothermal annealing, or pearlitizing, differs from marquenching and austempering in that the bath temperature is sufficiently high that isothermal transformation into pearlite occurs (Figure 8d). Pearlite has a high toughness and sufficient strength to be the optimal structure for parts such as wires or cables and railroad rails.

Spray quenching with a liquid quenchant, generally water or an aqueous polymer solution, at sufficiently high pressures on the surface of the workpiece produces fast cooling rates because the liquid droplets impact the surface and cause a high rate of heat transfer. The rate of heat extraction can be varied over a wide range by varying the quantity of the sprayed liquid or by spraying a mixture of water and air (fog quenching). Compared to immersion quenching, spray quenching thus allows better control in cooling the workpiece.

In gas quenching, the heat removal is achieved by blowing a stream of gas over the workpiece, sometimes after austenitizing it in a vacuum furnace. Usually, the cooling rate is faster than that obtained in still air but slower than that achieved in oil and is controlled by the type, pressure, and velocity of the cooling gas. Inert gases including helium, argon, and nitrogen are most commonly used.

IV. WETTING KINEMATICS

During quenching in liquid media with boiling temperatures far below the initial temperature of the body, three stages of heat removal occur. These are referred to as (1) the film boiling or vapor blanket stage, (2) the nucleate boiling stage, and (3) the convection stage.

In the film boiling stage the surface temperature of the workpiece is sufficiently high to vaporize the quenching liquid and form a stable film around the part. The vapor film has an insulating effect; therefore the cooling rate during film boiling is relatively slow. The temperature above which the stable vapor blanket occurs is called the Leidenfrost temperature after Johann Gottlieb Leidenfrost.

When surface temperature is less than the Leidenfrost temperature, the vapor film collapses and the nucleate boiling begins [11]. In this stage, the liquid in contact with the hot surface evaporates immediately the vapor bubbles leave the surface. This causes strong convection, which results in a high rate of heat transfer from the metal to the fluid.

Upon further cooling, the surface temperature becomes less than the boiling point of the liquid, and the surface is permanently wetted by the fluid. The cooling rate is low and is determined mainly be the rate of convection and the viscosity of the liquid quenchant.

Some typical examples of the wetting sequences on steel and silver samples quenched in water, oil, and aqueous polymer solutions are depicted in Figures 9–12. Quenching in water and oil usually results in "slow wetting" with a clearly visible wetting front. Figures 9 and 10 show two wetting processes that were observed during quenching of a cylindrical CrNi steel specimen in water at 30°C (86°F) and oil at 60°C (140°F) with an agitation rate of 0.3 m/s [12,13]. Wetting begins at the lower edge of the sample, and the wetting front, which is the interface between film boiling and nucleate boiling, ascends to the top in an almost annular manner. An additional descending wetting front can be observed from the upper edge that is more characteristic for quenching in oil. The time required for completely wetting the sample is about 14 s for the water quench and about 13 s for the oil quench. Thus, the three phases of cooling, with their widely varying heat transfer coefficients, are simultaneously present on the sample surface. Therefore, the cooling of the sample is subject to substantial variations.

In the quenching of prismatic cylinders, the wetting front is parabolically shaped as shown in Figure 11 [14,15]. The wetting along the cylinder edges accelerates the wetting front of the flanks and reduces the time interval of simultaneous presence of film boiling and nucleate boiling.

The wetting process can be strongly influenced by the addition of additives [12,14–18]. Figures 10a and 10b show two wetting sequences that were obtained in water with different chemical admixtures. On the surface of the sample, a polymer film forms that provides a uniform breakdown of the vapor blanket and reduces heat transfer in the lower temperature range. When the polymer film has completely redissolved, heat transfer is achieved entirely by convection.

The velocity of the spreading wetting front and the time interval of the simultaneous presence of film boiling and nucleate boiling can be strongly influenced by changing the physical properties of the quenchant and the sample. The items varied are

1. The type of quenchant as described by its boiling temperature, viscosity, thermal capacity, and surface tension
2. Additives to the quenchant, and their concentration

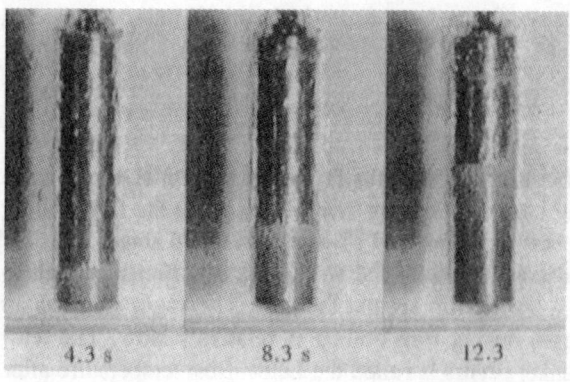

(a)

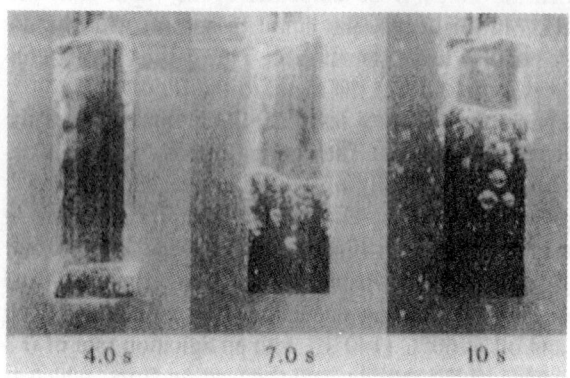

(b)

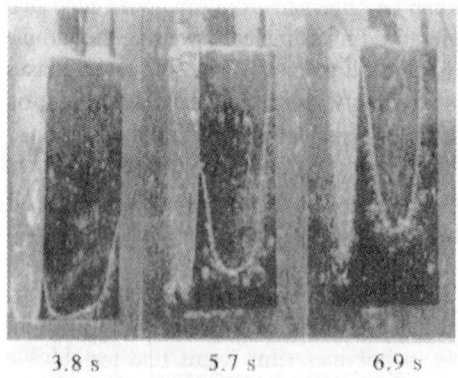

(c)

Figure 9 Wetting process on the surface of CrNi steel specimens quenched from 850°C into water and oil. (a) Cylinder (25 mm diameter × 100 mm) in water at 30°C flowing at 0.3 m/s; (b) cylinder (25 mm diameter × 100 mm) in oil at 60°C flowing at 0.3 m/s; (c) prismatic cylinder (15 ×15 × 45 mm) in water at 60°C without forced convection.

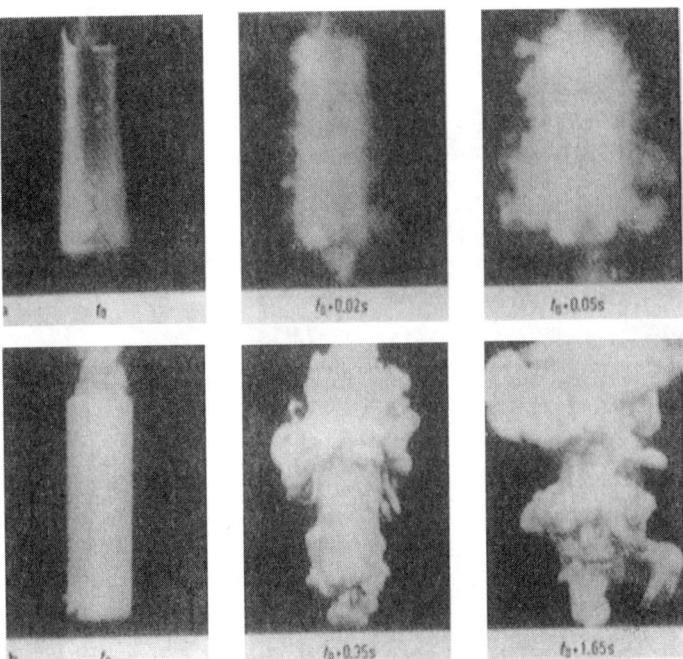

Figure 10 Wetting process on the surface of cylindrical samples being quenched in water with different admixtures. (a) CrNi steel cylinder (25 mm diameter × 100 mm) quenched in 5% aqueous polymer solution at 30°C without forced convection. (b) Silver cylinder (15 mm diameter × 45 mm) quenched in a 10% aqueous polymer solution with a chemical additive at 25°C without forced convection.

3. The temperature and agitation rate of the quenchant
4. The thermal characteristics of the body and its transformation behavior
5. The surface roughness of the body and surface layers
6. The geometry and initial temperature distribution of the sample

An overview of the effect that varying these properties has on the wetting kinematics is extensively described in Reference 19. The influence of some selected properties on the time when wetting starts t_s, the time when wetting is finished t_f, the time interval of wetting $\Delta t_w = t_f - t_s$, and the heat transfer coefficient α is summarized in Table 1.

The great importance of the wetting kinematics for steel hardening can be explained with respect to the wetting process and the corresponding surface hardness along the length of a cylinder of 1040 steel being quenched in water with different agitation rates (Figures 11 and 12). Agitation increases the wetting speed and reduces the time during which the different cooling phases are simultaneously present. In Figure 11 the finishing time of wetting is reduced from 9 s to 2.5 s by agitation at 1 m/s. Considering that the heat transfer from the surface to the fluid dramatically increases at the transition from film boiling to nucleate boiling (see Figure 22), it becomes apparent that a change of the wetting process has a significant effect on the cooling behavior and the hardness distribution. Figure 12 compares the hardness profile and the wetting time related to the distance z from the sample bottom for cooling in still and agitated water [20]. The wetting time t_k increases toward the top of the sample in accordance with the migration of the wetting front (see Figure 12). With increasing wetting time, a distinct drop of surface and core hardness becomes apparent for the still water quench. With

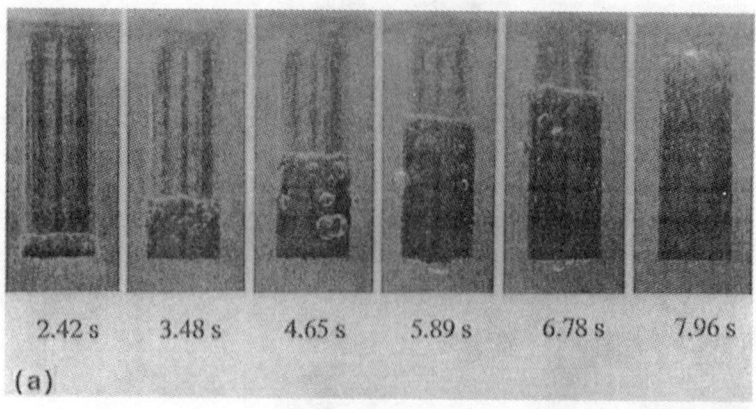

2.42 s   3.48 s   4.65 s   5.89 s   6.78 s   7.96 s

(a)

0.89 s   1.58 s   2.35 s   2.89 s   3.88 s

(b)

Figure 11 Wetting process on the surface of a 1040 steel (15 mm diameter × 45 mm) quenched from 850°C in water at 50°C (a) without forced convection and (b) with an agitation rate of 1 m/s.

water agitated at 1 m/s, the wetting front proceeds to the top in approximately 2 s, resulting in a constant hardness along the cylinder length. Other examples of the effect of wetting kinematics on steel hardening are given in References 12, 15, 21, and 22.

V.   DETERMINATION OF COOLING CHARACTERISTICS

A.   Acquisition of Cooling Curves with Thermocouples

The most common way to describe the complex mechanism of quenching is to use cooling curves. In almost all heat treatment processes, cooling curves are measured with metal test pieces instrumented with thermocouples [23]. A thermocouple consists of two metallic wires of different chemical composition that are brought into contact at their ends; usually the wires are welded. When it is connected to a measurement apparatus and the two metal junctions have different temperatures, a voltage can be measured that is independent of the absolute tempera-

Table 1 Influence of Some Physical Properties of the Fluid and the Body on the Characteristic Data of Wetting and the Heat Transfer Coefficient for Immersion Quenching

Properties of the fluid	t_s, t_f	Δt_w	α
Type of quenchant	↑↓	↑↓	↑↓
Additives to the fluid	↑↓	↑↓	↑↓
Rate of agitation v (↑)	↓	↓	↑
Bath temperature T_b (↑)	↑	↑	↓
Properties of the sample			
Thermal diffusivity a (↑)	↑	↑	↑
Sample diameter (↑)	↑	↑	
Surface roughness (↑)	↓	↓	↑
Surface oxidation (↑)	↓	↓	↓

ture but dependent on the temperature difference between the two ends of the thermocouple. Therefore, one end, called the cold junction or reference junction, is immersed in melting ice at the constant reference temperature of 0°C (32°F). The temperature at the other end, called the hot or measuring junction, can then be determined. In practice, instead of melting ice, electronic temperature compensation is used.

For many applications, especially very fast cooling processes, it is important to identify the exact response time of the thermocouples. Response time is primarily a function of the thermocouple dimension and its construction, thermal properties of the thermocouple materials, the method of installation, and heat exchange conditions within the surrounding metal.

Commercial thermocouples are constructed of two thermoelectric wires, insulation material, and a surrounding sheath that protects the thermoelectric couple against high-temperature corrosion and oxidation (Figure 13). The most common type of hot junction is the insulated version (Figure 13a). Despite the delayed heat conductance due to the electrical insulation of the hot junction, a short response time can be achieved by choosing small diameters down to 0.25 mm. Figure 14 compares the response time of two sheathed thermocouples with outside diameters of 0.5 and 1.0 mm [24]. The thermocouples were installed into the test piece as shown in Figure 13a. For exactly the same quenching process, the maximum cooling rate and the temperature/time gradient are significantly greater for the smaller dimension. The response time can also be slightly reduced by welding the hot-junction wires to the metal sheath (Figure 13b) provided the sample does not contain an electric charge that can influence the measuring process.

An elegant method to determine the true change of temperature during quenching is shown in Figures 13c and 13d. In Figure 13c, the metal sheath of the thermocouple is opened and the thermoelectric wires are welded to the sample material. The advantage of this installation is that the response time is minimized and the integral temperature of the crosshatched region is measured. An error in thermoelectric voltage may be due to the inserted sample volume, which is not known.

Figure 13d describes a similar thermocouple installation where the hot-junction wires are insulated with a ceramic tube. Since the thermophysical properties of ceramics and steels dif-

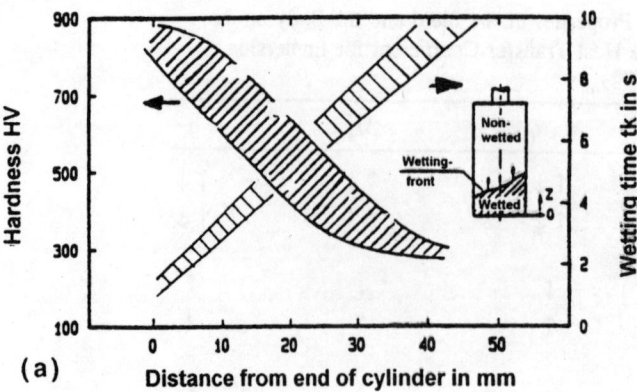

(a)

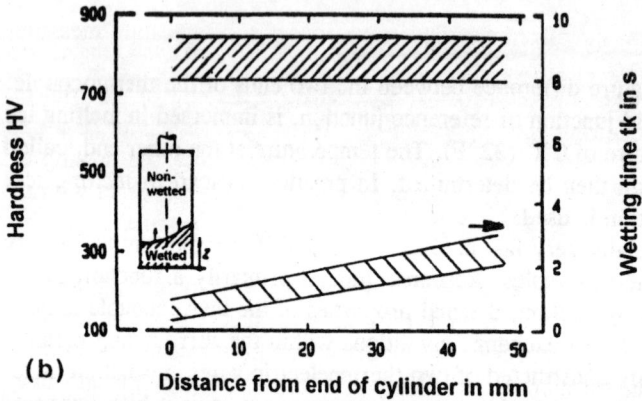

(b)

Figure 12 Wetting time t_b and surface hardness HV 2 along the length of a cylindrical 1040 steel (15 mm diameter × 45 mm) being quenched in water of 50°C (a) without forced convection and (b) with an agitation rate of 1 m/s.

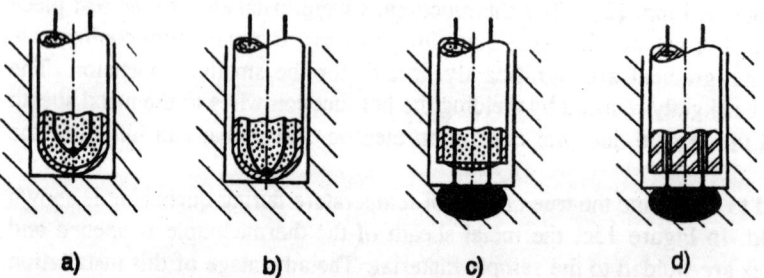

a)   b)   c)   d)

Figure 13 Construction of different thermocouples. (a) Mantle-sheathed thermocouple with insulated hot junction; (b) mantle-sheathed thermocouple with sheath-welded hot junction; (c) mantle-sheathed thermocouple with open sheath and thermoelectric wires are welded to the sample material, and (d) thermocouple in which the thermoelectric wires are welded to the sample material and insulated with a ceramic tube.

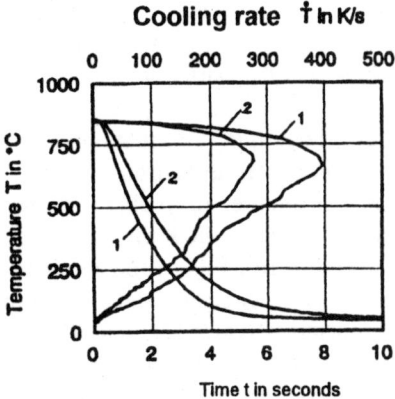

Figure 14 Response sensitivity of mantle-sheathed thermocouples with different outer diameters. The temperature–time cycles were measured in the center of a cylindrical CrNi steel sample of 8 mm diameter during spraying with water of room temperature. *1*, 0.5 mm diameter; *2*, 1.0 mm diameter.

fer, this thermocouple construction may produce significant errors in temperature measurement.

When measuring cooling curves, it is important to minimize the amount of material removed from the workpiece when drilling the thermocouple. The cooling behavior of a part especially a part with a small cross section, changes dramatically with increased drilling diameter. Therefore, all investigations concerning the cooling behavior of workpieces with respect to temperature measurement require good thermocouple-to-specimen contact and minimal thermocouple diameter and material removal.

Cooling curves are usually obtained by quenching a test specimen and measuring the temperature as a function of time at a specified point within the specimen. From the temperature–time curve a temperature–cooling rate curve can be calculated. The test pieces used are cylinders, plates, and spheres. The most common test piece is a cylinder with a length three to four times its diameter and with a thermocouple located at its geometric center (see diagram at left of Figure 15) [25,26].

Two characteristic parameters that can be obtained from cooling curves are the maximum cooling rate $\dot{T}_{max}$ and the cooling rate at 300°C (572°F) ($\dot{T}_{300°C}$). When hardening steel, the

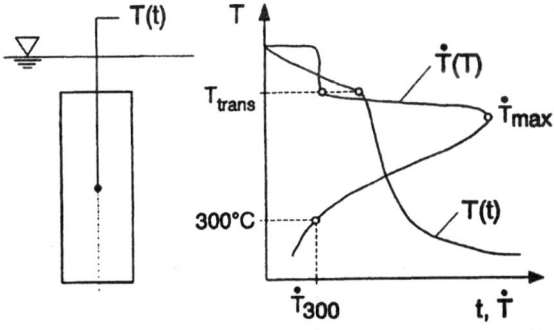

Figure 15 Determination of cooling characteristics by measuring temperature–time and temperature–cooling rate curves. (Schematic.)

maximum cooling rate and the temperature at which it occurs must be sufficiently high to minimize ferrite and pearlite transformation. The cooling rate at 300°C (572°F) can be used to indicate the probability of distortion and cracking because it is near the martensite temperature of most carbon and low-alloy steels.

The main limitation of cooling curve analysis is that relatively little information can be derived about the wetting process. The temperature in the sample center at the transition point from lower to higher cooling velocity (the so-called transition temperature, T_{trans}) only describes the wetting of the corresponding surface point and gives no information about the time when wetting starts (t_s) and the time when wetting is finished (t_f) (Figure 16). Figure 16a compares cooling curves that were measured in the center of a cylindrical CrNi steel sample (25 mm diameter × 100 mm) being quenched in water and a 10% aqueous polymer solution [27,28]. The bath temperature is 30°C (86°F), and the rate of agitation is 0.3 m/s. Though the sample center cools in almost the same way, the process of wetting differs considerably, as evidenced by the near-surface temperatures measured at three heights of the sample (Figure 16b). Quenching in water causes a slow wetting with a total wetting time of about 14 s. Wetting begins at the lower edge of the sample, and the wetting front ascends to the upper edge. This results in high localized temperature differences (continuous lines in Figure 16b). When the sample is quenched in the polymer solution, "explosive" wetting occurs after 6 s of stable film boiling. The temperature–time curves for the three locations near the sample surface are almost congruent (broken lines in Figure 16b). Since the wetting process controls the temperature distribution in the quenched part, it directly influences the hardness distribution. Whereas the slowly wetted sample will have a characteristic axial hardness gradient, similar hardness values along a centerline can be expected in the suddenly wetted sample (with the exception of a small range near the ends).

B. Measurement of Wetting Kinematics

1. Conductance Measurement

A method better suited to determining the wetting process is to measure the electrical conductance between sample and quenchant and between the sample and a counter electrode and measure the temperature change at the specimen center [21,29,30] (Figure 17a). When a va-

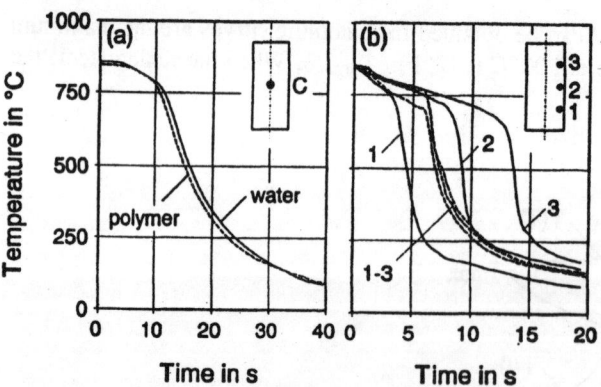

Figure 16 Comparison of cooling curves measured at different points in a cylindrical CrNi steel sample (25 mm diameter × 100 mm) during slow wetting (water) and during sudden wetting (10% aqueous polymer solution). Points of measurement: (a) center, (b) close to the specimen surface at three heights.

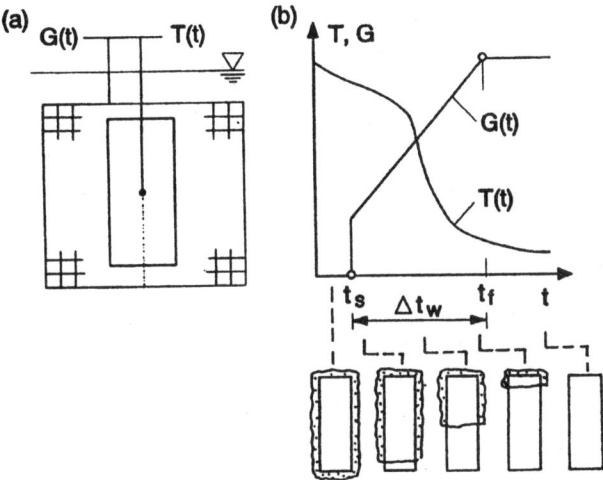

Figure 17 Determination of the percentage of the wetted surface area. (a) Measuring principle; (b) temperature–time $T(t)$ and conductance–time $G(t)$ curves with corresponding wetting state of the quenched sample. (Schematic.)

por blanket forms around the entire sample surface, the conductance between the sample and the counter electrode is low. When the vapor blanket collapses at an arbitrary position on the sample surface, conductance increases (Figure 17b). The increase in conductance is approximately proportional to the wetted portion of the sample surface. Conductance is at its highest value when the surface is completely wetted. (Note that the formation of bubbles after wetting can reduce conductance. This may result in a failure to identify the end of wetting). The main parameters provided from conductance–time curves are the time when wetting starts (t_s) and the time when wetting is finished (t_f).

In Figure 18a, wetting starts at time t_s, and is completed at time t_f (usually for quenching in water, oil, and some aqueous polymer solutions; see Figures 9–11). In Figure 18b, the transition from film boiling to nucleate boiling is "explosive" (see Figure 10). In Figure 18c, the wetting time is also infinitely short, but bubbles or polymer separations remain on the surface, which reduces conductance. In Figure 18d, sudden wetting occurs, but immediately a new vapor blanket forms, when then collapses again. This sequence may be repeated several times [13].

2. Temperature Measurement

A second method to ascertain the wetting process is by inserting thermocouples in near-surface positions along the cylindrical sample [31]. Figure 19 shows the basic information that can be derived from the near-surface cooling curves when a wetting front migrates from the lower end of the sample to its top. Each temperature cycle defines a period of slow cooling and a subsequent period of fast cooling. The point of transition indicates the transition from film boiling to nucleate boiling at the thermocouple location. The onset and conclusion of wetting, and thus the velocity of the wetting edge, can be calculated by extrapolating the local transition times to the front ends of the sample. The cooling curves in Figure 19 were obtained during quenching of a CrNi steel cylinder (15 mm diameter × 100 mm) in water at 50°C (120°F) without forced convection.

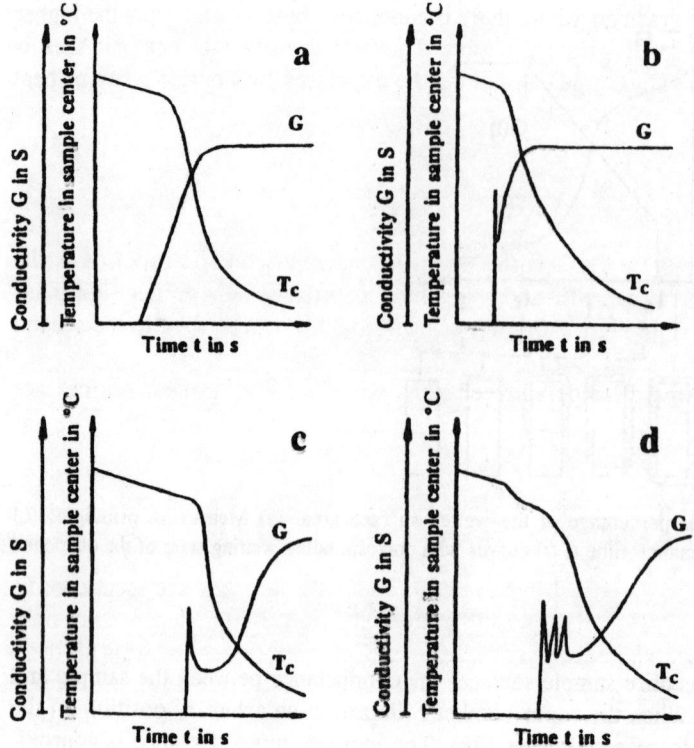

Figure 18 Schematic temperature–time and conductance–time curves for different wetting processes. (a) Slow wetting; (b) "explosive" wetting; (c) "explosive" wetting with large bubbles remaining on the surface; (d) repeatedly "explosive" wetting with large bubbles remaining on the surface.

VI. QUENCHING AS A HEAT TRANSFER PROBLEM

A. Heat Transfer in a Solid

When a part is heated to a specified temperature, heat is transferred to it by the furnace. Conversely, when the part is quenched, heat is transferred to the surrounding medium. This

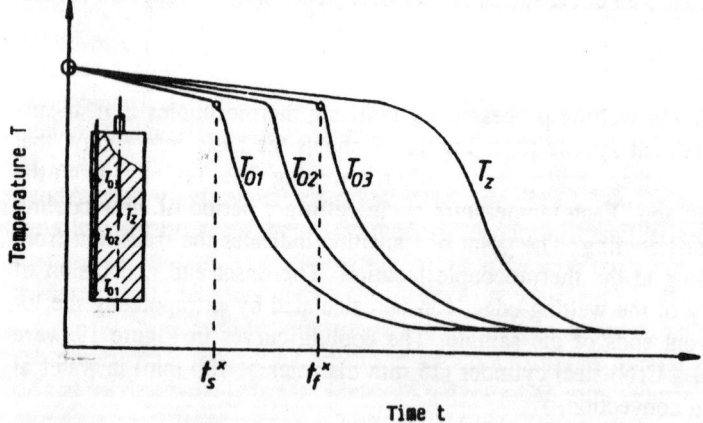

Figure 19 Determination of wetting kinematics by means of three thermocouples close to the sample surface.

produces localized temperature gradients where there is conductive heat transfer from the higher temperature region to the lower temperature region. The heat transfer rate per unit area is proportional to the local temperature gradient and can be expressed by Fourier's law of heat transfer,

$$Q = -\lambda A \frac{\delta T}{\delta x} \tag{1}$$

where Q is the heat transfer rate in J/s, λ is the thermal conductivity in J/(s·m·K), A is the unit area in m^2, T is the temperature in K, and x is a local coordinate. The minus sign is inserted because the heat flows from higher to lower temperature areas according to the second law of thermodynamics.

Heat transfer in a solid where temperature changes with time and no heat sources are present within the body is

$$\frac{\partial T}{\partial t} = a \left(\frac{\delta^2 T}{\delta x^2} + \frac{\delta^2 T}{\delta y^2} + \frac{\delta^2 T}{\delta z^2} \right) \tag{2}$$

where t is the time in s, a is the thermal diffusivity in m^2/s, and x, y, and z are local coordinates. According to Equation (2) the temperature distribution within a body depends not only on the local temperature gradients but also on the thermal diffusivity, which includes all thermodynamic parameters of the material. The thermal diffusivity a is defined as

$$a = \frac{\lambda}{\rho C_p} \tag{3}$$

where ρ is the density in kg/m^3 and C_p is the specific heat capacity under constant pressure in J/(kg·K). The larger the value of a, the more quickly heat will diffuse through the material. A high value of a can be achieved by having either high thermal conductivity λ or low heat capacity ρC_p, which means that less of the heat moving through the material is absorbed and thus will raise the temperature of the material. This fact is illustrated by faster cooling of silver relative to austenitic steels. In the simple form of the Fourier equation (Equation 2), thermal diffusivity is assumed to be constant, but in reality λ, ρ, and C_p, and therefore a, depend on temperature.

The heat capacity ρC_p of steel does not vary considerably with the chemical composition. However, the thermal conductivity (λ) of iron alloys does vary with temperature as illustrated in Figure 20 [26]. At room temperature, the value of λ is drastically reduced with increasing carbon content and to a lesser extent with the content of other alloying elements. The decrease of λ is caused by the distortion of the iron lattice due to insertion of foreign atoms. The thermal conductivities of pure iron, plain carbon steel with 0.45% carbon (1045 steel), and an austenitized (CrNi) stainless steel are 75, 55, and 1415 W/(m·K), respectively [27]. Above 800°C (1472°F), there is no significant influence of chemical composition on thermal conductivity. Figure 20 shows that λ decreases with increasing temperature for pure iron, plain carbon steels, and low-alloy steels. Low-alloy steels exhibit a maximum value that increases with higher amounts of alloying elements and shifts to a higher temperature.

High-alloy steels such as ferritic and austenitic steels exhibit an inverse temperature dependence [32]. In addition to temperature and alloying elements, thermal conductivity is also influenced by internal stresses and structural changes. During transformation from austenite to ferrite, λ increases discontinuously by approximately 7% for pure iron [32]. Every phase transformation is also accompanied by a change of latent heat and specific volume. Since exact

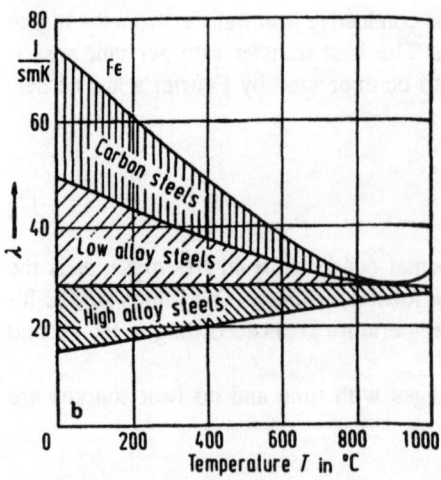

Figure 20 Influence of temperature and alloying elements on the thermal conductivity of steels.

information about the influence of temperature, chemical composition, structural changes, and stresses on the thermodynamic data is not available, the accuracy of calculations of the temperature distribution during heating or cooling of steels is limited. For examination of the cooling characteristics of quenchants, test pieces (probes) constructed from silver, nickel-base alloys such as Inconel, and stainless steel (1040 steel) are often used. The thermodynamic data for these materials are compared to those of austenitic and ferritic steels in Table 2.

B. Heat Transfer Across the Surface of a Body

Another important problem for the determination of the cooling behavior of steels concerns heat transfer across the surface of the body to the surrounding medium. This is mathematically described in terms of the (interfacial) heat transfer coefficient α:

$$\alpha = \frac{Q}{A(T_1 - T_2)} \tag{4}$$

where the units of α are $J/(s \cdot m^2 \cdot K)^8$, T_1 is the surface temperature of the body, and T_2 is the temperature of the medium. Determination of the heat transfer coefficient is based on Fourier's

Table 2 Approximate Values of Thermal Conductivity λ and Thermal Diffusivity α of Selected Materials at Room Temperature

Material	Thermal conductivity [$J/(s \cdot m \cdot K)$]	Thermal diffusivity (10^6 m²/s)
Silver	407	165
Inconel		
Austenitic steel	15	3.8
Ferritic steel	19	5.1
1040 steel	55	14.3
Iron	75	21

law of heat transfer (Equation 1), which states that heat flow across the surface of a body is proportional to the temperature gradient at the surface. In practice, the surface temperature of a body can only be determined with high-precision measuring equipment. Therefore α is often calculated from cooling curves that are recorded just beneath the surface.

During quenching, the heat transfer coefficient of steel is strongly dependent on the surface temperature and may vary by more than an order of magnitude between the austenitizing and bath temperatures due to variations in the interfacial film such as film boiling and nucleate boiling. The temperature dependence of α during quenching of an austenitic steel in water and oil is shown in Figure 21 [12].

Three cooling stages can be present simultaneously on the surface of an immersion-quenched part for a significant period of time. Therefore, the heat transfer coefficient not only varies with temperature but also varies with surface location. Figure 22 shows the wetting state of a cylinder of 1040 steel during quenching in water for a constant cooling time and the corresponding variation of the heat transfer coefficient. Stable film boiling produces a low heat transfer coefficient due to the insulating properties of the surrounding vapor blanket [$\alpha_{FB} \approx$ 400 W/(m²·K)]. Just below the wetting front, nucleate boiling occurs. The heat transfer coefficient α reaches its maximum value of approximately 15,000 W/(m²·K) and subsequently decreases. When the surface temperature decreases to less than the boiling point of the fluid, heat transfer occurs by convection and conduction, with α reaching even lower values, or α_{conv} ≈ 1500 W/(m²·K). This example demonstrates that heat transfer conditions during quenching can be complicated, especially when considering wetting behavior variation throughout the quench.

An example of the influence of wetting kinematics on heat removal during immersion quenching is illustrated in Figure 23a, which shows the change of wetting time and Leidenfrost temperature T_L, i.e., the surface temperature of the body at the transition from film boiling to nucleate boiling, along the length of a cylindrical CrNi steel during quenching in water. Wetting starts at the lower edge of the cylinder, and the wetting front ascends from the bottom to the top (see also Figure 9). This results in an almost linear decrease of Leidenfrost temperature with increasing wetting time. The change in the corresponding heat transfer coefficient α at half-height of the cylindrical sample ($z = 50$ mm) with the migrating wetting front is given in Figure 23b. Here, α is plotted as a function of the distance x from the actual

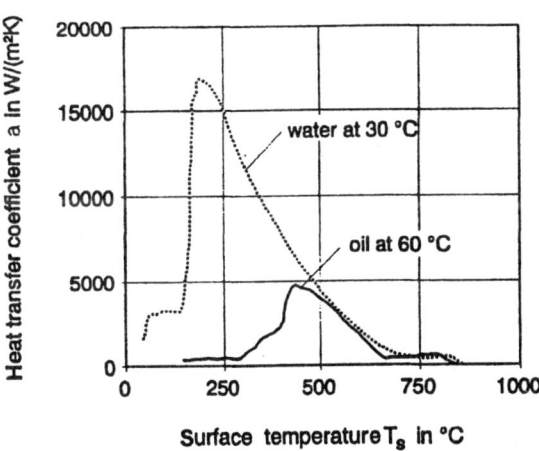

Figure 21   Heat transfer coefficient versus surface temperature of an austenitic steel cylinder (25 mm diameter × 100 mm) quenched into water at 30°C and into a fast oil at 60°C flowing at 0.3 m/s.

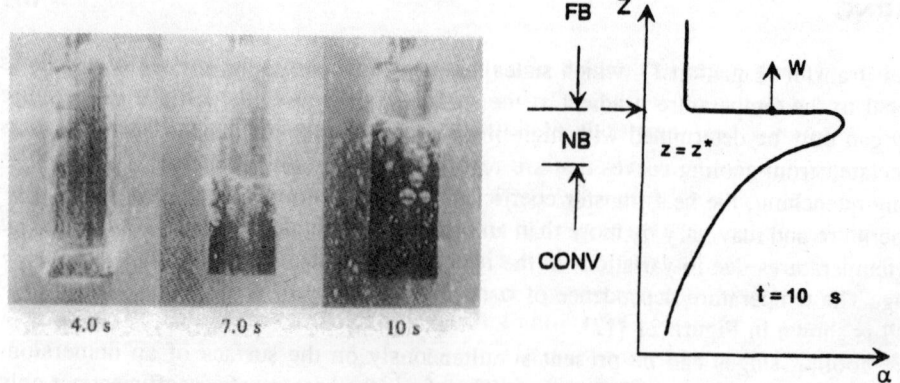

Figure 22 Wetting state on the surface of a cylindrical sample quenched into water at 30°C and schematic of variation of the corresponding heat transfer coefficient along the length of the cylinder for a given cooling time.

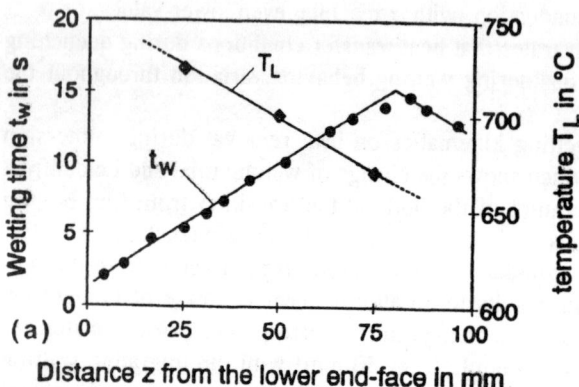

(a)

Distance z from the lower end-face in mm

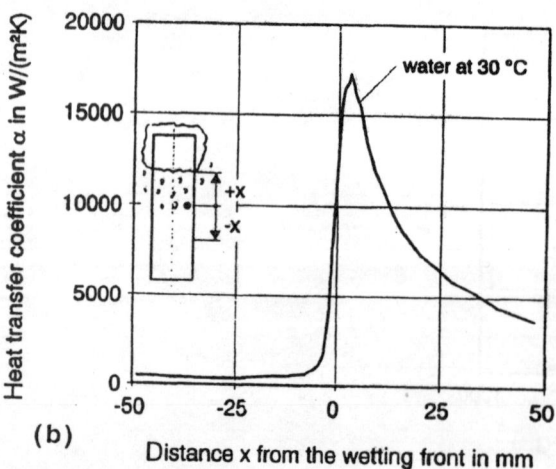

(b)

Distance x from the wetting front in mm

Figure 23 (a) Wetting time t_w and Leidenfrost temperature T_L along the length of an austenitic steel cylinder (25 mm diameter × 100 mm) quenched into water at 30°C flowing at 0.3 m/s. (b) Correspondingly changing heat transfer coefficient at half-height for varying positions of the wetting front.

position of the wetting edge. In the case of a negative value of x, the surface point at half-height is covered with the vapor film, whereas for positive values of x, nucleate boiling and subsequently convective heat transfer occur. When the edge of nucleate boiling passes the considered surface point ($x = 0$), the heat transfer coefficient rapidly increases to its maximum value and then decreases with decreasing surface temperature.

Heat transfer coefficients are required for finite-element or finite-difference heat transfer calculations of temperature distribution during quenching. Calculated cooling curves may be used to determine as-quenched microstructures and mechanical properties such as the hardness and yield strength of steel. The wetting kinematics of the quenching process, expressed by the nonconstant Leidenfrost temperature, must also be incorporated in these calculations. Figure 24 compares calculations achieved with different heat transfer coefficients [33,34]. In Figure 24a, the heat transfer coefficient was assumed to be equal for all surface locations. This would correspond to a very fast velocity of the wetting front (see Figure 10). It is apparent that the lines of constant temperature are parallel along the length of the cylinder and parallel to the cylinder end face in the region of the lower front end. However, high axial temperature differences actually occur when the wetting front migrates from the bottom to the top of the cylinder with an infinite wetting speed.

Figure 24b illustrates the calculated temperature distribution during cooling, at four points in time, for a wetting velocity of 9 mm/s. This calculation procedure is identical with the assumption that the heat transfer coefficient α varies with the surface temperature and the distance from the lower front end. From these figures, it can be seen that the axial temperature difference is greater than 600 K even for the small sample used. As already shown, this wetting process may have a substantial impact on hardening results.

Thermal stresses are dependent on temperature gradients within the sample. Figures 25 and 26 are for austenitic steel cylinders 15 mm in diameter × 45 mm. Figure 25 shows the distributions of radial and axial heat flux for the relatively slow wetting speed of 9 mm/s. The lines of constant radial and axial heat flux q_r and q_a expand in the opposite direction to the ascending wetting front. Distributions of axial heat flux are determined by relatively high heat

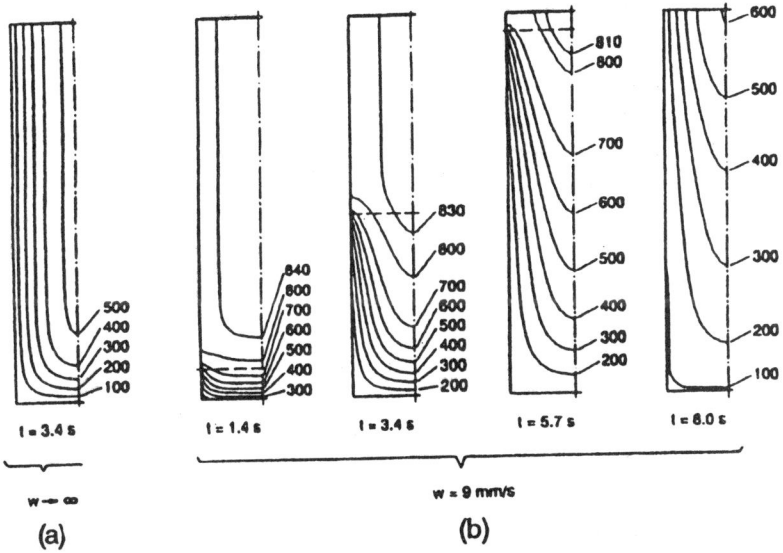

Figure 24 Calculated temperature distribution in an austenitic steel cylinder (15 mm diameter × 45 mm) during immersion quenching in water with different velocities of the ascending wetting front. (a) Sudden wetting; (b) slow setting with $w = 9$ mm/s.

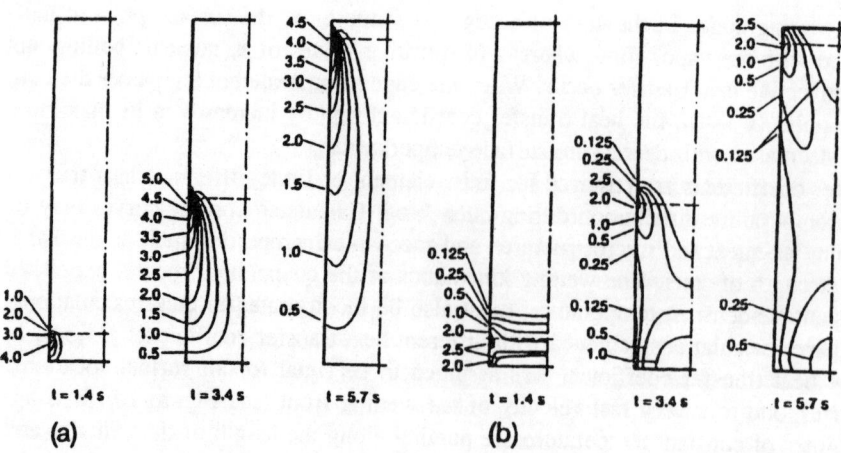

Figure 25 Calculated distributions of (a) radial and (b) axial heat flux in an austenitic steel cylinder (15 mm diameter × 45 mm) during immersion quenching in water with a wetting velocity of 9 mm/s.

removal in the region below the wetting front and by heat removal at the cylinder end. The highest values of radial and axial heat flux occur along the wetting front. With increasing wetting velocity, the maximum value of radial heat flux increases, while the axial heat flux decreases after it has reached a maximum at a wetting velocity of approximately 3 mm/s are shown in Figure 26. Therefore, high wetting velocities reduce thermal gradients along the length of the cylinder, which may decrease the distortion and cracking. In quench hardening, heat removal is often expressed by the Grossmann number or H factor instead of the heat transfer coefficient α. The Grossmann number [35] is defined as

$$H = \alpha/2\lambda \tag{5}$$

Thus, the H factor is equal to the interfacial heat transfer coefficient divided by twice the thermal conductivity. When the H factor is multiplied by the diameter of the body D, their product corresponds to the well-known dimensionless BiOT number (Bi):

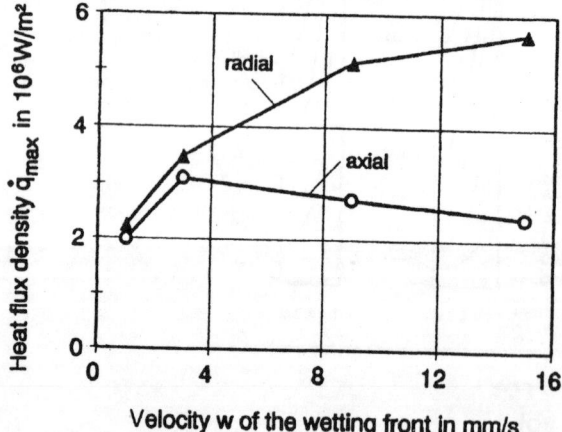

Figure 26 Influence of the velocity of an ascending wetting front on the maximum values of radial and axial heat flux for an austenitic steel specimen (15 mm diameter × 45 mm) quenched into water at 20°C.

$$HD = \text{Bi} \frac{\alpha}{\lambda} R \qquad (6)$$

The BiOT number is important because all bodies with the same BiOT number exhibit similar heat flow. If thermal conductivity λ is constant, although it is dependent on the chemical composition of the steel, the structural changes during quenching and temperature are as shown in Figure 20, and the value of H depends only on the heat transfer at the surface and thus on the cooling capacity of the quenchant.

The most significant deficiency of the H factor is the incorrect assumption that cooling occurs at a constant rate during quenching. The heat transfer coefficient and therefore the H value vary throughout the three stages of cooling and also with the surface wetting process as shown in Figures 21 and 22. Even with the well-known limitation, the H factor has been widely accepted by heat treaters and is often required as input data in software packages for predicting the hardness distribution of quenched components. Table 3 provides some illustrative H factor data along with the corresponding interfacial heat transfer coefficients for selected quenchants [36].

VII. PROCESS VARIABLES AFFECTING COOLING BEHAVIOR AND HEAT TRANSFER

Steel hardening requires a wide variation and sufficient reproducibility of heat transfer across the cooling surface to achieve the required cooling rate (which is dependent on the hardenability of the steel and the section size of the sample) and the desired mechanical properties. The influence of the cooling behavior during various quenching processes will now be described.

Table 3 Grossmann H Factor and Corresponding Heat Transfer Coefficient for Selected Quenchants

Quenchant	Quenchant temperature (°C)	Quenchant velocity (m/s)	H value (in.$^{-1}$)	Heat transfer coefficient [W/(m²·K)]
Air	27	0.0	0.05	35
		5.1	0.08	62
Conventional oil	65	0.51	0.7	3000
Fast oil	60	0.00	0.5	2000
		0.25	1.0	4500
		0.51	1.1	5000
		0.76	1.5	6500
Water	32	0.00	1.1	5000
		0.25	2.1	9000
		0.51	2.7	12000
		0.76	2.8	12000
	55	0.00	0.2	1000
		0.25	0.6	2500
		0.51	1.5	6500
		0.76	2.4	10500

A. Immersion Quenching

Heat transfer during immersion quenching is influenced by many factors such as the dimensions and shape of the part being quenched, the quenchant, and the quenching facility. In the heat treatment shop only a few of these parameters can be realistically varied, including bath temperature, agitation rate, and the quantity and racking arrangement of the parts. Of these, only agitation is readily varied, because rapid bath temperature changes and variation of the quenchant (and concentration, if an aqueous polymer solution is used) cannot realistically be accomplished during the quenching process itself.

1. Bath Temperature

The principal mechanism of heat transfer for vapor blanket cooling during immersion quenching is illustrated in Figure 27. Heat is transported across the surface through the vapor blanket by conduction ($\dot{q}_\lambda$) and radiation ($\dot{q}_\varepsilon$). Only a farction of the heat is released to the liquid by convection ($\dot{q}_\alpha$). The remainder ($\dot{q}_v$) vaporizes and stabilizes the fluid into the vapor blanket. The hot vapor flows upward, and at the vapor/liquid interface bubbles pass from the vapor film into the fluid, especially at the top of the cooling workpiece. A local decrease of the thickness δ of the vapor film immediately raises the heat flow by conduction ($\dot{q}_\lambda$), resulting in additional vaporization of the fluid, thus sustaining the vapor film.

When the surface temperature drops, the thickness δ of the vapor film is reduced until the fluid contacts the hot metal, which is the "start of wetting." Heat is no longer passed through the vapor film, and fluid is evaporated upon direct contact with the sample surface, dramatically increasing the rate of heat flow. With increasing liquid temperature T_1, the energy required for fluid evaporation is reduced and is proportional to the difference between the boiling point and the liquid temperature. The thickness δ of the vapor film is increased, and film boiling occurs to lower surface temperatures. Figure 27 shows that within the vapor film the

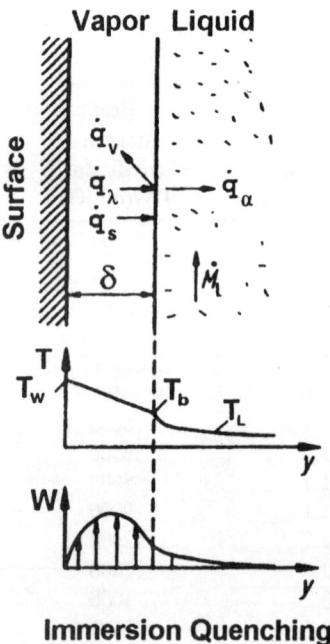

Immersion Quenching

Figure 27   Mechanism of heat transfer during film boiling and distribution of temperature and flow velocity in the bordering layer (schematic).

temperature drops from the surface temperature T_s to the boiling temperature T_b of the fluid and further decreases to the liquid temperature T_l.

The effect of water temperature on cooling curves in the center of an Inconel 600 probe (12.5 mm diameter × 60 mm) is illustrated in Figure 28 [37]. With increasing water temperature the duration of film boiling increases, which is indicated by the delayed transition from low cooling to fast cooling. In addition, the cooling rates of the three different stages of heat transfer are reduced.

2.   Effect of Agitation

In addition to bath temperature, the stability of the vapor film is greatly influenced by the "velocity profile" in the liquid. As illustrated in the lower part of Figure 27, the velocity profile is caused by the bouyancy-driven vapor flow and possibly by agitation (forced convection). A high flow velocity increases the heat transfer by convection ($\dot{q}_\alpha$) and reduces δ, therefore reducing the duration of film boiling. With agitation, heat transfer during the three stages of cooling is increased.

Figure 29 shows the effect of selected liquid velocities (0, 0.3, and 0.6 m/s) on cooling curves measured in the center of an austenitic stainless steel specimen quenched in 60°C (140°F) water [13]. In the range of stable film boiling, the temperature of the cylinder decreases slowly, almost independently of the liquid flow velocity. As the vapor film collapses, the steel temperature drops rapidly. Thus, the transition from lower to higher cooling rates may be strongly influenced by the liquid flow velocity.

3.   Effect of Quenchant Selection

The quenchant also exhibits dramatic effects on cooling behavior. The cooling curves in Figure 30a were obtained with a 25 mm diameter probe quenched in water at two temperatures, a 10% aqueous polymer solution, and a "fast" quenching oil. All quenchants were evaluated at 0.6 m/s. Using the same probe size (cross section) and flow velocity, different cooling curves were obtained for each quenchant.

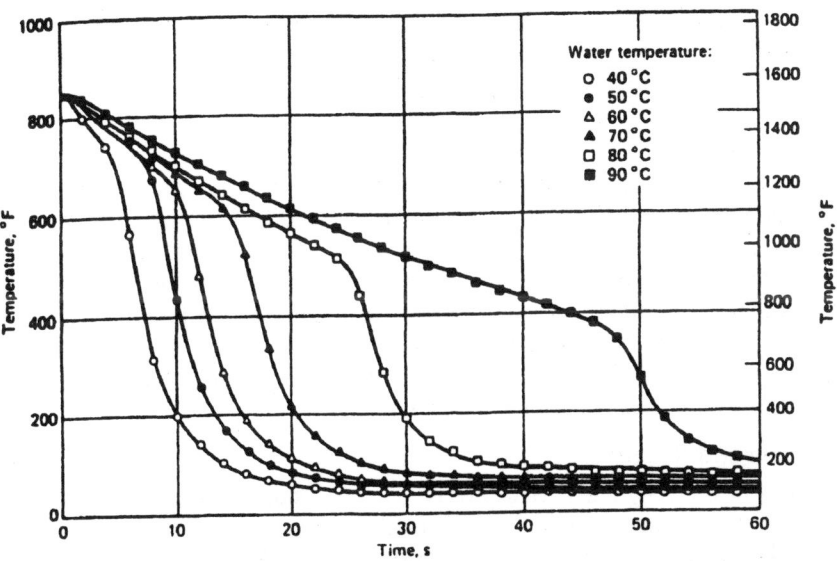

Figure 28   Effect of bath temperature on cooling curves measured in the center of an Inconel 600 probe (12.5 mm diameter × 60 mm) quenched into water flowing at 0.25 m/s.

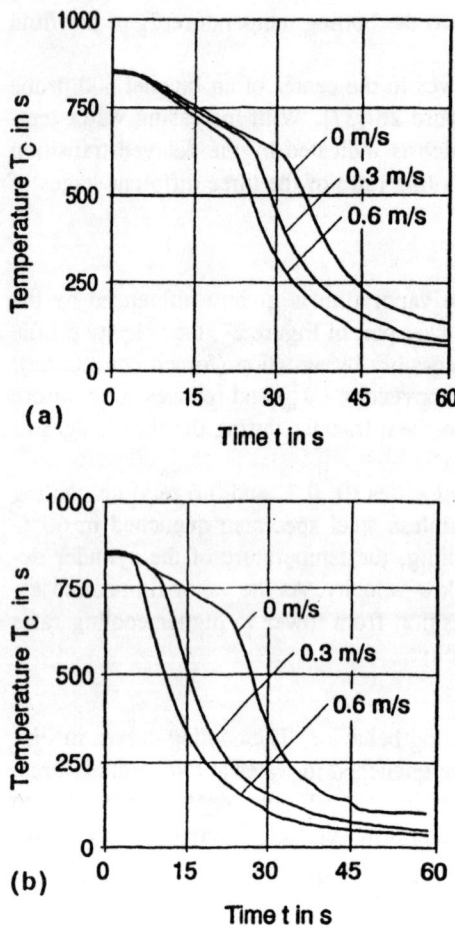

Figure 29 Effect of quenchant velocity on cooling curves in the center of an austenitic stainless steel specimen (25 mm diameter × 100 mm) quenched (a) into water at 60°C and (b) into a 10% aqueous polymer solution at 30°C.

The effect of the quenchant and flow velocity becomes even more apparent when the time $t_{A/5}$ from austenitizing temperature [for this example 850°C (1562°F)] to 500°C (932°F) is considered as shown in Figure 30b [12,13]. While the values of $t_{A/5}$ strongly differ for the quenchants shown, only a small influence of the flow velocity can be observed in water at 30°C (86°F) and oil at 60°C (140°F) for the cross-sectional size used (25 mm diamte). The $t_{A/5}$ time can be drastically reduced with increasing slow velocity in water at 60°C (140°F) and the 10% aqueous polymer solution at 30°C (86°F). As expected, the effect of the flow velocity becomes clearer when smaller cross sections are used.

4. Surface Oxidation and Roughness Effects

The surface roughness of the body and surface layers, such as oxides or organic substances, also strongly influences the cooling process. Chromium-alloy steels are oxidation-resistant due to concentration of chromium oxide at the surface. Oxide layers have greater surface roughness and a lower heat conductivity.

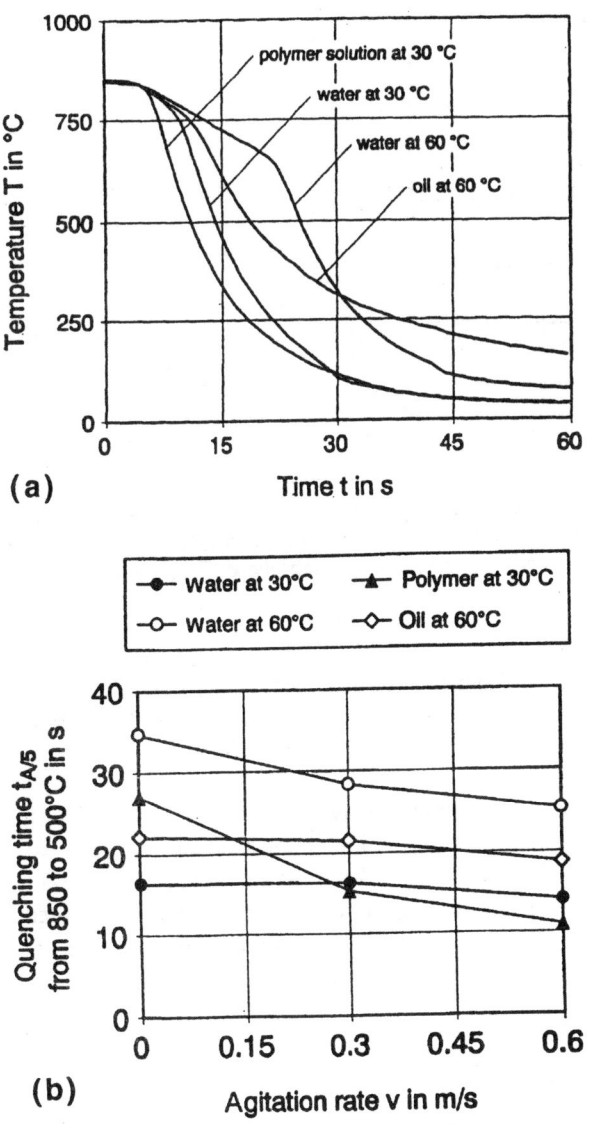

(a)

(b)

Figure 30 Influence of quenching media and agitation on the cooling behavior of a cylindrical specimen (25 mm (dia.) × 100 mm (length)). (a) Influence of different quenchants, some with different bath
temperature, on cooling curve behavior. (Agitation velocity of 0.6 m/s); (b) Influence of agitation rate
on the cooling time from 850°C to 500°C using the same probe with different quenchants at different
temperatures.

In Figure 30a, the time interval of wetting $(t_f - t_s)$ of austenitic stainless steel cylinders
heated to 850°C (1562°F) in an oxidizing atmosphere after annealing at 820°C (1508°F) for
20 h in air is compared to that of similar cylinders heated to the same temperature in a protective, reducing atmosphere [20]. Water at 20 and 50°C (68 and 122°F) flowing at various
flow velocities was used on the quenchant. The wetting time is shorter for the oxidized surface and is further reduced with increasing liquid flow velocity sustained by a decreasing bath

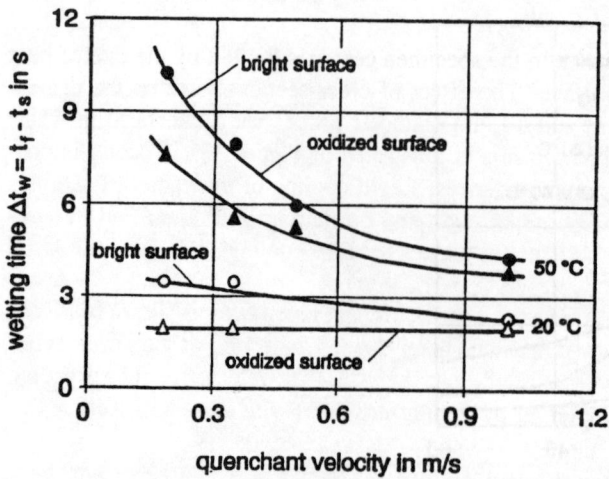

Figure 31 Influence of surface oxidation on the time interval of wetting ($t_f - t_s$) of an austenitic stainless steel specimen (15 mm diameter × 45 mm) quenched into water at two bath temperatures, 20 and 50°C. Oxidized surface: after annealing 20 h at 820°C in oxidizing atmosphere. Bright surface: after heating in protective atmosphere.

temperature. When the oxidized specimen was quenched in an aqueous polymer solution, the rough oxide layer facilitated immediate contact between the specimen surface and the liquid, while an identical specimen that had been heated in a protective atmosphere was surrounded by a vapor envelope for many seconds.

A more distinct effect of surface oxidation on wetting behavior is given in Figure 32 [13]. In addition to the oxidized and bright surfaces, the austenitic stainless steel specimen was passivated by etching with 1.5% and 3% nitric acid over 10 min in all. There is no observable influence on t_s, the time when wetting starts, but t_f, the time when wetting is finished, decreases and the time interval of wetting $t_f - t_s$ becomes shorter with increasing surface oxidation. In addition, the variance of the t_f values rises extremely.

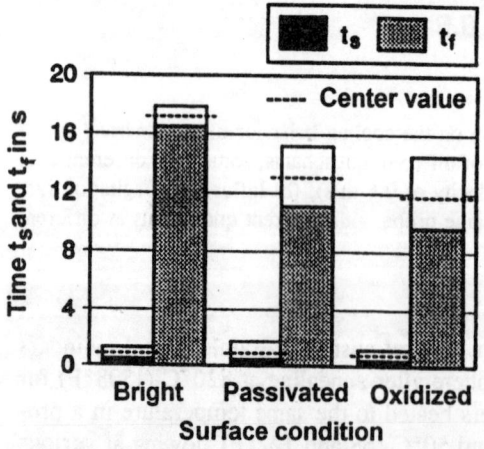

Figure 32 Influence of surface oxidation on the time when wetting starts (t_s) and the time when wetting is finished (t_f) of an austenitic stainless steel specimen (25 mm diameter × 100 mm) quenched into water at 30°C without forced convection.

5. *Effect of Cross-Sectional Size on Cooling*

With large cross sections, the cooling rate in the specimen center is limited by the rate of heat conduction from the interior to the surface. The effect of cross-sectional size on the center-cooling curves produced when 15, 25, and 40 mm diameter probes were quenched in water at 30°C (86°F) flowing at a velocity of 0.3 m/s is illustrated in Figure 33a. Cooling is proportionately slower with increasing sample thickness. Rapid cooling of the center of samples with large cross sections is therefore impossible with any quenching technique. Full hardening of large sections requires a high-hardenability steel because cooling rates are limited by the thermal diffusion of the steel.

Section size also exhibits a dramatic influence on the wetting kinematics as shown in Figure 33b. In water at 30°C (86°F), wetting starts immediately upon immersion at the lower cylinder edge and the time t_s remains unaffected with varying size. The velocity of the spreading wetting front is reduced and the time interval of wetting increases with increased section size [12].

The wetting kinematics vary with water temperature. As the water temperature increases, the vapor blanket surrounds the surface for a longer time and the time (t_s) of first contact between the specimen surface and the water increases substantially with increasing diameter.

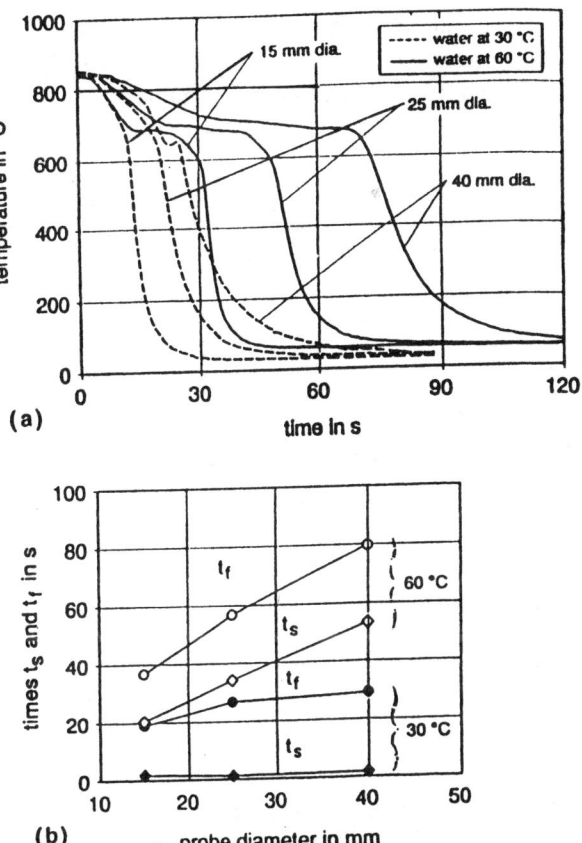

Figure 33 Influence of sample diameter on cooling behavior and wetting kinematics of a 25 mm diameter × 100 mm CrNi steel specimen quenched into 30 and 60°C water flowing at 0.3 m/s. (a) Cooling curve in the center of the specimen; (b) characteristic values of wetting t_s and t_f.

Practically, this means that water temperature must be carefully controlled to get reproducible hardness results, as increasing the wetting times results in decreasing the cooling rates, producing soft spots on parts. For example, hardness values of the 1045 steel bars quenched in water at 60°C (140°F) (Figure 33b) are less than $R_c = 30$ (HRC 30) and are independent of the section size.

6. Effects of Cooling Edge Geometry

Since the initial rupture of the vapor blanket is always related to the decrease of surface temperatures below the wetting temperature (Leidenfrost temperature), wetting behavior is strongly influenced by the radius of the lower edge of a cylindrical steel sample. Figure 34 shows the influence of different types of lower-edge geometry on the starting temperature T_s durign wetting of austenitic stainless steel specimens in water with different bath temperatures [14,20].

A sharp edge will cause high heat removal rates across the surface and a premature breakdown of the vapor blanket. Increasingly, the lower surface radius from a rounded edge to a radius of 2.5 mm reduces the initial wetting temperature, and the influence of bath temperature is diminished due to the greater thickness of the vapor blanket.

7. Effects of Steel Composition

The chemical composition of steel determines the thermodynamic parameters of the material, the transformation behavior of austenite, and oxidation of the surface and therefore influences the wetting and cooling behavior during quenching. The effect of chemical composition on cooling curves and cooling rate curves produced in the center of steel cylinders with 25 mm diameter quenched in water at 60°C (140°F) and 0.3 m/s agitation is illustrated in Figure 35a [12]. All samples were austenitized at 850°C (1562°F), except the 20MnCr5 steel, which was austenitized at 870°C (1600°F).

Cooling curves with different alloys were obtained with the same sample size and cooling conditions. High cooling rates for 1045 carbon steel are related directly to its high thermal conductivity.

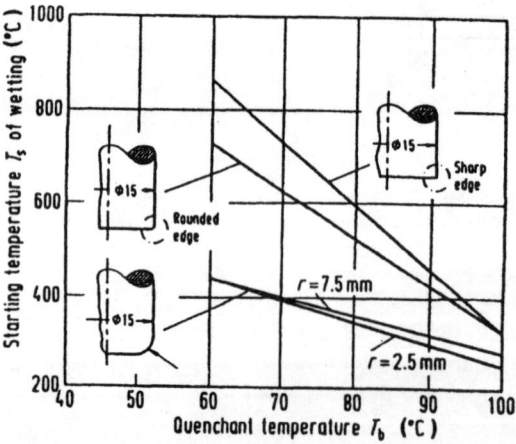

Figure 34 Influence of quenchant temperature T_b on wetting kinematics, here on the temperature T_s when wetting starts, of a CrNi steel specimen (15 mm diameter × 45 mm) with different kinds of lower-edge geometry during quenching in water. Sharp edge according to C1 DIN 6784; rounded edge according to D2 DIN 6784.

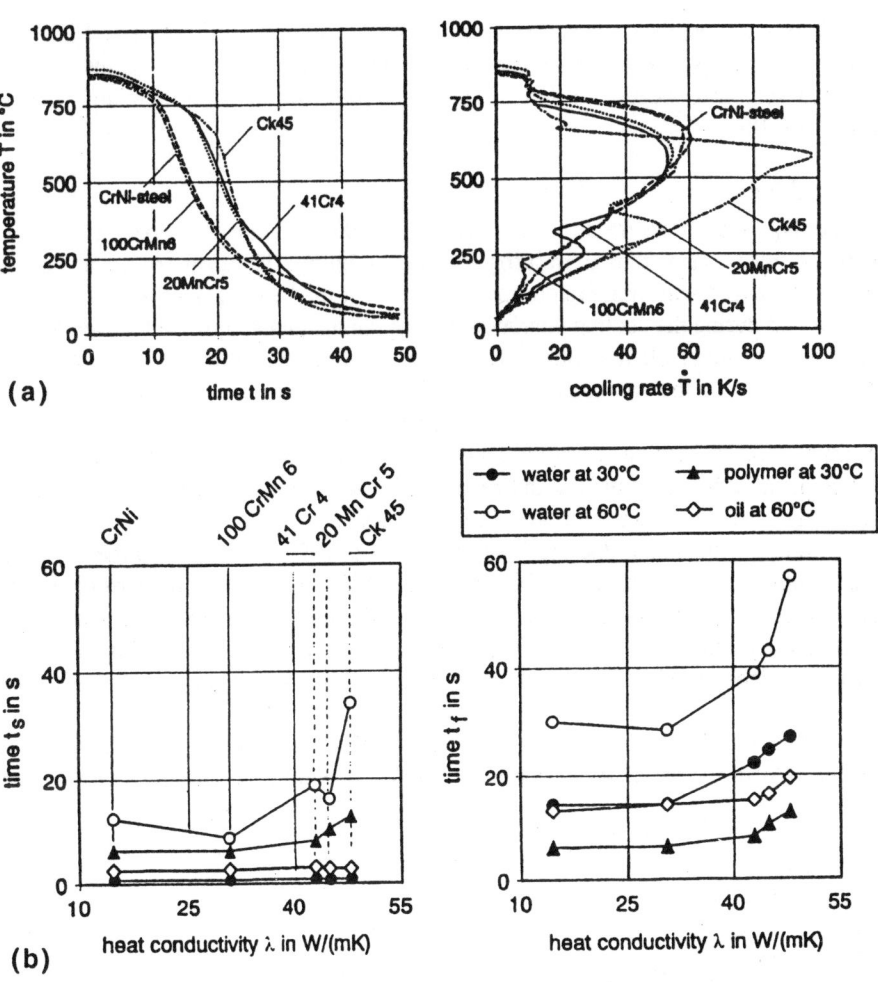

Figure 35 Influence of the chemical composition of steels on the cooling behavior and wetting kinematics of 25 × 100 mm steel bars quenched into different fluids flowing at 0.3 m/s. (a) Center cooling curves and cooling rate curves for 30°C water; (b) characteristic values of wetting t_s and t_f for water at 30 and 60°C, a 10% aqueous polymer solution at 30°C, and a fast oil at 60°C.

The minima in the cooling rate curves are caused by the latent heat of transformation of austenite into ferrite/pearlite, bainite, or martensite. The transformation temperature and the amount of latent heat depend on the hardenability of the steel grade and the cooling rate.

The thermodynamic properties of the material also exhibit a marked effect on the wetting kinematics as indicated in Figure 35b, which shows the time when wetting starts (t_s) and the time when wetting is finished (t_f) as a function of the heat conductivity λ at 20°C (68°F). Samples were quenched into different fluids agitated at 0.3 m/s. In 30°C (86°F) water and 60°C (140°F) oil, the start of wetting remains unaffected by thermal effects from the heat conductivity. However, water at 60°C (140°F) and the 10% aqueous polymer solution at 30°C (86°F) exhibit delayed wetting start time for λ values greater than about 30 W/(m·K). The time when wettnig is finished increases with increasing thermal conductivity for all fluids.

Surface effects were neglected because the probes were heated and transferred into the quenching bath in a protective, reducing atmosphere. For practical applications, unalloyed steels must be quenched very fast because of their transformation sensitivity and their unfavorable wetting process [12,28].

B. Spray Quenching

For spray quenching, high-pressure streams of liquid are directed through nozzles onto selected parts of the hot surface, thus permitting localized cooling. The primary advantage of spray quenching compared with immersion quenching is that the heat transfer across the surface can be controlled during the cooling operation by varying the quantity of the sprayed liquid, thus allowing computer-controlled cooling. However, careful adaptation of the quenching nozzles to the geometry of the workpiece is necessary because the cooling operation is extremely sensitive to any variations of the spraying action [38].

As with immersion quenching, a stable vapor film initially forms over the hot surface, reducing heat removal across the surface and resulting in slow cooling. Subsequently, nucleate boiling from the surface allows high heat flux densities.

The initial wetting process and the mechanism of heat transfer from steel to the liquid in the film boiling stage are shown in Figure 36. The sprayed liquid may flow upward or downward. Vapor flow is also affected by liquid flow. Since heat transfer is defined primarily by the velocity distribution within the liquid, it depends on the amount of sprayed fluid and thus on the impingement density, which is the amount of liquid at the metal surface per unit time per unit area.

Figure 37 shows the dependence of the heat transfer coefficient during film boiling on the impingement density for water spray cooling [39–41]. It is evident that the heat transfer

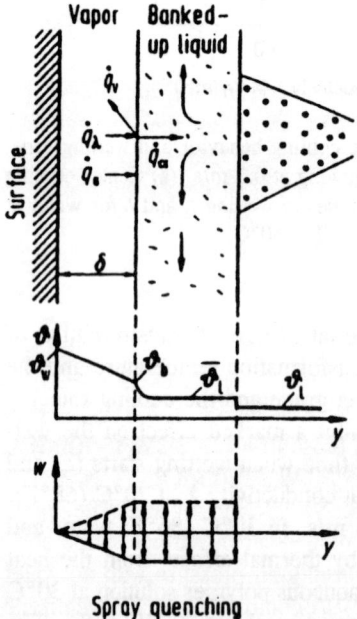

Spray quenching

Figure 36 Heat balance during film boiling in the boundary region of a spray-cooled surface and distribution of temperature and flow velocity.

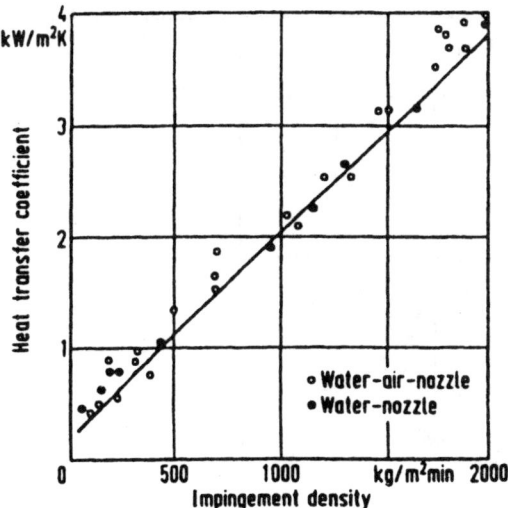

Figure 37 Effect of the impingement density on the heat transfer coefficient during film boiling for spraying with water.

coefficient values of film bonding (α_{FB}) may be much higher than with immersion quenching (see Figure 21) because of the higher density of sprayed water on the cooling surface.

The addition of air to the water spray, as shown in Figure 37, influences local distribution of the impingement density but does not increase heat transfer. It should be noted that heat transfer coefficients may also be influenced by the liquid temperature. For water, this influence increases with increasing impingement densities at temperatures above 20°C (68°F) [42].

In Figure 38a, the total heat transfer coefficient for water is given as a function of the surface temperature with various impingement densities [43]. Increasing the impingement densities increases heat transfer across the surface and shifts the wetting temperature to higher values. In the range of stable film boiling, the heat transfer coefficient does not depend on the surface temperature; however, significant variation is observed at lower temperatures.

Figure 38b indicates that the heat transfer coefficient depends not only on the impingement density but also on heat flow within the specimen and thus on the type of metal. This influence can be described by the coefficient of heat penetration b.

$$b = (\lambda \rho C_p)^{1/2}$$

The maximum heat transfer coefficient increases with increasing values of b. The value of b in Figure 38b varies from nickel [$b = 14.3$ kW·s$^{1/2}$/(m·K)], brass [$b = 21.3$ kW·s$^{1/2}$/(m·K)], and aluminum [$b = 23.6$ kW·s$^{1/2}$/(m·K)] to copper [$b = 35.5$ kW·s$^{1/2}$/(m·K)].

When a mixture of water and air is sprayed, heat transfer varies with air and water pressure as illustrated in Figure 39 [44] where heat flux density is plotted versus surface temperature for three sets of air and water supply conditions (air only, varying air pressure with no water pressure, varying water pressure with maximum air pressure). A control system for water and air pressure provides continual variation of the heat flux across the surface following a predetermined cooling curve [45].

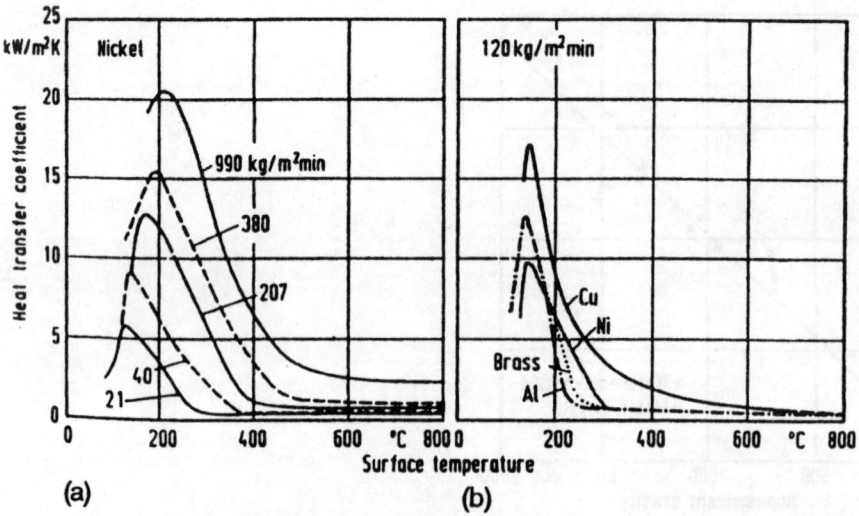

Figure 38 Heat transfer coefficient as a function of the surface temperature for spray quenching with water. (a) Influence of the impingement density; (b) influence of the chemical composition of the metal.

In Figure 40a, an isothermal pearlitic transformation of an AISI 1034 steel was obtained by real-time controlled spray quenching. The desired set-point curve (curve 2) is nearly completely reproduced by the regulated spray-quench operation (curve 3). Since the isothermally formed pearlite is very thin, the mechanical properties are higher compared to these in the continuous transformation (curve 1).

A second example, shown in Figure 40b, illustrates the effect of surface heat treatment of carbon steel. Induction heating of cylindrical parts was coupled with computer-controlled cooling, providing a fully automated heat treatment process. Here too, the real-time cooling control (irregular curve) allows the extraction of heat according to the desired temperature–time curve (straight-line segments).

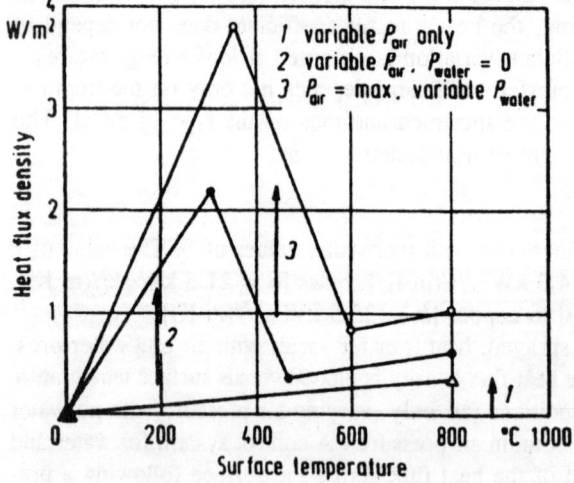

Figure 39 Heat flux density versus surface temperature for different, but constant, water and air pressures. (Δ) Variable P_{air}, no water; ($\bullet$) variable P_{air}, $P_{water} = 0$; ($\bigcirc$) P_{air} = max, variable P_{water}.

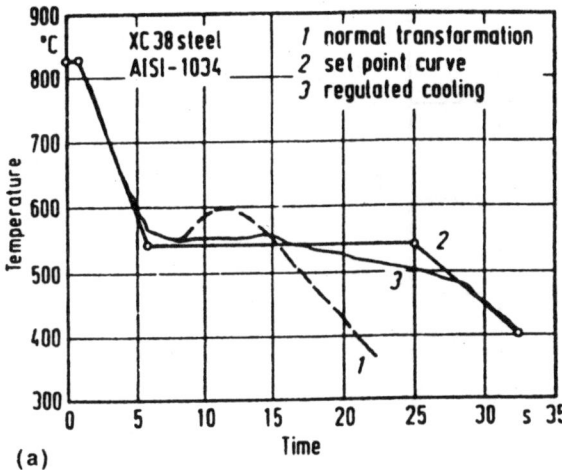

(a)

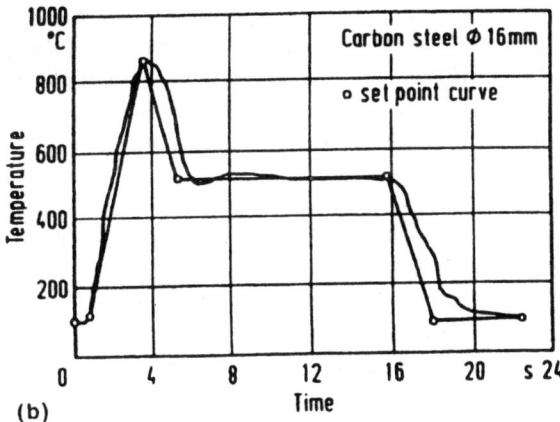

(b)

Figure 40 (a) Isothermal pearlitic transformation of a 16 mm diameter 1034 steel. The real time-controlled spray cooling (curve 3) allows the extraction of heat due to the desired transformation. (b) Surface heat treatment of a 16 mm diameter carbon steel. The set-point curve is achieved by coupling the heating and spray-cooling process.

C. Gas Quenching

In gas quenching, the hot workpiece is cooled by blowing gas over the surface. Gas quenching is generally used when cooling rates obtained in still air are too slow to produce the required hardness and oil quenching is undesirable because of residual stresses, cracking distortion, or handling (ecological, toxicological, or safety) problems.

Figure 41 compares cooling curves that were obtained during quenching of AISI 4130 steel tubing with an outer diameter of 31.7 mm and a wall thickness of 1.6 mm in still air, gas (the specific gas quenchant was not identified), and oil [37]. It is evident that the cooling rate achieved in gas is lower than that obtained in oil. This example illustrates that the favorable use of gas quenching depends on hardenability of the steel and the size of the workpiece.

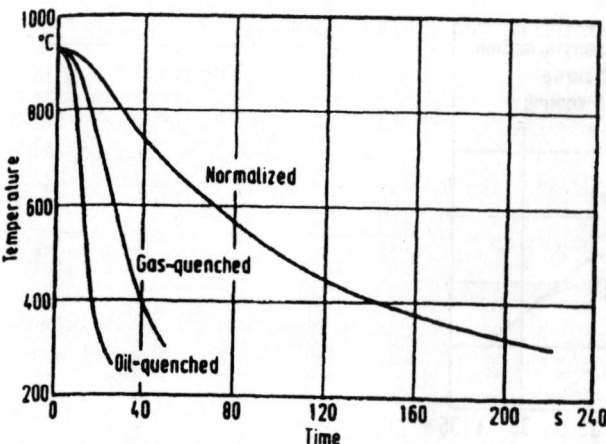

Figure 41 Cooling curve of a 4130 steel tube 31.7 mm diameter × 1.6 mm quenched into oil, gas, and still air (normalizing).

An advantage of gas quenching compared to liquid quenchants (oil, water, and aqueous polymers) is that quenching with gas proceeds more uniformly, minimizing residual stresses and distortion. Vaporizable liquids such as oils may exhibit an extended film boiling phase, and the beginning of nucleate boiling is variable. (see Section IV). This nonuniform heat removal from the surface may produce a nonuniform hardness (soft spots) and an unfavorable stress distribution leading to cracking and distortion.

In addition to potentially improved quench uniformity, gas quenching is a "clean" process, eliminating the need for the vapor degreasing step often used for oil quenching processes. Potential fire hazards and disposal problems are also eliminated.

The cooling operation can be controlled precisely by varying the type, pressure, velocity, and temperature of the gas, thus providing uniform heat transfer across the surface and a variety of possible cooling rates. The use of inert gases avoids chemical reactions with the steel surface, eliminating the need for any surface cleaning or finishing treatment after quenching.

Gas quenching is usually used after austenitization in vacuum furnaces. During quenching, gases are blown through nozzles or vanes toward the workload. After absorbing heat from the load, the gas is cooled by passage through water-cooled heat exchangers and then recycled to the nozzles for subsequent reuse.

Cooling rates obtainable for gas quenching are limited by the type, pressure, velocity, and temperature of the cooling gas and by the surface conditions, geometry, and thermal properties of the material [46]. The most common cooling gases that can be used in vacuum furnaces are argon, nitrogen, helium, and hydrogen. The physical properties of these gases are summarized in Table 4 [47]. The thermal conductivity, specific heat, density, and dynamic viscosity influence the heat transfer coefficient. Figure 42 shows that the greatest heat transfer coefficient is achieved with hydrogen, followed by helium, nitrogen, and argon, in that order. The heat transfer coefficients are directly proportional to the thermal conductivity and specific heat of these gases.

Hydrogen is explosive and exhibits decarburizing properties above 1000°C (1832°F). Helium is relatively expensive and is therefore used only with a recycling facility. Argon exhibits relatively low cooling rates. Nitrogen is the gas most commonly used in vacuum fur-

Table 4 Physical Properties of Hydrogen, Helium, Nitrogen, and Argon

Property	Hydrogen	Helium	Nitrogen	Argon
Density, kg/m³	0.303	0.601	4.207	6.008
Specific heat, J/(kg·K)	14450	5200	1050	520
Heat conductivity, 10^4 W/(m·K)	2256	1901	326	222
Dynamic viscosity, 10^6 N·s/m²	10.8	24.4	21.6	28.2

naces today. Both nitrogen and argon pressure are limited to approximately 10^6 Pa because of the high flow resistance of these gases, which is due to their high density and dynamic viscosity.

Figure 42 shows that the heat transfer coefficient increases with gas pressure [48]. The heat transfer coefficient was calculated for a solid cylinder that was radially quenched with a velocity of 20 m/s and a temperature of 200°C (392°F). In practice, gas pressures vary between 1 and 10 bar for argon and nitrogen and between 1 and 20 bar for helium and hydrogen. At these gas pressures, heat transfer coefficients are considerably less than those for oil quenchants. However, oil quenching produces less uniform surfaces during heat transfer because of the widely varying heat transfer coefficients exhibited by the three different cooling stages. (Increasing the flow velocity of the gas also increases the heat transfer coefficient.)

Figure 43 shows the effect of the bar diameter and the type of gas on the center cooling curves of a DIN 1.2080 steel cylinder with a length of 200 mm, which was quenched, after vacuum heat treatment, in nitrogen and hydrogen at a pressure of 5 bars [49]. The cooling rate decreases with increased section size, and greater cooling rates will be achieved with hydrogen than with nitrogen under the same quenching conditions. The center cooling curve of the 80 mm diameter sample quenched in nitrogen is similar to that obtained in the center of the 120 mm diameter sample quenched in hydrogen. Corresponding hardness values in the core are almost identical: HRC = 63 vs 62. For an 80 mm diameter, the core hardnesses can be improved to HRC = 66 by quenching with hydrogen.

Figure 44 compares the cooling curves produced when steel cylinders with a diameter of 20 mm and a length of 500 mm, arranged in a 520 kg net mass workload, were quenched in

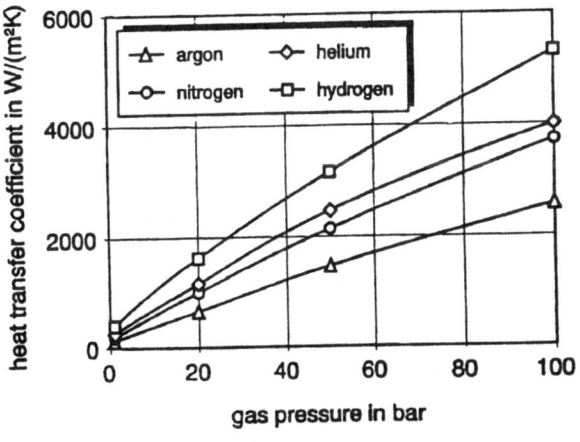

Figure 42 Heat transfer coefficients of inert gases (hydrogen, helium, nitrogen and argon) as a function of the gas pressure, calculated for the case of a cylinder with transverse gas flow.

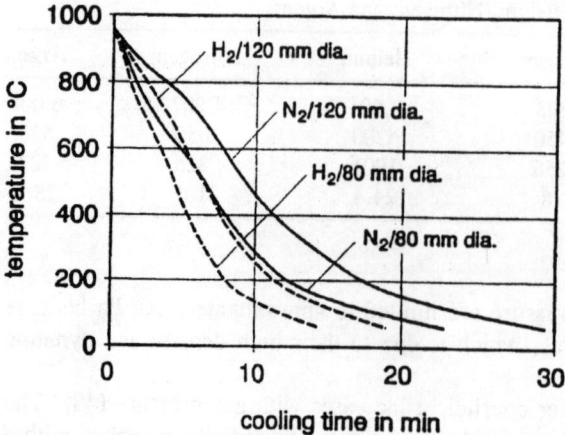

Figure 43   Cooling curves in the center of a DIN 1.2080 steel with 80 and 120 mm diameter and 200 mm length during cooling in hydrogen and nitrogen at a pressure of 5×10^5 Pa.

nitrogen at 6 bar and in hydrogen at 6 and 20 bar [49]. It is evident that cooling times can be decreased by 30% with the use of hydrogen instead of nitrogen at the same gas pressure. Increasing the hydrogen pressure from 6 bar to 20 bar further increases the cooling rate by about 75% compared to the 6 bar nitrogen quench.

Figure 45 compares the hardness values in the core and in the boundary region of a case-hardened DIN 1.7131 steel that was obtained when 20, 40, and 60 mm diameter probes were quenched in helium at a pressure of 17.5 bar and in oil [48]. Examination of the 60 mm diameter probe shows that the oil quench can be replaced by the high-pressure helium quench (see Figure 45a) and that the surface and core hardnesses are slightly higher for the helium quench.

Figure 45b illustrates that when the section size is reduced from 40 to 20 mm diameter, core hardness increases and the hardness increase is higher for helium than oil. These results show that the more uniform quenching obtained by gas quenching is advantageous for large

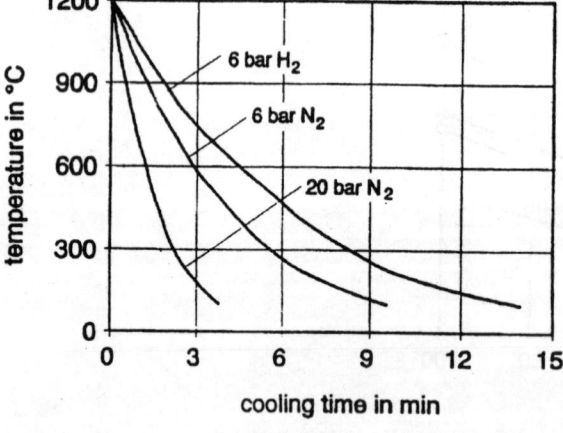

Figure 44   Influence of the type of gas, hydrogen and nitrogen, and of hydrogen pressure on the cooling curve during quenching of a batch of cylinders (30 mm diameter × 500 mm).

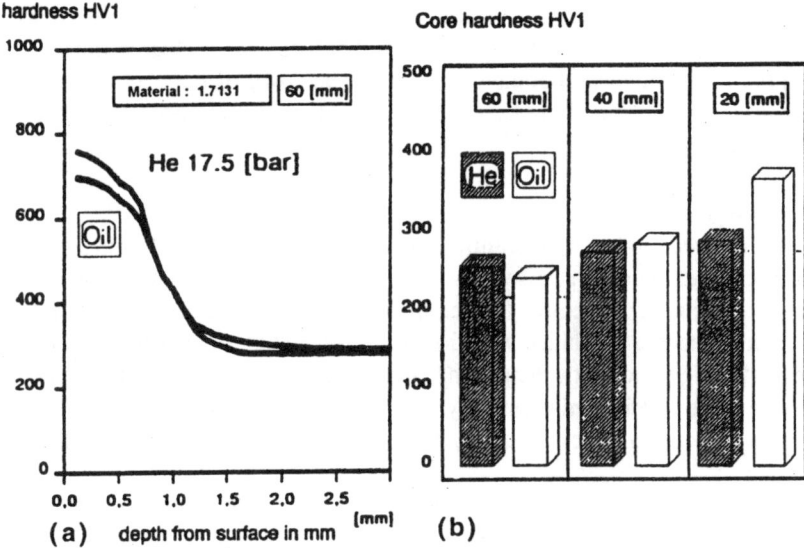

Figure 45 Hardness of case-hardened DIN 1.7131 steel bars of 20, 40, and 60 mm diameter during quenching in oil and in helium at a pressure of 17.5 x 10⁵ Pa. (a) Hardness distribution in the boundary region of the 60 mm steel bar; (b) hardness in the center of the 20, 40, and 60 mm bars.

sections. Conversely, for smaller diameter sections, oil quenching is preferred because the film boiling stage decreases with decreasing section size and nucleate boiling begins at a higher temperature, resulting in higher heat transfer coefficients at higher temperatures.

A comparison of nitrogen at 5 bar and molten salt for high-speed quenching of steel after vacuum heat treatment has also been reported [46,47]. Cooling curve and hardness results showed that the two processes were nearly identical. An example of an automated gas quenching facility for parts is presented in Figure 46 [50–52]. The hot ball-bearing races fall with the help of a cone in the quenching device, where they are multidirectionally sprayed with streams of gas. The pieces are turned during the quenching operation to minimize the differences in heat transfer in the region of a single gas stream. The primary advantage of this quenching

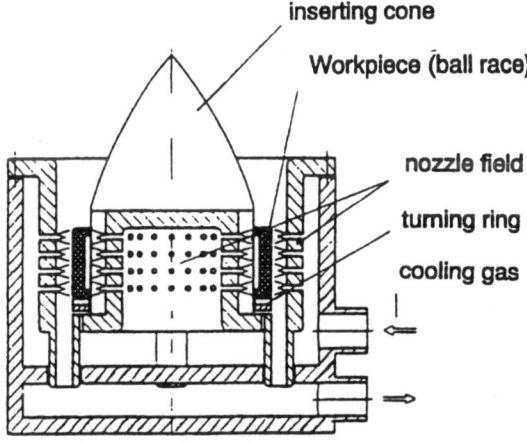

Figure 46 Gas quenching device with standard nozzle length for bearing shells.

device is the uniformity of heat transfer across the surface, which cannot be achieved with other quenching methods. In an oil or water quench, the heat transfer increases during the transition from film boiling to nucleate boiling. Since the breakdown of the vapor film does not occur simultaneously around the circumference of the ball race, this process results in distortion and high residual stresses.

D. Intensive Quenching

During quenching from austenitizing temperature, localized temperature differences occur within the workpiece, creating thermal stresses. These are due to the temperature dependence of the specific volume of the microstructures formed in the steel during quenching. The transformation-induced volume changes combined with the transformation plasticity cause additional transformation stresses that interact with the thermal stress state. Both thermal and transformational stresses cause plastic deformations if they are in excess of the yield strength of the material. This may result in quench cracks if the equivalent stresses exceed the tensile strength of the hot steel.

Traditionally, heat treaters have striven to achieve the critical cooling rate necessary to form 100% martensite. (This "critical cooling rate" is the cooling rate at the initial transition from austenite, pearlite, and bainite.) However, maximizing cooling rates may increase the risk of quench crack formation. Numerous experimental studies have shown that as cooling rates within the martensitic transformation range increase, the probability of quench crack formation first increases to a maximum, then reduces to zero as shown in Figure 47 [10]. This was confirmed by calculation of the thermal and transformational quenching processes using finite-element methods and a mathematical model that included the equation of nonstationary heat conduction and equations for thermoelastic-plastic flow with kinematic strengthening under the appropriate boundary conditions. It was found that the tensile residual stresses first increase to a maximum value with increasing cooling rate, then decrease and then change to compressive stresses. The probability of quench cracking decreases with increasing cooling rate due to the presence of high compressive stresses at the cooling metal surface. These results were validated by measurement of the surface residual stress of quenched parts.

The development of stresses during rapid cooling of steel cylinders is schematically illustrated in Figure 48. (Reference 4 provides an excellent overview on the formation of quench

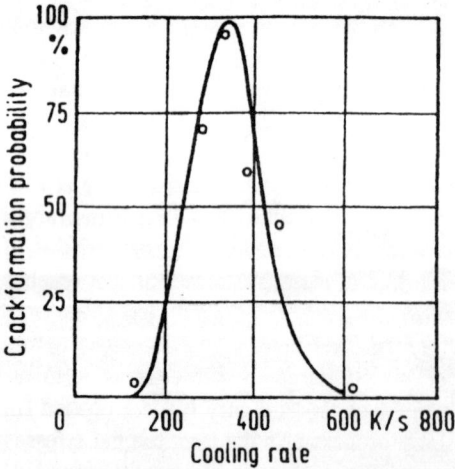

Figure 47 Probability of crack formation versus cooling rate within the martensite formation range of a 6 mm diameter steel cylinder (41Cr4 according to DIN).

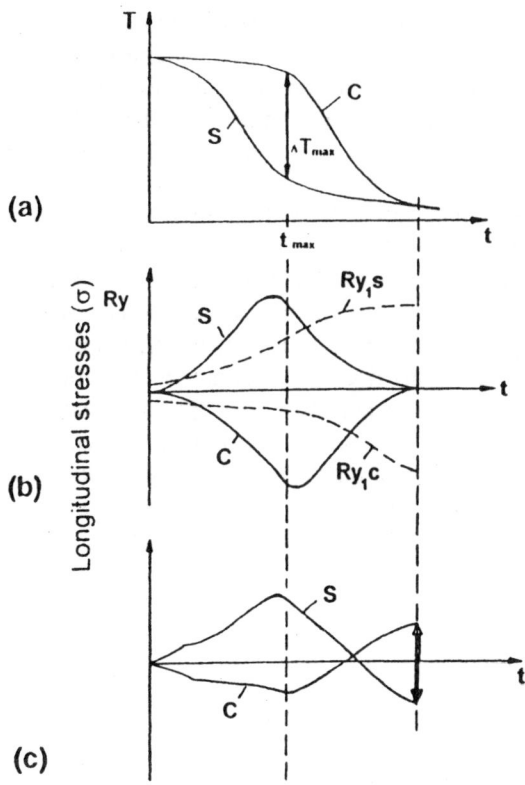

Figure 48 (a) Development of temperature and (b,c) formation of longitudinal stresses in the core and at the surface of a steel cylinder during transformation free quenching considering (b) ideal elastic and (c) elastic plastic deformation behavior (schematic). S, surface; C, core.

stresses.) The temperatures at the surface and core of the cylinder and the corresponding stress components will be considered as functions of the cooling time. For simplification, ideal elastic deformation behavior and transformation-free quenching will be assumed.

Initially, the surface temperature decreases faster than the core temperature, as shown in Figure 48a. This causes the surface of the cylinder to contract more rapidly than the core, resulting in tensile residual stresses in the longitudinal and tangential directions. These stresses are balanced by compressive stresses in the core of the cylinder, as shown in Figure 48b. Maximum thermal stresses develop when the maximum temperature difference between surface and core (T_{max}) is attained at cooling time t_{max}. Maximum surface stresses occur at $t < t_{max}$, and maximum core stresses occur at a later time, $t > t_{max}$. At $t = t_{max}$, the temperature of the core decreases faster than that of the surface, leading to a reduction of the stresses in both regions. If these stresses are elastic, the cylinder is free of residual stresses after it reaches thermal equilibrium.

In reality, metals always exhibit elastic-plastic deformation behavior. The elastic deformation is limited by the yield strength of the material, which increases with decreasing material temperature. In Figure 48b, the yield strength of surface (S) and core (C) are plotted (in broken lines) assuming a transformation-free quenching process. As the longitudinal stresses approach the yield strength, plastic deformation begins. The surface is plastically extended, and the core is plastically compressed. Both effects reduce the stresses in the core and surface regions compared to the ideal elastic deformation behavior shown in Figure 48c.

At $t > t_{max}$, the temperature difference between the core and surface decreases and the "shrinking" stresses in both reginos are reduced. As a result of the plastic compression in the core and the plastic extension at the surface, the stresses of the core and surface become zero before the temperature balance is reached. At the end of the cooling process, the plastic deformations of core and surface produce tensile residual stresses at the core and compressive stresses at the surface.

If phase transformations occur during quenching, they will result in thermal transformation stresses. For simplicity, it will be assumed that austenite is completely transformed into martensite and that the coefficient of thermal expansion is zero. This condition will produce no thermal stresses. Temperature–time curves for the surface and core of such a material are illustrated in Figure 49a. The surface of the cylinder begins transformation at t_1, at or below the martensite start temperature M_s. Volume expansions due to the formation of martensite produce compressive stresses at the surface, which are compensated for by tensile stresses within the core as shown in Figure 49b. Both stress components increase until the temperature of the core reaches the martensite start temperature, at time $t = t_2$. The volume increase of the core transformation immediately reduces the tensile stresses at the core and compressive stresses at the surface. Below the martensite finish temperature (M_f) in the core, the same

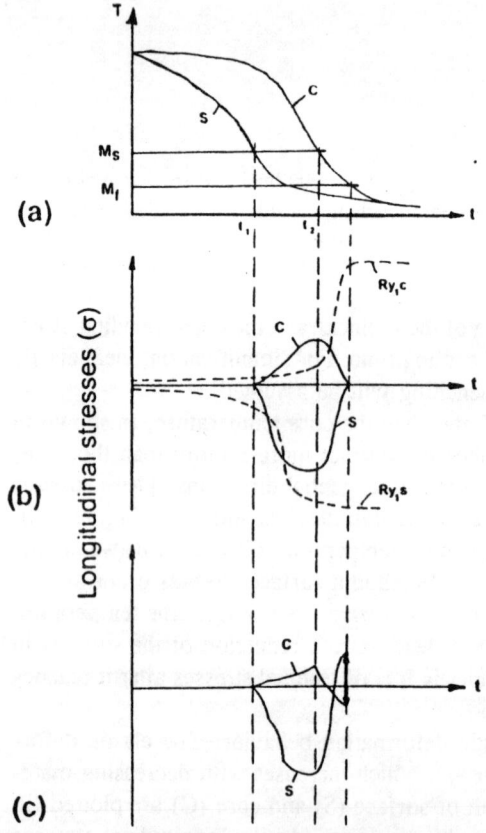

Figure 49 (a) Development of temperature and (b,c) formation of longitudinal stresses in the core and at the surface of a steel cylinder during with (b) ideal elastic and (c) elastic plastic deformation behavior, considering only austenite–martensite transformation and neglecting thermal expansion (schematic.)

amounts of martensite are formed within the cylinder and no residual stresses will occur if ideal elastic transformation behavior is assumed.

In elastic-plastic deformation, the temperature-dependent yield strength must be considered. Figure 49b shows yield strength variation at the surface and core with respect to cooling time. Martensitic formation strongly increases the yield strength at $T < M_s$. As the stresses approach the yield strength, the surface and core are plastically deformed as shown in Figure 49c. Upon further cooling, the stress values increase until the core temperature reaches M_s. The corresponding volume increase in the core reduces the stress values of both areas, and at temperature equilibrium, compressive stresses remain in the core and tensile stresses at the surface.

During quenching of hardenable steels, thermal and transformational stresses occur simultaneously. Residual stress after cooling is primarily determined by the time when transformation begins at the surface and in the core as expressed by the times t_1 and t_2, respectively, in relation to the time t_0 when the thermal stresses of surface and core change sign as shown in Figure 50a. Generally, transformations occuring in compressively stressed material increase the stresses, whereas transformations occurring in tensile-stressed areas reduce stresses. Figures 50b and 50c illustrate the simultaneous formation of both thermal and transformational stresses. Figure 50b illustrates that both surface and core transformation start after time t_0. In this particular case, surface compressive stresses and tensile stresses occur in the core after

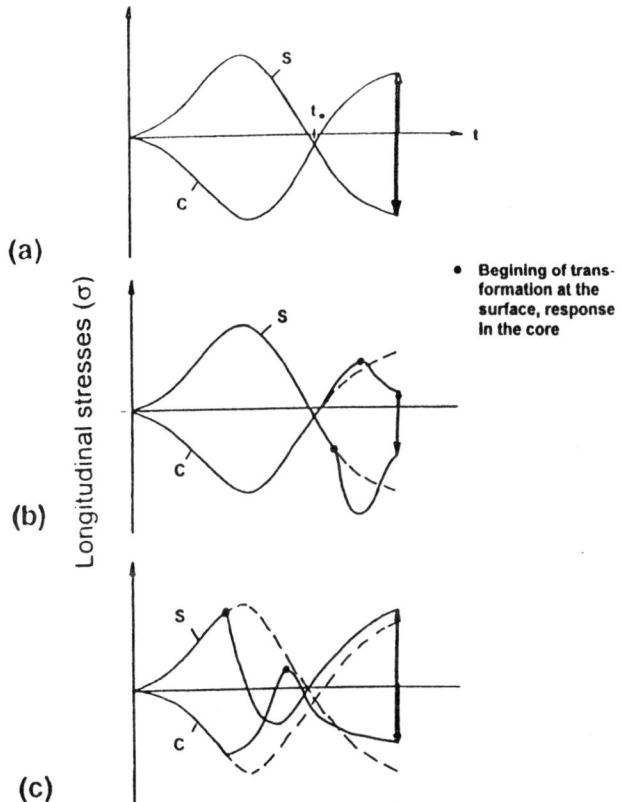

● Begining of trans-
formation at the
surface, response
in the core

Figure 50   (a) Development of temperature and (b,c) formation of longitudinal stresses in the core and at the surface of a steel cylinder during transformation free quenching considering thermal expansion and martensite transformation (schematic).

quenching. Figure 50c shows the development of stresses when surface and core begin to transform before t_0. The increase of the specific volume at the surface after transformation is initiated reduces tensile stresses and cause surface compressive stresses upon further cooling.
Tensile stresses occur when martensitic transformation begins in the core. This inverts the stress
state in both areas, leading to compressive residual stresses in the core and tensile residual
stresses at the surface after thermal equilibrium.

Figure 50 shows that the final stress state after quenching depends on the temperature
gradient within the sample and on the transformation behavior of the material. However,
measurements of residual stresses at the surface of quenched cylinders have shown that the
M_s temperature strongly influences the sign of the stress values. Increasing M_s temperature
decreases yield strength, and the tensile residual stresses change to transformation stresses as
shown in Figure 51 [9]. Therefore, the effects of transformation plasticity must also be considered. (See Section II.D.)

Development of residual stresses during intensive quenching are caused by transformation plasticity and changes in specific volume due to the transformation of austenite into martensite. With rapid cooling rates, the surface temperature is cooled immediately to the bath
temperature, while the core temperature is nearly unaffected. Rapid cooling causes surface
layer contraction and high tensile stresses, which are balanced by stresses in the core. Increasing
temperature gradients produce increasing tensile stresses at the beginning of martensite transformation, and increasing M_s temperatures produce surface layer expansions due to transformation plasticity as shown in Figure 7. If high M_s temperatures accompany high volume expansion during martensite transformation, surface tensile stresses are significantly reduced,
changing to compressive stresses. The amount of surface compression is proportional to the
amount of surface martensite formed. These surface compressive stresses dictate whether the

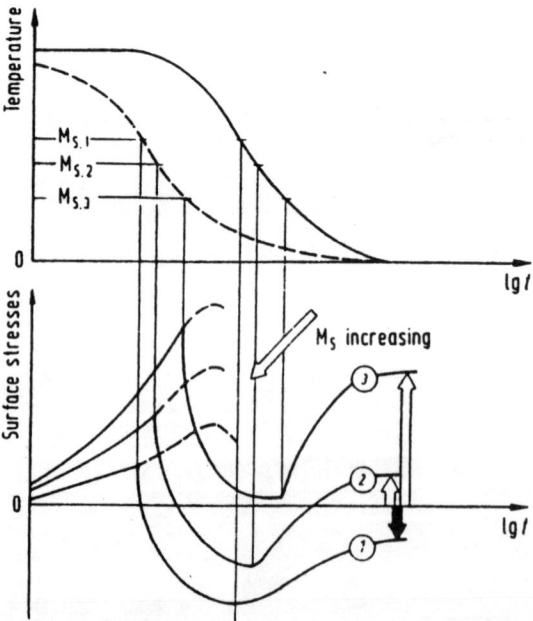

Figure 51 Effect of martensite start temperature M_s on the formation of longitudinal stresses at the
surface of a cylinder (schematic).

martensitic transformation at the core occurs under compression or forces a tensile stress inversion at the surface during further cooling. If the volume increase of the core upon martensite transformation is sufficiently large and if the surface martensite layer is very hard and brittle, destruction of the surface layer may occur due to stress inversion. Therefore, to ensure surface compressive stresses, martensite formation in the core must be retarded [10].

These results indicate that the intensive quenching process is stopped when surface maximum compressive stresses are formed. The isothermal cooling process is then held at approximately the M_s temperature. This will delay cooling at the core, thus retarding martensite formation, which will produce high surface compressive stresses. The intensive quenching process is completed when an optimal depth of the surface layer is achieved corresponding to maximum compressive stresses.

A second method of minimizing quench crack formation is to ensure material plasticity, which will occur if there is less than about 30% newly formed martensite in the supercooled austenite [10,53]. Initially, intensive quenching occurs until the surface temperature of the part is maintained at such a level that not more than 30% martensite can be formed in the subcooled austenite. Then the intensive quenching process is interrupted and the part is cooled in air until temperature equilibrium is achieved over the cross section. The newly formed martensite in the surface zone is self-tempered and quench crack formation is hindered because stresses are decreased. Subsequently, intensive cooling to room temperature is applied to transform the rest of the austenite into martensite. The intensification of cooling within the martensite range improves the plastic characteristics of the material and increases its strength properties.

The advantage of intensive quenching methods is that high residual surface compressive stresses are created, reducing the probability of quench crack formation, with a corresponding improvement in hardness and strength values. Microstructures consisting of 100% martensite can be achieved at the surface, and a maximum hardening depth can be obtained for a given steel grade. In this way, an unalloyed carbon steel may be used instead of a more expensive alloy steel. Intensive quenching also produces more uniform mechanical properties and minimal distortion. Furthermore, it has been demonstrated that the service life under cyclic loads may be increased by approximately one order of magnitude [54].

Intensive quenching requires appropriate quenching facilities and quenching media. Quenching media include pressurized water streams, water containing various additives, and liquid nitrogen. Figure 52 shows a quenching chamber for the intensive cooling of an automobile semiaxis using pressurized water flow.

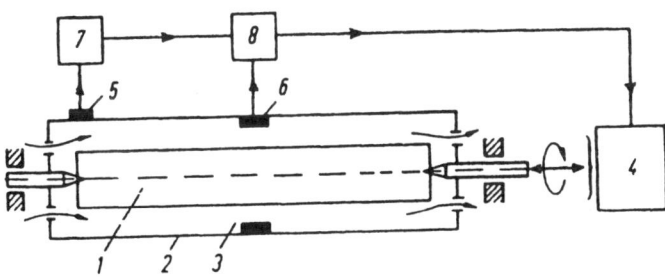

Figure 52 Quenching chamber for intensive quenching of semiaxes in a pressurized water flow. *1*, semiaxis; *2*, quenching chamber; *3*, water flow; *4*, mechanical drive for the semiaxis; *5*, sensor for analyzing the process of film boiling and nucleate boiling; *6*, sensor for analyzing the portion of formed structures; *7, 8*, amplifiers.

The water supply to the chamber and the charging and discharging of the axles are controlled by two sensors. The first sensor (5 in Figure 52) analyzes the process of film boiling and nucleate boiling, and the second, 6, describes the transformation of austenite into martensite by the change of the ferromagnetic state of the material. One method of intensive quenching has been used that achieves maximum compressive stresses at the surface when sensor 6 indicates a specific magnetic phase transformation. In this case, sensor 5 is used to minimize the duration of film boiling by regulating the water flow velocity. A second method has also been used, with sensor 5 indicating the beginning and completion of nucleate boiling, while sensor 6 controls the water pressure and determines the end of intensive quenching, so that no more than 30% martensite is formed.

Intensive quenching methods offer many possibilities for the successful cooling of parts with optimized strength properties and improved service life. However, a precondition for the use of this technology is the development of appropriate quenching equipment that enables precise control of the quenching performance.

VIII. PROPERTY PREDICTION METHODS

There are increasing demands on the heat treater to achieve as-quenched properties while simultaneously reducing heat treatment costs. To achieve these goals it is becoming increasingly important that experimentally or mathematically based methods to predetermine the as-quenched strength and/or hardness properties be applied with sufficient accuracy. Currently, a computer-based selection of steels and optimization of quenching conditions according to the desired service properties are generally possible. Hardenability is one of the most important properties to be predicted because it determines as-quenched microstructure formation. The ability to predict hardenability curves from chemical composition has already been described in Section IV of Chapter 3. However, these hardenability curves provide only limited information about the distribution of mechanical properties in the quenched part. It is necessary to correlate steel chemical composition, cooling rates during quenching, metallurgical transformation behavior, and the final physical properties. These correlations are often complex.

This section discusses various methods, ranging from relatively simple to more complicated methods, for determining as-quenched hardness distribution. Traditionally only hardness value has been effectively predicted. However, with the application of computer technology, additional calculations of residual stresses and distortion are becoming increasingly possible, at least for simple shapes. A detailed overview of published methods for prediction of hardness is given in Reference 55.

A. Potential Limitations to Hardness Prediction

The final hardness of a steel is determined by the amount of different microstructural constituents formed during cooling and by their individual hardnesses. The microstructures that develop with increasing cooling rate from the austenitizing temperature are ferrite-pearlite, bainite, and martensite, probably containing retained austenite. The individual hardnesses of these microstructures depend primarily on chemical composition of the steel, especially the carbon content, and the cooling rate and formation temperature. For example, in a 1040 steel, austenite fully transforms into martensite with a Vickers hardness of approximately 850 (HRC 66) at high cooling rates. At low cooling rates, ferrite and pearlite are formed with Vickers hardness of about 200 (HRA 57). At medium cooling rates, microstructures consisting of ferrite, pearlite, bainite, and martensite are produced. To render hardness prediction more difficult, different microstructural combinations can have the same hardness, so that there is a one-way relationship between microstructural combinations and final hardness [56,57]:

$$\begin{bmatrix} \text{Microstructure and} \\ \text{microstructural} \\ \text{combinations} \end{bmatrix} \rightarrow \begin{bmatrix} \text{final} \\ \text{hardness} \end{bmatrix} \tag{7}$$

A two-way relationship exists only between the course of cooling and microstructure formation:

$$\begin{bmatrix} T(t) \text{ curve} \\ \text{during} \\ \text{cooling} \end{bmatrix} \leftrightarrow \begin{bmatrix} \text{microstructure and} \\ \text{microstructural} \\ \text{combinations} \end{bmatrix} \tag{8}$$

Therefore, accurate prediction of the final hardness from the course of cooling always requires a quantification of the transformation structure:

$$\begin{bmatrix} T(t) \text{ curve} \\ \text{during} \\ \text{cooling} \end{bmatrix} \rightarrow \begin{bmatrix} \text{microstructure and} \\ \text{microstructural} \\ \text{combinations} \end{bmatrix} \rightarrow \begin{bmatrix} \text{final} \\ \text{hardness} \end{bmatrix} \tag{9}$$

Even with these restrictions, several methods of hardness prediction have been developed that use the simplified, but physically incorrect, assignment of hardness from cooling curves:

$$\begin{bmatrix} T(t) \text{ curve} \\ \text{during} \\ \text{cooling} \end{bmatrix} \rightarrow \begin{bmatrix} \text{final} \\ \text{hardness} \end{bmatrix} \tag{10}$$

Cooling curves obtained at different points over the cross section of real workpieces are compared to temperature–time relationships obtained from Jominy end-quench tests or CCT diagrams. However, these data describe only those transformations of austenite that occur along the cooling curves of the specimens used for their construction. Therefore, the accuracy of the predicted hardness values depends on how well cooling curves in real workpieces correlate to the cooling of the steel that was used for the construction of the CCT diagrams or Jominy end-quench hardness curves.

An example of insufficient conformity between these data is provided in Figure 53, which compares cooling curves that were obtained at different points in an austenitic stainless steel cylinder that was immersion-quenched in 60°C (140°F) water flowing at 0.3 m/s (solid lines) and spray quenched with 12°C (54°F) water at the lower end face according to the Jominy end-quench test (dashed lines) [12]. Cooling curves obtained in the end-quench test have progressively lower cooling rates with increasing distance from the water-quenched end face and differ completely from those produced with the immersion-quenched cylinder, which have low cooling rates in the vapor blanket stage and highest cooling rates in the nucleate boiling stage. Therefore, for the same cooling time $t_{8/5}$, from 800 to 500°C (1472 to 932°F), the microstructure formed may be quite different if different temperature–time cycles are applied.

In addition to proper modeling of the cooling process, exact prediction of the material properties also requires that all metallurgical and thermal boundary condition reference curves

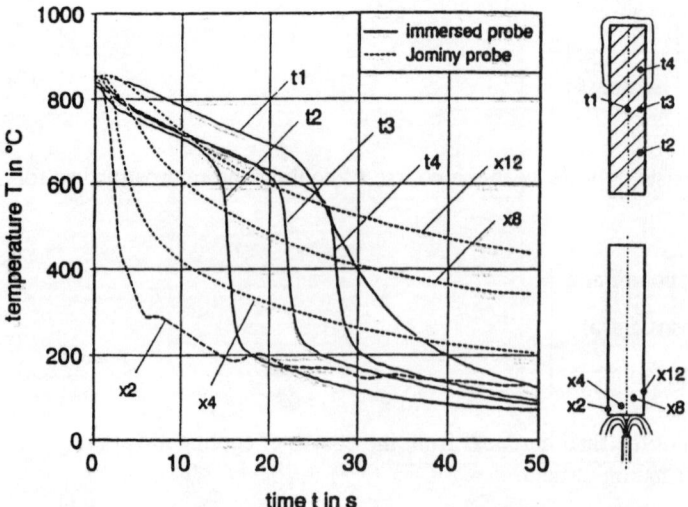

Figure 53 Cooling curves for the surface of a Jominy probe 2, 4, 8, and 12 mm from the quenched end (dashed lines) and for an identical cylinder quenched in 60°C water flowing at 0.3 m/s for the points 1 mm below the surface at 25, 50, and 75 mm height and for the probe center (full lines). Material: austenitic stainless steel. Geometry: 25 × 100 mm. (From Ref. 11.)

and the production quenching process be similar. These are the material composition and initial microstructure before austenitizing, the austenitizing conditions (temperature and time), and the chemical properties of the quenchant and its physical state (temperature and flow conditions).

B. Grossmann *H* Values

A practical method to predict the hardness distribution in a steel cylinder after quenching is based on the Grossmann *H* value [35]. The *H* value reflects the ability of a quenchant to remove heat from the surface of a hot workpiece during quenching and is defined by the heat transfer coefficient divided by twice the thermal conductivity (see Equation 5). The *H* value of still water at 18°C (64°F) is usually taken as 1.0. The *H* values of oil, water, and brine with different bath temperatures and flow velocities are presented in Table 3 (see Section VI.B) [58]. They are commonly determined by recording a cooling curve in the center of a test probe, determining the average cooling rate within a temperature region [e.g., from 700°C to 600°C (1292 to 1112°F) and comparing it with the cooling rate at standard conditions, i.e., for still water at 18°C (64°F). A polynomial expression has been published that allows a direct estimation of the *H* value from the center cooling rate at 700°C (1292°F) [59,60].

When using *H* values one has to be aware that *H* is a constant and does not describe the heat transfer conditions at the surface of a workpiece quenched in a vaporizable fluid. It is well known that the heat transfer coefficient α varies substantially with the three different stages of cooling—film boiling, nucleate boiling, and convection—and also varies with the wetting sequence, i.e., with the actual position of the wetting front (see Section VI). The thermal conductivity (λ) of steel depends on temperature and possible microstructural changes.

Figure 54 compares the temperature–time curve and the temperature–heat flux density curve of an actual cooling process with values calculated using a constant *H* value [61]. The solid line was obtained in the center of an 8 mm diameter silver cylinder quenched in oil; the dashed line was calculated according to Newton's law of cooling with a constant *H* value of

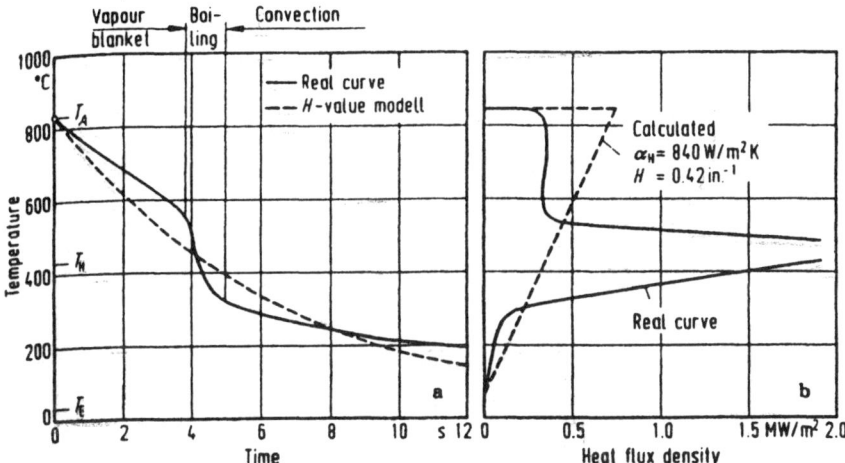

Figure 54 (a) Temperature–time curves and (b) temperature–heat flux density curves for an 8 mm diameter silver cylinder quenched in oil. The full lines are the real curves, the dashed lines are calculated using an H value of 0.42 in.$^{-1}$. T_A: temperature at start of cooling, T_E: temperature at the end of cooling, T_H: half-temperature. $T_H = (T_A - T_E)/2 + T_E$. (From Ref. 25.)

0.42 in.$^{-1}$. As Figure 54a illustrates, the calculated cooling curve does not show the rapid decrease in temperature that occurs in the nucleate boiling stage and intersects the actual cooling curve at temperature T_H (temperature halfway between austenitizing temperature and bath temperature).

Another limitation of H values is that sample size is not accounted for when the H value is determined from core cooling curves. Cooling curves measured in the core of steel cylinders do not reflect the heat flux at the surface due to the damping effect of thermal conductivity (λ).

Figure 55 illustrates that the temperature difference between surface and core increases with increasing bar diameter [9]. For example, after 40 s, when the temperature within the cylinder of smallest cross section has dropped below 180°C (356°F) and the convection stage begins, the surface temperature of the 100 mm diameter cylinder has only cooled to approximately 380°C (716°F), and the temperature in its core is 700°C (1292°F). This shows that the heat transfer conditions at the surface differ substantially from those at the core. Depending on sample dimensions and the cooling conditions, heat extraction from the surface can occur during the film boiling, nucleate boiling, or convection stage. Therefore, application of H values determined from a relatively small cross section (a diameter of 12.7 mm is often used) to larger cross sections is fundamentally erroneous.

For hardness predictions, reference H values are typically taken from a reference table (see Table 3), accounting for the particular quenchant bath temperature and agitation. (Another fundamental limitation of this approach is that the reference data do not adequately quantify either flow rates or turbulence.) Temperature–time cycles at different locations over the cross section are calculated from the solution of the heat transfer equation (Equation 2) using the determined H value. The calculation of hardness is based on the supposition that specimens that were made of the same steel grade have followed the same law of cooling if they have the same hardness. That means that if a point on the cross section of a quenched part has the same hardness as the point shown, e.g., at a given distance from the quenched end of a Jominy specimen made from the steel being quenched, both points have undergone identi-

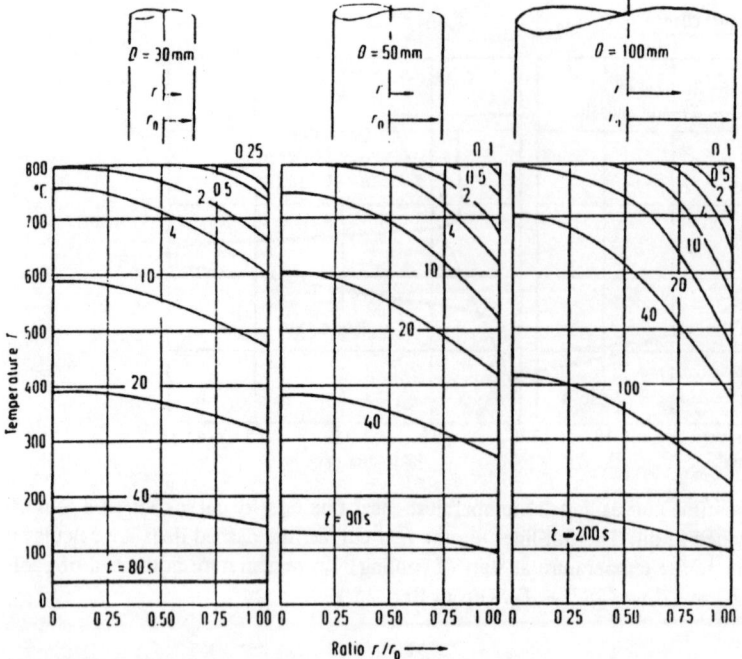

Figure 55 Calculated temperature distributions at selected times as a function of the radius ratio for 30, 50, and 100 mm diameter cylinders (34Cr4 steel) quenched in water at 20°C. (From Ref. 4.)

cal cooling processes. Therefore, a calculated cooling curve at a point on the cross section of the quenched part is related to an equivalent Jominy distance whose cooling curve shows the identical half-temperature time t_H (time until T_H is reached, as shown in Figure 54a). The corresponding hardness of the Jominy hardenability curve yields the hardness at the cross-sectional location in question. Of course, this supposition is not necessarily correct and can be taken only as an approximation, because the shape of the cooling curve obtained in the actual quenching process and that achieved in the end-quench test may differ considerably (see Figure 53)

Grossmann [62] or Lamont [63] diagrams interrelate the H value, the bar diameter, and the hardenability of the steel and have been used to calculate hardness. The Lamont diagram shown in Figure 56 is applicable for the center of round bars. Lamont developed these diagrams for different radius ratios in the range $0 \leq r/R \leq 1$, in steps of 0.1. In this diagram H values are plotted as follows. If a 25 mm bar is quenched in a "strongly" agitated oil quench, which is assigned an H value of ≈ 0.7 in.$^{-1}$, the value of 6.35 mm on the Jominy hardenability specimen is obtained by reading the diagram across the 25 mm horizontal line to the line of $H = 0.7$ in.$^{-1}$ and then down as shown. This means that the core of the bar will have the indicated hardness at 6.35 mm from the quenched end of the Jominy hardenability specimen made of the steel being quenched.

C. The QTA Method

As opposed to the Grossmann approach, the QTA method developed by Wünning [61,64,65] does account for the three stages of cooling—film boiling, nucleate boiling, and convective heat transfer—which are described by the three model parameters Q, T, and A.

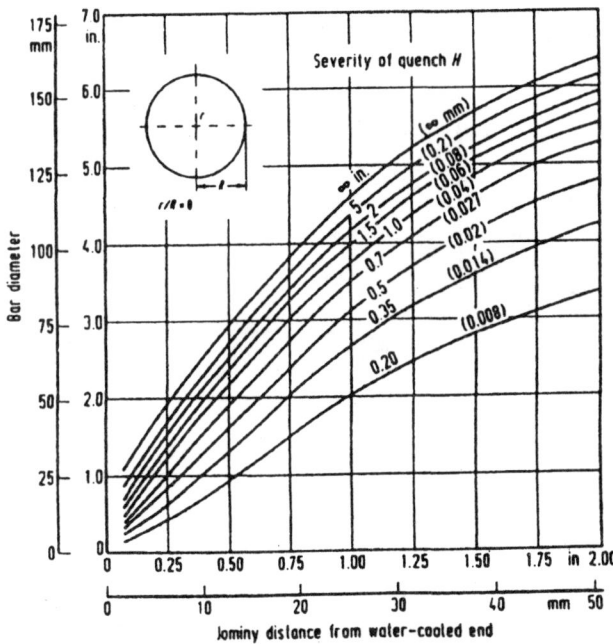

Figure 56 Lamont diagram for the center of round bars: the relation between bar diameter and distance from the quenched end of a Jominy probe for different H values. (From Ref. 26.)

Q is the average heat flux density on the surface of the quenched part until it cools to the surface temperature at which the vapor envelope collapses, i.e., the wetting temperature T_L. Since T_L values are generally unavailable (the wetting temperature is influenced by many factors, some of which cannot be exactly quantified), Wünning recommends the use of a constant T_L value of 500°C (932°F). To determine the value of Q for each particular quenching process, a standard steel probe with known hardenability is used.

T is the temperature at which the extrapolated line of the heat flux density in the nucleate boiling stage intersects the zero temperature axis as shown in Figure 57. This temperature approximately corresponds to the boiling temperature of the quenchant and therefore does not depend on the cooling conditions ($T \approx T_b$).

A is the average heat transfer coefficient of the convection phase and can be calculated using known heat transfer relationships.

Surface heat transfer described by the QTA model is illustrated in Figure 57. During the film boiling stage, the temperature variations are calculated using the heat flux density Q. The parameter Q depends on the properties of the fluid (heat conductivity, viscosity, etc.), the subcooling and flow conditions, and the geometry of the part. The assumption of constant average heat flux density is valid because the cooling course in the film boiling stage is usually linear. At the wetting temperature T_L [500°C (932°F)], transition from the film boiling to the nucleate boiling occurs.

The heat transfer coefficient increases by up to two orders of magnitude so that the heat extraction from the surface can be calculated using the first boundary condition assuming that the surface temperature drops immediately to temperature T. To dampen this infinitely high

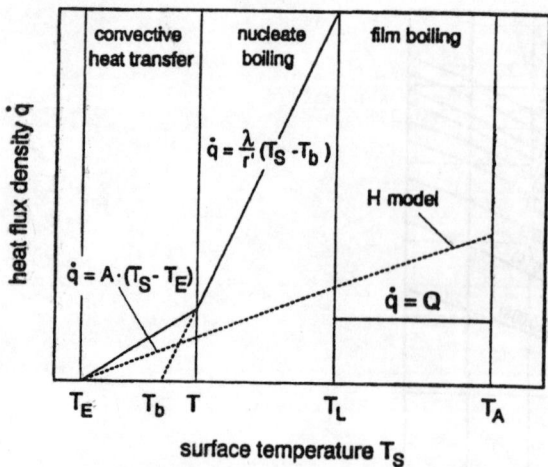

Figure 57   Parameters of the QTA model.

heat transfer rate, the temperature T is arbitrarily assigned the distance r' on the sample surface. According to Wünning, r' is 1 mm for quenching in water and 2 mm for quenching in oil.

The transition from nucleate boiling to convective cooling occurs when the heat flux densities become equal. This temperature (T') is approximately the boiling temperature of the quenchant, $T_b = T$.

The value A depends on the agitation rate and on the dimensions of the workpieces. Although A values can be calculated from the physical conditions of the quenchant and the T value can be determined from a laboratory test, the Q value must be measured under workshop conditions. Wünning claims that only one hardness measurement is necessary to define the Q value of a steel part quenched under known conditions, provided that a calibration curve is available describing the relationship between different Q values and the resulting as-quenched hardness. The calibration curve is determined using a standard "Q-probe" made of low-alloy steel (Wünning suggests DIN 46Cr2 steel) with 25–30 mm diameter and a length of 100 mm. Using different quenching conditions, the temperature–time cooling curve at radius ratio $r/R = 0.7$ and the corresponding hardness value are measured. Calculations have shown that for most applications the average cooling curve follows the cooling curve at $r = 0.7R$. The Q value is then inversely proportional to the cooling time down to 500°C (932°F) (t 500°C) and depends on the thermodynamic properties of the material and the volume/surface relation:

$$Q = \rho C_p \frac{V}{A} \left(\frac{T_A - T_{500°C}}{t_{500°C}} \right) \tag{11}$$

The Q value for a batch of workpieces under particular quenching conditions can be determined by measuring the hardness at the point $0.7R$ of the Q-probe quenched in exactly the same manner. Figure 58 illustrates the calibration curve for the Q-probe and interrelates the Q values to the appropriate Jominy hardenability curve. If, after quenching, a hardness of 44 HRC is measured at point $0.7R$, Q yields a value of 1.05 MW/m² while the same hardness is found at a distance of 6 mm from the water-cooled end of the Jominy sample.

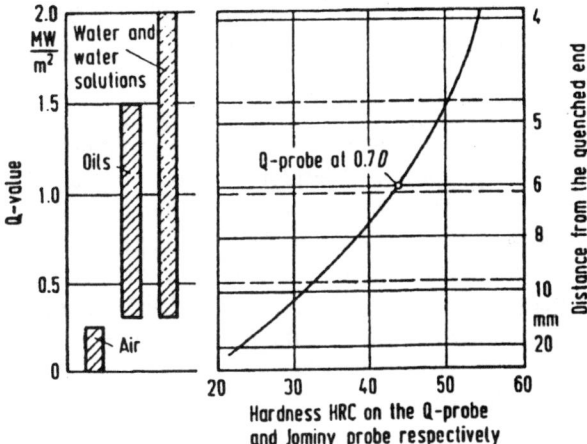

Figure 58 Calibration curve for the Q-probe, relation between Q value at $0.7R$ and hardness on the Q-probe. The hardness is also related to the distance of the quenched end of the Jominy specimen. (From Ref. 27.)

Figure 59 validates that the rapid decline of temperature during the nucleate boiling stage can be modeled by calculations based on the QTA model but that cooling to 500°C (932°F) is less accurate relative to the actual cooling curve. A more precise calculation can be obtained using the actual value of wetting temperature. According to Lübben et al. [66], the QTA model is limited by the assumption that Q values and hardness strongly depend on the boiling temperature of the fluid. The use of a single Q-probe made of a particular steel is not sufficient for establishing the calibration curve because the hardness at point $0.7R$ of the standard Q-probe (steel 46Cr2) is occasionally insensitive to variations of Q, especially for hot oils. However, the QTA model can still produce reasonably accurate calculations of cooling curves at

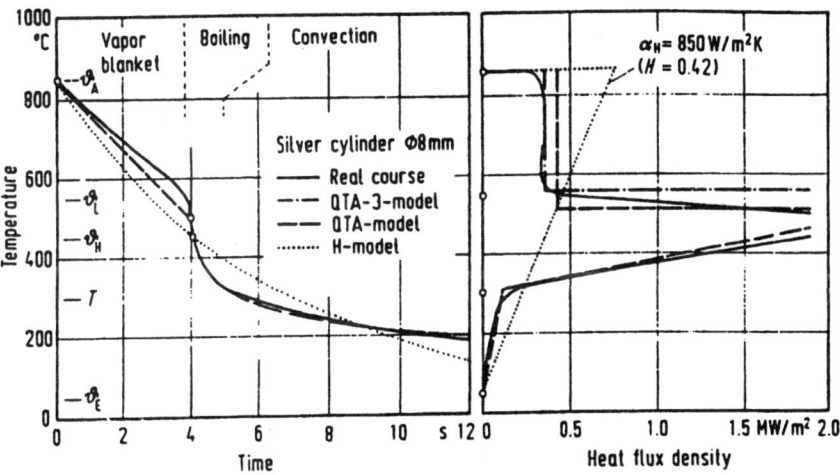

Figure 59 (a) Temperature–time curves and (b) temperature–heat flux density curves for a 9 mm diameter silver cylinder quenched in oil. Solid lines are the real curves; the dashed lines were calculated using a constant H value, the QTA model, and the QTA-3 model. (From Ref. 25.)

different points of a workpiece taking into account boiling effects. Thus the results can be used as input data in property prediction methods such as those described next.

D. Correlation Between Hardness and Wetting Kinematics

One limitation of property prediction methods is that no correlations between wetting kinematics and metallurgical properties have been performed. Therefore, correlation diagrams between surface hardness and wetting time have recently been developed with various steels. Figure 60 shows that a wetting time of more than 25 s was observed for a 1040 steel quenched in 30°C (86°F) water flowing at 0.3 m/s. The vapor envelope first collapses at the bottom of the cylinder, and the wetting time increases toward the cylinder top. The hardness values measured along the cylinder length decrease in accordance with the progression of the wetting front. From this plot, a correlation between surface hardness and wettnig time can be derived that may be used as a calibration curve for hardness prediction. The prediction method itself is based on the fact that the heat extraction on the surface of a quenched part varies with the position of the wetting front but varies only slightly with the quenching conditions in identical quenching liquids. Therefore, the same hardness can be expected at points showing the same wetting time [15,22,57,67].

The correlations between surface hardness and wetting time shown in Figure 61 were developed from a series of calibration tests in water with five different steel cylindrical specimens (100 mm long and 25 mm in diameter) [12]. The hardness increased with carbon concentration and decreased with decreasing concentration of other alloying elements. The higher "hardenability" of the 1045 steel compared to the 1040 steel (nominally both steels have the same concentration of carbon and alloying elements) depends on deviations in the chemical composition and the higher austenitizing temperature (880°C (1616°F) and 850°C (1562°F), respectively). This result shows that the relationship between surface hardness and wetting time is valid only for the initial state of the material, metallurgically defined by alloy composition

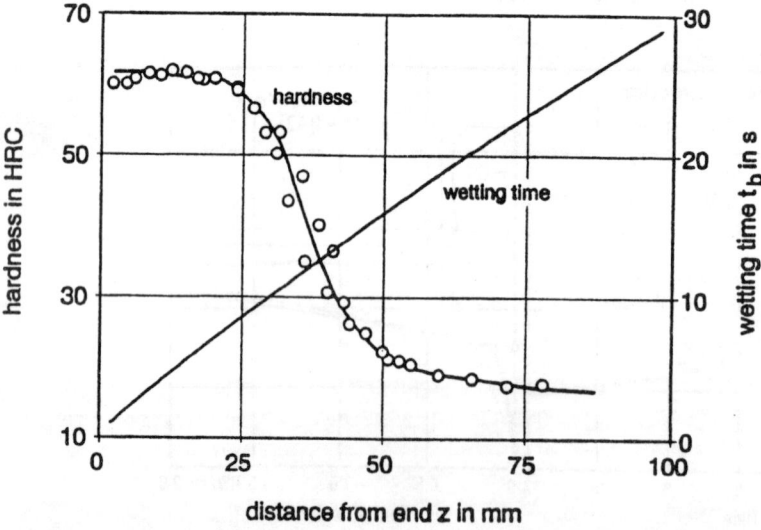

Figure 60 Surface hardness (HV) and wetting time t_b as a function of the distance from the lower end of a 1040 steel bar of 25 mm diameter and 100 mm length quenched from 850°C in 30°C water flowing at 0.5 m/s.

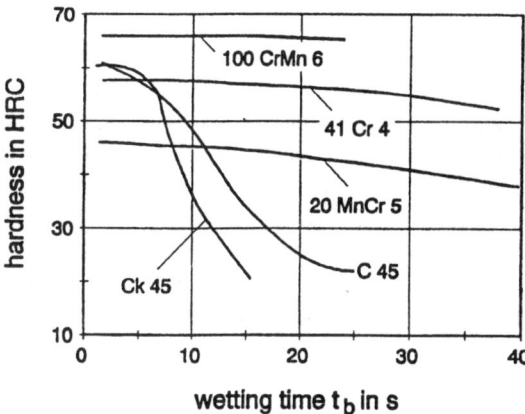

Figure 61 Calibration curves for selected steels. Relation between surface hardness and wetting time for quenching in water and an initial state of the material defined by the steel composition and the austenitizing conditions. (From Ref. 11.)

and the austenitizing temperature. Similar calibration curves must be developed for other quenchants.

Calibration curves permit reliable prediction of surface hardness obtained with varying wetting conditions such as bath temperature and agitation rate. Two procedures have been developed for obtaining wetting times, one by recording cooling curves at different heights of a cylinder with multiple thermocouples as described previously in Section V.B.2 and the other by measuring the change of conductance between quenched workpiece and a counter electrode, described previously in Section V.B.1. The hardness at a certain point of the workpiece surface is then derived from the calibration curve and wetting time. Interestingly, no effect on section size was found for the calibration curves when 15, 25, and 40 mm diameter probes were quenched in water [12]. These calibration curves are used to correlate the critical wetting time with hardness. A 100% martensitic microstructure can be achieved only when the time t_f at the conclusion of wetting is less that the upper critical cooling time where martensite is formed.

For hardness determination across the cylinder cross section, a method is suggested that is based on the Grossmann H value and the Jominy hardenability curve for the steel being quenched [15,57,67]. However, the H values are not taken from a table but are determined from the Lamont diagram of the surface $r/R = 1$. The equivalent distance from the quenched end of the Jominy sample is determined from the surface hardness of the cylindrical section desired, which is obtained from the calibration and wetting time. Then, from the Lamont diagram for the surface, the intersection of the horizontal line of the bar diameter and the line of the determined Jominy distance provides the H value. This H value is then used to calculate the Jominy distances corresponding to different radius ratios r/R of the quenched sample. From the Jominy hardenability curve, each Jominy distance is assigned to a specific hardness value. If these hardness values are transferred to the immersion-cooled cylinder, the radial hardness profile is determined.

If the three determined phases of cooling vapor blanket, nucleate boiling, and convective heat transfer are simultaneously present on the sample surface, large variations in axial cooling occur, which will produce axial hardness gradients as shown in Figure 60. In this case, different hardnesses along the cylinder length produce different H values for each distance z from the lower cylinder end face.

E. Computer-Based Calculation of Hardness Profile

Finite-element modeling has been used to compute the transient temperature fields in a quenched part and to predict microstructure and properties such as hardness or residual stresses. In this method, the temperature and corresponding phase transformation are predicted, and in the second step the corresponding mechanical properties are calculated. These microstructural calculations assume that the transformation of austenite during continuous cooling is predicted by dividing a cooling curve into discrete temperature–time increments and comparing the amount of time at each horizontal step with the transformations that occur at that temperature in the isothermal TTT diagram [68–71] (Figure 62). The transfer from isothermal to continuous cooling conditions at a certain temperature is carried out using the Avrami equation

$$w_{\text{flp}/b} = 1 - \exp\left[- b(t/t_0)^n\right] \tag{12}$$

where $w_{\text{f/p/b}}$ is the volume fraction of austenite transformed into ferrite/pearlite or bainite, t is the transformation time, t_0 is the reference time, b is the coefficient of the austenite transformation kinetics, and n is the exponent of the austenite transformation kinetics. The values of

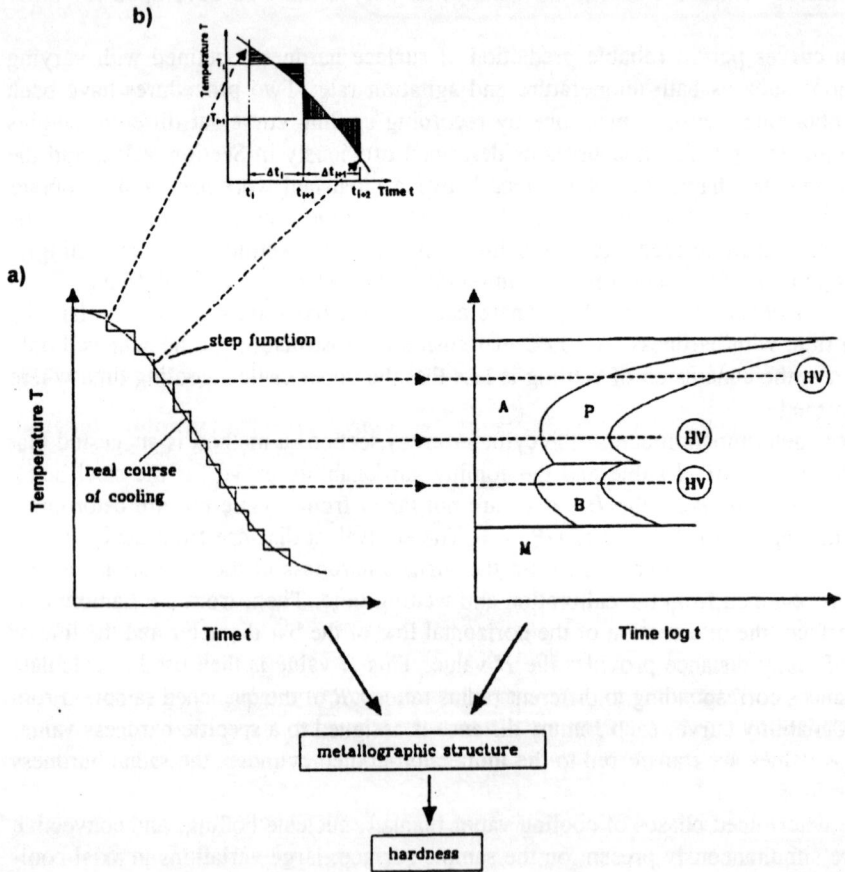

Figure 62 Prediction of microstructure and mechanical properties by dividing a cooling curve into a step function and comparing the horizontal steps to the transformations in the isothermal TTT diagram. (Schematic.)

b and n depend on the transformation temperature and are fitted for each individual steel grade from the transformation curves plotted in the isothermal TTT diagram.

The microstructural computation sequence is illustrated in Figure 63. Austenite transformation begins when the actual time step represents the line for 1% volume fraction transformed. If at temperature T_i the volume fraction of transformed austenite is w_i, then according to Equation (11) the fictitious time $t_{i+1,\text{fict}}$ is calculated where the same volume fraction would transform at temperature T_{i+1}. This time is added to the time interval Δt_{i+1} and then the total transformed volume fraction of austenite w_{i+1} at temperature T_{i+1} is calculated. The new transformed volume fraction is $w_{i+1} - w_i$.

If in a known temperature range the austenite transforms into two microstructures, e.g., into microstructures a and b, then it is assumed that microstructure a ceases to grow when microstructure b has increased to 1%. If correct data are available, an overlapping transformation can likewise be represented.

The athermic transformation of austenite into martensite below the martensite start temperature M_s is described by the equations

$$w_M = 1 - \exp\left[c(M_s - T)^m\right] \tag{13}$$

or

$$w_M = 1 - \left(\frac{T - M_f}{M_s - M_f}\right)^m \tag{14}$$

where w_m is the volume fraction of austenite transformed into martensite, M_s is the martensite start temperature, M_f is the martensite finish temperature, T is the transformation temperature, c is the coefficient of austenite transformation, and m is the exponent of austenite transformation.

The final hardness after quenching can be determined by applying an additivity rule provided that the hardness values of the different microstructures formed at the individual temperatures are known [72]:

$$HV = \Sigma\{\Sigma[\Delta w_k(T_i)\, HV_k(T_i)]\} \tag{15}$$

where HV is the final hardness at a defined location, w_k is the volume fraction of microstructure k formed at temperature T_i and HV_k is the microhardness of microstructure k formed at temperature T_i.

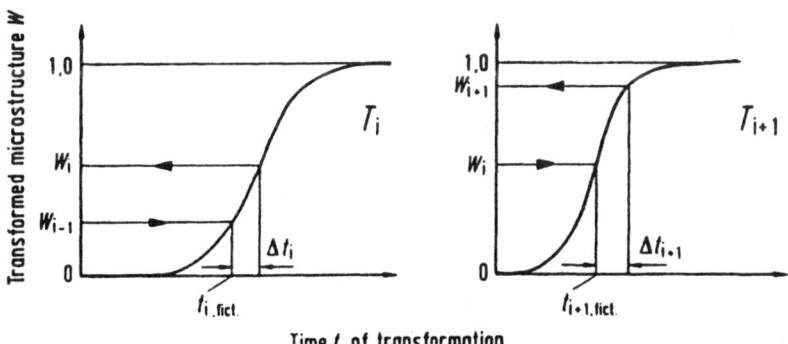

Figure 63 Sequence of the microstructural computation for iteration steps i and $i + 1$. (From Ref. 28.)

This method can be used to predict the transformation behavior of austenite during continuous cooling and then to calculate the mechanical properties of the final product based on the predicted microstructure. The effect of latent heat evolution on the temperature–time cycle and the transformation kinetics can be taken into account. But this method is limited by the isothermal TTT diagram, which usually fails to give reliable information on the transformation behavior of an actual quenched part because of deviations in the chemical composition within the standard range, segregations within the workpiece, and variations in the initial microstructures prior to austenitizing. In addition, the effects of transformation plasticity are not quantified. Another limitation is that Equation (11) is based only on empirical data for the transformation behavior. Theoretical approaches to nucleation and nuclear growth are more suitable and provide a greater range of application. Finally, it is questionable whether the application of isothermal transformation kinetics of austenite to a continuous cooling curve that is divided into a step function is even permissible.

IX. QUENCHANTS

From the standpoint of availability, perhaps the most common quenchants are air and water. Air is one of the slowest quenching media and controls heat transfer by a vapor blanket cooling mechanism. As with the liquid quenchants described earlier, cooling rates obtained with air may be accelerated by increasing the flow rate by the hot metal surface as shown in Figure 64 [73]. The cooling rates obtained with air, however, are usually insufficient to harden medium- and high-hardenability steels. For these steel grades, faster cooling rates are required.

Cooling rates intermediate between those of atmospheric air and liquid quench media such as molten salt, oil, and water are obtained by using gases such as hydrogen, helium, nitrogen, and argon under pressure (see Section VII.C). Convective heat transfer properties vary with the particular gas or gas mixture and can be predicted using Equation (16) and known values of a reference gas under the same conditions [74].

$$\alpha_g = \alpha_{Ref} \left[\frac{\lambda_g}{\gamma_{Ref}} \right]^{0.67} \left[\frac{\lambda_g}{\gamma_{Ref}} \right]^{0.70} \left[\frac{\eta_g}{\eta_{Ref}} \right]^{-0.37} \left[\frac{C_{pg}}{C_{pRef}} \right]^{0.33} \qquad (16)$$

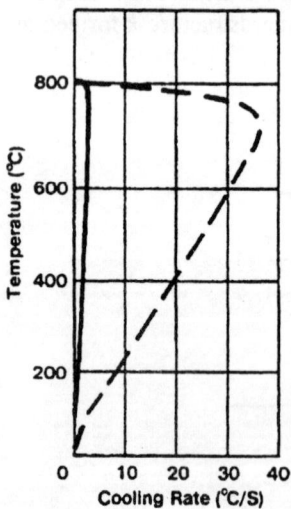

Figure 64   Effect of flow rate on air quenching. (——) Still air; (----) compressed air.

where α_g is the heat transfer coefficient of the gas being calculated, α_{ref} is the heat transfer coefficient of the reference gas, λ_g is the thermal conductivity of the gas, λ_{ref} is the thermal conductivity of the reference gas, η_g is the viscosity of the gas, η_{ref} is the viscosity of the reference gas, γ_g is the density of the gas, and γ_{ref} is the density of the reference gas.

Water exhibits some of the fastest cooling rates. Quenching properties of water as a function of water temperature were discussed previously and are summarized in Figure 28. Water, although usually readily available and relatively inexpensive, frequently produces high thermal gradients, resulting in high residual stresses or distortion due to the nonuniform quenching behavior described in Section IV and illustrated in Figures 9a and 9c or unacceptable transformational stresses that are due to excessive cooling rates, particularly near the M_s temperature [75, pp. 293–304].

Quenchants that exhibit cooling rates intermediate between those of air cooling and lower pressure (6 bar) gas quenching and water are very important, especially for quenching medium- and high-carbon steels. Such quenchants include oil, aqueous polymer solutions, molten salt, and fluidized beds. In some heat treatment applications, cooling rates significantly greater than that of water are required. In these situations, aqueous salts (brine) are used. The remainder of this discussion will focus on these quenchants. This discussion will provide a description of these quenchants, typical cooling rates, and quenchant bath maintenance procedures.

A. Oil Quenchants

1. *Oil Chemistry*

Most quenching oils, with the exception of some recently developed products that are based on synthetic fluids such as poly(α-olefins) and more recently developed quenchants based on vegetable oil and polyol esters are derived from petroleum base stock. Petroleum base stock is composed of hundreds of components, and the composition of the oil will vary, often dramatically, over time for oil from a single well and between different oil fields [75, pp. 129–159]. This is critically important because oil composition will significantly affect a number of performance properties including viscosity–temperature behavior, thermal stability, ash and stain formation, and wetting properties [75, pp. 129–159]. The impact of varying oil composition on quenching behavior is illustrated in Figure 65 [76].

The interfacial cooling process during oil quenching was illustrated earlier in Figure 9b. One of the properties that affect heat transfer from the hot metal to the cooler oil is the viscosity of the oil at the metal/oil interface throughout the quenching process. Typically, heat transfer decreases exponentially with increasing viscosity. Therefore, the viscosity of the oil as a function of temperature is critically important, and minimal viscosities at higher temperatures will facilitate heat transfer during the early stages of the quench. However, upon cooling, it is desirable for the oil to exhibit higher viscosity as the M_s transformation temperature is approached to minimize transformational stresses upon martensite formation. Therefore, the viscosity–temperature relationship of the oil is a fundamentally important physical property of the oil, and this behavior will be affected by the chemical composition.

The boiling point of the oil will also affect heat transfer rates during quenching. It is desirable for the oil to exhibit nucleate boiling, the fastest boiling mechanism, at higher temperatures to minimize the potential for the formation of undesirable transformation products. However, nucleate boiling should be minimized during martensite transformation to minimize transformation stresses. Oil compositions with components with higher boiling points will produce the desired effect.

Finally, to enhance the uniformity of the wetting process, it is desirable for the oil to wet the surface of the metal. This is also affected by the chemical composition of the oil. Wetting

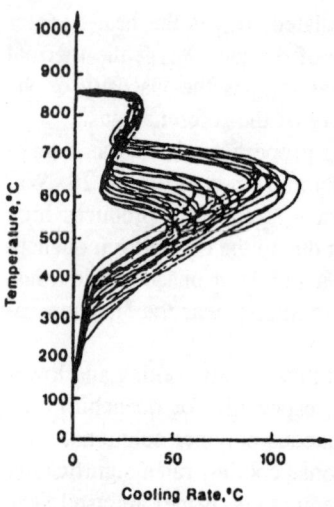

Figure 65 Illustration of cooling rate variability attainable with various quench oils. Each curve is for a different oil.

agents are actually added to some oils (accelerated or fast oils) to enhance the wetting process and reduce the duration of the relatively slow vapor blanket cooling process.

The chemical components typically present in petroleum oil include aliphatic, naphthenic, and aromatic derivatives and heterocyclic derivatives of nitrogen, oxygen, and sulfur. The amounts of these components will vary with the source of the oil and, as described above, affect quenching performance. Figure 66 shows the effect of paraffinic and naphthenic relative to aromatic hydrocarbon composition on cooling rate behavior [77]. The aromatic hydrocarbon content of the oils increases in the order MZM-16 < I-20A < I-20AR1 < I-20ARI2. The composition of the oils along with the viscosity and the contact angle, which is a quantitative measure of the ability of the oil to wet the surface of steel, are summarized in Table 5.

Increasing oil viscosity produces an exponential decrease in the maximum cooling rate as shown in Figure 67 [78]. Unfortunately, the viscosity–temperature relationships of the oils used to generate this curve were not reported. Figure 67 similarly illustrates the increasing wettability, as reflected by contact angle measurements, with increasing cooling rates [74,78].

2. Additive Effects—Accelerated Quench Oils

In some steel hardening applications, such as low-hardenability steels, it is critically important to cool the steel as fast as possible until the M_s temperature is approached. Since most quench oils exhibit substantial vapor blanket cooling properties, it is often difficult if not impossible to adequately harden steels with the oils whose quenching curves are depicted in Figure 68 [79]. This problem has been resolved by the addition of rate-accelerating additives such as rosins, sulfonates, and other surface-active compositions. However, it is important to note that "rate acceleration" actually refers to cooling times, not necessarily cooling rates in the convective region of the quench oil where martensitic transformation typically occurs. The so-called rate acceleration process occurs by facilitating the rupture of the vapor blanket, cooling by using additives to provide nucleation sites to initiate faster nucleate boiling processes, or enhancing the wetting of the steel by increasing the contact angle as shown in Figure 69 [78].

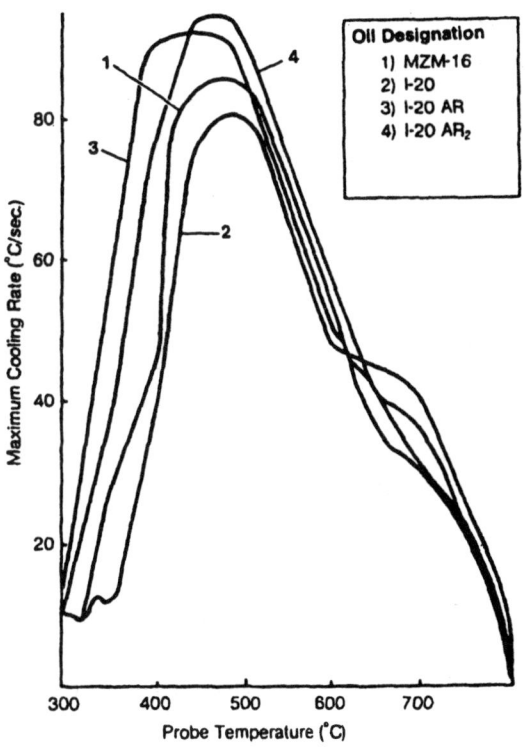

Figure 66 Effect of quench oil composition on cooling rates.

Additive performance in quench oils is concentration-dependent as shown in Table 6 [80]. This is important because additives operate by an adsorptive mechanism on the steel surface, whether they provide increased nucleation sites to enhance boiling or increase surface wetting. Quenching rates will decrease with use due to additive drag-out as shown in Figure 70

Table 5 Impact of Quench Oil Composition on Physical Properties

| | Oil I.D. No. | | | |
Property	MZM-16	I-20A	I-20AR1	I-20AR2
Hydrocarbon composition, %				
Paraffin-naphthene	73.70	71.48	67.04	62.25
Aromatic groups				
I	11.30	11.96	11.60	16.00
II	6.70	7.80	3.16	10.58
III	3.80	4.00	2.80	4.20
IV	3.50	2.00	3.84	1.40
Resins	0.50	2.10	4.00	4.10
Losses	0.40	0.66	2.96	1.48
Viscosity at 50°C, m²/s	17.5	18.3	24.8	25.6
Oxidation test, wt % decrease	1.54	1.8	2.2	3.0

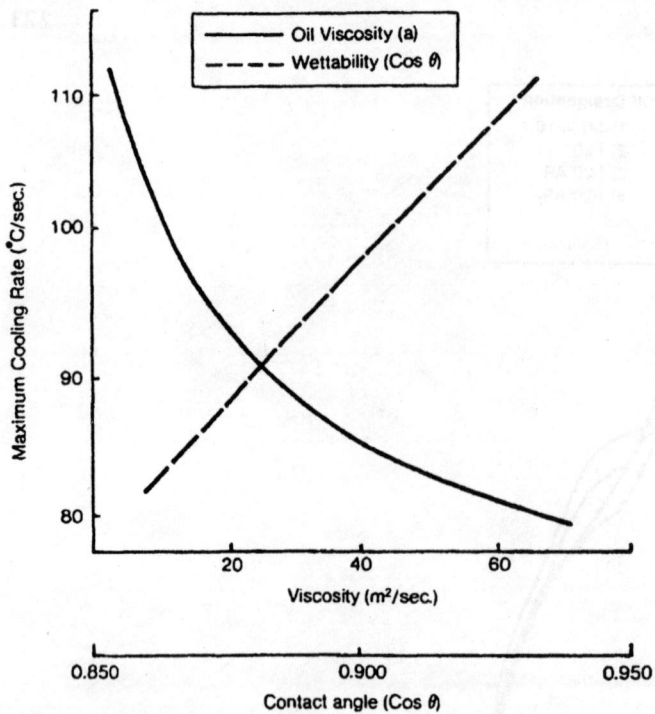

Figure 67 Relationship of viscosity and contact angle (wettability) on the maximum cooling rate of different quench oils.

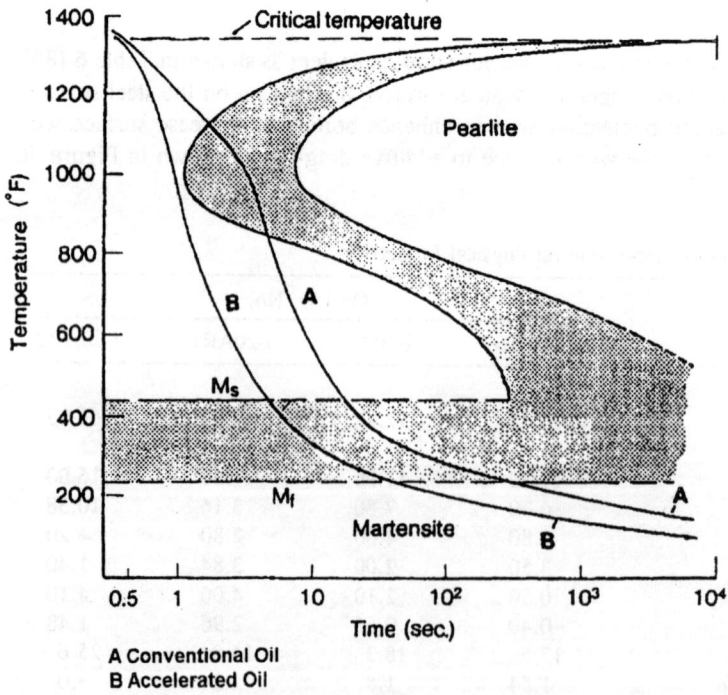

A Conventional Oil
B Accelerated Oil

Figure 68 Effect of quench oil additives on vapor blanket cooling and steel transformation behavior.

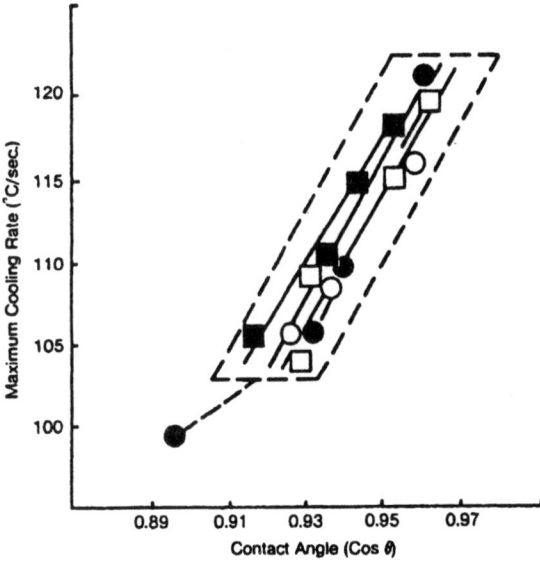

Figure 69 Effect of quench oil wettability on the maximum cooling rate.

[81]. Therefore, it is important to monitor the cooling rate performance of accelerated quench oils during use to ensure consistency in quenching performance.

3. Degradation and Contamination

Additive drag-out is not the only long-term concern that accompanies the use of quench oils. One of the most significant problems with any organic quenching medium, including petroleum oil, is degradation. Typically, gradual rate acceleration with increasing time in use is observed for conventional (non-rate-accelerated) quench oils as shown in Figure 71 [82]. Oil

Table 6 Effect of Addition of Sodium Sulfonate on the Properties of a Conventional Quench Oil[a]

Cooling property	Sodium sulfonate (wt %)		
	0%	1.5%	3.0%
Vapor blanket cooling, s	36.5	21.5	16.5
Average temperature after vapor blanket cooling, °C	600	650	650
Mean nucleate boiling rate, °C/s	10.4	14.2	20.5
Maximum surface heat transfer coefficient, W/(m²·K)	921	1779	2592
Temperature at maximum surface heat transfer, °C	476	425	400
Viscosity, cSt at 40°C)	32	33	34
Surface tension, N/m	0.0388	0.0334	0.0326

[a]Still-quench properties. Cooling rates were determined using a 120 × 120 × 20 mm (5 × 5 × 0.8 in.) stainless steel plate (0.29Si-1.16Mn-17.43Cr-2.29Mo-10.56Ni).

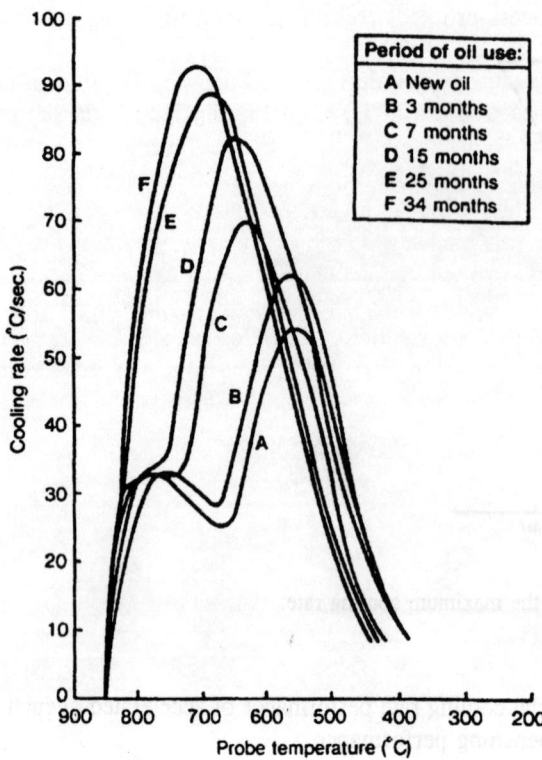

Figure 70 Cooling rate variation with time for an accelerated quench oil.

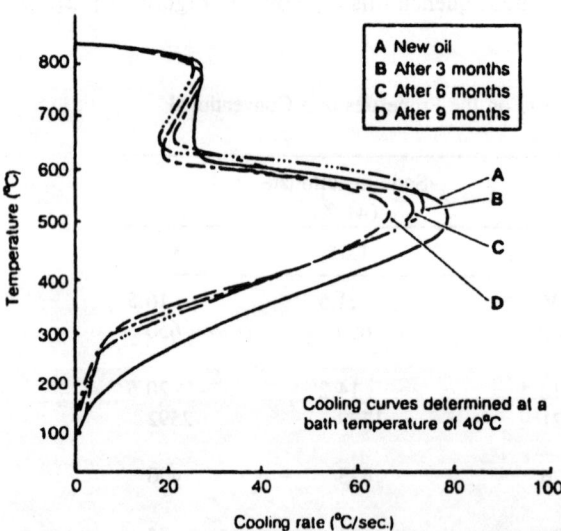

Figure 71 Cooling rate variation with time for a nonaccelerated quench oil.

viscosity decreases as the degradation process proceeds, resulting in corresponding cooling rate increases.

Oil degradation may increase the potential fire risk during use. For safe operation, it is recommended that the quench bath be at least 50°F (28°C) below the flash point of the oil [83, p. 169].

Cooling rates of both conventional and accelerated quench oils may be dramatically affected by contamination. One of the most common contaminants encountered in the heat treatment shop is water, which may arise from heat exchanger leakage and other sources. Water typically increases cooling rates for conventional oils and decreases cooling rates for accelerated quench oils as shown in Figure 72 [81]. This is due to the water insolubility of the rate-accelerating additive, which therefore separates from solution. In addition to these cooling rate effects, the presence of water may lead to increased staining [84], uneven hardness, and soft spots on the workpiece [85]. Most important, the presence of water, particularly at concen-

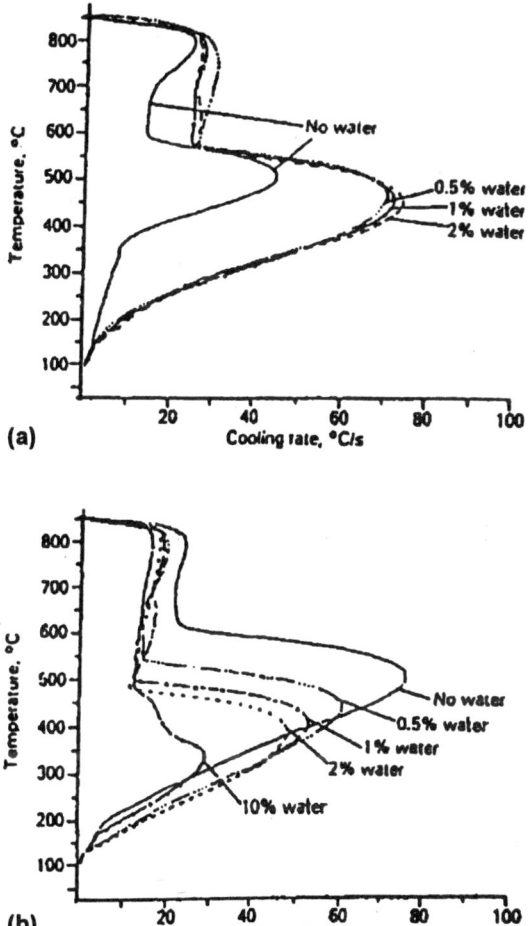

Figure 72 Effect of water contamination on the cooling curve behavior of (a) a conventional and (b) an accelerated quench oil.

trations above approximately 0.05%, may lead to foaming and, potentially, fire or explosion [83, pp. 44–45].

4. Quench Oil Characterization

Cooling curve analysis alone is insufficient to provide the needed characterization of a quench oil during use. Chemical and physical property characterization are required to explain cooling curve variation if observed. It may be recommended that chemical and physical property characterization precede cooling curve analysis. The most commonly recommended quench oil tests and methods are summarized in Table 7 [85].

B. Polymer Quenchants

1. Polymer Quenching Mechanism

Although quench oils are among the most common vaporizable liquid quenchants, they exhibit a number of significant disadvantages including limited process variability due to an inability to produce significant cooling rate variation with a single oil, fire hazards, smoke emissions, and disposal problems [86]. These disadvantages have led to progressive growth in the use of aqueous polymer solutions as quenchants. In this section, an overview of the most common polymer quenchants in use today, their performance, and bath maintenance procedures will be provided.

Aqueous polymer quenchants are typically dilute aqueous solutions of a polymer and additives, primarily corrosion inhibitors. Corrosion inhibitors include sodium nitrite, mixtures of aromatic carboxylic acids, and amine salts of various aliphatic carboxylic acids.

It is necessary for the polymer film to surround the cooling metal as uniformly as possible. The formation of a uniform polymer film is favored by optimizing the uniformity of agitation of the quenchant solution in the quenching zone and by ensuring that sufficient quantities of the quenchant are available to provide the required film around the workpiece.

2. Polymer Quenching Process Variables

Examples of polymer quenching processes are shown in Section IV in Figure 10. This figure shows that heat transfer during quenching is mediated by the formation and stability of the polymer "film" that initially surrounds the hot metal upon immersion, its subsequent breakage (or rupture), and finally the timing of the redissolution of the polymer into the surrounding aqueous solution. To optimize the uniformity of this process and minimize unacceptable

Table 7 Recommended Test Methods for Quench Oils

Property	ASTM method
Viscosity	D445, D92, D93
Flash point	D1310
Fire point	D92, D95, D1744
Water content	D4007
Neutralization number	D974
Saponification number	D94
Ash	D482
Conradson carbon residue	D189
Precipitation number or sludge	D91, D2773
Specific gravity	D287

thermal and transformational gradients, it is important that the initial polymer deposition and subsequent film formation be as uniform as possible.

Heat transfer rates will be determined by both the thickness of the polymer film and the viscosity of the hydrated polymer film. Slower cooling rates will be favored by increasing polymer molecular weight (polymer viscosity exponentially increases with molecular weight) and the concentration of the polymer in solution. Increasing concentration increases film thickness. Therefore, polymer concentration is an important process variable.

Figure 29b (see Section VII.A) shows that cooling times decreases (cooling rates increase) with increasing agitation. This is due to increased destabilization of the polymer film around the hot metal part as shown in Figure 10. This process variable may be readily applied, occasionally in combination with a "time quenching" process, to provide widely varying quench severities using a single quench bath. In some cases, computer-controlled agitation rates are varied throughout the quenching process [87].

A third process variable that can be employed is bath temperature. Bath temperature effects, along with agitation and concentration, are illustrated in the Grossmann H factor plots in Figure 73 [75, p. 180]. Typically, increasing the bath temperature decreases quench severity.

The effect of the three process variables concentration, agitation, and bath temperature are summarized in Table 1. The impact of these variables on quench severity is dependent on the polymer used to formulate the quenchant. This is illustrated in Figures 73a and 73b, where it is observed that the quench severity response to concentration, agitation, and bath temperature is dramatically different for the two polymer quenchants shown [75, p. 180]. This also illustrates the importance of controlling these parameters throughout the quench zone both to obtain reproducibility from quench to quench and to promote uniformity during each quench.

3. Quenchant Polymer Drag-Out

As with oils, the quenchant may be removed from the tank by a drag-out process. The amount of drag-out is dependent on a number of factors, including the polymer type, degree of agitation, and concentration. (Increasing aqueous polymer concentration results in increased viscosity.) These effects are illustrated in Figures 74a and 74b [88]. The drag-out values shown in Figure 74a are absolute values. It would be of ever greater value to show the drag-out as a relative loss with respect to the amount of polymer remaining, which, as described earlier, is necessary to provide uniform part coverage during the quench. It would be expected that higher molecular weight polymers present in solution at lower concentrations would exhibit even higher relative drag-out (as percent drag-out) than the values shown. Unfortunately, these data were not provided [88].

4. Mechanodegradation

In order to provide a relatively stable quenchant medium, it is important that any variation of the solution viscosity, at a particular concentration, be minimized. One potential source of viscosity instability is "mechanodegradation," where the mechanical energy introduced by agitating the quenchant would be sufficient to cause polymer chain scission, resulting in a viscosity decrease. The effect of polymer structure on mechanodegradation using a Waring blender shear test is illustrated in Table 8 [75, pp. 182–184].

5. Quenchant Polymers

The most common aqueous polymer quenchants in use today [89] are derived from poly-(alkylene glycol) copolymers [90], poly(vinylpyrrolidone) [91], poly(sodium acrylate) [92], and poly(ethyl oxazoline) [93].

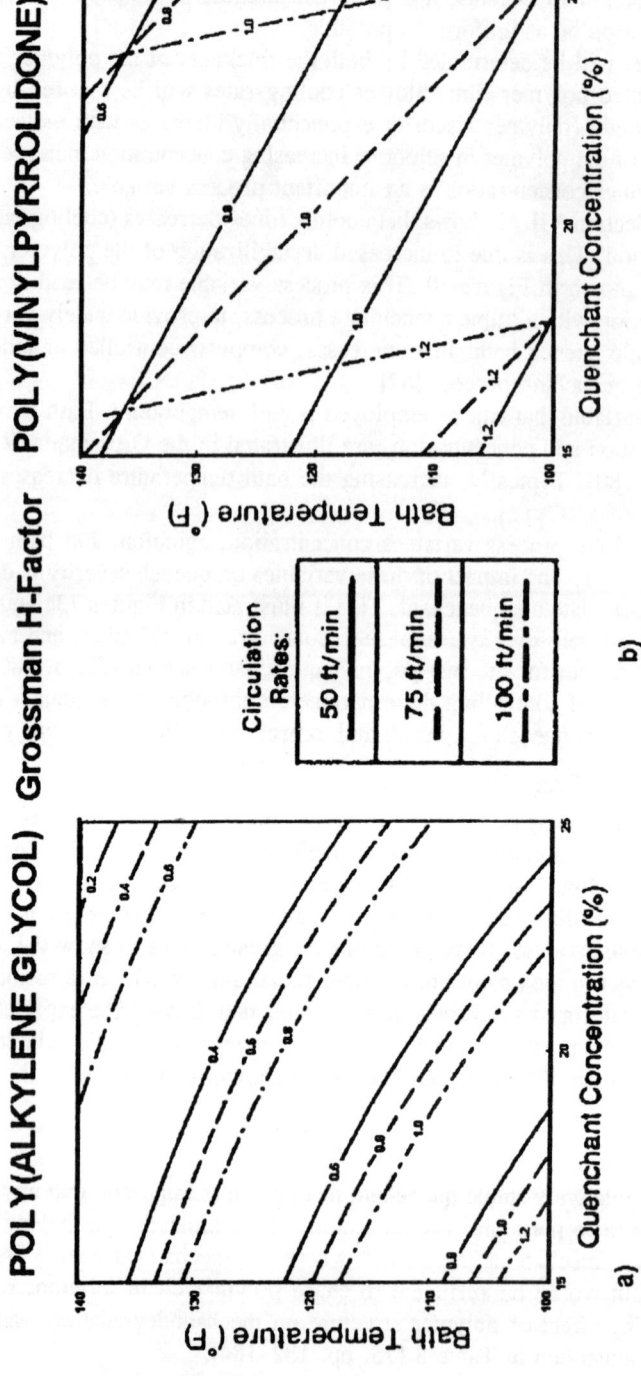

Figure 73 Comparison of Grossmann hardenability of two aqueous polymer quenchants as a function of varying concentration, agitation and bath temperature. (——) 50 ft/min; (---) 75 ft/min; (-·-) 100 ft/min. (a) Polyalkylene glycol; (b) polyvinylpyrrolidone.

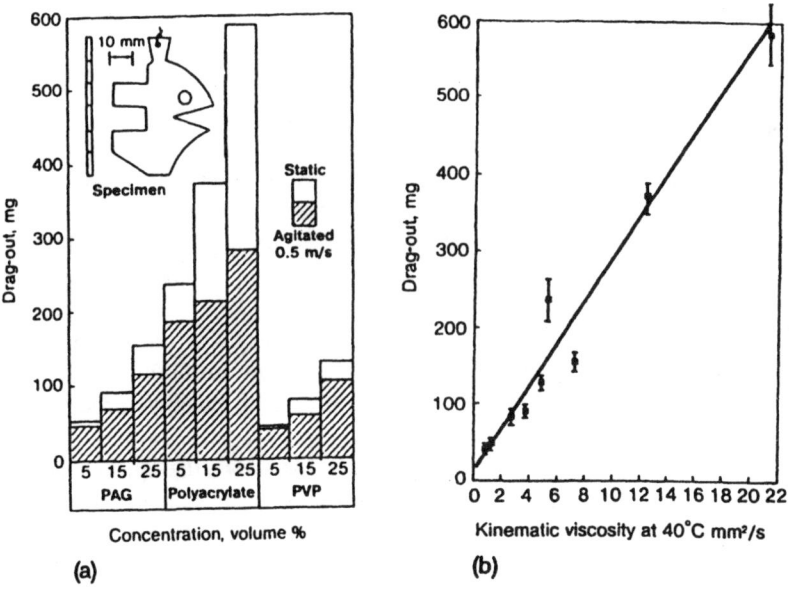

Figure 74 Comparison of polymer drag-out (a) as a function of polymer structure and concentration and (b) as a function of solution viscosity.

Poly(alkylene glycol) copolymers are the most commonly encountered polymer quenchants. Poly(alkylene glycol) copolymers are derived from the coreaction of ethylene oxide and propylene oxide as shown below:

$$nCH_2 - CH_2 + mCH_2 - CH \longrightarrow [(CH_2CH_2O)_n(CH_2CHO)_m]-$$

$$\overset{|}{CH_3} \qquad\qquad\qquad \overset{|}{CH_3}$$

Ethylene   Propylene   PAG
oxide   oxide

Table 8 Comparison of Shear Stability of a Polyacrylamide and a Poly(alkylene glycol) Copolymer (Waring Blender Shear Stability Test)

	Polymer concentration (wt %)	Viscosity at 40°C[a] (cSt)			
		Shear time (min)			
		1	2	3	ΔcSt
Polyacrylamide					
20% hydrolyzed	2.5	6.3	6.2	6.0	−0.3
90% hydrolyzed	0.25	6.3	5.9	5.6	−0.7
PAG copolymer	7.6	6.0	6.0	6.1	+0.1

[a]Sheared at maximum blender rate for time indicated.

Poly(alkylene glycol) copolymers are often incorrectly referred to as glycols. These polymers are not glycols. Examples of glycols are ethylene glycol, diethylene glycol, and propylene glycol, which are not polymers. Glycols are used to formulate antifreeze compositions that have generally unacceptable quenching behavior, typically little different from that of water itself. Furthermore, in view of their volatility, glycols should not be used as quenchants for safety, toxicological, and environmental reasons. The term "glycol" appears to have arisen from the unfortunate shortening of the common name for these copolymers, "polyalkylene glycol" (often abbreviated as PAG), which was further shortened to "polyglycol." Although the term "polyglycol" is technically incorrect and should not be used, "polyalkylene glycol" is acceptable.

PAG quenchants are stable to mechanodegradation, exhibit fairly low relative drag-out rates and oxidative degradation, and are hydrolytically stable. Small variations in polymer concentration exhibit relatively little impact on cooling rates, providing excellent process latitude.

The Grossmann H-factor contour plot illustrated in Figure 73a shows that quench severities ranging from those attainable with conventional oil to quench severities greater than those of water (0.20 to >1.0) are achievable with the appropriate selection of polymer concentration, agitation, and bath temperature [75, p. 180].

Aqueous PAG quenchants exhibit an *inverse solubility* property; an insoluble polymer hydrate separates from solution at a characteristic temperature that is dependent on the polymer structure, degree of degradation, and any additives that may be present [94]. Thermal separation temperatures for PAG quenchants typically vary from 65 to 85°C (149 to 185°F), depending on the particular quenchant. This reversible process is illustrated in Figure 75.

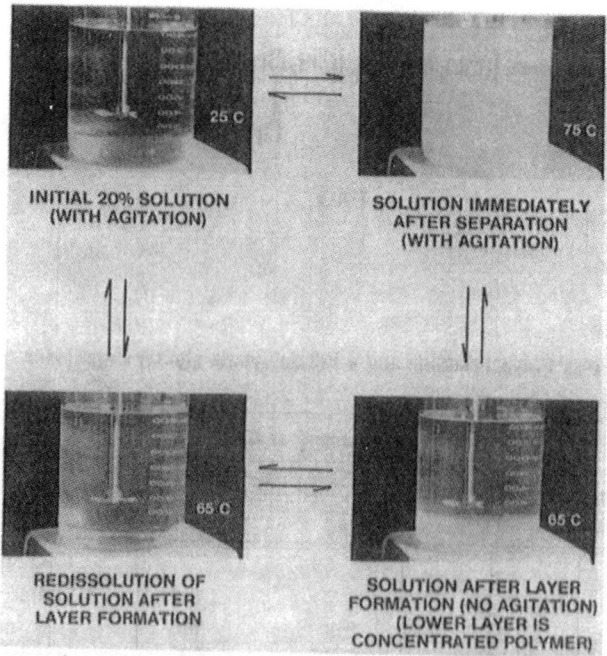

INITIAL 20% SOLUTION
(WITH AGITATION)

SOLUTION IMMEDIATELY
AFTER SEPARATION
(WITH AGITATION)

REDISSOLUTION OF
SOLUTION AFTER
LAYER FORMATION

SOLUTION AFTER LAYER
FORMATION (NO AGITATION)
(LOWER LAYER IS
CONCENTRATED POLYMER)

Figure 75   Illustration of the reversible thermal separation of a PAG polymer quenchant.

Thermal separation processes may be used to remove undesirable salt buildup from PAG quenchants if the quench tank is not too large [94]. Salt buildup may occur from the evaporative concentration of hard metal ions if distilled or deionized water is not used or the parts are preheated in a salt bath and the salt is dragged out on the parts just prior to the quench. Ionic salts are typically removed with the aqueous layer during thermal separation. Typically the "aqeuous layer" is the upper layer unless concentrations of the salt are sufficient to make the aqueous layer more dense than the "hydrated polymer layer." One can readily determine the difference between the two layers because the polymer layer will exhibit a thermal separation temperature when rediluted. For larger quench systems (and non-PAG polymer quenchants), salt removal by reverse osmosis (RO) is recommended [95].

Aqueous solutions of poly(vinyl pyrrolidone) (PVP) have also been used as quenchants [91]. PVP is prepared by homopolymerization of vinyl pyrrolidone as follows:

$$nCH_2 = CH \longrightarrow -[CH_2CH]_n -$$

Vinyl pyrrolidone   PVP

Aqueous solutions of PVP do not exhibit a thermal separation temperature and may be used at temperatures up to the boiling point of water. However, although such high bath temperatues are possible, the quenching performance of PVP polymer quenchants is relatively sensitive to small changes in concentration, bath temperature, and agitation as shown in Figure 73b.

Poly(sodium acrylate) (PSA) is used to formulate polymer quenchants [92]. PSA is prepared by the homopolymerization of sodium acrylate (shown below) or by the alkaline hydrolysis of a polyacrylate ester.

$$CH_2 = CH \quad \xrightarrow{\text{Catalyst}} \quad -[CH_2CH]_n$$
$$C=O \qquad\qquad\qquad C=O$$
$$ONa \qquad\qquad\qquad ONa$$

Sodium acrylate   Poly(sodium acrylate)
monomer   polymer

PSA quenchants are not widely used in the heat treatment industry; however, they are used to harden high-hardenability steels that are particularly prone to quench cracking. Applications include the quenching of deep-carburized parts such as bearing races and rollers, direct quenching of plain carbon steels such as SAE 1070 or 1090 automotive sway bars, and rod or wire patenting. The relatively long vapor blanket cooling stages encountered with PSA quenchants have favored its use in the quenching of railway rails, where pearlite or fine bainite transformation products are desired.

Aqueous PSA quenchants suffer from a number of disadvantages precluding broader use in the heat treatment industry. However, with the possible exception of sensitivity to mechanodegradation, perhaps the most significant disadvantage is the propensity of polyanionic polymers to form insoluble complexes, coacervates, and precipitates [96] with polyvalent metals such as calcium and iron. Polyvalent metals may be present from hard water or corrosion metals.

Recently, aqueous solutions of another polymer, poly(ethyl oxazoline) (PEOX) [93], have been used to formulate "nontacky" quenchants [97]. Nontacky quenchants were developed for use in induction heat treatment systems where the potential formation of residual tacky residues after quenching may interfere with robotic operations. PEOX is prepared by the homopolymerization of ethyl oxazoline.

$$nC_2H_5C \overset{\displaystyle N}{\underset{\displaystyle O}{\Big\langle}} \quad \longrightarrow \quad [NCH_2CH_2]_n$$
$$\overset{|}{\underset{\displaystyle C_2H_5}{C=O}}$$

Ethyl oxazoline   PEOX

Although aqueous solutions of PEOX, like PAG-based quenchants, exhibit inverse solubility temperatures, there have been no reports of this technology for quench bath purification. This may be due to the inherent hydrolytic instability of the amide linkages present in the PEOX polymer [75, pp. 187–188].

6. Polymer Quenchant Bath Maintenance

Various authors [98–101] have used cooling curve analysis to illustrate that, as with all organic compounds, polymer quenchants will undergo degradation via oxidative/thermal processes, resulting in an overall cooling rate acceleration as shown in Figure 76 [102]. Cooling rates increase because oxidative and thermal degradation reduce the polymer viscosity. According to the quenching process depicted in Figure 10, the lower film viscosity of the degraded polymer surrounding the cooling workpiece will exhibit higher heat transfer rates than the higher viscosity, undegraded polymer.

Another potential cause of increased cooling rates in used polymer quenchants is salt contamination. Generally, salt contamination produces an overall cooling rate increase as shown in Table 9. Although cooling curve analysis will detect cooling rate increases, it will not identify the cause. Therefore, cooling curve analysis must be accompanied by physical property characterization in order to assess the root cause of the cooling curve variation observed [103].

Perhaps the most commonly encountered method of tankside determination of quenchant concentration uses the refractive index of the bath. This is most typically performed with a temperature-compensated, handheld refractometer as shown in Figure 77. If a temperature-compensated instrument is not used, a constant-temperature bath must be used because refractive index is temperature-dependent. The quenchant concentration can then be determined from a refractive index vs concentration calibration chart or by the use of a linear regression equation of the form $y = mx + b$ because the refractive index–quenchant concentration relationship is linear. (The y intercept, b, is the refractive index of the water used for dilution of the quenchant.)

Although refractive index is easily measured, this method is susceptible to erroneous results due to both polymer degradation and contamination. In some cases, the refractive index obtained is more indicative of the contaminant than of the polymer used to formulate the quenchant, whether undegraded or degraded. Therefore, periodic calibration by viscosity is recommended.

Refractive index is most applicable to the use of PAG quenchants. For polymer quenchants based on PVP, PSA, and PEOX, viscosity is the preferred indicator for determining quenchant

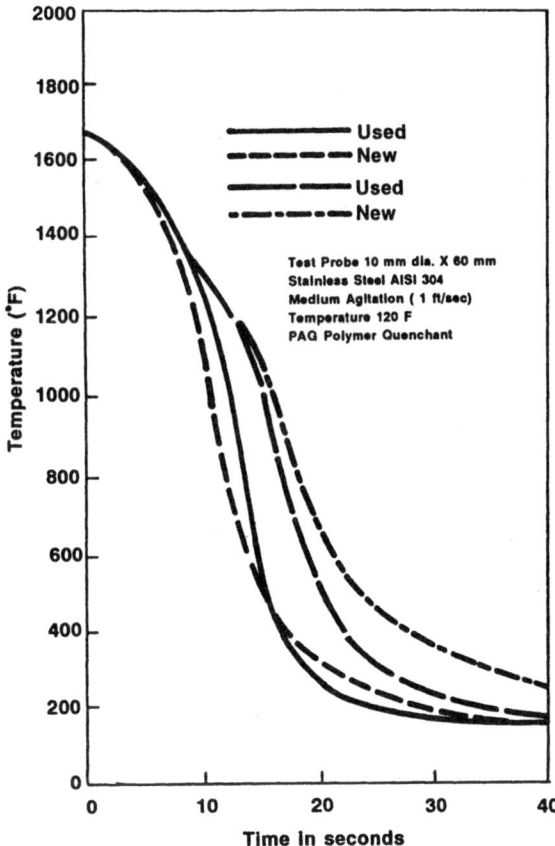

Figure 76 Illustration of the relative cooling properties of a fresh and used PAG polymer quenchant.

concentration. Viscosity is typically measured at constant temperature and compared to a calibration curve such as that shown in Figure 78. Viscosity is most sensitive to polymer concentration and degradation and is generally not significantly affected by low concentrations of a contaminant.

Table 9 Effect of Salt on Cooling Rates of a 20% Aqueous Solution of Poly(alkylene glycol)[a]

	Maximum cooling rate (°C/s)	Cooling rate at 343°C (°C/s)
Sodium nitrite		
0.0 wt %	39.0	25.3
3.0 wt %	54.0	31.8
6.0 wt %	65.8	35.0
Water[b]	61.8	33.0

[a]Data were obtained using a 25 × 50 mm (1 × 2 in.) cylindrical type 304 stainless steel probe instrumented with a type K thermocouple inserted at the geometric center. Axial flow by the probe surface was 23 L/min at 40°C.
[b]Distilled water.

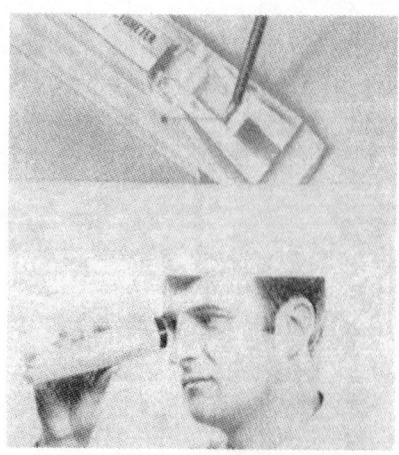

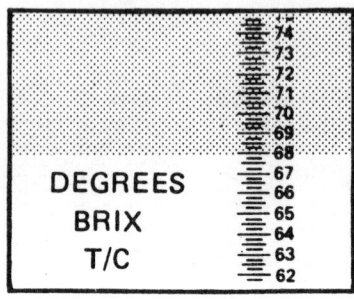

10432

Figure 77 Use of the hand-held refractomer to measure polymer quenchant concentration.

Polymer concentrations obtained by measuring refractive index and viscosity should be compared. If a discrepancy exists, either polymer degradation or contamination is indicated. Ionic contamination can be readily detected by measuring solution conductance. If the conductance is higher than that expected by comparison to the conductance of a fresh solution at the same nominal concentration, then salt contamination, perhaps from hard water metal ions, is indicated.

If polymer degradation has occurred (without accompanying salt contamination), this may be detected from the thermal separation temperature, which increases with the amount of polymer degradation. Salt contamination may either increase or decrease the thermal separation temperature. However, the presence of salts will be detected by measuring conductance.

All polymer quenchants, because they contain water, must also contain a corrosion inhibitor. One of the most common corrosion inhibitors is sodium nitrite. Sodium nitrite may be easily determined with the use of readily available colorimetric reagents. The analysis of other commonly used corrosion inhibitors such as amine/carboxylic acid salts or the sodium salts of mixed organofunctional aromatic acids is more complex and is usually performed free of charge by the polymer quenchant supplier.

Some polymer quenchants are susceptible to biological degradation [104]. Where it occurs, it is usually anaerobic degradation, which may be caused by leaving the quench tank

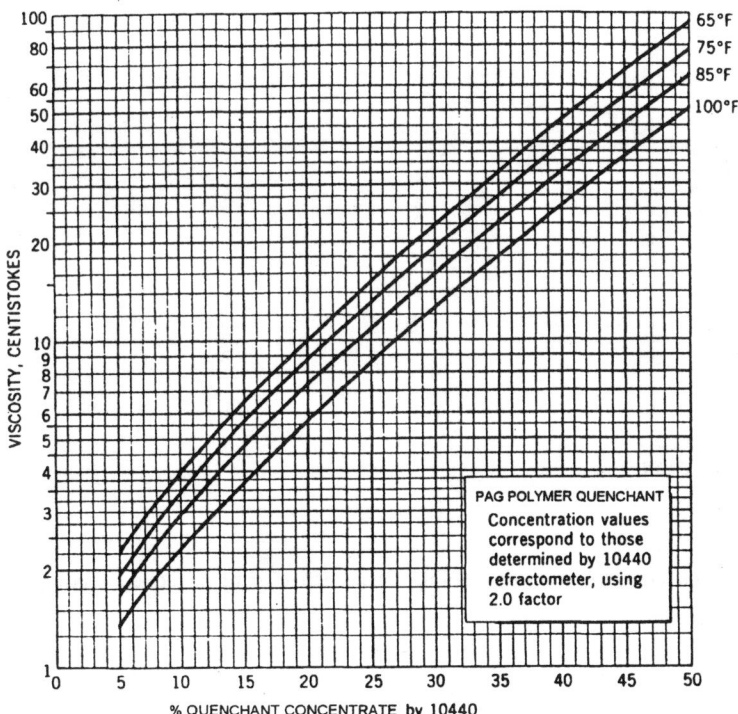

Figure 78 Viscosity–temperature calibration curve for a PAG polymer quenchant. (These curves vary with the individual polymer quenchant being used.)

unagitated for prolonged periods of time. Therefore, it is recommended that quench tanks be agitated for approximately 30–60 min daily. When bacterial contamination occurs, it may be readily monitored with commercially availabe dipsticks. Analytical procedures and subsequent inhibitor recommendations are available from the quenchant supplier.

The performance of aqueous polymer quenchants is affected by the presence of oil contaminants such as hydraulic, metalworking, and quench oils [105]. The effect of oil contamination on cooling curve performance is illustrated in Figure 79. As oil and aqueous polymers do not form compatible mixtures, they will form a discontinuous film around the hot metal part, which will produce substantial thermal gradients, which in turn may result in increased thermal and transformational stresses, possibly producing cracking or increasing distortion.

Gaseous contaminants such as ammonia from carbonitriding processes may produce significant extension of vapor blanket cooling behavior as shown in Figure 80. Although a technical solution is possible, this is the reason that aqueous polymer quenchants are not widely used in nitrocarburizing and nitriding processes.

In some cases, a contaminant may be introduced into an aqueous polymer quenchant that will cause foaming. Foam formation around the workpiece during quenching is always undesirable because it enhances nonuniform cooling. When foaming occurs, which can be detected either in the tank during agitation, by laboratory bottle shaking tests, or by Waring blender test, an antifoam agent should be added. The most preferable are cloud point antifoam agents, which are usually based on a poly(alkylene glycol) copolymer with a separation temperature of 30–55°C (86–131°F). Although silicone antifoam agents can be used, their cost is relatively high due to drag-out on the quenched parts.

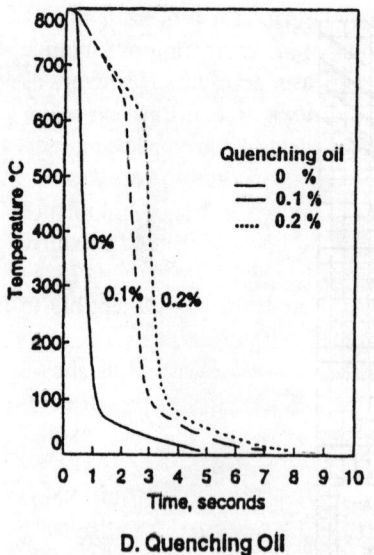

D. Quenching Oil

Figure 79 Effect of quench oil contamination on cooling curve behavior of an aqueous polymer quenchant.

C. Molten Salt Quenchants

Two high-temperature quenching processes discussed in Section III and illustrated schematically in Figure 8 are marquenching and austempering. Specially formulated marquenching oils, which may contain quench rate accelerator additives for certain applications, or molten salts may be used depending on the desired operating temperature. Marquenching oils are typically

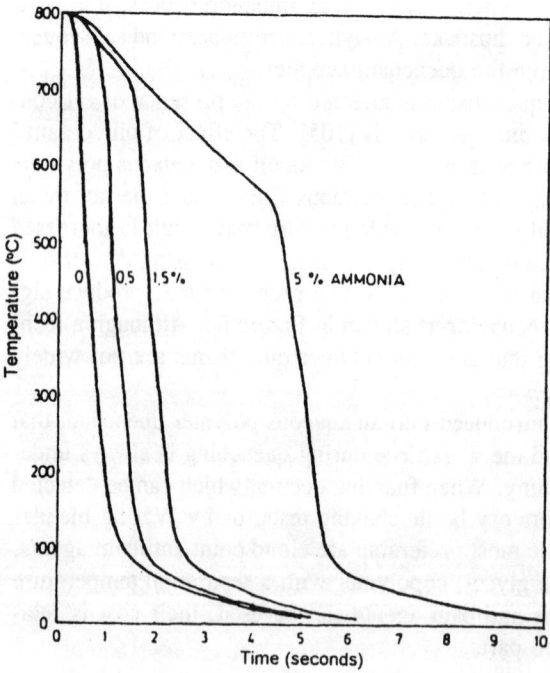

Figure 80 Effect of ammonia on cooling curve performance of an aqueous polymer quenchant.

used for processes up to 205°C (400°F), and molten salts may be used for process temperatures of 160–400°C (320–750°F) [106]. Some of the advantages of martempering oils include lower use temperatures, ease of low-temperature handling, and less drag-out. However, as illustrated in Table 10, relatively high flash and fire points are required and must be maintained to minimize fire hazards. For this reason and becasue they afford higher operating temperatures, molten salts are usually the medium of choice for high-temperature processes.

Molten salt baths used for quenching are usually mixtures of sodium or potassium nitrites and nitrates. The use temperature is dependent on the melting point of the particular mixture. A few typical examples are provided in Table 11 [107]. Melting points as low as 80°C (175°F) may be obtained with the addition of up to 10% water [108]. Salt baths are susceptible to potential degradation at temperatures above 600°C (1110°F) [109,110].

Salt bath operating temperature exhibits little if any effect on quench severity, as shown in Table 12 [111]. Agitation, however, exhibits a very substantial effect on maximum cooling rates, as illustrated in Figure 81 [112]. Figure 81 also shows that cooling rates exhibit significant effects on cooling rates and may affect the depth of hardening.

In addition to water content, salt baths are also affected by various contaminants. Some of the most common contaminants are salts, typically chlorides or carbonates (which arise from the thermal decomposition of sodium nitrite). The solubility of chlorides is temperature-dependent, and these salts may be removed by filtration from the molten salt mixture at temperatures 30–50°C (55–90°F) below the precipitation temperature. When chloride salts are present above their saturation temperature and there is some moisture present, they have been reported to decrease the quench severity of ternary nitrite–nitrate mixtures such as those listed in Table 11 [113]. However, cooling rate increases are observed if there is no water present in the molten salt [114].

D. Fluidized Bed Quenching

Fluidized bed quenching provides a number of advantages relative to quenching in molten metal [115] and molten salt [116] baths. In addition to improved process safety and reduced ecological hazards, these advantages include improved process control, flexibility, and cleanliness. Cost is a significant disadvantage if the fluidizing gas is not recycled.

Vaporizable liquid quench systems such as oil exhibit relatively complex vapor blanket cooling, nucleate boiling, and convective cooling mechanisms during the quench. Fluidized

Table 10   Typical Physical Properties of Two Marquenching Oils

Property	Operating temperature	
	95–150°C (200–300°F)	150–250°C (300–450°F)
Min. flash point, °C (°F)	210 (410)	275 (525)
Min. fire point, °C (°F)	245 (470)	310 (595)
Viscosity, SUS, at		
38°C (100°F)	235–575	—
100°C (210°F)	50.5–51	118–122
150°C (300°F)	36.5–37.5	51–52
175°C (350°F)	—	42–43
205°C (400°F)	—	38–39
230°C (450°F)	—	35–36
Viscosity index (min)	95	95
Acid number	0.00	0.00
Carbon residue	0.05	0.45

Table 11 Composition and Operating Temperature for Some Common Salt Mixtures

	Mixture			
	I	II	III	IV
Composition, %				
Sodium nitrate	45–55	15–25	57	—
Sodium nitrite	—	23–55	43	53
Potassium nitrate	45–55	45–55	—	47
Operating temperature, °C	260–595	175–590	230–550	160–550

Table 12 Effect of Operating Temperature on Quench Severity[a]

Operating temperature		Grossman H factor
°C	°F	(in.$^{-1}$)
195	385	0.40
200	390	0.45
230	450	0.40
270	515	0.45
295	560	0.41
350	660	0.43

[a]Potassium nitrate–sodium nitrite salt with a melting point of 135°C (275°F) with no agitation.

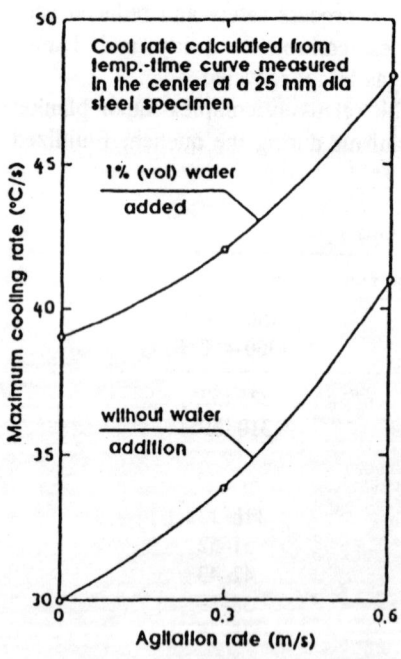

Figure 81 Effect of water and agitation rate on the maximum cooling rate for a molten salt quenchant.

bed quenching, which is controlled by the gas phase, proceeds by a convective cooling mechanism and does not vary throughout the quenching process [117].

A typical fluidized bed is shown in Figure 82. It includes a furnace system (shell, heaters, and insulation) and a quench system (gas diffusion assembly, fluidized bed support, and gas) [118]. The fluidized bed is generated by blowing a gas such as nitrogen, argon, helium, or hydrogen through a solid support such as aluminum oxide [119].

Heat transfer in a fluidized bed is affected by the particle size of the solid support, volumetric heat capacity, thermal conductivity of the gas, and gas flow rate through the bed. The impact of varying these parameters on the heat transfer coefficient is summarized in Table 13 [110].

The cooling behavior during fluidized bed quenching is described by the equation [113]

$$\frac{T - T_f}{T_i - T_f} = \exp\left[\frac{- Aht}{C_p V \rho}\right] \tag{17}$$

where T_f is the temperature of the fluidized bed, T_i is the initial temperature of the metal being quenched, A is the surface area of the cooling surface, V is the volume of the metal, C_p is the heat capacity of the metal, t is the cooling time, and h is the heat transfer coefficient.

The two variables that most strongly affect heat transfer are the thermal conductivity and the flow rate of the gas as shown in Figure 83 [112]. The most common gases used for fluidized beds have been nitrogen and argon. However, as Figure 83 shows, other gases provide much greater potential for cooling rate variation. Reynoldson recently provided an illustrative summary of the range of cooling rates possible relative to other quenching processes as shown in Figure 84 [120].

E. Aqueous Salt (Brine) Quenchants

It is very difficult to through-harden many low-hardenability steels even with a relatively severe water quench. This is due to the vapor blanket cooling properties exhibited by water, which may extend cooling into the pearlitic transformation region of these steels. In such cases, it is necessary to eliminate, or at least minimize, vapor blanket cooling to obtain the desired hardening.

Another problem with water quenching is the relative stability of the vapor blanket cooling process. During water quenching, it is not unusual for all three cooling processes to oc-

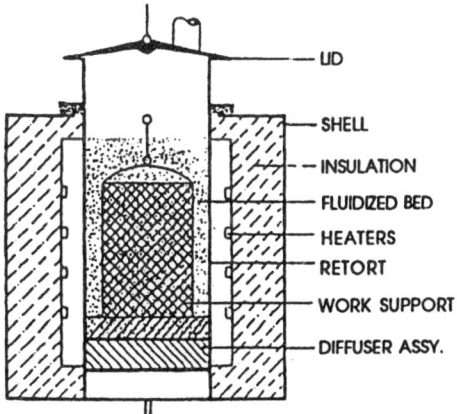

Figure 82 Illustration of an electrically heated fluidized bed furnace.

Table 13 Fluidized Bed Variables and Their Effect on Heat Transfer

Parameter	Effect	Comment
Particle size d	$d \uparrow, \downarrow$	Valid for $d > d_{crit}$
Volumetric heat capacity C_v	$C_v \uparrow, \uparrow$	Al_2O_3
Gas conductivity k	$k \uparrow, \uparrow$	He, H_2
Gas flow rate U_g	α_{max} for $U_g \approx 5U_{mff}$	

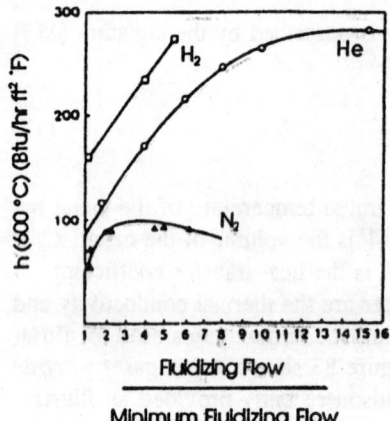

Figure 83 Effect of gas selection on heat transfer properties in a fluidized bed.

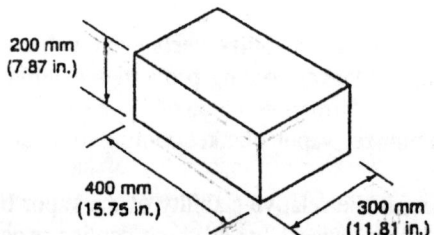

Figure 84 Comparison of cooling rates achievable with different quenchant media relative to fluidized bed quenching.

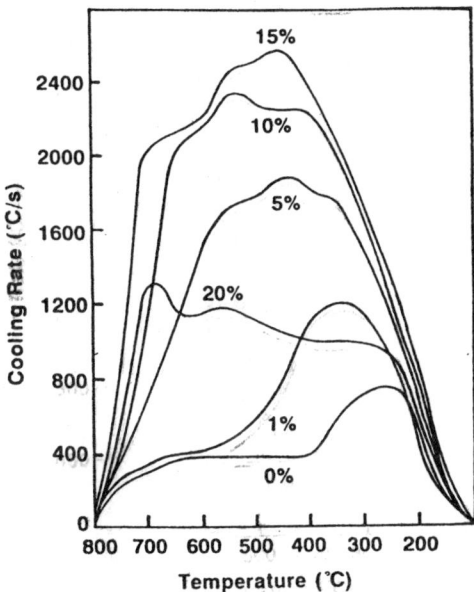

Figure 85 Comparison of cooling rates and vapor blanket cooling elimination as a function of NaCl concentration.

cur on the cooling surface simultaneously as illustrated in Figure 9. This will lead to increased thermal gradients during cooling, possibly resulting in increased cracking and distortion. One way to minimize this problem is to introduce an additive that will provide increased nucleation sites to initiate nucleate boiling, destabilizing vapor blanket cooling. This is done by the addition of salts.

Rose [121] was the first to report the use of sodium chloride (NaCl) and sodium hydroxide (NaOH) to increase the quench severity of water and minimize quench distortion and cracking by destabilizing the vapor blanket cooling.

Petrash [122,123] subsequently showed that approximately 5% of sodium chloride eliminated vapor blanket cooling as shown in Figure 85. This behavior was not dramatically affected by quenchant bath temperature over the range of 20–60°C (68–140°F). Kobasko [124] showed that the maximum cooling rate occurred for both salts at approximately the same concentration, as shown in Figure 86. Although sodium hydroxide (caustic) brines exhibit cool-

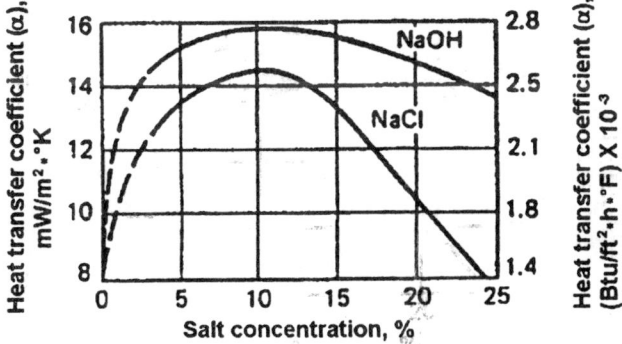

Figure 86 Comparison of the maximum heat transfer coefficients achievable with NaCl and NaOH brines.

Table 14 Relationship of Brine Density and Brine Concentration

Solution conc.	Specific gravity	Degrees Baumé	Density	
			g/L	lb/gal
NaCl				
4%	1.0268	3.8	41.1	0.343
6%	1.0413	5.8	62.4	0.521
8%	1.0559	7.7	84.5	0.705
9%	1.0559	7.7	84.5	0.705
10%	1.0707	9.6	107.1	0.894
12%	1.0857	11.5	130.3	1.087
NaOH				
1%	1.0095	1.4	10.1	0.842
2%	1.0207	2.9	20.4	0.1704
3%	1.0318	4.5	31.0	0.2583
4%	1.0428	6.0	41.7	0.3481
5%	1.0538	7.4	52.7	0.4397

Table 15 Comparison of Heat Fluxes of Some Inorganic Salts

Salt	Conc (%)	Density		Maximum heat flux	
		kg/m^3	lb/ft^3	mW/m^2	$mW/in.^2$
NaCl	10	1070	67	13	0.0084
LiCl	23	1138	71	9.5	0.0061
$MgCl_2$	14	1119	70	13	0.0084
$CaCl_2$	10–12	1083.5–1101.5	67.6–68.8	14	0.0090
NaOH	8–10	1087–1109	67.9–69.2	15	0.0097

ing rates similar to that of NaCl at high surface temperatures, they do produce slower cooling rates near the M_s transformation region of 350°C (660°F), reducing the potential for cracking [122].

Since the performance of a brine quenchant is concentration-dependent and since the salt can be removed by adsorptive drag-out, it is important that the concentration be maintained to reduce process variability. This is normally done by monitoring the specific gravity of the quenchant soltion. Typical specific gravity values for different concentrations of NaCl and NaOH solutions are provided in Table 14.

Although the most commonly encountered brine solutions are derived from NaCl and NaOH, numerous other salts can potentially be used, although each salt will exhibit its own particular quenching behavior. The maximum heat fluxes provided by aqueous solutions of these salts at similar densities are provided in Table 15 [125].

LIST OF SYMBOLS

A	surface, m^2
A	elongation, %

A	parameter of the QTA model, average heat transfer coefficient of the convection stage, W/(m² K)
A_b	temperature of beginning bainite formation dependent on cooling rate, °C
A_{r1}	temperature of beginning pearlite formation dependent on cooling rate, eutectoid temperature, °C
A_{r3}	temperature of proeutectoid ferrite precipitation dependent on cooling rate, °C
$A_{r'}$	temperature of beginning pearlite transformation and supression of ferrite precipitation dependent on cooling rate, °C
Bi	Biot number, dimensionless
G	electrical conductance, S
H	heat transfer equivalent, m⁻¹ or in.⁻¹
HV	Vickers hardness
M_f	temperature for completing the martensite formation, °C
M_s	temperature at beginning of martensite formation, °C
Q	parameter of the QTA model, average heat flux density of the vapor blanket stage from austenitizing temperature to 500°C (932°F), W/m²
R	probe radius, m
R_m	tensile strength, Pa
R_p	yield strength, Pa
T	temperature, °C
T	parameter of the QTA model, temperature at which the extrapolated line of heat flux density of the nucleate boiling stage intersects the zero temperature axis, °C
T_A	starting temperature in sample (austenitizing temperature), °C
T_E	temperature at the end of cooling, °C
T_H	half-temperature $[= (T_A - T_E)/2 + T_E]$, °C
T_b	boiling temperature of the quenchant, °C
T_s	core temperature at which wetting begins, °C
T_s	surface temperature of the probe, °C
T_{trans}	transition temperature, core temperature at the point of transition from lower to higher cooling rate, °C
$\dot{T}_{max}$	highest core cooling rate, K/s
$\dot{T}_{300°C}$	cooling rate at 300°C (572°F), K/s
V	probe volume, m³
a	thermal conductivity $(= \lambda/\rho C_p)$, m²/s
b	coefficient of heat penetration, W·s^{1/2}/(m·K)
C_p	specific heat capacity, J/(kg·K)
$\dot{q}$	heat flux density, W/m²
$\dot{q}_\lambda$	heat transfer by conduction, W/m²
$\dot{q}_\varepsilon$	heat transfer by radiation, W/m²
$\dot{q}_v$	heat transfer by vaporization of the fluid, W/m²
t	time, s
$t_{A/5}$	time from austenitizing temperature down to 500°C (932°F), s
t_b	wetting time, s
t_s	time at which wetting starts, s
t_f	time at which wetting is finished, s
w	average velocity of the ascending wetting front, mm/s
$w_{f/p/b}$	volume fraction of austenite transformed into ferrite/pearlite or bainite, %
w_M	volume fraction of austenite transformed into martensite, %

α heat transfer coefficient, $W/(m^2 \cdot K)$

α_{conv} heat transfer coefficient of the convection stage, $W/(m^2 \cdot K)$

α_{FB} heat transfer coefficient of the film boiling stage, $W/(m^2 \cdot K)$

α_{NB} heat transfer coefficient of the nucleate boiling stage, $W/(m^2 \cdot K)$

δ thickness of film boiling phase, m

λ heat conductivity, $W/(m \cdot K)$

ρ density, kg/m^3

Δt_w time interval of simultaneous presence of film boiling and nucleate boiling ($=$ $t_f - t_s$), s

REFERENCES

1. D. Hortsmann, Das Zustandsschaubild Eisen-Kohlenstoff, in Bericht Nr. 180 des Werkstoffausschusses des Vereins Deutscher Eisenhüttenleute, 6 Aufl., 1985, Verlag Stahleisen m.b.H., Düsseldorf.
2. *Metals Handbook*, Vol. 8, *Metallography, Structures, and Phase Diagrams*, 8th ed., ASM Int., Cleveland, OH, 1973.
3. H. P. Hougardy, Description and control of transformation in technical applications, in *Steel: A Handbook for Material Research and Engineering*, Vols. 1 and 2, Springer-Verlag, New York, 1991, 1992.
4. W. Pitsch and S. Sauthof, Microstructure of steels, in *Steel: A Handbook for Material Research and Engineering*, Vols. 1 and 2, Springer-Verlag, New York, 1991, 1992.
5. F. Wever, A. Rose, W. Peter, W. Strassburg, and L. Rademacher, *Atlas zur Wärmebehandlung der Stähle*, Max-Planck-Institut für Eisenforschung in Zusammenarbeit mit dem Werkstoffausschuss des Vereins deutscher Eisenhüttenleute, Vols. 1–4, 1954–1956.
6. A. Kulmberg, F. Kornteuer, and E. Kaiser, *HTM 42*: 69–74 (1987).
7. U. Wyss, Die wichtigsten Gesetzmässigkeiten des Verzugs bei der Wärmebehandlung von Stählen, in *Wärmebehandlung der Bau- und Werkzeugstähle*, W. Benninghoff, Ed., BAZ-Verlag, Basel, 1978.
8. H.-J. Eckstein, *Wärmebehandlung von Stahl*, VEB Deutscher Verlag für Grundstoffindustrie, 1971.
9. E. Macherauch and O. Vöhringer, Residual stresses after cooling, in *Theory and Technology of Quenching*, Springer-Verlag, New York, 1992, B. Liscic, H. M. Tensi, and W. Luty, Eds., pp. 117–181.
10. N. I. Kobasko, Intensive steel quenching methods, in *Theory and Technology of Quenching*, B. Liscic, H. M. Tensi, and W. Luty, Eds., Springer-Verlag, New York, 1992, pp. 366–389.
11. J. G. Leidenfrost, De aqua communis nonnullis qualitatibus tractatus, 1756, *J. Heat Mass Transfer 9*: 1153–1166 (1966), translated by C. Waves.
12. A. Stitch, Wechselwirkung zwischen Bauteil, Wärmeübergang und Fluid beim Abschreckhärten, Doctoral Thesis, Tech. Univ. Munich, Faculty Mech. Eng., Munich, 1994.
13. H. M. Tensi, H.-J. Spies, A. Spengler, and A. Stich, Wechselwirkung zwischen Benetzungskinematik und Stahlhärtung beim Tauchkühlen, DFG Rep., Deutsche Forschungsgemeinschaft, Bonn, Contracts Te 65/35-1, 2 and Sp 367/1-1, 2, 1994.
14. H. M. Tensi, P. Stitzelberger-Jakob, T. Künzel, and A. Stich, Gefügebeeinflussung/Benetzungskinetik, DFG Rep., Deutsche Forschungsgemeinschaft, Bonn, Contract number Te 65/27-1, 2, 1989.
15. P. Stitzelberger-Jakob, Härtevorherbestimmung mit Hilfe des Benetzungsablaufs beim Tauchkühler von Stählen, Doctoral Thesis, Tech. Univ. Munich, Faculty of Mech. Eng.
16. H. M. Tensi and A. Stich, Characterization of polymer quenchants, *J. Heat Treat. 5*: 25–29 (1993).
17. E. Steffen, Untersuchung von Abkühlmedien auf Polymerbasis für die Stahlhärtung, Doctoral Thesis, Tech. Univ. Munich, Faculty of Mech. Eng., 1986.
18. H. M. Tensi and M. Schwalm, Wirkung von Abschreckflüssigkeiten unter Berücksichtigung spezieller wässriger Kunststofflösungen (Polyäthylenoxide), *HTM 34*: 122–131 (1980).
19. H. M. Tensi, Wetting kinematics, in *Theory and Technology of Quenching*, B. Liscic, H. M. Tensi, and W. Luty, Eds., Springer-Verlag, New York, 1993, pp. 93–116.
20. T. Künzel, Eingluss der Wiederbenetzung auf die allotrope Modifikationsänderung tauchgekühlter Metallkörper, Doctoral Thesis, Tech. Univ. Munich, Faculty of Mech. Eng., 1986.
21. H. M. Tensi and P. Stitzelberger-Jakob, Influence of wetting kinematics on quenching and hardening in water based polymers with forced convection, *Proc. 6th Int. Conf. Heat Treatment of Metals*, 28–30 Sept. 1988, Chicago, published by ASM Int.
22. H. M. Tensi, P. Stitzelberger-Jakob, and A. Stich, Steuern und Berechnen der Abkühlung zylinderfömiger Körper und Vorherbestimmung der Härteverteilung, *HTM 45*: 145–153 (1990).
23. L. V. Körtvelyessy, *Thermoelement Praxis*, Vulkan-Verlag, Essen, 1987.
24. H. M. Tensi and A. Stich, Anzeigeträgheit unterschiedlicher Thermoelemente für Temperaturmessungen beim Abschrecken, *HTM 49*: 26–30 (1994).

25. S. Segerberg, Draft Int. Standard ISO/DIS 9950 (VDC 621.78.084: 621.78.065.2), Industrial Quenching Oils—Determination of Cooling Characteristics—Laboratory Test Method, 1988.

26. H. M. Tensi, P. Stitzelberger-Jakob, and A. Stich, A Proposal for Standardization: Industrial Polymer Quenchants—Determination of Cooling Characteristics—Laboratory Test Method, 1992.

27. H. M. Tensi, A. Stich, and G. E. Totten, Fundamentals about quenching by submerging, *Proc. 1st Int. Conf. Equipment and Processes*, 18–20 Apr. 1994, Schaumburg, IL, published by ASM Int.

28. H. M. Tensi, A. Stich, H.-J. Spies, and A. Spengler, Grundlagen des Abschreckens durch Tauchkühlen, *Proc. ATTT/AWT Tagung Abkühlen/Abschrecken*, 20–21 Apr. 1994, Strasbourg, France, pp. 43–50.

29. H. M. Tensi, G. Welzel, and T. Künzel, Problems of getting characteristics out of quenching experiments useful to control the heat treatment of metallic materials, *Proc. 8th Int. Conf. Heat Transfer*, August 1986, San Francisco, pp. 3031–3035.

30. H. M. Tensi and P. Stitzelberger-Jakob, Evaluation of apparatus for assessing effect of forced convection on quenching characteristics, *Mater. Sci. Technol. 5*: 718–724 (1989).

31. H. M. Tensi, Methods and standards for laboratory tests of liquid quenchants, in *Theory and Technology of Quenching*, B. Liscic, H. M. Tensi, and W. Luty, Eds., Springer-Verlag, New York, 1992, pp. 208–219.

32. F. Richter, Die wichtigsten physikalischen Eigenschaften von 52 Eisenwerkstoffen, *Stahl-Eisen-Sonderber. 8*: (1973).

33. V. E. Loshkaroev, H. M. Tensi, H. Gese, and A. Stich, Calculation of temperature and heat flux in quenched cylinders for different wetting processes, *Steel Res. 9*: (1994).

34. H. M. Tensi, A. Stich, and V. E. Loshkaroev, Influence of wetting kinematics on temperature and heat flux, *Proc. 1st Int. Conf. Equipment and Processes*, 18–20 Apr. 1994, Schaumburg, IL, published by ASM Int.

35. M. A. Grossmann and M. Asimow, Hardenability and quenching, *Iron Age*, May 2, 1940.

36. C. E. Bates, *J. Heat Treat. 6*: 27–45 (1988).

37. C. E. Bates, G. E. Totten, and R. L. Brennan, Quenching of steel, in *ASM Handbook*, Vol. 4, *Heat Treating*, ASM Int., 1991, pp. 67–120.

38. R. Jeschar and E. Specht, Abschreckwirkung durch Aufspritzen von Gas-Wasser-Gemischen, *Proc. ATTT-AWT-Tagung Abkühlen/Abschrecken*, Strasbourg, France, 20–21 Apr. 1994, pp. 161–167.

39. R. Jeschar, U. Reiners, and R. Scholz, Heat transfer during water and water-air spray cooling in the secondary cooling zone of continuous casting plants, *Proc 69th Conf. Steelmaking*, Washington, 1986, Vol. 69, pp. 511–521.

40. R. Jeschar, U. Reiners, and R. Scholz, Wärmeübergang bei der zweiphasigen Spritzwasserkühlung, *Gaswärme Int. 33*: 299–308 (1984).

41. R. Jeschar, E. Specht, and C. Köhler, Heat transfer during cooling of heated metallic objects with evaporating liquids, in *Theory and Technology of Quenching*, B. Liscic, H. M. Tensi, and W. Luty, Eds., Springer-Verlag, New York, 1992, pp. 73–92.

42. U. Reiners, Wärmeübertragung durch Spritzwasserkühlung heisser Oberflächen im Bereich der stabilen Filmverdampfung, Doctoral Thesis, Tech. Univ. Clausthal, 1987.

43. H. R. Müller and R. Jeschar, Wärmeübergang bei der Spritzwasserkühlung von Nichteisenmetallen, *Metallkunde 74*: 257–264 (1983).

44. G. Didier and F. Moreaux, Automatisation dùn système de refroidissement par pulvèrisation pneumatique, *Rev. Gén. Therm. 256*: 333–339 (April 1983).

45. P. Archambault and F. Moreaux, Computer controlled spray quenching, in *Theory and Technology of Quenching*, B. Liscic, H. M. Tensi, and W. Luty, Eds., Springer-Verlag, New York, 1992, pp. 360–367.

46. E. J. Radcliffe, Gas quenching in vacuum furnaces—a review of fundamentals, *Ind. Heat.*, November 1987, pp. 34–39.

47. K. Ahlberg, *AGA Gas Handbook*, Almquist & Wiksell, Sweden, 1985.

48. G. Schmitt and P. Heilmann, Die Vorteile der Hochdruckgasabschreckung, *Proc. ATTT-AWT-Tagung Abschrecken/Abkühlen*, 20–21 Apr. 1994, Strasbourg, France, pp. 115–125.

49. H. Altena, Hochdruck-Wasserstoff-Abschreckung, *Proc. ATTT-AWT-Tagung Abschrecken/Abkühlen*, 20–21 Apr. 1994, Strasbourg, France, pp. 127–136.

50. F. Limque and F. Bless, Härten von Werkzeugstählen in Vakuumöfen mit Hochdruck-Gasabschreckung, *Z. Wirtschaftl. Fertigung 9*: 1–4 (1982).

51. J. W. Bauwmann, Erfahrungen mit der Hochdruckgasabschreckung in Vakuumöfen und desses Weiterentwicklung, *HTM 39*: 1–5 (1984).

52. J. Wünning, Gasabschrecken von Serienbauteilen, *Proc. ATTT-AWT-Tagung Abschrecken/Abkühlen*, 20–21 Apr. 1994, Strasbourg, France, pp. 195–204.

53. J. R. Lyman, High carbon steel microcracking control during hardening, U.S. Pat. 4,523,965; Patent 472656 (1983).

54. R. Kern, Intense quenching, *Heat Treat. 9*: 19–23 (1986).

55. B. Liscic, H. M. Tensi, and W. Luty (Eds.), *Theory and Technology of Quenching*, Springer-Verlag, New York, 1992.

56. H. M. Tensi and A. Stich, Possibilities and limits to predict the quench hardening of steels, *Proc. 1st Int. Conf. Quenching and Distortion Control*, 22–25 Sept. 1992, Chicago, pp. 27–32.

57. H. M. Tensi, Prediction of hardness profile in workpiece, based on characteristic cooling parameters and material behaviour during cooling—prediction of hardening behaviour from wetting kinematics, in *Theory and Technology of Quenching*, B. Liscic, H. M. Tensi, and W. Luty, Eds. Springer-Verlag, New York, 1992, pp. 390–408.

58. C. E. Bates, *J. Heat Treat. 6*: 27–45 (1988).

59. R. W. Monroe and C. E. Bates, Evaluating quenchants and facilities for hardening steel, *J. Heat Treat. 3*(2): 83–99 (1983).

60. G. E. Totten, C. E. Bates, and N. A. Clinton, *Handbook of Quenchants and Quenching Technology*, ASM Int., Materials Park, OH, pp. 94–96.

61. J. Wünning, Berechnung und Steuerung des Temperaturverlaufs beim Abschrecken von Stahl in Wasser und Öl, *HTM 36*(5): 231–241 (1981).

62. M. A. Grossmann, M. A. Asimov, and S. F. Urban, Hardenability, its relationship to quenching and some quantitative data, *Hardenability of Alloy Steels*, ASM, Cleveland, OH, 1939, 237–249.

63. J. L. Lamont, How to estimate hardening depth in bars, *Iron Age 142*(14): 64–70 (1943).

64. J. Wünning and D. Liedtke, Versuche zum Ermitteln der Wärmestromdichte beim Abschrecken von Stahl in flüssigen Abschreckmitteln nach der QTA-Methode, *HTM 38*(5): 149–155 (1993).

65. B. Liscic, Predetermination of hardness results—the QTA method, in *Theory and Technology of Quenching*, B. Liscic, H. M. Tensi, W. Luty, Eds., Springer-Verlag, New York, 1992, pp. 409–419.

66. T. Lübben, H. Bomas, H. P. Hougardy and P. Mayr, Beschreibung der Abschreckwirkung flüssiger Abschreckmittel am Beispiel zweier Härteöle, *HTM 46*(1): 24–34 (1991); *HTM 46*(3):155–170 (1991).

67. H. M. Tensi and P. Stitzelberger-Jakob, Bedeutung des H-Wertes für die Bestimmung der Härteverteilung, *HTM 44*: 99–106 (1989).

68. H. P. Hougardy and K. Yamazaki, An improved calculation of the transformation of steels, *Steel Res. 57*(9):466–471 (1986).

69. M. Umemeto, N. Nishioka, and I. Tamura, Prediction of hardenability from isothermal transformation diagrams, *J. Heat Treat. 2*: 130–138 (1981).

70. I. Tzitzelkov, Eine mathematische Methode zur Beschreibung des Umwandlungsverhaltens eutektoider Stähle, Doctoral Thesis, RWTH Aachen, 1973.

71. I. Tzitzelkov, H. P. Hougardy, and H. P. Rose, Mathematische Beschreibung des ZTU-Schaubildes für isotherme Umwandlung und kontinuierliche Abkühlung, *Arch. Eisenhüttenwesen 45*: 525–532 (1974).

72. S. Somogyi and M. Gergely, Prediction of macrohardness by the help of the individual hardness of the microstructural elements, *Proc. 4th Int. Conf. Heat Treatment of Materials*, Berlin, 1985, pp. 84–90.

73. K. Speith and H. Lange, *Mitt. Kaiser Wilhelm Inst. Eisenforsch. 17*: 175–184 (1935).

74. P. Heilmann, Universal and Economical—The New Vacuum Furnace with Convective Heating and Gas High-Pressure Quenching, Leybold Durferrit Tech. Inf. Bull. No. 184-1-3-390 Ki.

75. G. E. Totten, C. E. Bates, and N. A. Clinton, *Handbook of Quenchants and Quenching Technology*, ASM Int., Materials Park, OH.

76. S. O. Segerberg, *Heat Treat.*, December 1988, pp. 30–33.

77. P. S. Protsidim, N. Ya Rudakova, and B. K. Sheremeta, *Metalloved. Term. Obrab. Met.*, February 1988, pp. 5–7.

78. T. I. Tkachuk, N. Ya Rukakova, B. Shermeta, and M. A. Aleshuler, *Metalloved. Term. Obrab. Met.*, October 1986, pp. 45–47.

79. Gulf Oil Co. Gulf Super-Quench 70: The Revolutionary Dual-Action Quench Oil.

80. F. S. Allen, A. J. Fletcher, and A. Mills, *Steel Res. 60*: 522–530 (1989).

81. W. Hewitt, *Heat Treat. Met. 13*: 9–14 (1986).

82. J. M. Hampshire, *Heat Treat. Met. 11*: 15–20 (1984).

83. H. E. Boyer, and Cary, P. R., *Quenching and Control of Distortion*, ASM Int., Materials Park, OH, 1988.

84. G. R. Furman, *Lubrication 57*: 25–36 (1971).

85. J. A. Hasson, *Ind. Heat.*, September 1981, pp. 21–23.

86. R. K. Singh, and C. R. Chakravorty, *Tool Alloy Steels 21*(5): 145–147 (1987).

87. S. G. Yun and S.-W. Han, *Ind. Heat.*, January 1994, pp. 35–38.

88. N. A. Hilder, Ph.D. Thesis, Univ. Birmingham (Aston), 1988.

89. G. E. Totten, *Adv. Mater. Process 2*: 51–53 (1990).

90. R. R. Blackwood, and W. D. Cheesman, U.S. Patent 3,220,893 (1965).

91. A. G. Meszaros, U.S. Patent 3,902,929 (1975).

92. K.-H. Kopietz, and F. S. Munjat, U.S. Patent 4,078,290 (1978).

93. J. F. Warchol, U.S. Patent 4,486,246 (1984).
94. L. M. Jarvis, R. R. Blackwood, and G. E. Totten, *Ind. Heat.*, November 1989, pp. 23-24.
95. R. D. Howard, and G. E. Totten, *Met. Heat Treat. 1*(5): 22-24 (1994).
96. G. E. Totten, E. D. Goddard, G. H. Matteson, and M. L. Wanschisen, *J. Am. Oil Chem. Soc., 63*: 1586-1589 (1986).
97. Brennan, B., *Ind. Heat.*, September 1988, pp. 34-36.
98. S.O. Segerberg, 4th Int. Congr. Heat Treat. Mater., Berlin, 3-7 June 1985, Vol. II, p. 1252-1265.
99. T. Hibi, *Netsu Shori 25*: 46-50 (1985).
100. N. Kobayashi, *Netsu Shori 25*: 51-54 (1985).
101. E. Varela and P. Cordoba, Proc. First Int. Conf. on Quench and Control of Distortion, Chicago, IL, 22-25 Sept. 1992, ASM Int., p. 63.
102. K.-H. Kopietz, *Heat Treat.*, September 1984, pp. 20-26.
103. G. E. Totten, R. R. Blackwood, and L. M. Jarvis, *Heat Treat.*, March 1991, pp. 16-18.
104. J. F. Kramer, *Proc. First Int. Conf. on Quench. and Control of Distortion*, Chicago, IL, 22-25 Sept. 1992, ASM Int. p. 141-145.
105. E. H. Burgdorf, *Ind. Heat.*, October 1981, pp. 18-25.
106. H. Webster, and J. W. Laird, Martempering of steel, in *ASM Handbook*, Vol. 4, *Heat Treating*, ASM Int., 1991, pp. 137-151.
107. L. Shu-Zhong, *Jinshu Rechuli 5*: 59-61 (1991).
108. C. Skidmore, *Heat Treat. Met. 2*: 34-36 (1986).
109. M. A. H. Howes, Ph.D. Thesis, London University, 1959.
110. R. W. Foreman, *Heat Treat.*, October 1980, pp. 26-29.
111. M. J. Sinnott and J. C. Shyne, *Trans. ASM 44*: 758-774 (1952).
112. B. Liscic, in *Quenching and Carburizing*, 3rd Int. Seminar, Int. Fed. Heat Treat. Surface Eng., Melbourne, 1991, pp. 1-27.
113. L. Rosseau, *Metallurgia, 49*: 27-33 (1954).
114. R. W. Foreman, paper presented orally at ASM Int. Nat. Heat Treat. Conf., Chicago, IL, September 1988.
115. R. Branders, *Wire Ind.*, February 1990, pp. 89-91.
116. W. Krebs, *Giesserei 77*: 337-344 (1990).
117. A. Dinunzi, *Heat Treat.*, March 1993, pp. 42-46.
118. Product brochure, Fluidtherm Technology Ltd., Madras, India.
119. M. A. Delano, and J. Van den Sype, *Heat Treat.*, December 1988, pp. 34-37.
120. R. W. Reynoldson, *Heat Treatment in Fluidized Bed Furnaces*, ASM Int., Materials Park, OH, 1993, p. 47.
121. A. Rose, *Arch. Eisenhuttenwes. 13*: 410-415 (1940).
122. L. V. Petrash, *Izv. V.U.Z. Chern. Metall. 1*: 153-159 (1958).
123. L. V. Petrash, *Metalloved. Term. Obrab. Met. 3* 56-61 (1958).
124. N. I. Kobasko, *Metalloved. Term. Obrab. Met. 3*: 2-6 (1968).
125. M. P. Mukhina, N. I. Kobasko, and L. V. Gordeeva, *Metalloved. Term. Obrab. Met. 9*: 32-36 (1989).

5

DISTORTION OF HEAT-TREATED COMPONENTS

George E. Totten
Union Carbide Corporation, Tarrytown, New York

Maurice A. H. Howes
IIT Research Institute, Chicago, Illinois

I. INTRODUCTION

In manufacturing processes, a leading cause of quality problems, scrap, and rework is the shape changes caused by heat treatment distortion. The cost impact of heat treatment distortion on manufacturing in the U.S. automotive industry alone has been estimated by Dodds of the Ford Motor Company [1] to be $1.5 billion in facilities and $230 million in annual ongoing costs. Other industries affected include manufacturers of

> Other automotive parts such as steering gears, rack-and-pinion assemblies; U-joints, CV joints, bearings, etc.
>
> Aerospace parts such as for engines, transmissions, and other drives
>
> Marine engines and drives
>
> Agriculture and earthmoving equipment
>
> Construction equipment
>
> Oil and chemical processing (pumps and gear boxes)
>
> Military vehicles
>
> Mining equipment
>
> Appliances (compressors, motors, gear boxes, shafts)
>
> Power tools

If all these industries are taken into account, it is easy to see that heat treatment distortion is a multibillion-dollar problem.

As long as parts have been heat treated, distortion has been a concern. As greater dimensional accuracy is required for components, distortion becomes even more of a problem. Recent studies and contacts with industry have often highlighted the frustrations experienced by manufacturers trying to control dimensions consistently. An often-asked question is, What is meant by distortion?, and the clue is in the word *consistently*. It is well known that parts change size due to the formation of metallurgical phases that have different volumes. Parts deform due to differential heating and cooling. Parts also deform to accommodate the relief of residual stresses.

251

It is known that almost every step in the process can affect the final shape of the part, but should this be a concern? If it could be accurately predicted what the new shape of a part would be after heat treatment, then this could be allowed for during manufacture. But at this stage of knowledge, as Walton [2] has depicted (Figure 1), there are so many variables interacting in so many ways that the problem is often beyond our present capacity to analyze, and thus distortion cannot be accurately predicted. This leads to a definition of heat treatment distortion:

> Distortion is the unexpected or inconsistent change in size or shape caused by variations in manufacturing process conditions.

It will be noted that heat treatment is not directly mentioned in the definition. This is because although distortion may become noticeable after heat treatment, the root cause may lie in another part of the manufacturing process that is contributing to variability, such as variable residual stress, due to differences in machining.

However, surface hardening and heat treatment of steel often require that the steel be heated to high temperatures, held at temperature for long periods, and then rapidly cooled by quenching. These drastic measures are necessary to generate high mechanical properties, but they can also cause the part to change shape in unpredictable ways unless conditions are closely controlled.

II. THE BASIC DISTORTION MECHANISMS

Shape and volume changes of a part during manufacture can be attributed to three fundamental causes:

1. Residual stresses that cause shape change when they exceed material yield strength. This will occur on heating when strength properties decline.
2. Stresses caused by differential expansion due to temperature gradients. These stresses will increase with the temperature gradient and will cause plastic deformation as the yield strength is exceeded.
3. Volume changes due to phase changes as transformations occur. These volume changes will be contained as residual stress systems until yield is exceeded.

A. Relief of Residual Stresses

If a part has locked-in residual stresses, these stresses can be relieved by heating the part until the locked-in stresses exceed the yield strength of the material. A typical stress-strain curve obtained from a tension test is shown in Figure 2. The initial changes in shape are elastic, but under increased stress they occur in the plastic zone and are permanent. On heating the part, the stresses are gradually relieved by changes in the shape of the part due to plastic flow. This is a continuous process, and as the temperature of the part is increased the material yield stress decreases (see Figure 3). It is a function not only of temperature but also of time, as the material will creep under lower applied stresses. It is apparent that the stresses can never be reduced to zero, for the material will always possess some level of yield strength below which residual stresses cannot be reduced.

B. Material Movement Due to Temperature Gradients During Heating and Cooling

When parts are heated to perform a heat treatment, a temperature gradient exists across the cross section of the component. If a section heats under restraint, i.e., if part of the compo-

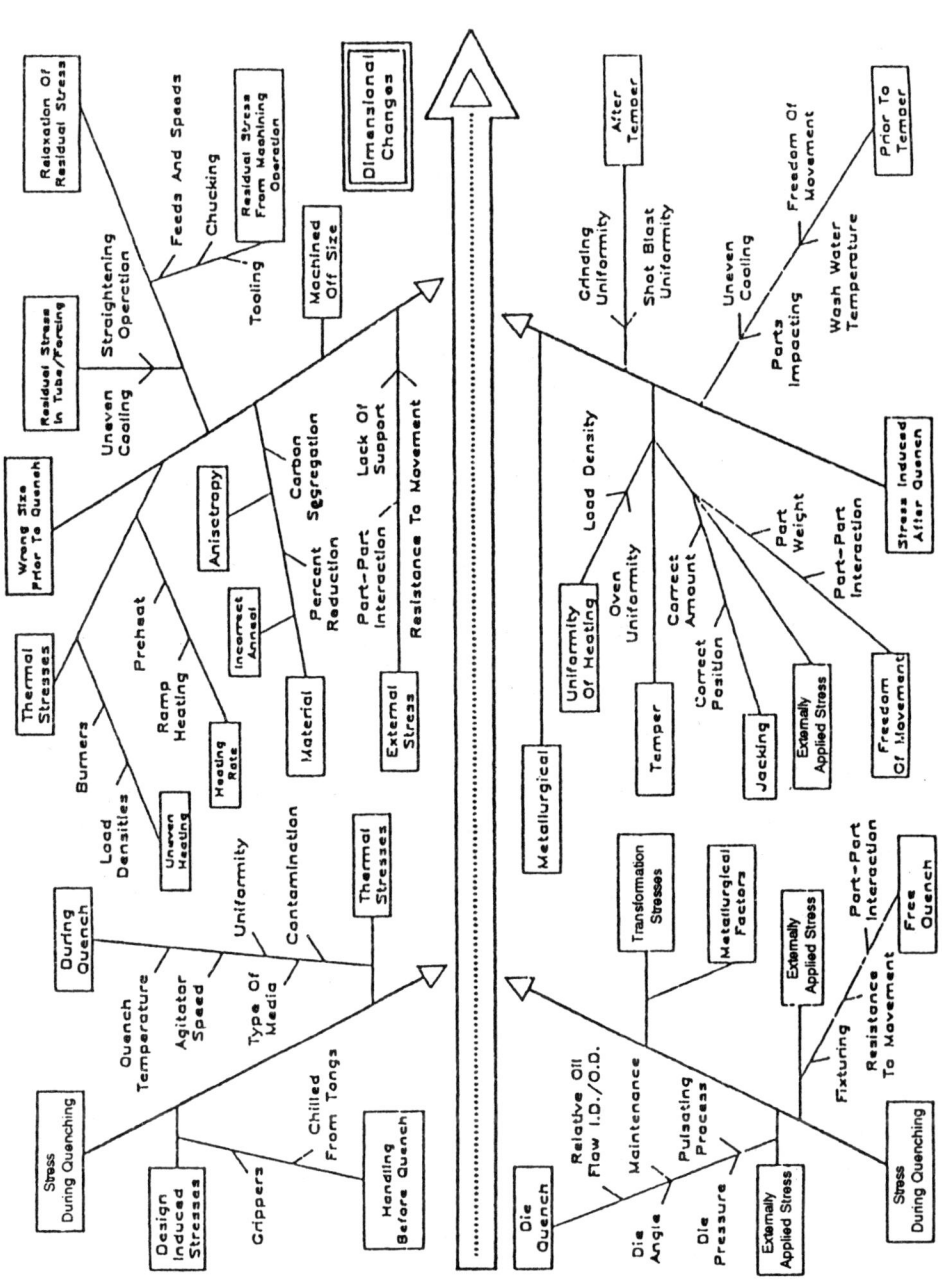

Figure 1 Summary of the many causes of dimensional changes in a quenched and tempered steel component. (From Ref. 2.)

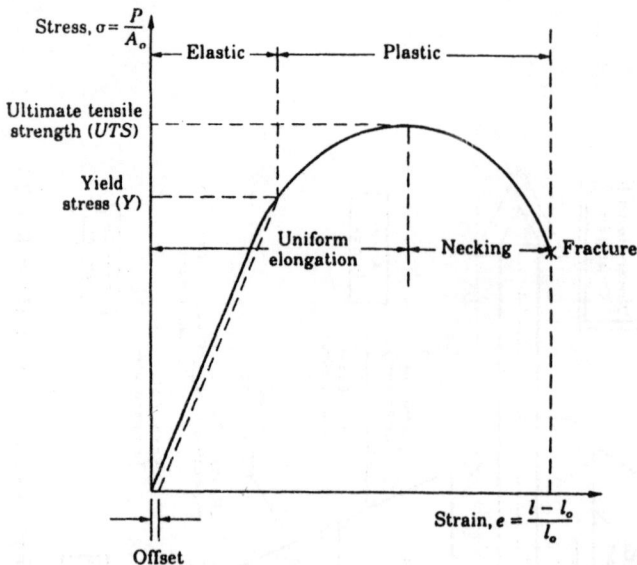

Figure 2 Typical stress–strain curve obtained from a tension test, showing various features.

nent becomes hotter than the surrounding material, then this material expands and occupies a greater volume than the adjacent material and thus will be exposed to applied stresses that will cause a shape change when they exceed the material strength. These movements can be related to the heating rate and the section thicknesses of the component.

C. Volume Changes During Phase Transformations

When a steel part is heated it transforms to austenite with an accompanying reduction in volume (Figure 4). When it is quenched, the structure transforms from austenite to martensite and its volume increases. If these volume changes cause stresses to be set up that are contained within the strength of the material, then a residual stress system is created. If the stresses cannot be contained, then material movement will occur, which under extreme conditions will cause cracking. The expansion is related to the composition of the steel, and Figure 5 shows

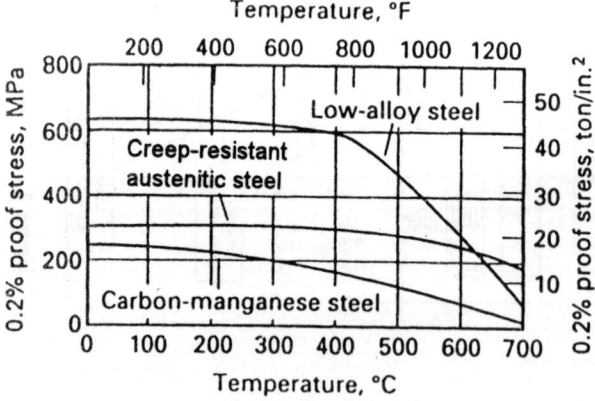

Figure 3 Variation of yield strength with temperature for three generic classes of steel. (From Ref. 65.)

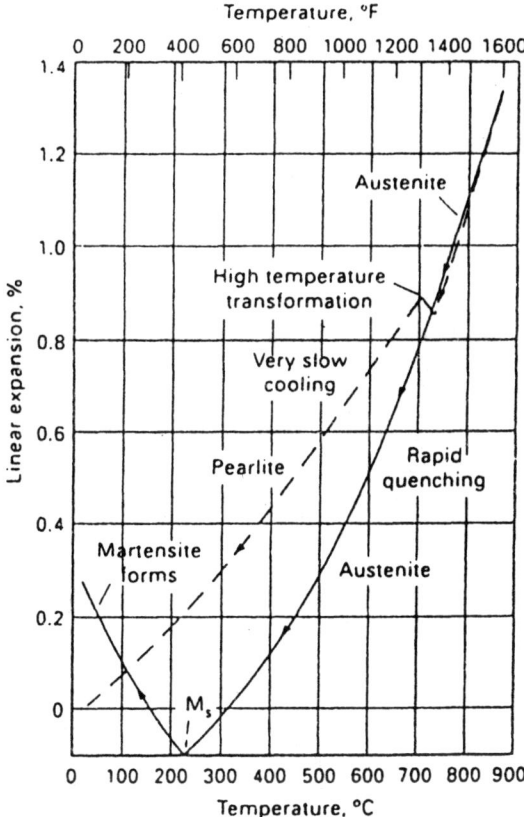

Figure 4 Steel expansion and contraction upon heating and cooling. (From Ref. 66.)

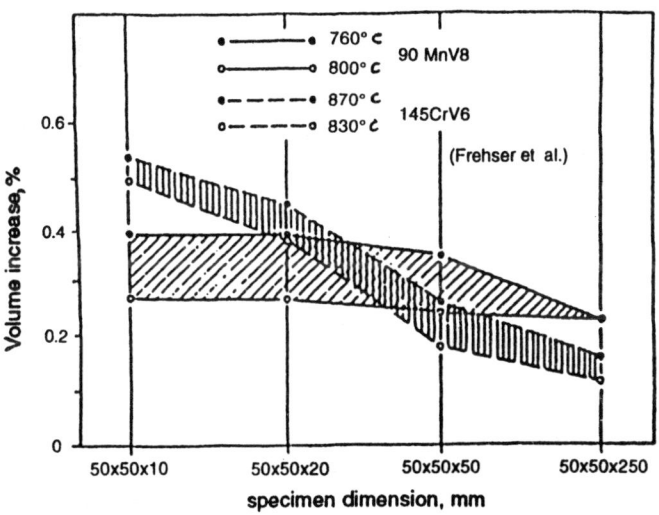

Figure 5 Volume increase of the steels 90MnV8 and 145CrV6 as a function of austenitizing temperature and specimen dimensions.

the relative volume increase of two steels as a function of the austenitizing temperature and the specimen dimensions.

While each of these phenomena is a well-known physical change, the situation is made more complex when all three events occur at the same time. In addition, other variables such as rate of heating, quenching, and inconsistent material composition cause further complications that are discussed later.

III. RESIDUAL STRESSES

A. Residual Stress in Components

Residual stresses are present in parts after any process that strains the material. Heavy metalworking such as forging, rolling, and extrusion causes stresses that remain in the metal if the working is performed below the hot-working temperature. If a material is hot-worked, the stresses are continually removed. Processes such as machining, shot peening, and grinding also cause residual stress formation but to a much shallower depth. While compressive residual stresses are desirable in a finished component to enable it to resist applied stress systems, stresses that exist during manufacture will be relieved during heating with consequent movement in the material as the stress system readjusts.

Residual stresses result not only from heat treatment but also from cold-working steel through metalworking, machining processes, etc. Within any steel part there is a balanced stress system consisting of tensile and compressive residual stresses. If the finished part has the compressive stresses at the surface, under normal tensile loading these stresses increase the strength of the part and are thus beneficial. Processes like shot peening also are used to increase surface compressive stresses to improve performance and compensate for structural defects. This type of residual stress is intentional and is part of the design. The problem arises when a metal part has a residual stress system prior to heat treatment. Then a shape change will occur that is predictable only if the magnitude and distribution of the stress system is known.

B. Residual Stresses Prior to Heat Treatment

The parts presented for heat treatment should not only have the correct dimensions but should also have a consistent residual stress pattern. Ideally, the part should be absolutely stress-free so that movement due to this cause can be disregarded, but in practice some final machining passes are necessary before heat treatment. The best compromise is to completely stress-relieve the part before the final machining cuts are made. With care, the stresses caused by the initial machining stages will not cause the part to go out of tolerance before final machining and the final operations can be done relatively gently and thus cause a minimum of stresses to be retained in the part. Several stress-relieving treatments may be necessary during initial machining to prevent dimensions from going out of control. If a part with a preexisting stress system is machined and has thus had some of the stresses removed, the system will constantly rebalance itself by changing its stress pattern. However a part is made, if it is made consistently the shape changes are predictable.

Parts harden because of phase transformations that usually occur with accompanying volume changes. These are predictable, however, if all parts undergo the same transformation sequence. Quenching is often carried out as a batch process, leading to variations in response because the quench severity will often not be the same in all parts of the batch. Components are almost always made by processes that leave residual stresses in the material. Any forming or machining process leaves stress systems that will be relieved by a dimensional change during heat treatment. Thus, if the part is heavily stressed prior to heat treatment, the shape will change due to this factor alone. Processing should be arranged so that virtually stress-free parts are heat treated. Variations in heat treatment parameters such as case carbon level and processing temperatures will also cause final shape and size differences.

C. Measurement of Residual Stresses

In production, residual stresses are rarely measured, but when they are, X-ray diffraction techniques are usually employed. There are, however, a range of applicable measurement methods.

1. Bending and Deflection Methods

Bending and deflection methods involve the measurement of a change in the diameter of a slit tube or the curvature of an otherwise flat plate [3]. The use of such methods requires knowledge of the interrelationship of stress and the amount of deflection observed. Although these methods are not applicable to the determination of radial stresses, with appropriate procedural adaptation they may be low-cost options for the determination of a systematic distribution of residual stress and uniform biaxial stresses in bars, tubes, sheets, and plates [4,5]. An illustration of the directionality of radial, longitudinal, and tangential stresses is provided in Figure 6.

2. Circumferential (Hoop) Stress

Circumferential (hoop) stress (σ_c) is determined by cutting a slit in the tube and measuring the change in diameter due to stress relief as shown in Figure 7 [3].

$$\sigma_c = E't\left(\frac{1}{D_0} - \frac{1}{D_1}\right)$$

where

E' = effective modulus of elasticity

t = thickness

D_0 = initial diameter of the tube

D_1 = diameter of the tube after opening

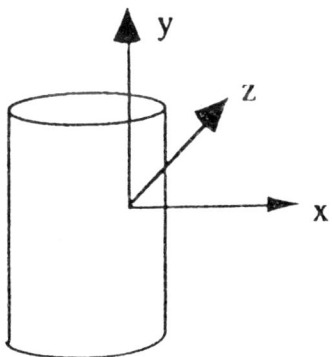

Figure 6 Directionality of tangential (x), longitudinal (y), and radial (z) stresses.

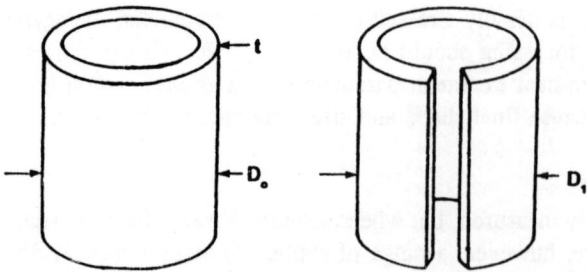

Figure 7 Circumferential (hoop) stress causing a change in tube diameter after slitting. (From Ref. 3.)

The size of the slit opening in the tubes will vary with the length of the tube. Generally, maximum deflection is obtained when the tube length is greater than 3 times the diameter. However, optimal lengths can be determined only by experiment [6]. Some illustrative values of modulus at different temperatures are shown in Table 1.

3. Longitudinal Stress

Longitudinal stress at the surface of a tube may be determined by the deflection method shown in Figure 8. A tongue is cut into the tube, the deflection (δ) of the tongue with respect to initial curvature and change in curvature after the tongue is removed is measured, and the longitudinal stress is calculated from

$$\sigma_L = \frac{E't\delta}{L^2}$$

where E is the modulus of elasticity, v is Poisson's ratio of transverse strain to longitudinal strain, t is the thickness of the tube, δ is the amount of deflection of the tongue as shown in Figure 8, and L is the length of the tube. E' is the effective modulus of elasticity and is calculated from

$$E' = \frac{E}{1 - v^2}$$

If circumferential stresses are present, they may produce significant errors in longitudinal stress. Slitting the tube prior to cutting the tongue into the tube may even further exaggerate the error. Therefore, it is recommended that the circumferential stress (σ_c) be determined on a long tube and a short tube (σ'). Since all longitudinal stress has been released from the short tube,

Table 1 Typical Values of Modulus of Elasticity (E) at Selected Temperatures

Material	Modulus of elasticity (psi $\times$ 10^{-6})				
	Room temp.	400°F	800°F	1000°F	1200°F
Carbon steel	30	27.0	22.5	19.5	18.0
Austenitic steel	28	25.5	23.0	22.5	21.0
Titanium alloys	16.5	14.0	10.7	10.1	—
Aluminum alloys	10.5	9.5	7.8	—	—

Source: Ref. 3.

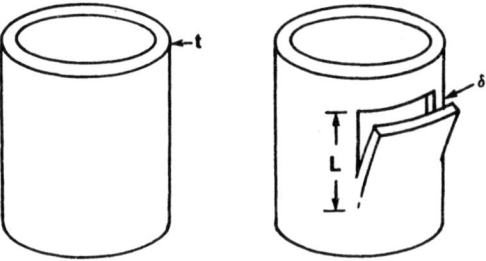

Figure 8 Longitudinal stresses as determined by the deflection method. (From Ref. 3.)

$$\sigma_c - \sigma' = \nu \sigma_L$$

When a force or load is applied to a material, a dimensional change results. This is illustrated by the load–elongation curve and the derived stress-strain curve in Figure 9. Engineering stress (σ) is defined as the total load (P) divided by the original cross-sectional area (A_0):

$$\sigma = P/A_0$$

Strain (ε) is obtained by measuring the ratio of the change in length (ΔL) of the material relative to the original length (L_0) when load P is applied.

$$\varepsilon = \frac{\Delta L}{L}$$

Stress is related to the strain by the elastic modulus (E) if the elastic limit of the material is not exceeded:

$$E = \frac{P}{A_0}\left(\frac{L}{L_0}\right)$$

When a tensile force is applied in the x direction, the resulting stress (σ_x) produces a strain (ε_x) not only in the same direction but also in the transverse direction. The ratio of transverse strain to longitudinal strain in Poisson's ratio (ν), with

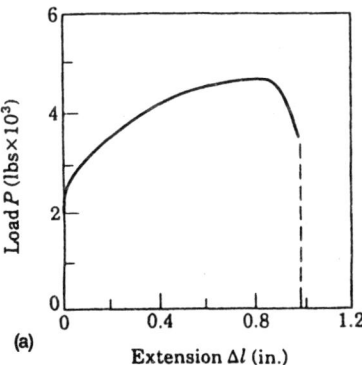

(a)

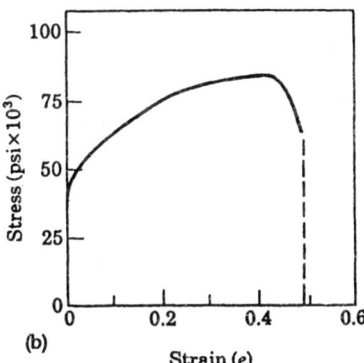

(b)

Figure 9 (a) Load–elongation curve in tension testing; (b) engineering stress–engineering strain curve obtained from data in (a).

$$\varepsilon_y = \frac{-v\sigma_x}{E}$$

Typical values of the elastic modulus are provided in Table 2.

For best results, to properly account for material variation, the elastic modulus should be determined experimentally instead of by using reference book values. The modulus is determined by attaching a strain gauge to the test specimen and then measuring the corresponding deflection with the application of different loads. The elastic modulus is determined graphically from the slope as shown in Figure 9 or by regression analysis:

$$Slope = E\frac{K}{1+v}$$

where K is a stress factor that is dependent on the particular metal being tested. For steel, K is approximately 1.7×10^{-4}.

4. Stress in Round Bars

Localized longitudinal stresses in round bars that vary over the cross section but not along the length may be measured by the deflection technique as shown in Figure 10. This is done by etching grooves in the bars and recording the resulting change in curvature [7]. The moment (M) due to relief of longitudinal stress (σ_z) in a circular cross section of thickness h and subtending an angle ϕ is

$$M = \int_0^h \int_{-\Phi z}^{+\Phi z} \sigma_z[(R-h)\cos\phi + C](R-h)\cdot d\phi\, dh$$

where R is the outer radius of the shaft and C is the centerline displacement of the neutral axis.

The surface longitudinal stress of the round bar may also be expressed as

$$\sigma_L = \frac{1.65E'\delta r}{L^2}$$

where E' is the elastic modulus, δ is the deflection of the bar, L is the length of the bar, and r is the radius of the bar.

5. Stress in Flat Plates

The bending moment (M) of a flat plate (or sheet) is

$$M = E'I/R$$

Table 2 Typical Room Temperature Values of Elastic Modulus for Isotropic Materials

Material	Modulus of elasticity (10^{-6} psi)	Shear modulus (10^{-6} psi)	Poisson's ratio
Aluminum alloys	10.5	4.0	0.31
Copper	16.0	6.0	0.33
Steel (plain carbon and low-alloy)	29.0	11.0	0.33
Stainless steel	28.0	9.5	0.28
Titanium	17.0	6.5	0.31
Tungsten	58.0	22.8	0.27

Source: Ref. 3, p. 38.

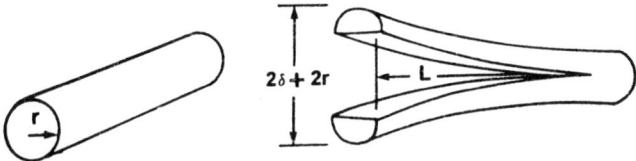

Figure 10 Localized longitudinal stresses in round cylindrical bars. (From Ref. 3.)

where I is the moment of inertia of the split section (see Figure 11), R is the radius of curvature, and E and E' are defined as above.

If it is assumed that the residual stress distribution varies linearly over the thickness t, the maximum longitudinal stress (σ_L) becomes

$$\sigma_L = Mc/I$$

where c is the distance from the neutral axis to the outer surface ($t/4$). Rearrangement of the above equations produces

$$\sigma_L = \frac{E't}{4R}$$

The radius of curvature R is related to the deflection δ by

$$R = \frac{L^2}{2\delta}$$

Combining the above relationships produces the equation for longitudinal residual stress for a flat plate,

$$\sigma_L = \frac{E't\delta}{2L^2}$$

6. Strain Gauge Technology

Strain gauges have been used to measure material strain since 1856 [8]. Since that time various strain gauge designs have been developed, including carbon, bonded wire, foil, and semiconductor strain gauges [9]. Of these, foil strain gauges are the most commonly used today. However, wire strain gauges are still used for high temperature (400–110°C; 750–230°F) strain measurement [8].

The basic concept is to attach an electrically conductive material such as a metal foil to the material of interest. The resistance of the foil is measured before and after the material is

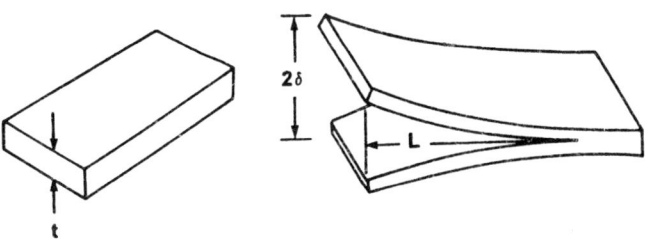

Figure 11 Stresses in a flat plate. (From Ref. 3.)

loaded. The strain in the material is proportional to the change in resistance. This is shown numerically by

$$F_S = \frac{\Delta R / R_0}{\Delta L / L_0}$$

where F_S is the strain sensitivity factor, ΔR is the change in resistance (ohms), R_0 is the initial resistance of the conductor (ohms), ΔL is the change in length (inches or millimeters), and L_0 is the initial length of the conductor (inches or millimeters). The $\Delta L/L$ term is often referred to as engineering strain or microstrain and is usually reported in microinches per inch.

Although the above relationship implies that the strain sensitivity factor of a metallic conductor (F_S) is linear, this is not always true. Generally, only strain gauge alloys that exhibit a strain sensitivity of 2.0 in the elastic region will exhibit linear behavior over large strain ranges [9]. Values of strain sensitivity for various metals are shown in Table 3.

The most commonly used strain gauges are those with resistance values of 120, 350, or 1000 ohms. For specific information on strain gauge selection, the reader is referred to Reference 10. Commonly available strain gauge rosettes are illustrated in Figure 12.

7. Hole-Drilling Methods

Residual stress may be measured by a method in which a hole is drilled into the material being tested and the change in strain is measured, usually with a strain gauge. Residual stress is then calculated from the magnitude and direction of this strain, hole size, and material properties.

There are many hole-drilling methods. The more common ones are summarized in Table 4. However, one of the most commonly used methods is the classical Sachs bore-out method [12], which is shown in Figure 13. Changes in residual stress with depth can be determined by incrementally drilling the hole and measuring the changes in stress with depth [14].

The sensitivity of the strain measurement and the accuracy of the stress calculation may be improved by using the reverse taper hole-drilling method that is illustrated in Figure 14 [13]. The reverse taper hole is cut by using an inverted cone drill bit and a high-speed air turbine drive. The tapered hole provides a greater hole size in the region of measurement and an increase in flexibility of material close to the hole, which provides greater strain relief and greater measurement sensitivity.

The Sachs bore-out method involves four assumptions [4]:

1. The metal is effectively isotropic and has constant Young's modulus and Poisson's ratio.
2. Residual stresses are distributed with rotational symmetry about the axis of the bar.
3. The tube formed by boring the bar is circular in cross section, and its inner and outer walls are concentric.
4. The specimen is sufficiently long to prevent bending.

The stress equations are [12]

$$\sigma_z = E' \left[(A_1 - A) \frac{d\Lambda}{dA} - \Lambda \right]$$

$$\sigma_\theta = E' \left[(A_1 - A) \frac{d\theta}{dA} - \frac{(A_1 + A)}{2A} \theta \right]$$

Table 3 Strain Sensitivities of Metals

Metal or alloy	Trade name	Typical strain sensitivity (elastic range)	Comments
Iron	—	+4.0	
Copper	—	+2.6	
Silver	—	+3.0	
Nickel	—	+12.0	Approximately value at low strains (<500 με)
Platinum	—	+6.0	
Titanium (commercially pure)		−1.1	
Titanium (6Al4V alloy)	—	−0.2	
Aluminum	—	+0.85	At strain levels below 750 με
Copper-nickel 55–45 (constantan)	Advance Cupron, Copel, etc.	+2.1	
Nickel-chromium 80–20	Nichrome V, Tophel A, etc.	+2.2	
Copper-nickel-manganese (manganin)	Manganin	+0.6	
Nickel-chromium 75–20 + additions	Karma, Evanohm, Chromel R, etc.	+2.1	In precipitation-treated condition; varies with cold work and specific composition
Iron-nickel-chromium	18–8 Stainless	+2.6	Varies with cold work and specific composition; precipitation-transformation
Iron-chromium-molybdenum (isoelastic)		+3.5	In highly cold-worked condition at low strain levels, in ordered condition
Gold-chromium	—	0	At low strain levels, in ordered condition
Iron-chromium-aluminum	Armour D, Alloy 815, Alchrome special, etc.	+2.4	Varies widely with composition
Copper-gold		−80 to −120	
Platinum-tungsten 92–8	Alloy 479, etc.	+4.1	Highly ordered condition, low strains

Source: Ref. 9, p. 5.

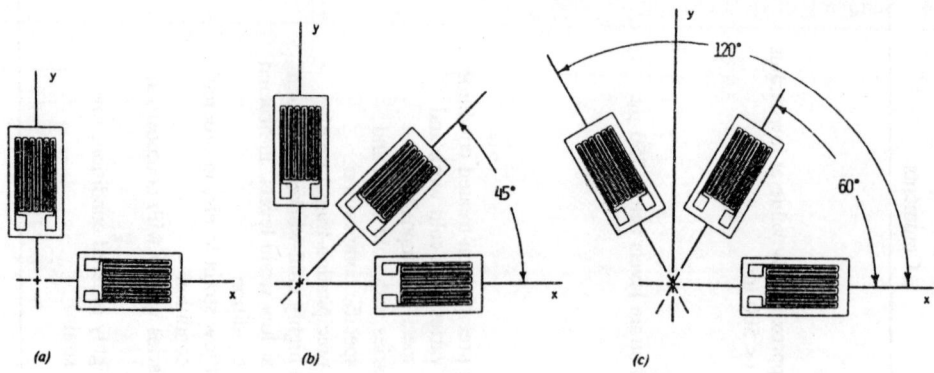

Figure 12 Strain gauge configurations. (a) Biaxial rosette; (b) rectangular rosette; (c) delta rosette. (From Ref. 8, p. 20.)

$$\sigma_r = E'\left[\frac{(A_1 - A)}{2A}\theta\right]$$

where

$$E' = \frac{E}{1 - v^2} \qquad \text{and} \qquad A_1 = \pi r_1^2$$

where r_1 is the external radius of the cylinder,

$$A = \pi r^2$$

where r is the current internal radius of the cylinder,

$$\Lambda = \lambda + r\theta$$

where λ is the change in longitudinal surface strain due to removal of the inner layer and θ is the corresponding change in circumferential strain, and finally Θ deformation is:

$$\Theta = \theta + r\lambda$$

The final stress distribution must satisfy the equilibrium conditions for the initial state:

$$\int_{A_0}^{A} \sigma_z dA = 0, \qquad [\sigma_r]A = A_1 = 0$$

and

$$\int_{r_0}^{r_1} \sigma_\theta dr = 0, \qquad [\sigma_r]_A = A_0 = 0$$

where $A_0 = \pi r_0^2$, r_0 being the initial radius of the bore.

When strain gauges are used to measure the residual stress of the drilled hole, for either of the methods illustrated by Figures 13 and 14 the residual stress equations are [13]

$$\varepsilon = \bar{A}(\sigma_{max} + \sigma_{min}) + \bar{B}(\sigma_{max} - \sigma_{min}) \cos 2\phi$$

Table 4 Summary of Hole-Drilling Methods

Method	Date	Ref.	Comments
Sachs	1927	12	A hole in a cylinder or tube is bored out in stages, and longitudinal and circumferential stresses are measured.
Mathar	1934	15	Measure stress-dependent changes in shape of hole using a mirror extensometer.
Soete and VanCrombrugge	1950	16,17	Simliar to Mathar's except strain is measured with electrical resistance strain gauge. Structure assumed to be infinite, homogeneous isotropic elastic plate under biaxial stress.
Riparbelli	1950	18	Effective use of electrical resistance strain gauge connected to Wheatstone bridge.
Bolten and Ten Gate	1952	19	Improved on Soete and VanCrombrugge and Riparbelli by double integration of varying strain area length and width of gauge.
Kelsey	1956	20	Proposed a method to measure residual stresses that vary with depth below the surface. Provides both magnitude and direction of principal residual stress.
Nisida and Takabayashi	1965	21	Used birefingent coating to measure residual stress by hole-drilling method.
Rendler and Vigness; Gupta	1966	22,23	Improved on Kelsey's method by eliminating the need for individual, complementary calibration tests.
Nathgate	1968	24	Developed method relating residual stresses and orientation at base of a blind hole to measured surface strain relaxation around drilled hole.
Shewchuck et al.	1976	25,26	Method requiring no supplemental calibration specimen.

Source: Ref. 11

where

ε = relieved strain measured by the strain gauge

σ_{max} = maximum principal stress

σ_{min} = minimum principal stress

ϕ = angular coordinate of the radial axis of the strain gauge measured counterclockwise from the maximum principal stress

$\overline{A}, \overline{B}$ = calibration coefficients

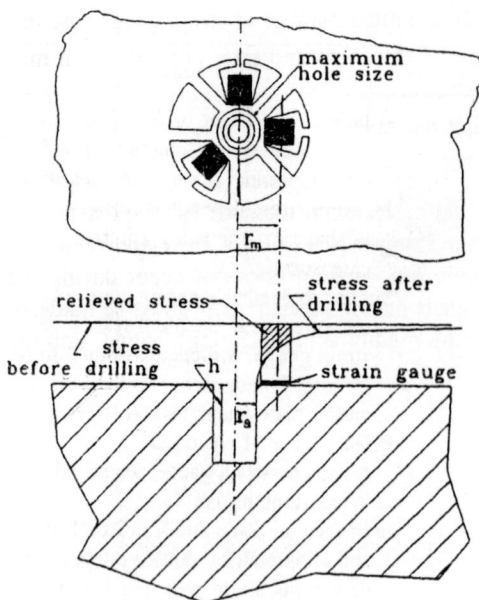

Figure 13 Hole drilling and strain gauge arrangements. (From Ref. 13.)

The principal stresses and their orientation can be calculated from the following equations. *For each strain gauge,*

$$\sigma_{max}, \sigma_{min} = \frac{\varepsilon_1 + \varepsilon_3}{4\bar{A}} \pm \frac{[(2\varepsilon_2 - \varepsilon_1 - \varepsilon_3)^2 + (\varepsilon_1 - \varepsilon_3)^2]^{1/2}}{4\bar{B}}$$

$$\phi - \frac{1}{2}\arctan \frac{2\varepsilon_2 - \varepsilon_1 - \varepsilon_3}{\varepsilon_1 - \varepsilon_3}$$

where the epsilons are the strains measured by three strain gauges and ϕ is the clockwise angle of gauge 1 in the direction of ε_{max}.

The particular variation of the hole-drilling method currently most often used in industry is ASTM E837. Although commonly used, this method suffers from a number of sources of potential errors [18]:

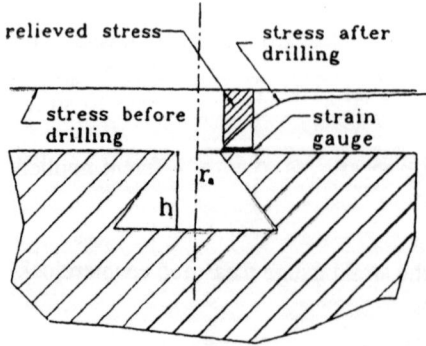

Figure 14 Cross section of a tapered hole. (From Ref. 13.)

1. Calibration constants are available for only a limited number of strain gauge rosettes.
2. The calculation does not account for material thickness and size, both of which may affect the accuracy of the measurement.
3. There may be stress gradients surrounding the hole.
4. Residual stress may vary at larger interfaces.

The Sachs bore-out procedure, while one of the best known methods for the determination of residual stress, also has a number of disadvantages [4,27]: It is slow and relatively expensive, care must be taken to ensure that plastic deformation does not occur during the hole-drilling process, strain gauge corrections for drift during measurement must be made, it can only measure final stress and therefore cannot be readily applied during cooling, and the hole-drilling process damages the test specimen.

$$\delta_z = E^1 \left[(A_1 - A)d\frac{\Lambda}{dA} - \Lambda \right]$$

8. X-Ray Diffraction

A nondestructive alternative to hole-drilling methods for the measurement of surface residual stresses is X-ray diffraction analysis [28]. When steel is irradiated with X-rays, a characteristic diffraction pattern that is dependent on the crystal structure of the iron and alloying elements persists. Examples of possible crystal structures for steel are provided in Figure 15.

The interplanar spacing for any set of parallel planes passing through the crystal lattice is called d spacing. Bragg's equation permits the calculation of the interplanar d spacing from the X-ray diffraction data:

$$n\lambda = 2d \sin \theta$$

where n is an integer, λ is the wavelength of the X-ray beam, d is the spacing between reflecting planes, and θ is the angle of incidence of the beam with the sample. This relationship is illustrated in Figure 16 [29].

When a load is applied to the sample, there will be a perturbation in the d spacing (Δd). From this perturbation, the strain ($\Delta d/d$) can be calculated as

$$\frac{\Delta d}{d} = -\cot \theta \left[\frac{\Delta 2\theta}{2} \right]$$

There are a number of methods for calculating residual stress from diffraction data. The most common is the $\sin^2 \psi$ method, in which the steel specimen is irradiated and changes in the diffraction angle (ψ) pattern are related to the interplanar spacing (d) and thus to strain ($\Delta d/d$).

The change in interplanar spacing is determined by measuring d with different applied stresses. The stress (σ) is calculated from

$$\sigma = \frac{d - d_0}{d_0} \left(\frac{E}{1 + v} \right) \left(\frac{1}{\sin^2 \psi} \right)$$

The d spacing is determined from the Bragg equation [30,31].

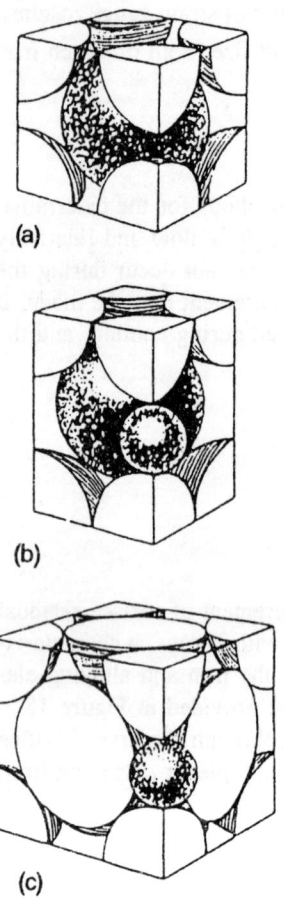

(a)

(b)

(c)

Figure 15 Possible crystal structures for steel. (a) Ferrite: body-centered cubic (bcc): (b) martensite: body-centered tetragonal (bct); (c) austenite: face-centered cubic (fcc). (From Ref. 30, p. 447.)

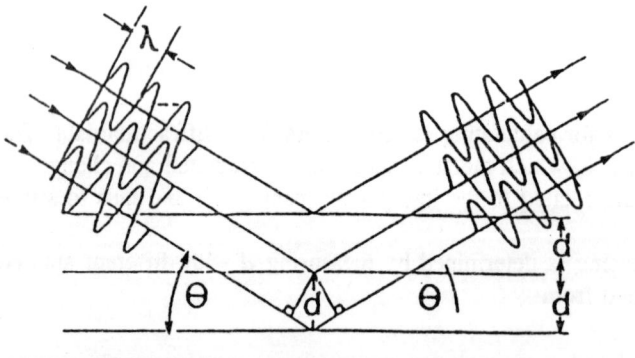

Figure 16 Illustration of the Bragg relation. (From Ref. 29.)

A $\sin^2 \psi$ plot is constructed as shown in Figure 17 [5]. The residual stress can be calculated from the slope:

$$\text{Slope} = \frac{\sigma(1 - v)}{E}$$

which can be rearranged to yield

$$\sigma = \text{Slope} \times \frac{E}{1 + v}$$

The stress (σ) can be readily calculated because the elastic modulus (E) and Poisson's ratio (v) are known.

Possible sources of X-ray measurement errors include [30]

1. Error in peak position
2. Stress relief by aging
3. Sample anisotropy

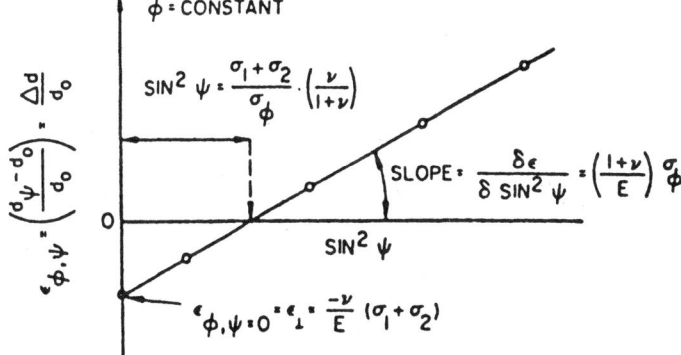

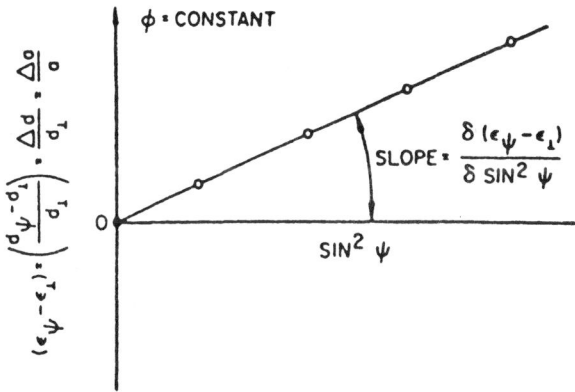

Figure 17 Plots of measured lattice strain relationships. (From Ref. 5.)

4. Grain size
5. Round surfaces (vs. flat)

Because of these and other potential errors, it is recommended that multiple measurements be made at distances at least twice the diameter from the end of the bar stock.

9. Other Residual Stress Measurement Methods

There are a number of other less commonly used but valuable experimental methods that have been used for residual stress measurement. These include Barkhausen noise [31,32], ultrasonics [33–35], and neutron diffraction [36]. The reader is referred to the indicated references for a detailed description of these methods and their use.

D. Heat Treatment Prior to the Hardening Process

Steel supplied by the steel processor to the machining operation is often required to be in a stress-free condition. When a steel is hot-worked, the material cannot work harden and the steel can be processed easily until its temperature falls into the cold-working range where work hardening stresses start to build up and increase the hardness. After metalworking, the forgings or rolled products are often given an annealing or normalizing heat treatment to reduce hardness so that the steel may be in the best condition for machining. These processes also reduce the residual stresses present in the steel.

Annealing and normalizing are terms used somewhat interchangeably, but they do have specific meaning. Both terms imply heating the steel above the transformation range. The difference lies in the cooling method. Annealing requires a slow cooling rate, the thought being that equilibrium conditions are approached; whereas normalized parts are cooled faster in still, room-temperature air. Annealing can be a lengthy process but produces relatively consistent results, whereas normalizing is much faster (and therefore favored from a cost point of view) but can lead to variable results depending on the position of the part in the batch and the variation of section thickness in the part being stress-relieved.

1. Normalizing Type Treatments

Normalizing always involves transforming the steel to the austenitic condition by heating to about 100°F above the A_{c3} temperature as defined in the iron–iron carbide phase diagram. A homogenous face-centered cubic structure should be produced prior to cooling. Cooling then usually takes place in air, and the actual cooling rate depends on the mass being cooled.

These treatments have three main purposes:

1. To control hardness for machinability purposes. Metal that is not too soft is often believed to cut cleaner with less tool wear.
2. To control structure. Heating to above the austenitizing temperature range will allow the material to recrystallize on cooling and to form a fine grain structure having superior mechanical properties.
3. To remove residual stresses on heating. However, if cooling is not controlled, a new stress system may result after cooling.

2. Stress Relief Treatments

Stress relief is used to remove stresses residual from other manufacturing steps. The treatment involves controlled heating to a temperature below A_{c1}, holding for the required time, and then cooling at a rate to avoid the introduction of thermal stresses. The stress relaxation involves microscopic creep. Creep-resistant materials will be more difficult to stress relieve, and the results will be dependent on both time and temperature as correlated by the Larson–Miller equation,

Thermal effect = $T(\log t + 20) /1000$

where T is temperature in Rankine degrees and t is time in hours.

Resistance of a steel to stress relief is related to the yield strength at the treatment temperature. The temperature selected should be at the point where the material yield strength corresponds to an acceptable level of residual stress remaining in the part. Contrary to popular belief, residual stresses cannot be reduced to zero. Figure 18 shows the yield strength of SAE 9310 steel at elevated temperatures, where it can be seen that residual stress levels will decrease very slowly at temperatures above 300°C (572°F). After treatment, uniform cooling is vital. Otherwise the thermal stresses can cause a new system of residual stresses that may be greater than those the treatment was intended to relieve.

Subcritical treatment can also reduce stresses by overtempering previously hardened parts. This type of treatment is also favored because lower temperatures are used and the results can be more uniform, but grain-refined structures are not produced.

3. Terminology

The functions and use of terms such as hardening, normalizing, annealing, stress relieving, and tempering often overlap and may even be confusing at times. For example, a stress-relieving treatment may appear exactly the same as a high temperature tempering treatment, and in fact they can be identical. To maintain clarity, a treatment should always be referred to by the name that indicates its *primary purpose*.

IV. DISTORTION DURING MANUFACTURING

The causes of distortion will be considered during five separate stages of manufacturing and processing:

1. Prior to heat treatment
2. During heat-up for heat treatment
3. At treatment temperature, i.e. during carburizing, nitriding, etc.

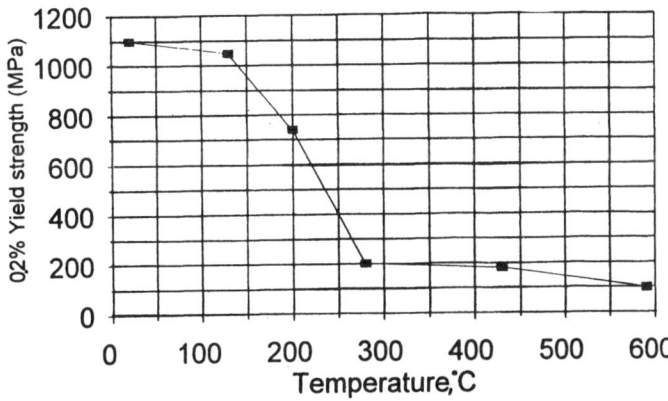

Figure 18   Decrease in yield strength with temperature for SAE 9310 steel. (INFAC DATA)

4. During quenching and cooling
5. During postquench processing

A. Manufacturing and Design Factors Prior to Heat Treatment that Affect Distortion

Manufacturing and design factors that will affect distortion prior to heat treatment may be summarized as

1. Material properties
2. Homogeneity of properties across the cross section of the material
3. The residual stress system magnitude and distribution
4. The part geometry

1. Material Properties

Material properties affect distortion response in several ways. As discussed above, the strength properties have important effects on the response to stress-relieving treatments and also on movement during differential thermal expansion and on the residual stresses caused during quenching. The composition also is related to the hardenability, which determines the phase changes during quenching. These properties can vary according to the actual composition of the steel used. The composition specification allows a range for each element, which means, in practice, that each batch of steel is unique and will respond slightly differently.

2. Homogeneity of Material

The first variable that must be considered is the material source, starting with the steel supplier. Compositional variations across the section of the cast ingot can cause different responses during heat treatment. The processing of the steel into the form required by the manufacturing process (forgings, bar stock, plate, tubes, etc.) can cause further variations and may leave high levels of residual stress, which may be removed partially by normalizing or another stress-relieving process. Since these heat treatments are usually carried out on large batches, they produce variable results from part to part, which causes different responses in subsequent processing. Steel supplied to the manufacturers of precision parts is typically either forgings or rolled products, which are made from ingots or continuously cast products. In rolling and forging, the steel is heated into the 1050–1200°C (1900–2200°F) range and then worked by hammering, pressing, or rolling to break down the cast structure and produce a homogeneous cross section in both composition and structure. However, the effects of earlier processing are never totally eradicated, and careful examination shows that they persist into the finished components.

Figure 19 shows typical solidification patterns that occur when steel solidifies in an ingot. Solidification starts at the walls and finishes at the center of the ingot. Note the geometrical effect of processing a square ingot to finish with a cylindrical component. This causes the possibility of different structures occurring at 90° positions. It is also common to observe differences between the top and bottom of an ingot and also between the first ingots to be cast from a heat of steel and the last ingots to be poured. As an ingot solidifies, a compositional gradient also arises for each element in the alloy. For example, manganese, in a nominal 1.0% manganese steel could vary from 0.8% at the surface to 1.2% in the center. These variations cause variable responses in hardenability, microstructure after heat treatment, residual stress levels, and consequently distortion.

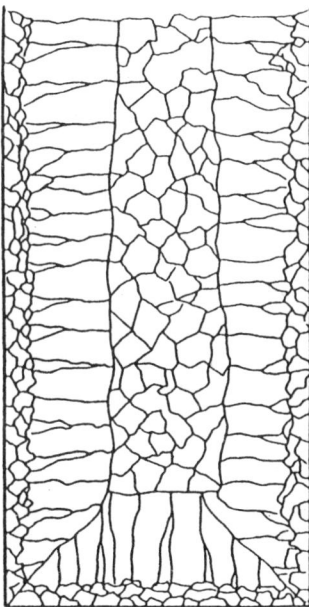

Figure 19 Ingot solidification pattern. (From Ref. 67.)

3. Distribution of Residual Stress System

If the source of steel supply is consistent and the steel is processed the same way every time, these effects cause consistent, predictable residual stress behavior that is acceptable. However, if the steel is coming from different melt shops, rolling mills, and forgers with different processing schedules, heat treatment and residual stress responses can vary, often without apparent explanation. Most steel is hot rolled, and after rolling it is allowed to cool in air on a "hot bed." This causes a difference in cooling due to conduction of heat from the bottom of the bar and convection cooling from the top. If the bar is allowed to cool completely in this position, the top of the bar will have residual tensile stresses that will tend to bend the bar and make straightening necessary. Straightening can produce very high levels of residual stresses, and a stress-relieving treatment must be carried out. For critical components a complete understanding of the processing that the steel has undergone must be obtained by discussion between suppliers and users.

4. The Part Geometry

Nonuniform heating and quenching can be caused by changes in section thickness in the same component. When a part is designed, most enlightened designers recognize the need to keep the section sizes as uniform as possible to minimize temperature gradients and the tendency to produce high stresses due to differential expansion and contraction during heating and quenching. If a part is made with features such as gear teeth, however, it is unavoidable that these areas will have higher surface-to-volume ratios than the rest of the parts and that gear teeth will often tend to heat and cool faster than the rest of the section. Due to the coefficient of expansion, the base of the tooth will be restrained by the rim, and this area will tend to go into compression during heating and into tension during cooling. Similar effects will take place elsewhere in the part wherever there is a change of section.

B. Distortion During Component Heating

The major effects during heat-up are initiated by three distortion-causing mechanisms acting at the same time:

1. Shape change due to relief of residual stress
2. Shape change due to thermal stresses causing plastic flow
3. Volume change due to phase change on heating

1. Shape Change Due to Relief of Residual Stress

As discussed earlier, the presence of residual stresses will cause shape changes to occur if the stresses are relieved by heating the part to a point where the yield strength of the material decreases below the residual stress level in the material. The extent of the resulting plastic deformation will therefore depend on the magnitude and distribution of the stress fields in the part.

2. Shape Change Due to Thermal Stresses

If a simple form could be heated at the same rate throughout the section, it would expand uniformly at a rate determined by the coefficient of thermal expansion but maintain the same shape. In actual practice, as the part heats up, the surface will heat first and will expand or try to occupy a greater volume than the colder internal material. Expansion of the outer layers is therefore constrained by the colder, stronger inner layers of material. Compressive stresses will be present in the outer layers on heating, balanced by tensile stresses in the interior of the component. Shape change depends on the geometry of the part, the heating rate, the coefficient of thermal expansion, the material properties, and the fixturing of the part.

3. Volume Change Due to Phase Change on Heating

The shape and volume changes as transformation occurs, depending on the heating rate, the part geometry, and the phase volume change (i.e., on heating, the pearlite-to-austenite phase change causes approximately a 4% contraction; see Figure 4).

A study of these factors shows that the major control of distortion during heat-up is the heating rate, and differences in heat-up rate (due, for example, to position in load) will lead to inconsistent distortion. It is also evident that the slower the heating rate the less shape change is likely. Unfortunately this is in direct conflict with normal practice, since in the interest of production rate the parts are usually heated as fast as possible to the treatment temperature.

C. Distortion During High Temperature Processing

Once the parts are at a constant temperature there are some minor factors that will cause shape change, but the major changes will occur on cooling from processing. Carburized parts can be direct quenched from carburizing or just below carburizing temperature, or they can be slow cooled, given an optional temper, reheated to austenitizing temperature, and quenched. The latter treatment is used to give optimum case properties. The factors to be considered during and after high temperature processing are

1. Volume expansion during diffusion treatments during processing
2. Distortion due to creep
3. Cooling from processing, which can set up a residual stress system

1. Volume Expansion During Case Diffusion

There is a measurable volume expansion during diffusion treatments, the extent depending on the diffused element (carbon, nitrogen, etc.), the depth of diffusion, the concentration profile, and the furnace temperature and atmosphere uniformity. The major heat treatment used for high quality parts is a case hardening process designed to form a hard surface layer on the gear surface. This layer gives the part a hard, wear-resistant finish but also sets up a compressive stress system at the surface that helps to resist fatigue failures. In components having gear teeth and bearing races, the type of fatigue encountered is usually pitting fatigue in areas of high loading and slipping under concentrated contact.

The conventional process is carburizing, although nitriding is used for parts particularly susceptible to distortion. Carburizing involves the diffusion of carbon from a gaseous atmosphere while the part is heated to about 925°C (1700°F) in an atmosphere or vacuum furnace. Carbon is introduced to a level of 0.70–1.00% at the surface. After carburizing, the part is quenched, usually in oil to produce a hard martensitic layer on the surface. The diffusion times used are usually in the range of 4–20 h depending on the temperature of treatment and the case depth required. The case depth required by the designer is related to the size of the part and is often greater the larger the part to produce the correct residual stress pattern. Nitriding involves the diffusion of nitrogen from a gaseous atmosphere in the temperature range of 495–565°C (925–1050°F). It may be done in an atmosphere furnace or in vacuum ion-nitriding equipment. After nitriding, the parts are hard without quenching, and they have some degree of compressive stresses due to compound formation in the surface layers. Nitriding takes anywhere from one day to one week because of the slow diffusion rates. Since nitriding is performed at relatively low temperatures and quenching is unnecessary, distortion is a minor problem; accordingly, nitriding will not be considered any further in this discussion. Another diffusion process sometimes used is carbonitriding, the simultaneous diffusion of carbon and nitrogen, generally for lower cost parts.

2. Distortion Caused by Metal Creep

Distortion due to creep will depend on the geometry of the part and the support during processing, the temperature and time of treatment, and the creep strength of the material. Any heat treatment process that requires heating the part to above the transformation temperature causes the part to lose most of the strength it possesses at room temperature. A part subjected to elevated temperatures for extended times (as in carburizing) could creep under its own weight unless it is properly fixtured and supported. Long slender parts are best suspended vertically. If this is not practical, the support should have the same contour as the component resting on it.

3. Residual Stresses After Cooling

Cooling after heat treatment will cause temperature gradients that are proportional to the rate of cooling. The differential expansion across the section will cause shape changes that will result in the retention of stresses after the part has cooled. Slow cooling will minimize the residual stresses retained in the part.

D. Distortion During Component Hardening

Distortion during component hardening is related to four factors:

1. Properties of phases present in heat-treated steels
2. Homogeneity of properties across the section of the material
3. The residual stress system magnitude and distribution
4. The part geometry

1. Properties of Heat-Treated Materials

Parts harden because of phase transformations that usually occur with accompanying volume changes. These are predictable, however, if all parts undergo the same transformation sequence.

Quenching is often carried out as a batch process, leading to variations in response because the quench severity will often not be the same for all parts of the batch. Components are almost always made by processes that leave residual stresses in the material. Any forming or machining process leaves stress systems that will be relieved by a dimensional change during heat treatment. Thus, if the part is heavily stressed prior to heat treatment, the shape will change due to this factor alone. Processing should be arranged so that virtually stress-free parts are heat treated. Variations in heat treatment parameters such as case carbon level and processing temperatures will also cause differences in final shape and size.

Whichever method of surface hardening is selected (except for nitriding), the mechanism of hardening remains the same. For simplicity, only three phases are considered in the following discussion:

1. Pearlite, a lamellar eutectoid mixture of body-centered cubic ferrite and orthorhombic cementite (Fe_3C)
2. Austenite, a face-centered cubic structure, relatively soft and stable
3. Martensite, a hard, body-centered tetragonal metastable phase

When pearlite, which probably exists within the microstructure of a part prior to heat treatment, is heated to carburizing temperatures, the dense austenite phase is formed, so the percentage volume will decrease by 4.64 – 2.21 (%C). There is 0.10–0.25% C in the core. If the steel contains 0.10% C, on heating its volume will decrease by [4.64 – 2.21 (0.1)]%, or 4.4%. Since the heating for carburizing is relatively slow and both phases are relatively soft, the part will accommodate this change. The linear changes are approximately one-third of the volume change or, in this example, 1.4%. For comparison, if a 25 mm (1 in.) long part is being machined to a tolerance of 0.13 mm (0.005 in.) (a wide tolerance), this represents an allowable deviation of only 0.5% in the length of the part. In general, therefore, linear changes due to phase transformations cannot be ignored.

2. Material Properties Across Heat-Treated Sections

There are further complications. An implicit assumption in the above discussion is that the carbon content of the steel is uniform across the cross section. This may not be true because (1) segregation occurs and (2) some heat treatments call for two case depths on a part. Thus, when the part is reheated to diffuse in the second case, the carbon content can vary from 0.1% to 1.0%, leading to a range of linear change of 0.8–1.4% in different areas of the part.

Carburizing is usually carried out using the "boost–diffuse" method: The furnace atmosphere is first controlled at a high carbon potential, and later in the cycle the atmospheric carbon potential is reduced to the desired level. Depending on how the boost–diffuse cycle is controlled, different carbon profiles are produced, the result being different transformation temperature gradients with different phase expansion characteristics. Changing the type of carburizing cycle will change the distortion behavior.

The most difficult volume changes to understand are those caused by quenching. Part of the effect is due to the phase changes on cooling. In general, shape and volume changes depend on the quenching rate, the part geometry, and the phase volume changes (i.e., the transformation of austenite to martensite causes a 4% expansion).

If a component is quenched under conditions of uniform heat abstraction, the situation may be represented by Figure 20. This figure represents a cylinder that has been carburized and heated to a temperature above the austenitizing temperature and then quenched. The isochronal lines show how the surface cools faster than the center of the section because heat is abstracted from the surface by the quenching medium. This trend continues throughout the quenching process. Figure 20 also shows lines representing the start of martensitic transfor-

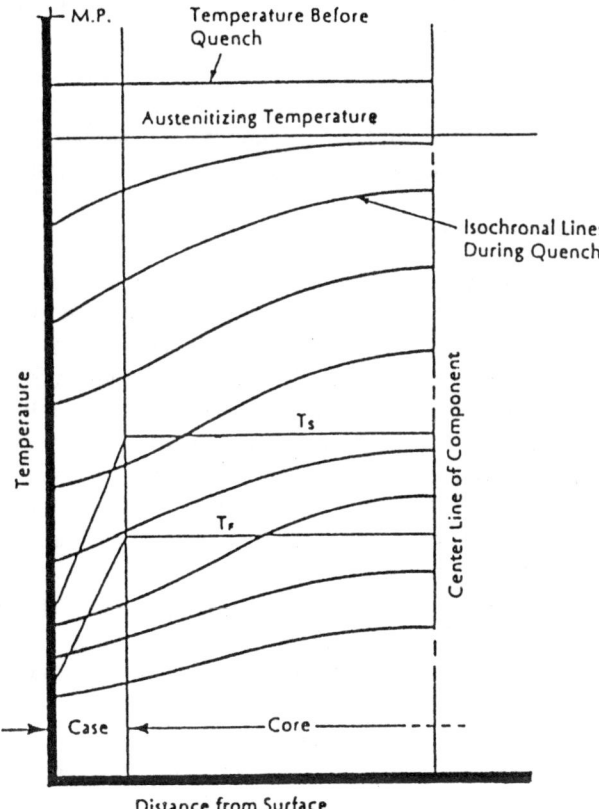

Figure 20 Temperature distribution in steel parts during the quench after carburizing. (From Ref. 67, p. 254.)

mation (T_S) and the finish of transformation (T_F). It will be seen that these temperatures are depressed as the case carbon increases. The net result is that the transformation of austenite to martensite starts at the case/core interface with an expansion as martensite is formed. The case is the last material to transform, and the expansion to martensite causes compressive stresses because the core is already transformed and restrains the tendency of the case to expand.

If the T_F temperature is below room temperature, austenite will be retained. This lowers the surface hardness and will cause dimensional instability if very low temperatures are encountered in service because more austenite will transform, resulting in a subsequent volume increase. Usually a subzero treatment is performed after carburizing and quenching if the carburized case is likely to retain excessive austenite. In practice, a small amount of retained austenite (about 15%) is often thought to be desirable and can improve the performance of the part. It is normal to temper the freshly formed martensite.

Although Figure 20 assumes "conditions of uniform heat abstraction" at the surface of the part being quenched, this is a most unlikely occurrence in practice. One of the major difficulties in distortion calculations is knowing what the heat abstraction is at all parts of the surface of the quenched body. The work of Pilling and Lynch [37] showed that the mechanism of heat transfer during quenching may be considered in three stages. Quenchants typically go through three stages of cooling as the part cools. During the first or "vapor" stage, cooling is slow due to an insulating blanket of gas. This gradually breaks down, and quenching enters the second or "agitated boiling" stage, where heat abstraction is the highest. The final stage is one of convection cooling in liquid, which is entered below the boiling point of the quenching fluid.

What actually happens during quenching is that two or even three of these stages can be occurring in different areas of the part surface at the same time. Phase transformations then take place over a wide time interval, with the resulting volume expansions and the rigid martensite phase occurring at different times.

Consider a part being quenched. The following factors will affect heat removal:

1. The importance of mechanical agitation during quenching has been stressed by many studies, including the earliest comprehensive account of quenching by French [38]. Some natural movement of the quenching medium always occurs when hot parts are quenched by virtue of the convection currents set up in the liquid. This liquid movement past the hot part will slow down as the temperature difference between the liquid adjacent to the metal surface and the bulk of the medium decreases. It is therefore desirable that the quench bath be agitated, but the quenchant cannot be directed at all areas of the surface at the same time. The part has to be in a flowing stream of quenchant, which has the effect of having maximum cooling on the surface first impacted but has minimum effect on the downstream surfaces.

2. Geometry of the part will affect cooling according to the direction of the quenchant stream relative to major areas of flat surface. Russell [39] showed that during the cooling of a sphere without agitation of the quenchant, the bottom cools at a slower rate than the top because the convection currents in the liquid are concentrated around the top of the sphere (Figure 21a). Russell's work extended to cylinders shows that a long cylinder (Figure 21c) has a greater proportion of effective vertical cooling surface than a short cylinder (Figure 21b), and if the volume-to-surface area ratios are equal the long cylinder will cool faster. Howes [40] constructed a set of three rectangular specimens as shown in Figure 22. According to the direction of quenching, the 40%, 70%, or 90% vertical surface could be exposed to the quenchant. After

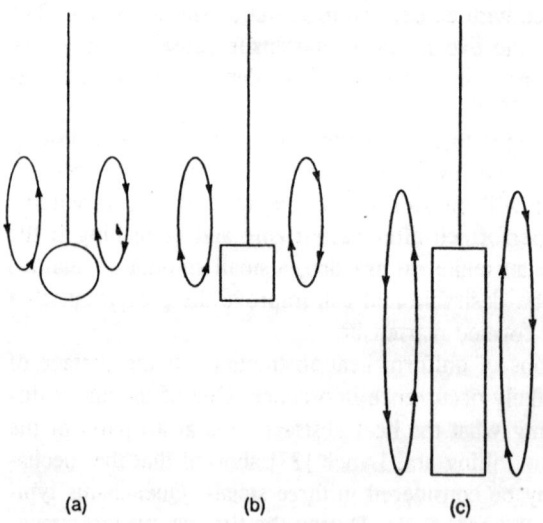

(a)   (b)   (c)

Figure 21 Convection currents existing around (a) a sphere, (b) a short cylinder, and (c) a long cylinder. (From Ref. 40 and Ref. 67, p. 255.)

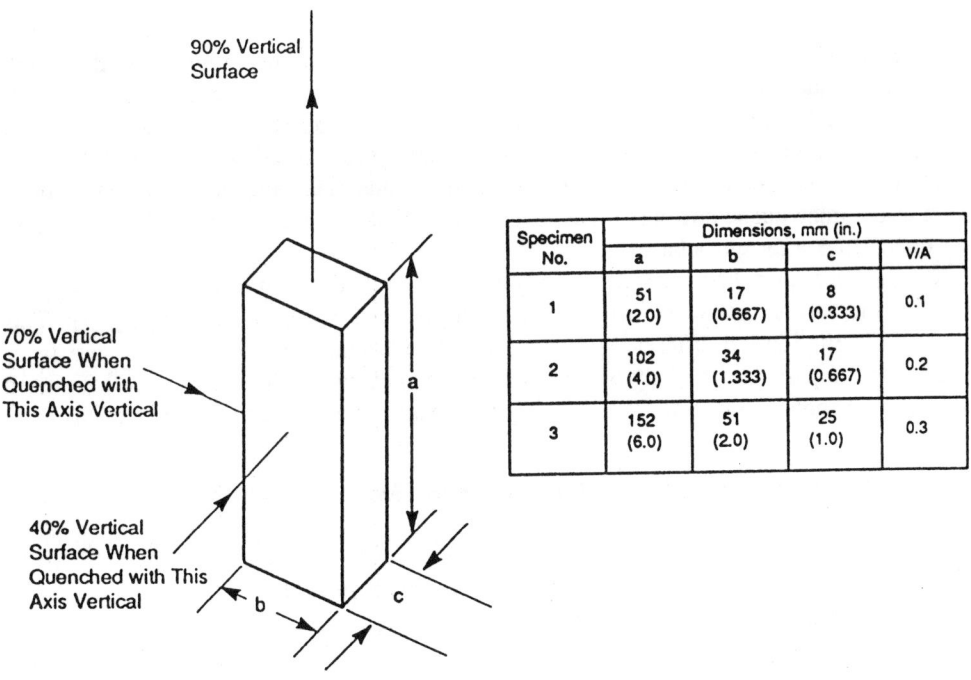

90% Vertical Surface

70% Vertical Surface When Quenched with This Axis Vertical

a

40% Vertical Surface When Quenched with This Axis Vertical

b

c

Specimen No.	Dimensions, mm (in.)			
	a	b	c	V/A
1	51 (2.0)	17 (0.667)	8 (0.333)	0.1
2	102 (4.0)	34 (1.333)	17 (0.667)	0.2
3	152 (6.0)	51 (2.0)	25 (1.0)	0.3

Figure 22   Dimensions of three rectangular specimens that were quenched in three different directions. (From Ref. 40 and Ref. 67, p. 255.)

quenching, center cooling with the 40% vertical surface exposed took about one-third longer than with the 90% vertical surface exposed. (Figure 23).

Deterioration of the quenchant may also affect heat abstraction rates. It may occur after long service and must be taken into account if distortion is to remain under control. Quenchants should be tested at intervals dictated by experience and should be replaced or reconditioned as necessary.

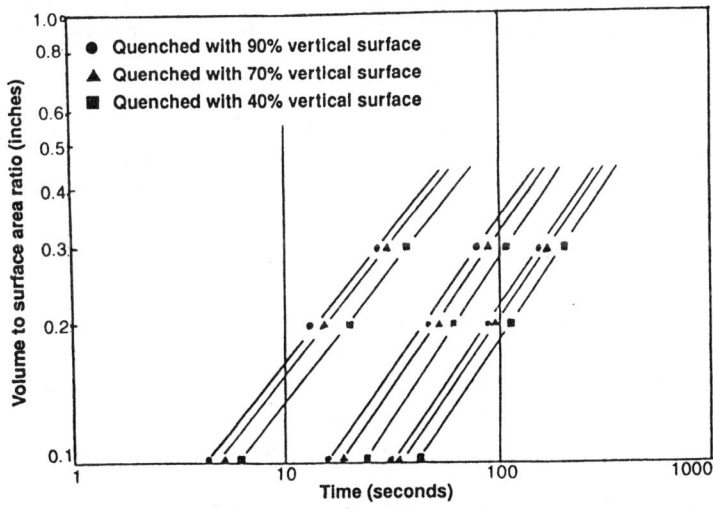

● Quenched with 90% vertical surface
▲ Quenched with 70% vertical surface
■ Quenched with 40% vertical surface

Volume to surface area ratio (inches)

Time (seconds)

Figure 23   Center cooling times for the three rectangular shapes described in Figure 22. Quenched in molten salt at 200°C. (From Ref. 40.)

3. *Press Quenching*

Press quenching involves quenching usually one part at a time by restraining it in dies while controlling the quenchant flow to different parts of the surface until the part is fully cooled to a predetermined temperature. Since a batch of parts is not involved and quenchant flows are controllable within a wide range, this quenching method should offer the best possibilities for control. It is an expensive quenching method, and considerable experimental work is often needed to optimize quenching conditions. It is widely used for precision parts that must have exact dimensions after quenching.

Although the quenched parts are under closely controlled conditions while in the press, very often the part is transferred manually from the reheat furnace to the quench press. This leads to variability due to differences in operators. The operator may have power assistance for handling the part, but the only way to ensure documented consistency is a well-designed and controlled robotized transfer system with the reheat furnace under computer control.

V. DISTORTION DURING POSTQUENCH PROCESSING

There are many possible treatments that can be carried out after quenching. Four typical operations are

1. Stabilization with a low temperature temper treatment
2. Subzero treatment
3. Retempering after subzero treatment
4. Metal removal by grinding, etc.

A. Tempering Treatments

Tempering treatments relieve stresses to an extent that depends on the treatment temperature. This may cause shape change on heating, depending on the steel properties and the treatment.

B. Subzero Treatments

Subzero treatment can cause a size change by further austenite-to-martensite transformation resulting in further expansion. If the size change is restrained, then additional stresses will be locked in. This effect depends on the M_s–M_f temperature range, the temperature and time of subzero treatment, and the creep strength of the material.

C. Metal Removal After Heat Treatment

A finishing process such as grinding or hard finishing is often required to correct dimensional changes caused by heat treatment. The tendency is to try to use parts as heat treated without touching the surface again (except perhaps for lapping), because in this condition the part can have a much higher fatigue strength [41]. This is particularly true for parts loaded under concentrated contact such as bearings or gears.

For parts with close tolerances, however, the component size must be brought under control in the finish grinding or hard turning stage. This leads to a dilemma: if excess material is left on the part prior to heat treatment, there will be enough stock to enable the size to be brought under control. However, if too much is taken off, the most effective regions of the carburized (or nitrided) case are removed. Figure 24 shows a gear with excessive material being removed from a tooth after heat treatment.

In the example shown in Figure 25, the tooth has distorted to the right, and to correct the profile, excess stock has to be ground from the right side of the tooth. This has several

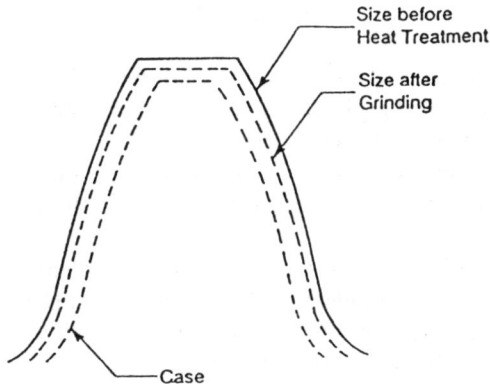

Figure 24   Schematic of material ground from a gear tooth after heat treatment. (From Ref. 67.)

serious consequences. First, the lack of uniformity in case depth leads to uneven residual stress distribution. Second, and worse, is that the gear appears satisfactory in a nondestructive inspection even though its performance will be less than optimum. Third, a considerable thickness of material has to be removed during grinding, increasing the probability of grinding burns. Some problems that are blamed on grinding can doubtless be traced back to badly controlled heat treatment and the prior machining processes.

Thus, the effects of the heat treatment process have to be considered before and after the process in both the soft machining and hard finishing stages if the final dimensions and structure are to be successfully controlled.

VI. CARBURIZING VS. DIRECT HARDENING AND SELECTIVE HARDENING

As an alternative to carburizing and nitriding, a hard case may be produced by selective direct hardening. Instead of increasing the carbon content of the steel surface by diffusing extra carbon into the case, a medium-carbon steel composition that already contains 0.4–0.6% carbon is selected. This steel could then be through-hardened and tempered in a furnace, but a hard case and tough core can also be produced by selectively heating the surface where the case is required to a controlled depth and leaving the core in the original hardened and tem-

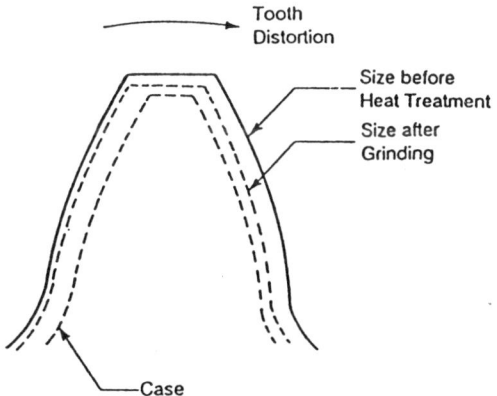

Figure 25   Schematic of material ground from a distorted gear tooth after heat treatment. (From Ref. 67.)

pered condition. Both methods produce a case, one by controlling depth of carbon or nitrogen diffusion and the other by the depth of heating.

Through-hardened components are not carburized and hardened but are made from a medium-carbon steel. These steels can be through-heated to above their austenitizing temperature and then quenched and tempered to produce a hard tempered martensitic structure. This treatment causes transformation to proceed from the surface to the center of the section and often results in unfavorable tensile residual stresses at the surface and increased tendency to crack during quenching or later during grinding.

An alternative method of direct hardening medium-carbon steels is selective hardening. Energy is transmitted quickly into the surface by applying induction, laser, or electron beam heating on the specific areas where surface hardening is required. The surface layer is thereby heated to above the austenitizing temperature and will later become the hardened case. When the energy is turned off, rapid cooling progresses (due to self-quenching) and, as in carburizing, the case is the last to transform; the restraint induces residual compressive stresses as the surface expands during transformation from austenite to martensite. This situation is illustrated in Figure 26, which shows how the final results are similar to those obtained with a carburized case.

The advantages of selective hardening using high energy sources are that the process avoids the lengthy carburizing period and can thus be completed in seconds or minutes. Also, since only a small portion of the part is heated at any time, most of the structure is cold and rigid

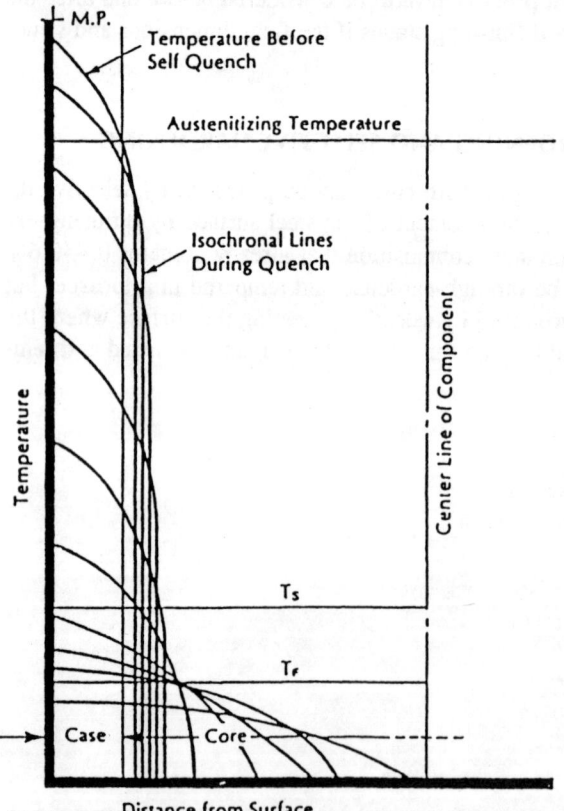

Figure 26   Isochronal plots of section after high-energy surface hardening. (From Ref. 67.)

enough to preclude distortion. From a practical point of view, these processes can often be in the manufacturing line, avoiding the batching used for carburizing and other diffusion processes, eliminating part transportation, and making inspection easier. Part-to-part consistency is also likely to be better. Consistency of part structure will only be achieved, however, if processing conditions are kept constant. The energy applied to the work surface must be carefully controlled as to power level and time, as too little energy results in a shallow case, if any, whereas too much can cause surface melting. Both conditions can affect final dimensions.

The nonuniform heating can cause high stresses due to the coefficient of expansion of the steel. This may be enough to upset thin sections and cause distortion after the self-quench or applied liquid quench. Another difficulty is that sometimes parts are "scanned" by induction, flame, or laser, and where the scan ends, an area with tensile residual stresses can exist. The hardening process should be carefully designed to prevent or minimize this effect.

VII. TESTS FOR PROPENSITY FOR DISTORTION AND CRACKING

Numerous tests have been applied to evaluate the potential of a steel to undergo distortion or cracking upon heat treatment. Test specimens range from simple balls or bars to actual parts with intricate designs. This section provides an overview of test specimens that have been reported for this purpose. In most cases, the test specimens are manufactured specifically for these test procedures. One of the most difficult challenges is to devise a testing procedure that accounts for the statistical nature of the occurrence of cracking, because it is seldom that 100% of all parts actually undergo cracking in the heat treatment process.

A. Cylindrical, Tubular, and Solid Block Test Specimens

Many workers have simply quenched cylindrical test specimens and observed them for cracking and/or volumetric changes. For example, Moreaux [42] used round bar test specimens of 0.60% C, 1.6% Si, and 0.5% Cr steel whose length was three times their diameter. These studies showed that the transition temperature from film to nucleate boiling contributes primarily to thermal stress. The nucleate boiling to convective cooling transition will primarily affect the formation of transformational stresses. Beck [43] studied the effect of the interrelationship between quench severity and average cooling rates on the severity of crack formation. The work was performed using 34 mm (1.3 in.) diameter 60SC7 steel cylindrical test specimens, and the results are shown in Figure 27.

Owaku [44] identified two forms of quench cracking using both round bars and solid blocks. In general, two kinds of cracking were identified for nonuniform cooling as shown in Figure 28. Pull cracking is caused by nonuniform surface cooling during quenching. Push

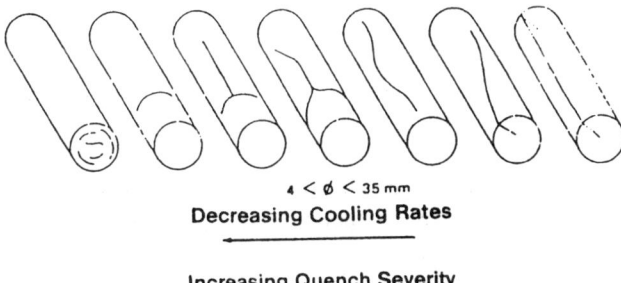

$4 < \phi < 35\,mm$
Decreasing Cooling Rates

Increasing Quench Severity

Figure 27 Quench cracking morphology of a 34 mm diameter cylinder of 60SC7 steel quenched in water. (From Ref. 43.)

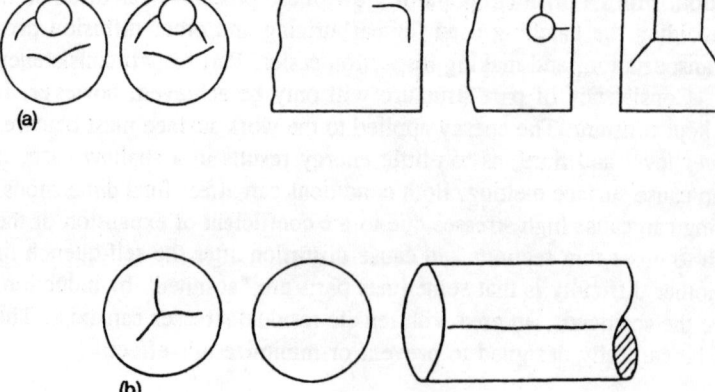

Figure 28 Two forms of quench cracking. (a) Pull cracking; (b) push cracking. (From Ref. 44.)

cracking is caused by nonuniform cooling within the body of the metal during the quench and occurs in rapidly cooled areas.

Mamatkulov and Domnina [45] used the following test pieces to measure the potential distortion of tools and machine parts during heat treating:

1. Cylinders 10 mm in diameter and 100 mm long, to evaluate changes in length
2. Tubular pieces 40, 65, and 75 mm in diameter and 20 and 25 mm thick, to determine change in diameter
3. Rings with an outside diameters of 40, 65, and 75 mm, inside diameters of 20 and 25 mm, and thickness of 20 and 25 mm, to determine changes in inside and outside diameter
4. Long parts of small cross section (5 × 100 × 140 mm), to determine change in flexure

Rectangular solid specimens of 16 × 16 × 40 mm or 20 × 20 × 40 mm were used by Kubota [46] to study the potential for quench cracking with a variety of quenchants. Three test specimens were used. Two were quenched from the optimum austenitizing condition, and one from 900°C, which represented an overheated condition. After quenching, the test specimens (Figure 28) were inspected for cracks. The possibility of quench cracking is determined by whether the sum of thermal and transformational stress is compressive or tensile. In general, no cracking occurs if compressive stress remains. However, cracking will occur if tensile stress remains [46].

B. Plates with Holes

In addition to solid blocks as shown in Figure 29, Owaku [44] used blocks with holes to examine cracking mechanisms. Hirose et al. [47] also used plates with holes or slots to evaluate the potential for distortion to occur. Changes in both inside hole and outside plate (block) dimensions were measured because the particular stresses operating on each of these dimensions are different. These results may be used for more complex shapes.

C. Step Bar Test

A "step bar" test using the test specimen shown in Figure 30 has been used to measure the effect of cross-sectional size on distortion [48]. In this test, the "running" distortion of a long bar per unit length is measured.

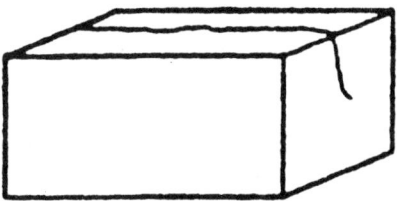

Figure 29 Appearance of quench cracking. (From Ref. 44.)

D. Navy C-Ring and Slotted Disk Test

One of the oldest standard tests for evaluating quench distortion is the so-called Navy C-ring test [49-51]. Examples of various reported test specimens are shown in Figure 31.

A modified Navy C-ring test (see Figure 32) has also been reported [51]. The notched test specimen imparts greater crack sensitivity.

A variant of the Navy C-ring test employs the slotted disk used by Kakuchi [52]. This test specimen is a relatively thin 8 mm (0.3 in.) disk with a 3 mm (0.12 in.) outside diameter, a 15.5 mm (0.6 in.) outside diameter, and a 1 mm (0.04 in.) slot or gap.

E. Finned Tubes

The distortion of hollow tubes has been studied by Mikita and coworkers [53,54] using the "finned tube" test specimen shown in Figure 33, where a large hole is drilled into a solid block, effectively creating a "fin" on one end. In this test, two types of cracks; fin and vertical, were quantified. The susceptibility for fin cracking can be increased by cutting a V-notch in the fin [53]. Mikita et al. [54] determined the relative cracking frequency using a limited number n of test specimens by quenching each specimen repeatedly until a crack appeared. Sensitivity for quench cracking was designated as the $1/n$ value.

VIII. ESTIMATION OF DISTORTION AND CRACKING BY CALCULATION

Kunitake and Susiyawa [55] found that there was no, or very poor, correlation of the occurrence of quench cracking with M_s (martensite start temperature), D_I, carbon content, or grain size. However, they did report excellent correlation of quench cracking with the calculated alloy-dependent value designated the "carbon equivalent" (C_{eq}), which is defined as

$$C_{eq} = \%C + \frac{Mn}{5} + \frac{Mo}{5} + \frac{Cr}{10} + \frac{Ni}{50}$$

where %C is the percent carbon content and Mn, Mo, Cr, Ni are the alloy contents in percent. Correlation of C_{eq} with cracking frequency is shown in Figure 34. Generally, steels are classified as crack-sensitive when the C_{eq} value is greater than 0.52–0.55 [55].

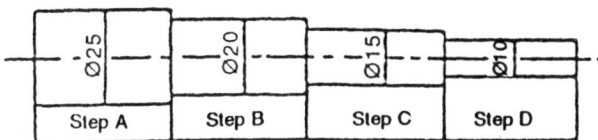

Stepped Bar Test

Figure 30 Stepped bar test.

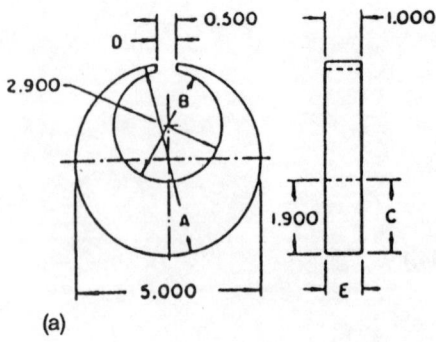

(a)

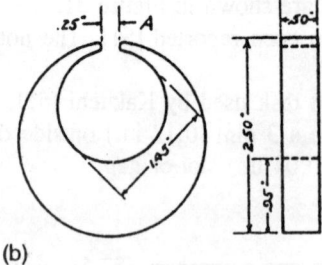

(b)

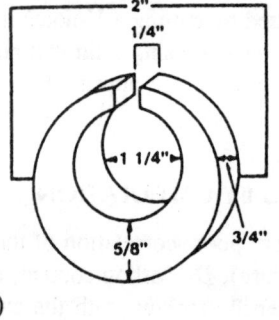

(c)

Figure 31 Examples of C-ring test specimens used for quench distortion studies. ((a) from Ref. 49; (b) from Ref. 38; (c) from Ref. 50.)

Thelning reported that volumetric expansion occurring during quenching can be calculated from

$$\frac{\Delta V}{V} \times 100 = (100 - V_c - V_a) \times 1.68(\%C) + V_a[-4.64 + 2.21(\%C)]$$

where $\Delta V/V$ equals the percent change in volume, V_c equals the percent by volume of undissolved cementite, V_a equals the percent by volume of austenite, and $100 - V_c - V_a$ equals the percent by volume of martensite, and C equals the percent by weight of carbon dissolved in austenite and martensite, respectively. The volume of typical ferrous structures is shown in Table 5 [56].

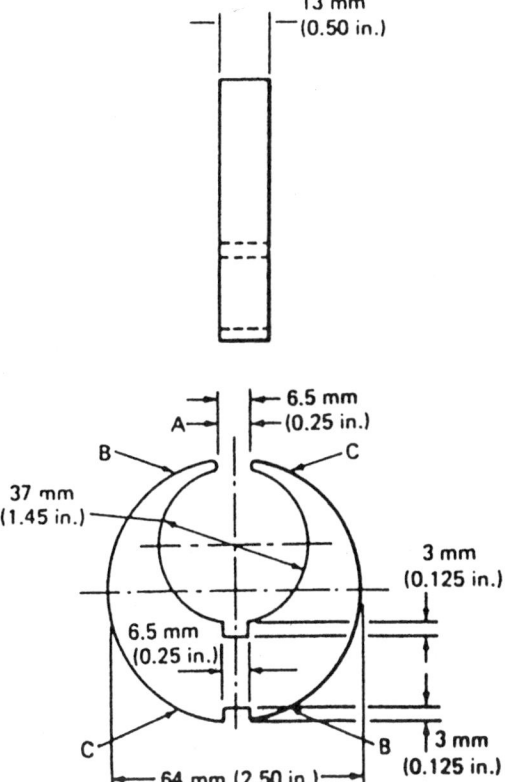

Figure 32 Modified Navy C-ring distortion test specimen. (From Ref. 51.)

Bavaro [56] showed that

1. Pearlite does not involve any variation in volume.
2. Bainite causes an increase in volume.
3. Martensite increases in volume with increasing carbon content.
4. Austenite causes a redirection in volume that decreases with decreasing carbon content.

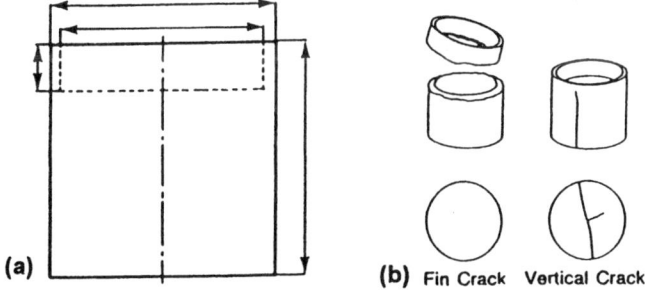

Figure 33 (a) Finned tube specimen; (b) types of cracks. (From Ref. 53.)

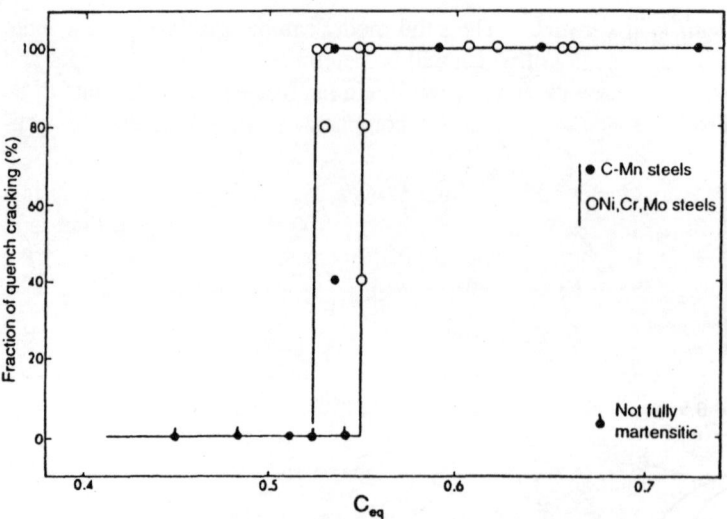

Figure 34 Relation between fraction of quench cracking and carbon equivalent (C_{eq}). (From Ref. 55.)

IX. PREDICTION OF DISTORTION

A. Factors for Consideration

Many attempts are being made to predict distortion in simple shapes by first considering coolant characteristics, heat transfer coefficients, flow rates, agitation, surface geometry, material conductivity, latent heats, etc. This information is then combined with phase transformation data and deformation and stress calculations using Poisson's ratio, yield strength, work hardening modulus, and density changes. To be of practical use these predictions should recognize the variability occurs in parts during processing. Once manufacturers recognize that variability in materials and processes does have important effects on precision parts at later stages in manufacturing, better process controls and the tracking of quality will be accepted as being beneficial and cost-effective.

B. Current State of the Art

The many factors described demonstrate that prediction of distortion is no easy task, but some first attempts can be made. A suggested approach would be as follows.

1. *Define the shape.*   The dimensions of the part can be defined, preferably as a three-dimensional CAD presentation of the part as a solid model.

Table 5 Volume of the Structure Constituents of
Ferrous Alloys

Phase	Apparent atomic volume (Å^3)
Ferrite	11.789
Cementite	12.769
Ferrite + carbides	$11.786 + 0.163(\%C)$
Pearlite	11.916
Austenite	$11.401 + 0.329\ (\%C)$
Martensite	$11.789 + 0.370\ (\%C)$

Source: Ref. 56.

2. *Construct a finite element network.* The solid model can be translated into a finite element network, the complexity of which will be determined by the size, shape, and form of the part. Experience will show the optimum number of mesh elements to be used. This will also be related to the available computer power and the desired accuracy.

3. *Perform quenching calculations.* The cooling of every network cell during the quench can be calculated based on the known characteristics of the quenching medium and the part. These will include the heat transfer coefficient of the quenchant under the conditions specified by the designer and those naturally present in the quenching system, as well as the following items:

Initial temperature of the part

Temperature of the quenchant

Cooling characteristics of the quenchant

Agitation of the quench medium

Orientation of quenchant flow to surface being cooled

Complexity of surface features

Conductivity of the part

Latent heat data

4. *Calculate the phase transformations.* Phase transformations can be calculated using time–temperature–transformation data available on the source material for the part. For carburized and nitrided steels the differences in composition and thus hardenability and phase boundary temperatures must be allowed for. Continuous cooling data will be the preferred source.

5. *Calculate deformation and stress.* These calculations can be made with available data on Poisson's ratio, yield strength, work hardening modulus, and density changes due to phase changes. All these data are obviously temperature-dependent. The output of these calculations will be a predicted final shape and the level of residual stresses.

As part of the predictive methodology, it is necessary to be able to calculate how every element of the part being quenched will respond in the chosen quenchant. The vast literature dealing with quenching shows that many early attempts were made to develop a method of predicting cooling curves [57–64]. These methods are often unsatisfactory because of the difficulty in obtaining the data on which some of the mathematical analyses are based. The following assumptions are often made.

1. That Newton's law of cooling is true under all quenching conditions. This law is usually expressed in the form

$$\frac{dQ}{dt} = C(T_s - T_m)$$

where C is assumed constant, T_s = surface temperature of the part, and T_m = temperature of the quenchant. Heindlhofer [63] showed that this law is not strictly true unless $T_s - T_m$ is small, which is not a valid assumption under normal quenching conditions.

2. That physical properties of the material are constant over the entire temperature range considered, which is another invalid assumption.

3. That the heat transfer coefficient of the quenchant is constant. However, this factor has been shown to vary during quenching [38]. It will also be dependent on quench bath temperature and degree of agitation.

The complexity of issues and assumptions makes it necessary to have complete data available on the geometry of the part being quenched, on the material used for the part, and on the quenchant. Most methods of calculation have been applied only to simple symmetrical shapes, which are useful as exercises but are rarely reproduced in engineering practice.

Although many factors must be considered in predicting distortion, a level of understanding has been reached to enable development of the data for building the relationships between the variables from which to construct predictive equations. A major difficulty may be in calculating heat abstraction during the quench because of the many variables involved. Validation of proposed distortion prediction methods will be a complex process because the number and range of variables will be extensive. The benefits of a successful predictive method would be most useful to the manufacturing and heat treatment industries, and even a more comprehensive understanding of the interaction between manufacturing and processing factors would be a big step forward.

C. Summary

Production of quality components with distortion under control can be considered as a value-added chain, and everyone in the chain from the material maker onward should have a good working knowledge of how the chain links together to produce the final components. Too often each link operates in almost total ignorance of even the adjacent links, much less the entire chain. For example, the steelmaker should know that how he makes his product can cause problems in controlling part dimensions later in the manufacturing sequence. The machinist should know that variation in production methods can change final dimensions. The furnace maker should know how variations in both atmosphere flow and quench fluid flow will affect consistency of product. This awareness by all manufacturing personnel is vital to ensuring final quality.

REFERENCES

1. D. Dodds, Gear Research Institute meeting, Chicago, 1992.
2. H. Walton, Dimensional changes during hardening and tempering, in *Proceedings of the First International Conference on Quenching and the Control of Distortion* (G. E. Totten, ed.), ASM Int., Materials Park, OH, 1992, p. 265–275.
3. G. E. Dieter, *Mechanical Metallurgy*, McGraw-Hill, New York, 1961, pp. 405–407.
4. A. A. Denton, Determination of residual stress, *Metall. Rev. 11*: 1–23 (1966).
5. A. A. Denton and J. M. Alexander, *J. Mech. Eng. Soc. 5*(1): 75 (1963).
6. E. M. Loxley, British Iron and Steel Research Assoc. Rep. MW/E 148/51.
7. G. Z. Serebrennikov, *Ind. Lab. 28*: 1177 (1963).
8. M. E. Tuttle, Fundamental strain-gauge technology, in *Manual on Experimental Methods for Mechanical Testing of Composites* (R. L. Pendleton and M. E. Tuttle, eds.), Soc. Exp. Mech., Bethel, CT, 1989, p. 17.
9. J. E. Starr, Basic strain gage characteristics, in *Strain Gage Users's Handbook* (R. L. Hannah and S. E. Reed, eds.), Soc. Exp Mech., Bethel, CT, 1992, pp. 1–77.
10. Anon., Optimizing Strain Gage Excitation Levels, M-M Tech. Note TN-502, Measurements Group, Inc., Micro-Measurements Div., Raleigh, NC.
11. R. E. Rowlands, Residual stress, in *Handbook on Experimental Mechanics* (A. S. Kobayashi, ed.), Prentice-Hall, Englewood Cliffs, NJ, 1987, pp. 768–813.
12. G. Sachs, The determination of internal stresses in rods and tube, *Z. Metalk. 19* (9): 352–357 (1927).
13. M. Tootoonian and G. S. Schafer, Enforced sensitivity residual stress management using taper hole drilling, *Proc. Fourth Int. Conf. Residual Stress*, June 8–10, 1994, Soc. Exp. Mech., Bethel, CT, pp. 52–62.

14. E. Proctor, Principles and practice of the center-hole residual stress measurements, in *Import Surface Treatment* (Proc. Conf. Bedford, UK, Sept. 22–26, 1986), Elsevier Appl. Sci., London, 1986.

15. J. Mathar, Determination of initial stress by measuring the deformation around drilled holes, *Trans. ASME 56*: 249–254 (1934).

16. W. Soete and R. VanCrombrugge, An industrial method for the determination of residual stress, *Proc. SESA VIII* (1):17–28 (1950).

17. W. Soete, Measurement and relaxation of residual stress, *Sheet Met. Ind. 26* (266): 1269–1281 (1949).

18. C. R. Riparbelli, A method for the determination of critical stress, *Proc. SESA VIII* (1): 173–196 (1950).

19. C. R. Bolten and W. Ten Gate, A routine method for the measuremenet of residual stress in plates, *Appl. Sci. Res.* (Sec. A) *3*: 317–343 (1952).

20. R. A. Kelsey, Non-uniform residual stress by the hole-drilling method, *Proc. SESA XIV* (1): 181–194 (1956).

21. M. Nisida and H. Takabayashi, Thickness effects in hole method and application of method to residual stress measurement, *Sci. Pap. IPCR 59* (2): 78–86 (1965).

22. N. J. Rendler and I. Vigness, Hole-drilling strain-gage method of measuring residual stress, *Exp. Mech. 6* (12): 577–586 (1966).

23. B. Gupta, Hole-drilling technique; Modifications in the analysis of residual stresses, *Exp. Mech 13*: 45–58 (1973).

24. R. G. Nathgate, Measurement of non-uniform biaxial stresses by the hole-drilling method, *Strain 4* (2), 1968.

25. A. M. Newar, K. McLachlan, and J. Shewchuck, A modified hole-drilling technique for determining residual stresses in thin plates, *Exp. Mech. 16* (6): 226–232 (1976).

26. A. M. Newar and J. Shewchuck, On the measurement of residual stress gradients in aluminum alloy specimens, *Exp. Mech, 18* (7): 269–276 (1978).

27. S. Nishimura, Y. Moorita, and K. Tokimasa, *Bull. Jpn. Soc. Mech. Eng. 18* (116): 116–122 (1975).

28. Residual Stress Measurement by X-Ray Diffraction, SAE J784a.

29. C. O. Ruud, Residual stresses and their measurement, *Proceedings of the First International Conference on Quenching and Control of Distortion* (G. E. Totten, ed.), ASM Int., Materials Park, OH, 1992, pp. 193–198.

30. G. E. Totten, C. E. Bates, and N. A. Clinton, *Handbook of Quenchants and Quenching Technology*, ASM Int., Materials Park, OH, 1993, pp. 483–486.

31. T. Piech and K. Pomorski, Barkhausen noise method for estimation of direction of residual stresses in surface layers, in *Residual Stress* (V. Hauk, H. P. Hougardy, E. Macherauck, and H.-D. Tietz, eds.), DGM Informations Gmbh, 1993, pp. 333–340.

32. K. Titto, Use of Barkhausen effect in testing for residual stresses and material defects, in *Residual Stress in Design, Process and Material Selection* (W. B. Young, ed.), ASM Int., Materials Park, OH, 1978, pp. 27–36.

33. R. E. Schromm, A. V. Clock, D. V. Mitrakovic, and S. R. Schaps, Ultrasonic measurements of residual stress in railroad wheels, in *Practical Application of Residual Stress Technology* (C. Ruud, ed.), ASM Int., Materials Park, OH, 1991, pp. 61–68.

34. R. B. Thompson, An overview of ultrasonic measurement techniques, *Proc. Fourth Int. Conf. on Residual Stresses*, Soc. Exp. Mech., 1994, pp. 97–111.

35. D. R. Allen and C. M. Sayers, The measurements of residual stress in textured steel using ultrasonic velocity combination technique, *Ultrasonics*, July 1984, pp. 179–188.

36. H. J. Praskard and C. S. Choi, Residual stress measurements in system components by means of neutron diffraction, in *Residual Stress in Design Process and Materials* (W. B. Young, ed.), ASM Int., Materials Park, OH, 1987, pp. 21–26.

37. N. B. Pilling and T. D. Lynch, *Trans. Am. Inst. Min. Metall. Eng. 62*: 669 (1920).

38. H. J. French, *The Quenching of Steels*, Am. Soc. for Steel Treating, 1930, p. 133.

39. T. F. Russell, *Iron Steel Inst. (Lond.) Spec. Rep. 24*: 283 (1939).

40. M. A. H. Howes, Ph.D. Dissertation, London University, 1959, p. 38.

41. M. A. H. Howes and J. P. Sheehan, SAE (Soc. Automot. Eng.) Tech. Paper 740222, 1974.

42. F. Moreaux, A. Simon, and G. Beck, *J. Heat Treat. 1* (3): 50–56 (1980).

43. G. Beck, *Mem. Sci. Rev. Metall.*, June 1985, pp. 269–282. Etud.

44. S. Owaku, Criterion of quench cracking: its sources and prevention, *Netsu Shori 30* (2): 63–67 (1990).

45. D. D. Mamatkulov and N. V. Domnina, Heat treatment to reduce distortion, *Izv. Vyssh. Uchebn. Zaved., Chern. Metall.* (8): 44–46 (August 1991).

46. T. Kubota, Quenching crack test for polymer quenchants, *Netsu Shori 25* (1): 34–37 (1985).

47. N. Hirose, S. Yamamoto, and K. Shinji, Quenching deformation in 13Cr steels, *Netsu Shori 26* (4): 268–273 (1987).

48. W. A. J. Moerdijk, *Adv. Mater. Process. 137* (3): 19–28 (1990).

49. *ASM Metals Handbook*, 8th ed., Vol. 2, *Heat Treating, Cleaning and Finishing*, American Society for Metals, Metals Park, OH, 1964, p. 41.

50. Tenaxal, *Ucon Quench A—The Fast Safe Way to Quench Steel Alloys*, Product Inf. Bull., Tenaxal, Inc., Milwaukee, WI, revised 1972.
51. H. E. Boyer and P. R. Cary, *Quenching and Distortion*, ASM International, Metals Park, OH, 1988, p. 39.
52. K. Kakuchi, Distortion by quenching SNCM 439 steel in circular ring with slit, *Netsu Shori 30C* (1): 40–44 (1990).
53. Y. Mikita, I. Nakabayaski, and K. Sakamaki, *Heat Treat.* December 1989, pp. 21–24.
54. Y. Mikita and I. Nakabayashi, *Nipon Kikai Gakkai, Ronbunshu (A-hen) 53* (489): 884–889 (1987).
55. T. Kunatake and S. Sugisawa, The quench-cracking susceptibility of steel, *Sumitomo Search 5*: 16–25 (1971).
56. A. Bavaro, Heat treatments and deformation, *Traitement Therm. 20*: 37–41 (1990).
57. M. J. Sinnot and J. C. Shyne, *Trans. Am. Soc. Met. 44*: 758 (1952).
58. H. J. French and O. Z. Klopsch, *Trans. Am. Soc. Steel Treat. 9*: 857 (1926).
59. T. F. Russell, *Iron Steel Inst. (Lond.) Spec. Rep. 14*: 149 (1936).
60. C. B. Post and W. H. Fenstermacher, *Trans. Am. Soc. Met. 33*: 19 (1944).
61. R. Jackson and R. J. Sarjant, J. B. Wagstaff, N. R. Eyres
62. D. R. Hartree and J. Ingham, *J. Iron Steel Inst. (Lond.) 150*: 211 (1944).
63. K. Heindlhofer, *Phys. Rev. (Ser. 2) 20* (3): 221 (1922).
64. V. Paschkis, *Trans. Am. Soc. Met. 37*: 216 (1946).
65. D. A. Canonico, Stress and relief heat treating of steel, in *ASM Handbook*, Vol. 4, ASM Int., Materials Park, OH, 1991.
66. Bates, Totem and Brennan, Quenching of Steel, in *ASM Handbook*, Vol. 4, ASM Int., Materials Park, OH, 1991, p. 67–120.
67. M. A. H. Howes, Factors affecting distortion in hardened steel components, in *Quenching and Control Distortion* (G. E. Totten, ed.), ASM Int., Materials Park, OH, 1992, p. 251–258.

6

HEAT TREATMENT EQUIPMENT

George E. Totten
Union Carbide Corporation, Tarrytown, New York

Gary R. Garsombke
ITW Shakeproof, Milwaukee, Wisconsin

David Pye
Pye Metallurgical Consulting Inc., Meadville, Pennsylvania

Ray W. Reynoldson
Quality Heat Treatment Ltd., North Bayswater, Victoria, Australia

I. INTRODUCTION

Some of the most important decisions that the heat treater will make are related to the selection of furnaces and ancillary equipment. These decisions involve selection of the energy source, gas or electricity, which is vital to the overall profitability of the heat treatment process. Another is the selection of the furnace transfer mode, batch or continuous, and the particular furnace type. If a furnace is being rebuilt, the proper choice and installation of the refractory material is vital.

In this chapter, the selection and operation of heat treatment equipment are addressed. The focus of the discussion is on the furnace, furnace atmosphere generation, and ancillary equipment. Discussion subjects include

Furnace part transfer mechanisms.

Furnace heating. Heat transfer principles and application to furnace calculations.

Atmosphere generation. Atmosphere sensors; thermocouples.

Refractory materials.

Fans.

Fixtures.

Parts washing.

Quenching systems.

Furnace safety.

Salt bath furnaces.

Fluidized bed furnaces.

II. FURNACE TRANSFER MECHANISMS

Heat treatment furnaces are classified as batch, semicontinuous, or continuous. In batch furnaces, which are the most common and most versatile in the heat treatment industry, the work is typically held stationary in the furnace vestibule. The furnace is loaded or unloaded in a single (batch mode) operation.

In continuous furnaces, the load moves through different zones, usually with varying temperatures as shown in Figure 1. Parts being heat treated moved through the furnace in a continuous process.

In semicontinuous furnaces, parts move through in a continuous but stepwise manner. For example, they may move through the furnace in a tray or basket. As the tray or basket is charged from the furnace, it is quenched. After the quench cycle is completed, the tray or basket moves on, the next one is discharged from the furnace and quenched, and the process continues in a stepwise manner.

A comparison of the features of selected examples of each furnace is provided in Table 1.

For the purposes of this discussion, semicontinuous furnaces are considered to be continuous. Batch-type furnaces discussed include box, pit, integral quench (IQ), and tip-up (both circular and car-bottom). Continuous-type furnaces discussed here include walking beam, rotary hearth, pusher, roller hearth, conveyor, shaker hearth, screw conveyor, and rotary retort. Table 2 provides a selection guide for these types of furnaces as a function of the heat treatment process. A comprehensive review of furnace applications and a summary of suppliers are available in Reference 4.

A. Batch Furnaces

1. *Box Furnace*

The box furnace, such as the one shown in Figure 2, is the simplest heat treating furnace. It is used for tempering, annealing, normalizing, stress relieving, and pack-carburizing. It is capable of operating over a wide range of temperatures, 95–1095°C (200–2000°F).

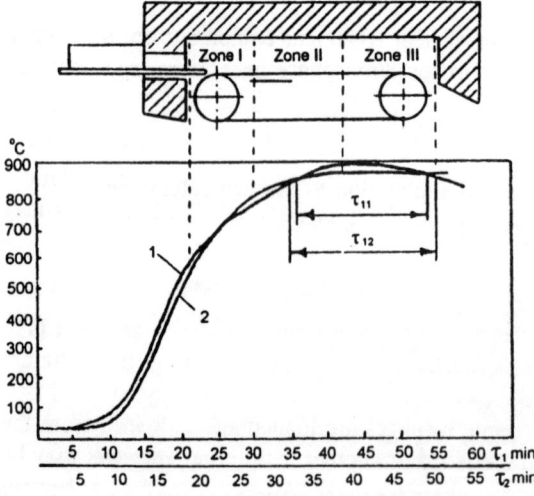

Figure 1   Temperature variation with distance through a continuous furnace (from Ref. 1.)

Table 1 Comparison of Types of Heat Treating Furnaces for Small Parts

Class of equipment	Furnace type	Versatility in use	Labor needs	Atmosphere quality	Quenchability
Batch	Integral quench	E	F	E	G
Batch	Salt pot	F	H	F	LD
Batch	Rotary drum	F	F	L	G
Semicontinuous	Pan conveyor	G	F	F	F
Semicontinuous	Tray pusher	G	F	G	F
Continuous	Belt shaker	F	L	F	E
Continuous	Shaker hearth	F	L	F	E
Continuous	Rotary retort	LD	L	F	E

E = excellent; G = good; F = fair; H = high; L = low; LD = limited.
Source: Ref. 2.

2. *Integral Quench (Sealed Quench) Furnace*

Integral quench (IQ) furnaces are among the most commonly used and most flexible furnaces for processing small parts. They can be used for either neutral or atmosphere hardening processes in addition to normalizing and stress relief.

Integral quench furnaces are similar to box furnaces except that the quench tank is located at the discharge end of the furnace as shown in Figure 3. As the parts are removed from the furnace, the baskets used to hold the parts being heat treated are lowered into the quenchant with an elevator.

The attached quench tank can be a disdvantage because the bath loading and height of the baskets restricts the size of the parts that can be quenched [2]. Another limiting factor is the size of the heating chamber.

Economics typically dictate that the largest furnace available with respect to the amount of work to be treated be used. Furnace design developments have led to ever greater automation, reduced cycle times, and greater fuel efficiency of these furnaces [5]. Cycle times at temperature are reduced even with newer furnace designs. High temperature carburizing is not practiced because of the limitations of materials engineering.

3. *Pit Furnaces*

Pit furnaces are circular furnaces that can be either floor- or pit-mounted. They are used for such processes as annealing, carburizing, tempering, normalizing, and stress relieving. Some of the advantages of the circular furnace are the smaller heat release areas in comparison to box or rectangular furnaces, more uniform atmosphere, improved temperature distribution, and smaller furnace body weight [6]. A pit furnace is shown in Figure 4. Such furnaces may also be used as "pit carburizers" or nitriders.

4. *Car-Bottom Furnaces*

Car-bottom furnaces are used for thermal processing of very large parts such as gears [11–13] and forgings [5]. They may be used for carburizing, annealing, hardening, normalizing, stress relieving, and tempering [3]. The bottom of the furnace is a refractory-covered flatbed railcar that moves on rails in the shop. Some furnaces are loaded and unloaded from the same end. Others are loaded at one end and unloaded at another. Still other furnaces may be loaded from the side to permit the use of more than one railcar. An example of a car-bottom furnace is shown in Figure 5.

Table 2 Furnace Selection Guide

Function	Production process	Atmosphere	Furnace types
	Continuous		Conveyor; pusher; rotary hearth; shaker hearth; roller hearth cast-link belt
Annealing steel	Batch		Car type
Normalizing steel	Batch		Semi type look-up
Spheroidizing steel	Batch		Semi-muffle oven
		Air	Full muffle oven
			Vertical muffle oven[a]
		Salt	Round pot
			Rectangular pot
Blueing steel	Batch	Steam or air	Round pot
			Rectangular pot
Bright annealing steel	Continuous	Air	Conveyor atmosphere controlled
(also copper, brass, etc.)		Air	Pusher atmosphere controlled
	Batch	Air	Full muffle oven atmosphere controlled
Carburizing steel	Continuous		Pusher
	Batch		Car type
			Semi-muffle oven
Cyaniding steel		Cyanide	Round pot
(or liquid carburizing)		or salt	
Forming and forging	Continuous		
steels (steel, brass,	1. Slugs, billets		Rotary hearth
copper, etc.)	2. Billets		Pusher type
	Batch		
	1. End or center heating of slugs		Open slot
	2. Large billets, heavy forgings, plates structural steel shapes, rods, bars, etc.	Oven type, direct fired	
Hardening steel	Continuous		Conveyor; pusher; rotary hearth; shaker hearth; roller heart cast-link belt
	Batch	Air	Car type; semi-muffle oven; full-muffle oven; verticaler muffle oven
Stress-relieving steel	Continuous		Conveyor air-recirculating; pusher air-recirculating
	Batch		Car-type air-recirculating
Tempering/drawing	Continuous		Conveyor air-recirculating; conveyor; pusher air-recirculating; Pusher
	Batch		Car type air-recirculating[b]; basket air-recirculating[c]

[a]The vertical muffle oven furnace is not commonly encountered.
[b]Used for high temperature stress relieving and subcritical annealing.
[c]Used for large quantities.
Source: Ref. 3.

Figure 2 Box furnace with electric rod overbend heating elements. (Courtesy of Lindberg—A General Signal Company.)

5. *Tip-Up (Lift-Off) Furnaces*

Some heat treatment furnaces are designed so that the top can be hydraulically lifted over the load to facilitate removal by a forklift or removal from a car-bottom railcar. These furnaces may be circular [6,7] or a variation of the car-bottom furnace.

B. Continuous Furnaces

Continuous furnaces are particularly suited to continuous heat treatment processing of parts that are the same or at least similar. A nomograph is provided in Figure 6 that interrelates heating time, furnace load, furnace length, and production rate for a continuous furnace [8]. In this section, an overview of the features of different furnace designs.

1. *Walking-Beam Furnaces*

The movement of parts through a walking beam furnace, illustrated in Figure 7, is by repeated lifting, moving, and lowering of the parts in the furnace in a "walking" action [4]. Walking beam furnaces are particularly suitable for continuous thermal processing of large parts and heavy loads.

2. *Roller Hearth Furnace*

In a roller hearth furnace, the load is moved through on externally driven heat-resistant alloy rollers. This furnace is best suited for continuous heat treatment of large parts and plates [4] and is illustrated in Figures 8 and 9.

(a)

(b)

Figure 3 Examples of integral quench (IQ) furnaces. [Photos courtesy of (a) Lindberg—A General Signal Company; (b) Abar Ipsen Industries; (c) Atmosphere Furnace Company.]

3. Pusher Furnaces

Pusher furnaces have one of the simplest furnace designs. The load is hydraulically or pneumatically mechanically pushed on rollers or skid rails inside the furnace as shown in Figure 10. One variant of this furnace design is the tray pusher furnace [2], which can be used for both neutral carburizing and carbonitriding in addition to annealing, normalizing, tempering, and stress relieving [9].

4. Mesh Belt Conveyor Furnaces

Belt conveyor furnaces may vary in size from very small to very large. They are similar to roller hearth furnaces except that the load is continuously moved through the furnace on a metal mesh, roller chain, or link chain as shown in Figure 11 [10]. The load may be moved continuously through other process lines such as quenching and tempering. Maintenance costs

(c)

Figure 3   Continued

may be high due to frequent belt repair or replacement, and since these furnaces are difficult to seal there may be significant leakage of protective atmospheres [2]. Figure 12 illustrates some of the designs of mesh belts that are available.

5.   Shaker Hearth Furnaces

The shaker hearth furnace (Figure 13) is another example of a continuous furnace [2,10] that is used primarily for heat treatment processing of small parts. Parts are moved through the furnace by a mechanically induced vibrating mechanism that has the advantage of not requiring a belt or chain to move back to the beginning.

6.   Screw Conveyor Furnace

In the screw conveyor furnace, parts proceed through the furnace on one or more screw mechanisms [2,10] as shown in Figure 14. These furnaces are suitable for hardening, tempering, annealing, and stress relieving long, thin parts, which require careful handling during heating.

7.   Rotary Hearth Furnace

The rotary hearth furnace differs from previously discussed furnaces in that the load is transferred through the heating zone in a rotary motion on a moving furnace hearth. This is illustrated in Figure 15, where the work to be heat treated is loaded into the furnace near the dis-

Figure 4   Pit furnace with retort for gas nitriding. (Courtesy of Lindberg—A General Signal Company.)

Figure 5   Car-bottom furnace for annealing and/or normalizing. (Courtesy of Lindberg—A General Signal Company.)

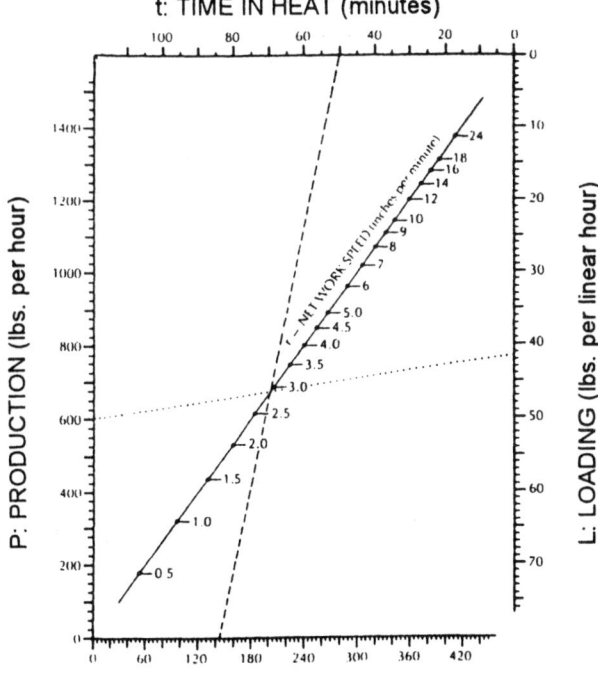

t: TIME IN HEAT (minutes)

P: PRODUCTION (lbs. per hour)

L: LOADING (lbs. per linear hour)

ℓ: LENGTH OF HEATING CHAMBER (inches)

EXAMPLE

———— 1. Tie line between t (150 minutes) and ℓ (145 inches) will indicate net work speed r (2.9 inches per minute)

............2. Tie line from P desired production (600 lbs.) through r (2.9 inches per minute) will indicate required loading L (41.4 lbs. per linear foot)

$$\frac{\ell}{t} = t = \frac{P}{5t}$$

Figure 6 Production rate nomograph for continuous furnaces. (From Ref. 8.)

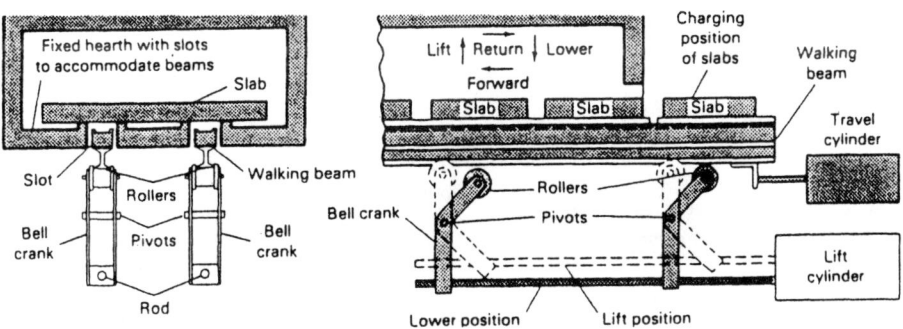

Fixed hearth with slots to accommodate beams

Slab

Slot

Rollers

Walking beam

Bell crank

Pivots

Bell crank

Rod

Lift ↑ Return ↓ Lower

Forward

Slab   Slab   Slab

Bell crank

Rollers

Pivots

Lower position   Lift position

Charging position of slabs

Walking beam

Travel cylinder

Lift cylinder

Figure 7 Illustration of a walking beam furnace mechanism.

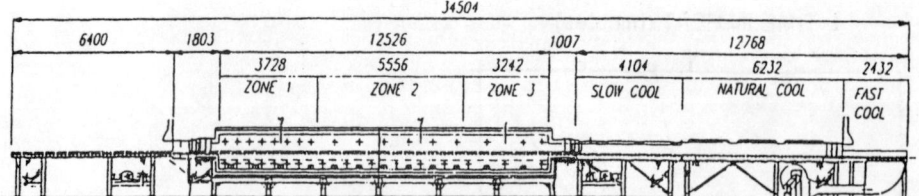

			J4504				
6400	1803		12526	1007		12768	
		3728	5556	3242	4104	6232	2432
		ZONE 1	ZONE 2	ZONE 3	SLOW COOL	NATURAL COOL	FAST COOL

Figure 8 Heating/cooling zones and transfer mechanism in a roller hearth furnace.

charge point. An example of a rotary hearth furnace is provided in Figure 16. In the commercial design illustrated in Figure 17 several rotary hearth furnaces are connected in sequence to obtain even greater furnace and floor-space flexibility. This type of furnace can be used for a variety of processing methods such as atmosphere hardening or carburizing.

III. FURNACE HEATING: ELECTRICITY OR GAS

One of the first steps in furnace design is to select the energy source, typically electricity or gas, that will be used to heat the furnace. The economics of gas and electricity vary with

(a)

Figure 9 Roller hearth furnaces. [Photos courtesy of (a) Lindberg—A General Signal Company); (b,c) Abar Ipsen Industries.]

(b)

(c)

Figure 9 Continued

(a)

(b)

Figure 10 Pusher furnaces. (a). Twin-lane pusher furnace. (Photos courtesy of (a) Abar Ipsen Industries). (b) Lindberg—A General Signal Company.]

availability and cost. In this section, the calculation of furnace heating economics is discussed, followed by a brief overview of electric and gas furnace heating designs.

A. Furnace Heating Economics

The first step in determining the economics of heating a furnace is to perform an energy balance [14]. The heat loss factors to be considered relative to total heat input [15,16] are

1. Wall losses
2. Heat to atmosphere

3. Heat to trays and fixtures
4. Heat to stock

Heat losses to the atmosphere such as fresh air, infiltration, and exhaust are calculated from

$$\text{Btu/h} = \text{scfm} \times 1.08 \; \Delta T$$

where
Btu/h = heat transfer rate
scfm = standard cubic feet of air per minute
1.08 = conversion constant
ΔT = change in temperature of air (°F)

Heat losses to materials passing through the furnace such as trays, fixtures, and steel are calculated from [15]

$$\text{Btu/h} = W \, C_p \, \Delta T$$

where
W = weight of materials passing through the furnace in lb/h
C_p = specific heat of the material [Btu/(lb · °F)]
ΔT = change in temperature between entry and exit of the furnace (°F)

The heat loss through the furnace wall as shown in Figure 2 is calculated from [15]

$$\text{Btu/h} = A \, U \, \Delta T$$

where A is the surface area of the furnace (ft^2) and U is the overall heat transfer coefficient [Btu/(h · ft^2 · °F)] and is calculated from

$$U = \frac{1}{1/h_1 + x/k = 1/h_2}$$

where h_1 = inner film coefficient (Btu/hr·ft^2·°F), h_2 = outer film coefficient (Btu/hr·ft·°F), k = panel thermal conductivity (Btu/hr·ft^2·°F), and x = panel thickness (ft).
The heat input is calculated from the percent energy available, both latent and sensible:

$$\text{Gross Btu/h} = \frac{\text{Btu/h}}{\% \text{ energy available}}$$

The total energy available from the fuel gas or oil will decrease with increasing furnace exhaust temperature, although the amount of decrease depends on the combustion system and raw material source for gas. This is illustrated in Figure 18.
The first step in the selection of a gas or electric furnace can be made once these calculations are performed for both energy sources.

B. Electric Element Furnace Heating

Electrical heat has a number of advantages over gas. These include lower furnace cost and higher furnace efficiency because there are no exhaust losses, advantageous regulations that permit unattended operation, lower maintenance costs, improved temperature uniformity, the relative ease of replacing electric elements, and wider operating temperature range [17,18].
In this section, design recommendations and general properties of the most commonly used heating elements are discussed.

(a)

(b)

Figure 11   Mesh belt furnaces. (a) Loading end; (b,c) unloading end. [Photos courtesy of (a,b) Abar Ipsen Industries; (c) C.I. Hayes, Inc.; (d) Lindberg—A General Signal Company.]

(c)

(d)

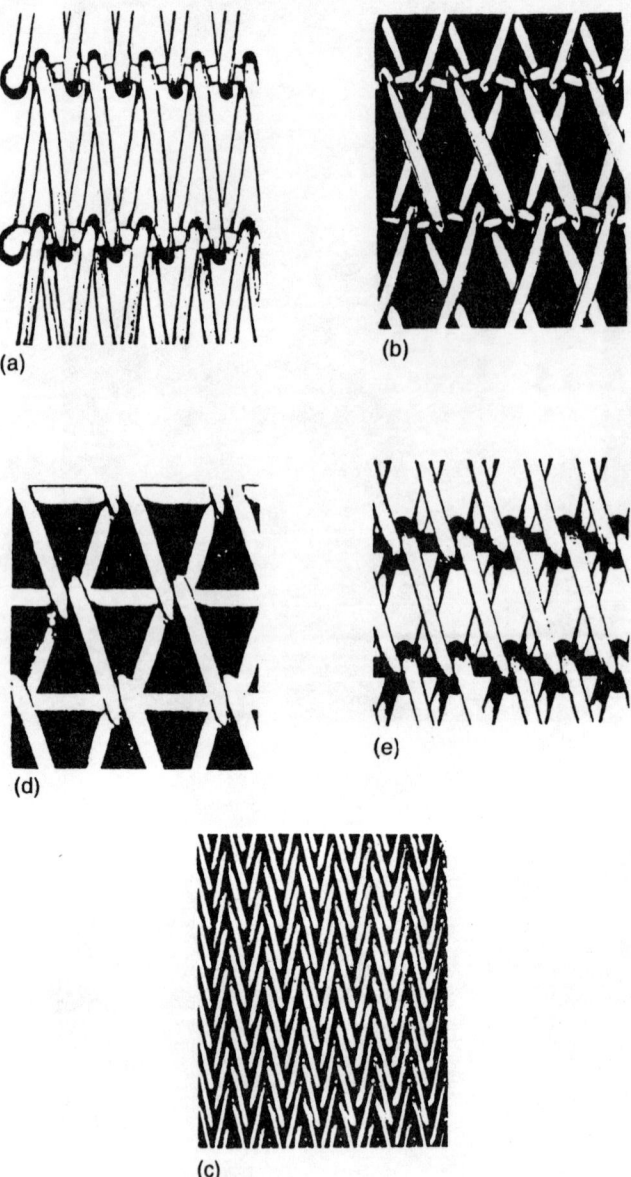

Figure 12 Examples of mesh belt types. (a) Balanced weave; (b) double balanced weave; (c) compound balanced weave; (d) rod-reinforced; (e) double rod reinforced. (Courtesy of The Furnace Belt Company Ltd.)

Electrical heating is dependent on the ability of electric current to flow through a conductor, which may be either metallic or nonmetallic. The heating rate of a material depends on the current density and the specific resistance of the material. (Resistance is inversely related to conductance.) The basic electricity equations that illustrate these concepts are provided in Table 3.

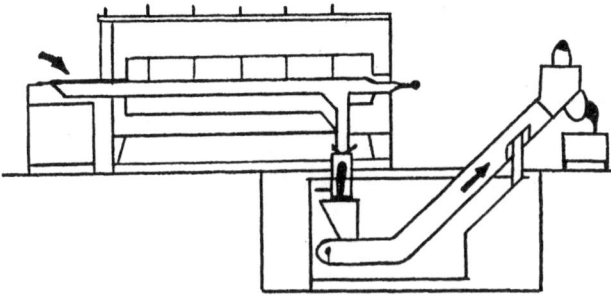

Figure 13   Schematic of a vibratory retort continuous furnace.

When designing an electrically heated furnace, three variables that affect element performance must be considered [20]:

1. Electrical characteristics of the element material
2. Watt loading of the elements
3. The furnace atmosphere

Currently, there are primarily four materials used for heating element construction: nickel-chromium (80 Ni, 20 Cr) [18,21], iron-chromium-aluminum [18], silicon carbide [20,21], and molybdenum disilicide ($MoSi_2$) [22].

Except in reducing atmospheres, plain carbon steel elements cannot be used at temperatures above 800°F and therefore are rarely used in heat treatment operations. In reducing atmospheres, they can be used up to 1200°F [21]. Because of these limitations, elements constructed from plain carbon steels are not used in heat treatment furnaces.

Electrical resistance alloys based on nickel-chromium (Ni-Cr) were developed for high temperature furnace heating [18]. Alloys of this type are suitable for use in furnaces at temperatures up to 2190°F (1200°C) [18].

To obtain even greater lifetimes and higher maximum operating temperatures, alloys based on iron, chromium, and aluminum (Fe-Cr-Al) were developed and are marketed under the trade name Kanthal [18,21]. The aluminum alloying element is used to form a chemically resistant protective layer in the furnaces [21]. Kanthal elements are the most commonly used elements in electric furnaces employing reactive atmospheres. Table 4 lists the maximum recommended operating temperature for Kanthal elements in various environments.

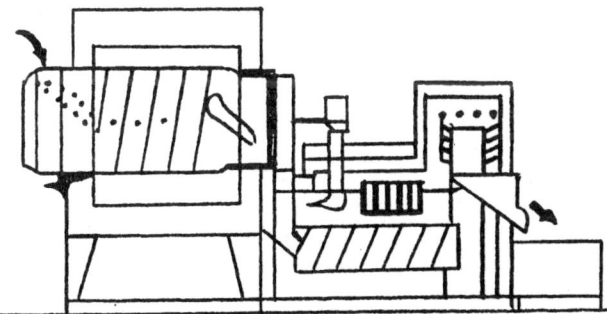

Figure 14   Schematic of a rotary drum screw conveyor continuous furnace.

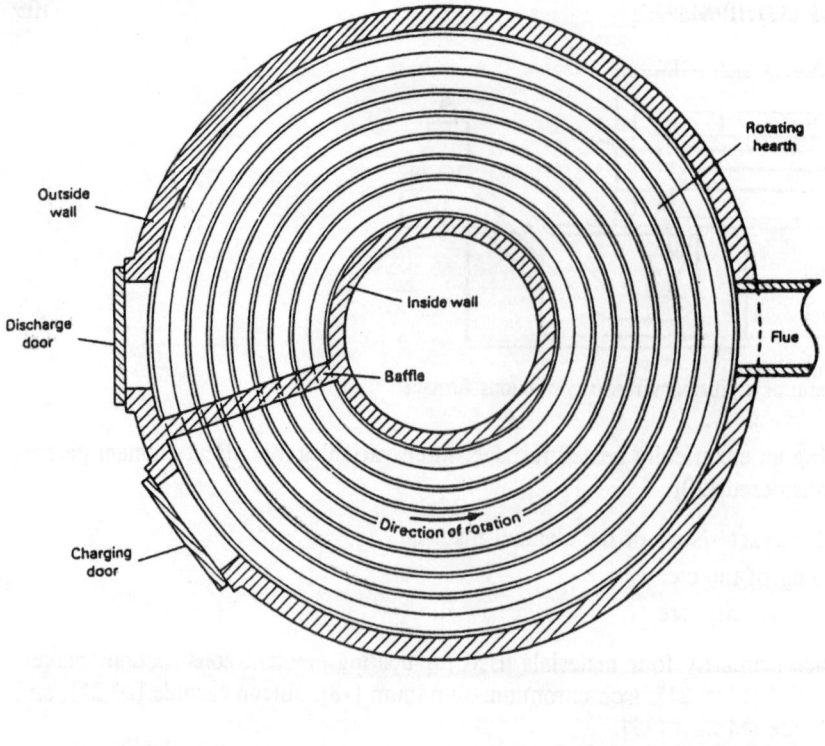

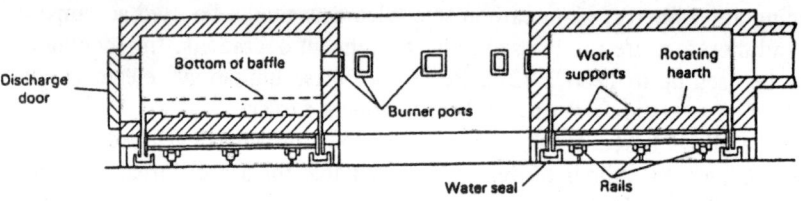

Figure 15   A rotary hearth continuous furnace mechanism.

Figure 16   Rotary hearth furnace. (Courtesy of Abar Ipsen Industries.)

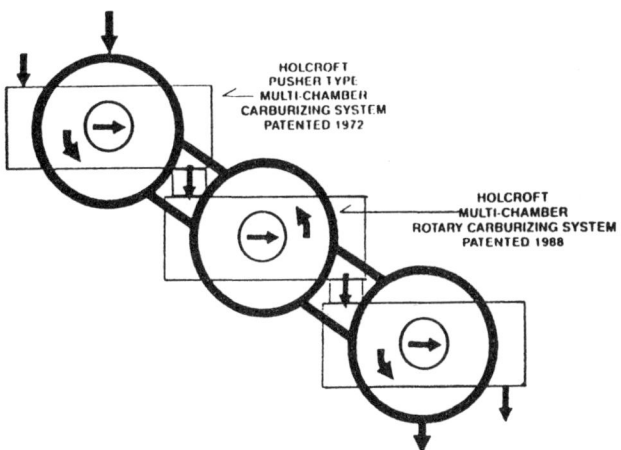

Figure 17 The Holcroft multichamber rotary carburizing system. (Courtesy of Holcroft—A Division of Thermal Process Systems Inc.)

Kanthal elements can be used at temperatures up to 2550°F (1400°C) [18]. A comparison of the use of Ni-Cr and Kanthal electric heating elements is provided in Table 5. There are positive features to both materials. Kanthal provides greater high temperature creep strength and longer life at high temperatures and is more chemically resistant.

Watt loading increases with the cross-sectional area of the element. This is illustrated in Figure 19 for a Kanthal element at 900°C (1650°F) furnace temperature.

The resistivity of a heating element increases with temperature, as illustrated in Figure 20 for a Kanthal element. These curves may vary with the material and even within a class of materials and should be obtained from the manufacturer of the specific heating element being considered.

Some materials such as graphite [17] and silicon carbide [18] that are used as refractories become electrical conductors at high temperatures [15]. Although they are subject to embrittlement, silicon carbide elements may be used in heat treatment furnaces [21].

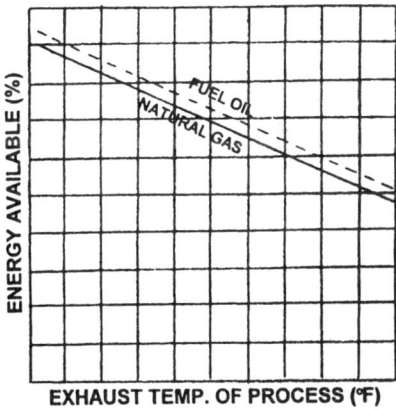

Figure 18 Dependence of available heat energy on exhaust temperature for fuel oil and natural gas. (From Ref. 15.)

Table 3 Electricity Equations

$$W = EI = I^2R = \frac{E^2}{R}$$

$$E = IR = \sqrt{WR} = \frac{W}{I}$$

$$I = \frac{E}{R} = \frac{\sqrt{W}}{R} = \frac{W}{E}$$

$$R = \frac{E}{I} = \frac{E^2}{W} = \frac{W}{I^2}$$

Btu = kW × 3412

$$kW = \frac{BTU}{3412}$$

W = heat flow rate (W); E = electromotive force (V);
I = electric current (A): R = electrical resistance (Ω).
Source: Ref. 19.

Silicon carbide elements are known by the trade name Globar [19] and are constructed from high density silicon carbide crystals that are relatively chemically resistant. However, Ni-Cr exhibits superior high temperature mechanical properties [18]. Advantages of silicon carbide elements include slowness of resistance changes with aging, wide use temperature range (up to 3000°F), suitability for use with high wattage density, and good lifetimes (6 months to 2 years) [20].

The watt loading properties of Globar heating elements are summarized in Table 6. Figure 21 shows that Globar elements exhibit higher watt loadings in oxidizing air environments. The Globar element may accumulate carbon when used with some endothermic atmospheres. Excessive carbon can be removed by preventing replenishment of the atmospheric gas and

Table 4 Maximum Recommended Fe-Cr-Al (Kanthal) Element Temperatures with Different Atmospheres[a]

Atmosphere	Temperature	
	°F	°C
Air	3090	1700
Nitrogen	2910	1600
Argon, helium	2910	1600
Hydrogen[b]	2000–2640	1100–1450
Nitrogen/hydrogen (95/5)[b]	2280–2910	1250–1600
Exogas (10% CO_2, 5% CO, 15% H_2)	2910	1600
Endogas (40% H_2, 20% CO)	2550	1400
Cracked and partially burned NH_3 (8% H_2)	2550	1400

[a]The values shown are for Kanthal Super 1700 and will vary with manufacturer and grade.
[b]The useful element temperature varies with the dew point of the atmosphere.
Source: Ref. 23.

Table 5 Comparison of Ni-Cr and Fe-Cr-Al Elements[a]

Process Variable	Element	
	Ni-Cr	Fe-Cr-Al[b]
Furnace temperature, °C (°F)	1000 (1830)	1000 (1830)
Element temperature, °C (°F)	1068 (1955)	1106 (2025)
Hot resistance Rw	3.61	3.61
Temperature factor Ct	1.05	1.06
Cold resistance, R_{20}	3.44	3.41
Wire diameter, mm (in.)	5.5 (0.217)	5.5 (0.217)
Surface load, W/cm² (W/in.)	3.09 (19.9)	3.98 (25.7)
Wire length, m (ft), for three elements	224.9 (738)	174.6 (573)
Wire weight, kg (lb), for three elements	44.4 (98)	29.6 (65)

[a]120 kW furnace. Three elements at 40 kW each, 380 V.
[b]These are Kanthal AF elements.
Source: Ref. 8.

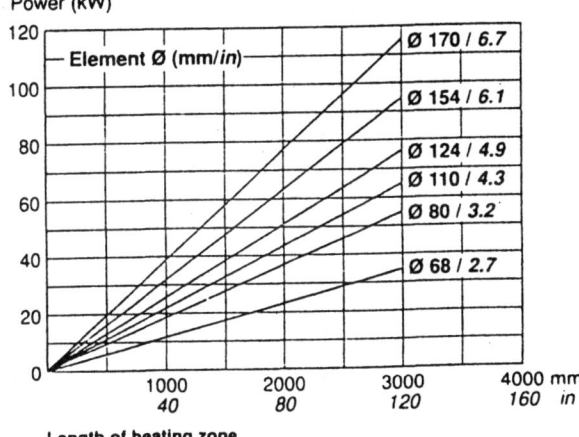

Figure 19 Possible loading at 900°C (1650°F) furnace temperature. (From Ref. 18.)

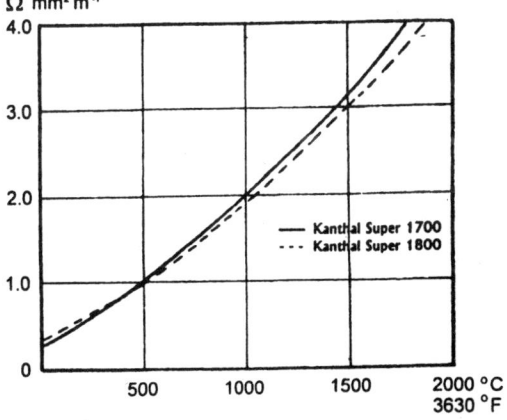

Figure 20 Resistivity of Kanthal Super 1700 and 1800. (From Ref. 20.)

Table 6 Recommended Operating Limits of Globar Elements and Effect of Various Atmospheres

Atmosphere	Recommended operating limits		Effect on elements
	Temperature (°F)	Watt loading	
Ammonia	2370	25–30	Reduces silicon film; forms met.
Argon	Max	Max	No effect.
Carbon dioxide	2730	20–25	Attacks silicon carbide.
Carbon monoxide	2800	25	Attacks silicon Carbide.
Endothermic			
18° CO	Max	Max	No effect.
20° CO	2500	25	Carbon pick-up.
Exothermic	Max	Max	No effect.
Halogens	1300	25	Attacks silicon carbide.
Helium	Max	Max	No effect.
Hydrocarbons	2400	20	Hot spotting from carbon pick-up.
Hydrogen	2370	25–30	Reduces silicon film.
Methane	2400	20	Hot spotting from carbon pick-up.
Nitrogen	2500	20–30	Forms insulating silicon nitrides.
Oxygen	2400	25	Oxidizes silicon carbide.
Sulfur dioxide	2400	25	Attacks silicon carbide.
Vacuum	2200	25	Below 7 millions, vaporizes silicon carbide.
Water			Reacts with silicon carbide to form silicon hydrates.
Dew Point	60	2000	20–30
	50	2200	25–35
	0	2500	30–40
	–50	2800	25–45

Source: Ref. 20.

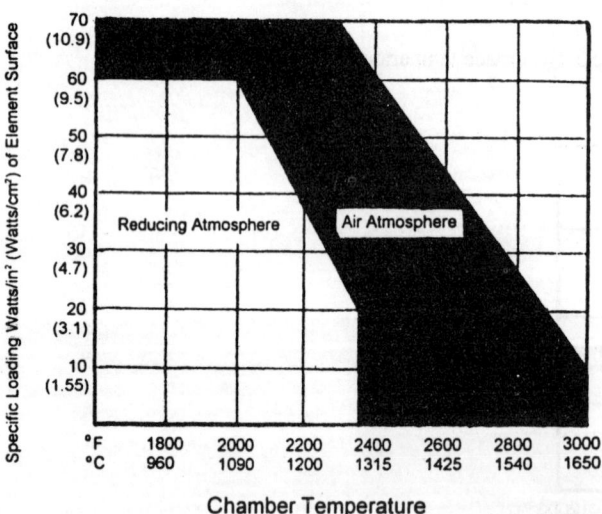

Figure 21 Recommended loading for Globar heating elements.

introducing air periodically to burn off residual carbon [20]. The high temperature zone should be kept free of excessive moisture and carbonaceous vapors [20].

Electrical resistance properties of Globar elements are summarized in Figure 22. Electrical resistance decreases with increasing temperature up to approximately 1200°F (650°C). Above this temperature it increases with temperature. The different negative resistance values shown in Figure 22 are due to the effect of trace impurities in the silicon carbide [20].

Another, less often used, heating element material is molybdenum disilicide, $MoSi_2$. These elements typically contain 90% $MoSi_2$ and 10% of metallic and ceramic additions [22]. They are used in furnaces for both high and low temperature processes and are suitable for use with atmospheres of pure hydrogen and cracked ammonia (with a very low dew point) [17].

An electrical resistance curve for an $MoSi_2$ element is provided in Figure 23. As with the Kanthal and Globar elements shown earlier, electrical resistance generally increases with increasing temperature [22].

Element watt loading with respect to temperature is plotted in Figure 24, which shows that relatively high loadings at high temperature are possible [22].

In a furnace atmosphere, $MoSi_2$ reacts with oxygen above 1800°F (980°C) to form a layer of silicon dioxide (SiO_2), which protects these elements against further chemical attack.

The selection of electric elements in furnace design is carried out in 13 steps [20]. The necessary equation for each step is given in Table 7. Although optimal furnace design is considerably more complicated than this process and should be reviewed with an appropriate engineering consultant, these calculations do provide a first approximation of the likely design requirements for an electric furnace.

Thomander [24] reviewed the design and construction recommendations for electric furnace design using Kanthal elements. Some of these recommendations follow.

1. The proper refractory material should be used for each part of the furnace. Generally a low iron content brick or low fiber modulus (1 mcf) refractory material is used for an electric furnace. Brick refractory should exhibit an electrical resistance of at least $4 \times 10^4 \ \Omega \cdot cm$ at 1200°C. The voltage drop through the brick section should be less than 25 V/cm, if possible.

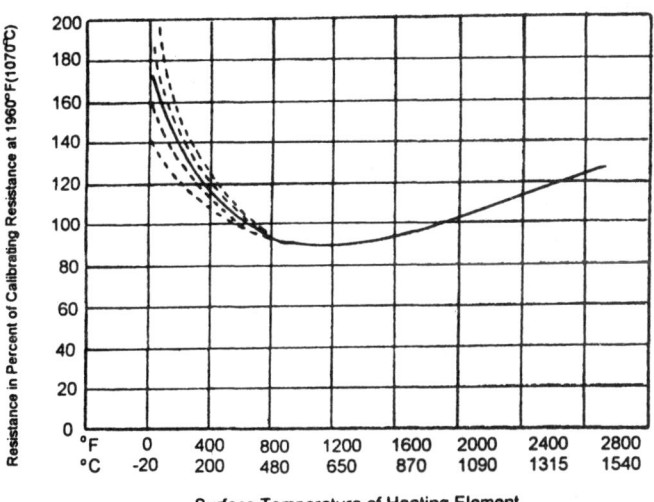

Figure 22 Electrical resistance versus surface temperature for Globar heating elements. (From Ref. 20.)

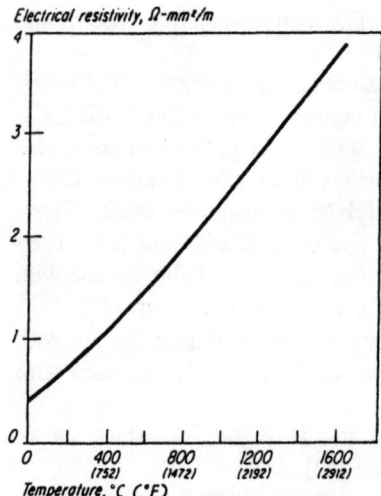

Figure 23   Electrical resistance curve for molybdenum silicide heating elements.

2. Wire element size depends on the watt loading, and the proper size should be used to obtain optimal lifetime from the element. A wire thickness of at least 3 mm (0.12 in.) is often used.

3. Wire is generally used for relatively low amperage applications.

4. The spiral diameter should be four to six times the wire diameter for furnace temperatures > 1000°C (1830°F) and four to seven times the wire diameter at furnace temperatures below 1000°C (1830°F).

5. For strip elements, the thickness should be at least 1.5–2.5 mm (0.59–0.09 in.).

6. To minimize the potential of wires breaking off, the terminal area should be approximately three times the area of the heating zone. The area of the wall should be at least as large as the heating zone.

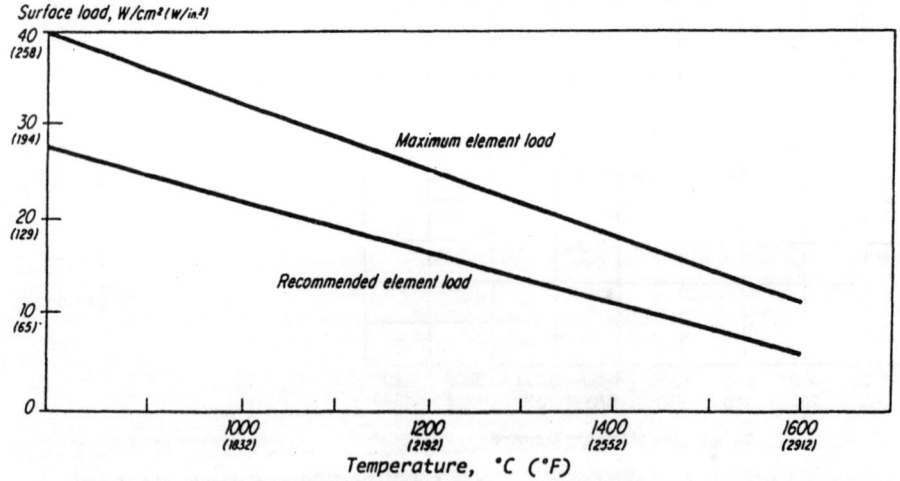

Figure 24   Recommended loading for molybdenum silicide heating elements.

Table 7 Electric Element Furnace Design Equations

1. Process heat (kW) $= \dfrac{W \times C_p \times \Delta T}{3412}$ (1)

   where W = load weight, ΔT = °F heat-up/h, C_p = specific heat.

2. Heat loss (kW) $\dfrac{(\text{heat loss/ft}^2) \times A}{3412}$ (2)

   where A = total furnace area.

3. Heat Storage (kW) $= \dfrac{(\text{heat storage/ft}^2) \times A}{3412}$ (3)

4. Total power requirement = process heat + heat loss + heat storage (4)

5. Watts/element = heating section area (in.^{2}) × watts/in.^{2} (5)

6. Number of elements $= \dfrac{\text{total power requirement (W)}}{\text{W/element}}$ (6)

7. Volts/element $= [(\text{W/element}) \times (\text{resistance/element})]^{1/2}$ (7)

8. Total volts = (volts/element) × (number of elements in series) (8)

9. Maximum Amperes/element $= \dfrac{\text{volts/element}}{\text{resistance/element}} \times 1.56$ (9)

10. Total Amperes $= \dfrac{\text{amperes}}{\text{element}} \times$ number of elements in parallel circuit (10)

11. Delta circuit:

    Three-phase volts = single-phase volts (11a)

    Three-phase amperes = 1.73 × single-phase amperes (11b)

12. Wye circuit:

    Three-phase volts = 1.73 × single-phase volts (12a)

    Three-phase amperes = single-phase amperes (12b)

Transformers with Taps:

13. Total voltage = lowest tap (13a)

    Voltage × 1.0 = nominal tap (13b)

    Voltage × 2.0 = high tap (13c)

Source: Ref. 20.

7. Electric furnace efficiency can generally be assumed to be 50–80%. However, this approximation may be insufficient for calculating power requirements with respect to a small load and large chamber. In such cases, Figure 25 may be used to approximate furnace power requirements [24].

8. Element loading is dependent on furnace wall construction, atmosphere, temperature, and load capacity (throughput per hour). Figure 26 may be used to determine element temperature [24]. (This figure should be used only for unrestricted radiation.)

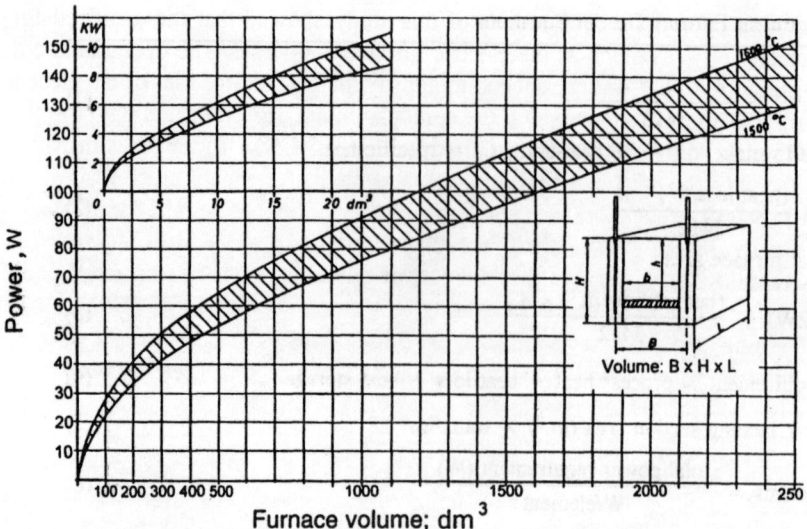

Figure 25 Determination of approximate furnace power based on furnace volume.

C. Gas-Fired Furnaces

Although electrically heated furnaces may be much more efficient (>85%), they may also be significantly more expensive to operate than a less efficient (50–70%) gas-fired furnace [16,25]. However, this depends on the cost of gas and on whether the natural gas is "spiked."

For example, in a 1981 paper, the cost of operating a gas-fired Surface Combustion Super All Case furnace was compared with the cost of operating an electric furnace performing

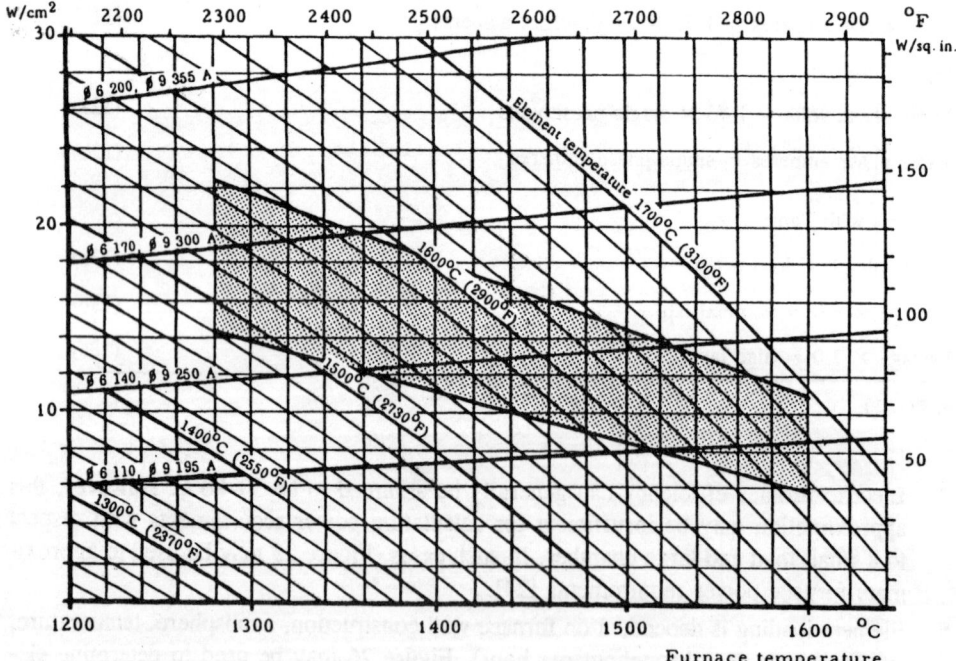

Figure 26 Surface loading graph for Kanthal heating elements.

the same heat treatment operations. The results of this study showed that the gas-fired furnace cost approximately 80% as much as the electric furnace to operate. The operational cost of the gas-fired furnace could be further reduced to approximately 30% of that of the electric furnace with a heat recovery process in which the combustion air was preheated with the flue gases. Although this may be practical and cost-effective for large four-row pusher furnaces, that would not be the case for smaller batch, temper, and medium temperature furnaces. An illustration of fuel savings of this process is shown in Figure 27. A nomogram for the calculation of fuel savings by the combustion air preheating process is provided in Figure 28.

In a more recent study, it was shown that the cost of operating a gas-fired furnace could be reduced to approximately 8% of that of operating an electric furnace [26]. Furthermore, the use of gas had a number of additional advantages. For example: (1) It is possible to use a more useful heat input into the heating process; (2) the use of gas increases the heat treatment process rate; (3) natural gas is reliable and burns cleanly; (4) furnace conversions are fast and low cost, and gas-to-electric conversions of any size of heat treatment equipment are readily performed [26]. Disadvantages of this conversion included: (1) Conversion downtime causes production loss; (2) it is necessary to install flue ducts; (3) temperatures are higher around equipment; and (4) it is necessary to install flame safety controls and train operators [26].

In this section, methods of improving gas-fired furnace efficiency through combustion control and waste heat recovery are discussed.

1. Gas Combustion

Combustion is an oxidative chemical reaction between a hydrocarbon fuel source such as methane (CH_4) and oxygen. If sufficient oxygen is available, there are no impurities in the gas, and the reaction is complete, then the sole reaction products are carbon dioxide (CO_2) and water (H_2O) [26]. Figure 29 shows that increasing the combustion temperature increases the degree of completion of the combustion reaction (increasing CO_2) [28].

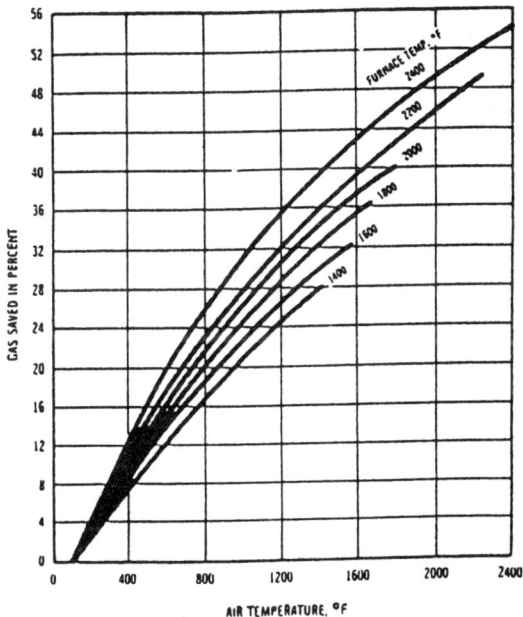

Figure 27 Potential fuel savings available with exhaust gas recovery. (From Ref. 26.)

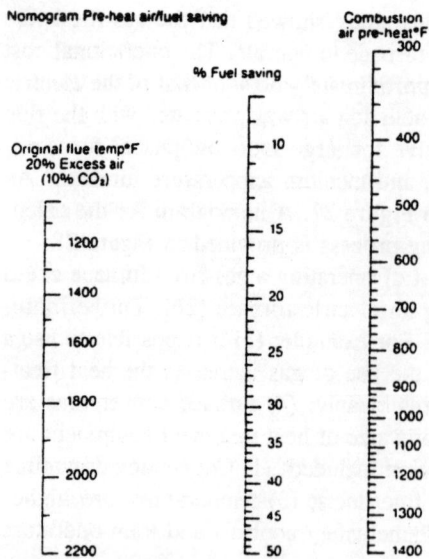

Figure 28 Nomogram to determine expected fuel savings from preheating combustion air. (From Ref. 27.)

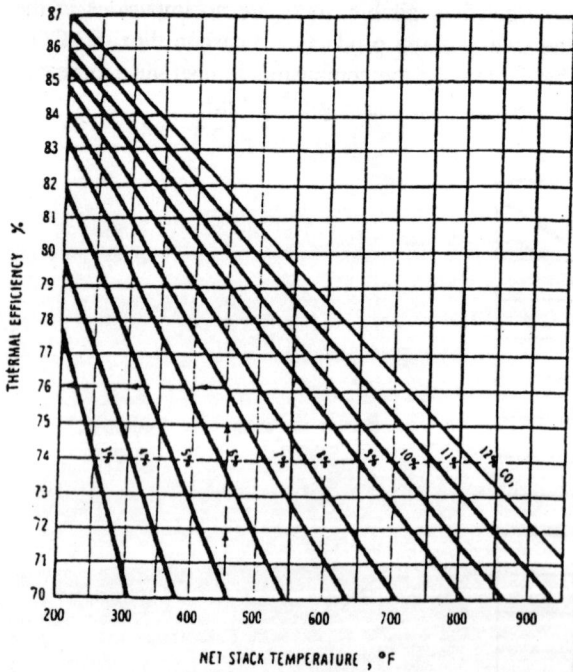

Figure 29 Thermal efficiency as a function of exhaust temperature. Note: This chart applies only to cases where the percentage of CO_2 is less than ultimate because the fuel-air ratio is leaner than that required for perfect combustion.

$$CH_4 + 2O_2 \rightarrow CO_2 + 2H_2O$$

Incomplete combustion will occur is insufficient oxygen is available, and carbon monoxide (CO) will be produced.

$$2CH_4 + O_2 \rightarrow 2CO + 4H_2$$

If additional oxygen is added, the overall reaction can be driven to produce the products of complete combustion [29].

$$2CO + O_2 \rightarrow 2CO_2$$

$$2H_2 + O_2 \rightarrow 2H_2O$$

To achieve the desired degree of combustion and combustion efficiency, excess air is used. The completion of combustion and the percent available gross fuel input depend on the amount of excess air (combustion ratio) as shown in Figure 6.30 [30,31] (see also Fig. [32]).

The effect of the combustion ratio on the amount of heat available from the combustion process is determined from Figure 31 [30]. For example, if the excess air increases from 5% to 30% with a flue temperature of 1500°F, available heat decreases from 58% to 50% [30]. The percent decrease in available heat is

$$\frac{\text{Available heat at 5\% excess air}}{\text{Available heat at 30\%}} = \frac{58 - 50}{50} \times 100 = 16\%$$

The decrease in available heat is due to the energy required to heat the additional air. Figure 32 provides an estimate of energy savings attainable by minimizing the excess of air used for combustion [30]. It should be noted that any opening in the furnace wall produces a

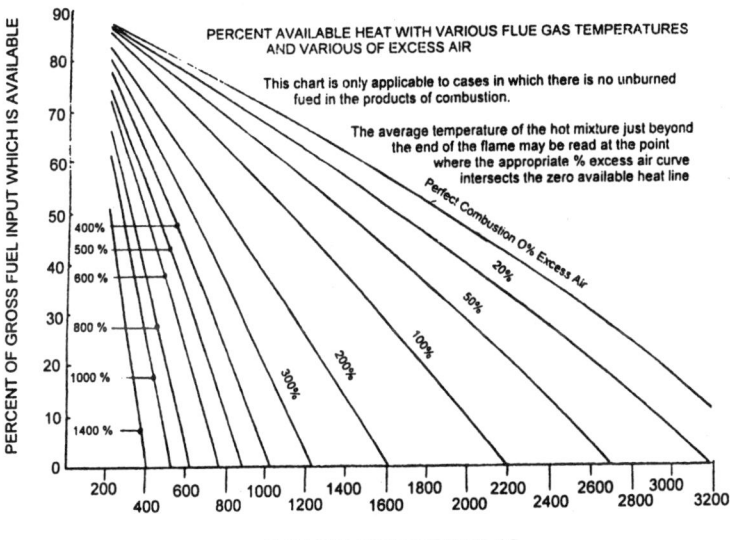

Figure 30 Available heat vs. excess air for flue temperatures of 200–3200°F.

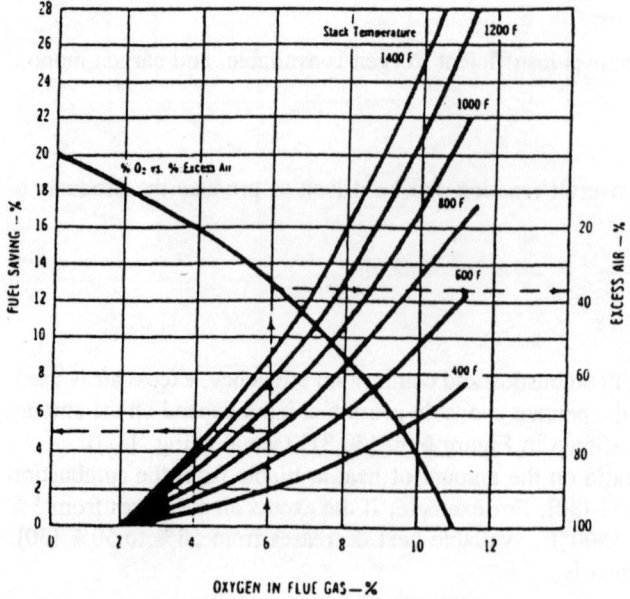

OXYGEN IN FLUE GAS—%

Figure 31 Fuel savings obtainable by controlling excess air.

"chimney effect," which results in heat loss that may significantly perturb the air–fuel balance as shown in Figure 33 [31].

The exit gas temperature will increase with the furnace temperature because the theoretical combustion temperature increases with the amount of air (oxygen) as shown in Figure 34 [33]. Figure 35 shows that combustion efficiency is significantly improved by oxygen addition (enrichment) into the combustion mixture [33].

Common fuel sources are city gas, natural gas ($\sim 85\%$ CH$_4$), propane, and butane. Table 8 provides burner combustion data for these gases.

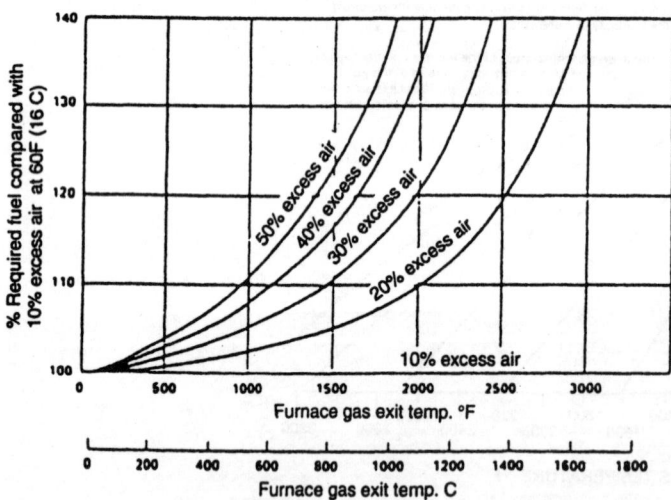

Figure 32 Fuel requirements vs. excess air for furnace gas exit temperatures of 32–3000°F.

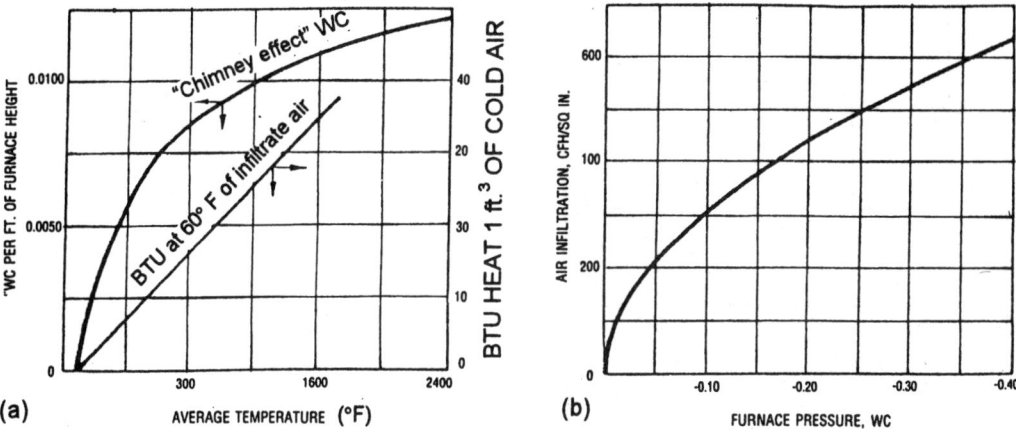

Figure 33 Calculation of "chimney effect." (a) is used to determine the negative pressure of the hearth; (b) is used to calculate the flow rate due to negative pressure.

Specific gravity [(density gas)/(density air)] is used to calculate flow and combustion products in gas mixing.

Air requirement is used for the calculation of total air required for the combustion of a particular gas.

Flame temperature reflects the energy available in the combustion process. The flame temperature is dependent on the amount of oxygen available during combustion as shown in Figure 36 [34].

Flame propagation speed reflects the ability to obtain a stable flame.

Limits of inflammability are used to determine the safety of the use of a particular gas–air mixture.

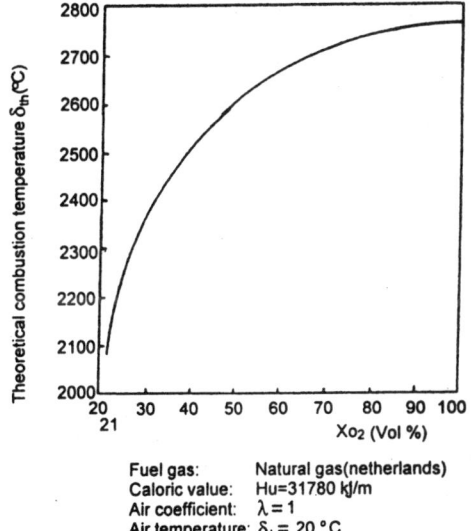

Figure 34 Theoretical combustion temperature as a function of oxygen content in combustion air.

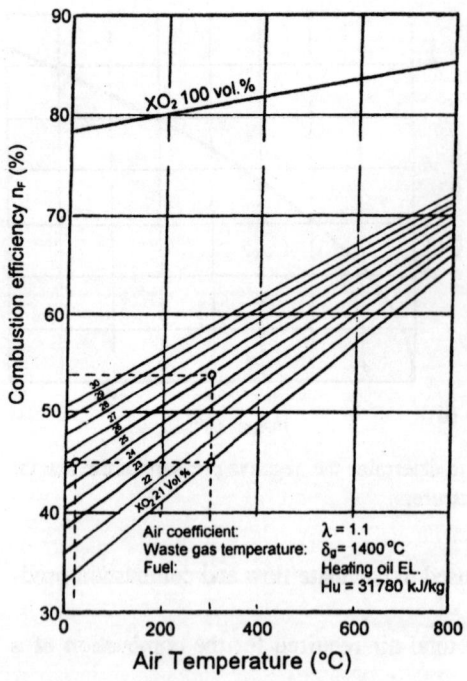

Figure 35 Influence of combustion air temperature and oxygen content on combustion efficiency.

The combustion air/fuel ratio can be controlled either by precise metering of the fuel and air entering the burner or by flue gas analysis. The preferred method of flue gas analysis is usually to use an oxygen sensor [35].

Although perfect combustion is achieved by mixing the exact quantities of fuel and air to produce only CO_2 and water, this is often not practical. Typically the fuel–air mixture if either "rich" or "lean." A rich mixture contains excess fuel, and since CO and H_2 are produced, it produces a reducing atmosphere. If excess air is used, an oxidizing atmosphere will result. In addition to the proper fuel/air ratio, it is important to provide sufficient time at temperature for complete combustion of the fuel to occur, as shown in Figure 37 [36].

As an approximation for furnace calculations, it is often assumed that approximately 1 ft^3 of air is required for each 100 Btu of heating value [37]. For 1 gal of fuel oil, approximately 1500 ft^3 of air is required [37].

Table 8 Burner Combustion Data for Common Fuel Sources

Gas	Spec. gravity[a]	(Btu/ft^3)	Air reqd. (ft^3 air/ ft^3 gas)	Flame temp. (°F)	Flame speed (in./s)	Limits of inflammabil., % gas in air Low	High
City gas	0.5	500	5	3600	60	5	40
Natural gas	0.6	1000	10	3550	25	4	13
Propane	1.52	2500	25	3650	30	2	10
Butane	1.95	3200	32	3660	30	2	9

[a]Specific gravity = (density of gas)/(density of air).
Source: Ref. 29.

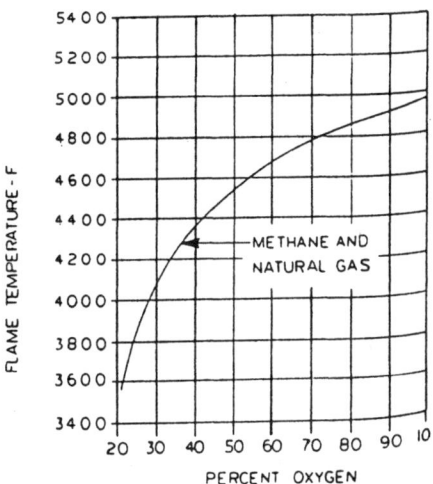

Figure 36   Effect of oxygen content on the flame temperature of methane and natural gas.

2.   Burner Selection

The two most common types of burners for heat treatment furnaces are direct-fired, high velocity (Figure 38) and indirect-fired, radiant tube (Figure 39). With direct-fired burners, fuel combustion occurs in the furnace vestibule. The circulation of the hot gases, which may be oxidizing or reducing, depending on the air/fuel ratio, provides the temperature uniformity within the furnace.

Direct-fired burners are not favored in heat treatment processes such as carburizing where atmosphere control is critical, and also the generation of reducing gases is not a very efficient furnace heating process. In these cases, more efficient combustion may be attained inside a protective radiant tube.

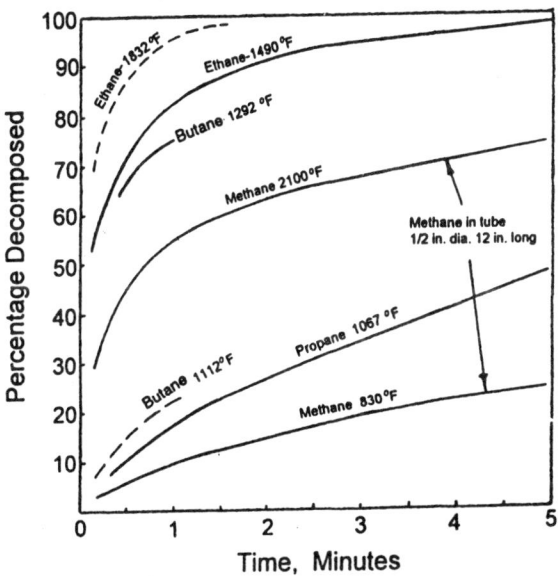

Figure 37   Effect of fuel composition and time at combustion temperature on combustion efficiency.

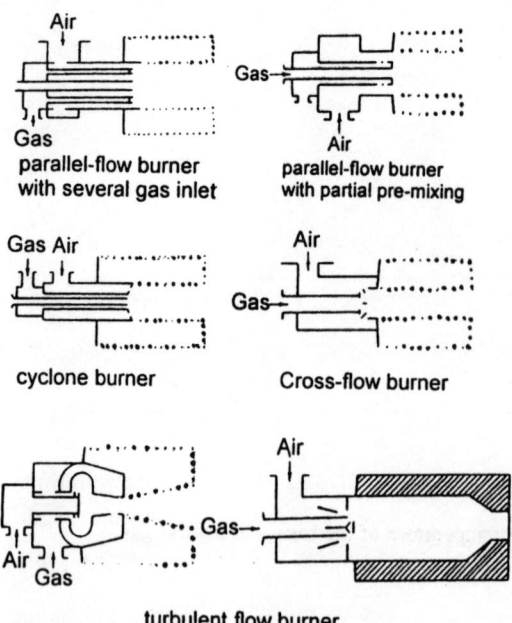

Figure 38 Several types of direct-fired high velocity burners. (From Ref. 38.)

High Velocity Burners To obtain uniform microstructure and properties and to minimize undesirable residual stresses that arise from thermal gradients, it is important to facilitate uniform temperature distribution within the furnace. Some methods that have been used to accomplish this are [38,40]

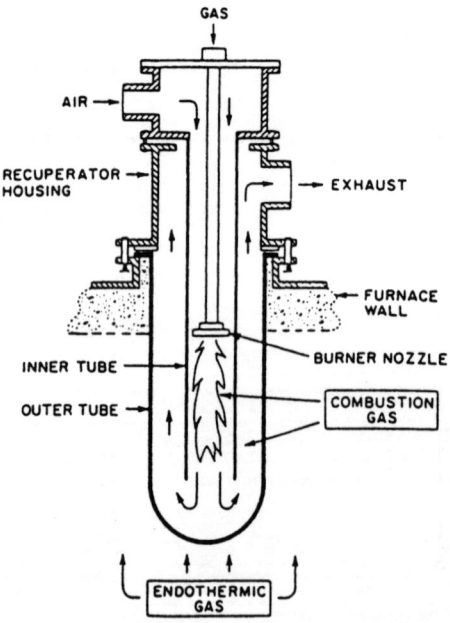

Figure 39 A single-ended radient (SER) tube burner. (From Ref. 39.)

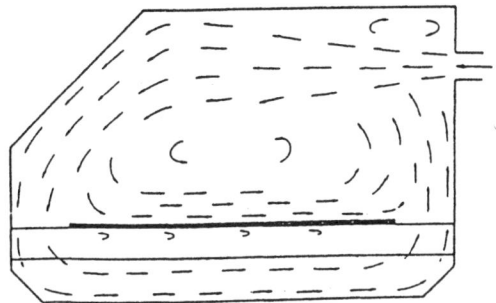

Figure 40 Furnace gas flow patterns from one-sided burner firing arrangement.

1. Use of high temperature fans.
2. Use of baffled walls in both the upper and lower parts of the furnace.
3. Utilization of fewer heating zones. The use of fewer burners instead of the traditional array of multiple burners may produce excellent thermal uniformity as shown in Figure 40 [38,41].
4. Hot-charging and preheating steel with flue gases.
5. Utilization of high velocity burners.

In high velocity burners, fuel is burned either completely or partially in a ceramic-lined combustion chamber (see Figure 41). The exit velocity of the combustion products is 50–200 m/s. The high speed jet promotes gas mixing and temperature uniformity in the vestibule. Depending on the design and type of gas–air mixing, high velocity burners are classified as parallel-flow, cross-flow, cyclone, or turbulent jet designs. Examples of each are illustrated in Figure 38.

Combustion efficiency varies with the rated furnace load of the combustion chamber, Q/V, where Q combustion is the total quantity of heat released in the burner chamber and V is the volume of the burner combustion chamber. In a recent study, Keller [38] reported that

1. As the specific load increases, combustion shifts from the burner toward the furnace.
2. Combustion efficiency decreases as the rated furnace load increases (see Figure 43a).
3. The specific load of the combustion chamber increases with fuel flow rate (see Figure 42b).
4. The burner with the largest rated load and largest inside diameter produces the lowest specific load (see Figure 43).
5. Combustion efficiency and the mean load on the combustion chamber vary inversely with the diameter of the combustion chamber (see Figure 43).

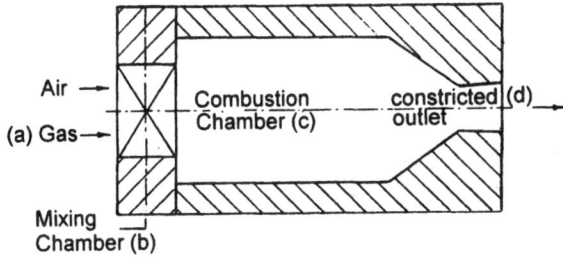

Figure 41 Schematic of a high speed burner. (From Ref. 38.)

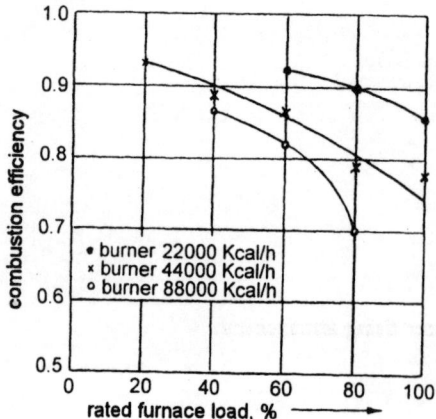

(a)

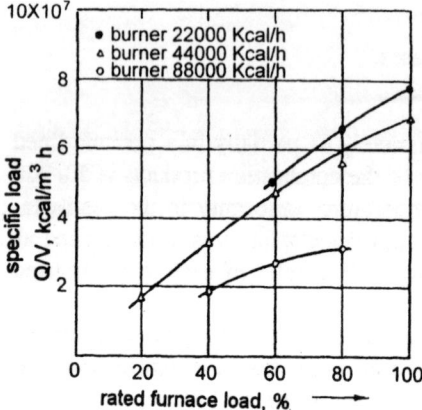

(b)

Figure 42 Variation of the specific load of the combustion chamber with fuel flow rate. (From Ref. 38.)

High velocity burners have a number of advantages for use in heat treatment furnace applications [43]:

1. The velocity of the exiting combustion products re-entrains the existing furnace atmosphere, maximizing the available heat.
2. Depending on the type of furnace, the relatively short flame of a high velocity burner reduces the risk of flame impingement on the load.
3. The turbulence created by the burner combustion products facilitates temperature uniformity.
4. The turbulence of the furnace gases facilitates heat transfer from the atmosphere to the load.

Some general rules that have been proposed for a heat treatment combustion system are [44]

1. There should be a maximum of two burners, one at each end of the furnace, with firing occurring at a high position along the long walls.

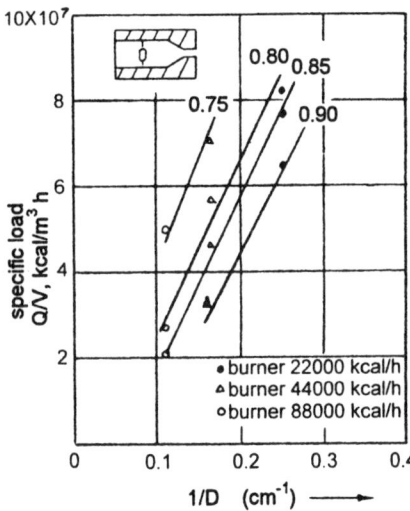

Figure 43 Variation of combustion efficiency and mean load with chamber diameter.

2. All flues should be placed under the center of the load.
3. The high velocity discharge should be sufficient to provide temperature uniformity.

Note: The validity of these "general rules" is dependent on the particular furnace being designed. For example, exceptions to these rules include large furnace systems such as continuous carburizing and continuous tempering furnaces.

When lower heating temperatures [<700°C (<1300°F)] are required, it is often difficult to obtain the necessary furnace uniformity with a low burner velocity. One method of improving furnace uniformity in such cases is to use "excess air" [44]. A variable excess air furnace combustion system has been described that achieves superior temperature control and greater combustion efficiency. A 42% energy savings was reported [45].

Another method of increasing the heating performance and fuel efficiency of a high velocity burner at lower temperature is to use a computer-controlled pulse-firing system [43,46]. Since high velocity burners are more efficient when they are on, the idea is to turn them on or off with a timed "pulse." The burner firing frequency may vary from 3 to 6 s between pulses [35].

Radiant Tube Burners When heat treatment processes requiring exclusion of combustion gases from the furnace load and also requiring precise temperature control are being carried out; the use of direct-fired burners is unacceptable. These furnace applications generally require isolation of the burner combustion process. This is accomplished by encasing the burner, often pulse-fired, in a "radiant" tube such as the single-ended radiant (SER) tube design shown in Figure 39 [25]. Currently, many manufacturers favor radiant tubes constructed of ceramic materials such as reaction-bonded silicon carbide [39,42].

D. Heat Recovery

Combustion efficiency can be significantly improved by preheating the incoming cold air with the hot flue gas. The effect of the temperature of the flue gas and percentage of excess combustion air on gas efficiency is illustrated in Figure 30 [40]. Alternatively, fuel savings may be calculated using the nomogram provided in Figure 28 [27].

There are two principal methods for recovering and reusing heat that is normally lost through flue gas emission: recuperation and regeneration. Recuperation uses a heat exchanger

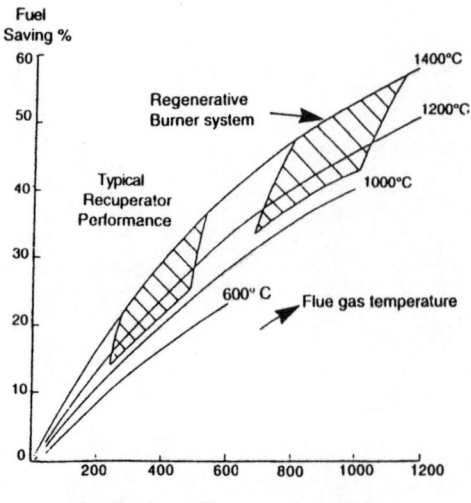

Air Preheat Temperature (°C)

Figure 44 Relative impacts of recuperators and regenerative burner systems on fuel savings. (From Ref. 40.)

to transfer heat from the hot flue gas to the incoming cold combustion air. Regeneration increases combustion efficiency by using the hot flue gas to both preheat combustion air and further increase the burner flame temperature. Figure 44 shows the enhancement in combustion efficiency gained by recuperation and regeneration. Overviews of both processes follow.

1. Recuperation

The use of recuperator systems to provide substantial improvement in both batch and continuous heat treatment processes [14,47] has been reported for both new furnace systems [25] and retrofitted older systems [48,49].

The improvement of combustion efficiency by preheating air for natural gas combustion is illustrated in Figure 45. The recuperative process that increases combustion air temperature through heat exchange with the hot exiting flue gas is illustrated in Figure 46.

There are three types of heat exchangers for gases: continuous flow, parallel-flow, and cross-flow recuperators. These heat exchanger flue gas and air flow patterns are illustrated in Figure 47.

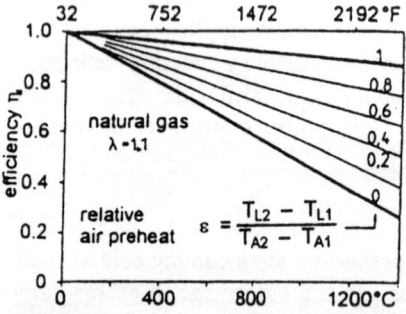

Figure 45 Effect of temperature of furnace exhaust gas entering recuperator on combustion efficiency. (From Ref. 50.)

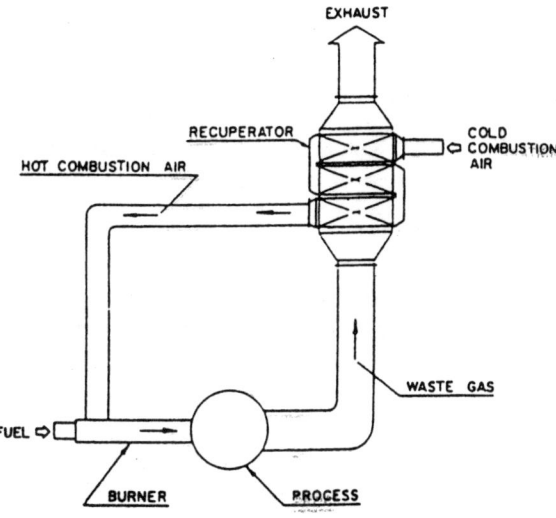

Figure 46 Schematic of a recuperator process. (From Ref. 47.)

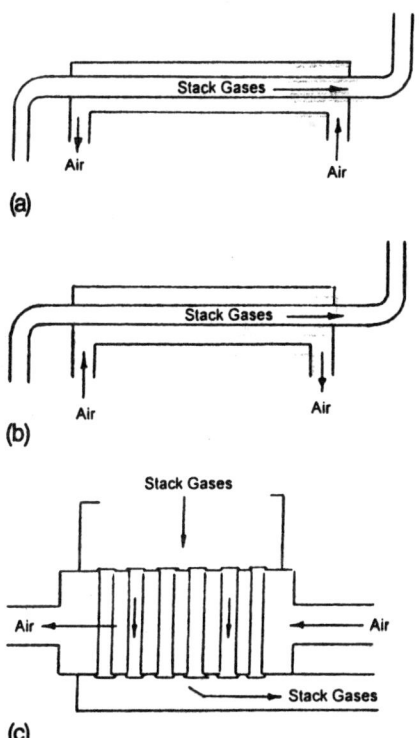

Figure 47 Recuperator heat exchanger processes. (a) Counterflow; (b) parallel-flow; (c) crossflow. (From Ref. 51.)

Examples of commercial recuperator systems are provided in Figures 48 and 49. Convective heat transfer efficiency increases with flow rate through the heat exchanger (Figure 50) and is nearly independent of gas temperature [51]. Larger heat exchanger surface areas are required with increasing air preheat temperature as shown in Figure 51 [51].

Recuperators may be constructed from either high temperature, corrosion-resistant metallic materials or nonmetallic ceramics. However, the use of ceramic materials is much less dependent on service temperature and contact with corrosive flue gases [52].

2. Regeneration

The performance of recuperator systems is generally limited by the surface area of the heat exchanger [22] and the upper temperature of the preheated air due to potential oxidation of the preheated recuperator surfaces [53]. These and other limitations are circumvented by the more efficient regenerative combustion system.

Numerous publications have discussed the use of regenerators in both batch [53–55] and continuous [41,56,57] systems to achieve substantial energy cost reductions [58]. This section provides an overview of the operation of regenerator burner combustion systems.

A regenerative combustion system generally consists of two regenerators, two burners, a flow reversal valve, and the necessary control system [59]. (*Note:* Systems such as these are

Figure 48 Comparison of (left) a recuperator and (right) a typical exhaust stack (Courtesy of Holcroft—A division of Thermo Process Systems Inc.)

Figure 49   Side-mounted recuperators illustrate air flow both parallel and counter to the exhaust. (Courtesy of Holcroft—A Division of Thermo Process Systems Inc.)

not often encountered in heat treatment furnaces.) A schematic of a typical system is shown in Figure 52. Pulse-firing burners are often selected to minimize the potential for radiant tube burn-out [35]. Temperature fluctuations are controlled by the mass of the radiant tube.

The regenerator is a two-chamber system that contains firebrick [51] or a ceramic material for heat storage. These materials may be in the form of balls [59], honeycomb [51,56], or even a granular refractory [58].

The performance of a regenerator depends on [37,48]

The size of the furnace and regenerator

Reversal time

Thickness of the firebrick or other refractory material

Conductivity of the refractory

Heat storage ratio

Geometry of the regenerator

Temperature and flow rate of the gases

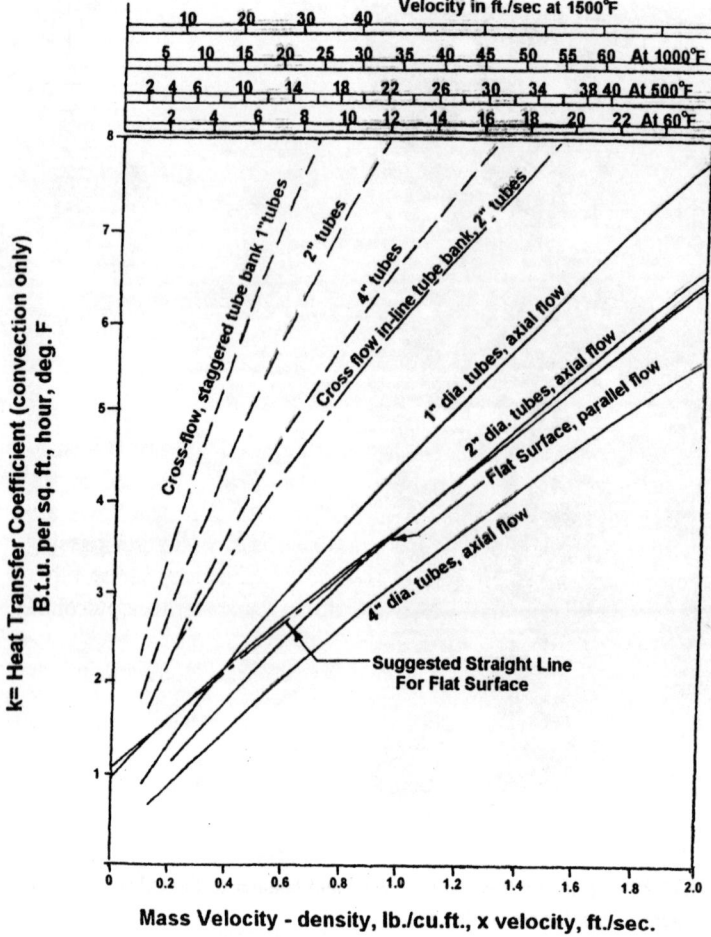

Figure 50 Convective heat transfer for different recuperator sizes and flow patterns.

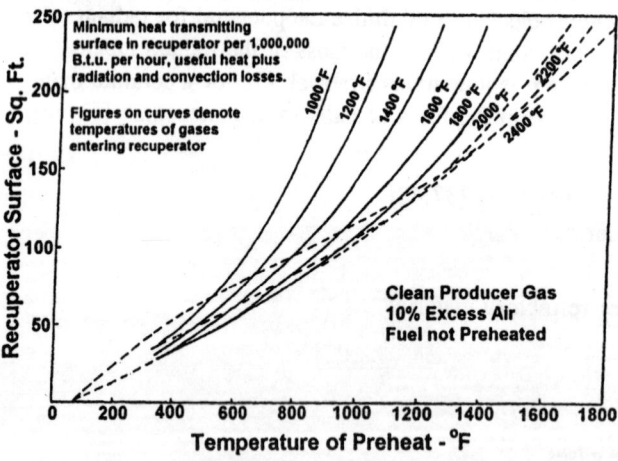

Figure 51 Recuperator surface requirements for natural gas.

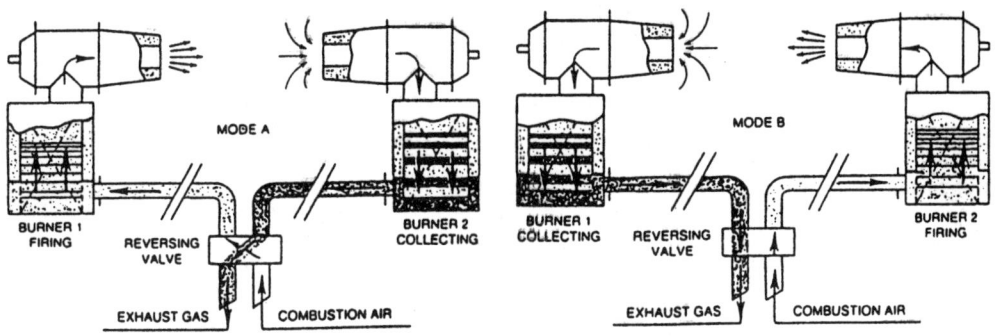

MODE A   MODE B

BURNER 1   BURNER 2   BURNER 1   BURNER 2
FIRING   COLLECTING   COLLECTING   FIRING
        REVERSING   REVERSING
        VALVE   VALVE

EXHAUST GAS   COMBUSTION AIR   EXHAUST GAS   COMBUSTION AIR

Figure 52 Schematic of a regenerative burner process. (From Ref. 59.)

The basic principle of operation of a regenerator system is that flue gas gives up its heat to the refractory in one of the regenerators. At the same time, combustion air is heated by the hot refractory material in the other regenerator. After approximately 20 min, the flows are reversed, and the hot refractory in the regenerator that previously had hot flue gases passing through it is used to heat the combustion air. Conversely, the cooled refractory in the other regenerator chamber is then reheated by hot flue gases. Regenerator microprocessor-controlled cycle times of 20 s have been reported [58].

The burner typically used for regenerator heat treatment furnace applications is of the radiant tube type shown in Figure 53.

3. Rapid Heating

Rapid heating is any heating method that accelerates conventional furnace heating. It may be accomplished in gas-fired furnaces by [60]

1. Lifting the stock off the furnace hearth
2. Rotating stock to eliminate "cold" surfaces
3. Separating stock in the furnace
4. Increasing the heat flux on the metal surface by increasing the furnace temperature
5. Increasing the flow velocity in the furnace heated by high velocity burners by (a) matching the internal shape of the furnace to the shape of the stock or (b) having the burner jet impinge on the stock

The heat transfer rates attainable by these methods are summarized in Table 9.

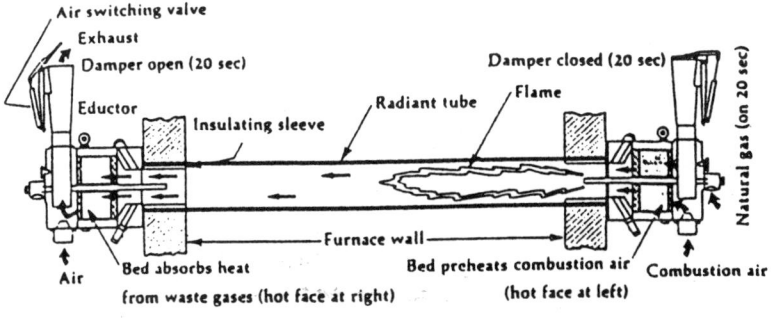

Figure 53 Schematic of regenerative ceramic radiant tube burner. (From Ref. 58.)

Table 9 Typical Gas-Fired Furnace Rapid Heating Convective Heat Transfer Coefficients

Heating method	Gas velocity		Convective heat transfer coefficient	
	m/s	ft/s	W/(m² · K)	Btu/(ft² · h · °F)
Jet impingement	150	500	250–500	50–100
Furnace stock matching	50	150	100–250	20–50
Conventional furnace	<5	<15	<25	<5

Source: Ref. 60.

The nomogram in Figure 54 is provided for the estimation of soaking times for steel >75 mm (3 in.) in diameter in the temperature range of 1000–1250°C (1830–2280°F) following rapid heating to the soaking temperature [60].

Although rapid heating is not often used in heat treating furnaces, it is used in forge heating furnaces.

IV. HEAT TRANSFER

The heat transfer process that occurs when a part is heated in a furnace is depicted in Figure 55. Typically the heat transfer rate is rapid initially and decreases as the temperature of the center of the part approaches the surface temperature, which achieves the furnace temperature more rapidly. Ideal furnace design permits thermal equilibrium to be reached as quickly as possible while minimizing thermal gradients within the past during the heating process.

Heat transfer in furnaces occurs by convection, radiation, and conduction. The application of these modes of heat transfer in furnace heating is illustrated in Figure 56.

A. Convective Heat Transfer

Heat flux (q) in a furnace is dependent on the change in temperature (dT) with an incremental change in distance (dx):

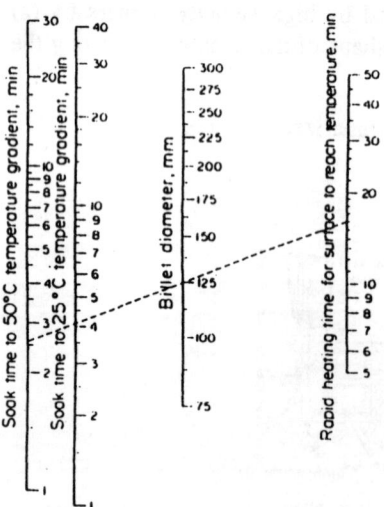

Figure 54 Nomogram for rapidly heating steel billets to 1000–1250°C.

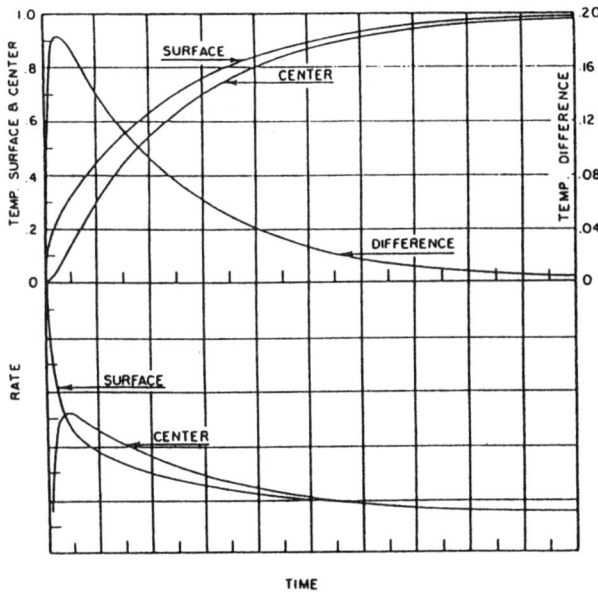

Figure 55 Relationship of temperature rise and heating rates with respect to time. (From Ref. 61.)

$$q = -K \frac{dt}{dx}$$

Heat flux q is related to total power Q by

$$Q = Aq$$

where A is the total area of the part. Total power (Q) is often preferred because it accounts for the variation of heat flux with total area. Figure 57 illustrates the calculation of total power

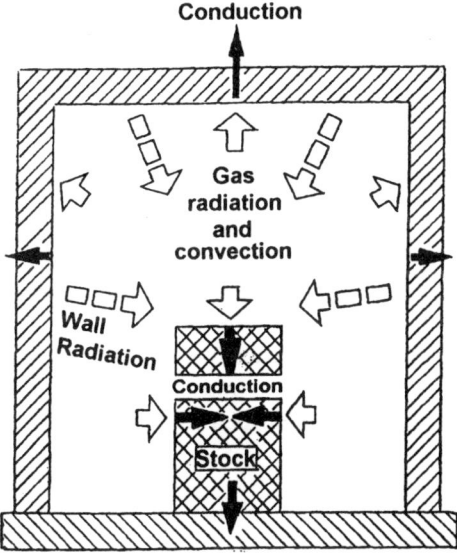

Figure 56 Furnace heat transfer processes. (From Ref. 62.)

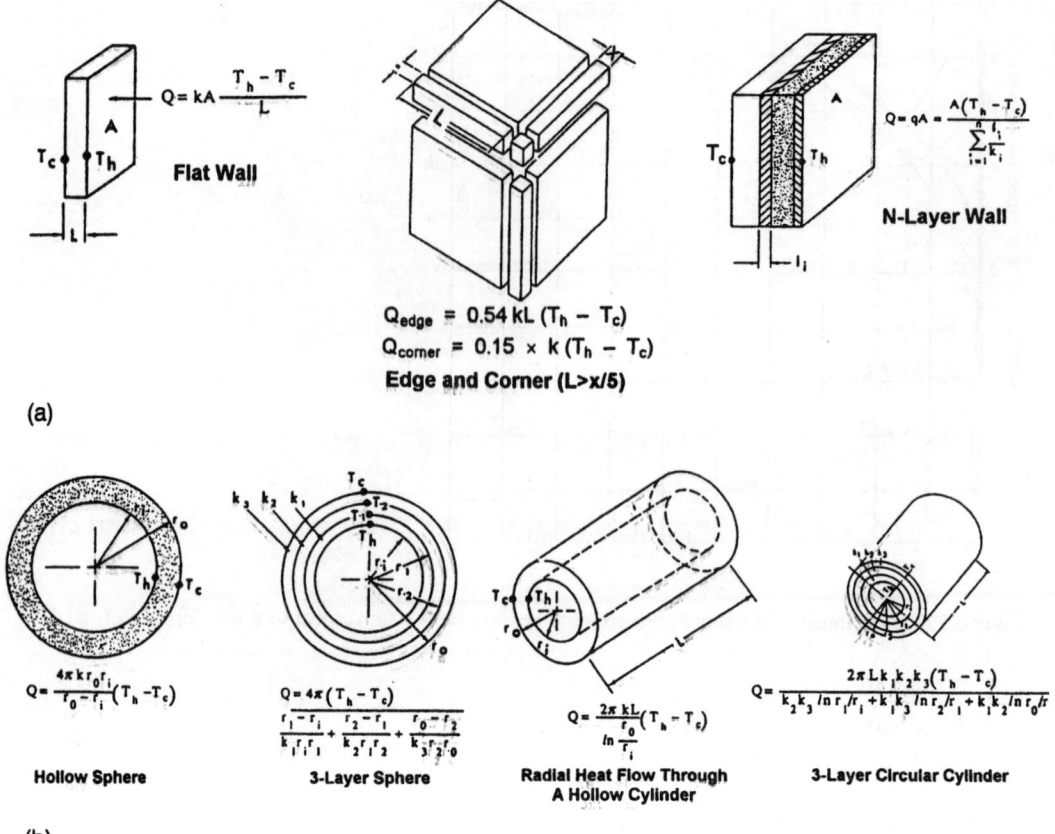

Figure 57   Heat transfer equations for commonly encountered shapes in heat treatment. (From Ref. 63.)

transferred through members of simple shapes that can be combined to approximate power losses in actual equipment.

The variation of thermal conductivity with temperature is shown in Figure 58. The variation within the area bounded by *abcda* in Figure 58 can be approximated as

$$\int_{T_c}^{T_h} k(T)dT$$

Once the variation of thermal conductivity with temperature is known, the heat transferred through a flat wall can be approximated as shown in Figure 59. Heat transfer through a curved wall may be approximated from the flat wall expression as shown in Table 10.

The thermal conductivity of various metals and refractory materials with respect to temperature are shown in Figure 60 and 61, respectively. The mean heat transfer coefficient accounting for fluid flow properties in natural convection can be calculated from [63,64]

$$\bar{h} = \frac{k}{L} C \left[\frac{g\beta(T_s - T_b)L^3}{\nu^2} N_{Pr} \right]^m = \frac{k}{L} C (N_{Gr} N_{Pr})^m$$

where
  k = thermal conductivity

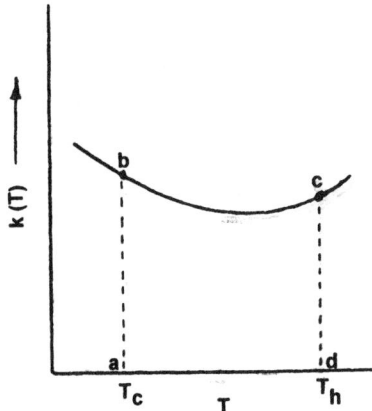

Figure 58 Variation of thermal conductivity with temperature.

L = length (for vertical planes and cylinders, L = height of surface; for horizontal cylinder,

L = diameter; for horizontal squares, L = length of a side)

g = gravitational constant

β = coefficient of volumetric thermal expansion

T_s = surface temperature

T_b = temperature of the boundary layer

v = kinematic viscosity

N_{Pr} = Prandtl number, $N_{Pr} = v/a$

N_{Gr} = Grashof number, $N_{Gr} = g \beta (T_s - T_b) x^3/v^2$

C and m are constants determined from Table 11 and Figure 62 for vertical planes and cylinders, horizontal cylinders, and square surfaces [63]. When the mean value of the heat transfer coefficient is calculated, it is assumed that

$$T_f = \frac{(T_s - T_b)}{2} \qquad \beta = \frac{1}{T}$$

where T is in kelvins.

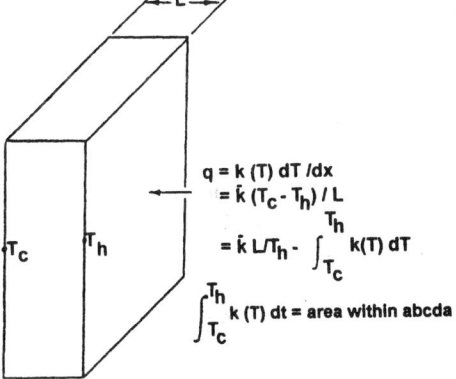

Figure 59 Calculation of heat flux for a flat wall.

Table 10 Flat Wall Thickness Equivalents for Curved Surfaces

Shape of wall	Assume flat wall thickness of	Assume flat wall area of
Cylindrical	$r_o - r_i$	$\pi (r_o + r_i)$
Spherical	$r_o - r_i$	$4 \pi r_o r_i$

Source: Ref. 63.

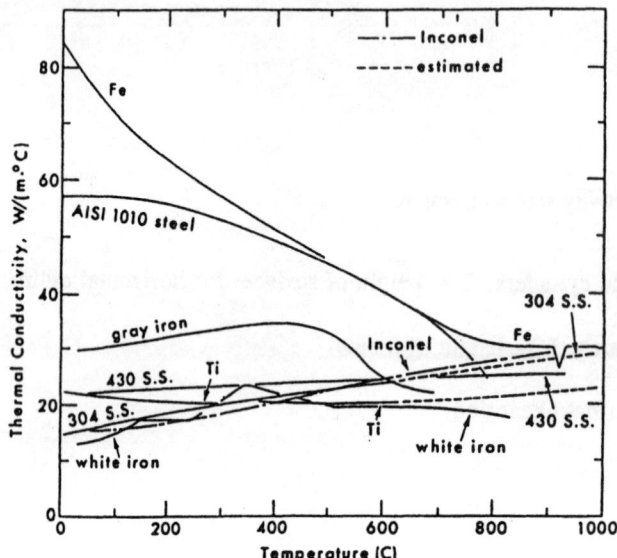

Figure 60 Thermal conductivity of various ferrous metals and Inconel.

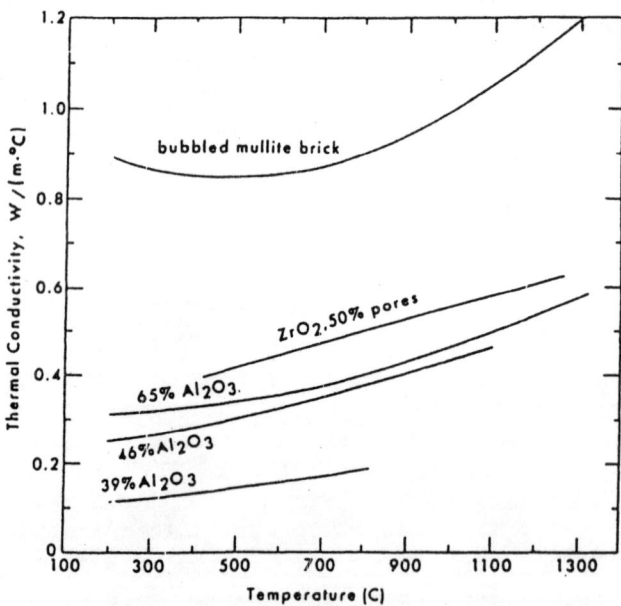

Figure 61 Thermal conductivity of various refractory materials.

Table 11 Equation Constants for Natural Convection

Surface	$N_{Gr}N_{Pr}$	C	M	Note
Vertical	$10^{-1}-10^4$			Use Figure 57.
	10^4-10^9	0.59	1/4	
	10^9-10^{12}	0.13	1/3	
Horizontal cylinder	$0-10^{-5}$	0.40	0	
	$10^{-5}-10^4$			Use Figure 57.
	10^4-10^9	0.53	1/4	
	10^9-10^{12}	0.13	1/3	
Horizontal square surface	$10^5-2 \times 10^7$	0.54	1/4	Upper surface if heated, lower surface if cooled
	2×10^7 to -3×10^{10}	0.14	1/3	
	3×10^5 to -3×10^{10}	0.27	1/4	Lower surface if heated plate; upper surface if cooled plate

Source: Ref. 51.

For laminar ($N_{Pr} < 2300$) flow through smooth tubes, the equation for the mean heat transfer coefficient is [63]

$$\bar{h} = \frac{k}{d}\left[3.66 + \frac{0.0668 \,(d/L)\, N_{Re}N_{Pr}^{\;3}}{1 + 0.04\,[(d/L)\, N_{Re}N_{Pr}]^{2/3}}\right]$$

where L = tube length and d = inside diameter for circular tubes or $d = 4 \times$ (area of cross section)/(inside circumference) for noncircular tubes.

For turbulent flow through smooth tubes [63],

$$\bar{h} = 0.023\frac{k}{d}\, N_{Re}^{0.8}\, N_{Pr}^{n}$$

where $N_{Re} > 4000$ and $0.6 \leq N_{Pr} \leq 100$; $n = 0.4$ for heating and 0.3 for cooling.

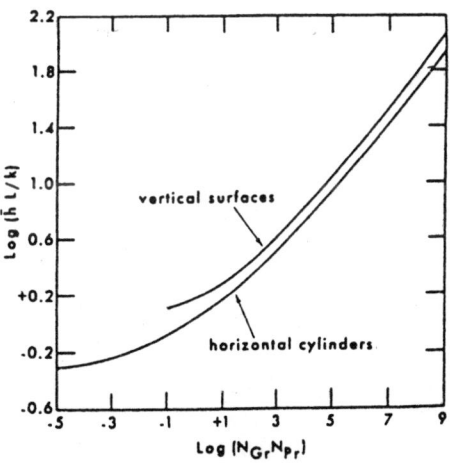

Figure 62 Correlation of natural convection for vertical and horizontal surfaces.

The equation for the mean heat transfer coefficient for gas flow across a cylinder is [65]

$$\bar{h} = C\frac{k}{d}N_{Re}^n \quad C\frac{k}{d}\left(\frac{vd}{v}\right)^n$$

where the mean film temperature (T_f) is

$$T_f = \frac{T_s + T_b}{2}$$

The values for C and n are given in Table 12 [65].

The constants for calculating the mean heat transfer coefficient for flow across noncylindrical shapes are provided in Table 13 [65].

Heat transfer coefficient for various furnace heating media used for heat treating are provided in Table 14.

B. Radiant Heat Transfer

Radiant heat transfer is dependent on the amount of radiation emitted or absorbed, the wavelength of the radiation, and the temperature and physical condition of the surface. The rate of radiant heat transfer (Q) between two surfaces at T_1 and T_2 is [65]

$$Q = A_1 F_{12} \delta (T_1^4 - T_2^4)$$

where A is the area, δ is the Stefan–Boltzmann constant, and F_{12} is a constant that depends on emissivity and geometry and is usually determined empirically.

The δ radiative heat transfer coefficient (h_r) can be calculated from

$$h_r = F_{12} \delta (T_1^3 + T_1^2 T_2 + T_1 T_2^2 + T_2^3)$$

Table 12   Constants for Gaseous Cylindrical Crossflow

N_{Re}	C	n
0.4–4	0.891	0.330
4–40	0.821	0.385
40–4000	0.615	0.466
4000–40,000	0.174	0.618
40,000–400,000	0.0239	0.805

Table 13   Constants for Calculation of Gaseous Crossflow of Noncylindrical Tubes

Tube shapes	N_{Re}	C	n
	$5 \times 10^3 - 1 \times 10^5$	0.222	0.588
	$5 \times 10^5 - 1 \times 10^5$	0.092	0.675
	$5 \times 10^3 - 1.95 \times 10^4$	0.144	0.638
	$1.95 \times 10^4 - 1 \times 10^5$	0.0347	0.782
	$5 \times 10^3 - 1 \times 10^5$	0.138	0.638
	$4 \times 10^3 - 1.5 \times 10^4$	0.205	0.731

Table 14 Heat Transfer Coefficients for Various Furnace Heating Media

Medium	Heat transfer coefficient Btu/(ft$^2 \cdot$ h $\cdot$ °F)
Air circulation furnace	2–8
Jet heating/cooling	20–50
Batch and pusher furnaces[a]	15–80
Gaseous fluidized bed	50–110
Stirred salt bath	200–600
Liquid fluidized bed	1300
Lead bath	1000–6000

[a]Convection/radiation.
Source: Ref. 65.

One of the greatest sources of radiative heat loss is when the furnace door is opened and the rate of heat loss can be calculated from [66]

$$q = 0.173 \ Ae\left[\left(\frac{T_0}{100}\right)^4 - \left(\frac{T_a}{100}\right)^4\right]$$

where A is the effective area of the door opening, e is the emissivity, and T_0 is the absolute temperature of the air.

Some illustrative values of surface emissivity are provided in Table 15.

C. Conductive Heat Transfer

The rate of conductive heat transfer (q) depends on thermal conductivity k, temperature T, and the distance from the heat source X:

$$q = k\frac{dT}{dX}$$

For plane surfaces or thin layers,

$$q = \frac{k}{t}(T_1 - T_2)$$

where t is the thickness of the material.

Table 15 Total Emissivities of Selected Surfaces

Material	Emissivity
Refractory	0.8
Carbon	0.9
Steel plate	0.95
Oxidized aluminum	0.15

Source: Ref. 67.

Figure 63 provides a illustrative summary of these heat transfer processes for an electrically heated furnace.

Heat transfer within the part being heated may be modeled from the differential equation [62]

$$\rho C \frac{\partial \theta}{\partial t} = k \frac{\partial^2 \theta}{\partial x^2}$$

where ρ = density, C = specific heat, θ = temperature, k = thermal conductivity, and x = distance.

It is important to note that these data are temperature-dependent. This dependence should be accounted for when conducting any furnace thermal modeling.

Furnace designers use computerized computational methods for the solution of these and related equations. A detailed discussion of these methods is beyond the scope of this text. However, the methods discussed here provide an excellent approximation to the solution of many routine heat transfer problems encountered in heat treatment shops.

The temperature rise of a simple shape can be estimated using Heisler charts, which are constructed with the following information [61].

1. Diffusivity (α):

    $$\alpha = k/C_\rho$$

    where k = conductivity, C = specific heat, and ρ = density of the material.
2. Fourier number (N_{Fo}):

    $$N_{Fo} = \alpha T/L^2$$

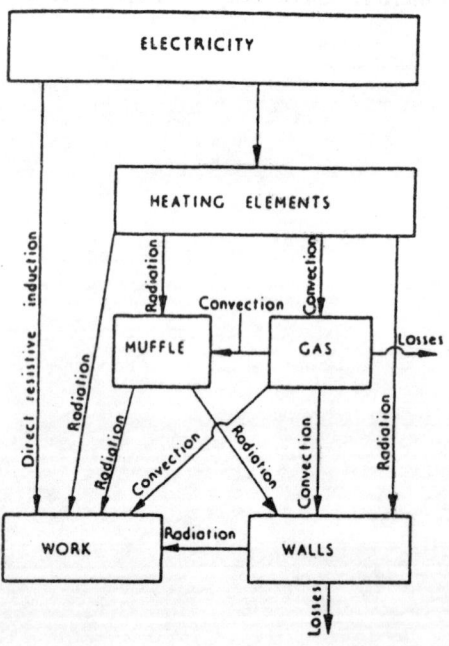

Figure 63 Summary of heat transfer processes for various furnace components. (From Ref. 68.)

where T is the heating time and L is the critical dimension of the shape. The critical dimension for slabs, cylinders, and and spheres (assuming no end effects) is determined as follows:

Slab	Half the thickness
Cylinder	Radius
Sphere	Radius

3. Relative boundary resistance (m):

$$m = k/hL$$

where $h =$ boundary conductance.

4. Temperature function (TF)

$$TF = (t_f - t)/(t_f - t_i)$$

where $t_i =$ initial temperature of the part, $t_f =$ constant furnace temperature, and $t =$ temperature at any point in the part.

The Heisler charts for spheres, cylinders, and slabs are shown in Figure 64 for "short times" where $N_{Fo} = 0$–0.2 and $m < 100$ and in Figure 65 for "long times," where $N_{Fo} > 0.2$ and $m < 100$. When $m > 100$, Figure 66 should be used. The relationship between the ratio of TF on the surface (TF_s) and in the center (TF_c) of the heated object and m is shown in Figure 67 [61].

Furnace temperature and heating time can be interrelated using the uniformity factor U, which is defined as [61]

$$U = \frac{t_s - t_c}{t_s - t_i}$$

where t_s is the temperature at the surface of the part, t_c is the temperature at its center, and t_i is its initial temperature. Uniformity factors for spheres, cylinders, and slabs are shown in Figure 68.

D. Furnace Temperature Uniformity

Characteristic thermal flow patterns in a furnace may be significant and will cause nonuniform heating of parts. This nonuniformity may be due to various factors, including [62]

1. Interaction between burners
2. Unstable flow
3. Variation of mass circulation rates and thermal distribution within the load
4. Stagnant regions of high or low temperature
5. Combustion patterns

Two important criteria in considering appropriate furnace design are the variation of size and type of the materials being heated (see Figure 69) [66,67], if more than one, and the required production rate [68]. Another factor that must be considered is the necessity for various heat treatment processes such as austenitizing, normalizing, and stress relief in a single furnace.

The first step in determining the best furnace design is to conduct an energy balance to determine the relative efficiencies of different furnace designs being considered. One method

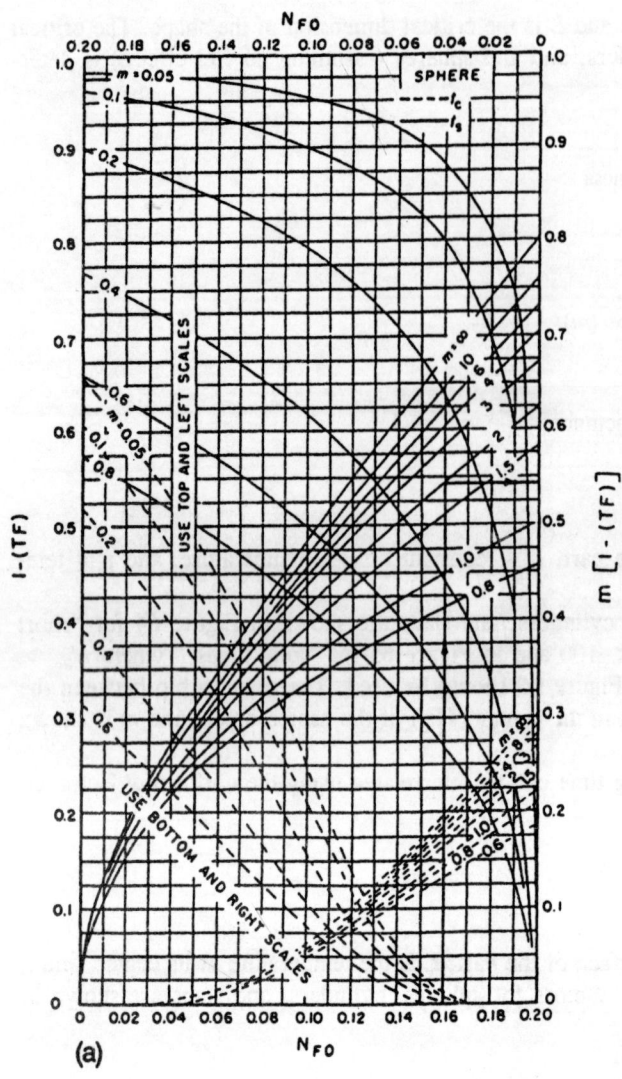

(a)

Figure 64 Short-time temperature function (TF) relationships for (a) spheres, (b) cylinders, (c) slabs.

of conducting this assessment is to model the various heat transfer processes in the furnace and conduct an energy balance [69]. The energy balance may be illustrated using a Sankey diagram such as the chart depicted in Figure 70, which shows that furnace efficiency [(useful output)/(fuel input) × 100] is dependent on

1. Thermal energy in the material
2. Structural losses
3. Waste gas losses
4. Heat recovery
5. Unaccounted losses

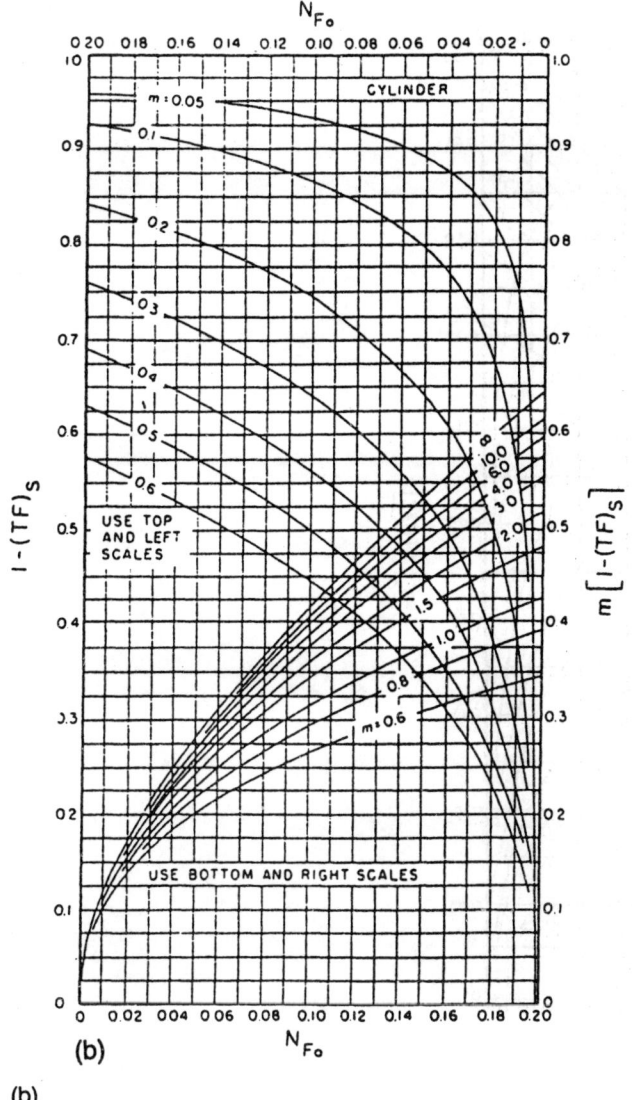

(b)

Figure 64 Continued

One method of improving furnace temperature uniformity is to use forced circulation of the heated gaseous atmosphere [66,67]. The effect of increased flow velocity on furnace temperature uniformity is shown in Figure 71. The required air flow to maintain a given temperature tolerance, typically 5–15°F, may be calculated from [66,67]

$$\text{Air flow (cfm)} = \frac{HA}{625.7U} \times T_A$$

where
H = heat loss in Btu through 1 ft^2 of furnace wall per hour
A = furnace wall area (ft^2)

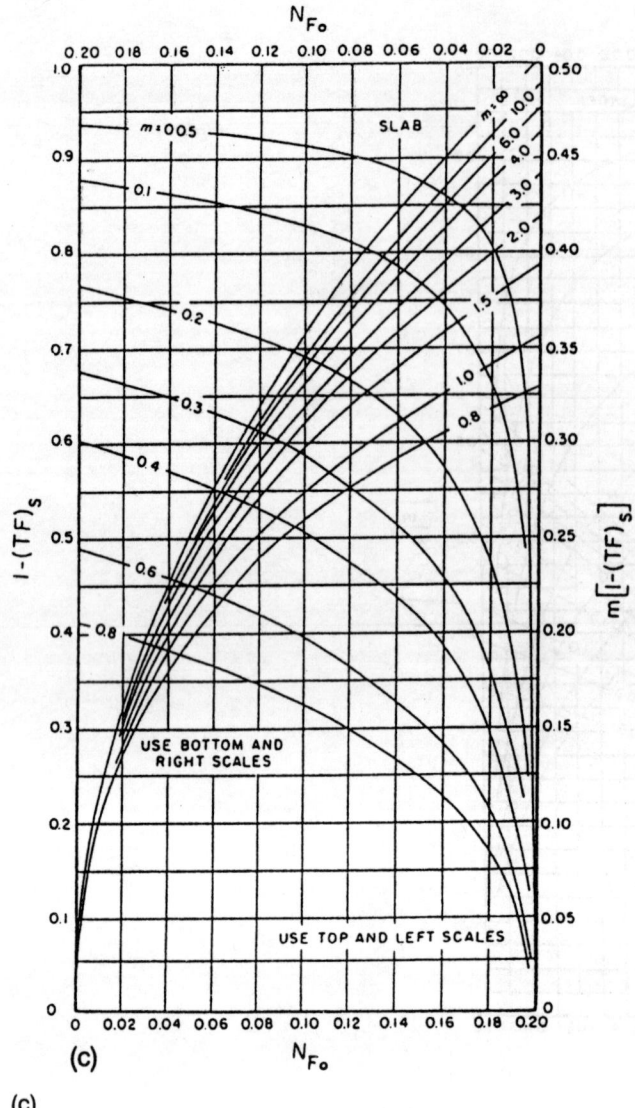

(c)

Figure 64 Continued

T_A = absolute furnace temperature (460° + °F)
U = maximum allowable variation in furnace temperature (°F)

This calculation assumes that the heated air has the necessary Btu content to heat the load for the furnace cycles being calculated and that the heat losses through the furnace walls are included [66].

E. Soaking Time

Soaking time is dependent on (1) gas metrical factors relating to the particular furnace and load, (2) type of load, (3) type of steel, (4) thermal properties of the load, (5) load and fur-

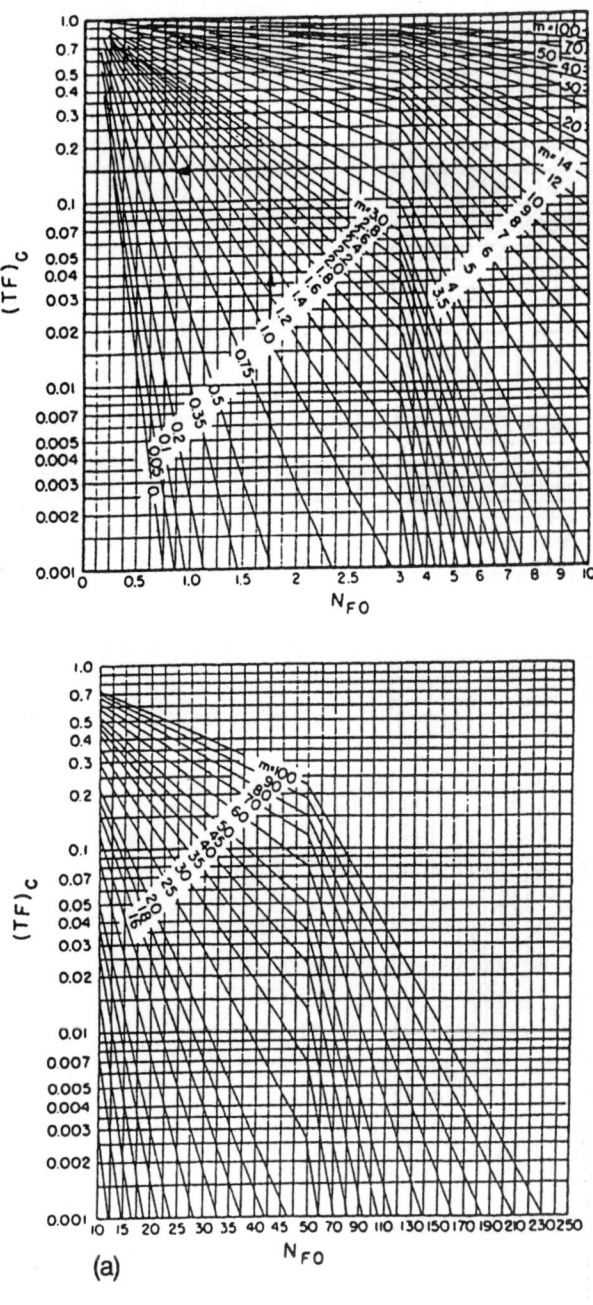

$(TF)_c$

N_{FO}

$(TF)_c$

N_{FO}

(a)

(a)

Figure 65 Long-time temperature function (TF) relationships for (a) spheres, (b) cylinders, (c) slabs.

nace emissivities, (6) initial furnace and load temperatures, (7) characteristic fan curves, and (8) the chemical composition of the atmosphere [70,71].

   Aronov et al. [70,71] modeled soaking times as a function of these parameters and developed menu-driven software for predicting furnace soaking times based on "load character-

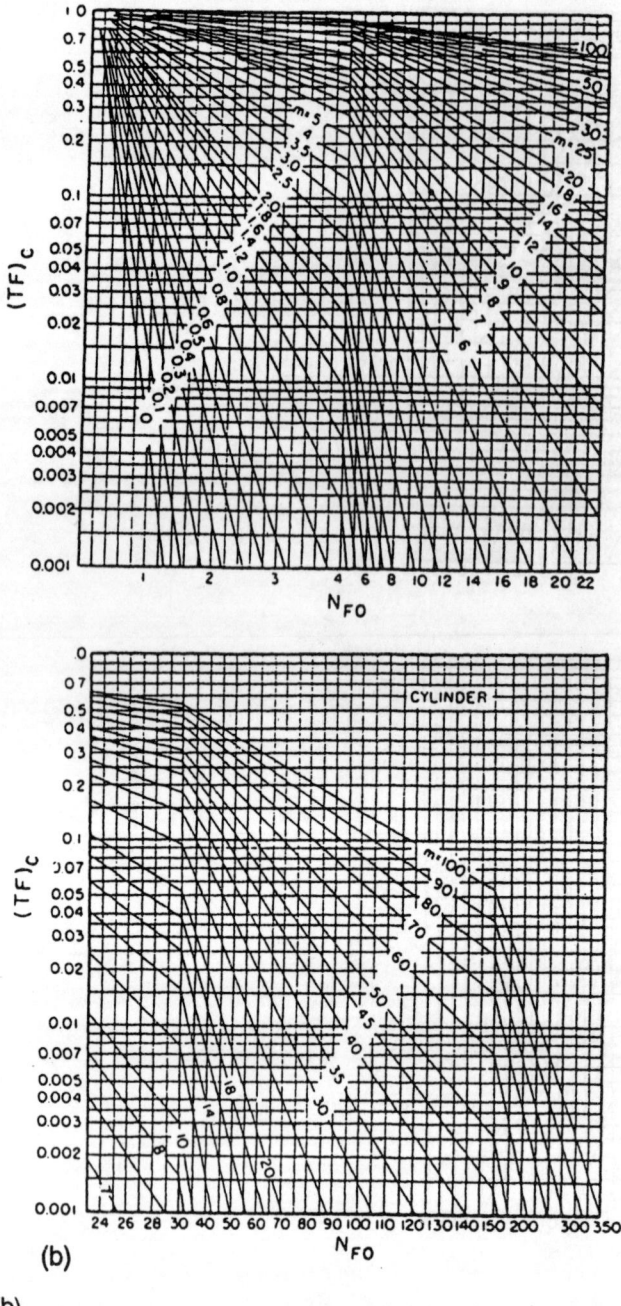

(b)

Figure 65 Continued

ization." Load characterization diagrams are provided in Figure 72. These models are based on the generalized equation for soaking time (T_s),

$$T_s = T_{sb} K$$

where

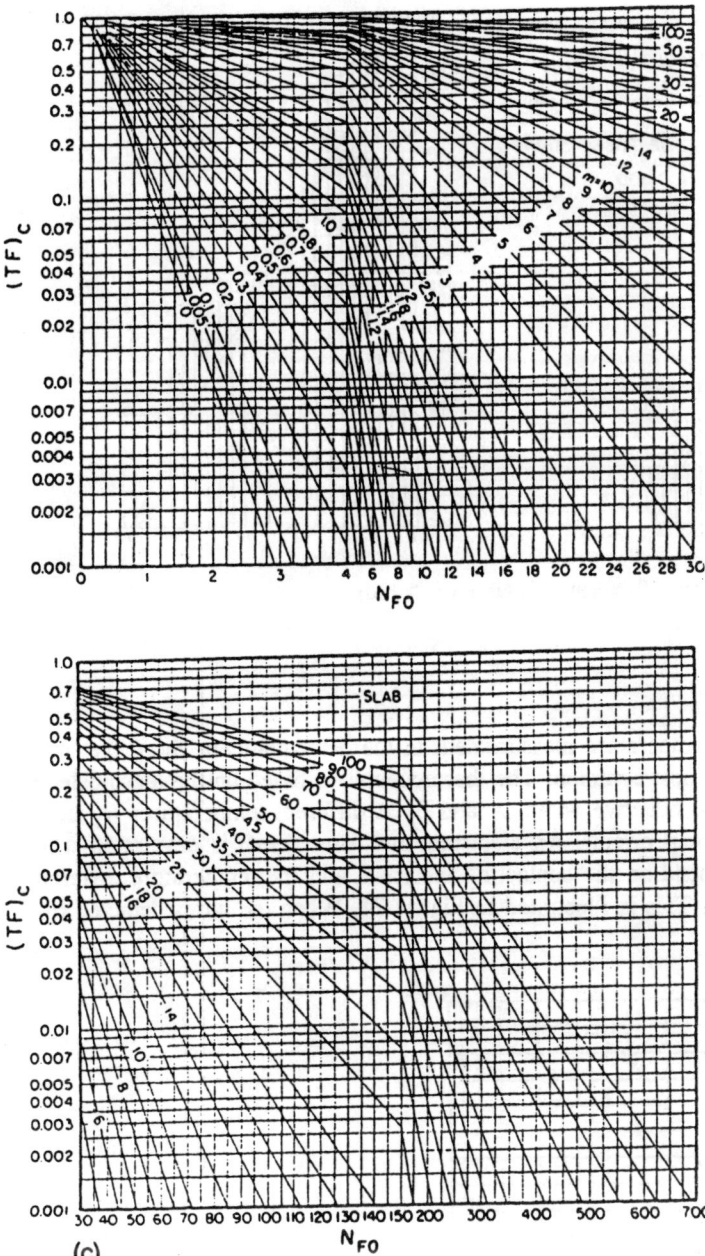

(c)

Figure 65 Continued

T_s = calculated soaking time (min)

T_{sb} = baseline soak temperature condition selected from graphs such as those in Figures 73a–73d [70,71]

K = correction factor for the type of steel ($K = 1$ for low alloy steel and 0.85 for high alloy steel)

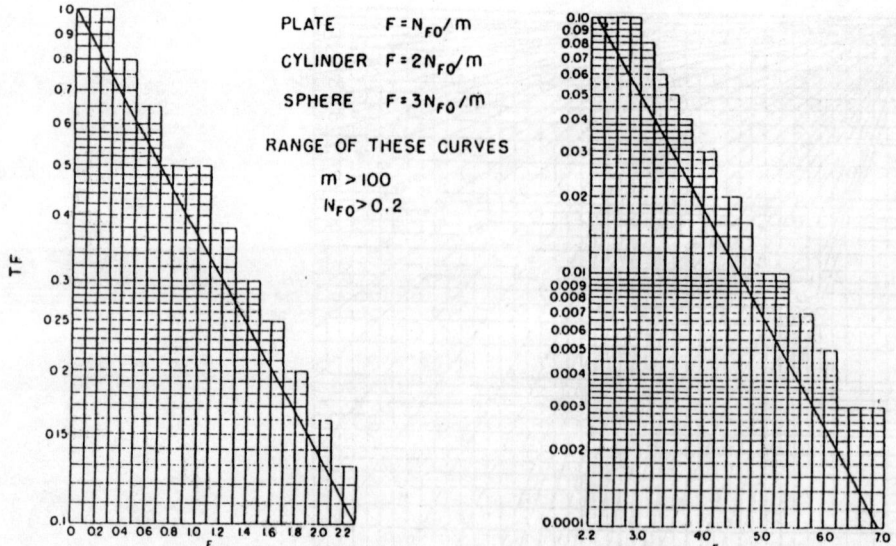

PLATE $F = N_{FO}/m$

CYLINDER $F = 2N_{FO}/m$

SPHERE $F = 3N_{FO}/m$

RANGE OF THESE CURVES

$m > 100$

$N_{FO} > 0.2$

Figure 66   Temperature function (TF) when m is large ($m > 100$).

V.   THERMOCOUPLES

Of the various methods of temperature measurement in the heat treatment shop, the use of thermocouples is one of the most common. The thermocouple is based on the thermoelectric effect that exists when two conductive wires (A and B) at different temperatures (t_1 and t_2) are connected to form a closed circuit. An electromotive force (emf) is developed whose magnitude and direction depend on the contacting materials and the temperature difference between the two points [72]. This is illustrated in Figure 74. When the wires depicted by A

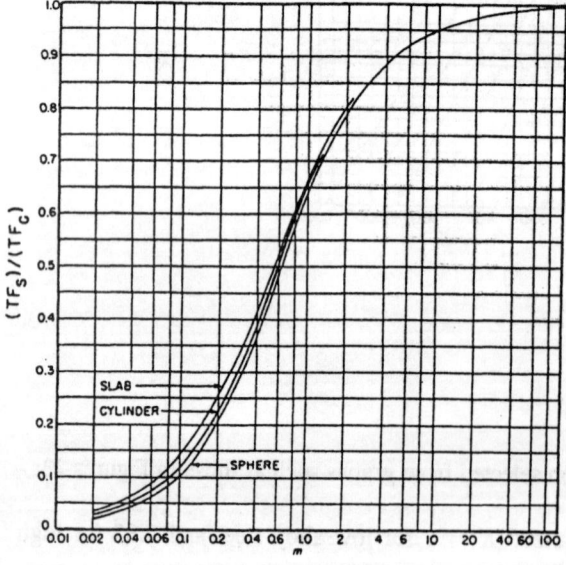

Figure 67   Surface to center temperature function ratio for spheres, cylinders, and slabs. (From Ref. 61.)

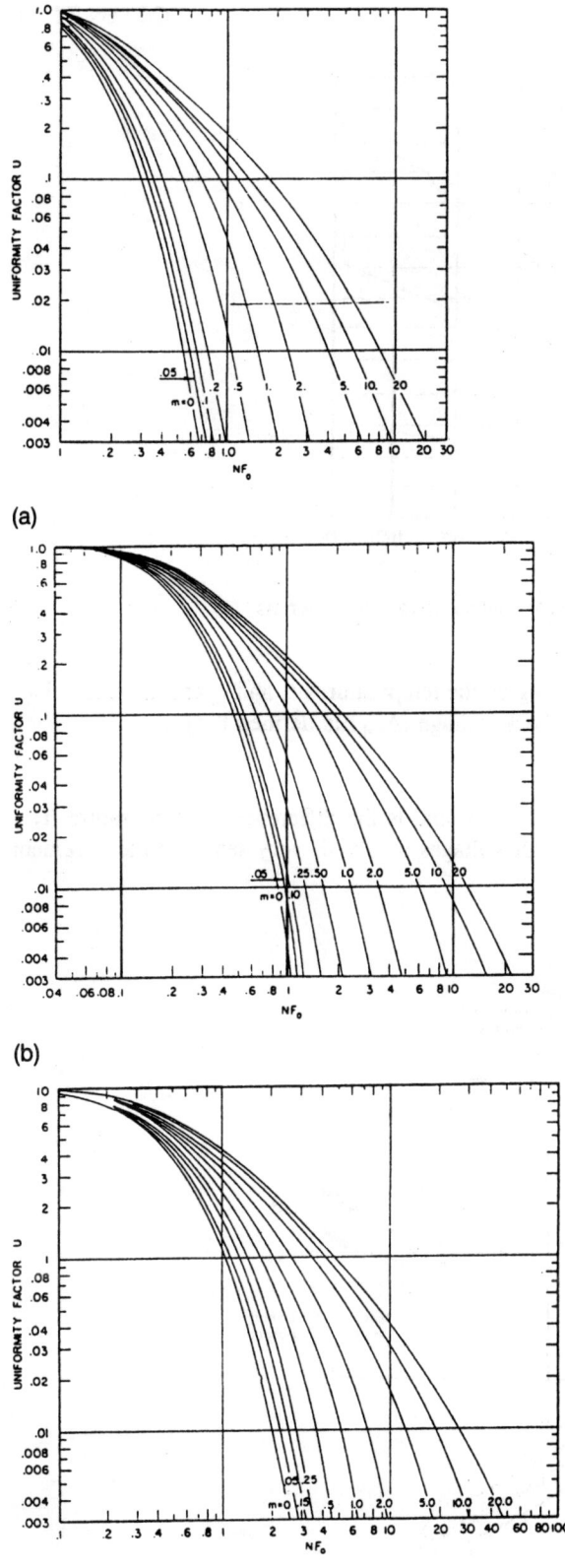

Figure 68 Uniformity factors for (a) spheres, (b) cylinders, (c) slabs.

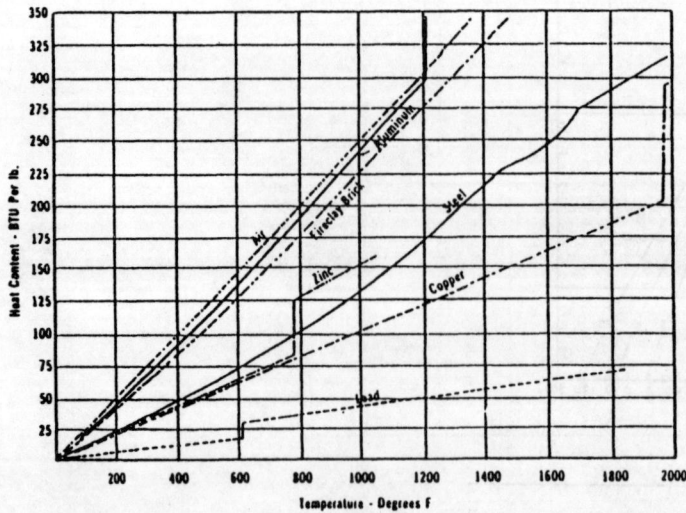

Figure 69 Temperature dependence of the heat content of various materials.

and B are different, current will flow as long as the temperatures t_1 and t_2 are different. This is called the Seebeck effect [74]. The Seebeck voltage (Δe_{ab}) is defined [75] as

$$\alpha e_{ab} = \alpha \, \Delta T$$

where α is the Seebeck coefficient (see Table 16) and is the difference in temperature ($t_2 - t_1$). Table 17 shows that in view of the small voltages involved, very sensitive measurement instruments are required [75].

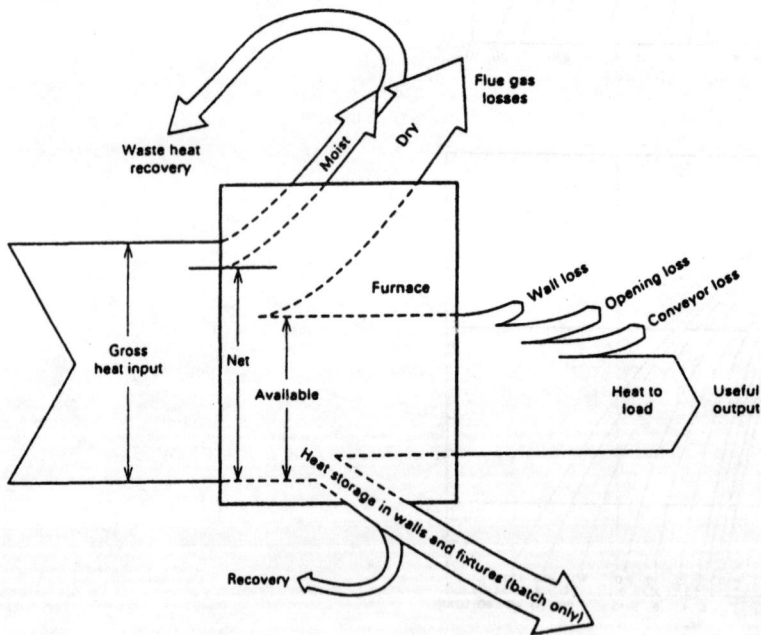

Figure 70 Illustration of the use of a Sankey diagram to track furnace heat losses.

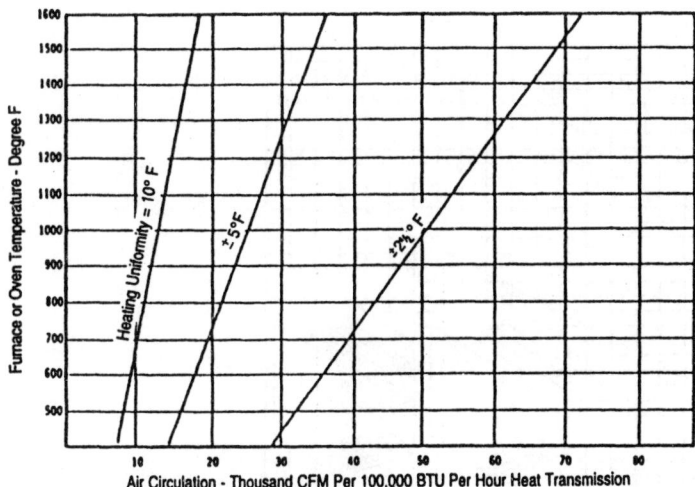

Figure 71 Effect of air circulation and oven temperature on furnace temperature uniformity.

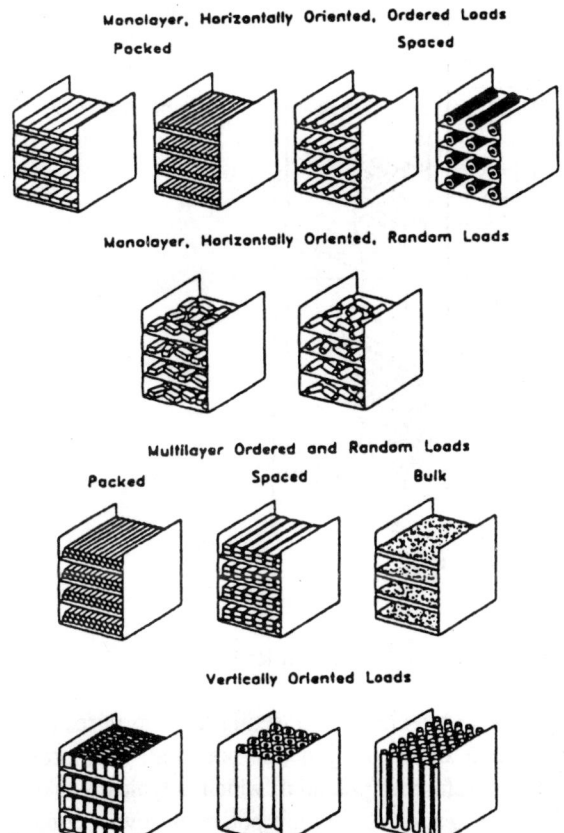

Figure 72 Aronov load characterization diagram for soaking time calculations.

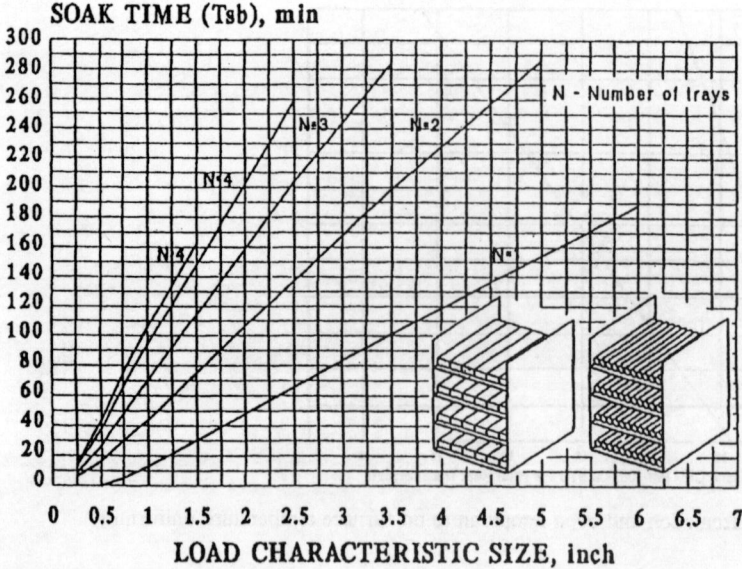

(a)

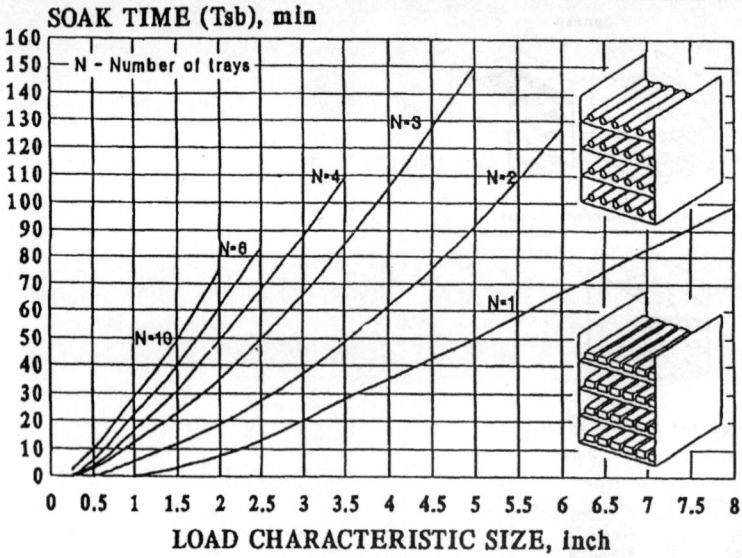

(b)

Figure 73 Soaking times for (a) packed load, (b) spaced load, (c) vertical load, (d) disks.

It is not possible to measure the voltage (Δe_{AB}) by simply connecting the two wires to a voltmeter because the voltmeter itself introduces a significant junction potential. This problem is solved by adding the junction potential to a reference potential, which is typically taken as the freezing point of water ($T_{ref} = 0°C$). The measured junction potential now becomes [75]

$$V = \alpha \, (T_1 - T_{ref})$$

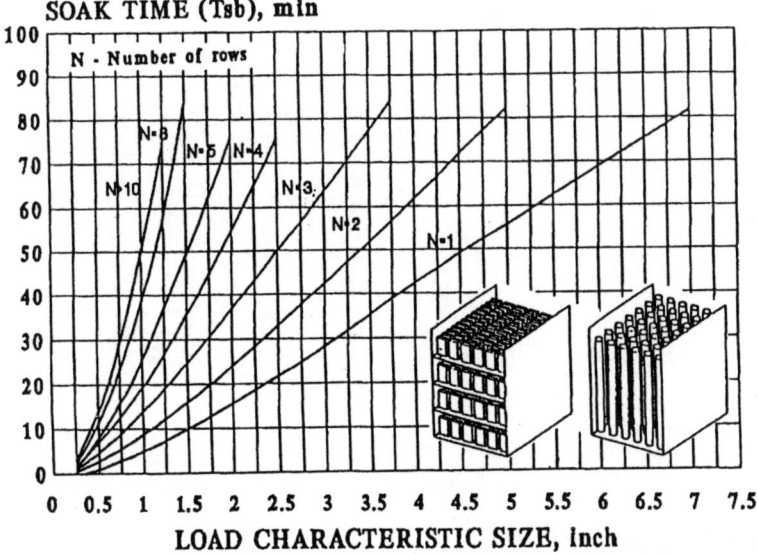

(c)

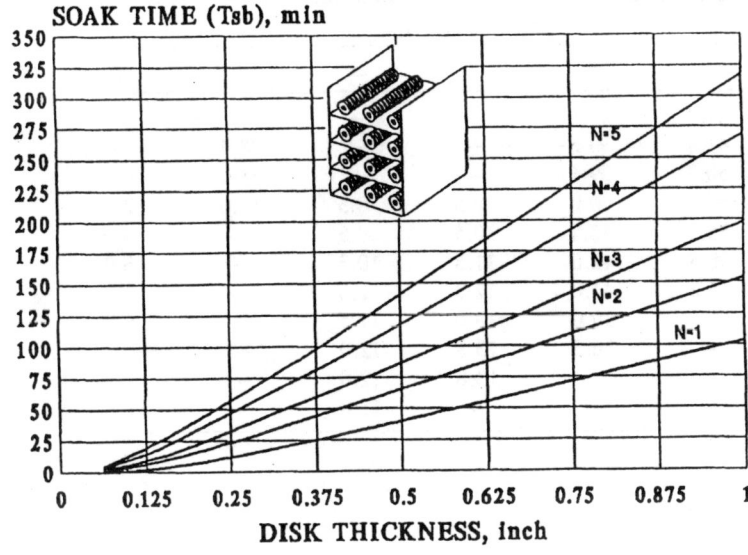

(d)

Figure 73 Continued

Most thermocouples use an external reference junction compensation with either hardware or software compensation instead of an ice-water bath [75].

Table 18 summarizes the composition and maximum use temperature for various standard thermocouples [76]. However, the maximum recommended use temperature is dependent on the size of the thermocouple wire as shown in Figure 75 [77].

The voltage for a thermocouple may be read directly from voltmeter using either hardware or software compensation or calculated from [75]

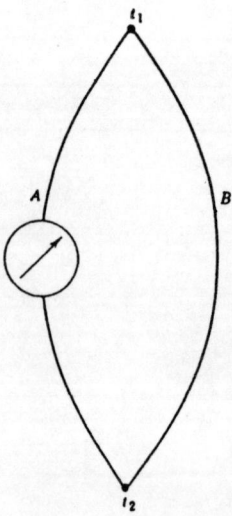

Figure 74 A thermocouple circuit. (From Ref. 73.)

Table 16 Nominal Seebeck Coefficient (μV/°C) for Standard Thermocouples

Temp. (°C)	Thermocouple type						
	E	J	K	R	S	T	B
−190	27.3	24.2	17.1	—	—	17.1	—
−100	44.8	41.4	30.6	—	—	28.4	—
0	58.5	50.2	39.4	—	—	38.0	—
200	74.5	55.8	40.0	8.8	8.5	53.0	2.0
400	80.0	55.3	42.3	10.5	9.5	—	4.0
600	81.0	58.5	42.6	11.5	10.3	—	6.0
800	78.5	64.3	41.0	12.3	11.0	—	7.7
1000	—	—	39.0	13.0	11.5	—	9.2
1200	—	—	36.5	13.8	12.0	—	90.3
1400	—	—	—	13.8	12.0	—	11.3
1600	—	—	—	—	11.8	—	11.6

Table 17 Required Voltmeter Sensitivity

Thermocouple type	Seebeck coefficient (μV/°C at 20°C)	Voltmeter sensitivity for 0.1°C (μV)
E	62	6.2
J	51	5.1
K	40	4.0
R	7	0.7
S	7	0.7
T	40	4.0

Source: Ref. 75.

Table 18 Standard Letter-Designated Thermocouples

Type J	Thermo-elements	Typical alloy	Base Composition	Application Atmosphere	Max temp.
J	JP	Iron	Fe	Oxidizing	760°C
	JN	Constantan (J)	44Ni/55Cu	Reducing	1000°F
K	KP	Chromel	90N/9Cr	Oxidizing	1260°C
	KN	Alumel	94Ni/Al, Mn, Si	Inert	2300°F
T	TP	Copper	OFHC/Cu	Oxidizing	370°C
	TN	Constantan (T)	44Ni/55Cr	Reducing	700°F
E	EP	Chromel	90Ni/9Cr	Oxidizing	870°C
	EN	Constantan (T)	44Ni/55Cr	Inert	1600°F
N	NP	Nicrosil	Ni/14.2Cr 1.4Si	Oxidizing	1260°C
	NN	Nisil	Ni/4Si/15Mg	Inert	
R	RP	Pt/Rh	87Pt/13Rh	Oxidizing	1480°C
	RN	Pt	Pt	Inert	2700°F
S	SP	Pt/Rh	90Pt/10Rh	Oxidizing	1480°C
	SN	Pt	Pt	Inert	2700°F
B	BP	Pt/Rh	70Pt/30Rh	Oxidizing	1700°C
	BN	Pt/Rh	94Pt/6Rh	Inert & Vacuum	3100°F

$$T = a_0 + a_1x + a_2x^2 + a_3x^3 + \cdots + a_nx^n$$

Where T is temperature, x is the thermocouple voltage, the a's are thermocouple-dependent polynomial coefficients (Table 19), and n is the maximum order of the polynomial; as n increases, accuracy increases, e.g., when $n = a$, the accuracy is $\pm 1°C$.

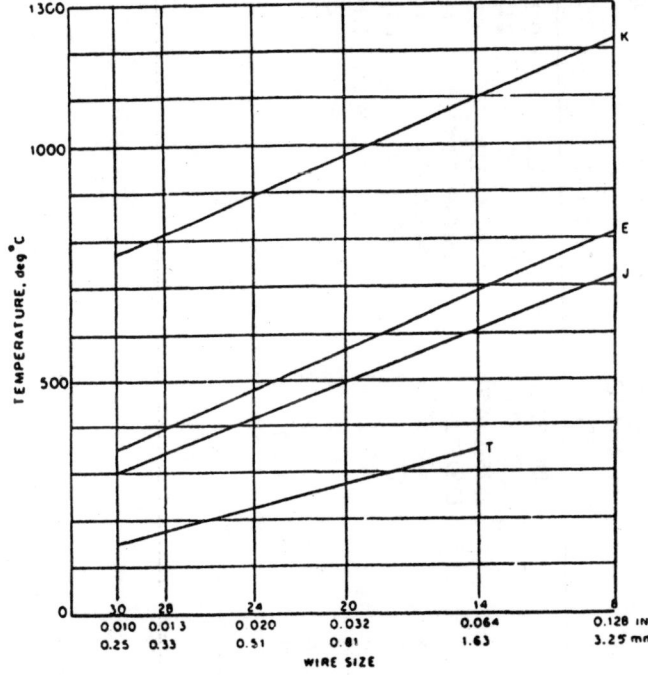

Figure 75 Recommended upper temperature limits for Type K, E, J, and T thermocouples of various sizes. (From Ref. 77.)

Table 19 NBS Thermocouple Polynomial Coefficients (av)

n	Type E Ni-10%Cr (+) vs. Constantan (−) −100–1000°C ±0.5°C	Type J Fe (+) vs. Constantan (−) 0°–760°C ±0.1°C	Type K Ni-10%Cr (+) vs. Ni-5% (−) 0°–1370°C ±0.7°C	Type R Pt-Rh (+) vs. Pt (−) 0°–1000°C ±0.5°C	Type S Pt-10% Rh (+) vs. Pt (−) 0°–1750°C ±1°C	Type T Cu (+) vs. Constantan (−) −160–+400°C ±0.5°C
0	0.104967248	−0.048868252	0.226584602	0.26362917	0.927763167	0.100860910
1	17189.45282	19873.14503	24152.10900	179075.491	169526.5150	25727.94369
2	−282639.0850	−218614.5353	67233.4248	−48840341.37	−31568363.94	−767345.8295
3	12695339.5	11569199.78	2210340.682	1.90002E+10	8990730663	78025595.81
4	−448703084.6	−264917531.4	−860963914.9	−4.82704E+12	−1.63565E+12	−9247486589
5	1.10866 E+10	2018441314	4.83506E+10	7.62091E+14	1.88027E+14	6.97688E+11
6	−1.76807 E+11		−1.18452E+12	−7.20026E+16	−1.37241E+16	−2.66192E+13
7	1.71842 E+12		1.38690E+13	3.71496E+18	6.17501E+17	3.94078E+14
8	−9.19278 E+12		−6.37708E+13	−8.03104E+19	−1.56105E+19	
9	2.06132 E+13				1.69535E+20	

Table 20 Thermocouple Response Time (s) Variation with Wire and
Heat Transfer Medium

Wire size (in.)	Still air (800°F/100°F)	Air at 60 ft/s (800°F/100°F)	Still water (200°F/100°F)
0.001	0.05	0.004	0.002
0.005	1.0	0.08	0.04
0.015	10.0	0.80	0.40
0.032	40.0	3.2	1.6

Source: Ref. 75.

The thermocouple response times are also dependent on the size of the wire as shown in Figure 75 [78] and the heat transfer medium as shown in Table 20.

There are three conventional styles of thermocouples as shown in Figure 76 [78].

1. *Exposed.* These thermocouples are used when very fast response times are necessary. They are characterized by their exposed thermocouple junctions.
2. *Grounded.* These thermocouples are characterized by grounding to the thermocouple sheath, which provides both excellent response time and protection of the thermocouple junction.
3. *Ungrounded* (insulated). The thermocouple junction is electrically isolated from the protective sheath (which is usually stainless steel or Inconel) and is used when electrical noise hinders measurement of the thermocouple voltage. These thermocouples typically have somewhat slower response times than grounded thermocouples.

Thermocouple assemblies can be constructed in various ways such as those depicted in Figure 77. The most common types of insulation for high temperature applications are fiberglass, fibrous silica, and asbestos. Asbestos, because of its toxicity, is no longer commonly used.

It is important that thermocouple measurement be both accurate and precise. It is possible, as illustrated in Figure 78, for thermocouple readings to be reproducible and precise but still wrong. Therefore, the temperature reading of a thermocouple should be traceable to an NIST standard [76]. The uncertainty of the thermocouple calibration is calculated from [76]

$$U_{TE} = [(U_{NIST})^2 - (R_{TE})^2]^{1/2}$$

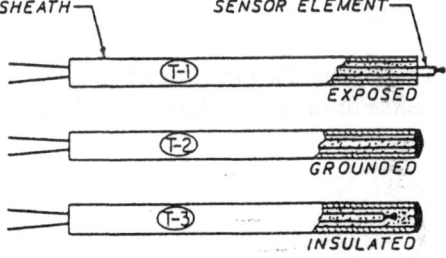

Figure 76 Common thermocouple junction styles.

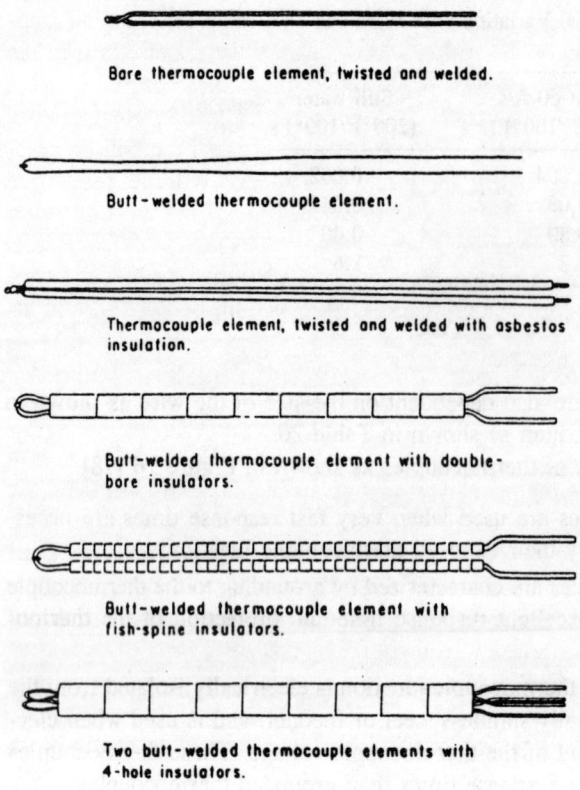

Bore thermocouple element, twisted and welded.

Butt-welded thermocouple element.

Thermocouple element, twisted and welded with asbestos insulation.

Butt-welded thermocouple element with double-bore insulators.

Butt-welded thermocouple element with fish-spine insulators.

Two butt-welded thermocouple elements with 4-hole insulators.

Figure 77   Thermocouple element assemblies. (From Ref. 77.)

where

U_{TE} = uncertainty of calibration in the user's laboratory
U_{NIST} = uncertainty of calibration at NIST
R = reproducibility in the user's laboratory

Ideally, thermocouples will be calibrated at temperatures between the use temperature and room temperature. Thermocouples of types J, K, E, and N may be calibrated at temperatures between ice water (32°F; 0°C), boiling water (212°F; 100°C), the melting point of tin (449°F; 232°C), and the melting point of zinc (787°F; 420°C) [74].

VI. ATMOSPHERES

Furnace atmosphere selection, generation, and control are among the most important steps in the heat treatment operation. For example, control of oxide formation, facilitation of the for-

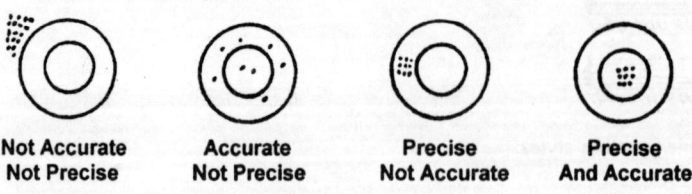

Not Accurate   Accurate   Precise   Precise
Not Precise   Not Precise   Not Accurate   And Accurate

Figure 78   Illustration of accuracy versus precision. (From Ref. 76.)

mation of the desired steel surface chemistry, and prevention of decarburization are all critically important to ensure overall success of the process, and all are integrally related to the proper selection and operation of the heat treatment equipment.

Furnace atmosphere surface reactions will vary with steel chemistry, process temperature and time, and the purity of the atmosphere itself. In some cases an atmosphere will be reactive with a steel surface, and in other cases the same atmosphere will be protective (nonreactive). Tables 21–24 and Figure 79 provide a summary of some of the most common furnace atmospheres used in heat treatment and their properties [71].

In this section, an overview of furnace atmospheres, both primary furnace gases and controlled atmospheres, is provided, followed by classification, composition, properties, and atmosphere generation.

A. Primary Furnace Gases

1. Nitrogen

Nitrogen (N_2), an inert diatomic gas,[1] is the primary component (78.1%) of atmospheric air as shown in Table 25. The remaining components of air are oxygen (20.9%) and other gases in much lower concentrations (<1%). The physical properties of nitrogen and other gases used in furnace atmosphere preparation are summarized in Tables 26A–26D.

Nitrogen is considered to be chemically inert and is used as a carrier gas for reactive furnace atmospheres, for purging furnaces, and in other processes requiring inert gases. However, at high temperature, nitrogen may not be compatible with certain metals such as molybdenum, chromium, titanium, and columbium [82].

Nitrogen of high purity (<1 ppm O_2) can be produced by oxidation of ammonia or in somewhat less pure form by the combustion of hydrocarbons in insufficient air (which removes O_2) and subsequent purification of residual hydrogen and methane [83]. Table 27 illustrates the purity of nitrogen obtained by these processes.

High purity nitrogen and other gases are also obtained by air liquefaction. A schematic for this process is shown in Figure 80. The nitrogen thus obtained can be subsequently purified by palladium-catalyzed hydrogen reduction of the residual oxygen, typically present at a level of ~0.2%. Alternatively, residual oxygen can be removed by high temperature reaction with copper to form copper oxide, which eliminates the possibility of water vapor contamination of the combustion process [82].

Residual water vapor, which is present as a by-product of this process, must be removed both to minimize vaporization (flash-off) upon expansion, as shown in Figure 81, and to ensure subsequent control of reactive gas chemistry in processes such as carburization.

Nitrogen can also be obtained by membrane separation from air based on the selective permeability of a composite membrane fiber. In this process, atmospheric air is filtered, compressed, and cooled and then passed through an air-separation membrane as shown in Figure 82 [84,85]. Oxygen, carbon dioxide, and water vapor permeate the hollow membrane fibers and are then vented at low pressure to the atmosphere. Nitrogen is then stored in the desired form.

---

[1]The term "inert gas" is often misleading to both heat treaters and furnace designers. It is incorrectly assumed that furnace atmospheres of nitrogen will not undergo any reactions with steel surfaces. However, steel will be decarburized if it is held for extended periods of time at elevated temperature with a nitrogen atmosphere. To make atmospheres "nonreactive" and "neutral" to steel, an enriching gas such as methane or propane that will generate the same C_p as steel must be used with nitrogen. (These will be discussed subsequently.) If nonferrous materials are used, then a reducing gas such as hydrogen is required, but not more than 10%.

Table 21   Heat Content of Various Gases Above 77°F (Btu/ft³ at 60°F and 30 in. Hg)

Gas temp. (°F)	CO	CO_2	CH_4	H_2	N_2	O_2	H_2O	AX^a	DX^b	DX^c
100	.4	.6	.5	.4	.4	.4	.5	.4	.4	.4
200	2.3	3.0	2.9	2.3	2.3	2.3	2.6	2.3	2.3	2.3
300	4.1	5.6	5.4	4.1	4.1	4.2	4.8	4.1	4.2	4.3
400	6.0	8.3	8.2	6.0	6.0	6.1	7.0	6.0	6.1	6.2
500	7.9	11.2	11.1	7.8	7.9	8.1	9.2	7.8	8.0	8.2
600	9.8	14.1	14.3	9.7	9.7	10.1	11.5	9.7	10.0	10.2
700	11.7	17.1	17.7	11.5	11.7	12.2	13.8	11.6	11.9	12.2
800	13.7	20.2	21.3	13.4	13.6	14.2	16.1	13.4	13.9	14.2
900	15.7	23.4	25.0	15.2	15.5	16.3	18.5	15.3	15.9	16.3
1000	17.7	26.6	29.0	17.1	17.5	18.5	21.0	17.2	18.0	18.4
1100	19.7	29.9								
1200	21.8	33.2	37.4	20.9	21.6	22.8	26.0	21.0	22.1	22.7
1300	23.8	36.6	41.9	22.8	23.6	25.0	28.5	23.0	24.2	24.9
1400	25.9	40.1	46.5	24.7	25.7	27.2	31.2	24.9	26.3	27.1
1500	28.1	43.5	51.2	26.6	27.8	29.4	33.8	26.9	28.5	29.3
1600	30.2	47.1	56.1	28.5	29.9	31.6	36.5	28.9	30.7	31.6
1700	32.3	50.6	61.1	30.5	32.0	33.9	39.3	30.9	32.8	33.8
1800	34.5	54.2	66.2	32.5	34.1	36.1	42.1	32.9	35.0	36.1
1900	36.7	57.8	71.4	34.5	36.3	38.4	44.9	34.9	37.2	38.4
2000	38.9	61.4	76.7	36.5	38.4	40.7	47.8	37.0	39.5	40.7
2100	41.1	65.0	82.1	38.5	40.6	43.0	50.7	39.0	41.7	43.0
2200	43.3	68.7	87.6	40.5	42.8	45.3	53.6	41.1	44.0	45.4
2300	45.5	72.4	93.2	42.5	45.0	47.6	56.6	43.2	46.2	47.7
2400	47.8	76.1	98.9	44.6	47.2	49.9	59.6	45.3	48.5	50.1
2500	50.0	79.8	104.7	46.7	49.4	52.2	62.7	47.4	50.8	52.5
2600	52.3	83.6	110.5	48.7	51.7	54.6	65.7	49.5	53.1	54.8

2700	54.5	87.3	116.4	50.8	53.9	56.9	68.9	51.6	55.4	57.2
2800	56.8	91.1	122.4	53.0	56.2	59.3	72.0	53.8	57.7	59.6
2900	59.1	94.9	128.4	55.1	58.4	61.6	75.2	55.9	60.0	62.0
3000	61.3	98.7	134.5	57.2	60.7	64.0	78.3	58.1	62.4	64.5
3100	63.6	102.5	140.7	59.4	63.0	66.4	81.6	60.3	64.7	66.9
3200	65.9	106.3	146.9	61.5	65.2	68.8	84.8	62.4	67.1	69.3
3300	68.2	110.1	153.1	63.7	67.5	71.2	88.1	64.6	69.4	71.7
3400	70.5	114.0	159.4	65.9	69.8	73.6	91.4	66.8	71.8	74.2
3500	72.8	117.8	165.7	68.0	72.1	76.0	94.7	69.1	74.2	76.6
3600	75.2	121.7	172.1	70.2	74.4	78.4	98.0	71.3	76.5	79.1
3700	77.5	125.6	178.5	72.5	76.7	80.9	101.4	73.5	78.9	81.5
3800	79.8	129.4	185.0	74.7	79.0	83.3	104.7	75.8	81.3	84.0
3900	82.1	133.3	191.5	76.9	81.3	85.8	108.1	78.0	83.7	86.5
4000	84.4	137.2	198.0	79.1	83.6	88.2	111.5	80.3	86.1	88.9
4100	86.8	141.1	204.6	81.4	85.9	90.7	115.0	82.5	88.5	91.4
4200	89.1	145.0	211.1	83.7	88.3	93.2	118.4	84.8	90.9	93.9
4300	91.4	148.9	217.8	85.9	90.6	95.6	121.9	87.1	93.3	96.4
4400	93.8	152.8	224.4	88.2	92.9	98.1	125.3	89.4	95.7	98.9
4500	96.1	156.7	231.1	90.5	95.2	100.6	128.8	91.7	98.1	101.3
4600	98.5	160.6	237.8	92.8	97.6	103.1	132.3	94.0	100.5	103.8
4700	100.8	164.5	244.5	95.1	99.9	105.6	135.9	96.3	102.9	106.3
4800	103.2	168.5	251.2	97.4	102.2	108.1	139.4	98.6	105.4	108.8
4900	105.5	172.4	258.0	99.7	104.6	110.7	142.9	100.9	107.8	111.3
5000	107.9	176.3	254.8	102.0	106.9	113.2	146.5	103.2	110.2	113.8

[a] AX = 75.0% H_2, 25.0% N_2.
[b] 12.0% H_2, 72.8% N_2.
[c] 1.0% H_2, 88.0% N_2.
Source: Ref. 79.

Table 22 Characteristics of Simple Gases

Simple gases and compound	Critical temp. (°F)	Critical pressure (psia)	Limits of inflammability Lower	Limits of inflammability Upper	Ignition temp. (°F)	Combustion velocity of max. speed mixture (ft/s)	Toxicity — Maximum amt. inhaled for 1 h without serious disturbance (ppm)	Toxicity — Dangerous in 30 min 1 hr (ppm)	Toxicity — Rapidly fatal (ppm)	Solubility H₂O at 60°F, 30 in. Hg	Thermal conductivity [Btu/(ft²·°F·in.·s)]	Specific gravity of the liquid at 60°F	Heat of vaporization at 60°F (Btu/lb)
H₂	−400	188	4.1	74	1076–1094	8.2	—	Simple asphyxiant	—	0.0_167	3.05×10^{-4}	—	—
O₂	−181	731	—	—	—	—	—	—	—	0.049	4.47×10^{-5}	—	—
N₂	−233	492	—	—	—	—	—	Simple asphyxiant	—	0.0_22	4.38×10^{-5}	—	—
CO	−218	515	12.5	74	1191–1216	1.6	1000–1200	1500–2000	4,000	Very slight	4.12×10^{-5}	—	—
CO₂	88	1073	—	—	—	—	—	5 to 7% respiratory stimulant	—	0.090	2.63×10^{-5}	—	—
CH₄	−116	673	5.3	14.0	1200–1382	1.2	—	Simple asphyxiant	—	—	5.64×10^{-5}	—	—
C₂H₆	90	717	3.2	12.5	968–1166	—	—	Simple asphyxiant	—	—	3.46×10^{-5}	0.38	223
C₃H₈	204	632	2.4	9.5	~965	1.03	—	Anesthetic	—	—	—	0.51	210
n-C₄H₁₀	308	529	1.9	8.5	~930	—	—	Anesthetic	—	—	—	0.58	183
iso-C₄H₁₀	273	544	1.4	8.0	~890	—	—	—	—	—	—	0.56	166
n-C₅H₁₂	387	485	—	—	—	—	—	Anesthetic, convulsive, irritant	—	—	—	0.63	159
iso-C₅H₁₂	370	482	—	—	—	—	—	Anesthetic, convulsive, irritant	—	—	—	—	153
C₆H₁₄	455	434	3.3	34	1000–1020	2.1	—	Anesthetic, convulsive, irritant	—	—	223×10^{-5}	0.66	143
C₂H₄	49	748	2.2	10	—	—	—	Simple asphyxiant and anesthetic	—	—	1.98×10^{-5}	—	—
C₃H₆	198	662	1.7	9	—	—	—	Anesthetic	—	—	3.14×10^{-5}	—	—
C₄H₈	—	—	2.5	80	763–824	4.1	—	Anesthetic	—	—	—	—	—
C₂H₂	97	911	1.4	8.0	1364	—	3100–4700	Simple asphyxiant and anesthetic	19,000	—	332×10^{-5}	—	—
C₆H₆	551	701	1.3	6.75	1490	—	3100–4700	—	19,000	—	1.60×10^{-5}	0.88	—
C₇H₈	609	612	—	—	—	—	3100–4700	—	19,000	—	—	—	—
C₁₀H₈	—	—	—	—	—	—	—	—	—	—	—	—	—
NH₃	270	1639	16	27	—	—	300–500	—	5,000–10,000	0.612	384×10^{-5}	—	—
H₂S	212	1307	—	—	—	—	200–300	500–700	1000–3000	0.00466	2.30×10^{-5}	—	—
H₂O	706	3226	—	—	—	—	—	—	—	—	4.17×10^{-5}	—	1058
Air	−285	547	—	—	—	—	—	—	—	—	4.28×10^{-5}	—	—

Source: Ref. 80.

Table 23 Properties of Typical Commercial Gases

No.	Gas	Constituents of gas (% v/v) CO_2	O_2	N_2	CO	H_2	CH_4	C_2H_4	Illuminants C_2H_4	C_2H_4	C_2H_4	Spec. grav.	Air ref. for comb. of 1 ft³ gas (ft³)	Btu/ft³ Gross	Btu/ft³ Net	Products of combustion per cu. ft. of gas (ft³) H_2O	CO_2	N_2	Total	Ultimate % CO_2	Btu (net)/ft³ prod. of comb.	Flame temp., no excess air (°F)
1	Natural gas (Birmingham)	—	—	5.0	—	—	90.0	5.0	—	—	—	0.60	9.41	1002	904	2.02	1.00	7.48	10.50	11.8	86.0	3565
2	Natural gas (Pittsburgh)	—	—	0.8	—	—	83.4	15.8	—	—	—	0.61	10.58	1129	1021	2.22	1.15	8.37	11.73	12.1	87.0	3562
3	Natural gas (So. California)	0.7	—	0.5	—	—	84.0	14.8	—	—	—	0.64	10.47	1116	1009	2.20	1.14	8.28	11.62	12.1	87.0	3550
4	Natural gas (Los Angeles)	6.5	—	—	—	—	77.5	16.0	—	—	—	0.70	10.05	1073	971	2.10	1.16	7.94	11.20	12.7	86.7	3550
5	Natural gas (Kansas City)	0.8	—	8.4	—	—	84.1	6.7	—	—	—	0.63	9.13	974	879	1.95	0.98	7.30	10.23	11.9	86.0	3535
6	Reformed natural gas	1.4	0.2	2.9	9.7	46.6	37.1	—	1.3		(C_3H_6 0.8)	0.41	5.22	599	536	1.30	0.53	4.16	5.99	11.3	89.6	3615
7	Mixed natural and water gas	4.4	2.1	4.7	25.5	35.1	23.1	4.7	0.2	0.2	—	0.61	4.43	525	477	1.01	0.64	3.55	5.20	15.3	91.7	3630
8	Coke oven gas	2.2	0.8	8.1	6.3	46.5	32.1	—	3.5	0.5	—	0.44	4.99	574	514	1.25	0.51	4.02	5.78	11.2	87.0	3610
9	Coal gas (continuous verticals)	3.0	0.2	4.4	10.9	54.5	24.2	—	1.5	1.3	—	0.42	4.53	532	477	1.15	0.49	3.62	5.26	11.9	90.7	3645
10	Coal gas (inclined retorts)	1.7	0.8	8.1	7.3	49.5	29.2	—	0.4	—	—	0.47	5.23	599	540	1.23	0.57	4.21	6.01	11.9	89.9	3660
11	Coal gas (intermittent verticals)	1.7	0.5	8.2	6.9	49.7	29.9	—	3.0	0.1	—	0.41	4.64	540	482	1.21	0.45	3.75	5.41	10.7	89.0	3610
12	Coal gas (horizontal retorts)	2.4	0.75	11.35	7.35	47.95	27.15	—	1.32	1.73	—	0.47	4.68	542	486	1.15	0.50	3.81	5.46	11.6	89.0	3600
13	Mixed coke oven and carburetted water gas	3.4	0.3	12.0	17.4	36.8	24.9	—	3.7	1.5	—	0.58	4.71	545	495	1.04	0.62	3.85	5.51	13.9	90.0	3630
14	Mixed coal, coke oven, and carburetted water gas	1.8	1.6	13.6	9.0	42.6	28.0	—	2.4	1.0	—	0.50	4.52	528	475	1.11	0.50	3.71	5.32	11.8	89.3	3640
15	Carburetted water gas	3.0	0.5	2.9	34.0	40.5	10.2	—	6.1	2.8	—	0.63	4.60	550	508	0.87	0.76	3.66	5.29	17.2	96.2	3725
16	Carburetted water gas	4.3	0.7	6.5	32.0	34.0	15.5	—	4.7	2.3	—	0.67	4.51	534	493	0.75	0.86	3.63	5.24	17.1	94.2	3700
17	Carburetted water gas (low gravity)	2.8	1.0	5.1	21.0	47.5	15.0	—	5.2	2.4	—	0.54	4.61	549	501	0.98	0.64	3.70	5.31	14.7	94.3	3690
18	Water gas (coke)	5.4	0.7	8.3	37.0	47.3	1.3	—	—	—	—	0.57	2.10	287	262	0.53	0.44	1.74	2.71	20.1	96.6	3670
19	Water gas (bituminous)	5.5	0.9	27.6	28.2	32.5	4.6	—	0.4	0.3	—	0.70	2.01	261	239	0.47	0.41	1.86	2.74	18.0	87.2	3510
20	Oil gas (Pacific Coast)	4.7	0.3	3.6	12.7	48.6	26.3	—	2.7	1.1	—	0.47	4.73	551	496	1.15	0.56	3.77	5.48	12.9	90.5	3630
21	Producer gas (buckwheat anthracite)	8.0	0.1	50.0	23.2	17.7	1.0	—	—	—	—	0.86	1.06	143	133	0.22	0.32	1.34	1.88	19.4	70.5	3040
22	Producer gas (bituminous)	4.5	0.6	50.9	27.0	14.0	3.0	—	—	—	—	0.86	1.23	163	153	0.23	0.35	1.48	2.06	18.9	74.6	3175
23	Producer gas (0.6 lb steam/lb coke)	6.4	—	52.8	27.1	13.3	0.4	—	—	—	—	0.88	1.00	135	128	0.17	0.34	1.32	1.82	20.5	70.3	3010
24	Blast furnace gas	11.5	—	60.0	27.5	1.0	—	—	—	—	—	1.02	0.68	92	92	0.02	0.39	1.14	1.54	25.5	59.5	2650
25	Commercial butane	—	(C_4H_{10} 93.0)				(C_3H_8 7.0)					1.95	30.47	3225	2977	4.93	3.93	24.07	32.93	14.0	90.5	3640
26	Commercial propane	—	(C_3H_8 100.0)									1.52	23.82	2572	2371	4.17	3.00	18.82	25.99	13.7	91.2	3660

Source: Ref. 80.

Table 24 Gas Combustion Constants

Column groups: **Heat of combustion** = Btu/ft^3 (Gross, Net) and Btu/lb (Gross, Net). **Vol gas (ft^3/ft^3)** = Required for combustion (O$_2$, N$_2$, Air) and Flue products (CO$_2$, H$_2$O, N$_2$). **Mass gas (lb/lb)** = Required for combustion (O$_2$, N$_2$, Air) and Flue products (CO$_2$, H$_2$O, N$_2$).

No.	Gas	Formula	Mol. wt.	lb/ft^3	ft^3/lb	Sp. gr. (Air = 1.000)	Btu/ft^3 Gross	Btu/ft^3 Net	Btu/lb Gross	Btu/lb Net	Vol O$_2$	Vol N$_2$	Vol Air	Vol CO$_2$	Vol H$_2$O	Vol N$_2$	Mass O$_2$	Mass N$_2$	Mass Air	Mass CO$_2$	Mass H$_2$O	Mass N$_2$
1	Carbon	C	12.000	—	—	—	—	—	14,140	14,140	—	—	—	—	—	—	2.667	8.873	11.540	3.667	—	8.873
2	Hydrogen	H_2	2.015	0.005327	187.723	0.06959	325	275.1	61,100	51,643	0.5	1.882	2.382	—	1.0	1.882	7.939	26.414	34.353	—	8.939	26.414
3	Oxygen	O_2	32.000	0.08461	11.819	1.1053	—	—	—	—	—	—	—	—	—	—	—	—	—	—	—	—
4	Nitrogen (atmos.)	N_2	28.016	0.07439	13.443	0.9718	—	—	—	—	—	—	—	—	—	—	—	—	—	—	—	—
5	Carbon monoxide	CO	28.000	0.07404	13.506	0.9672	323.5	323.5	4,369	4,369	0.5	1.882	2.382	1.0	—	1.882	0.571	1.900	2.471	1.571	—	1.900
6	Carbon dioxide	CO_2	44.000	0.1170	8.548	1.5282	—	—	—	—	—	—	—	—	—	—	—	—	—	—	—	—
	Paraffin series C_nH_{2n+2}																					
7	Methane	CH_4	16.031	0.04243	23.565	0.5543	1014.7	913.8	23,912	21,533	2.0	7.528	9.528	1.0	2.0	7.528	3.992	13.282	17.274	2.745	2.248	13.282
8	Ethane	C_2H_6	30.046	0.08029	12.455	1.04882	1781	1631	22,215	20,312	3.5	13.175	16.675	2.0	3.0	13.175	3.728	12.404	16.132	2.929	1.799	12.404
9	Propane	C_3H_8	44.062	0.1196	8.365	1.5617	2572	2371	21,564	19,834	5.0	18.821	23.821	3.0	4.0	18.821	3.631	12.081	15.712	2.996	1.635	12.081
10	Isobutane	C_4H_{10}	58.077	0.1582	6.321	2.06654	3251	2999	21,247	19,606	6.5	24.467	30.967	4.0	5.0	24.467	3.581	11.914	15.495	3.030	1.551	11.914
11	n-Butane	C_4H_{10}	58.077	0.1582	6.321	2.06654	3353	3102	21,247	19,606	6.5	24.467	30.967	4.0	5.0	24.467	3.581	11.914	15.495	3.030	1.551	11.914
12	n-Pentane	C_5H_{12}	72.092	0.1904	5.252	2.4872	3981	3679	20,908	19,322	8.0	30.114	38.114	5.0	6.0	30.114	3.551	11.815	15.366	3.052	1.499	11.815
13	n-Hexane	C_6H_{14}	86.107	0.2274	4.398	2.9704	4667	4315	20,526	18,976	9.5	35.760	45.260	6.0	7.0	35.760	3.530	11.745	15.275	3.067	1.465	11.745
	Olefin series C_nH_{2n}																					
14	Ethylene	C_2H_4	28.031	0.07456	13.412	0.9740	1631	1530	21,884	20,525	3.0	11.293	14.293	2.0	2.0	11.293	3.425	11.395	14.820	3.139	1.285	11.395
15	Propylene	C_3H_6	42.046	0.1110	9.007	1.4504	2336	2185	21,042	19,683	4.5	16.939	21.439	3.0	3.0	16.939	3.425	11.395	14.820	3.139	1.285	11.395
16	Butylene	C_4H_8	56.062	0.1480	6.756	1.9336	3135	2884	20,840	19,481	6.0	22.585	28.585	4.0	4.0	22.585	3.425	11.395	14.820	3.139	1.285	11.395
17	Acetylene	C_2H_2	26.015	0.06971	14.344	0.9107	1503	1453	21,572	20,840	2.5	9.411	11.911	2.0	1.0	9.411	3.075	10.231	13.306	3.383	0.692	10.231
	Aromatic series C_nH_{2n-6}																					
18	Benzene	C_6H_6	78.046	0.2060	4.852	2.6920	3741	3590	18,150	17,418	7.5	28.232	35.732	6.0	3.0	28.232	3.075	10.231	13.306	3.383	0.692	10.231
19	Toluene	C_7H_8	92.062	0.2431	4.113	3.1760	4408	4206	18,129	17,301	9.0	33.878	42.878	7.0	4.0	33.878	3.128	10.407	13.535	3.346	0.783	10.407
20	Xylene	C_8H_{10}	106.077	—	—	—	—	—	18,410	—	10.5	39.524	50.024	8.0	5.0	39.524	3.168	10.540	13.708	3.318	0.849	10.540
21	Naphthalene	$C_{10}H_8$	128.062	—	—	—	—	—	17,298	—	12.0	45.170	57.170	10.0	4.0	45.170	2.999	9.978	12.977	3.436	0.563	9.978
	Miscellaneous gases																					
22	Ammonia	NH_3	17.031	0.04563	21.914	0.5961	433	—	9,598	—	—	—	—	—	—	—	—	—	—	—	—	—
23	Hydrogen Sulfide	H_2S	34.080	0.09109	10.979	1.189	672	—	7,479	—	1.5	5.646	7.146	SO_2=1.0	1.0	5.646	1.408	4.685	6.093	SO_2=1.880	0.529	4.085
24	Sulfur dioxide	SO_2	64.06	0.1733	5.770	2.264	—	—	—	—	—	—	—	—	—	—	—	—	—	—	—	—
25	Water vapor	H_2O	18.015	0.04758	21.017	0.6215	—	—	—	—	—	—	—	—	—	—	—	—	—	—	—	—
26	Air	—	28.9	0.07655	13.063	1.0000	—	—	—	—	—	—	—	—	—	—	—	—	—	—	—	—

Source: Ref. 80.

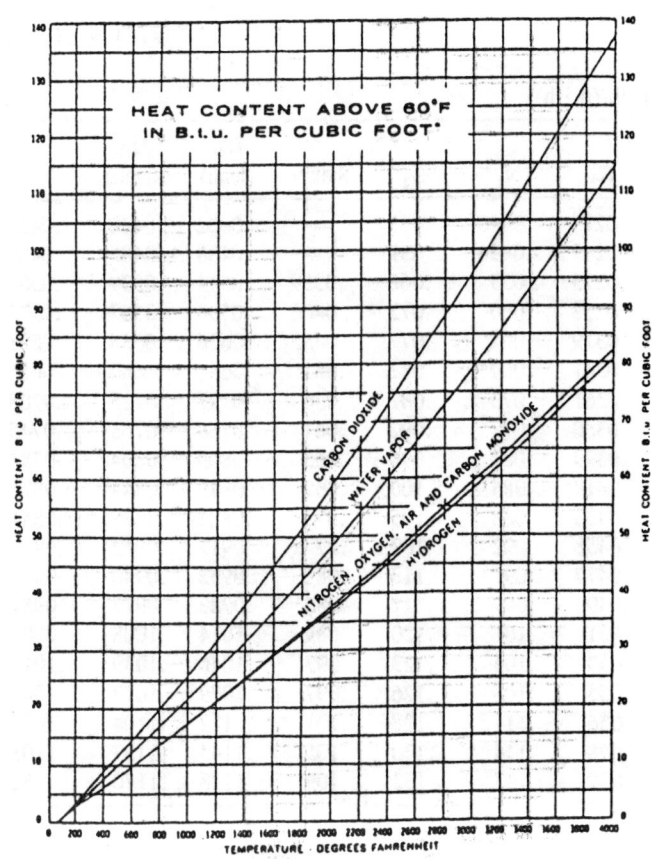

Figure 79 Temperature dependence of heat content for various gases. (From Ref. 14.)

Table 25 Composition of Atmospheric Air

	Concn., dry basis	
Gaseous component	Vol %	ppm
Fixed		
Nitrogen (N_2)	78.084	—
Oxygen (O_2)	20.9476	—
Argon (Ar)	0.934	—
Neon (Ne)	—	18.18
Helium (He)	—	5.24
Krypton (Kr)	—	1.14
Xenon (Xe)	—	0.087
Variable		
Carbon dioxide (CO_2)	—	30–400
Nitrous oxide (N_2O)	—	0.5
Nitrogen dioxide (NO_2)	—	0–0.02
Water (H_2O)	1.25[a]	—
Hydrogen (H_2)	—	0.5 Type
Carbon monoxide (CO)	—	1 Type
Methane (CH_4)	—	2 Type
Ethane (C_2H_6)	—	<0.1 Type
Other hydrocarbons (C_nH_{2n+2})	—	<0.1 Type

Source: Ref. 81.

Table 26A Viscosity (μ) of Gases[a] [μlb/(ft · h)]

Temp. (°F)	Air	CO	CO$_2$	H$_2$	N$_2$	O$_2$	Steam	DX (lean)	DX (rich)	RX
100	.0462	.0432	.0379	.0223	.0440	.052	—	.0432	.0430	.0427
200	.0520	.0487	.0439	.0249	.0496	.059	—	.0488	.0485	.0480
300	.0575	.054	.0497	.0273	.055	.066	.0359	.0544	.0539	.0533
400	.0626	.059	.055	.0297	.060	.072	.0408	.0590	.0587	.0577
500	.0675	.063	.060	.0319	.064	.077	.0455	.0636	.0635	.0621
600	.0722	.068	.065	.0341	.069	.082	.051	.0681	.0680	.0665
700	.0767	.072	.070	.0361	.073	.087	.056	.0727	.0725	.0708
800	.0810	.076	.075	.0380	.077	.092	.061	.0768	.0765	.0745
900	.0852	.080	.079	.0399	.081	.096	.065	.0809	.0804	.0784
1000	.0892	.083	.083	.0419	.085	.100	.069	.0849	.0844	.0821
1100	.0932	.086	.087	.0438	.088	.105	.073	.0883	.0878	.0854
1200	.0970	.089	.091	.0458	.092	.109	.077	.0917	.0912	.0887
1300	.101	.093	.095	.0477	.095	.113	.081	.0952	.0946	.0921
1400	.104	.096	.099	.0496	.099	.117	.085	.0983	.0977	.0951
1500	.108	.099	.103	.051	.101	.120	.089	.1014	.1008	.0981
1600	.111	.102	.107	.053	.104	.123	.093	.1046	.1039	.1011
1700	.115	.105	.110	.055	.108	.126	.097	.1079	.1072	.1045
1800	.118	.108	.113	.056	.111	.128	.101	.1112	.1104	.1071
1900	.121	.111	.116	.058	.114	.130	.105	.1145	.1137	.1105
2000	.124	.114	.119	.059	.117	.132	.109	.1178	.1170	.1138

[a]DX, RX defined in Table 33.
Source: Ref. 79.

Table 26B Thermal Conductivity (k) of Gases [klb/(ft · h)]

Temp. (°F)	Air	CO	CO$_2$	H$_2$	N$_2$	O$_2$	Steam	DX (lean)	DX (rich)	RX
100	.0154	.0142	.0101	.109	.0151	.0157	—	.0150	.0212	.0404
200	.0174	.0160	.0125	.122	.0170	.0180	—	.0171	.0239	.0455
300	.0193	.0178	.0150	.135	.0189	.0203	.0171	.0191	.0267	.0506
400	.0212	.0196	.0174	.146	.0207	.0225	.0200	.0210	.0292	.0549
500	.0231	.0214	.0198	.157	.0225	.0246	.0228	.0229	.0318	.0592
600	.0250	.0231	.0222	.168	.0242	.0265	.0257	.0247	.0342	.0634
700	.0268	.0248	.0246	.178	.0259	.0283	.0288	.0266	.0366	.0677
800	.0286	.0264	.0270	.188	.0275	.0301	.0321	.0283	.0388	.0716
900	.0303	.0279	.0294	.198	.0290	.0319	.0355	.0300	.0411	.0754
1000	.0319	.0294	.0317	.208	.0305	.0337	.0388	.0317	.0433	.0793
1100	.0336	.0309	.0339	.219	.0319	.0354	.0422	.0333	.0454	.0832
1200	.0353	.0324	.0360	.229	.0334	.0370	.0457	.0349	.0475	.0870
1300	.0369	.0339	.0380	.240	.0349	.0386	.0494	.0365	.0497	.0909
1400	.0385	.0353	.0399	.250	.0364	.0401	.053	.0381	.0518	.0946
1500	.0400	.0367	.0418	.260	.0379	.0414	.057	.0397	.0538	.0983
1600	.0415	.0381	.0436	.270	.0394	.0425	.061	.0413	.0559	.1020
1700	.0430	.0395	.0453	.280	.0409	.0436	.064	.0428	.0579	.1055
1800	.0444	.0408	.0469	.289	.0423	.0446	.068	.0443	.0599	.1090
1900	.0458	.0420	.0484	.298	.0437	.0456	.072	.0458	.0619	.1126
2000	.0471	.0431	.050	.307	.0450	.0466	.076	.0473	.0638	.1161

Source: Ref. 79.

Table 26C Specific Heat (Cp) of Gases [Btu / (lb/ft · °F)]

Temp. (°F)	Air	CO	CO_2	H_2	N_2	O_2	DX Steam	DX (lean)	(rich)	RX
100	.240	.249	.203	3.42	.249	.220	—	.2449	.2756	.3938
200	.241	.250	.216	3.44	.249	.223	—	.2468	.2770	.3948
300	.243	.251	.227	3.45	.250	.226	.456	.2487	.2784	.3957
400	.245	.253	.237	3.46	.252	.230	.462	.2515	.2808	.3977
500	.247	.256	.247	3.47	.254	.234	.470	.2542	.2832	.3996
600	.250	.259	.256	3.48	.256	.239	.477	.2572	.2862	.4032
700	.253	.263	.263	3.49	.259	.243	.485	.2602	.2893	.4069
800	.256	.266	.269	3.49	.262	.246	.494	.2637	.2927	.4101
900	.259	.270	.275	3.50	.265	.249	.50	.2672	.2962	.4133
1000	.262	.273	.280	3.51	.269	.252	.51	.2708	.2996	.4164
1100	.265	.276	.284	3.53	.272	.255	.52	.2739	.3028	.4201
1200	.268	.279	.288	3.55	.275	.257	.53	.2770	.3060	.4238
1300	.271	.282	.292	3.57	.278	.259	.54	.2800	.3091	.4276
1400	.274	.284	.295	3.59	.280	.261	.55	.2831	.3124	.4314
1500	.276	.287	.298	3.62	.282	.263	.56	.2862	.3157	.4352
1600	.278	.290	.301	3.64	.285	.265	.56	.2893	.3190	.4390
1700	.280	.292	.303	3.67	.288	.266	.57	.2915	.3214	.4424
1800	.282	.294	.305	3.70	.290	.268	.58	.2938	.3238	.4457
1900	.284	.296	.307	3.73	.292	.269	.59	.2960	.3263	.4491
2000	.286	.298	.309	3.76	.294	.270	.60	.2982	.3287	.4524

Source: Ref. 79.

Table 26D Prandtl Number of Gases

Temp. (°F)	Air	CO	CO_2	H_2	N_2	O_2	DX Steam	DX (lean)	(rich)	RX
100	0.72	0.76	0.76	0.70	0.73	0.73	—	0.71	0.56	0.42
200	0.72	0.76	0.76	0.70	0.73	0.73	—	0.71	0.56	0.42
300	0.72	0.76	0.75	0.70	0.73	0.73	0.95	0.71	0.56	0.42
400	0.72	0.76	0.75	0.70	0.73	0.74	0.94	0.71	0.56	0.42
500	0.72	0.76	0.75	0.70	0.73	0.74	0.94	0.71	0.57	0.42
600	0.72	0.76	0.75	0.70	0.73	0.74	0.94	0.71	0.57	0.42
700	0.72	0.76	0.75	0.71	0.73	0.75	0.93	0.71	0.57	0.43
800	0.72	0.77	0.75	0.71	0.73	0.75	0.92	0.72	0.58	0.43
900	0.72	0.77	0.74	0.71	0.75	0.75	0.91	0.72	0.58	0.43
1000	0.73	0.77	0.73	0.71	0.74	0.75	0.91	0.73	0.58	0.43
1100	0.73	0.77	0.73	0.71	0.75	0.75	0.90	0.73	0.59	0.43
1200	0.74	0.77	0.73	0.71	0.75	0.75	0.88	0.73	0.59	0.43
1300	0.74	0.77	0.73	0.71	0.75	0.75	0.88	0.73	0.59	0.43
1400	0.74	0.77	0.73	0.72	0.76	0.75	0.87	0.73	0.59	0.43
1500	0.74	0.77	0.73	0.72	0.75	0.75	0.87	0.73	0.59	0.43
1600	0.74	0.78	0.74	0.72	0.75	0.76	0.87	0.73	0.59	0.44
1700	0.75	0.78	0.73	0.72	0.76	0.76	0.87	0.74	0.59	0.44
1800	0.75	0.78	0.73	0.72	0.76	0.77	0.87	0.74	0.60	0.44
1900	0.75	0.78	0.73	0.72	0.76	0.77	0.87	0.74	0.60	0.44
2000	0.75	0.79	0.74	0.72	0.77	0.77	0.87	0.74	0.60	0.44

Source: Ref. 79.

Table 27 Impurities in "Pure" Nitrogen

Source of nitrogen	Impurity					Dew point (°C)
	O_2	CO_2	CO	H_2	CH_4	
Decomposition of air	0.0001	0.0005	0.001	0.001	0.001	−70
Hydrocarbon combustion	0.0001	0.005	0.001	0.002	0.002	−65

Liquid nitrogen is also of value for its refrigeration or cooling capabilities as shown in Figure 83 [81]. Liquid nitrogen (LN$_2$) processes are summarized in Table 28. Liquid nitrogen is subsequently vaporized to a gas before use [86]. On-site membrane separation system are available from various suppliers in various gas separation capacities.

2. Hydrogen

In addition to being used as a quenchant, hydrogen is a highly reducing atmosphere that is used both for preventing steel oxidation and for oxide reduction according to the surface reactions

$$H_2 + FeO \rightarrow H_2O + 3Fe$$

and

$$H_2 + Fe_3O_4 \rightarrow H_2O + 3FeO$$

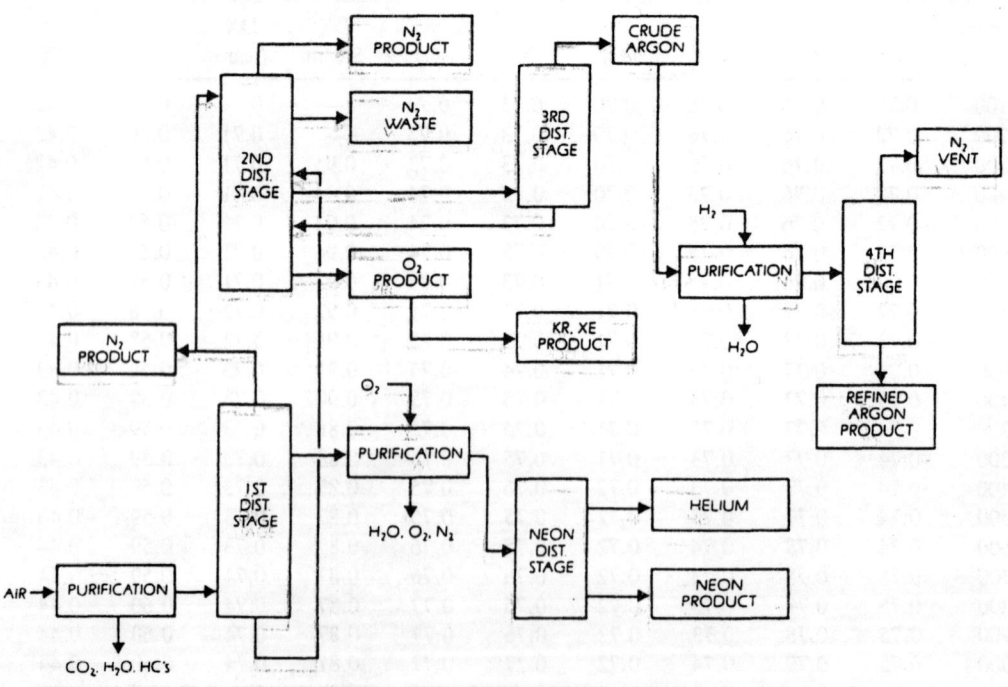

Figure 80 Process for gas production by air separation. (From Ref. 81.)

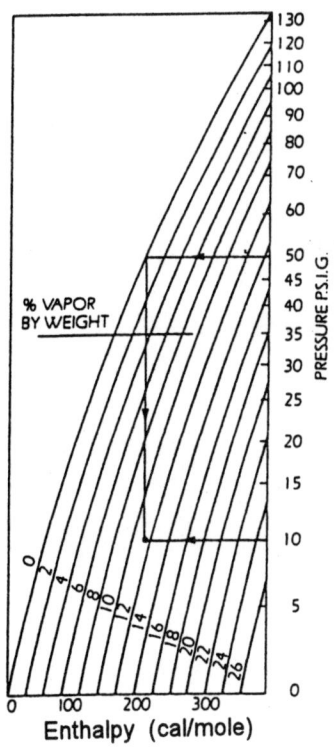

Figure 81 Example of "flash-off" with constant-enthalpy throttling of liquid nitrogen. (From Ref. 81.)

The thermal properties of hydrogen, relative to other heat treatment atmospheres, are summarized in Tables 26A–26D. At temperatures greater than 1300°F (700°C), hydrogen may cause decarburization to form methane by reaction with carbon:

$$C + 2H_2 \rightleftharpoons CH_4$$

In some cases, hydrogen may be adsorbed by the metal at elevated temperatures, causing hydrogen embrittlement [83].

Hydrogen is potentially an extremely explosive and flammable gas. However, if proper safety precautions are followed, it can be used safely in heat treatment.

Hydrogen can be produced by electrolysis of water:

$$2H_2O \rightarrow 2H_2 + O_2$$

Hydrogen produced in this way must be further purified to remove traces of contaminants, CO_2 and especially O_2, that may be present. The dew point of hydrogen is dependent on the degree of contamination as shown in Table 29.

The highest purity hydrogen is prepared by ammonia dissociation:

$$2NH_3 \rightarrow 3H_2 + N_2$$

The most common impurities in hydrogen are O_2 and H_2O. Purification is performed by filtration over palladium, which traps all gases except hydrogen, the smallest molecule. Residual water vapor can be removed by passing hydrogen through either silica gel or a molecular sieve column [83].

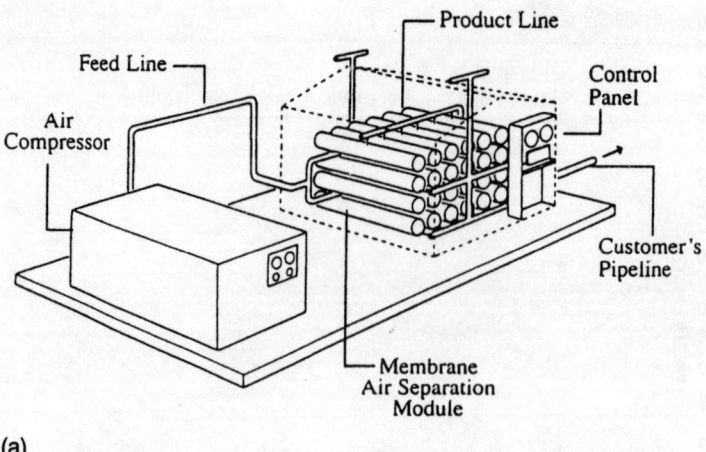

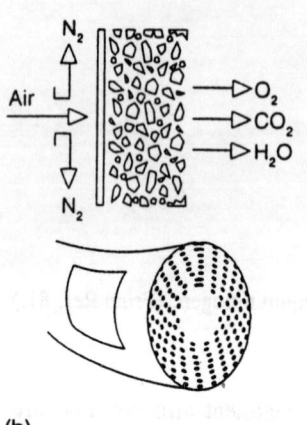

Figure 82 Gas production by membrane separation. (a) Commercial membrane separation unit; (b) schematic of membrane separation. (Courtesy of Praxair, Inc.)

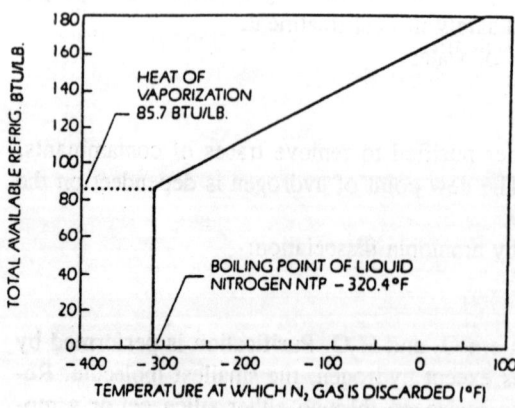

Figure 83 Refrigeration capacity in liquid nitrogen.

Table 28 Cooling Chamber Selection Chart for LN_2 Processes

Cooling application	LN_2 immersion bath	LN_2 cooled glycol bath	LN_2 fluid bed	Indirect cold gas cooling	Direct cold gas cooling
Rapid cooling to 40°F		X	X		
Slow cooling to 40°C			X	X	X
Rapid cooling to 200°F			X		X
Slow cooling to 200°F			X	X	X
Rapid cooling to 320°F	X				X
Slow cooling to 320°F					X
Constant-temperature soaking to –40°F		X	X	X	
Constant soaking to –200°F			X	X	X
Shrink fitting (cool to –320°F)	X		X		X
Metallurgical treatment (–140°F)			X	X	X
Metallurgical treatment (–320°F)	X				X

Source: Ref. 81.

3. Carbon Monoxide

Carbon monoxide is also considered to be a reducing gas as it may reduce iron oxide:

$$CO + FeO \rightleftharpoons Fe + CO_2$$

Although CO is a reducing atmosphere, it is not as good a reducing agent as hydrogen. Thermal properties of CO are given in Tables 26A–26D [79]. The preparation of CO is discussed later.

4. Carbon Dioxide

Carbon dioxide is a mildly oxidizing gas. It will form oxides upon reaction with iron at elevated temperatures. When the temperature is $>1030°F$ ($540°C$), FeO is formed [87]:

Table 29 Impurities of "Pure" Hydrogen

Method	Impurities (%)			Dew point (°C)
	CO_2	O_2	$CO_2 + O_2$	
Electrolysis of H_2	<0.003	<0.0001	<0.0004	−65
Electrolysis of H_2	<0.03	<0.0001	<0.0301	−18
Diffusing through palladium	—	—	<0.0001	−80

Source: Ref. 83.

$$Fe + CO_2 \rightleftharpoons FeO + CO$$

When the temperature is $< 1030°F$ (540°C),

$$3\,FeO + CO_2 \rightleftharpoons Fe_3O_4 + CO$$

Decarburization may also result from the reaction of CO_2 with carbides of iron or free carbon [87]:

$$Fe_3C + CO_2 \rightleftharpoons 3Fe + 2CO$$

$$C + CO_2 \rightleftharpoons 2CO$$

The properties of CO_2 are listed in Tables 26. The generation of CO_2 is discussed in a subsequent section.

5. Argon and Helium

Helium and argon are also considered to be inert gases for heat treatment processes because they will not undergo gas–solid reactions, even at high temperatures.

Argon is a significant component of air as shown in Table 25 and is obtained in high purity by an air separation process. High purity argon ($>99.999\%$ Ar) contains <2 ppm O_2 and <10 ppm N_2 and has a dew point of $-110°C$.

Helium concentration in air is insignificant, too little for air to be a commercial source of this gas. Instead, natural gas, which contains 5–8% helium, is the commercial source of helium. Air liquefaction is used to separate the helium, which has a dew point of $-100°C$.

Physical properties of argon and helium are provided in Table 30.

6. Dissociated Ammonia

From ammonia dissociation, which occurs at temperatures of $>300°C$ in the presence of a catalyst such as Fe or Ni, it is possible to obtain mixtures of hydrogen and nitrogen that are free of oxygen contamination (see Figure 84) [80]. This may be a critically important requirement for some heat treatment processes.

$$2NH_3 \rightarrow N_2 + 3H_2$$

The relative concentrations of nitrogen and hydrogen may be varied by subsequent burning of the hydrogen. The most common atmospheres obtained by ammonia dissociation are summarized in Table 31.

Residual ammonia, typically less than 0.05%, is the primary impurity in dissociated ammonia atmospheres [82]. Residual ammonia is removed using the same methods as those used for water vapor removal. Refrigeration or adsorption in regenerative dryers are the most efficient.

Table 30 Physical Properties of Inert Gas Protective Atmospheres

Gas	Specific gravity	Thermal conductivity relative to air	Thermal content (Btu/ft³)
Nitrogen	0.972	0.999	0
Argon	1.379	0.745	0
Helium	1.137	6.217	0

Source: Ref. 81.

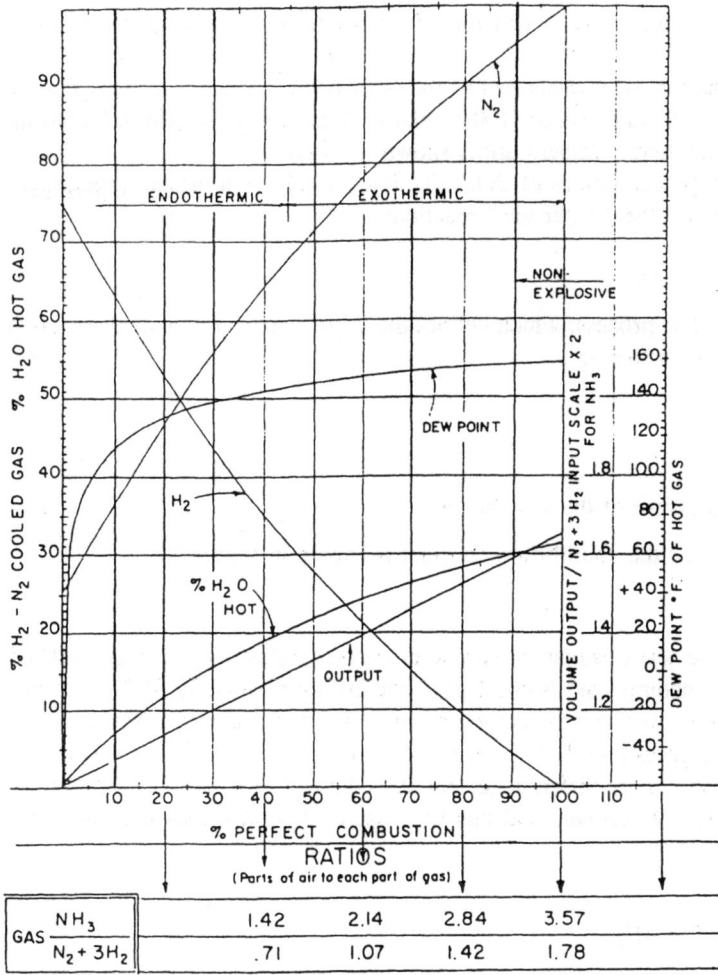

Figure 84   Generation of nitrogen base atmospheres (From Ref. 80.)

7.   Steam

Water vapor (steam) is also an important component in heat treating. As originally reported by Barff [88], steam will react with steel at 650–1200°F (343–650°C) to produce a "blueing" effect, which imparts a wear-resistant and oxidation-resistant surface furnish. This is due to the formation of either Fe_2O_3, Fe_3O_4, or FeO, depending on the surface temperature of the

Table 31   Atmospheres Produced by Dissociation of Ammonia

AGA		Composition (%)	
class	Atmosphere	H_2	N_2
601	Dissociated ammonia	75	25
621	Substantial combustion	1	99
622	Partial combustion	1–20	99–80

Source: Ref. 83.

steel and the ratio of water vapor pressure to hydrogen pressure in the atmosphere as illustrated in Figure 85.

The concentration of water vapor is quantified by the "dew point," which is the temperature at which a gas is saturated with water vapor (100% relative humidity) [87]. The relationship between dew point and atmospheric temperature is shown in Figure 86.

As discussed previously, the concentrations of CO, CO_2, H_2O, and H_2 in a furnace atmosphere are interdependent as shown by the "water gas" reaction:

$$CO + H_2O \rightleftharpoons CO_2 + H_2$$

The equilibrium constant for this process, which defines the actual concentration of these gases in the furnace atmosphere, is written as

$$K = \frac{pH_2 \times pCO_2}{pCO \times pH_2O}$$

where p is the partial pressure of the gas shown.

The temperature dependence of the equilibrium constant is shown in Figure 87.

8. Hydrocarbons

Hydrocarbons used to generate heat treatment atmospheres are most often derived from methane (CH_4), propane (C_3H_8), and natural gas. Natural gas contains approximately 85% methane. The combustion of these gases provides the carbon required for heat treatment processes.

The combustion of these gases and that of other potential sources of carbon result in different furnace efficiencies and require different control procedures. For example, consider the stoichiometry of the oxidation of methane and that of propane, hydrocarbons that differ by only two carbons.

For methane:

$$2CH_4 + O_2 \rightleftharpoons 2CO + 4H_2$$

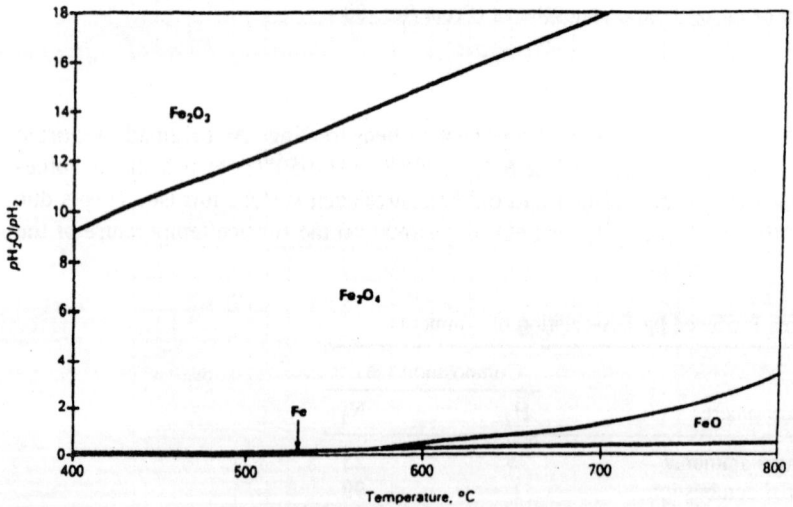

Figure 85 Effect of hydrogen/water ratio as a function of temperature on the oxidation of iron. (From Ref. 89.)

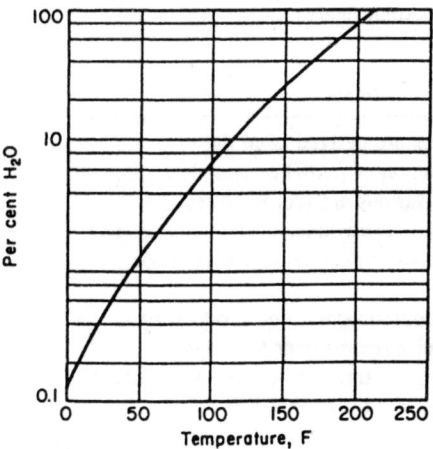

Figure 86 Dew point curve. (From Ref. 87.)

For propane:

$$2C_3H_8 + 3O_2 \rightleftharpoons 6CO + 8H_2$$

Propane provides more CO and less H_2 than methane. In addition, propane may contain small quantities of propylene or butylene, which may lead to soot formation [103].

Properties of various hydrocarbons are provided in Tables 22 and 23, and gas combustion constants are given in Table 24 [80].

B. Classification

Heat treatment furnace atmospheres have been classified by the American Gas Association (AGA) and are summarized in Table 32. Selected AGA and European gas classifications are

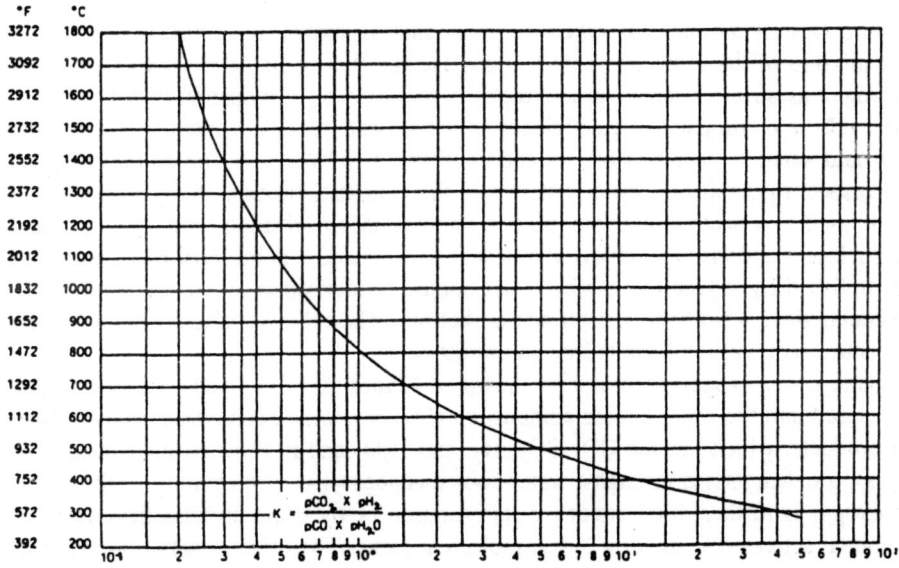

Figure 87 Water-gas equilibrium temperature as a function of temperature. (From Ref. 87.)

Table 32 AGA Classification of Heat Treatment Furnace Atmospheres

AGA class	Atmosphere designation	Notes
100	Exothermic base	These atmospheres are derived from the exothermic reaction of partial or complete combustion of various air–fuel mixtures and may contain various amounts of ammonia or H_2O.
200	Prepared nitrogen base	Class 100 with most of the CO_2 and H_2O vapor removed.
300	Endothermic base	These are the endothermic reaction products of various gas–fuel mixtures over a catalyst.
400	Charcoal base	Formed by passing air through heated, incandescent charcoal. The desired gases are removed at the maximum temperature zone and purified.
500	Exothermic–endothermic	A mixture of air and fuel undergoes complete combustion, and most of the water vapor is removed. The CO is then endothermically formed from the CO_2.
600	Ammonia base	Any atmosphere derived from ammonia; may include ammonia or various forms of dissociated ammonia with water vapor and residual ammonia removed.

Source: Ref. 90.

compared in Table 33. The applicability of some of these gases to particular types of heat treatment processes is illustrated in Tables 34 and 35.

1. *Protective Atmospheres and Gas Generation*

Exothermic Gas Generators Exothermic (exo) gas, as shown in Tables 32 and 33, is essentially a mixture of the reducing gases CO and H_2 and the oxidizing gases CO_2 and water, with nitrogen as an inert carrier gas. Exogas is prepared by combustion of a hydrocarbon such as methane or propane in a deficiency of air. "Lean" exogas contains larger quantities of CO_2 and smaller quantities of CO and H_2 than "rich" exogas. The composition of exogas with respect to the air/gas ratio is shown in the combustion chart of Figure 88.

Exogas is used as a protective atmosphere to prevent decarburization, scaling, and other undesirable surface reactions. To achieve the best results, the dew point must be minimized. This is accomplished by drying the gas by water-cooled condensation or refrigeration or by using a molecular sieve adsorbent (see discussion below). The relative effectiveness of these forms of moisture removal is illustrated in Figure 89.

A schematic of an exogas generator is shown in Figure 90. The hydrocarbon–air fuel mixture is metered into the water-cooled combustion chamber, which contains a catalyst. The combustion products, which include CO, CO_2, H_2, and H_2O (and in some cases residual hydrocarbon gas), are then passed through a water-cooled condenser to remove most of the water vapor. The exogas is further dried either by refrigeration or by using an adsorbent such as activated alumina, activated silica, or a 3–4 Å molecular sieve. The dried exogas is then piped directly to the furnace for use. An example of a commercial exogas generator is shown in Figure 91.

Table 33 European and AGA Classification of Protective Atmospheres

Atmosphere	Classification European	AGA	Dew point (°F)	Typical gas composition (% v/v)					
				CO_2	CO	H_2	CH_4	H_2O	N_2
Exothermic	DX, inert	101	100	10.4	0.5	0.5	0.0	6.5	82.1
			40	11.0	0.5	0.5	0.0	0.8	87.2
			100	4.7	9.4	9.4	0.4	6.5	69.6
	DX, rich	102	40	5.0	10.0	10.0	0.4	0.8	73.8
	NX	201	-40	0.05	0.5	0.5	0.0	0.0	98.95
	HNX	223	-40	0.05	0.05	3.0	0.0	0.0	96.9
		224				to			to
						10.0			89.8
Endothermic	RX	302	30-45	0.1	20.7	40.6	0.4	0.3	37.9
	SRX	323	100	5.2	16.8	71.1	0.4	6.5	0.0
	ASRX	323	100	3.3	18.7	65.4	0.4	6.5	5.7
	HX	325	-40	0.05	4.55	95.0	0.4	0.0	0.0
				to	to	to	to		
				0.0	0.0	100.0	0.0		
Dissociated ammonia	AX	601	-40	0.0	0.0	75.0	0.0	0.0	25.0

Source: Ref. 90.

Table 34 Composition and Dew Point of Selected Protective Atmospheres

AGA class	Typical atmosphere	Composition (%)					Dew point (°F)
		CO	CO_2	H_2	N_2	CH_4	
101	Exothermic (lean)	1.5	10.5	1.2	86.8	—	40[a]
102	Exothermic (rich)	10.5	5.0	12.5	71.5	0.5	40[a]
201	Prepared nitrogen (lean)	1.7	—	1.2	97.1	—	−40
202	Prepared nitrogen (rich)	11.0	—	13.2	75.3	—	−40
301	Endothermic (lean)	19.6	0.4	34.6	45.1	0.3	+50
302	Endothermic (rich)	20.7	—	38.7	39.8	0.8	0 to −5
601	Dissociated ammonia	—	—	25	75	—	−60

[a]If tap water cooling, dew point is room temperature and is reduced to 40°F using −50°F refrigeration.

Molecular Sieves Activated alumina, activated silica, and activated carbon are common adsorbents with high surface areas. The term "activation" refers to either vacuum or thermal surface desorption freeing the surface-active adsorption sites for subsequent adsorption. Although these materials possess porosity, it is not uniform.

Molecular sieves are aluminosilicates of elements of Group I (potassium and sodium) and Group II (magnesium, calcium, and barium). These crystalline structures of Al_3O_4 and SiO_4 which are interconnected through oxygen atoms through shared bonding to the metal cations. They possess uniform microporosity with pore sizes varying from 3 to 10 Å, depending on the particular zeolite. Table 36 lists commonly available commercial zeolites and their pore sizes.

Table 35 Selected Heat Treatment Processes Using AGA Classified Atmosphere

AGA class	Atmosphere	Heat treatment process
101	Lean exothermic	Forms an oxide coating on steel.
201	Lean prepared nitrogen	Neutral heating can cause decarburization of steel.
202	Rich prepared nitrogen	Annealing and brazing of stainless steel.
301	Lean endothermic	Clean hardening.
302	Rich endothermic	Carburizing (rich endogas is not usually used; lean endogas and enrichment gas addition is the preferred method.)
501	Lean exothermic-endothermic	Clean hardening
502	Rich exothermic-endothermic	Carburizing
601	Dissociated ammonia	Brazing; sintering
621	Lean combusted ammonia	Neutral hardening

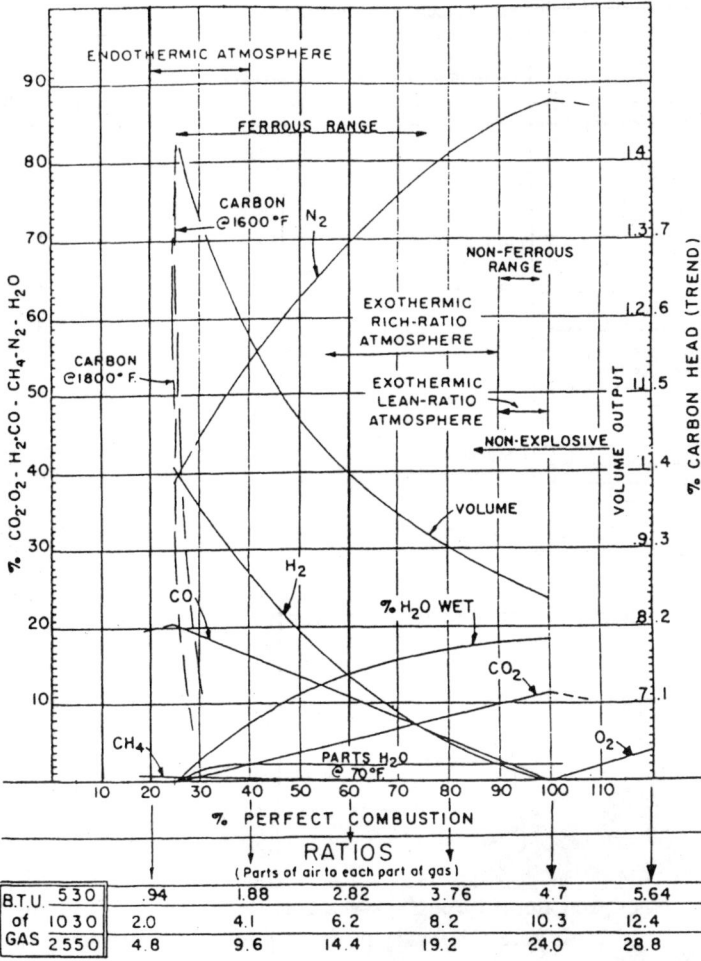

Figure 88 Generation of exothermic and endothermic atmospheres. (From Ref. 80.)

Zeolites characteristically possess very high internal surface area relative to their outside surface area. Molecular separation is based primarily on zeolite pore size; molecules of dimensions larger than the zeolite pore size are excluded. Tables 37 and 38 provide various physical constants, including molecular size, for various adsorbate gases.

Molecular sieves provide excellent alternatives to other activated solid substrates such as silica gel and activated alumina for atmosphere dehydration. Although all of the Type A molecular sieves (Table 36) readily absorb water, Type 3A (3 Å) molecular sieves are the preferred zeolite for dehydration of hydrocarbon gases [91]. The 3 Å pore size excludes other hydrocarbons such as ethylene and propylene that may catalytically undergo secondary polymerization reactions on the zeolite bed.

Gas adsorption and desorption on molecular sieves are both pressure- and temperature-sensitive and are the basis for the commercial highly efficient engineering practices called pressure swing and temperature (or thermal) swing separations. These separation practices are illustrated schematically in Figure 92. For temperature swing processes, adsorption occurs at lower temperatures and desorption at higher temperatures. Pressure swing adsorption is as-

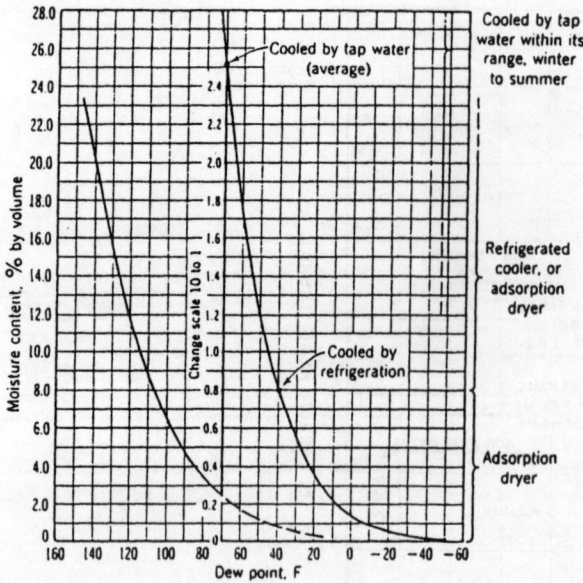

Figure 89 Effect of cooling temperature on moisture removal. (From Ref. 82.)

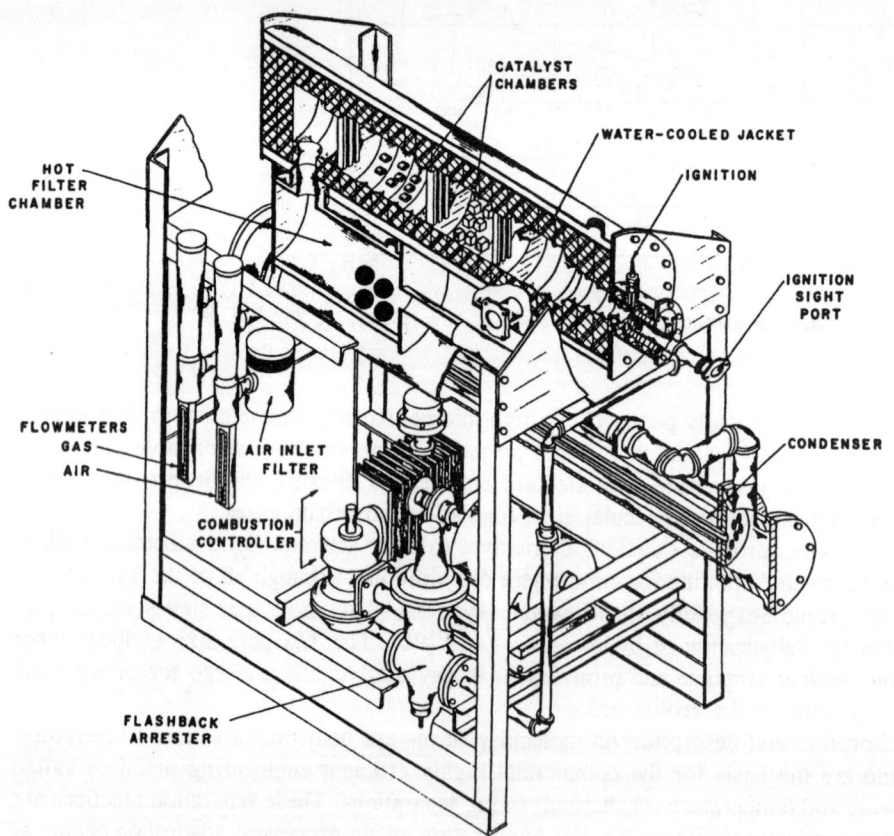

Figure 90 Schematic illustration of an exothermic generator. (Courtesy of C.I. Hayes, Inc.)

Figure 91 Exothermic generator. (Courtesy of Lindberg—A General Signal Company.)

sumed to be an isothermal process, and adsorption occurs at higher pressures and desorption at lower pressures.

Monogas (Prepared Nitrogen) Generators Monogas or prepared nitrogen atmospheres are nitrogen-rich gases (>90%) such as those listed in Table 34 that are obtained by the combustion of a hydrocarbon in the presence of a slight deficiency of air using the exogas generator discussed above. The residual CO_2 may be removed (to <0.05%) using a 4–5 Å molecular sieve [27,28]. A molecular sieve of this pore size will remove both residual water vapor and CO_2. (Previously, methanolamine gas scrubbers, which may reduce CO_2 content to 0.05%, were used, but this technology is rarely if ever used today.)

For optimal results in preventing decarburization, it is often important to use a dry, CO_2-free monogas protective atmosphere. This is illustrated using Gonser's curves in Figure 93, where increasing CO_2 concentration in the exogas resulted in a corresponding increase in decarburization.

Endothermic Gas Generators As shown in the the combustion chart of Figure 88, endothermic atmospheres are prepared by combustion of richer mixtures of hydrocarbons in air

Table 36 Molecular Sieves

Product type[a]	Major cation	Pore size (Å)	Water capacity (wt %)
3A	K^+	3	20
4A	Na^+	4	22
5A	Ca^{2+}	5	21.5
10X	Ca^{2+}	8	31.6

[a]*Source*: Ref. 91.

Table 37 Physical Constants of Adsorbate Gases

Gas	Boiling point (°C)	Critical temp. (°C)	Polarizability (A³)	Ionization pot. (V)	Length (Å)	Width (Å)	Kinetic diameter[a]
Argon	-187.8	-122.4	1.6	15.7	1.92	1.92	3.4
Oxygen	-183.0	-118.8	1.2	12.5	2.0	1.4	3.46
Nitrogen	-195.8	-147.1	1.4	15.5	2.1	1.5	3.64
Methane	-161.4	-82.5	2.6	14.5	2.0	2.0	3.8
CO	-192.0	-139	1.6	14.3	2.1	1.8	3.76
Ethylene	-103.7	9.7	3.5	12.2	2.5	2.2	3.9
Ethane	-88.6	32.1	3.9	12.8	2.6	2.5	3.8
CO_2	-78.5	31.1	1.9	14.4	2.6	1.8	3.3
Propylene	-47.6	92.0	3.5	12.2	3.4	2.2	—
Propane	-42.3	96.8	5.0	12.8	3.3	2.5	4.3

[a]The kinetic diameter is the intermolecular distance of closest approach for two molecules colliding with zero potential energy.
Source: Ref. 91.

than those used for the preparation of exogases [80]. Generally, the air/gas ratio is selected to favor the formation of CO + H_2 and to be insufficient to form large amounts of CO_2 and H_2O. Endogases contain much higher concentrations of CO and H_2 than exogases and exhibit higher dew points, as shown in Table 33. The relationship between the air/gas ratio and dew point for endogas is shown in Figure 94. Nitrogen is used as the inert carrier gas.

Since very small amounts of air are used, a catalyst and heat are required to facilitate combustion as shown in the schematic of an endogas generator in Figure 95. The endogas is cooled immediately upon departure from the externally heated combustion chamber and be-

Table 38 Pressure and Temperature Sensitivity of Gas Adsorption on Molecular Sieve[a]

Gas	Temp. (K)	Pressure (torr)	x/m	Pressure (torr)	x/m	Pressure (torr)	x/m
Argon	77	100	<0.01				
	195	100	5	300	15	700	30
	273	100	1	300	1.5	700	3.7
Oxygen	90	0.2	0.11	1	0.17	700	0.26
	195	40	0.003	150	0.01	700	0.044
	195	100	6	300	18	700	34
Nitrogen	97	700	<0.01				
	195	100	0.05	300	0.085	700	0.115
	195	100	30	300	42	700	49
Water	298	0.025	0.16	0.1	0.20	4	0.25
	373	1	0.06	4	0.13	12	0.17
Ammonia	298	3	0.090	10	0.11	100	0.15
CO_2	198	10	0.25	700	0.30		
	298	2	0.070	10	0.12	100	0.165
	423	100	0.034	700	0.105		
CO	198	15	0.070	100	0.091	700	0.11
	273	150	0.024	700	0.055		
	273	150	0.007	700	0.02		

[a]x/m is grams of gas per gram of dehydrated crystalline zeolite.
Source: Ref. 91.

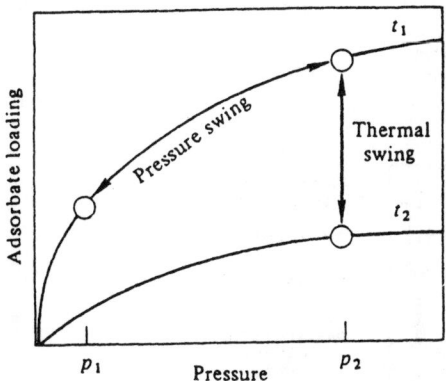

Figure 92 Illustration of temperature swing and pressure swing separation cycles. (From Ref. 91.)

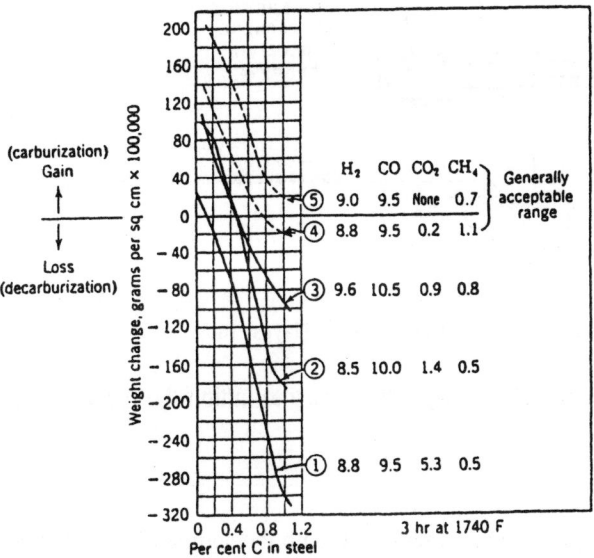

Figure 93 Gonser's curves illustrating decarburization with increasing CO_2 content of furnace atmosphere. (From Ref. 92.)

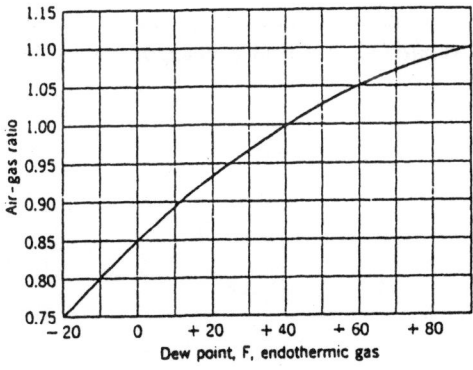

Figure 94 Effect of air/gas ratio on dew point. (From Ref. 82.)

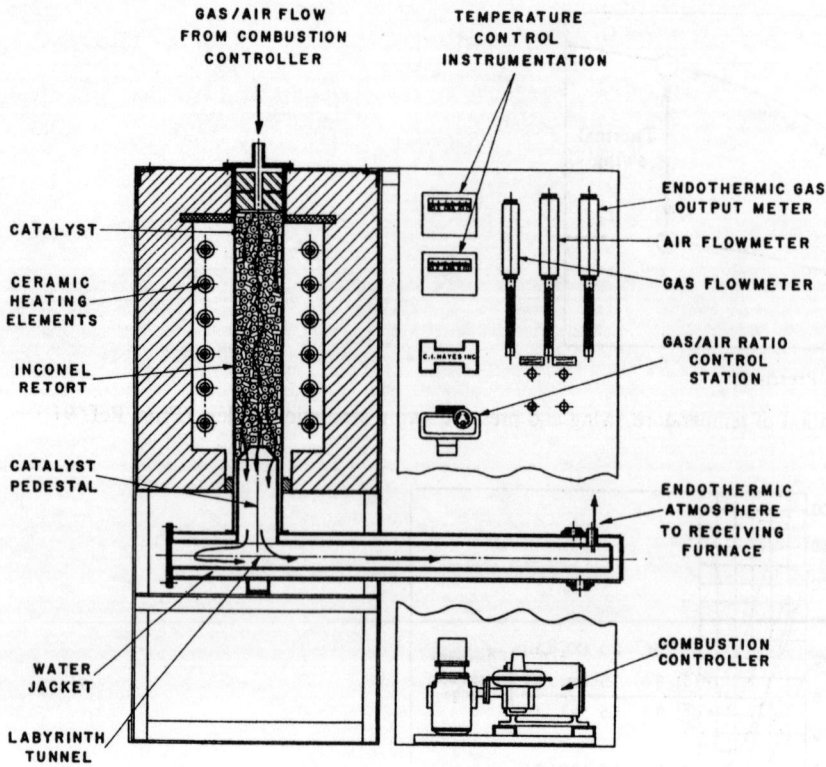

Figure 95   Schematic of an endothermic generator. (Courtesy of C.I. Hayes, Inc.)

fore being piped to the furnace for use. Rapid cooling is necessary to inhibit the formation of CO_2 and soot (carbon):

$$2CO \rightleftharpoons CO_2 + C$$

The air/gas ratio required for endogas formation will vary with the source of combustion gas as shown in Table 39. An example of an endogas generator is provided in Figure 96.

Ammonia Dissociators (Crackers)   Ammonia dissociation (cracking) is the preferred method for preparing the highest purity, oxygen-free nitrogen. This reaction is conducted in an "ammonia dissociator" such as the one illustrated in Figure 97.

Ammonia dissociators typically have three essential parts: an instrument panel, a vaporizer, and a reaction chamber. Liquid ammonia is vaporized in a heat exchanger and is fed

Table 39   Air/Gas Ratio for Various Hydrocarbon Sources

Fuel source	Air/gas ratio
Propane	7.0
Butane	9.5
Natural gas[a]	2.5

[a]Natural gas typically contains 85% methane.
Source: Ref. 83.

Figure 96 Endothermic generator. (Courtesy of Lindberg—A General Signal Company.)

Figure 97 Drever nitrogen generator. (Courtesy of Drever Company.)

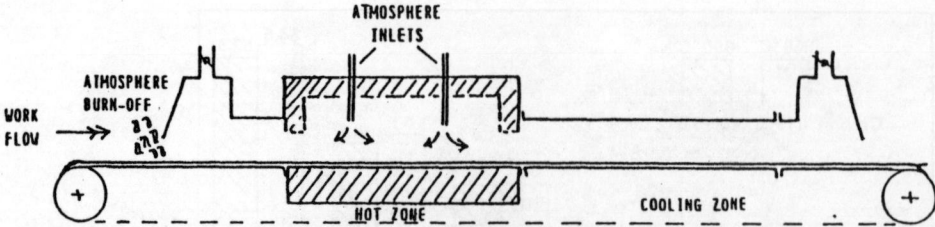

Figure 98 Illustration of the use of a conventional atmosphere in a continuous furnace. (From Ref. 93.)

directly to the dissociator, which is a heat-resistant, electrically heated coiled pipe surrounded by refractory material. Hot dissociated ammonia exiting the dissociator is used to heat liquid ammonia entering the heat exchanger and is then piped to the furnace if no further purification is required. The control panel permits control of the gas flows. Gaseous ammonia from cylinders can also be used.

C. Furnace Zoning

Traditionally, continuous heat treatment furnaces that use protective atmospheres such as exogas, endogas, and dissociated ammonia were open-ended with an inlet, hot zone, cooling zone, and exit (as shown in Figure 98) [93]. This was a relatively inefficient and limiting process because a single reducing or nonoxidizing atmosphere must be used, the atmosphere is typically fed into the hot zone, flammable atmospheres are burned at the ends of the furnace, and a positive pressure is used to prevent atmospheric air from entering the furnace [93].

There is an increasing trend to use zoning to optimize atmosphere in different zones within the furnace rather than using a control in a continuous furnace. Zoning permits the controlled use of different atmospheres. Some advantages of zoning are that it improves atmosphere control throughout the furnace, reduces the amount of gas readded, and improves atmospheric flow. A schematic of a typical continuous furnace using zoning is shown in Figure 99.

Significant improvements in zoned temperature control and energy savings are achieved with the use of furnace curtains. Furnace curtains can be fabricated to withstand temperatures up to 2000°F. Suitable materials include steel or silica-based or non-asbestos textiles [94].

D. In Situ Atmosphere Generation

Volatile organic liquids or hydrocarbons such as methanol, acetone, isopropanol, and ethyl acetate are used for in situ furnace atmosphere generation [95–99]. Other liquids that have

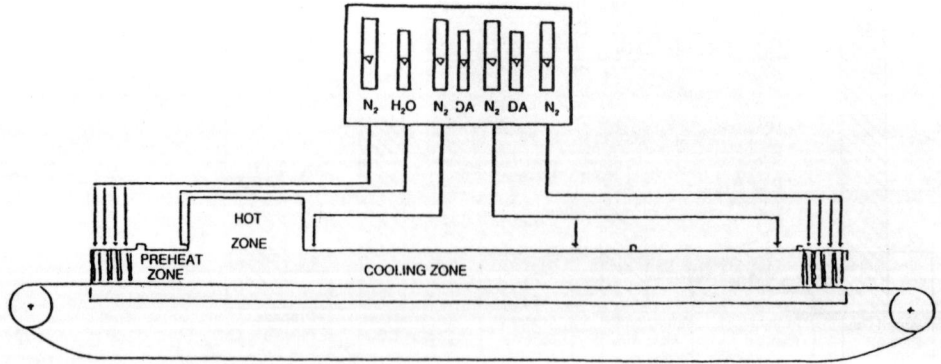

Figure 99 Illustration of atmosphere zoning.

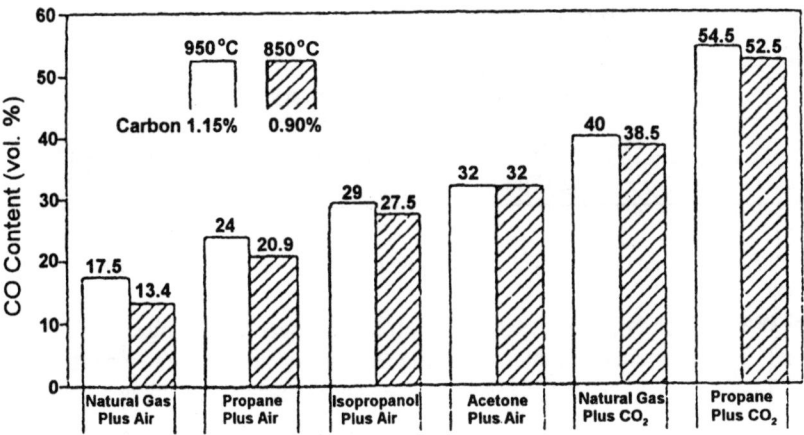

Figure 100   CO content of selected gas mixtures at 950°C and 850°C. (From Ref. 98.)

been tried include higher alcohols, butane, diesel oil, and kerosene [98]. Comparisons of the CO and CH_4 content of in situ generated atmospheres from selected volatile organic liquids are shown in Figures 100 and 101, and the reactive carbon availability of various fuels is summarized in Table 40.

Atmosphere generation occurs by either direct injection of the liquid into the furnace, usually onto a hot surface such as electric heating elements or gas-fired radiant tube heaters, or by external furnace atmosphere preparation. An illustration of an external nitrogen–methanol generator is provided in Figure 102. Furnace requirements for this process are: [98]

1. The furnace must be able to supply the additional energy required for in situ atmosphere generation.
2. The gas reactions with air must be exothermic.
3. The flow must be strong and directed.
4. Gases entering the furnace must be thoroughly mixed and directed to the hot dissociating surface.

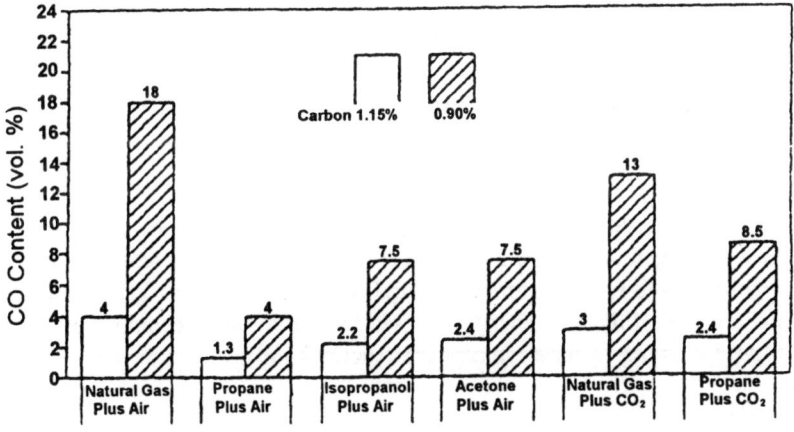

Figure 101   Methane content of selected gas mixtures at 950°C and 850°. (From Ref. 98.)

Table 40 Primary and Secondary Carbon Availability and the Carbon Transfer Coefficient for Various Carburizing Methods at 925°C

Carburizing method	Primary carbon availability[a] (g C/nm³)	Secondary carbon availability[b] (g C/nm³)	Carbon Transfer coefficient[c] (N · m/s)
Endothermic generator, propane	27	0.5	120
Methanol, generator gas	28	1.1	280
Endo/exothermic processor gas	27	0.5	120
Propane processor gas	≤400	0.5–1.5	—
In situ air–propane	25–30	0.5	120
In situ methanol–acetone	3–268	0.8–1.1	250
In situ methanol–ethyl acetate	3–178	1.1–1.2	280
Methanol–water–propane	~28	1.1	280
Nitrogen–methanol	~27	0.5	130

[a]Primary carbon availability is the amount of carbon that 1 m³ (NTP) of a gas carburizer can supply before the carbon potential is reduced to 1.0% C.
[b]Secondary carbon availability is the amount of carbon absorbed at the steel surface by 1 m³/(NTP) of a gas carburizer while its carbon potential is reduced from 0.9 to 1.0% C.
[c]Carbon transfer coefficient describes the kinetics of carbon flow from the gas phase near the steel surface to the steel surface.

Similarly, there are some general guidelines for selection of the fuel source for in situ atmosphere generation [83]:

1. The fuel cannot contain solid material.
2. The components of the fuel mixture must be mutually compatible, and the mixture must have a freezing point above even the coldest use temperature.
3. Liquids with flash points greater than methanol should be avoided.
4. The use of water-immiscible liquids may be undesirable from the standpoint of fire safety.
5. Thermal decomposition should not produce tars or soot.
6. The liquids should not be corrosive or toxic.
7. No liquid should be introduced into the furnace at a temperature below 750°C because of explosion hazards.

1. Nitrogen-Methanol

The most common fuel mixture used for in situ generation of furnace atmospheres is nitrogen–methanol. High purity nitrogen may be generated on site with a membrane separator such as the one shown in Figure 82 and may be stored as liquid nitrogen or used as generated.

Figure 103 presents a graph of the free energy released by the thermal decomposition of methanol according to the following furnace atmosphere reactions [99]:

$$CH_3OH \rightleftharpoons + 2H_2 \tag{1}$$

$$2CH_3OH \rightleftharpoons + CH_4 + C + 2H_2O \tag{2}$$

(a)

(b)

Figure 102   Methanol dissociator. (Courtesy of Surface Combustion, Inc.)

$$CH_3OH \rightleftharpoons C + H_2 + H_2O \tag{3}$$

$$2CH_3OH \rightleftharpoons CH_4 + CO_2 + 2H_2 \tag{4}$$

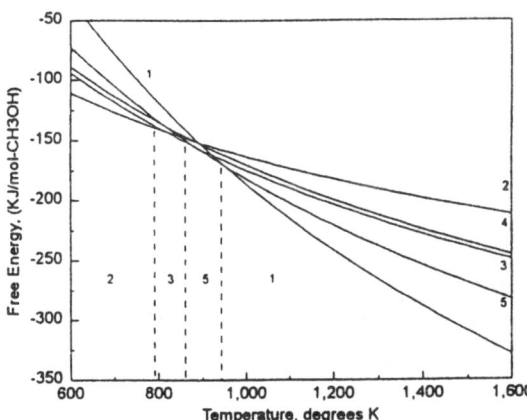

Figure 103   Free energy released by methanol dissociation reactions. (See page 392 for Equations 1 and 2, above for Equations 3 and 4, and page 394 for Equation 5.)

$$2CH_3OH \rightleftharpoons C + CO_2 + 4H_2 \tag{5}$$

The overall furnace atmosphere composition is determined by the water–gas equilibrium

$$CO_2 + H_2 \rightleftharpoons CO + H_2O$$

The temperature dependence of methanol dissociation is shown in Figure 104 [97,100]. Methanol dissociation may be performed either in the furnace chamber or in an external dissociator. To facilitate more complete decomposition, external liquid methanol dissociators may be used that employ catalysts such as copper and zinc and somewhat lower dissociation temperatures [100].

A comparison of the "typical" composition of atmospheres attainable with nitrogen–methanol mixes and endogases from natural gas (which contains approximately 85% methane) or propane is shown in Table 41.

The delivery of a nitrogen–methanol mixture to the furnace in a direct feed system is shown in Figure 105. Proper flow control is critical to the success of the process.

Conversion from conventional endothermic gas to nitrogen–methanol is relatively straightforward. The nitrogen–methanol injector pipe is introduced directly into the endogas entry port [96]. The injector atomizes the nitrogen–methanol mixture and directs it to the heated dissociation surface. When high dew points and poor control are encountered, it is usually due to improper injection. Praxair [96] gives some additional guidelines:

1. To ensure safe operation, the inert gas purge line should enter the furnace through a separate furnace port to minimize any possibility of plugging.

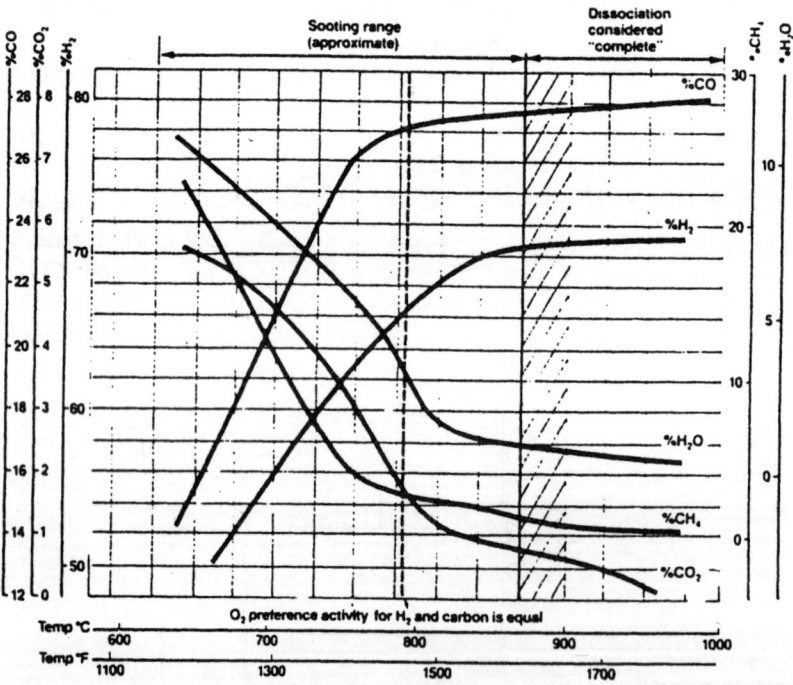

Figure 104 Dissociation of methanol at various temperatures.

Table 41 Comparison of the Composition Ranges Attainable with Nitrogen–Methanol and Endothermic Gas

Analysis	Nitrogen methanol	Endothermic gas from natural gas	Endothermic gas from propane
% CO	15–20	19.8	23.8
% H_2	35–45	40.4	31.2
% CO_2	0.4	0.3	0.3
% CH_4	0.3	0.5	0.1
N_2	Remainder	Remainder	Remainder

Source: Ref. 101.

2. A methanol pipe that is oversized or excessively long increases the possibility of liquid gas separation (Figure 106) and will lead to furnace "puffing" as shown in Figure 107.

VII. ATMOSPHERE SENSORS

The primary constituents of furnace atmospheres are carbon monoxide (CO), carbon dioxide (CO_2), nitrogen (N_2), hydrogen (H_2), water (H_2O), and hydrocarbon gases such as methane (CH_4). The ratio of these gases with respect to each other is critical to prevent decarburization, oxidation, hydrogen embrittlement, surface blueing due to the low temperature (400–700°F; 205–370°C) reaction of water vapor with steel, and soot formation. Therefore, it is necessary to use various sensing devices to accurately monitor and control the concentration of these gases throughout the heat treatment cycle. A summary of the various options to analyze heat treating atmospheres is provided in Table 42.

The objective of the remainder of this section is to provide an overview of the various measurement techniques that can be used to determine gas composition.

A. Orsat Analyzer

The analytical procedure followed with the Orsat analyzer is based on the relative gas adsorptivity of different gases in selective reagents. A sample of the gaseous atmosphere is passed through a series of liquids, each specific to one of the various gases being analyzed: CO_2, CO, O_2, and H_2 (see Table 43). The concentration of nitrogen is obtained by difference [83]. The increase in volume is proportional to the relative volume of the gas in the atmosphere being analyzed.

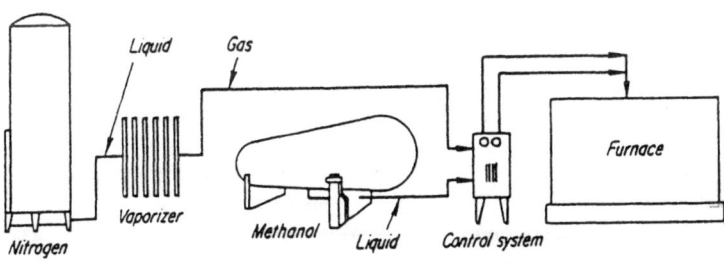

Figure 105 Schematic of a nitrogen–methanol atmosphere supply system. (From Ref. 101.)

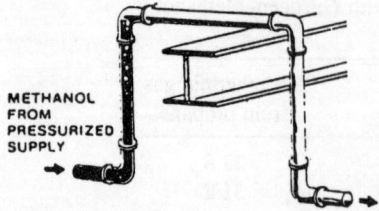

Figure 106 Elevated piping handling methanol from pressurized supply can accumulate nitrogen with resultant flow irregularities. (From Ref. 96.)

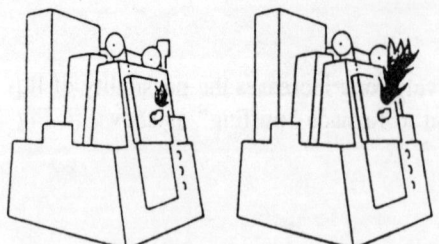

Figure 107 Furnace puffing due to segregation of liquid and gas in piping. (From Ref. 96.)

Orsat analysis is not commonly used today. It has been replaced by significantly faster and more sensitive methods such as gas chromatography, and dew-point and infrared analyzers [83].

B. Gas Chromatography

Gas chromatography is a relatively fast and accurate method for measuring the concentration of the gaseous components in a heat treatment atmosphere [83,104–108]. Gas chromatographic

Table 42 Gas Composition Measurement Options

Measurement system	Advantages	Disadvantages
O_2 only	Most applicable because only one instrument is needed and low O_2 content indicates proper air/fuel ratio.	Substoichiometric conditions are not defined. Inefficient combustion cannot be detected.
O_2 + total combustibles	Preferable because there is an instrument that will measure both O_2 and total combustibles. This defines total range.	
O_2 + CO_2	Defines entire combustion range.	Two instruments required.
O_2 + CO	Defines entire combustion range.	Two instruments required.
CO_2 only	One one instrument required.	Particular information on equilibrium flue gas products should be available.
Complete gas analysis	Stoichiometric composition gives better information for troubleshooting.	More complex. Requires carrier gas and standard gas.

Source: Ref. 104.

Table 43 Liquid Adsorbents for Heat
Treatment Atmosphere for Orsat Analysis

Gas	Liquid adsorbent
CO_2	Aqueous potassium hydroxide
CO	Ammoniacal cupric chloride
O_2	Alkaline pyrogallol

analysis is conducted by injecting the gas to be analyzed onto a column with a carrier gas, typically helium or argon [109]. The separation is primarily dependent on how strongly the adsorbed gases adhere to the column solid support. Gases, as they elute from the column, are detected by flame ionization, thermal conductivity, or some other suitable detector. The more weakly adsorbed gases will elute first, and the most strongly adsorbed gases will elute last. Care should be taken when molecular sieves are used as the solid support because an error due to coelution of argon, which is present in combustion air, with oxygen may inflate the oxygen concentration by approximately 2% [104]. An illustrative gas chromatogram is provided in Figure 108.

C. Thermal Conductivity
The thermal conductivity of a gas is dependent on its composition, as illustrated by the data presented in Table 44. The thermal conductivity of a gas is measured electrically using a Pt or Pt/Rh wire in a heated chamber. The thermal conductivity is determined by the amount the wire is cooled in the presence of the gas relative to the cooling provided by a known gas at the same temperature.

D. Oxygen Sensors

1. Paramagnetic Oxygen Analyzers
A unique feature of oxygen, relative to the other gaseous constituents of heat treatment atmosphere gases, is its magnetic susceptibility [83,110]. This is illustrated in Table 45.

Three types of instruments are used to measure the paramagnetic properties of oxygen, magnetodynamic (paramagnetic), thermomagnetic, and magnetopneumatic oxygen analyzers. A schematic of the magnetodynamic oxygen analyzer is shown in Figure 109. When the gas

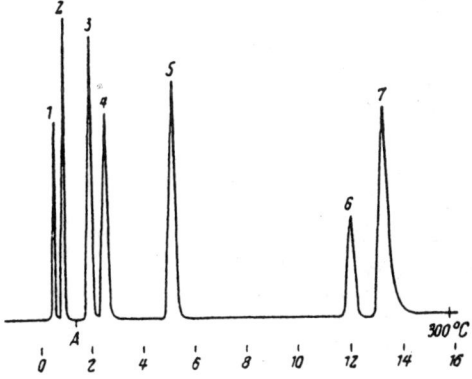

Figure 108 Gas chromatographic analysis of furnace atmospheres. (From Ref. 14.)

Table 44 Thermal Conductivity of Heat Treatment Gases

Gas	Relative thermal conductivity	Gas	Relative thermal conductivity
Air (100°C)	100	CO	97
H_2	715	NH_3	90
CH_4	150	CO_2	68
O_2	102	H_2S	54
N_2	99	SO_2	35

Source: Ref. 83.

Table 45 Relative Magnetic Susceptibility of Various Gases

Gas	Relative magnetic susceptibility
Oxygen (O_2)	100
Nitrogen oxide (NO)	43
Nitrogen dioxide (NO_2)	28
Hydrogen (H_2)	0.24
Carbon monoxide (CO)	0.01
Nitrogen (N_2)	0.00
Methane (CH_4)	−0.20
Argon (Ar)	−0.22
Ethylene ($C_2\dot{H}_4$)	−0.26
Ammonia (NH_3)	−0.26
Carbon dioxide (CO_2)	−0.27

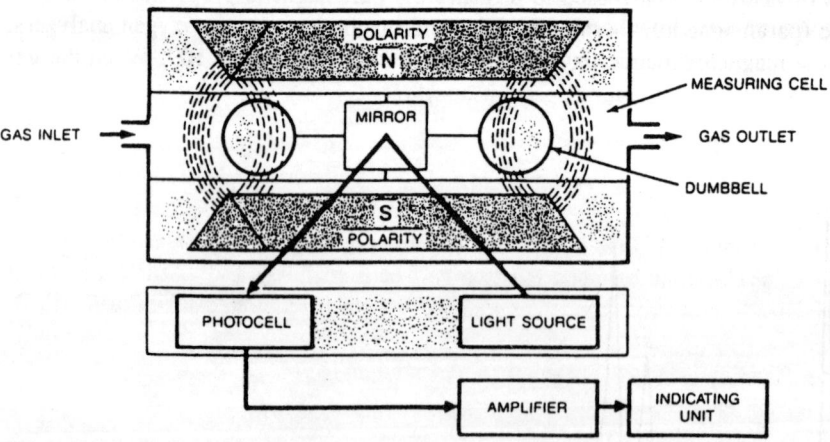

Figure 109 Schematic illustration of a paramagnetic oxygen analyzer.

to be analyzed is introduced into the instrument, oxygen is attracted to the point of greatest magnetic susceptibility, the "dumbbell" area. The magnitude of displacement of the dumbbell by the oxygen gas is proportional to the amount of oxygen present in the gas.

The operation of the thermomagnetic analyzer is based on the temperature dependence of the magnetic susceptibility c of oxygen [83]:

$$cT = Z$$

where T is the absolute temperature and Z is the Curie constant.

The thermomagnetic analyzer consists of two tubes for gas entry that are connected by a heated crossover tube with an electric heating filament at each end. Each filament is attached to a Wheatstone bridge, so there is a magnetic field surrounding one of the heating filaments. The gas is then introduced equally to both tubes. Because of its magnetic susceptibility, oxygen in the gas is attracted to the magnet where it is heated by the electric filament. As the sample is heated, the magnetic susceptibility decreases until it is displaced by cooler, more magnetic oxygen from the gas.

This mechanism creates a "magnetic wind," which cools the filaments. The difference in the resistance of the two wire filaments is proportional to the amount of oxygen gas that is present [110].

The third type of instrumentation that may be used, which is based on the paramagnetic properties of oxygen, is the magnetopneumatic oxygen analyzer. This instrument uses a nonhomogeneous magnetic field, and the oxygen in the gas is attracted toward the pole of greatest magnetic strength. A reference gas of known oxygen content is introduced into the unknown sample and is also analyzed separately, creating a pressure differential. The pressure differential is balanced by a measured flow from the unknown sample gas that is proportional to the differential pressure and is also proportional to the oxygen content of the unknown gas [104].

These types of instruments have not found widespread use in the heat treatment industry because [104]

1. The magnetodynamic oxygen analyzer is too delicate an instrument for use in rugged heat treatment shop environments.
2. The accuracy of the thermomagnetic oxygen analyzer may be unstable and unreliable when used with high temperature gases.
3. The magnetopneumatic oxygen analyzer is very sensitive to vibrations, which are often present in the heat treatment shop environment.

2.   Electrochemical Oxygen Analyzers

It is also possible to measure the oxygen content of a gas by measuring the electrochemical potential (or voltage, emf) for electrons to flow between platinum electrodes in a known reference cell and an unknown cell [110–116]. The two cells are separated by an electrolyte that permits transfer of the electrons between the two half-cells (cathode and anode). The most common electrolyte used for oxygen ions is zirconium oxide stabilized with calcium oxide (ZrO_2 + 4% CaO).

The electrochemical cell that is formed is represented as [83]

$Pt:O_2$	ZrO_2	$O_2:Pt$
(unknown)	(electrolyte for oxygen ions)	(unknown)

The emf potential (or voltage) between the two half-cell reactions is measured and is proportional to the electron transfer between the two cells. The potential can be calculated by the
well-known Nernst equation [83]

$$E = \frac{RT}{nF} \ln\left[\frac{O_2 I}{O_2 II}\right]$$

where R is the gas constant, T is the absolute temperature, n is the number of electrons transferred ($n = 4$ for $O_2 \rightarrow 2O^{-2}$), F is the Faraday constant, and O_2I and O_2II are the reactivi-
ties of oxygen at either side of the electrolyte.

By combining constants, the Nernst equation for the oxygen probe can be rewritten as
[104]

$$\text{emf} = 0.0496T \log\left[\frac{P_{O_2}(\text{ref})}{P_{O_2}}\right]$$

A schematic of the electrochemical oxygen probe is provided in Figure 110. The corresponding chemical reactions are

Anode:

$$2O^{2-} \rightarrow 4e + O_2 \ (P_{O_2} \text{ furnace})$$

Cathode:

$$O_2 \ (P_{O_2} \text{ air}) + 4e^- \rightarrow 2O^{2-}$$

Sum:

$$O_2 \ (P_{O_2} \text{ air}) \rightarrow O_2 \ (P_{O_2} \text{ furnace})$$

The relationship between the emf for the oxygen probe and the carbon potential is shown in
Figure 111.

A alternative electrochemical oxygen sensor is the nondepleting coulometric sensor illustrated in Figure 112. The advantage of the coulometric sensor is that the sensor does not require replacement. Analysis of the oxygen concentration occurs in an electrochemical cell where
potassium hydroxide is used as the electrolyte. The half-cell reactions, which are driven by
an external emf of 1.3 V, are [110]

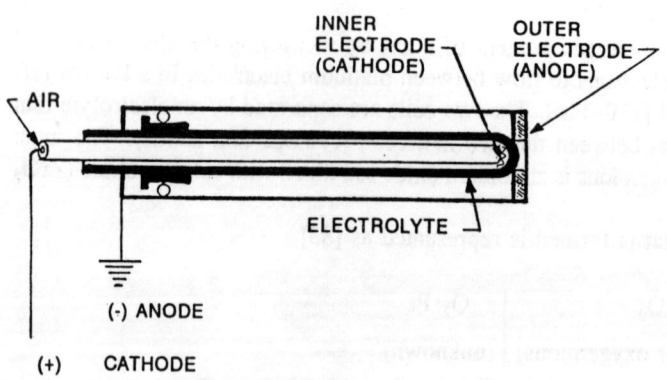

Figure 110 Schematic of an oxygen probe.

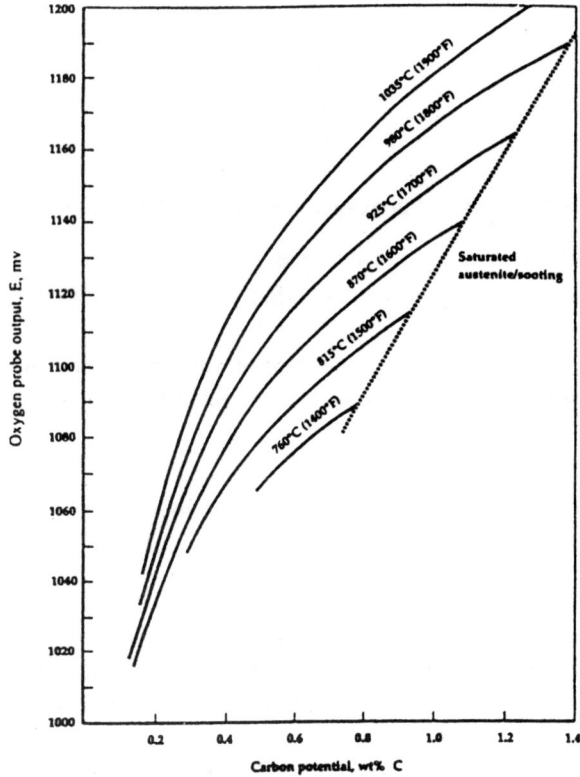

Figure 111 Isothermal relationship between oxygen probe output and carbon potential. (From Ref. 113.)

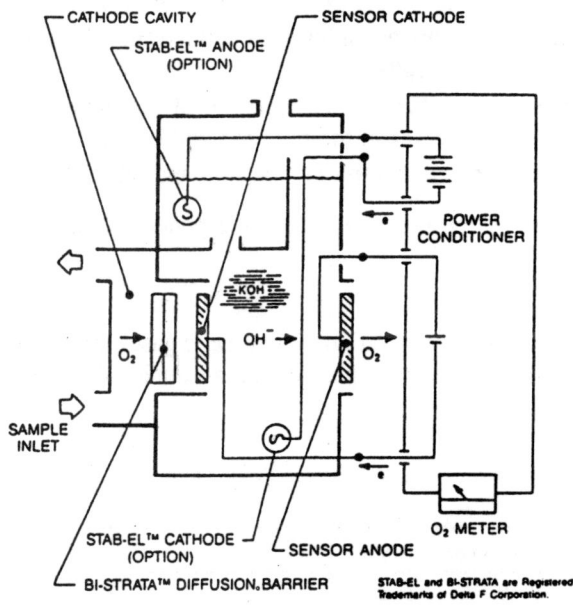

Figure 112 Coulometric sensor for oxygen probe. (From Ref. 110.)

Cathode:

$$O_2 + 2H_2O + 4e^- \rightarrow 4OH^-$$

Anode:

$$4OH^- \rightarrow O_2 + 2H_2O + 4e^-$$

The following recommendations have been made for oxygen sensor installation [111]:

1. The sensor should be installed in a separate protective tube.
2. The sensor should be installed to monitor the gas stream after it has passed through the load.
3. It is suggested that the sensor be located in the top one-third of the heated zone and away from the fan, halfway between the front and back of the heated zone and close to the furnace thermocouple.
4. The sensor should be mounted to permit easy access.
5. A good source of "reference" air must be available, preferably at a flow rate of 0.5–2.0 scfh.

Some of the most common sources of oxygen probe errors are [113,114]

1. Improper operating conditions due to probe deterioration. For example, soot formation, which is due to the operation of the furnace outside of the austenite phase field shown in Figure 111, will cause carbon to be deposited on the probe, leading to an incorrect response.
2. Alteration of $CO + CO_2$ content in the furnace atmosphere. As shown in Table 46, there will be a significant error in carbon content if CO varies outside the target concentration.
3. Electrolyte failure may develop from cracks within the tube, making it possible for reference air to leak into the sample, leading to an incorrect probe response.
4. The platinum or nickel wire used in the probe assembly may catalytically degrade the CH_4 gas into CO, producing erroneous results.
5. If the reference air sample that is necessary for correct operation of the probe becomes contaminated with furnace atmosphere gases, the response will be erroneous.
6. The reference air may leak out of the probe due to probe cracking and other flaws.

Detection of most of these problems is aided by the use of a second independent measurement of the furnace atmosphere [114].

3. *Infrared Sensors*

Infrared spectrophotometry provides an indirect measure of various atmospheric gases including CH_4, CO, and CO_2. This method involves passing infrared light through a sample of the gas

Table 46   Effect of CO Content on Carbon Potential for Constant Oxygen Probe Output of 1149 mV

	CO content (%)					
	15	17	19	20	21	23
Wt % carbon	0.80	0.88	0.95	1.0	1.05	1.11

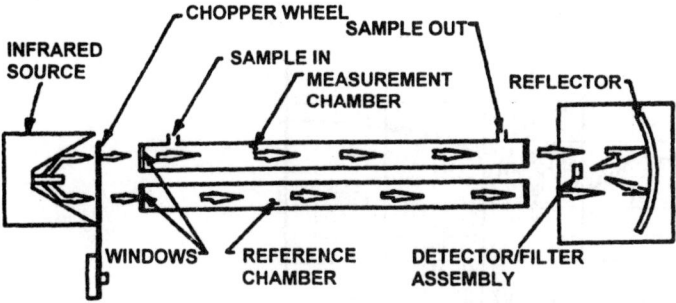

Figure 113 Nondispersive infrared detector (NDIR) for gas analysis. (From Ref. 117.)

as shown in Figure 113. The wavelength of infrared light characteristically varies from 0.75 to 300 μm, although most applications are performed in the region of 2.5–5 μm. The particular wavelength of infrared light selected depends on the gas being analyzed. Figure 114 illustrates the infrared spectrum of a CO–CO_2–H_2O mix. Figure 115 illustrates the infrared spectrum for CO, CO_2, NH_3, and CH_4.

When the proper wavelength of light is passed through the gas, it is absorbed, and the amount of absorption is defined by Beer's law,

$$A = abc$$

where
A = absorption of light
a = absorptivity, a characteristic value for each gas
b = path length of light through the sample
c = concentration of the infrared-active material

Since the pathlength b is fixed and a is a constant for the gas being analyzed, the amount of light absorbed is proportional to the concentration.

Infrared analyzers are used continuously in the system with gas sampling rates of approximately 1 min [83,109,119,120]. They have the advantage of being accurate and fast. However, infrared analyzers are relatively expensive, and significiant damage to the optics in the infrared detector may be caused by contamination.

4. *Dew Point Analyzers*

Another indirect measure of the carbon content of an atmospheric gas is the dew point. The dew point is defined as the temperature at which as gas is saturated with water vapor, that is,

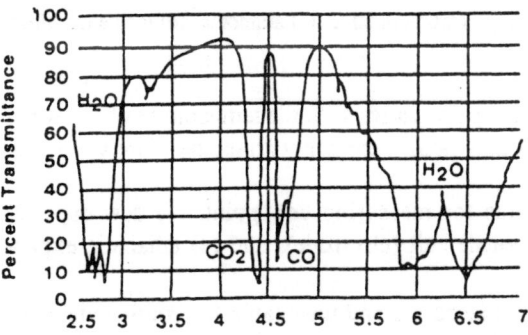

Figure 114 Infrared spectral frequencies for heat treatment gases. (From Ref. 114.)

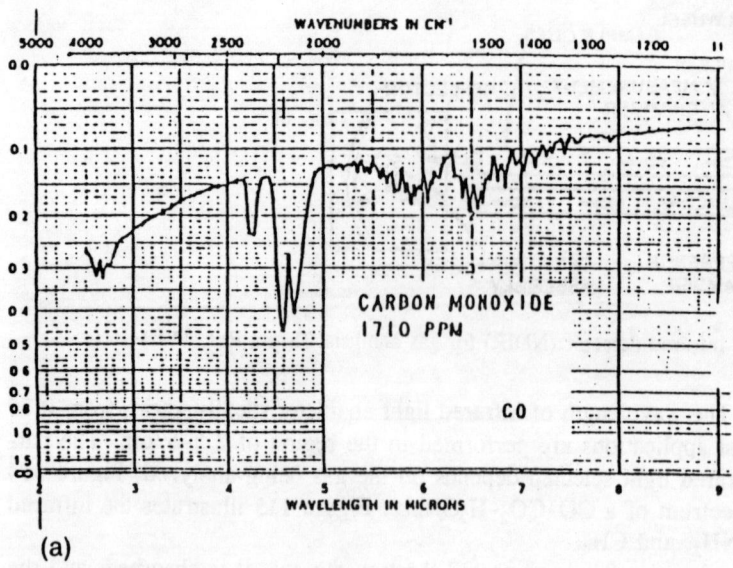

(a)

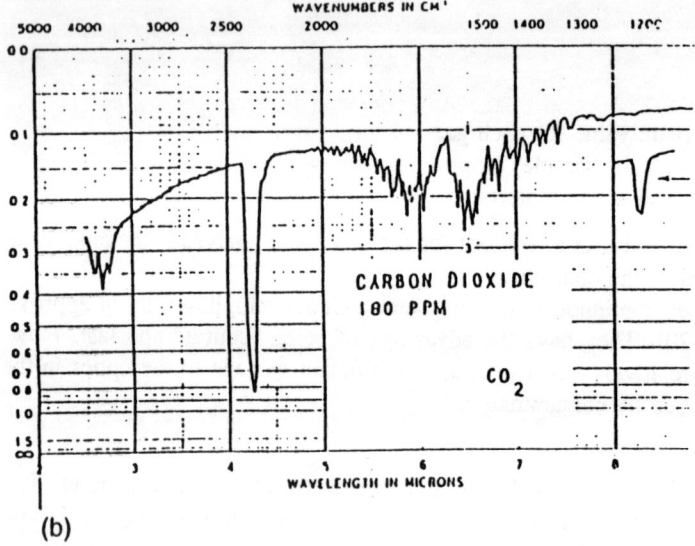

(b)

Figure 115   Infrared spectra. (a) Carbon monoxide; (b) carbon dioxide; (c) ammonia; (d) methane. (From Ref. 115.)

it has a relative humidity of 100% [87]. The dew point varies with the temperature of the water vapor as shown in Figure 116. A summary of dew-point variation with temperature and moisture is provided in Table 47.

There are four principal types of apparatus for measuring dew point: (1) dew cup, (2) chilled mirror (Peltier effect), (3) lithium chloride cell (electrode), and (4) adiabatic expansion [83,109].

Dew Cup Apparatus   Although relatively inaccurate and unsuitable for continuous automated atmosphere control, the simplest method of measuring the dew point is with the "dew cup" apparatus shown in Figure 117 [83]. This method involves cooling a chromium-plated

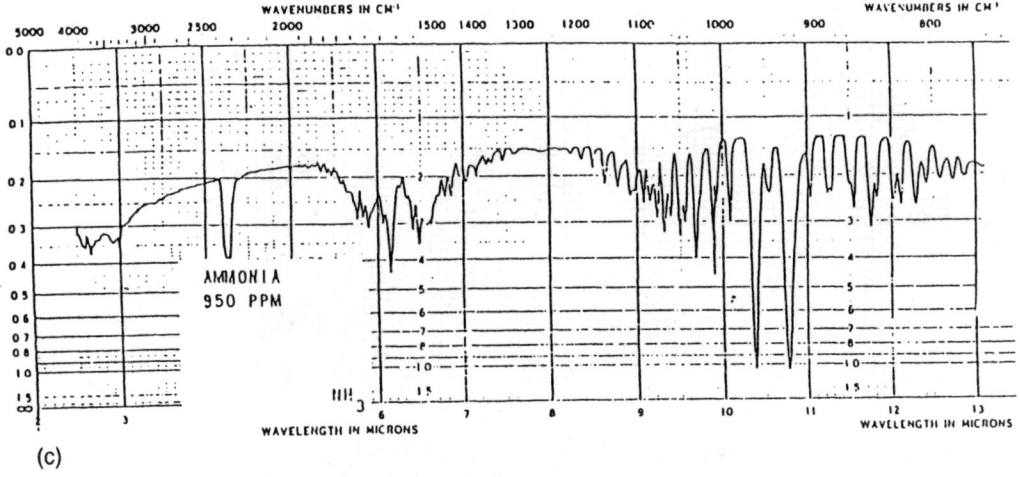

(c)

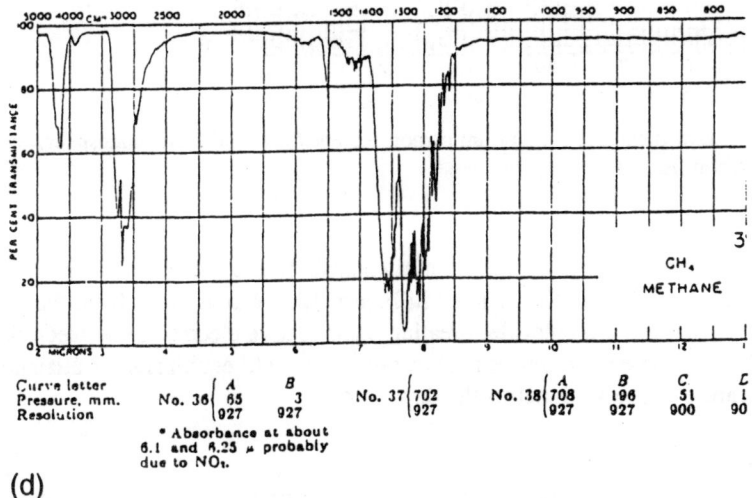

Curve letter		A	B			A	B	C	L
Pressure, mm.	No. 36	65	3	No. 37	702	No. 38	708	196	51	1
Resolution		927	927		927		927	927	900	90

* Absorbance at about
6.1 and 6.25 μ probably
due to NO₂.

(d)

Figure 115 Continued

copper cup within a glass enclosure containing a dry ice–alcohol bath. The gas to be measured enters the instrument, and the cup is cooled until condensation is observed on its surface. The temperature at which this occurs is the dew point.

Chilled Mirror—The Peltier Effect A method of measuring the dew point that is based
on the Peltier effect is shown schematically in Figure 118. The Peltier effect refers to the difference in current intensity between two ends of a wire where one end is at the temperature
of a reference cold junction and the other is at a higher temperature.

When a chilled mirror is used, the surface of the metal mirror is scanned photoelectri-
cally while it is cooled in the gas to be analyzed. At the dew point, water vapor forms on the
mirror. A photoelectric cell will direct the intensity of light reflected from the surface of the
mirror, which is dependent on the amount of moisture in the gas. A schematic of a continuous chilled mirror detector is shown in Figure 119.

Measurement by Lithium Chloride Cell (Electrode) The lithium chloride cell is the most
widely used method for automatic measurement of dew point [83]. This method is based on

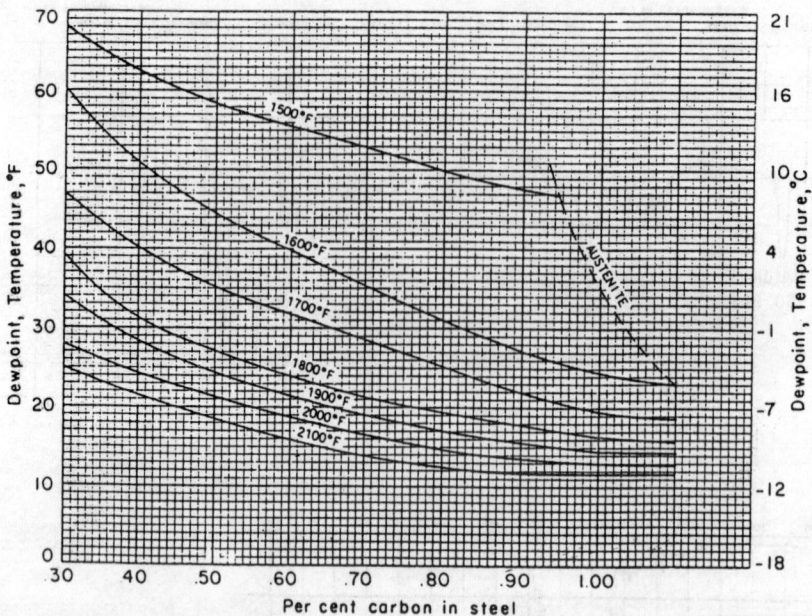

Figure 116 Correlation of dew point and equilibrium carbon content for endothermic atmosphere as a function of temperature. (From Ref. 87.)

the variation in the conductivity of hydroscopic lithium chloride in the presence of water vapor. An example of a lithium chloride cell is shown in Figure 120. Water vapor from the gas atmosphere is absorbed by the lithium chloride, creating a conductive electrolyte between the platinum wire electrodes. This process produces a temperature rise until equilibrium is attained. The equilibrium temperature is proportional to the dew point.

E. Adiabatic Expansion

When a gas cools adiabatically, the dew point is achieved only for a particular pressure drop, temperature, and moisture content. The temperature where this occurs corresponds to the dew point.

F. Carbon Resistance Gauge

The electrical resistance of steel varies with carbon content as shown in Figure 121. Instruments based on the measurements of continuous iron-nickel alloy wire expanded to the atmosphere gases available.

G. Weight Measurement of Equilibrium Shim Stock

Blumenthal and Hlasny [123] describe a method for calibration of carbon sensors that uses a test for "true carbon potential," which is measured by equilibrating an AISI 1010 steel shim sample, 0.003 in. thick and 2.5 × 3 in. in area in a furnace atmosphere and then determining the carbon content of the shim by weight gain (or by chemical analysis). The shim-holding device is illustrated in Figure 122. The test procedure is as follows [123]

　　1.　Wearing rubber gloves, clean the shim test specimen with acetone and weigh it on
　　　　a balance to the nearest 0.1 mg.

Table 47 Moisture Conversion Table

Dew Point		Vapor pressure (water/ice in equilibrium) (mm Hg)	Water content (ppm) on volume basis at 760 mm Hg	Relative humidity at 70°F (%)	Moisture (ppm) on weight basis in air
°C	°F				
−110	−166	.0000010	.00132	0.0000053	0.00082
−108	−162	.0000018	.00237	0.0000096	0.0015
−106	−159	.0000028	.00368	0.000015	0.0023
−104	−155	.0000043	.00566	0.000023	0.0035
−102	−152	.0000065	.00855	0.000035	0.0053
−100	−148	.0000099	.0130	0.000053	0.0081
−98	−144	.000015	.0197	0.000080	0.012
−96	−141	.00002	.0289	0.00012	0.018
−94	−137	.000033	.0434	0.0018	0.027
−92	−134	.00048	.0632	0.00026	0.039
−90	−130	.000070	.0921	0.00037	0.057
−88	−126	.00010	.132	0.00054	0.082
−86	−123	.00014	.184	0.00075	0.11
−84	−119	.00020	.263	0.00107	0.16
−82	−116	.00029	.382	0.00155	0.24
−80	−112	.00040	.562	0.00214	0.33
−78	−108	.00056	.737	0.00300	0.46
−76	−105	.00077	1.01	0.00410	0.63
−74	−101	.00105	1.38	0.00559	0.86
−72	−98	.00143	1.88	0.00762	1.17
−70	−94	.00194	2.55	0.0104	1.58
−68	−90	.00261	3.43	0.0140	2.13
−66	−87	.00349	4.59	0.0187	2.84
−64	−83	.00464	6.11	0.0248	3.79

(continued)

Table 47 Continued

Dew Point		Vapor pressure (water/ice in equilibrium) (mm Hg)	Water content (ppm) on volume basis at 760 mm Hg	Relative humidity at 70°F (%)	Moisture (ppm) on weight basis in air
°C	°F				
-62	-80	.00614	8.08	0.0328	5.01
-60	-76	.00808	10.6	0.0430	6.59
-58	-72	.0106	13.9	0.0565	8.63
-56	-69	.0138	18.2	0.0735	11.3
-54	-65	.0178	23.4	0.0948	14.5
-52	-62	.0230	30.3	0.123	18.8
-50	-58	.0296	38.8	0.157	24.1
-48	-54	.0378	49.7	0.202	30.9
-46	-51	.0481	63.3	0.257	39.3
-44	-47	.0609	80.0	0.325	49.7
-42	-44	.0768	101	0.410	62.7
-40	-40	.0966	127	0.516	78.9
-38	-36	.1209	159	0.644	98.6
-36	-33	.1507	198	0.804	122.9
-34	-29	.1873	246	1.00	152
-32	-26	.2318	305	1.24	189
-30	-22	.2859	376	1.52	234
-28	-18	.351	462	1.88	287
-26	-15	.430	566	2.30	351
-24	-11	.526	692	2.81	430
-22	-8	.640	842	3.41	523
-20	-4	.776	1020	4.13	633
-18	-0	.939	1240	5.00	770

-16	+3	1.132	1490	6.03	925
-14	+7	1.361	1790	7.25	1110
-12	+10	1.632	2150	8.69	1335
-10	+14	1.950	2570	10.4	1596
-8	+18	2.326	3060	12.4	1900
-6	+21	2.765	3640	14.7	2260
-4	+25	3.280	4320	17.5	2680
-2	+28	3.880	5100	20.7	3170
0	+32	4.579	6020	24.4	3640
+2	+36	5.294	6970	28.2	4330
+4	+39	6.101	8030	2.5	4990
+6	+43	7.013	9230	37.4	5730
+8	+46	8.045	10590	42.9	6580
+10	+50	9.029	12120	49.1	7530
+12	+54	10.52	13840	56.1	8600
+14	+57	11.99	15780	63.9	9800
+16	+61	13.63	17930	72.6	11140
+18	+64	15.48	20370	82.5	12650
+20	+68	17.54	23080	93.5	14330
+22	+71.5	19.827	26088		16699
+24	+75	22.377	29443		18847
+26	+79	25.209	33169		21232
+28	+82	28.349	37301		23877
+30	+86	31.824	41874		26804

Source: Ref. 116.

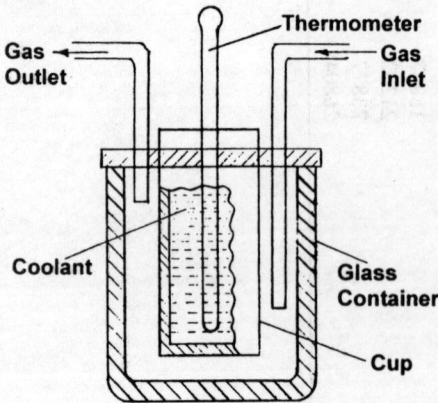

Figure 117 Dew cup instrument. (From Ref. 83.)

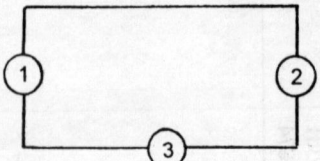

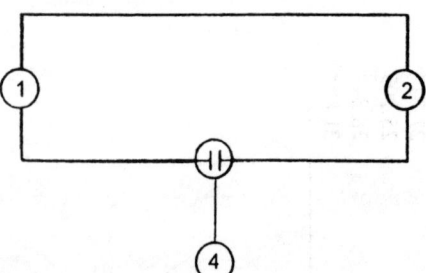

Figure 118 Dew-point measurement by the Peltier effect. (From Ref. 83.)

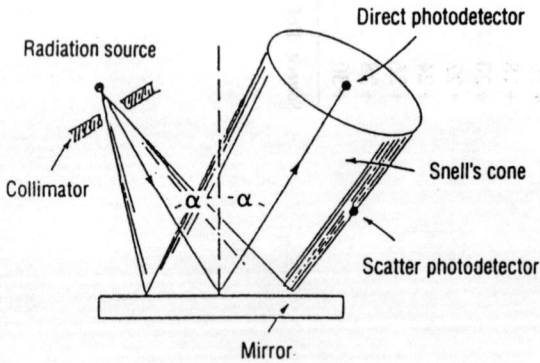

Figure 119 Continuous on-line dew-point monitor. (From Ref. 122.)

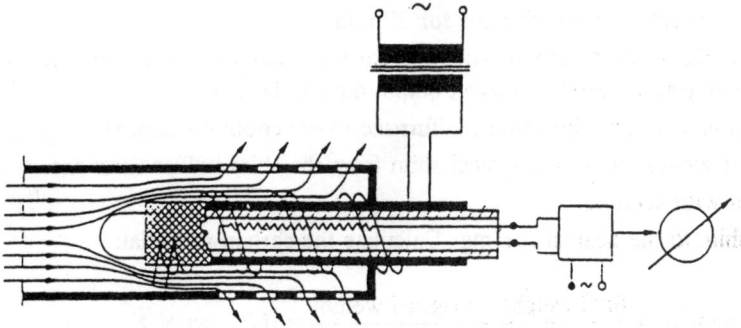

Figure 120 Lithium chloride dew-point probe. (From Ref. 93.)

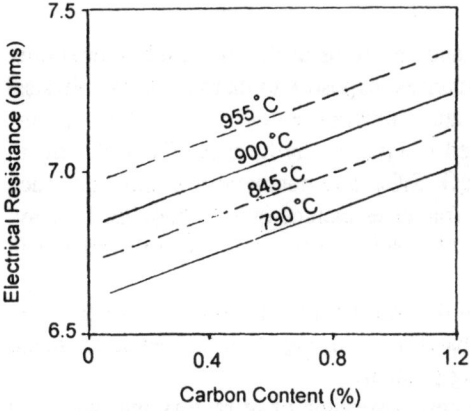

Figure 121 Electrical resistance of steel vs. carbon content at various temperatures. (From Ref. 83.)

2. Roll the specimen into a cylinder approximately 3/4 in. in diameter, and insert it into the shim holder as shown in Figure 122.

3. Be sure the furnace is operating above 1600°F (871°C) and that there is a load in it. Close the doors with the atmosphere circulation fan running; usually 30 min is sufficient to attain equilibrium.

4. Remove the plug from the sample side of the furnace, and open the gate valve (see Figure 122). Insert the shim holder to the same depth as the carbon/oxygen sensor, usually approximately 20 in.

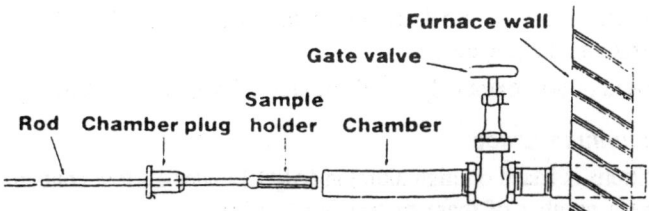

Figure 122 Apparatus for determination of carbon content by shim-stock exposure and measurement. (From Ref. 123.)

5. Leave the shim specimen in the furnace for 30 min.
6. Record the weight percent carbon from the carbon controller, average furnace temperature, and average probe millivolt output during the test.
7. Move the shim-holder cartridge from the furnace to the cooling chamber.
8. Wearing rubber gloves, remove the steel shim from the shim-holder cartridge.
9. Clean the shim with acetone.
10. Reweigh the shim to the nearest 0.1 mg. Calculate the carbon potential:

$$\text{Carbon potential} = \frac{\text{final weight} - \text{original weight}}{\text{final weight}} \times 100 + \text{wt \% C}$$

where wt % C is the original weight percent carbon content.

VIII. REFRACTORY MATERIALS

Furnace refractories are materials that must withstand severe or destructive high temperature service conditions and resist attack by chemical substances, exposure to thermal shock, physical impact, and other abuse [124]. Heating efficiency and temperature uniformity within the furnace depend on the type of refractory material used to insulate the furnace. Examples of refractory materials include metal oxides such as MgO, SiO_2, and nonmetallic compounds such as SiC and SiN. Other materials that are used as refractories include quartz, bauxite, fireclay, chromite ore, magnesite, and zirconium oxide [124]. Refractory materials are available in various shapes such as brick and in bulk form.

Traditionally, the most common lining used in the heat treatment furnaces has been firebrick. Although firebrick continues to be used, materials with significantly better chemical properties such as ceramics and fiber are becoming popular.

The objective of this section is to provide a general overview of refractory materials and their classification.

A. Refractory Classifications

1. Magnesium Compositions

Refractory materials are available in five classes of acid-resistant composition [124]:

1. Magnesia or magnesite (MgO).
2. Magnesia in combination with chromium-containing materials such as Cr_2O_3 [63]. If the chromium is the major component, then it is referred to as chrom-magnesite. If magnesia is the predominant component, then it is referred to as magnesite-chrome.
3. Magnesia in combination with spinel (magnesium aluminum silicate).
4. Magnesia in combination with 2.5% or 4.5% carbon, where carbon is in the form of pitch, which bonds refractory aggregates.
5. Dolomite, which is composed of approximately equal amounts of $MgCO_3$ and $CaCO_3$.

2. Compositions Containing Aluminum Oxide

Another class of refractory compositions is based on high alumina (Al_2O_3) materials that contain more than 47.5% Al_2O_3. There are a number of these materials [83,124].

1. Mullite brick, which is compositionally 71.8% Al_2O_3 and 28% SiO_2. This material has excellent volume stability and strength at high temperatures.

2. Chemically bonded, normally phosphate-bonded, brick that reacts to form aluminum orthophosphate ($AlPO_4$).

3. Alumina-chrome brick, which is composed of a solid solution of chromium oxide (Cr_2O_3) and a high purity alumina.

4. Alumina-carbon brick, which is a resin-bonded high alumina graphite-containing composition.

3. Fireclay Compositions

Compositionally, fireclay is a hydrated aluminum silicate ($Al_2O_3 \cdot 2SiO_2 \cdot 2H_2O$). After dehydration at elevated temperatures, the residual material should contain 45.9% alumina and 54.1% silica. There are five standard ASTM classifications of firebrick [124]:

1. Super duty fireclay contains 40–45% Al_2O_3. This is the refractory classification with the greatest high temperature volume stability and spalling resistance.

2. High duty fireclay is slightly less refractory than the super duty type but is still spalling- or slag-resistant.

3. Medium duty fireclay is suitable for moderately severe applications.

4. Low duty fireclay is suitable for moderate temperature application only.

5. Semisilica brick contains only 18–15% alumina. It exhibits good high temperature load-bearing capacity and volume stability.

4. Silica Refractories

Silica refractories are used for high temperature (3080–3110°F) melting applications. They are capable of withstanding pressures up to 50 psi and are chemically resistant to acid slag. They do not undergo spalling at temperatures above 1200°F. There are two ASTM classifications of silica firebrick: Type A and Type B.

5. Monolithic Refractories

Monolithic refractories are special mixes or blends of dry granular or cohesive plastic materials that form a virtually joint-free lining [124]. They are classified as plastic refractories [125]; ramming mixes, which are denser and stronger than plastic refractories; gunning mixes, which may be pneumatically bonded to a furnace wall; and castables, or "refractory concrete." Plastic refractories are prepared from fireclay, high alumina graphite, alumina-chrome, and other materials.

Numerous other materials are being developed, for example high alumina fiber [126], glass fiber [127], and microporous insulating [128] refractories, to provide greater resistance to slag attack, low thermal expansion, increased resistance to spalling, reduced impurities, improved hot strength, acceptable creep resistance, and improved resistance to thermal shock and corrosion [124].

B. Design Properties

Dimensional change of a material with respect to changes in temperature is described by

$$\frac{\Delta L_0}{L_0} = \alpha(T - T_0)$$

where α is the coefficient of linear expansion. Table 48 provides viral coefficients to calculate the percent of linear expansion with respect to its length at 20°C (293 K) for various materials.

Table 48 Virial Coefficients for Linear Thermal Expansion of Selected Solid Materials

$$100 \Delta L_t/L_{293} = A + B (10^{-4}T) + C (10^{-4}T)^2 + D (10^{-4}T)^3 \quad (T \text{ in kelvins})$$

Material	MP (K)	A	B	C	D	Note
Al_2O_3 (hex.)	2327	-0.180	+5.494	+22.520	-22.940	1
C_2O	3200	-0.321	+10.590	+13.100	-14.050	
Cr_2O_3 (hex.)	2603	-0.280	+10.380	-31.220	+106.200	2
Fe_2O_3 (trig.)	1838	-2.537	+7.300	+49.640	-114.000	3
MgO	3125	-0.326	+10.400	+25.810	-28.340	
SiO_2 (lo qtz.)	~873 (tr.)	-0.236	+6.912	+0.556	+1312.00	4
SiO_2 (hi qtz.)	1743 (tr.)	+1.040	+0.068	+11.660	+18.000	Est.
SiO_2 (vitr.)	~1273 (cryst.)	-0.015	+0.397	+4.666	-34.460	
ZrO_2 (monocl.)	2988	-0.314	+13.040	-90.920	+408.400	5
$Al_6Si_2O_{13}$	2193	-0.0929	+2.580	+21.530	-45.720	
$CaAl_2O_4$	1873	-0.107	+2.578	+39.680	-90.770	
Ca_2SiO_4	2403	-0.345	+11.260	+16.560	+27.330	
$MgAl_2O_4$	2408	-0.183	+5.456	+28.060	-41.810	
$Mg_2Al_4Si_5O_{18}$	~1773	+0.00911	-0.912	+20.640	-3.921	6

						Ref.
MgCr$_2$O$_4$	2673	-0.176	+5.822	+55.80	+23.360	
MgFe$_2$O$_4$	2023	-0.218	+6.003	+52.560	-94.040	
Mg$_2$SiO$_4$	2183	-0.238	+7.166	+33.810	-37.970	
Mg$_2$TiO$_4$	~2100	-0.249	+8.294	+4.074	+94.300	
ZrSiO$_4$	2673	-0.136	+5.337	-30.420	+209.400	
AlN (hex.)	~2500	-0.0809	+1.806	+31.760	-72.560	
B$_4$C (rhomb.)	2623	-0.114	+3.523	+12.660	-5.085	8
BN	~3273 (sub.)	-0.00133	-1.278	+49.110	-86.350	
SiC	~2923 (dec.)	-0.0991	+2.970	+13.880	-15.480	
TiC	~3410	-0.177	+5.710	+11.740	+2.412	
C (graphite)	~3900	-0.0550	+1.552	+12.050	-10.330	9
C (graphite)	~3900	-0.1580	+5.561	-8.850	+35.550	9
C (vir.)	~2700 (cryst.)	-0.890	+3.015	+1.285	+17.240	
Fe	1800	-0.289	+7.350	+93.300	-314.000	10

Notes:
1. Cryst. exp. c/a ~1.1.
2. Cryst. exp. a/c ~1.3.
3. Cryst. exp. a/c ~1.26.
4. Cryst. exp. a/c ~1.58.
5. Cryst. exp. c/b ~2.5.
6. Cordierite refractory.
7. Cryst. exp. a/c ~1.18.
8. Cryst. exp. ~ isotrop.
9. Grade ATJ, parallel and perpendicular to the textural "grain," respectively. Cryst. exp. c/a ~10.
10. Numerous steels and SS agree with Fe within ~ ±15%.

The value of $\Delta L_0/L_0$ is of considerable value when estimating the impact of a known temperature change (ΔT) across a wall [(hot face temperature)–(cold face temperature)] and refractory deformation. If the modulus of elasticity (E) is known, it is possible to calculate the stress σ from

$$\sigma = E\,\alpha(T_2 - T_1)$$

Thermal conductivity δ is calculated from the heat capacity C_p:

$$\delta = k/C_p\rho$$

where k is the thermal conductivity, δ is thermal diffusivity, and ρ is 123 density.

Figure 123 and Table 49 provide heat capacity data as a function of temperature for various simple refractory compositions [129]. Thermal conductivity as a function of temperature is illustrated in Figure 124 [129]. Thermal conductivity data for firebrick and alumina are provided in Figures 125 and 126, respectively [129].

For firebrick, the thermal conductivity k is dependent on bulk density BD [129]:

$$k_{500°F}\ [Btu \cdot in./(h \cdot ft^2 \cdot °F)] = 0.03455\ BD\ (lb/ft^3) - 0.2545$$

$$k_{260°C}\ [W/(m \cdot K)] = 0.3110\rho_b\ (g/cm^3) - 0.0367$$

For insulating castables, the thermal conductivity may be calculated from [129]

$$k_{500°F}\ [Btu \cdot in./(h \cdot ft^2 \cdot °F)] = 0.04375\ BD\ (lb/ft^3) - 0.4250$$

$$k_{260°C}\ [W/(m \cdot K)] = 0.3939\rho_b\ (g/cm^3) - 0.0613$$

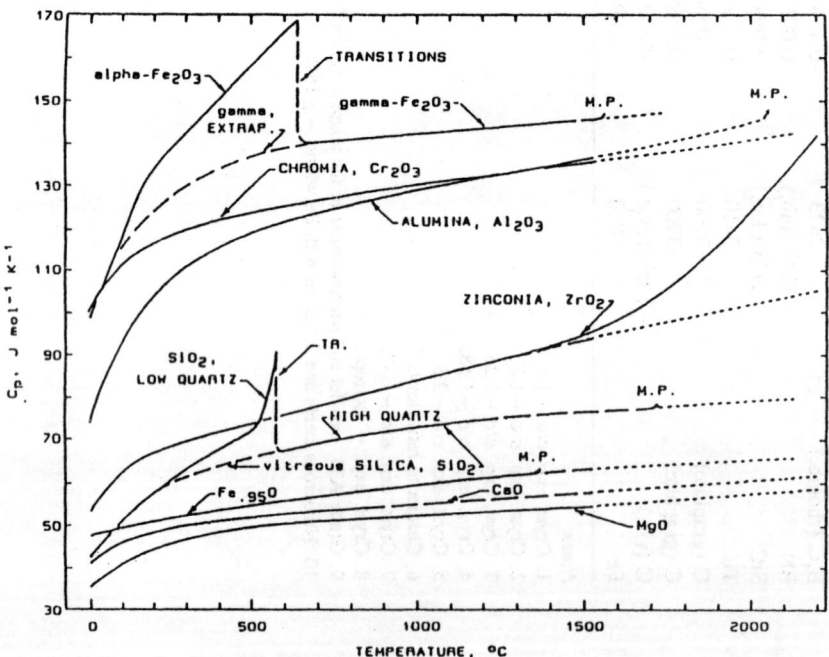

Figure 123 Heat capacity of various refractory materials.

Table 49 Molal Heat Capacity of Solid Substances [C_p in J/(mol · K); T in K]

$$C_p = a + b(10^{-3}T) + c(10^{-3}T)^2 + d/(10^{-3}T)$$

	Formula	a	b	c	d	Note
$Al_2O_3$101.96	+154.96	-16.168	+7.120	-20.817		
CaO	56.08	+57.68	-1.324	+1.560	-4.418	
Cr_2O_3	152.00	+137.14	-3.568	+3.120	-9.585	
Fe_2O_3	159.70	+176.70	-24.000	+7.200	-19.843	1
$Fe_{0.95}O$	69.06	+45.31	+14.900	-2.600	-0.374	2
MgO	40.31	+63.24	-7.632	+2.880	-7.263	
SiO_2	60.09	+77.09	+3.384	-0.160	-10.558	3
ZrO_2	123.32	+60.88	+22.320	-1.600	-3.370	
$Al_6Si_2O_{13}$	426.06	+607.83	-42.752	+22.880	-81.020	4
$CaAl_2O_4$	158.04	+236.00	-39.240	+14.360	-32.120	
Ca_2SiO_4	172.25	+182.73	+3.768	+11.580	-17.372	
$CaTiO_3$	135.96	+149.08	-2.916	-0.360	-15.051	
$MgAl_2O_4$	142.27	+233.16	-39.244	+14.360	-32.124	4
$MgCr_2O_4$	192.31	+212.76	-18.404	+8.360	-24.261	
Mg_2SiO_4	140.71	+191.48	-4.236	+8.040	-21.790	5
$MgTiO_3$	120.19	+145.39	-3.128	+3.520	-15.949	
$ZrSiO_4$	183.31	+179.94	-21.380	+12.200	-22.838	
AlN	40.99	+61.68	-7.344	+2.560	-8.911	
B_4C	55.25	+35.55	+86.508	-16.920	-3.145	
BN	24.82	+37.02	+19.324	-5.560	-6.814	
SiC	40.10	+55.08	-0.876	+2.040	-8.312	
TiC	59.89	+65.44	-7.964	+2.760	-8.911	
C (graph.)	12.01	+18.37	+8.592	-2.480	-3.969	6

Notes:
1. Constants fit γ - Fe_2O_3 · Tr. ~630°C; mp 1565°C.
2. Melting point 1369°C.
3. Constants fit high-quartz. Tr. ~577°C; mp 1723°C.
4. Constants estimated as sum of $CaO+Al_2O_3$ and $MgO+Al_2O_3$.
5. Melting point 1910°C.
6. No "reference" nongraphitic carbon exists.

C. Furnace Refractory Installation

Traditionally, firebrick was used to reline heat treatment furnaces. Today, refractory materials can be installed in a number of ways. Furnaces may be lined with brick or refractory blankets, or they may be lined with refractory modules as shown in Figure 127.

IX. FANS

A fan, blower, or compressor may be selected to move air or gas. They differ from each other in their operational pressure as shown in Table 50 [130].

This section presents an overview of the selection and operation of a fan that can be used in heat treatment equipment.

A. Calculation of Fan Performance (The Fan Laws)

The work done by a fan is related to the volume change of the gas being moved due to compression (see Figure 128) and is calculated from

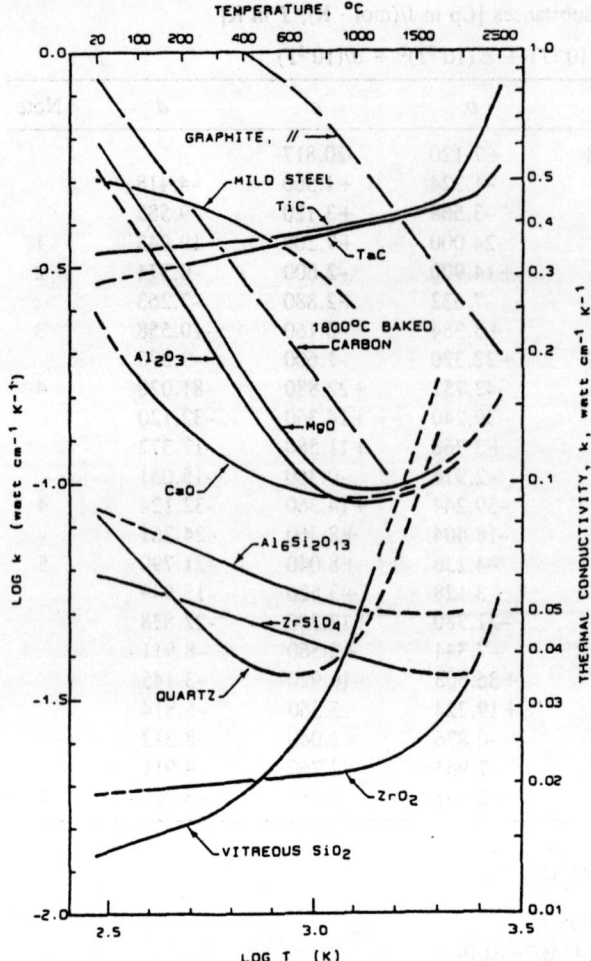

Figure 124 Thermal conductivity for various refractory materials.

$$\text{Work} = \Delta p \times V$$

where Δp is the pressure increase across the fan and V is the volume of gas moved. The pressure difference (Δp) is referred to as "head" and is usually quantified in inches of water.

Static pressure is the pressure on the walls, ducts, and piping. The velocity pressure required to move the gas is the difference between total pressure and the static head.

Air horsepower output (Ahp) is defined as

$$\text{Ahp} = \frac{VH}{6356}$$

where V is the volumetric flow through the fan (ft³/min) and H is the head (Δp) across the fan. The head may be static (H_s) or total (H_t):

$$H_t = H_s + H_s$$

Static air horsepower (Ahp$_s$) is the resistance that must be overcome and is related to static head. Total air horsepower (Ahp$_t$) or power output is based on total head.

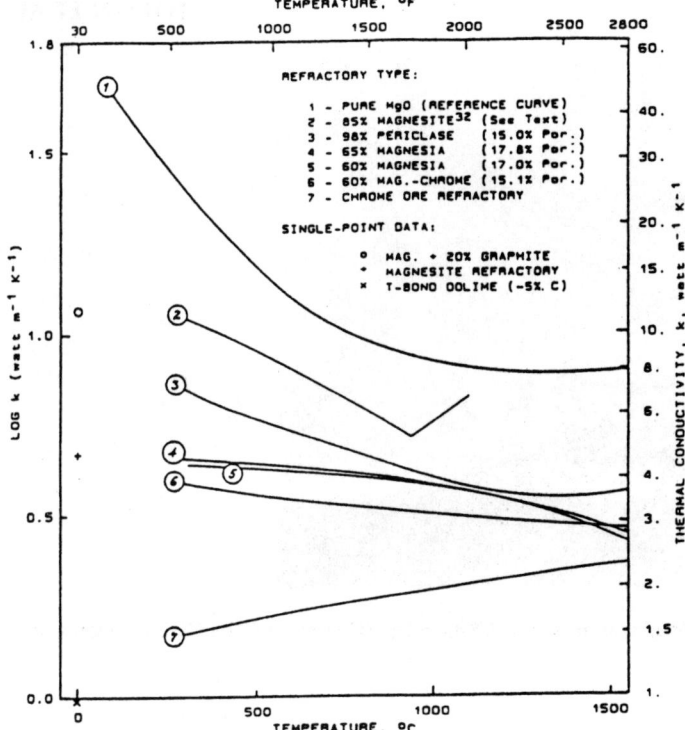

Figure 125 Thermal conductivity of firebrick.

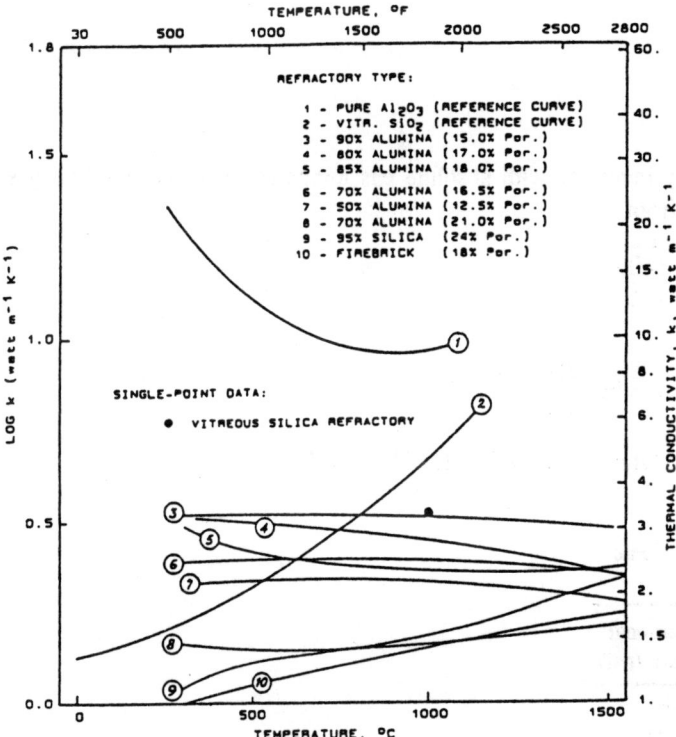

Figure 126 Thermal conductivity of alumina.

Figure 127 Car-bottom furnace lined with power-lock® modules. (Courtesy of The Carborundum Company, Fibers Division.)

The static and mechanical efficiency is calculated from

$$E_s = \frac{Ahp_s}{power\ input}$$

$$E_t = \frac{Ahp_t}{power\ input}$$

Head and horsepower vary inversely with absolute gas temperature T and absolute gas pressure P. This is calculated from

$$\text{Corrected head } (H_a) = H_b \frac{P_a T_b}{P_b T_a}$$

$$\text{Corrected horsepower } (hp_a) = hp_b \frac{P_a T_b}{P_b T_a}$$

where the subscript a indicates "after" and b indicates "before."

Table 50 Operational Pressures of Fans, Blowers, and Compressors

Machine	Operational pressure (psi)
Fan	<1
Blower	<50
Compressor	>35

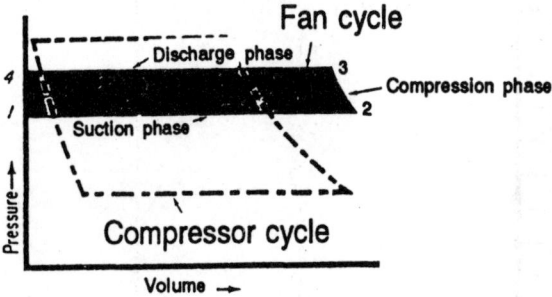

Figure 128   Illustration of a fan pressure cycle.

The interrelationships of head, efficiency, volume, and horsepower are shown in Figure 129. A system resistance curve relates fan pressure and horsepower to flow as shown in Figure 130. The fan supply equals the demand at the intersection point. To conserve energy, fan speed can be varied to match output pressure and system resistance for a given flow as shown in Figure 131. From these curves, it is evident that fan capacity Q, pressure P, and horsepower (hp) vary with speed and can be calculated as

$$\frac{Q_1}{Q_2} = \frac{\text{rpm}_1}{\text{rpm}_2}$$

$$\frac{P_1}{P_2} = \left(\frac{\text{rpm}_1}{\text{rpm}_2}\right)^2$$

$$\frac{\text{hp}_1}{\text{hp}_2} = \left(\frac{\text{rpm}_1}{\text{rpm}_2}\right)^3$$

Fan capacity and speed vary with the square root of pressure:

$$\frac{\text{rpm}_1}{\text{rpm}_2} = \frac{Q_1}{Q_2} = \left(\frac{P_1}{P_2}\right)^{1/2}$$

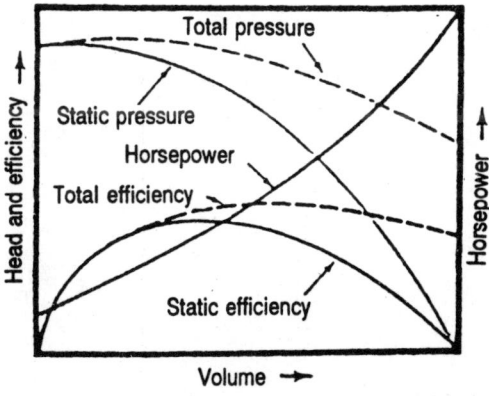

Figure 129   Fan performance curve. (From Ref. 130.)

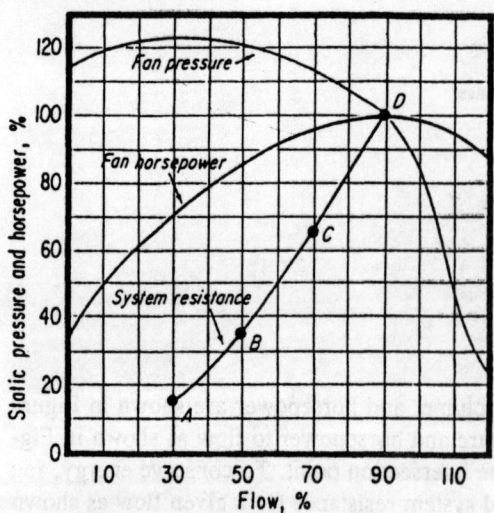

Figure 130 System resistance curve.

Horsepower varies with pressure:

$$\frac{hp_1}{hp_2} = \left(\frac{P_1}{P_2}\right)^{3/2}$$

At constant pressure and density, for geometrically similar fans the capacity power and speed are calculated from the wheel diameter (dia):

$$\frac{Q_1}{Q_2} = \frac{hp_1}{hp_2} = \left(\frac{dia_1}{dia_2}\right)^2$$

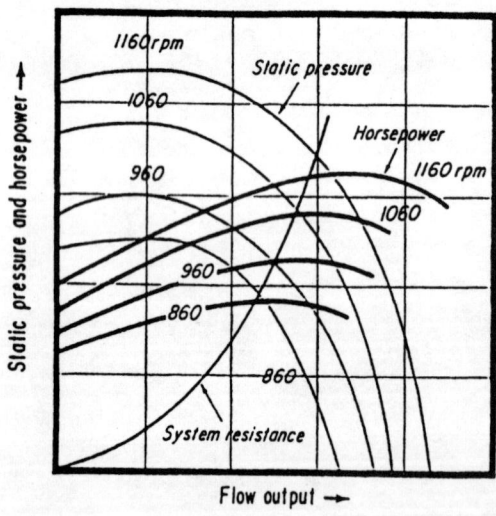

Figure 131 System resistance curve variation with fan speed.

$$\frac{rpm_1}{rpm_2} = \frac{dia_1}{dia_2}$$

If both the speed and diameter change, then

$$\frac{Q_1}{Q_2} = \frac{rpm_1}{rpm_2} \times \left(\frac{dia_1}{dia_2}\right)^3$$

$$\frac{P_1}{P_2} = \left(\frac{rpm_1}{rpm_2}\right)^2 \times \left(\frac{dia_1}{dia_2}\right)^2$$

$$\frac{hp_1}{hp_2} = \left(\frac{rpm_1}{rpm_2}\right)^3 \times \left(\frac{dia_1}{dia_2}\right)^5$$

$$\frac{hp_1}{hp_2} = \frac{Q_1}{Q_2} \times \frac{P_1}{P_2}$$

$$\frac{rpm_1}{rpm_2} = \frac{Q_1}{Q_2} = \frac{hp_1}{hp_2} = \left(\frac{d_2}{d_1}\right)^{1/2}$$

and

$$\frac{rpm_1}{rpm_2} = \frac{Q_1}{Q_2} = \frac{hp_1}{hp_2} = \left(\frac{b_2}{b_1}\right)^{1/2} = \left(\frac{T_1}{T_2}\right)^{1/2}$$

where d is the density of the gas being transferred, b is the barometric pressure, and T is the absolute temperature.

At constant speed and capacity,

$$\frac{hp_1}{hp_2} = \frac{P_1}{P_2} = \frac{d_1}{d_2} = \frac{b_1}{b_2} = \frac{T_2}{T_1}$$

Transferring a constant amount of gas results in

$$\frac{Q_1}{Q_2} = \frac{rpm_1}{rpm_2} = \frac{P_1}{P_2} = \frac{d_2}{d_1} = \frac{b_2}{b_1} = \frac{T_1}{T_2}$$

$$\frac{hp_1}{hp_2} = \left(\frac{d_2}{d_1}\right)^2$$

$$\frac{hp_1}{hp_2} = \left(\frac{b_2}{b_1}\right)^2 = \left(\frac{T_1}{T_2}\right)^2$$

If both the temperature and pressure vary, then

$$\frac{Q_1}{Q_2} = \frac{\text{rpm}_1}{\text{rpm}_2} = \left(\frac{P_1}{P_2} \times \frac{T_1}{T_2}\right)^{1/2}$$

$$\frac{\text{hp}_1}{\text{hp}_2} = \left(\frac{P_1^3}{P_2^3} \times \frac{T_1}{T_2}\right)^{1/2}$$

For geometrically similar fans, the flow, speed, and head are related by specific speed and specific diameter [130]. Specific speed (N_s) is that rpm at which a fan would operate if reduced proportionately in size so that it delivers 1 ft^3/min of air at standard conditions against a 1 in. H_2O static pressure (SP) [130].

$$N_s = \frac{\text{rpm (ft }^3\text{/min)}^{1/2}}{\text{SP}^{3/4}}$$

The specific diameter D_s is the fan diameter D required to deliver 1 ft^3/min standard air against 1 in. H_2O static pressure at a given speed.

$$D_s = \frac{D \text{ (SP)}^{1/4}}{(\text{ft}^3\text{/min})^{1/2}}$$

B. Fan Selection

The first step in selecting a fan is to determine the system requirements. First determine the amount of gas to be moved, making appropriate corrections for temperature, density, and barometric pressure. If multiple fans are to be used, divide the total volume by the number of fans.

The second step is to determine static pressure where the fan will operate. This is done by plotting specific speed versus specific diameter and static efficiency for the number of fans to be considered as shown in Figure 132 [130].

The flow (ft^3/min), static pressure (SP), and specific speed (N_s) can be calculated from the nomogram shown in Figure 133 from the N_s. The specific diameter (D_s) is determined by calculation, and the type of fan required is identified from a graph such as Figure 132.

When selecting fans, consideration must be given to whether they will be installed in series or in parallel. When installed in series, each fan handles the same amount of gas and the volumetric capacity varies with the density of the gas. When installed in parallel, each fan handles only part of the total volume of gas.

When determining the system resistance, one must account for flow losses due to duct resistance and diameter (Figure 134), shape (Figure 135), and frictional losses in elbows in the duct system (Figure 136) [130].

C. Flow Calibration

The flow rate of the fan system can be determined after initial installation or at any time by placing a pitot tube in a straight length of duct at approximately 10 diameter lengths from either the inlet or outlet side of the fan. Velocity pressure (H_v) is then determined at 20 different positions in the duct as shown in Figure 137. The square root of each velocity is determined and then averaged for the 20 readings.

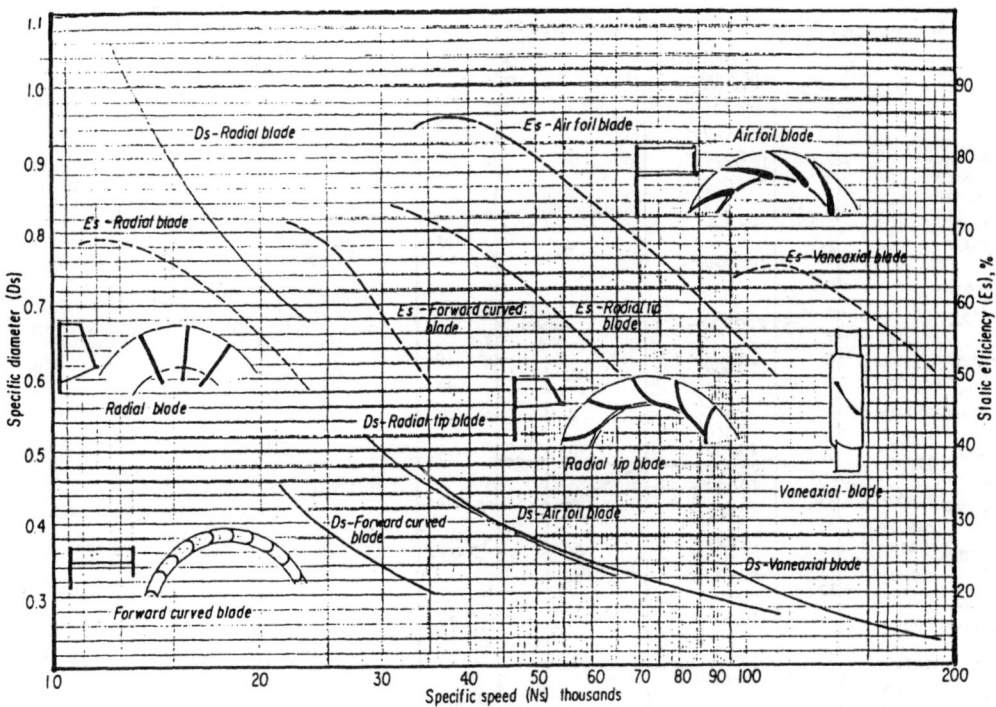

Figure 132 Static efficiency curves for various fan blade configurations. (From Ref. 130.)

The gas density (D) is determined by correcting for temperature and pressure;

$$D \ (\mathrm{lb/ft^3}) = 0.075 \left[\frac{530}{460 + \text{local temp. } (°F)} \right] \left[\frac{\text{barometer reading}}{29.92} \right]$$

The ratio of standard to actual gas (air) density (K) is determined as

$$K = \frac{0.075}{D}$$

The average air velocity V (ft/min) is calculated from

$$V = 4000 \left(\sqrt{H_v} \right)_{\text{ave}}$$

The volume flow rate is determined by multiplying the linear flow rate by the cross-sectional area of the duct.

X. FIXTURE MATERIALS

Heat treatment conditions such as corrosive and reactive atmospheres and high temperatures provide an especially harsh environment for furnace fixtures: baskets, trays, chains, conveyor

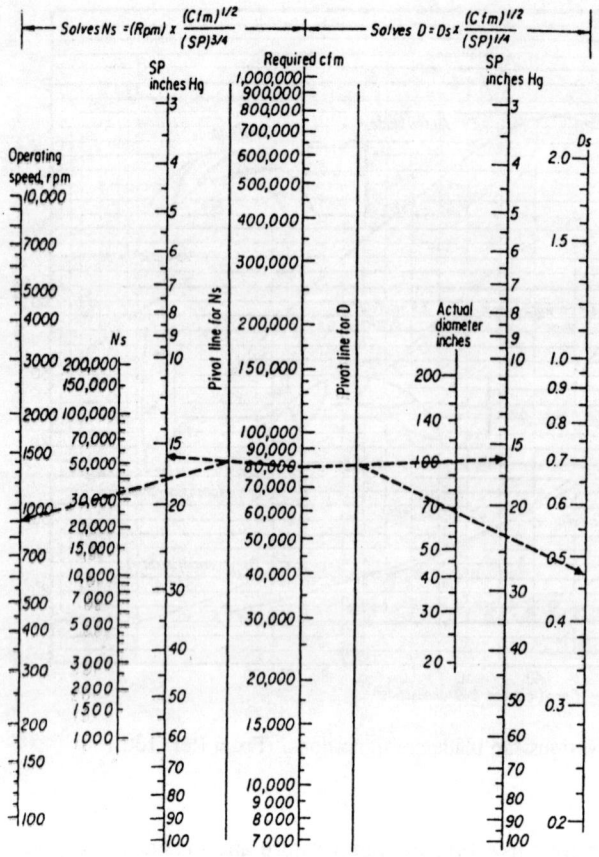

Figure 133   Nomogram for calculating fan performance parameters.

belts, and radiant tubes. This section provides an overview of typical heat-resistant alloys that are used for fixture construction.

A.   Common High-Temperature Alloys

The most common heat-resistant alloy materials used for furnace fixture construction are wrought and cast materials of Fe-Cr-Ni, Fe-Ni-Cr, and Ni-based alloys. Data on the composition of selected alloys are provided in Table 51.

A number of alloying elements are used for heat-resistant materials [132]:

Chromium (Cr) improves oxidation resistance at temperatures below 950°C, improves resistance to sulfidation and carburization, exhibits poor nitriding resistance, increases high-temperature resistance.

Silicon (Si) improves resistance to sulfidation, nitriding, oxidation, and carburizing. In synergy with Cr, Si improves scale resilience.

Molybdenum (Mo) improves high-temperature resistance and creep strength and reduces oxidation resistance.

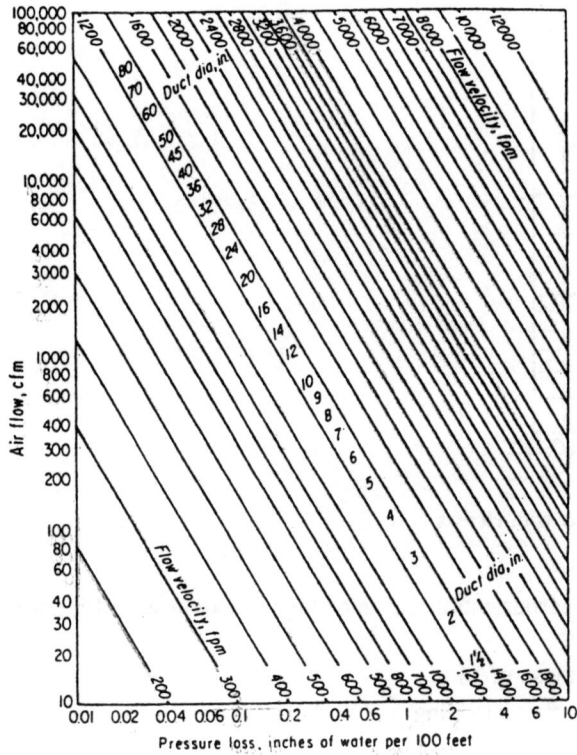

Figure 134 Fan pressure loss with varying flow rates and duct diameters.

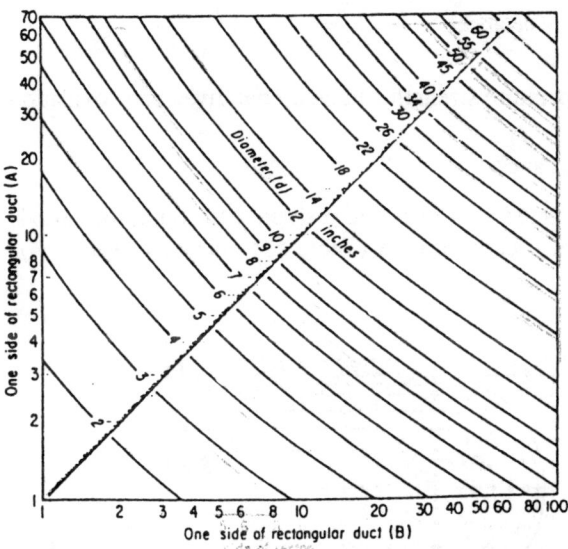

Figure 135 Equivalent round duct size conversion chart.

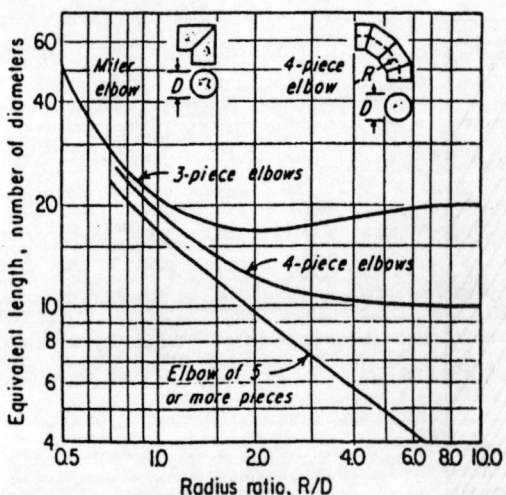

Figure 136 Equivalent length chart for ducts with elbows.

Nickel (Ni) improves resistance to carburization and nitriding and reduces sulfidation resistance.

Tungsten (W) has effects similar to those of Mo.

Carbon (C) improves strength, improves nitriding and carburizing resistance, reduces resistance to oxidation.

Yttrium (Y) and rhenium (Re) improve resistance to spalling, oxidation, carburizing, and sulfidation.

Aluminum (Al) acts independently of and synergistically with Cr to improve oxidation and sulfidation resistance and reduces nitriding resistance.

Titanium (Ti) imparts poor nitriding resistance.

Niobium (Nb) improves creep strength.

Manganese (Mn) improves high temperature strength and creep but gives poor oxidation and nitriding resistance.

Cobalt (Co) improves sulfidation resistance.

High-temperature alloys should be selected with regard to resistance to combustion products (see Figure 139), oxidation (Table 52), carburization (Table 53), nitriding (Table 54),

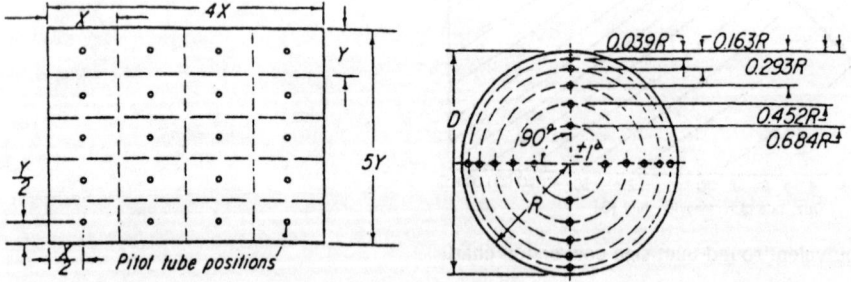

Figure 137 Recommended pitot placement positions for fan flow calibrations. (a) In rectangular duct; (b) in round duct, keep pitot tube within ±0.5% of R and ±1° of indicated positions.

Table 51 Nominal Compositions of Alloys Used in Laboratory and Field Tests[a]

Alloy (UNS No.)	Composition (%)										
	Ni	Fe	Co	Cr	Mo	W	Al	Ti	Si	C	Other
Cabot No. 214	75	2.5	2.0	16.0	0.5	0.5	4.5	—	—	—	Y (present)
Hastelloy X (N06002)	—	47	—	18.5	1.5	22.0	9.0	0.6	—	—	—
Alloy 600 (N06600)	72	8.0	1.0	15.5	—	—	0.35[b]	0.3[b]	—	—	—
Inconel 601 (N06601)	Bal	14.0	—	25	—	—	1.4	0.3	—	—	—
Cabot No. 625 (N06625)	Bal	5.0	1.5	9.0	—	0.4	0.4[b]	—	—	—	Cb + Ta = 3.5
Haynes No. 230	57	3.0	3.0	22.0	2.0	14.0	0.3	—	—	—	0.03 La
Hastelloy S	67	3.0	2.0	5.5	14.5	1.0	0.25	—	—	—	0.05 La
Waspaloy (N07001)	58	2.0	13.5	19.0	4.3	—	1.5	3.0	—	—	0.05 Zr
RA 333 (N06333)	45	18.0	3.0	25.0	3.0	3.0	—	—	—	—	1.25 Si
Haynes No. 188 (R30188)	22	3.0	39.0	22.0	—	14.0	—	—	—	—	0.07 La
Alloy 800H (N08810)	32.5	44	2.0[b]	21	—	—	0.4	0.4	—	—	—
Haynes No. 556	20.0	31	18.0	22	3.0	2.5	0.2	—	—	—	0.8 Ta, 0.02 La
Multimet (R30155)	20.0	30	20.0	21	3.0	25	—	—	—	—	Cb + Ta = 1.0
RA330 (N08330)	35.0	43	—	19	—	—	—	—	—	—	1.25 Si
Type 304 SS (S30400)	9.0	Bal	—	—	—	—	—	—	—	—	2.0 Mn, 1.09 Si
Type 310 SS (S31000)	20	—	—	25	—	—	—	—	—	—	2.0 Mn, 1.5 Si
Type 316 SS (S31600)	12	—	—	17	2.5	—	—	—	—	—	2.0 Mn, 1.0 Si
Type 446 SS (S44600)	—	—	—	25	—	—	—	—	—	—	1.5 Mn, 1.0 Si
RA 85 H (S30615)	14.5	—	—	18.5	—	—	1.0	—	3.5	0.2	—
HR 120	37	—	—	25	—	—	0.1	—	0.6	0.05	0.7 Cb 0.2 N
RA 309 (S30908)	13	—	—	23	—	—	—	—	0.8	0.05	—
RA 310 (S31008)	20	—	—	25	—	—	—	—	0.5	0.05	—
RA 446 (S44600)	—	—	—	25	—	—	—	—	0.5	0.05	0.7 Mn 0.1 N

(continued)

Table 51 Continued

Alloy (UNS No.)	Ni	Fe	Co	Cr	Mo	W	Al	Ti	Si	C	Other
253 MA (S30815)	11	—	—	21	—	—	—	—	1.7	0.08	0.04 Ce 17N
314 (S31400)	20	—	—	25	—	—	—	—	2.2	0.1	—
Alloy 800 (N08800)	31.8	Bal	—	21.4	—	—	0.35	—	0.35	0.04	0.79 Mn, 0.44 Ti
Alloy 520	35.0	Bal	—	21.0	—	—	NA	—	2.0	—	1 Cb
Alloy DS	34.3	Bal	—	18.0	—	—	<0.1	—	2.20	0.03	1.3 Mn, <0.05 Ti
253	11	Bal	—	21	—	—	—	—	1.7	—	N, Ce
DS	36	Bal	—	18	—	—	—	—	2.2	0.06	—
45 TM	Bal	23	—	27	—	—	—	—	2.7	0.08	—
602 CA	Bal	9.5	—	25	—	—	2.1	—	—	0.18	Y, Zr, Ti
X	Bal	18	—	22	—	—	—	—	—	0.10	9 Mo, W, Co
625	Bal	3	—	22	—	—	—	—	—	0.10	9 Mo, 3.5 Cb
617	Bal	1.5	—	22	—	—	1.2	—	—	0.06	9 Mo, 12 Co

SS = stainless steel.

[a]Cabot, Hastelloy, Haynes, and Multimel are registered trademarks of Cabot Corp; Waspaloy is a trademark of United Technologies. RA is a registered trademark of Rolled Alloys. Inconel is a registered trademark of the INCO family of companies.

[b]Maximum.

Source: Refs. 131 and 134.

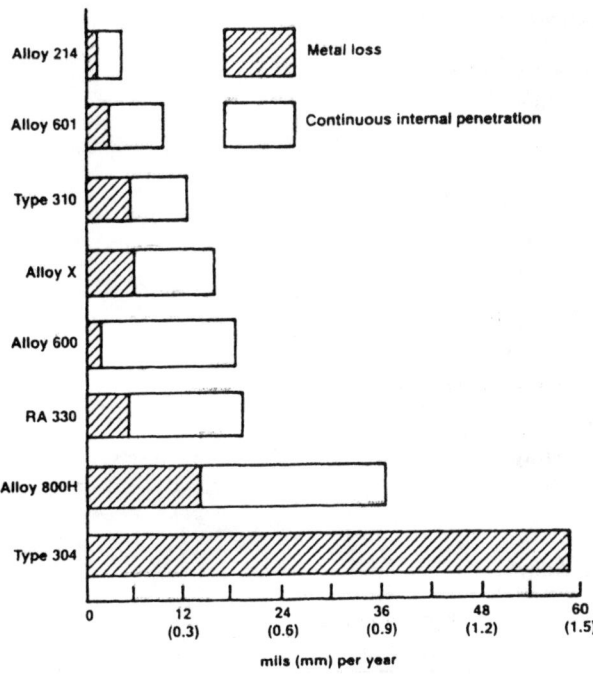

Figure 138   Alloy attack by combustion by-products.

molten salts (Table 55), and other severe environments encountered in the particular heat treatment of interest [131].

XI. PARTS WASHING

A. Washing Processes

Many parts enter the heat treatment shop with residual lubricants, coolants, or corrosion inhibitor films on the surface. These surface contaminants may contain graphite, molybdenum,

Table 52   Static Oxidation Behavior of Various Alloys—Metal Loss and Continuous Internal Penetration, mils (μm) per side[a]

Alloy	1800°F (980°C)		2000°F (1095°C)		2200°F (1205°C)
Alloy 214	0.2	(5)	0.1	(3)	0.7 (18)
Alloy 600	0.9	(23)	1.6	(41)	8.4 (213)
Alloy X	0.9	(23)	2.7	(69)	(672 h)[b]
Alloy 556	1.1	(28)	2.6	(66)	(168 h)[b]
Type 310	1.1	(28)	2.3	(58)	10.3 (262)
Alloy 800H	1.8	(46)	7.4	(188)	13.6 (345)
Type 446	2.3	(58)	14.5	(368)	(1000 h)[b]
RA 330	4.3	(109)	6.7	(170)	8.3 (211)
Type 304	8.1	(206)	>23.0	(584)	(336 h)[b]
Type 316	14.3	(363)	>69.0	(1753)	(168 h)[b]

[a]Exposure at 1008 h in air. Cooled to room temperature weekly.
[b]Alloy element consumed in test.

Table 53 Carburization Resistance of Various Alloys[a]

Alloy	Carbon absorption (g/cm^2)
Alloy 214	0.6
RA 333	1.0
Alloy 800H	1.0
Multimelt	1.3
Alloy 556	1.3
Alloy X	2.5
Alloy 600	2.8
Alloy 625	5.3
Type 310 stainless steel	Heavy localized attack

[a]Carburization conditions: In Ar, 5% H_2, and 5% CH_4 at 1800°F (980°C) for 55 h. $P_{O_2} = 9 \times 10^{-22}$ atm, carbon activity 1.0.

Table 54 Nitriding Resistance of Selected Alloy[a]

Alloy	Nitrogen absorption (mg/cm^2)	Depth of nitrided layer (μm)
Alloy 230	0.7	30
Alloy 600	0.8	33
Alloy 188	1.2	15
Alloy 214	1.5	38
Alloy X	1.7	38
RA 330	3.9	97
Alloy 800H	4.3	104
Alloy 556	4.9	89
Type 310	7.4	152
Type 304	9.8	213

[a]Tested in ammonia at 1200°F for 1 week.

Table 55 Resistance to Molten Chloride Salts at 1550°F (845°C)

Alloy	Metal loss and continuous internal penetration (mm/side)
Alloy 188	0.69
Multimet alloy	0.76
Alloy X	0.97
Alloy s	1.0
Alloy 556	1.1
Waspaloy alloy	1.7
Alloy 214	1.8
Type 304	1.9
Type 310	2.0
Alloy 600	2.4

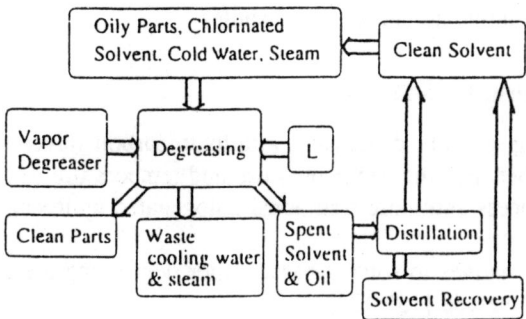

Figure 139 Anderson diagram for vapor degreasing (From Ref. 136.)

sulfide (MoS_2), or compounds of chlorine, sulfur, silicon, phosphorus, or boron. Their presence on the surface of parts to be heat treated may prevent uniform atmosphere diffusion and/or heat transfer (on heating or cooling) and lead to "soft spotting" or increased distortion prior to the heat treatment operation [135]. Degreasing may be performed before or after the heat treatment process, or both. A schematic illustrating the process is shown in Figure 139.

Until recently, one of the most common parts cleaning operations was vapor degreasing. The solvents that have been used for vapor degreasing include trichloroethane, carbon tetrachoride, trichloroethylene, perchloroethylene, and tetrachloroethane (see Figure 140) [135,137]. Cleaning with such solvents is expensive, in many cases toxic; they are often "pollutants"; and their use is accompanied by a high disposal cost.

An alternative to vapor degreasing using a chlorohydrocarbon is to use a hydrocarbon solvent. Hydrocarbon solvents for these processes typically exhibit flash points of $>70°C$. Benefits of hydrocarbon-based parts cleaning processes include the following [138].

1. They have no detrimental impact on the ozone layer.
2. They provide high solvent efficiency.
3. Petroleum solvents are noncorrosive.

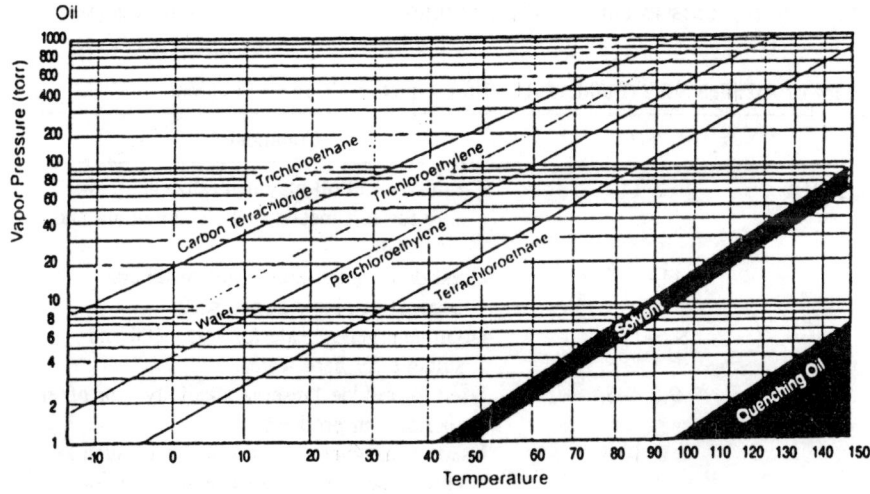

Figure 140 Vapor pressure–temperature relationship of various cleaning solvents.

4. Hydrocarbon solvents are relatively nontoxic and easy to handle.
5. The solvents are reusable.
6. There is no need for drainage or exhaust gas facilities.

Currently, systems employ a combination of vapor degreasing with hydrocarbon solvents and vacuum drying [138]. The relationship between the vapor pressure and temperature of a typical hydrocarbon solvent used for this process and more traditional halogenated hydrocarbons is shown in Figure 140.

Currently, the most commonly used alternatives to solvent-based vapor degreasing are a water-based washing system. These systems include [135,137]

Emulsion cleaners.   An organic solvent is emulsified in water that contains various additives such as surfactants to aid in the surface soil removal process.

Semiaqueous cleaners.   Parts may be cleaned using solvents such as terpenes, esters, or a blend of hydrocarbons. Residual solvents are removed as an emulsion by subsequent cleaning with water and/or surfactant.

Alkaline cleaners.   This process uses alkaline metal salts (also known as detergent builders) (see Table 56) that include a mixture of silicates, phosphates, and surfactant. These cleaning solutions are suitable for automated systems and heavily soiled parts.

Neutral cleaners.   These cleaning solutions are composed of water and are usually nontoxic surfactants. They are suitable for lightly soiled surfaces and are considerably less alkaline (neutral) than the alkaline cleaners described above.

A schematic of a water-based cleaning process is presented in Figure 141.

The mechanistic cleaning process of alkaline and neutral cleaners is illustrated in Figure 142. The aqueous surfactant solution penetrates the oil film on the metal surface and solubilizes it by a micellation process. The micellized oil is then removed to the bulk of the aqueous cleaning fluid. The detergent cleaning action is affected by time, temperature, concentration, contamination, and additional additives, including surfactants and defoamants.

B. Equipment

Most parts washers used in heat treatment operations are classified as either "dunk" or "spray-and-dunk" washers. Both processes enhance the mixing of the cleaning fluid with the part and

Table 56 Summary of Alkaline Salts (Builders) Used for Aqueous Cleaning Solutions

Salt class	pH	Comment
Caustics (NaOH, KOH)	2–14	Clean fats and oil may neutralize acid. *Not safe* for soft metals such as aluminum and zinc.
Silicates	11–12.5	Good detergency; attack soft metal and become insoluble at pH < 10.
Phosphates	9.5–11.5	Good detergency; safe for soft metal and suitable for hard water.
Carbonates	9.0–9.5	Used to provide "reserve alkalinity" for other builders in products.
Borates	8.0–10.5	Used when other more alkaline salts such as phosphates and silicates cannot be used. Provides some rust protection.

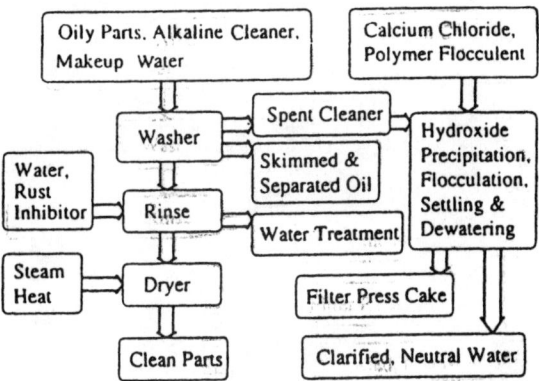

Figure 141 Schematic of aqueous alkaline washing process. (From Ref. 136.)

facilitate subsequent removal of the micellized soil from the part surface. Examples of these processes are shown in Figure 143. A photo of a commercial washer is shown in Figure 144.

The washing process should [137,139]

Use hot water, 160°F minimum.

Continually remove residual oil throughout the cycle.

Employ a drying cycle to eliminate the possibility of water contamination of the furnace.

Use the correct cleaner for the job and use it correctly.

XII. QUENCH SYSTEM DESIGN

Quenching is one of the most critical processes in the overall heat treatment operation. It is essential to achieve the necessary heat transfer rates and optimal uniformity of the heat transfer process if the desired steel transformation products are to be obtained and if thermal gradients are to be minimized [140–142]. Optimizing the overall quenching process is interdependent with quench system design, especially the direction, turbulence, and velocity of quenchant flow [142]. This is true for all quench systems, whether gas or liquid, including high pressure gas quenching and molten salt, fluidized-bed, brine, oil, and aqueous polymer quenchants.

In this section, an introduction is provided to quench system design criteria with particular focus on vaporizable liquids. The design criteria include (1) tank sizing, (2) heat exchanger selection, (3) agitator selection, (4) chute design, (5) flood quench systems, and (6) filtration.

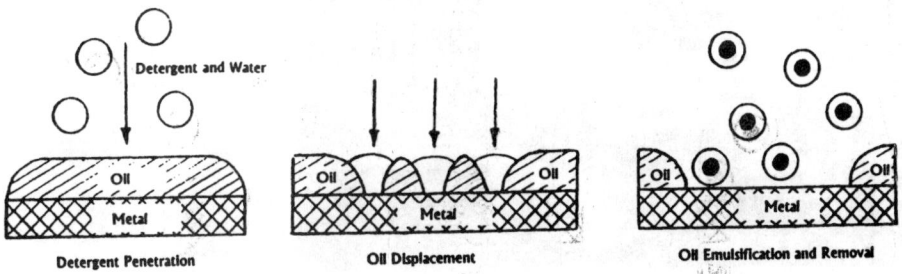

Figure 142 Detergency mechanism for alkaline washing process. (From Ref. 37.)

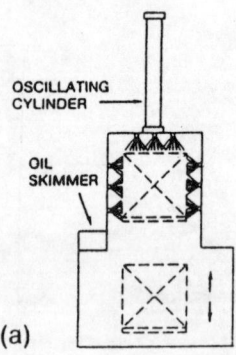

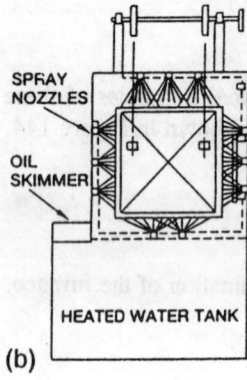

Figure 143 Schematics of (a) spray-and-dunk washer; (b) spray washer. (Courtesy of BeaverMatic, Inc.)

Figure 144 A spray washer. (Courtesy of Abor-Ispen Industries.)

A. Quench Tank Sizing

The sizing of the quench system depends on the thermal properties of the metal (Figure 145) and on those of the quenchant. Heat (q) transfer rates depend on the heat transfer coefficient of the quenchant film surrounding the heated metal (h), surface area (A), and the difference between the temperature of the metal before quenching (T_i) and the bath temperature (T_b).

$$q = hA\,(T_i - T_b)$$

The amount of heat transferred from the hot metal to the cooler quenchant can be estimated as [143]

$$Q = C_{pm}\,W_m\,(T_i - T_b)$$

where C_{pm} = specific heat of the metal, W_m = weight of the metal, including fixtures such as trays, and T_i and T_b are the initial and final temperatures, respectively, of the metal.

The temperature rise (T_r) of the quenchant can be estimated from [143]

$$T_r = (T_i - T_b) = QW_q C_{pq}$$

where C_{pq} = heat capacity of the quenchant, W_q = weight of the quenchant, and T_i and T_b are as defined above.

Clearly the temperature increase of the quenchant is dependent on the initial temperature of the metal; the volume, temperature, and heat capacity of the quenchant; and the heat capacity and size of the load being quenched.

Two general rules have been traditionally and successfully applied to oil and polymer quenchant systems [144,145]:

1. The size of the load (including fixtures) being quenched should not exceed 1 lb/gal (0.11 kg/L).
2. The temperature rise of the quenchant should not exceed 10°F (5°C).

It is recommended that mild steel be used for quench tank construction. Cast iron may be used for pumps and agitators [146]. Soft metals such as zinc, lead, magnesium, and

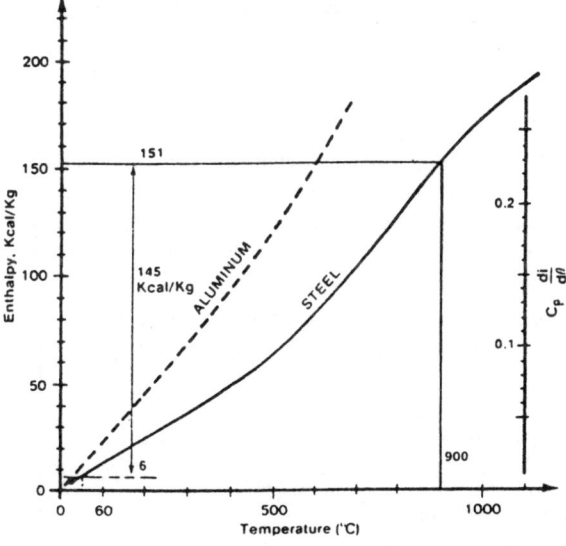

Figure 145 Heat capacity curves for aluminum and steel.

galvanized steel should not be used for aqueous polymer quenchants that have a basic pH (pH > 7.0). Copper oil and brass should not be used for mineral oil systems because they promote oxidation [146].

B.   Heat Exchanger Selection

There are four primary types of heat exchangers used in quenching systems: mechanical refrigeration, shell-and-tube (or plate-and-frame), evaporative water-cooled spray towers, and air-cooled radiators [140]. For small systems, submerged water cooling pipes and cooling jackets around the quench tank may also be used [146].

Selection of the proper heat exchanger depends on the final temperature of the fluid. The selection criteria are summarized Table 57. Examples of air-cooled and evaporative heat exchangers, two of the more commonly used heat exchangers in larger heat treatment processes, are shown schematically in Figures 146 and 147.

Shell-and-tube heat exchangers typically contain copper cooling surfaces and are subject to fouling and plugging, particularly when mineral oil quenchants are used. Plate-and-frame heat exchangers have stainless steel plates on the heat-exchange surface. Water contamination problems may occur with both of these types of heat exchangers. These can be disastrous for oil quenchants, ultimately leading to fires. This problem can be avoided by pressurizing the quench oil side to prevent water from flowing into the oil. To avoid potential problems in the event of pressurizing pump failure, it is recommended that a pressure-sensing device (140–280 kPa) be installed [140].

Centrifugal pumps are generally recommended. The appropriate pipe size is dependent on the required flow rate as shown in Table 58. Correct pipe size is essential to ensure the necessary turbulent flow and to reduce fouling potential. Maximum efficiency is obtained by removing the heated quenchant from the top of the tank and returing the cooled fluid to the bottom of the tank [146].

C.   Agitator Selection

There are numerous possible design alternatives to provide the necessary agitation to optimize the heat transfer rates and the uniformity of the quenching process. They may be used either individually or in combination with each other. Some of the more common agitator design options are briefly reviewed here.

1.   Sparging

Bubbling air or inert gases into the quenchant, a process known as sparging, may be used either as the sole source of agitation [147,148] or to supplement other forms of agitation such as pump agitation [149]. Air sparging can be readily used for water and brine. It is a relatively poor choice for other vaporizable quenchants such as oil because the increased presence of air will facilitate oxidative degradation. Although an inert gas such as nitrogen should be used, it is significantly more expensive than air.

Agitation rates provided by gas sparging are dependent on the gas pressure. The pressure required to provide the desired amount of agitation depends on the head of the fluid,

Table 57   Heat Exchanger Selection Guidelines

Final fluid temperature	Heat exchanger type
<24°C (<75°F)	Mechanical refrigeration
35–45°C (95–110°F)	Evaporation or water-cooled
>45°C (>110°F)	Air-cooled

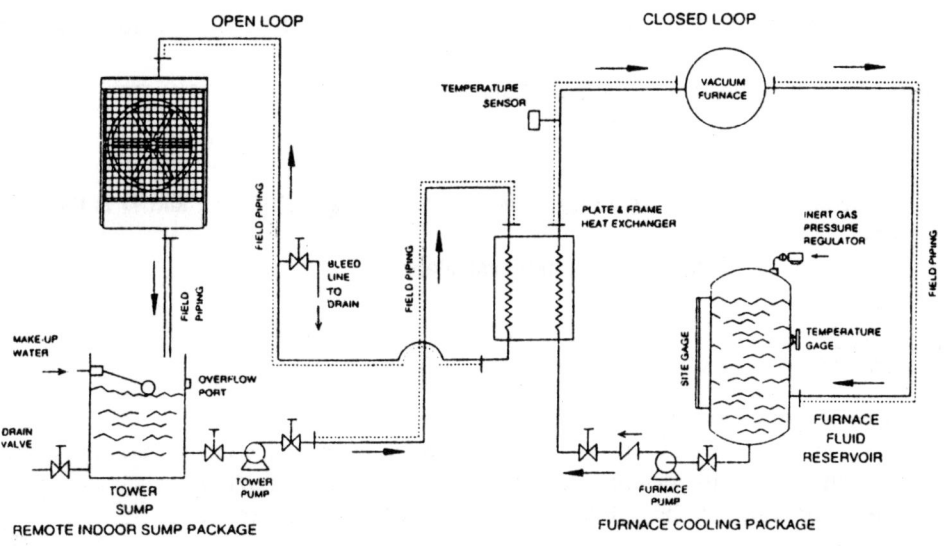

Figure 146 Example of an evaporative heat exchanger. (Courtesy of Industrial Heat Recovery Equipment.)

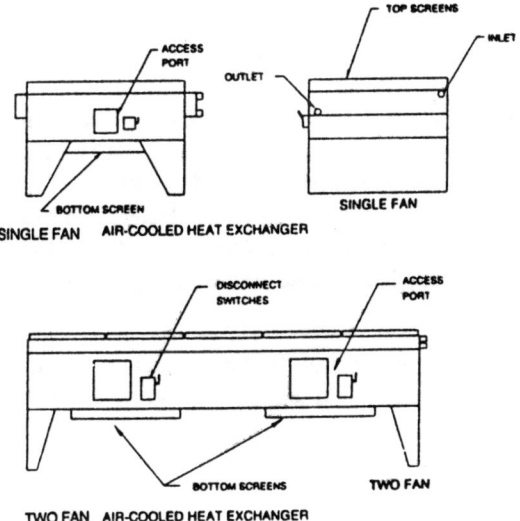

SINGLE FAN   AIR-COOLED HEAT EXCHANGER

TWO FAN   AIR-COOLED HEAT EXCHANGER

Figure 147 Example of an air-cooled heat exchanger. (Courtesy of Industrial Heat Recovery Equipment.)

Table 58 Heat Exchanger Pipe Sizing Recommendations

Flow rate		Pipe size	
gal/min	L/min	in.	mm
50–90	190–340	2	50
90–180	340–680	2.5	63
180–250	680–950	3	75

frictional losses in the delivery pipe, and the pressure difference required to force the gas through the pipe orifice [148].

Air flow rate is a function of the pipe diameter:

$$Q = 0.327VD^2$$

where Q is the air flow volume (ft^3/min), V is the linear velocity (ft/min), and D is the pipe diameter (ft).

Pressure losses due to friction (P) can be calculated as

$$P = \frac{Q^2L}{2690D^5}$$

where L is the length of the pipe (ft).

To determine pressure loss due to friction in flow through a pipe elbow, the equivalent length of straight pipe is first determined by using a table such as Table 59.

The rate of air flow (Q) through a pipe orifice of known diameter (in.) is

$$Q = 21.7 \, D^2C\sqrt{H}$$

where H = head pressure (in. H$_2$O) and C is an orifice-dependent constant selected from Table 60.

2. Centrifugal Pumps

Centrifugal pump systems can be used to provide quenchant bath agitation and to transfer the fluid to and from the heat exchanger. However, if centrifugal pumps are used, particular attention must be given to the position of the sparge manifolds around the part, or load, being quenched. This is especially significant because the flow velocity drops rapidly at distances greater than 10 diameters from the discharge point, as shown in Figure 148 [140].

A detailed description of centrifugal pump selection, use, and system design is beyond the scope of this text. However, excellent information is provided in References 152–154 and other texts. Reference 153, which is particularly good, is available free of charge.

3. Impeller Agitation

Perhaps the most commonly used source of agitation, especially for larger quench tanks, is the impeller. One of the earliest references to the selection and use of impeller agitation for

Table 59 Equivalent Straight Pipe Length for Elbows

Pipe elbow radius		Equiv. length of straight pipe	
in.	mm	in.	mm.
0.5	12.7	121.0	3070
0.75	19.0	35.0	889
1.0	25.4	17.5	445
1.25	31.8	12.7	323
1.5	38.1	10.3	262
2.0	50.8	9.0	229
3.0	76.2	8.4	213
5.0	127.0	7.8	198

Source: Ref. 147.

Table 60   Gas Injection Orifice Constant[a]

Orifice type	Constant C
Conoidal mouthpiece (contracted vein)	0.97–0.99
Conoidal (converging mouthpiece)	0.90–0.99
Short (cylindrical mouthpiece rounded at inner end)	0.92–0.93
Short (cylindrical mouthpiece)	0.81–0.84
Thin circular plate	0.56–0.59

[a]For equation $Q = 21.7D^2C \sqrt{H}$; see text.

quench systems was a now-classic paper published by U. S. Steel [150]. The newer impeller mixer technology is discussed in References 151 and 155.

Many quench tanks are agitated using one or more open-impeller mixers or an impeller encased in a tube for directing the flow. If a draft tube is not used, then flow must be directed by the impeller itself or by flow-directing baffles in the quenching region of the tank.

A common impeller used for open systems is an axial flow impeller such as the marine impeller shown in Figure 149b. An axial flow impeller directs flow parallel to the impeller shaft. Axial flow impellers may be top-entering, side-entering, or angled top-entering as shown diagramatically in Figure 150.

Three-blade marine impeller mixers have traditionally been used for quench tank agitation. The mixer power requirements for a marine impeller with a 1.0 pitch ratio operating at 420 rpm are summarized in Table 61 [150].

The impeller currently most often recommended for angled top entry or vertical top entry is the airfoil type illustrated in Figure 150a. A comparison of the power requirements for a marine and an airflow-type impeller is given in Table 62 [151]. The recommended power requirements for an airfoil impeller in both open and draft tube configurations are provided in Table 63.

The required power P is proportional to speed N and can be calculated for other output speeds from

$$P \propto N^{4/3}$$

The impeller blades used for a side-entry mixer (Figure 151c) have a different shape than those used for top-entry mixers (Figure 151b) to balance the mechanical forces acting on the impeller in the side-entry mode.

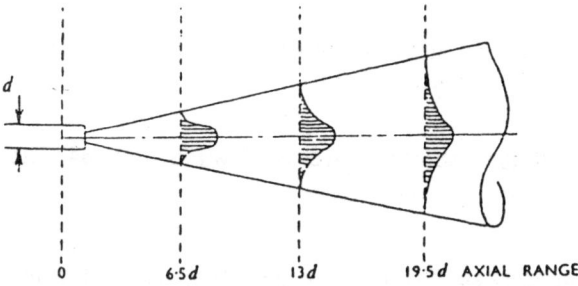

Figure 148   Flow pattern as a function of nozzle diameter and distance from nozzle for a jet mixer. (From Ref. 140.)

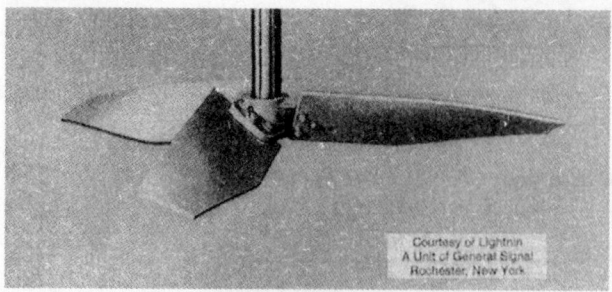

(a)

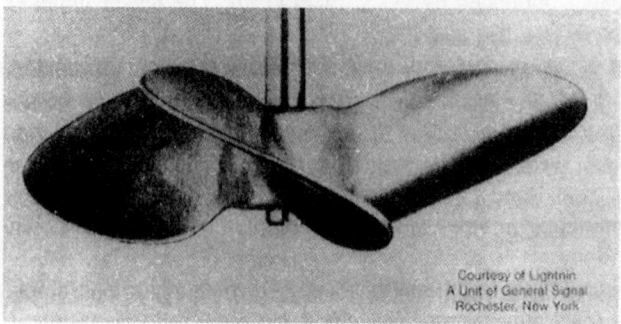

(b)

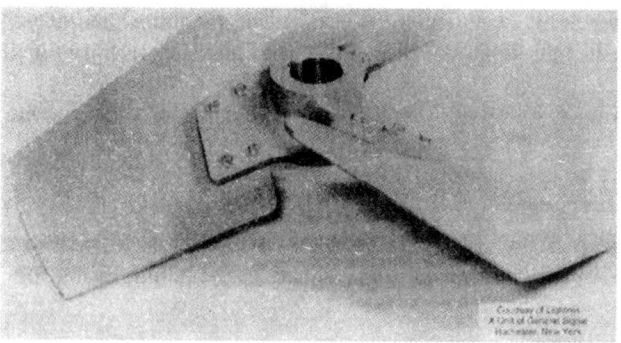

(c)

Figure 149 Examples of impellers. (a) High-efficiency airfoil-type impeller; (b) marine propeller; (c) side-entry impeller. (Courtesy of Lightnin—A Unit of General Signal.)

The sizing of the impeller diameter with the respect to power is shown in Table 64. The required power is taken from Table 63.

4. Draft Tubes

A draft tube agitator is used to provide directed air flow in many quenching system designs. A schematic of a draft tube agitator is provided in Figure 151.

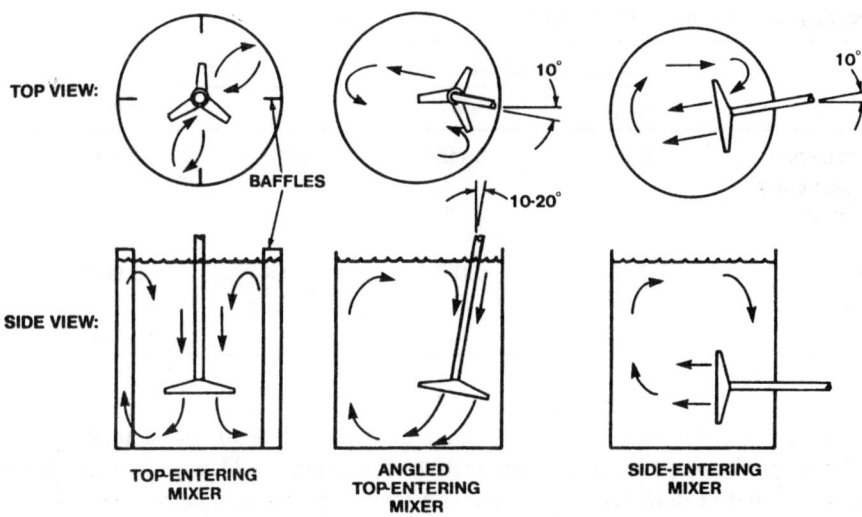

Figure 150 Examples of flow patterns and impeller arrangements for top-entering, angled top-entering, and side-entering mixers.

Table 61 Power Requirements for Propeller Agitation[a]

Tank volume		Power required			
		Standard quench oil		Water or brine	
Gal.	Liters	hp/gal	kW/L	hp/gal	kW/L
50–800	2000–3200	0.005	0.001	0.004	0.0008
800–2000	3200–8000	0.006	0.0012	0.004	0.0008
2000-3000	8000-12,000	0.006	0.0012	0.005	0.001
>3000	>12,000	0.007	0.0014	0.005	0.001

[a]Agitation at 420 rpm. Marine propeller with 1.0 pitch ratio.
Source: Ref. 150.

Table 62 Equivalent Quench Tank Mixer Size[a]

Impeller	rpm[b]	Standard quench oil		Water or brine	
		hp/gal	kW/L	hp/gal	kW/L
Marine	480	0.007	0.0014	0.005	0.001
Marine	280	0.004	0.0008	0.003	0.0006
Air foil	280	0.003	0.0006	0.002	0.0004

[a]Based on a 3000 gal tank with an open-impeller mixer providing "violent" circulation. The recommended power requirement for an airfoil impeller used in a draft tube application is 0.006 hp/gal for a quench oil and 0.0045 hp/gal for water or brine.
[b]The power levels for other output speeds are adjusted using $P \propto N^{4/3}$, where P = required power level and N = speed.

Table 63 Recommended Quench Tank Mixer Sizes

Mixer type	Standard quench oil		Water or brine	
	hp/gal	kW/L	hp/gal	kW/L
Open-impeller mixers— top-entry or side-entry airfoil impeller at 280 rpm	0.004	0.0008	0.003	0.0006
Draft-tube mixer— airfoil impeller at 280 rpm	0.006	0.0012	0.0045	0.0009

Marine impellers are normally used for smaller draft-tube systems (<24 in.). Airfoil impellers are better performing, lower cost alternatives for larger (>24 in.) systems. When airfoil impellers are used in draft tube applications, a lower tip-chord angle is used than for side-entering mixers to avoid stalling under conditions of high head resistance.

Table 64 Size of Impeller Mixers

Motor[a,b]		Impeller size[c,d]	
hp	kW	in.	cm.
0.25	0.19	13	33.0
0.33	0.25	14	35.6
0.50	0.37	15	38.1
0.75	0.56	16	40.6
1.0	0.75	17	43.2
2.0	1.49	20	50.8
3.0	2.34	22	55.9
5.0	3.73	24	61.0
7.5	5.59	26	66.0
10.0	7.46	28	71.1
15.0	11.19	30	76.2
20.0	14.92	32	81.3
25	18.65	33	83.8

[a]The power requirements were calculated assuming 280 rpm, specific gravity 1.0, and airfoil impeller with Np 0.33. (Airfoil and marine propeller power numbers are nearly identical.)
[b]The shaft horsepower (hp$_s$ is equal to 80% of the motor horsepower (hp$_m$) (0.8 × hp$_m$ – hp$_s$).
[c]These are the power requirements for an open impeller operating at 280 rpm.
[d]When used in a draft tube, the impeller size should be reduced by 3%. Axial flow impellers are used in draft tubes to more closely control the direction of the flow pattern. Draft tube circulators have a higher resistance head that the impeller must pump against, which is due to the fluid friction losses in the draft tube pipe. Velocity head losses also occur at the entrance, exit, and at any angles in the draft tube. The higher head conditions require a sligtly different impeller for optimum pumping performance. Impellers in both open and draft tube configurations are provided in Table 6.58.

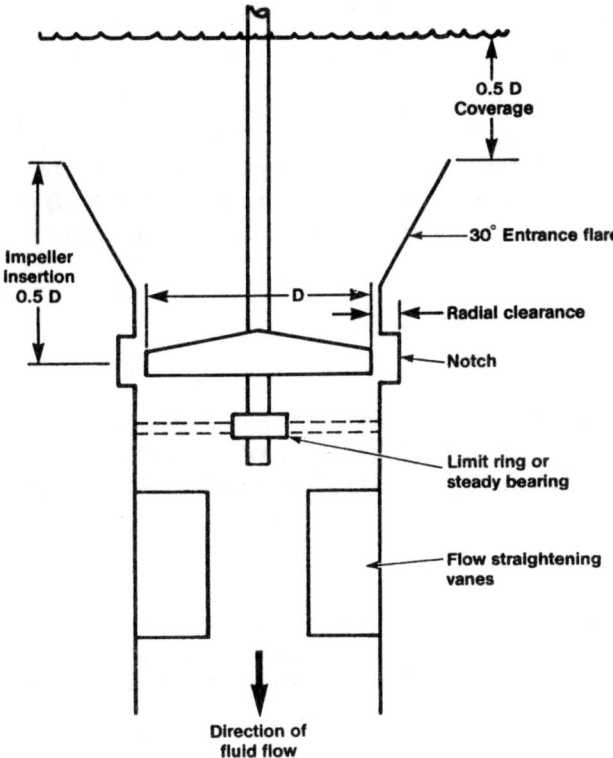

0.5 D Coverage

30° Entrance flare

Impeller Insertion 0.5 D

D

Radial clearance

Notch

Limit ring or steady bearing

Flow straightening vanes

Direction of fluid flow

Figure 151 Draft tube impeller design.

A draft tube should have the following characteristics [151,155]:

1. A down-pumping operation is used to take advantage of the tank bottom as a flow-directing device.

2. A 30° entrance flow on the draft tube minimizes the entrance head losses and ensures a uniform velocity profile at the inlet.

3. Liquid depth over the draft tube should be at least one-half of the tube diameter to avoid flow loss due to disruption of the impeller inlet velocity profile.

4. Internal flow straightening vanes are used to prevent fluid swirl.

5. The impeller should be inserted into the draft tube a distance equal to at least one-half of the tube diameter.

6. A steady bearing or limit ring is used to protect the impeller from occasional high deflection. A steady bearing is the lower cost alternative but requires maintenance.

7. The impeller requires 1–2 in. (25–50 mm) of radial clearance between the blade tips and the draft tube. When the draft tube must be minimized, an external notch can be used to reduce the draft tube dimensions by 2–3 in. (50–75 mm).

5. *Multiple Mixers*

Ensuring uniform heat transfer throughout the quench zone often requires multiple mixers. Although there are no simple quantitative prediction methods to determine the necessity of using multiple mixers, there are two general guidelines:

1. If the length/width ratio is 2, a single properly design mixer is usually sufficient.
2. If the length/width ratio is greater than 2, then multiple mixers are recommended.

Possible draft tube and side-entry mixer arrangements for rectangular tanks are shown in Figure 152. Top-entry mixers may be arranged differently.

To size a mixer used in a multiple arrangement, it is first necessary to determine the total power requirement for the tank using Table 63. The impeller sizing for each mixer is determined from Table 64, and the power per mixer is determined from the relationship

$$\text{Power per mixer} = \frac{\text{total power}}{\text{number of mixers}}$$

6. Cavitation

Cavitation occurs when the pumping action of the impeller creates localized zones of low pressure, pressure below the vapor pressure of the fluid. Vaporization can lead to erosion of the impeller and unstable operating performance. This should not occur with properly designed systems.

D. Computational Fluid Dynamics

Computational fluid dynamics (CFD) may provide greater understanding of the details of fluid flow in the quench system and greater insights into how to determine the optimium location and orientation of the part in the quench zone [156,157]. It can also assist in troubleshooting fluid flow problems that can produce distortion or cracking [158].

An example of the use of CFD to model flow in a simple quench tank configuration using a draft tube [155] is shown in Figure 153. The CFD model illustration shows that there

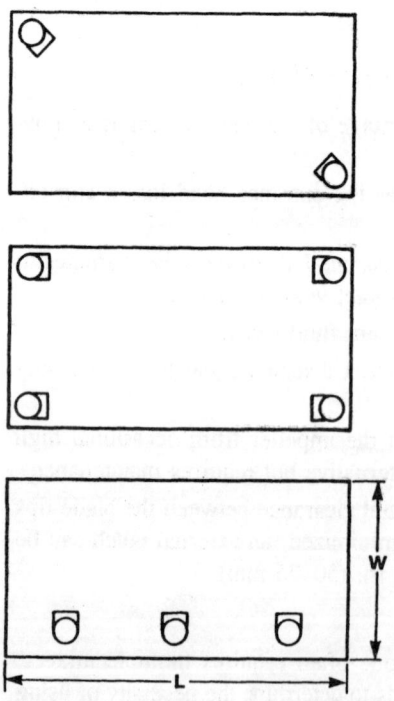

Figure 152   Examples of draft tube mixer placement for tanks with $L/W > 2$.

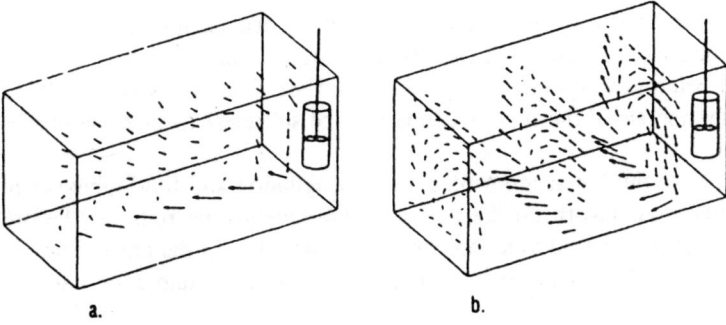

a.   b.

Figure 153 Example of the utility of CFD analysis to model agitation uniformity in quench tanks. (a) Vectors in xz planes; (b) vectors in yz planes.

is a general flow pattern that sweeps across the bottom of the tank from left to right with a rotational element around the axis of the general flow pattern. The flow is relatively quiet in the corners of the tank except where the draft tube is located.

Similar studies were performed by Garwood et al. [156] and Wallis et al. [157]. Their work also showed that even with considerable surface flow, there were substantial velocity gradients within the tank around the mixers positioned at all four corners of the tank.

These results show that CFD calculations have tremendous potential in the study of existing quench tank flow problems and in assisting in the design of new systems.

E.   Chute-Quench Design

Illgner [159] conducted a study of chute quenching in the continuous heat treatment process illustrated in Figure 154. The variables affecting quenching in the chute zone include effect of the time to drop through the chute zone, cross-sectional size, and quenchant viscosity. The conclusions of this study were that [159,160]

1.   The quenchant viscosity has only minimal effect on sinking speeds.
2.   The position of the part as it enters the quenchant will significantly affect the sinking speed.
3.   Even small (16 × 100 mm) parts do not cool to the M_s transformation temperature in the chute zone.

For optimal quench uniformity, both vigorous agitation and adequate quenchant turnover in the chute zone are necessary. If the temperature is allowed to vary, it may not be possible to control hardness because quench severity depends on the quenchant temperature.

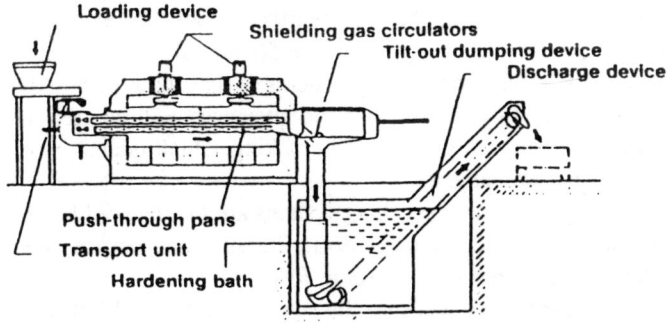

Figure 154   Illgner continuous chute-quench design, (From Ref. 159.)

Quenchant temperature control in the chute zone is especially critical for aqueous poly-alkylene glycol polymer quenchants, which will thermally separate from solution at elevated temperatures, producing a heterogeneous quenching medium. These conditions will result in nonuniform cooling, and the increased thermal gradients may be sufficient to increase distortion or cracking.

If thermal separation of the fluid is accompanied by poor agitation and fluid turnover in the chute zone, the polymer may adsorb on the part and subsequently be removed before polymer dissolution occurs. This will lead to a depletion of the polymer in the chute zone. If sufficient polymer is removed with replenishment, the quenchant in the chute zone will become almost entirely water.

Various chute-quench designs have been used to achieve the appropriate agitation and quenchant turnover in the chute zone. Illustrative examples are shown in Figure 155 [160]. The systems include (a–c) impeller agitation in the chute zone, (d–f) up-flow and submerged spray-quench designs, (h) the SECO-Warwick Whirl-Away countercurrent pump flow design, and (i) a special roller-track design (which must also be accompanied by some form of agitation such as the submerged spray). A rotating chute design has also been reported (Figure 155j) [162]. The dual-belt chute-quench design (Immersion Time Quenching System; ITQS) shown in Figure 155g permits variable cooling rates during the quench [161].

A properly designed chute-quench system should incorporate the following features [145]:

1. Sufficient agitation and turnover in the chute zone to provide adequate and uniform heat transfer.
2. A cooling jacket for the chute above the quench zone to prevent oil and water vapor from entering the furnace vestibule. Cooling can be achieved by routing the quenchant returning from the heat exchanger through the chute zone cooling jacket.
3. A fume eductor located in the chute zone above the cooling jacket to prevent vapor contamination of the furnace atmosphere.
4. A perforated or screened opening in the chute area to allow heated quenchant to escape during the quench. Solid chutes should never be used.
5. A mesh belt of sufficient porosity and length to permit quenchant agitation around the part to facilitate completion of the quench.

F. Flood Quench Systems

Typically either spray or "dunk" quenching is used for induction hardening [142,144,163]. Polymer quenchants are the most commonly used quenchants for either process [163].

For dunk quenching, it is important that the reservoir be sufficiently large to allow the foam head to dissipate before the quenchant is pumped back into the system [145]. Therefore, the reservoir volume should be at least 5–8 times the volume rate of flow. For example, if the flow rate is 10 gal/min, then the reservoir capacity should be 50–80 gal.

One of the most common problems with dunk quenching systems is that the reservoir is undersized. If the reservoir is too small, a mixture of the foam and quenchant will be used to quench the part. This will often lead to increased distortion and cracking. Dunk quench systems also require the use of filters and heat exchangers.

Major factors in the selection and design of spray quenching systems are the material being induction hardened, the area to be quenched, and the quenchant. Pressure and orifice size recommendations for polymer quenchants are presented in Table 65.

The total flow of the quenchant onto the part depends on the size and number of holes in the spray ring. Generally, the total quench hole area should be a minimum of 5–10% of the

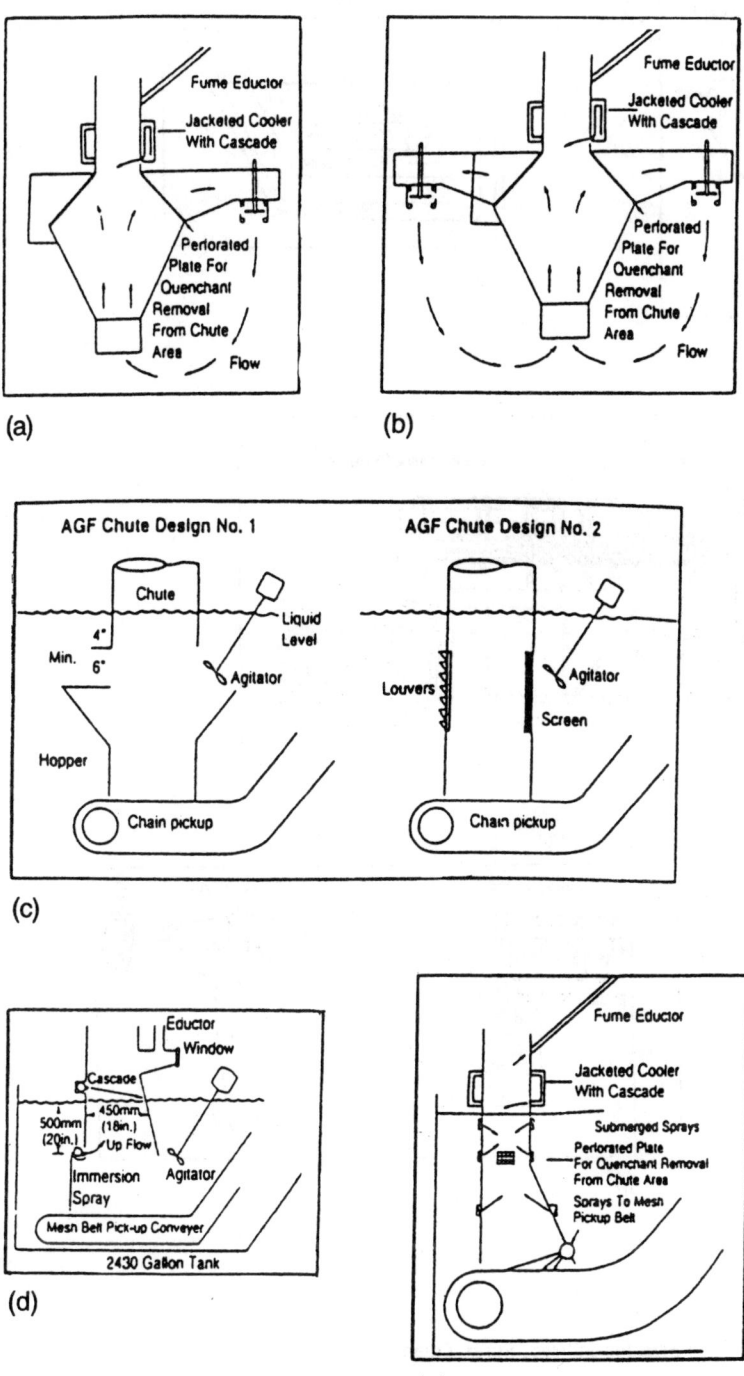

Figure 155   Examples of possible chute-quench designs.

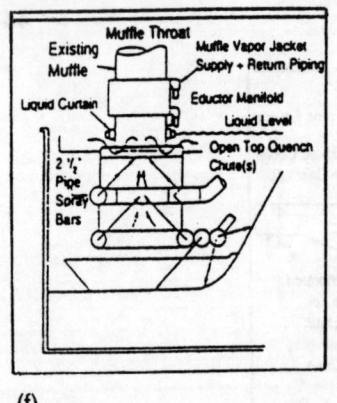

(f)

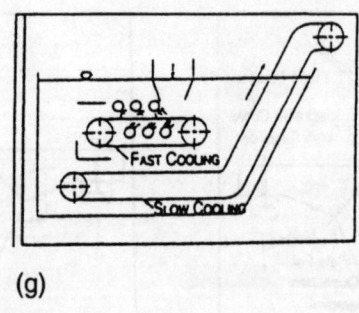

(g)

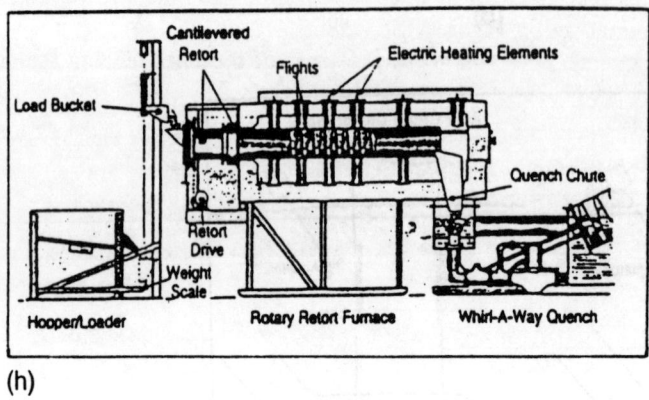

(h)

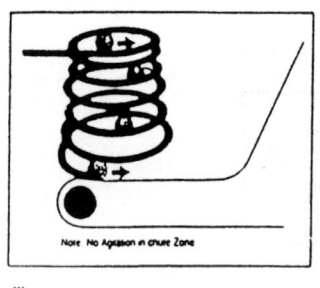

(i)

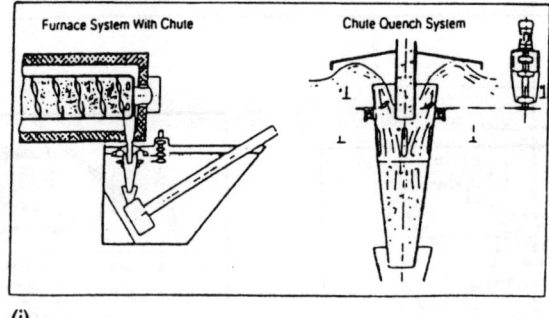

(j)

Figure 155 Continued

Table 65 Pressure and Orifice Size for Indirect Spray Quench Systems

Type of spray	Pressure (psig)	Orifice size	
		in.	mm
Open	<20	1/8	3.18
Submerged	>40	1/4	6.35

Table 66 Interrelationship Between Hole Cross Section and Flow Rate

Cross section		Hole diameter		Flow rate at 20 psi
in.	mm	in.	mm	(gpm/hole)
0.5	12.7	1/16	1.58	0.33
1.0	25.4	1/8	3.18	1.5

Source: Ref. 163.

area being quenched [163]. The size of the holes in the quench ring is a function of the cross-sectional size of the part (Table 66).

Mass flow versus pipe size, effect of inlet/outlet ratio, and hole size on quenching size versus pressure are summarized in Figures 156–158.

Ideally the rate of quench hole area to surface area of the lines feeding the coil will be 1:1 and no greater than 2:1 [163].

G. Filtration

Quenchants may contain various types of solid contaminants such as sludge and carbon. In addition to causing excessive wear of pumps and seals, scale, and heat exchange fouling, the

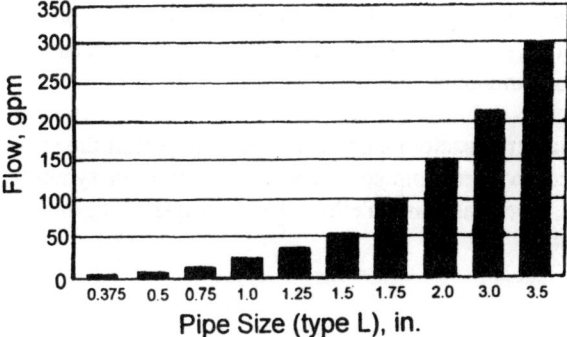

Figure 156 Correlation of volume flow rate and pipe size at quench flow of 10 ft/s. (From Ref. 163.)

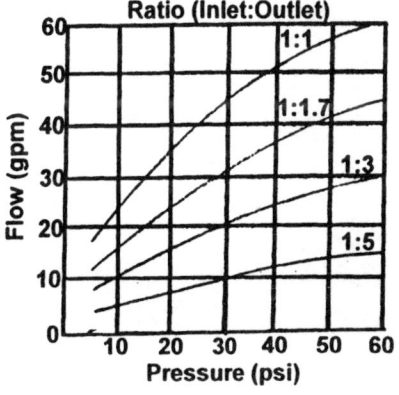

Figure 157 Correlation of pipe inlet/outlet ratio with flow rate and pressure. (From Ref. 163.)

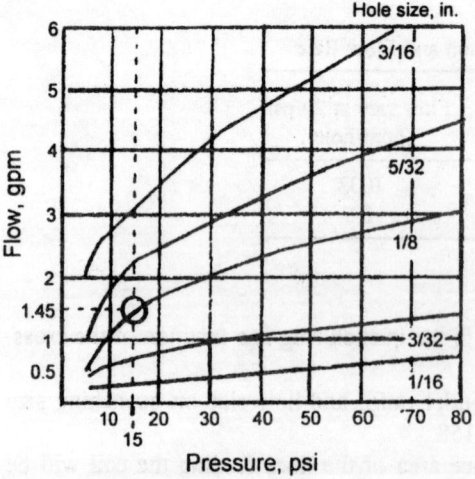

Figure 158 Correlation of orifice size with flow rate and pressure. (From Ref. 163.)

presence of contaminants may affect the uniformity of the quenching process, resulting in increased distortion and cracking frequency for the lifetime of the bath.

One method of solid contaminant removal is centrifugation. An example of a centrifugal separator is shown in Figure 159. Solid contaminants may also be removed by filtration. Filters can be divided into two classes; fixed-pore surface filters and non-fixed-pore depth filters. Fixed-pore surfaces such as mesh or screen filters are suitable for removal of relatively large solids (or small parts). Dimensions of commonly available mesh filters are given in Table 67.

Non-fixed-pore depth filters have a higher capacity than mesh filters and permit the separation of smaller particles. They may be constructed from cotton, wool, cellulose, or synthetic materials such as polyester, polyethylene, fiberglass, and Teflon. The best quenching results are achieved using 3–5 μm filtration [164].

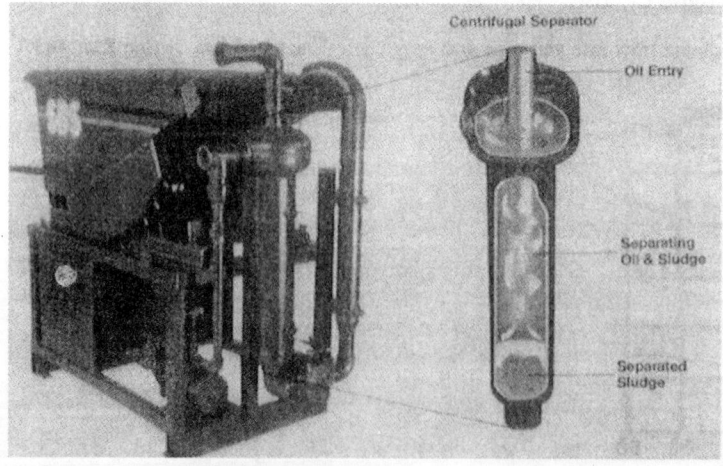

Figure 159 Continuous centrifugal oil sludge separator. (Courtesy of SBS Corporation.)

Table 67   Standard Mesh Screen Sizes

Mesh per linear inch	U.S. mesh no.	Sieve opening	
		in.	μm
52.36	50	0.0017	297
72.45	70	0.0083	210
101.01	100	0.0059	149
142.86	140	0.0041	105
200.00	200	0.0029	74
270.26	270	0.0021	53
323.00	325	0.0017	44

Filters can be used in either full-flow or proportional flow mode. For proportional flow filtration, only a portion of the pumped fluid stream is passed through the filter. For full-flow filtration, all of the fluid stream being pumped is passed through the filter.

Filter selection is dependent on flow rate, fluid composition, and temperature. The maximum flow rate and allowable pressure will dictate the filter element size that can be used. The filter selected must be compatible with the fluid type at the operating temperature. *Note*: Cellulosic filters should never be used with aqueous quenchants.

1.   *Membrane Separation*

As illustrated in Figure 160, microporous membranes are capable of removing various organic and inorganic materials from water [165]. The porosity of these membranes, which are commonly manufactured from cellulose acetate, is selected to separate large molecules such as polymer from smaller molecules such as salt and water. If the porosity is sufficiently small, everything larger than a water molecule can be removed.

Two of the most common membrane separation techniques for the removal of quenchant polymers, salts, and additives from water are reverse osmosis (RO) and ultrafiltration (UF). The primary differences between these methods are the system pressure and molecular size being separated (see Table 68). Reverse osmosis is more common because it provides the most complete separation.

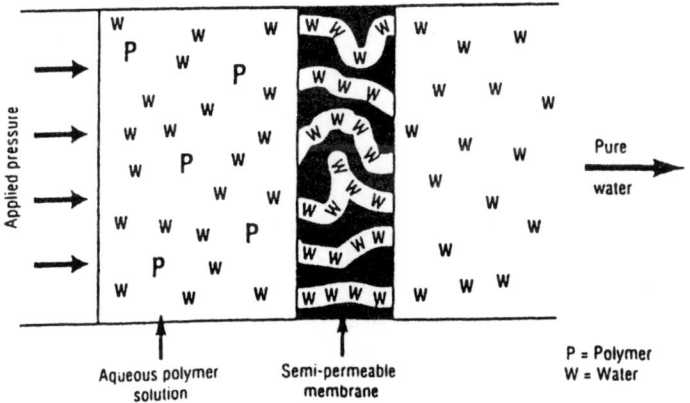

Figure 160   The semipermeable membrane separation process. P = polymer; W = water.

Table 68 System Pressure and Solute Size Membrane Separation Techniques

Technique	Size of solute separated (μm)	Pressure (psig)
Ultrafiltration	0.001–2	—
Reverse osmosis	0.0001–0.001	200–1000

The separation process is carried out by pumping the quenchant solution through an array of membrane cartridges such as the one illustrated in Figure 161. Pure water passes into the core of the filter cartridge, from which it is released to the environment or delivered to a storage tank. The separated polymer is pumped to a storage tank, rediluted to the proper concentration, and reused.

If the aqueous quenchant is contaminated with oil, the oil must be removed prior to membrane separation. This can be accomplished by liquid–liquid surface coalescence of the micrometer-size oil droplets, which are then separated by density difference as illustrated in Figure 162. The agglomerated oil is then collected in a settling tank.

H.   Press Die Quenching

Many batch and continuous processes call for the heat treatment of parts, such as gears and rings, for which the maintenance of dimensional tolerances is critically necessary. Various metallurgical and heat treatment process variables will affect the final part dimensions [166,167]:

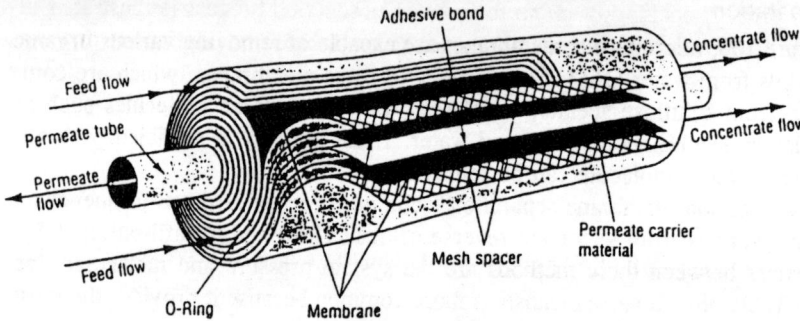

Figure 161 Typical membrane assembly. (Courtesy of Despatch Industries.)

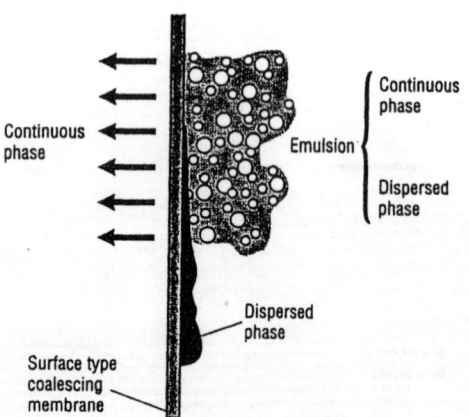

Figure 162 Schematic of separation by coalescence. (Courtesy of Despatch Industries.)

Materials: Irregular grain structure (texture, grain size), hardenability, chemical composition

Workpieces: Geometry or thickness variation, internal stress (machining/forming)

Heat treatment: Thermal stress during heating and/or cooling, texture transformation (austenitizing/hardening), nonuniform carburizing, exterior stress (loading configuration, part weight)

Variables that may potentially affect heat treatment, particularly quenching, are often the processes that receive the most attention. (For a more complete discussion of this subject, see Chapter 5.)

Previously, it was shown that to minimize thermal and transformational gradients, optimization of quench uniformity is critical. However, this may be insufficient to obtain the desired dimensional tolerances. In these cases, the use of press (die) quenching may be required.

1. Press Quenching Machines

A simplified schematic of a press quenching machine is shown in Figure 163 [168,169]. The part to be quenched is placed on a die that is custom manufactured for the particular part being processed. The press, which is either hydraulically or pneumatically actuated, closes over the part and die. The part, pressed in the die at the minimum required pressure, is then either submerged in and/or sprayed with the quenchant. After the quench, the press is opened and the part is removed. (The die-pressing operation also holds the part to prevent distortion during quenching.)

An example of a press quenching machine for a batch production process is illustrated in Figure 164. However, press quenching may be part of either a batch or continuous process [28].

Symmetrical parts may be manufactured to the straightness relationship [166]

$$\text{TIR} = k\frac{l}{d}$$

where TIR = total indicator reading or straightness, l = length (in.), d = diameter (in.), and $k = 10^{-4}$.

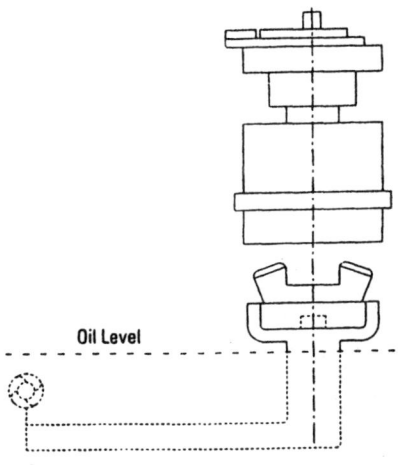

Oil Level

Figure 163   Schematic of a press quench process.

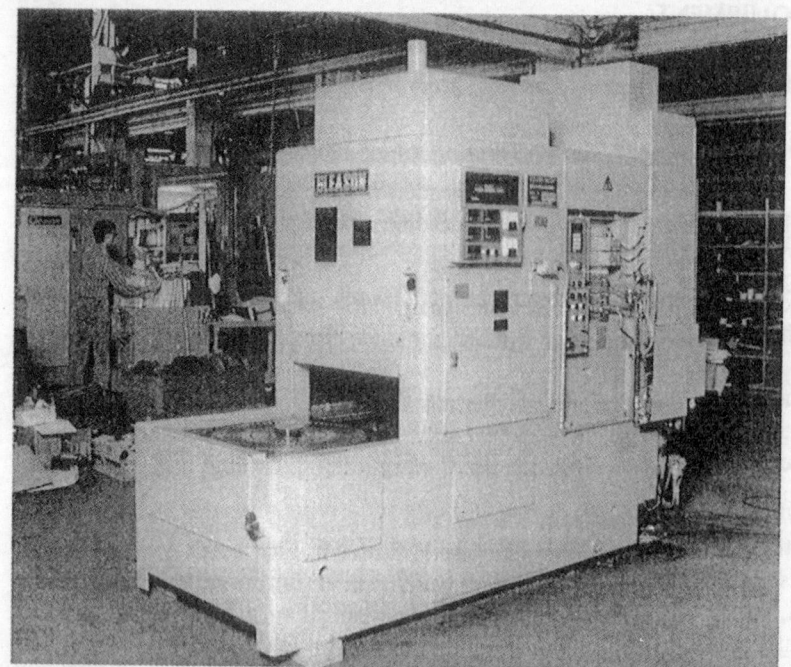

(a)

(b)

Figure 164 (a) A press quench machine. (b) Die installation into press quench machine.

XIII. FURNACE SAFETY

There are numerous examples of industrial accidents that have been caused by improper and unsafe furnace operation [172]. Many of these incidents are related to the unsafe use of furnaces with controlled atmospheres. For example, furnace explosions may occur with sufficient force to literally blow the door off the furnace and propel it across the shop. Unfortunately these incidents occur with such frequency that it is commonly understood that one should never stand in front of the furnace door.

The objective in this section is to provide an overview and summary of furnace safety precautions. *Note: The National Fire Protection Agency and the U.S. Occupational Safety and Health Administration, insurance underwriters, and equipment manufacturers all provide guidelines for safe operation of industrial furnaces. The furnace operator should be well versed in these regulations before operating any piece of equipment.*

A. Explosive Mixtures

When combustible gas and air are mixed and the necessary source of ignition, such as a spark, is present, combustion of the mixture will occur. There will be a corresponding volumetric expansion due to the temperature rise of the gaseous combustion by-products that may lead to the rupture of the container in which they are confined (an explosion).

Many furnace atmospheres, when combined with the proper amount of air, are explosive. Some of those mixtures have been noted in Table 21. One of the first steps of safety is for the furnace operator to be familiar with the explosive limits of all the gases being used in the shop!

When positive ignition is necessary to facilitate combustion of a mixture containing >4% combustibles, the minimum temperature for safe operations is 1400°F (760°C).

B. Purging

The furnace must be purged of air with an inert gas, usually nitrogen, before a flammable mixture is introduced. It is generally true that there is no residual oxygen if the furnace is purged with a volume of N_2 equal to at least five times the volume of the furnace [171].

C. Safety of Operation Temperature

There are 4 variations of furnace temperatures. They include [171]

1. Entire furnace temperature is above 1400°F (760°C)
2. Part of the furnace is above 1400°F with short cooler zones.
3. Part of the furnace is above 1400°F (760°C) with longer cold zones.
4. The entire furnace is below 1400°F (760°C).

When the furnace temperature is above 1400°F (760°C), combustible gas ignites immediately and is therefore relatively safe. However, if the furnace temperature is <1400°F (760°C), an explosive situation may result. In such cases, it is recommended that the furnace be purged with an inert gas to remove the combustible gas.

If there are long hot zones, especially at the ends of the furnace, and relatively short cooler zones, the inert purge gas need be added only in the cold zones. Care should be taken to ensure that the atmosphere is added only to the hot zones. Batch furnaces below 1400°F (760°C) should be purged with the introduction of air when starting cold and at the end of the cooling cycle, only after it is ensured that there can be no ignition source [171].

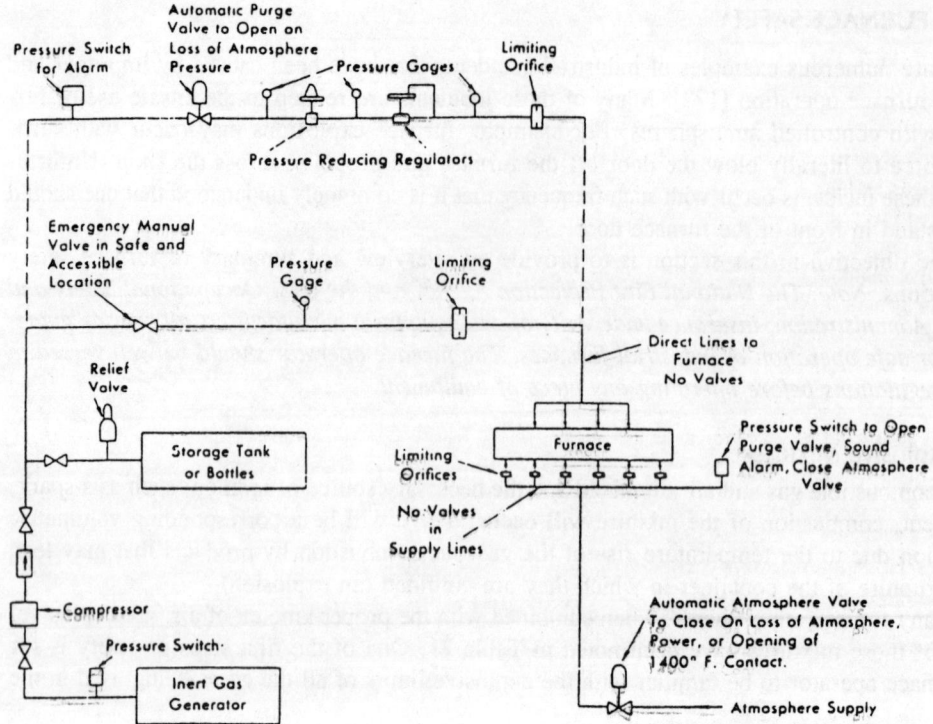

Figure 165   Inert gas purging system.

Large cold zones with short hot zones represents a potentially unsafe condition, and the furnace must be operated only with special precautions such as with the use of an inert gas production and storage system (Figure 165).

D. Power Failures

If a power failure occurs, the atmosphere and heat will be lost, creating unsafe <1400°F (760°C) conditions. In such cases [171]

1. The flammable mixture should be "burned out" immediately if the burn-out procedure is safe.
2. If the furnace is operated below 1400°F, standby storage of an inert gas purge is recommended.

Some common protective systems are summarized in Table 69 [171].

The following protective controls should be installed and interlocked for all atmosphere furnace systems [172]:

1. The atmosphere supply line should have a safety shut-off valve.
2. The operator should be able to visually monitor the atmosphere gas supply to verify gas flow.
3. The temperature of all zones in the furnace should be adequately monitored and interlocked to the atmosphere supply safety shut-off at (or below) 1400°F (760°C).
4. An automatic safety shut-off valve for the flame curtain should be installed and interlocked to prevent opening below 1400°F (760°C).

Table 69   Prevention of Explosions of Furnace Atmospheres During Emergencies

	Emergency interruption		
	Power	Fuel Gas	Atmosphere
Protective system	X	X	X
Inert gas storage	X	X	X
Bottled inert gas	X	X	X
Steam purging system	X	X	X
Preventive method			
Motor generator	X		
Manifold generator			X
Natural gas[a]	X		X
Bottle LP fuel gas		X	

[a]Natural gas should be used only at temperatures $> 1400°F$ (760°C).

5. Both audible and visual temperature atmosphere and gas flow alarms should be installed.
6. It should be possible to open the door manually in the event of power failure.

XIV.   SALT BATH FURNACE

There are three components to a salt bath furnace: the salt, hardware to melt the salt and control temperature, and a means of reclaiming the salt. This section first provides an overview of the use of molten salts in heat treatment and their selection. The second subsection reviews the furnaces (or "salt pots") that are used to provide the necessary temperature control. The third and fourth subsections deal with salt bath furnace safety and salt contamination. Finally, salt reclamation is briefly discussed.

A.   Salt Baths

A alternative to heating a metal part in a conventional furnace is to heat it in a liquid such as a molten salt. Advantages of using moltent salts for heating include rapid heating, superior temperature uniformity and control, and protection from air, thus eliminating surface oxidation and scale formation. Heating processes where molten salt heating is used [174–177] include neutral hardening, isothermal heat treating, annealing, tempering, descaling, cyanide hardening, carburizing, and brazing.

Hardening of some distortion- and crack-sensitive steels requires cooling the steel at a temperature at or near its martensitic temperature. The high temperature cooling process (martempering or austempering) can also be performed in a molten salt [175–179]. Figure 166 provides typical heat-up times for steel rods of various cross sections.

The use temperature of a salt bath is dependent on the melting point of the salt. Melting points are controlled by blending mixtures, either binary or ternary, of different salts. Table 70 lists some typical salt mixtures used for preheating, heating (austenitizing), and cooling [175]. (Preheating of crack- or distortion-sensitive parts may be required to minimize thermal shock and to minimize heating time in the higher temperature furnace [175,180].) Generally, higher melting chlorides are used for heating. Lower melting nitrates (and nitrite ternary mixtures discussed in Chapter 4) are used for martempering and austempering.

B. Furnace (Salt Pot) Design

1. Gas- or Oil-Fired Furnaces

Molten salt furnaces and salt pots can be heated with a gas-or oil-fired radiant tube combustion system as shown in Figure 167. The schematic of the heating system illustrated in Figure 168 indicates that the pressurized burner fires into a radiant tube, which heats the salt inside the baffled area. The molten salt is subsequently recirculated throughout the furnace zone using mechanical agitators.

Gas- or oil-fired burners provide significant energy savings in comparison with electrical heating. Further savings can be achieved by recirculating the hot off-gases to provide additional thermal energy to heat the salt [180].

2. Electrically Heated Furnaces

Molten salt furnaces may also be heated electrically. There are a number of advantages to electrical heating: excellent temperature control ($\pm 5°C$ compared to $\pm 10°C$ for gas-fired heating), ease of changing electrodes, faster heating, and more furnace flexibility.

There are three common designs used for electrical heating of molten salts: "over-the-top" (see Figure 169), "through-the-wall" (see Figure 170), and submerged electrode (Calrod) heating (see Figure 171).

C. Salt Bath Furnace Safety

In addition to the potential hazards of overheating a salt, a number of additional furnace safety concerns have been noted by Laird [180] and Becherer [175].

1. Some mixtures are subject to potential explosive degradation at temperatures above 600°C (1110°F) [181]. However, molten salts can be used safely if the manufacturer's use recommendations and hazard warnings, which are available on the product material safety data sheets (MSDSs), are followed.

2. Care should be taken not to overheat the side walls or bottom during start-up to prevent the salt from flowing from the pot.

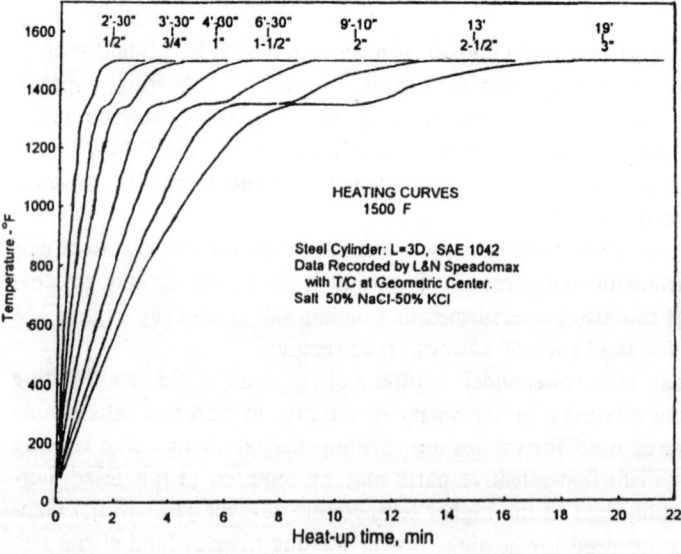

Figure 166 Steel heating curves with respect to section size. (Courtesy of Ajax Electric Company.)

Table 70 Composition and Recommended Use Temperature Range for Steel Heat Treating Salt Mixture

Process	Composition (%)						Melting pt		Use temp.	
	$BaCl_2$	$NaCl_2$	KCl	$CaCl_2$	$NaNO_3$	KNO_3	°C	°F	°C	°F
Austenitizing										
	98–100	—	—	—	—	—	950	1742	1035–1300	1895–2370
	80–90	10–20	—	—	—	—	870	1598	930–1300	1705–2370
Preheating salts										
	70	30	—	—	—	—	335	635	700–1035	1290–1895
	55	20	25	—	—	—	550	1022	590–925	1095–1700
Quench and temper										
	30	20	—	50	—	—	450	842	500–675	930–1250
	—	—	—	—	55–80	20–45	250	482	285–575	545–1065

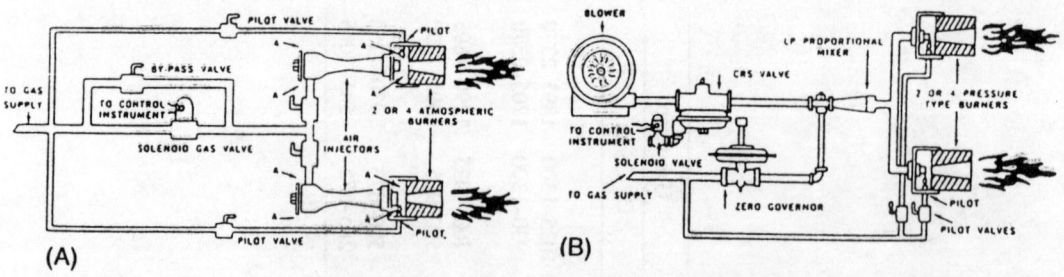

Figure 167 A gas combustion heating systems. (a) Atmospheric combustion system (A = air); (b) blast-type combustion system. (Courtesy of Ajax Electric Company.)

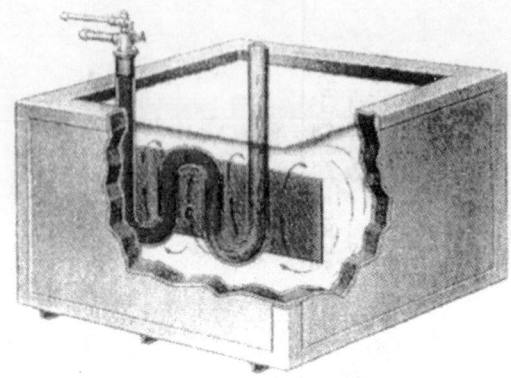

Figure 168 A gas-fired radiant tube heating/circulation system. (Courtesy of Ajax Electric Company.)

Figure 169 An over-the-top electrical heating system. (Courtesy of Ajax Electric Company.)

Figure 170 A through-the-wall heating system. (Courtesy of Ajax Electric Company.)

3. A nitrate–nitrite salt mixture "dripping" on a hot refractory may present a fire hazard.
4. Chloride salts used at temperatures in excess of 650°C (1200°F), although nontoxic, may produce fumes that require venting from the work area.
5. Flue gases must be vented from the work area.

D. Salt Contamination

A part may be austenitized in a chloride salt pot, withdrawn, and then austempered in molten nitrite salt. In this situation, it is likely that some of the chloride salt will drag out on the heated part and subsequently contaminate the nitrate salt. Since chloride salts are insoluble, a "chloride sludge" will be formed that must be removed by filtration. Conversely, contamination of

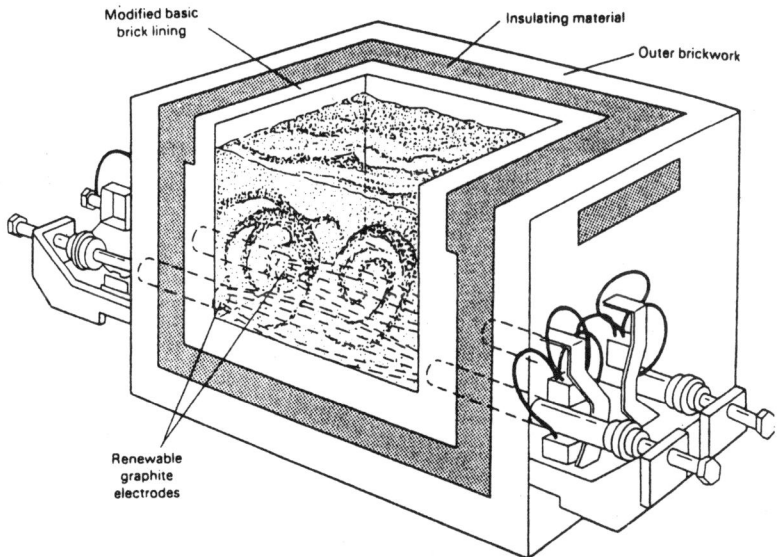

Figure 171 An internal Calrod heating system. (Courtesy of Ajax Electric Company.)

a chloride salt solution with even 600 ppm of a nitrate salt can result in severe damage to the surface of the heated part.

Chloride salts may become contaminated with soluble oxides and dissolved metals. This will produce an oxidizing environment that may lead to decarburization. When this occurs, the salt solutions must be "rectified," sometimes daily. Rectification can be accomplished in one of two ways. The first method involves the reaction of the soluble oxide to form an insoluble compound. For example, silica or silicon carbide will react with a soluble metal oxide to form an insoluble silicate that can be removed by filtration. A second method of rectification is to react the soluble metal oxides with methyl chloride gas or ammonium chloride, a solid. Either process results in the re-formation of the original chloride salt.

Soluble metals will be reduced at the surface of a graphite rod, thus facilitating their removal.

E. Salt Reclamation

Increasingly stringent environmental regulations reflect worldwide demand that salt contaminants not be released to the environment. This problem has been successfully addressed by the use of an automatic surface evaporation scheme that permits salvaging the salt from the waste cleaning water and its subsequent reuse [182]. One process used to accomplish this is shown schematically in Figure 172 [174,182]. Figure 173 is a schematic of a continuous salt recovery system for use with a conveyorized wash system.

XV. FLUIDIZED BED FURNACES

Fluidized bed furnaces employ one of the newer techniques for heat treating. They were first developed in the early 1970s when the oil crisis caused designers to look at more energy-efficient techniques for heating metals. Fluidized beds are formed by passing a gas, often air or nitrogen, through solid particles such as aluminum oxide and silica sand, causing the particles to become microscopically separated from each other and to behave like a bubbling fluid analogous to a boiling liquid. Typical properties are shown in Figure 174. The particles in general are inert (with a high melting point) and do not react with metal parts, but act purely as the mechanism to improve heat transfer between the fluidizing gas and the part being processed.

A broad range of heat transfer rates are possible, depending on bed particle size (particle composition is relatively insignificant), fluidizing gas velocity, choice of fluidizing gas, and bed temperature.

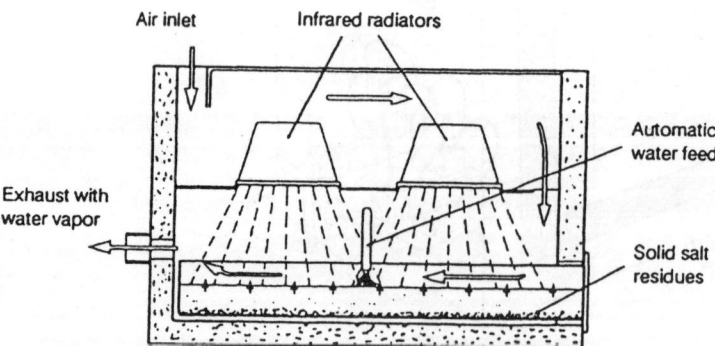

Figure 172 Schematic illustration of a salt recovery system.

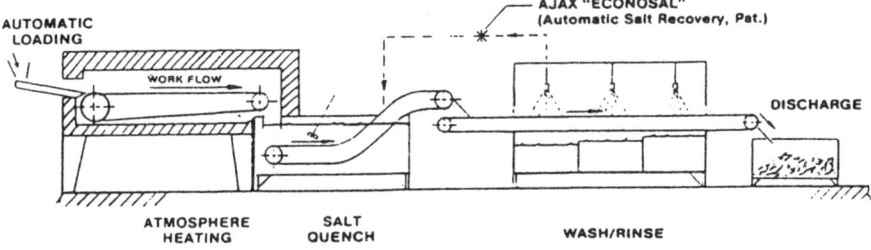

Figure 173   Example of a continuous salt recovery system. (Courtesy of Ajax Electric Company.)

Apart from being more energy-efficient than conventional furnaces [183], it was found that the natural physical properties of the basic fluidized bed technique such as temperature uniformity and heat transfer rates similar to those of salt and lead made it a safe and ecologically sound alternatives to molten metal [184,185] and salt baths [186].

It is possible using fluidizing gases such as argon and nitrogen to perform conventional neutral hardening processes such as austenitizing of all ferrous and nonferrous alloys [183]. In addition, the natural charateristics of the fluidized bed allow the normal thermochemical processes to proceed as well as newly developed processes, e.g., carbide dispersed carburizing [187], which cannot be performed in other atmosphere-based furnaces. The basic carburizing and nitriding [186] surface treatments can be performed as well as hard surface coating [183] of ferrous materials. The relatively high heat transfer characteristics of the fluidized bed can also be used to cool or quench metal parts after austenitizing or solution treatment [5,6] under certain circumstances [188[, providing significant advantages over the more established quenching techniques.

Heating and cooling rates approaching those for salt and lead baths as shown in Figure 175 can be achieved depending on temperature and choice of gas. For comparison, recommended heat-up times for various types of heat transfer media are shown in Figure 176. Similarly, when the fluidized bed is used for quenching, a broad range of cooling rates are also possible, from gas quenching to those attainable with the slower quench oils. The primary focus of this section is placed on the fluidized bed furnace rather than on the associated process technology.

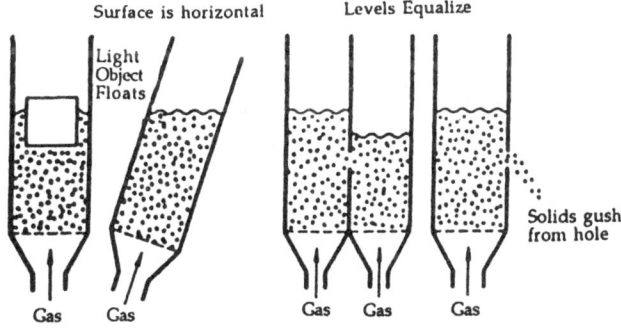

Figure 174   Typical properties of a fluidized bed mixture. (Courtesy of Quality Heat Treatment Pty. Ltd.)

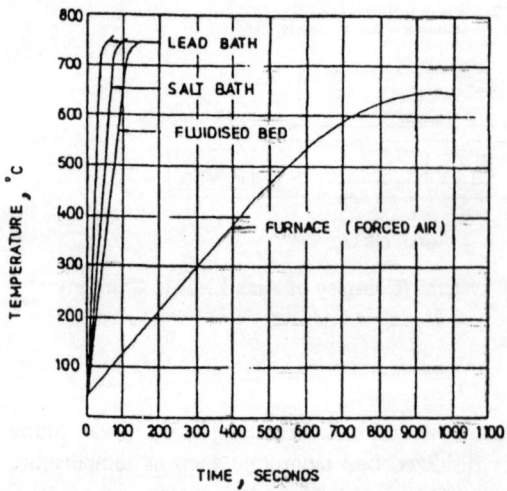

Figure 175 Comparison of cooling rates achievable with salt and lead baths and fluidized beds for a steel bar 16 mm in diameter. (From Ref. 189.)

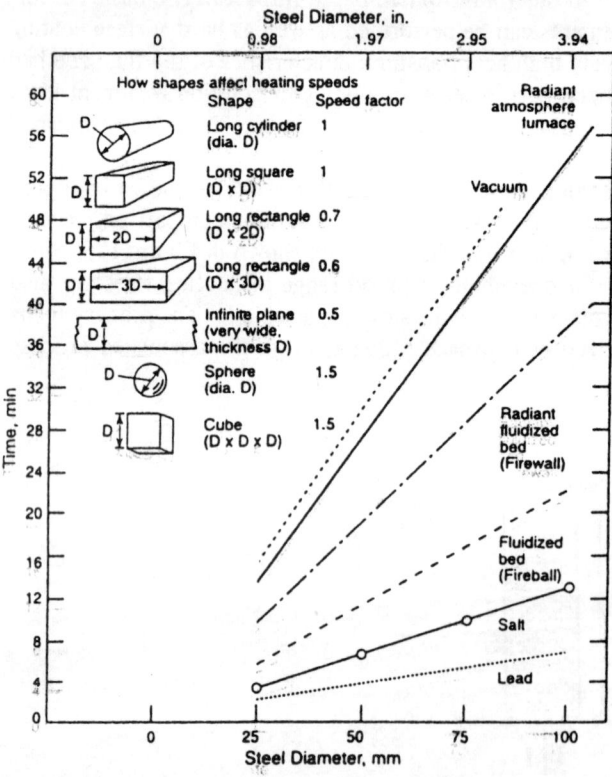

Figure 176 Recommended heat-up times for various heating media. (From Ref. 183.)

A. Design of Fluidized Bed Furnaces

An illustrative schematic of a fluidized bed furnace is provided in Figure 177. This section discusses the main components of a fluidized bed furnace.

1. Heat Transfer Particles

The characteristic properties of fluidized bed media are based on their position in the Geldart chart as shown in Figure 178. The Geldart chart permits characterization of solid media based on their ability to form gas solid fluid. The greatest heat transfer rates occur with Type B media, which include aluminum oxide. Of these media, aluminum oxide provides the best heat transfer capacity, thermal stability, and uniformity [191]

In general, the optimum heat transfer on heating and cooling is achieved with particles in the range of 100–125 μm. If the particle size is less than 100 μm, the uniformity of fluidizing deteriorates, and if the particle size becomes greater than 250 μm, the use of gas becomes uneconomical.

The exception to the choice of aluminum oxide for the fluidized particles is where the particles play a significant role in the thermochemical treatment. For example, ferrovanadium particles are used in the hard surface or vanadium carbide coating of ferrous metals [183]. The choice is based on the coating required and its reaction with the fluidizing gas.

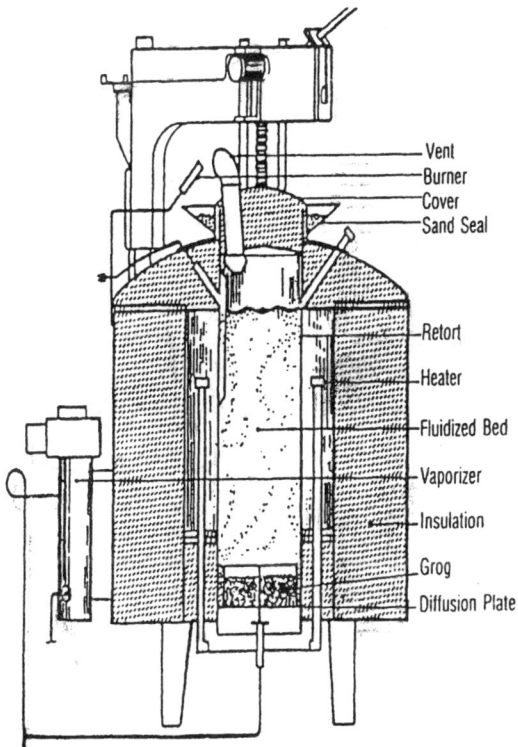

Vent
Burner
Cover
Sand Seal

Retort

Heater

Fluidized Bed

Vaporizer

Insulation

Grog

Diffusion Plate

Figure 177 Schematic illustration of a fluidized bed furnace for methanol carburizing. (Courtesy of Procedyne Corp.)

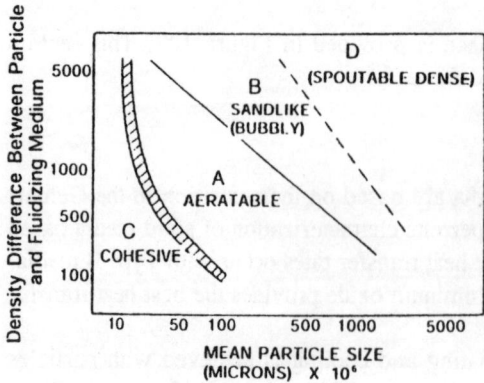

Figure 178 Geldart chart illustration of effect of selection of solid on fluidized bed properties. (From Ref. 190.)

2. Retort

The retort, which contains the Particles, is generally fabricated from a high melting point metallic alloy which determines the operating temperature limitations on the bed (i.e., safe working temperature). For some heating methods, e.g., internal gas combustion, refractory containers have been used.

3. Fluidizing Gas Distributor

The design of the distributor, which controls fluidization characteristics, is critical to the function of the bed in that it controls the temperature uniformity and distribution of the gas phase. Many types of distributors have been used, ranging from perforated metal to ceramic filters.

4. Fluidizing Gas

The type of gas used for fluidizing has an effect on the heat transfer rates, particularly at temperatures below 600°C (where radiation effects are minimal). Hydrogen and helium give the highest rates of heat transfer (see Figure 179). A mixture of gases used for thermochemical treatments can play a significant part not only in heat transfer but also in controlling the rate of thermochemical reactions.

5. Heating Systems

External Heating—Electricity or Gas Externally resistance-heated systems are the most commonly used. For gas, natural draft and flat flame burners with and without heat recovery systems have been developed (Figure 180). One recent development has been the use of an externally heated fluidized bed as shown in Figure 181, in which internal gas combustion is used to replace burners, resulting in significant increases in heating energy utilization. Externally heated fluidized beds are limited to maximum operating temperatures of 1200°C, which in fact is the safe operating temperature for most heat-resistant alloys. With the development of alloys with higher operating temperatures, fluidized bed furnaces could operate at much higher temperatures.

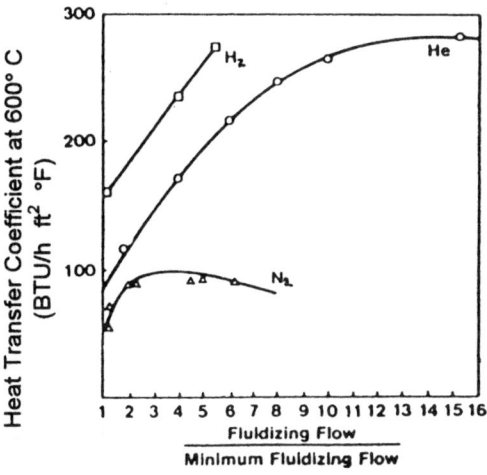

Figure 179   Effect of fluidizing gas on heat transfer.

Internal Heating—Electricity or Gas   The use of internal or submerged combustion techniques using stoichiometric mixtures of gas and air are very energy-efficient, as is direct resistance heating. However, separation of the control of the fluidizing atmosphere and velocity from the control of heat input makes these techniques more difficult to use than external heating.

6.   *Retort Support and Casing*

The method of construction for supporting the fluidized bed uses conventional high performance insulation and heating element design.

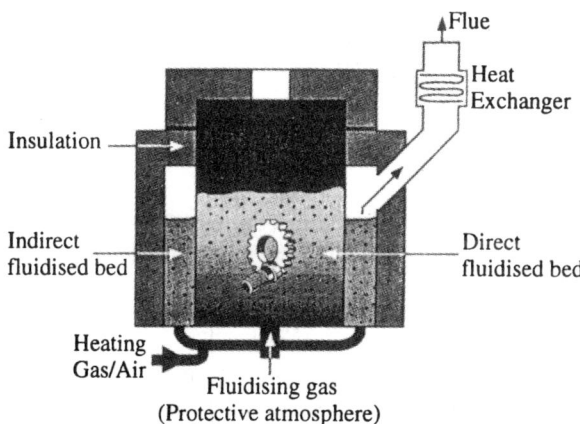

Figure 180   A gas (or electric) fluidized bed furnace.

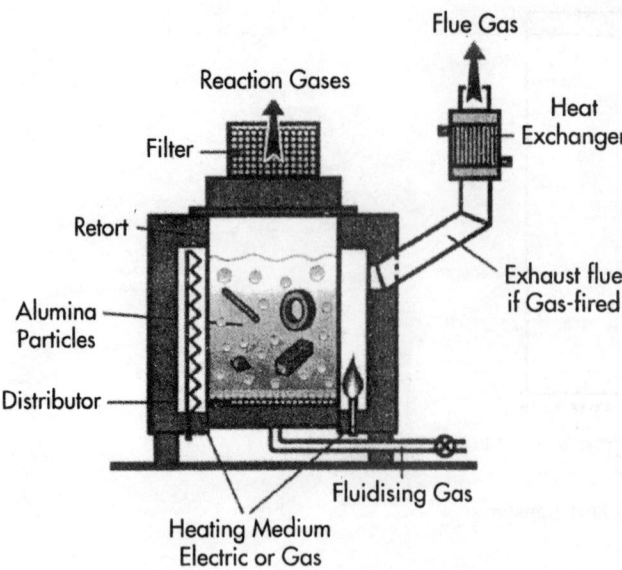

Figure 181   An externally heated fluidized bed furnace.

B.   Development of Fluidized Bed Furnaces – Energy and Fluidizing Gas Utilization

As with all furnace designs, the development of the fluidized bed has occurred in discrete stages since the early 1970s. After the initial signs with relatively large particles of aluminium oxide, it was recognized that the use of a column of fluidizing gas was uneconomical. This led to the development of various techniques.

Fluidizing gas recirculation.

Smaller particle size (Fluidizing gas use is proportional to the square of the diameter of the particle.)

Slump techniques combined with controlled fluidization.

Computer control.

Today these techniques can be combined, making the furnaces (see Figure 182) not only energy efficient in the transfer of heat to the metal but also efficient in the use of fluidizing gases.

C.   Applications of Fluidized Beds for Metal Processing

The applications of fluidized bed furnaces are rapidly increasing in number as the technology becomes more widely accepted. The flowchart in Figure 183 can be used in choosing the most suitable type of equipment. In continuous processing, such as wire heating and patenting, fluidized bed furnaces have been used to replace lead and salt baths and should be considered as a viable alternative processing technique for many types of materials.

Figures 184–189 illustrate some typical commercial installations of fluidized bed furnaces.

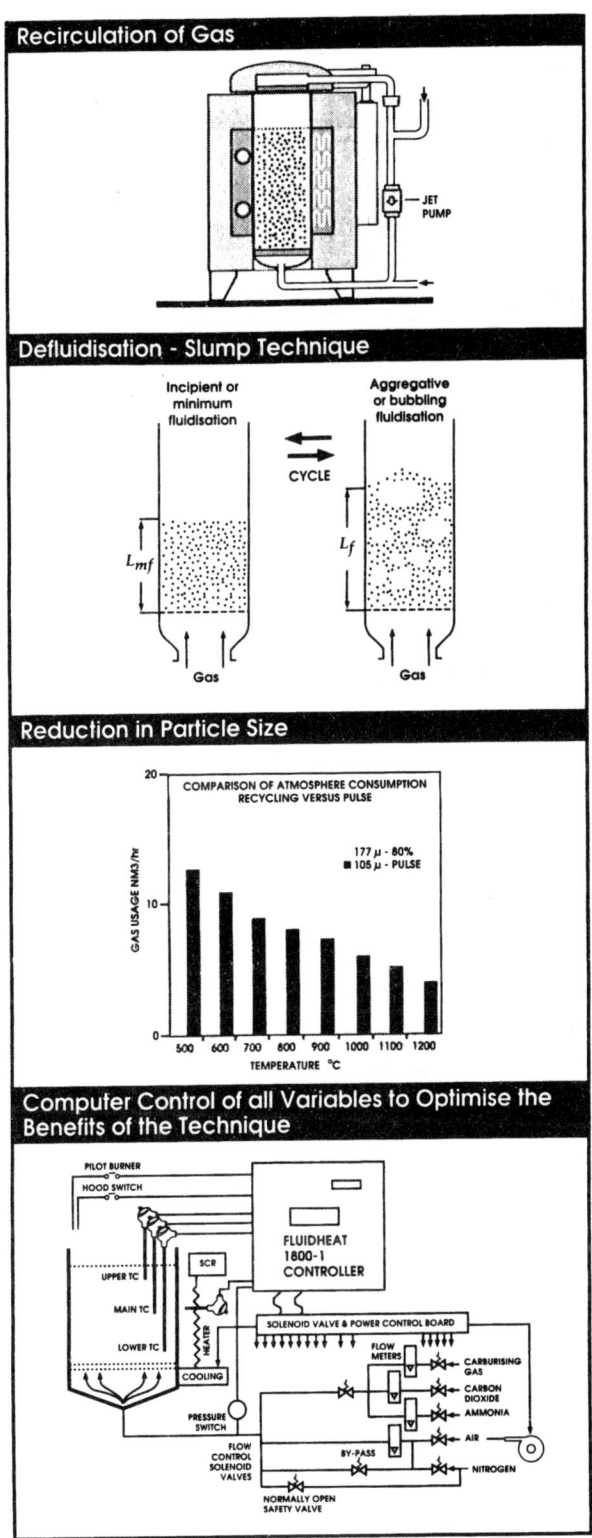

Figure 182   Procedures to reduce fluidization gas use. (Courtesy of Quality Heat Treatment Pty. Ltd.)

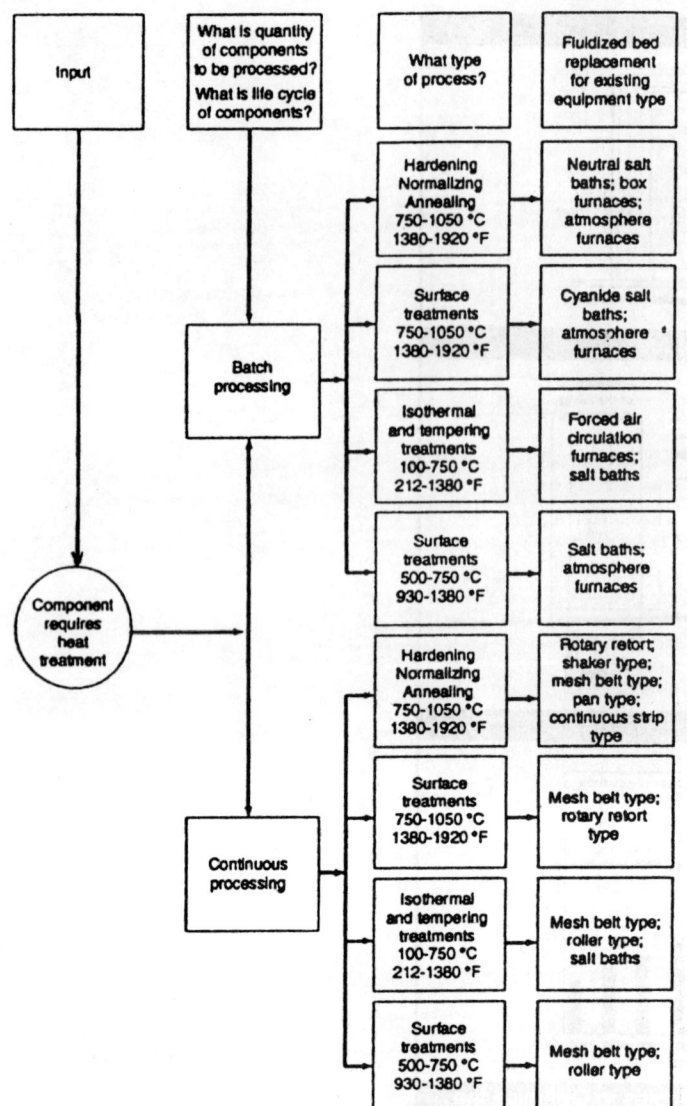

Figure 183   Decision model for applicability of fluidized bed furnaces. (Courtesy of Quality Heat Treatment Pty. Ltd.)

Figure 184 High speed hardening furnace. (Courtesy of Procedyne Corp.)

Figure 185 Fluidized bed heat treating line with fluidized bed tempering furnace, fluidized carburizing furnace, polymer quench, and box tempering furnace. (Courtesy of Procedyne Corp.)

Figure 186 Fluidized bed furnace shown in covered position to conserve energy when not in use. (Courtesy of Fluidtherm Technologies Pty. Ltd.)

Figure 187 Fluidized bed heat treating line with control system. (Courtesy of Fluidtherm Technologies Pty. Ltd.)

Figure 188 Batch fluidized furnace installation for general-purpose heat treatment (600 mm diameter × 900 mm deep including washing and quenching accessories). (Courtesy of Quality Heat Treatment Pty. Ltd.)

Figure 189 Continuous fluidized bed line for hardening and tempering of metal parts. (Courtesy of Quality Heat Treatment Pty. Ltd.)

REFERENCES

1. A. N. Kulakov, V. Ya. Lipov, A. P. Potapov, G. K. Rubin, and I. I. Trusova, Modernization of the electric furnace SKZ-8.40.1/9, *Met. Sci. Heat Treat. Met.* 8:551–553 (1981).
2. T. W. Ruffler, Bulk heat treatment of small components, 2nd Int. Congr. Heat Treat. Mater.: 1st Natl. Conf. Metall. Coatings, Florence, Italy, Sept. 20–24, 1982, pp. 597–608.
3. *Heat Treating Furnacers and Ovens*, Brochure, K. H. Hupper & Co. (KHH), South Holland, MI.
4. J. H. Greenberg, *Industrial Thermal Processing Equipment Handbook*, ASM Int., Materials Park, OH, 1994.
5. Anon., New heat-treating furnace sets high-performance standards, *Nat. Gas. Ind. Technol.*, Spring 1988, pp. 4–5.
6. Anon., Furnace technology applied to the production of wire, *Wire Ind.*, November 1994, pp. 744–745.
7. E. Ford, The new efficiencies of anti-pollutant furnaces, *Eng. Dig.* 38(1):11–12 (1992).
8. *Sintering Systems*, Product Bull. No. 901, C. I. Hayes, Cranston, RI.
9. J. P. Hutchinson and S. T. Passano, Control of Lukens 110-in. Plate mill reheat furnaces at Conshohocken, *Iron Steel Eng.*, August 1990, pp. 29–32.
10. K. H. Illgner, Furnace plant for the heat treatment of small mass-produced components, *Wire* 39:246–250 (1989).
11. H. Trojan, Car-bottom furnace of versatile design built in co-operative venture to heat treat gears, *Ind. Heat.*, October 1992, pp. 25–27.
12. J. Jirka, Annealing car-hearth for foundries, *Skoda Rev. 2*.
13. T. Banno, Heat treating technology present states and challenges, *Heat and Surface '92*, Kyoto, Japan, November 1992, pp. 449–454.
14. D. Schwalm, Energy saving design concepts for heat treating type furnace, in *The Directory of Industrial Heat Processing and Combustion Equipment: United States Manufacturers, 1981–1982*, Energy ed., Published for Ind. Heating Equipment Assoc. by Information Clearing House, pp. 147–153.
15. S. N. Piwtorak, Energy conservation in low temperature oven, in *The Directory of Industrial Heat Processing and Combustion Equipment: United States Manufacturers, 1981–1982*, Energy ed., Published for Ind. Heating Equipment Assoc. by Information Clearing House,
16. P. Shefsiek, Gas vs. electricity: another look at relative efficiencies, *Heat Treating*, September 1980, pp. 40–41.
17. M. D. Bullen and J. Bacon, Advances towards the electric heat treatment shop, *Metallurgia* 49(11):518–519, 523–524 (1982).
18. *Kanthal (R) Handbook—Resistance Heating Alloys and Elements for Industrial Furnaces*, Brochure, Kanthal Corporation Heating Systems, Bethel, CT.
19. *Hot Tips for Maximum Performance and Service—Globar Silicon Carbide Electric Heating Elements*, Brochure, The Carborundun Company, Niagara Falls, NY.
20. G. C. Schwartz and R. L. Hexemer, Designing for most effective electric element furnace operation, *Ind. Heat*, March 1995, pp. 69–72.
21. W. Trinks and M. H. Mawhinney, *Industrial Furnaces*, 4th ed., Vol. 2, Wiley, New York, 1967, pp. 93–117.
22. J. Oare and G. Eklund, Gas-to-electric furnace conversion hikes production, *Metal Prog.*, February 1978, pp. 32–35.
23. *Kanthal Super Electric Heating elements for Use Up to 1900°C*, Brochure, Kanthal Furnace Products, Hallstahammer, Sweden.
24. T. Thomonder, *Some View Points Concerning the Design of Electric Resistance Furnaces*, Brochure, Aktiebolaget Kanthal, Hallstahammar, Sweden.
25. B. Barton, Case study showns natural gas to be cheaper than electricity, *Heat Treating*, February 1981, pp. 36–38.
26. R. A. Andrews, Convert electric furnaces to gas, *Heat Treating*, February 1993, pp. 2–23.
27. *Single-End Recuperative Radiant Tube Combustion Systems*, Brochure, Pyronics Inc., Cleveland, OH.
28. *Fundamentals of Gas Combustion*, 10th ed. 1994, Brochure, Catalog No. XH 03737, American Gas Association, Arlingon, VA. 22209.
29. R. G. Martinek, *Eclipse Industrial Process Heating Guide*, Brochure, Eclipse Fired Engineering Co. of Canada, Ltd., Dan Mills, ON, Canada.
30. The Prudent Use of Gas—An Industrial Guide to Energy Conservation, Brochure No. 7782-573, Catalog No. C30070, American Gas Association, Arlington, VA.
31. R. J. Reed, Maintaining and adjusting combustion systems for fuel economy, *J. Heat Treat.* 1(1):93–95 (1979).
32. *Handbook Supplement 6–57*, Chart 40, North American Mfg. Co., Cleveland, OH.

33. G. Gross, The use of oxygen in industrial furnace, Int. Foundry Heat Treat. Conf., 1985, Johannesburg, South Africa, Sept. 30–October. 4, Vol. 4, Paper No. 6.
34. L. A. Weaver, Systems control for energy conservation, in *The Directory of Industrial Heat Processing and Combustion Equipment: United States Manufacturers, 1981–1982*, Energy ed., Published for Industrial Heating Equipment Association by Information Clearing House, pp. 160–168.
35. S. Lampman, Energy-efficient furnace design and operation, in *ASM Handbook*, Vol. 4, *Heat Treating*, ASM Int., Materials Park, OH, 1991, pp. 519–526.
36. W. Trinks and M. H. Mawhinney, *Industrial Furnaces*, 4th Ed., Vol. 2, Wiley, New York, 1967, pp. 29–73.
37. *Heat Treating Handbook*, Seco/Warwick Corp., Meadville, PA.
38. K. Keller, Use high-speed burners in heating and heat treating furnaces", *Formage Trait. Met.*, November 1977, pp. 39–44.
39. T. Darroudi, J. R. Hellmann, R. E. Tressler, and L. Gorski, Strength evaluation of reaction-bonded silicon carbide radiant tubes, *J. Am. Ceram. Soc.* 75(12):3445–3451 (1992).
40. D. Hibberd, Recent developments in reheating and heat-treatment furnaces, *Metallurgia*, February 1968, pp. 52–58.
41. D. Hibberd and T. Hallatt, Regenerative firing in the steel industry, *Metallurgia* 58(12): FUS 3-Fus (1991).
42. B. Vinton, Ceramic radiant tube system speeds batch furnace recovery, *Heat Treat.*, February 1989, pp. 24–27.
43. Anon., Furnace design moves into the digital age with pulse-firing burner technology, *Nat. Gas. Ind. Technol.*, Summer 1988, pp. 1–2.
44. Anon., High velocity combustion in a heat treatment furnace, Steelmaker, September 1968, pp. 28–30.
45. R. A. Wallis, B. C. McMillian, and A. C. Sanderfer, Variable excess air system for controlled firing saves gas while maintaining temperature uniformity in a batch heat treat furnace, *Ind. Heat.*, June 1995, pp. 53–55.
46. M. Shay, Pulse firing—how this "new" combustion technique works, *Heat Treat.*, February 1987, pp. 24–25.
47. F. M. Heyn, Heat recovery from waste gases with recuperators, in *The Directory of Industrial Heat Processing and Combustion Equipment: United States Manufacturers, 1981–1982*, Energy ed., Published for the Industrial Heating Equipment Association by Information Clearing House.
48. R. J. Evans, New recuperative-burner system gives MOI a better workhorse, *Heat Treat.*, 1984, pp. 18–20.
49. F. J. Bartkowski and K. H. Kohnken, Carbottom furnace retrofitted with burner—ceramic recuperative system to save energy in heat treating casting, *Ind. Heat.*, 1982, pp. 48–50.
50. J. A. Wünning and J. G. Wünning, Burner design for flameless oxidation with low NO-formation even at maximum preheat, *Ind. Heat.*, January 1995, pp. 24–28.
51. W. Trinks, *Industrial Furnaces*, 4th ed., Vol. 1, Wiley, New York, 1950, pp. 220–262.
52. T. Ward and R. J. Webb, Regenerative burners for use in high temperature furnaces, *Inst. Gas Eng. Commun. 1273*, 1985.
53. D. O. Swinder, Gas—the natural winner for heat treatment, *Heat Treat. Steel*, September 1989, pp. 6–25.
54. Anon., The Performance of Impulse-Fired Regenerative Burners on a Small Batch Heat Treatment Furnace, Br. Energy Efficiency Office, Dept. of Energy, Rep. ED/220/232.
55. D. Lupton, Regenerative burners in bloom reheating furnace, *Steel Times Int.*, November 1989, pp. 22–25.
56. Anon., New generation, regenerative burner for continuous slab reheating furnace in Japan to reduce NO_x emission and save energy, *Ind. Heat.*, March 1996, p. 112.
57. Anon., The Application of Regenerative Burners to a Continuous Heat Treatment Furnace, Br. Energy Efficiency Office, Dept. of Energy, Rep. ED/166/191.
58. J. D. Bowers, Regenerative burners slash fuel consumption, *Adv. Mater. Proc.*, March 1990, pp. 63–64.
59. T. Martin, Regenerative ceramic burner technology and utilization, *Ind. Heat.*, November 1988, pp. 12–15.
60. N. Fricker, K. F. Pomfret, and J. D. Waddington, Commun. 1072, Inst. Gas Eng., 44th Annu. Meeting, London, November 1978.
61. V. Paschkis and J. Persson, *Industrial Electric Furnaces and Appliances*, Interscience, New York, 1960, pp. 14–25.
62. D. F. Hibbard, Modern approaches to heat treatment furnace design, *Melt. Mater.* 3(1):22–27 (1987).
63. AFS, *Refractories Manual*, 2nd ed., American Foundrymen's Society, Des Plaines, IL, 1989.
64. W. H. McAdams, *Heat Transmission*, 3rd ed., McGraw-Hill, New York, 1954.

65. J. P. Holman, *Heat Transfer*, 2nd ed., McGraw-Hill, New York, 1968.

66. A. J. Beck, *Heat Treat.*, May 1973, pp. 29–32.

67. R. N. Britz, Convection heat treating. Part II. Furnace considerations, *Ind. Heat.*, January 1975, pp. 39–47.

68. D. Nicholson, S. Ruhemann, and R. J. Wingrove, Heat transmission in reheating and heat treatment furnace: some recent developments, in *Heat Treatment of Metals*, Spec. Rep 95, Iron and Steel Inst., 1966, pp. 173–182.

69. S. Lampman, Energy-efficient furnace design and operation, in *ASM Handbook*, Vol. 4, *Heat Treating*, ASM Int., Materials Park, OH, pp. 519–526.

70. M. A. Aronov, J. F. Wallace, and M. A. Ordillas, System for prediction of heat-up and soak times for bulk heat treatment processes, in *Heat Treating: Equipment and Processes* (G. E. Totten and R. A. Wallis, Eds.), Proc. 1994 Conf., ASM Int., Materials Park, OH, 1994, pp. 55–61.

71. M. Aronov, J. Wallace, and M. Ordillas, Development of validated system for prediction of heat-up and soak times for bulk heat treatment process for materials, in Soak Time Determination Manual, Appendix to Find Report HTN/CT-020112/TR93, Heat Treating Network, Cleveland, OH.

72. Z. S. Tian, M. S. Xu, Y. H. Chui, and J. F. Sui, The thermoelectric effect and theory of the thermocouple, *J. Heat Treat. Met. (China)* 6:42–44 (1994).

73. H. D. Baker, E. A. Ryder, and N. H. Baker, *Temperature Measurement in Engineering*, Vol. 1, Omega Press, Stamford, CT, 1975.

74. Anon., *The Theory and Properties of Thermocouple Elements*, ASTM STP 492, Am. Soc. Testing and Materials, Philadelphia, PA, 1971.

75. Anon., *The Temperature Handbook*, Vol. 28, Omega Engineering Corp., Stamford, CT.

76. T. P. Wang, Thermocouples for special applications, in *Heat Treating: Equipment and Process*, (G. E. Totten and R. A. Wallis, Eds.), Proc. 1994 Conf., ASM Int. Materials Park, OH, 1994, pp. 171–174.

77. Anon., *Manual on the Use of Thermocuples in Temperature Measurement*, ASTM STP 470 B, Am. Soc. Testing and Materials, Philadelphia, PA, 1968.

78. J. Nanigian, Improving Accuracy and response of thermocouples in ovens and furnace, in *Heat Treating: Equipment and Processes* (G. E. Totten and R. A. Wallis, Eds.), Proc. 1994 Conf., ASM Int., Materials Park, OH, 1994, pp. 171–174.

79. *Atmospheres for Heat Treating Equipment*, Brochure, Surface Combustion, Inc., Maumee, OH.

80. *Protective Atmospheres and Analysis Curves*, Brochure, Electric Furnace Company, Salem,, OH.

81. *Bulk Gases for the Electronics Industry*, Brochure, Praxair Inc., Chicago, IL.

82. A. G. Hotchkiss and H. M. Weber, *Protective Atmospheres*, Wiley, New York, 1963.

83. R. Nemenyi and G. Bennett, *Controlled Atmospheres for Heat Treatment*, Franklin Book Co., 1995, pp. 22–1022.

84. *Praxair Nitrogen® Membrane System*, Brochure, Praxair Inc., Danbury, CT.

85. *Nitrogen Supply Systems to Meet Every Requirement*, Brochure, Praxair Inc., Danbury, CT.

86. *Liquid Nitrogen Cooling Systems for Heat Treaters*, Brochure, Praxair Inc., Chicago, IL.

87. J. A. Zahniser, *Furnace Atmospheres*, ASM Int. Materials Park, OH.

88. F. S. Barff, Zinc white as paint, and the treatment of iron for the prevention of corrosion, *J. Soc. Arts* 25:254–260 (1877).

89. J. Morris, The use of water in furnace atmospheres, *Heat Treat. Met.* 2:33–37 (1989).

90. F. E. Vandaveer and C. G. Segeley, Prepared atmospheres, in *Gas Engineers Handbook*, Industrial Press, New York, 1965, pp. 12/278–12/289.

91. D. W. Breck, *Zeolite Molecular Sieves—Structure, Chemistry and Use*, Krieger, Malabar, FL, 1984.

92. B. W. Gonser, The status of prepared atmospheres in the heat treatment of steel, *Ind. Heat.* 6(12):1123–1134 (1939).

93. M. J. Hill, The efficient use of atmospheres in a continuous furnace using the concept of zoning, Int. Foundry Heat Treat. Conf., Johannesburg, S. Africa; 1985, Vol 2, pp. 1–30.

94. Anon., Furnace curtains save energy, *Metal Heat Treat.* November/December 1994, p. 40.

95. W. Olszanski, T. Sobusiak, and T. Trzcialkowski, Carbomix system of controlled carburizing and carbonitriding in pit-type furnaces, 5th Int. Cong. Heat Treat. Mater., Budapest, Hungary. Oct. 20–24, 1986, Vol. II, pp. 1276–1285.

96. *Installation of Nitrogen Methanol Atmosphere Systems*, Brochure, Praxair Inc., Chicago, IL.

97. Data available from Airco Industrial Gases, Murray Hill, NJ.

98. B. Edenhofer, Progress in the technology and applications of in-situ atmosphere production in hardening and case-hardening furnaces, Proc. 2nd Int. Conf. on Carburizing and Nitriding with Atmospheres, 6–8 Dec. 1995, Cleveland, pp. 37–42.

99. D. S. Mackenzie, The dissociation of methanol used for neutral hardening of steel, in *Heat Treating:*

Equipment and Processes (G. E. Totten and R. A. Wallis, Eds.), Proc. 1994 Conf., ASM Int., Materials Park, OH, 1994.

100. M. J. Huber, High temperature methanol dissociation, in *Heat Treating: Equipment and Processes* (G. E. Totten and R. A. Wallis, Eds.), Proc. 1994 Conf., ASM Int. Materials Park, OH, 1994, pp. 437–440.

102. D. H. Herring, Nitrogen Based Atmospheres: A 1982 Status Report, Lindberg Furnace Company, WI.

103. H. Walton, Atmospheres for the hardening of steel, in *Heat Treating: Equipment and Processes*, (G. E. Totten and R. A. Wallis, Eds.), Proc. 1994 Conf., ASM Int., Materials Park, OH, 1994, pp. 441–447.

104. T. J. Schultz, Portable flue gas analyzers for industrial furnaces, Inf. Letter No. 153, Catalog No. C10877, American Gas Association, Arlington, VA.

105. S. Yasui, Automatic gas chromatography, *Netsu Shori 26*(3):231–237 (1986).

106. M. Okumiya, Y. Tsunekawa, I. Niimi, M. Hamada, and M. Mabe, Control of the surface carbon content by measurement of retained methane in plasma-carborizing, *Nippon Kinzoku Gakkaishi 55*(1):981–985 (1991).

107. F. Trombetta and M. Caon, Confronto ta analizzatore a raggi infraosse e la sonda ord ossido di friconio nel trattamento. Termico di cementazione e tempra di pari en sinterizzato, *Metall. Stalenana 2*:65–74 (1981).

108. K. Derge, Uberwacbury du Often Atmosphaere mit Hilfe de Gas-Chromatographie, *Giesserei-Praxis 21*:433–440 (1965).

109. R. K. Singh and C. R. Chakrovorty, Method for atmosphere control of heat treatment furnaces, *Tool Alloy Steels*, April/May 1986, pp. 141–144.

110. *Guide to the Selection of Oxygen Analyzers*, Brochure, Delta F Corporation, Woburn, MA.

111. H. W. Bond, Atmosphere control of heat treating furnace using O_2 sensors: Current standing and the future, *Heat and Surface '92*, Kyoto, Japan, Nov. 17–20, 1992, pp. 479–482.

112. Y.-C. Chen, Automatic control of carbon potential of furnace atmospheres without adding enriched gas, *Metall. Trans. B 24B*:881–888 (Nov. 5., 1993).

113. R. W. Blumenthal, A technical presentation of the factors affecting the accuracy of carbon/oxygen probes, Proc. 2nd Int. Conf. on Carbonizing and Nitriding with Atmospheres, Dec. 6–8, 1995, Cleveland, OH, pp. 17–22.

114. D. W. McCurdy, Improving the accuracy of oxygen probe control system, Heat Treating: Equipment and Processes (G. E. Totten and R. A. Willis, Eds.), Proc. 1994 Conf., ASM, Int., Materials Park, OH, 1994, pp. 117–121.

115. H. W. Bond, Oxygen sensors—A review of their impact on the heat treating industry, *Mat. Sci. Forum, 102–104*:831–838 (1992).

116. R. W. Blumenthal and A. T. Melville, Hot gas measuring probe, U.S. Patent 4,588,493 (May 13, 1986).

117. Data provided by Teledyne Brown Engineering Analytical Instruments, City of Industry, CA.

118. Data provided by Horiba Instruments Inc., Irvine, CA.

119. P. C. Prasannan, Carburization of steels—An overview, *Indian J. Eng. Mater. Sci. 1*:221–228 (August 1994).

120. M. J. Fischer, Distributed infrared atmosphere monitoring system, Heat Treating: Equipment and Processes. (G. E. Totten and R. A. Wallis Eds.), Proc. 1994 Conf., ASM Int., Materials Park, OH, 1994, pp. 167–169.

121. Chart available from Endress and Hauser, 2350 Endress Place, Greenwood, IN 46143.

122. W. Cole, Continuous dew point monitor increases accuracy, *Heat Treating*, March 1993, pp. 19–20.

123. R. N. Blumenthal and R. Hlasny, Check out carbon control systems—step by step, *Heat Treating*, August 1991.

124. H. Heine, Refractories revisited: a review and outlook, *Met. Heat Treat.*, March/April 1996, pp. 87–93.

125. R. C. Johnson, Evaluation of state-of-the-art refractory system, *Ind. Heating*, February 1994, pp. 39–43.

126. J. Dinwoodie, High alumina fiber: manufacture, properties and application in high temperature furnaces, *Ind. Heating*, February 1996, pp. 40–46.

127. G. Deren and M. A. Rhoa, Favorable properties of high temperature glass fiber insulating material by improved chemistry, *Ind. Heating*, November 1993, pp. 46–49.

128. Anon., Characteristics and applicability of specially designed microporous insulating refractory, *Ind. Heating*, August 1995, pp. 47–51.

129. S. C. Carniglia and G. L. Barna, *Handbook of Industrial Refractories Technology*, Noyes, Park Ridge, NJ, 1992.

130. R. J. Aberbach Fans—A Special Report, *Power*, New York, NY.

131. D. E. Fluck, R. B. Herchenroeder, G. Y. Lai, and M. F. Rothman, Selecting alloys for heat treatment equipment, *Met. Prog. 128*(4):35–40 (1985).

132. D. C. Agarwal and U. Brill, Material degradation problems in high temperature environments (alloys—alloying elements—solutions), *Ind. Heat.*, October 1994, pp. 55–60.

133. G. D. Smith and R. A. Smith, Characteristics of current and advanced wire mesh belt alloys for sintering furnaces, *Inc. Heat.*, October 1995, pp. 61–63.

134. J. Kelly, Heat resistant alloy performance, *Heat Treat.*, July 1993, pp. 24–27.

135. T. Cronan, Parts cleaning and its integration into heat treating, *Heat Treating: Equipment and Processes* (G. E. Totten and R. A. Wallis, Eds.), Proc. 1994 Conf., ASM Int., Materials Park, OH, 1994, pp. 311–315.

136. W. H. Michels, Pollution prevention analysis of oil and polymer quenching in the heat treatment of steel, *Heat Treating: Equipment and Processes* (G. E. Totten and R. A. Wallis, Eds.), Proc. 1994 Conf., ASM Int., Materials Park, OH, 1994, pp. 383–388.

137. D. B. Lebart, Parts cleaning: alternatives for the heat treat shop, *Metal Heat Treating*, January/February 1995, pp. 21–24.

138. M. Sugiyama and N. Hiramoto, Vacuum vapor solvent degreasing: an effective alternative for pollution control in thermal processing, *Ind. Heat.*, November 1994, pp. 36–39.

139. L. E. Jones and B. Strebing, Heat Treating: Equipment Maintenance, Safety and Equipment Operations, ASM Int. Short Course, presented April 16–17, 1994, Schaumberg, IL.

140. G. E. Totten, C. E. Bates, and N. A. Clinton, in *Handbook of Quenchants and Quenching Technology*, ASM Int., Materials Park, OH, 1993, Chap. 9.

141. F. Mayinger, Thermo and fluid-dynamic principles of heat transfer during cooling, in *Theory and Technology of Quenching: A Handbook* (B. Liscic, H. M. Tensi, and W. Luty, Eds.), Springer-Verlag, New York, Chap. 3.

142. G. E. Totten, G. M. Webster, and N. Gopinath, Quenching fundamentals—effect of agitation, *Adv. Mater. Process.* 149(2):73–76 (1996).

143. C. E. Bates, G. E. Totten, and R. L. Bremman, in *ASM Handbook*, Vol. 4, *Heat Treating*, ASM Int., Materials Park, OH, 1991, pp. 67–120.

144. J. J. Lakin, *Heat Treat. Met.* 3:73–76 (1982).

145. G. E. Totten K. B. Orszak, L. M. Jarvis, and R. R. Blackwood, *Ind. Heat.*, October 1991, pp. 37–41.

146. J. Hasson, Quench system design factors, *Adv. Mater. Process.* 148(3):425–424 (1995).

147. G. B. Tatterson, *Fluid Mixing and Gas Dispersion in Agitated Tanks*, McGraw-Hill, New York, 1991.

148. H. L. Kauffman, *Chem. Metall. Eng.* 37:177–180 (1930).

149. V. G. Stognei and A. T. Kruk, *Met. Sci. Heat Treat.* 31(1–2):69–71 (1989).

150. U.S. Steel, *Improved Quenching of Steel by Propeller Agitation*, 1954.

151. G. E. Totten and K. S. Lally, Proper agitation dictates quench success. Part I, *Heat Treat.*, September 1992, pp. 12–17.

152. I. J. Krassik, W. C. Krutzch, W. H. Fraser, and J. P. Messina, *Pump Handbook*, 2nd ed., McGraw-Hill, New York, 1985.

153. E. E. Ludwig, *Applied Process Design for Chemical and Petrochemical Plants*, 2nd ed., Vol. 1., Gulf Publishing, 1984.

154. *Goulds Pump Manual*, Goulds Pumps, Inc., Industrial Products Group, Seneca Falls, NY.

155. K. S. Lally and G. E. Totten, Proper agitation dictates quench success. Part 2, *Heat Treat.*, October 1992, pp. 28–31.

156. D. R. Garwood, J. D. Lucas, R. A. Walls, and J. Ward, Modeling of flow distribution in an oil quench tank, *J. Mater. Eng. Perf.*, 1(6):781–787 (1992).

157. R. A. Wallis, D. R. Garwood, and J. Ward, The use of modeling techniques and improved quenching of components, in *Heat Treating: Equipment and Processes* (G. E. Totten and R. A. Wallis, Eds.), Proc. 1994 Conf., ASM Int., Materials Park, OH, 1994, pp. 51–54.

158. N. Bogh, Quench tank agitation design using flow modeling, in *Heat Treating: Equipment and Processes* (G. E. Totten and R. A. Wallis, Eds.), Proc. 1994 Conf., ASM Int., Materials Park, OH, 1994, pp. 51–54.

159. K. Illgner, Quenching of small parts, *Harterei-Tech. Mitt.* 41(2):113–120 (1987).

160. G. E. Totten, G. M. Webster, R. R. Blackwood, L. M. Jarvis, and T. Narumi, Designing chute quench for continuous furnace heat treating effectively, *Ind. Heating*, November 1995, pp. 49–52.

161. S. W. Han, S. H. Kang, G. E. Totten, and G. M. Webster, Immersion time quenching, *Adv. Mater. Process.*, September 1995, pp. 42AA–42DD.

162. A. P. Petrukhin, U.S.S.R. Patent 1,247,424 (1986).

163. D. J. Williams, Quench system for industrious hardening, *Met. Heat Treat.*, July/August 1995, pp. 33–37.

164. V. Srimongkolkul, Is there a need for really clean oil in quenching operations?, *Heat Treat.*, December 1990, pp. 27–28.

165. R. D. Howard and G. E. Totten, Membrane separation of polymer quenchants, *Met. Heat Treat.*, Sept./Oct. 1994, pp 22–24.
166. R. Kern, Distortion and cracking. II. Distortion from quenching, *Heat Treat.*, March 1985, pp. 41–45.
167. D. Grassl, Heat treating furnace system with integrated single-part press quenching, *Heat and Surface '92*, Conf., Proc., Kyoto, Japan, Nov. 1992, pp. 625–628.
168. L. E. Jones, The fundamentals of gear press quenching, *Ind. Heat.*, April 1995, pp. 54–58.
169. L. E. Jones, The fundamentals of gear press quenching, *Gear Technol.* March/April 1994, pp. 32–40.
170. X. Ping and L. Gang, The cause of an explosion in a gas-tight box-type furnace with a controllable atmosphere and related preventative measures, *Jinshu Rechuli 3*:43–45 (1994).
171. Anon., The safe operation of atmosphere furnace, *Metal Prog.*, December 1956, pp. 1–7.
172. R. Ostrowski, Furnace safety, in *ASM Handbook*, Vol. 4, *Heat Treating*, ASM Int., Materials Park, OH, 1991, pp. 657–663.
173. L. E. Jones and B. Strebing, Heat treating: equipment maintenance, safety and equipment operations, *Equip. Safety*, Sec. VI, pp. 1–19.
174. G. Wahl, Development and application of salt baths in the heat treatment of case-hardened steels, Proc. ASM Heat Treating Conf. Carburizing, Processing and Performance, ASM Int., Materials Park, OH, 1989, pp. 41–56.
175. B. A. Becherer, Processes and furnace equipment for heat treating of tool steels, in *ASM Handbook*, Vol. 4, *Heat Treating*, ASM Int., Materials Park, OH, 1991, pp. 726–733.
176. Q. D. Mehrkam, Salt bath heat treating, Part I, *Tooling Prod.*, June 1967, pp.
177. Q. D. Mehrkam, Salt bath heat treating, Part II, *Tooling Prod.*, July 1967, pp.
178. R. W. Foreman, Salt-bath quenching, in *Quenching and Distortion Control* (G. E. Totten, Ed.), ASM Int., Materials Park, OH, 1992, pp. 87–94.
179. K. S. Sreenivasa Marthy and K. S. Shamanna, Heat treatment salts, *Toal and Alloy Steels*, May 1992, pp. 115–117.
180. W. J. Laird, Salt bath equipment, in *ASM Handbook*, Vol. 4: *Heat Treating*, ASM Int., Materials Park, OH, 1991, pp. 475–483.
181. M. A. H. Howes, The cooling of steel shapes in molten salt and hot oil, Ph.D. Thesis, London Univ., 1959.
182. E. H. Burgdorf, Use and disposal of quenching media—recent developments with respect to environmental regulations, *Quenching and Carburizing*, 3rd Int. IFHT Seminar, Melbourne, Australia, September 1991, pp. 66–77.
183. R. W. Reynoldson, *Heat Treatment in Fluidized Bed Furnace*, ASM Int., Materials Park, OH, 1993.
184. R. Branders, Patenting in a fluidized bed, *Wire Ind.*, February 1990, pp. 89–91.
185. H. Lochner, Molten-metal and hydrogen quenching technologies for steel and strip, *Rev. Fr. Metall.*, February 1993, pp. 65–73.
186. W. Krebs, Fluidized-bed furnaces instead of salt baths for the heat treatment of tool steels, *Giesserei 77*(10):337–334.
187. A. Killian, Fluidized bed and QCD carburizing, Materials Research Conf., 1996.
188. R. W. Reynoldson and E. Ninham, Optimizing the performance of die casting tools manufactured from H13 hot work tool steels, Tech. Paper, 1996.
189. *An Introduction to Fluidized Bed Heat Treating*, Brochure, Fluidtherm Technology, Madras, India.
190. M. A. Delano and J. Van den Sype, Fluid bed quenching of steels: applications are widening, *Heat Treat.*, December 1988.
191. P. Sommer, Quenching in fluidized beds, *Heat Treat. Met. 2*:39–44 (1986).

7

VACUUM HEAT TREATMENT

Bernd Edenhofer and Jan W. Bouwman
Ipsen Industries International GmbH, Kleve, Germany

I. INTRODUCTION

The term *vacuum heat treatment* refers to heat treatment processes in which steel components are subjected to heat in a vacuum.

What is a vacuum? The word *vacuum* originates from the Latin and means empty or empty space. By empty space one understands a space entirely devoid of matter. Such a space is produced by removing the air (or other gases) from a gastight container with, for example, a pump. In technical terms a vacuum is not a space totally devoid of matter, as such a space does not exist nor can it be produced. Thus a vacuum can be considered a space with highly reduced gas density.

The quality of a vacuum is described by the degree of reduction of gas density, i.e., gas pressure. Gas pressure is measured in pascals (Pa), where $1 \text{ Pa} = 1 \text{ N/m}^2$. The conversions between the units Pa, bar, torr, and others are shown in Table 1.

One distinguishes four different vacuum levels or qualities as shown in Table 2. The heat treatment of steel is carried out in three qualities of vacuum—rough, fine, and high. The majority of applications are processed in the fine vacuum range.

What is the purpose of utilizing vacua for heat treatment? The heat treatment of steel components in air leads to surface oxidation. The type and thickness of oxide layers produced are dependent on the temperature of the heat treatment, the duration of exposure, and, naturally, also on the type of steel.

To avoid surface oxidation there are several possibilities. One is to replace the air in the heat treatment furnace by an atmosphere that does not contain oxygen, a so-called protective atmosphere. Another is to reduce the amount of air surrounding the workpieces by evacuation to such a low level that the remaining oxygen is below the material's oxidation level. Replacing the air in a furnace by a protective atmosphere such as pure nitrogen or nitrogen with reducing elements like hydrogen or carbon monoxide requires purging the furnace with this atmosphere. Figure 1 shows the reduction of the oxygen level, through purging, in a gastight furnace with increasing numbers of volume changes.

From the known temperature dependence of formation energies of oxides [1], the oxidation boundaries of iron and typical alloying elements have been calculated (Figure 2) [2]. These curves show that to avoid the oxidation of iron at, for example, 1000°C (1832 °F), the par-

Table 1 Conversion of Pressure Units

Pressure unit	Pa	Bar	Torr
1 Pa = 1 N/m²	1	10^{-5}	7.5006×10^{-3}
1 bar	10^5	1	750.06
1 torr = 1 mm Hg	133.32	1.3332×10^{-3}	1
1 m/H$_2$O	9807	9.807×10^{-2}	73.56
1 atm	1.0133×10^5	1.0133	760.0
1 μm	0.13332	1.3332×10^{-6}	10^{-3}
1 in. Hg	3386.4	3.3864×10^{-2}	25.4
1 lb/ft²	47.88	4.7880×10^{-4}	0.3591

Table 2 Classification of Vacua

Quality of vacuum	Pressure range (hPa)[a]
Rough	$1 - 10^3$
Fine	$10^{-3} - 10^0$
High	$10^{-7} - 10^{-3}$
Ultrahigh	$< 10^{-7}$

[a]1 hPa = 1×10^2 Pa $\approx$ 1 mbar.

tial pressure of oxygen (which in air is 0.207 bar) has to be reduced below 10^{-15} bar. This is not only impossible, it is also not necessary.

Practical experience shows that fine vacua of 10^{-2}–10^{-3} mbar suffice to produce bright surfaces on most steels. Thus, even though the oxygen in the fine vacuum oxidizes the steel surface, the degree is insufficient to produce a visible effect. Also, certain furnace materials getter oxygen (by producing oxides themselves) and thus reduce the partial pressure of oxy-

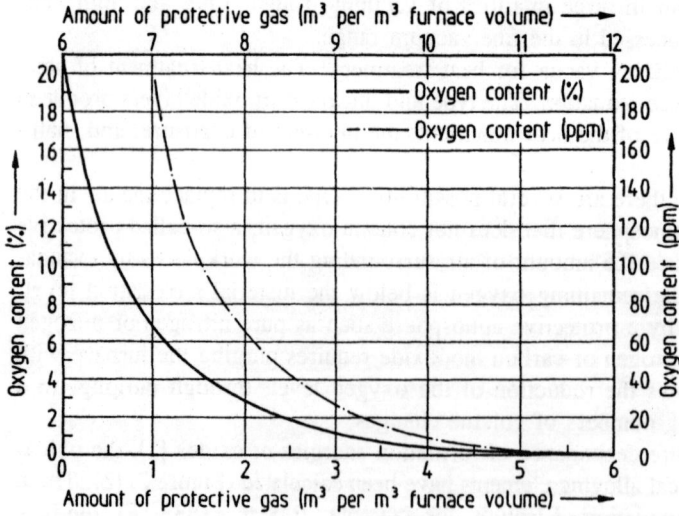

Figure 1 Reduction of oxygen level in a furnance through purging with a protective atmosphere.

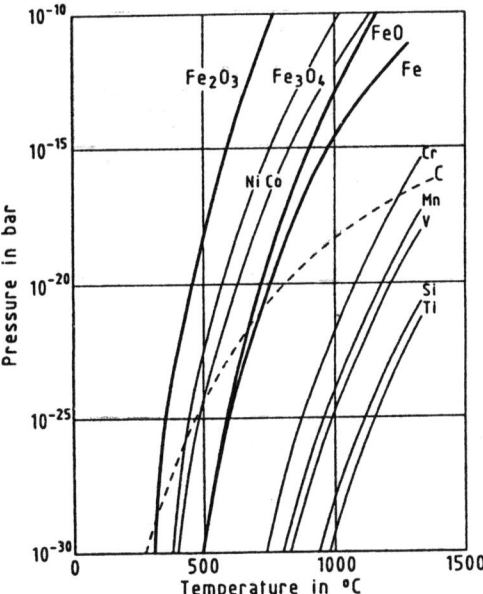

Figure 2 Partial pressure of oxygen for the formation of iron oxides and oxides of other elements.

gen within the furnace even further. All materials that, like carbon, chromium, and manganese, etc., have lower oxide formation energies than iron contribute to this effect if they are sufficiently hot.

II. COMPARISON TO ATMOSPHERIC PROCESSES

Almost all heat treatment processes carried out at normal pressure in protective atmospheres have an equivalent counterpart in vacuum processing. Table 3 gives an overview.

If atmosphere furnaces have no separate purge chambers, a neutral atmosphere like nitrogen is not sufficient to produce bright surfaces. Therefore, protective atmospheres usually contain certain amounts of reducing gases, like H_2 and/or CO, to counter the oxidizing and decarburizing effects of constituents such as H_2O and O_2 that enter the furnace during door openings. In the case of vacuum processing of high-alloy steels, it may also be advisable to feed hydrogen at low partial pressures into the vacuum furnace to counter the effect of water vapor being adsorbed on the inside of the cold furnace walls during door openings.

Carrying out thermochemical processes like carburizing and nitriding in vacuum furnaces at low pressures usually produces little mass transfer, as the thermal dissociation of low-density process gases yields only small amounts of reactive elements. In addition, these elements will react with the steel surface directly on first contact, resulting in a higher mass transfer on the outside of a dense load than in the load center. Activating and ionizing the process gas molecules by applying an electric field of high voltage will overcome this weakness by substantially increasing the number of reactive gas species and distributing them uniformly throughout the load. Processes that include electric activation of the gases are called plasma processes, as shown in Table 3.

A general difference between atmospheric and vacuum processes is given by the cleanliness and the substantially reduced environmental loading of vacuum heat treatments. Vacuum furnaces produce no fumes or exhaust gases, possess no flames, and are usually cold. They

Table 3   Comparison of Equivalent Atmospheric[a] and Vacuum Heat Treatment Processes

| Atmospheric processes | | Vacuum processes | | |
Process	Gas constituents	Process	Pressure (mbar)	Gas type
Annealing	N_2 $N_2 + CH_4$ (C_3H_8) $N_2 + CH_3OH$, etc. Endothermic gas, etc.	Annealing	10^{-3}–10^{-1}	None N_2 $N_2 + H_2$
Hardening	See Annealing, above	Hardening	10^{-3}–10^{-1}	See above
Tempering	See Annealing, above	Tempering	10^{-2}–10^3	See under Annealing, above
Carburizing	$N_2 + CH_3OH + CH_4$(C_3H_8) Endothermic Gas + CH_4 etc.	Low-pressure carburizing	4–400	C_3H_8(CH_4)
		Plasma carburizing	2–10	CH_4($+H_2$), $C_3H_8+H_2$($+Ar$)
Carbonitriding	As above + NH_3	Plasma carbonitriding	2–10	As above $+N_2$
Nitriding	NH_3 or $NH_3 + N_2(H_2)$ or $NH_3 + O_2$($+N_2$)	Plasma nitriding	0.5–10	$N_2 + H_2$
Nitrocarburizing	NH_3 + endothermic gas or + CO_2	Plasma nitrocarburizing	0.5–10	$N_2 + H_2 + CH_4$ (CO_2)
Brazing	H_2 or $N_2 + H_2$ or endothermic gas	Brazing	10^{-4}–10^{-2}	None or H_2
Sintering	H_2 or $N_2 + H_2$ or endothermic gas	Sintering	10^{-3}–10^{-2}	None or N_2($+H_2$)

[a]At normal pressure.

behave more like machines than like furnaces. In addition, the optical appearance of the treated steel workpieces differs. Pieces treated in atmospheric furnaces usually achieve a light gray color even in reducing atmospheres, whereas the surfaces of vacuum heat treated components remain bright and shiny. At elevated temperatures the surfaces become shinier than they were before treatment. This is due to the evaporation effect of residues and the reduction of oxides on the surfaces at low pressure levels in vacuum.

Low pressures, i.e., high vacua, will not only evaporate adsorbed material from the surface, they also cause the volatilization of those alloyed elements in the steels that have a high vapor pressure.

III. VOLATILIZATION, DISSOCIATION, AND DEGASSING

During vacuum heat treatment of steel it is always necessary to consider the evaporation (sublimation) of the alloying constituents. In particular, manganese and chromium have relatively high vapor pressures (Table 4). Here, the vaporization rate is the determining factor, and this is a direct function of the furnace pressure. The higher the pressure, the more frequent the collisions of gas molecules; hence, fewer metal atoms escape from the metal surface.

Elements with very high vapor pressure, such as zinc, will evaporate very rapidly when heated at relatively high temperatures, even at low vacuum levels. Alloys with high concentrations of volatile elements, such as brass, are therefore not heat treated in vacuum because of the risk of dealloying the brass and contaminating the cold furnace parts with condensed zinc.

Another example, a 14% chromium steel, shows no chromium loss at 990°C and 10^{-2} mbar after 2.5 h. However, at 10^{-4} mbar, with identical time and temperature conditions, a surface chromium loss of 0.5% was measured.

Alloyed steels can still be processed in vacuum by using a backfill pressure that at the temperature involved exceeds the vapor pressure of the volatile alloying elements. It is common practice to use a backfill pressure of inert gas between 10^{-1} and 10 mbar at temperatures above 800°C (1472°F) to preclude the vaporization of elements such as chromium, copper, and manganese from steels processed at higher temperatures.

Like the pure metals, metal–hydrogen, –oxygen, and –nitrogen compounds will also volatilize when heated to sufficiently high temperatures at correspondingly low pressures. For this reason, vacuum treating can be used both to dissociate these compounds and to evacuate the evolved gases, leaving an undisturbed and clean base metal behind.

Gas incorporated in steel during manufacturing or posttreatment such as pickling and welding can also be removed in vacuum. To degas a metal, the pressure in the furnace vessel must be lower than the pressure of the gas in the metal. Under these conditions, the gas will diffuse out of the metal into the vacuum. As the outer layer of the metal outgasses, a gradient is set up effecting gas desorption from the interior to the surface. The desorption rate is accelerated by higher temperatures. In the degassing of thicker parts, the gas diffusion rate through the mass becomes a limiting factor. Gases are removed from the metal surface into the vacuum surrounding the charge and are finally trapped by the vacuum pumps and exhausted from the system.

The degassing of oxygen, nitrogen, and hydrogen from refractory metals is particularly important in view of the improvements in properties, including ductility and fatigue, that can be achieved as a result of reducing the interstitial alloy content in these materials.

Table 4 Temperatures at Which Common Elements Exhibit Specific Vapor Pressures

Element	Temperature (°C) at which vapor pressure is				
	1.3×10^{-4} mbar	1.3×10^{-3} mbar	1.3×10^{-2} mbar	1.3×10^{-1} mbar	1000 mbar
Aluminum	808	890	997	1124	2058
Antimony	525	595	678	780	1441
Arsenic	—	220	—	310	610
Barium	544	626	717	830	1404
Beryllium	1030	1131	1247	1396	—
Bismuth	537	609	699	721	1421
Boron	1141	1240	1356	1490	—
Cadmium	180	220	264	321	766
Calcium	463	528	605	701	1488
Carbon	2290	2473	2683	2928	4831
Cerium	1092	1191	1306	1440	—
Cesium	74	110	153	207	691
Chromium	993	1091	1206	1343	2484
Cobalt	1363	1495	1650	1834	—
Copper	1036	1142	1274	1433	2764
Gallium	860	966	1094	1249	—
Germanium	997	1113	1252	1422	—
Gold	1191	1317	1466	1647	2999
Indium	747	841	953	1089	—
Iridium	2156	2342	2558	2813	—
Iron	1196	1311	1448	1604	2737
Lanthanum	1126	1243	1382	1550	—
Lead	548	620	718	821	1745
Lithium	378	439	514	608	1373
Magnesium	331	380	443	515	1108
Manganese	792	878	981	1021	2153
Molybdenum	2097	2297	2535	3011	5573
Nickel	1258	1372	1511	1680	2734
Niobium	2357	2541	—	—	—
Osmium	2266	2453	2669	2922	—
Palladium	1272	1406	1567	1760	—
Platinum	1745	1905	2092	2295	4411
Potassium	123	161	207	265	643
Rhodium	1816	1973	2151	2359	—
Rubidium	88	123	165	217	679
Ruthenium	2040	2232	2433	2668	—
Scandium	1162	1283	1424	1596	—
Silicon	1117	1224	1344	1486	2289
Silver	848	921	1048	1161	2214
Sodium	195	238	291	356	893
Strontium	413	475	549	639	1385
Tantalum	2601	2822	—	—	—
Thallium	461	500	607	661	1485
Thorium	1833	2000	2198	2433	—
Tin	923	1011	1190	1271	2272
Titanium	1250	1385	1547	1725	—
Tungsten	2769	3019	3312	—	5932
Uranium	1586	1731	1899	2099	—

Table 4   Continued

Element	Temperature (°C) at which vapour pressure is				
	1.3×10^{-4} mbar	1.3×10^{-3} mbar	1.3×10^{-2} mbar	1.3×10^{-1} mbar	1000 mbar
Vanadium	1587	1726	1889	2081	—
Yttrium	1363	1495	1630	1834	—
Zinc	248	290	343	405	908
Zirconium	1661	1818	2003	2214	—

IV.   FURNACE EQUIPMENT

A.   Vacuum Furnaces

Heat treatment in vacuum furnaces is characterized by special conditions with regard to the design of the furnaces as well as the control of temperature and vacuum during the heat treatment. The design of the furnaces generally depends on the size of the load, the pressure and temperature to be attained, and the medium to be used in cooling the load.

The main parts of a vacuum furnace are the recipient, pumping system, heating chamber, temperature-measuring and control system, and cooling system.

1.   The Recipient

Recipient designs can be grouped into so-called hot-wall and cold-wall furnaces. The hot-wall furnace has a ceramic or metallic retort (Figure 3a). The heating system is usually located outside of the retort and consists of resistive heating elements or an induction coil. Limitations of this retort-type furnace are the restricted dimensions of the heating zones and the restricted temperature range of the metallic retort [maximum 1100°C (2012°F)], arising from low mechanical strength, gas permeability, and evaporation of alloying elements.

Because of the demand of the heat treatment industry for higher temperatures, lower pressures, rapid heating and cooling capabilities, and higher production rates, the cold-wall vacuum furnace has become the dominant design for high-temperature furnaces since the late 1960s.

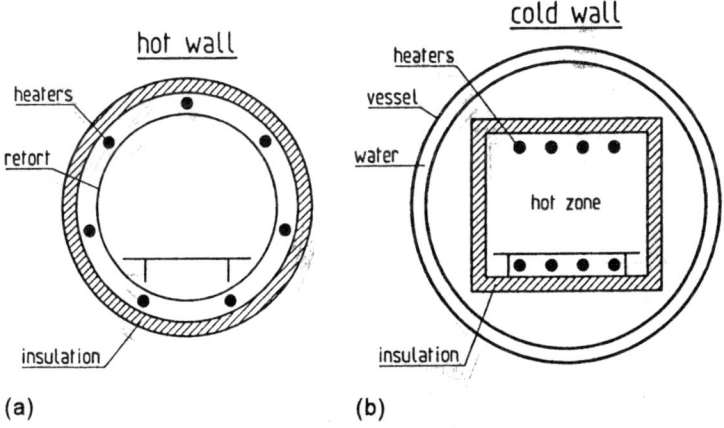

(a)   (b)

Figure 3   Vessel design for (a) hot-wall and (b) cold-wall vacuum furnaces.

With cold-wall furnaces, the vacuum vessel is cooled with a cooling medium (usually water) and is kept near ambient temperature during high-temperature operations (Figure 3b). In comparison to the hot-wall furnace, the features of the cold-wall furnace are

The higher temperature range (1350°C (2462°F) or higher)

The small heat accumulation

The faster heating and cooling performance

A disadvantage is the greater adsorption of gases and water vapors on the cooled furnace walls and in the insulation after opening of the furnace.

2. Pumping System

The construction of the pumping system depends on the following factors:

1. The volume of the recipient
2. The surface of the recipient and the type of furnace parts inside
3. The degassing of the charge
4. The time required for evacuation down to the final pressure

Pumping systems are usually divided into two subsystems, the roughing pump and the high-vacuum pump, although for certain requirements a single pumping system can handle the entire range and cycle (Figure 4). Pumps are usually classified in two general categories, mechanical pumps and diffusion pumps, although there are other specialized types such as ejectors, ion pumps cryo pumps, and "chemical getter" pumps.

Table 5 gives a survey of the types of pumps, or pump combinations, for achieving various final pressures. The various vacuum levels are grouped in Table 6 together with their specific application in the heat treatment of steel in vacuum.

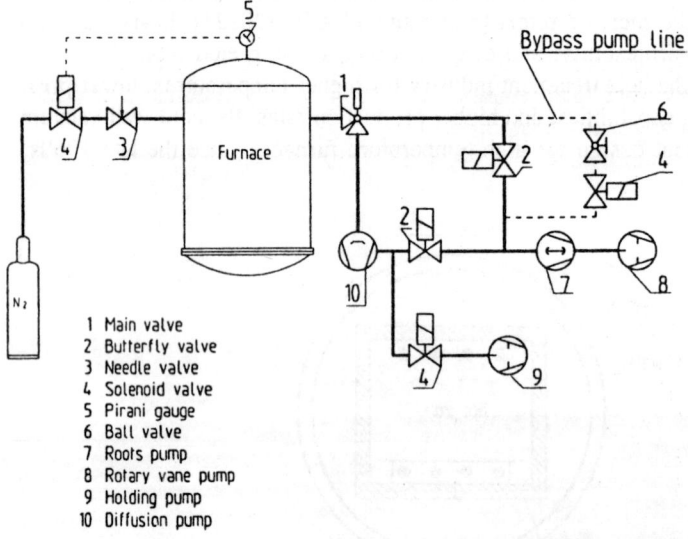

1 Main valve
2 Butterfly valve
3 Needle valve
4 Solenoid valve
5 Pirani gauge
6 Ball valve
7 Roots pump
8 Rotary vane pump
9 Holding pump
10 Diffusion pump

Figure 4 Typical vacuum system with mechanical and diffusion pump, connection valves, and lines. *1*, Main valve; *2*, butterfly valve; *3*, needle valve; *4*, solenoid valve; *5*, Pirani gauge; *6*, ball valve; *7*, roots pump; *8*, rotary vane pump; *9*, holding pump; *10*, diffusion pump.

Table 5 Final Vacuum Levels for Certain Pump Combinations

Pump	Pressure (mbar)			
	$1-10^{-2}$	$10^{-2}-10^{-3}$	$10^{-3}-10^{-6}$	$< 10^{-6}$
Rotary vane pump, rotary piston pump	Req	Req	Req	Req
Roots pump	Rec	Req	Rec	Req
Oil diffusion pump		Rec	Req	
Turbomolec. pump				Req

Req = required; Rec = recommended.

3. Heating Chamber

For the insulation of the hot zone the following designs and materials are in use (see Figure 5):

Radiation shields

A combination of radiation shields and ceramic insulating material

Multiple-layer insulation (sandwich)

Graphite insulation

Radiation shields are manufactured from tungsten or tantalum [maximum operating temperature 2400°C (4350°F)], molybdenum [1700°C (3092°F)], or stainless steel or nickel alloys [1150°C (2100°F)]. Radiation shields adsorb only small amounts of gases and water vapors during opening of the furnace. They are, however, expensive to purchase and maintain. Compared with other types of insulation, their heat losses are high and become higher with the gradual contamination of the shields.

A sandwich insulation is composed of one or more radiation shields with ceramic insulation wool between them. Combinations of graphite fiber sheets and ceramic insulation wool are also used. These versions are cheaper to buy and maintain but adsorb much more gases and water vapor (due to the very large surface area of the insulation wool). Their heat losses are considerably lower than those of radiation shields.

Graphite fiber sheets cost somewhat more than sandwich insulation. However, as their heat losses are lower, a smaller thicknesses is sufficient. In this way, the adsorption of gases and water vapor is considerably reduced. Furthermore, the heating costs are lower, and the lifetime of this type of insulation is much longer. The maximum operating temperature is 2000°C (3632°F). The lifetime depends strongly on the purity of the graphite.

For most heat treatments in vacuum furnaces, graphite insulation is used.

Table 6 Pressure Ranges Used in Vacuum Heat Treatment of Steel

Pressure range		Applications
Rough vacuum,	1000–1 mbar	Unalloyed steel
Medium vacuum,	$1-10^{-3}$ mbar	Alloyed tool steels; stainless steel (not Ti-alloyed)
High vacuum,	$10^{-3}-10^{-6}$ mbar	Stainless steel (Ti-alloyed); CVD-coated tools; critical brazing treatments (e.g., with Ag-base filler metal)

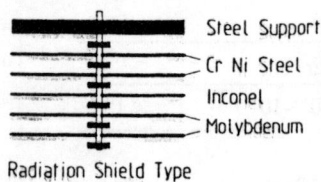

Steel Support
Cr Ni Steel
Inconel
Molybdenum

Radiation Shield Type

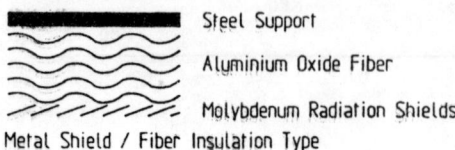

Steel Support

Aluminium Oxide Fiber

Molybdenum Radiation Shields

Metal Shield / Fiber Insulation Type

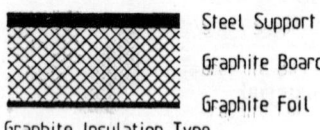

Steel Support

Graphite Board

Graphite Foil

Graphite Insulation Type

Figure 5 Vacuum insulation.

4. Heating System

In general, the heating elements for heating systems in vacuum furnaces are made from one of the following materials:

Chromium-nickel alloy, which can be used up to 1150°C (2100°F). Above 800°C (1472°F) there is a risk of evaporation of chromium.

Silicon carbide with a maximum operating temperature of 1200°C (2192°F). There is a risk of evaporation of silicon at higher temperatures.

Molybdenum, maximum operating temperature 1700°C (3092°F), becomes brittle at high temperature and is sensitive to leaks (change in emissivity).

Graphite can be used up to 2000°C (3632°F) but is sensitive to leaks (heavy wear due to the formation of CO, which will be evacuated by the pumps). The strength of graphite increases with temperature.

Tantalum, maximum operating temperature 2400°C (4350°F), becomes brittle at high temperatures and is sensitive to leaks (change in emissivity).

Uniformity of temperature is of great importance to heat treatment results. The construction of the heating system should be such that temperature uniformity in the load during heating is optimal; it should be better than ±5°C (±9°F) after temperature equalization. This is realized with two, or multiple-side, temperature control zones and a continuously adjustable supply of heating power (using thyristor or transductor control) for each zone.

In the lower temperature range [below 850°C (1562°F)], the heat transfer (by radiation) is low and can be increased by convection-assisted heating. For this purpose, after evacuation the furnace is backfilled with inert gas up to an overpressure of 1–2 bar, and a built-in convection fan circulates the gas around the heating elements and the load. In this way, the time to heat different loads to moderate temperatures [e.g., 550°C (1022°F)], can be reduced by 30–40%. At the same time the temperature uniformity during convection-assisted heating is much better, resulting in less distortion of the heat-treated part.

Figure 6 shows the design of a vacuum furnace with an integrated convection heating system (see Section IV.A.9).

5. Temperature Measurement and Control System

One of the requirements of industrial furnaces is to quickly and easily achieve various processing criteria such as a specified hardness or hardness profile with optimum energy consumption and by a method that is exactly reproducible from cycle to cycle. Microprocessor-controlled programming systems have become indispensable for satisfying this requirement.

A temperature sensor attached to the workpiece is an additional safeguard against unacceptable deviations between the workpiece and the furnace temperature. The fundamental advantage of a process control system with a load thermocouple is that the user, when preparing the program, does not need to consider any temperature/time lags between furnace and load and need only input the duration of soaking times for the load at the various stages. Also, the varying mass of different loads is rendered unimportant even if it significantly influences the actual heating up rate.

Thermocouple breakage can also be detected by computer program control followed by an automatic thermocouple changeover, so that the continuity of the automatic cycle is not interrupted.

Hence, microprocessor-controlled programming systems make it possible to prevent intolerable temperature differences between furnace and workpiece and also within the workpiece during the heat-up phase. This feature, coupled with uniform gas quenching, enables virtually distortion-free heat treatment.

6. Thermocouples

The thermocouples usually used in vacuum applications are nickel-nickel/chromium [up to 1175°C (2147°F)], nickel-nickel/molybdenum [up to 1350°C (2462°F)], platinum-platinum/

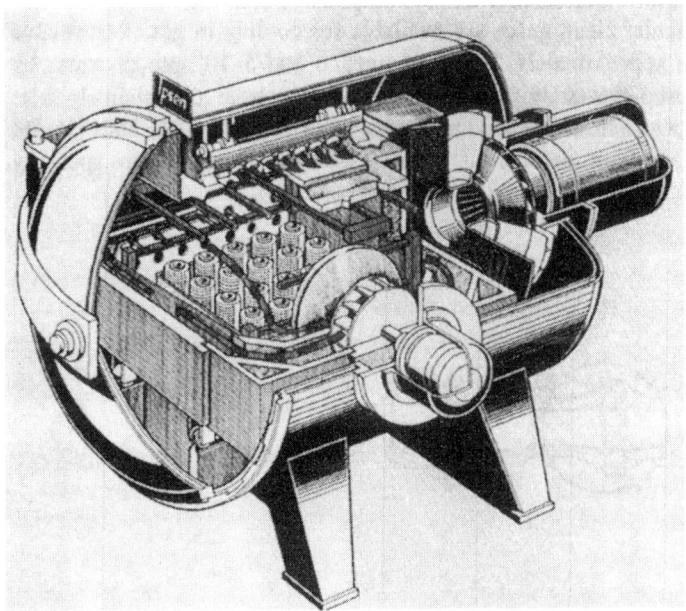

Figure 6   Schematic of a vacuum furnace with convective heating unit.

rhodium [up to 1600°C (2912°F)], and tungsten-tungsten/rhenium [up to 2700°C (4892°F)]. At higher temperatures the platinum-platinum/rhodium can only be used in the form of a compacted ceramic thermocouple because the rhodium would evaporate at these temperatures. These elements are also sensitive to the adsorption of contaminants, resulting in deviations.

A special version of a temperature-measuring system is the so-called triple-thermocouple assembly (Figure 7). Three very thin ceramic protective tubes are fitted into a conventional thermocouple junction head. Two protective tubes hold platinum-platinum/rhodium thermocouples. The third protective tube is empty. The two thermocouples can be connected with two digital controllers whose outgoing signals are compared. If the difference between these two outgoing signals exceeds a preset value, an alarm is activated [3]. Using a temperature controller, the deviation of these thermocouples from a standard thermocouple can be detected. For this purpose a standard thermocouple is inserted in the third protective tube. In this way, any deviations of the thermocouples are indicated and can be corrected, even during operation. The lifetime of these thermocouples is longer than that of other thermocouples.

More and more microprocessor-based digital programmers are being used for temperature control. In this way, in combination with a multizone temperature control, a most exact and uniform heating of the load is achieved.

7. Cooling System

The following media (listed in order of increasing intensity of heat transfer) are used for the cooling of components in vacuum furnaces:

1. Vacuum
2. Unagitated inert gas (Ar, N_2) at normal conditions
3. Agitated, recirculated gas (Ar, N_2) at normal conditions
4. Agitated, recirculated gas (Ar, N_2, He, H_2) at overpressure up to 20 bar or more
5. Agitated oil

After heating in vacuum, the bright surface of the components must be maintained during the cooling. Today, sufficiently clean gases are available for cooling in gas. Permissible levels of impurities amount to approximately 5 ppm of oxygen and 5–10 ppm of water by volume. Normally nitrogen is used as a cooling medium because it is cheap and relatively safe.

With multichamber furnaces such as a vacuum furnace with integrated oil bath (sealed quench), an additional cooling medium, namely oil, is also available. These oils are specially selected (evaporation-resistant) bright quenching oils (see Section VI).

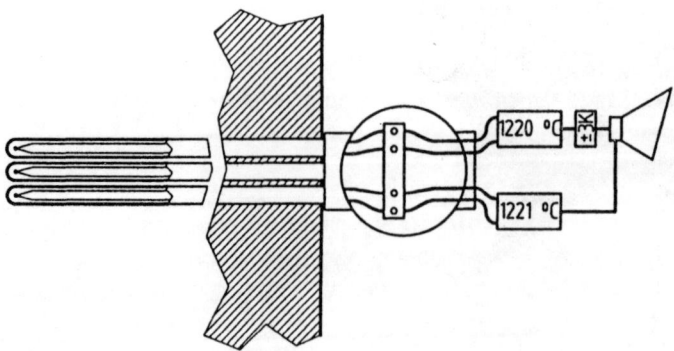

Figure 7   Triple-thermocouple assembly.

8. *Workload Support*

Materials like Inconel, molybdenum, and graphite are very often used for baskets, trays, and fixtures. Under vacuum conditions and at high temperatures, these materials might react with the workload; for example, graphite and stainless steel form a eutectic that melts at 1125°C (2057°F).

Table 7 shows the maximum admissible temperature for selected materials in mutual contact under vacuum. These materials must be separated (by using ceramic materials like paper, cloth, or rods) if the working temperature in the furnace exceeds these maximum admissible temperatures.

9. *Types of Vacuum Furnaces*

Vacuum furnaces can be classified, according to the mode of loading, into horizontal and vertical furnaces. A large number of configurations exist that are described in detail in the literature [4]. Here the description of these furnaces is restricted to some of the latest developments in cold-wall furnaces.

Horizontal Furnaces. Figure 8 shows a classical horizontal box-type vacuum furnace with heating on two sides and gas cooling from bottom to top. The maximum cooling gas pressure is 1.5 bar. Because of the simple design and the mode of loading, this type of furnace finds many applications in heat treatment.

To increase the speed of cooling, vacuum furnaces with gas cooling at higher pressures were developed. The realization of high cooling rates in vacuum hardening furnaces with a high-pressure cooling system is no longer a problem for practical applications. Vacuum furnaces with gas quenching facilities of up to 20 bar gas pressure and more exist today. With high-pressure gas cooling, the influencing factors are well known, enabling the cooling rate to be controlled.

It is important to achieve the required hardness by employing a sufficiently high rate of cooling, but the minimization of distortion of the quenched components is of at least equal importance. In the first place, the development of distortion is not governed by the speed of cooling itself, but by the uniformity of the cooling of the components. The uniformity of cooling in every section of the components depends on the design of the furnace, i.e., the design of the cooling system and the way the components are positioned in the charge. Figure 9 shows several cooling systems in use with today's vacuum hardening furnaces.

The all-around jet system (Figure 9a) directs streams of gas onto the charge from all sides. With large, heavy workpieces, this leads to uniform cooling. However, with a dense load of many small components, such a jet system creates a zone of less movement of the cooling gas in the center of the load, consequently with a reduced cooling rate in this area.

Cooling systems with straight-through gas flow (Figures 9b–9d) realized by the opening of two opposite dampers in the heating chamber provide a good cooling rate in the center of a load of many small components. The disadvantage is the distortion of large components, which receive the flow from one side only. However, by alternating the direction of flow (e.g., 10 s from above, 10 s from below) and oscillating the jet stream horizontally (Figures 9e–9f), this disadvantage is reduced to a minimum.

Figure 10 shows a schematic of a horizontal vacuum furnace with a dual dynamic gas cooling system (see also Fig. 9e-9f).

The heating rate and temperature uniformity in the lower temperature range [below 850°C (1562°F)] are low in vacuum furnaces and can be increased by convection-assisted heating (see also Section IV.A.4).

Multichamber Furnaces. To increase the throughput, multichamber furnaces can be used. Figure 11 shows a three-chamber furnace. The heating chamber is in the middle and is sepa-

Table 7 Maximum Temperature (°C) of Materials in Mutual Contact Under Vacuum

	W	Mo	Al_2O_3	BeO	MgO	SiO_2	ThO_2	ZrO_2	Ta	Ti	Ni[a]	Fe[a]	C
W	2540	1930	1825	1760	1370	1370	2205	1595	1930		1260	1205	1480
Mo		1930	1825	1760	1370	1370	1900	1900	1825	1260	1260	1205	1480
Al_2O_3			1825										
BeO				1760	1370		1760	1760	1595				1760
MgO					1370		1370	1370	1370				1370
SiO_2						1370							1370
ThO_2							1980	2205	1900				1980
ZrO_2								2040	1595				1595
Ta									2340	1260	1260	1205	1930
Ti										1260	930	1040	1260
Ni											1260	1205	1260
Fe												1205	1095
C													2205

[a] Also for Ni, Fe, Cr alloys.

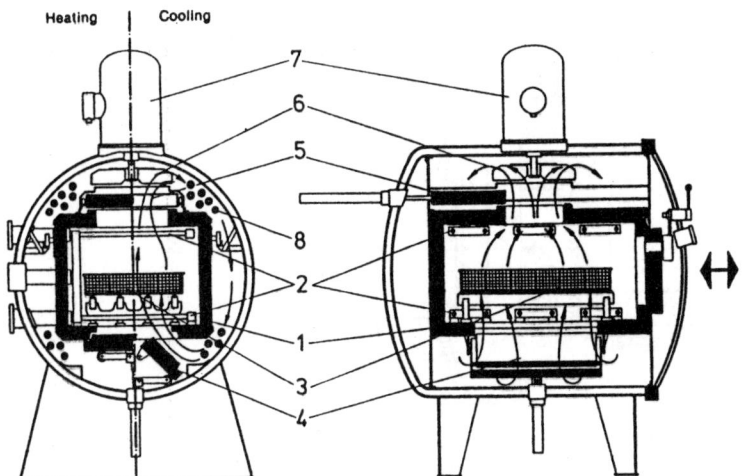

Heating | Cooling

Figure 8 Horizontal vaccum furnace. *1*, Heating chamber; *2*, heating elements; *3*, charge carrier; *4*, bottom bung; *5*, top bung; *6*, fan; *7*; fan motor; *8*, heat exchanger.

rated from the loading chamber and gas/oil quench chamber by vacuum-sealed doors. The heating chamber remains constantly evacuated and can be kept at temperature.

Furnaces like this can also manage to successfully quench components with low hardenability that cannot be fully hardened in single-chamber vacuum furnaces by gas quenching, e.g., ball bearing rings made from AISI 52100 steel. Oil or gas can be chosen as a cooling

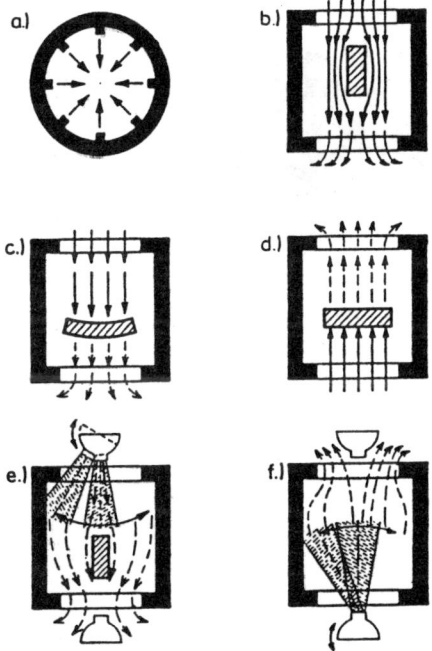

Figure 9 Three types of gas quenching systems in vacuum furnaces; all-around jet system (a); static straight-through (b-d); and dynamic straight-through (e-f).

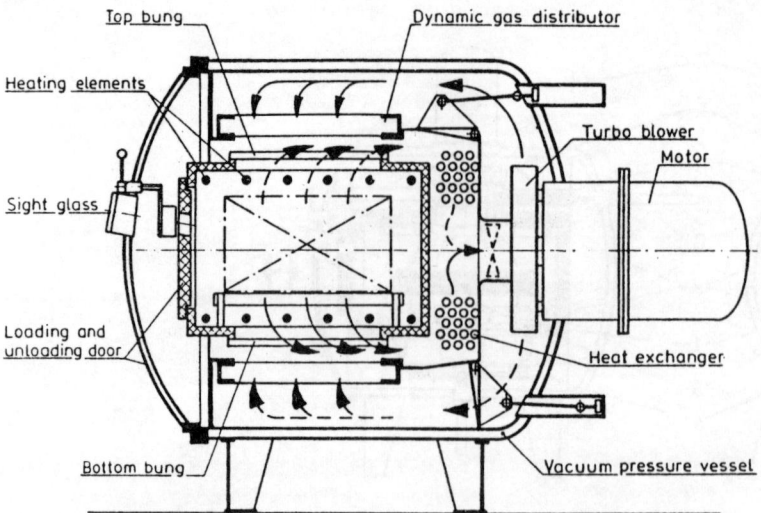

Figure 10 Horizontal vacuum furnace with dual dynamic gas cooling system.

medium. In this way, the furnace combines the features of vacuum heat treatment with oil hardening.

Vertical Furnaces.   The previously described high-pressure gas quench technology of horizontal furnaces has also been applied to vertical vacuum furnaces. Figure 12 shows schematically a vertical vacuum furnace for the hardening of long and large tools (e.g., broaches,

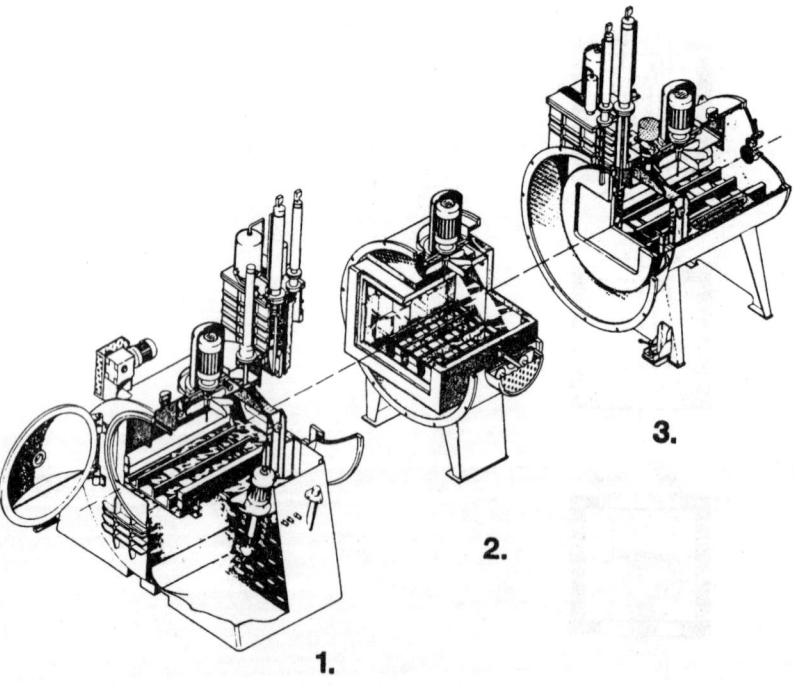

Figure 11 Three-chamber vacuum furnace with integrated gas and oil quench.

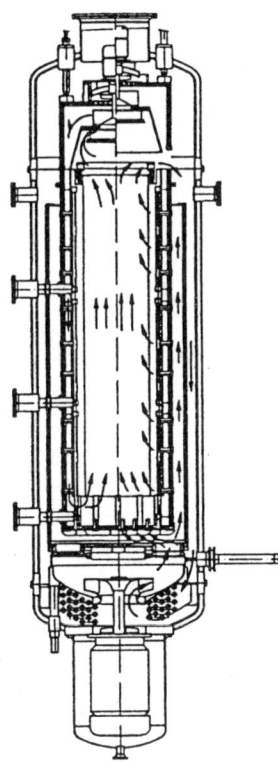

Figure 12 Cross section of a vertical vacuum hardening furnace with convection heating and overpressure gas quenching.

with weights up to 1000 kg and length up to 3 m) in a high-speed stream of nitrogen at pressures up to 6 bar. The furnace is equipped with a convection-assisted heating system to reduce heat-up time and increase temperature uniformity during heating, thereby minimizing distortion.

B. Plasma Furnaces

Plasma furnaces are of very similar design to normal vacuum heat treatment furnaces. The basic differences, as shown in Figure 13, are the electrical isolation of the load from the furnace vessel via load support isolators; the plasma current feed-through; the high-voltage generator, which creates the plasma; and the gas dosage and distribution system. Plasma furnaces also have the normal vacuum furnace chamber and pumping station.

Plasma furnaces exist in all variations (just like standard vacuum furnaces)—as horizontal single- or multiple-chamber furnaces, car bottom furnaces, vertical bell furnaces, and pit and bottom loader furnaces. Depending on the specific application, they are either low-temperature furnaces [up to 650 or 750°C (1200 or 1382°F)] for plasma nitriding or high-temperature furnaces [up to 1100 or 1300°C (2012 or 2372°F)] for plasma carburizing. High-temperature furnaces are usually cold-wall furnaces with water-cooled double walls. They are equipped either with a high-pressure gas quench system or an integrated oil quench tank [5].

Figure 14 shows the schematic of a horizontal chamber furnace with incorporated oil quench tank. The hot zone is usually made up of graphite heating elements and graphite lining but can also be fully metallic. Figure 15 depicts such a plasma carburizer with a load size

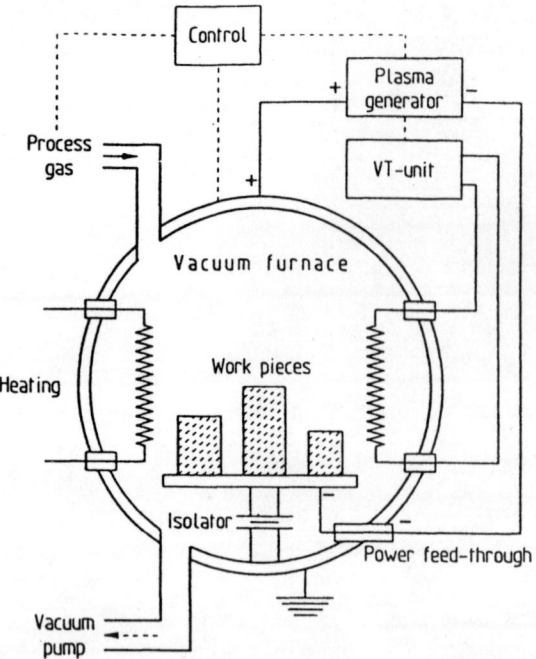

Figure 13   Schematic of a plasma heat treatment installation and components.

of 910 × 760 × 1220 mm (36 × 30 × 48 in) being used for plasma case-hardening of gears.

Low-temperature furnaces for plasma nitriding are constructed as cold-wall or hot-wall furnaces. The cold-wall design is very similar to those of the high-temperature furnaces with a hot zone inside a double-walled and water-cooled vacuum vessel. The hot-wall design con-

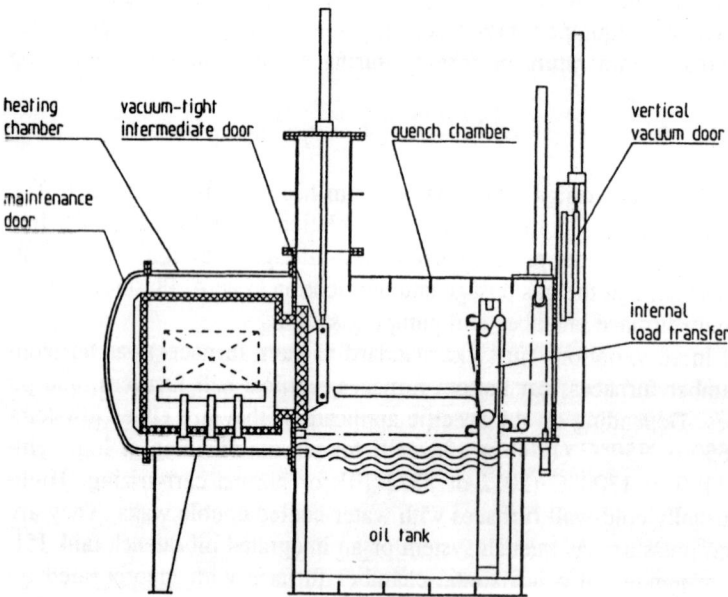

Figure 14   Schematic of a horizontal chamber plasma carburizer with integrated oil quench.

Figure 15   View of the front side (loading side) of a horizontal plasma carburizer with oil quench, for loads up to 910 × 760 × 1220 mm (w × h × l).

sists of a retort that is heated from the outside and has fiber insulation around it (Figure 16). The heating of the load is usually carried out with a convection system with a convection fan inside the retort. After evacuation and backfilling with nitrogen, the convective heating system will bring the load to nitriding temperature, where the furnace is evacuated once more

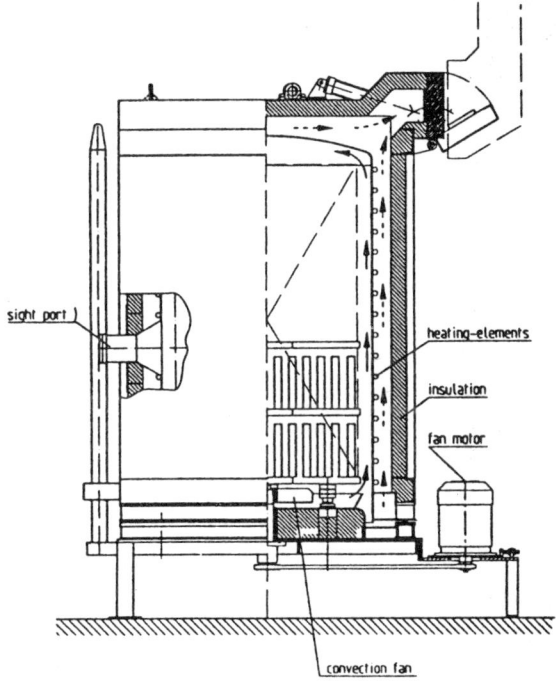

sight port

heating-elements

insulation

fan motor

convection fan

Figure 16   Schematic of a vertical (bell-type) plasma nitrider of the hot-wall design (with heat exchanger).

for the start of the plasma process. Cooling is carried out by air being blown around the retort and/or by recirculation of nitrogen gas through the retort via an external heat exchanger.

Figure 17 shows a bell-type plasma nitrider (a cold-wall furnace) for load sizes of 850 mm diameter × 1,500 mm (33.5 × 59 in.) being used for treating hydraulic pump components.

The generator needed to create a plasma inside a plasma furnace has to be a high-voltage dc generator (up to 1000 V). Currently there are two types of generators in use; one type has continuous-current outputs and the other has pulsed current output. Continuous-current generators have essentially only one advantage; they can be built with very high power and current outputs (up to 450 A at 1000 V).

The advantages of pulsed current generators are manifold. These generators

1. Permit the use of better insulated furnaces, thus saving energy and improving the temperature uniformity within the load
2. Stabilize the plasma and suppress the occurrence of arc discharges
3. Support plasmas at higher gas pressures (even above 10 mbar), thus penetrating much better into small holes and slits
4. Prevent the occurrence of cathode hole discharges in holes and slits with their unwanted overheating effects

Generators with fixed and variable frequencies exist, all in the kilohertz range. It has not

Figure 17 View of a vertical bell-type plasma nitrider with two bases for load sizes of 850 mm diameter × 1500 mm.

yet been established whether variable-frequency generators have advantages over fixed-frequency generators.

V. ANNEALING

Annealing treatments are undertaken primarily to soften a material, to relieve internal stresses, and to modify the grain structure. These operations are carried out by heating to the required temperature and soaking at this temperature for sufficient time to allow the required changes to take place, usually followed by a slow cooling at a predetermined rate. A metallurgical decision to vacuum-anneal is primarily influenced by the cleanliness and high quality of surface finish that can be obtained relatively easily compared to controlled atmosphere heat treatment operations.

A. Stainless Steel

Although many grades of stainless steel have been successfully annealed in low-dew-point hydrogen atmospheres for many years, there is now a considerable quantity of all grades of stainless steel being annealed in vacuum furnaces. Processing of stainless steel components in vacuum furnaces is often specified not only because of the cleanliness of the finished product, but also because the fast inert gas quench capability supports a high production rate. Austenitic and ferritic stainless steels are usually gas quenched in nitrogen for general commercial applications. However, austenitic steel grades stabilized with titanium and columbium are argon quenched, particularly for nuclear energy and aerospace applications.

Some chromium evaporation can take place during the annealing of stainless steels, but normally the amount lost is not significant, and seldom even measurable, because of the short time at heat [~1000°C (1832°F) temperature] and the slow diffusion rates of chromium in steel. The annealing parameters for a range of stainless steels are presented in Table 8.

B. Carbon and Low-Alloy Steels

Increasing use is being made of vacuum annealing of carbon and low-alloy steels where it is economically justifiable because of the cleanliness of the products and the prevention of carburization or decarburization.

C. Tool Steels

Because vacuum furnaces do not have any reaction with the materials being processed, it is possible to anneal tools that have already been hardened, modify their design to meet part changes, and reharden them in vacuum. This is impractical with other types of furnace equipment, as all working surfaces of the tools could be affected to the extent that they might have

Table 8 Annealing Parameters for Stainless Steels

Type	Typical analysis (%)	Annealing temp. range (°C)	Vacuum range (mbar)
Ferritic	12/27% Cr, 0.08% C max.	630–870	10^{-2}–10^{-3}
Martensitic	12/14% Cr, 0.2% C and 16/18% Cr, 0.9% C	830–900	10^{-2}–10^{-3}
Austenitic			
Unstabilized	18% Cr, 8% Ni	1010–1120	10^{-2}–10^{-3}
Stabilized	18% Cr, 8% Ni, 1% Nb or Ti	950–1120	10^{-4}–10^{-5}

to be reground, thus losing the dimensional precision required.

VI. HARDENING BY LIQUID QUENCHING

Liquid quenching in vacuum furnaces is usually done in integrated oil quench tanks. Only a few cases are known of large dies being transferred, via the air, into a salt quench tank sitting next to a vacuum furnace. Today these applications are normally executed in vacuum furnaces with high-pressure gas quenching.

The design of an oil quench tank in a vacuum furnace is very similar to its equivalent in an atmosphere furnace. There are oil circulation fans on both sides of the tank and baffles to guide the respective oil flow below the load (Figure 18). The oil is heated and its temperature is controlled. It is cooled via an external oil cooler, usually employing air as cooling medium (for safety reasons).

The peculiarity of quenching in vacuum furnaces is the low pressure above the oil, which causes standard quench oils to degas violently. The duration of this degassing process depends on the amount of air absorbed by the oil during the loading and unloading of the furnace. Vacuum oils are better with respect to this gas take-up and their readiness to degas. Oils that are not degassed properly have a worse quenching severity and produce discolored components. Vacuum quench oils are distilled and fractionated to a higher purity than normal oils, which is important in producing bright and shiny quenched parts.

Table 9 shows some physical data of typical vacuum quench oils and normal quench oils.

In practice, the quenching in vacuum furnaces is frequently done with a partial pressure of nitrogen above the oil. Pressures between 50 and 200 mbar are very common. This pressure increase just before initiating the quench serves mainly to reduce the evaporation of the oil. It is well known, however, that such a pressure increase also changes the oil cooling char-

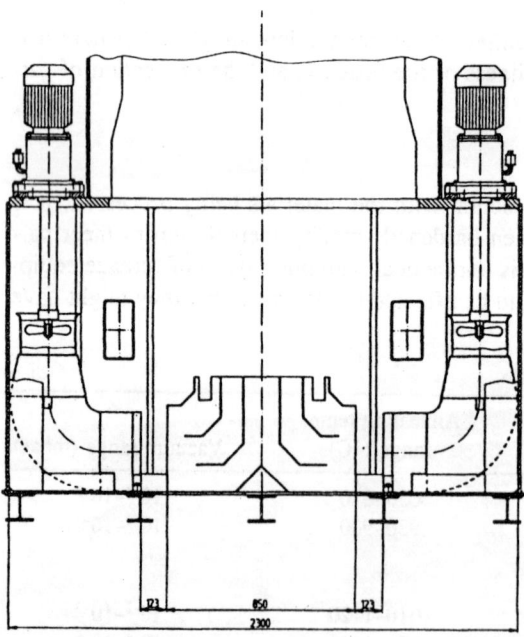

Figure 18   Schematic of an oil quench tank in a horizontal vacuum chamber furnace.

Table 9 Examples of Physical Data of Quenching Oils for Atmosphere and Vacuum Furnaces

Physical property	Severe high-duty quench oils for air and atmosphere furnaces		Hot quench oil for atmosphere furnaces	Vacuum quench oils	
	A	B	C	D	E
Density at 20°C, g/cm³	0.879	0.864	0.878	0.852	0.868
Viscosity, mm²/s					
at 40°C	46	21	120	20	50
at 50°C	31.6	16	81	12	30
Flame point, °C	228	200	264	196	230
CCT, wt %	0.17	~0.2	~0.2	0.04	0.05
Temperature range, °C	50–150	50–130	90–180	40–60	60–100
Vapor pressure at 75°C, mbar	$\sim 2 \times 10^{-3}$	$\sim 10^{-2}$	$\sim 10^{-4}$	$\sim 10^{-2}$	$\sim 2 \times 10^{-2}$
Sulfur content, wt %	<0.3	<0.3	<0.3	0.04	0.05

acteristics of the quench oils. As Figure 19 [6] demonstrates, the pressure increase shortens the vapor blanket phase, thus increasing the quench severity at high temperatures (in the pearlite–ferrite transformation). On the other hand, it lowers the quench rate in the convective cooling phase, i.e., in the lower temperature range of the bainite or martensite transformation. Thus, high gas pressures above the oil can be advantageous in producing full hardness on unalloyed or very low alloy materials, whereas low nitrogen pressures above the oil produce higher hardnesses and lower distortions on components of alloyed steels.

Very low pressures above the oil (<50 mbar) and very high quenching temperatures [like 1200°C (2200°F)] can lead to carbon pick-up of the surface of the workpieces as has been experienced in hardening components of high-speed steels [7]. The carbon originates from the oil vapor in contact with the hot surface of the load in the initial phase of the quench process. High nitrogen pressures (>200 mbar) reduce or eliminate this effect.

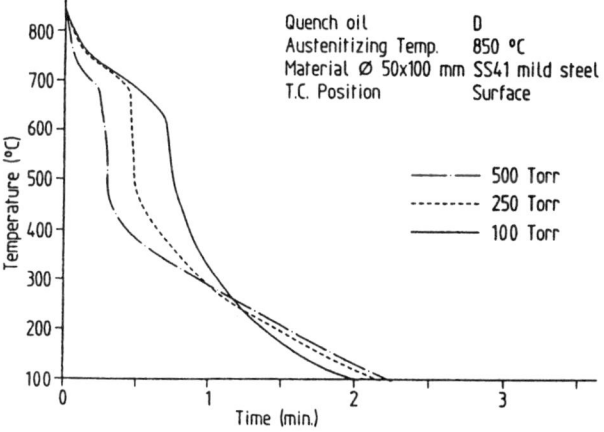

Figure 19 Cooling curves of a quench oil with three different nitrogen pressures (100, 250, 500 torr) above the oil. (From Ref. 4.)

VII.   HARDENING BY GAS QUENCHING

The capabilities of vacuum furnaces are continually being examined to improve their productivity by reducing cycle time and to improve the metallurgical applications by improving the quenching capability. In recent years, great attention has been paid to obtaining more rapid gas quenches [8].

A.   Cooling Properties of Gases

With the advent of rapid gas quenching systems in vacuum furnaces, it was recognized that the flow rate and density of the cooling gas blown onto the surface of the load were the decisive factors for reaching high heat transfer, i.e., high cooling rates. In addition to high gas velocities, high gas pressures are needed to through-harden a wide variety of steel parts with appreciable dimensions. The first vacuum quenching furnace to operate at a pressure of 2 bar was developed in 1975, and the first 5 bar furnace in 1977.

   Calculations of heat transfer show that the heat transfer coefficient α is proportional to the product of gas velocity and gas pressure:

$$\alpha \approx (vP)^n \tag{1}$$

where v is the gas velocity and P the pressure of the gas. The exponent n depends on the furnace design, the load, and the properties of gas. It lies typically in the range of 0.6–0.8. The exponential behavior of the heat transfer makes it clear that the difference in the increase of heat transfer is considerable with the first few bars of pressure but decreases with increasing pressure.

   This behavior is demonstrated in Figure 20, which shows cooling curves from 1200 to 550°C (2200 to 1022°F). An increase of the pressure from 1 to 2 bar reduces the cooling time from 120 to 60 s, i.e., a reduction of 60 s. An increase from 4 to 5 bar reduces the cooling time from 30 to 24 s, i.e., a reduction of 6 s. It is obvious that any further increase of the pressure above 5 or 6 bar gives very little increase in heat transfer per bar of pressure in-

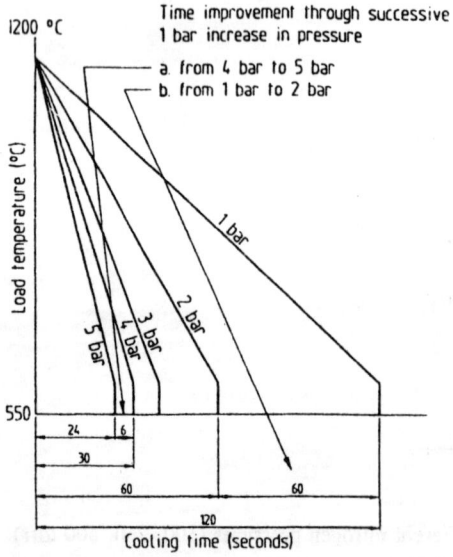

Figure 20   Dependence of cooling time on gas pressure.

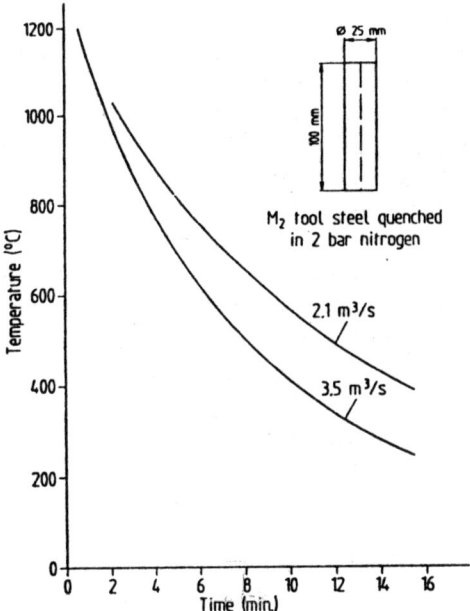

Figure 21 Effect of gas velocity on cooling speed.

crease. Therefore, a 5 or 6 bar furnace represents an economic optimum. It gives very high quenching rates at reasonable cost.

The increase of cooling speed with higher gas velocity (higher flow rate) is shown in Figure 21.

Figure 22 shows, for different types of cooling gas, that argon gives the slowest cooling, followed by nitrogen, helium, and hydrogen, in the order of increasing cooling rate.

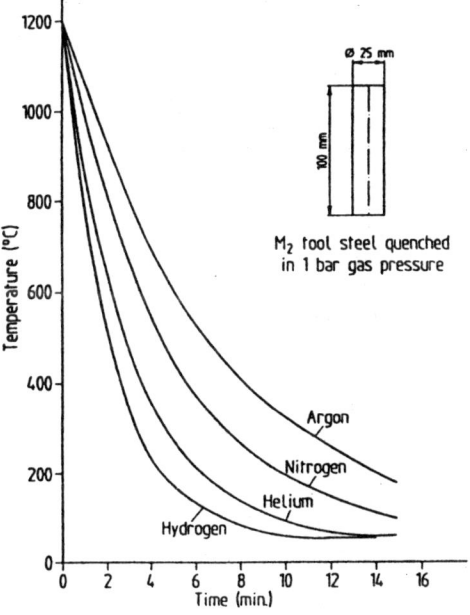

Figure 22 Effect of type of gas on cooling speed during gas quenching.

Theoretically, there is no limit to the improvement in cooling rate that can be obtained by increasing gas velocity and pressure. Practically, however, very high pressure and very high velocity systems are difficult and costly to construct. In particular, the power required for nitrogen gas recirculation increases faster than benefits accrue. On the other hand, there are pressure–gas combinations that provide heat transfer coefficients within the range of those produced by agitated oil quenchants (Table 10) [9]. Naturally, the power requirements for gas circulation decrease dramatically when less dense gases like helium or hydrogen are being used. Thus, gas quenching can be used to produce full hardness in many oil-quenching steels.

Gas quenching has certain advantages over liquid (oil and salt bath) quenching. Because the cooling rate can be easily changed by altering gas velocity or pressure, the same vacuum furnace and gaseous cooling medium can be used to quench a wide variety of materials. A furnace using nitrogen at 6 bar, for example, can quench a range of cold-work, hot-work, high-speed, stainless, and bearing steels.

It also is possible in gas quenching to change the cooling rate during a single cycle, whereas multiple baths are required in liquid quenching systems. Figure 23 shows the controlled cooling with an isothermal hold in vacuum furnaces. The isothermal hold requires two load thermocouples. The cooling speed of the workpiece is controlled by variation in gas speed and pressure and by reheating of the furnace if necessary. This ability to perform interrupted quenching is particularly important in applications where parts can distort or crack if cooling occurs too rapidly over the entire quenching range.

The effect of load weight on the resultant cooling speed during gas quenching is more pronounced than, for example, in liquid or salt bath quenching. Figure 24 shows the cooling speed for different load weights obtained with different part dimensions during gas quenching with 6 bar nitrogen (λ is the cooling time of the core in seconds from 800°C to 500°C (1472 to 932°F) divided by 100). This type of data will vary from furnace to furnace depending on design and loading capacity. However, the data provide an indication of the magnitude of the parameters involved. The determination of such data for any particular furnace is of considerable use to a heat treater who needs to obtain the maximum performance from a furnace. The heat treater should also consider the largest section size that can be through-hardened and

Table 10   Effects of Quenching Parameters on Heat Transfer Coefficient

Medium and quenching parameters	Heat transfer coefficient [W/(m²·K)]
Air, no forced flow	50–80
Nitrogen, 6 bar, fast[a]	300–400
Nitrogen, 10 bar, fast[a]	400–500
Helium, 6 bar, fast[a]	400–500
Helium, 10 bar, fast[a]	550–650
Helium, 20 bar, fast[a]	900–1000
Hydrogen, 6 bar, fast[a]	450–600
Hydrogen, 10 bar, fast	750[b]
Hydrogen, 20 bar, fast	1300[b]
Hydrogen, 40 bar, fast	2200[b]
Oil, 20–80°C, no flow	1000–1500
Oil, 20–80°C, agitated	1800–2200
Water, 15–25°C, agitated	3000–3500

[a]Fan speed of 3000 rpm.
[b]Calculated values.

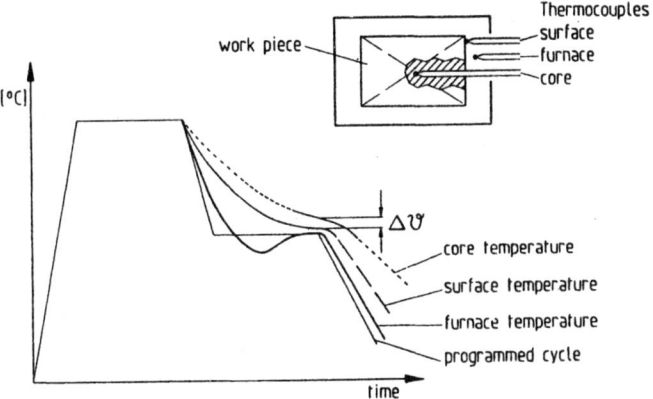

Figure 23   Controlled cooling in vacuum furnaces during gas quenching.

compare this to the appropriate continuous cooling diagram for the material to be processed, with respect to the data obtained above.

B.   Tool Steels

The advantages of vacuum furnaces for the heat treatment of tool steels are well known to-day. Vacuum furnaces provide unreactive conditions for the materials being treated that ensure development of the optimum properties desired. Parts processed in this way have surfaces that are neither carburized nor decarburized and consequently exhibit superior performance. Such a condition is practically impossible to obtain when using protective atmospheres generated by cracking some types of hydrocarbons, because their carbon potential cannot be kept in balance with the carbon content of the material being processed over the total range of temperatures experienced in the heat treatment cycle. Even dry hydrogen atmospheres present problems in maintaining the dew points necessary to prevent some degree of decarburization in the furnace.

The bright hardening of tool steels from the air-hardening category has proven in many situations to be economically viable relative to salt bath treatments and is the most important

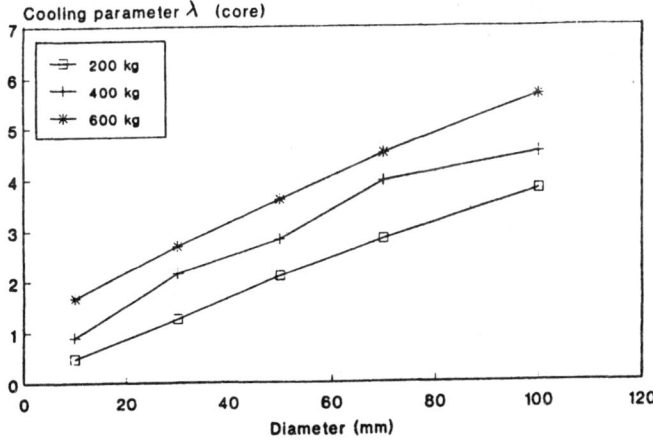

Figure 24   Effect of load weight on cooling speed.

application of gas quench hardening in vacuum furnaces. The ability to minimize shape distortion and in many situations to virtually eliminate finishing costs on complex shapes through the use of vacuum furnaces is particularly advantageous.

Steels in the following group designations based on the AISI classification of tool steels have been successfully treated in vacuum for many years now.

A Series	Air-hardening medium-alloy cold-work tool steels
D Series	High-carbon, high-chromium cold-work steels
H Series	Hot-work tool steels of three types: chromium, tungsten, and molybdenum
T Series	Tungsten high-speed steels
M Series	Molybdenum high-speed steels

In general, the air-hardening tool steel parts are hardened in much the same way as in atmosphere furnaces. They are preheated, heated to a high austenitizing temperature, and cooled at a moderate rate.

The medium-alloy air-hardening steels in the A series and the high-carbon, high-chromium steels in the D series are regularly hardened in gas quench furnaces (nitrogen up to 6 bar). The notable exception is the widely used wear-resistant 12% Cr, 2.5% C tool steel D3, which requires oil quenching (or higher pressures of helium gas) if full core hardness is to be attained for massive tools.

Also, it is recommended that some of the grades of hot-work tool steels, e.g., H23, be oil quenched if full hardness is to be achieved. In general, there is a limit to the section size of these materials. However, this limitation of section size is similar to that for tools hardened in a salt bath, marquenched, and air-cooled.

The tungsten and molybdenum high-speed steels can be hardened up to moderate sizes [~120 mm (4.72 in)] by a high-pressure nitrogen quench without any loss of essential properties in comparison to salt bath heat treatment [10].

It should be noted that only a rough vacuum of 10^{-1}–1 mbar should be used during the heat treatment of tool steels. This level of vacuum is required mainly because of the relatively high vapor pressures of chromium, manganese, and sometimes other easily vaporized elements in the tool steel (see Section III). The values of the vapor pressures of these alloying elements in pure form can be seen as a function of temperature in Table 4. The vapor pressures of the elements in solid solution should be somewhat lower than these values, but the actual values will depend on the alloy concentration in the steel.

C. Martensitic Stainless Steels

The martensitic stainless steels can be adequately hardened in vacuum furnaces. Vacuum furnace processing has eliminated many heat treatment variables, particularly with atmosphere control, which previously often resulted in high rejection rates with these heat treatment–sensitive materials.

All grades of martensitic stainless steels have been processed in vacuum furnaces using the same austenitizing temperatures and considerations as those used in atmosphere furnaces. Since the austenitizing temperatures are usually below 1090°C (2000°F), vacuum levels in the range of 10^{-3} mbar are very often employed, which result in clean and bright parts upon unloading from the vacuum furnace. To avoid evaporation of alloying elements, processing is also done at vacuum levels ranging from 10^{-3} up to 10^{-1} mbar, but the lower pressure levels result in cleaner and brighter workpieces. Because of the differences in the hardenability of the various martensitic stainless alloys, there is a limitation on the section sizes that can be fully hardened by recirculated nitrogen gas quenching. The actual values of section size limits depend on the type of cooling system and the capability of the specific furnace employed.

D. Heat Treatable Construction Steels

Parts for the aircraft industry are very often made of heat-treatable steels that are normally oil hardened. Table 11 shows that at least some of these steels can also be successfully hardened with 6 bar nitrogen gas pressure [11]. Normal low-alloy case-hardening steels, however, cannot be hardened sufficiently in 6 bar nitrogen gas. With higher alloy steels, this is possible up to certain section sizes.

Figure 25 shows the hardness profile of plasma-carburized shafts made from AFNOR 16NCD13 constructional steel. These shafts, with diameters of 10, 20, and 30 mm (0.4, 0.8, and 1.4 in) were gas quenched with 6 bar N_2. If larger loads of low-hardenability steels are to be fully martensitically hardened, the 6 bar nitrogen gas quenching reaches its limit, and other gases and higher pressures have to be used.

Figure 26 shows the hardness profile of a plasma-carburized ball bearing cage made from 17NiCrMo2 ball bearing steel that was quenched with 10 bar N_2 and 10 bar He, respectively. It is obvious that the use of the helium gas gives a higher case depth and a higher core hardness than nitrogen with the same gas pressure. Further applications of gas quenching with 10 bar N_2 and 10 bar He are (plasma) carburized gear wheels made of case-hardening steels 17NiCrMo6 and 16MnCr5 [11].

It can be summarized that in vacuum heat treatment the 6 bar nitrogen gas quenching has become a standard treatment in batch furnaces. It can be applied to a lot of cold-working steels, hot-working steels, high-speed steels, stainless steels, and even high-alloy heat-treatable steels and case-hardening steels. However, for some of these steels there are, of course, limitations with respect to the maximum section size.

A further increase in quenching speed is obtained by using a higher gas pressure (up to 20 bar) and gases other than nitrogen (e.g., helium). This enables some low- to medium-alloy tool steels and case-hardening steels to be hardened.

Table 11   Gas Quenching of Some Oil Hardening Steels in Vacuum Furnaces with 6 Bar Nitrogen

SAE/ AISI	DIN		AFNOR	Hardness HRC		
				Max value att.	20 mm diam.	40 mm diam.
4135	34CrMo4	1.7220	35CD4	55	39	38
	36NiCr6	1.5710	35NC6	56	45	44
4130	25CrMo4	1.7218	25CD4S	50	28	
	100Cr6	1.3505	100C6	65	34	34
	32NiCrMo125	1.6655	30NCD16	55	54	54
	45SiCrV6	1.2249	45SCD6	60	58	58
	32CrMo12	1.7361	30CD12	55	48	47
440B	X90CrMoV18	1.4112	Z100CD17	64	63	63
431	X22CrNi17	1.4057	Z15CN17	51	51	51
4340	40NiCrMo73	1.6562		58	57	56
300M				59	59	59

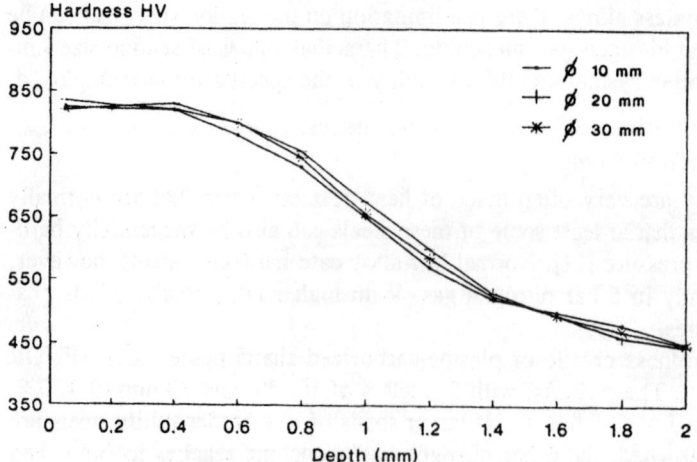

Figure 25 Hardness profile of plasma-carburized 16NCD13 shafts gas quenched with 6 bar N_2.

VIII. CARBURIZING

With normal atmosphere carburizing in atmospheres containing CO and H_2, the carbon transfer is dominated by the CO adsorption reaction

$$CO \rightarrow CO_{ad} \rightarrow [C] + O_{ad} \tag{2}$$

and the oxygen desorption reaction

$$O_{ad} + H_2 \rightarrow H_2O \tag{3}$$

which together yield the well-known heterogeneous water gas reaction

$$CO + H_2 \rightleftharpoons [C] + H_2O \tag{4}$$

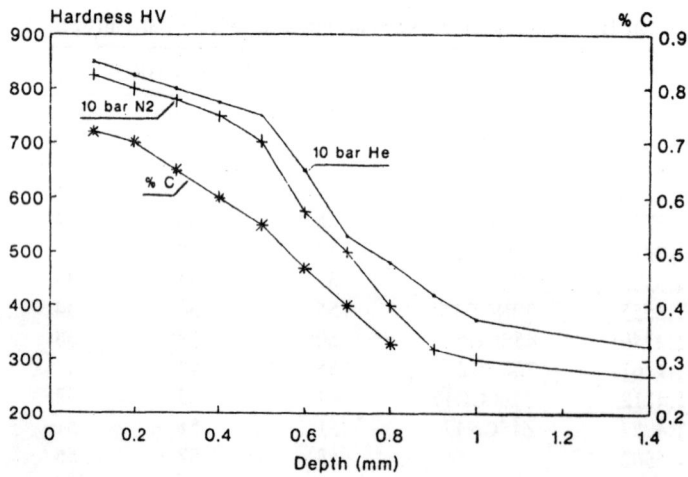

Figure 26 Hardness profile of a plasma-carburized 17NiCrMo2 ball bearing cage gas quenched with 10 bar N_2 and 10 bar He.

Thus he transfer of carbon in atmospheres containing CO and H_2 is connected with a transfer of oxygen, giving rise to an oxidation effect of alloying elements such as Si, Cr, and Mn. This phenomenon is known as internal oxidation of steel (see Chapter 12).

Carburizing in pure hydrocarbons at normal pressures delivers carbon via reactions such as

$$CH_4 \rightleftharpoons [C] + 2H_2 \tag{5}$$

and

$$C_3H_8 \rightleftharpoons 3\,[C] + 4H_2 \tag{6}$$

At normal pressure, these reactions come into equilibrium only at very high hydrogen levels. Thus undiluted hydrocarbons will produce carbon in excess and will therefore quickly soot up the furnaces. To lower their carbon availability, one would have to either dilute the hydrocarbons with nitrogen (or hydrogen), thus reducing their partial pressure, or lower the total pressure in the furnace. The latter is known as the low-pressure carburizing process.

A. Low-Pressure Carburizing

In the 1970s, the opportunities of low-pressure carburizing were extensively studied. It was found that the carbon transfer depends to a large extent on the partial pressure of the hydrocarbon, as shown for the case of methane in Figure 27, and increases with rising partial pressure [12]. At low partial pressures of methane (<100 mbar), the carbon transfer is negligibly small. Propane carburizes at lower pressures than methane because of its higher carbon availability.

The problem with the process using methane is that in order to achieve sufficient carbon transfer the process has to be carried out at higher gas pressures, which leads to overcarburized edges of the workpieces, nonuniform carburizing on single components and within a load, and massive soot formation in the furnace. To reduce these adverse effects, basically a pulsation of the pressure of the carburizing gas was introduced by cycling the mechanical pump either in a time-controlled manner [12,13] or by using a measuring device such as a soot sensor [12]. This cycling of the pressure of the carburizing gas (pressure pulsing) did improve the situation. It reduced the soot formation and gave better case uniformity on complex shapes as shown

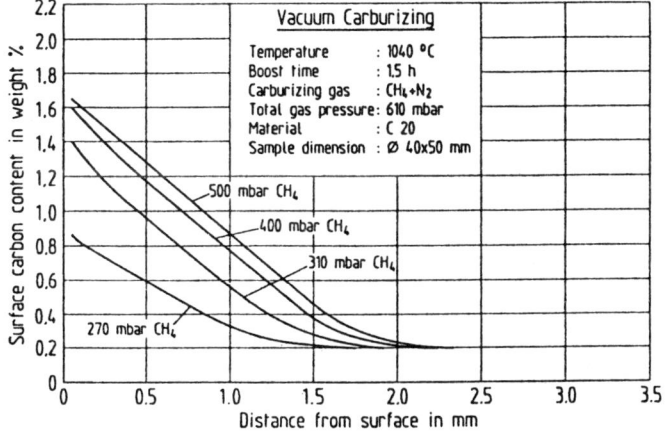

Figure 27 Carbon transfer in CH_4 plus N_2 (at 610 mbar total pressure) after 90 min of carburizing at 1040°C with different partial pressures of CH_4. (From Ref. 12.)

in Figure 28. As it never totally solved the problems with respect to soot formation and carburizing uniformity within the load, this process was never really accepted by industry.

It was not until some French researchers found out in 1985 that, under certain conditions of gas flow, high rates of carbon transfer are possible at pressures of propane below 10 mbar that a revival of the low-pressure carburizing process took place. The idea of Pourprix and Naudot [14] was to inject the propane into the vacuum furnace at such a high rate, i.e., high gas velocity, that the thermal dissociation of propane into methane and ethylene and further into methane, carbon, and hydrogen,

$$C_3H_8 \rightarrow CH_4 + C_2H_4 \rightarrow CH_4 + 2C + 2H_2 \tag{7}$$

would not occur before the propane gas reached the workpieces of the load. This is achieved by feeding sufficiently high propane gas flows, via an adapted nozzle system, onto all sides of loads of limited size and density. The limitation in load size and density is necessary for the propane gas to reach the center part of the load in sufficient time and quantity.

Figure 29 [14] shows the hardness profiles achieved within a load of 80 pinions of 27MnCr5 steel after 15 min carburizing at 910°C (1670°F) (and 5 mbar) and 20 min of subsequent diffusion followed by a quench in nitrogen gas (at 3.5 bar). The control of such a low-pressure carburizing process in hydrocarbons can be realized by time-controlling the carbon transfer and the diffusion phases. As no workable sensors exist for the time being that allow an on-line measurement of the carbon transfer, the transfer rates are usually evaluated empirically as a function of temperature, gas pressure, and flow rate and applied to the respective kinetic and diffusion models (see Chapter 12) to determine the boost and diffuse times of the cycle.

Even though this new low-pressure carburizing process yields largely improved case uniformity on single parts and whole loads (Figure 29) and thus has found some industrial utilization in Europe, it is still not completely satisfactory for parts with deeps small slits and holes or for large dense loads. The process called plasma carburizing overcomes this weakness by using an electric field to enhance the carbon transfer.

B. Plasma Carburizing

Plasma carburizing also works with hydrocarbon gases at pressures below 10 mbar. It was recognized at a very early stage that an electric field greatly enhances the carbon transfer [15].

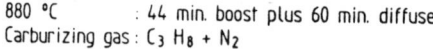

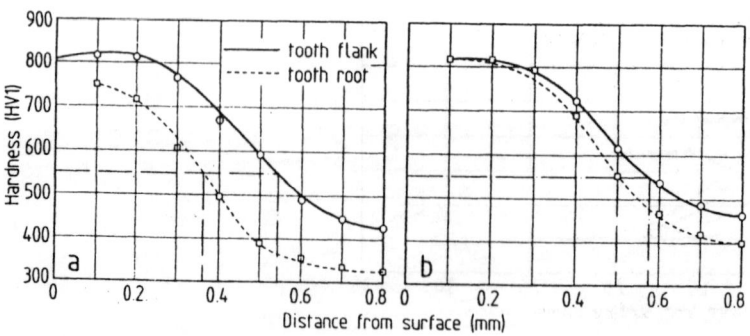

Figure 28 Comparison of case uniformity on a drive shaft with (a) unpulsed and (b) pulsed pressure of the carburizing gas ($C_3H_8 + N_2$). Carburized at 880°C with 44 min boost and 60 min diffuse. (From Ref. 12.)

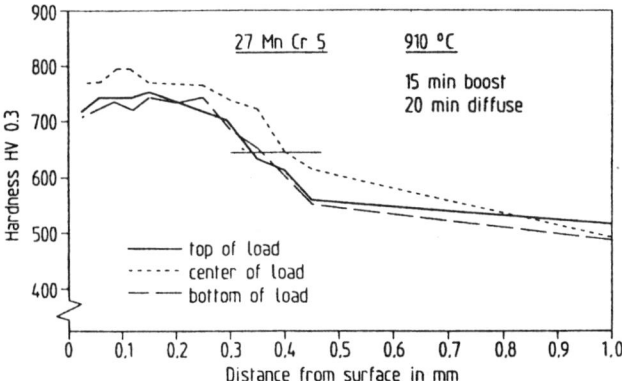

Figure 29   Uniformity of carburizing response in a load of 80 pinions of 27MnCr5 steel after low-pressure carburizing in propane at 5 mbar. (From Ref. 14.)

Grube and Gay [16] demonstrated in 1978 that the enhanced carbon transfer takes place even at low levels of plasma energy, thus relieving the plasma energy source from the task of heating the load. This was the major breakthrough for industrial use of this technology. It meant that basically standard vacuum furnaces with some modifications (see Section IV.B) could be used for the plasma carburizing process.

Today, two plasma carburizing technologies exist. One uses methane gas, the other propane gas (usually diluted with hydrogen). The carbon transfer of the two gases differs under equivalent conditions (same temperature, gas pressure, and electrical conditions). Figure 30 shows that at 900°C (1650°F) in a propane plasma the surface carbon content rises much faster than in a methane plasma [17].

Plasma carburizing in methane is usually done in a two-stage cycle with one boost and one diffuse period [18]. The boost period with the plasma ignited takes about one-third of the total cycle time, and the diffuse period takes place in vacuum.

The carbon transfer in propane is so high that the carbon content at the surface of the workload reaches the point of austenite saturation within 10–15 min. This calls for a different cycle technology. The process is usually run in a multistage cycle with many boost and

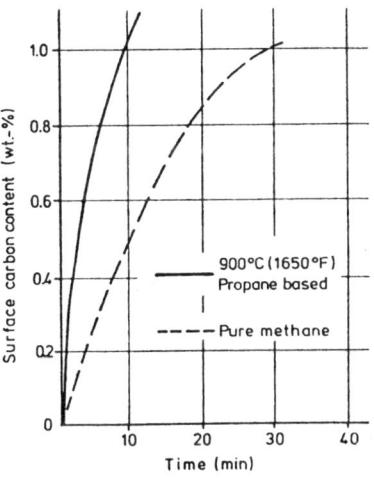

Figure 30   Effect of type of gas on plasma carburizing rate. (From Ref. 17.)

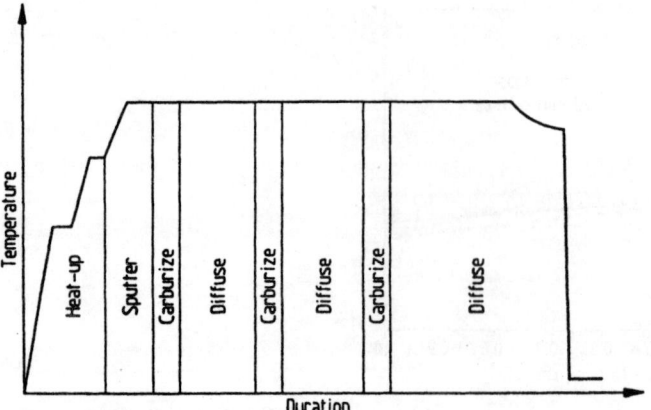

Figure 31 Schematic of a multistage cycle used for plasma carburizing in propane-base gases. (From Ref. 19.)

diffuse periods as demonstrated in Figure 31 [19]. The high carbon transfer with propane gas yields a time gain in the first phase of the cycle, which is roughly 1 h, compared to the best atmospheric carburizing processes [19]. This is true for small and large case depths as, with increasing carburizing time, the diffusion is the time-controlling step, whereas the carbon transfer is the time-controlling step during the initial phase. Figure 32 shows two identical carbon profiles produced at 930°C (1700°F), one after gas carburizing in 186 min and one after plasma carburizing in propane in 130 min.

Another control variable of the plasma carburizing process is the plasma energy or, more precisely, the plasma current density (at constant voltage) [20].

Plasma carburizing is used industrially today for constant-velocity joints [21], diesel engine components [18], hydraulic valve components [22,23], gears [19,24], and clutch com-

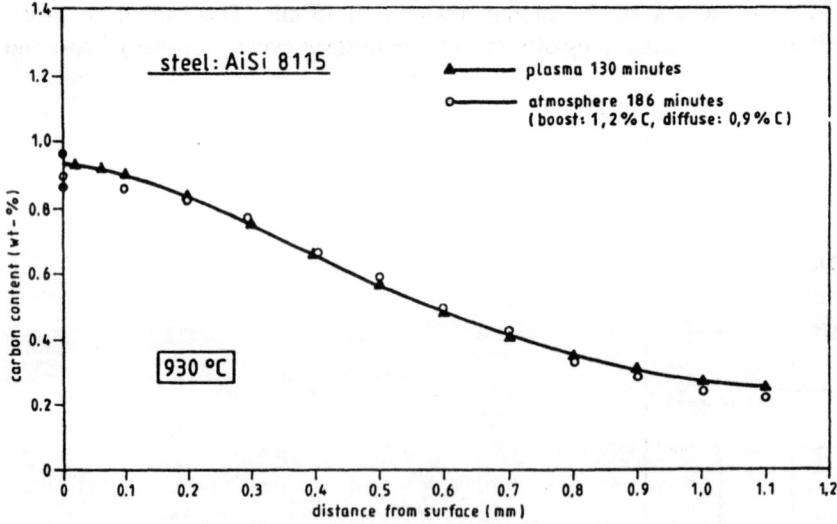

Figure 32 Comparison of carbon profiles produced at 930°C by plasma and gas carburizing cycles (steel: AISI 8115).

ponents. Further applications in the gear box and bearing industry can be envisioned. Currently these industrial applications use mostly standard carburizing temperatures between 850 and 980°C (1562 and 1800°F). Plasma carburizing furnaces, however, lend themselves to high-temperature treatments in the 1050–1100°C (1922 to 2012°F) temperature range, giving rise to dramatic reductions in treatment time. Also, the plasma carburizing of high-alloy tool or stainless steels with the aim of producing carbide-rich and extremely wear-resistant surface layers is another future area of utilization of this process [25].

The plasma carburizing process yields the following benefits compared to gaseous carburizing:

1. Improved quality of the surface layer with no internal oxidation and no soft pearlite structures and, as a consequence, higher fatigue strength
2. Better case uniformity on complex shapes (in holes and recessions)
3. Better case uniformity within a load, especially for shallow case depths
4. Shorter cycle times (by increased carbon transfer)
5. Possibility of higher carburizing temperatures
6. Excellent process reproducibility because of controlled carbon transfer
7. Simple masking methods for selective carburizing
8. No need for furnace conditioning
9. Applicability to high-alloy steels (stainless, etc.)
10. No environmental pollution
11. Excellent working conditions
12. Integrable into manufacturing lines
13. Lower running costs

The main disadvantages are [19]

1. Not applicable to bulk loads (individual loading of components necessary).
2. Carburizing of all surfaces of one component is not possible (because of necessary cathode contact).
3. Higher capital costs (counterbalanced by much reduced running costs).

IX. PLASMA NITRIDING

Plasma nitriding is today a widely utilized process, with approximately 400 industrial furnaces in operation in Europe and about 150 in North America. The acceptance of the process started roughly 25 years ago with a better understanding of the process [26] and with the availability of large production units. The process received a second push in the 1980s with the advent of pulsed power generators and insulated furnaces (see Section IV.B) [27]. This improved the process's range of application and especially its economics.

Plasma nitriding uses nitrogen gas at low pressures (1–10 mbar) as the source for nitrogen transfer. Nitrogen, being a neutral gas that does not react with steel surfaces at temperatures below 1000°C (1832°F), becomes reactive when an electric field of a few hundred volts (300–1200 V) is applied. The electric field is established in such a way that the workload is at the negative potential (cathode) and the furnace wall is at ground potential (anode) (Figure 33).

The nitrogen transfer is caused by the attraction of the positively charged nitrogen ions to the cathode (workpieces), with the ionization and excitation processes taking place in the

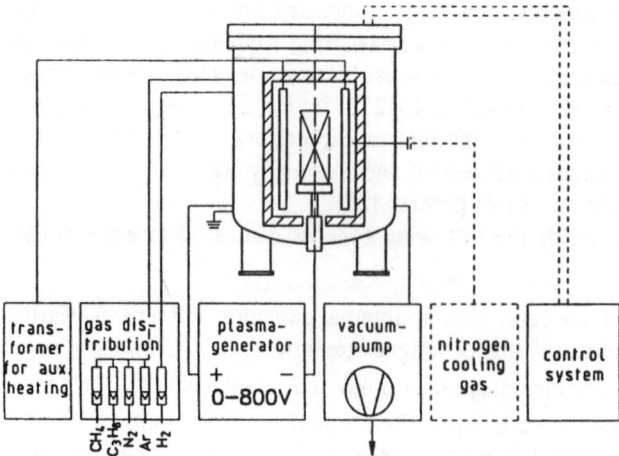

Figure 33 Schematic of a plasma nitriding system.

glow discharge near the cathode's surface (Figure 34). This region where the main potential drop takes place is called the cathode fall [26]. The rate of nitrogen transfer can be adjusted by diluting the nitrogen gas with hydrogen. The addition of hydrogen to the gas mixture has to be above 75 vol % in order for a reduction of the nitrogen transfer rate to take place (Figure 35). This influences the thickness of the compound layer ("white layer") as Table 12 shows. The higher the nitrogen concentration, the thicker the compound layer. A nitrogen gas concentration of 2% does not produce a compound layer.

The compound layer consists of iron nitrides that develop in the outer region of the diffusion layer after saturation with nitrogen. According to the iron–nitrogen phase diagram (see Chapter 12), basically two iron nitrides are possible, the nitrogen-poor γ' phase (Fe_4N) and the nitrogen-rich ε phase ($Fe_{2-3}N$). Figure 36 shows the microstructure of these different nitride layers.

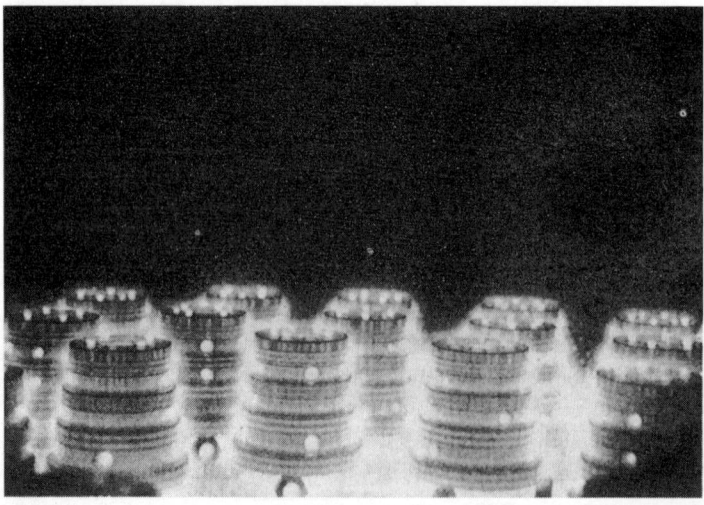

Figure 34 View into a plasma nitrider onto the workpieces showing the plasma glow during plasma nitriding.

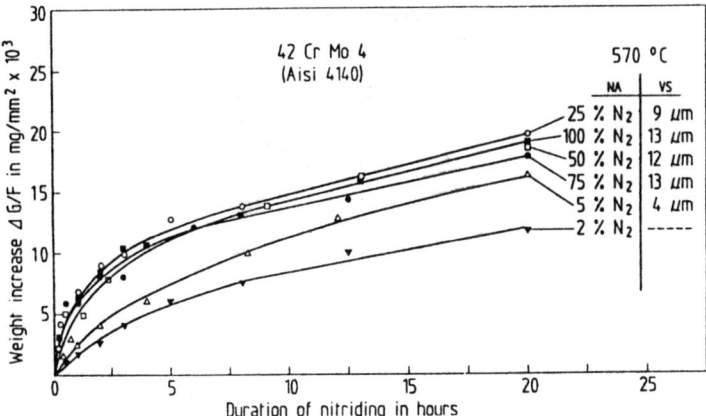

Figure 35 Nitrogen transfer, as measured by the weight increase per surface area, in different nitrogen–hydrogen gas mixtures at 570°C for 42CrMo4 (AISI 4140) steel.

The temperature of the workpiece is another important control variable. Figure 37 shows how the nitrogen transfer increases with the nitriding temperature for type 42CrMo4 (AISI 4140) steel.

The depth of the diffusion layer also depends strongly on the nitriding temperature and time (Figure 38). For a given temperature the case depth increases with the square root of time [26].

A third process variable is the plasma power or current density, which has a certain influence on the thickness of the compound layer [20]. Adding small amounts (1–3 vol %) of methane or carbon dioxide gas to the nitrogen–hydrogen gas mixture will produce a carbon-containing ε compound layer ($Fe_{2-3}C_xN_y$). This process is called plasma nitrocarburizing. It is normally used only for unalloyed steels and cast irons.

The different possible layer structures and large number of process variables make it difficult to make the correct choice for each application. Table 13 gives some indications of which layer structure to choose for certain applications and steels [28]. Hints for case depths, expected surface hardnesses, white layer thickness, and applicable nitriding temperatures for some typical steels and cast irons are presented in Table 14 [28]. Table 15 presents typical treatment cycles and results for a variety of steels.

Table 12 Thickness and Structure of the Compound Layer after 2 and 5 h of Plasma Nitriding of 42CrMo4 or AISI 4140 Steel at 570°C in N_2–H_2 Gas Mixtures with Different N_2 Contents

Amount of nitrogen in the N_2–H_2 gas (vol %)	Thickness of compound layer (μm) after plasmanitriding at 570°C for		Structure of compound layer
	2 h	5 h	
2	0	0	none
5	2.5	3	γ'
25	6.5	8	γ'
50	7	9	γ' ($+\varepsilon$)
100	7	9.5	$\gamma' +\varepsilon$

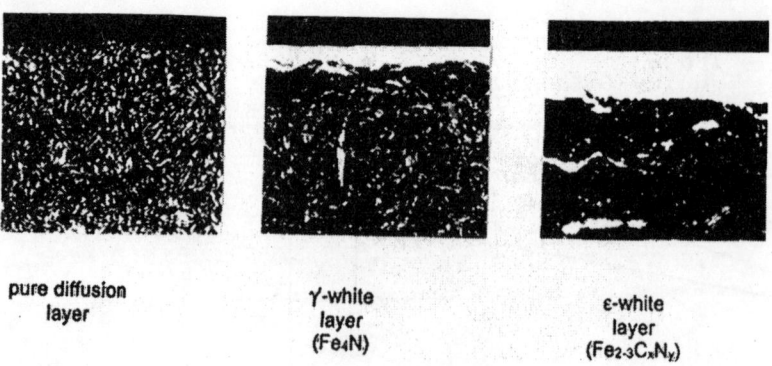

pure diffusion
layer

γ-white
layer
(Fe₄N)

ε-white
layer
(Fe₂₋₃CₓNᵧ)

Figure 36 Microsections of (a) a pure diffusion layer, (b) a thin γ' compound layer of approximately 6 μm plus diffusion layer, and (c) an ε-compound layer of approximately 15 μm plus diffusion layer.

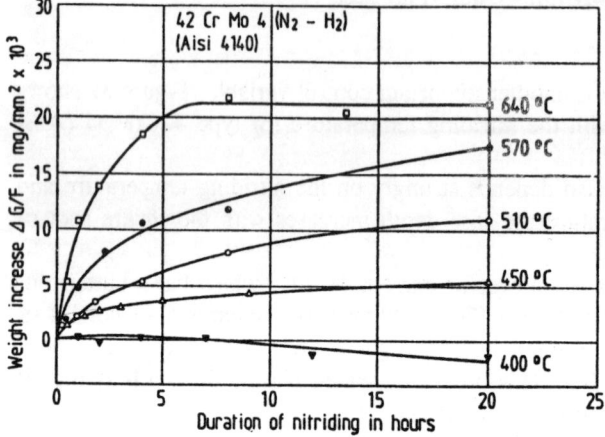

Figure 37 Increase of nitrogen transfer (weight gain) with increasing plasma nitriding temperature for 42CrMo4 (AISI 4140) steel and gas composition with 25% N_2 (the balance H_2).

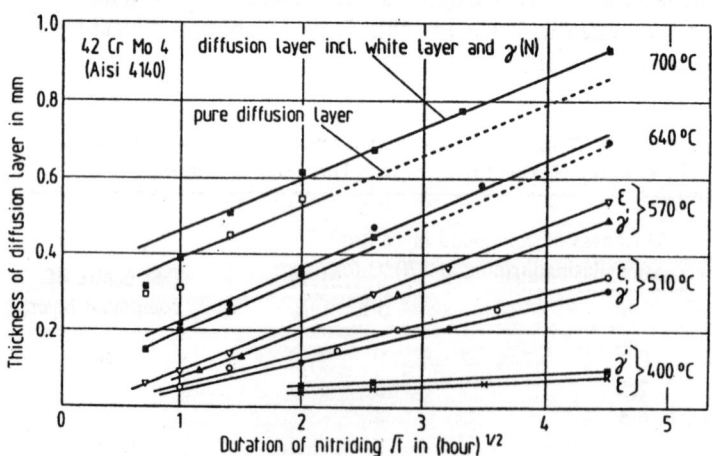

Figure 38 Growth of the diffusion layer in plasma nitriding 42CrMo4 (AISI 4140) steel at different temperatures.

Table 13 Choice of Surface Layers for Different Steels and Applied Stresses

	Unalloyed steels, cast irons, annealed alloy steels	Alloy steels (hardened), nitriding steels, hot-work steels	Stainless, heat-resistant, precipitation hardening	High-chromium, high-speed tool steels
Adhesive wear				
Low load	ε	γ′	γ′	γ′
High load	N.R.	γ′	γ′ or Diffusion	Diffusion
Abrasive wear				
Low load	ε	γ′	Diffusion	Diffusion
High load	N.R.	γ′	Diffusion	Diffusion
Pitting or impact				
Low load	N.R.	Diffusion	Diffusion	Diffusion
High load	N.R.	Diffusion	N.R.	Diffusion

N.R. = not relevant.

Plasma nitriding increases the wear resistance of components, their gliding properties, load-carrying capacity, fatigue strength, and corrosion resistance. Corrosion resistance can be especially enhanced by a plasma postoxidation. Dimensional changes are minimal; the masking method is simple and effective.

The process is nonpolluting and creates excellent working conditions. It is being used in many applications in the automotive industry, gearing, machinery (plastic extrusion, agricultural, food, etc.), chemical apparatus, and tooling (dies, molds, drills, punches, etc.). Also, the plasma nitriding of sintered iron materials is very effective [29]. The nitriding effect is limited to the surface area, keeping the growth of the components very small.

Table 14 Typical Property Ranges for Plasma-Nitrided Materials

Steel	Nitriding temperature range (°C)	Surface hardness (HV)	Case depth (mm)	Compound layer thickness (μm)
Alloy				
AISI 1045	550–580	350–500	0.3–0.8	4–15
AISI 4140	500–550	550–650	0.2–0.6	4–8
AISI 6150	500–550	550–650	0.2–0.6	4–8
AISI 9310	520–550	500–650	0.3–0.7	4–8
Nitralloy 135	510–550	900–1100	0.2–0.5	2–10
Tool steel				
H13	450–570	900–1150	0.1–0.3	2–6
A2	450–550	800–1050	0.1–0.3	2–7
D2	480–550	900–1250	0.1–0.2	0–3
M2	480–510	1000–1250	0.03–0.05	none
Stainless steel				
18–8	550–580	900–1200	0.05–0.2	none
430	550–570	900–1200	0.10–0.2	none

Table 15 Representative Treatment Cycles and Results for Plasma Nitriding

Steel	Temp/time (C°/h)	Gas[a]	Core hardness (HV)	Surface hardness (HV)	Case depth (mm)	Compound layer (μm)
Duct. iron	540/6	N-H-CH	290	370	0.14	6-11 (ε)
AISI 1045	560/6	N-H-CH	230	350	—	8-10 (ε)
AISI 4140	550/15	N-H	300	600	0.4	4-8 (γ')
AISI 9310	530/14	N-H	—	640	—	—
Nitral. 135	510/10	N-H	300	1100	0.33	3-5 (γ')
H13	530/10	N-H	480	970	0.22	2 (γ')
D2	550/10	N-H	—	930	0.07	1-3 (γ')
M2	500/20 min	N-H	700	840	—	none
18-8	550/16	N-H	210	1180	0.14	none

[a]N-H-Nitrogen/hydrogen, N-H-CH-nitrogen/hydrogen/methane.

X. VACUUM BRAZING

The dollar volume of work produced by vacuum brazing probably far exceeds that of any other process in which vacuum furnaces are being used. The development of heat-resisting alloys containing aluminum and titanium, which are difficult if not impossible to braze in very dry hydrogen gas atmospheres, promoted vacuum brazing as the first large commercial application of vacuum furnaces. The transportation industry has provided the impetus for the increasing use of vacuum furnaces for brazing, as the use of lightweight, high-strength materials such as titanium has expanded. It also provided the impetus that led to the development of the semicontinuous vacuum furnace for high-production, fluxless brazing of aluminum heat exchangers.

Vacuum furnaces have many advantages, some of which are as follows:

The process permits brazing of complex, dense assemblies with blind passages that would be almost impossible to braze and adequately clean using atmospheric flux brazing techniques.

Vacuum furnace processing at 10^{-4}–10^{-5} mbar removes essentially all gases that could inhibit the flow of brazing alloy, prevents the development of tenacious oxide films, and promotes the wetting and flow of the braze alloy over the vacuum conditioned surfaces.

Properly processed parts are unloaded in a clean and bright condition that often requires no additional processing.

Today a wide variety of steels, cast irons, stainless steel, aluminum, titanium, nickel alloys, and cobalt-base heat-resisting alloys are brazed successfully in vacuum furnaces without the use of any flux. In fact, flux would contaminate the vacuum furnace and degrade the work results.

Many different types of nickel-base, cooper, copper-base, gold-base, palladium-base, aluminum-base, and some silver-base brazing alloys are used for the filler metal. Generally, alloys that contain easily vaporized elements for lowering the melting points are avoided. With respect to the heat treatment of steel, the copper and the nickel-base brazing alloys are the most widely used filler metals.

A. Brazing with Copper

During vacuum brazing with copper filler metal, the high vapor pressure of copper at its melting point causes some evaporation and undesirable contamination of the furnace parts. To prevent this action, the furnace is first evacuated to a low pressure ($\sim 10^{-2}$–10^{-4} mbar) to remove residual air. The temperature is then raised to approximately 955°C (1750°F) to outgas and to remove any surface contamination. Finally, the furnace is heated, usually to about 1120°C (2050°F) brazing temperature, with a partial pressure of up to 1 mbar to inhibit evaporation of the copper. When brazing is completed at 1120°C, usually within minutes, the work is allowed to cool undisturbed to approximately 982°C (1800°F), then cooled by recirculated gas quenching.

B. Brazing with Nickel-Base Alloys

Brazing with nickel-base alloys is usually done without any partial pressure at the vacuum levels attainable by the furnace ($\sim 10^{-3}$–10^{-4} mbar). Normally a preheat soak at 920–980°C (1690–1800°F) is used to ensure that large workloads are uniformly heated through. After brazing, the furnace temperature is sometimes lowered for additional solution or hardening heat treatments before gas fan-cooling and unloading.

XI. SINTERING

A continuous increase has been experienced in the use of vacuum furnaces for a wide range of sintering applications. This transition from the more traditional methods of controlled atmosphere sintering arises not only from the increased volume of production of sintered parts but also from

The overall benefits to be derived from the purity of the vacuum environment

The efficiency of some sintering reactions (notably with difficult materials) that are assisted by subatmospheric pressures

The ability of the vacuum process to reduce pore size and improve pore size distribution

The higher furnace temperature capabilities that permit faster reactions from sinter processing carried out much closer to the melting point and with alloys of higher melting point interstitial elements

The limitation on the application of sintering in vacuum furnaces is the vapor pressure of the metals being processed at the sintering temperature. If the vapor pressure is comparable with the working pressure in the vacuum furnace, there will be considerable loss of metal by vaporization unless a sufficiently high partial pressure of a pure inert gas is used.

A. Stainless Steel

Vacuum sintering of stainless steel powder metal parts is a common process, employed for the AISI 410, 420, 303, 316, and 17-4PH grades of stainless steel. These products are very often superior to those sintered in hydrogen or cracked ammonia atmospheres with respect to their corrosion resistance and mechanical properties.

B. High-Speed Steel

Powder metallurgy manufacturing methods have been developed for producing finished and full-density cutting tools of high-speed tool steel. For complex geometries—hobs, pipe taps, reamers, etc.—special isostatic compacting techniques have been developed that use neither lubricants nor binders. The pressed compacts are sintered in vacuum furnaces under precise control of heating rate, sintering time, temperature, and vacuum pressure in order to eliminate porosity. The result is predictable densification of the pressed compact with final size tolerances of ± 0.5–1.0%. Full-density, sintered high-speed steel tools have been shown to be at least equivalent to conventional wrought material in cutting properties. Grindability is dramatically improved, in particular for the high alloy grades such as M4 and T15. This is attributed to the finer and more uniform carbide distribution.

XII. TEMPERING AND STRESS RELIEVING

Where surface finish is very critical and clean parts are desired to avoid any additional processing, heat treaters now employ vacuum furnaces for tempering and stress relief. Below temperatures at which radiant energy is an efficient method for heating, the furnace is evacuated and backfilled with an inert gas such as dry nitrogen or argon. A fan in the furnace circulates this atmosphere, and parts are heated by convection and conduction. This process is used when tempering high-speed steel components and a variety of other items made from many types of steels. Also, special furnaces have been designed to combine the tempering treatment with a plasma or gas nitriding process in the same installation (see Section IX).

REFERENCES

1. F. D. Richardson and J. H. Jeffes, The thermodynamics of substances of interest in iron and steel making III, *J. Iron Steel Inst.* *171*:165–175 (June 1952).
2. R. Hoffmann, Oxidation events during heat treatments *Härterei-Techn. Mitt.* *39* (2): 61–70 (1984) (in German).
3. German patent DE-P3032010, L. Körtvelyessy, Aug. 1980.
4. ASM, *Metals Handbook, 9th ed.*, Vol. 4, *Heat Treating*, Am. Soc. Materials Park, Cleveland, OH, 1991, pp. 307–324.
5. G. Legge, Plasma carburizing facility design and operating data, *Ind. Heating*, March 1988, pp. 26–30.
6. D. H. Herring, M. Sugiyama, and M. Uchigaito, Vacuum furnace oil quenching influence of oil surface pressure on steel hardness and distortion, *Ind. Heating,* June 1986, pp. 14–17.
7. R. W. Reynoldson and K. C. Harris, The vacuum heat treatment of tool steels, *Metal Treat.* Oct.–Nov. 1970, pp. 3–24.
8. B. Edenhofer and J. W. Bouwman, Progress in design and use of vacuum furnaces with high pressure gas quenching systems, *Ind. Heat.*, February 1988, pp. 12–16.
9. J. G. Conybear, High pressure gas quenching, *Adv. Mater. Process.*, February 1993.
10. H. J. Becker and E. Haberling, Comparison between hardenability and toughness of high speed tool steels after hardening in a salt bath and in a vacuum furnace, *Proc. First Int. High Speed Steel Conf.*, Leoben, Austria, Mar. 26–28, 1990.
11. B. Edenhofer and J. W. Bouwman, Innovative quenching methods in vacuum heat treating for improving product quality and furnace productivity, Presented at Heat Treatment Conference, Indianapolis, March 1993.
12. C. H. Luiten, F. Limque, and F. Bless, Carburizing in vacuum furnaces, Heat Treatment '79, Birmingham, U.K., May 22–24, 1979.
13. J. A. Coffey, Supercell-carburizing, *Adv. Mater. Process. 1989*: 81–83.
14. Y. Pourprix and J. Naudot, Carburizing under reduced pressure, *Trait. Therm. 197*:51–58 (1985) (in French).
15. B. Edenhofer, Carbonitriding in the plasma of a glow discharge, *Härterei-Techn. Mitt. 28* (3): 165–172 (1973) (in German).
16. W. L. Grube and J. G. Gay, High rate carburizing in a glow-discharge methane plasma, *Metl. Trans. A, 9A*:1421–1424 (1978).
17. J. G. Conybear, The role of process variables in plasma carburizing, *Heat Treat. 3*: 24–27 (1988).
18. F. Hombeck and W. Rembges, User experience with an integral quench plasma (ion) carburizing furnace, Heat Treatment '94, May 2–4, 1984, The Metals Society, London.

19. B. Edenhofer, J. G. Conybear, and G. T. Legge, Opportunities and limitations of plasma carburizing, *Heat Treat. Met. 1*:6–12 (1991).
20. J. G. Conybear and B. Edenhofer, Progress in the control of plasma nitriding and carburizing for better layer consistency and reproducibility, 6th Int. Congr. Heat Treatment of Materials, Chicago, Sept. 28–30, 1988, ASM, Materials Park, Ohio.
21. T. J. Schulte, T. J. Kuhn and D. E. Goodman, Carburizing—stepping into the future, *Proc. Carburizing: Processing and Performance*, July 12–14, 1989, Lakewood, Colorado, ASM Intern., Materials Park, OH.
22. S. McFaul, Case uniformity in plasma carbonitriding using Taguchi method, ASM 14th Int. Heat Treatment Conf., Indianapolis, Mar. 16–18, 1993.
23. B. Edenhofer and J. G. Conybear, Experience with a flexible plasma-carburizing cell, *Härterei-Techn. Mitt. 48* (3):191–198 (1993).
24. M. H. Jacobs, T. J. Law, and F. Ribet, Plasma carburizing: theory, industrial benefits and practices, *Surf. Eng. 1* (2):105–114 (1985).
25. K. Akutsu and M. Nakamura, Practice and experience with plasma carburizing furnace, Ionitriding and Ion Carburizing, Sept. 18–20, 1989, Cincinnati, ASM Int., Materials Park, OH.
26. B. Edenhofer, Physical and metallurgical aspects of ionitriding, *Heat Treat. Met. 1*:23–28 (1974); 2:59–67 (1974).
27. M. H. Jacobs, Technological, industrial trends and new applications, *Heat Treat. 1*:26–29 (1986).
28. J. G. Conybear, Guidelines for choosing the ionitriding process, *Heat Treat. 19* (3):33–37 (1987).
29. W. Rembges, Ionitriding applications grow for automotive components, *Heat Treat. 22*: 3 (1990).

8

STEEL HEAT TREATMENT

Božidar Liščić
Faculty of Mechanical Engineering and Naval Architecture,
University of Zagreb, Zagreb, Croatia

I. FUNDAMENTALS OF HEAT TREATMENT

A. Heat Transfer

Heat treatment operations require a direct or indirect supply of energy onto the treated workpieces and its subsequent removal to realize their heating and cooling, respectively. Because this chapter deals only with heat treatment operations involving the whole volume of treated workpieces, let us consider only the relevant heat transfer problems, not taking into account other heating methods connected to surface heat treatment operations. As an example, Figure 1 shows the temperature distribution on the cross section of a plate during heating (Figure 1a) and during cooling (Figure 1b).

In heat treatment operations, when heating or cooling the treated workpieces, nonstationary temperature fields develop in which the temperature distribution changes with time. Through the surface F of the plate of thickness s (Figure 1), the heat flux Q is supplied (during heating) or extracted (during cooling):

$$Q = \frac{dQ}{dt} = -\lambda F \frac{dT}{dx}, \qquad x = 0,\ldots,s/2 \tag{1}$$

where T is temperature, K; t is time, s; λ is heat conductivity, W/(m·K); F is surface area, m²; and dT/dx is the temperature gradient, K/m.

From the other side, based on the first law of thermodynamics,

$$dQ = \rho \, dx \, F C_p \, dT \tag{2}$$

where ρ is the density of the workpiece, kg/m³; and C_p is the specific heat capacity of the workpiece at constant pressure, J/(kg·K).

From Equation (1) it follows that

$$\frac{d}{dx}\left(\frac{dQ}{dt}\right) = -\lambda F \frac{d^2T}{dx^2} = \frac{d}{dt}\left(\frac{dQ}{dx}\right) \tag{3}$$

From Equation (2) it follows that

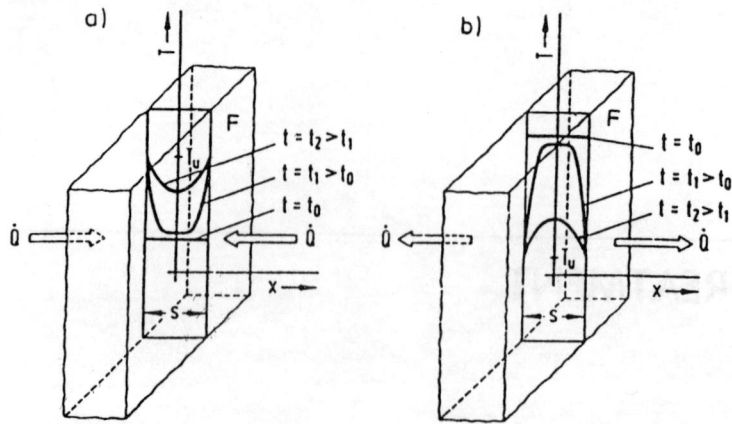

Figure 1 Temperature distribution on the cross section of a plate (a) during heating and (b) during cooling. t_0 = beginning of temperature change; T_u = surrounding temperature; Q = heat flux; s = thickness of the plate. (From Ref. 1.)

$$\frac{d}{dt}\left(\frac{dQ}{dx}\right) = -\rho F C_p \frac{dT}{dt} \tag{4}$$

and for the one-dimensional heat flux the time-dependent temperature distribution inside the workpiece is

$$\frac{dT}{dt} = \frac{\lambda}{\rho C_p}\frac{d^2T}{dx^2} = a\frac{d^2T}{dx^2} \tag{5}$$

where $a = \lambda/\rho C_p$ m²/s is the temperature diffusivity.

In the case of three-dimensional heat flux, Equation (5) reads

$$\frac{dT}{dt} = a\left(\frac{d^2T}{dx^2} + \frac{d^2T}{dy^2} + \frac{d^2T}{dz^2}\right) = a\,\nabla\,T \tag{6}$$

where

$$\nabla = \frac{d^2}{dx^2} + \frac{d^2}{dy^2} + \frac{d^2}{dz^2}$$

is the Laplace operator

Equations (5) and (6) are the *temperature conduction equations* in which the *temperature diffusivity* represents the amount of the time-dependent temperature change of a workpiece because of nonstationary heat conduction.

A heat flux dQ flowing through a surface of area F is, according to Fourier's heat conduction law, proportional to the temperature gradient at the relevant position:

Table 1 Approximate Values for Heat Conductivity λ, W/(m·K)

Metals		Stiff Anorganic Materials	
Copper	380	Chamotte	0.5–1.2
Aluminum	170–230	Brick	0.8
Brass	80–115	Concrete	0.8–1.4
Gray iron	58	Mineral wool	0.05–0.1
Steel	40		
Liquids		Gases	
Water	0.6–0.7	Air (20–2000°C)	0.026–0.11
Oil	0.1–0.2	H_2 (20–2000°C)	0.18–0.75

Source: Ref. 1.

$$dQ = -\lambda \frac{dT}{dx} F \, dt \tag{7}$$

or, expressed as heat flux density per unit time (s) and unit surface (m²),

$$q = -\lambda \frac{dT}{dx} = -\lambda \, \mathrm{grad}T \qquad [\mathrm{W/m^2}] \tag{8}$$

Equation (8) clearly shows that the temperature gradient is the driving force of the heat flux. The heat conductivity λ, as a proportional factor in this *heat conduction equation*, represents the influence of the material's properties on the heat transport. Table 1 gives approximate values for heat conductivity λ [in W/(m·K)] for selected materials.

In the above equations, the heat conductivity λ is assumed to be a constant value, but in reality it depends on temperature. Figure 2 shows the temperature dependence of heat conductivity for groups of steel. As can be seen, the biggest differences in heat conductivity among different steel grades are at room temperature. Whereas for unalloyed steels the heat conductivity decreases with increasing temperature, for high-alloy steels it slightly increases with

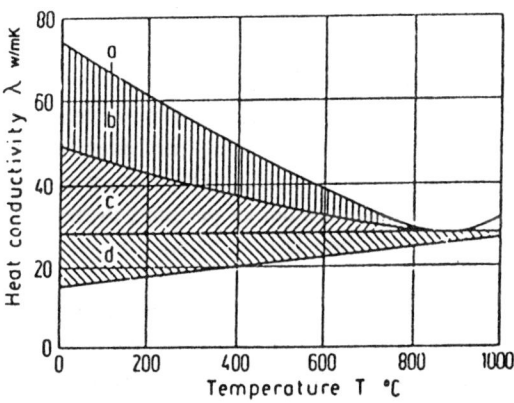

Figure 2 Temperature dependence of the heat conductivity λ for selected steel groups. (a) Pure iron; (b) unalloyed steels; (c) low-alloy steels; (d) high-alloy steels. (From Ref. 1.)

increasing temperature. At about 900°C (1652°F) the value of λ is almost the same for all steel grades. The specific heat capacity C_p depends also on temperature.

The transport of thermal energy through a solid body, described by the heat conduction equation (Equation 8), extends naturally beyond the body surface; i.e., heat transfer takes place between the body and its environment. This heat transfer is expressed as the amount of heat exchanged between the surface of the body and its environment per unit surface area and per unit time. According to Newton's law of cooling, the amount of heat exchanged between a body and its environment depends on the difference between the body surface temperature and the temperature of its environment. The relevant heat flux density is

$$q = \frac{dQ}{dF\,dt} = \alpha(T_K - T_U) \qquad \text{for } T_K > T_U \tag{9}$$

where T_K is the body surface temperature, T_U is the temperature of the environment, and α is the heat transfer coefficient, $W/(m^2 \cdot K)$.

The actual conditions of the heat transfer in each case are represented by the relevant heat transfer coefficient α, which depends on

1. The shape and cross-sectional size of the body
2. The position of the body (standing or lying)
3. The surface condition of the body
4. The physical properties of the body's material
5. The physical properties of the surrounding fluid (density, specific heat capacity, dynamic viscosity)
6. The agitation (flow) rate of the surrounding fluid

During every heating or cooling process, the temperature difference between the body surface and its environment becomes smaller with time, i.e., the exchanged heat quantity becomes smaller. The heat transfer coefficient α is therefore not a constant but varies with the body surface temperature.

Gases, liquids, and vacuums are the environments in which heat transfer occurs during heat treatment operations. Heat can be transferred by three different heat transfer mechanisms: heat conduction, heat convection, and heat radiation.

Heat conduction (in fluids) is the heat transfer that occurs in a nonagitated liquid or gaseous medium between directly adjoining particles owing to a temperature gradient.

Heat convection is directly connected to the movement (flow or agitation) of the heat-carrying fluid, because it is realized through movement of the fluid particles from one place to another. Therefore heat convection is possible only in liquids and gases. The amount of heat transferred by heat convection in a gas also depends on the number of particles in the gas. Because this number depends on gas pressure, heat convection is proportional to gas pressure.

If the only cause of particle movement is the difference in density caused by the temperature difference, the movement is called free or natural convection. When the movement of particles of the fluid is caused by an outside force, it is called forced convection. Generally, free and forced convection take place simultaneously. The amount of free convection contributing to the heat transfer depends on the temperature gradient within the fluid, and the contribution of the forced convection depends on the flow velocity, i.e., on the agitation rate.

When an air stream passes toward a cylinder, the convective heat transfer coefficient α_K can be calculated, according to Reference 2, by using the formula

$$\alpha_K = (4.64 + 3.49 \times 10^{-3} \, \Delta T) \frac{v^{0.61}}{D^{0.39}} \qquad [W/(m^2 \cdot K)] \tag{10}$$

where D is the diameter, m; ΔT is the temperature difference between air and cylinder surface; and v is the air velocity, m/s.

The third heat transfer mechanism is heat radiation. Solid bodies, liquids, and gases can all transfer heat in the form of radiation. This kind of heat transfer does not depend on any heat transfer carrier; therefore it can take place in vacuum also. Heat radiation is in the form of electromagnetic waves whose length is in the range of 0.3 to 500 μm. When radiation strikes the surface of a body, part of it will be absorbed, part will be reflected, and the rest may pass through the body. Every body emits radiation corresponding to its temperature. The body that, at a certain temperature, emits or absorbs the largest amount of radiation is called a *blackbody*. All other bodies emit or absorb less radiation than the blackbody. The ratio of the radiation of a body to that of a blackbody is called the emission-relation coefficient ε.

The total heat flux density emitted by radiation from a body can be calculated according to the Stefan–Boltzmann law,

$$q = \varepsilon \sigma T^4 \tag{11}$$

where ε is the emission-relation coefficient, $\varepsilon < 1.0$; σ is the Stefan–Boltzmann constant, $\sigma = 5.67 \times 10^{-8} \, W/(m^2 \cdot K^4)$; and T is the absolute temperature, K.

If two bodies mutually exchange radiant heat, then not only is the warmer body emitting heat to the colder one, but the colder body is also emitting heat to the warmer body, so that the transferred heat consists of the difference of the amounts of heat absorbed by the two bodies. The total heat transferred by radiation from one body having a surface area F_1 to another solid body of any surface area can be calculated according to

$$Q = \varepsilon_{1,2} \sigma F_1 (T_1^4 - T_2^4) \tag{12}$$

where $\varepsilon_{1,2}$ is the emission-relation coefficient, which depends on the emission-relation coefficients of both bodies, their surface relation, and their mutual position in space; T_1 is absolute temperature of the emitting body; and T_2 is absolute temperature of the absorbing body.

In industrial furnaces, heat is transferred substantially by simultaneous heat convection and heat radiation. The heat transferred by heat conduction (in fluids) is negligible compared to the heat transferred by convection and radiation. When calculating the total heat transferred by both mechanisms, it is appropriate to express the heat transferred by radiation with a formula similar to Newton's law,

$$q = \alpha_\varepsilon (T_1 - T_2) \tag{13}$$

The heat transfer coefficient for radiation α_ε can be calculated by combining formulas (12) and (13):

$$\alpha_\varepsilon = \varepsilon_{1,2} \, \sigma \frac{T_1^4 - T_2^4}{T_1 - T_2} \tag{14}$$

The total heat transfer coefficient is then

$$\alpha = \alpha_K + \alpha_\varepsilon \tag{15}$$

where α_K is the heat transfer coefficient for convection.

Table 2 gives, according to Reference 2, the average values of the heat transfer coefficient for cooling or quenching in liquid or gaseous media. This complex heat transfer coeffi-

Table 2 Average Values of the Heat
Transfer Coefficient α When Cooling or
Quenching in Liquid or Gaseous Media

Medium	$\alpha \ [W/(m^2 \cdot K)]$
Furnace	15
Still air	30
Moving air	40
Compressed air	70
Air–water mixture	520
Hardening oil	580
Liquid lead	1200
Water	3500

Source: Ref. 2.

cient depends in each case on many specific factors, discussed earlier, but also depends strongly on the surface temperature of the workpiece. It is a temperature-dependent and location-dependent value that changes during heat transfer as the body surface temperature equalizes to the environment's temperature. According to Reference 2, the complex heat transfer coefficient can increase thirty to fifty times between 50 and 1500°C (122 and 2732°F). At temperatures below 300°C (572°F), heat transfer by convection is predominant. With increasing temperature, heat transfer by radiation becomes more important, and at about 800°C (1472°F) it reaches 80% of the total heat transfer.

Especially in operations that employ immersion quenching in liquids, where two-phase heat transfer takes place with high heat flux densities under nonstationary conditions, the heat transfer coefficient value changes very much. Therefore nowadays when heat transfer calculations are carried out by computer a temperature-dependent function of the heat transfer coefficient instead of an average value should be used. One practical way to obtain this function in each actual case is to measure the surface temperature of an adequately instrumented probe (cylinder or plate of adequate dimensions) placed appropriately in the quenching tank and, from the measured surface temperature vs. time history, calculate the corresponding heat flux density and heat transfer coefficient vs. temperature functions. Figure 3 shows, as an example, the heat transfer coefficient vs. surface temperature for quenching a stainless steel cylinder of 50 mm diameter × 200 mm in still oil of 20°C (68°F) [3]. If there is no adequate thermocouple to measure the surface temperature of the probe, the temperature can be measured near the surface and, by using the inverse heat conduction method and an adequate mathematical procedure (FEM), the surface temperature of the probe can be calculated.

To explain the dependence between the heat transfer conditions and the temperature fields in solid bodies, let us consider the heating of a plate of thickness s (see Figure 4). At the beginning of heating ($t = 0$) the plate has a temperature $T_K = 0$ and is suddenly transferred in standing position into a furnace, where the environmental temperature is T_U. Equal amounts of heat are transferred from both sides of the plate. Because boundary conditions of the third kind exist, the ratio between the heat conductivity and the heat transfer coefficient (λ/α) gives a point at temperature T_U outside the plate. The straight line connecting this point and the relevant surface temperature is the tangent on the temperature distribution curve at the body surface. As time progresses, both the surface temperature and the temperature in the middle of the plate increase. The temperature gradient inside the body is different for different λ/α ratios and changes over time. If the heat conductivity λ of the body's material is small or the

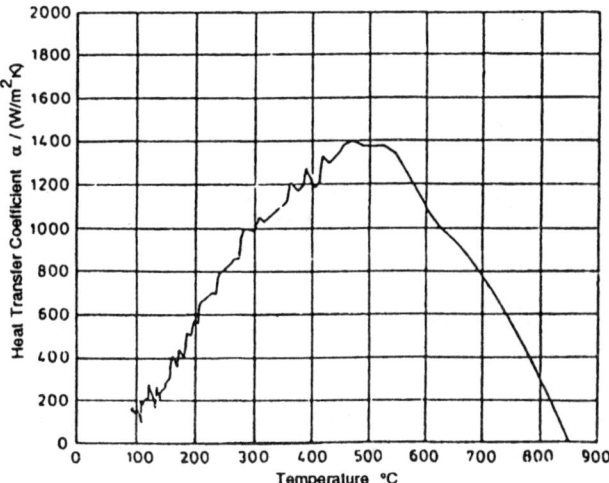

Figure 3 Heat transfer coefficient vs. surface temperature when quenching a stainless steel cylinder of 50 mm diameter × 200 mm in still oil of 20°C, calculated from the measured surface temperature–time history. (From Ref. 3.)

heat transfer between the environment and the body surface is large, the ratio λ/α is small and heat accumulates in the surface region because the amount of heat transferred is greater than the amount transported by conduction into the body's interior. The smaller the ratio λ/α, the faster the surface temperature equalizes to the temperature of the environment. The relevant changes of temperature at points x_1 and x_2 and the value and the change in the temperature gradient over time are also greater. This can be seen in Figure 5 when comparing the curves for $(\lambda/\alpha)_1$ small to $(\lambda/\alpha)_2$ big. If the heat conductivity λ is big or the heat transfer coefficient α is small, i.e., the heat is transported from the surface to the core of the body faster than it is transferred from the environment to the body's surface, the λ/α ratio becomes

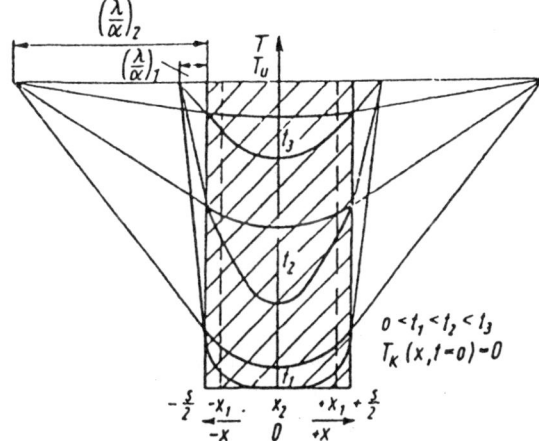

Figure 4 Change of the temperature distribution with time when heating a plate of thickness s, depending on different heat transfer conditions expressed by the ratio λ/α. (From Ref. 2.)

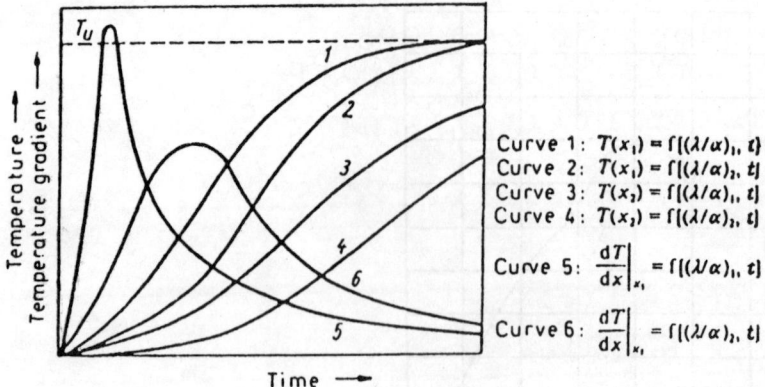

Curve 1: $T(x_1) = f[(\lambda/\alpha)_1, t]$
Curve 2: $T(x_1) = f[(\lambda/\alpha)_2, t]$
Curve 3: $T(x_2) = f[(\lambda/\alpha)_1, t]$
Curve 4: $T(x_2) = f[(\lambda/\alpha)_2, t]$

Curve 5: $\dfrac{dT}{dx}\Big|_{x_1} = f[(\lambda/\alpha)_1, t]$

Curve 6: $\dfrac{dT}{dx}\Big|_{x_1} = f[(\lambda/\alpha)_2, t]$

Figure 5 Change of the plate temperatures at points x_1 and x_2 and temperature gradients at point x_1 in the cross section of the plate shown in Figure 4 when different heat transfer conditions $(\lambda/\alpha)_1$ and $(\lambda/\alpha)_2$, respectively, exist. (From Ref. 2.)

big, and the temperature in the interior of the body increases relatively faster than the surface temperature.

Other factors that should be taken into account when analyzing such heat transfer problems are the shape and cross-sectional size of the body. As to the influence of different shapes of workpieces it should be borne in mind that at constant heat transfer conditions and constant thermal properties of the material and equal temperature of the environment, the temperature change with time depends on the surface-to-volume ratio of the body. The greater this ratio is, the greater is the temperature change over time.

B.   Lattice Defects

Generally the lattice of a metal crystal contains imperfections, i.e., aberrations from a perfect atomic arrangement. These imperfections may be divided, from the geometrical standpoint, into the categories

Zero-dimensional or point imperfections
One-dimensional or linear
Two-dimensional or superficial
Three-dimensional or spatial

The most important lattice defects that occur in metals are shown schematically in Figure 6. Figure 6a shows a plane consisting of equal regular atoms *a* of which the spatial lattice is built, with four different types of lattice point defects. At position *b* one atom is missing; this defect is called a *vacancy*. Atom *c* occupies a place between the regular places in the lattice and it is called an *interlattice atom*; in this case *c* is the same kind of atom as the regular atoms. In position *d*, a strange atom (of an alloying element) with a larger diameter has taken the place of a regular atom; therefore it is called a *substitutional atom*. A practical example of this is a manganese atom dissolved in iron. In position *e*, a strange atom with a much smaller diameter than the regular atoms of the lattice is inserted between regular atoms in a position that is not occupied by regular atoms. It is called an *interstitial atom*. A practical example of this is a carbon atom dissolved in iron. Both substitutional and interstitial atoms cause local deformations and microstresses of the crystal lattice.

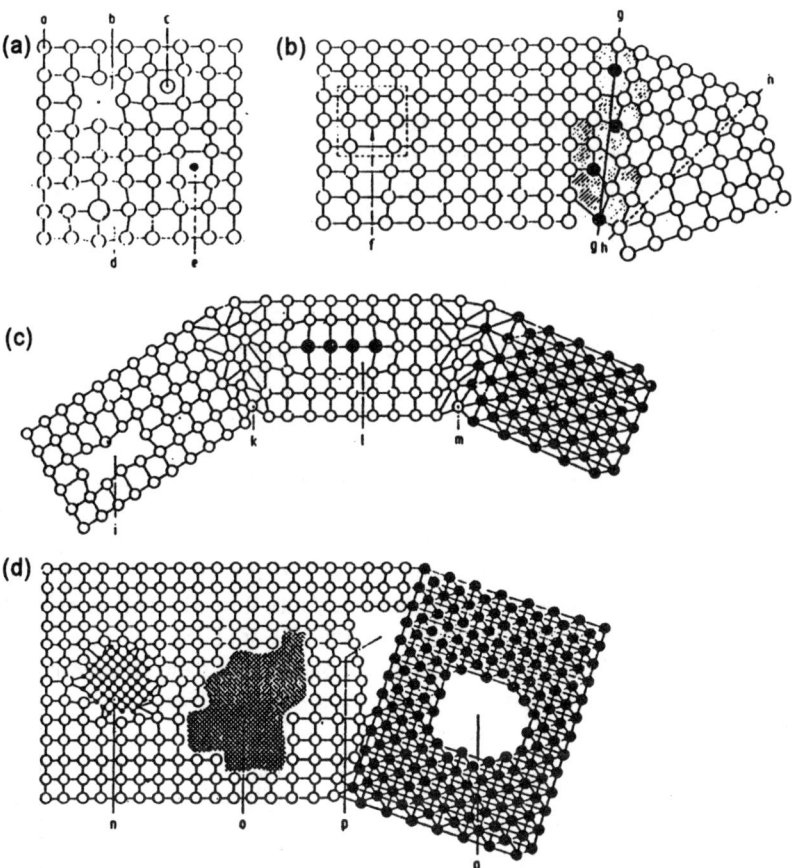

Figure 6 Lattice defects. (a) Lattice point defects; (b) linear and superficial lattice defects; (c) superficial lattice defects; (d) spatial lattice defects. *a*, Regular lattice atom; *b*, vacancy; *c*, interlattice atom; *d*, substitutional atom; *e*, interstitial atom; *f*, edge dislocation; *g*, low-angle grain boundary; *h*, twinning boundary; *i*, vacancy zone; *k*, high-angle grain boundary; *l*, strange atoms zone; *m*, phase boundary; *n*, precipitate; *o*, inclusion; *p*, microcrack; *q*, micropore. (From Ref. 1.)

Figure 6b shows at *f* a linear lattice defect. A row of atoms in the outlined atomic plane terminates at this point. If we imagine the outlined atomic plane as a section through a crystal that stretches perpendicular to the plane shown, then the row of atoms terminating at *f* becomes a half-plane of atoms that has been inserted between regular planes of atoms and ends inside the crystal. The boundary line of the inserted half-plane of atoms that stretches through the greater lattice region, perpendicular to the plane shown, is a linear lattice defect called an *edge dislocation*. Every edge dislocation is connected with characteristic deformations and microstresses of the lattice.

The lattice defects *g* and *h* are superficial defects. The line *g–g* represents schematically a *low-angle grain boundary* that consists of edge dislocations arranged regularly one under the other. The inserted half-plane of each edge dislocation terminates at the associated atom shown in black. The dashed area is a section through a low-angle grain boundary between neighboring parts of the crystal lattice that are inclined at a low angle to each other. The line *h–h* represents a *twinning boundary*. It is characterized by the fact that the atoms on both sides of the boundary are symmetrically distributed, and therefore neighboring parts of the crystal lattice are completely equal, looking like twins in a mirror.

Figure 6c shows at *i* a superficial imperfection (in the outlined plane) where a group of atoms is missing. This zone of missing atoms could have developed by way of an accumulation of vacancies. It can be stretched to other planes of atoms perpendicular to the outlined one. The imperfection *k* is a more or less irregular distribution of atoms between two neighboring parts of the crystal lattice with big differences in orientation, which interrupts the continuity of the lattice. It is called a *high-angle grain boundary*, or simply a grain boundary. The superficial imperfection at *l* is a section through a zone of strange atoms that stretches in two dimensions perpendicular to the plane shown. The boundary plane *m* between two different lattices is called a *phase boundary*, which is also a two-dimensional lattice defect.

Figure 6d shows schematically the characteristic three-dimensional lattice defects. In many metal alloys, within the lattice of grains under specific thermodynamic conditions, new lattice regions with changed structure are formed. Such a lattice defect shown at *n* is called a *precipitate*. The spatial imperfection at *o* is an *inclusion*. Such inclusions, which develop unfailingly during the production of alloys, are nonmetallic or intermetallic compounds. Like precipitates, inclusions have their own structure and/or phase and are separated by a phase boundary from the surrounding lattice. *p* denotes a *microcrack*, a spatial imperfection that is created by three edge dislocations that came to a phase boundary and formed a hollow among the three half-planes of the lattice. The hollow stretches perpendicular or at a slope to the plane shown. At *q* a sphere-like hollow inside the crystal's lattice is shown; this is called a *micropore*. Such defects can originate from the accumulation of either vacancies or gases.

Of all the lattice defects discussed above, vacancies and edge dislocations are especially important in the heat treatment of metals. Vacancies enable neighboring atoms or substitutional atoms of alloying elements to change their positions and thus enable diffusion processes. The diffusion of interstitial atoms (e.g., a carburizing process) is possible without vacancies. Dislocations can move, increase in number, and accumulate. By lowering the share force (as a consequence of the intermittent movement of the atoms), compared to the case of a perfect iron crystal (whisker), dislocations facilitate the plastic deformation of the material.

C.   Application of TTT (IT) and CCT Diagrams

Time–temperature–transformation diagrams for isothermal transformation (IT diagrams), and for continuous cooling transformation (CCT diagrams) are used to predict the microstructure and hardness after a heat treatment process or to specify the heat treatment process that will yield the desired microstructure and hardness. The use of the either type of diagram requires that the user be acquainted with its specific features, possibilities, and limitations.

1.   Isothermal Transformation (IT) Diagram

Figure 7 shows an isothermal transformation diagram of the low-alloy steel DIN 50CrV4. The regions of transformation of the structural phases ferrite (F), pearlite (P), and bainite (B) as positioned in the time–temperature diagram (the abscissa of which is always in logarithmic scale) are valid only under conditions of fast quenching from the austenitizing temperature to the chosen transformation temperature and subsequent holding at that temperature. This is the way the diagram itself was developed. Therefore the IT diagram may be read only along the isotherms. The beginning and end of transformation of ferrite, pearlite, and bainite in isothermal processes take place according to the function

$$M = 1 - \exp(-bt^n) \tag{16}$$

where M is the fraction of phase transformed; t is time, s; $b = 2 \times 10^{-9}$; and $n = 3$.

%C	%Si	%Mn	%P	%S	%Cr	%Cu	%Ni	%V	Method of melting	Mc Quaid-Ehn
0,43	0,47	0,82	0,041	0,015	1,22	0,14	0,04	0,11	b.S.-M.	4

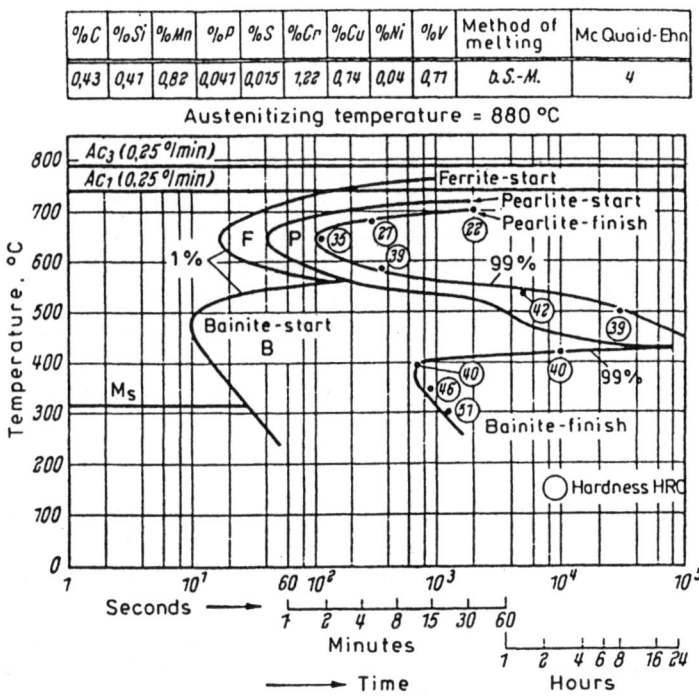

Austenitizing temperature = 880 °C

Figure 7 Isothermal transformation (IT) diagram of DIN 50CrV4 steel. (From Ref. 4.)

Because, as shown in Figure 8, this function starts and ends very flat, the actual beginning and end of transformation are difficult to determine exactly. Therefore, an agreement is reached, according to which the curve marking the beginning of transformation denotes 1% of relevant phase originated, and the curve marking the end of transformation denotes 99% of the austenite transformed.

Only the formation of martensite takes place without diffusion, instantly, depending only on the temperature below the M_s point. Hougardy [5] gave the following formula (valid for structural steels for hardening and tempering) for this transformation:

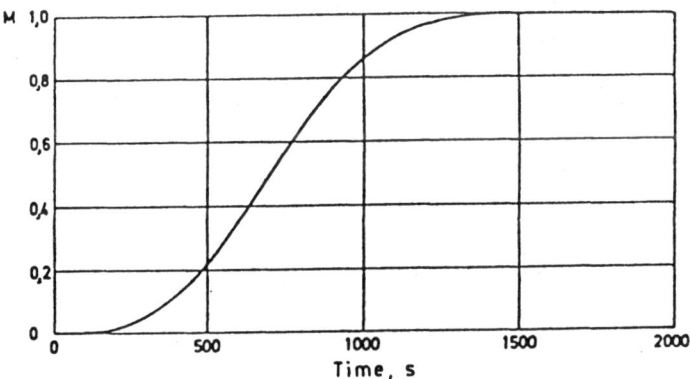

Figure 8 Relation between the amount of transformed structure (M) and time in IT diagrams, according to Equation (16). (From Ref. 5.)

$$M_a = 1 - 0.929 \exp[-0.976 \times 10^{-2}(M_s - T)^{1.07}] \qquad (17)$$

where M_a is the amount of martensite; M_s is the martensite start temperature; and T is a temperature below M_s.

Some IT diagrams, when read along the isotherms, enable the user to determine the percentages of phases transformed and the hardness achieved. Figure 9, for example, shows that when the DIN 41Cr4 steel (austenitized at 840°C (1544°F) with 5 min holding time) is fast quenched to 650°C (1200°F) and held at this temperature, after 12 s ferrite starts to form. After 30 s the formation of pearlite begins. After 160 s the transformation is completed with 5 vol % of ferrite and 95 vol % of pearlite formed. The hardness achieved is about 20 HRC. If a specimen of this steel is quenched to 300°C (572°F), instantly, 50% (v/v) of martensite will be formed.

The accuracy of an IT diagram with respect to the position of isotherms can generally be taken as ±10°C (50°F), and with respect to the time ordinates, as ±10%.

2. Continuous Cooling Transformation (CCT) Diagram

Figure 10 shows the continuous cooling transformation (CCT) diagram of the same heat (as Figure 7) of the low-alloy DIN 50CrV4 steel.

When comparing the curves for the start of transformation in CCT and IT diagrams for the same heat and steel grade (Figures 10 and 7), we found that in the CCT diagram the curves are slightly shifted to longer times and lower temperatures. For example, in the IT diagram of Figure 7, the shortest time to start the transformation of ferrite is 16 s at 650°C (1200°F) and the corresponding time for bainite is 9 s at 480°C (900°F). In the CCT diagram of Figure 10, however, the shortest transformation start time for ferrite is 32 s at 620°C (1150°F) and the corresponding time for bainite is 20 s at 380°C (716°F). This indicates that in continuous cooling transformation processes the transformation starts later than in isothermal transformation processes. The basis of this phenomenon is related to the incubation time and can be found in a hypothesis of E. Scheil [6].

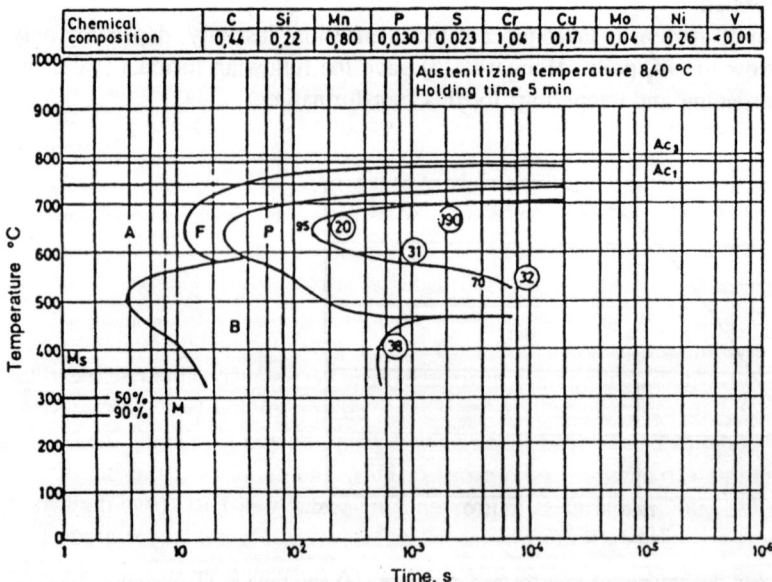

Figure 9   Isothermal transformation (IT) diagram of DIN 41Cr4 steel. (From Ref. 5.)

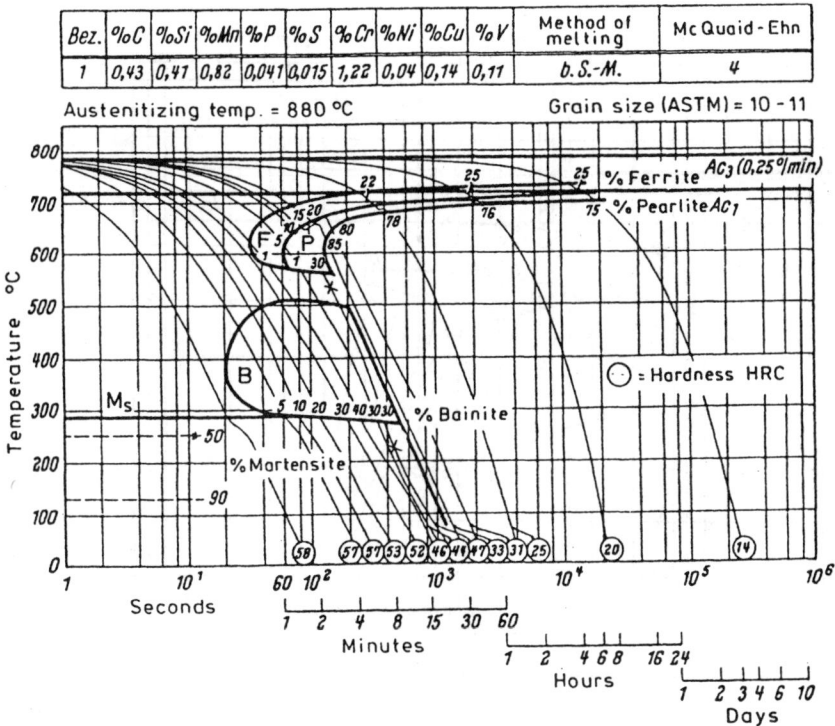

Bez.	%C	%Si	%Mn	%P	%S	%Cr	%Ni	%Cu	%V	Method of melting	McQuaid-Ehn
1	0,43	0,47	0,82	0,041	0,015	1,22	0,04	0,14	0,11	b.S.-M.	4

Austenitizing temp. = 880 °C   Grain size (ASTM) = 10 – 11

Figure 10 Continuous cooling transformation (CCT) diagram of DIN 50CrV4 steel. (From Ref. 4.)

It should also be noted that with higher austenitizing temperature the curves denoting the start of transformation of a particular phase can be shifted to longer times. Figure 11 shows the CCT diagram of DIN 16MnCr5 steel after austenitizing at 870°C (1600°F) (a), and after austenitizing at 1050°C (1922°F) (b). In the latter case the regions of ferrite and pearlite are shifted to longer times. It is necessary, therefore, when using a CCT diagram, to ascertain that the austenitizing temperature used to develop the CCT diagram corresponds to the austenitizing temperature of the parts treated.

A CCT diagram is developed in the following way. Many small specimens (e.g., 4 mm diameter × 2 mm for high cooling rates, and 4.5 mm diameter × 15 mm for medium and low cooling rates) are austenitized and cooled within a dilatometer with different cooling rates. Start and finish of transformation of relevant phases with each cooling curve are recorded and these points are connected to obtain the regions of transformation for the relevant phases (see Figure 10). Therefore, a CCT diagram can be read only in the way in which it was developed, i.e., along the cooling curves. As can be seen from Figure 10, a single phase structure occurs only in cases of very high cooling rates (martensite) and very slow cooling rate (pearlite). In all other cooling regimes a mixture of more structural phases results. How much of each phase such a mixture contains can be read in percentage from the numbers along the cooling curve (usually marked in CCT diagrams of German origin). The numbers at the end of each cooling curve denote the relevant hardness after quenching [usually in HRC (two-digit numbers) or in HV (three-digit numbers)]. For example, as shown in Figure 10 for grade DIN 50CrV4 steel, if cooling proceeds at the rate marked with ×, a mixture of 10% ferrite, 30% pearlite, 30% bainite, and (the rest) 30% martensite will result at room temperature, and the

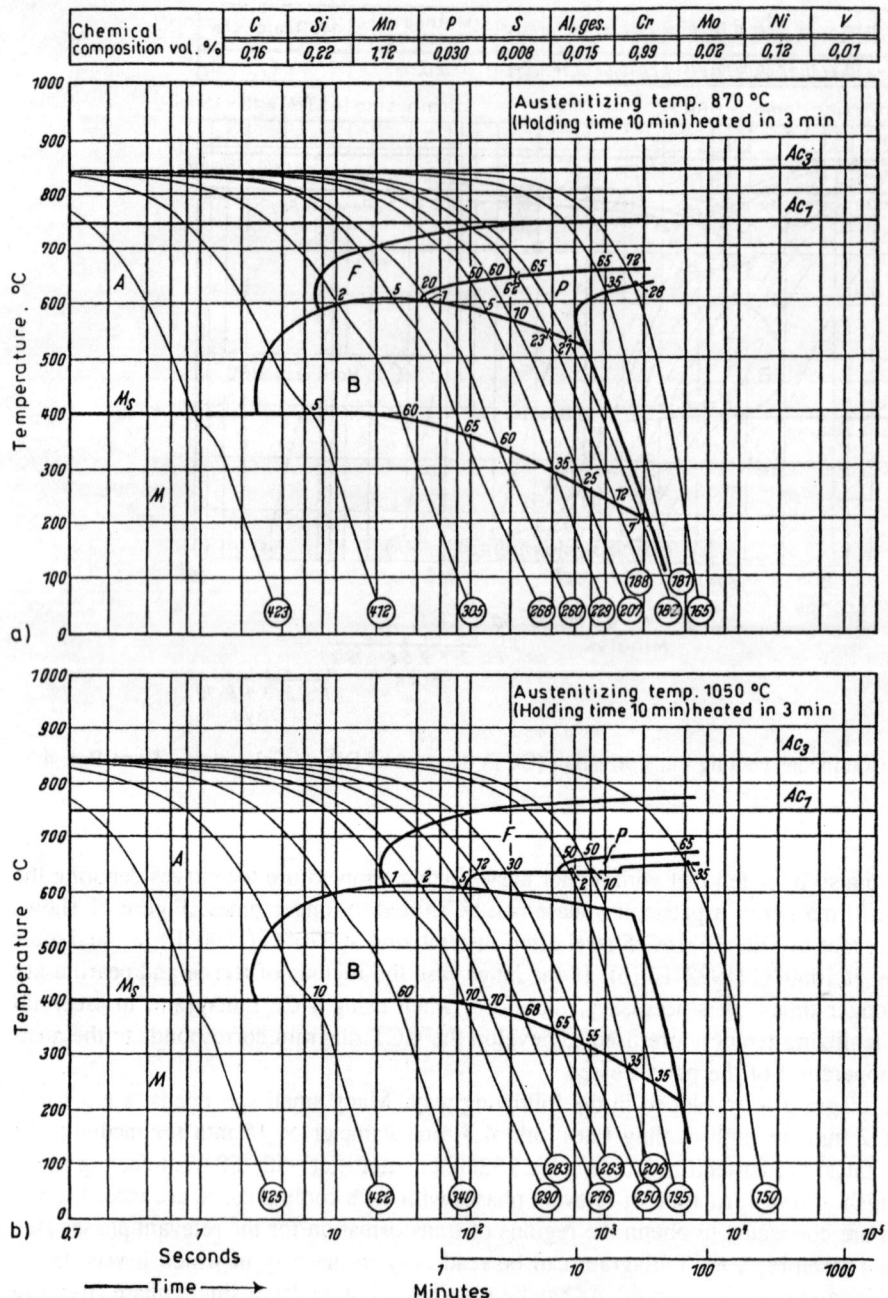

Chemical composition vol. %	C	Si	Mn	P	S	Al, ges.	Cr	Mo	Ni	V
	0,16	0,22	1,12	0,030	0,008	0,015	0,99	0,02	0,12	0,01

Figure 11 CCT diagrams of DIN 16MnCr5 steel (a) when austenitizing temperature is 870°C and (b) when austenitizing temperature is 1050°C. (From Ref. 7.)

hardness after quenching will by 47 HRC. It should be noted that the part of the area (region) of a phase that the cooling curve intersects is by no means a measure of the amount of transformed phase!

Sometimes a CCT diagram can be supplemented with a diagram showing portions of each structural phase and hardness after quenching; see the lower part of Figure 12. The abscissa

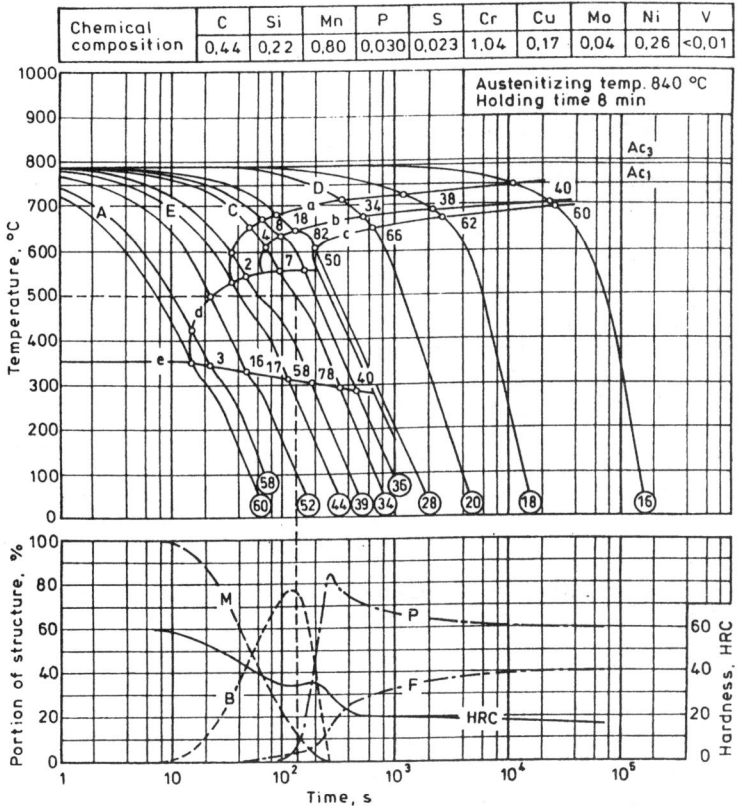

Figure 12 CCT diagram of 41Cr4 steel (top) with the diagram at bottom showing portions of each structural phase and hardness after quenching depending on the cooling time to 500°C. (From Ref. 5.)

of this diagram denoting time enables the cooling time to 500°C (932°F) to be determined for every cooling curve. To determine the portions of structural phases and hardness after quenching, one should follow the relevant cooling curve until its intersection with the 500°C (932°F) isotherm and from this point down along the vertical line read the phase portions and hardness after quenching. For example, for cooling curve C, which intersects the 500°C (932°F) isotherm at 135 s, the readings are 4% ferrite, 7% pearlite, 78% bainite, and 11% martensite and a hardness of 34 HRC.

It should be noted that every CCT diagram is exactly valid only for that heat of a steel that was used for its construction. The influence of different heats (having slightly different compositions) of the same grade of steel on the position of transformation curves in the relevant CCT diagram, as an example, is shown in Figure 13.

As for IT diagrams, the accuracy of a CCT diagram, according to Reference 5, with respect to the position of isotherms is ±10°C (50°F) and with respect to time ordinates ±10% of the relevant time.

3. Heat Treatment Processes for Which an IT or CCT Diagram May Be Used

Taking into account what was explained above about how IT and CCT diagrams can be read, Figure 14 shows the isothermal heat treatment processes for which only IT diagrams may be used. The first is *isothermal annealing* to obtain a coarse ferritic-pearlitic structure, for better machinability (Figure 14a). In this case the IT diagram gives the crucial information, the

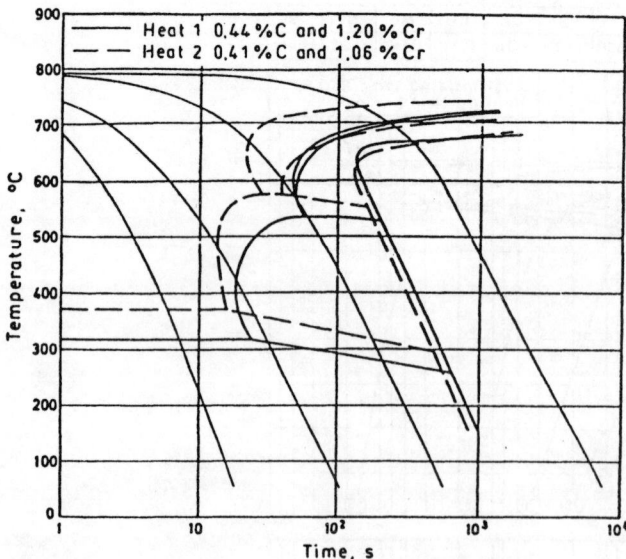

Figure 13 Influence of the difference in composition between two heats of DIN 41Cr4 steel on the position of transformation curves in the relevant CCT diagram. (From Ref. 5.)

optimum temperature at which annealing should take place to achieve the full transformation in the shortest possible time.

The second process is isothermal transformation to bainite, i.e., the *austempering process* (Figure 14b). In this case the IT diagram is used first of all to ascertain that the steel in question is applicable for this process, i.e., that is has enough hardenability (which means that its start of transformation curves are not too close to the ordinate). If this condition is fulfilled, the diagram enables the heat treater to select the appropriate temperature according to the hardness desired and read the minimum time needed at this temperature to achieve the full transformation to bainite.

The third process is the *martempering process* (Figure 14c), an interrupted quenching in a hot bath to obtain the martensite structure with minimum stress and distortion. The applicability of a steel for the martempering process may be checked in the same way as above. In this case the diagram gives information about the necessary temperature of the hot bath and the maximum time the parts can be immersed in it (in order to obtain only martensite) before they are taken out to be cooled in air.

Figure 15 shows, as an example, the only three cases of continuous cooling for which only a CCT diagram may be used. The first case (Figure 15a) is direct quenching to obtain full martensitic structure. In this case the diagram enables the user to determine the critical cooling rate for the steel in question. The second case (Figure 15b) is a continuous slow cooling process, e.g., cooling in air after normalizing annealing. In this case the diagram enables the user to select the cooling rate required to yield the desired hardness of the ferritic-pearlitic structure at room temperature. The percentage of ferrite and pearlite can be read as described above if the diagram allows it. The third case (Figure 15c) represents any continuous cooling regime that results in more than two structural phases. In any of these cases the diagram enables the user to determine the portion of each phase and the hardness after quenching.

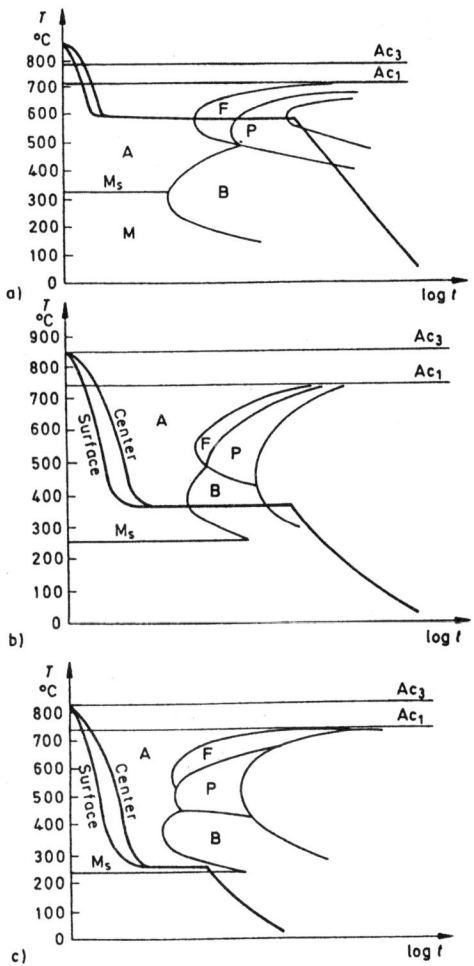

Figure 14 Isothermal heat treatment processes for which only IT diagrams may be used. (a) Isothermal annealing; (b) austempering; (c) martempering.

4. *Using the CCT Diagram to Predict Structural Constituents and Hardness upon Quenching Real Workpieces*

Each CCT diagram describes only those transformations of the structure that occur along the cooling curves of specimens used for its construction. This means that a CCT diagram is valid only for the cooling conditions under which it was constructed. The cooling law for the specimens of small diameter and small volume that were used in constructing the CCT diagram can, according to Rose and Strassburg [4], be described by the exponential function

$$T = T_0 e^{-\alpha t} \tag{18}$$

where T_0 is the austenitizing temperature, α is the heat transfer coefficient, and t is time.

The exactness of the predictions of structural constituents (phases) and hardness values upon quenching depends on the extent to which the cooling law at different points in the cross

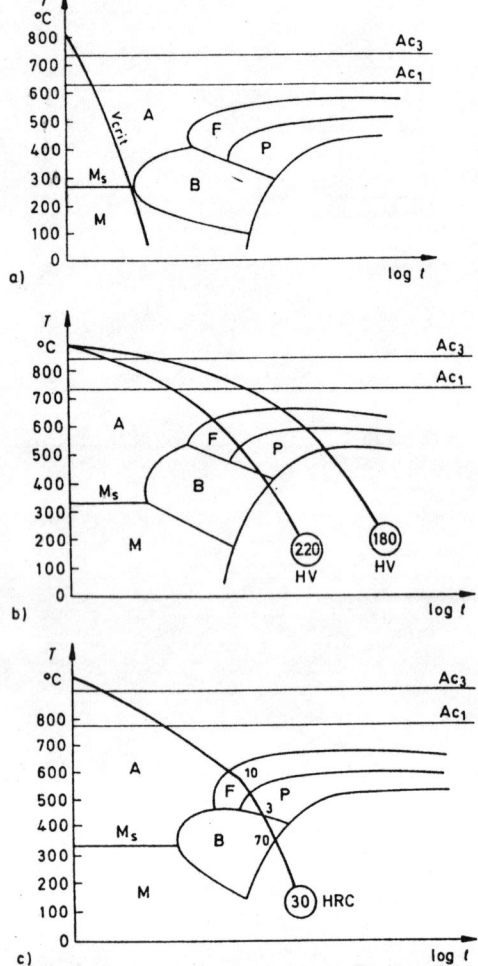

Figure 15 Heat treatment processes with continuous cooling for which only CCT diagrams may be used. (a) Direct quenching to obtain full martensitic structure; (b) slow cooling to obtain a ferrite-pearlite structure of required hardness; (c) continuous cooling regime where a mixed structure is obtained.

section of real workpieces corresponds to the cooling curves of specimens drawn in the CCT diagram. Experimental work [4] using round bars of 50 mm diameter × 300 mm with thermocouples placed 1, 5, 10, 15, and 25 mm below the surface showed that the cooling curves in different points of a round bar's cross section correspond in form to the cooling curves in CCT diagrams to the extent that the structural transformation, i.e., the resulting structural constituents and hardness values, can be compared. Figure 16 shows how a hardness distribution can be predicted by using this correspondence.

If the temperature–time scales of the measured cooling curves and the CCT diagram are the same (as in Figures 16a and 16b), then, by using a transparent sheet of paper the measured cooling curves on CCT diagrams of different steel grades can be superimposed, and in this way the steel grade can be selected that will develop the required structure and hardness at the desired point of the cross section. The accuracy of such prediction from a CCT diagram decreases as the radius of the workpiece's cross section increases. According to Peter

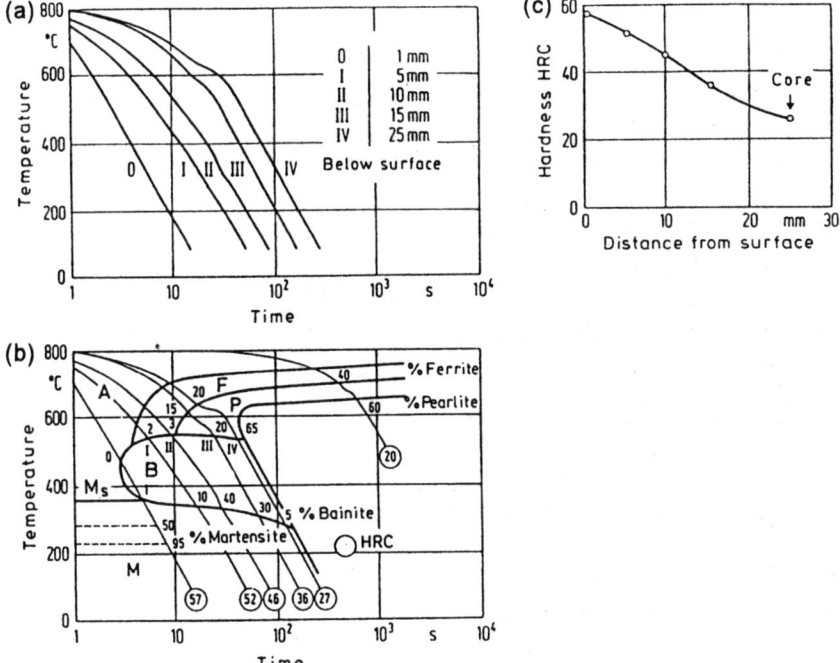

Figure 16 Prediction of hardness distribution on a round bar cross section. (a) Cooling curves measured at different points below surface, as indicated. (b) CCT diagram of the relevant steel with superimposed cooling curves from (a). (c) Hardness distribution on the bars' cross section upon quenching, obtained by reading the hardness values from (b).

and Hassdenteufel [8], sufficiently exact predictions are possible using CCT diagrams for round bars up to 100 mm in diameter when quenching in oil and up to 150 mm in diameter when quenching in water.

It appears that the main problem in the practical use of CCT diagrams for prediction of structural constituents and hardness upon quenching is to establish exactly the cooling curve for the specified point on the workpiece's cross section. This can be done either by calculation (if symmetrical parts and one-dimensional heat flow are involved and the boundary conditions are known) or experimentally (for asymmetric parts) by measuring the temperature–time history with a thermoelement. The correspondence between cooling curves of real workpieces and cooling curves drawn on CCT diagrams also enables the reverse, to draw conclusions about the cooling history (curve) at a specified point of the cross section of a workpiece of any shape and size based on metallographic analysis of the structure and measured hardness upon quenching.

When CCT diagrams (of American origin) are used, the manner of predicting structural constituents and hardness is slightly different. For example, in Figure 17, instead of dilatometric cooling curves, cooling curves taken at different distances from the quenched end of the Jominy test specimen are superimposed.

If one follows one of these cooling curves, e.g., the one for 19.1 mm (3/4 in.) from the quenched end (the heavier line in the diagram), one can read that after 25 s of cooling the Jominy specimen made of AISI 3140 steel at this distance from the quenched end starts to develop ferrite, after 30 s, pearlite; and after 45 s bainite. After 90 s of cooling, 50% of the

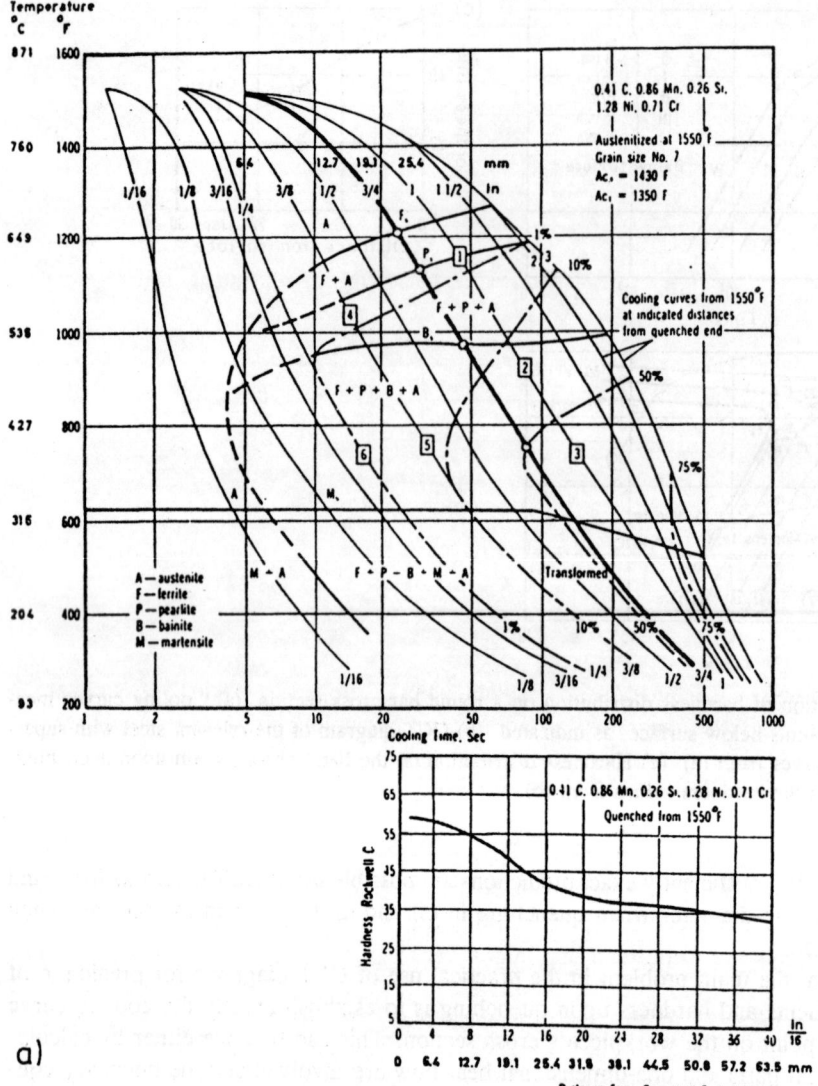

a)

Figure 17 (a) CCT diagram and Jominy hardenability curve for AISI 3140 (From Ref. 33). (b) Chart showing relationship between rate of cooling at different Jominy distances and rate of cooling in moderately agitated oil of round bars of 12.5–100 mm diameter. (From Ref. 14.)

austenite is already transformed. After 140 s, when the temperature at this point has fallen to 315°C (599°F), formation of martensite begins.

The corresponding Jominy curve at the bottom of Figure 17a shows that this cooling curve (at this Jominy distance) with the steel in question will yield a hardness of 48 HRC. To correlate this hardness to different points of the round bars' cross sections of different diameters, an auxiliary diagram (valid in this case only for quenching in moderately agitated oil) such as that shown in Figure 17b should be used. From this diagram one can see that the same hardness of 48 HRC can be met, after quenching in moderately agitated oil, 9 mm below the surface of a round bar of 75 mm diameter.

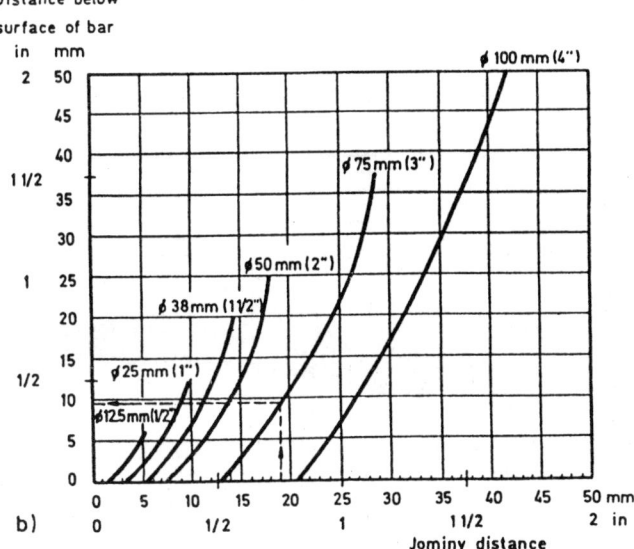

Figure 17 Continued

There also exist CCT diagrams of another type, developed by Atkins [9]; an example is given in Figure 18. These diagrams were developed by cooling/quenching round bars of different diameters in air, oil, and water, recording their cooling curves in the center of the bar, and later simulating these cooling curves in a dilatometric test in order to identify the transformation temperatures, microstructures, and hardness. These diagrams therefore refer only to the center of a bar. Instead of a time scale on the abscissa, these diagrams have three parallel scales, denoting bar diameters cooled in air, quenched in oil, and quenched in water. A scale of cooling rates [usually at 700°C (1292°F)] in °C/min is added.

These diagrams are to be read only along vertical lines (from top to bottom), denoting different cooling rates. For example, to determine the microstructure developed and resulting hardness in the center of a 10 mm bar of the steel in question when cooling it in air, one takes the vertical line at 10 mm diameter on the scale for air cool (see Figure 18), starts in the austenite region and proceeds downward. Transformation in this case (unalloyed steel grade with 0.38% C) starts at 700°C (1292°F) with the formation of ferrite, continuing to nearly 50% transformation at 640°C (1184°F) when pearlite begins to form. At 580°C (1076°F), a trace of bainite is indicated before transformation is complete.

If oil quenching of a 10 mm bar is now considered, the 10 mm position should be located on the oil-quenched bar diameter scale in Figure 18. Again starting in the top region and following the vertical line down, it is seen that in this case bainite is the first phase to form from austenite at 560°C (1040°F). At 330°C (626°F), after about 40% transformation, the remaining austenite transforms to martensite until the reaction is complete at 150°C (300°F). Similarly, the center of a water-quenched 10 mm diameter bar will transform to martensite starting at 360°C (680°F) and finishing at 150°C (300°F).

Relevant hardness values after quenching (and in some cases after tempering to different specified temperatures) can be read following the same vertical line further down, into the "hardness after transformation" diagram.

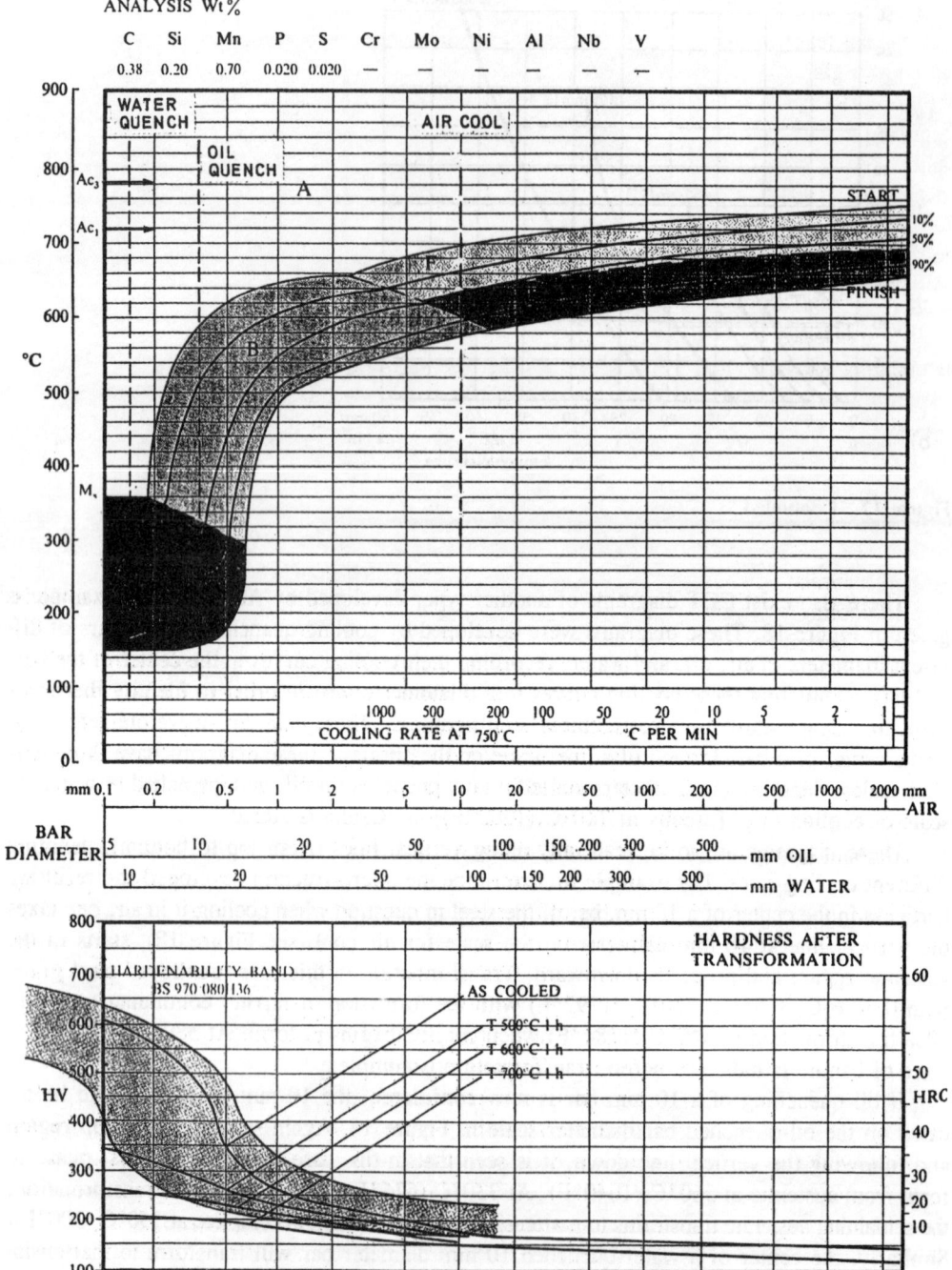

Figure 18   CCT diagram for rolled steel austenitized at 860°C. (From Ref. 9.)

Examination of the left-hand side of the upper diagram in Figure 18 for the steel in question shows that martensite will form on air cooling with bars up to 0.18 mm diameter, on oil quenching up to 8 mm diameter, and on water quenching up to 13 mm diameter.

A special feature of this type of CCT diagram is that the hardenability of the steel can be assessed at a glance.

Figure 19a is a CCT diagram for a very low hardenability steel previously rolled and austenitized at 950°C (1742°F). It shows early transformation to ferrite and pearlite (even with oil and water quenching of smallest diameters). Figure 19b shows a similar diagram for a high-hardenability steel previously rolled, softened at 600°C (1112°F) and austenitized at 830°C (1526°F). In this case the austenite changed predominantly to martensite and bainite over a wide range of bar diameters and quenching rates. Diagrams of this type representing 172 steel grades have been published in the BSC atlas [9].

5. Special Cases and Limitations in the Use of CCT Diagrams

When dealing with carburized steels, one should be aware that, because of the big difference in carbon content between the core ($\approx 0.2\%$) and the case ($\approx 0.8\%$), the CCT diagram for the case will be totally different from the one for the core of the same steel, as shown by Figures 20 and 21. The increased carbon content in the case increased the hardenability and caused the pearlite and bainite regions to be shifted to much longer times. The ferrite region disappeared, and the M_S point was lowered. Cooling at the same rate results in different portions of structural constituents and substantially different hardness values.

Another limitation in the use of CCT diagrams concerns cooling regimes with discontinuous change in cooling rate, as for example a delayed quenching in air followed by water or oil quenching. The left-hand part (a) of Figure 22 shows the start of transformation as in the conventional CCT diagram for the steel in question. The right-hand part (b) of this diagram holds for air cooling to approximately A_{c1}, followed by water quenching (a delayed quenching process). It shows a significant displacement of the ferrite and bainite regions to longer times. Such a cooling mode enhances hardenability and results in higher hardnesses than expected from the conventional CCT diagram for the same steel. The effect of discontinuous change in cooling rate is based on nucleation and on incubation time before the change in cooling rate occurs and is theoretically explained by Shimizu and Tamura [11].

D. Oxidation

Oxidation takes place as an undesirable accompanying phenomenon during every heat treatment of metals in a noninert atmosphere. Chemical reactions that occur during the oxidation of a metal are generally expressed by the formula

$$\frac{2x}{y} \text{Me} + O_2 \rightleftharpoons \frac{2}{y} \text{Me}_x O_y \qquad (19)$$

where x and y denote integer numbers.

The oxidation process proceeds at a set temperature spontaneously from left to right (according to formula 19). During this process the free enthalpy of the reaction products (G_R) becomes smaller than the enthalpy of the original materials (G_A), i.e., the difference can be expressed as

$$\Delta G_0 = G_R - G_A \qquad \text{for } G_A > G_R \qquad (20)$$

If $\Delta G_0 > 0$, reaction (19) will take place from right to left, i.e., the metal oxide will be reduced.

AUSTENITISED AT 950°C

PREVIOUS TREATMENT   ROLLED

ANALYSIS Wt%

C	Si	Mn	P	S	Cr	Mo	Ni	Al	Nb	V
0.06	—	0.30	..	..	..	—	—	—	—	—

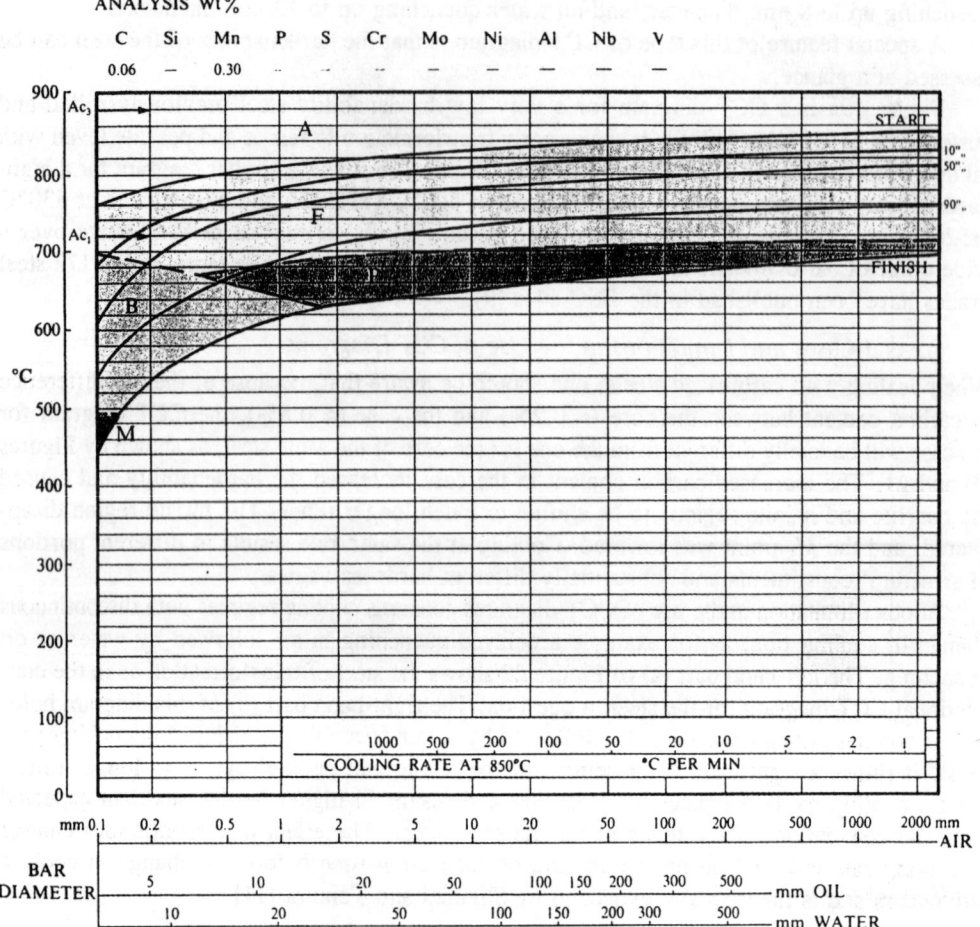

Figure 19 Examples of (a) a low-hardenability and (b) a high-hardenability steel as depicted in CCT diagrams. (According to Atkins [9].)

When the oxygen pressure (p_{O_2}) equals 1 bar, ΔG_0 is called the "free standard creating enthalpy." Figure 23 shows the temperature dependence of the free standard creating enthalpy (ΔG_0) for oxidation reactions of some metals.

If (p_{O_2}) at temperature T differs from 1 bar, then the characteristic change of the free creating enthalpy may be calculated as follows:

$$\Delta G = \Delta G_0 - RT \ln p_{O_2} \qquad (21)$$

where R is the universal gas constant and T is absolute temperature.

As can be seen from Figure 23 for all of the metals represented except silver, the values of ΔG_0 are negative with an increasing trend at higher temperatures. In the case of silver, $\Delta G_0 = 0$ at 190°C (324°F). At this temperature, equilibrium exists between Ag, O_2, and Ag_2O, i.e., the disintegration pressure of Ag_2O has reached the oxygen pressure of 1 bar that was

AUSTENITISED AT 830°C

PREVIOUS TREATMENT ROLLED. SOFTENED 600°C 1 h

ANALYSIS Wt%

C	Si	Mn	P	S	Cr	Mo	Ni	Al	Nb	V
0.40	0.25	0.60	0.020	0.020	0.65	0.55	2.55	—	—	—

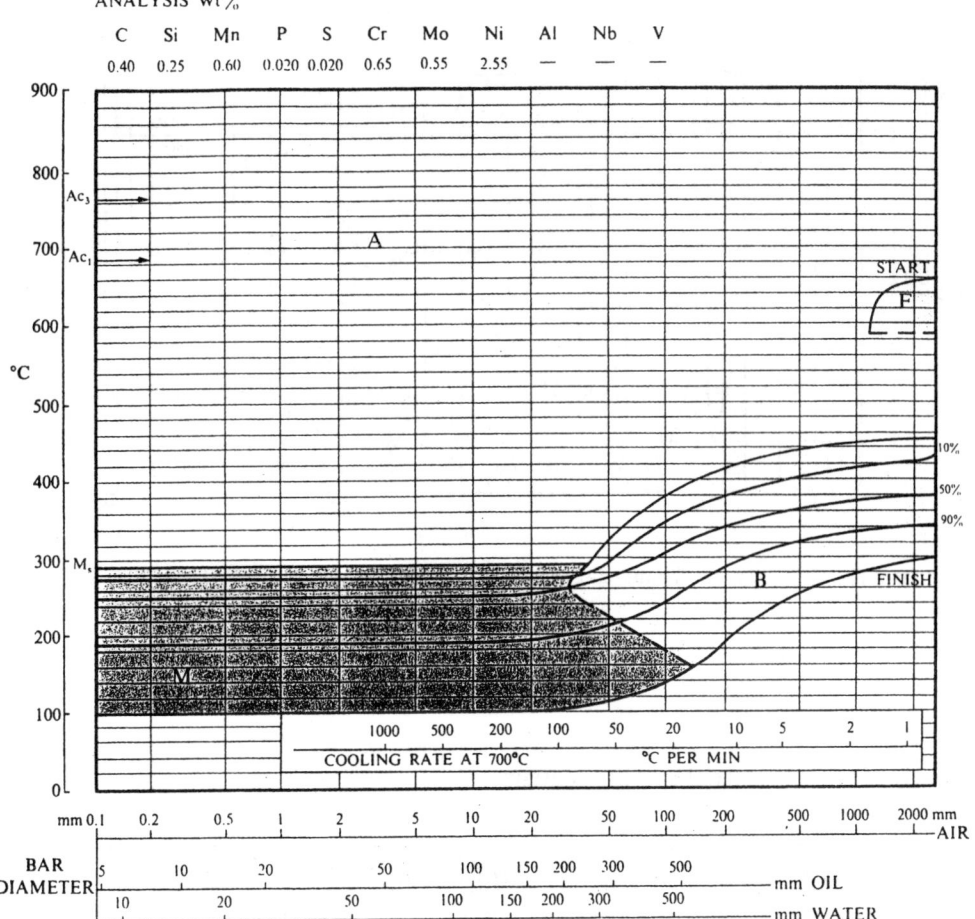

Figure 19 Continued

taken as the basis. At higher temperatures the disintegration pressure of Ag_2O becomes higher and the metal oxide (Ag_2O) will be disintegrated.

From Figure 23 it can be concluded that the chosen metals, with the exception of silver, within the shown temperature range would form oxides. Because the oxidation takes place on the surface, the oxide layer that is formed separates the two reaction partners, i.e., the metal and the oxygen. This oxide layer, which is material-specific, becomes thicker with time. There are several formulas expressing the dependence of the oxide layer thickness on time. For higher temperatures a parabolic law is usually used:

$$y = \sqrt{A_1 t} \tag{22}$$

where y is the oxide layer thickness, A_1 is a material-specific constant, and t is time. This parabolic law is valid for oxidation processes when the rate of oxidation depends on the diffusion of metal ions and oxygen ions through the oxide layer.

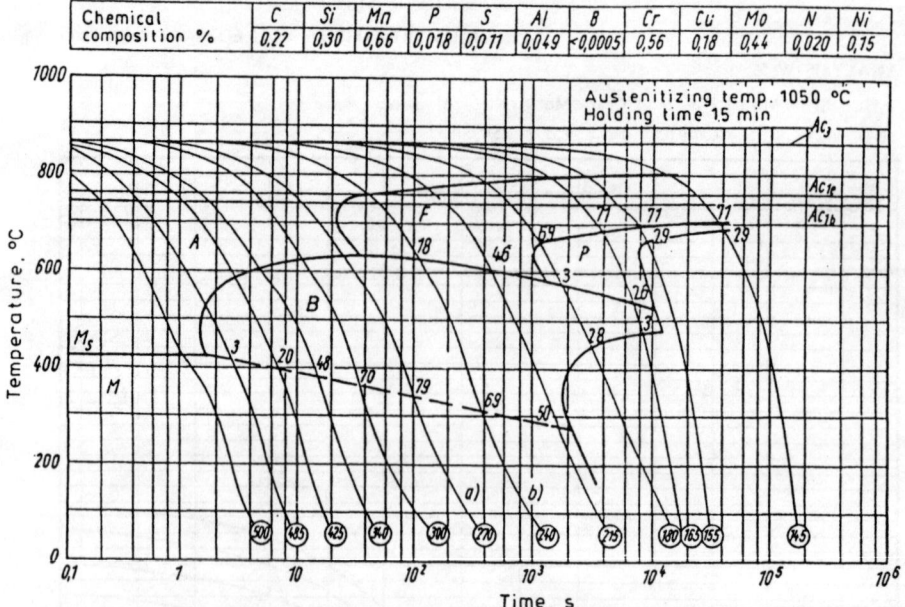

Figure 20 The CCT diagram for the core (0.22% C) of DIN 20MoCr5 steel. (From Ref. 2.)

When the oxide layer is porous, i.e., permeable for the gas, and therefore the metal and oxygen are not separated, build-up of the layer follows a linear law:

$$y = A_2 t \tag{23}$$

where A_2 is another material-specific constant.

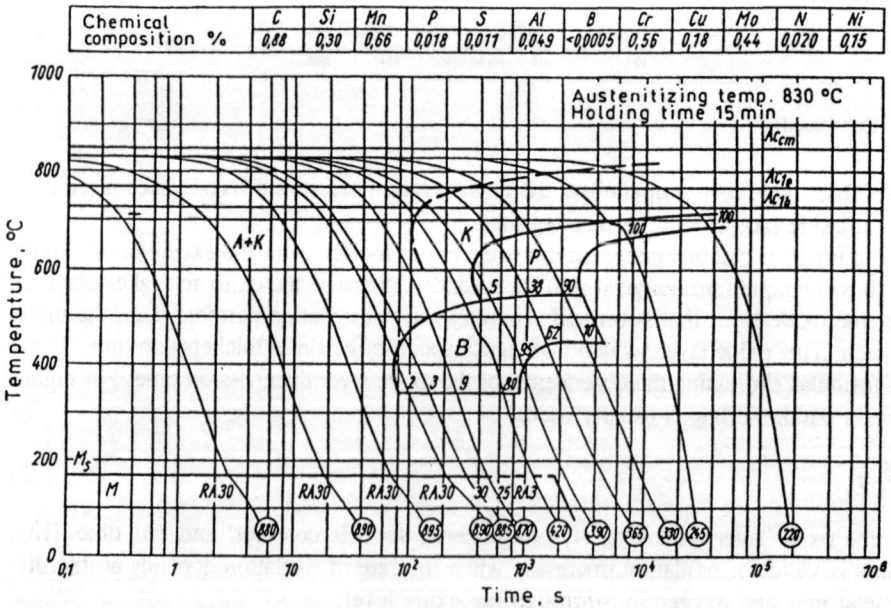

Figure 21 The CCT diagram for the case (0.88% C) of DIN 20MoCr5 steel. (From Ref. 2.)

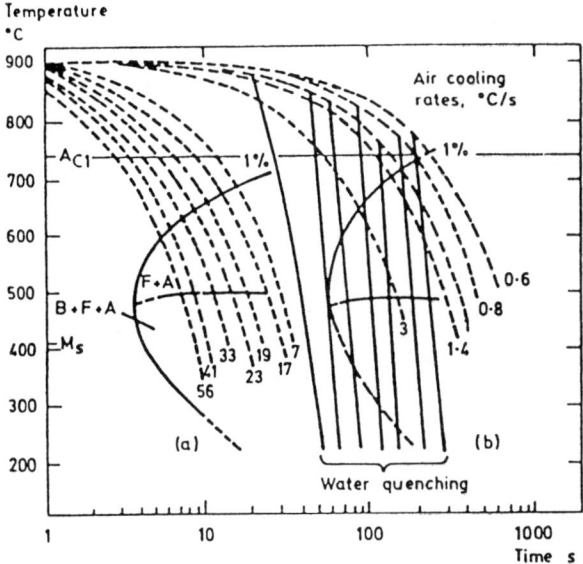

Figure 22 Cooling curves and CCT diagrams for a steel containing 0.20% C, 0.78% Mn, 0.60% Cr, 0.52% Ni, and 0.16% Mo after austenitizing at 900°C. (a) Conventional CCT diagram; (b) various air cooling rates to approximately A_{c1} followed by water quenching. (From Ref. 10.)

An oxide layer of a pure metal is constituted of uniform chemical compound if a single valency is involved, for example FeO. If more valencies are involved, the oxide layer consists of sublayers with oxygen valencies increasing from the inside to the outside, for example: FeO, Fe_3O_4, and Fe_2O_3, as shown in Figure 24. In Figure 24A an oxide layer of pure iron, created during a 5 h annealing at 1000°C (1832°F), is shown. Relevant processes during development and build-up of the layer are schematically shown in Figure 24B.

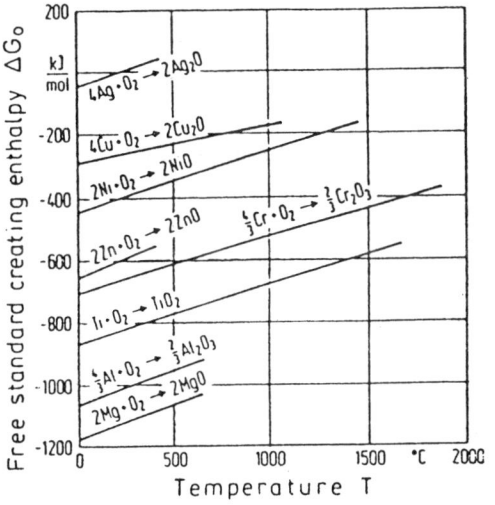

Figure 23 Temperature dependence of the "free standard creating enthalpy" of some oxidation reactions between 0°C and the melting point of the relevant metal. (From Ref. 1.)

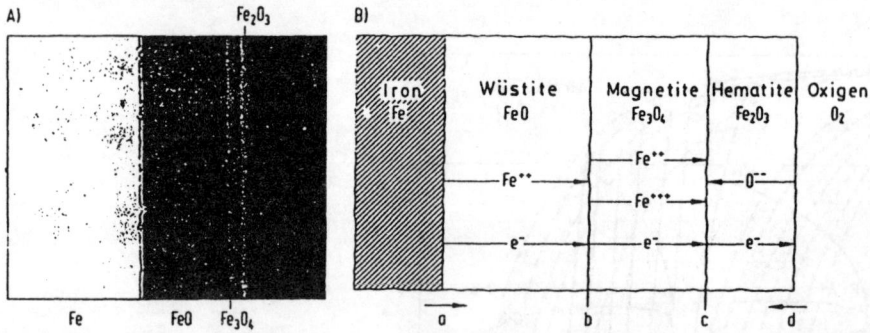

Figure 24   Oxidation of pure iron. (A) Oxide layer. (B) Processes during the build-up of the oxide layer; a, transition of $Fe^{2+} + 2e^-$ from the metal into FeO; b, creation of FeO; c, creation of Fe_3O_4 and Fe_2O_3; d, input of oxygen. (From. Ref. 1.)

1. Scaling of Steel

When metallic parts are heated above 560°C (1040°F) (this is the temperature at which the creation of wüstite or FeO begins), after creation of the first part of the layer in the starting phase, the reaction follows by diffusion of Fe^{2+} ions from the steel toward the outside and the diffusion of oxygen ions at the scale/metal interface toward the inside. As time passes, the linear law of oxidation valid for the starting phase changes to a parabolic law.

The growth of the oxide layer (scale) depends very much on the chemical composition of the steel. Different alloying elements, having different diffusion abilities, have different influences on the oxidation process and the build-up of scale. The chemical composition of the original material on the surface is subject to change.

According to their affinity for oxygen, the alloying elements in steel can be divided into three groups with respect to their influence on the scaling process [2]:

> Group I contains those elements whose oxygen affinity is less than the affinity of the richest oxide compound wüstite (FeO), for example, Ni and Co. After saturation of the basic metal with oxygen, the outer oxidation of iron begins with the creation of wüstite. The alloying elements become richer at the scale/metal interface.
>
> Group II contains those elements whose oxygen affinity is greater than that of iron (Cr, Si, V, Al). After saturation of the basic metal with oxygen, inner oxidation begins. Because of the creation of internal oxides of alloying elements, a diffusion barrier builds up against the diffusion of metal and oxygen ions, hampering the development of scale.
>
> Group III contains those elements whose oxygen affinity is similar to that of wüstite (Mo, W). No inner oxidation takes place. The alloying elements become richer in the basic metal at the scale/metal interface.

Figure 25 shows the influence of Cr additions to a steel on the depth of scale (mm/y) at temperatures of 600, 700, and 800°C (1112, 1292, 1472°F). A particularly high oxidation resistance of steels may be achieved by Cr additions of 6–30% by mass.

E.   Decarburization

Under conditions that cause the oxidation of iron, the oxidation of carbon is also to be expected. Decarburization of a metal is based on the oxidation at the surface of carbon that is dissolved in the metal lattice. It should be noted that, depending on the carbon potential of

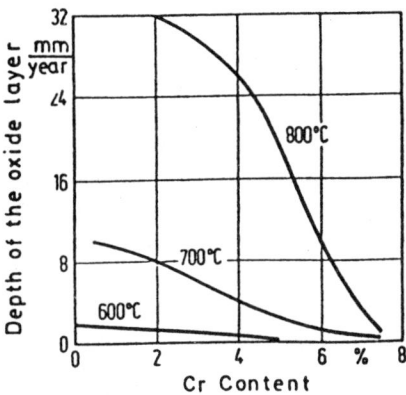

Figure 25 Influence of Cr on the oxidization of a steel at temperatures of 600, 700, and 800°C. (From Ref. 12.)

the surrounding atmosphere, decarburization can take place independently of scaling. However, in heat treatment processes iron and carbon usually oxidize simultaneously. During the oxidation of carbon, gaseous products (CO and CO_2) develop. In the case of a scale layer, substantial decarburization is possible only when the gaseous products can escape, i.e., when the equilibrium pressures of the carbon oxides are high enough to break the scale layer or when the scale layer is porous.

The carbon consumed on the surface has to be replaced by diffusion from the inside. Hence the process of decarburization consists of three steps:

1. Oxygen transport within the gas to the metal surface
2. Carbon exchange at the gas/metal interface
3. Diffusion of carbon within the metal

Generally the diffusion of carbon within the metal is the most important factor in controlling the rate of decarburization, which after a short starting period follows a parabolic time law. When a mild steel is heated below 910°C (1670°F), a surface layer of ferrite is formed that acts as a barrier to carbon transport owing to the very low solubility of carbon in ferrite. Above 910°C (1670°F) the steel remains austenitic throughout, and decarburization becomes severe. The model used to represent decarburization in the fully austenitic condition is shown in Figure 26. The steel surface is continually oxidized to form a scale, while the carbon is oxidized

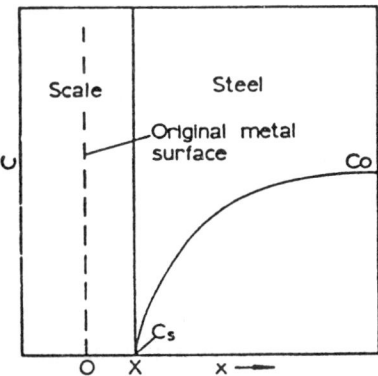

Figure 26 Model for decarburization in fully austenitic condition. (From Ref. 12.)

to form the gases CO and CO_2. The scale is assumed to be permeable to these gases, which escape to the atmosphere.

The carbon content at a scale/metal interface is assumed to be in equilibrium with the oxygen potential of the scale, which at that position corresponds to the equilibrium between iron and wüstite. The carbon concentration profile in the metal varies from the low surface concentration to the original carbon content within the metal, as shown in Figure 26. In using this model, distances are measured from the original metal surface; the instantaneous scale/metal interface lies at the position $x = X$ at time t. This means that scaling has consumed a thickness X of metal during time t.

To calculate the depth of decarburization, the distribution of carbon in the metal is calculated by solving Fick's second law for the relevant boundary conditions:

$$\frac{\delta C}{\delta t} = D \frac{\delta^2 C}{\delta x^2} \qquad \text{for } x > X \tag{24}$$

$$C = C_0 \qquad x > 0; \; t = 0 \tag{25}$$

$$C = C_s \qquad x = X; \; t > 0 \tag{26}$$

Equation (25) indicates that initially the carbon concentration was uniform throughout the specimen; Equation (26) indicates that the carbon concentration at the metal/scale interface is constant (in equilibrium with the scale). It is assumed that decarburization does not extend to the center of the specimen and that the diffusion coefficient of carbon in austenite is independent of composition; enhanced diffusion down grain boundaries is neglected. Under these conditions the solution at a constant temperature (for which the diffusion coefficient is valid) reads

$$\frac{C_0 - C}{C_0 - C_s} = \frac{\text{erfc}(x/2\sqrt{2Dt})}{\text{erfc}(k_c/2D)^{1/2}} \tag{27}$$

where C_0 is the original carbon content of the metal, C_s is the carbon concentration at the metal/scale interface, D is the carbon diffusion coefficient, t is time, erfc $= 1 - \text{erf}$ (here erf is the error function), and k_c is the corrosion constant of the metal ($k_c = X^2/2t$).

Equation (27) gives the carbon content within the metal for $x > X$ as a function of time and position.

The value of k_c for the relevant steel in the relevant atmosphere is expressed as

$$k_c = 0.571 \; \exp \; (-43,238/RT) \; \text{cm}^2/\text{s} \tag{28}$$

Although the variation of the diffusion coefficient of carbon in austenite was ignored in solving Equation (24), it was found that the best agreement between calculated and measured carbon profiles was obtained when values relating to very low carbon content were used. Therefore in this calculation a diffusion coefficient for zero carbon content was used, which reads

$$D_{(C=0)} = 0.246 \; \exp(-34,900/RT) \; \text{cm}^2/\text{s} \tag{29}$$

A comparison of measured decarburization depths with values calculated using these data showed that with 12 measurements of isothermal treatments between 1050 and 1250°C (1922 and 2282°F) for times between 900 and 10,800 s, the mean prediction was 97% of the measured value [12]. It was found that the inner limit of the decarburized zone is placed, by

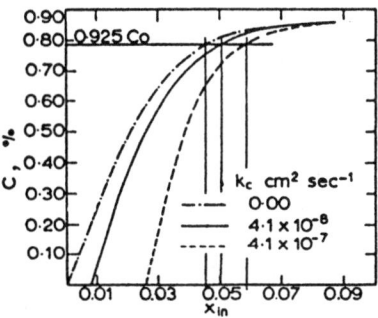

Figure 27 Effect of scaling rate on decarburization of an 0.85% C steel after 1.5 h at 1050°C. The position where the carbon profile cuts the x axis indicates the position of the scale/metal interface for the three k_c values. (From Ref. 12.)

metallographic examination, at the position where the carbon content is 92.5% of the original carbon content.

To ascertain the effect of scaling rate on decarburization, it may seem logical to try to reduce decarburization by reducing the oxidizing potential of the atmosphere. This is a fallacy, as the carbon concentration at the metal/scale interface is constant in equilibrium with iron oxide as long as scale is present. However, the scaling rate can be affected by changing the atmosphere, and this will affect the observed depth of decarburization.

To illustrate this, carbon profiles have been calculated and plotted in Figure 27 for a 0.85% C steel heated for 1.5 h at 1050°C (1922°F). The carbon profiles are plotted relative to the original metal surface, while the point at which a carbon profile cuts the x axis indicates the position of the scale/metal interface for the related conditions. The curves refer to k_c values of 0, 4.1 × 10^{-8}, and 4.1 × 10^{-7} cm²/s. The depth of decarburization is determined by the position at which $C = 0.925 C_o$ (see the horizontal line drawn of Figure 27). From Figure 27 it appears that with increased values of k_c the scale/metal interface on the x axis shifts progressively toward the inside of the metal, while the depth of decarburization (which is the horizontal distance between the intersection of a carbon profile curve with the horizontal $0.925 C_o$ line and its intersection with the x axis) is found to decrease as k_c increases. These results are shown in Table 3. This reveals an interesting situation where, by reducing the oxidation rate, the depth of decarburization is increased yet less metal is wasted.

When scaling and decarburization take place simultaneously, decarburization is prevented during the starting phase of scaling. It takes place substantially only after the equilibrium pressures of CO and CO_2 increase at increased temperatures and the adhesion strength of the scale (because of its increased thickness) diminishes or the scale becomes porous.

Table 3 Effect of Scaling Rate on Decarburization

Scaling rate k. (cm²/s)	Total depth of metal affected		Depth of decarburized layer	
	cm	in.	cm	in.
0	0.119	0.047	0.119	0.047
4.1×10^{-8}	0.130	0.051	0.109	0.043
4.1×10^{-7}	0.150	0.059	0.084	0.033

Source: Ref. 12.

1. *The Effect of Alloying Elements on Decarburization*

Alloying elements may affect decarburization due to their effect on

1. The ferrite-austenite transformation temperature
2. The activity of carbon in solution
3. The diffusion coefficient of carbon in solution
4. The scaling characteristics of iron

Although this subject is a complex one, even when considered only qualitatively, the following statement is generally valid. Decarburization increases with (1) increased rate of carbon diffusion, (2) increased carbon activity, and (3) increased ferrite-austenite transformation temperature.

A complication arises due to the fact that during scaling alloying elements tend to concentrate either in the scale or in the metal at the scale/metal interface. In special cases, when strong carbide-forming elements are involved, decarburization may also be influenced by the rate of dissolution of carbides in the matrix. When the alloying elements are less valuable than iron, the possibility of internal oxidation arises; an external oxide layer may also be formed under circumstances that are normally protective to iron. If either type of oxide formation occurs, the concentration of the alloying element in solution is reduced at the metal surface, and so the effect on carbon behavior will be altered correspondingly.

When an external scale is formed, the effect of the alloying element on the scaling rate must also be considered. If the scaling rate is increasing, then, in the absence of other factors, the observed depth of decarburization will be reduced.

Although quantitative predictions are not possible, it is instructive to predict what the effects of a few common alloying elements may be.

Nickel will concentrate at the scale/metal interface and, although the scaling rate may not be greatly affected, the solubility of carbon in the surface layers may be reduced, thus restricting carbon diffusion outward and reducing the depth of decarburization.

Manganese is taken into the scale in solid solution in the wüstite and magnetite layers. Scaling rates are hardly affected, and any effect on decarburization will be restricted to its effects on carbon activity and the diffusion coefficient. Since manganese is denuded in the surface layers of the metal, however, the effect may be only slight.

Silicon also concentrates in the scale and forms fayalite, which reduces the scaling rate. This should lead to deeper observed decarburization. Silicon also increases the activity of carbon and therefore increases the tendency of carbon to diffuse out to the scale/metal interface. Thus the general effect expected of silicon is to increase decarburization.

Chromium concentrates in the scale, forming spinel, depending on its concentration. In general, scaling rates are reduced. The formation of stable carbides introduces the possibility of a slow carbide decomposition step into the mechanism. At the usual reheating temperatures and times, however, the chromium carbides may dissolve completely. In this case the effect of chromium would be to reduce the activity of carbon in solution, thus tending to reduce the rate of migration to the surface. There are therefore two conflicting factors. The lower scaling rate would tend to increase the observed decarburization, whereas the reduction of carbon activity would tend to reduce it. The later factor may be expected to predominate and reduce decarburization.

2. Definitions and Measurement of Decarburization

The strength of a steel depends on the presence of carbides in its structure; therefore, loss of carbon from the surface softens and weakens the surface layers. In such a case the wear resistance is obviously decreased, and in many circumstances there can be a serious drop in fatigue resistance.

To avoid the real risk of failure or inferior performance of engineering components, it is essential to minimize decarburization at all stages in the processing of steel. This requires inspection and the laying down of specifications for decarburization in various components and semiproducts. The decarburization limits in such specifications must be related to the function of the component and must enable checking by the use of agreed-upon and preferably standardized measuring techniques.

There are several different definitions of decarburization. A rigorous definition is that the depth of decarburization is the thickness of the layer in which the carbon content is less than that of the core, i.e., the distance from the surface to a boundary at which the carbon content of the core is reached. This boundary corresponds to the asymptote of the graph of carbon content versus distance from the surface and is therefore somewhat diffuse and difficult to locate with precision. Thus the depth of decarburization according to this definition is difficult to measure reproducibly. A functional definition is that the depth of decarburization is the thickness of the layer in which loss of carbon has a significant effect on properties that affect the functioning of the final component. The limit of this layer can be expressed as a carbon content or a hardness level. Finally, a practical definition of the depth of decarburization is that it is the thickness of the layer in which the structure differs significantly from that of the core. This definition is suitable when a metallographic examination is undertaken.

There is a distinction between complete decarburization and partial decarburization. Complete decarburization leaves the surface layer entirely ferritic, which can be clearly distinguished under the microscope. The depth of complete decarburization is the thickness of the ferrite layer, i.e., the distance from the surface to the first particle of a second phase. In the zone of partial decarburization the carbon content increases progressively from the ferrite layer to the core and approaches the core composition asymptotically.

The total thickness of the decarburization layer, i.e., the distance from the surface to the inner boundary of the core, is the total depth of decarburization—the sum of the depths of complete and partial decarburization. It should be noted also that the depth of decarburization may vary around the circumference of a component.

A number of methods for measuring decarburization are available. The requirements that such techniques have to meet are

1. The ability to measure a clearly defined depth of decarburization, e.g., compatibility with the functional definition of the depth of decarburization
2. Reproducibility of measurement
3. Ease and convenience of measurement

Optical metallography is the most useful and convenient method. A cross section of the component or sample around the periphery is examined, and the depth of decarburization is measured from the surface to the "practical" boundaries of complete and partial decarburization. This method is suitable for ferrite-pearlite structures only.

For the metallographic examination of high-speed steels, a method has been established that depends on color staining by means of etching in alcoholic nitric acid (nital). A polished cross section of annealed high-speed steel is etched in 4% nital. During the first 30 s in the etchant the specimen surface progresses through a gray color to a purplish-blue, which changes suddenly after about 60 s to a blue-green. Where the functional definition of decarburization calls for the development of the full hardness in the surface layers, the practical boundary is the start of the general core structure, i.e., the edge of the blue-green zone.

The *arrest quench method* consists of austenitizing a very thin specimen in a neutral atmosphere or salt bath and quenching it in a hot bath held at an appropriate temperature. This is the M_s temperature corresponding to the carbon content at which it is desired to place the boundary. The specimen is held at that temperature for about 5 s and is then water-quenched. During this short arrest the decarburized zone, which has an M_s temperature above the temperature of the bath, will partly transform to martensite; the core will remain austenitic. As soon as martensite has formed in the decarburized zone, the martensite needles will begin to temper slightly. Thus, after water quenching the core will consist of fresh light-etching martensite, while the decarburized zone will contain dark-etching tempered martensite needles. A very sharp contrast is achieved at the boundary between the decarburized zone and the core, and the boundary can be located with considerable accuracy at any desired carbon content below the original carbon content of the core.

Figure 28 shows specimens taken from the same hot-rolled rod that have been arrest quenched at different temperatures to place the decarburization boundary at different carbon contents. The micrographs are placed on a graph of carbon content versus depth of decarburization, and the carbon profile has been drawn through the microstructures. This technique is compatible with the functional definition of the depth of decarburization based on a particular carbon content and gives very reproducible results.

Microhardness measurement is a fairly convenient method for quantitatively accurate determinations of a functional decarburization limit by determining the variations of hardness with distance from the surface of the test piece. Since this involves polishing a cross section, it is invariably preceded by a metallographic scan that facilitates the location of the best area for the hardness survey.

A graph of hardness versus distance from the surface is plotted, and the deviation from the core hardness can be detected.

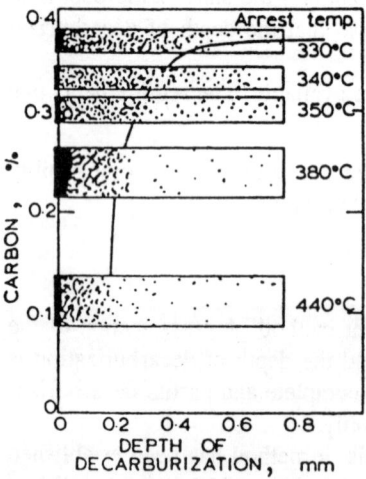

Figure 28 Depth of decarburization to various carbon contents, established by the arrest-quench method. (From Ref. 12.)

Chemical analysis of successive surface layers is the classical referee method for measuring decarburization. The sample has to be large enough to permit accurate chemical analysis, and yet each surface layer must be fairly thin in order to give an adequate number of points on a graph of carbon content versus distance from the surface. The graph of carbon content against distance can be used to indicate the first deviation from the core composition or to locate any decarburization boundary. Complete decarburization is not very easy to locate on this graph, because the carbon content of ferrite is too low for very accurate chemical analysis.

Chemical analysis can be replaced by the carbon determination with a vacuum spectrograph. This has several advantages, particularly in speed and convenience and also because the sample size required is much smaller than for other methods. The only limitation is the need to place the spark accurately on a flat area parallel to the original surface and at least 15 mm in diameter. Successive layers have to be exposed by grinding, because the maximum depth measured in one exposure is limited to about 500 μm.

F. Residual Stresses, Dimensional Changes, and Distortion

Residual stresses are stresses in a body that is not externally loaded by forces and/or moments. They are in mechanical equilibrium within the body, and consequently the resultant force and the resultant moment produced by residual stresses must be zero. Residual stresses are classified, according to the area within which they are constant in magnitude and direction (i.e., in which they are homogeneous), into three categories:

> Residual stresses of the first kind are those stresses that are homogeneous across large areas of the material, i.e., across several grains. Internal forces resulting from these stresses are in equilibrium with respect to any cross section, and mechanical moments resulting from these stresses are in equilibrium with respect to any axis. Any intervention in the equilibrium of forces and moments of a volume element containing such residual stresses will change the element's macroscopic dimensions.

> Residual stresses of the second kind are those stresses that are homogeneous across microscopically small areas (one grain or subgrain region) and are in equilibrium across a sufficient number of grains. Macroscopic changes in the dimensions of a volume element possessing these stresses may become apparent only if distinct disturbances of this equilibrium occur.

> Residual stresses of the third kind are those stresses that are inhomogeneous across microscopically small areas (within several atomic distances of single grains) and are in equilibrium across subgrain regions). No macroscopic changes of the dimensions of the stressed material will result when such equilibria are disturbed.

Residual stresses of the first kind are called *macro residual stresses,* and those of the second and third kinds are called *micro residual stresses.*

Typical residual stresses of the third kind are stresses connected with dislocations and other lattice defects. An example of residual stresses of the second kind are stresses within grains of a material consisting of two structural phases with different expansion coefficients.

In practice, only residual stresses of the first kind are considered, and they are characterized by the technological processes by which they originate. The main groups of residual stresses are

Casting residual stresses

Forming residual stresses

Working-out residual stresses
Heat treatment residual stresses
Joining residual stresses
Coating residual stresses

In every stressed workpiece all three kinds of residual stresses are present.

Figure 29 is a schematic presentation of all three kinds of residual stresses and their superposition in a two-phase material after quenching. (RS I–III denote residual stresses of the first to third kinds, respectively.)

Estimation of residual stresses in a workpiece is very important because they represent a preloading of the material. There is always a linear superposition of internal (residual) and external stresses, and the resulting stress affects the strength of the material and its deformation behavior.

In the case of a dynamic loading on a component, the residual stresses act as a constant preloading. Tensile stresses decrease the fatigue strength, and compression stresses increase it. The fatigue strength of a component depends not only on the resulting stresses on the surface but also on the distribution of stresses across the section. Figure 30 shows schematically two cases with the same external stress (straight line c), and same fatigue strength (straight line a). The only difference is in the distribution of residual stresses (curve b). In case I, a

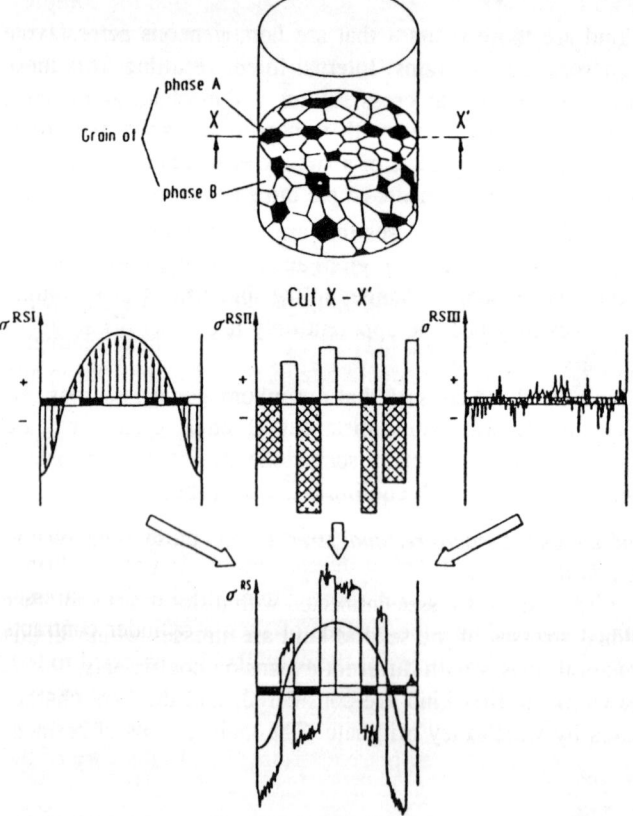

Figure 29 All three kinds of residual stresses in a two-phase material after quenching, and their superposition (shown schematically). (From Ref. 13.)

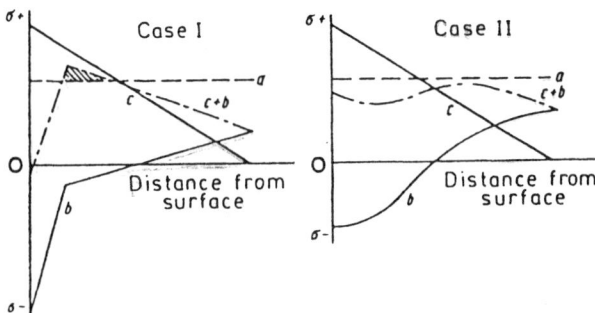

Figure 30 Schematic presentation of superposition of the external load and residual stresses at a fatigue test. (From Ref. 2.)

high residual stress (compressive) on the surface rapidly decreases below the surface, while in case II the residual stress, although smaller at the surface, decreases more slowly below the surface. The component can withstand the applied load only when the curve $c + b$ representing the sum of the external and residual stresses does not intersect the fatigue strength (straight line a). In case I, in spite of higher compressive residual stresses at the surface, at a distance below the surface the sum of external and residual stresses is higher than the fatigue strength, and a crack can be expected to form at this point. In case II, although the compressive residual stress is lower at the surface, its distribution below the surface is more favorable and the sum of external and residual stresses does not intersect the fatigue strength curve at any point.

There is a further point to consider when dealing with resulting stresses. This is their multiaxis nature. In practice, the estimation of the sum of external and residual stresses is complicated by the difficulty of determining the direction of the stresses at the critical point of the workpiece.

1. Thermal Stresses in the Case of Ideal Linear-Elastic Deformation Behavior

When a metallic body is heated or cooled, as soon as a temperature difference between the surface and the core is established, residual stresses of the first kind occur. In heat treatment, quenching processes usually produce the biggest temperature gradients across the section and hence the greatest residual stresses. Let us therefore discuss thermal stresses due to local and temporal differences in shrinking during quenching of ideal linear-elastic cylinders in which no plastic deformation can arise.

Transformation-free cooling of cylinders is accomplished by the development of a sequence of inhomogeneous temperature distributions, which, as a consequence of the thermal shrinking behavior, in turn cause locally and temporally different thermal strains and hence shrinking stresses. It is assumed that linear-elastic cylinders can elastically accommodate these stresses at all temperatures. At the beginning of quenching, the surface of such a cylinder contracts more rapidly than its core. As a result, the surface zones of the cylinder are subjected to tensile stresses in the longitudinal and tangential directions, while radially compressive stresses are created, as shown in Figure 31. In order to establish equilibrium, these stresses are counterbalanced by longitudinal, tangential, and radial compressive stresses within the core of the cylinder.

Figure 32 shows the temperature–time history at the very surface and at the core of the cylinder, the temperature difference between surface and core, and the development of longitudinal stresses during quenching of an ideal linear-elastic cylinder. The largest temperature

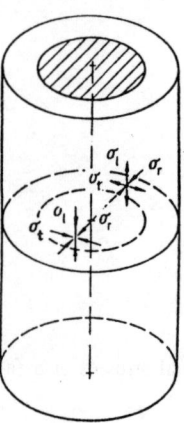

Figure 31 Thermal stresses in the surface zone and core of an ideal linear-elastic cylinder during rapid cooling (quenching). (From Ref. 13.)

difference ΔT_{max} is attained at $t = t_{max}$, where the slopes of temperature–time curves are identical for the core and the surface. Obviously, the surface reaches its maximum thermal stress before $t = t_{max}$; the core, however, reaches its maximum later than $t = t_{max}$.

The magnitude of the developed longitudinal stresses depends on cylinder diameter, as shown in Figure 33 for cylinder diameters of 30, 50, and 100 mm, when the cylinders were quenched from 800°C (1472°F) in water at 20°C (68°F). Because the maximum tempera-

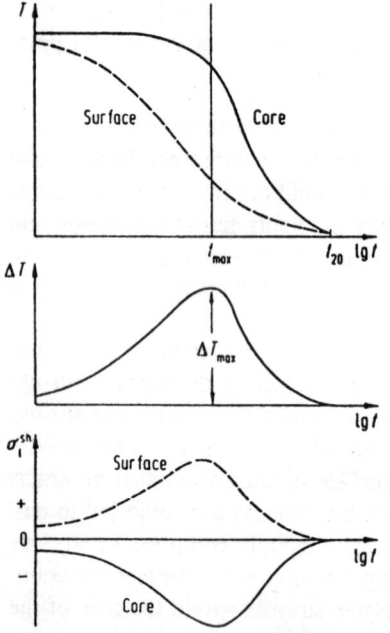

Figure 32 Top to bottom: Time–temperature history, temperature difference between surface and core, and development of longitudinal stresses during transformation-free quenching of an ideal linear-elastic cylinder. (From Ref. 13.)

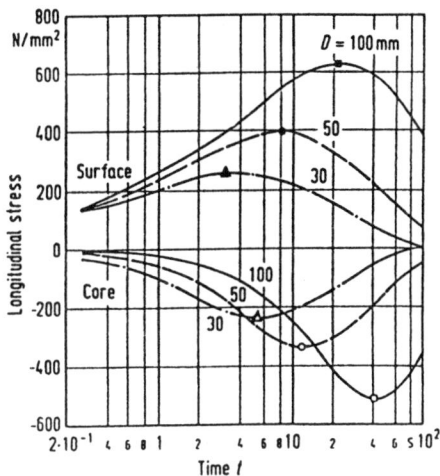

Figure 33 Dependence of longitudinal stresses at surface and core of ideal linear-elastic cylinders on their diameters, when quenched in water from 800°C to 20°C. Calculated for unalloyed steel with medium carbon content. (From Ref. 13.)

ture difference between surface and core occurs later for the larger diameter cylinders, the maximum stresses also occur later for larger diameters. The longitudinal surface stress maxima always occur at $t < t_{max}$, whereas those of the core occur later than t_{max}. At $t < t_{max}$, steep temperature gradients are present near the cylinder surface (see Figure 32), which cause high tensile stresses. In contrast, at $t > t_{max}$, relatively steep core temperature gradients are established, which cause large compressive stresses in the core. Upon reaching the temperature balance at 20°C (68°F) ($t = t_{20}$), the ideal linear-elastic cylinders are free of residual stresses.

2. Transformational Stresses

Let us consider the development of pure transformational stresses in a material whose coefficient of thermal expansion is zero. Furthermore, assume that if in the course of quenching the martensite start temperature M_s is passed, complete martensitic transformation occurs, with corresponding volume increase.

The temperature–time curves for the surface and core of a cylinder of such a material are shown in Figure 34. After passing the M_s temperature at time $t = t_1$, as a consequence of transformation-induced volume increase, compressive transformational stresses develop at the surface. These stresses within the surface zone must be compensated for by tensile transformational stresses within the core of the cylinder. The magnitudes of both stresses increase in the course of further surface cooling.

When the core temperature reaches M_S at time $t = t_2$, a transformation-induced volume increase occurs in the core, which leads to a reduction of the tensile stresses present there. The surface compressive stresses are correspondingly reduced. After reaching temperature equalization at $t = t_{20}$, the same amounts of martensite are present across the whole cylinder, so that finally a residual stress-free state is established. If, however, different amounts of martensite are formed within distinct areas, also under the idealized assumptions made here, some transformational residual stresses will remain.

In addition to the longitudinal stresses, tangential and radial residual stresses are caused by structural transformation. Within the surface zone, tangential compressive and radial tensile stresses are to be expected, while in the core all components should be tensile stresses.

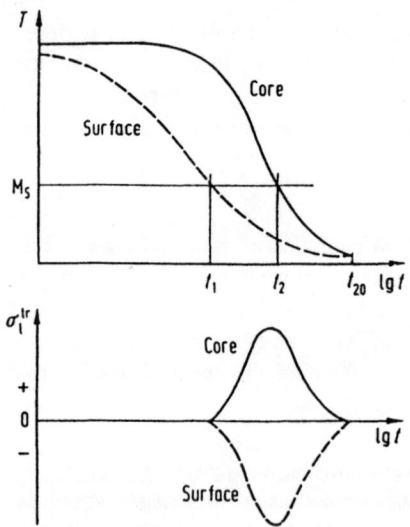

Figure 34 Temperature–time history and development of longitudinal transformation stresses, when quenching an ideal linear-elastic cylinder that transforms only to martensite. (From Ref. 13.)

When thermal (shrinking) and transformational stresses act simultaneously during quenching of an ideal linear-elastic cylinder that transforms from austenite to martensite, superposition of the two types of stresses occurs as shown in Figure 35. The upper graph shows the time dependence of the longitudinal components of thermal and transformational stresses at surface and core. The lower graph shows the time dependence of the total stress after the formal superposition of the two. The initiation of martensitic transformation immediately reduces the absolute stress value within both core and surface.

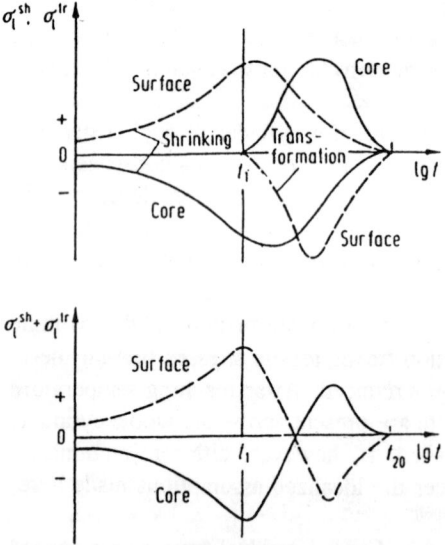

Figure 35 Combined thermal (shrinking) and transformation stresses during quenching of an ideal linear-elastic cylinder that transforms from austenite to martensite. (From Ref. 13.)

Further increasing martensitic transformation causes a stress inversion in both regions. Provided that the transformation occurs uniformly across the whole cylinder, at $t = t_{20}$ the tensile core stresses and the compressive surface stresses approach zero. Hence when temperature equalization is achieved, in an ideal linear-elastic cylinder no residual stresses remain.

3. *Residual Stresses when Quenching Cylinders with Real Elastic-Plastic Deformation Behavior*

In real practice there is no ideal linear-elastic deformation behavior as assumed above. The yield strength (R_y) of metallic materials, which limits the elastic deformation range, is strongly temperature-dependent and decreases with increasing temperature.

At any temperature, plastic deformations will develop when stresses surpass the corresponding yield strength. The ultimate tensile strength, which limits the uniaxial loading capacity of the material, is also temperature-dependent as shown in Figure 36 for two low-alloy steels. During quenching of a cylinder, biaxial longitudinal and tangential stresses develop in its surface zone, whereas triaxial longitudinal, tangential, and radial stresses develop in the cylinder core. Plastic deformations can occur only if the local equivalent stresses equal or exceed the yield strength of the material at the corresponding temperature.

Equivalent stresses can be calculated according to various hypotheses. Assuming the validity of the van Mises criterion, the equivalent stress of a triaxial stress state, given by the principal stresses σ_1, σ_2, σ_3, is

$$\sigma_{eq} = \frac{1}{\sqrt{2}}[(\sigma_1 - \sigma_2)^2 + (\sigma_2 - \sigma_3)^2 + (\sigma_3 - \sigma_1)^2]^{1/2} \tag{30}$$

During quenching of a cylinder in its surface zone, $\sigma_1 = \sigma_l$ and $\sigma_2 = \sigma_t$, while in its core $\sigma_1 = \sigma_l$, $\sigma_2 = \sigma_t$, and $\sigma_3 = \sigma_r$.

The condition for the onset of plastic deformation will be fulfilled when $\sigma_{eq} = R_y$. The local shrinking and transformational stress components and consequently the equivalent stress (σ_{eq}) depend on temperature, cooling conditions, geometry, and the mechanical and thermal properties of the material, and the yield strength (R_y) depends on temperature and the structure of the material.

The temperature dependence of the yield strength is obviously of particular importance for the stresses that result upon quenching. Figure 37 shows the temperature–time history and

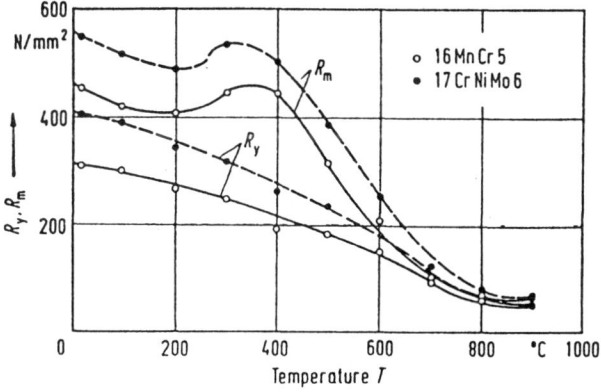

Figure 36 Yield strength (R_y) and tensile strength (R_m) of the steels DIN 16 MnCr5 and DIN 17CrNiMo6 as a function of temperature. (From Ref. 13.)

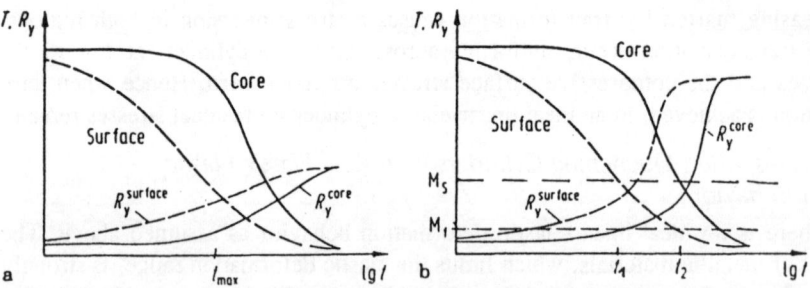

Figure 37 Temperature–time history and development of yield strength for surface and core during quenching of a cylinder (a) without and (b) with martensitic transformation. (From Ref. 13.)

development of yield strength for surface and core of a cylinder during quenching. Figure 37a depicts the case of transformation-free cooling, and Figure 37b is valid for cooling with martensitic transformation. To determine the occurrence of plastic deformations at any instant, the local yield strength must be compared with the local equivalent stress. Because plastic deformations never occur homogeneously over the whole cross section of the cylinder, residual stresses always remain after temperature equalization. Plastic deformations can be caused by either thermal (shrinking) stresses or transformational stresses or by a combination of the two.

   Thermal (Shrinking) Residual Stresses   Figure 38 shows the cooling curves for the surface and core of a cylinder during quenching without martensitic transformation and the temperature- (and time-) dependent yield strengths, which at the same temperature are assumed to be identical for tensile and compressive loading. At the start of quenching, the surface temperature decreases faster than the core temperature (Figure 38a). As a result, longitudinal tensile stresses develop at the surface and compressive stresses develop at the core. If they were elastically accommodated, their development would be as shown in Figure 38b. However, because of the temperature dependence of the yield strengths for surface and core, neither the surface nor the core can withstand these stresses without plastic deformation, and so the surface zone is plastically extended and the core is plastically compressed. After the time $t = t_{max}$, the temperature of the core decreases faster than that of the surface, leading to a reduction of the magnitudes of shrinking stresses in both regions. However, the stress values of core and surface reach zero at different instants, since they can no longer coexist at the same time in a stress-free state because of plastic extension at the surface and plastic compression in the core. Upon further cooling, this extension and compression causes compressive and tensile stresses, respectively, which are opposed by those due to the temperature differences still existing between core and surface. These latter stresses ultimately vanish after reaching the temperature equalization at the end of quenching, and hence thermal (shrinking) residual stresses remain that are compressive at the surface and tensile in the core, as depicted in Figure 38c.

   Transformational Residual Stresses   Figure 39 shows cooling curves for surface and core when quenching a cylinder that, upon cooling below the M_s temperature, transforms completely to martensite. For simplicity, it is assumed that no thermal (shrinking) stresses occur. Figure 39b shows the yield strengths for surface and core, showing their strong increase with the onset of martensitic transformations.

   The surface of the cylinder starts to transform to martensite at $t = t_1$. At that time the volume expansion of the surface zone is impeded by the core being not yet transformed. As a result, compressive transformational stresses are established at the surface that are compen-

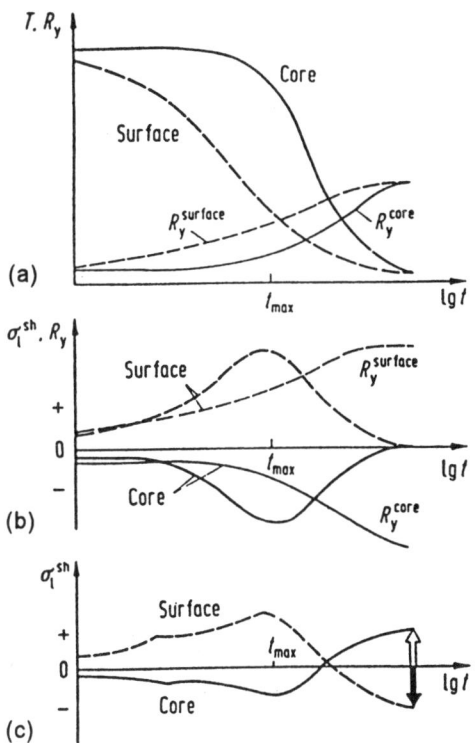

Figure 38 Longitudinal thermal (shrinking) residual stresses when quenching a cylinder. (From Ref. 13.)

sated for by tensile stresses at the core. From Figure 39b it can be concluded that both areas plastically deform. In the course of further cooling, the tensile stressed core reaches M_S at $t = t_2$. The immediate volume increase reduces both the tensile stresses of the core and the compressive stresses of the surface. Due to the differently sized and opposing plastic deformations generated, the stresses at surface and core pass zero values at different times. Upon further cooling the still existing volume incompatibilities between surface and core create transformational stresses of opposite sign to those that are produced by the plastic deformations. After reaching temperature equalization, compressive residual stresses remain in the core and tensile residual stresses remain at the surface, as shown in Figure 39c.

It should be noted also that transformation-induced plastic deformations that occur under local tensile or compressive stresses may enhance the local strains.

Hardening Residual Stresses When austenitized steel cylinders are quenched to room temperature, both thermal (shrinking) and transformational stresses develop, causing hardening residual stresses, which cannot be described by simply superimposing the shrinking and transformational stresses. Of fundamental importance is the fact that any local martensitic transformation coupled with a volume increase always shifts the existing stress (irrespective of its sign) to more negative values. As a reaction, for reasons of equilibrium, the unaffected material zones react with positive stress changes. Structural transformations that occur in tensile-stressed material regions therefore inevitable reduce the stresses, while transformations that take place in compressively stressed zones always enhance the (negative) values of the stresses. Consequently, since the thermal (shrinking) stresses of core and surface change sign in the

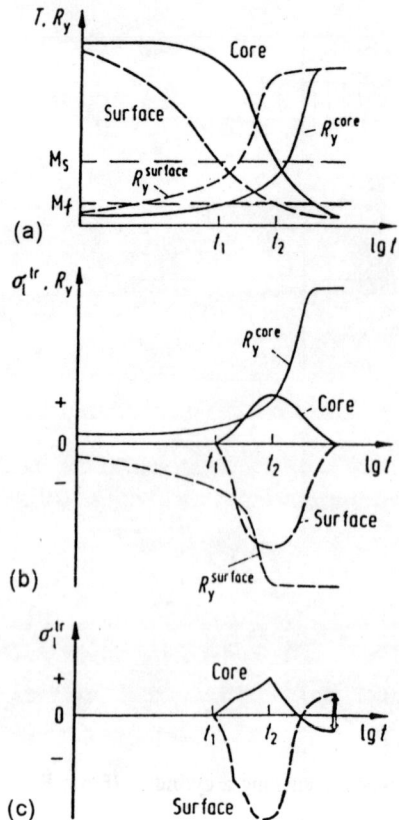

Figure 39 Longitudinal transformation residual stresses when quenching a cylinder. (From Ref. 13.)

course of cooling during the time interval $t_{c,o} - t_{s,o}$ as depicted in Figure 40a, the positions of the initiation time of transformation at the surface $(t_{s,i})$ and in the core $(t_{c,i})$ relative to this time interval are of key importance for the hardening residual stresses that will remain at the end of quenching.

The average time that passes before the quenching stresses invert is

$$t_o = (1/2)(t_{s,o} + t_{c,o}) \tag{31}$$

Since for full-hardening steel cylinders the time for surface transformation $(t_{s,i})$ always occurs earlier than the time for core transformation $(t_{c,i})$, it is appropriate to distinguish between the following cases:

$t_o < t_{s,i} < t_{c,i}$ (Figure 40b)

$t_o \approx t_{s,i} < t_{c,i}$ (Figure 40c)

$t_{s,i} < t_{c,i} \approx t_o$ (Figure 40d)

$t_{s,i} < t_{c,i} < t_o$ (Figure 40e)

Figure 40 shows schematically the development of longitudinal stresses as a function of time and remaining longitudinal residual stress distributions across the section of cylinder speci-

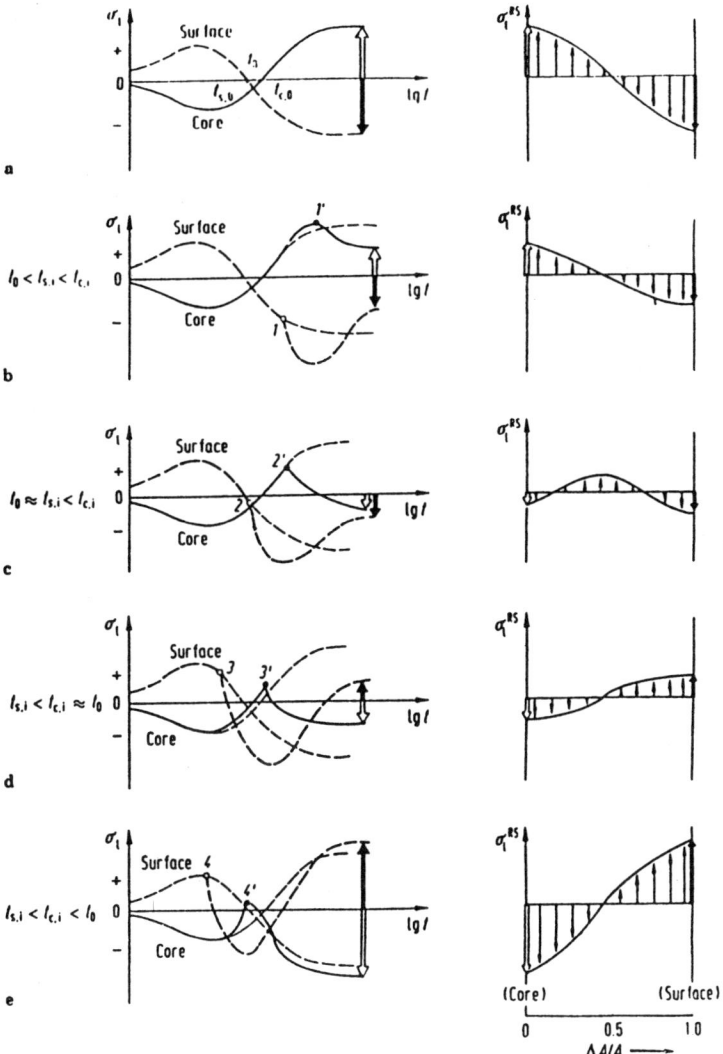

Figure 40 Different possibilities of generation and development of hardening residual stresses (b–e) compared to pure thermal (shrinking) residual stresses (a), when quenching a cylinder with real elastic-plastic deformation behavior. (From Ref. 13.)

mens with real elastic-plastic deformation behavior after complete temperature equalization at the end of the quenching process.

Figure 40a shows a transformation-free quenching, and Figures 40b–40e demonstrate the combined effects of thermal (shrinking) and transformational processes. The numbers 1–4 depict the initiation of transformation at the surface, while 1'–4' represent that of the core. Figure 40b illustrates the case when both surface and core transform after t_0. At the end of this cooling process, compressive stresses at the surface and tensile stresses in the core remain.

Figure 40c illustrates the stress development in the case when the surface transforms slightly before t_0 and the core transforms later. At the end of this cooling process, both core and surface remain under compressive residual stresses, while the regions in between are subjected to tensile residual stresses.

Figure 40d illustrates the case when the surface transforms before t_0 and the core at about t_0. At the end of this cooling process, tensile surface residual stresses and compressive core residual stresses remain.

Figure 40e illustrates the case when both surface and core transform before t_0. In this case the start of transformation at the surface caused a rapid reduction of the tensile stresses. For reasons of equilibrium, the longitudinal stresses at the core must also change during further cooling. Martensitic transformation in the core takes place when tensile stresses are acting there. This again causes stress inversions in the surface zone and in the core. At the end of this cooling process, tensile stresses at the surface and compressive stresses in the core remain.

When full-hardening of equal-sized cylinders with different M_s temperatures is compared with respect to residual stress distributions, one finds that cylinders made of steels with low M_s temperatures show tensile surface residual stresses, whereas cylinders made of steels with high M_s temperatures give compressive surface residual stresses, as schematically illustrated in Figure 41.

Because the high-temperature yield strength usually increases with decreasing M_s temperature, the largest tensile shrinking stresses develop at the surface of the steel with $M_{s,3}$ and the smallest at the surface of the steel with $M_{s,1}$. The martensitic transformation, however, begins earliest for the steel with the highest and latest for the steel with the lowest M_s temperature.

When high shrinking stresses and high M_s values act together, no secondary stress inversion occurs during further cooling, and ultimately compressive residual stresses remain within the surface zone.

On the basis of the preceding discussion, the whole range of expected hardening residual stress distributions in quenched steel cylinders can be divided into three main groups, as schematically illustrated in Figure 42. The arrows indicate how local transformations under existing stress states will affect the residual stress distribution.

It should be emphasized that the residual stress distributions that are created during quenching of cylinders with different diameters but made of the same steel can be shifted from the

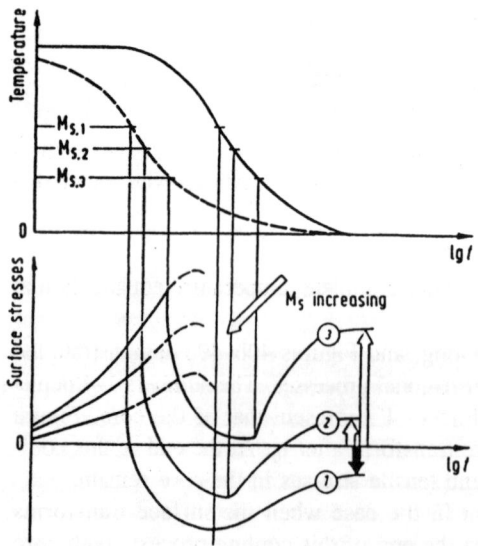

Figure 41 Influence of different M_s temperatures on the development of surface residual stresses. (From Ref. 13.)

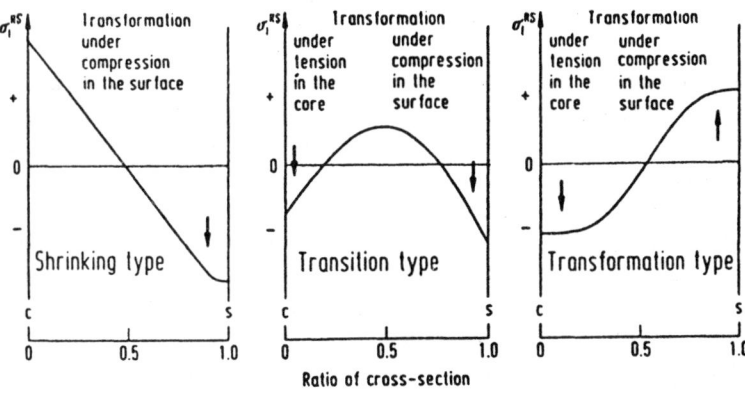

Steel	Quenching process	Cylinder diameters in mm for residual stresses of		
		shrinking type	transition type	transformation type
Ck 45	850 °C — 20 °C, H₂0	100...30	15	5
	850 °C — 60 °C, oil	50	30	10

Figure 42 Basic types of hardening residual stresses. (From Ref. 13.)

transformation type to the shrinking type with increasing cylinder diameter as well as with the higher quenching intensity, i.e., higher cooling rates. Some cylinder diameters and quenchants are specified for the unalloyed steel DIN Ck45 where the basic residual stress types occur.

The above statements and specifically the principle that local stresses are shifted to more negative values due to transformation-induced volume increase also hold for all nonmartensitic transformations that are accompanied by volume changes. In the individual case, the effect of volume changes on the final residual stress state depends on when the transformations start at the core and surface relative to time t_0.

4. Dimensional Changes and Distortion During Hardening and Tempering

As a consequence of thermal (shrinking) stresses and transformational stresses, changes occur in both the shape and size of workpieces during hardening and tempering. Because there are many factors that influence dimensional changes and distortion, the most difficult problem in practice is to predict the amount of dimensional changes and distortion. It is likely that computer modeling of the quenching process, which can account for the influence of all relevant factors, will in the future enable more precise predictions. Let us therefore discuss only some basic mechanisms of dimensional changes and distortion during hardening and tempering.

Influence of Thermal (Shrinking) Stresses Because of thermal (shrinking) stresses during quenching, generally all bodies whose shape is different from a sphere tend by deformation to assume a spherical shape, this being the shape that offers the least resistance during deformation. This means, at the practical level, that bodies that have the shape of a cube will assume spherically distorted sides, bodies with the shape of a prism will become thicker and shorter, and plates will shrink in area and become thicker. These deformations are greater with greater temperature differences between the surface and the core, i.e., with higher quenching intensity (which also corresponds to bigger differences between the austenitizing temperature and the temperature of the quenchant); with greater cross-sectional size of the workpiece; and with smaller heat conductivity and smaller high-temperature strength of the material.

The effect of thermal (shrinking) stresses can be studied in a low-carbon steel or an austenitic steel, in which the martensitic transformation can be disregarded. Figure 43 shows, according to Frehser and Lowitzer [15], the effect of different quenching intensities on dimensional changes and distortion of plates made of low-carbon steel (0.10% C) after quenching in water, oil, molten salt bath, and air. In Figure 43a the plate is solid, and in Figure 43b the plate has an inner square hole of 100 × 100 mm. The outer full lines denote the original size of each plate. To illustrate the dimensional changes more clearly, they have been drawn to a larger scale (see the 0.4 mm scale). From this figure it is evident that the more drastic the quench, the greater are the dimensional changes and distortion.

Figure 44 shows that a greater difference between the austenitizing temperature and the temperature of the quenchant causes greater dimensional changes and distortion.

Figure 45 shows the effect of the high-temperature strength of the material. The steel having the greatest high-temperature strength (18/8 steel) exhibits the highest dimensional stability.

Influence of Transformation Stresses   During heating and cooling, steels pass through various structural transformations accompanied by volume changes. These changes are usually studied by using a dilatometer and are registered as changes in length of the specimen, as shown, for example, in Figure 46 for eutectoid steel. During heating a continuous increase in length occurs up to A_{c1}, where the steel shrinks as it transforms to austenite. After the aus-

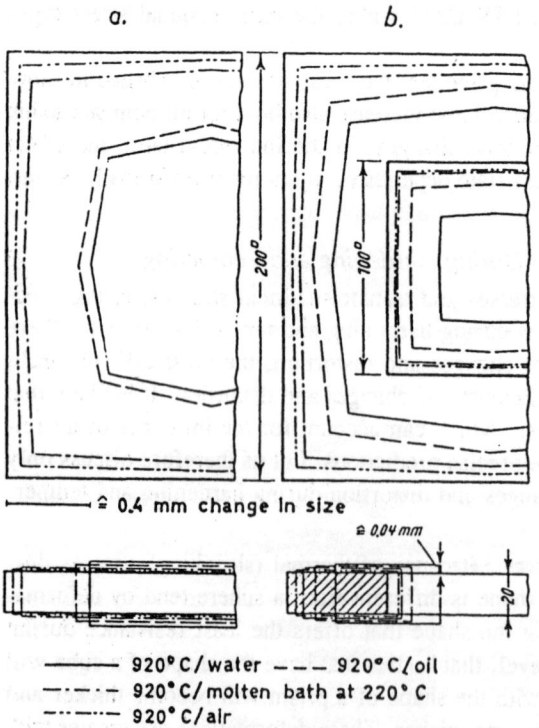

a.   b.

├──── ≙ 0.4 mm change In size

≙ 0,04 mm

─────── 920°C/water   ─ ─ ─ 920°C/oil
─·─··─ 920°C/molten bath at 220°C
─··─···─ 920°C/air

Figure 43 Dimensional changes and distortion of plates made from low-carbon steel (0.10% C) after cooling in water, oil, molten salt bath, and air. (From Ref. 14.)

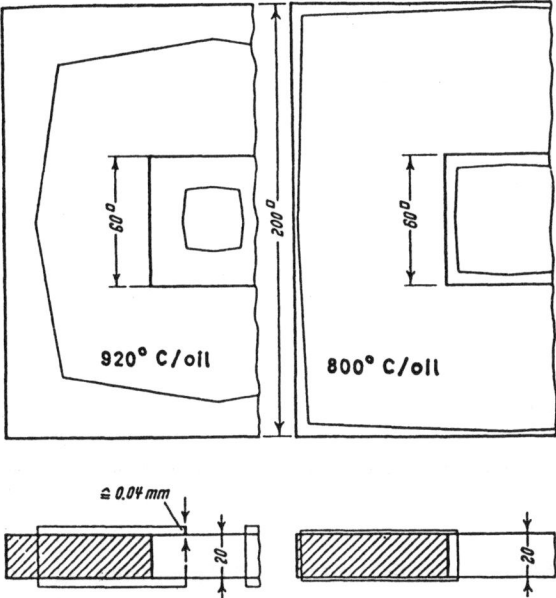

Figure 44 Effect of difference between the austenitizing temperature and the temperature of the quenchant on dimensional changes after quenching plates of low-carbon steel in oil. (From Ref. 14.)

tenite formation is completed, the length increases again. However, the expansion coefficient for austenite is not the same as the expansion coefficient for ferrite.

On cooling, thermal contraction takes place, and when martensite starts to form at the M_s temperature, the volume increases and the length of the specimen therefore increases. After cooling to room temperature, most martensitic steels contain some retained austenite, the amount of which increases with increased carbon content, with higher austenitizing temperature, and with the amount of some alloying elements dissolved during austenitization. The larger the quantity of retained austenite contained in the steel after hardening, the smaller the increase in volume and in length of specimen.

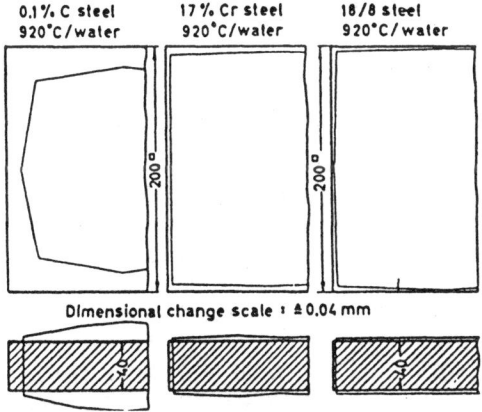

Figure 45 Dimensional changes and distortion after quenching steel plates of different composition from 920°C in water. (From Ref. 14.)

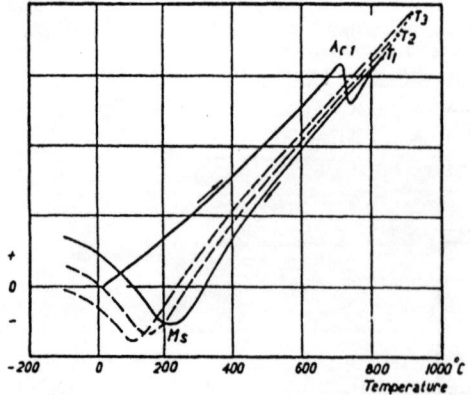

Figure 46 Dilatometer curves showing change in length during heating and rapid cooling of a eutectoid (0.8% C) steel. (From Ref. 14.)

Various structural constituents have different densities and hence different values of specific volume, as shown in Table 4. The amount of carbon dissolved in austenite, in martensite, or in different carbides has a relatively strong effect on the specific volume as the formulas for calculating specific volume in this table indicate. When calculating the changes in volume that take place during the transformations of different structural phases, the carbon content must be taken into account, as shown in Table 5.

Taking as a basis the proportions of martensite and austenite, together with the amount of carbon dissolved therein, and using the data from Table 5, one can calculate the changes in volume that occur during hardening. If the steel contains undissolved cementite, this volume has to be deducted during the calculation. The following equation should be used:

$$\frac{\Delta V}{V} = \frac{100 - V_c - V_a}{100} \times 1.68C + \frac{V_a}{100}(-4.64 + 2.21C) \tag{32}$$

where $\Delta V/V$ is the change in volume in %, V_c is the amount of undissolved cementite in vol %, V_a is the amount of austenite in vol %, $100 - V_c - V_a$ is the amount of martensite in vol %, and C is the carbon dissolved in austenite and martensite, respectively, in % by weight.

Table 4 Specific Volume of Phases Present in Carbon Tool Steels

Phase or phase mixture	Range of carbon (%)	Calculated specific volume at 20°C (cm³/g)
Austenite	0–2	$0.1212 + 0.0033 \times (\%C)$
Martensite	0–2	$0.1271 + 0.0025 \times (\%C)$
Ferrite	0–0.02	0.1271
Cementite	6.7 ± 0.2	0.130 ± 0.001
ε Carbide	8.5 ± 0.7	0.140 ± 0.002
Graphite	100	0.451
Ferrite+cementite	0–2	$0.271 + 0.0005 \times (\%C)$
Low-carbon content martensite ε-carbide	0.25–2	$0.1277 + 0.0015 \times (\%C - 0.25)$
Ferrite + ε-carbide	0–2	$0.1271 + 0.0015 \times (\%C)$

Source: Ref. 14.

Table 5 Changes in Volume During Transformation to Different Phases

Transformation	Change in volume (%)
Spheroidized pearlite → austenite	-4.64 + 2.21 × (%C)
Austenite → martensite	4.64 – 0.53 × (%C)
Spheroidized pearlite → martensite	1.68 × (%C)
Austenite → lower bainite	4.64 – 1.43 × (%C)
Spheroidized pearlite → lower bainite	0.78 × (%C)
Austenite → upper bainite	4.64 – 2.21 × (%C)
Spheroidized pearlite → upper bainite	0

Source: Ref. 14.

The increase in volume during martensitic transformation depends not only on the carbon content but also on the kind and amount of alloying elements in the steel. Consequently, different groups of steels undergo different changes in volume during hardening. The unalloyed water-hardening steels experience the greatest volume changes, followed by low-alloy oil-hardening steels, while the high-alloy ledeburitic Cr alloy steels show the least volume increase during hardening, as shown in Figure 47. The austenitizing temperature, as mentioned earlier, has an influence on the amount of retained austenite after hardening. Because the retained austenite producing volume contraction (compared to the original volume) counteracts the volume increase caused by martensitic transformation, the austenitizing temperature may influence the volume changes during hardening.

It should also be noted that engineering steels are not isotropic materials (because of the rolling process they have undergone), which means that the linear change occurring during hardening will not be the same in the direction of rolling as in the direction perpendicular to it.

Dimensional Changes During Tempering During tempering, relaxation as well as structural transformations occur, which change the volume of the hardened steel and its state of stress. Martensite decomposes to form ferrite and cementite, which implies that there is a continuous decrease in volume. The continuous decomposition of martensite during temper-

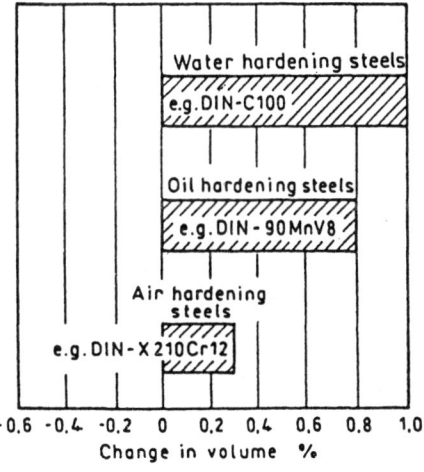

Figure 47 Volume changes of different steels during hardening when martensite is formed across the whole section. (From Ref. 2.)

Increase
in volume

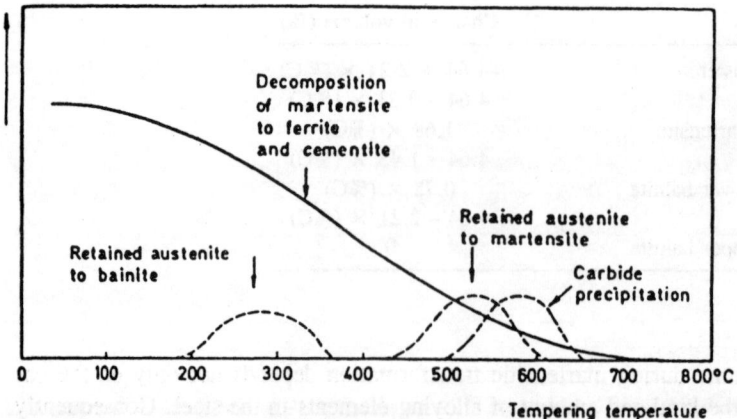

Figure 48 Schematic presentation of the effect of changes of structural constituents on volume changes during tempering of hardened steel. (From Ref. 14.)

ing causes at the same time a continuous reduction in the state of stress. Figure 48 is a schematic presentation of the effect of changes of structural constituents on the volume changes during tempering of a hardened steel. The dashed curves represent increases in volume during different tempering stages. The retained austenite, which in carbon steels and low-alloy steels is transformed to bainite in the second stage of tempering at about 300°C (572°F), results in an increase in volume.

When high-alloy tool steels are tempered at 500–600°C (932–1112°F), very finely distributed carbides are precipitated. This gives rise to a stress condition that results in increased hardness and greater volume. Simultaneously with the precipitation of carbides the alloy content of the matrix is reduced, which implies that the M_s point of the retained austenite will be raised to higher temperatures. During cooldown from the tempering temperature, the retained austenite will transform to martensite, which also results in an increase in volume.

Figure 49 shows changes in length for different steels as a function of tempering temperature. For low-alloy steels (105WCr6, see curve 1 of Figure 49), one can easily recognize the particular tempering stages. At low tempering temperatures (first tempering stage) a volume contraction takes place as a consequence of ε-carbide precipitation. At higher tempering temperatures (second tempering stage), transformation of the retained austenite again causes a certain volume increase, and in the third tempering stage the progressive decomposition of martensite leads to the volume decrease.

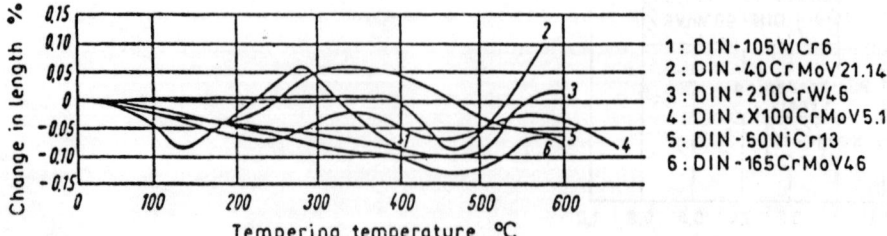

1 : DIN -105WCr6
2 : DIN -40CrMoV21.14
3 : DIN -210CrW46
4 : DIN -X100CrMoV5.1
5 : DIN -50NiCr13
6 : DIN -165CrMoV46

Figure 49 Change in length of different steels during tempering as a function of tempering temperature. (Designation of steels according to DIN.) (From Ref. 2.)

For high-alloy tool steels (e.g., 210CrW46, curve 3 of Figure 49), a stabilization of austenite is evident, so that the effect of the volume increase (due to austenite–bainite or austenite–martensite transformation) takes place only at higher temperatures.

In most cases, as can be seen from Figure 49, a reduction in length, i.e., a volume decrease, can be found after tempering.

It should be noted that the changes in length shown in Figure 49 represent only the order of magnitude of the expected changes, because the actual values depend in each case on the specific heat treatment conditions. The austenitizing temperature, which determines the amount of carbon dissolved and the amount of retained austenite, has a strong influence on expected volume changes.

II. ANNEALING PROCESSES

A. Stress-Relief Annealing

Stress-relief annealing is an annealing process below the transformation temperature A_{c1}, with subsequent slow cooling, the aim of which is to reduce the internal residual stresses in a workpiece without intentionally changing its structure and mechanical properties.

Residual stresses in a workpiece may be caused by

1. Thermal factors (e.g., thermal stresses caused by temperature gradients within the workpiece during heating or cooling)
2. Mechanical factors (e.g., cold-working)
3. Metallurgical factors (e.g., transformation of the microstructure)

In processes that involve heat, residual stresses are usually caused by the simultaneous existence of thermal and transformational stresses (e.g., during the solidification of liquid metals, hot forming, hardening, or welding). Thermal stresses are always directly proportional to the existing temperature gradient, which further depends on the cross-sectional size and on the heating or cooling rate.

In workpieces made of steel, for the above reasons, local residual stresses may amount to between about 10 N/mm^2 and values close to the yield strength at room temperature. The consequences of residual stresses may include

1. Dimensional changes and warpage of the workpiece
2. The formation of macroscopic and microscopic cracks
3. Asymmetric rotation of shafts
4. Impairment of the fatigue strength of engineering components

Residual stresses in a workpiece can be reduced only by a plastic deformation in the microstructure. This requires that the yield strength of the material be lowered below the value of the residual stresses. The more the yield strength is lowered, the greater the plastic deformation and correspondingly the greater the possibility of reducing the residual stresses. The yield strength and the ultimate tensile strength of the steel both decrease with increasing temperature, as shown in Figure 50 for a low-carbon unalloyed steel. Because of this, stress-relief annealing means a through-heating process at a correspondingly high temperature. For plain carbon and low-alloy steels this temperature is usually between 450 and 650°C (842 and 1200°F), whereas for hot-working tool steels and high-speed steels it is between 600 and 750°C (1112 and 1382°F). This treatment will not cause any phase changes, but recrystallization may take place.

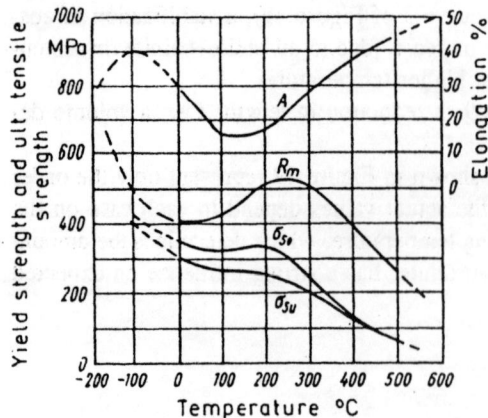

Figure 50 Change in some mechanical properties of low-carbon unalloyed steel with increasing temperature, according to Christen. A = elongation; R_m = ultimate tensile strength; σ_{so} = upper yield strength; σ_{su} = lower yield strength. (From Ref. 2.)

Tools and machine components that are to be subjected to stress-relief annealing should be left with a machining allowance sufficient to compensate for any warping resulting from stress relief.

When dealing with hardened and tempered steel, the temperature of stress-relief annealing should be about 25°C (77°F) below that used for tempering. If the tempering temperature was quite low, after stress-relief annealing quite a high level of residual stresses will remain. In some other cases, for instance with a gray iron, the maximum temperature of the stress-relief annealing should be limited because of possible strength loss. Therefore gray iron must not be stress-relief-annealed about 550°C (1022°F).

In the heat treatment of metals, quenching or rapid cooling is the cause of the greatest residual stresses. A high level of residual stress is generally to be expected with workpieces that have a large cross section, are quenched at a high cooling rate, and are made of a steel of low hardenability. In such a case high temperature gradients will arise on one side, and on the other side structural transformations will occur at different points of the cross section at different temperatures and different times.

In contrast to heat treatment processes with continuous cooling, processes with isothermal transformation (e.g., austempering) result in a low level of residual stresses.

To activate plastic deformations, the local residual stresses must be above the yield strength of the material. Because of this fact, steels that have a high yield strength at elevated temperatures can withstand higher levels of residual stress than those that have a low yield strength at elevated temperatures.

The level of yield strength at elevated temperatures depends on the alloying elements in the steel. Figure 51 shows the increase in yield strength at temperatures of 300–550°C (572–1022°F) when 0.5% of each element was added to an unalloyed steel. It can be seen from this diagram that additions of Mo and V are most effective in increasing the yield strength at elevated temperatures.

To reduce residual stresses in a workpiece by stress-relief annealing, a temperature must be reached above the temperature corresponding to the yield strength that is adequate to the maximum of the residual stresses present. In other words, every level of residual stress in a workpiece corresponds to a yield strength that in turn depends on temperature.

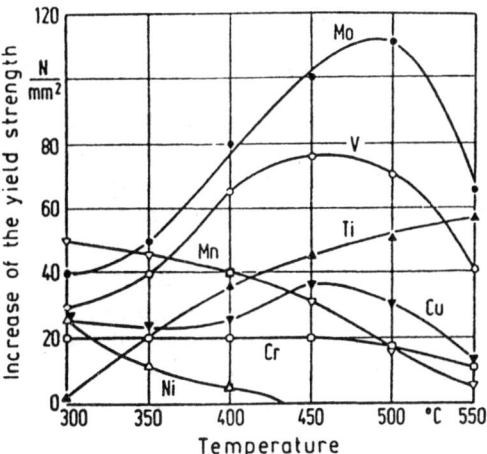

Figure 51 Increase in yield strength at elevated temperatures when 0.5% of each alloying element indicated is added to an unalloyed steel. (From Ref. 1.)

In addition to temperature, soaking time also has an influence on the effect of stress-relief annealing, i.e., on the reduction of residual stresses, as shown in Figure 52.

The relation between temperature and soaking time during stress-relief annealing can be described by Hollomon's parameter:

$$P = T(C + \log t) \tag{33}$$

where P is Hollomon's parameter (heat treatment processes with the same Hollomon parameter value have the same effect), C is the Hollomon–Jaffe constant, T is temperature in K, and t is time in hours.

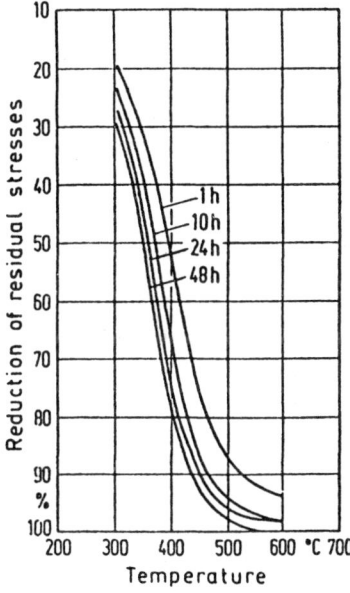

Figure 52 Effect of soaking time (at different temperatures) of stress-relief annealing on the reduction of residual stresses for hardening and tempering steels. (From Ref. 1.)

The Hollomon–Jaffe constant can be calculated as

$$C = 21.3 - (5.8 \times \% \text{ carbon})$$ (34)

Figure 53 shows (according to Larson–Miller method) calculated values of the yield strength at elevated temperatures (for 0.2% strain) for three grades of alloyed structural steels for hardening and tempering (designations according to DIN). Using this diagram, the abscissa of which represents the actual Hollomon parameter P, knowing the temperature and time of the stress-relief annealing, one can read off the level of residual stresses that will remain in the workpiece after this annealing process, i.e., the level up to which the residual stresses will be reduced by this stress-relief annealing. If, for instance, for DIN 24CrMoV5.5 steel a temperature of 600°C (1112°F) and a soaking time of 10 h are chosen for stress-relief annealing, the residual stresses will, after this annealing, be reduced to a maximum of 70 N/mm². Higher temperatures and longer times of annealing may reduce residual stresses to lower levels, as can be seen from Figure 53.

As in all heat treatment processes where Hollomon's parameter is involved, selection of a higher temperature may dramatically shorten the soaking time and contribute substantially to the economy of the annealing process.

Dealing with structural steels for hardening and tempering, the stress-relief process and the tempering process can be performed simultaneously as one operation, because Hollomon's parameter is also applicable to tempering. In such a case the stress-relief diagram may be used

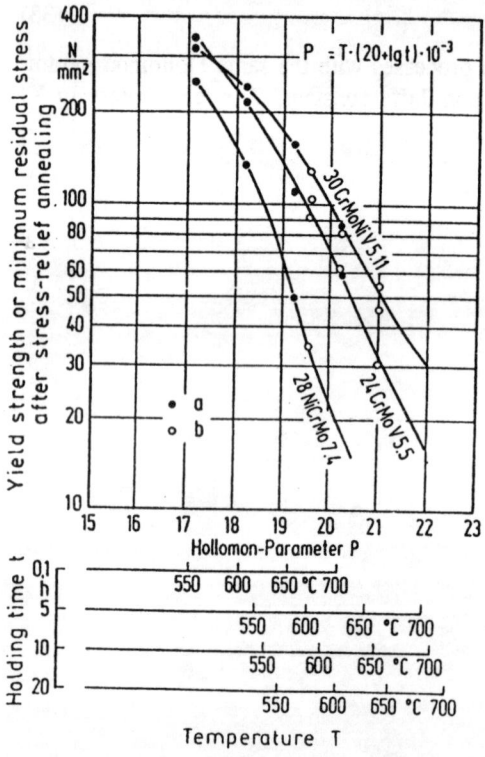

Figure 53 Yield strength at elevated temperatures (for 0.2% strain) calculated according to the Larson–Miller method for three grades of alloyed structural steels for hardening and tempering (designations according to DIN). (a) Calculated values; (b) experimentally obtained values. (From Ref. 1.)

in combination with the tempering diagram to optimize both the hardness and the level of reduced residual stresses.

The residual stress level after stress-relief annealing will be maintained only if the cooldown from the annealing temperature is controlled and slow enough that no new internal stresses arise. New stresses that may be induced during cooling depend on the cooling rate, on the cross-sectional size of the workpiece, and on the composition of the steel. Figure 54 shows the effect of cooling rate and cross-sectional diameter of forgings made of a CrMoNiV steel on the level of tangential residual stresses after stress-relief annealing.

A general conclusion about stress-relief annealing is the following: In the temperature range 450–650°C (842–1200°F) the yield strength of unalloyed and low-alloyed steels is lowered so much that a great deal of residual stress may be reduced by plastic deformation. The influence of the steel composition on the level of residual stresses after annealing can be considerable. While unalloyed and low-alloy steels with Ni, Mn, and Cr after stress-relief annealing above 500°C (932°F) may get the residual stresses reduced to a low level, steels alloyed with Mo or Mo+V will retain a much higher level of the residual stresses after stress-relief annealing at the same temperature because of their much higher yield strength at elevated temperature.

B. Normalizing

Normalizing or normalizing annealing is a heat treatment process consisting of austenitizing at temperatures of 30–80°C (86–176°F) above the A_{c3} transformation temperature (for hypoeutectoid steels) followed by slow cooling (usually in air), the aim of which is to obtain a fine-grained, uniformly distributed, ferrite-pearlite structure.

Normalizing is applied mainly to unalloyed and low-alloy hypoeutectoid steels. For hypereutectoid steels normalizing is performed only in special cases, and for these steels the austenitizing temperature is 30–80°C (86–176°F) above the A_{c1} transformation temperature.

Figure 55 shows the thermal cycle of a normalizing process, and Figure 56 shows the range of austenitizing temperatures for normalizing unalloyed steels depending on their carbon content.

The parameters of a normalizing process are the heating rate, the austenitizing temperature, the holding time at austenitizing temperature, and the cooling rate.

Normalizing treatment refines the grain of a steel that has become coarse-grained as a result of having been heated to a high temperature, e.g., for forging or welding. Figure 57 shows

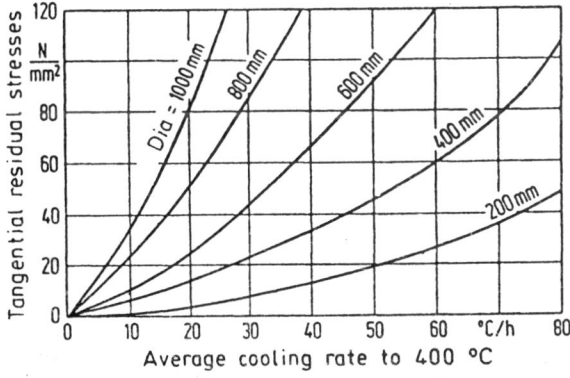

Figure 54 Tangential residual stresses in a CrMoNiV alloy steel depending on the cooling rate and cross-section diameter. (From Ref. 1.)

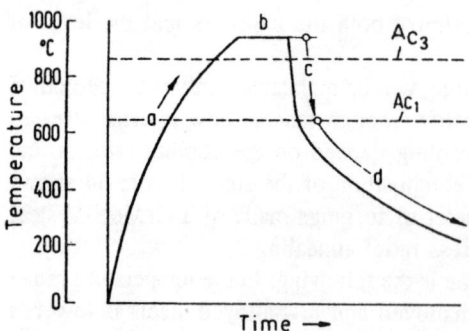

Figure 55 Time–temperature regime of normalizing. *a*, Heating; *b*, holding at austenitizing temperature; *c*, air cooling; *d* air or furnace cooling. (From Ref. 1.)

the effect of grain refining by normalizing a carbon steel of 0.5% C. Such grain refinement and homogenization of the structure by normalizing is usually performed either to improve the mechanical properties of the workpiece or (previous to hardening) to obtain better and more uniform results after hardening. Better machinability of low-carbon steels is also sometimes a reason to apply normalizing.

A special need for normalizing exists with steel castings because, due to slow cooling after casting, a coarse-grained structure develops that usually contains needle-like ferrite (Widmann-stätten's structure), as shown in Figure 58. A normalizing treatment at 780–950°C (1436–1742°F) (depending on chemical composition) removes this undesirable structure of unalloyed and alloyed steel castings having 0.3–0.6% C.

After hot rolling, the structure of steel is usually oriented in the rolling direction, as shown in Figure 59. In such a case, of course, mechanical properties differ between the rolling di-

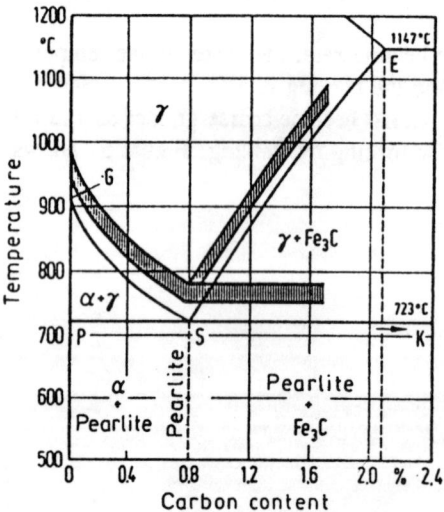

Figure 56 Austenitizing temperatures for normalizing unalloyed steels depending on their carbon content. (Temperature range above the line S–E is used for dissolution of secondary carbides). α = ferrite; γ = austenite; Fe₃C = cementite. (From Ref. 1.)

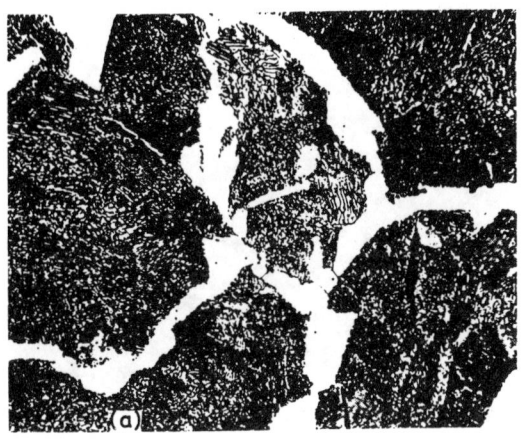

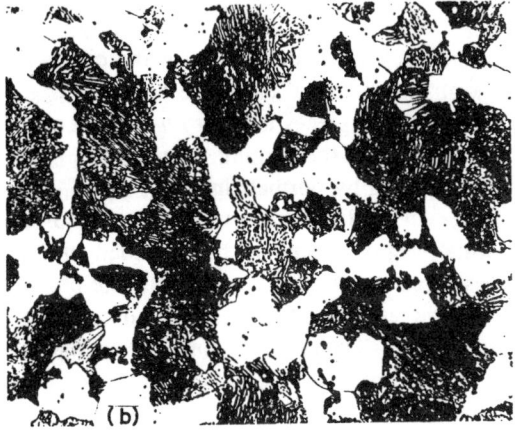

Figure 57 Effect of grain refining by normalizing a carbon steel of 0.5% C. (a) As-rolled or forged, grain size ASTM 3; (b) normalized, grain size ASTM 6. Magnification 500×. (From Ref. 14.)

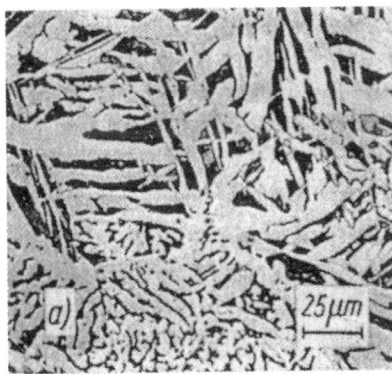 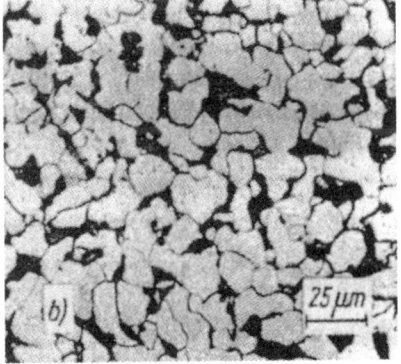

Figure 58 Structure of a steel casting (a) before normalizing; (b) after normalizing. (From Ref. 2.)

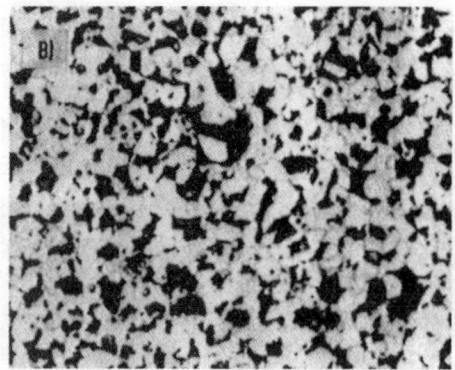

Figure 59 Structure of DIN 20MnCr 5 steel (A) after hot rolling; (B) after normalizing at 880°C. Magnification 100×. (From Ref. 1.)

rection and the direction perpendicular to it. To remove the oriented structure and obtain the same mechanical properties in all directions, a normalizing anneal has to be performed.

After forging at high temperatures, especially with workpieces that vary widely in cross-sectional size, because of the different rates of cooling from the forging temperature, a heterogeneous structure is obtained that can be made uniform by normalizing.

From the metallurgical aspect the grain refinement and the uniform distribution of the newly formed ferrite-pearlite structure during normalizing treatment can be explained with the following mechanism. At normalizing, the steel is subjected first to a $\alpha \rightarrow \gamma$ (ferrite-pearlite to austenite) transformation, and after the holding time at austenitizing temperature, to a recurring $\gamma \rightarrow \alpha$ (austenite to ferrite-pearlite) transformation. The effect of normalizing depends on both what happens during austenitizing and what happens during cooling from the austenitizing temperature.

During austenitizing a far-reaching dissolution of carbides is aimed at, but this process competes with the growth of austenite grains after complete carbide dissolution, which is not desirable. Besides the carbide dissolution, the degree of homogenization within the austenite matrix is important for obtaining a new arrangement of ferrite and pearlite constituents in the structure after normalizing. Both dissolution and homogenizing are time- and temperature-dependent diffusion processes that are slower when the diffusion paths are longer (higher local differences in carbon concentration) and the diffusion rates are smaller (e.g., increasing amounts of alloying elements). Therefore, especially with alloyed steels, lower austenitizing temperatures and longer holding times for normalizing give advantages taking into account the austenite grain growth. As shown in Figure 60, high austenitizing temperatures result in a coarse-grained austenite structure, which yields a coarse structure after normalizing.

Holding time at austenitizing temperature may be calculated using the empirical formula

$$t = 60 + D \tag{35}$$

where t is the holding time in minutes and D is the maximum diameter of the workpiece in millimeters.

When normalizing hypoeutectoid steels (i.e., steels with less than 0.8% C), during cooling from the austenitizing temperature, first a pre-eutectoid precipitation of ferrite takes place. With a lower cooling rate, the precipitation of ferrite increases along the austenite grain boundaries. For the desired uniform distribution of ferrite and pearlite after normalizing, however,

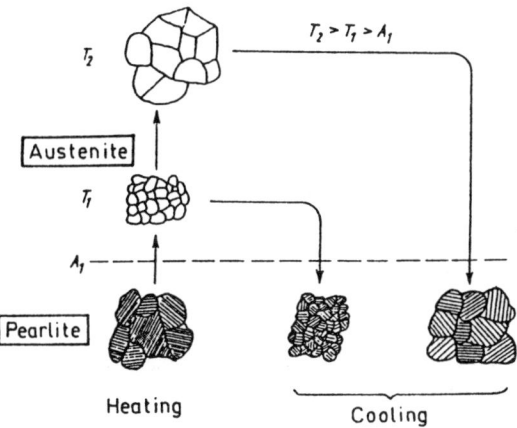

Figure 60 Schematic presentation of the influence of austenitizing temperature on the grain size of the structure of a eutectoid steel after normalizing. (From Ref. 2.)

a possibly simultaneous formation of ferrite and pearlite is necessary. Steels having carbon contents between 0.35 and 0.55% C, especially, tend to develop nonuniform ferrite distributions as shown in Figure 61. The structure in this figure indicates overly slow cooling in the temperature range of pre-eutectoid ferrite precipitation between A_{r3} and A_{r1}. On the other hand, if the cooling through this temperature region takes place too fast, with steels having carbon contents between 0.2 and 0.5%, formation of an undesirable needle-like ferrite (oriented at austenite grain boundaries), the so-called Widmannstätten's structure, may result as shown in Figure 62. Formation of pearlite follows only after complete precipitation of ferrite by transformation of the remaining austenite structure at temperature A_{r1}. It starts first at the boundaries of ferrite and austenite and spreads to the interior of the austenite grains. The greater the number of the pearlitic regions formed, the more mutually hindered the pearlite grains are in their growth, and consequently the finer the grains of the normalized structure. The influence of alloying elements on the austenite to ferrite and pearlite transformation may be read off from the relevant CCT diagram.

Care should be taken to ensure that the cooling rate within the workpiece is in a range corresponding to the transformation behavior of the steel in question that results in a pure ferrite-pearlite structure. If, for round bars of different diameters cooled in air, the cooling

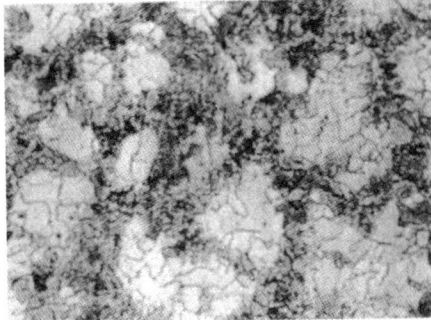

Figure 61 Nonuniform distribution of ferrite and pearlite as a consequence of unfavorable temperature control during normalizing of unalloyed DIN C35 steel. Magnification 100×. (From Ref. 1.)

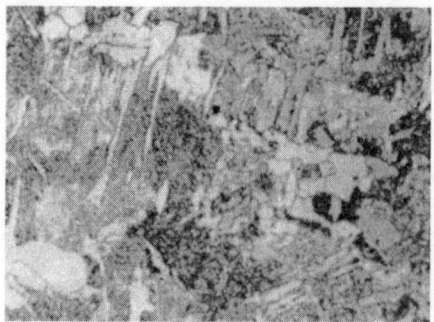

Figure 62 Formation of needle-like ferrite at grain boundaries after normalizing of the unalloyed steel DIN C35, because of too fast a cooling rate. Magnification 500×. (From Ref. 1.)

curves in the core have been experimentally measured and recorded, then by using the appropriate CCT diagram for the steel grade in question, it is possible to predict the structure and hardness after normalizing. To superimpose the recorded cooling curves onto the CCT diagram, the time–temperature scales must be equal to those of the CCT diagram.

Figure 63 shows, for example, that the unalloyed steel DIN Ck45 will attain the desired ferrite-pearlite structure in the core of all investigated bars of different diameters cooled in air. On the other hand, as shown in Figure 64, the alloyed steel DIN 55NiCrMoV6 cooled in the same way in air will transform to martensite and bainite. In this case, to obtain a desired structure and hardness after normalizing, a much slower cooling of about 10°C/h (50°F/h), ie., furnace cooling, has to be applied from the austenitizing temperature to the temperature at which the formation of pearlite is finished [≈ 600°C (≈ 1100°F)].

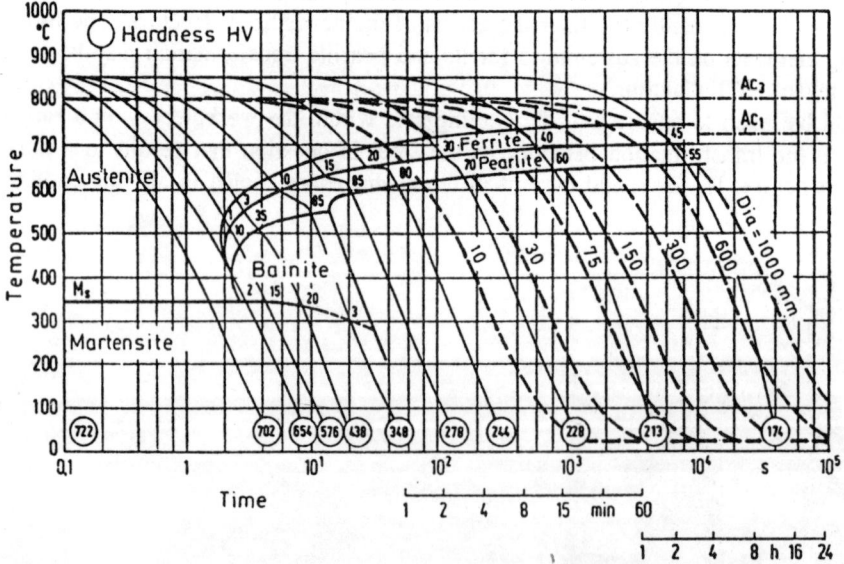

Figure 63 CCT diagram of the unalloyed steel DIN Ck45 (austenitizing temperature 850°C), with superimposed cooling curves measured in the core of round bars of different diameters cooled in air. (From Ref. 1.)

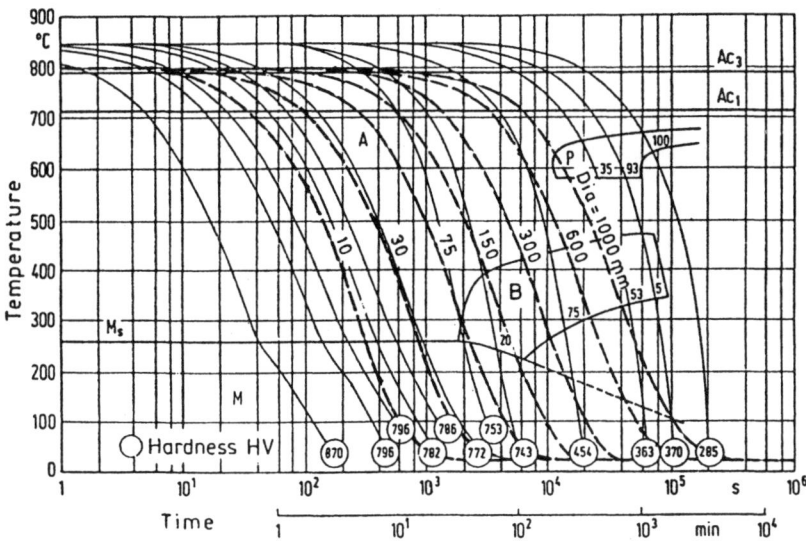

Figure 64 CCT diagram of the alloyed steel DIN 55NiCrMoV6 (austenitizing temperature 950°C), with superimposed cooling curves measured in the core of round bars of different diameters cooled in air. (From Ref. 1.)

C. Isothermal Annealing

Hypoeutectoid low-carbon steels for carburizing as well as medium-carbon structural steels for hardening and tempering are often isothermally annealed, for best machinability, because a well-differentiated, nontextured ferrite-pearlite structure is the optimum structure for machinability of these steels. If low-carbon steels are soft-annealed they give long shavings when turned and a bad surface appearance (sometimes called "smearing" or "tearing") because of the accumulation of the material on the tool's cutting edge. On the other hand, nonannealed workpieces, having harder structural constituents like bainite, results in heavy wear of the cutting edge when machined.

An isothermally annealed structure should have the following characteristics:

1. High proportion of ferrite
2. Uniformly distributed pearlite grains
3. Fine lamellar pearlite grains
4. Short pearlite lamellae
5. Coarse ferrite grains

Figure 65 shows the structure of a thin-wall die forging made of low-alloy steel for carburizing (DIN 16MnCr5) after a normalizing anneal (A) and after an isothermal anneal process (B). The desired ferrite-pearlite structure originates during an isothermal annealing, the principle of which is explained by Figure 66. This figure shows an isothermal transformation (IT) diagram of a low-alloy steel for carburizing (DIN 15CrNi6) with superimposed cooling curves for different cooling rates at continuous cooling. The slowest cooling rate of 3 K/min relates to a furnace cooling, and the fastest cooling rate of 3000 K/min relates to a quenching process. From the diagram in Figure 66 it can be clearly seen that bainite formation can be avoided only by very slow continuous cooling, but with such a slow cooling a textured (elongated ferrite) structure results (hatched area in Figure 66). There is only one way to avoid both the

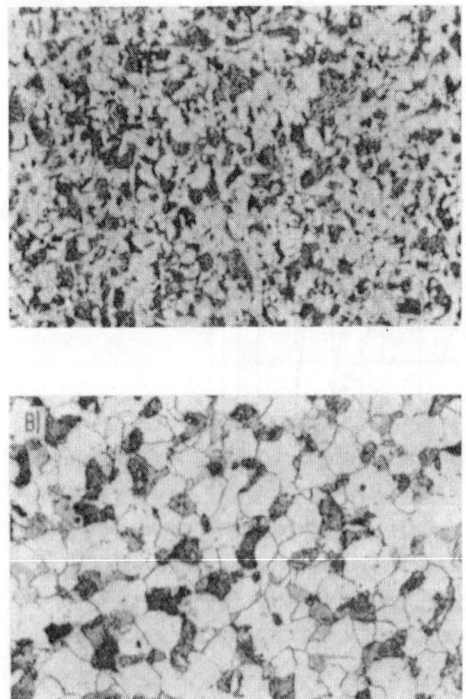

Figure 65 Structure of a forging made of low-carbon steel for carburizing (DIN 16MnCr5) (A) after normalizing; (B) after isothermal annealing. Magnification 200×. (From Ref. 1.)

formation of bainite and the formation of a textured structure (see the open arrow in Figure 66), and this is the isothermal annealing process, which consists of austenitizing followed by a fast cooling to the temperature range of pearlite formation [usually about 650°C (1200°F)], holding at this temperature until the complete transformation of pearlite, and cooling to room

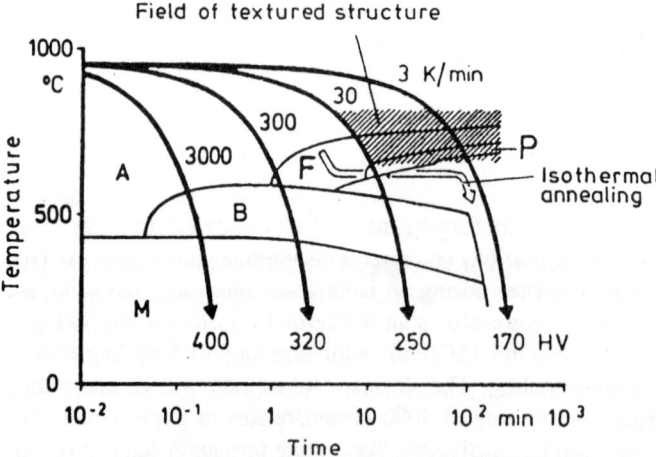

Figure 66 The principle of isothermal annealing. TTT diagram of the low-alloy steel for carburizing DIN 15CrNi6. (From Ref. 16.)

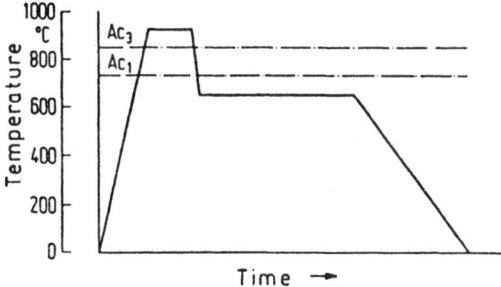

Figure 67 Temperature–time cycle of isothermal annealing. (From Ref. 1.)

temperature at an arbitrary cooling rate. The temperature–time diagram of an isothermal annealing is given in Figure 67. The metallurgical mechanism of a good isothermally annealed structure depends on the austenitizing conditions as well as on the temperature and time of the isothermal transformation and on cooling from the austenitizing temperature to the isothermal transformation temperature.

The austenitizing temperature and time should be high enough to completely dissolve all carbides, to homogenize the austenite matrix, to stabilize the austenite structure, and achieve a coarse-grained ferrite-pearlite structure after cooling. The undesired textured structure originates by pre-eutectoid ferrite precipitation along stretched phases acting as germs, for instance manganese sulfides, carbon segregations, or aluminum nitride precipitations. These phases have been stretched as a consequence of a preliminary hot-forming process.

To avoid the textured structure the steel has to contain as little sulfur, nitrogen, and aluminum as possible, and during austenitizing a complete dissolution of nitride precipitations and carbides should be achieved. Therefore the austenitizing temperature is adequately high, i.e., about 100°C (212°F) above A_{c3}, and the holding times are usually about 2 h.

Another very important condition to avoid a textured structure is to realize a minimum cooling rate between the austenitizing temperature [≈ 950°C (≈ 1750°F)] and the isothermal transformation temperature [≈ 650°C (1200°F)]. This, about a 300°C (572°F)-wide temperature range should be passed through at a minimum cooling rate of 20–40 K/min. This means that the whole batch of treated workpieces should be cooled from about 950°C (1750°F) to about 650°C (1200°F) in less than 10 min. During this cooling process an undercooling below the chosen isothermal transformation temperature must be avoided to prevent the formation of bainite.

The physical mechanism that accounts for the manner and magnitude of ferrite precipitation is the carbon diffusion during cooling from the austenitizing temperature. To achieve a good structure after isothermal annealing, all measures that reduce the carbon diffusion rate or restrict the diffusion time for carbon atoms during cooling are useful.

Figure 68 shows three structures after isothermal annealing of the low-alloy steel DIN 16MnCr5. It can be seen that cooling too slowly from the austenitizing temperature to the transformation temperature results in an undesirable textured structure of ferrite and pearlite, and if during this cooling process an undercooling takes place (i.e., the transformation temperature has been chosen too low) before the pearlite formation, then bainite will be present in the structure, which is not allowed.

Big companies usually have internal standards to estimate the allowable degree of texturing of the isothermally annealed structures, with respect to machinability, as shown in Figure 69. The transformation temperature and the necessary transformation time for the steel in

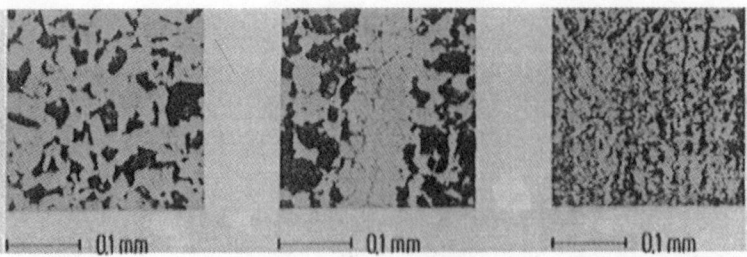

Figure 68 Different structures after isothermal annealing of the low-alloy steel DIN 16MnCr 5. (left) Well-distributed ferrite-pearlite; correct annealing. (center) Textured ferrite-pearlite structure; too slow cooling from the austenitizing to the transformation temperature. (right) Ferrite-pearlite + bainite; undercooling before pearlite transformation. (From Ref. 16.)

question may be determined by means of the appropriate isothermal transformation (IT) diagram. Figure 70 shows such a diagram for the steel DIN 17CrNiMo6. As can be seen, the lower the transformation temperature chosen, the sooner the transformation starts, up to a temperature (the so-called pearlite nose) at which the shortest time to start the transformation is achieved. Below this temperature, longer times are again necessary to start the transformation. In the range of the pearlite nose temperature, fine lamellar pearlite will be formed, and the time to complete pearlite transformation is the shortest.

For unalloyed steels, the pearlite nose temperatures are between 550 and 580°C (1022 and 1076°F), while for alloyed steels they are between 640 and 680°C (1184 and 1256°F).

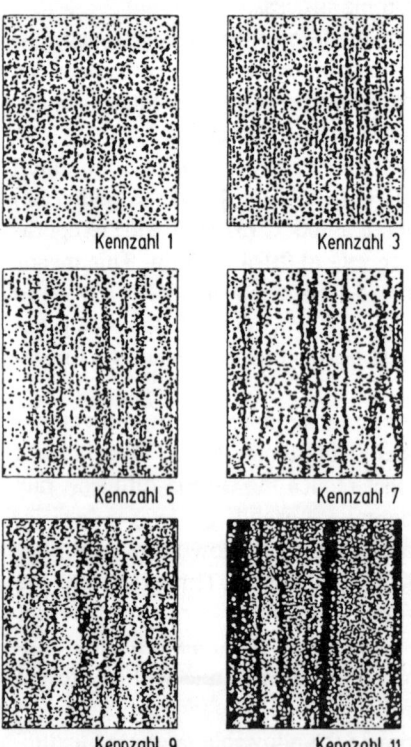

Kennzahl 1   Kennzahl 3

Kennzahl 5   Kennzahl 7

Kennzahl 9   Kennzahl 11

Figure 69 Internal standard of the German company Edelstahlwerke Buderus A.G.-Wetzlar for estimation of the allowable degree of texturing of the structure after isothermal annealing. Magnification 100×. (From Ref. 1.)

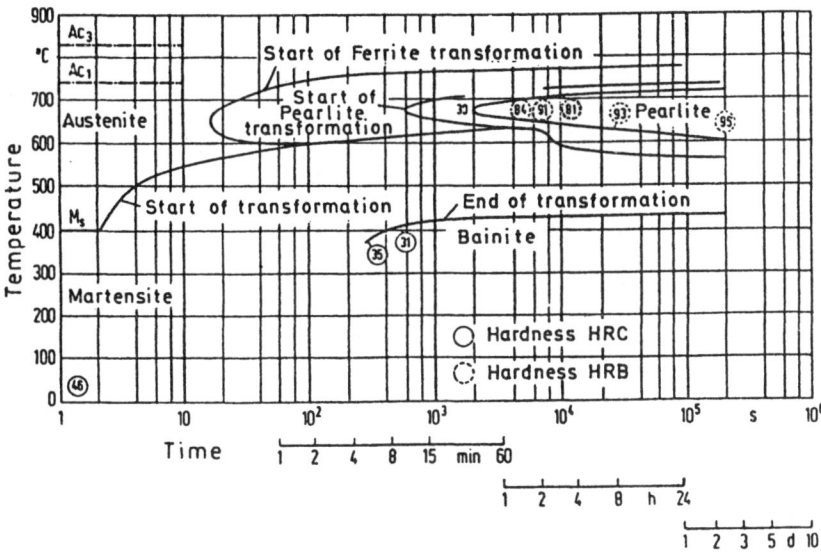

Figure 70 Isothermal transformation (IT) diagram of the steel DIN 17CrNiMo6. Austenitizing temperature 870°C. (From Ref. 1.)

The optimum isothermal annealing temperature is 10–20°C (50–68°F) above the pearlite nose temperature.

The necessary transformation time depends on the alloying elements in the steel. In the practice of isothermal annealing the holding time at the transformation temperature includes an adequate reserve because of compositional tolerances in different steel heats. Usually for low-alloy steels for carburizing and structural steels for hardening and tempering the transformation times are below 2 h.

From the technical standpoint, when a batch of workpieces has to be isothermally annealed, the biggest problem is to realize sufficiently fast cooling from the austenitizing temperature to the chosen transformation temperature without any undercooling. This cooling process depends on several factors, the main ones being the workpiece cross-sectional size, the loading arrangement, the temperature difference between the austenitizing temperature and the temperature of the cooling medium, and the heat transfer coefficient between the workpieces' surface and the ambient.

D. Soft Annealing (Spheroidizing Annealing)

Soft or spheroidizing annealing is an annealing process at temperatures close below or close above the A_{c1} temperature, with subsequent slow cooling. The microstructure of steel before soft annealing is either ferrite-pearlite (hypoeutectoid steels), pearlite (eutectoid steels), or cementite-pearlite (hypereutectoid steels). Sometimes a previously hardened structure exists before soft annealing. The aim of soft annealing is to produce a soft structure by changing all hard constituents like pearlite, bainite, and martensite (especially in steels with carbon contents above 0.5% and in tool steels) into a structure of spheroidized carbides in a ferritic matrix.

Figure 71 shows the structure with spheroidized carbides (a) after soft annealing of a medium-carbon low-alloy steel and (b) after soft annealing of a high-speed steel. Such a soft structure is required for good machinability of steels having more than 0.6% C and for all cold-working processes that include plastic deformation. Whereas for cold-working processes

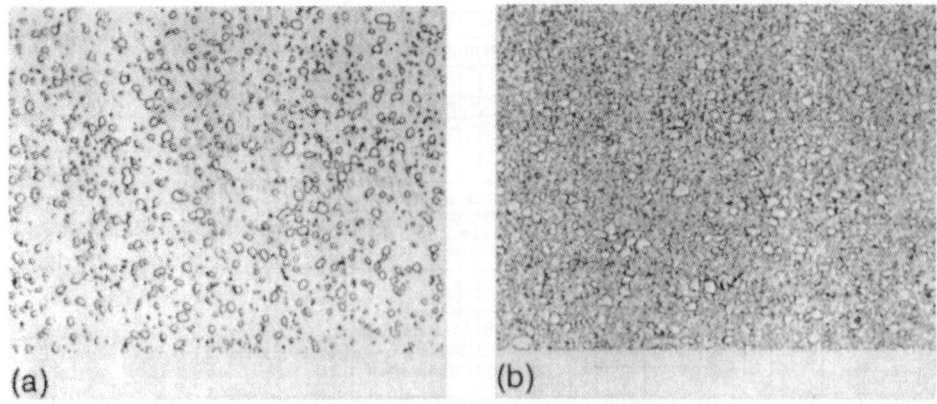

Figure 71 Structures of (a) a medium-carbon low-alloy steel DIN 50CrMoV4 after soft annealing at 720–740°C and (b) a high-speed steel annealed at 820°C. Magnification 500×. (From Ref. 1.)

the strength and hardness of the material should be as low as possible, for good machinability medium strength or hardness values are required. Therefore, for instance, when ball bearing steels are soft annealed, a hardness tolerance is usually specified. In the production sequence, soft annealing is usually performed with a semiproduct (after rolling or forging), and the sequence of operations is hot working, soft annealing, cold forming, hardening, tempering.

The required degree of spheroidization (i.e., 80–90% of globular cementite or carbides) is sometimes specified. To evaluate the structure after soft annealing, there are sometimes internal standards, for a particular steel grade, showing the percentage of achieved globular cementite, as shown in Figure 72 for the ball bearing steel DIN 100Cr6. The degree of spheroidization is expressed in this case as percentage of remaining lamellar pearlite.

The physical mechanism of soft annealing is based on the coagulation of cementite particles within the ferrite matrix, for which the diffusion of carbon is decisive. Globular cementite within the ferritic matrix is the structure having the lowest energy content of all structures in the iron–carbon system. The carbon diffusion depends on temperature, time, and the kind and amount of alloying elements in the steel. The solubility of carbon in ferrite, which is very low at room temperature (0.02% C), increases considerably up to the A_{c1} temperature. At temperatures close to A_{c1}, the diffusion of carbon, iron, and alloying atoms is so great that it is possible to change the structure in the direction of minimizing its energy content.

The degree of coagulation as well as the size of carbides after soft annealing is dependent also on the starting structure before annealing. If the starting structure is pearlite, the spheroidization of carbides takes place by the coagulation of the cementite lamellae. This process can be formally divided in two stages. At first the cementite lamellae assume a "knucklebone" shape, as shown in Figure 73. As annealing continues, the lamellae form globules at their ends and, by means of boundary surface energy, split up into spheroids, hence the name spheroidizing. During the second stage, some cementite (carbide) globules grow at the cost of fine carbide particles, which disappear. In both stages the rate of this process is controlled by diffusion. The thicker the cementite lamellae, the more energy necessary for this process. A fine lamellar pearlite structure may more easily be transformed to a globular form.

In establishing the process parameters for a soft (spheroidizing) annealing, a distinction should be drawn among hypoeutectoid carbon steels, hypereutectoid carbon steels, and alloyed steels. In any case the value of the relevant A_{c1} temperature must be known. It can be taken from the relevant IT or CCT diagram or calculated according to the formula

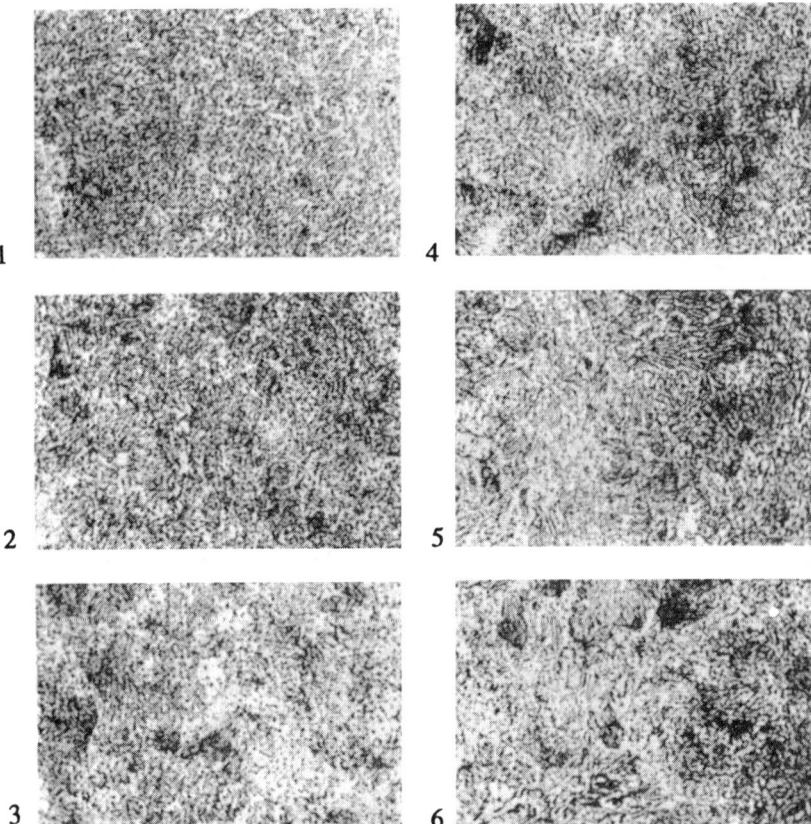

Figure 72 Internal standard of the German company Edelstahlwerke Buderus A.G.-Wetzlar for evaluation of the degree of spheroidization after soft annealing of grade DIN 100Cr6 steel. Magnification 500×. (From Ref. 1.) Amount of lamellar pearlite remaining *1*, 0%; *2*, 8%; *3*, 20%; *4*, 35%; *5*, 60%; *6*, 80%.

$$A_{c1} = 739 - 22(\%C) + 2(\%Si) - 7(\%Mn) + 14(\%Cr) + 13(\%Mo)$$
$$+ 13(\%Ni) + 20(\%V) \ [°C] \tag{36}$$

The temperature range for soft annealing of unalloyed carbon steels may be taken from the iron-carbon diagram as shown in Figure 74. The holding time at the selected temperature is approximately 1 min per millimeter of the workpiece cross section.

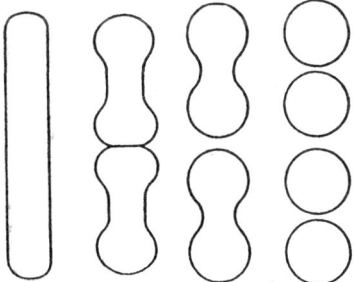

Figure 73 Schematic presentation of the process of transforming cementite lamella to spheroids during soft annealing. (From Ref. 14.)

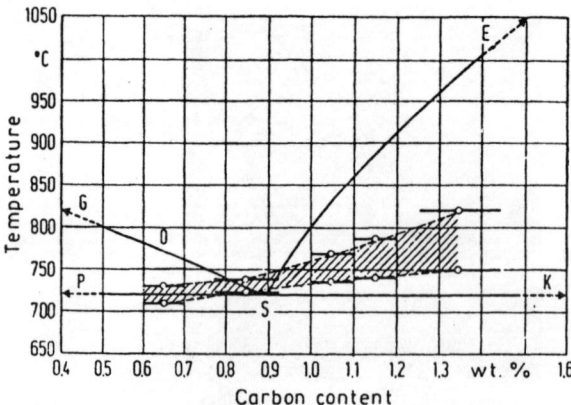

Figure 74 Temperature range for soft annealing of unalloyed steels having carbon contents of 0.6–1.35% C. (From Ref. 1.)

For alloyed steels, the soft annealing temperature may be calculated according to the empirical formula

$$T = 705 + 20(\%\text{Si} - \%\text{Mn} + \%\text{Cr} - \%\text{Mo} - \%\text{Ni} + \%\text{W}) + 100(\%\text{V}) \quad [°\text{C}] \quad (37)$$

This formula is valid only up to the following values of the alloying elements:
0.9% C; 1.8% Si; 1.1% Mn; 1.8% Cr; 0.5% Mo; 5% Ni; 0.5% W; 0.25% V. If the steel has higher amounts of alloying elements, only these indicated maximum values are to be taken into account.

Figure 75 shows possible temperature–time regimes for soft annealing. The swinging regime (Figure 75c) is used to accelerate the transformation of cementite lamellae to globular form. Increasing the temperature above A_{c1} facilitates the dissolution of cementite lamellae. At subsequent cooling below A_{c1} this dissolution process is interrupted and the parts broken off (which have less resistance to boundary surface energy) coagulate more easily and quickly.

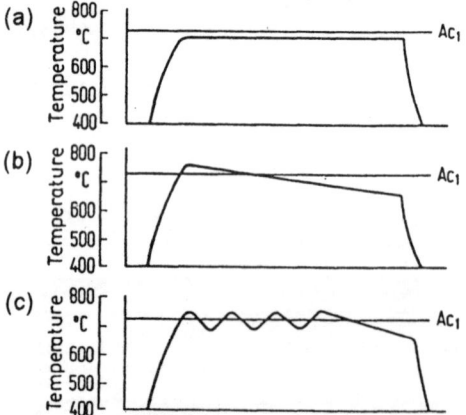

Figure 75 Temperature–time regimes at soft annealing. (a) Annealing at 20°C below A_{c1}, for unalloyed steels and for alloyed steels with bainitic or martensitic starting structure. (b) Annealing at 10°C above A_{c1} (start) and decreasing temperature to 30°C below A_{c1} for alloyed steels. (c) Swinging annealing ±5°C around A_{c1} for hypereutectoid steels. (From Ref. 1.)

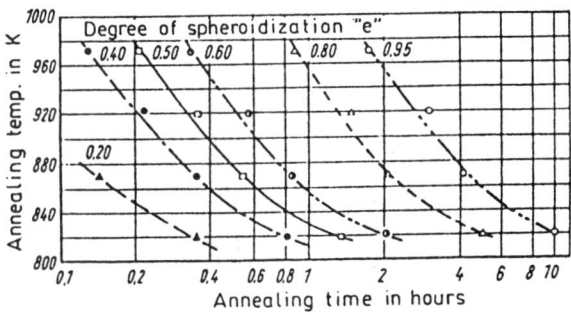

Figure 76 Time–temperature diagram for soft annealing of the unalloyed steel DIN C35 (previously deformed 50%), to achieve the required degree of spheroidization. (After Köstler; see Ref. 2.)

On the basis of the investigations of Köstler, a degree of spheroidization e has been established that gives the amount of globular cementite compared to the total amount of cementite in a steel after soft annealing. $e = 1$ means that 100% of the globular cementite (i.e., no lamellar cementite) has remained. Because the degree of spheroidization depends on the time and temperature of the soft annealing process, diagrams may be established that correlate the degree of spheroidization with the time and temperature of soft annealing. Figure 76 shows such a diagram for the unalloyed steel DIN C35.

The degree of spheroidization, especially above 80% ($e = 0.8$), has considerable influence on ultimate tensile strength, yield strength, and elongation, as shown in Figure 77 for the unalloyed eutectoid steel DIN C75. The hardness after soft annealing depends on the time and temperature of spheroidization, as shown in Figure 78 for an unalloyed steel with 0.89% C.

The machinability of steels with more than 0.6% C can be increased by soft annealing as shown in Figure 79, from which it can be seen that decreasing tensile strength and increasing the degree of spheroidization allows a higher turning speed (v_{60}) in m/min.

The cooling after soft annealing should generally be slow. Depending on the steel grade, the cooling should be performed as follows:

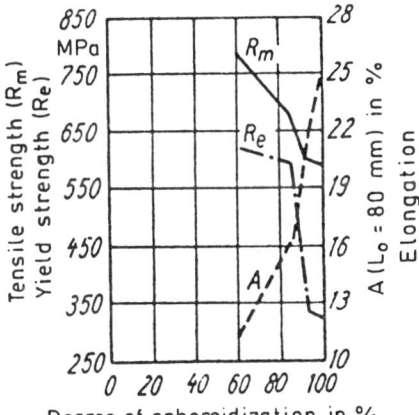

Figure 77 Change of ultimate tensile strength, yield strength, and elongation with increasing spheroidization of an unalloyed eutectoid steel, DIN C75. (From Ref. 2.)

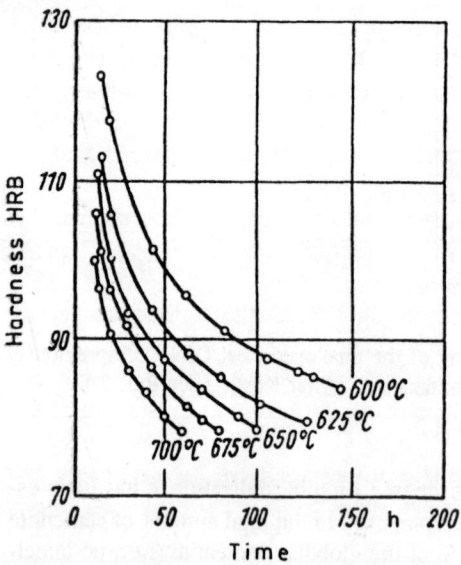

Figure 78 Hardness of an unalloyed steel with 0.89% C after soft annealing, depending on the spheroidization time and temperature. (From Ref. 2.)

For carbon and low-alloy steels up to 650°C (1200°F), with a cooling rate of 20–25 K/h (furnace cooling)

For medium-alloy steels up to 630°C (1166°F), with a cooling rate of 15–20 K/h (furnace cooling)

For high-alloy steels up to 600°C (1112°F), with a cooling rate of 10–15 K/h (furnace cooling)

Further cooling below the temperatures indicated is usually performed in air.

E. Recrystallization Annealing

Recrystallization annealing is an annealing process at temperatures above the recrystallization temperature of the cold-worked material, without phase transformation, that aims at regen-

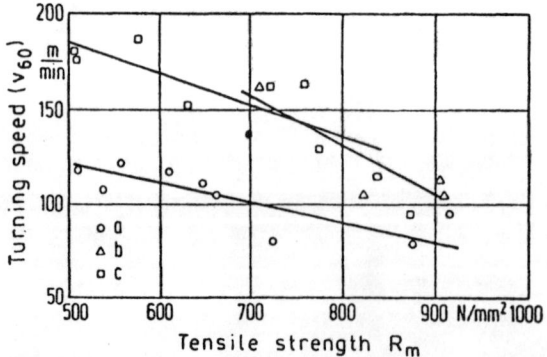

Figure 79 Influence of the ultimate tensile strength and degree of spheroidization on machinability of steels for carburizing and structural steels for hardening and tempering, expressed as 1 h turning speed (v_{60}) in m/min. (a) Spheroidization degree < 30%; (b) spheroidization degree between 40 and 60%; (c) spheroidization degree > 70%. (From Ref. 1.)

eration of properties and changes in the structure that exists after a cold-forming process such as cold rolling, deep drawing, or wire drawing. Materials that are to be subjected to a cold-forming process and subsequent recrystallization annealing must possess good cold-forming ability. These materials include soft unalloyed steels, microalloyed steels for deep drawing, microalloyed high-strength steels, unalloyed and alloyed carbon steels, stainless steels, and soft magnetic steels.

The prerequisite to recrystallization on annealing is that the degree of deformation during cold working has been large enough to produce the required number of defects in the crystals to initiate nucleation, which is then followed by grain growth. Figure 80 shows the microstructure of a low-carbon steel (a) after cold working and (b) after subsequent recrystallization annealing. During cold working of metallic materials, by far the greatest amount of the energy applied for deformation is transformed into heat, but a relatively small part (<5%) of it remains accumulated in the material due to the formation of crystal lattice defects. It is a known fact that every cold-working process (i.e., plastic deformation of the material) increases the dislocation density by some orders of magnitude. Because every dislocation is a crystal defect associated with internal stresses, the increase in the dislocation density causes the accumulation of internal stresses (i.e., of internal energy) and thereby increases the free enthalpy. Such a thermodynamically unstable material condition tends, at increased temperatures, to decrease the free enthalpy by rearranging and demolishing lattice defects. The greater the plastic deformation in a cold-forming process, the greater the strengthening of the material, which is characterized by an increase in tensile strength and yield strength and a decrease in elongation as shown in Figure 81. The material becomes harder and more brittle, so that in some cases a further step in the forming process cannot be applied without a recrystallization annealing. Also the anisotropy of the material, i.e., the dependence of mechanical properties on the direction of the cold-forming process, can be annulated by recrystallization annealing, by bringing the oriented grains that are deformed in one direction back to the original globular form.

Thermic activation, i.e., increasing the temperature at recrystallization annealing, can be used to reestablish the original structure (before cold working) with the original density of dislocations, which results in decreased hardness and strength and increased ductility and formability. The recrystallization annealing process includes the following phenomena: grain recovery, polygonization, recrystallization, and grain growth.

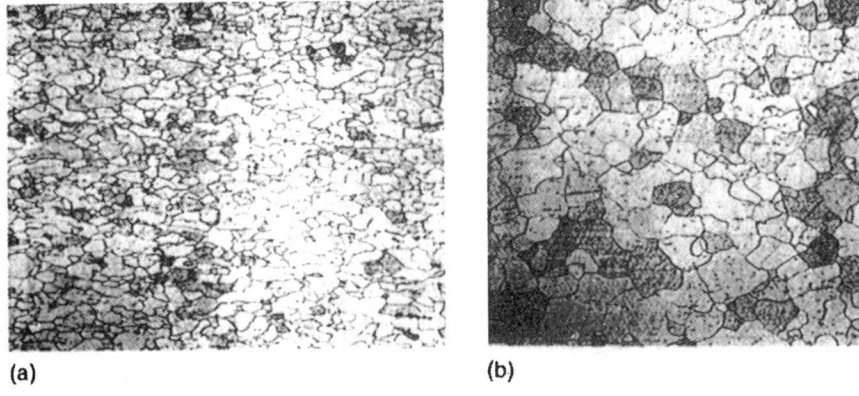

(a)   (b)

Figure 80 Low-carbon steel with 0.05% C (a) after cold working with 20% reduction (hardness 135 HV); (b) after subsequent recyrstallization annealing at 750°C (hardness 75 HV). Magnification 200×. (From Ref. 14.)

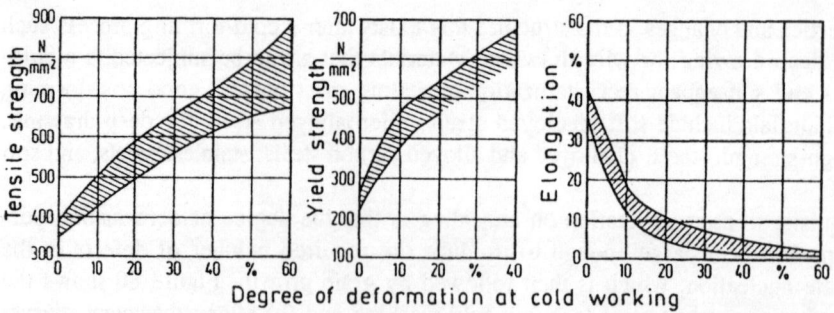

Figure 81 Strengthening of a low-carbon steel by the cold-rolling process. (From Ref. 1.)

1.   Grain Recovery

Grain recovery is a process of tempering a cold-worked metallic structure at low temperatures [150–350°C (300–662°F)] without causing any discernible changes in the microstructure. It results only in decreasing the internal stresses without substantially decreasing the
strength of the material. However, during this process characteristic changes occur in the electrical resistance and its temperature coefficient of the cold-worked material. The activation energy needed for grain recovery depends on the degree of cold working. The higher the degree (i.e., the greater the deformation), the less the activation energy required. The temperature
of grain recovery correlates with the recrystallization temperature of the same material according
to the formula

$$T_{GR} = T_R - 300 \quad [°C] \tag{38}$$

2.   Polygonization

Polygonization of a cold-worked structure is the creation of a new polygonal arrangement of
edge dislocations in the metallic crystal lattice that takes place at temperatures close above the
grain recovery temperature. As shown in Figure 82, in such a case the applied thermal energy is sufficient to rearrange the edge dislocations. In this case the originally bent sliding
planes take a polygonal shape, forming segments within a grain called subgrains. The angles
between subgrains are very small (about 1°). As a consequence of a substantial energy discharge by discharge of internal stresses, material strength is decreased. Polygonization takes
place primarily in heavily cold-worked structures, especially in ferritic matrices, below the
recrystallization temperature.

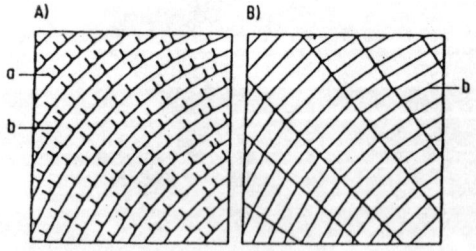

Figure 82 Schematic presentation of polygonization. Arrangement of edge dislocations and sliding planes
(A) before polygonization and (B) after polygonization. a, Edge dislocations; b, sliding planes. (From
Ref. 1.)

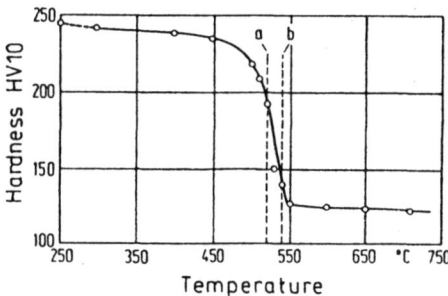

Figure 83 Decrease in hardness during recrystallization of a steel having 0.03% C, 0.54% Si, and 0.20% Mn that was cold rolled (80% deformation), as a function of annealing temperature. (Heating rate 20°C/h.) a, Begin formation of new grains; b, end formation of new grains. (From Ref. 1.)

3. Recrystallization and Grain Growth

The process of recrystallization begins when the recrystallization temperature is overstepped. The recrystallization temperature of a material is the temperature at which the formation of new grains begins within a cold-worked microstructure, as shown in Figure 83. From this figure one can conclude that for the steel in question the recrystallization temperature is 520°C (968°F). During recrystallization, as can be seen from Figure 83, hardness and strength decrease substantially while ductility increases. In practice, the recrystallization temperature T_R is often considered the temperature of a cold-worked material at which recrystallization is completed after 1 h of annealing. There is a correlation between the recrystallization temperature (T_R) and the melting temperature (T_M) of the material, which reads

$$T_R = 0.4 \ T_M \tag{39}$$

Figure 84 shows that this correlation holds for practically all pure metals if both temperatures T_R and T_M are taken in kelvins. The recrystallization temperature can be influenced by the degree of deformation during cold working, the heating rate, and the starting microstructure.

In contrast to the grain recovery process (which follows a parabolic law), the recrystallization process begins only after an incubation period (because of nucleation), starting slowly,

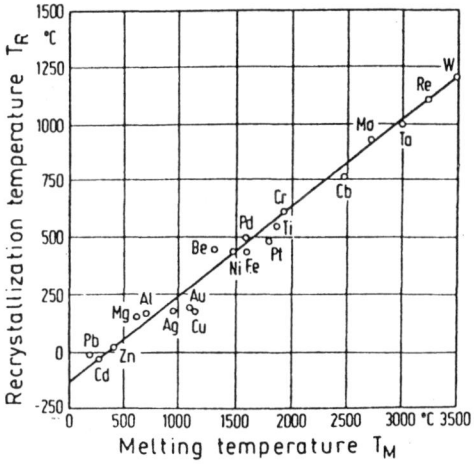

Figure 84 Correlation between the recrystallization temperature and the melting temperature for pure metals. (From Ref. 1.)

reaching a maximum rate, and finishing slowly. The nuclei from which new grains grow are situated preferably at the grain boundaries of compressed cold-worked grains. New grains grow from these nuclei until they meet up with other grains. Recrystallization brings about the formation and movement of large-angle grain boundaries.

Figure 85 is a schematic presentation of new grain formation and growth during the recrystallization process as a function of annealing time. As time passes, the new grains, starting from nuclei, grow unhindered within the cold-worked grains. Simultaneously, new nuclei are formed. At the movement of big-angle grain boundaries, new grains consume the previously deformed grains. The recrystallization process is locally finished when new neighboring grains collide wih each other. The size, form, and orientation of the new structure, as well as the condition of the lattice defects in it, differ substantially from those of the previous structure. The recrystallization process itself can be hindered by precipitations, dispersions, and a second phase.

The most important technological parameters of recrystallization annealing that influence the rate of recrystallization and the material properties after recrystallization are

1. Material-dependent parameters. The chemical composition and the starting structure (including the degree of deformation)
2. Process-dependent parameters. Annealing temperature, annealing time, and heating and cooling rates

The course of a recrystallization process can be presented in an isothermal time-temperature–recrystallization diagram as shown in Figure 86. As can be seen from this diagram, the higher the temperature of recrystallization annealing, the shorter the necessary annealing time. The lower the degree of deformation at cold working, the higher the required recrystallization temperature, as shown in Figure 87. The higher the heating rate, the higher the recrystallization temperature. It can be concluded from Figure 86 that with substantially longer annealing times, a full recrystallization can be achieved at relatively low temperatures.

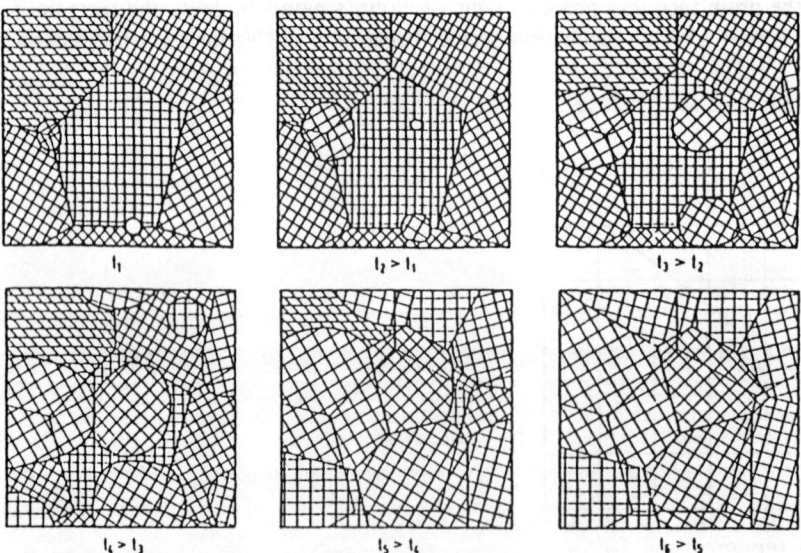

Figure 85 Schematic presentation of new grain formation and growth during the recrystallization process as a function of annealing time t. (From Ref. 1.)

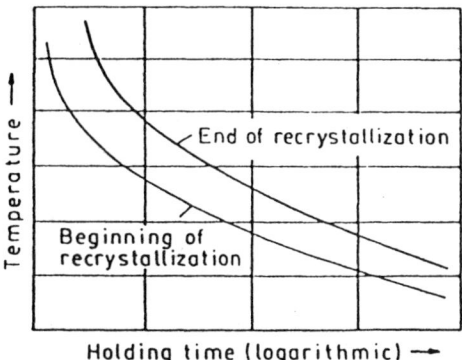

Figure 86   An isothermal time–temperature–recrystallization diagram. (From Ref. 2.)

The degree of deformation at cold working has a very important influence on the size of newly formed grains during recrystallization. If the cold working is carried out with a very low degree of deformation but without sufficient strengthening of the material to enable the process of recrystallization, a decrease in stresses can still be achieved by movement of the deformed grain boundaries. In this case grains with low dislocation density grow (because there are only a few nuclei) and a coarse-grained structure develops as shown in Figure 88. Consequently, there is a critical degree of deformation at cold working that at subsequent recrystallization annealing leads to sudden grain growth, as shown in Figure 89 for a low-carbon steel. With an increase in the carbon content of the steel, this critical degree of deformation shifts from about 8% to 20% of deformation at cold working.

III.   HARDENING BY FORMATION OF MARTENSITE

A.   Austenitizing

Austenitizing is the first operation in many of the most important heat treatment processes (hardening, carburizing, normalizing) on which the properties of heat-treated parts depend. Let us assume the bulk heat treatment of real batches of workpieces and consider the metallurgical and technological aspects of austenitizing.

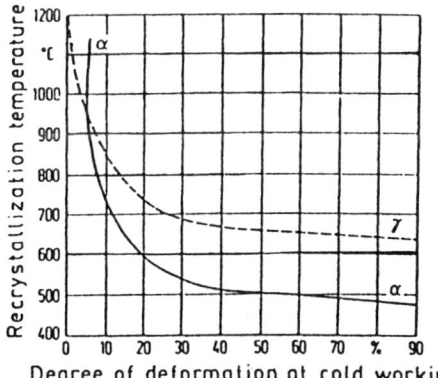

Figure 87   Recrystallization temperature of α- and γ-iron as a function of the degree of deformation at cold working. (From Ref. 1.)

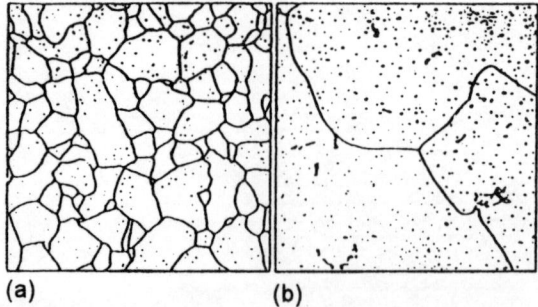

(a)   (b)

Figure 88   Development of coarse-grained structure during recrystallization of soft iron. (a) Microstructure before cold working; (b) microstructure after cold working with very low degree of deformation (10%) and subsequent recrystallization annealing at 700°C. Magnification 500×. (From Ref. 1.)

1.   Metallurgical Aspects of Austenitizing

The way austenite is formed when a certain steel is heated depends very much on the steel's starting microstructure. Let us take as an example an unalloyed eutectoid steel with 0.8% C and follow the process of its austenitization using the schemes shown in Figure 90. At room temperature the cementite (Fe$_3$C) plates of the pearlite are in direct contact with ferrite (α-Fe, see Figure 90a). The carbon atoms from cementite have a tendency to diffuse into the ferrite lattice. The higher the temperature, the greater this tendency is. When, upon heating, the A_{c1} temperature [723°C (1333°F)] has been crossed, the transformation of ferrite into austenite (γ-Fe) starts immediately adjacent to the cementite plates (see Figure 90b). After that the cementite plates start to dissolve within the newly formed austenite, becoming thinner and thinner (Figures 90c and 90d). So two processes are taking place at the same time: the formation of austenite grains from ferrite and the dissolution of cementite plates in the austenite lattice. Experiments have shown that the process of ferrite-to-austenite transformation ends before all the cementite has been dissolved. This means that after all the ferrite has transformed into austenite, small particles of cementite will remain within the austenite grains (Figure 90e). Figure 91 shows the formation of austenite in a microstructure of eutectoid steel. Areas of austenite formation are visible as white patches within the lamellar pearlitic structure. Some of the cementite persists in the form of spheroidized particles (the small dark spots in the white

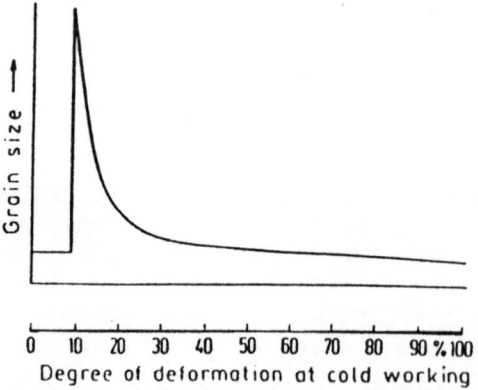

Figure 89   Grain growth in the range of the critical degree of deformation (at 10%) for a steel with 0.06% C. Recrystallization temperature, 700°C. (From Ref. 1.)

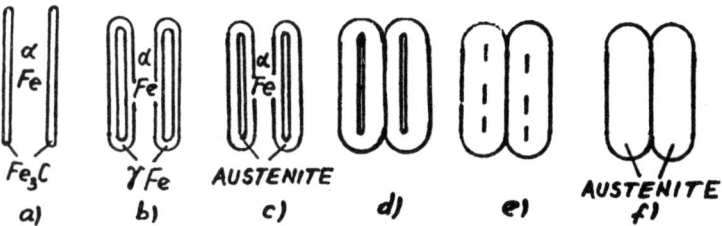

Figure 90 Transformation of a pearlitic structure to austenite when heating an unalloyed eutectoid steel of 0.8% C.

areas). They dissolve only with longer holding times at temperature. Once these cementite particles completely dissolve, the structure consists of only one phase—austenite (see Figure 90f). In this state, however, there are still differences in carbon concentration among particular austenite grains. In spots where cementite plates were previously to be found, the carbon concentration is high, while in other spots far from cementite plates it is low.

Equalizing of the carbon concentration proceeds gradually by diffusion, giving, at the end of this process, a homogeneous austenite structure. The holding time at austenitizing temperature necessary for this process is called the homogenization time. At pearlite-austenite transformation several austenite grains are formed from one pearlite grain, i.e., the newly formed austenite is fine-grained.

Nucleation sites for austenite formation depend on the starting microstructure as shown in Figure 92. In ferrite the nucleation sites are situated primarily at grain boundaries. In

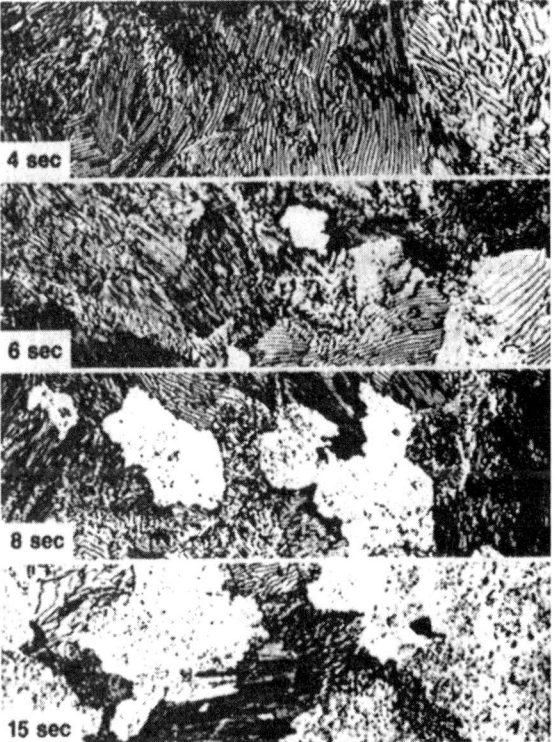

Figure 91 Formation of austenite (light patches) from pearlite as a function of time. (From Ref. 17.)

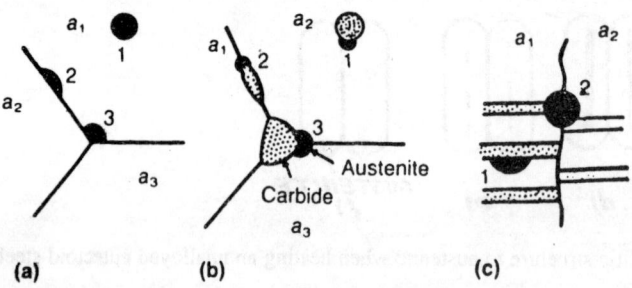

Figure 92 Nucleation sites for austenite formation in microstructures of (a) ferrite, (b) spheroidite, (c) pearlite. (From Ref. 17.)

spheroidized structures nucleation starts on carbide particles, whereas in pearlitic structures it starts primarily at the intersection of pearlite colonies but also at cementite lamellae.

Kinetics of Transformation During Austenitizing   Figure 93 shows the volume percent of austenite formed from pearlite in eutectoid steel as a function of time at a constant austenitizing temperature. From the beginning of austenitization a certain incubation time is necessary to form the first nuclei, and then the process proceeds at a more rapid rate as more nuclei develop and grow. At higher temperatures the diffusion rate increases and austenite forms more rapidly, as shown in Figure 94.

The duration of an austenitizing process depends on the austenitizing temperature and the steel composition. The influence of time at austenitization can best be explained by the diagrammatic illustrations shown in Figure 95. From Figures 95a and 95b, which apply to eutectoid carbon steel of 0.8% C, one can see that if an austenitizing temperature of 730°C (1346°F) is maintained (after a rapid heating to this temperature), the transformation will start in about 30 s. If instead an austenitizing temperature of 750°C (1382°F) is chosen, the transformation will begin in 10 s, and if a temperature of 810°C (1490°F) is selected, in about 1 s. The transformation of pearlite to austenite and cementite is in this case completed in about 6 s. If the steel is to be fully austenitic (all carbides dissolved, hatched area), it must be held at this temperature for about 2 h (7×10^3 s).

Figures 95c and 95d apply to a hypoeutectoid plain carbon steel of 0.45% C. They show that in this case at an austenitizing temperature of 810°C (1490°F) the transformation from pearlite to austenite starts in about 1 s. In about 5 s the pearlite has been transformed and the

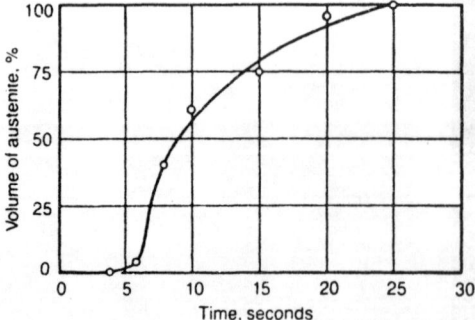

Figure 93 Volume percent austenite formed from pearlite in eutectoid steel as a function of time at a constant austenitizing temperature. (From Ref. 17.)

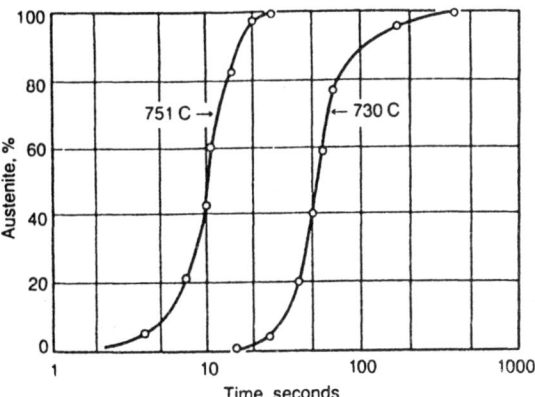

Figure 94   Effect of austenitizing temperature on the rate of austenite formation from pearlite in a eutectoid steel. (From Ref. 17.)

structure consists of ferrite, austenite, and cementite. About 1 min later the carbon has diffused to the ferrite, which has thereby been transformed to austenite. Residual particles of cementite remain, however, and it takes about 5 h at this temperature to dissolve them completely.

Figures 95e and 95f apply to a hypereutectoid steel containing 1.2% C. If this steel is austenitized at 810°C (1490°F), the pearlite starts to transform in about 2 s, and in about 5 s the structure consists only of austenite and cementite. It is not possible for the cementite to be completely dissolved at this temperature. To achieve complete solution of the cementite, the temperature must be increased above A_{cm}, in this case to at least 860°C (1580°F).

The holding time at austenitizing (hardening) temperature depends on the desired degree of carbide dissolution and acceptable grain size, taking into account that the grain growth increases with higher austenitizing temperatures and longer holding times. Since the amount of carbide is different for different types of steel, the holding time (from the metallurgical point of view) depends on the grade of steel. However, carbide dissolution and the holding time is dependent not only on the austenitizing temperature but also the rate of heating to this temperature. Varying the rate of heating to this temperature will have an effect on the rate of transformation and dissolution of the constituents.

The influence of the rate of heating (and correspondingly of the holding time) on carbide dissolution, grain growth, and hardness after hardening for various grades of steel has been studied in detail and published in References 18 and 19. These time–temperature–austenitizing diagrams (*Zeit–Temperatur–Austenitisierung Schaubilder* in German) have been produced either as isothermal diagrams (the steel specimens were heated rapidly at the rate of 130°C/s (266°F/s) to the temperature in question and held there for a certain predetermined time) or as continuous heating diagrams (the steel specimens were heated continuously at different heating rates). Consequently, isothermal diagrams may be read only along the isotherms, and the continuous heating diagrams may be read only along the heating rate lines.

Figure 96 shows an isothermal type of time–temperature–austenitizing diagram of grade DIN 50CrV4 steel. From this type of diagram one can read off, for instance, that if the steel is held at 830°C (1526°F), after about 1 s pearlite and ferrite will be transformed to austenite, but more than 1000 s is necessary to completely dissolve the carbides and achieve a homogeneous austenite.

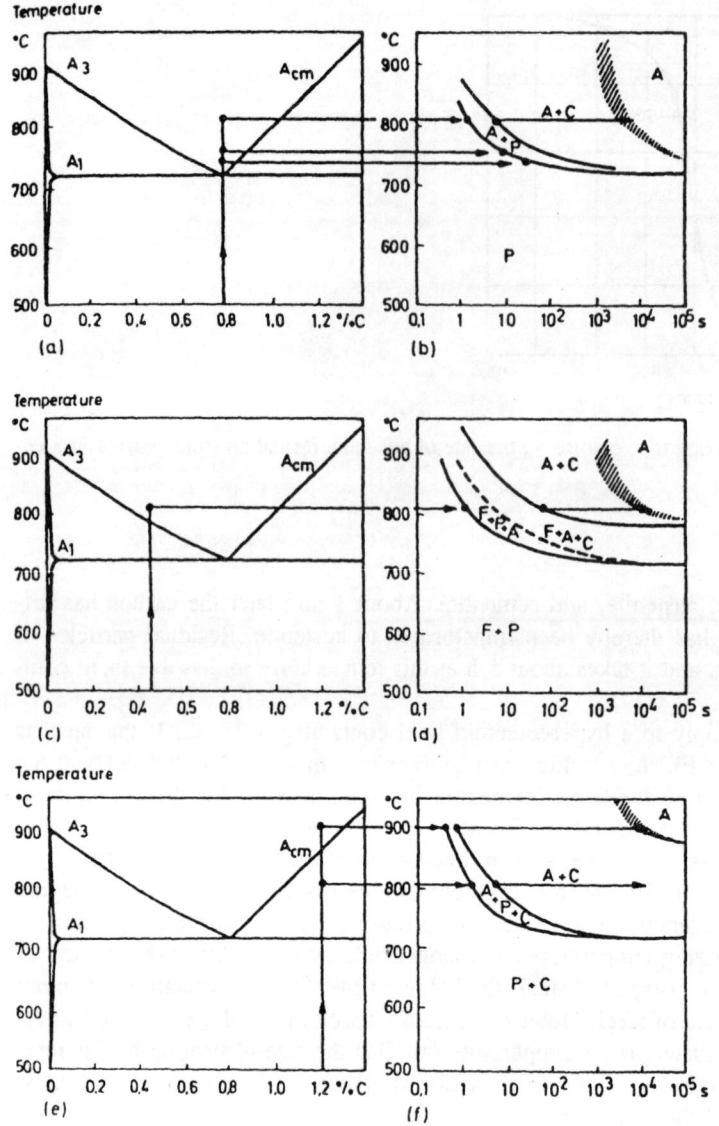

Figure 95   Structural transformations during austenitizing steels containing (a, b) 0.8% C; (c, d) 0.45% C; (e, f) 1.2% C. A = austenite, C = cementite, F = ferrite, P = pearlite. (From Ref. 14.)

In practice, the continuous heating diagrams are much more important because every austenitizing process is carried out at a specified heating rate. Figure 97 shows a time–temperature–austenitizing diagram of the continuous heating type for grade DIN Ck45 steel. The continuous heating was carried out at various constant rates ranging from 0.05 to 2400°C/s (32.09 to 4352°F). If the heating rate was extremely slow [e.g., 0.22°C/s (32.4°F/s)] to about 775°C (1427°F), on crossing the A_{c3} temperature after about 1 h all pearlite and ferrite would have been transformed to inhomogeneous austenite. At a heating rate of 10°C/s (50°F/s) the pearlite and ferrite would have been transformed to inhomogeneous austenite after crossing the A_{c3} temperature at about 800°C (1472°F) after only 80 s.

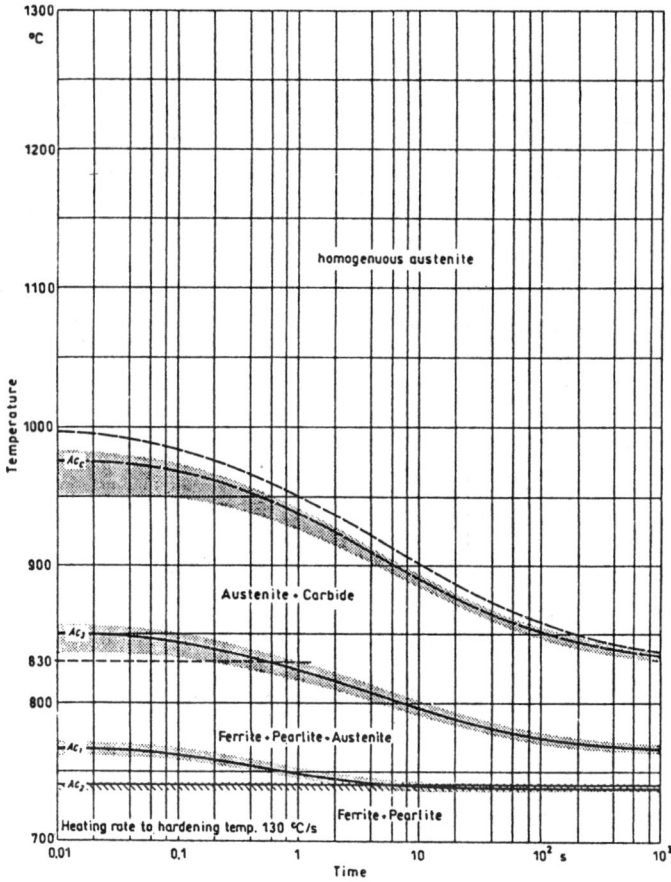

Figure 96 Isothermal time–temperature–austenitizing diagram of the steel grade DIN 50CrV4 (0.47% C, 0.27% Si, 0.90% Mn, 1.10% Cr). (From Ref. 19.)

A remarkable feature of such diagrams is that they show precisely the increase of A_{c1} and A_{c3} transformation temperatures with increasing heating rates. This is especially important when short-time heating processes like induction hardening or laser beam hardening, with heating rates ranging to about 1000°C/s (1832°F/s), are applied for surface hardening. In such a case this diagram should be consulted to determine the required austenitizing temperature, which is much higher than in conventional hardening of the same grade of steel. For the steel in question, for example, the conventional hardening temperature would be in the range of 830–850°C (1526–1562°F), but for induction or laser beam hardening processes the hardening temperatures required are between 950 and 1000°C (1742 and 1832°F). When heating at a rate of 1000°C/s (1832°F/s) to the austenitizing temperature of 1000°C (1832°F), only 1 s is necessary, and the above-mentioned short heating time processes operate in approximately this time range.

As Figure 97 shows, much higher temperatures are necessary to achieve the homogeneous austenite structure. In such a case one is, of course, concerned with the grain growth.

Figure 98 shows the grain growth (according to ASTM) when grade DIN Ck45 steel is continuously heated at different heating rates to different austenitizing temperatures. Figure

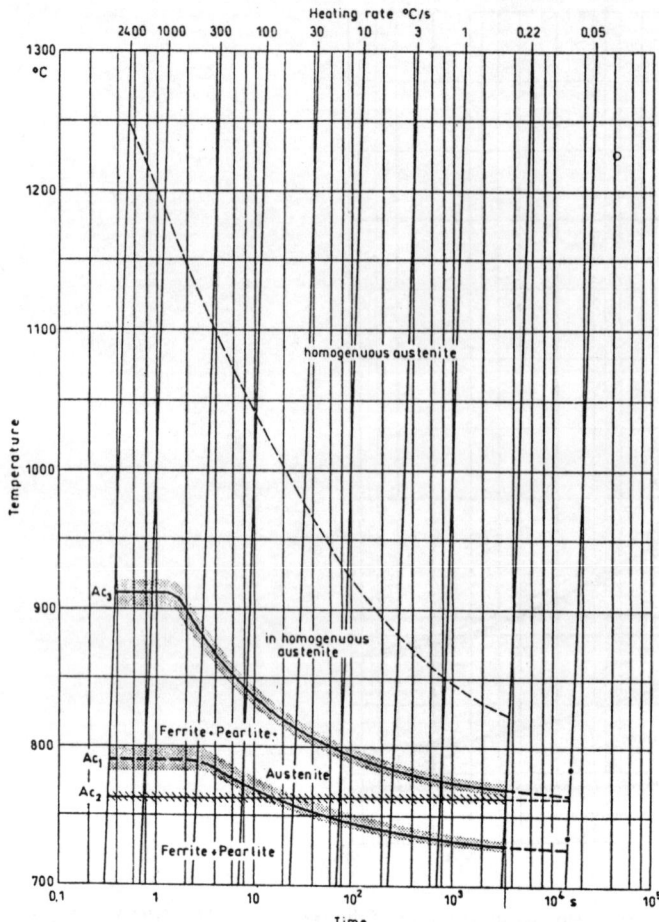

Figure 97 Time–temperature–austenitizing diagram for continuous heating of the steel grade DIN Ck45 (0.49% C, 0.26% Si, 0.74% Mn). (From Ref. 18.)

99 shows the achievable Vickers hardness after hardening for grade DIN Ck45 steel austenitized at various heating rates to various temperatures. It shows, for example, that maximum hardness would be achieved upon austenitizing the steel at 850°C (1562°F) for about 900 s [or heating at a heating rate of 1°C/s (33.8°F/s), which corresponds to the field of homogeneous austenite (see Figure 97).

The hardness after quenching, which depends on the amount of carbide dissolution, is also dependent on the initial structure of the steel. This is illustrated in Figure 100. Figure 100a shows that a structure of spheroidized cementite (after soft annealing) of the hypoeutectoid DIN Cf53 carbon steel will attain the maximum hardness of 770 HV when heated at a rate of 1°C/s (33.8°F/s) to 875°C (1609°F) (holding time 855 s or 14 min). The hardened and tempered structure (tempered martensite) of the same steel, as shown in Figure 100b, will attain the maximum hardness of 770 HV, however, if heated to 875°C (1609°F) at the rate of 1000°C/s (1832°F/s) (holding time less than 1 s). For this reason, when short-time heating processes are used, the best results are achieved with hardened and tempered initial structures.

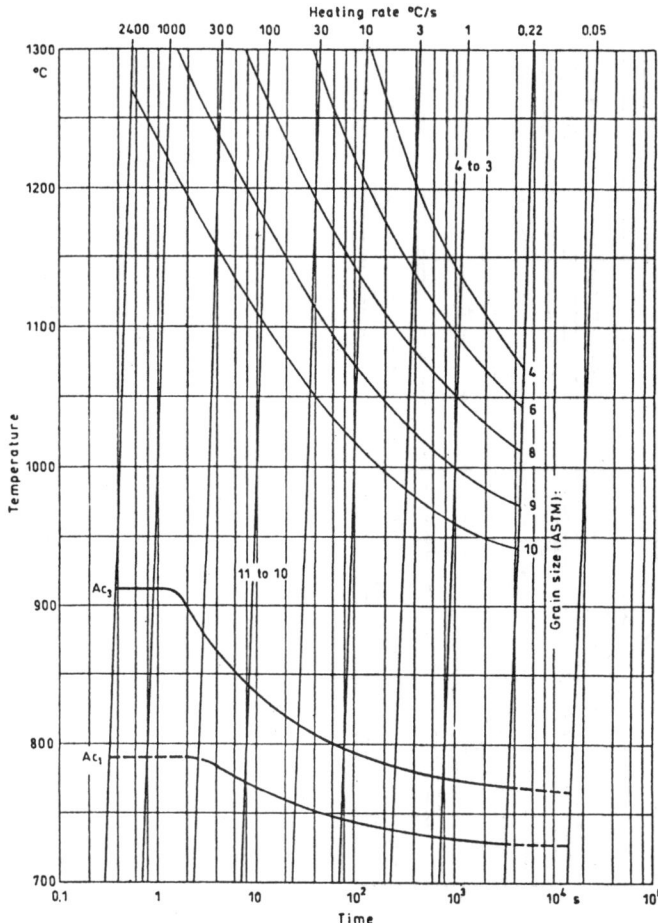

Figure 98 Time-temperature-austenitizing diagram for continuous heating showing the grain growth of steel grade DIN Ck45. (From Ref. 18.)

For eutectoid and hypereutectoid steel grades, which after quenching develop substantial amounts of retained austenite, the attainment of maximum hardness after quenching is more complicated. Figure 101 shows the hardness after quenching for the ball bearing hypereutectoid grade DIN 100Cr6 steel (1.0% C, 0.22% Si, 0.24% Mn, and 1.52% Cr). The maximum hardness of 900 HV after quenching is attained on heating to a very narrow temperature range, and furthermore this temperature range is displaced toward higher temperatures as the heating rate is increased. If this steel is quenched from temperatures that exceed the optimum range, the resulting hardness is reduced owing the presence of an increasing amount of retained austenite.

For plain carbon and low-alloy structural steels, which contain easily dissolved carbides, a holding time of 5–15 min after they have reached the hardening temperature is quite enough to make certain that there has been sufficient carbide dissolution. For medium-alloy structural steels this holding time is about 15–25 min. For low-alloy tool steels, it is between 10 and 30 min; and for high-alloy Cr steels, between 10 min and 1 h.

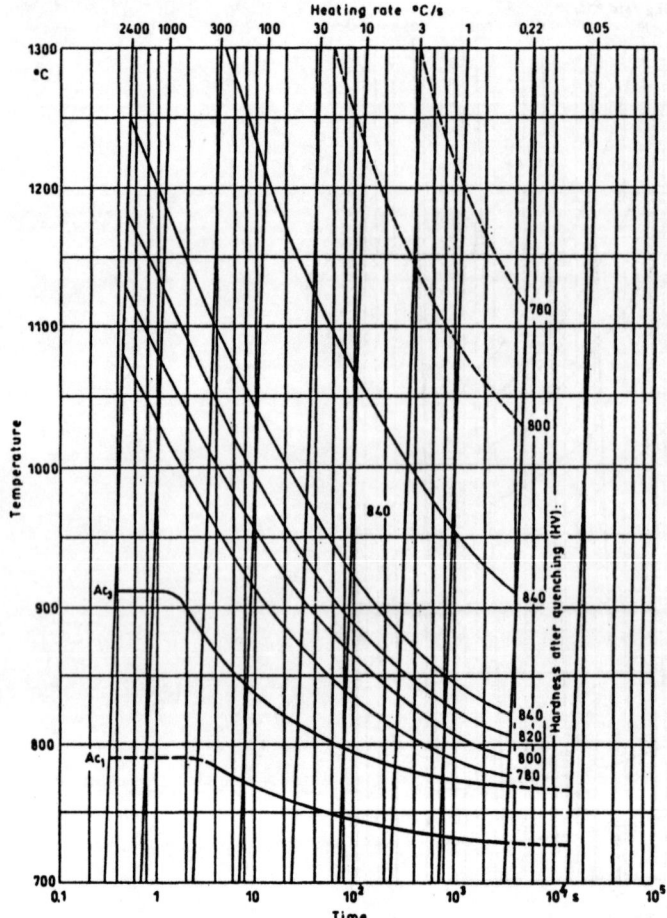

Figure 99 Time–temperature–austenitizing diagram for continuous heating showing the achievable hardness after hardening steel grade DIN Ck45. (From Ref. 18.)

2. Technological Aspects of Austenitizing

In heating metallic objects to their austenitizing (hardening) temperature, there are two kinds of heating rates to be distinguished: those that are technically possible and those that are technologically allowed.

The technically possible heating rate is the heating rate the heating equipment could realize in actual use. It depends on

1. The installed heating capacity of the equipment
2. The heat transfer medium (gas, liquid, vacuum)
3. The temperature difference between the heat source and the surface of the heated objects (workpieces put in a hot or cold furnace)
4. The mass and shape of the workpiece (the ratio between its volume and superficial area)
5. The number of workpieces in a batch and their loading arrangement

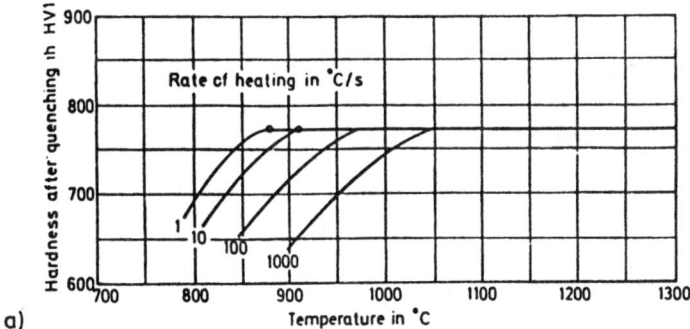

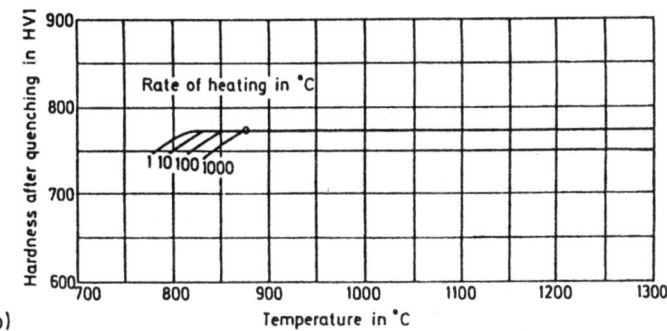

Figure 100 Hardness after quenching as a function of the rate of heating and austenitizing temperature for grade DIN Cf53 steel (hypoeutectoid carbon steel) (a) for soft-annealed condition, (b) for hardened and tempered condition. (From Ref. 14.)

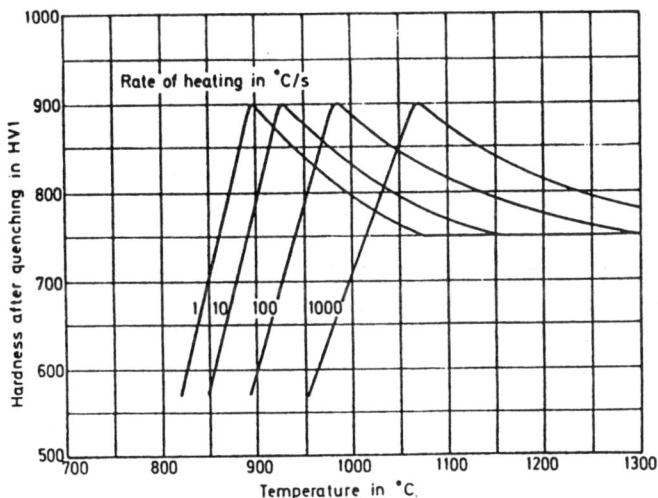

Figure 101 Hardness after quenching as a function of the rate of heating and austenitizing temperature for grade DIN 100Cr6 steel initially soft annealed. (From Ref. 14.)

The technologically allowed heating rate is the maximum heating rate that can be applied in actual circumstances, taking into account the fact that thermal stresses that develop within the workpiece must not exceed the critical value because this could cause warping or cracking, since sections having different dimensions heat up at different speeds and large temperature gradients can arise between the surface and the core of the workpiece. This heating rate depends on

1. The mass and shape of the workpiece (the ratio between its volume and superficial area)
2. The chemical composition of the material
3. The initial microstructure

When workpieces of heavy sections or of complicated shapes are heated, temperatures between 250 and 600°C (482 and 1112°F) are particularly dangerous, because in this temperature range the steel does not have enough plasticity to compensate for thermal stresses. If the heating of an object is asymmetrical, the object will warp. If thermal stresses are developed that overstep the strength of the material (which is substantially lower at higher temperatures), cracks will result.

If the heating rate is too high through the transformation temperature range (between A_{c1} and A_{c3}), warping may occur because of volume change of the structure lattice. The tendency of a steel to crack during heating depends on its chemical composition. Carbon content has the decisive influence. The higher the carbon content, the greater the sensitivity to cracking. The complex influence of carbon and other alloying elements is expressed by the following empirical formula termed the C equivalent (C_{ekv}):

$$C_{ekv} = C + \frac{Mn}{5} + \frac{Cr}{4} + \frac{Mo}{3} + \frac{Ni}{10} + \frac{V}{5} + \frac{Si - 0.5}{5} + \frac{Ti}{5} + \frac{W}{10} + \frac{Al}{10} \qquad (40)$$

where the element symbols represent weight percent content. This formula is valid up to the following maximum values of alloying elements:

C ≤ 0.9%	V ≤ 0.25%
Mn ≤ 1.1%	Si ≤ 1.8%
Cr ≤ 1.8%	Ti ≤ 0.5%
Mo ≤ 0.5%	W ≤ 2.0%
Ni ≤ 5.0%	Al ≤ 2.0%

The values of the alloying elements actually present are put into the formula in weight percent. If the amount of an alloying element exceeds the limit given above, then the indicated maximum value should be put into the formula.

The higher the calculated C_{ekv} value, the greater the sensitivity of the steel to cracking. For instance,

$C_{ekv} \leq 0.4$:   The steel is not sensitive to cracking (it may be heated quite rapidly)

$C_{ekv} = 0.4$-0.7:   The steel is medium sensitive to cracking

$C_{ekv} \geq 0.7$:   The steel is very sensitive to cracking (when heating up a preheating operation should be included)

The initial microstructure also has some influence on the technologically allowed heating rate. A steel with a homogeneous microstructure of low hardness may be heated more rapidly than a steel of high hardness with inhomogeneous microstructure.

The thermal gradients and consequently the thermal stresses developed when heating to austenitizing temperature can usually be diminished by preheating the workpiece to a temperature lying close below the transformation temperature A_{c1} and holding it there until temperature equalizes throughout the cross section.

The theoretical time–temperature diagram of the austenitizing process is shown in Figure 102. Practically, however, there is no such strict distinction between the heating and soaking times. Contrary to the generally widespread belief that the surface of the steel reaches the preset temperature considerably earlier than the center, the closer the temperature of the steel approaches the preset temperature, the smaller the temperature difference between surface and core, as shown in Figure 103. It can therefore be assumed that when the surface has reached the preset temperature, part of the soaking time (depending on the cross-sectional size) has already been accomplished. Certainly, one has to be aware of the "corner effect"—corners, sharp edges, and thin sections reach the preset temperature much earlier than the core of the workpiece.

The most important parameters of every austenitizing process are

1. The austenitizing temperature
2. The heat-up and soak time at austenitizing temperature

For each grade of steel there is an optimum austenitizing (hardening) temperature range. This temperature range is chosen so as to give maximum hardness after quenching and maintain a fine-grained microstructure. It can be determined experimentally as shown in Figures 104 and 105. From Figure 104 it is clear that the lowest possible hardening temperature for the steel

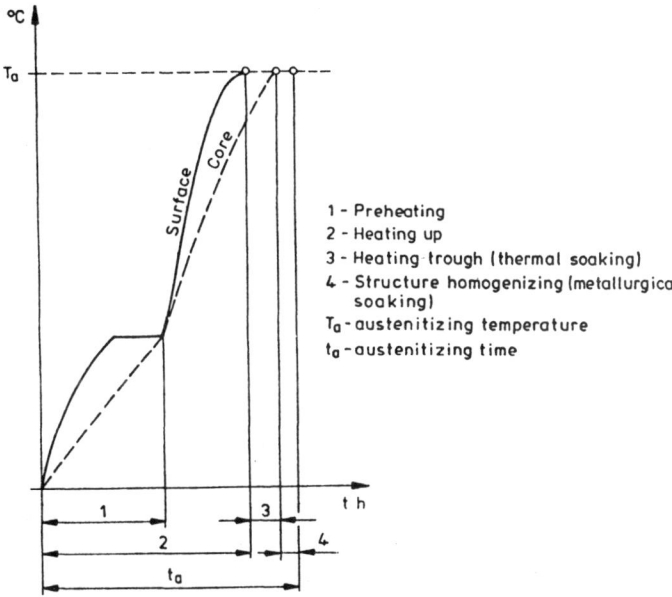

Figure 102   Austenitizing process (theoretically).

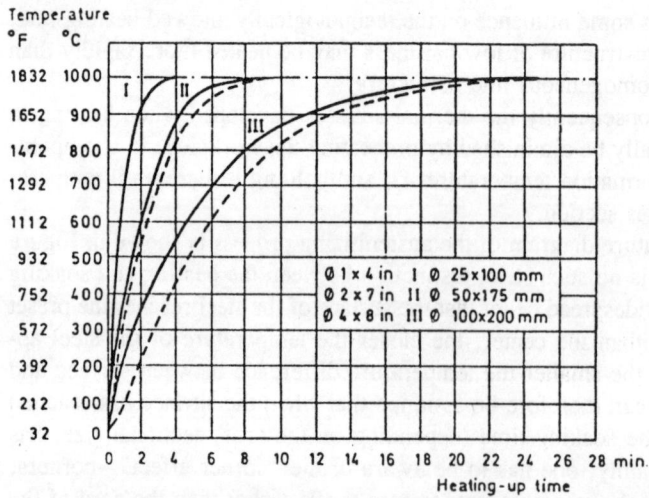

Figure 103 Time–temperature curves for steel bars of different diameters heated in a salt bath at 1000°C. Full line, measured temperature at surface; dashed line, measured temperature at center. (From Ref. 14.)

in question is 850°C (1562°F). A lower hardening temperature would result in the formation of bainite and even pearlite with inadequate hardness.

When the hardening temperature is increased (see Figure 105), the grain size and the amount of retained austenite increase. At 920 and 970°C (1688 and 1778°F) the retained austenite may be discerned as light angular areas. On the basis of these experiments the optimum hardening temperature range for the steel in question has been fixed at 850–880°C (1562–1616°F). The optimum hardening temperature range for unalloyed steels can be determined from the iron–carbon equilibrium diagram according to the carbon content of the steel. This range is 30–50°C (86–122°F) above the A_{c3} temperature for hypoeutectoid steels and 30–50°C (86–122°F) above A_{c1} for hypereutectoid steels. as shown in Figure 106. Because the curve S–E in this diagram denotes also the maximum solubility of carbon in austenite, it is clear that the higher the austenitizing (hardening) temperature, the more carbon can be dissolved in austenite. For alloyed steels the optimum austenitizing (hardening) temperature range depends on the chemical composition, because different alloying elements shift the A_1 temperature to either higher or lower temperatures. For these steels, therefore, data from the literature on the optimum hardening temperature range have to be consulted.

It should also be mentioned that increasing the austenitizing temperature causes the following effects:

1. It increases the hardenability of the steel because of the greater amount of carbide going into solution and the increased grain size.
2. It lowers the martensite start temperature (M_s). Owing to the more complete carbide dissolution, the austenite becomes more stable and starts to transform upon quenching at lower temperature.
3. It increases the incubation time, i.e., the time until the isothermal transformation to pearlite or bainite starts. This is expressed as a shift in the start of transformation curves in an isothermal transformation (IT) diagram to later times.

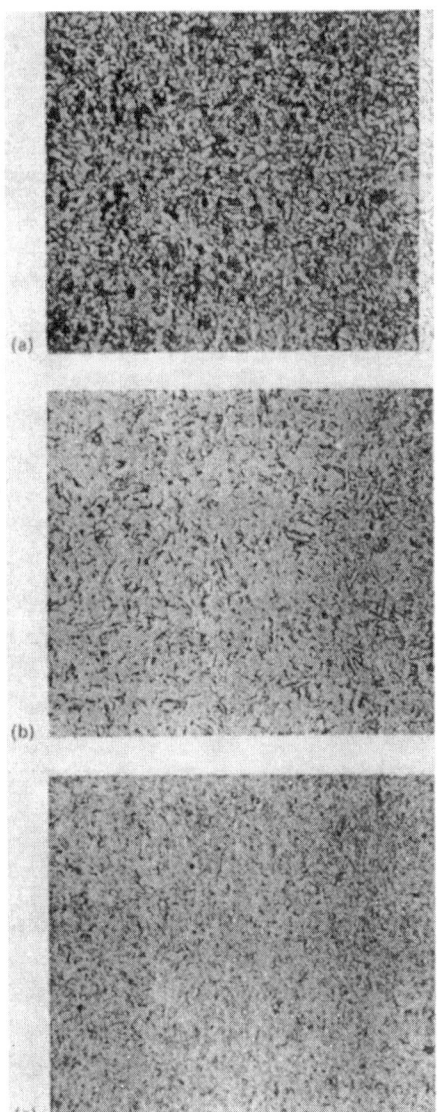

Figure 104 Microstructures of a steel having 1% C, 1.5% Si, 0.8% Mn, and 1% Cr, hardened from hardening temperatures between 800 and 850°C. Dimensions of test pieces: 30 mm diameter × 100 mm. Magnification 400×. (a) Hardening temperature 800°C, hardness 55 HRC. (b) Hardening temperature 825°C, hardness 61.5 HRC. (c) Hardening temperature 850°C, hardness 66 HRC. (From Ref. 14.)

4. It increases the amount of retained austenite after quenching due to stabilization of the austenite, which at higher temperatures is more saturated with carbon from dissolved carbides.

Heat-up and soak time at austenitizing temperature is a very important parameter for bulk heat treatment because it not only determines the furnace productivity and economy (consumption of energy) but may also affect the properties of the treated workpieces. Until recently

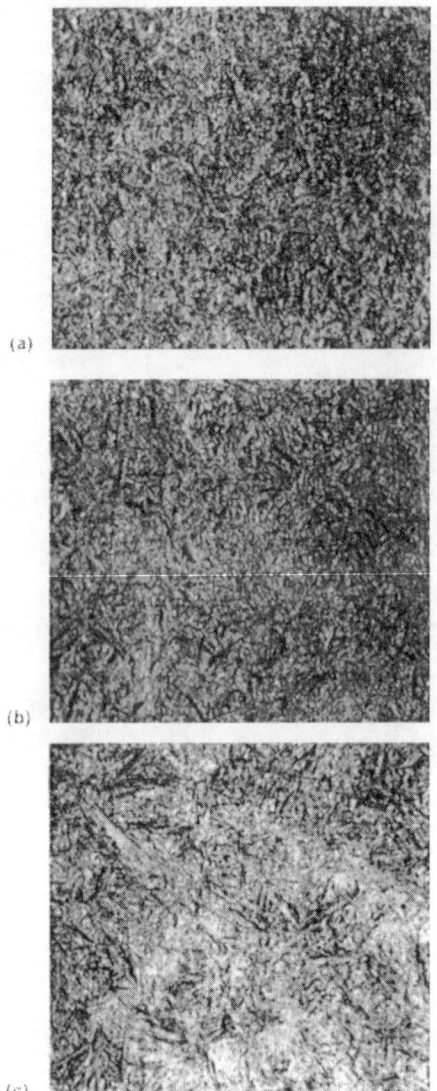

(a)

(b)

(c)

Figure 105 Microstructures of steel having 1% C, 1.5% Si, 0.8% Mn, and 1% Cr, hardened from hardening temperatures between 870 and 970°C. Dimensions of test pieces: 30 mm diameter × 100 mm. Magnification 400×. (a) Hardening temperature 870°C, hardness 62.5 HRC, retained austenite 12%. (b) Hardening temperature 920°C, hardness 62 HRC, retained austenite 20%. (c) Hardening temperature 970°C, hardness 61 HRC, retained austenite 28%. (From Ref. 14.)

there was no reliable, objective method for accurately predicting heat-up and soak times for heat treatment cycles that took into account all workpiece characteristics, variations in furnace design, and load arrangement. Current determinations of heat up and soak time are based on either a very conservative and general rule (e.g., 1 h per inch of cross section) or some empirical method, the results of which [20] differ substantially.

By heat-up and soak time we mean the time it takes for the heated workpiece to go from starting (room) temperature to the preset temperature in its core. The many factors that influence heat-up and soak time are diagrammed in Figure 107.

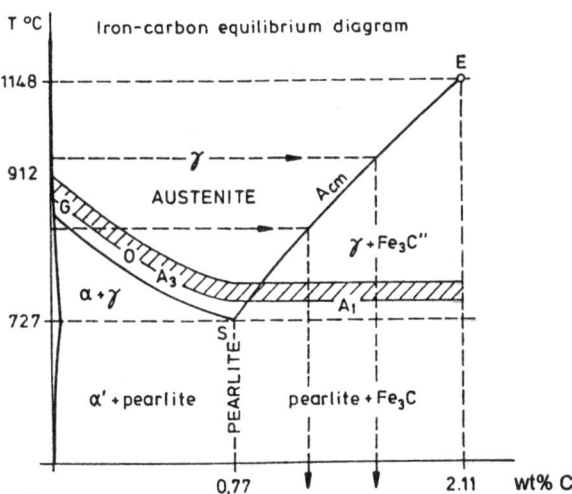

T °C | Iron-carbon equilibrium diagram

Figure 106 Optimum hardening temperature range for unalloyed steels, depending on the carbon content.

On the basis of experiments with 26 specimens (cylinders, round plates, and rings of various dimensions) made of unalloyed and low-alloy structural steels, Jost et al. [20] found from core temperature measurements that the heat-up and soak time depends substantially on the geometry of the heated workpiece and its mass. They found the heat-up and soak time to be directly proportional to the mass/surface area (m/A, kg/m^2) ratio, as shown in Figure 108. By regression analysis for their conditions (the specimens were heated in an electrically heated chamber furnace of 8 kW capacity and 240×240×400 mm working space, to the hardening temperature, 870°C (1598°F), they found that the heat-up and soak time (t) can be calculated using the equation

$$t = 0.42 (m/A) - 3.7 \tag{41}$$

The regression coefficients 0.42 and 3.7 are, of course, valid for their experimental conditions only. Comparison with their experimentally obtained results (see the points in Figure

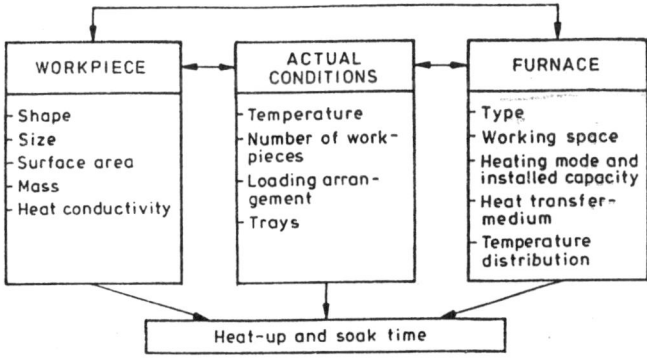

Figure 107 The main factors that influence the heat-up and soak time. (From Ref. 20.)

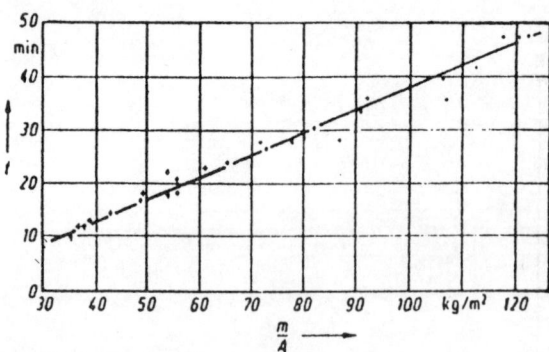

Figure 108 Dependence of the heat-up and soak time on the mass/surface area ratio (m/A). (From Ref. 20.)

108) showed a standard deviation of $s^2 = 1.4$ min^2, or $s = \pm 1.2$ min, indicating that this way of predicting heat-up and soak time in specific circumstances may be quite precise.

The Jost et al. [20] approach may be used generally for prediction of heat-up and soak times according to the general expression

$$t = a(m/A) + b \qquad (42)$$

provided that for a given situation the straight line of regression and relevant values of the regression coefficients a and b are fixed by means of some preliminary experiments. It should be stressed, however, that the described results of this investigation are valid for single workpieces only.

In another investigation [21] a method enabling heat treaters to accurately determine the heat-up and soak times for different loads treated in batch-type indirect fired furnaces was developed. To develop the method, a statistical and experimental investigation of load temperature conditions was performed. A computer-aided mathematical model of heat and mass transfer throughout the furnace and load was developed. The computer model accurately predicts the suitable heat-up and soak times for various types of furnace loads, load arrangements, workpiece shapes, and thermal properties. The treated loads were divided into several groups in terms of workpiece allocation and aerodynamic patterns of the furnace atmosphere, as shown in Figure 109.

The experiments with six different loads were conducted in indirectly fired batch furnaces, the working space of which was length 915–1680 mm, width 610–1420 mm, and height 610–1270 mm. The furnaces were equipped with four burners firing into the trident burner tubes located on the side walls, with a circulating fan located on top of the furnace as shown in Figure 110. The thermocouples were located in different parts of the load (measuring always the surface temperature of the workpieces)—on the top and bottom, in the core, at the corners, and on the surfaces facing radiant tubes—to determine temperature variations across the load.

As can be seen from Figure 110, the heat and mass transfer in the furnace and load are very complicated and are characterized by nonlinear three-dimensional radiation and convection and by nonlinear heat conduction within the workpieces. In this case the mathematical model to describe the heat and mass exchange is a system of integral and differential nonlinear equations. The input parameters to the computer program were as follows:

Geometrical data of the furnace and load: Furnace working space dimensions, radiant tube diameter and layout in the furnace, dimensions of the baskets, number of trays in the basket, workpiece characteristic size

(a)   Monolayer, Horizontally Oriented, Ordered Loads
        Packed   Spaced

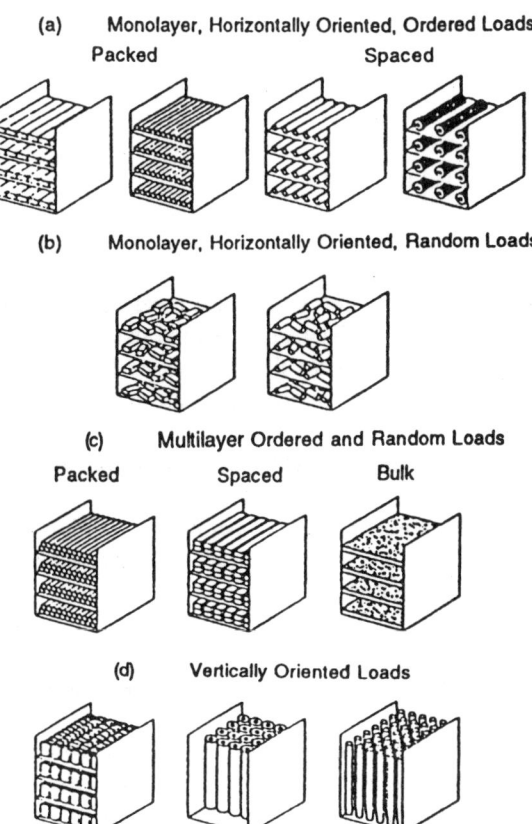

(b)   Monolayer, Horizontally Oriented, Random Loads

(c)   Multilayer Ordered and Random Loads
    Packed   Spaced   Bulk

(d)   Vertically Oriented Loads

Figure 109   Load characterization. (From Ref. 21.) (a) Monolayer, horizontally oriented, ordered loads. (b) Monolayer, horizontally oriented, random loads.(c) Multilayer ordered and random loads. (d) Vertically oriented loads.

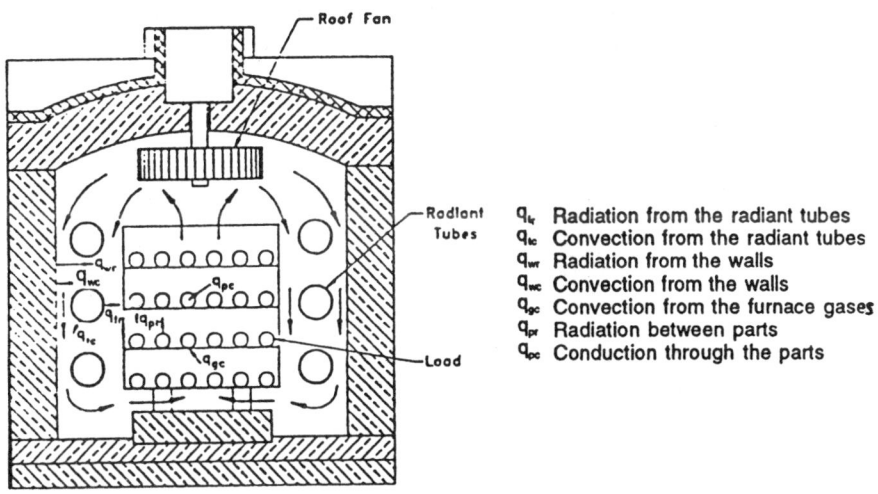

q_{tr}	Radiation from the radiant tubes
q_{tc}	Convection from the radiant tubes
q_{wr}	Radiation from the walls
q_{wc}	Convection from the walls
q_{gc}	Convection from the furnace gases
q_{pr}	Radiation between parts
q_{pc}	Conduction through the parts

Figure 110   Heat transfer in the used furnace and load. (From Ref. 21.)

Type of load (according to load characterization, see Figure 109).

Type of steel (carbon, alloyed, high-alloy)

Load thermal properties

Load and furnace emissivities

Temperature conditions (initial furnace and load temperature)

Fan characteristic curve parameters

Composition of protective atmosphere

As an example, maximum and minimum steel part temperatures for a test (heating of shafts) together with the calculated data are shown in Figure 111. The experimental data show that the temperature curve of the load thermocouple usually reaches the set furnace temperature well within the soak time requirements. The experimentally determined soak time is seen to be considerably shorter than the soak time defined by the heat treater. It was found that the discrepancy between soak times determined from the test data and calculations does not exceed 8%, which is acceptable for workshop practice.

The developed computer model was used for simulation of temperature conditions for different load configurations, and a generalized formula and set of graphs were developed. The generalized equation for the soak time determination is

$$t_s = t_{sb}k \tag{43}$$

where t_s is the calculated soak time, min; t_{sb} is soak time for baseline temperature conditions, min; and k is a correction factor for the type of steel.

The basic soak time (t_{sb}) is obtained from graphs derived from the computer simulation. Such a graph for packed loads is shown in Figure 112. Other load shapes and configurations require different graphs. The correction factor k depends on the type of steel. The generalized equation (Equation 43) for the heat-up and soak time determination was set into a user-friendly computer package that incorporates charts for the calculation. This resulted in a straightforward way of determining the soak time without the use of charts while allowing for a quick and accurate soak time calculation.

B. Quenching Intensity Measurement and Evaluation Based on Heat Flux Density

In designing the method for practical measurement, recording, and evaluation of the quenching/cooling intensity in workshop conditions, in contrast to the Grossmann H value concept,

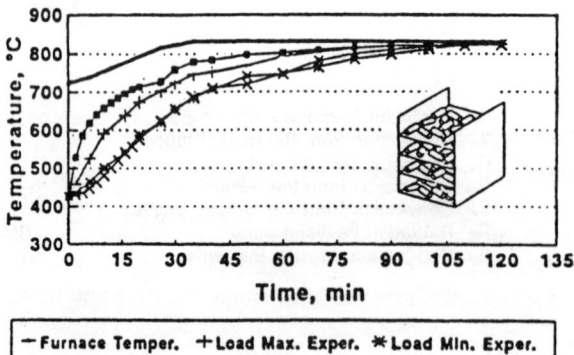

Figure 111   Computer simulation for heating of shafts. (From Ref. 21.)

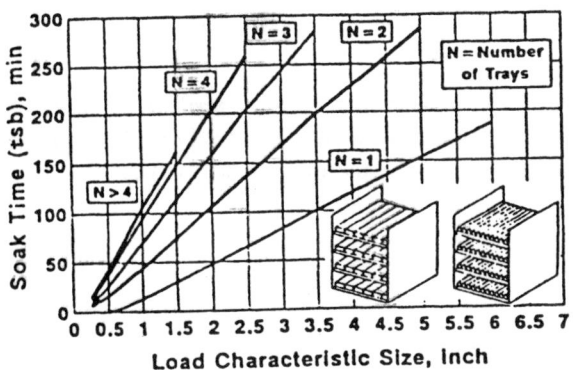

Figure 112 Thermal soak time for a packed load. (From Ref. 21.)

which expresses quenching intensity by a single number, the main idea of Liščić was to express the quenching intensity by continuous change of relevant thermodynamic functions during the whole quenching process. Instead of recording only one cooling curve (as in laboratory-designed tests) in the center of a small (usually 1/2 in.) cylindrical specimen, the heat flux density at the surface of a standard-size probe becomes the main feature in measuring, recording, and evaluating the quenching intensity.

The first substantial difference between using the small laboratory specimen and using the probe applied in the method described below is that when quenching, for example, in an oil quenching bath, because of its small mass and small heat capacity the former will cool down in 15–30 s, whereas the latter will take 500–600 s to cool to the bath temperature, allowing the entire quenching process of real components to be followed.

This workshop-designed method is applicable to

1. All kinds of quenchants (water and brine, oils, polymer solutions, salt baths, fluid beds, circulated gases)
2. A variety of quenching conditions (different bath temperatures, different agitation rates, different fluid pressures)
3. All quenching techniques (direct immersion quenching, interrupted quenching, martempering, austempering, spray quenching)

The method is sufficiently sensitive to reflect changes in each of the important quenching parameters (specific character of the quenchant, its temperature, and mode and degree of agitation).

This method

Enables a real comparison of the quenching intensity among different quenchants, different quenching conditions, and different quenching techniques.

Provides an unambiguous conclusion as to which of two quenchants will give a greater depth of hardening (in the case of the same workpiece and same steel grade) and enables the exact calculation of cooling curves for an arbitrary point on a round bar cross section of a specified diameter, to predict the resulting microstructure and hardness. (An exception is the case of delayed quenching, where the cooling rate is discontinuously changed; for an explanation see Reference 23.)

Furnishes information about thermal stresses and possible superposition of structural transformation stresses that will occur during a quenching process.

Provides the basis for automatic control of the quenching intensity during the quenching
process.

The method itself, known in the literature as the *temperature gradient method*, is based on
the known physical rule that the heat flux at the surface of a body is directly proportional to
the temperature gradient at the surface multiplied by the thermal conductivity of the body
material:

$$q = \lambda \frac{\delta T}{\delta x} \tag{44}$$

where q is the heat flux density, W/m^2 (i.e., quantity of heat transferred through a surface
unit perpendicular to the surface per unit time); λ is thermal conductivity of the body material, W/(m·K), and $\delta T / \delta x$ is the temperature gradient at the probe surface perpendicular to
the surface, K/m.

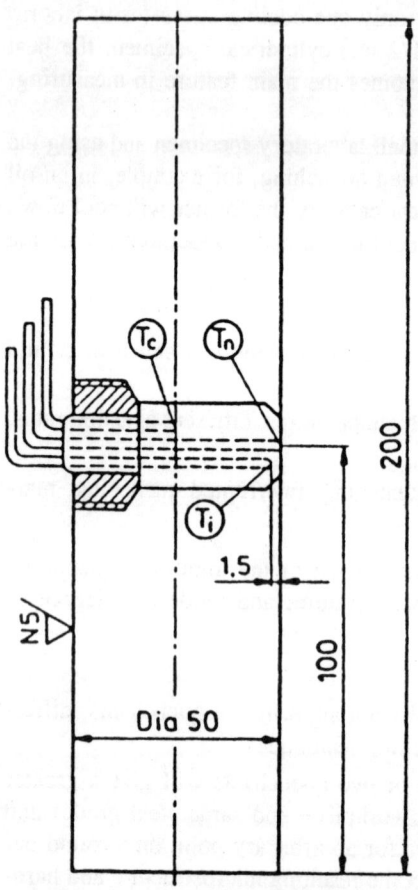

All dimensions in mm

Figure 113 The Liščić–NANMAC probe (made by the NANMAC Corp., Framingham Center, MA)
for measurement of the temperature gradient on the surface.

The essential feature of the method is a cylindrical probe [32] of 50 mm diameter × 200 mm, instrumented with three thermocouples placed along the same radius at the half-length cross section as shown in Figure 113. As can be seen, the thermocouple at the surface reproducibly measures the real surface temperature of the probe (T_n), which is important to register all the phenomena that are taking place on the surface during quenching. The intermediate thermocouple (T_i) measures the temperature at a point 1.5 mm below the surface. The readings of T_n and T_i enable the heat treater to easily calculate the temperature gradient near the surface of the probe at each moment of cooling. The central thermocouple (T_c), placed at the center of the cross section, indicates how long it takes to extract heat from the core and provides at every moment the temperature difference between the surface and the core, which is essential for the calculation of thermal stresses.

Specific features of probe are the following.

1. The response time of the surface thermocouple is 10^{-5} s; the fastest transient temperature changes can be recorded.
2. The intermediate thermocouple can be positioned with an accuracy of ± 0.025 mm.
3. The surface condition of the probe can be maintained by polishing the sensing tip before each measurement (self-renewable thermocouple).
4. The body of the probe, made of an austenitic stainless steel, does not change in structure during the heating/quenching process, nor does it evolve or absorb heat because of structural transformation.
5. The size of the probe and its mass ensure sufficient heat capacity and symmetrical radial heat flow in the cross-sectional plane where the thermocouples are located.
6. The heat transfer coefficient during the boiling stage, which, according to Kobasko [22], depends on bar diameter, becomes independent of diameter for diameters > 50 mm.

When a test of the quenching intensity is performed, the probe is heated to 850°C (1562°F) in a suitable furnace and transferred quickly to the quenching bath and immersed. The probe is connected to the temperature acquisition unit and a PC. Adequate software enables the storage of the temperature–time data for all three thermocouples and the calculation and graphical display of relevant functions. The software package consists of three modules:

Module I.   TEMP-GRAD (temperature gradient method)

Module II.   HEAT-TRANSF (calculation of heat transfer coefficient and cooling curves)

Module III. CCT-DIAGR (prediction of microstructure and hardness after quenching).

As an example let us compare two different quenching cases:

Case A.   Quenching in a mineral oil bath at 20°C (68°F) without agitation (Figures 114a–114f)

Case B.   Quenching in a 25% PAG polymer solution at 40°C (104°F) and 0.8 m/s agitation rate (Figures 115a–115f).

By comparing Figures 114b and 115b it is clear that case B involves delayed quenching with a discontinuous change in cooling rate, because in case A the time when maximum heat flux density occurs (t_{qmax}) is 15 s whereas in case B it is 72 s.

In case A (oil quenching), by 20 s after immersion (see Figure 114e) 34 MJ/m² of heat has already been extracted, and by 50 s, 50 MJ/m², whereas in case B (high concentration polymer solution quenching; see Figure 115e) by 20 s only 5 MJ/m² and by 50 s only 20 MJ/

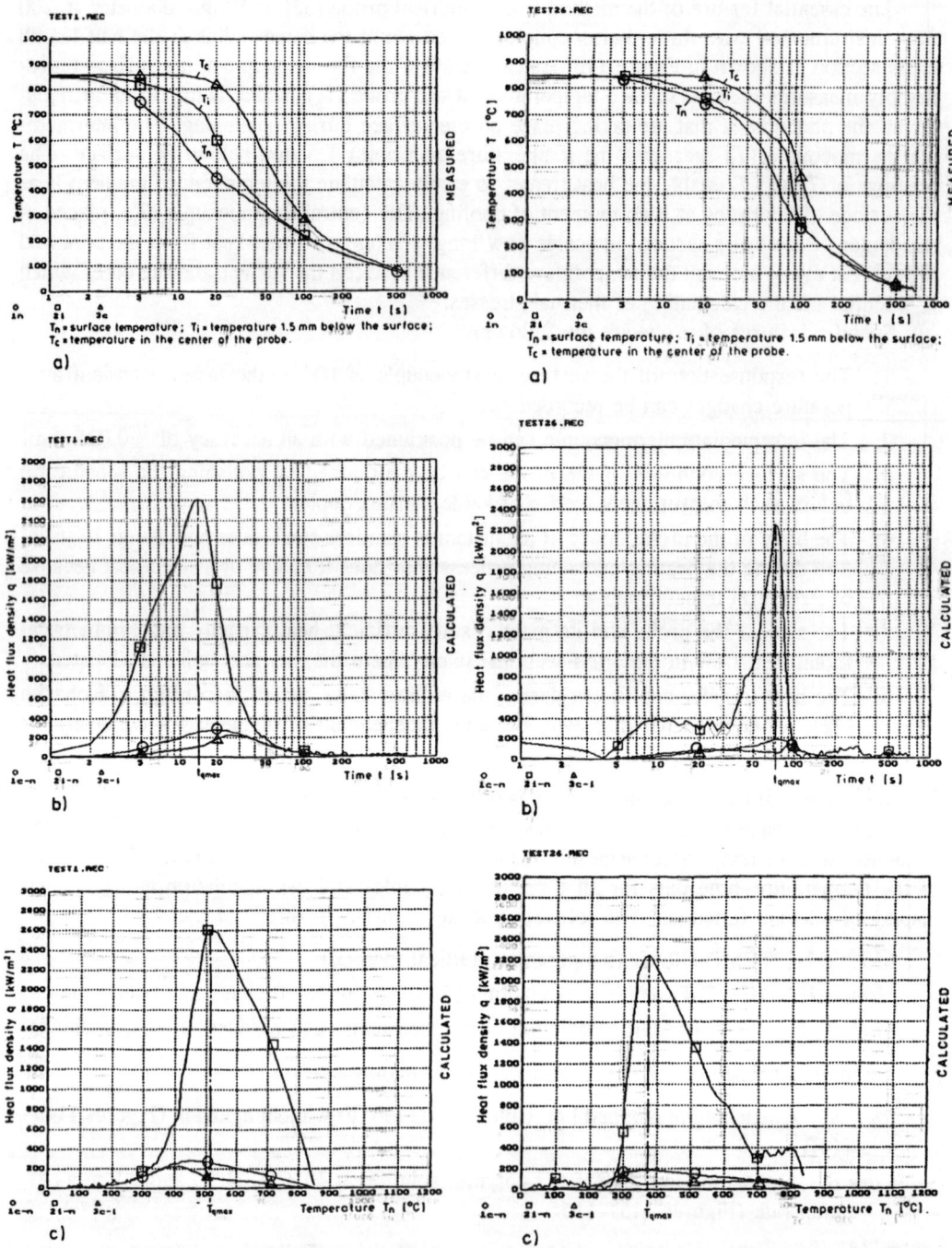

Figure 114 Graphical display from module I, TEMP-GRAD, when quenching the Liščić-NANMAC probe in a 20°C mineral oil bath without agitation. (a) Measured and recorded temperature vs. time, $T = f(t)$. (b) Calculated heat flux density vs. time, $q = f(t)$. (c) Calculated heat flux density

Figure 115 Graphical display from module I, TEMP-GRAD, when quenching the Liščić-NANMAC probe in a PAG polymer solution of 25% concentration, 40°C bath temperature, and 0.8 m/s agitation rate. (a) Measured and recorded temperature vs. time data, $T = f(t)$. (b) Calculated heat flux

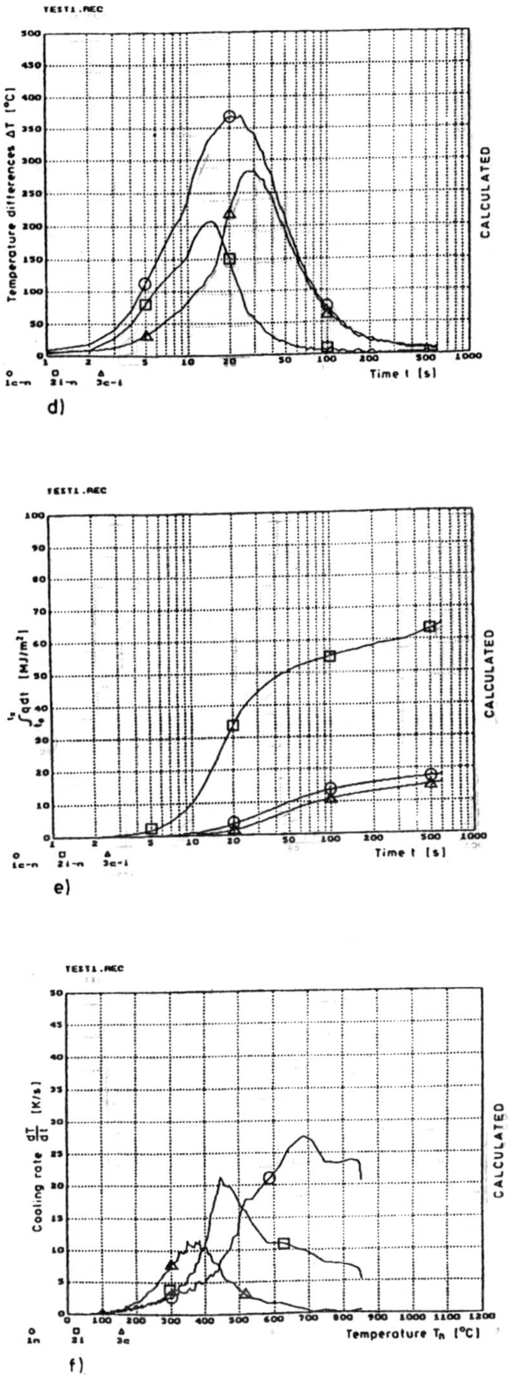

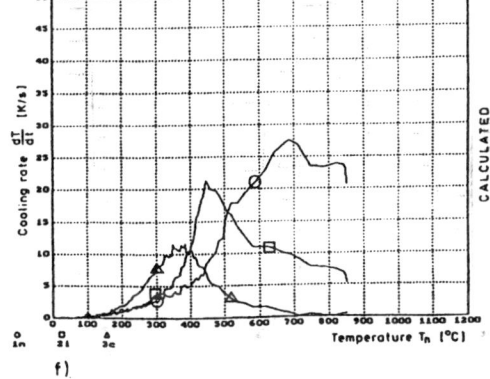

Figure 114 (Continued)
vs. surface temperature, $q = f(T_n)$. (d) Calculated temperature differences vs. time, $\Delta T = f(t)$. (e) Calculated integral $\int q\,dt$ = heat extracted vs. time. (f) Calculated cooling rates vs. surface temperature $dT/dt = f(T_n)$.

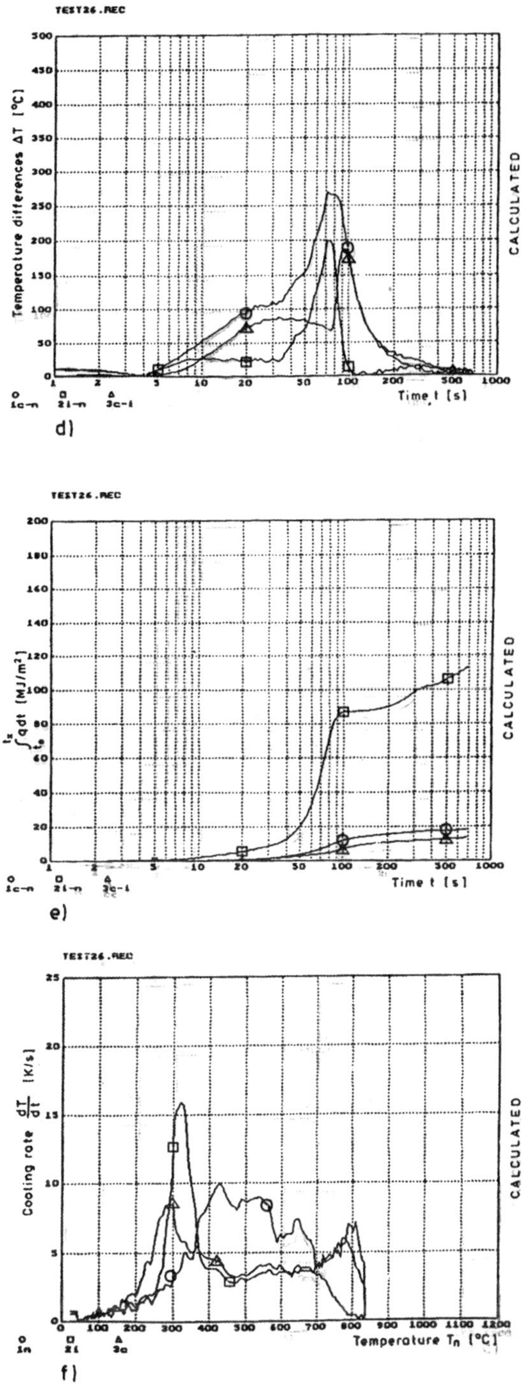

Figure 115 (Continued)
density vs. time, $q = f(t)$. (c) Calculated heat flux density vs. surface temperature, $q = f(T_n)$. (d) Calculated temperature differences vs. time, $\Delta T = f(t)$. (e) Calculated integral $\int q\,dt$ = heat extracted vs. time. (f) Calculated cooling rates vs. surface temperature, $dT/dt = f(T_n)$.

m^2 of heat was extracted. However, immediately after that in the period between 50 and 100 s, in case A the extracted amount of heat has increased from 50 to only 55 MJ/m^2, whereas in case B it has increased from 20 to 86 MJ/m^2. Both of the calculated integral ($\int q dt$) curves, designated with the open square symbols in Figures 114e and 115e, have been calculated as the area below the heat flux density vs. time curves, designated similarly in Figures 114b and 115b. That is, they represent the heat extracted only through the surface region between the point 1.5 mm below the surface and the surface itself.

Comparing the time required to decrease the heat flux density from its maximum to a low value of, e.g., 100 kW/m^2 (see Figures 114b and 115b), one can see that in case A 45 s is necessary, whereas in case B only 28 s is necessary. This analysis certifies that case B (quenching in PAG polymer solution of high concentration) is a quenching process with delayed burst of the thick polymer film.

Discontinuous change in cooling rate is inherent to this quenching regime. In this respect it is interesting to analyze the cooling rate vs. surface temperature diagrams of Figures 114f and 115f. While in oil quenching (case A), the cooling rate at the surface of the probe (O) has a higher maximum than the cooling rates at 1.5 mm below the surface (□) and at the center (Δ), in case B the maximum cooling rate at 1.5 mm below surface (during a certain short period between 350 and 300°C (662 and 572°F) surface temperature) is higher than the maximum cooling rate at the surface itself! This can also be seen in Figure 115a, which shows that the slope of the cooling curve T_i between 500 and 300°C (932 and 842°F) is greater than the slope of the cooling curve for the very surface (T_n). This is another experimental proof that in delayed quenching cooling rates below the workpiece surface can be higher than at the surface itself.

Another analysis, with respect to thermal stresses during quenching (on which the residual stresses and possible distortion depend), is possible by comparing Figures 114d and 115d. This comparison shows that quenching in a PAG polymer solution of high concentration (case B), compared to oil quenching (case A), resulted in 27% lower maximum temperature difference (read thermal stress) between the center and surface of the probe (O) or 36% lower maximum temperature difference between the center and the point 1.5 mm below the surface (Δ), contributing to lower distortion than in oil quenching. Whereas with oil quenching the maximum temperature difference between the center and the point 1.5 mm below the surface (Δ) is higher than the maximum temperature difference between the point 1.5 mm below the surface and the surface itself (□), in the case of delayed quenching (case B), the maximum temperature difference between the point 1.5 mm below the surface and the surface itself (□ in Figure 115d) is slightly higher than the maximum temperature difference between the center and the point 1.5 mm below the surface (Δ), which is reached about 20 s later. This also shows an abrupt heat extraction when the polymer film bursts.

On the other hand, Figure 114d shows that in oil quenching the maximum temperature difference between the center and surface (O) occurs 20 s after immersion, when the surface temperature is 450° (842°F) (see Figure 114a), i.e., above the temperature of the M_s point. In PAG polymer solution quenching (Figure 115d), the maximum temperature difference between the center and the surface (O) occurs much later, when the surface temperature has already fallen to 360°C (680°F) (see Figure 115a). In this respect, dealing with steels that have a high M_s temperature, water-based polymer solutions always run a higher risk of overlapping thermal stresses with those created due to austenite-to-martensite transformation.

The probability of crack formation can be seen at a glance by comparing the surface temperature of the probe at the moment the maximum heat flux density occurs (T_{qmax}). As seen in Figure 114c, T_{qmax} is 515°C (959°F) for oil quenching (case A), while for water-based polymer solution (case B), T_{qmax} is 380°C (716°F) (see Figure 115c). The lower the value of T_{qmax}, the higher is the risk of crack formation, especially with steel grades having high M_S temperature.

When direct immersion quenching is involved with continuous cooling (not delayed quenching with discontinuous cooling), the depth of hardening, when comparing two quenching processes, is determined as follows: The larger the values of q_{max} and $\int q dt$ and the shorter the time t_{qmax}, the greater will be the depth of hardening.

Module II of the software package, HEAT-TRANSF, makes it possible (based on the input of measured surface temperatures and calculated heat flux density on the very surface) to calculate (by a numerically solved method of control volumes) and graphically present

1. The heat transfer coefficient between the probe's surface and the surrounding fluid vs. time, $\alpha = f(t)$ (Figure 116a)
2. The heat transfer coefficient between the probe's surface and the surrounding fluid vs. surface temperature, $\alpha = f(T_n)$ (Figure 116b)

Using the calculated values of α, the software program enables the calculation of cooling curves at any arbitrary point of the round bar cross section of different diameters, as shown in Figures 117a and 117b.

The Module III of the software package, CCT-DIAGR, is used to predict the resulting microstructure and hardness after quenching of round bar cross sections of different diameters. This module contains an open data file of CCT (continuous cooling transformation) diagrams in which the user can store up to 100 CCT diagrams of his own choice. The program enables the user to superimpose every calculated cooling curve on the CCT diagram of the steel in question. From the superimposed cooling curves shown on the computer screen, the user can read off the percentage of structural phases transformed and the hardness value at the selected point after hardening as shown by Figure 118.

If for a round bar cross section of the chosen diameter the cooling curves are calculated at three or five characteristic points [surface, $(3/4)R$, $(1/2)R$, $(1/4)R$, center], using the HEAT-TRANSF module, the CCT-DIAGR module enables the user to read off the hardness values after quenching at those points and to obtain the hardness distribution curve displayed graphically on the computer screen. In the case of delayed quenching with discontinuous change of

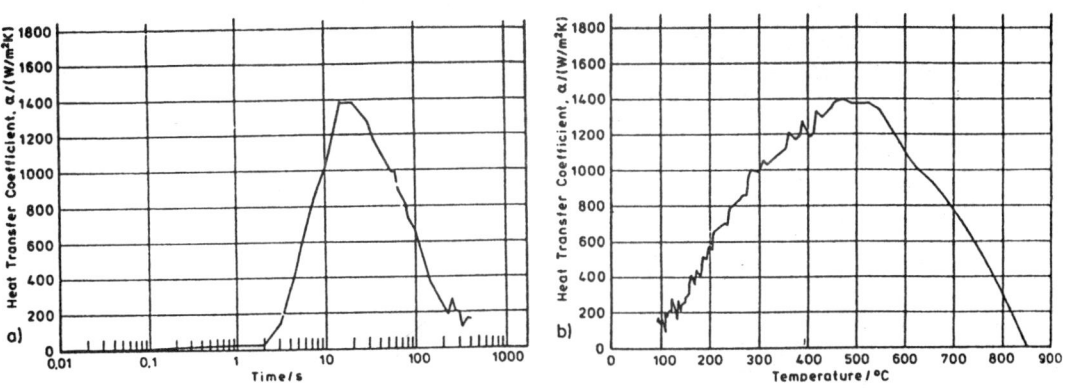

Figure 116 Heat transfer coefficient (a) vs. time and (b) vs. surface temperature when quenching the Liščić–NANMAC probe (50 mm diameter × 200 mm) in a 20°C mineral oil bath without agitation.

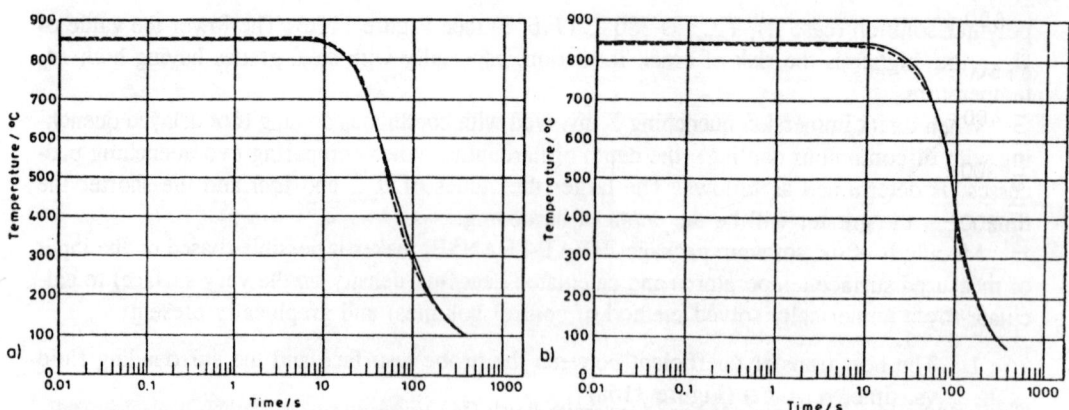

Figure 117 Comparison of measured (---) and calculated (—) cooling curves for the center of a 50 mm diameter bar quenched in (a) mineral oil at 20°C, without agitation; (b) 25% PAG polymer solution, 40°C bath temperature, and 0.8 m/s agitation rate.

cooling rate, the prediction of structural transformations and hardness values after quenching from an ordinary CCT diagram is not correct because the incubation time consumed (at any point of the cross section) before the cooling rate abruptly changes has not been taken into account.

For a detailed explanation see Shimizu and Tamura [11].

C. Retained Austenite and Cryogenic Treatment

The martensite start (M_s) and martensite finish (M_f) temperatures for unalloyed steels depend on their carbon content, as shown in Figure 119. As can be seen from this diagram, when steels of more than 0.65% C are quenched the austenite-to-martensite transformation does not

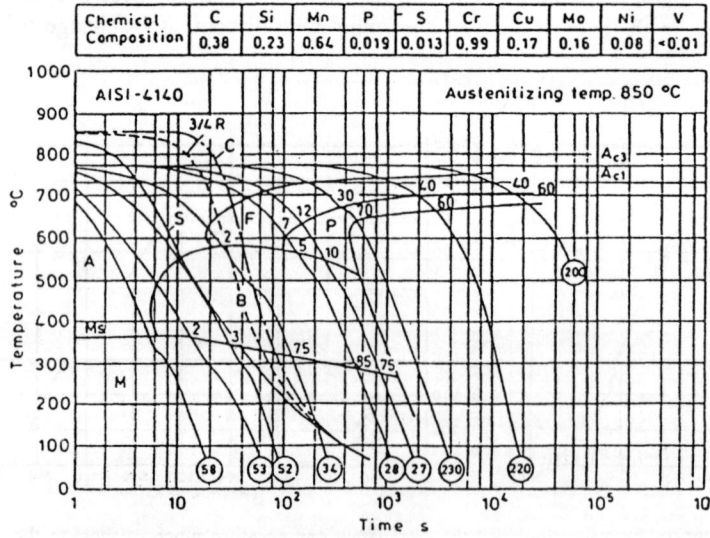

Figure 118 CCT diagram of AISI 4140 steel with superimposed calculated cooling curves for surface (S), three-quarter radius (3/4 R) and center (C) of a round bar of 50 mm diameter.

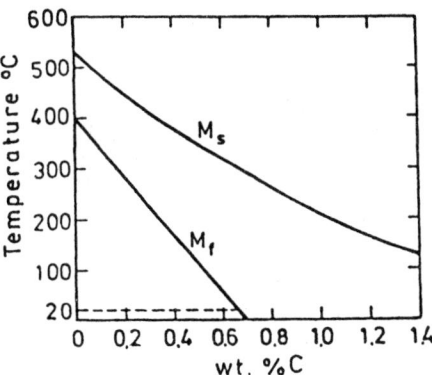

Figure 119 Martensite start (M_s) and martensite finish (M_f) temperatures vs. carbon content in unalloyed steels.

end at room temperature [20°C (68°F)] but at some lower temperature, even at temperatures much lower than 0°C (32°F). Consequently, after these steels are quenched to room temperature, a portion of austenite will remain untransformed; this is referred to as *retained austenite*. The greater the amount of carbon in the steel, the greater the amount of retained austenite after quenching, as shown in Figure 120c.

Retained austenite, being a softer constituent of the structure, decreases the steel's hardness after quenching. If present in amounts of more than 10%, a substantial decrease in the hardness of the quenched steel may result (see curve *a* of Figure 120a).

When quenching hypereutectoid steels from the usual hardening temperature (Figure 120b) i.e., from the $\gamma + Fe_3C$ region, the same hardness will result independently of carbon content (curve *b* in Figure 120a), because the hardness of martensite depends only on the carbon dissolved in austenite (γ), which further depends (according to the solubility limit, line S–E) on the hardening temperature. The structure of hardened hypereutectoid steels therefore consists of martensite + Fe_3C + retained austenite.

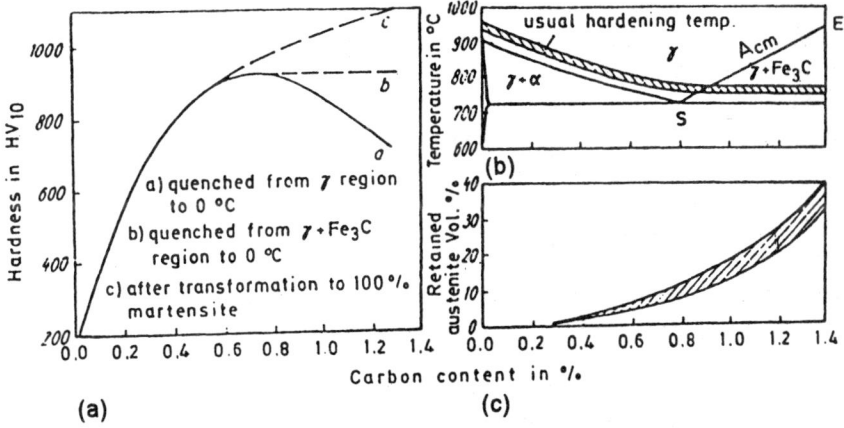

(a)   (c)

Figure 120 (a) Hardness of carbon (unalloyed) steels depending on carbon content and austenitizing temperature. (b) The range of usual hardening temperatures. (c) Volume percent of retained austenite. (From Ref. 2.)

When quenching hypereutectoid steels from the region of pure austenite (γ), i.e., from above the A_{cm} temperature (which is not usual), the structure after hardening consists only of martensite and retained austenite, and the hardness decreases with carbon content as shown by curve a of Figure 120a. If the retained austenite is transformed (e.g., by subsequent cryogenic treatment) to 100% martensite, the hardness would follow curve c in Figure 120a.

When after hardening the steel is kept at room temperature for some time or is heated to the temperature range corresponding to the first tempering stage, the retained austenite is stabilized, which implies that it has become more difficult to transform when subjected to cryogenic treatment. The stabilization of retained austenite is assumed to be due to the dissolution, at the arrest temperature, of the martensite nuclei formed during cooling from the austenitizing temperature.

When martempering is performed, i.e., the quenching process is interrupted somewhere around the M_s temperature, a similar stabilization of retained austenite occurs. When the cooling to room temperature is then continued, the same effect, in principle, results as that obtained by the subzero cryogenic treatment with respect to the transformation of retained austenite to martensite.

The initial amount of retained austenite (after quenching) is dependent to a very large extent on the austenitizing (hardening) temperature. The higher the hardening temperature, the greater the amount of retained austenite, but greater amounts of retained austenite may also be transformed to martensite by subzero cryogenic treatment for the same stabilizing temperature and same stabilizing time as shown in Figure 121. The stabilizing effect increases as the stabilizing temperature and time increase. After quenching from, say, 840°C (1544°F) (Figure 121b), there is 18% retained austenite. If the subzero treatment is carried out within 5 min after the temperature of the steel has reached 20°C (68°F), about 70% of the retained austenite will be transformed. If 40 min is allowed to pass before the subzero treatment, 60% will be transformed, and after 50 h holding at 20°C (68°F) only 30% of the retained austenite will respond to the subzero cryogenic treatment.

If the steel is held after quenching at a higher temperature, e.g., at 120°C (248°F), for only 10 min before subzero treatment, only 30% of the retained austenite will be transformed. In order to transform the greatest possible amount of retained austenite, the subzero cryogenic treatment should be performed immediately after quenching prior to tempering.

The question of whether the retained austenite in the structure is always detrimental or whether in some cases it can be advantageous has still not been answered unambiguously. When dealing with carburized and case-hardened components, because of the high carbon content in the case, the problem of retained austenite is always a real one, especially with steels containing nickel. The higher alloy nickel steels, such as types AISI 4620, 4820, and 9310, are particularly likely to have retained austenite in their microstructures after heat treatment because nickel acts as an austenite stabilizer.

Tests performed by M. Shea (cited in [24]), showed marked improvement in tensile bending strain values when retained austenite was present in the 20–30% range for AISI 8620 and 4620 steels and up to 40% for AISI 3310 steel. This report indicated that the transformation of retained austenite in the range of >20% allows more plastic strain to be accommodated prior to crack initiation because the austenite deforms and subsequently transforms to martensite.

The graph in Figure 122, taken from work done by C. Razim, shows where large quantities of retained austenite (in the range of 50%) improve contact fatigue of carburized and case-hardened steels. In another publication [24], several grades of carburized and case-hardened steels were compared (both before and after subzero treatment), and a clear improvement in bend ductility is reported for those having retained austenite.

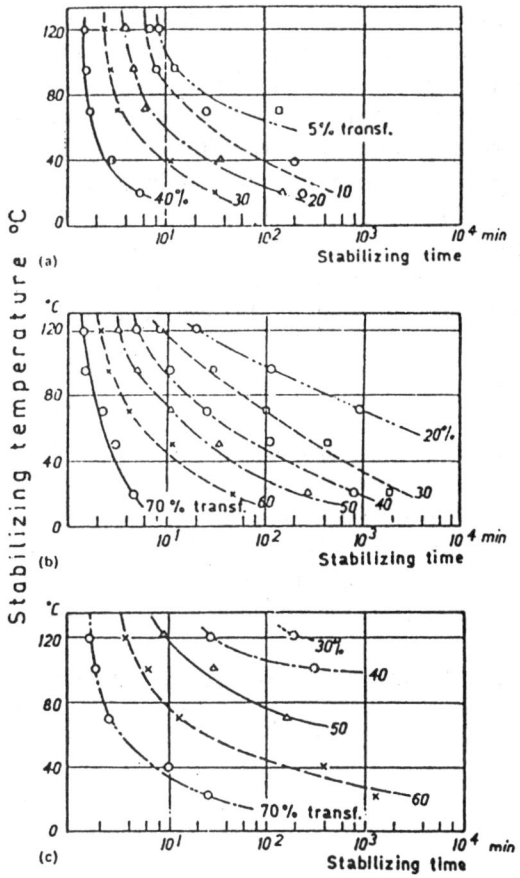

Figure 121 Influence of stabilizing temperature and time on the amount of retained austenite that transforms on being subzero treated at $-180°C$ for the ball bearing steel AISI 52100. (a) Austenitizing temperature 780°C; 9.4% retained austenite after quenching. (b) Austenitizing temperature 840°C; 18% retained austenite after quenching. (c) Austenitizing temperature 900°C; 27% retained austenite after quenching. (From Ref. 14.)

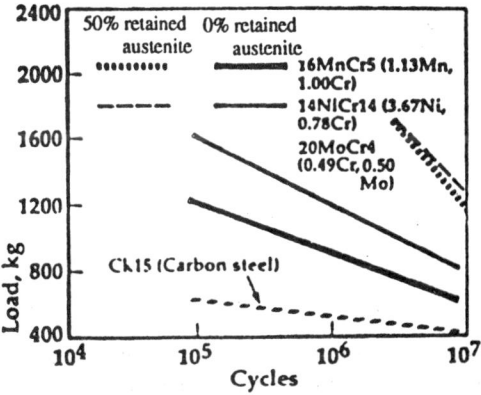

Figure 122 Improvement of contact fatigue of carburized and case-hardened steels containing 50% retained austenite, according to C. Razim. (From Ref. 24.)

As a result of the above-mentioned investigations, when dealing with carburized and case-hardened gears, an amount of 10–20%, and in some instances up to 25%, of retained austenite is not objectionable for most applications and may be beneficial. On the other hand, retained austenite can be detrimental, causing premature wear of sliding on the components' surface or of sliding and rolling of gear teeth, because it is a softer constituent of the microstructure. The presence of retained austenite in cases of carburized and case-hardened components that are to be subsequently ground is definitely detrimental because under certain grinding conditions it causes severe grinding burns and cracking. The susceptibility of carburized and case-hardened components to cracking during grinding becomes greater the greater the amount of retained austenite, this amount further depending on the steel grade and carburizing temperature as shown in Figure 123.

1. Transforming the Retained Austenite

When steels are tempered, retained austenite decomposes to bainite during the second tempering stage [230–280°C (446–536°F)]. For high-alloy chromium steels, hot-work steels, and high-speed steels, the range of decomposition of retained austenite is shifted toward higher temperatures. The product of decomposition, i.e., whether it is bainite or martensite, depends on the tempering temperature and time. Bainite formation occurs isothermally, i.e., at constant temperature during the tempering process, whereas martensite forms as the steel is cooling down from the tempering temperature.

Subzero cryogenic treatment may be applied to transform the retained austenite to martensite, substantially lowering its amount, sometimes to as little as about 1 vol %, which cannot be detected metallographicaly but only by X-ray diffraction.

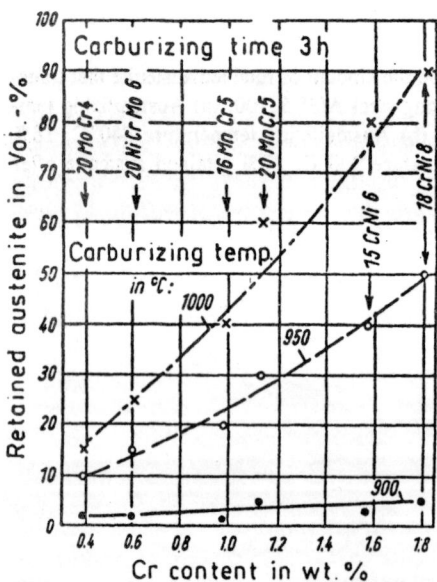

Figure 123 Influence of the Cr content of low-alloy steels for carburizing on the amount of retained austenite at carburizing. The amount of retained austenite was determined metallographically 0.05 mm below the specimen's surface. (From Ref. 2.)

Decreasing the amount of retained austenite achieves

1. An increase in hardness and consequently in wear resistance
2. More dimensional stability in the finished part (smaller change in dimensions due to structural volume change in use)
3. Less susceptibility to the development of cracks at grinding

Figure 124 shows a heat treatment cycle that includes subzero treatment. The most important parameters of the treatment are (1) the temperature below 0°C (32°F) that should be attained and (2) the cooling capacity of the equipment. In some cases, temperatures of –80 to –100°C (–112 to –148°F) are sufficient, but for other steels, especially high-alloy ones, lower temperatures of –140°C (–280°F) or even –180°C (–292°F) are necessary. Holding time at low temperature is unimportant, because the transformation of retained austenite to martensite does not depend on time, but only on the temperature to which the metal has been cooled. Only that portion of the retained austenite will be transformed to martensite that corresponds to the cooling temperature realized. Further transformation will take place only if the temperature is lowered further.

There are four methods using different types of equipment for the subzero treatment:

1. Evaporation of dry ice (CO_2 in the solid state).   This method is capable of reaching at most –75°C (–103°F) or –78°C (–108°F) and is used for small quantities and small mass of parts.
2. Circulating air that has been cooled in a heat exchanger.   This low-temperature cascade system (Figure 125) cools the parts put in a basket with air circulated by a fan. The air flows from top to bottom, extracting the heat of the parts, exiting through a grate at the bottom of the basket and flowing further through the heat exchanger, which is cooled by two or four compressors. Such metal-treating freezers have been built with a capacity to cool a mass of 270–680 kg of parts to –85°C (–121°F) in about 2 h.
3. Evaporation of liquid nitrogen.   For subzero treatment of relatively small quantities of parts down to –180°C (–292°F), equipment such as that shown in Figure 126 is used. The parts to be cooled are put in the working space 1, and the liquid nitrogen is in the container 4. Because of heat coming through the walls, the pressure in the container 4 increases. Using this pressure, when the valve 6 is opened, liquid nitro-

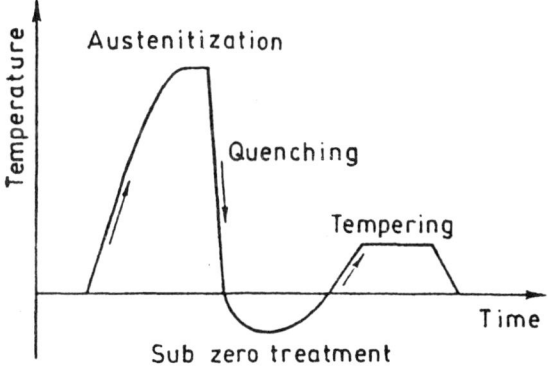

Figure 124   A heat treatment cycle including subzero treatment.

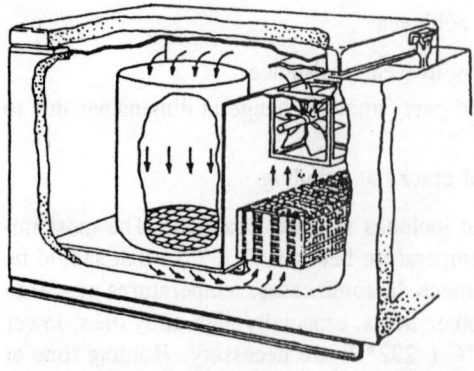

Figure 125   Low-temperature cascade system for subzero cooling by circulating air that has been cooled in a heat exchanger.

gen is injected into the working space, where it evaporates instantly. A fan *7* circulates the evaporated nitrogen through the working space, taking the heat out of parts and lowering their temperature. The amount of the injected liquid nitrogen and consequently the temperature of cooling can be controlled by adjusting the valve *6*. The overpressure that develops in the working space because of constant nitrogen evaporation is let out through the exhaust valve *8*. A temperature of –180°C (–292°F) can be reached in less than 10 min.

4. In a container connected to a cryogenerator.   This method enables subzero treatment of large quantities of parts to as low as –190°C (–310). The cryogenerator powered by an electric motor works on the principle of the Stirling cycle. By continuous circulation of air the working space with parts is gradually cooled to desired temperature. Figure 127 shows the cooling curve from room temperature to –180°C (–292°F) and the natural reheating curve from –180 to 0°C (–292 to 32°F) for the empty container of 100 dm³, connected to a cryogenerator. It can be seen that after 1 h of cooling a temperature of –120°C (–184°F) has already been reached, but an additional 1.5 h is necessary to reach –180°C (–292°F). The natural reheating from –180 to 0°C (–292 to 32°F), as shown, takes much longer (about 20 h).

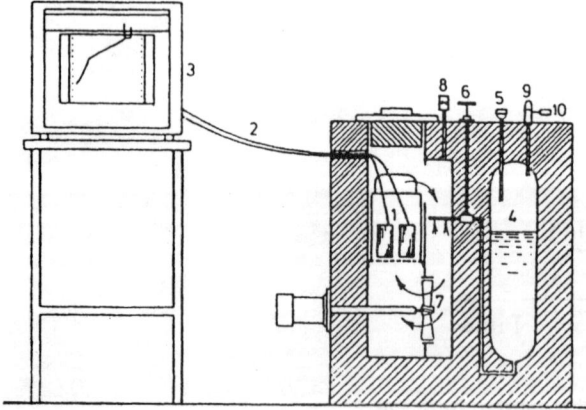

Figure 126   Subzero treatment equipment with evaporation of liquid nitrogen.

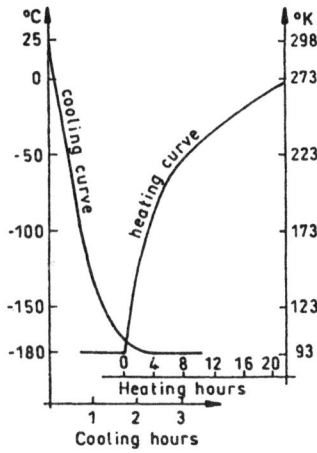

Figure 127 Cooling curve and natural heating back curve of an empty container connected to a cryogenerator.

IV. HARDENING AND TEMPERING OF STRUCTURAL STEELS

A. Mechanical Properties Required

A combined heat treatment process consisting of hardening plus tempering (to temperatures between 450 and 680°C (842 and 1256°F)) applied to structural steels (in German called *Vergütung*) is performed to achieve maximum toughness at a specified strength level. Toughness is a very important mechanical property, especially for components that must be able to withstand dynamic loading or impact. The aim of hardening and tempering structural steels will be better understood if one has a clear notion of the difference between toughness and ductility.

Ductility is the property denoting the deformability of a material and is measured in a tensile test as elongation (A in %) and reduction of area (Z in %). It is a one-dimensional property. *Toughness* of a material, however, is a two-dimensional property because it is an integral (or product) of strength and ductility, as schematically shown in Figure 128. Steels of the same ductility but different strength levels can differ in toughness. As Figure 128 shows, a normalized steel (N) having the same ductility as a hardened and tempered steel (V) will have lower toughness because of its lower strength level. Toughness is measured in separate tests as impact toughness (a_k, J/cm^2) or as fracture toughness (K_{IC}, N/mm$^{3/2}$).

The lower the ductility of a material, the more brittle it is. Total brittleness accordingly denotes zero ductility of the material.

The aim of the hardening and tempering process can also be explained by means of the stress–strain diagram schematically shown in Figure 129. As hardened, a steel has high yield strength but low ductility, and a small area below the stress–strain curve (curve 2) indicates low toughness. As-hardened and tempered (curve 3) steel has higher yield strength than in its normalized condition but also much higher ductility than in its hardened condition. The greatest area below the stress–strain curve indicates a substantial increase in toughness compared to either normalized or hardened conditions.

For a certain steel grade, the relation between mechanical properties and the tempering temperature can be read off from a diagram as shown in Figure 130 for the steel DIN 20 CrNi Mo 2 (0.15% C, 0.20% Si, 0.88% Mn, 0.53% Cr, 0.50% Mo, 0.86% Ni). It can be clearly seen from the lower part of this diagram how the impact toughness increases when the steel

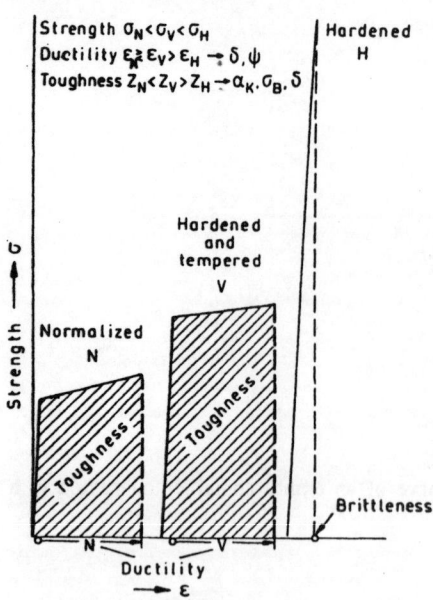

Figure 128   Schematic presentation of ductility, toughness, and brittleness. (From Ref. 25.)

is tempered to a temperature above 550°C (1022°F). Such diagrams enable precise optimization of the strength level and toughness by selection of the proper tempering temperature.

The properties of a hardened and tempered steel correlate to a high degree with the microstructure after hardening and tempering. Maximum toughness values are obtained when tempering a structure that after quenching consists of fine-grained martensite (having a grain size of ASTM ≥ 6 (see Figure 131). How different microstructures after different heat treatment processes influence the impact toughness of 3.5% Ni steel at low temperatures is shown in Figure 132. The maximum toughness is achieved after tempering water-quenched specimens (tempered martensite). When testing the impact toughness at low temperatures, the so-

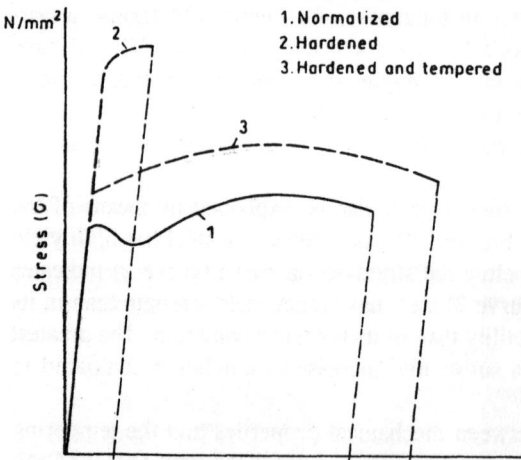

Figure 129   Stress-strain diagram of a steel after different heat treatments. *1*, Normalized; *2*, hardened; *3*, hardened and tempered.

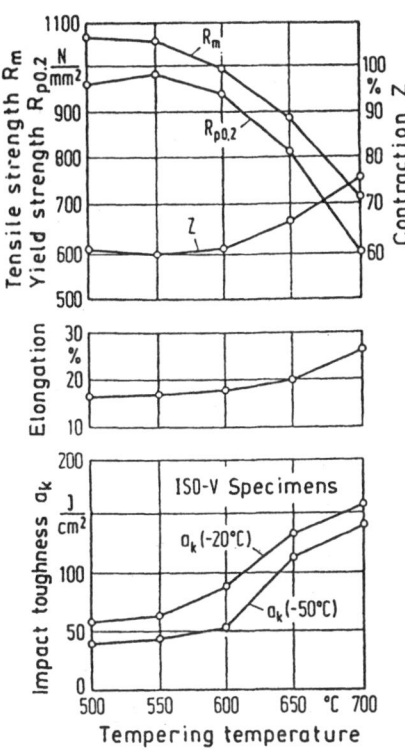

Figure 130 Hardening and tempering diagram of DIN 20CrNiMo2 steel. Hardening temperature 950°C; quenched in water. Specimen from a plate of 25 mm thickness; testing direction longitudinal. (From Ref. 1.)

called transition temperature (the temperature at which a substantial drop in impact toughness begins) is of special interest. The lower the transition temperature, the higher the toughness. Certainly, when hardening workpieces of big cross section, not only martensite is obtained, but also other constituents such as bainite, pearlite, and even pre-eutectoid ferrite, depending

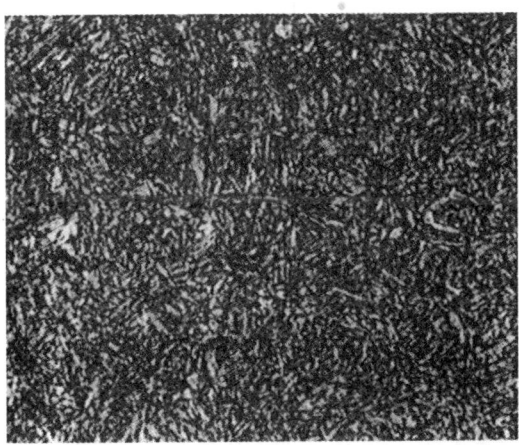

Figure 131 Microstructure of DIN 34CrNiMo6 steel after hardening and tempering. Tempered fine-grained martensite. Magnification 500×. (From Ref. 1.)

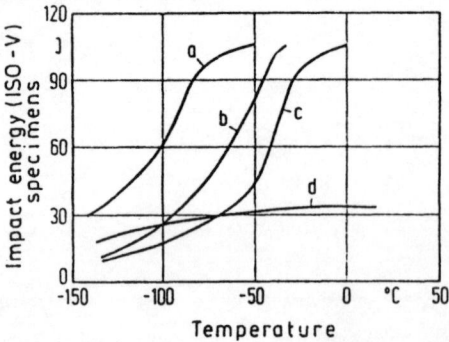

Figure 132 Influence of different microstructure and respective heat treatments on the impact tough-ness at low temperatures (ISO notch specimens) of a 3.5% Ni alloyed steel. *a*, Hardened by quenching in water and tempered; *b*, normalized and tempered; *c*, normalized only; *d*, hardened by quenching in water only. (From Ref. 1.)

on the decrease in cooling rate at quenching, below the surface toward the core of the work-piece. So after tempering, besides tempered martensite, other structural constituents having lower toughness are present.

Figure 133 shows the relationships among transition temperature, yield strength, and microstructure. For high strength values especially, the superiority of fine-grained martensite structure with respect to toughness is evident.

From a series of tests with hardened and tempered steels with about 0.4% C, Figure 134 shows a general relation between the structural constituents and the properties characterizing ductility (elongation and reduction of area) and impact toughness, respectively, for different levels of yield strength. It is clear that tempered martensite always gives the best ductility and toughness.

When comparing the impact toughness of tempered martensite at different strength levels (different hardness levels), one can perceive the influence of carbon content. As shown in Figure 135, of steels for hardening and tempering, those with 0.2-0.3% C have the best im-pact toughness. When testing the impact toughness of a steel, one should be aware that tough-

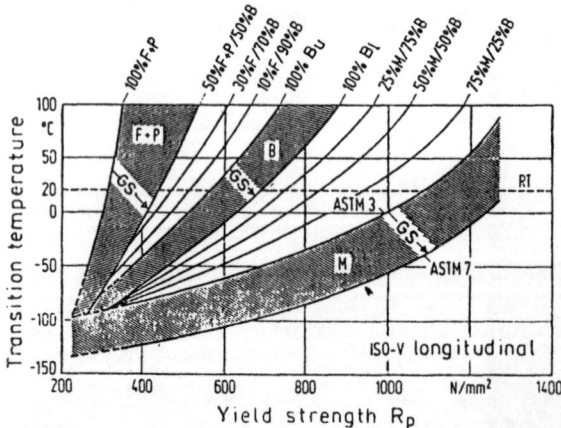

Figure 133 Transition temperature as a function of yield strength and microstructure. F = ferrite, P = pearlite, B = bainite, B_u = upper bainite, B_l = lower bainite, M = martensite, GS = grain size (ASTM). (From Ref. 1.)

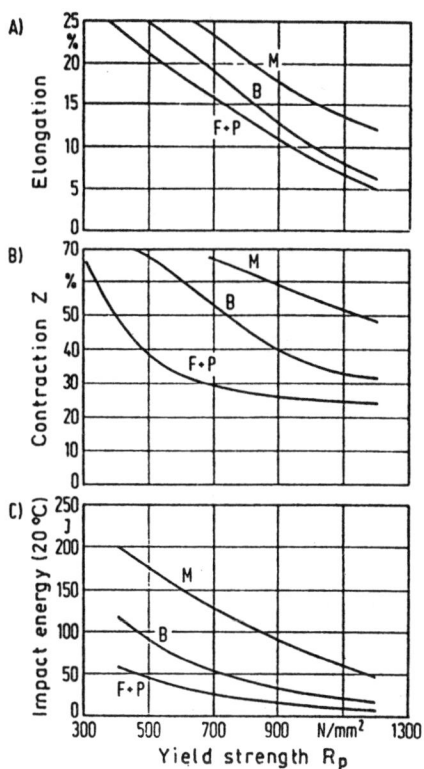

Figure 134   (A) Elongation, (B) reduction of area, and (C) impact toughness of hardened and tempered steels having about 0.4% C, as a function of structure constituents and yield strength. F = ferrite, P = pearlite, B = bainite, M = martensite. Grain size: ASTM 6–7. Impact toughness: ISO notch specimens. Testing direction: longitudinal. (From Ref. 1.)

ness is usually higher in the longitudinal direction (rolling direction) than in the transverse direction. That is because some phases or nonmetallic inclusions that are present in every steel (carbides, oxides, and sulfides) are stretched during rolling in the longitudinal direction. In this way a textured structure originates that has lower impact toughness in the transverse di-

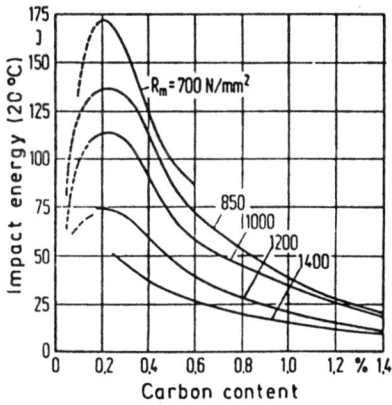

Figure 135   Impact toughness as a function of tensile strength and carbon content for the structure of tempered martensite. Grain size: ASTM 6–7. (From Ref. 1.)

rection than in the longitudinal direction. As a measure of this effect, the "factor of isotropy" (the ratio of transverse impact toughness to longitudinal impact toughness) is sometimes used.

The great influence of the microstructure after hardening (before tempering) on the impact toughness of a steel is evident from Figure 136. Appearance of pre-eutectoid ferrite or ferrite and pearlite in the structure results in a substantial decrease in the impact toughness.

When selecting a structural steel for hardening and tempering, in order to better adapt the mechanical properties to the requirements of the treated parts, the expected microstructure must be considered. To be able to reproducibly influence mechanical properties, one should know the relationships among the heat treatment regime, microstructure, and resulting mechanical properties.

Unalloyed steels for hardening and tempering, because of their low hardenability, exhibit a high degree of section sensitivity with respect to hardness distribution after hardening as shown in Figure 137. After quenching a bar specimen of 30 mm diameter of the steel in question in conventional quenching oil, a hardness of only 40 HRC was achieved at the surface. When specimens of the same diameter were quenched in fast quenching oil, the hardness was 45 HRC; when quenched in 10% Aquaquench solution the hardness was 56 HRC; and when quenched in water containing 5% Na_2CO_3, it was 58 HRC (see Figure 138).

This example leads to two important conclusions:

1. By using different quenchants and quenching conditions, different hardness distributions can be obtained with the same steel grade and same cross section size.
2. With an unalloyed steel (shallow-hardener), even when the most severe quenchant is used, for large cross-sectional sizes, the depth of hardening will be small and the core will remain unhardened.

Because of the second conclusion, when selecting a structural steel grade for hardening and tempering, its hardenability must always be adapted to the workpiece's cross-sectional size and the required strength level. Figure 139 shows the preferred fields for the application of some common steel grades for hardening and tempering according to the actual bar diameter and the strength level required. This recommendation is based on the assumption that a mini-

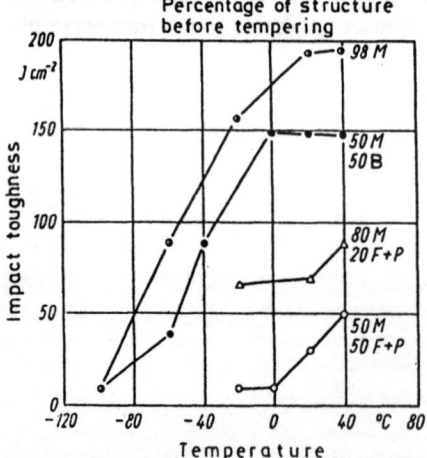

Figure 136 Influence of the microstructure after hardening (before tempering) on the impact toughness of DIN 16MnCr5 steel. (From Ref. 26.)

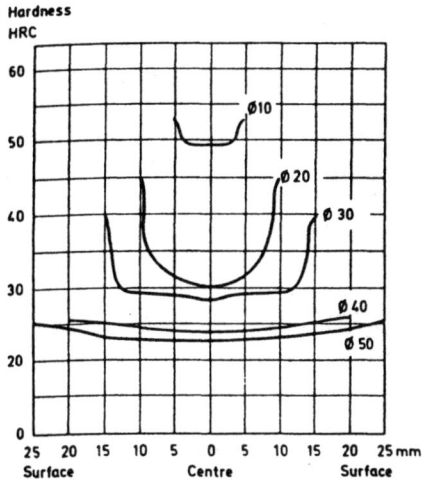

Figure 137 Hardness distribution (measured) on the cross section of bars of different diameters made of unalloyed steel (0.52% C, 0.24% Si, 0.90% Mn, 0.06% Cr) quenched in conventional hardening oil from 860°C. (From Ref. 14.)

mum impact toughness of about 50 J/cm^2 at room temperature will be achieved. As can be seen from Figure 139, for bigger cross-sectional sizes (bigger diameters) and higher strength levels, steels of higher hardenability (i.e., with more alloying elements) are required.

B. Technology of the Hardening and Tempering Process

Hardening, being the first operation of the hardening and tempering process, will yield a martensitic structure (provided a correct austenitization and quenching with a cooling rate greater

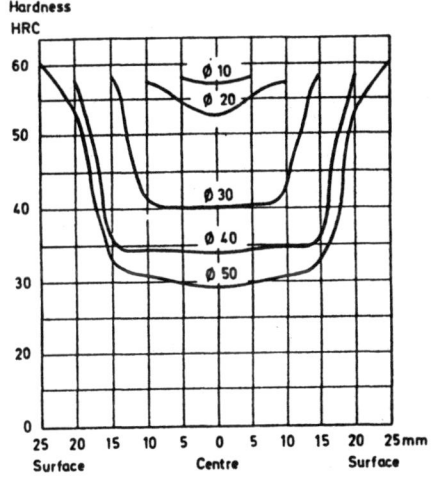

Figure 138 Hardness distribution (measured) on the cross section of bars of 10–50 mm diameters made of unalloyed steel (0.52% C, 0.24% Si, 0.90% Mn, 0.06% Cr) quenched from 860°C in water containing 5% Na$_2$CO$_3$. (From Ref. 14.)

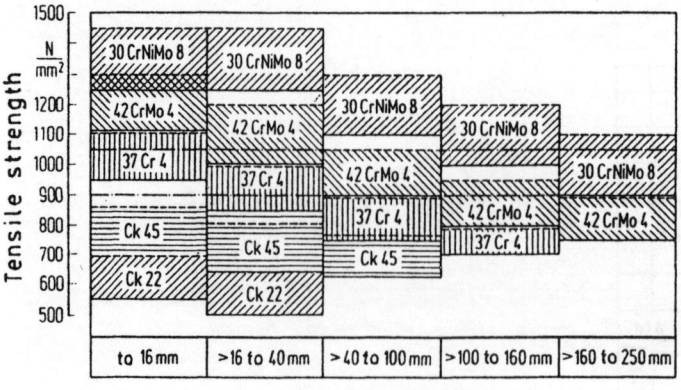

Bar diameter

Figure 139 Applicability of steel grades for hardening and tempering according to required strength level and bar diameter. Steel designations according to DIN. (From German standard DIN 17200.)

than the critical rate for the steel in question have been realized), the hardness of which depends on the dissolved carbon content, according to the empirical formula

$$H_{100\%mart} \approx 60\sqrt{C} + 20 \tag{45}$$

where C is the carbon content in wt%.

The required critical cooling rate for unalloyed steels is about 250°C/s (482°F/s). Alloyed steels have lower critical cooling rates or a higher hardenability, which means that the same quenching conditions yield a greater depth of hardening.

As a measure of hardenability, the ideal critical diameter D_I (see Chapter 3, Hardenability) can be applied. The actual depth of hardening, however, is influenced by, in addition to the alloying elements, the austenitizing temperature (especially for steels containing carbides difficult to dissolve) and quenching conditions. Consequently, two steels having the same D_I value may give different depths of hardening.

For a designer, therefore, information based only on the percentage of martensite doesn't seem practical, because even for the same D_I values he might get different depths of hardening. Besides, microstructures having the same amount of martensite don't always give the same hardness. The hardness of martensite depends on dissolved carbon content and may be calculated for 50% martensite according to the empirical formula

$$H_{50\% mart} \approx 44\sqrt{C} + 14 \quad \text{for } C < 0.7\% \tag{46}$$

More practical information for the designer, about expected depth of hardening, may be obtained for round bars from the correlation among the applied radius of the bar (R in mm), quenching intensity according to the Grossmann H factor (see Chapter 3), and the equivalent distance (E in mm) on the relevant Jominy curve. According to Just [25], this correlation for the surface of round bars reads

$$E_{surf} = \frac{R^{1/2}}{(3/4)H^{3/4}} - 3 \quad [mm] \qquad \text{for } R < 50 \text{ mm}; E < 30 \text{ mm} \tag{47}$$

and for the core of round bars:

$$E_{core} = \frac{R}{2H^{1/4}} \quad [\text{mm}] \tag{48}$$

After calculating the equivalent Jominy distance E, one can read off the hardness from the relevant Jominy curve. Figure 140 shows this correlation as a diagram (for radii from 0 to 50 mm and H values from 0.3 to 2) for convenient use.

As already explained, the properties of hardened and tempered parts depend first of all on how well the hardening operation has been performed. The higher the percentage of martensite at a specified point of the cross section after hardening, the better will be the properties after subsequent tempering. To check the quality of hardening achieved, the degree of hardening (S) can be used. It is the ratio between the achieved (measured) hardness and the maximum hardness attainable with the steel in question:

$$S = \frac{H}{H_{max}} = \frac{H}{60\sqrt{C} + 20} \tag{49}$$

where H is the actual hardness measured at a specified point of the cross section, in HRC, and H_{max} is the maximum attainable hardness in HRC. The degree of hardening is valid, or course, for the point of the cross section where the hardness was measured.

For highly stressed parts that are to be hardened and tempered to high strength levels, the required degree of hardening is $S > 0.95$, whereas for less stressed components values of $S > 0.7$ (corresponding to about 50% martensite) are satisfactory.

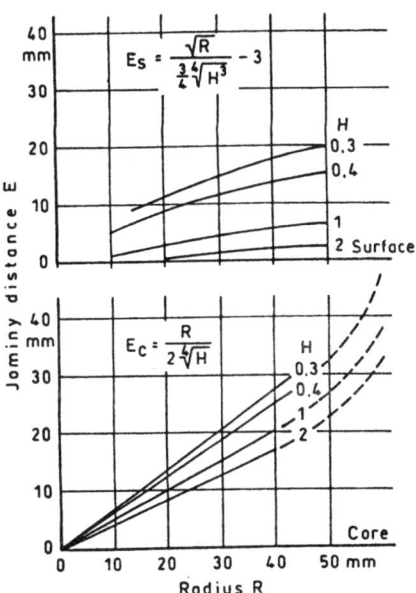

Figure 140 Correlation among radius of round bars (quenched by immersion), quenching intensity H and equivalent Jominy distance for the surface and core of the bars. (From Ref. 25.)

When hardening and tempering structural steels, the value of hardness after hardening and tempering is usually specified. The required degree of hardening can also be expressed as a function of the hardness after hardening and tempering (H_t):

$$S \geq \frac{1}{1 + 8e^{-H_t/8}} \tag{50}$$

Figure 141 shows the minimum values of the required degree of hardening as a function of hardness after hardening and tempering, limiting the allowed area.

By specifying the required degree of hardening, one can avoid the risk of an incorrect hardening and tempering. It is known that too low a value of hardness after hardening (not enough martensite) can be covered up by tempering intentionally at a lower temperature. Although in such a case the final hardness after hardening and tempering may correspond to the required value, the toughness and other mechanical properties important for dynamically stressed parts will be insufficient because of an inadequate microstructure. Such a risk can be avoided by specifying the minimum degree of hardness for the critical point of the cross section, which can easily be checked after hardening.

In selecting sufficiently severe quenchants to obtain a high percentage of martensite and great depth of hardening, one has to be aware of the risk of cracking. Hardening cracks are dependent on:

1. The shape of the workpiece. (Big differences in the size of the cross section, edges, and corners favor the formation of cracks.)
2. The heat treatment process. (High austenitizing temperatures and severe quenching conditions favor the formation of cracks.)
3. The steel grade itself. (The lower the M_s temperature of the steel, the greater the risk of cracking.) The M_s temperature can be calculated using the formula

$$M_s = 548 - 440 \text{ C} - 14 \text{ Si} - 26 \text{ Mn} - 11 \text{ Cr} - 9 \text{ Mo} - 14 \text{ Ni} + 2 \text{ V} \quad [°C] \tag{51}$$

where contents of alloying elements are in wt %.

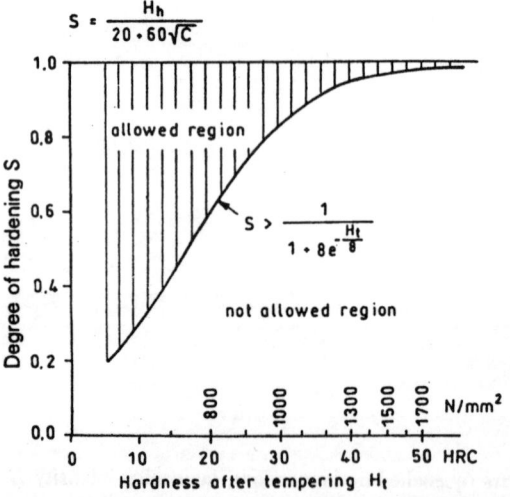

Figure 141 Minimum values of the degree of hardening required as a function of hardness after hardening and tempering. (From Ref. 25.)

The carbon content, as is known, has the greatest influence on the M_s temperature and on the risk of cracking.

Tempering, being the second important operation, decreases high hardnesses more than low hardnesses, as can be seen in Figure 142. This figure shows the Jominy curve of the steel DIN 28Cr4 (0.24–0.31% C, 0.15–0.36% Si, 0.62–0.78% Mn, 0.75–1.07% Cr) in hardened condition and after tempering the Jominy specimen to 500 and 600°C (932 and 1112°F). It can be seen that high hardness near the quenched end of the specimen has been decreased much more by tempering than low hardness values at greater distances from the quenched end. With respect to the cross section of hardened real components, this effect means that tempering more or less equalizes the hardness differences between surface and core. It is known that the hardness after tempering is a linear function of tempering temperature (in the range from about 320 to 720°C (608 to 1328°F)) and a logarithmic function of tempering time, according to the following formula [25], which is valid for a 100% martensite structure:

$$H_t = 102 - 5.7 \times 10^{-3} [T_t (12 + \log t)] \quad [\text{HRC}] \tag{52}$$

where T_t is the tempering temperature in kelvin and t is tempering time in seconds.

Tempering temperature and tempering time are consequently interchangeable with respect to resulting hardness; however, very short or very long tempering times do not yield optimum toughness. To obtain the optimum toughness for chromium steels, the tempering times should be between 1 and 5 h.

There is a firm relation between the hardness after tempering and the hardness after hardening. Spies et al. [26] have, by using multiple linear regression, quantified the influence of hardness after hardening, chemical composition, and tempering temperature on hardness after tempering and developed the formula

$$HB = 2.84\, H_h + 75(\%C) - 0.78(\%Si) + 14.24(\%Mn) + 14.77(\%Cr)$$
$$+ 128.22(\%Mo) - 54.0(\%V) - 0.55T_t + 435.66 \tag{53}$$

where HB is hardness after hardening and tempering (Brinell), H_h is hardness after hardening, (HRC); and T_t is tempering temperature, °C.

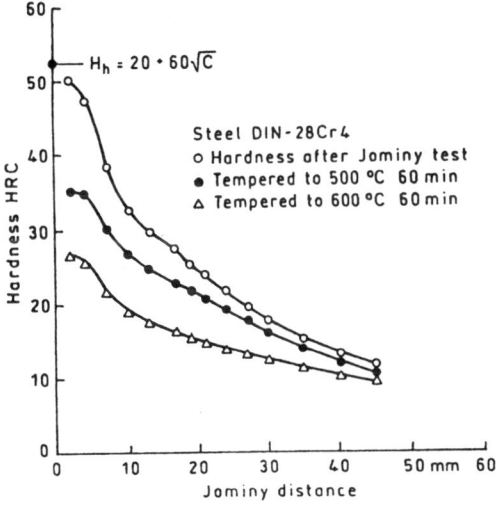

Figure 142   Influence of tempering temperature on level of hardness at various Jominy distances. (From Ref. 25.)

Equation (53) is valid for the following ranges:

H_h	20–65 HRC
C	0.20–0.54%
Si	0.17–1.40%
Mn	0.50–1.90%
Cr	0.03–1.20%
T_t	500–650°C (932–1202°F)

According to the German standard DIN 17021, an average relation between the hardness after hardening (H_h) and the hardness after hardening and tempering (H_t) reads

$$H_h = (T_t/167 - 1.2)H_t - 17 \quad [\text{HRC}] \tag{54}$$

where T_t is tempering temperature, °C; and H_t is hardness after hardening and tempering, (HRC). This formula is valid for 490°C (914°F) < T_t < 610°C (1130°F) and for tempering time of 1 h. Because, as already mentioned, high hardnesses decrease at tempering much more than low hardnesses, the prediction is more precise if the degree of hardening (S) is accounted for:

$$H_h = 8 + (H_t - 8) \exp [S(T_t/917)^6] \quad [\text{HRC}] \tag{55}$$

where S is the degree of hardening, $S < 1.0$; and T_t is tempering temperature, K.

Figure 143 shows a diagram from which it is possible to predict at a glance the hardness required after hardening for a desired hardness after tempering, taking into account the actual tempering temperature and the necessary degree of hardening.

It is also possible to calculate the necessary tempering temperature for a specified hardness after hardening and tempering when chemical composition and the degree of hardening are known:

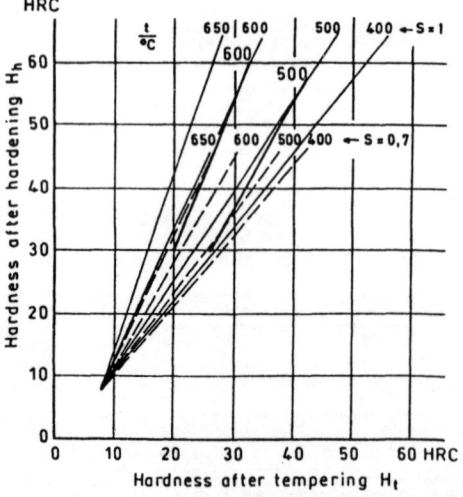

Figure 143 Relation among hardness after hardening, degree of hardening, tempering temperature, and hardness after tempering. (From Ref. 25.)

$$T_t = 647[S(60\sqrt{C} + 20)/H_t - 0.9]^{1/4} - 3.45SH_t + (537 - 561S)(\%C) + 505S(\%V)$$
$$+ 219S(\%Mo) + 75S(\%Cr) + 66S(\%Si) - 51 \quad [°C] \tag{56}$$

where H_t is hardness after hardening and tempering, [HRC]; S is the degree of hardening, $S \leq 1.0$; and alloying elements are given in weight percent. This formula is valid for a tempering time of 2 h.

There are also diagrams for practically every steel grade from which the tempering temperature may be determined if the ultimate tensile strength or yield strength required after hardening and tempering is known. Figures 144a and 144b show such diagrams for the unalloyed steel DIN Ck45 after quenching in water (25, 50, and 100 mm bar diameter) and in oil (25 and 50 mm bar diameter), respectively.

C. Computer-Aided Determination of Process Parameters

Increasingly, modern heat treatment equipment incorporates microprocessors for automatic control of temperature/time cycles, protective or reactive atmosphere, material handling, and,

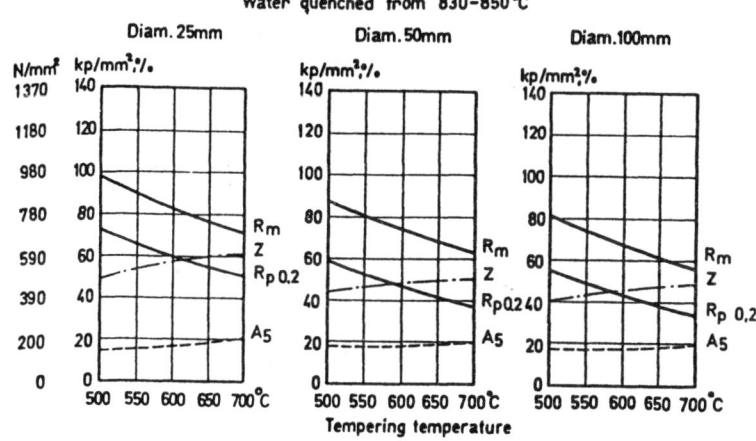

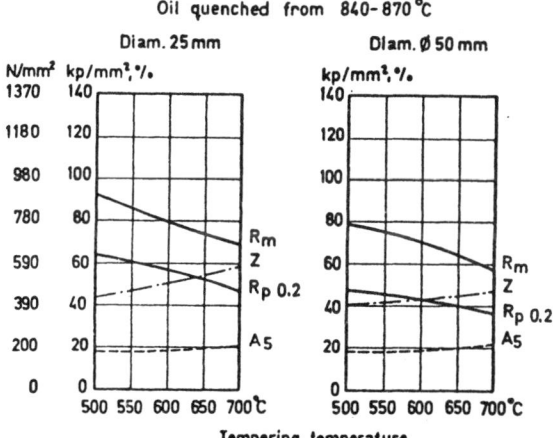

Figure 144 Tempering diagrams for the unalloyed steel DIN C45 when quenched (a) from 830–850°C in water, for bar diameters 25, 50, and 100 mm. and (b) from 840–870°C in oil, for bar diameters 25 and 50 mm. (From Ref. 14.)

to some extent, quenching operations. On the other hand, determination of the process parameters necessary to achieve the heat-treated properties required is normally based on empirical results.

For routine often-repeated heat treatment processes (e.g., carburizing, hardening, and tempering), computer programs can be written to establish treatment parameters provided that adequate data are available on workpiece geometry, material properties desired after heat treatment, steel grade used, and the actual heat treatment equipment itself. The aims of such an approach are to optimize the heat treatment operation by saving time and energy and to maintain close tolerances on the material properties imparted. The basic prerequisite is the availability of satisfactory mathematical models that enable the presentation and prediction of relevant metallurgical and physical phenomena.

As described by Liščić and Filetin [27], the process of hardening and tempering has been divided into operations of austenitization (A), quenching (Q), and tempering (T) as shown in Figure 145. Within the austenitization operation, the phases are (1) preheating, (2) heating to austenitization temperature with workpiece equalization at that temperature, and (3) homogenization of the structure. Within tempering, the phases are (5) heating to the tempering temperature and temperature equalization, (6) soaking at tempering temperature, and (7) cooling down from the tempering temperature.

Computer-aided determination of process parameters has a great number of advantages compared with earlier methods:

1. Planning of the process and preparation of the technological documentation are incomparably faster and simpler.
2. Since the computer program takes into account the influences of all relevant factors, provided that all necessary input data are used, the resulting parameters and time–temperature profiles can be determined more precisely.
3. The professional level of the technological documentation is always high and consistent, irrespective of the ability and experience of an individual technologist.
4. It is possible (by using appropriate subprograms and inserting the data for alternative equipment) to examine the potential energy savings or economy of using some other equipment for the same process.
5. If modern heat treatment equipment with microprocessor control is available, the resulting treatment parameters can be distributed directly (on-line) to all units where the process will be performed and controlled automatically.

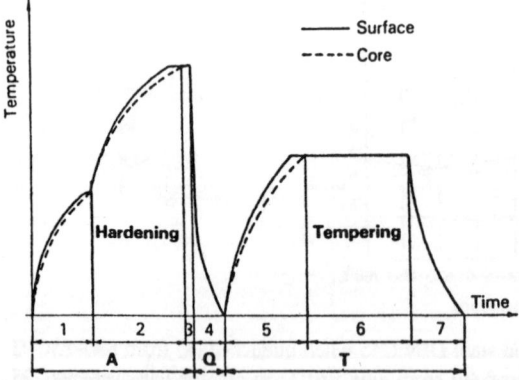

Figure 145   Operations and phases in the process of hardening and tempering.

The general scheme of computer-aided planning of the hardening and tempering process is shown in Figure 146. Use of a computer for this purpose requires

1. A database on characteristics of the steel grades treated
2. A database on the equipment employed (especially data on quenching severities available)
3. Subprograms stored in the computer memory for the necessary calculation of parameters.

The input data in a particular case are

1. Data on the steel grade in question
2. Data on the workpiece (shape, critical cross section, surface condition, number of pieces in a batch)
3. Data on mechanical properties required, after hardening and tempering, at a specified point of the cross section (hardness or ultimate tensile strength, yield strength, ductility, impact toughness, minimum grade of hardening or minimum percentage of martensite after quenching)
4. Data on the equipment used for all operations and phases of the process (preheating, austenitization, quenching, tempering).

The database on steels contains the following information for every specified steel grade or heat: chemical composition; carbon equivalent; austenitizing temperature; time for homogenization of the structure; M_s temperature; Jominy hardenability curve; holding time at tempering temperature; susceptibility to temper brittleness.

To determine the parameters of the hardening and tempering process, the following relationships must be known and stored in the computer memory in the form of adequate mathematical equations:

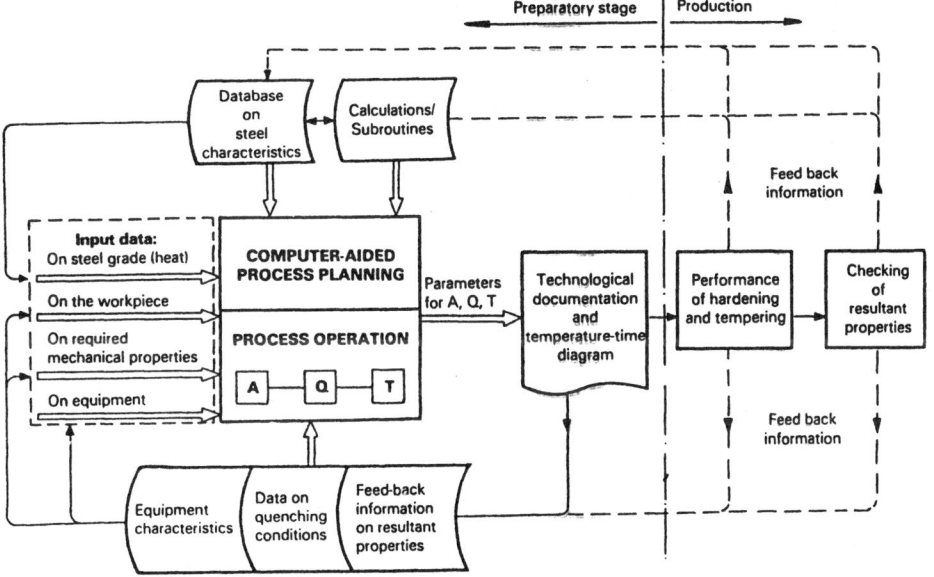

Figure 146   Scheme of computer-aided planning of the hardening and tempering process. (From Ref. 27.)

1. The effect of shape and cross-sectional size of the workpiece on the time necessary for heating and austenitization under the specific heat transfer conditions of the equipment employed. For the case of a 40 mm bolt made of grade BS 708A37 (En 19B) steel (see Figure 147), because of the high value of the carbon equivalent (0.82), a preheating stage at 450°C (842°F) was necessary. For calculation of preheating time, as well as the time for heating to austenitizing temperature and temperature equalization, formula (42) was used (see Section III.A) whereby the regression coefficients a and b were experimentally determined for the equipment used.

2. The influence of steel grade, cross-sectional size, and actual quenching conditions on the depth of hardening. This is necessary for selection of optimum quenching conditions (quenching medium, temperature, and agitation rate) to satisfy the required degree of hardening. The method by which this selection was performed is described in Chapter 3 (see Section VI.D of that chapter).

3. The relationship between hardness after tempering and tempering temperature for the steel in question.

The necessary tempering temperature (T_t) was calculated by means of the formula

$$T_t = 917\left[\frac{\ln[(H_h - 8) / (H_t - 8)]}{S}\right]^{1/6} - 273 \quad [°C] \tag{57}$$

where H_h is hardness after hardening, HRC; H_t is required hardness after tempering, HRC; and S is degree of hardening. This equation is valid for tempering temperatures between 390 and 660°C (734 and 1220°F).

The time necessary for heating up to the tempering temperature and for temperature equalization through the cross section was calculated in the same manner as for austenitization, taking into account data for the specific tempering furnace. The holding time at tempering temperature was taken from the database for the steel grade in question.

Cooling from the tempering temperature is carried out in air or inert gas in all cases where the steel is not prone to temper brittleness. If it is susceptible, faster cooling in oil or in air blast is necessary.

By using the described subprograms, the input data are processed by means of stored equations into the following output data or process parameters:

1. Temperature and time of preheating
2. Temperature and time for austenitization
3. Quenching conditions (quenching medium, its temperature, its agitation rate; time required for complete cooling of the workpiece)
4. Temperature and time for tempering
5. Mode of cooling from tempering temperature to room temperature.

Figure 147 shows an example of computer-generated documentation for hardening and tempering a 40 mm diameter bolt made of BS 708A37 (En 19B) steel.

Table 6 compares the required hardness (ultimate tensile strength) for the center of bars of 30, 50, and 80 mm diameters made of grade BS 708A37 steel with the measured hardness after hardening and tempering under computer-calculated conditions.

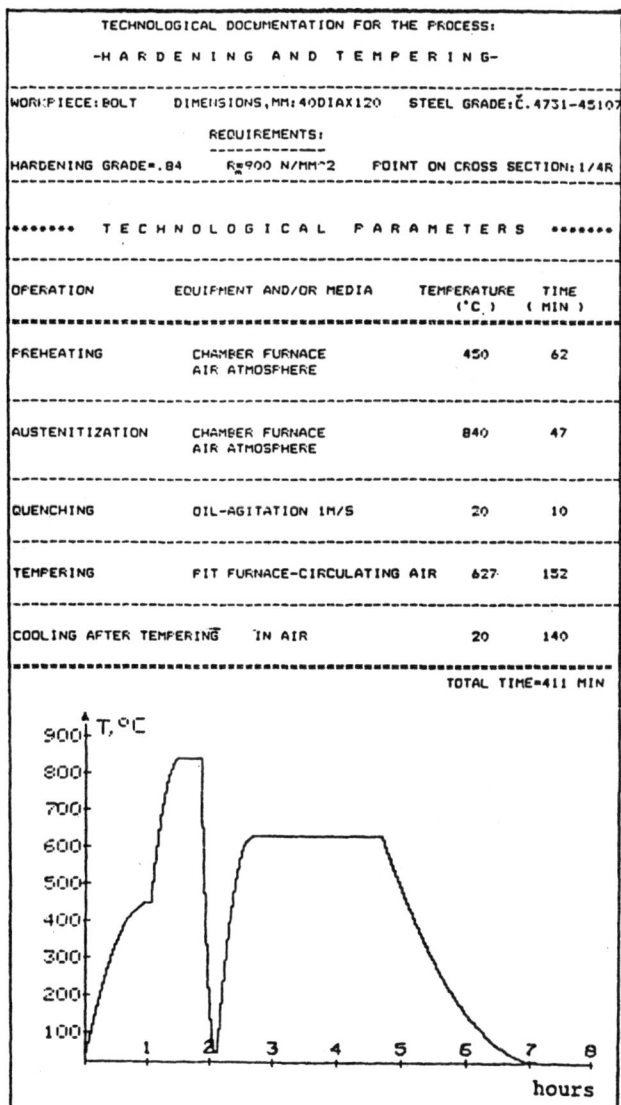

TECHNOLOGICAL DOCUMENTATION FOR THE PROCESS:

-H A R D E N I N G   A N D   T E M P E R I N G-

WORKPIECE: BOLT   DIMENSIONS,MM: 40DIAX120   STEEL GRADE: Č.4731-45107

REQUIREMENTS:

HARDENING GRADE=.84   Rm 900 N/MM^2   POINT ON CROSS SECTION: 1/4R

•••••••   T E C H N O L O G I C A L   P A R A M E T E R S   •••••••

OPERATION	EQUIPMENT AND/OR MEDIA	TEMPERATURE (°C)	TIME (MIN)
PREHEATING	CHAMBER FURNACE AIR ATMOSPHERE	450	62
AUSTENITIZATION	CHAMBER FURNACE AIR ATMOSPHERE	840	47
QUENCHING	OIL-AGITATION 1M/S	20	10
TEMPERING	PIT FURNACE-CIRCULATING AIR	627	152
COOLING AFTER TEMPERING	IN AIR	20	140

TOTAL TIME=411 MIN

Figure 147 An example of the computer-generated parameters and time–temperature cycle for hardening and tempering a 40 mm diameter bolt made of BS 708A37 (En 19B) steel. (From Ref. 27.)

V. AUSTEMPERING

The austempering process (see Figure 148) consists of austenitization, quenching into a hot bath maintained between 260 and 450°C (500 and 842°F), holding at this temperature until the transformation of austenite to bainite is complete, and cooling to room temperature at will. Compared with the process of hardening and tempering, there are the following substantial differences:

1. At austempering there is no austenite-to-martensite transformation, but the final structure (bainite) is obtained gradually during the isothermal transformation of austenite to bainite.

Table 6 Comparison of Required and Measured Hardness at the Center of Cylinders Made of Steel Grade BS 708A37 (En 19B), After Hardening and Tempering Under Computer-Calculated Conditions

Dimensions (mm)		Required values			Measured values		
Diameter	Length	Hardening grade (S)	Ultimate tensile strength[a] (N/mm²)	Hardness after tempering (HRC)	Tempering temperature (°C)	Ultimate tensile strength (N/mm²)	Hardness after tempering (HRC)
30	120	0.95	1240	38	519	1210	37.3
50	200	0.84	1000	31	583	1050	32.6
80	320	0.65	850	24.5	621	900	27.7

[a]Calculated from the hardness (DIN 50150).
Source: Ref. 27.

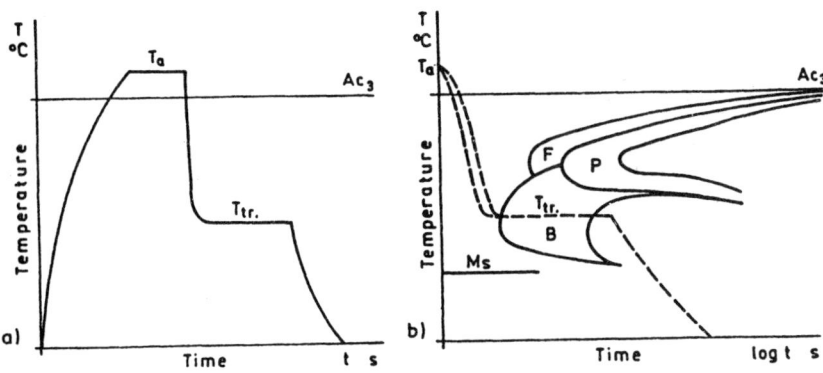

Figure 148 Scheme of an austempering process (a) in time–temperature diagram and (b) in isothermal transformation (IT) diagram.

2. After austempering there is no tempering.
3. While hardening and tempering is a two-operation process, austempering is performed in one cycle only, which is an advantage for the automation of the process.

Dealing with austempering one should use the isothermal transformation (IT) diagram of the steel in question to optimize the process parameters, among them first of all the transformation temperature (T_{tr}) and holding time at this temperature.

Austempering of steel offers two primary potential advantages:

1. Reduced distortion and less possibility of cracking
2. Increased ductility and toughness, especially in the range of high strength (hardness) values between 50 and 55 HRC.

Reduced distortion and less possibility of cracking are the result of lower thermal stresses, as well as lower transformational stresses compared to conventional hardening. Although at austempering there are also temperature differences between the surface and core of the workpiece, during quenching, these differences are substantially smaller, as shown in Figure 149, because the difference between the austenitizing temperature and the temperature of the quenching bath is much smaller [for 200–400°C (392–752°F)] than in conventional hardening. Smaller temperature gradients across the section mean smaller thermal stresses. On the other hand, at austempering there is no momentary austenite-to-martensite transformation, connected with the volume increase, taking place at different moments at different points of the cross section. Instead there is a gradual transformation from austenite to bainite, which takes place almost simultaneously in thin and thick cross sections. Both effects contribute to much lower risk of cracking and distortion, thereby minimizing the production of scraped parts and additional cost for straightening or grinding to repair the distortion.

Increased ductility and toughness as well as increased bendability and fatigue life are the strongest reasons to apply austempering instead of hardening and tempering. Figure 150 shows the relation of impact toughness and Brinell hardness (HB) of a Cr-Mn-Si steel after conventional hardening and tempering and after austempering, as a function of tempering temperature and austempering temperature, respectively. The most important difference is that a good combination of hardness and toughness after conventional hardening and tempering is possible only at high tempering temperatures, which means low hardness, whereas at austempering a good combination of hardness and impact toughness may be achieved at high hardness values.

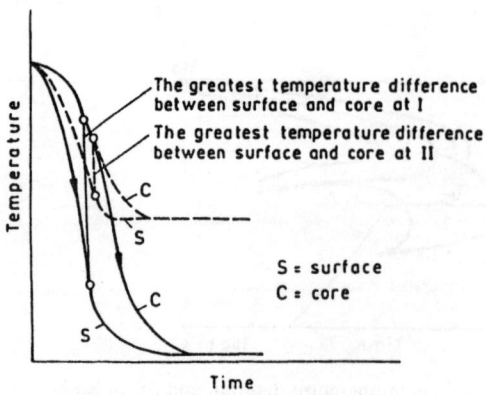

I   Conventional hardening and tempering
II Austempering

Figure 149 Temperature differences between surface and core of the workpiece in conventional hardening and in austempering. (From Ref. 28.)

Another comparison of impact toughness of a carbon steel after hardening and tempering and after austempering, as a function of hardness, is shown in Figure 151. It is evident that austempering yields much better impact toughness, especially at high hardness, around 50 HRC. It is necessary to emphasize that high toughness after austempering is possible only under conditions of complete transformation of austenite to bainite. Table 7 shows a comparison of some mechanical properties of austempered and of hardened and tempered bars made of AISI 1090 steel. In spite of having a little higher tensile strength and hardness, austempered specimens have had remarkably higher elongation, reduction of area, and fatigue life.

Figure 152 shows the fatigue diagram of DIN 30SiMnCr4 steel after conventional hardening and tempering and after austempering. The increase in fatigue resistance values is especially remarkable for notched specimens.

Regarding bendability, Figure 153, from an early work of Davenport, shows the results of bending a carbon steel wire austempered and hardened and tempered to 50 HRC.

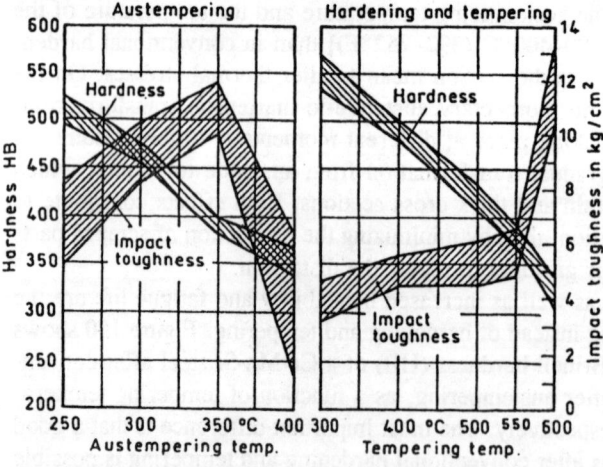

Figure 150 Impact toughness and hardness (HB) of five heats of a Cr-Mn-Si steel after conventional hardening and tempering and after austempering, as a function of tempering temperature and austempering temperature, respectively. (From Ref. 29.)

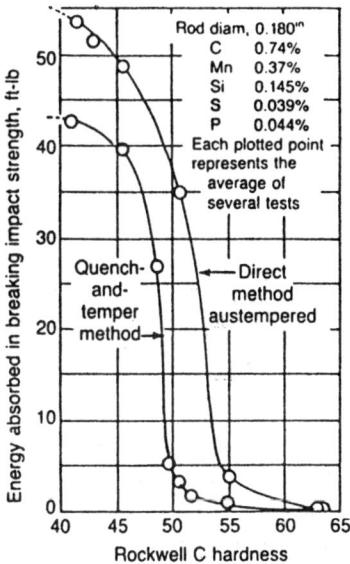

Figure 151 Comparison of impact toughness of a carbon steel after conventional hardening and tempering and after austempering, as a function of hardness. (From Ref. 17.)

When selecting a steel for austempering, IT diagrams should be consulted. The suitability of a steel for austempering is limited first of all with minimum incubation time (the distance of the transformation start curve from the ordinate). Another limitation may be the very long transformation time. Figure 154 shows the transformation characteristics of four AISI grades of steel in relation to their suitability for austempering. The AISI 1080 steel has only limited suitability for austempering (i.e., may be used only for very thin cross sections) because the pearlite reaction starts too soon near 540°C (1004°F). The AISI 5140 steel is well suited to austempering. It is impossible to austemper the AISI 1034 steel because of the extremely fast pearlite reaction at 540–595°C (1004–1103°F). The AISI 9261 steel is not suited to austempering because the reaction to form bainite is too slow (too long a transformation time) at 260–400°C (500–752°F).

Table 7 Comparison of Some Mechanical Properties of Austempered and of Hardened and Tempered Bars Made of AISI 1090 Steel

Property[a]	Austempered at 400°C (750°F)[b]	Quenched and tempered[c]
Tensile strength, MPa (ksi)	1415 (205)	1380 (200)
Yield strength, MPa (ksi)	1020 (148)	895 (130)
Elongation, %	11.5	6.0
Reduction of area, %	30	10.2
Hardness, HB	415	388
Fatigue cycles(d)	105,000[e]	58,600[f]

[a]Average values. [b]Six tests. [c]Two tests. [d]Fatigue specimens 21 mm (0.812 in.) in diameter. [e]Seven tests: range, 69,050–137,000. [f]Eight tests: range, 43,120–95,220.
Source: Ref. 31, p. 155.

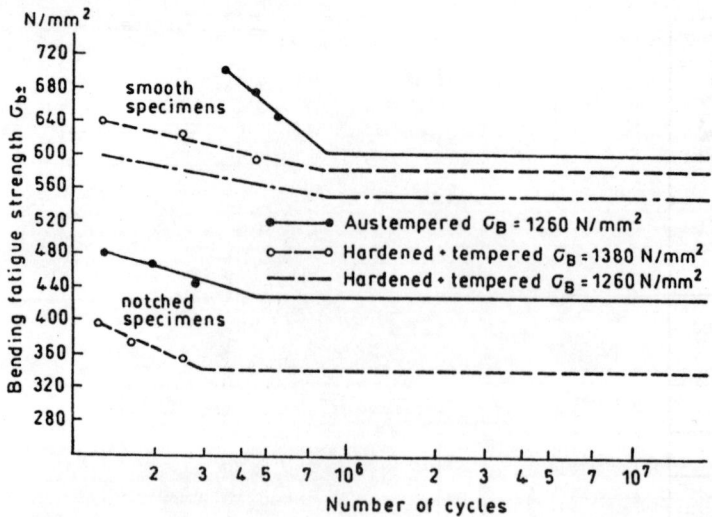

Figure 152 Fatigue of DIN 30SiMnCr4 steel after conventional hardening and tempering and after austempering. (From Ref. 29.)

The austempering process is limited to sections that can be cooled at a sufficient rate to prevent transformation to pearlite during quenching to the austempering bath temperature. Maximum section thickness is therefore important in determining whether or not a part can be successfully austempered. For eutectoid or hypereutectoid carbon steels like AISI 1080, a section thickness of about 5 mm is the maximum that can be austempered to a fully bainitic structure. Unalloyed steels of lower carbon content are restricted to a proportionately lesser thickness (except those containing boron). With increasing alloy content, heavier sections can be austempered, in some alloy steels up to 25 mm cross section. When some pearlite is permissible in the microstructure, even carbon steels can be austempered to sections significantly thicker than 5 mm. Table 8 lists section sizes and hardness values of austempered parts made of various steels.

Process parameters for the austempering process are

1. Austenitizing temperature and time
2. Quenching intensity when cooling from the austenitizing temperature to the austempering bath temperature

Hardness: Rockwell C 50

Austempered   Quenched and tempered

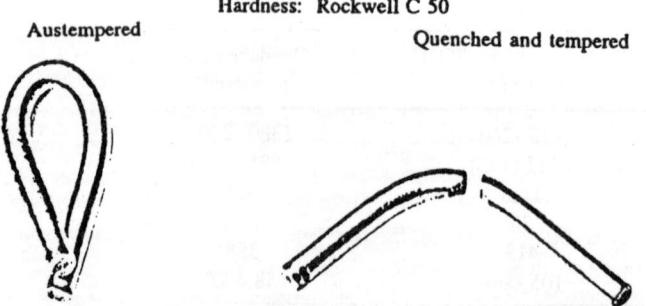

Figure 153 Carbon steel wire (0.78% C, 0.58% Mn) of 4.6 mm diameter, (left) austempered and (right) hardened and tempered to 50 HRC and bent under comparable conditions. (From Ref. 30.)

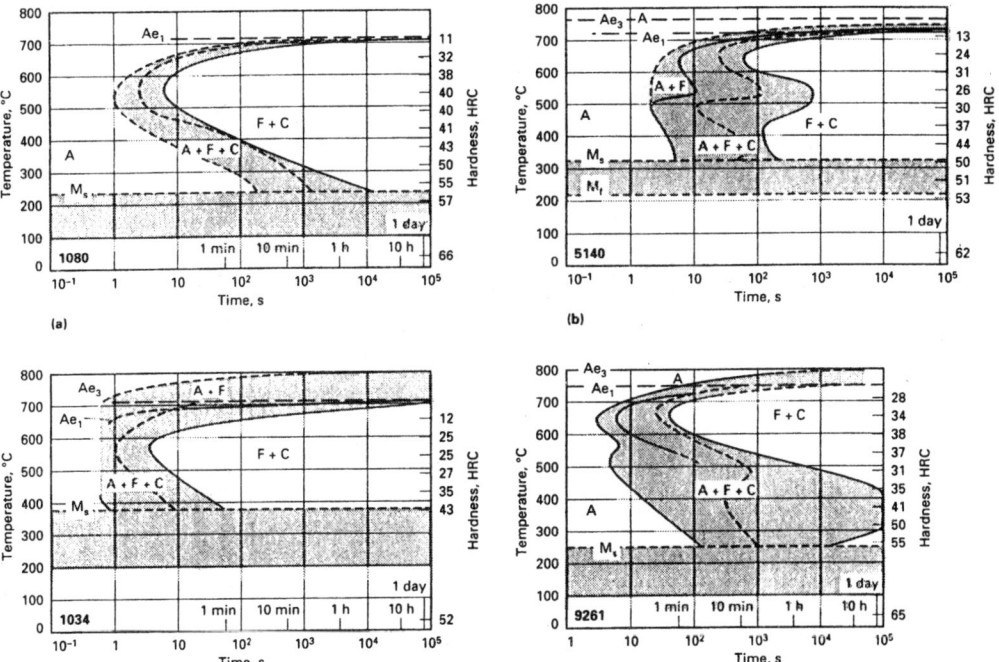

Figure 154 Transformation characteristics of steel of AISI grades (a) 1080, (b) 5140, (c) 1034, and (d) 9261 in relation to their suitability for austempering (see text). (From Ref. 31.)

3. Temperature of transformation, i.e., the austempering bath temperature
4. Holding time at austempering temperature

The austenitizing temperature and time (as in any hardening process) are responsible for carbide dissolution and homogenizing of the structure, which has a substantial influence on the impact toughness of treated parts.

Quenching must be severe enough to avoid any pearlite formation on cooling from the austenitizing temperature to the temperature of the austempering bath. Molten nitrite–nitrate salts are used as quenching media for austempering. To increase the quenching severity, agitation and sometimes the addition of some percentage of water is used. When adding water to a hot salt bath, care must be taken to prevent spattering. The higher the temperature of the salt bath, the less water should be added. Because of evaporation, the amount of water added must be controlled.

Transformation temperature, i.e, austempering bath temperature, is one of the two most important parameters as it directly influences the strength (hardness) level of the treated parts. The higher the austempering temperature, the lower the strength (hardness) of the austempered parts. The bainitic region can be divided according to austempering temperature into upper and lower bainite regions, the boundary between them being at about 350°C (662°F). The structure of upper bainite in steels (consisting of parallel plates of carbides and ferrite) is softer and tougher, whereas the structure of lower bainite (needlelike, with small carbides under 60° within the needles) is harder and more brittle.

Because not only the strength (hardness) level but also the impact toughness varies with austempering temperature, the temperature of the austempering bath must be kept within close tolerance [±6°C (±43°F)].

Table 8 Section Sizes and Hardness Values of Austempered Parts of Various Steel Grades

Steel	Section size		Salt temperature		M_s temperature[a]		Hardness (HRC)
	mm	in.	°C	°F	°C	°F	
1050	3[b]	0.125[b]	345	655	320	610	41–47
1065	5[c]	0.187[c]	—[d]	—[d]	275	525	53–56
1066	7[c]	0.281[c]	—[d]	—[d]	260	500	53–56
1084	6[c]	0.218[c]	—[d]	—[d]	200	395	55–58
1086	13[c]	0.516[c]	—[d]	—[d]	215	420	55–58
1090	5[c]	0.187[c]	—[d]	—[d]	—	—	57–60
1090[e]	20[c]	0.820[c]	315[f]	600[f]	—	—	44.5 (avg)
1095	4[c]	0.148[c]	—[d]	—[d]	210[g]	410[g]	57–60
1350	16[c]	0.625[c]	—[d]	—[d]	235	450	53–56
4063	16[c]	0.625[c]	—[d]	—[d]	245	475	53–56
4150	13[c]	0.500[c]	—[d]	—[d]	285	545	52 max
4365	25[c]	1.000[c]	—[d]	—[d]	210	410	54 max
5140	3[b]	0.125[b]	345	655	330	630	43–48
5160[e]	26[c]	1.035[c]	315[f]	600[f]	255	490	46.7 (avg)
8750	3[b]	0.125[b]	315	600	285	545	47–48
50100	8[c]	0.312[c]	—[d]	—[d]	—	—	57–60

[a]Calculated. [b]Sheet thickness. [c]Diameter of section. [d]Salt temperature adjusted to give maximum hardness and 100% bainite. [e]Modified austempering; microstructure contained pearlite as well as bainite. [f]Salt with water additions. [g]Experimental value.
Source: Ref. 31, p. 155.

The holding time at austempering temperature should be sufficient to allow complete transformation. Therefore the isothermal transformation (IT) diagram of the steel grade in question should be consulted. Allowing parts to remain in the bath for longer than the required time will increase the cost of treatment but it is not harmful to the mechanical properties of austempered parts.

REFERENCES

1. G. Spur and T. Stöferle (Eds.), *Handbuch der Fertigungstechnik, Vol 4/2, Wärmebehandeln,* Carl Hanser, Munich, 1987.
2. H. J. Eckstein (Ed.), *Technologie der Wärmebehandlung von Stahl,* 2nd ed., VEB Deutscher Verlag für Grundstoffindustrie, Leipzig, 1987.
3. B. Liščić, S. Švaić, and T. Filetin, Workshop designed system for quenching intensity evaluation and calculation of heat transfer data, *Proc. First Int. Conf. Quenching and Control of Distortion,* Chicago, IL, September 1992, pp. 17–26.
4. A. Rose and W. Strassburg, Anwendung des Zeit-Temperatur-Umwandlungs-Schaubildes für kontinuierliche Abkühlung auf Fragen der Wärmebehandlung, *Archiv. Eisenhüttenwes. 24*(11/12): 505–514 (1953) (in German).
5. H. P. Hougardy, Die Darstellung des Umwandlungsverhaltens von Stählen in den ZTU-Schaubildern, *Härterei-Tech. Mitt., 33*(2):63–70 (1978) (in German).
6. E. Scheil, *Arch. Eisenhüttenwes. 8*:565–567 (1934/35) (in German).
7. F. Wever aned A. Rose (Ed.), *Atlas zur Wärmebehandlung der Stähle,* Vols. I and II, Verlag Stahleisen, Düsseldorf, 1954/56/58.
8. W. Peter and H. Hassdenteufel, Aussagefähigkeit der Stirnabschreckprüfung und des Zeit-Temperatur-Umwandlungsschaubildes für das Ergebnis der Härtung von Rundstäben, *Stahl Eisen 87*(8):455–457 (1967) (in German).
9. M. Atkins, *Atlas of Continuous Transformation Diagrams for Engineering Steels,* British Steel Corporation, BSC Billet, Bar and Rod Product, Sheffield, U.K., 1977.
10. E. A. Loria, Transformation behaviour on air cooling steel in A_3–A_1 temperature range, *Met. Technol.,* October 1977, pp. 490–492.
11. N. Shimizu and I. Tamura, Effect of discontinuous change in cooling rate during continuous cooling on pearlitic transformation behaviour of steel, *Trans. ISIJ 17*:469–476 (1977).
12. ISI, *Decarburization,* ISI Publ. 133, Gresham Press, Old Woking, Surrey, England, 1970.
13. B. Liščić, H. M. Tensi, and W. Luty (Eds.), *Theory and Technology of Quenching,* Springer-Verlag, New York, 1992.
14. K. E. Thelning, *Steel and Its Heat Treatment,* 2nd ed., Butterworths, London, 1984.
15. J. Frehser and O. Lowitzer, The process of dimensional change during the heat treatment of tool steels, *Stahl Eisen 77*(18):1221–1233 (1957) (in German).
16. J. Wünning, Verfahrenstechnik des Isothermglühens, *Härterei-Tech. Mitt,* Vol. 32 (1977), pp. 43–49 (in German).
17. G. Krauss, *Steels: Heat Treatment and Processing Principles,* ASM Int., Materials Park, OH, 1990.
18. J. Orlich, A. Rose and P. Wiest (Eds.), *Atlas zur Wärmebehandlung der Stähle,* Vol. 3, *Zeit-Temperatur-Austenitisierung-Schaubilder,* Verlag Stahleisen, Düsseldorf, 1973 (in German).
19. J. Orlich and H. J. Pietrzenivk (Eds.), *Atlas zur Wärmebehandlung der Stähle,* Vol. 4, *Zeit-Temperatur-Austenitisierung-Schaubilder,* Part 2, Verlag Stahleisen, Düsseldorf, 1976 (in German).
20. S. Jost, H. Langer, D. Pietsch, and P. Uhlig, Rechnerische Ermittlung der Erwärmdauer bei der Wärmebehandlung von Stahl, *Fertigungstech. Betr 26*(5):298–301 (1976) (in German).
21. M. A. Aronov, J. F. Wallace, and M. A. Ordillas, System for prediction of heat-up and soak times for bulk heat treatment processes, *Proc. Int. Heat Treat. Conf. Equipment and Processes,* April, 18–20, 1994, Schaumburg, IL, pp. 55–61.
22. N. I. Kobasko, Teplovie procesi pri zakalke stali, *Mettaloved. Termiceskaja Obrab. Metalov. 3* (1968), pp. 2–6.
23. B. Liščić and G. E. Totten, Controllable delayed quenching, *Proc. Int. Conf. Equipm. Process.* April 18–20, 1994, Schaumburg, IL, pp. 253–262.
24. J. Parrish, Retained austenite: new look at an old debate, *Adv. Mater. Process. 3*:25–28 (1994).
25. E. Just, Vergüten-Werkstoffbeeinflussung durch Härten und Anlassen , *VDI-Ber. 256*:125–140 (1976) (in German).
26. H. J. Spies, G. Münch, and A. Prewitz, Möglichkeiten der Optimierung der Auswahl vergütbarer Baustähle durch Berechnung der Härt-und vergütbarkeit, *Neue Hütte 8*(22):443–445 (1977) (in German).

27. B. Liščić and T. Filetin, Computer-aided determination of the process parameters for hardening and tempering structural steels, *Heat Treat. Met. 3*:62–66 (1987).
28. K. H. Illgner, Qualitäts-und Kostenvorteile mit Zwischenstufenvergütungsanlagen im Vergleich zu normalen Vergütungsanlagen, *Fachber. Hüttenpraxis Metallweiterverarb. 17*(4):281–288 (1979) (in German).
29. F. W. Eysell, Die Zwischenstufenvergütung und ihre betriebliche Anwendung, Z. *TZ Prakt. Metallbearb. 66*:94–99 (1972) (in German).
30. E. S. Davenport, Heat treatment of steel by direct transformation from austenite, *Steel*, March 29, 1937.
31. *ASM Handbook,* 9th ed., Vol. 4, *Heat Treating*, ASM Int., Materials Park, OH, 1991.
32. U.S. Patent 2,829,185, regarding surface temperature measurement.
33. *Met. Prog.,* October 1963, p. 134.

9

HEAT TREATMENT WITH GASEOUS ATMOSPHERES

Johann Grosch
Institut für Werkstofftechnik, Technische Universität Berlin, Berlin, Germany

I. INTRODUCTION

Heat treatment of components is mostly accomplished in gaseous atmospheres, the more so if plasma and vacuum are regarded as special cases of gaseous atmospheres. In comparison, heat treatment in solid or liquid media is negligible. Heat treatment in gaseous atmospheres falls into two categories: processes with the aim of avoiding a mass transfer between gaseous atmosphere and material, and processes with the aim of achieving just such a transfer. Mass transfer occurs when there is a difference in the potential between the constituents of a gaseous atmosphere and those of the microstructure of a component. The direction of such a mass transfer is determined by the potential difference, which leaves two fundamental possibilities with regard to the component. One is the intake of elements of the gaseous atmosphere into the component, and the other is the emission of elements of the component into the gaseous atmosphere. This kind of heat treatment falls under the category of *heat treatment with gas*, the subject of this chapter. The deposition of constituents of a gaseous atmosphere onto the surface of a component (coating) that is not connected with the described mass transfer mechanism is therefore excluded from the subject of heat treatment with gas.

A protective gas is a gaseous atmosphere that is free of a potential difference with respect to those elements of both gas and steel that have the ability to transfer mass. A central matter of concern with all the homogeneity treatments (annealing, austenitizing, tempering) is to prevent oxidation. Gas compositions suitable for reducing oxidation may have a potential difference against carbon; furthermore, a reduction of oxide layers makes it always possible for carbon to get into the gas. Protective gases with a reducing effect must hence be adjusted to the carbon content of the steel to prevent decarburization. Inert gases such as a rare gas or pure nitrogen and high-quality vacua do not contain any reactant constituents and thus prevent mass transfer. In processes without mass transfer, the gaseous atmosphere as a protective gas is an important, basic requirement of heat treating but is not used as a parameter with which to attain or alter certain properties of the component. Processes like these are therefore referred to as *heat treatment in gas*.

Transport of matter and heat conduction can be formally calculated by applying the same rules. The heat conduction in steel, however, is of a much higher order than the transport of matter, which as a diffusion process causes a uniformly direct flow of atoms. Heat treatments with gases are therefore always isothermal processes. As a rule, the rather slow process of diffusion determines the time that is needed for the technical processing of a heat treatment with mass transfer. This in turn determines essential processing conditions.

During a technically and economically justifiable treatment time, only those atoms that are interstitially dissolved in iron are absorbed in adequate amounts and sufficiently deep to meet the given requirements. Elements that are used in heat treating are therefore carbon, nitrogen, oxygen, and hydrogen. With these interstitial atoms also, the exchange is limited to the case, with the exception of thin sheets and hydrogen as the smallest element that diffuses most easily, where it is possible to influence the bulk material. Heat treatment with gases is therefore mostly a surface process. The corresponding thermochemical surface processes with gases are carburizing and decarburizing, nitriding and denitriding, and soaking and the combined processes carbonitriding and nitrocarburizing. Treatments with oxygen as the reactant almost always cause the formation of an oxide layer (controlled oxidizing, blueing) or lead to a reduction of one or more oxide layers. Boriding with gaseous boron sources is seldom done in practice because the predominant medium, diborane (B_2H_6), is highly toxic. The boron halides BBr_3, BF_3, and BCl_3 are also seldom used due to their corrosive effect in humid conditions.

The industrial heat treatment of steel with gases consists primarily of carburizing, which is surface heat treatment in the austenite phase field, nitrocarburizing, and, to a lesser extent, nitriding, which is surface heat treatment in the ferrite phase field. Consequently, these heat treatment processes will be dealt with in detail as to the fundamental principles of introducing carbon, nitrogen, or a mixture of the two into the case of a steel, the characteristics of the heat-treated microstructures, and the properties of carburized or nitrocarburized components. Oxidation, reduction, and the effect and composition of protective gases are often connected with decarburizing processes. Hence this topic will be dealt with following the discussion on carburizing. Some conditions of a desired oxidation will also be treated in connection with oxynitriding.

A discussion of the fundamental principles of reactions in and with gaseous atmospheres and of diffusion in solid metals that the above-mentioned processes have in common [1–5] will precede the sections on the individual processes.

II. FUNDAMENTALS IN COMMON

The absorption of material from a gaseous atmosphere occurs in several steps [6,7]:

Processes in the Gaseous Atmosphere: Formation of transportable gas molecules and transfer of these molecules through the gas phase onto the surface of the metal with subsequent physical adsorption of the gas molecules

Processes in the Interface: Dissociation of the gas molecules and chemisorption of the gas atoms, penetration of the atoms through the surface of the metal with transition of the atoms from the state of chemisorption to the interstitial solute state in the solid solution

Diffusion of the atoms from the surface into the core of the material.

These steps are based on the premise that there is a potential difference between gas and steel. By analogy, the described steps are also valid for the emission of material. The atoms emitted from the solid solution recombine to form molecules at the surface of the material and penetrate into the gaseous atmosphere.

Independent of the composition of the initial gases, the gaseous atmospheres used in heat treating at processing temperatures consist of the elementary molecules carbon monoxide (CO), carbon dioxide (CO_2), hydrogen (H_2), water vapor (H_2O), oxygen (O_2), ammonia (NH_3), and sometimes also methane (CH_4), all of which are able to react with one another and with the catalyzing surface of the component (and the furnace wall), thus releasing or absorbing carbon, oxygen, nitrogen, and hydrogen. Reactions among the constituents of the gas are described as *homogeneous reactions*; reactions between elements of the gas and elements of the component surface are described as *heterogeneous reactions*. The heterogeneous reactions in the interface between gaseous atmosphere and component surface are chemophysical processes and are responsible for the mass transfer. The reactions, i.e., the transition from an initial state to a final state, are accompanied by a change in the energy, which is determined by the first law of thermodynamics:

$$du = q + w \tag{1}$$

where q is the amount of heat added during the change of states and w the work done by the system. In many cases it suffices to just consider the work against the surrounding pressure (volume work):

$$w = -P \, dV \tag{2}$$

Thus Equation (1) can be rewritten as

$$du = q - P \, dV \tag{3}$$

Combined with the entropy derived from the second law of thermodynamics,

$$dS = \frac{dq_{rev}}{T} \tag{4}$$

these give the basic equation for reversible thermodynamic processes at constant pressure and constant temperature:

$$du = T \, dS - p \, dV \tag{5}$$

This relation allows the derivation of thermodynamic potentials, one of which is the free enthalpy or Gibbs free energy

$$g = u + PV - TS \tag{6}$$

which is subsequently needed.

In a closed system the gas reactions approach a dynamic equilibrium state that is determined by pressure, temperature, gas composition, and material composition. In this dynamic equilibrium state, the Gibbs free energy is at its minimum, and reactions and reverse reactions, on average, take the same amount of time; i.e., the total of locally absorbed and emitted particles equals zero, thus causing the net flow to cease. The rate of a chemical reaction is in proportion to the active masses of the involved elements, which for gaseous material are described by their partial pressures p_A (volume of constituent A $\times$ total pressure in the system). From the fact that reaction and reverse reaction take the same amount of time, it follows that for the general reaction

$$aA + bB = cC + dD \tag{7}$$

(a–d are the stoichiometric factors of the reaction components A–D) the equilibrium constant K_p of the process is

$$K_p = \frac{p_C^c p_D^d}{p_A^a p_B^b} \tag{8}$$

where, by agreement, the reaction products C and D are placed over the reactants A and B. Thus, the gas composition is replaced by the equilibrium constant K_p, which is independent of pressure.

The Gibbs free energy for the standard state is thus

$$dg^0 = -RT \ln \frac{p_C^c p_D^d}{p_A^a p_B^b} = -RT \ln K_p \tag{9}$$

Values of the Gibbs free energy for standard-state reactions have been studied for many reactions and can be found in special tables [8,9].

The values of the partial pressures of the gaseous atmospheres discussed here vary between 10^{-17} and 10^{-25} bar at treatment temperature and are thus, for reasons of convenience, often replaced by the activity a_i of the gases by relating the partial pressure p to a standard-state pressure. It is most usual to choose as standard state the partial pressure $p_0 = 1$ of the pure component in the same phase at the same temperature, which for carbon, for instance, is the steam pressure p_0^C of graphite. That is, the activity of the graphite-saturated austenite is by definition $a_{carbon} = 1$ [10].

Equation (9) can thus be rewritten as

$$dg^0 = -RT \ln \frac{a_c a_d}{a_A a_B} \tag{10}$$

These deductions are valid only for reactions where substance is neither added to nor emitted from the system. On changing the amount of substance, the constituents must be taken into consideration by means of their chemical potential, which as the partial Gibbs free energy

$$\mu_i = \left(\frac{\partial g}{\partial n_i}\right)_{T,p,j} \left(\frac{\partial e}{\partial n_i}\right)_{S,V,j} \tag{11}$$

is defined with dn_i moles of substance i. In this case it is necessary to complete Equation (5),

$$du = T\, dS - p\, dV + \Sigma \mu_i\, dn_i \tag{12}$$

The mass transfer within the interface is technically described by the mass transfer coefficient β, which determines how fast the particles move, with the mass transfer equation

$$\dot{m} = \beta(a_{gas} - a_{surface}) \tag{13}$$

and is therefore also called the effective reaction rate constant. The direction of mass transfer is determined by the activity gradient between the gaseous atmosphere and the surface of the steel. The reaction rate constant indicates the total mass transfer in the interface and comprises as a global value the effects of material, the microgeometrical state of the surface, flow conditions, pressure, and temperature on the mass transfer. The individual physicochemical reactions that occur in the interface cannot be described by the mass transfer coefficient.

Single-phase systems are homogeneous when at thermodynamic equilibrium. Differences in the distribution of the involved atoms such as those caused during production are equalized by matter flowing from regions of higher concentration toward regions of lower concentration. The cause of this flux, which is called *diffusion*, are differences in the chemical potential μ_c of the diffusing substance. The partial molar Gibbs free energy, according to Equation (11), can be rewritten

$$\mu_c = \frac{dg}{dc} \qquad (14)$$

where c is the concentration of the diffusing substance.

The potential difference is equalized by the flux

$$\dot{m} = -\frac{D*}{RT} c \frac{\partial \mu_c}{\partial c} \, grad \ c \qquad (15)$$

where $\dot{m}$ is the number of atoms c that penetrate the area F in the time t, R is the universal gas constant [8.314 J/(mol·K)], and $D*$ is the diffusion coefficient (or diffusivity).

In multiphase systems with different chemical potentials it is likely that potential jumps occur at the phase boundaries. In this case it may happen that the flux is opposed to the concentration gradient (uphill diffusion). On carbide formation, for instance, carbon diffuses from the saturated austenite into carbide with a higher concentration of carbon, which, however, has a lower chemical potential in the carbide.

In dilute solutions, i.e., when the amount of the diffusing material is small, it is possible to use, with adequate precision, the more easily accessible concentration gradient as a driving force. This approach is valid for the thermochemical surface treatment and leads to Fick's first law of diffusion [11],

$$\frac{\partial c}{\partial t} = -D \frac{\partial c}{\partial x} \qquad (16)$$

according to which the variation of the concentration over time depends on the concentration gradient $\partial c/\partial x$ parallel to the x axis. The effective diffusion coefficient D has the units of area/time and is usually expressed in cm^2/s. Fick's first law is valid when there is no change over time in the concentration gradient and thus none in the flux. Frequently, diffusion causes a change in the concentration gradient and thus becomes dependent on time and location. This is covered by Fick's second law [11],

$$\frac{\partial c}{\partial t} = -\frac{\partial}{\partial x} D \frac{\partial c}{\partial x} \qquad \text{with } D = f(c) \qquad (17)$$

or, if D is independent of concentration and consequently of location,

$$\frac{\partial c}{\partial t} = -D \frac{\partial^2 c}{\partial x^2} \qquad \text{with } D \neq f(c) \qquad (18)$$

In the case of a semi-infinite system, that is, when the diffusion flow does not reach the end of the specimen as is the case in thermochemical surface treatment, Fick's second law as given in Equation (18), the van Orstrand–Dewey solution [12–14] applies:

$$c(x,t) - c_0 = (c_s - c_0)\left[1 - \mathrm{erf}\left(\frac{x}{2\sqrt{Dt}}\right)\right] \qquad (19)$$

where $c(x,t)$ is the concentration c at a distance x from the surface of a steel with the initial concentration c_0 following a diffusion time t, and c_s is the surface concentration of the diffusing element (erf is the Gaussian error function). According to this relation, the depth of penetration increases in proportion to the root of the diffusion time, which leads to the empirical rule that to get double the depth of penetration it is necessary to quadruple the diffusion time.

Fick's second law can also be resolved when substance is emitted, i.e., when c_s is smaller than c_0, in the form

$$c(x,t) - c_S = (c_0 - c_s)\,\mathrm{erf}\!\left(\frac{x}{2\sqrt{Dt}}\right) \tag{20}$$

The diffusion coefficient is given by the empirical relation

$$D = D_0 \exp\!\left(-\frac{Q}{RT}\right) \tag{21}$$

with the element-dependent constant D_0 and the activation energy Q of the diffusion. From Equations (16) and (19) it follows that the diffusion time that is needed for a specific depth of penetration can be reduced only by higher temperatures T and/or an increase in the concentration gradient $\partial c/\partial x$. The activation energy Q is dependent on the mechanisms of solid-state diffusion. The diffusion of the gases nitrogen, oxygen, and hydrogen and of carbon, which are located interstitially mainly on octahedral voids in the lattice, occurs primarily by the interstitial mechanism that is the cause of the already mentioned rather fast diffusion of the above elements, which can still occur at room temperature and even lower temperatures.

The diffusion coefficients D_H, D_N, D_C, and D_O have been thoroughly studied for α-iron. Figure 1 [15] offers a comparison of the magnitudes between hydrogen, carbon, oxygen, and

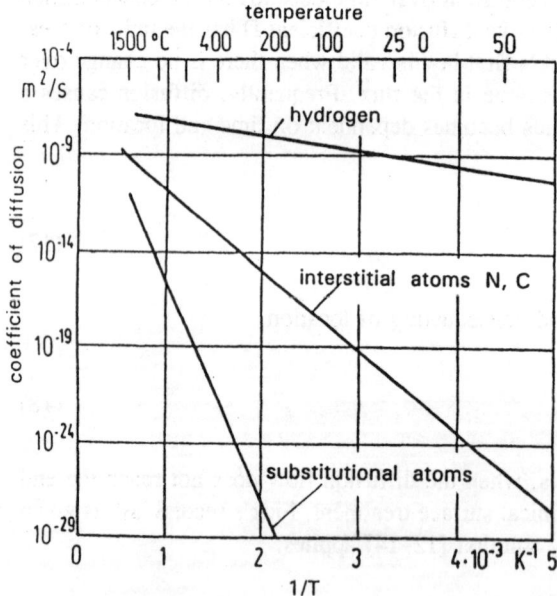

Figure 1 Diffusion coefficients of hydrogen and of interstitial and substitional elements in α-iron. (From Ref. 15.)

nitrogen and of substituted atoms. Figure 2 [4] shows a detailed plot of the diffusion coefficients of carbon, oxygen, and nitrogen. The diffusion coefficients in γ-iron are lower by approximately the second power of 10; details on this will be dealt with in Section III. It ought to be noted that diffusion proceeds faster alongside grain boundaries than in the matrix [3,4].

The directed exchange of matter requires a difference in potential or activity that is established and maintained by the gas composition. It is possible to relate processes near equilibrium to their final state, describe them by specifically derived and easily obtainable values such as the carbon potential, and control them accordingly. With processes far from equilibrium, where it is not possible to ascertain how great the differences in potential or activity are, it has proved helpful to use process characteristics that allow a required gas composition to be maintained provided that the processing conditions are fixed.

III. CARBURIZING

A. Introduction

Carburizing produces a hard and often relatively shallow surface on relatively soft components when the surface microstructure of steels with a (core) carbon content of usually 0.15–0.25% is carburized to carbon contents in the range of 0.7–0.9% (1.0 %) C and transformed to martensite. This treatment leads to the formation of a hardness gradient and to a distribution of residual stresses, with compressive stresses in the surface microstructure due to the changes in volume during the martensitic transformation. The combined effect of these two parameters causes the main properties of the components, i.e., fatigue strength, rolling contact fatigue strength, and toughness, to reach the highest values possible in the same part. Carburizing thus comprises the processing steps (1) carbon diffusion, leading to the forma-

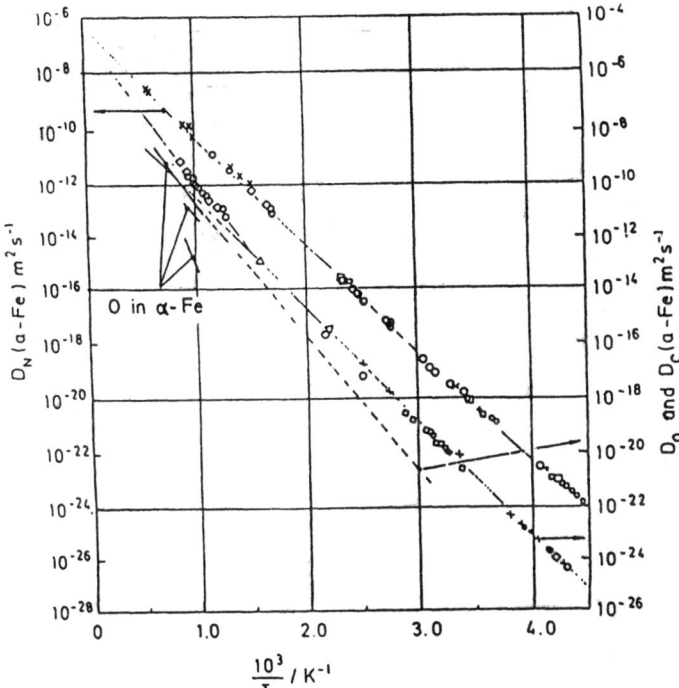

Figure 2 Diffusion coefficients of C, N, and O in α-iron. (From Ref. 4.)

tion of a carbon gradient, and (2) quenching, resulting in the hardness gradient and the distribution of residual stresses, where it is possible to establish equal carbon gradients in differently alloyed steels while the respective hardness gradients can differ because of the dependence of hardenability on the type of alloy. The remainder of Section III deals with the fundamentals of carburizing (carbon diffusion), hardenability as a criterion for the selection of suitable carburizing steels, and the microstructures resulting from hardening. The properties of carburized components will be described in Section VI.

B. Carburizing and Decarburizing with Gases

Carburizing is achieved by heating the steel at temperatures in the homogeneous austenite phase field in an environment of appropriate carbon sources. The carburizing time depends on the desired diffusion depth. The processes in use are classified according to their carbon sources into pack carburizing (solid compounds), salt bath carburizing (liquid carbon sources), and gas and plasma carburizing (gaseous carbon sources), the latter of which is subsequently described. Pack carburizing produces a gaseous atmosphere, which means that the phenomena described for gas carburizing are, in essence, also true for pack carburizing. The processes in the gas phase and in the interface can be described for the present as states of equilibrium [16–23]; the enforced transition to a nonequilibrium state leads to carburizing or decarburizing. This approach is valid because during the carburizing process the diffusion in the steel is considerably slower than the preceding steps described in Section II. Carburizing procedures for which this condition is no longer valid [24,25] and plasma carburizing [26–29] will not be covered in this context. (See Chapter 13 for plasma carburizing.)

1. Gas Equilibria

Gaseous carbon sources in wide use are hydrocarbon gases such as methane or propane [19–21], alcohols and alcohol derivatives (above all methanol [30]), and other organic carbon compounds [31]. These carbon sources decompose at carburizing temperature to the constituents carbon monoxide and hydrogen with small amounts of carbon dioxide, water vapor, oxygen, and methane. In general, certain constant proportions of carbon monoxide and hydrogen diluted with nitrogen are used as a carrier gas, which can be produced separately in a gas generator or in the furnace directly. Such a carrier gas has a carburizing or decarburizing effect due to the reactions

$$2\ CO \rightleftharpoons C(\gamma) + CO_2 \qquad \text{(Boudouard reaction)} \tag{22}$$

$$CO + H_2 \rightleftharpoons C(\gamma) + H_2O \qquad \text{(heterogeneous water gas reaction)} \tag{23}$$

$$CO \rightleftharpoons C(\gamma) + (1/2)O_2 \tag{24}$$

$$CH_4 \rightleftharpoons C(\gamma) + 2H_2 \tag{25}$$

and the homogeneous water gas equilibrium

$$CO_2 + H_2 \rightleftharpoons CO + H_2O \tag{26}$$

with which the reaction product CO_2 is decomposed again. This means that after a sufficiently long carburizing time an equilibrium between carburization and decarburization is established whose carbon content is described as the carbon potential (C potential C_p) [32] and can be measured by shim stock analysis [33]. (It is customary to define the carbon potential, in mass percent, as the carbon content of pure iron within the homogeneous austenite phase field at a given temperature that is in equilibrium with the furnace atmosphere.) Figure 3 [16,18] shows

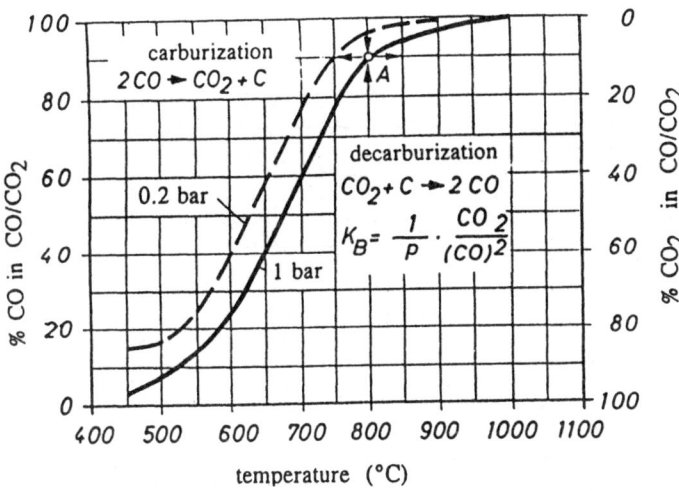

Figure 3 Boudouard reaction in equilibrium with pure carbon. (From Refs. 16 and 18.)

the equilibrium between carburization and decarburization for the Boudouard reaction in the presence of pure carbon dependent on pressure and temperature. The equilibrium can be described by the equilibrium constant

$$\log K_B = \log \frac{p_{CO}^2}{p_{CO_2} p_C^0} = \log \frac{p_{CO}^2}{p_{CO_2}} \tag{27}$$

with the concentration of the gas constituents as the respective partial pressures and the carbon concentration as vapor pressure p_C^0 of pure carbon (equal to unity). The equilibrium curve shown in Figure 3 at $p_{abs} = 0.2$ bar is equivalent to an amount of 20% carbon monoxide in the carrier gas.

In the carburizing of steel, the influence of the carbon content and the alloy composition of the steel on the carbon concentration must be considered by means of the activity of carbon, a_C [25], which is defined as the ratio of the vapor pressure p_C of carbon in the given condition to the vapor pressure p_C^0 of pure carbon:

$$a_C = p_C/p_C^0 \tag{28}$$

The activity of carbon for plain carbon steels can be calculated [19,23] according to

$$\log a_C = \frac{10{,}500}{4.575\ T} - \frac{3.95 - 0.69C}{4.575} + \frac{C}{0.785C + 21.5} \tag{29}$$

or, less complicated but sufficiently precise, as

$$\log a_C = \frac{2300}{T} - 2.21 + 0.15C + \log C \tag{30}$$

and is shown in Figure 4 [34] in the form of isoactivity plots [22,34].

The influence of the alloying elements on the activity of carbon is described by the ratio of the C potential c_p of the carburizing atmosphere (carbon content of pure iron at equilibrium) to the carbon content C_L of the alloyed steel at equilibrium [19,21,35,36]:

$$\lg a_C = \frac{2300}{T} - 2,21 + 0,15 \cdot (\%C) + \lg (\%C)$$

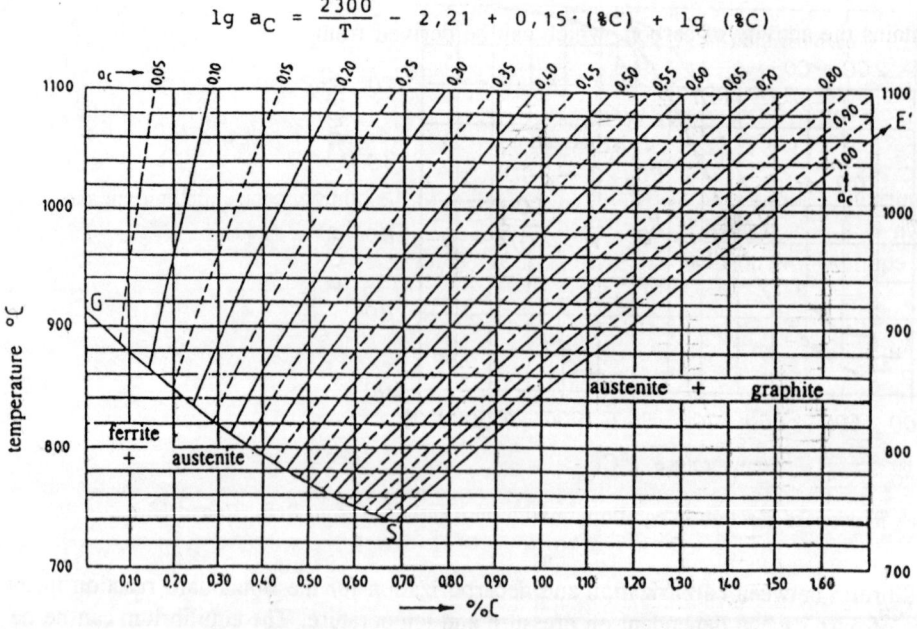

Figure 4   Isoactivity of carbon in the austenite phase field. (From Ref. 34.)

$$\log(C_P/C_L) = 0.055\ (\%Si) + 0.014\ (\%Ni) - 0.013\ (\%Mn)$$
$$-0.040\ (\%Cr) - 0.013\ (\%Mo) \qquad (\text{Ref. } 35) \tag{31}$$

$$\log(C_P/C_L) = 0.062\ (\%Si) + 0.014\ (\%Ni) - 0.016\ (\%Mn)$$
$$-0.057\ (\%Cr) - 0.015\ (\%Mo) - 0.102\ (\%V)$$
$$-0.014\ (\%Al) - 0.006\ (\%Cu) \qquad (\text{Ref. } 19) \tag{32}$$

$$\log(C_P/C_L) = 0.075\ (\%Si) + 0.02\ (\%Ni) - 0.04\ (\%Cr)$$
$$-0.01\ (\%Mn) - 0.01\ (\%Mo) \qquad (\text{Ref. } 36) \tag{33}$$

On calculating the value of a required carbon potential of a carburizing atmosphere, the influence of the alloying elements on the activity of carbon is taken into account by means of the alloy factor $f = C_L/C_P$ for a specific steel [37]. Nickel-alloyed steels have alloy factors smaller than 1; Cr alloys, Cr-Mn alloys, and Cr-Mo alloys have alloy factors higher than 1, which is why on carburizing these steels in a carburizing atmosphere with a C potential of 1%, the case carbon content of nickel-alloyed steels is smaller than 1% whereas the case carbon content of the other steels is higher than 1%.

The equilibrium constant of the Boudouard reaction in the carburizing process of steel [19,20,22],

$$\log K_B = \log \frac{p_{CO}^2}{p_{CO_2} a_C} - \frac{8817}{T} + 9.071 \tag{34}$$

thus contains the activity of carbon, which can be derived from

$$\log a_C = \log \frac{p_{CO}^2}{p_{CO_2}} - \frac{8817}{T} - 9.071 \tag{35}$$

The carburizing equilibria of the Boudouard reaction in the homogeneous austenite phase field are shown in Figure 5 [19].

The equilibrium constants of reactions (23)–(25) are

$$\log k_{H_2} = \log \frac{p_{H_2} p_{CO}}{p_{H_2O} a_C} - \frac{7100}{T} + 7.496 \tag{36}$$

$$\log k_{O_2} = \log \frac{p_{CO}}{p_{O_2} a_C} - \frac{5927}{T} + 4.545 \tag{37}$$

$$\log k_{CH_4} = \log \frac{p_{CH_4}}{p_{H_2} a_C} - \frac{4791}{T} + 5.789 \tag{38}$$

By analogy, it is possible to derive the corresponding relations valid for the activities from Equation (35).

Given a certain composition and volume of gas, the described processes cease eventually, having established an equilibrium with the corresponding carbon potential. With the reactions

$$CH_4 + CO_2 \rightarrow 2CO + 2H_2 \tag{39}$$

$$CH_4 + H_2O \rightarrow CO + 3H_2 \tag{40}$$

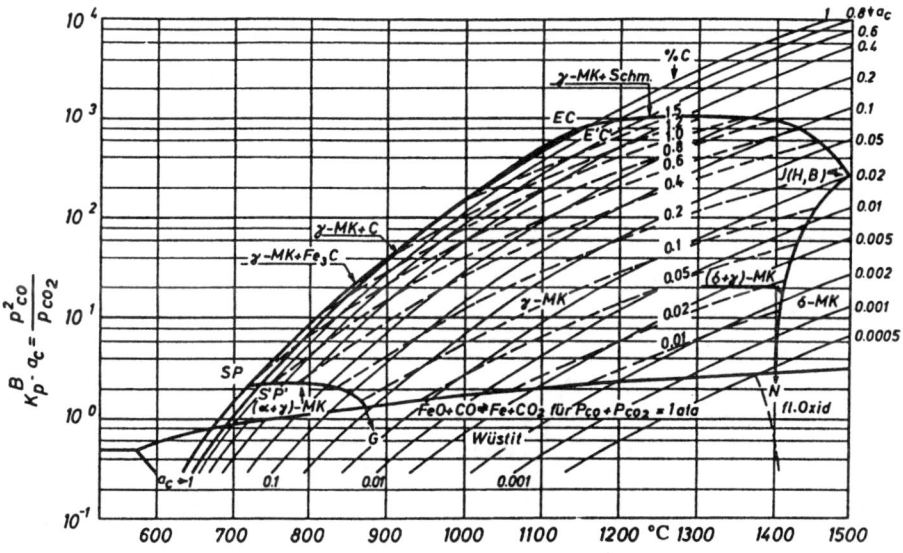

Figure 5 Boudouard reaction in equilibrium with steel. (From Ref. 19.)

methane decomposes the reaction products carbon dioxide and water vapor of the heterogeneous carburizing reactions (22) and (23), thus regenerating the carburizing atmosphere. Compared with the carburizing reactions (22)–(24), in particular with the heterogeneous water gas reaction, the reactions with methane are sluggish, which makes it necessary to add large quantities of methane (or propane, which decomposes to methane at carburizing temperature) to maintain the carburizing processes.

2. Kinetics of Carburizing

In the carburizing process, a carrier gas of a composition as constant as possible is enriched with a carburizing agent (e.g., propane) to establish and maintain, by means of further additions of the carburizing agent, a desired carbon potential that exceeds the carbon content of the steel to be carburized. In this nonequilibrium condition [38] the activity of carbon in the carburizing atmosphere $a_{C(gas)}$ is higher than the activity of carbon in the steel $a_{C(steel)}$. The difference in the activities leads to the desired carbon transfer into the steel, the flux $\dot{m}$ (number of atoms M that penetrate the area F in unit time) being proportional to the difference in activities [22],

$$\dot{m} = \frac{M}{F} dt = -\beta(a_{C(gas)} - a_{C(steel)}) \tag{41}$$

The surface reaction rate constant β is in particular dependent on the composition of the carburizing atmosphere and the carburizing temperature, as given in Figure 6 [20], with values for $\beta = 1.25 \times 10^{-5}$ cm/s for an endothermic gas consisting of methane/propane up to $\beta = 2.5 \times 10^{-5}$ cm/s for undiluted decomposed alcohols. The surface reaction rate constant is ascertained by means of thin iron foils (thickness of about 0.05 mm) [20,39] or thin iron wires [40]; the quantitative connections are, however, not yet sufficiently known in detail.

By combining Equations (16) (Fick's first law) and (41), a formula for the effect of time on the growth of the carbon diffusion depth can be derived [23,41,43]:

$$x = At = \frac{0.79\sqrt{Dt}}{0.24 + (C_{At} - C_0)/(C_P - C_0)} - 0.7\frac{d}{\beta} \tag{42}$$

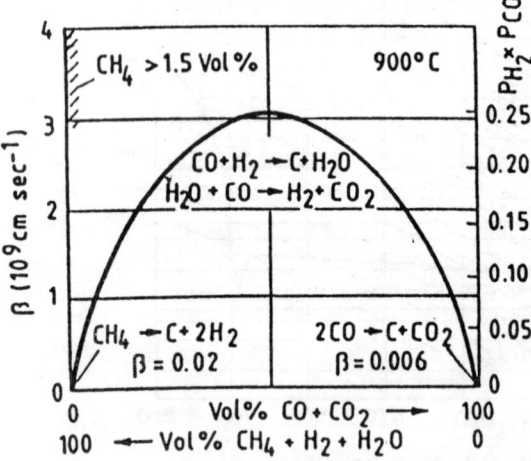

Figure 6   Surface rate constant β versus gas composition. (From Ref. 20.)

where the limiting carbon content C_{At} determines the depth At of carbon diffusion. For a limiting carbon content of 0.3% the depth $At_{0.3\%C}$ can be assessed [42] as

$$At_{0.3\%C} = k\sqrt{t} - D/\beta \qquad \text{[mm]} \qquad (43)$$

where the proportional value k depends on temperature, carbon potential, and the surface reaction rate constant. The $\sqrt{t}$ rule, as described in Section II, which states that to double the carburizing depth it is necessary to quadruple the carburizing time, is valid for $k = 0.5$ and of sufficient accuracy with carburizing depths of about 1 mm and more [38,42,43] (Figure 7). The formal meaning of the compensating factor D/β is that the point of intersection of the carbon diagram and the C potential is shifted from the surface into the gas atmosphere [38,42,43].

Effective decarburizing reactions are [6,44]

$$2C(\alpha,\gamma) + O_2 \rightarrow CO_2 \qquad (44)$$

$$C(\alpha,\gamma) + 2H_2 \rightarrow CH_4 \qquad (45)$$

$$C(\alpha,\gamma) + H_2O \rightarrow CO + H_2 \qquad (46)$$

where molecules from the gaseous atmosphere react with the carbon to form new molecules at the surface of the steel that in turn are emitted into the gaseous atmosphere. Reaction (46), occurring in several steps, is used for controlled decarburization; i.e., components such as electric sheets are annealed in moist hydrogen. Taking considerably more time, reaction (45), which is also involved, is of minor importance, whereas reaction (44) tends to have a slightly deleterious effect when oxygen-containing protective gases or annealing in air is used.

3. Control of Carburizing

The carbon gradient as the major aim of the carburizing process is determined by the carburizing depth At and the case carbon content C_s and can be controlled at a given temperature

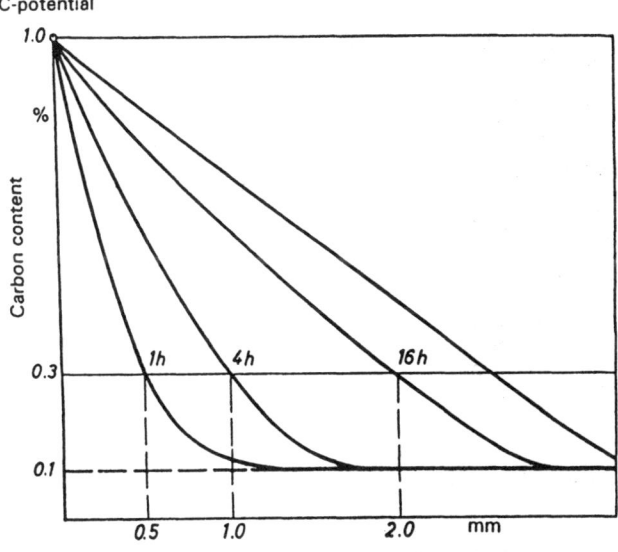

Figure 7 Effect of carburizing time on carbon gradients for $D/\beta = 0$. (From Ref. 42.)

by the carburizing time according to Equation (43) and by the carbon potential. In industrial carburizing the carbon potential is controlled by either using Equation (22), monitoring carbon dioxide content; Equation (23), ascertaining water vapor content; or Equation (24), monitoring oxygen content [37]. The partial pressure of CO_2 in the gaseous atmosphere is continually measured by infrared absorption, which is based on the differences in attenuation of an infrared beam due to the CO_2 content of the gas. The interrelationship between CO_2 content, C potential, and temperature,

$$\log p_{CO_2} = \frac{6552}{T} - 6.841 - 0.15C + \log\left[\frac{p_{CO}^2}{C}(0.785C + 21.5)\right] \qquad (47)$$

contains the CO content, which is generally held constant by means of an adequate carrier gas composition to avoid undesirable influences. The measuring accuracy, then, is about 0.005–0.01 relative to the C potential. Figure 8 shows an evaluation of Equation (47) for a carrier gas generated of propane with propane as the enriching gas [45].

The partial pressure of oxygen, which is about 10^{-17}–10^{-24} bar within the furnace, is measured by means of an oxygen probe [46] whose measuring element consists of a zirconia electrolytic solid that projects into the interior of the furnace in the form of a tube. The electrical potential, or electromotive force E, of such an electrolytic solid is, according to Nernst's law,

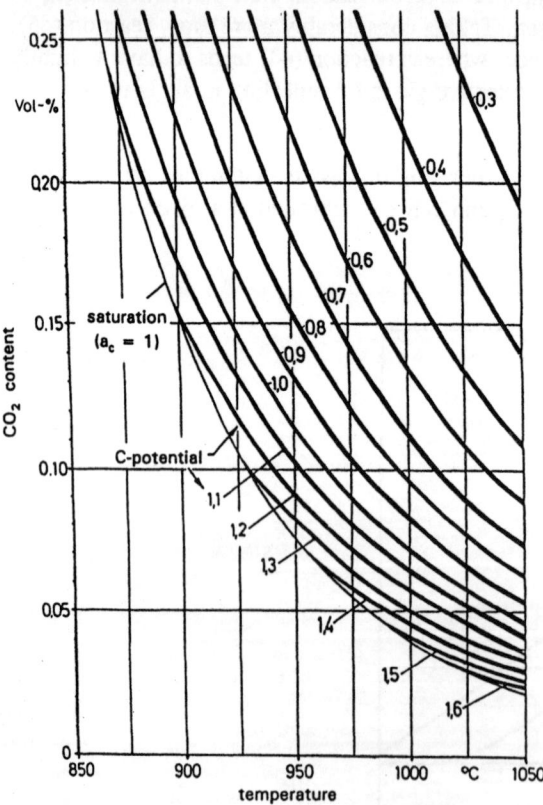

Figure 8 Controlling carbon potential by monitoring the carbon dioxide content (endo gas enriched with propane, 23.7 vol % CO). (From Ref. 45.)

$$E = \frac{RT}{4F} \ln \frac{p_{O_2}}{p_{O_2}^0} \tag{48}$$

or, with $R = 8.314$ J/(mol·K) and $F = 96.5727$ J/(mV·mol) (Faraday constant),

$$E = 0.0216T \ln \frac{p_{O_2}}{p_{O_2}^0} = 0.0497T \log \frac{p_{O_2}}{p_{O_2}^0} \quad [\text{mV}] \tag{49}$$

where $p_{O_2}^0$ is the partial pressure of the reference gas on the inside wall of the tube, usually surrounding air (=0.21 bar). By means of Equations (39) and (48), with air as the reference gas, it is possible to combine the measured electrical potential, which is about 1 V, with the C potential [22],

$$E = 0.09992T \ (\log p_{CO} - 1.995 - 0.15 \ C \ \log \ C) - 816 \tag{50}$$

or rewritten [45] as

$$E = 807 - \frac{T}{100}\left[36.86 + 1.5C + 4.31 \ln\left(\frac{0.457C}{p_{CO}(0.785C + 215}\right)\right] \tag{51}$$

The interrelationship between the electrical potential of the probe, temperature, and C potential according to Equation (51) is illustrated in Figure 9 [45] for an atmosphere consisting of

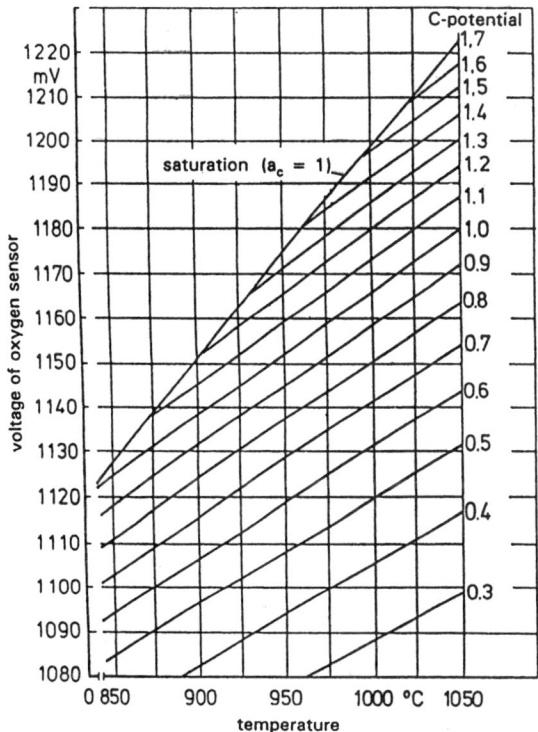

Figure 9   Controlling carbon potential by means of the oxygen sensor (endo gas enriched with propane). (From Ref. 45.)

an endothermic gas as a carrier gas and propane as enriching gas. The linear progression facilitates a high resolution that allows the C potential to be measured with an accuracy smaller than 0.005.

Methane reacts rather slowly according to Equation (25) but forms rapidly at carburizing temperature on cracking of the previously mentioned carburizing gases. That is why in actual practice carburizing atmospheres contain more CH_4 than would correspond to the state of equilibrium, thus making the content of methane, which can also be measured by infrared analysis, an unsuitable means of controlling carburizing potential. Dew-point analysis for measuring the water vapor content according to Equation (23) has lost its importance since the development of, above all, the oxygen probe. Thus, the C potential is always determined indirectly. Direct procedures such as measurement of the electrical potential of carbon [47] have so far not proved to be safe enough in processing.

With the development of computer techniques, the van Orstrand–Dewey solution [Equation (19), Section II] for carbon could be applied for controlling carbon on-line [43,48–53]. It suffices to calculate the diffusion coefficient D_C^γ step by step with appropriate formulas that have been derived from test results [54,55] by regression analysis:

$$D_C^\gamma = 0.47 \exp\left[-1.6C - \frac{37,000 - 6600C}{RT}\right] \qquad [cm^2/s] \quad \text{(Ref. 54)} \tag{52}$$

or

$$D_C^\gamma = 0.78 \exp\left[-\frac{18,900}{T} + \left(\frac{4300}{T} - 2.63\right)C^{1.5}\right] \quad \text{(Ref. 56)} \tag{53}$$

The exact and continuous quantitative determination of the C potential as a controlling factor and the on-line calculation of the carbon profile have facilitated computer-controlled carburizing, which is state of the art in heat treatment technique [57–59]. In particular, with the establishment of deeper carbon diffusion depths this technique is frequently applied in a two-stage process (boost-diffuse method of carburizing). In the first stage a C potential is established that can be held, by means of computer control, just below saturation, avoiding the formation of carbides or soot. The steep carbon concentration gradient, according to Equation (16), causes rapid carbon diffusion. In the second stage, the C potential is adjusted to the desired surface carbon content and the initial carbon gradient is equalized by some of the carbon diffusing deeper into the treated component and some of it diffusing into the gaseous atmosphere until the appropriate carbon gradient is reached [59–61].

4. Carbonitriding

In carbonitriding, ammonia (NH_3) is introduced into the carburizing atmosphere. Ammonia decomposes at carburizing temperature and releases nitrogen with the ability to be absorbed by the steel. The simultaneous and competing diffusion of carbon and nitrogen leads to some particularities [62–64] that are technically used. Solute nitrogen that occupies an interstitial position in the austenite lattice causes the GSE curve of the homogeneous austenite phase field to shift to lower temperatures and reduces the carbon content of the eutectoid composition. Nitrogen stabilizes the austenite, thus reducing the diffusion-controlled transformation of the austenite to ferrite and pearlite and lowering the martensite start temperature M_s. Martensite is also stabilized by dissolved nitrogen. The essential advantage of carbonitriding lies in the better hardenability (stabilized austenite), because this also allows better control of hardness profiles in steels that are not intended to be carburized as well as in sintered iron and/or the use of milder quenchants with less distortion. Moreover, C-N martensite has better tempering behavior. Whereas carburizing temperatures generally exceed 900°C (1652°F), carbonitriding is performed at lower temperatures in the range of 815–900°C (1499–1652°F) because at temperatures of more than 900°C (1652°F) ammonia decomposes too fast and the nitrogen potential [64] becomes too high, thus causing increased contents of retained austenite and even pores. Consequently, carbonitriding is usually applied to produce case depths of less than 0.5 mm, which can be better accomplished in carbonitriding than in carburizing.

C. Hardenability and Microstructures

After they have been carburized, the components are subjected to a hardening process in which they are quenched in oil [whose temperature is usually held at 60°C (140°F)], for economic reasons sometimes from their carburizing temperature [generally 930°C (1706°F)] but more often from a lower temperature [≈830–850°C (1526–1562°F)] that is adapted to the hardening temperature of their case carbon content (direct quenching). Reheating is a heat treatment in which a component is slowly cooled from its carburizing temperature to a temperature below Ac_3, most often to room temperature, and subsequently austenitized at hardening temperature, corresponding to the established case carbon content. Reheating is a suitable heat treatment for large or single components but otherwise not in wide use. Double quenching from austenitizing temperatures that are adapted to the carbon content of case and core, respectively, is seldom used in commercial practice, the ensuing distortions caused by two subsequent martensitic transformations being only tolerable with simple geometries. The carbon gradient (Figure 10 [65]) becomes the desired hardness gradient (Figure 11 [65]), which is determined according to DIN 50190, Part 1, by means of HV1 and contains the case (hardening) depth that is determined by the limiting hardness of 550 HV1. The carbon gradients in Figure 10 verify that it is possible to establish the same gradients for case carbon content and case depth for differently alloyed carburizing steels when drawing upon the discussed knowledge on carburizing and taking into consideration the influence of the alloy on carbon absorption by controlling the C potential. The corresponding hardness gradients depend on the hardenability [66], which has an obvious influence on the hardness of case and core, causing the case hardness values to drop and the core hardness values to rise with rising alloying content. The better hardenability of alloyed microstructures is understandable; maximum hard-

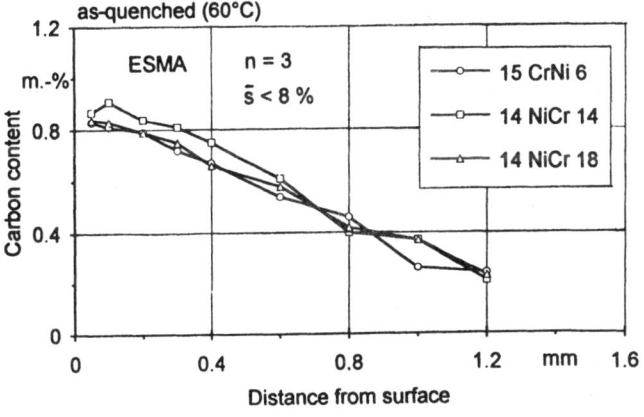

Figure 10 Carbon gradients of alloyed steels that have been carburized under computer control. (From Ref. 65.)

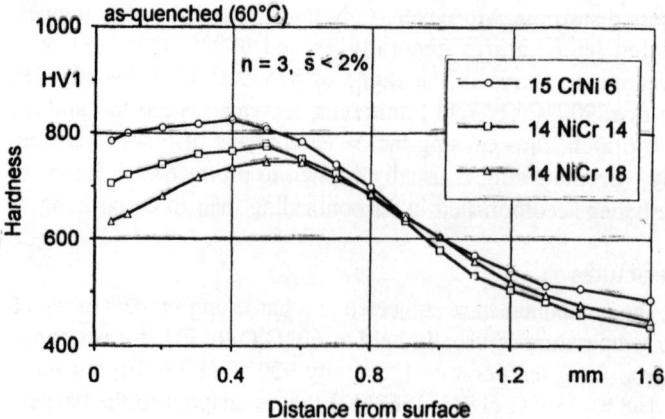

Figure 11 Hardness gradients of carburized microstructures with carbon gradients according to Figure 10. (From Ref. 65.)

ness is, however, generally attributed to the carbon content (which is equal in the above discussion). The influence of the alloy on the hardness of the case is an indirect one, as an increase in the alloying content (and in the carbon content) causes a drop in the martensite start temperature M_s [67,68], thus causing a drop in the martensite formation because the martensite finish temperature M_f is lower than the quench temperature [usually 60°C (140°F)]. The austenite that has not transformed to martensite is called retained austenite; its microstructure is soft and thus responsible for a drop in the hardness. The amount of retained austenite still existing after quenching (Figure 12 [65]) can be reduced by deep cooling, which causes further amounts of austenite to transform into martensite (Figure 13 [65]). The case hardness of deep-cooled microstructures does not differ within the measuring accuracy (Figure 14 [65]). Hence, the hardenability of the case [69] is primarily determined by the content of dissolved carbon. Maximum case hardness is determined by the amount of retained austenite in the case microstructure, which depends on the carbon content and the alloying elements. In contrast, the hardenability of the core is markedly dependent on the alloy (Figure 15 [70]) and serves as a criterion for the selection of suitable alloy steels. The preferred alloying elements are chromium, manganese, molybdenum, and nickel [71], and also, to a lesser extent, boron [72].

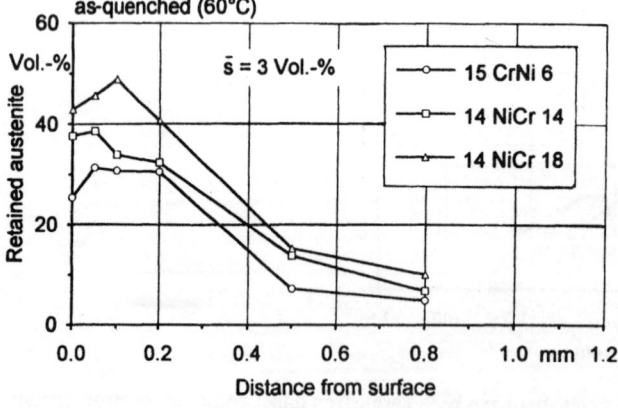

Figure 12 Amount of retained austenite, as-quenched conditions. (From Ref. 65.)

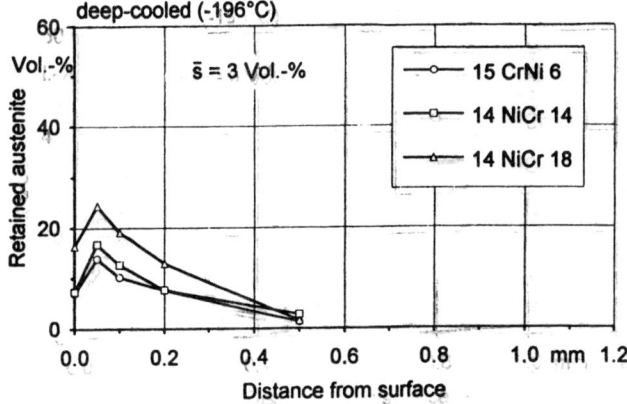

Figure 13 Amount of retained austenite, deep-cooled conditions. (From Ref. 65.)

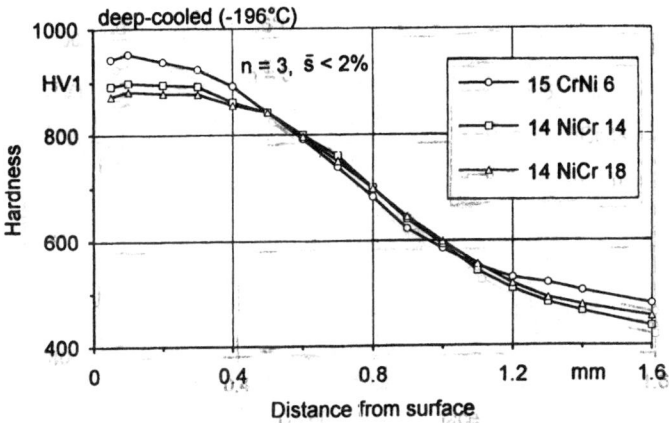

Figure 14 Hardness gradients, deep-cooled conditions. (From Ref. 65.)

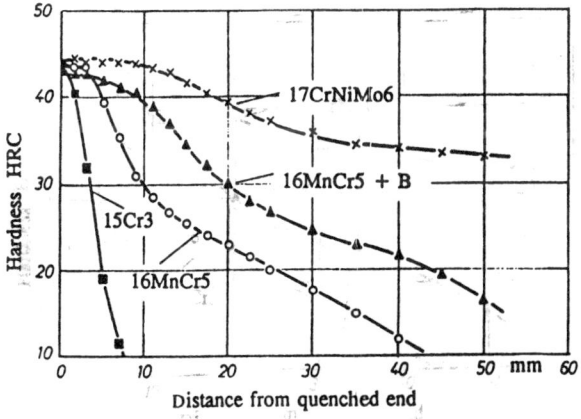

Figure 15 Jominy test results for some carburizing steels. (From Ref. 70.)

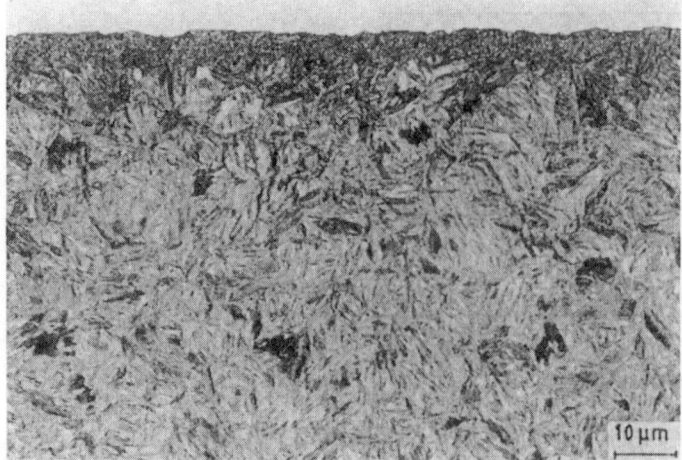

Figure 16   Plate martensite and retained austenite in the case (14NiCr18, 0.7% C, 25% retained austenite). (From Ref. 65.)

The case microstructure thus consists of (plate) martensite and retained austenite. The influence of the carbon content on the formation of the case microstructure is clearly visible in Figure 16 [65] (0.7% C, 25% retained austenite) and Figure 17 [65] (1.0% C, 60% retained austenite). Reliable values for the content of retained austenite can only be obtained by X-ray analysis. The core microstructure consists of (lath) martensite (Figure 18 [65]) or, given larger components, of bainite or ferrite and pearlite. The diffusionless martensitic transformation can occur only within the austenite grains. At low martensite start temperatures, i.e., when the carbon contents of the case microstructure are high, single martensite plates form successively within the austenite grain with $\{225\}_A$ and $\{259\}_A$ habit planes [73–77] originating from nuclei on the austenite grain boundaries [78]. The possible variants of the habit planes lead to the development of nonparallel, irregularly distributed martensite plates that, starting

Figure 17   Plate martensite and retained austenite in the case (14NiCr18, 1.0% C, 60% retained austenite). (From Ref. 65.)

Figure 18 Lath martensite in the core microstructure (14NiCr18, 0.15% C). (From Ref. 65.)

from grain diameter size, become successively smaller until they take up as much space as possible in the austenite grain. The irregularly distributed single plates are clearly visible in the scanning electron micrograph (Figure 19). Single needles that seem jagged, obtained from cross sections of the plates together with retained austenite, are shown in the light micrographs of Figures 16 and 17. This martensite morphology is called plate martensite and is characteristic of the carburized case microstructure. With carbon contents of up to 0.4% carbon in supersaturated solution, martensite consists of plates that are arranged in packets of different orientations within the former austenite grain boundaries (Figure 18). The single plates have a thickness of about 0.1–0.5 µm; their largest diameters are in the range of 1–5 µm. The single plates are clearly visible in SEM magnification (Figure 20), as is their arrangement in layered packets. Light micrographs often allow one to just discern cross sections of the plates. From studies using the transmission electron microscope it follows that the dislocation den-

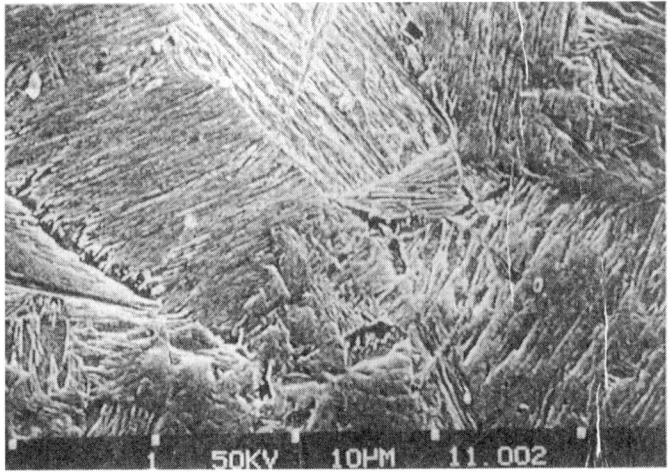

Figure 19 Lath martensite (16MnCr5), SEM micrograph.

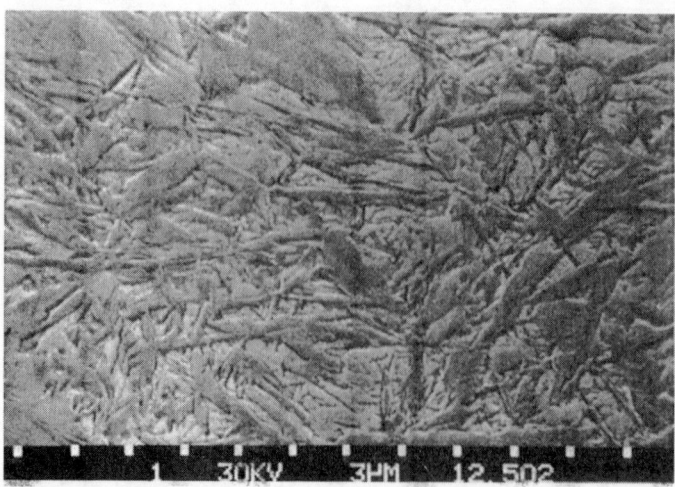

Figure 20   Plate martensite (16MnCr5), SEM micrograph.

sity of the single plates is high, in the range of 10^{11}–10^{12} dislocations per square centimeter [75].

It is possible to make the austenite grain boundaries visible in the carburized microstructure as the prior austenite grain size (Figure 21); they can usually only be etched separately from the matrix microstructure. The prior austenite grain size is a further characteristic of carburized microstructures, fine-grained microstructures having better behavior under all loading conditions. Martensitic transformations within large grains may cause microcracks in martensite plates (Figure 22) if a rapidly growing martensite plate (in the size range of 10^{-7} s [75,76]) impinges on an already existing plate, thus converting its kinetic energy into an impact that causes the plate to crack. Low transformation temperatures contribute to the development of microcracks [79,80].

In direct hardening, it is necessary to use steels whose austenite grain sizes do not increase much during carburizing. Suitable fine-grained steels are alloyed with aluminum and

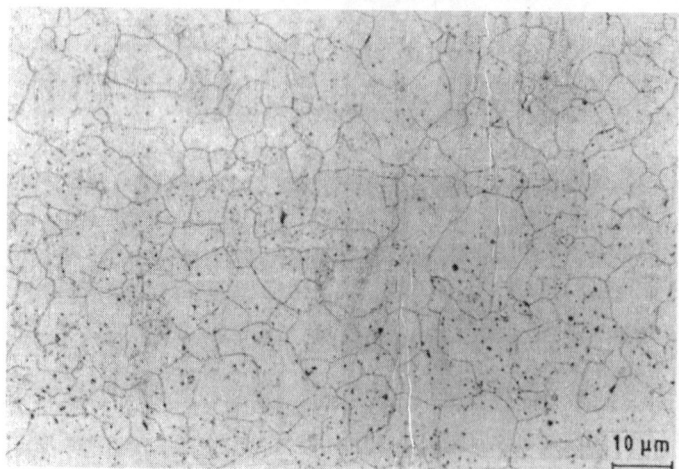

Figure 21   Austenite grain boundaries in carburized steel (15CrNi6).

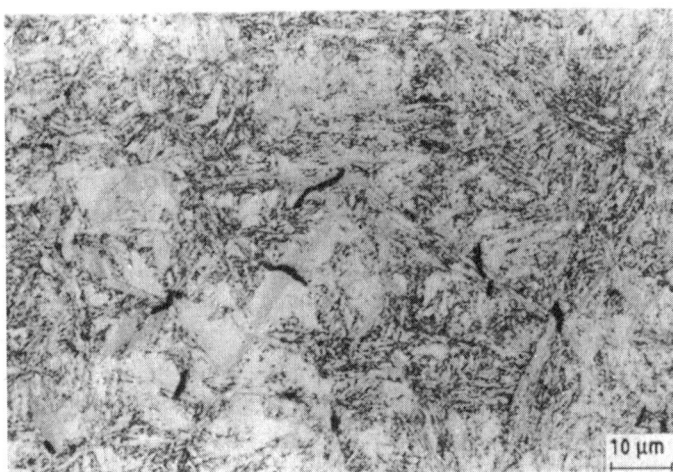

Figure 22 Microcracks in the plate martensite (16MnCr5).

nitrogen and subjected to a thermomechanical treatment in which a fine grain develops that is stabilitized by the precipitation of aluminum nitride at grain boundaries [81–84]. The ratio of aluminum to nitrogen ought to be about 3:5.

The absorption of carbon into the case microstructure and the martensitic transformation lead to an increase in volume with respect to the initial microstructure. In combination with the cooling and the transformation process, residual stresses develop in the carburized microstructure, with compressive residual stresses in the case that turn into tensile residual stresses upon reaching the case depth (Figure 23) [65].

Carburized microstructures are almost always tempered to transform the unstable and brittle as-quenched martensite into the more stable tempered martensite. This leads to an increase in ductility and thus minimizes the occurrence of delayed fracture [86]. The transformation of retained austenite tends to decrease distortion. With common carburizing steels the tempering temperatures are limited to 180–200°C (356–392°F)] to ensure that the usually required surface hardness of more than 60 HRC is still maintained, and for economic reasons the tempering time is almost always no more than 2 h. With high-alloy steels such as M50-NiL/AMS

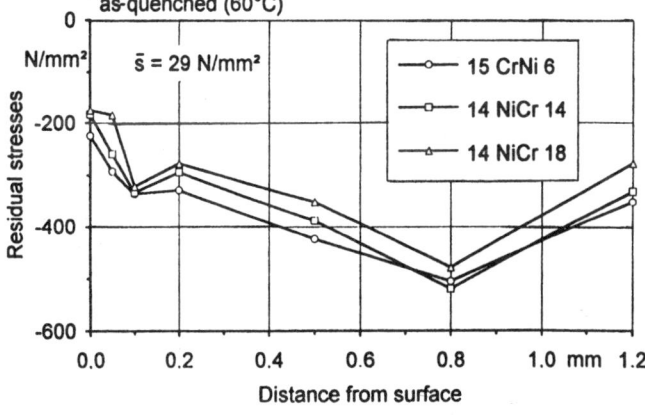

Figure 23 Residual stresses in the carburized case. (From Ref. 65.)

6278 [87,88], a change in tempering conditions may become necessary. In tempering, carbon diffuses from the supersaturated martensite lattice in several temperature-dependent stages and segregates at lattice defects or forms carbides with the iron atoms of the matrix [85,89–93]. In the temperature range between 100 and 200°C (212 and 392°F), i.e., in the temperature range of conventional tempering of carburized microstructures, transitional carbides are precipitated from the martensite in steels with a carbon content higher than 0.2% (stated are hexagonal ε-carbides Fe_2C and, particularly with higher carbon contents, orthorhombic η-carbides $Fe_{2.4}C$ and monoclinic γ-carbides $Fe_{2.5}C$) [91–97]. As a supersaturated α'-solid solution, the martensite contains up to 0.2% carbon segregated at lattice defects. The tetragonality and ensuing hardness of the martensite decrease [98]. With carbon contents lower than 0.2%, ε-carbide precipitation is highly improbable for energy-related reasons. Under the tempering conditions that are common practice in carburizing, only a small amount of retained austenite is transformed. The complete transformation of retained austenite would take 2 h of tempering at 300°C (572°F) [99] or more than 50 h of tempering at 180°C (356°F) [100], because the effects of tempering time and tempering temperature are interchangeable [101,102]. It is also possible to decrease the content of retained austenite by means of deep cooling, but only at the expense of mechanical properties, particularly fatigue resistance [65]. The transformations that occur in a microstructure subjected to 2 h of tempering at 180°C (356°F) cannot be resolved in the optical microscope.

Impurity atoms, phosphorus in particular, may segregate at the austenite grain boundaries during carbon diffusion. In the diffusionless transformation, this state of segregation is inherited from the martensite and can lead to its embrittlement if retained austenite is transformed in tempering and cementite is precipitated at the former austenite grain boundaries that are segregated with phosphorus [103–108]. This one-step temper embrittlement occurs at temperatures around 300°C (572°F) and is effective only in rapid loading. In common tempering practice, temper embrittlement of carburized components is hence very unlikely. Tempering causes a shift in the hardness gradient toward lower values and deeper case depth (Figure 24), and the residual stresses become smaller (Figure 25).

In all the oxygen-containing carburizing atmospheres, oxygen reacts with the elements of the case microstructure, with chromium, manganese, and silicon oxidizing under carburizing conditions and iron, molybdenum, and nickel being reduced [109]. As a consequence, surface oxidation occurs and is most prominent alongside the grain boundaries where diffusion

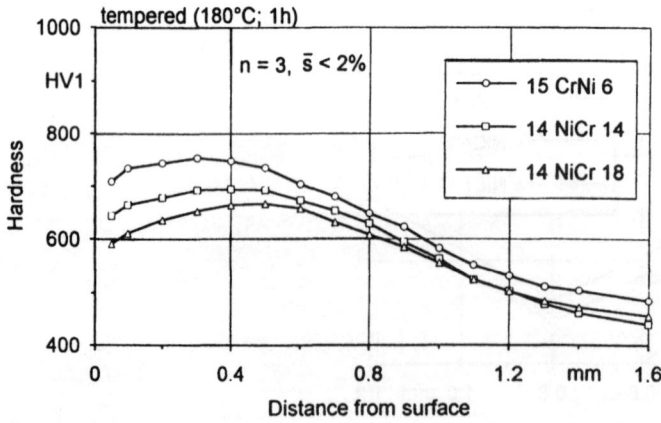

Figure 24   Effect of tempering on hardness gradients. (From Ref. 65.)

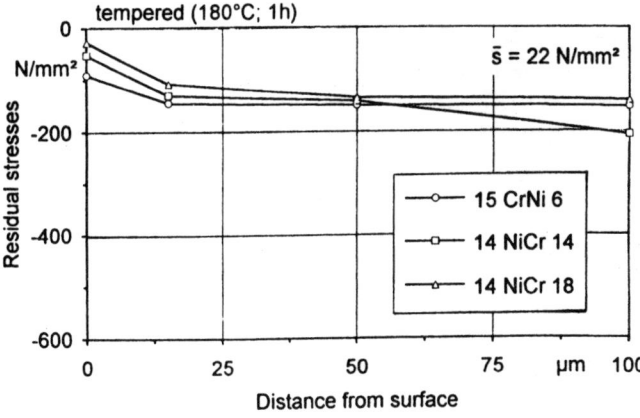

Figure 25 Effect of tempering on residual stresses. (From Ref. 65.)

is faster. Intergranular surface oxidation (Figure 26) is a characteristic of carburized micro-structures [110,111] that can be avoided only by oxygen-free carburizing atmospheres as are generally used in low-pressure carburizing or in plasma carburizing [26–29]. Carbide formation in carburized microstructures occurs when the C potential is in the two-phase austenite/cementite field or when a microstructure carburized in the homogeneous austenite phase field in a subsequent heat treatment has stayed in the two-phase field for a sufficiently long time. It should be taken into consideration that the carbon content of the eutectoid composition and the S–E line in the Fe–C diagram shift toward lower values by alloying elements, particularly by chromium [112]. With a C potential near the cementite point in a two-stage carburization, ledges carburized by two planes are especially vulnerable (Figure 27). Carbides are almost always unwelcome, the more so if precipitated on grain boundaries (Figure 28). Nonmetal inclusions in steel such as sulfides or oxides remain unchanged by carburizing (appearing as dark areas in Figure 27).

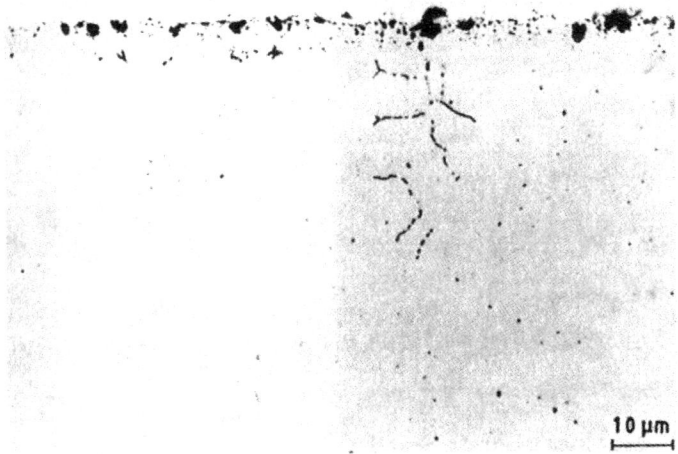

Figure 26 Surface intergranular oxidation (15CrNi6).

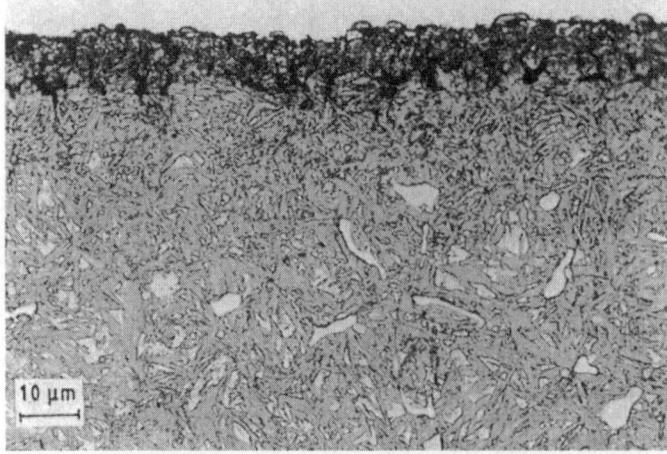

Figure 27   Globular carbides in the case microstructure (15CrNi6).

IV.   REACTIONS WITH HYDROGEN AND WITH OXYGEN

Atomic hydrogen with the ability to diffuse into steel develops in electrochemical processes [113], welding [114], and heat treatments such as carburizing [115]. Hydrogen in steel can recombine to form molecular hydrogen or cause failure by coupling with lattice defects, imperfections, and microcracks, the most severe failure being the hydrogen-induced embrittlement of high-strength steels [116]. The partial pressure of hydrogen, $p_{H2} = 5.3 \times 10^{-7}$, in an atmosphere for 1 bar is very low; therefore, it ought to be possible for hydrogen as a highly diffusible element in an interstitial position to effuse into the air, making effusion annealing (soaking) unnecessary. Hydrogen, however, besides being interstitially dissolved, is also attached to lattice defects [117–119] such as vacancies, dislocations, and grain boundaries, to elements of the microstructure [120] such as precipitates and inclusions, and to imperfections [121] such as pores and microcracks. These linkage conditions are zones of increased absorption

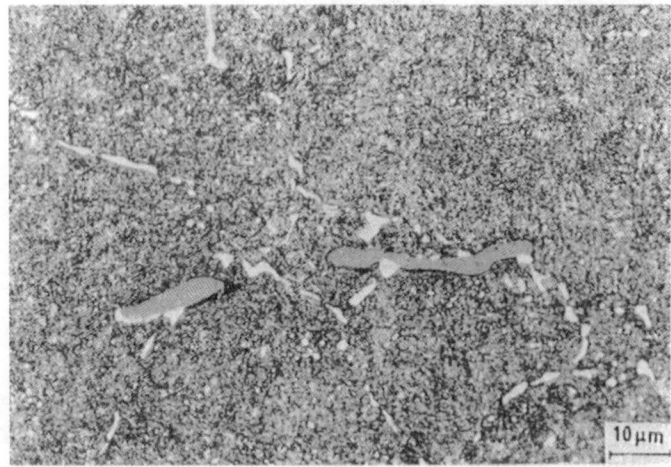

Figure 28   Carbides on grain boundaries in the carburized case (15CrNi6).

of hydrogen (hydrogen traps) during intake and act as hydrogen sources in degassing, which is why solubility and the diffusion coefficient are dependent on morphology at temperatures below about 400°C (752°F) [122,123]. Carburizing steels [123], for instance, are cited with diffusion coefficients of $D_H = 3.4 \times 10^{-7}$ cm^2/s for the core microstructure (0.2% carbon) and $D_H = 0.9 \times 10^{-7}$ cm^2/s for the case microstructure (0.8% carbon). Moreover, the combined hydrogen means that the maximum degassing rate [6],

$$f = \frac{c_t - c_0}{c_s - c_0}$$

(54)

is dependent on degassing time and temperature (Figure 29 [124]). In Equation (54), c_0 is the initial hydrogen concentration and c_t the concentration after the degassing time t in a component; c_s is the concentration at the surface. At room temperature, normalized steel has a hydrogen content of 0.05–0.1 µg H/g Fe [114], which may increase to 16 µg H/g Fe [125] in deformed microstructures that have a high lattice defect density. As oxide layers impede the diffusion of hydrogen into the atmosphere, degassing is usually carried out in an oxygen-reducing atmosphere.

Iron is produced from iron oxide (and iron sulfide) ores by supplied energy and therefore oxidizes in every atmosphere in which the oxygen partial pressure is higher than the oxygen partial pressure of iron oxide [32]. Figure 30 shows the dependence of the oxygen partial pressures of iron and its oxides on temperature [126–128]. Oxygen partial pressures above the Fe curve stimulate oxidation. The iron oxides are reduced on falling below their respective partial pressures. It follows that iron and steel with similar behavior oxidize in air ($p_{O_2} = 0.21$ bar). In technical terms, the surface of a steel is referred to as bright when the oxide layer, which inevitably forms in air, is only about 20 nm thick and thus not visible to the eye. Tarnished surfaces that are equivalent to oxide layers between ~ 50 nm (straw color) and ~ 70 nm (blue color) are regarded as technically clean. Layers of more than about 1 µm are discernible in the light microscope and develop up to about 5 µm during blueing, turning black. Oxide layers starting from about 10 µm thickness that form at temperatures exceeding 500°C (932°F) are no longer dense and scale off because of the different volumes of oxide and substrate.

Thick oxide layers may adversely affect the properties of a component, whereas thin oxide layers obstruct the mass transfer and must therefore be reduced, in particular prior to carburizing or nitriding. It was pointed out earlier that oxidation can be avoided either by lowering the (overall) pressure in the furnace (vacuum processing) or by displacing the air in the furnace with a protective gas. The partial pressure of wüstite (Fe$_{0.947}$O) is, according to Refer-

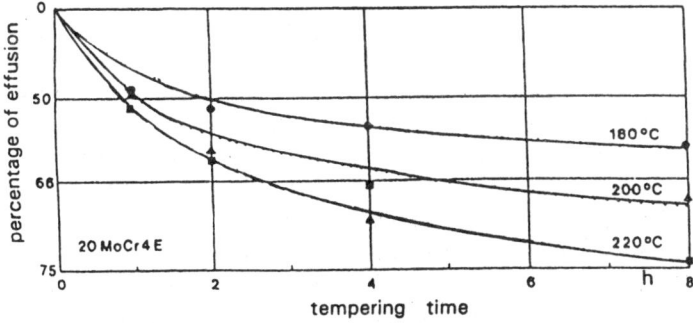

Figure 29 Effusion of hydrogen during tempering in air. (From Ref. 124.)

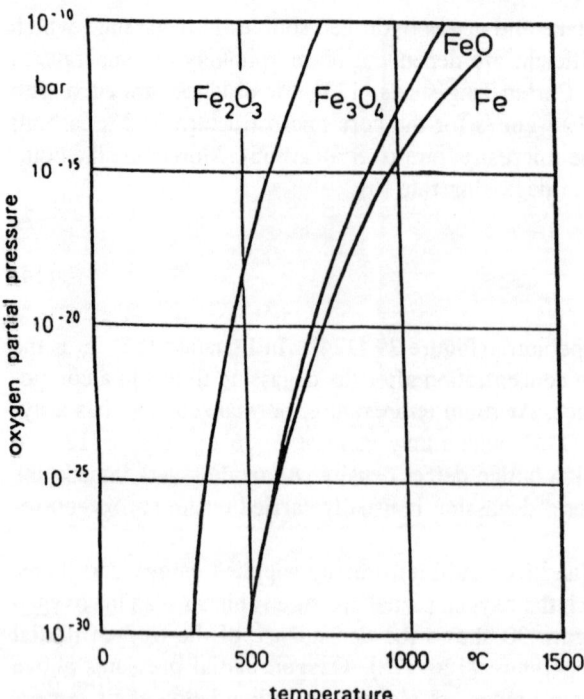

Figure 30 Oxygen partial pressure of iron and iron oxides. (Data from Refs. 126–128.)

ence 128, 1.07×10^{-19} bar at a temperature of, e.g., 800°C (1472°F). That is, to avoid oxidation caused by air ($p_{O_2} = 0.21$ bar) it would be necessary to decrease the pressure within the furnace to less than 10^{-18} bar. In vacuum furnaces, however, considerably smaller vacuum values in the range of 10^{-5} bar suffice to attain bright surfaces. During heating and evacuation, some of the trapped hydrogen will escape and react with the low, oxygen contents in the vacuum according to

$$2H_2 + O_2 \rightleftharpoons 2H_2O \tag{55}$$

to generate water vapor, whose oxygen partial pressure is considerably lower than that of free oxygen. The reduced oxygen partial pressure may result in the reduction of iron oxides. Oxygen molecules that have not transformed to water vapor react with the steel to form invisible oxide layers. This phenomenon becomes clear by studying how the leakage rate of the vacuum furnace affects the growth of the oxide layer [126]. Figure 31 [126] shows the relation between the maximum thickness of the layer (and the decarburizing depth of the case related to 1% carbon in the steel) and the leakage rate of a vacuum furnace with the parameters final pressure, furnace volume, and treatment time. With leakage rates smaller than 10^{-2} mbar·L/s, which can be obtained without much technical difficulty, the penetrating oxygen is of virtually no importance to the thickness of the oxide layer (and to the decarburizing depth). It follows that in industrial vacuum furnaces oxidation is avoided by gas reactions, and thus the vacuum (heat) treatment can be designated as heat treatment with gases.

Pure nitrogen is inert with respect to steel and is thus an economical protective gas. In a nitrogen atmosphere with an oxygen partial pressure of zero, iron oxide is reduced until the oxygen partial pressure of the atmosphere, which has developed from the emitted oxygen,

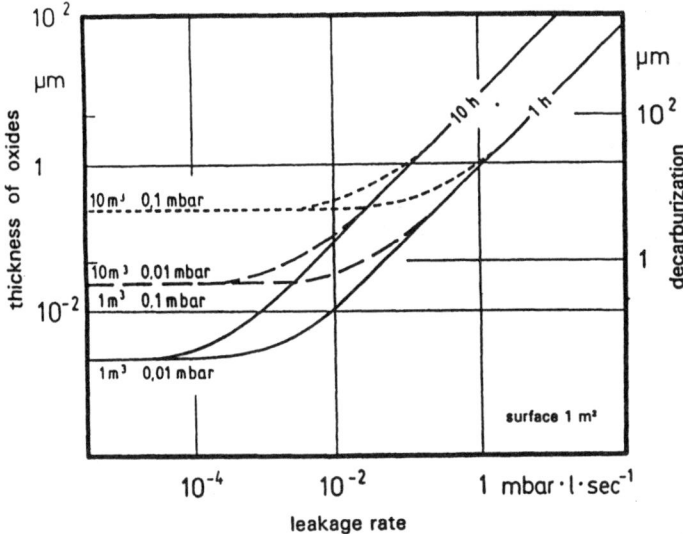

Figure 31 Leakage rate of a vacuum furnace and decarburization rate. (From Ref. 126.)

reaches the same level as the oxygen partial pressure of the oxide. The oxygen partial pressures of the iron oxides (Figure 30) being minimal, the nitrogen atmosphere scarcely reduces. Moreover, industrial nitrogen contains small amounts of oxygen that are equivalent to an oxygen partial pressure in the range of 10^{-5} bar, and even purified nitrogen still contains oxygen equivalent to a partial pressure of about 10^{-6} bar [129]. Therefore, at least 2 vol % hydrogen is generally added to a nitrogen atmosphere to ensure that, according to Equation (55), water vapor is generated via the transformation of the total of the existing or developing oxygen that with leaking furnaces diffuses into the furnace from the ambient air. The oxygen partial pressure of such an atmosphere can be calculated according to

$$p_{O_2} = \frac{k_p p_{H_2O}}{p_{H_2}} \tag{56}$$

with the equilibrium constant k_p listed in Reference 128. Atmospheres consisting of nitrogen, hydrogen, and water vapor can easily be assessed because the ratio of H_2O to H_2 of their respective partial pressures is independent of temperature. Figure 32 [126] shows the interrelationship between the ratio of partial pressures, oxygen potential, and temperature and also plots of the phase fields of iron and iron oxides. From these it is possible to ascertain gas compositions with neutral, reducing, or oxidizing effects. The HN gases, or mono gases composed of nitrogen and hydrogen, are generated by the complete decomposition of ammonia,

$$NH_3 \rightarrow 3H_2 + N_2 \tag{57}$$

or from a purified and dried generator gas. According to Equation (45), these gases have a decarburizing effect, with a rather sluggish reaction [44] that is effective only at treatment temperatures of about 500°C (932°F) and more. HN gases are therefore mainly used in the heat treatment of steels with a low carbon content or at temperatures below 500°C (932°F) such as for tempering. Pure hydrogen atmospheres have excellent heat conduction properties and are used as a protective gas, for instance in bell-type annealing furnaces. To avoid an

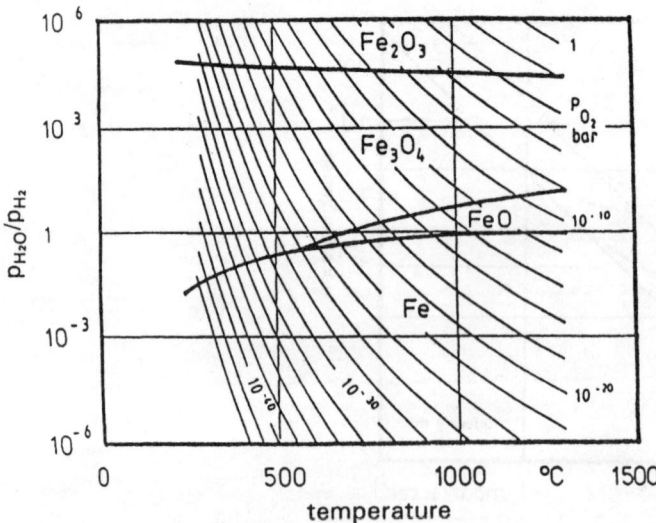

Figure 32 Partial pressure ratio p_{H_2O}/p_{H_2} versus temperature at various oxygen potentials. (From Ref. 126.)

explosive mixture of oxygen and hydrogen, it is necessary to control the oxygen content by means of the oxygen sensor.

In controlled oxidizing or blueing, annealing is accomplished in an atmosphere of saturated water vapor (99.9%), the reaction according to Equation (55) proceeding from H_2O, thereby forming a black oxide layer of about 1–5 μm thickness whose corrosion resistance is better than that of iron. This procedure is mainly applied for protecting tools against corrosion from handling.

Decarburization is avoided by using gases composed of CO and CO_2, H_2, and H_2O diluted with N_2 (referred to as CCHN gases, or exo gases) as protective gases, with their C potentials adjusted to the carbon content of the steel according to the interrelationships discussed in Section III.B. At thermodynamic equilibrium, which is established at higher temperatures according to the homogeneous water gas reaction, Equation (26), there is a correspondence among the oxygen partial pressures calculated from the H_2O/H_2 and CO_2/CO ratios and also with the free oxygen. This means that in addition to the interrelationship of the H_2O/H_2 ratio as shown in Figure 32, the corresponding interrelationship of the CO_2/CO ratio as shown in Figure 33 [126] has to be taken into consideration. As a rule, CCHN gases are generated as generator gases by the combustion of hydrocarbons such as methane (natural gas) or propane with air, the ratio of combustion gas to air determining the quantitative composition and thus the C potential.

V. NITRIDING AND NITROCARBURIZING

A. Introduction

The industrial application of nitriding started with studies by Machlet [130] and Fry [131,132], who introduced the classical gas nitriding in ammonia atmospheres at the temperature range of α-iron. This gas nitriding process as recommended by Machlet and Fry is still applied as a treatment that causes minimal distortion in components consisting primarily of nitriding steels

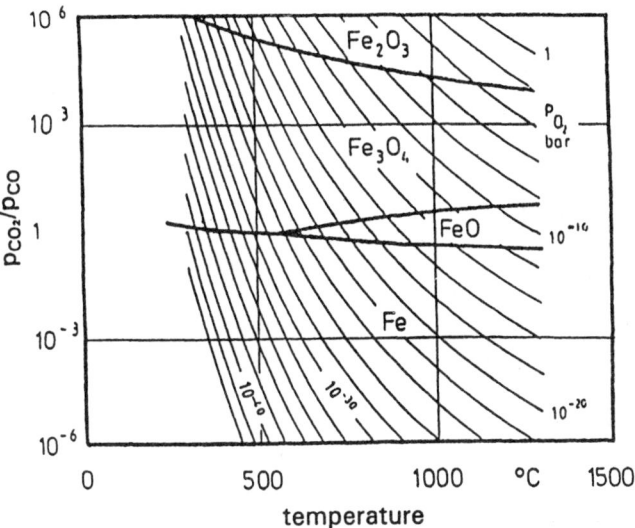

Figure 33 Partial pressure ratio p_{CO_2}/p_{CO} versus temperature at oxygen potentials of $1-10^{-40}$ bar. (From Ref. 126.)

that have to endure abrasion, rolling contact fatigue, and fatigue. Depending on the desired nitriding hardness depth, a treatment time of up to 120 h may be necessary. Based on studies by Kinzel and Egan [133], nitriding in cyanide-containing molten salts was subsequently developed [134–136]. In this procedure both carbon and nitrogen are introduced into the case, which is why it is called salt bath, or liquid, nitrocarburizing. Nitrocarburized microstructures develop also in gaseous atmospheres when ammonia is blended with a carbon-emitting gas [137–144]. Gas nitrocarburizing is widely applied, and the use of additional gases such as endo gas, exo gas, carbon dioxide, methylamine, and methanol has resulted in several processing variants. This section describes gas nitriding and gas nitrocarburizing.

At standard pressure, solid and liquid iron dissolve only a small amount of nitrogen, α-iron up to an optimum of 0.005 wt.% at 810°C (1490°F) on the transition to γ-iron, which at this temperature also has its maximum solubility of 0.035 wt.% for nitrogen (Figure 34 [145]), a binary phase diagram being of no interest. According to A. Sieverts, solubility can be improved by increasing the partial pressure of nitrogen in proportion to the square root of the partial pressure of nitrogen, $\sqrt{p_{N_2}}$. The main constituent of nitriding atmospheres is ammonia (NH_3), which on decomposing causes the partial pressures of nitrogen to increase considerably, thus being responsible for the formation of different solid phases at equilibrium [146,147] that vary in their respective pressure–temperature–concentration diagrams according to their nitrogen content and can be projected on a joint temperature–concentration surface plane in an Fe-N phase diagram (Figure 35) [146–148] that is consequently not at equilibrium at $p = 1$ bar. In technical application, processing is done mainly in the phase field of α-iron, which at 590°C dissolves a maximum of 0.115 wt.% nitrogen. The low solubility of nitrogen in α-iron together with the necessity of producing high partial pressures for nitriding has so far made it almost impossible to use nitrogen potentials that lead to nitrogen absorption only in the α-iron. The aim of the generous nitrogen supply is the formation of both nitrides and a microstructure that consists of a compound layer of nitrides and an adjoining diffusion zone. The nitriding atmospheres are at a nonequilibrium in this process, making a quantitative treatment such as carburizing impossible. In actual practice, a processing characteristic indicating

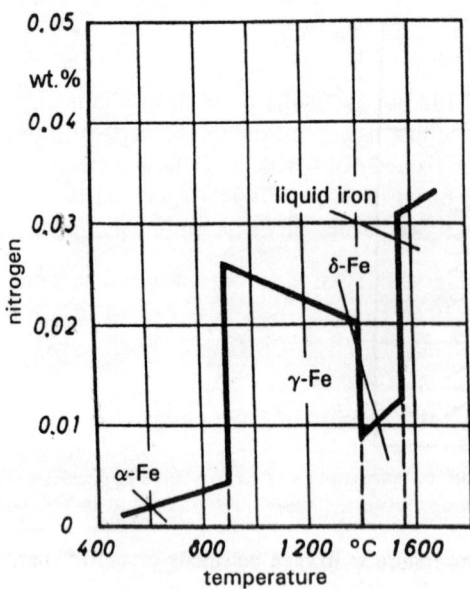

Figure 34 Solubility of nitrogen in steel at 1 bar. (From Ref. 145.)

the "nitriding power" or "nitriding potential" [149–152] or sensors [153–155] are therefore used, allowing the gas composition to be adjusted to the desired microstructure; i.e., characteristic and sensor signals are calibrated with regard to the nitriding microstructures. Hence, the following discussion starts with a description of nitriding and nitrocarburizing microstructures and subsequently deals with the processing.

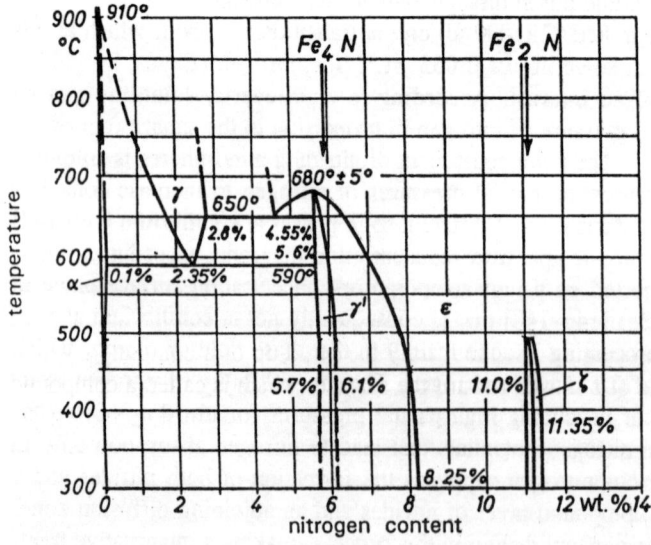

Figure 35 Binary Fe-N phase diagram. (From Ref. 146.)

B. Structural Data and Microstructures

1. *Structural Data*

Nitrogen has a (homopolar) atomic diameter of 0.142 nm and is dissolved in iron in interstitial positions in octahedral voids of the cubic lattice that have a maximum diameter of 0.038 nm in body-centered cubic α-iron and a maximum diameter of 0.104 nm in face-centered cubic γ-iron. Nitriding of pure iron at temperatures of up to 590°C (1094°F) with an increasing nitrogen content according to Figure 35 leads to the formation of the folllowing phases:

1. Body-centered cubic α-iron, which dissolves 0.001 wt.% nitrogen at room temperature and 0.115 wt.% nitrogen at 590°C (1094°F)

2. Face-centered cubic γ'-nitride, Fe_4N, which dissolves 5.7–6.1 wt.% nitrogen (stoichiometric 5.88 wt.%)

3. Hexahedral ε-nitride, $Fe_{2-3}N$, which exists in the range of 8–11 wt.% nitrogen.

Orthorhombic ζ-nitride (Fe_2N) forms at temperatures below about 500°C (932°F) and nitrogen contents exceeding 11 wt.%, i.e., at formation conditions that are not used in technical practice.

At temperatures above 590°C (1094°F), γ-iron is formed according to the binary Fe-N phase diagram (Figure 35); it dissolves a maximum of 2.8 wt.% nitrogen at 650°C (1202°F). At 590°C (1094°F) and 2.35 wt.% nitrogen, a eutectoid microstructure designated as braunite is produced that consists of α-iron and γ'-nitride. Nitrogen has a stabilizing effect on austenite, which therefore transforms slowly. Martensite formed on an Fe-N basis is moreover rather soft; therefore technical nitriding processes are usually limited to the ferrite field.

Strictly speaking, the structural data discussed above are valid only if pure iron is nitrided, that is to say they are valid only as an exception because generally the carbon present in the steel (in nitrocarburizing, also present in the nitriding medium) is incorporated into the case microstructure during the process and competes with the nitrogen. The nitrides absorb carbon, causing the range at which the carbonitrides $Fe_xN_yC_z$ exist to broaden relative to the binary system Fe-N, which can be inferred from the ternary system Fe-C-N (Figures 36 [156] and 37 [157]). Moreover, there is the possibility of θ-carbide (cementite) formation. In nitriding or nitrocarburizing alloyed steels, it should also be taken into account that nitrogen and carbon are prone to form nitrides or carbonitrides with alloying elements such as aluminum, chromium, and vanadium whose composition can only be given in the general form M_xN_y or $M_xN_yC_z$, where M stands for metal.

Present knowledge allows the application of the van Orstrand–Dewey solution (Equation 19) of the diffusion equation for nitrogen only in the α-iron [158]. The diffusion coefficients (as listed in Reference 153) are for α-iron, $D_N^\alpha = 6.6 \times 10^{-3} \exp(-77,820/RT)$; in γ'-nitride, $D_N^{\gamma'} \approx (1/25)D_N^\alpha$, and in ε-nitride, $D_N^\varepsilon \approx (1/60)D_N^\alpha$.

Nitrides develop by γ' nucleation at the interface between the nitriding atmosphere and the substrate and subsequent nucleation of ε-nitrides at the interface between the atmosphere and the already formed γ'-nitrides (Figure 38) [159] until the build-up of a complete layer of γ'-nitride toward the substrate and of ε-nitride toward the surface, which is called a compound layer. This compound layer allows controlled processing because of its defined concentration gradients. Nitrogen diffusion is, however, considerably slower in the compound layer than in the ferrite.

2. *Microstructures of Nitrided Iron*

The described microstructural formation is found in nitrided pure iron (Figure 39), and the sequence of nitride layers is clearly visible in SEM micrographs (Figure 40). The ε-nitride

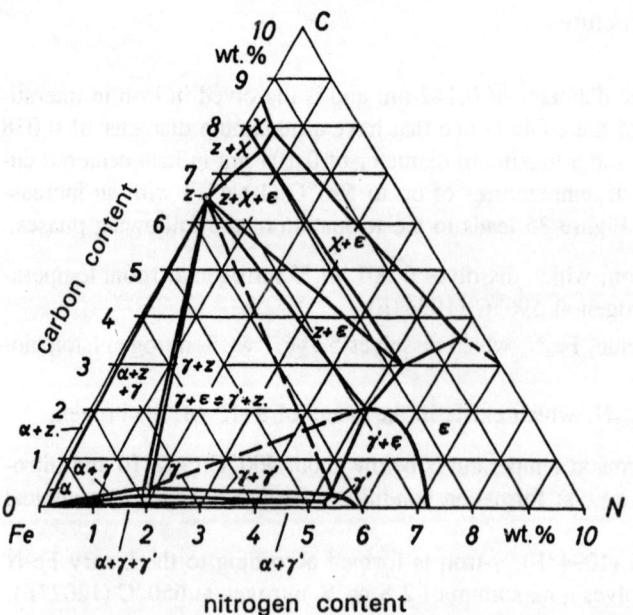

Figure 36 Ternary Fe-C-N phase diagram at 575°C. (From Ref. 156.)

layer shows pores that, according to the pressure theory [159], form when nitrogen atoms in the ε-nitride layer recombine to form nitrogen molecules, thus creating high pressures. A consistent theory on the causes of porosity has, however, not yet been offered [160]. Carbon changes the morphology of the compound layer, causing part of the ε-nitride and γ'-nitride to arrange side by side. The carbon may originate in the steel or in the nitriding atmosphere (Figure 41, nitrocarburized pure iron), in comparison with nitrided pure iron (Figure 42). The samples in Figures 41 and 42 are etched with Murakami solution, which has a stronger effect

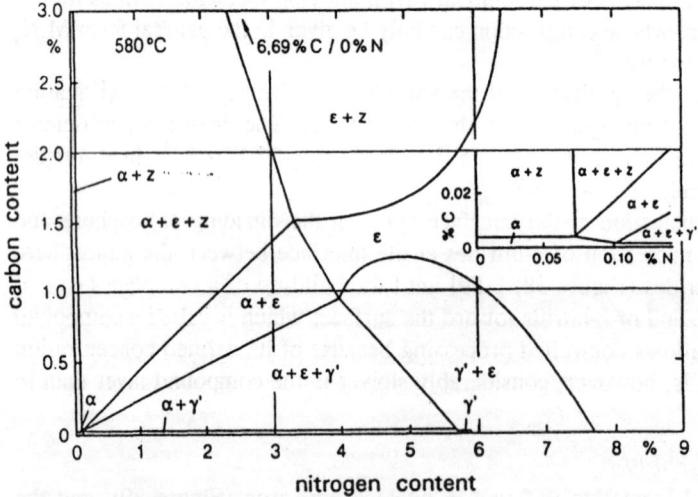

Figure 37 Part of ternary Fe-C-N phase diagram at 580°C. (From Ref. 157.)

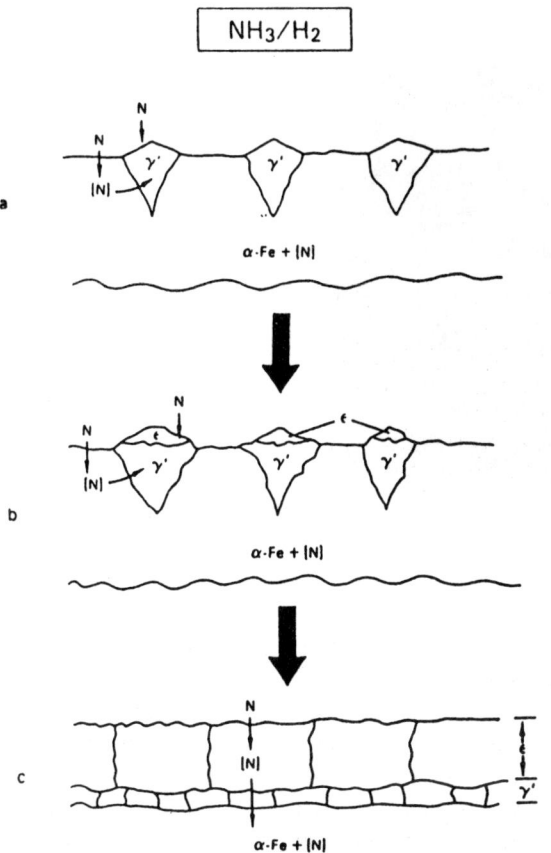

Figure 38 Scheme of nucleation of γ'- and ε-nitrides on iron. (From Ref. 159.)

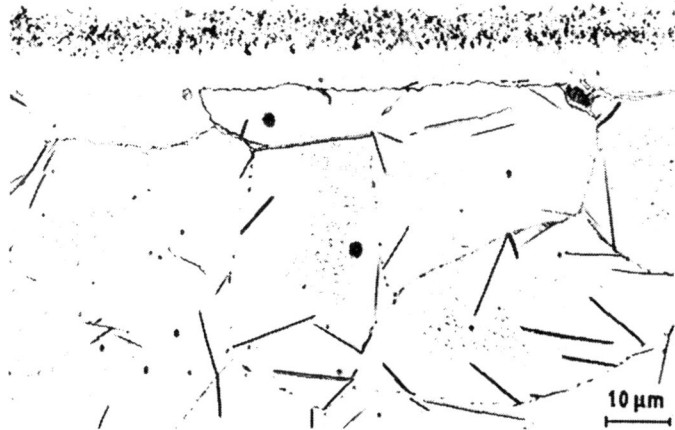

Figure 39 Pure iron nitrided, etched with Nital.

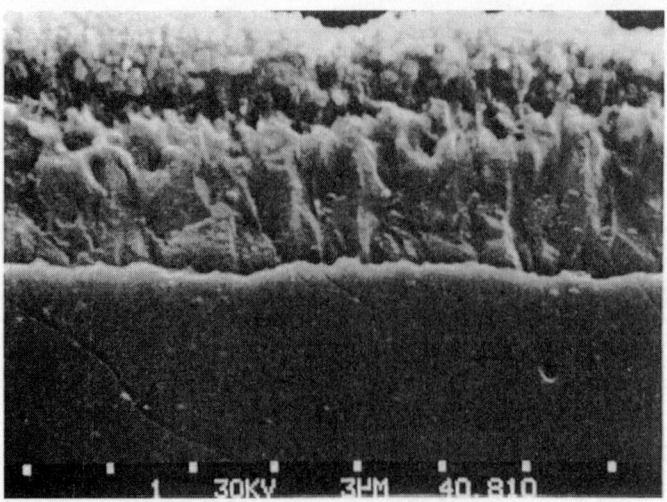

Figure 40 Pure iron nitrided, SEM micrograph.

on carbon-containing microstructure areas than on their surroundings, making them appear dark. The hardness of the compound layer is around 570 HV0.02 and is not significantly affected by the described morphological differences. Beneath the compound layer, nitrogen is dissolved in α-iron (diffusion zone) at nitriding temperature and segregates during cooling in correspondence with the decreasing solubility of nitrogen in α-iron as coarse hexahedral body-centered α''-nitrides (Fe_8N) (Figure 39). In quenched microstructures, nitrogen remains in supersaturated solution; subsequent aging leads to more uniformly dispersed finer α''-nitrides (Figure 43, aged 20 min at 325°C (617°F).

A considerable increase in the hardness of the diffusion zone is achieved only by increasing the amount of supersaturated dissolved nitrogen (solid solution strengthening) and is decreased again to the values characteristic of the slowly cooled diffusion zone by aging.

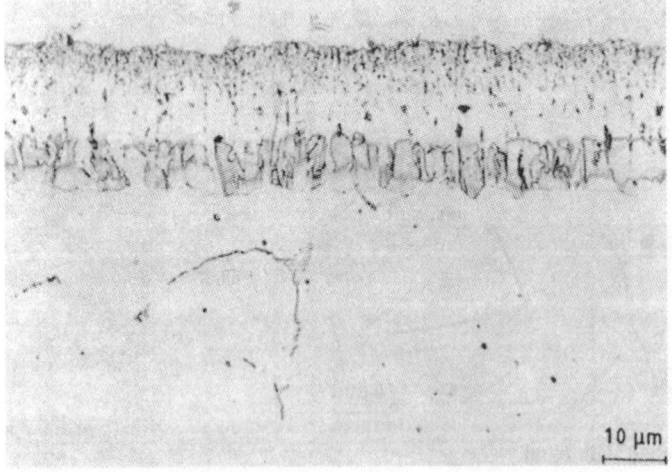

Figure 41 Pure iron nitrocarburized, etched with Murakami solution.

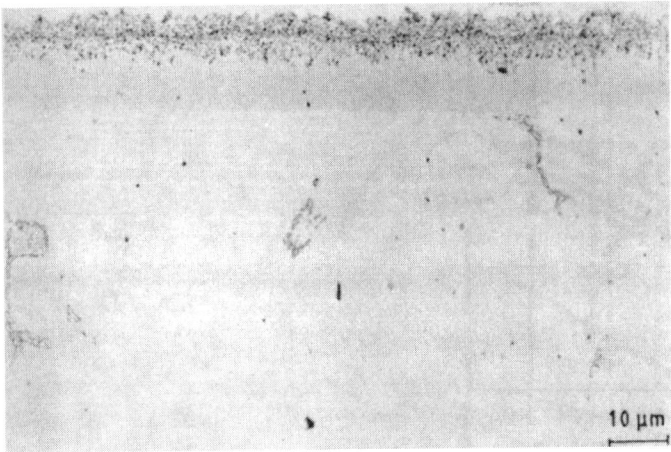

Figure 42 Pure iron nitrided, etched with Murakami.

3. Microstructures of Nitrided and Nitrocarburized Steels

With steels, there are no major changes in the described microstructures. Some specific characteristics, however, are altered by the existing carbon content and by nitride-forming alloying elements. With respect to the conditions at which those nitrides form it should noted that the temperatures and concentrations of the phase limits in Figures 35–37 are shifted by alloying elements.

Given equal nitriding conditions as to nitrogen supply and nitriding time, the compound layer of plain carbon steels is always thicker than that of steels alloyed with nitride-forming elements (Figure 44 [161]) because nitrides or carbonitrides formed with alloying elements contain more nitrogen than those formed with iron. The thickness of the compound layer, which is equivalent to the area of increased nitrogen content, is further dependent on the nitriding time (Figure 45), the nitriding temperature (at set nitriding time, Figure 46), and, most markedly, on its preliminary treatment (Figure 47, different processing). The share of the poros-

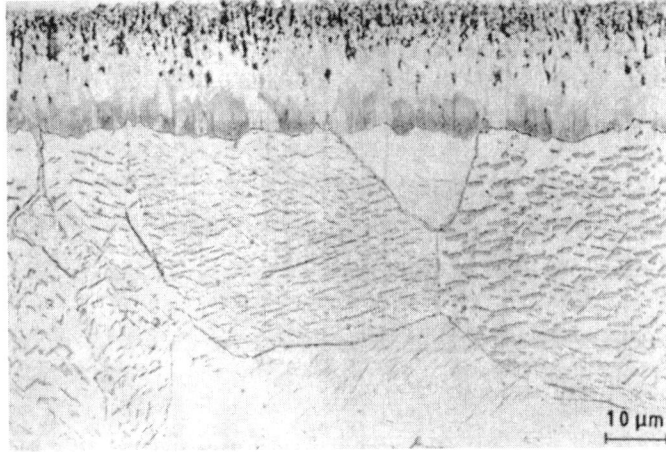

Figure 43 Pure iron, nitrided, quenched, and aged at 325°C for 20 min.

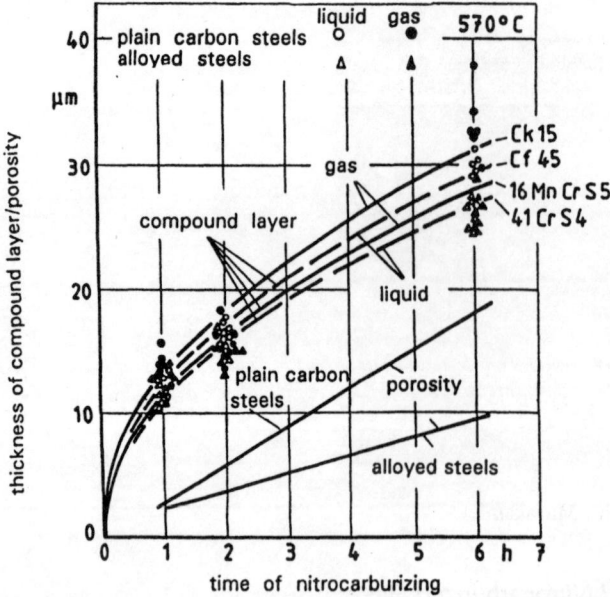

Figure 44 Thickness of compound layer versus treatment time. (From Ref. 161.)

ity fringe is lower with alloy steels than with plain carbon steels. The nitriding process, however, has little influence on the formation of the compound layer (Figure 44) or on the overall microstructure morphology provided that the respective nitriding conditions are equal. The
compound layer contains carbon, which is located as carbon enrichment in the form of a "carbon hill" at the inner interface between compound layer and diffusion zone and can be made
visible by means of an adequate etching technique (Figure 42, Murakami solution) as well as
by analytical methods (Figure 48). Within the diffusion zone of alloyed steels, the total of the
diffusing nitrogen is tied up in nitrides or carbonitrides that do not resolve under the light

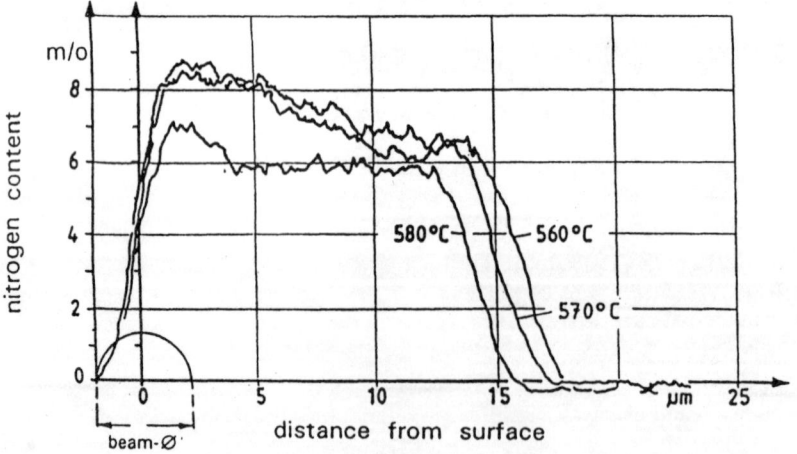

Figure 45 Influence of treatment temperature on thickness of compound layers. Ck15 steel gas
nitrocarburized at 570°C for 3 h.

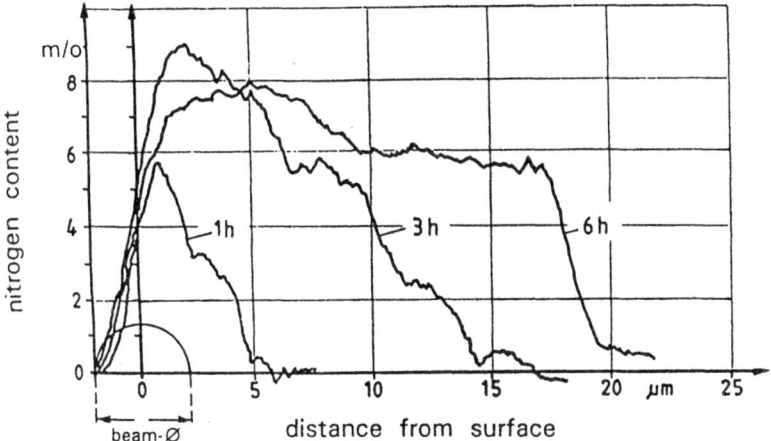

Figure 46 Influence of treatment time on thickness of compound layers. Ck15 steel gas nitrocarburized at 570°C for 3 h.

microscope. The formation of a supersaturated α-iron occurs only with plain carbon steels; it is possible for this α-iron to segregate iron nitrides by slow cooling from nitriding temperatures or by aging of the quenched microstructure (Figure 43).

The compound layers of plain carbon steels, which form iron nitrides or iron carbonitrides, reach a hardness of 500–800 HV0.02; those of alloyed steels in which the iron in the nitrides or the carbonitrides is replaced by alloying elements reach about 900–1000 HV0.02. Nitriding steels that are alloyed with aluminum or molybdenum can have a hardness of up to 1200 HV0.02. Denoting the increase in nitrogen and carbon atoms (dissolved or precipitated as nitrides or carbonitrides), the hardness of the diffusion zone must be related to the hardness of the material prior to its heat treatment. The initial hardness is dependent on the material proper, above all on its carbon content and on its microstructure; the higher hardness of quench and

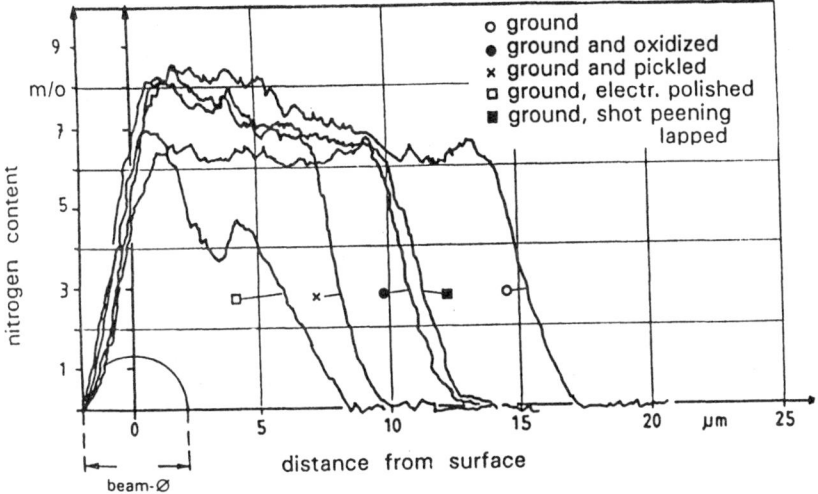

Figure 47 Influence of pretreatment preparation on thickness of compound layers. Ck15 steel gas nitrocarburized at 570°C for 3 h.

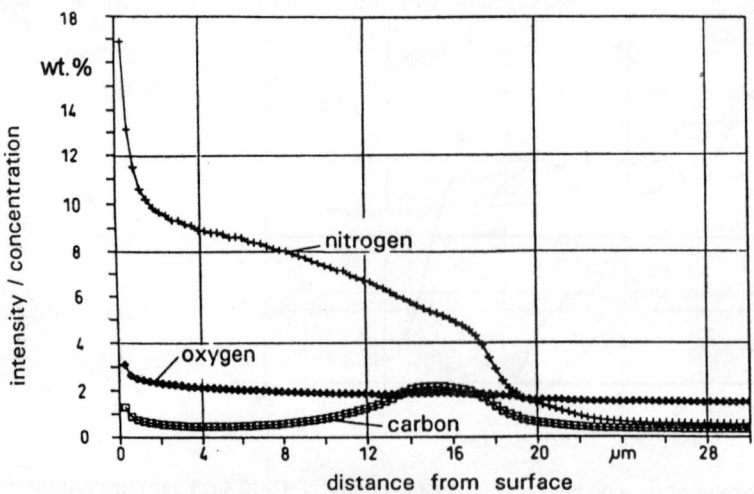

Figure 48 Gradients of nitrogen, carbon, and oxygen in the compound layer. C15 steel, gas nitro-carburized.

temper microstructures is maintained only if the material has good tempering properties at nitriding temperatures.

Figure 49 shows a result typical of a gas nitrocarburized quenched and tempered steel. The nitrides of nitriding steels contain more nitrogen, accounting for the formation of harder compound layers and more shallow nitriding case depths (Figure 50) at equal nitrogen supply.

4. *Microstructural Specialties*

On slow cooling, nitriding, and nitrocarburizing at temperatures above 590°C (1094°F), the eutectoid transformation of austenite to ferrite and γ'-nitride leads to a special microstructure (Figure 51) that is sometimes called braunite [132]. In quenched microstructures the austenite beneath the compound layer is retained (Figure 52). Carbides in the initial microstructure absorb nitrogen and are transformed to carbonitrides. With larger carbides and carbides with alloying elements, the nitriding time necessary to completely dissolve the carbides in the compound layer is often not long enough, thus explaining the existence of leftover carbides in the

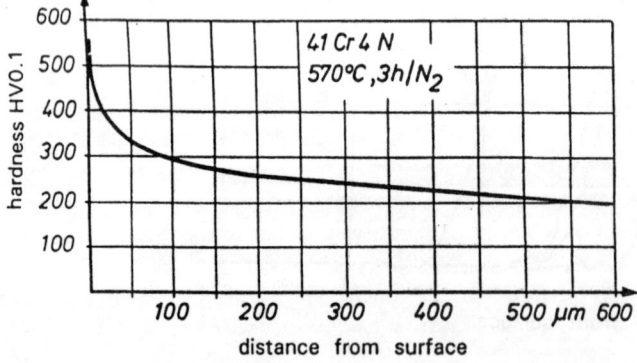

Figure 49 Hardness gradient of gas nitrocarburized steel 42CrS4. (Hardness of compound layer 928 HV0.015, surface hardness 352 HV10, core hardness 189 HV0.1.)

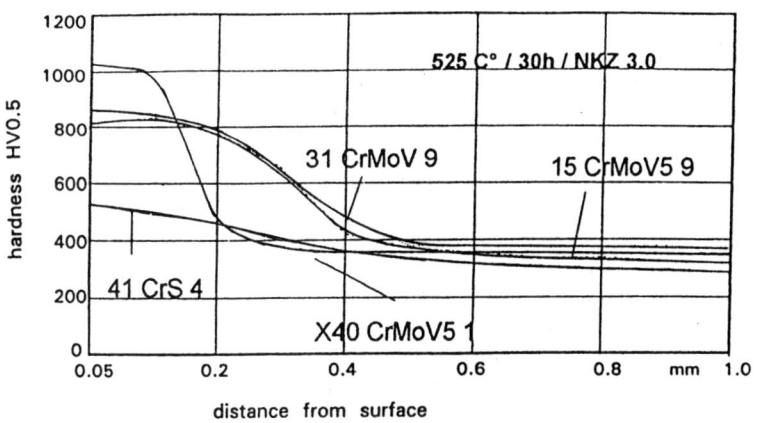

Figure 50 Hardness gradients of nitrided steels.

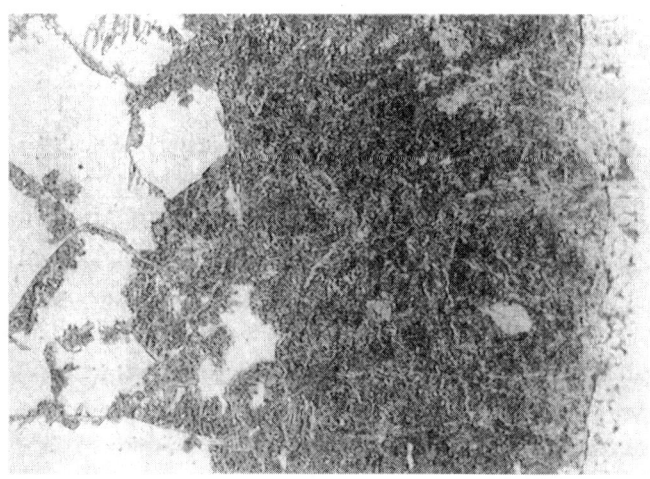

Figure 51 Braunite (St 37, nitrided at 640°C for 16 h).

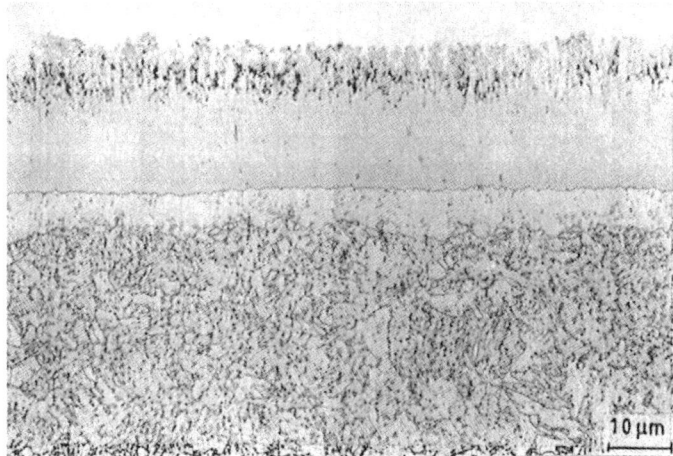

Figure 52 Nitrogen austenite (Ck35 steel nitrocarburized at 630°C for 40 min and quenched in water).

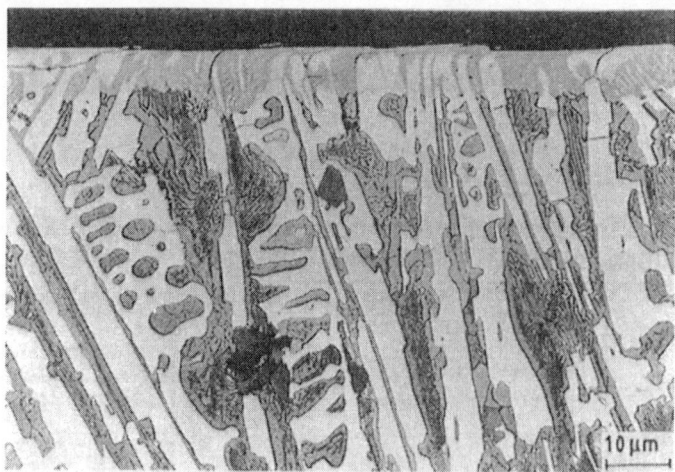

Figure 53 Carbides in the nitrocarburized case of chill-case iron

compound layer (Figure 53). The nitrogen absorption in the carbides is plainly visible from their changed etchability. Nonmetallic inclusions do not respond to nitrogen and remain in existence; examples are graphite in globular cast iron (Figure 54) and manganese sulfides in machining steels (Figure 55).

C. Nitriding and Nitrocarburizing Processes

1. Nitriding

A nitriding atmosphere consists always of ammonia sometimes diluted with additional gases such as nitrogen and hydrogen. At the usual nitriding temperatures of 500–570°C (932–1058°F), ammonia is in an unstable thermodynamic state and dissociates according to

$$2NH_3 \rightarrow 2N_{(ad)} + 6H_{(ad)} \rightarrow N_2 + 3H_2 \tag{58}$$

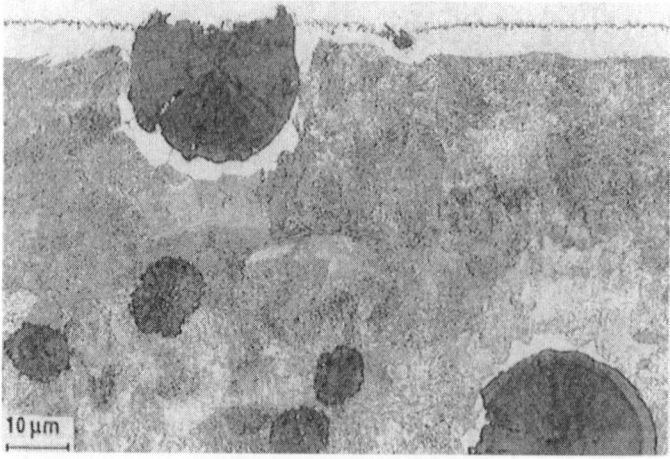

Figure 54 Graphite spheroids in the compound layer (GGG 70).

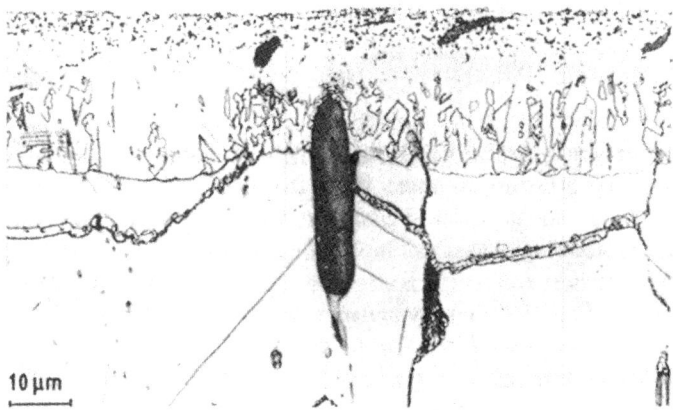

Figure 55   Manganese sulfide in the compound layer (9SMnPb28K).

This reaction, valid for thermal equilibrium, leads to degrees of dissociation higher than 99% and thus to the formation of a protective gas without any nitriding effect, as mentioned in Section IV (Equation 57). Despite the beneficial catalytic effect of the surfaces of the workpieces and the furnace wall, the dissociation of ammonia in the furnace is a very sluggish process, probably because the mean residence time of the gas in the furnace is too short for reaching an equilibrium by recombining to molecular nitrogen [152,163]. Therefore, ammonia-based nitriding atmospheres for steel treatment rarely contain less than 20%, frequently up to 50%, ammonia; hence their degree of dissociation is far from equilibrium. The remaining ammonia content is decisive for the effect of nitriding, during which nitrogen diffuses into the steel according to the reaction

$$NH_3 \rightarrow N(\alpha) + (3/2)H_2 \tag{59}$$

which occurs within the boundary layer. The nitrogen activity, which is the driving force in the mass transfer, can be calculated according to

$$a_N = K_N \frac{p_{NH_3}}{p_H^{3/2}} \tag{60}$$

The nitrogen transfer is comparatively low, and the release of hydrogen from the ammonia molecules determines the rate of the process [162,163]. Therefore, as mentioned earlier, nitriding times are rather long, up to 120 h [164].

The nitriding effect of a nitriding atmosphere is defined by its degree of dissociation which can be used for controlling the nitriding process, with high degrees of dissociation always indicating a near-equilibrium state where the nitriding effect is low. In processing, it is common practice to use a constant degree of dissociation, but sometimes a two-stage procedure involving variations of temperature and of dissociation degree (Floe process [165,166]) is followed. However, the measured ammonia content is not equivalent to the actual degree of dissociation. In ammonia dissociation according to Equation 58, two molecules of ammonia dissociate to one molecule of nitrogen and three molecules of hydrogen. This increase in volume dilutes the ammonia content, as do additional gases, and must be taken into account in ascertaining the actual degree of dissociation. The nitriding effect can be determined more easily by means of the nitriding characteristic NK ("nitriding power" or "nitriding potential") [149–152],

$$NK = \frac{p_{NH_3}}{p_H^{3/2}} \tag{61}$$

that, following Equation (60), is generally used for describing the nitridability of an ammonia atmosphere. The hydrogen partial pressure originates from the dissociation of ammonia and can be altered and controlled by additional gases. According to the Lehrer diagram (Figure 56) [167], the microstructure morphologies that can be obtained in a compound layer are determined by both nitriding characteristic and temperature. The validity of the diagram has, in essence, been corroborated in practice; the phase boundaries are valid for pure iron with respect to quantity, and those for steels are dependent on the alloy content. From Figure 56 it is discernible that at nitriding temperatures of more than 500°C (932°F) a nitriding characteristic of NK = 5 entails the formation of ε-nitride only; the necessary nitriding atmosphere would have to be composed of about 65 vol% NH_3 and 25 vol % H_2 at a standard pressure of 1 bar. The reaction

$$N_2 \rightleftharpoons 2N_{(ad)} \tag{62}$$

that occurs at the surface of the steel is also able to transfer nitrogen. The forward (left-to-right) reaction rate, however, is too slow to have any actual influence. The reverse reaction, from atomic to molecular nitrogen, is considerably faster and causes an increase in volume, which could account for the formation of pores [151].

2. Nitrocarburizing

In nitrocarburizing, an ammonia atmosphere is blended with carbon- and oxygen-containing gases that are usually based on carbon dioxide or allow carbon dioxide to be generated [139–

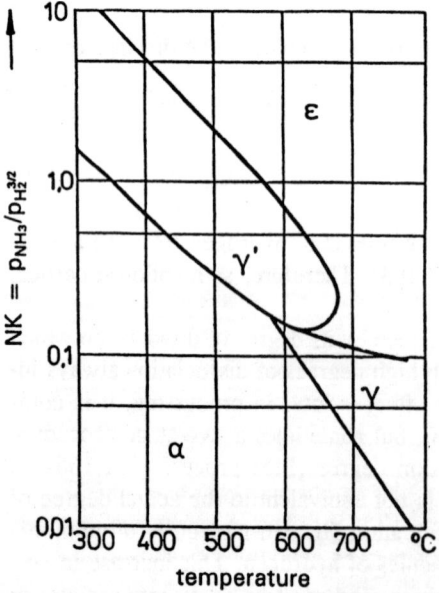

Figure 56   Lehrer diagram. (From Ref. 167.)

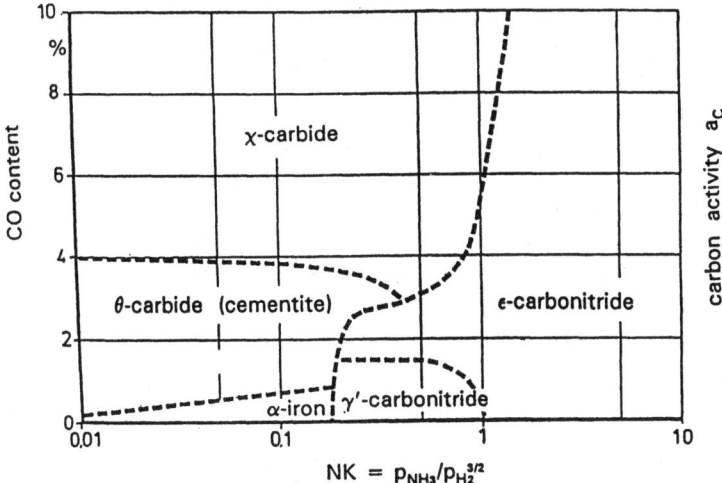

Figure 57 Influence of the gas composition on nitride formation in the compound layer. (From Ref. 156.)

144]. In accordance with the homogeneous water gas reaction (Equation 26, Section III.B.1), carbon dioxide reacts under near-equilibrium conditions with the hydrogen generated according to the reaction

$$CO_2 + H_2 \rightleftharpoons H_2O + CO \tag{22}$$

Given a constant ammonia partial pressure, this leads to a drop in the hydrogen partial pressure and a rise in the nitriding characteristic NK. A state of near-equilibrium is also maintained in the reaction of ammonia with carbon monoxide from the water gas reaction above or from an additional gas,

$$NH_3 + CO \rightleftharpoons HCN + H_2O \tag{63}$$

generating hydrogen cyanide (HCN) [149,151]. In addition to reaction (59), nitrogen transition is promoted by the reaction

$$HCN \rightarrow N(\alpha) + C(\alpha) + (1/2)H_2 \tag{64}$$

in which carbon also diffuses into the steel. Carbon, moreover, diffuses into the steel via the Boudouard reaction and the heterogeneous water gas reaction (Equations 22 and 23, Section III.B.1). Following Equation (61), the carburizing effect of a nitrocarburizing atmosphere is described by the carburizing characteristic CK,

$$CK = \frac{p_{CO}}{p_{CO_2}} \tag{65}$$

The interrelationship between nitriding characteristic NK, carburizing characteristic CK, and the nitrocarburized microstructures is shown in Figure 57 [156]. This diagram, too, has been corroborated in practice. As to quantity it is valid for the test conditions established by Langenscheidt [156]. More recent results are given in References 144, 157, 168, and 169. The addition of gases that release CO_2 at nitriding temperature to the original nitriding atmosphere consisting of NH_3 thus enhances the nitriding effect of a nitrocarburizing atmosphere because of the increased nitriding characteristic and the transformation of nitrogen from the decomposition of HCN. In such atmospheres, at the usual nitrocarburizing temperature of about 570°C (1058°F) for ferrous materials, processing times of 2–4 h are enough to lead to the formation of nitrocarburized microstructures consisting of a compound layer and a diffusion zone, whose properties are such as to greatly enhance the component behavior under many loading conditions. Above all, the economic advantage of a considerably shorter processing time and an easier solution to the problem of waste management than in salt bath nitrocarburizing have made gas nitrocarburizing the nitriding process most often used with ferrous materials to date. The advantage of nitriding with ammonia atmospheres lies above all in achieving deep nitriding depths, up to 1 mm, to meet the requirements of specific loading conditions [164].

3. Processing Effects on the Nitriding and Nitrocarburizing Results

By means of Figure 57 it is possible to assess the gas composition that has to be established and maintained for a specific microstructure at a given temperature. The thickness of the compound layer and the nitriding hardness depth are dependent on treatment time and temperature, on pretreatment (see Figures 45–47), and on the material. The rate at which the layer grows varies roughly with the square root of the treatment time (Figure 48) if processing conditions are such as to guarantee that in nitriding with 100% ammonia the rate of the gas exchange reaches a value that is at least five times higher than that of the furnace volume [170]. Nitrogen-diluted atmospheres have a less distinct nitriding effect, thus requiring increased rates of gas exchange. At an equal rate of gas exchange, additional gases yielding oxygen and carbon considerably accelerate the growth of the compound layer [171] by absorbing hydrogen and thus causing a rise in the nitriding characteristic because of the diminished hydrogen partial pressure. The content of ε-nitride in the compound layer can be controlled via the level of carbon dioxide (Figure 58) [172].

To control the nitriding and nitrocarburizing, it is common practice to determine the degree of dissociation and then calculate the nitriding characteristic NK. Values to be measured

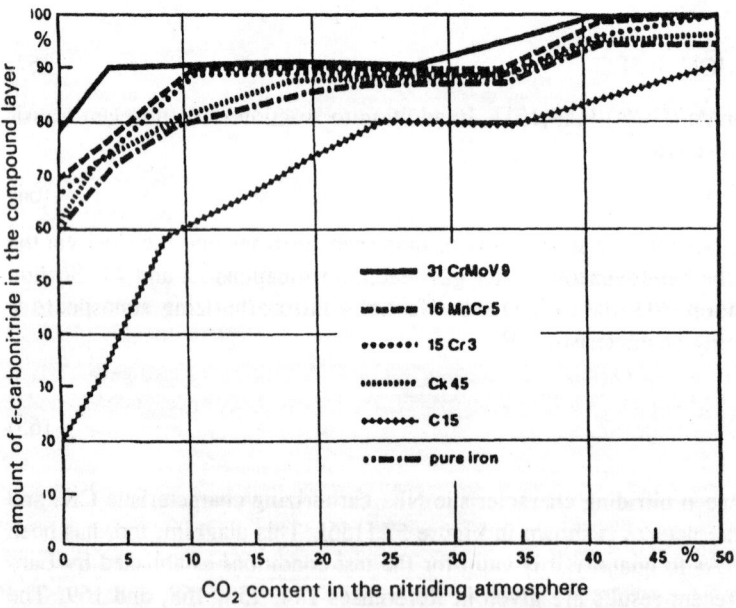

Figure 58 Influence of CO_2 content on the amount of ε-nitride in the compound layer. (From Ref. 172.)

include the volume percent of ammonia or hydrogen in the atmosphere. A more recent development is the utilization of nitriding sensors, which are designed either following a magnetoinductive measuring principle [153,154] or as a solid-state electrolyte [152,155]. Nitriding and nitrocarburizing cause the electrical and magnetic properties of the material to change. These changes are measured by special sensors [153,154] that are introduced into the furnace atmosphere and calibrated according to the nitriding result, the parameters being the material and the surface of the components. This measuring technique also allows separate determination of microstructure features such as thickness of compound layer or case depth. The oxygen partial pressure can be used as an indirect measure of the nitriding characteristic NK [152,155] if water vapor or oxygen or carbon dioxide are present in the nitriding atmosphere or are added in small amounts to make this measuring technique work. From the signal of the oxygen sensor it is possible to infer the oxygen partial pressure and thus the degree of dissociation and the nitriding characteristic NK.

Prior to gas nitriding and gas nitrocarburizing, the surfaces must be thoroughly cleaned, because adherent layers and impurities such as oils, forging scales, and preservatives are not entirely susceptible to the treatment at the comparatively low nitriding temperature, inhibiting adsorption and diffusion and diminishing the nitrogen transfer. The thickness of the compound layer and the case depth are then more shallow than would be expected from their nitriding time, or soft zones may develop. As cleaning systems containing halogenated hydrocarbons are no longer allowed, for environmental reasons, or are permitted only when certain expensive conditions are met, thorough cleaning is accomplished by use of hydrous cleaners [173,174] that sometimes are supported by ultrasonics. Layers still existing after cleaning are frequently further reduced by special preoxidation at about 350°C (662°F).

Oxidizing, frequently carried out in the form of a water vapor treatment at a temperature of at least 450°C (842°F) following nitriding [175–179], leads to the development on the compound layer of iron oxides FeO, $\alpha\text{-}Fe_2O_4$, or Fe_3O_4, depending on the processing mode. These oxide layers, above all the seemingly black magnetite layer Fe_3O_4, have a very beneficial effect on corrosion resistance. Magnetite adheres better on nitride layers than on iron because oxide layer and substrate reach a tighter fit (Pilling–Bedworth relationship) due to the intermediate nitride layer.

VI. PROPERTIES OF CARBURIZED AND NITRIDED OR NITROCARBURIZED COMPONENTS

The properties of a steel component [180] can be modified according to its loading requirements by means of heat treatment, the parameters and possibilities of which must be focused on this aim. Hence, this section deals with the essential properties that can be improved by carburizing and by nitriding or nitrocarburiznig. The properties of a steel component have their origin in the interrelationship between the microstructure properties within the component and the loading conditions the component has to endure.

One common characteristic of carburized and nitrided or nitrocarburized microstructures is the hardness gradient with its parameters case hardness, hardness difference between case and core, and case depth. Moreover, the heat treatment of the surface results in a specific residual stress distribution with compressive residual stresses in the case that turn into tensile residual stresses on reaching the case depth. Heavily loaded engineering components demand surface and substrate properties such as resistance to applied loads, to adhesive and abrasive wear, to rolling contact fatigue, and to corrosion, all of which must, of course, be provided by the same component. Most applied loads result in maximum stresses at the surface of the component or in an area near the surface. Corrosion and wear, by definition, affect component surfaces. Nearly all components are therefore stressed by a system of combined load-ing, wear, and corrosion with maximum effect at or near the surface. Frequently, the required material properties contradict each other, as can be observed, for instance, in the demand for a material that combines high hardness and enhanced abrasive wear resistance with ductility and enhanced toughness to withstand single impact loads. Surface hardening leads to a distri-bution of properties that fits the loading demands of most components excellently. In contrast to nontreated components, surface-hardened components therefore always have a much higher fatigue and rolling contact fatigue strength as well as better wear resistance in all those tribo-logical systems in which a high case hardness is beneficial. A composite with a hard case and a soft core is tougher than a through-hardened component (as to case behavior) but more brittle than a nonhardened microstructure (as to core behavior). Better corrosion resistance can de-velop only in nitrided or nitrocarburized components as their microstructures have a higher (more positive) corrosion potential.

The fatigue strength of case-hardened components is improved by the combined effect of higher case hardness and compressive residual stresses (Figure 59) [181], which results in a local endurance limit [181–183]. This has proved valid with carburized [184] and nitro-carburized [185] specimens. In carburizing, the highest values of fatigue resistance are ob-tained with case depths in the range of 10–20% of the component size (Figure 60) [186]. Given the same percentage of case depths, fine-grained microstructures [187,188] and a content of retained austenite as low as possible yield the best results [188,189] (Figure 61). However, the content of retained austenite must not be reduced by deep cooling, because deep cooling causes high tensile residual stresses to build up in the retained austenite at the surface [65], which may initiate local cracks. Reheated microstructures always have better fatigue resistance than as-quenched microstructures [190,191], where the detrimental effect of phosphorus seg-regation at the austenite grain boundaries is more distinct [103–107,191,192]. Nitriding and nitrocarburizing also lead to a substantial increase in fatigue resistance caused by the above-discussed mechanisms (Figure 62) [193]. The best effect can be expected starting from about 10% case depth relative to the component size [183]. At equal nitriding conditions, fatigue

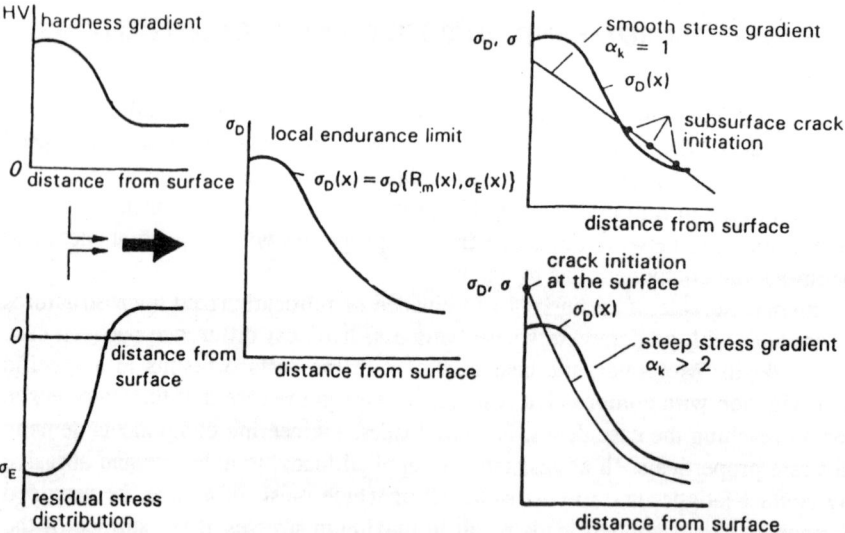

Figure 59 Local endurance limit of surface-hardened specimens. (From Ref. 181.)

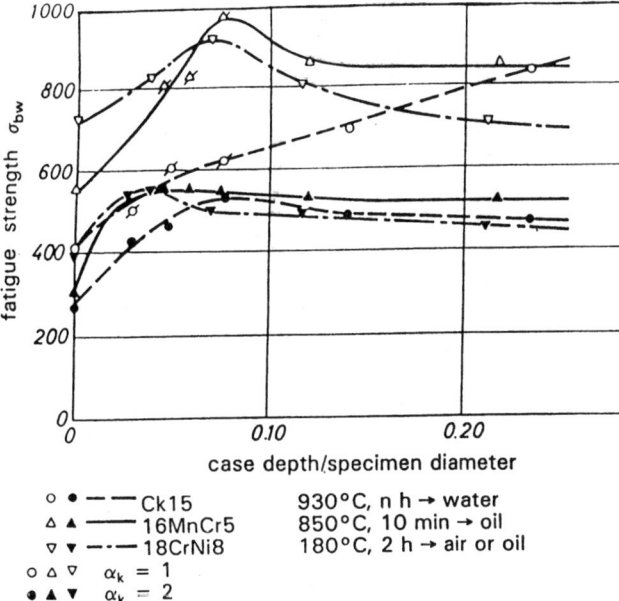

Figure 60 Influence of case depth on bending fatigue strength of carburized specimens. (From Ref. 186.)

resistance is better in microstructures with a high core strength (Figure 63); the nitrocarburizing process has no significant influence.

The case hardness of carburized and nitrided or nitrocarburized microstructures imparts better resistance to abrasive wear. To obtain maximum hardness, carburized microstructures

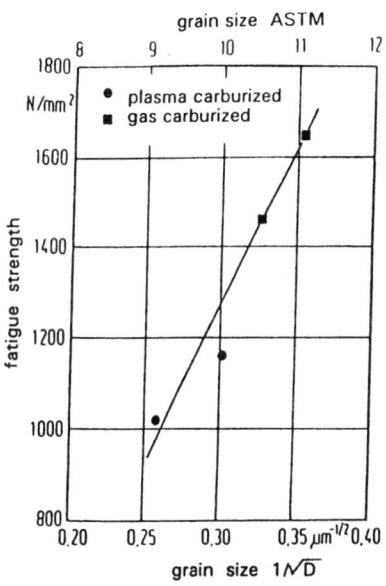

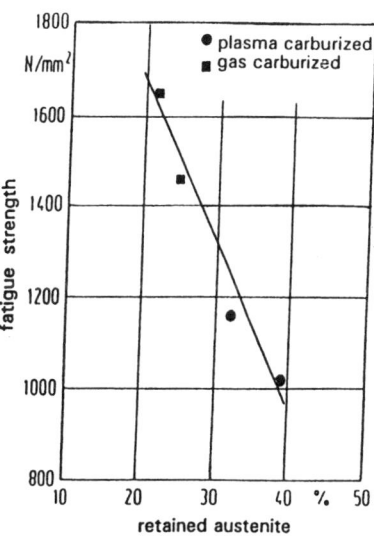

Figure 61 Effect of grain size and retained austenite content on bending fatigue of carburized steels. (From Ref. 188.)

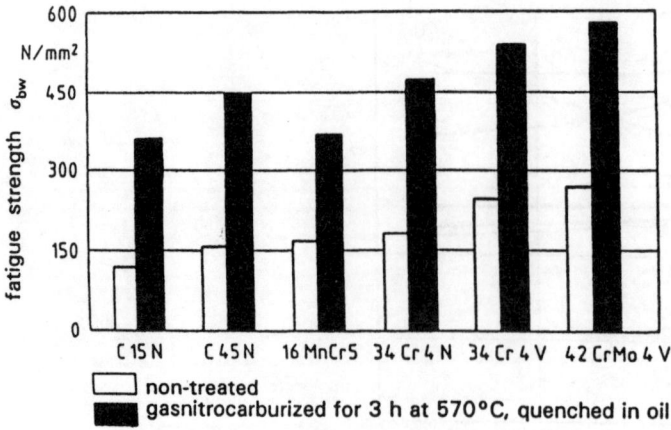

Figure 62 Bending fatigue strength of nitrocarburized steels. (From Ref. 193.)

ought to have carbon contents of about 0.7%. The case depth can be determined corresponding to the maximum value of tolerable wear depths. An additional effect of the compound layer
is to increase the resistance to adhesive wear, which is best achieved with the formation of ε-
nitride [194] and a small band of porosity. Nitride precipitations and a high strength, which
can be obtained by a quench-and-temper process, have an overall beneficial effect on the wear
resistance of the diffusion zone.

With the exception of stainless steels, nitrided microstructures have a higher chemical
potential than nonnitrided steels and thus better corrosion resistance [195,196], which can be
further improved by additional, passivating oxide layers [175–179]. This has been verified by
means of potential–current diagrams such as Figure 64 [179]. Porosity always accelerates the
corrosive effect.

The toughness of carburized components decreases with increasing relative case depth,
which describes the amount of brittle case microstructure in the composite. On reaching about

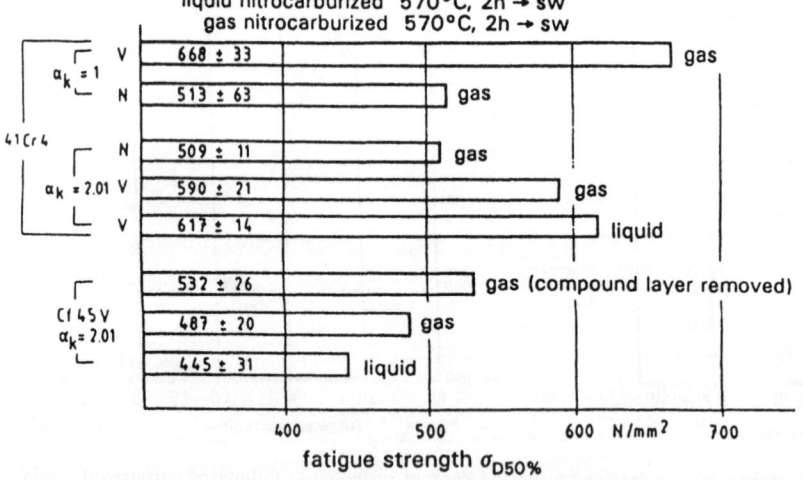

Figure 63 Bending fatigue strength of nitrocarburzied steels. sw = salt water.

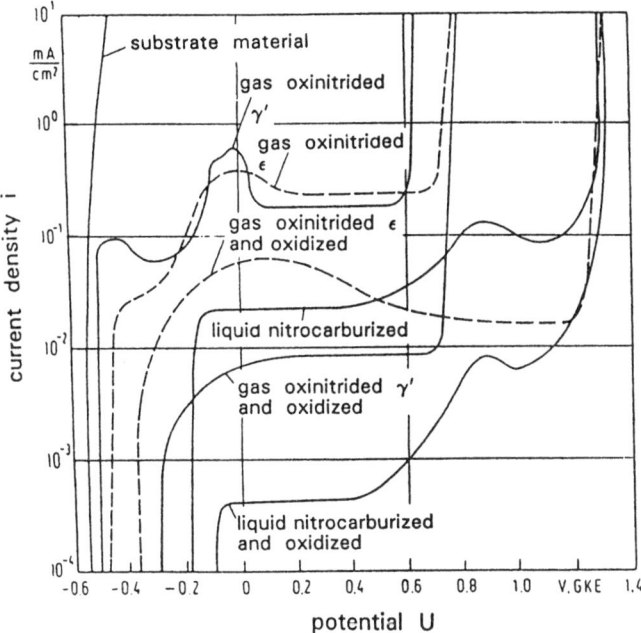

Figure 64 Current–potential curves of nitrocarburized and oxynitrocarburized steels. (From Ref. 179.)

30% relative case depth, the toughness has diminished to the value of the through-hardened microstructure (Figure 65) [197]. At equal relative case depth, toughness can be improved by using nickel-alloyed carburizing steels (Figure 66) [198], with tempered microstructures always being tougher than nontempered ones. In nitrocarburized microstructures, the influence of the core is dominant; microstructures consisting of ferrite and pearlite ought to be avoided (Figure 67) [199]. The different nitrocarburizing processes obviously have no effect.

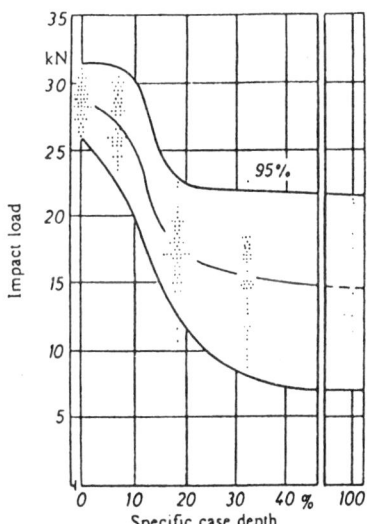
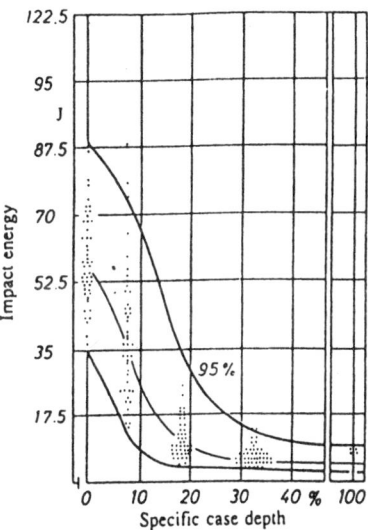

Figure 65 Influence of case depth on toughness of carburized specimens (gas carburizing 900°C, nh; quenched in oil at 180°C, 1 h. Data points: mean value of six specimens.) (From Ref. 197.)

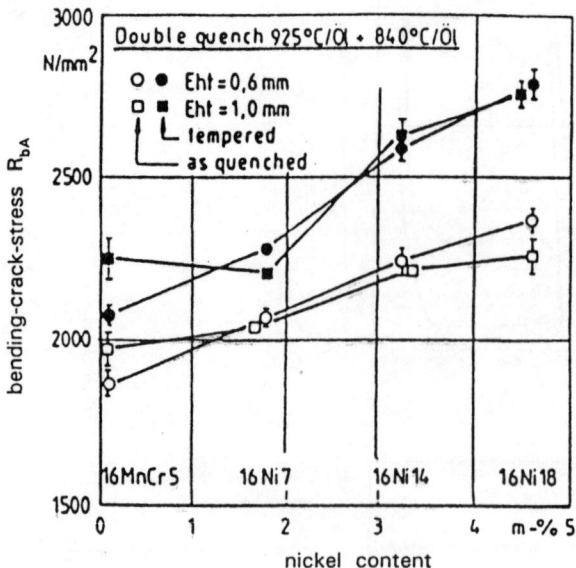

Figure 66 Effect of nickel content on bending crack stress of carburized specimens. (From Ref. 198.)

As for rolling contact fatigue, the acting stress according to Mises' criterion reaches its highest value beneath the surface at low friction coefficients. With increasing friction, the maximum value shifts toward the surface [200]. Depending on applied load, either case depth or surface condition (surface intergranular oxidation, porosity) must be fixed [201]. A high nitriding case depth, which can be produced in the gas nitriding of nitriding steels, are advantageous for gears. Their case depth should, however, not exceed 0.2 × modulus to avoid nitriding of the tooth tips that would entail embrittlement [202]. The maximum values that can

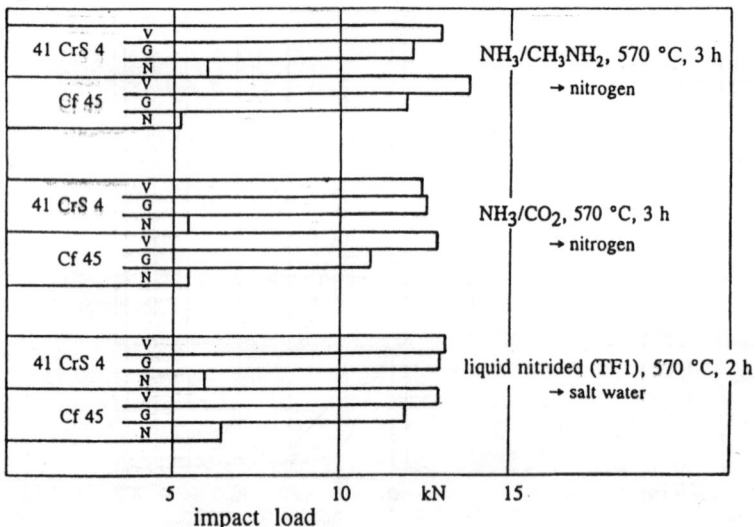

Figure 67 Influence of core strength on toughness of nitrocarburized specimens. (From Ref. 199.)

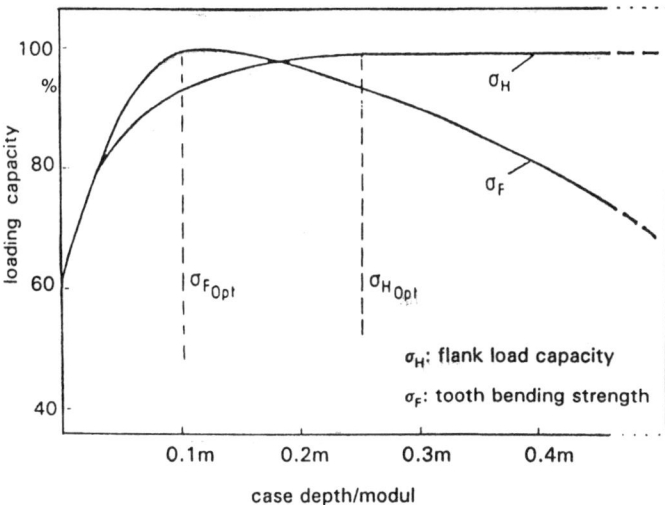

Figure 68 Influence of case depth on gear tooth bending endurance strength and pitting endurance strength of carburized steels. (From Ref. 203.)

be reached for the flank load capacity of carburized gears can be reached with case depths in the size range of $0.25 \times$ modulus (Figure 68) [203]. The optimal case depth for the gear tooth bending fatigue strength, which is also shown in Figure 68, illustrates that in determining the heat treatment parameters it is necessary to weigh optimal tooth-bending fatigue strength against optimal flank load capacity.

REFERENCES

1. P. G. Shewmon, Metallurgical thermodynamics, in *Physical Metallurgy*, 2nd ed., R. W. Cahn, Ed., North-Holland, Amsterdam, 1970, p. 281.
2. M. G. Frohberg, *Thermodynamik für Metallurgen und Werkstofftechniker*, VEB Deutscher Verlag für Grundstoffindustrie, Leipzig, 1980.
3. E. Fromm, *Thermodynamik*, Springer-Verlag, Berlin, 1970.
4. T. Heumann, *Diffusion in Metallen*, Springer-Verlag, Berlin, 1992.
5. P. G. Shewmon, *Diffusion in Solids*, 2nd ed., McGraw-Hill, New York, 1989.
6. G. Hörz, Kinetik und Mechanismen, in *Gase und Kohlenstoff in Metallen*, E. Fromm and E. Gebhardt, Eds., Springer-Verlag, Berlin, 1976, p. 84.
7. E. Doehlemeier, *Z. Elektrochem. 42*:561 (1936).
8. JANAF, *Thermochemical Tables*, 2nd ed., Natl. Bur. Stand. Ref. Ser., NBS 37, 1971.
9. J. F. Elliott and M. Gleiser, *Thermochemistry for Steelmaking*, Addison-Wesley, Reading, MA, 1960.
10. C. Wagner, *Z. Anorg. Chem. 199*:321 (1931).
11. A. Fick, *Poggendorf Ann. 94*:59 (1885).
12. C. E. van Orstrand and F. P. Dewey, U.S. Geol. Survey Paper 956, 1915; cited by W. Seith, *Ber. Naturforsch. Ges. 31*(1/2), Freiburg, 1931, 1–27.
13. A. Slattenscheck, *Härterei Tech. Mitt. 1*:85 (1942).
14. J. I. Goldstein and A. E. Moren, *Metall. Trans. 9A*:1515 (1978).
15. E. Hornbogen, *Werkstoffe*, 2nd ed., Springer-Verlag, Berlin, 1979.
16. F. E. Harris, *Met. Prog.*, Vol. 77, January 1945, p. 84.
17. R. W. Gurry, *Trans. AIME 188*:671 (April 1950).
18. Th. Schmidt, *Härterei Techn. Mitt. Sonderheft Gasaufkohlung 11*:1 (1952).
19. F. Neumann and B. Person, *Härterei Tech. Mitt. 23*:296 (1968).
20. F. Neumann and U. Wyss, *Härterei Tech. Mitt. 25*:253 (1970).
21. F. J. Harvey, *Met. Trans. 9A*:1507 (1978).
22. U. Wyss, *Härterei Tech. Mitt. 45*:44 (1990).

23. C. A. Stickels, in *Metals Handbook,* Vol. 4, *Heat Treating,* 1991, ASM Int., Materials Park, OH, p. 313.
24. W. Göhring and C. H. Luiten, *Härterei-Tech. Mitt. 43:*236 (1988).
25. H. J. Grabke, E. M. Müller, H. V. Speck, and G. Konczos, Steel Res. *56:*275 (1985).
26. W. Rembges, *Härterei-Techn. Mitt. 49:*112 (1994).
27. F. Hoffmann, S. Dorn, and P. Mayr, *Härterei Techn. Mitt. 49:*103 (1994).
28. M. H. Jacobs, T. J. Law, and F. Ribet, *Surf. Eng. 1:*105 (1985).
29. B. Edenhofer, *Härterei-Techn. Mitt. 45:*154 (1990).
30. U. Wyss, *Härterei Techn. Mitt. 17:*160 (1962).
31. H. V. Speck, H. J. Grabke, and E. M. Müller, *Härterei Techn. Mitt. 40:*92 (1985).
32. F. Neumann, *Härterei Techn. Mitt. 33:*192 (1978).
33. O. Schaaber and R. Fischer, *Industrieblatt 2:*89 (1956).
34. E. Schürmann, T. Schmidt, and H. Wagener, *Giesserei, 16:*91 (1964).
35. S. Gunnarson, *Härterei Techn. Mitt. 22:*292 (1967).
36. K. H. Sauer, M. Lucas, and H. J. Grabke, *Härterei Techn. Mitt. 43:*45 (1988).
37. F. Neumann, *Härterei Techn. Mitt. 44:*262 (1989).
38. H. J. Grabke, *Härterei Techn. Mitt. 45:*110 (1990).
39. J. Wünning, *Härterei Techn. Mitt. 23:*101 (1968).
40. K. H. Weissohn, *Härterei Techn. Mitt. 49:*118 (1994).
41. F. E. Harris, *Metal Progress 1943*, August, p. 265.
42. U. Wyss, in *Wärmebehandlung der Bau-und Weikzeugstähle,* H. Benninghoff, Ed., BAZ Buchverlag, Basel, 1978, p. 237.
43. J. Wünning, *Z. Wirtsch. Fertigung 77:*424 (1982).
44. H. J. Grabke and G. Tauber, *Arch. Eisenhüttenwes. 46:*215 (175).
45. D. Liedtke, Messen und Regeln beim Aufkohlen, in *Einsatzhärten,* J. Grosch, Ed., Expert-Verlag, Renningen, 1994, p. 16.
46. H. J. Grabke, *Härterei Techn. Mitt. 44:*270 (1989).
47. J. Wünning, *Härterei Techn. Mitt. 43:*266 (1988).
48. J. I. Goldstein and A. E. Moren, *Metall. Trans. 9A:*1515 (1978).
49. J. Pavlossoglou and H. Burkhard, *Härterei Techn. Mitt. 31:*209 (1976).
50. H. U. Fritsch and H. W. Bergmann, *Härterei Techn. Mitt. 41:*14 (1986).
51. A. R. Marder, S. M. Perpetpetua, J. A. Kowalik, and E. J. Stephenson, *Metall. Trans. 16A:*1160 (1985).
52. C. A. Stickels, in *Heat Treatment and Surface Engineering*, G. Krauss, Ed., ASM, Cleveland, 1988, p. 99.
53. T. Reti and M. Cseh, *Härterei Techn. Mitt. 42:*139 (1987).
54. C. Wells, W. Batz, and R. F. Mehl, *Trans. TMS-AIME 188:*553 (1950).
55. R. P. Smith, *Acta Metall. 1:*578 (1953).
56. G. Leyens, G. Woelk, and J. Wünning, *Arch. Eisenhüttenwes, 47:*385 (1976).
57. B. Thoden and J. Grosch, *Neue Hütte 34:*96 (1989).
58. B. Edenhofer and W. Lerche, *Härterei Techn. Mitt. 49:*88 (1994).
59. C. A. Stickels and C. M. Mack, in *Carburizing—Processing and Performance,* G. Krauss, Ed., ASM, Cleveland, 1989, p. 1.
60. B. Edenhofer and H. Pfau, in *Heat Treatment and Surface Engineering,* G. Krauss, Ed., ASM, Cleveland, 1988, p. 85.
61. T. Reti, M. Reger, and M. Gergely, in *Heat Treatment and Surface Engineering,* G. Krauss, Ed., ASM, Cleveland, 1988, p. 95.
62. B. Prenosil, *Härterei-Techn. Mitt. 21:*24, 124 (1966).
63. R. Chatterjee and O. Schaaber, *Härterei-Techn. Mitt. 24:*121, 292 (1969).
64. J. Slycke and T. Ericsson, *J. Heat Treat. 2:*3 (1981).
65. O. Schwarz, J. Grosch, C. Genzel, and W. Reimers, *Härterei Techn. Mitt. 49:*134 (1994).
66. U. Wyss, *Härterei-Techn. Mitt. 43:*27 (1988).
67. J. H. Hollomann and L. D. Jaffe, *Trans. AIME 180:*439 (1949).
68. J. Zhao, *Mat. Sci. Techn. 8:*997 (1992).
69. C. A .Siebert, D. V. Doane, and D. H. Breen, *The Hardenability of Steels,* ASM, Cleveland, 1977, p. 100.
70. H. Dietrich and W. Schmidt, *Thyssen Techn. Berichte 10:*105 (1984).
71. A. F. de Retana and D. V. Doane, *Met. Prog. 100* (3):65 (1971).
72. H. Dietrich, *Thyssen Edelst. Techn. Ber. 7:*14 (1981).
73. G. Krauss, *Härterei Techn. Mitt. 41:*56 (1986).
74. G. Krauss, *Steels—Heat Treatment and Processing Principles*, ASM, Cleveland, 1990, p. 43.
75. O. Vöhringer, in *Grundlagen der technischen Wärmebehandlung von Stahl,* J. Grosch, Ed., Werkstofftechn. Verlagsges, Karlsruhe, 1981, p. 75.

76. H.-J. Eckstein, *Wärmebehandlung von Stahl,* VEB-Verlag, Leipzig, 1971, p. 179.
77. Z. Nishiyama, *Martensitic Transformation,* Academic, New York, 1978, p. 236.
78. H. K. D. H. Bhadeshia, *J. Mat. Sci. 17:*383 (1982).
79. A. H. Rauch and W. R. Turtle, *Met. Prog. 69:*73 (1956).
80. R. P. Probst and G. Krauss, *Metall. Trans. 5:*457 (1974).
81. A. Randak and R. Eberbach, *Härterei Techn. Mitt. 32:*223 (1977).
82. W. Knorr, H.-J. Peters, and G. Tacke, *Härterei Techn. Mitt. 36:*129 (1981).
83. P. G. Dressel, R. Kohlmann, K.-J. Kremer, and A. Stanz, *Härterei Techn. Mitt. 39:*112 (1984).
84. V. Schüler, B. Huchtemann, and E. Wulfmeier, *Härterei Techn. Mitt. 45:*57 (1990).
85. L. Cheng, C. M. Brakman, B. M. Korevaar, and E. J. Mittemeijer, *Metall. Trans. 19A:*2415 (1988).
86. H. Streng, C. Razim, and J. Grosch, *Härterei Techn. Mitt. 43:*80 (1988).
87. H. J. Böhmer, H. W. Zoch, and H. Schlicht, *Härterei Techn. Mitt. 41:*258 (1986).
88. H. J. Böhmer, H. J. Ebert, and W. Trojahn, *Proc. 46th Annual Meeting STLE,* Montreal, Canada, 1991.
89. G. Krauss, *Steels—Heat Treatment and Processing Principles,* ASM, Cleveland, 1990, p. 205.
90. E. Macherauch and O. Vöhringer, *Härterei-Techn. Mitt. 41:*71 (1986).
91. Liu Cheng, A. Böttger, and E. J. Mittemeijer, *Metal. Trans. 22A:*1945 (1991).
92. G. R. Speich, *Trans. TMS-AIME 245:*2553 (1969).
93. D. Dengel, *Härterei-Techn. Mitt. 39:*182 (1984).
94. N. DeChristofaro and R. Kaplow, *Metall. Trans. 8A:*35 (1977).
95. N. DeChristofaro, R. Kaplow, and W. S. Owen, *Metall. Trans. 9A:*821 (1978).
96. S. Nagakura, T. Suzuki, and M. Kusonoki, *Trans. Jpn. Inst. Met. 22:*699 (1981).
97. S. Nagakura, Y. Hirotsu, M. Kusonoki, T. Suzuki, and Y. Nakamura, *Metall. Trans. 14A:*1025 (1983).
98. H. Faber, O. Vöhringer, and E. Macherauch, *Härterei Techn. Mitt. 34:*1 (1979).
99. M. Bacher, O. Vöhringer, and E. Macherauch, *Härterei Techn. Mitt. 45:*16 (1990).
100. T. A. Balliett and G. Krauss, *Metall. Trans. 7A:*81 (1976).
101. J. H. Hollomon and L. D. Jaffe, *Trans. AIME 162:*223 (1945).
102. R. A. Grange and R. W. Baughman, *Trans. ASM 48:*165 (1956).
103. H. K. Obermeyer and G. Krauss, *J. Heat. Treat. 1:*31 (1980).
104. J. Grosch, *VDI Ber. 318:*79 (1978).
105. T. Ando and G. Krauss, *Metall. Trans. 12A:*1283 (1981).
106. S. K. Banerjee, C. J. McMahon, Jr., and H. C. Feng, *Metall. Trans. 9A:*237 (1978).
107. G. Thomas, *Metall. Trans. 9A:*439 (1978).
108. F. Z. Ephraimi and G. Krauss, *Acta Met. 32:*1767 (1984).
109. I. S. Koslowski, *Met. Sci. Heat Treat. 25:*157 (1967).
110. R. Chatterjee, *Z. Wirtsch. Fert. 71:*367 (1976).
111. R. Chatterjee-Fischer, *Metall. Trans. 9A:*1553 (1978).
112. K. Bungardt, E. Kunze, and H. Brandis, *DEW-Techn. Ber. 5:*1 (1965).
113. J. C. Scully, *The Fundamentals of Corrosion,* 3rd ed., Pergamon, New York, 1990, p. 90.
114. W. Haumann, *Stah. Eisen 107:*585 (1987).
115. H. Streng, J. Grosch, and C. Razim, in *Heat Treatment and Surface Engineernig,* G. Krauss, Ed., ASM, Cleveland, 1988, p. 313.
116. H. G. Nelson, Hydrogen embrittlement, in *Treatise on Materials Science and Technology,* Vol. 25, C. L. Briant and S. K. Banerjee, Eds., New York, 1983, p. 275.
117. R. A. Oriani, *Acta Metall. 18:*147 (1970).
118. G. M. Pressouyre, *Acta Metall. 28:*895 (1980).
119. K. B. Kim and S. Pyun, *Arch. Eisenhüttenwes. 53:* 397 (1982).
120. G. M. Pressouyre, *Metall. Trans. 14A:*2189 (1983).
121. A. M. Brass, M. Aucouturier, and H. Barthelemy, *Mem. Sci. Rev. Metall. 78:*601 (1981).
122. H.-J. König and K. W. Lange, *Arch. Eisenhüttenwes. 46:*669 (1975).
123. G. Vibrans, *Arch. Eisenhüttenwes. 32:*667 (1961).
124. H. Streng, *Zähigkeitsoptimierung einsatzgehärteter Gefüge,* Dissertation, Techn. Univ., Berlin, 1986.
125. E. Riecke, *Arch. Eisenhüttenwes. 44:*647 (1973).
126. R. Hoffmann, *Härterei Techn. Mitt. 39:*61,99 (1984).
127. F. D. Richardson and J. H. Jeffes, *J. Iron Steel Inst. 171:*165 (1952).
128. I. Barin and O. Knacke, *Thermodynamical Properties of Inorganic Substances,* Springer-Verlag, Berlin, 1973.
129. G. Minkler, *Blech-Rohre-Profile 29:*484 (1982).
130. A. W. Machlet, U.S. Pat. 1,092,925 (Apr. 14, 1914).
131. A. Fry, *Stahl Eisen 23:*1271 (1923).
132. A. Fry, *Kruppsche Mh. 7:*17 (1926).
130. A. Fry, *Trans. Am. Soc. Steel Treat. 16:*111 (1929).
133. A. B. Kinzel, J. J. Egan, *Trans. Am. Soc. Steel Treat. 16:*175 (1929).

718   GROSCH

134. C. Albrecht, *TZ Prakt. Metallbearb. 52*:252 (1942).
135. J. G. Morrison, *Iron Coal Tr. Rev. 147*:135 (1943).
136. J. Müller, *Härterei Techn. Mitt. 3*:219 (1944).
137. F. K. Naumann and G. Langenscheidt, *Arch. Eisenhüttenwes. 36*:583 (1965).
138. B. Prenosil, *Härtenei Techn. Mitt. 20*:41 (1965).
139. C. Dawes and D. F. Trenter, *Metal Forming 40*:58 (1973).
140. C. Luiten, *Z. Wirtsch. Fert. 68*:482 (1973).
141. H. J. Grabke, *Scripta Met. 8*:121 (1974).
142. J. Wünning, *Härterei Techn. Mitt. 29*:42 (1974).
143. T. Bell, *Heat Treat. Met. 2*:39 (1975).
144. A. Wells and T. Bell, *Heat Treat. Met. 10*:39 (1983).
145. F. Leitner and E. Plöckinger, *Edelstahlerzeugung,* Springer-Verlag, Berlin, 1964, p. 259.
146. M. Hansen and K. Anderko, *Constitution of Binary Alloys,* McGraw-Hill, New York, 1958.
147. R. P. Elliott, *Constitution of Binary Alloys,* Suppl. 1, McGraw-Hill, New York, 1965.
148. H. A. Wriedt, N. A. Gokcen, and R. H. Nafziger, *Bull. Alloy Phase Diagrams 8*(4):355.
149. J. Slycke and L. Sproge, *Surf. Eng. 5*:125 (1989).
150. F. Hoffmann, R. Hoffmann, and E. J. Mittemeijer, *Härterei Techn. Mitt. 47*:365 (1992).
151. J. Slycke and L. Sproge, *Härterei Techn. Mitt. 47*:357 (1992).
152. R. Hoffmann, E. J. Mittemeijer, and M. A. J. Somers, *Härterei Techn. Mitt. 49*:177 (1994).
153. H. Klümper-Westkamp, *VDI-Fortschrittsberichte 5*(160):14 (1989).
154. H. Klümper-Westkamp, F. Hoffman, P. Mayr, and B. Edenhofer, *Härterei Techn. Mitt. 46*:367 (1991).
155. H.-J. Berg, H.-J. Spies and S. Böhmer, *Härterei Techn. Mitt. 46*:375 (1991).
156. G. Langenscheidt, Beitrag zum System Eisen-Stickstoff-Kohlenstoff, Dissertation BA, Clausthal, 1964.
157. U. Huchel and J. Kunze, *Härterei-Techn. Mitt. 46*:351 (1991).
158. H.-J. Spies and D. Bergner, *Härterei Techn. Mitt. 47*:346 (1992).
159. M. A. J. Somers and E. J. Mittemeijer, *Härterei-Techn. Mitt. 47*:5 (1992).
160. F. Hoffmann, in *Nitrieren und Nitrocarburieren,* E. J. Mittemeijer and J. Grosch, Eds., AWT, 1991, Darmstadt, p. 105.
161. D. Liedtke and J. Grosch, *Proc. 5th Int. Congr. Heat Treatment,* Budapest, 1986, pp. 8–16.
162. H. J. Grabke, *Ber. Bunsenges. 72*:533 (1968).
163. H. J. Grabke, *Arch. Eisenhüttenwes. 46*:75 (1975).
164. U. Huber-Gommann, *Heat Treating,* March 1991, p. 19; May 1991, p. 24.
165. C. F. Floe, *Trans. ASM 32*:134 (1944).
166. C. F. Floe, *Metal. Prog. 50*:1212 (Dec. 1946).
167. E. Lehrer, *Z. Elektrochemie 36*:3832 (1930).
168. J. Slycke, L. Sproge, and J. Agren, *Scand. J. Met. 17*:122 (1988).
169. E. J. Mittemeijer, *Härterei Techn. Mitt. 36*:57 (1981).
170. C. Dawes, D. F. Trantner, and C. G. Smith, *Met. Techn. 6*:345 (1979).
171. C. Dawes, D. F. Trantner, and C. G. Smith, *Heat Treat. Met. 7*:1 (1980).
172. W. Vogel, in *Heat Treatment and Surface Engineering,* G. Krauss, Ed., ASM, Cleveland, 1988, p. 337.
173. W. Burger, *Härterei Techn. Mitt. 47*:76 (1992).
174. B. Haase, K. Bauckhage, and A. Schreiner, *Härterei Techn. Mitt. 47*:67 (1992).
175. M. A. J. Somers and E. J. Mittemeijer, *Härterei Techn. Mitt. 47*:169 (1992).
176. D. J. Coates, B. Mortimer, and A. Hendry, *Corrosion Sci. 22*:951 (1982).
177. K. G. Schmitt-Thomas and B. Rauch, *Härterei Techn. Mitt. 42*:256 (1987).
178. H.-J. Spies, H.-P. Winkler, and B. Langenhan, *Härterei Techn. Mitt. 44*:75 (1989).
179. U. Ebersbach, S. Friedrich, T. Nghia, and H.-J. Spies, *Härterei Techn. Mitt. 46*:339 (1991).
180. J. Grosch, *Z. Wirtschaftl. Fertig. 69*:412 (1974).
181. K. H. Kloos, *Härterei-Techn. Mitt. 44*:157 (1989).
182. E. Velten, Entwicklung eines Schwingfestigkeitskonzeptes zur Berechnung der Dauerfestigkeit thermochemisch randschichtverfestigter bauteilähnlicher Proben, Dissertation, Techn. Hochschule, Darmstadt, 1984.
183. P. Starker, H. Wohlfahrt, and E. Macherauch, *Arch. Eisenhüttenwes. 51*:439 (1980).
184. T. Kuttner and H. Zieger, *Härterei Techn. Mitt. 47*:367 (1992).
185. H.-J. Spies, T. Kern, and D. N. Tan, *Mat. -Wiss. Werkstofftechn. 25*:191 (1994).
186. H. Wiegand and G. Tolasch, *Härterei-Techn. Mitt. 22*:213,230 (1967).
187. C. A. Apple and G. Krauss, *Metall. Trans. 4*:1195 (1973).
188. J. L. Pacheo and G. Krauss, *Härterei Techn. Mitt. 45*:77 (1990).
189. W. Beumelburg, Das Verhalten von einsatzgehärteten Proben mit verschiedenen Oberflächenzuständen und Randkohlenstoffgehalten im Umlaufbiege-, statischen Biege- und Schlagbiegeversuch, Dissertation, Univ. Karlsruhe, 1974.
190. G. Krauss, *Metall. Trans. 9A*:1527 (1978).

191. G. Krauss, *Härterei Techn. Mitt. 49*:157 (1994).
192. R. S. Hyde, G. Krauss, and D. K. Matlock, *Metall. Mat. Trans. 25A*:1229 (1994).
193. R. Chatterjee-Fischer, *Härterei Techn. Mitt. 38*:35 (1983).
194. H. Woska, *Härterei Techn. Mitt. 38*:10 (1983).
195. B. De Benedetti and E. Angelini, *Advances in Surface Treatments*, Vol. 5, Pergamon, New York, 1987, p. 3.
196. J. V. Scalera, Rep. 8894, U.S. Dept. Int., Bureau of Mines, Washington, D.C., 1984.
197. D. Wicke and J. Grosch, *Härterei-Techn. Mitt. 32*:223 (1977).
198. B. Thoden and J. Grosch, *Härterei-Techn. Mitt. 45*:7 (1990).
199. J. Grosch and D. Liedtke, in *Nitrieren und Nitrocarburieren,* E. J. Mittemeijer and J. Grosch, Eds., AWT, Darmstadt, 1991, p. 365.
200. O. Zwirlein and H. Schlicht, *Z. Werkstofftechn. 11*:1 (1980).
201. H. Stangner and H.-W. Zoch, *Härterei Techn. Mitt. 45*:223 (1990).
202. U. Huber-Gommann, H.-J. Grasemann, and G. Schulz, *Z. Wirtschaftl. Fertigung 75*:9 (1980).
203. J. Sauter, I. Schmidt, and U. Schulz, *Härterei-Techn. Mitt. 45*:98 (1990).

10

NITRIDING TECHNIQUES AND METHODS

David Pye
Pye Metallurgical Consulting Inc., Meadville, Pennsylvania

I. INTRODUCTION

The process of nitriding has been known and carried out for many years, almost since the beginning of the 20th century. It is, in essence, an extremely simple process and one for which it is not necessary to austenitize the steel being treated, nor is it necessary to consider critical cooling paths on the time–temperature–transformation diagrams. In general, the process involves raising the temperature of a steel to a suitable temperature below that of the Ac1 eutectoid line on the iron-cementite diagram (Figure 1) in a nitrogen-rich environment for a suitable period of time. Once the cycle has been completed, no further treatment is necessary save that of, perhaps, machining.

Nitrided case-hardened steels exhibit high surface hardness values at elevated temperatures as well as improved corrosion resistance under alkaline conditions (except in the case of stainless steels). Under acidic conditions, the nitrided case is not resistant to this type of corrosive environment. A large number of the alloy steels, including some of the cast irons and stainless steels, will nitride readily.

The nitrided case exhibits many useful physical properties that are attractive to the engineer. These include [1]

High surface hardness values
Increased wear resistance and antigalling properties
Improvement of fatigue life
Improvement of corrosion resistance (except for stainless steels)
High surface hardness, resistant to temperatures up to at least the nitriding temperature
Low distortion and movement due to low processing temperatures

This chapter examines the history of nitriding, the theory of the nitriding process, nitriding equipment, and the general practice of nitriding, embracing the gaseous, liquid, and pack methods.

II. DEVELOPMENT OF THE NITRIDING PROCESS

In the early years of this century, Dr. Adolph Machlet of the American Gas Company in Elizabeth, New Jersey [2], began work on a low-temperature heat treatment process to case harden

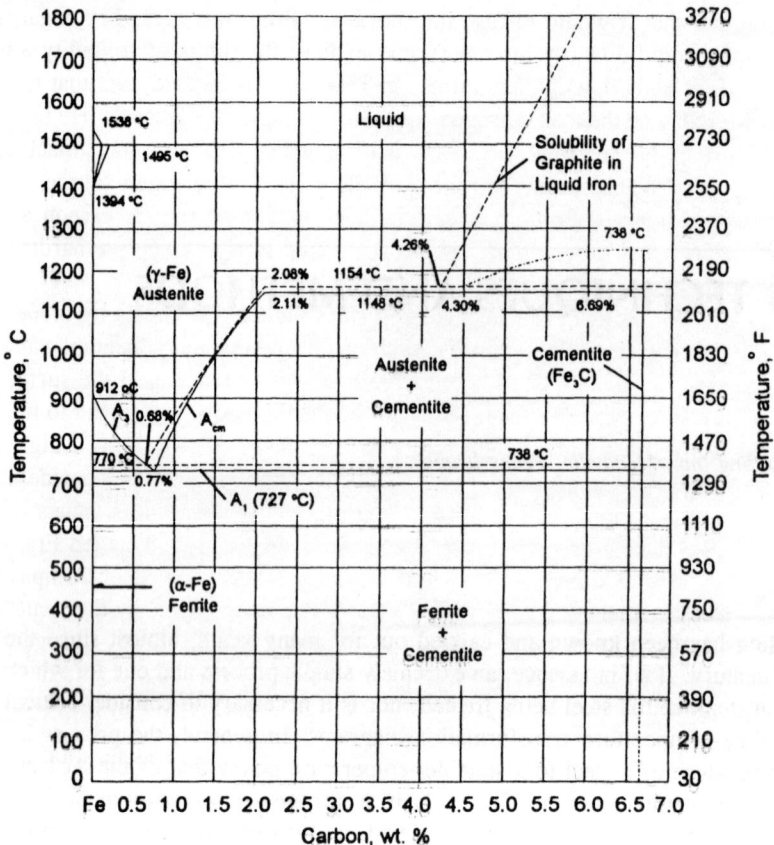

Figure 1   Iron–carbon equilibrium diagram.

low-alloy steels and cast irons. His theory was based on the solubility of nitrogen in cast iron and steel at low temperatures. For many years, he had been aware of the solubility of nitrogen in steel and cast iron, and as a result of this knowledge, he began a series of experiments to explore this phenomenon. At the conclusion of his experiments, he had proved that cast iron and steel did in fact absorb nitrogen in sufficient quantities to form a hard layer on the metal surface.

Machlet concluded that he had the basis for a new heat treatment process and applied for a patent in March 1908. Part of his patent read

> For the nitrogenization treatment of cast irons and steels, as well as other articles of some other compositions in a manner to produce upon the articles a skin, or shell, by heating them in a retort heated at a temperature of 480°C (900°F) to 982°C (1800°F) while ammonia gas, diluted with hydrogen, flows slowly through the retort.

Machlet's patent was granted on June 24, 1913 [2].

Although Machlet was the "inventor" of the nitriding process, he received very little recognition for it and American industry took no commercial advantage of Machlet's discovery and patent.

At the same period, in Europe, Dr. Adolph Fry of the Krupps Steel Company was also investigating the solubility of nitrogen in cast irons and steels. Fry had the idea that nitrogen had an affinity for iron and that it would diffuse into hot iron. From this idea, Fry began a series of experiments that would quantify his theory and explore the limits of solubility and diffusion of nitrogen in iron at various temperatures. In 1906, Fry presented data that contributed to what is known today as the iron–nitrogen equilibrium diagram (Figure 2). He noted that as nitrogen was absorbed into the steel surface, structural changes began to take place as well as a noticeable increase in the surface hardness of the steel. His nitrogen source was ammonia gas. Fry observed that a simple plain carbon steel containing 0.39% carbon and 2.85% chromium with an original hardness of 280 Brinell would exhibit a surface hardness of 470 Brinell after treatment with nitrogen.

Fry then began to look for ways of increasing the surface hardness of steels by investigating the effect of alloying the steel with elements such as chromium, molybdenum, vanadium, tungsten, manganese, and titanium and noted significant improvements in the surface hardness results. Some time after the conclusion of his alloying experiments, he began to note that steels that had been alloyed with aluminum up to 1% as "killed steels" exhibited higher than expected hardness values. On microscopic examination, he observed what he considered to be stable nitrides. The stable nitride not only produced high surface hardness values but also exhibited a resistance to nitride decomposition at elevated temperatures. This led Fry to apply for a patent for what is now known as the Nitralloy steels. The Krupp Steel Company then began the marketing and production of Nitralloy steels. Fry applied for another patent in 1921, this time for the process he called "a method of hardening alloy steels by nitrogenization." The patent was granted to Fry on March 18, 1924 [3]. Two of Fry's claims in the patent application were that his process was

1. A process for hardening a steel alloy by subjecting the alloy to nitrogenization at a temperature below the peritectoidal transformation temperature
2. A process for hardening steel alloys containing chromium, manganese, silicon, tungsten, vanadium, or aluminum in any desired combination, consisting of subjecting the alloy to a temperature below the temperature of peritectoidal transformation to the action of a substance that gives off nitrogen

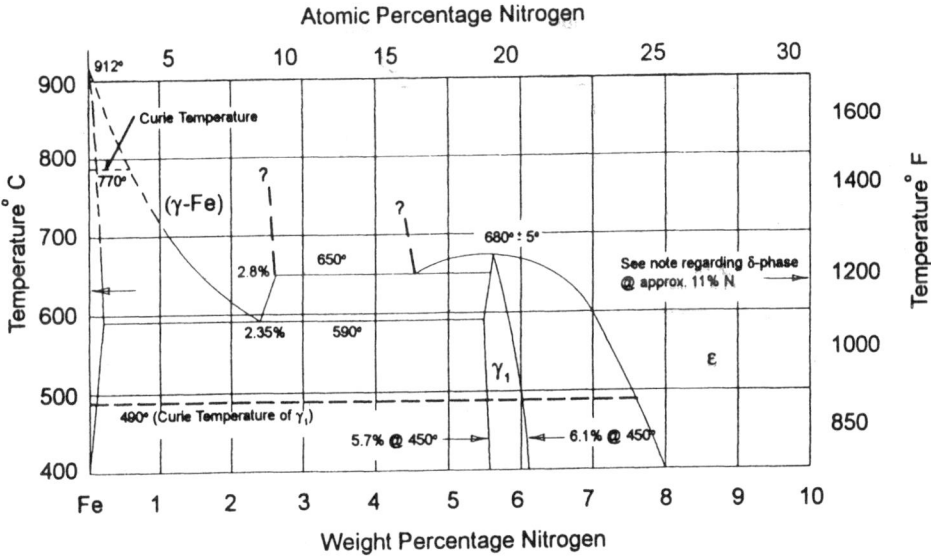

Figure 2 Iron–nitrogen equilibrium diagram.

A most useful feature of Fry's process, as noted by European engineers, was that of minimal distortion. This meant that scrap rates due to distortion were greatly reduced compared to those of other more conventional process methods such as carburizing. European engineers were quick to make use of this feature and thus began applying it to industries where distortion was critical, such as the aeronautical, railroad, and textile industries.

In 1926, information began to filter back to the United States of Fry's process. Jeffries, then president of the Society of Mechanical Engineers, was charged with the task of investigating Fry's work. Jeffries met with Fry toward the end of 1926 in Essen, Germany, to discuss the industrial applications of Fry's process. This meeting led to an invitation for Fry to present a paper on his work to the SME convention in 1927. Fry was unable to accept the invitation but recognized the importance of presenting his findings, so he asked his longstanding friend, Pierre Aubert, to make the presentation on his behalf. Aubert attended the convention and gave the presentation, which was centered on the following points:

1. The process results in high surface hardness with minimal distortion.
2. Core properties of treated materials are not diminished.
3. Wear resistance properties and hardness values are higher than those exhibited by other more conventional case hardening processes.
4. Core properties are not affected by nitriding at temperatures below the tempering temperature.
5. Nitrided parts are free of the internal stresses caused by the action of quenching when steels are processed by more conventional case hardening methods.
6. Alloy steels that have been nitrided exhibited a resistance to oxidation (or rust).
7. The best results of nitriding are obtained on steels that have been prehardened, tempered and machined before nitriding.

Aubert's presentation was received with great enthusiasm. Work was soon begun in earnest by American industrialists and metallurgists. Many of the advantages of the process described by Aubert hold true for today's industrial environment.

Although Machlet has been recognized as the "father" of the nitriding process, it was Fry who developed the process into a commercially viable process [4]. In 1929, intensive work was begun at MIT by a team headed by Victor Homerberg [4]. In 1928, work was undertaken on the process by McQuaid and Ketchum, who were at that time metallurgists at the Timken and Detroit Axle Company. Shortly after they had commenced their work, they presented a paper on the process as a commercially viable method of heat treatment, not only for their company but for the steel industry in general. At the same time, Robert Sergeson, employed in the laboratory of the Central Alloy Steel Corporation in Canton, Ohio, presented the results of his research program, which observed the surface hardness results of steels with varying aluminum content.

A gentlemen by the name of J. P. Walstead was, at that time, presenting his thesis for his Doctorate of Science in Metallurgy under the direction of Professor Homerberg, and jointly they presented a paper that reviewed the work of Dr. Fry. In addition, they commented on their own work, which studied the effects of temperature on the physical properties of the newly developed nitriding steels. They also reviewed their work on equipment development, the effects of decarburization and preheat treatments, and the first early comments on the phenomenon of "white layer" and its effect on the performance of the component. In 1943, Dr. Carl Floe, an associate professor of metallurgy at MIT, presented a paper on the composition of the white layer, its effects, and a method for reducing its formation that is now known as the two-stage process or the Floe process. It can be seen that considerable work was carried out on the nitriding process, much of which remains useful today.

III. THE PROCESS THEORY

Nitriding is thermochemical case hardening process that is carried out in the ferrite region of the iron-cementite diagram (Figure 3), which is located beneath the Ac1 horizontal eutectoid line (for ferritic steels). In order for the process to be carried out successfully, there are three primary requirements:

1. A source of heat. The heating method should be capable of raising the temperature to approximately 500°C (932°F). This will, of course, depend on the tempering temperature of the steel being treated.
2. A source of nitrogen (usually ammonia).
3. The steel product to be treated.

Gas nitriding is based on the decomposition of ammonia under heat to provide nascent or atomic nitrogen. The process is reliant on the diffusion of atomic nitrogen into the surface of the steel.

The crystalline structure of the steel at a temperature below the Ac1 line for ferritic steels is that of a body-centered cubic (bcc) lattice (as can be seen in Figure 4). The individual atoms of the crystalline structure begin to vibrate around their discrete lattice points, each one separated from its neighbors. The individual atoms that make up the bcc lattice are located one at each corner of the cube and one in the center of the cube. The nitrogen atom is small

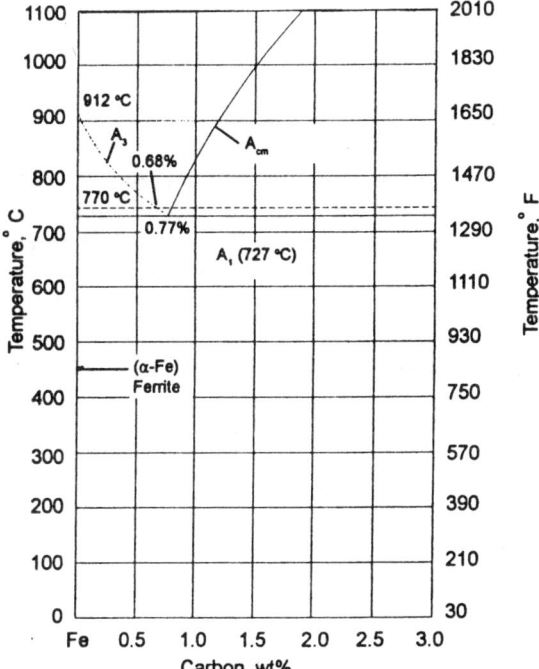

Figure 3   Iron–carbon equilibrium diagram and the Ac1 line.

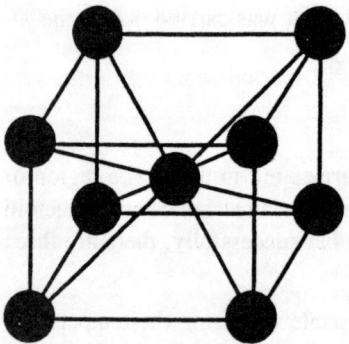

Figure 4 Body-centered cubic lattice structure.

enough to pass between the iron-based crystals as heat is applied up to a suitable process temperature. This is known as "interstitial diffusion." As the temperature increases, the lattice structure vibration increases accordingly, to such a point as to allow the nitrogen atoms to pass through the interstices and into the steel surface to form a case in which the nitrogen has not only diffused but has also reacted with other alloying elements to form nitrides in that surface area. The most common source of nitrogen for the nitriding process is ammonia (NH_3) supplied to the steel surface in gaseous form. There are other sources of nitrogen, but at this point we are only considering the gaseous method of nitriding.

When heat is applied to the gas, it will begin to decompose to its component parts of nitrogen and hydrogen. On cooling, it will recompose back to its original form. This reversible reaction is shown as

$$2NH_3 \rightleftharpoons N_2 + 3H_2$$

Because of the nature of the structure and the bonding mechanism of iron, certain other elements, such as carbon, nitrogen, sulfur, and other metals, dissolve readily in iron to form alloys of iron. In addition to this, elements such as carbon, nitrogen, sulfur, and oxygen can be introduced into the metallic surface in an easy manner to produce what can be considered to be a surface alloy of the steel, hence the principle of "surface treatment." As described previously, the nitrogen atom will take up a position within the interstices of the particular atomic phase structure of the steel at treatment temperature.

The steel parts to be treated are placed into a gastight container, usually made of a heat-resisting type of steel, and sealed into the container following an engineered sealing method. The container is then placed into a furnace, and the temperature is raised to the appropriate process temperature. At the process temperature, the following reactions take place at the surface of the steel:

$$NH_3 \rightarrow 3H + N \tag{1}$$

$$2N \rightarrow N_2 \tag{2}$$

$$2H \rightarrow H_2 \tag{3}$$

The atomic nitrogen and hydrogen components shown in Equation (1) are unstable and will unite with other like atoms to form molecules as shown in Equations (2) and (3). It is while they are in the atomic state that diffusion will take place. Nitriding atmospheres are in nonequilibrium while at the process temperature, which means that at the steel surface a high degree of nitrogen activity takes place from the ammonia. It is usual to have a 100% flow of ammonia into the container within the furnace, meaning that some of the ammonia will dissociate into nitrogen and hydrogen gases when subjected to the furnace temperature as shown earlier.

The ammonia that does not dissociate is known as "undissociated" ammonia. The undissociated ammonia is used to measure the nitriding activity taking place within the process container. The degree of decomposition can be expressed as [5]

$$\frac{\text{Amount of decomposed NH}_3}{\text{Added amount of NH}_3} \times 100\%$$

The decomposition rate of the ammonia gas is usually kept between 10% and 35%, depending on the steel being treated and the gas exchange rate [6, p. 407].

IV. COMPOSITION OF THE CASE

The formed nitrided case has two distinct zones:

1. The compound zone at the surface of the steel. This is known as the "white layer" because the layer is seen as a white zone when etched with Nital (3% nitric acid and alcohol). It is made up of two phases known as the epsilon and gamma prime phases. The concentration of each phase is dependent on the chemistry of the steel, primarily the carbon concentration and the nitriding activity. The compound zone composition will also determine the steel's surface resistance to corrosion and wear. The compound layer is usually up to 25–20 μm thick, the thickness varying in to time, temperature, and gas composition. A typical case formation can be seen in Figure 5.

2. The diffusion zone, the area immediately below the compound layer. This region is made up of stable nitrides formed by the reaction of the nitrogen with the nitride-forming elements, such as chromium, vanadium, molybdenum, tungsten, and aluminum. The region's thickness is again time- and temperature-dependent, ranging up to 0.040 in. It is from this region that the fatigue and load-bearing strengths are determined. This can be seen in Figure 6.

The area below the diffusion zone is the core of the steel and is unaffected by the surface nitriding activity. The core hardness is considered to be the hardness achieved by the prehardening and tempering operations. A typical case structure is shown in Figure 7.

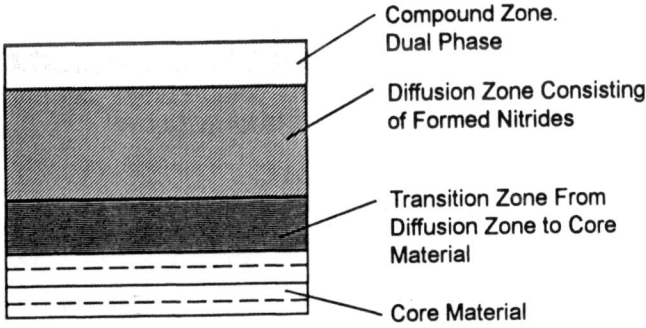

Compound Zone.
Dual Phase

Diffusion Zone Consisting
of Formed Nitrides

Transition Zone From
Diffusion Zone to Core
Material

Core Material

Figure 5 A typical nitrided case structure.

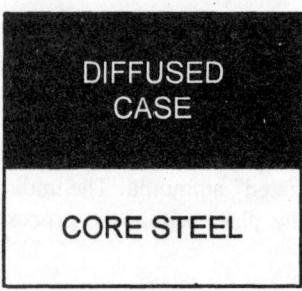

Figure 6   An illustration of the diffusion zone formation.

As a direct result of nitriding, the compound layer will always be present; it cannot be prevented. However, it can be minimized by a technique developed by Floe in the early 1940s [7]. The method involves nitriding at the normal process temperature and gas dissociation (surface nitriding activity) followed by raising the temperature to 570°C (1095°F) and a gas dissociation of 85% using dissociated ammonia. This method of nitriding will produce a compound zone of minimal thickness.

The thickness requirement of the compound layer is dependent on the operating conditions of the steel being treated. The compound layer normally formed does not display any ductility, is extremely brittle, has no impact value, and will normally exhibit some porosity depending on the process temperature and the degree of nitriding activity at the surface of the steel.

The preceding description is generalized to describe the defined zones of the formed case. The composition of the case is influenced by the steel composition and the process gas composition. Because of the use of a gaseous compound such as ammonia (NH_3), the resultant metallurgy of the compound zone will always be dual phase, both epsilon ($Fe_{2-3}N$) and gamma prime (Fe_2N). The formation of the epsilon phase can be directly influenced by (1) temperature, (2) percentage of carbon in the steel, (3) the introduction of a hydrocarbon gas such as

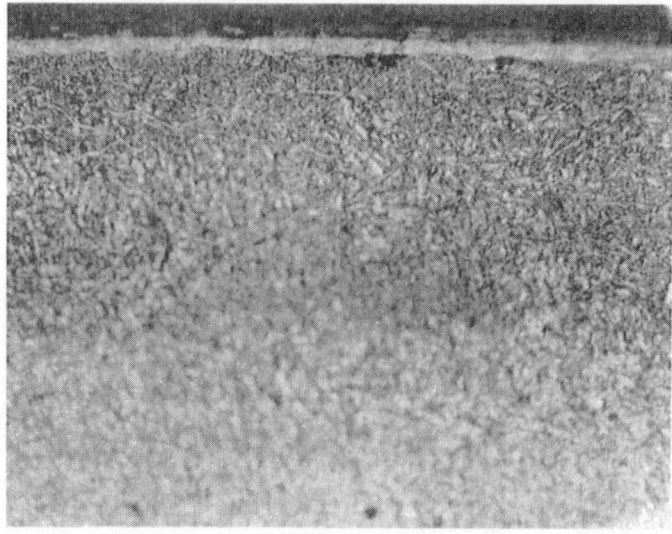

Figure 7   A typical case structure showing the core below the diffusion zone.

methane (CH_4) into the gas flow, and (4) the dilution of the process gas by the addition of a diluent gas such as hydrogen.

The action of the nitrided case formation in a molten salt bath can be represented by the general formulas

$$4NaCN + 2O_2 \rightarrow 4NaCNO$$

$$4NaCNO \rightarrow 4Na_2CO_3 + 2CN + CO + 2N$$

Salt manufacturers will vary the composition of the salts for specific purposes. It can be seen that the working liberates carbon monoxide (CO) and nascent nitrogen (N), which are responsible for the absorption of carbon and nitrogen by the treated parts. The nitrogen will react at the steel surface to form iron nitrides, while the carbon activity will lead to the formation of carbides and promote the nucleation of the epsilon phase within the compound zone. As the activity of the bath varies during operation, so too will the resulting surface metallurgy of processed steel parts.

Temperature selection of the nitriding process plays a very important role in the compound zone formation. As the process temperature increases, the amount of the epsilon phase will also increase up to a temperature of, say, 1100°F. Above this temperature, then, the complete activity changes because now the process is entering the range of nitrocarburizing treatments, in addition to which the thickness of the compound zone will increase with temperature and time as the epsilon phase increases.

V. GAS FLOW AND CONTROL OF DISSOCIATION

The control of the amount of ammonia dissociating throughout the process is a major consideration with respect to the compound layer and the diffusion zone. To measure the extent of dissociation, an adequate and continuous supply of ammonia gas is necessary, delivered at such a rate as to maintain a slight positive pressure in the process container in the furnace. The container will have an exhaust pipe that directs the ammonia exhaust gas out to the atmosphere or through a gas burnoff port. At a suitable point in the exhaust line, a measuring device will be fitted. This device can be in the form of a "water dissociation" unit or an electronic measuring instrument [6, p. 408].

Due to the temperature of nitriding and the catalytic effect of the steel being nitrided, the ammonia gas is decomposed into nitrogen and hydrogen, some of which then recombines to form ammonia. This means that the exhaust gases leaving the nitriding container and furnace are ammonia (NH_3), hydrogen (H_2), and nitrogen (N_2), of which only the ammonia is soluble in water. It is known that ammonia will absorb 70 times its own volume in water, and use is made of this fact in measuring the nitriding activity at the steel's surface. This is accomplished by passing the exhaust gases from the nitriding container through a calibrated glass buret with a glass reservoir filled with water above the buret. Once the buret has been completely purged by the exhaust gases, it is sealed from the exhaust gas flow. The exhaust gas in the buret is now sealed in, and the water reservoir located above the now sealed buret is opened to allow the water to flow directly into the gas-filled buret. As the water flows in, it will absorb the free ammonia, but not the free molecular nitrogen and hydrogen.

If the volume of the calibrated buret is 100 cm^3 and the water fills up to, say, the 70 cm^3 mark, then 30% of the exhaust gas is insoluble molecular nitrogen and hydrogen. This can be expressed as 30% dissociation in the nitriding container. Ideally, the amount of dissociation or nitriding activity should be around the 30% mark, depending on the composition of the steel being treated. If there is a deviation from the 30% mark, then the gas flow should be adjusted accordingly.

Consideration should be given to the amount of surface area being treated in the nitriding container. The internal surfaces of the container and any work support fixtures should also be considered when setting the gas flow for the nitriding process. If the same work support fixtures are used for each nitriding cycle, the fixture will also be nitrided, which will affect the gas flow setting from cycle to cycle.

An illustration of the complete furnace and dissociation system can be seen in Figure 8.

VI. THE TWO-STAGE (FLOE) PROCESS OF NITRIDING

Carl Floe was among the many early metallurgists to observe and identify the compound zone, or white layer, and its inherent problems in relation to the product application. The compound zone is very brittle and is prone to exfoliation and cracking. Floe developed a two-stage process, or double-stage process as it is sometimes known, in an attempt to eliminate the compound zone and patented the process in the early 1940s [7]. He was not successful in eliminating the layer, only in minimizing its formation. The process is still an accepted method for reducing the thickness of the layer.

The Floe process is carried out as two quite distinct events. The first portion of the cycle is accomplished as a normal nitriding cycle at a temperature of 495°C (925°F), the time having been determined by the case depth required, and a normal gas dissociation of between 15 and 30%. This will produce a nitrogen-rich compound layer at the steel surface. Once the cycle is complete, the furnace temperature is elevated to approximately 560°C (1030°F) and a gas dissociation of 75–80%. The gas is now supplied from an external ammonia dissociator, which is necessary for the higher dissociation required. The two-stage process is used to reduce the formation of the compound zone only; it serves no other purpose. It should be noted that its use will produce deeper overall case depths than can be obtained with the single-stage process because of the higher process temperatures employed. Also, slightly lower hardness values will be attained. When considering the use of the two-stage process, the tempering temperature of the steel part to be treated should be taken into account and the process temperature kept below that temperature. If the reduction of the compound layer is critical to the performance of the part, then consideration should be given to the single-stage process, which can be followed by grinding off the compound layer.

VII. SALT BATH NITRIDING

Shortly after the development of gas nitriding, alternative methods of nitriding were sought. One such method was the use of molten salt as a nitrogen source.

In early salt bath nitriding, the molten salts were prepared by melting and aging 96–98% sodium cyanide. Although these baths produced what would nowadays be considered a crudely nitrided case, it was a nitrided case nonetheless. The case uniformity was irregular, and the salt produced resulted in a poor surface finish. The nitriding salts of today are chemically blended to very close tolerances and produce a uniform and consistent case with a clean, bright gray surface finish. As with gas nitriding, the operating temperature range is in the ferrite range, below the Ac1 line on the iron-cemenite diagram. The liberated nitrogen enters solution in the ferrite and precipitates as nitrides without the need for quenching.

Again, the case depth is related to time, temperature, and steel chemistry. Nitriding in salt baths is comparable to gaseous nitriding in ammonia but is usually confined, although not limited, to shallow case depths of, say, 0.005 in. (0.13 mm).

The make-up of a new bath is accomplished simply by placing chemically blended salt into the steel nitriding pot and applying heat until the salt melts and reaches the appropriate

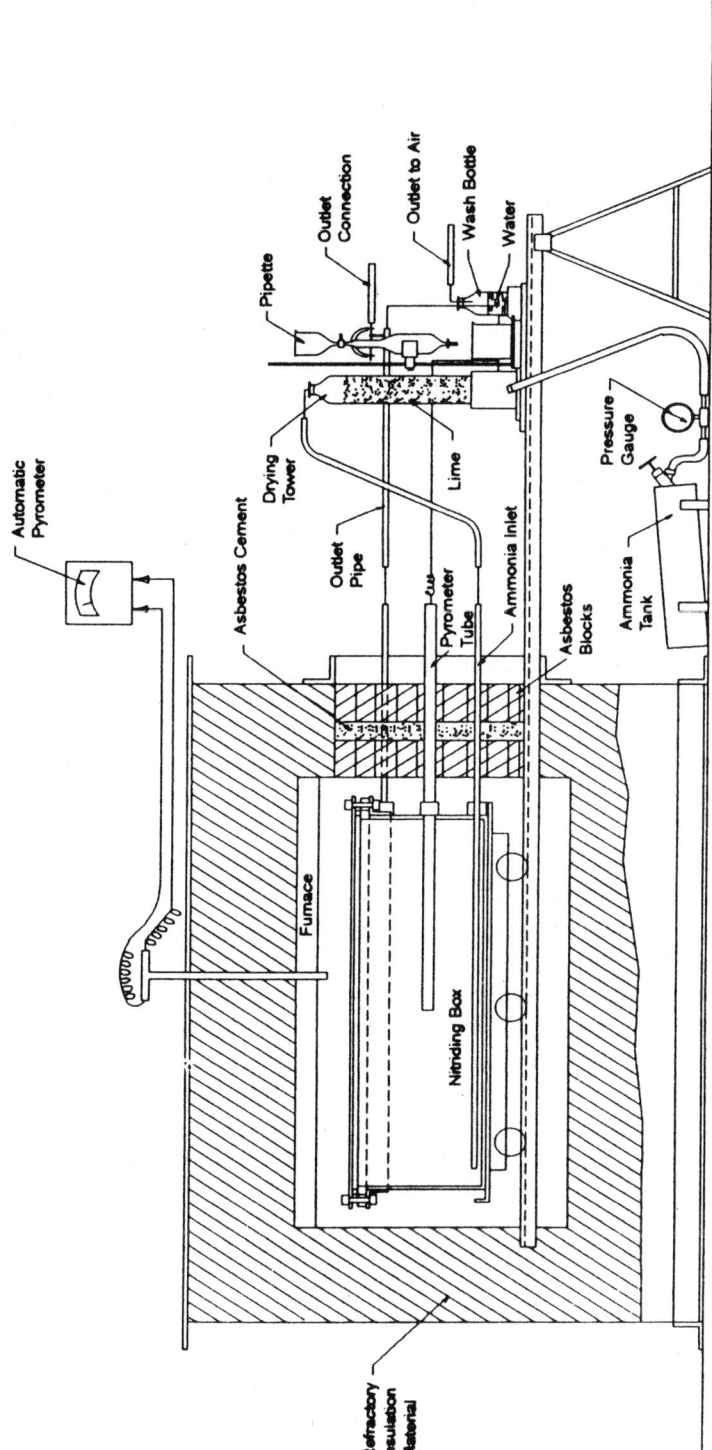

Figure 8 An early typical nitriding furnace construction, complete with anhydrous ammonia supply. (Courtesy *ASM International Heat-Treatment Handbook* Vol. 4.)

operating temperature. It is important and strongly recommended that the newly made-up bath not be used for at least 24 h. This is known as "aging" the bath [8]. In this condition, the salt activity at the steel surface is extremely aggressive and will cause serious pitting of the steel surface. The newly made-up bath will need to be analyzed by titration very carefully until a level of 5% sodium cyanate is reached in the bath. Once this level has been attained, it is safe to use the bath. With continual use of the bath, the cyanide level will continue to decrease. Conversely, the cyanate and carbonate levels will increase. A high cyanate level will cause a somewhat dark finish on the surface of the treated steel.

If the cyanate content of the bath is too high, it can be reduced simply by increasing the temperature of the bath to 650–700°C (1200–1300°F) for a period of approximately 1 h. It will be observed that a much higher than normal level of nitriding potential will occur at the higher cyanate levels. The bath should be analyzed at the commencement of each operating shift and the results recorded, and the amount of make-up salt added that is required to bring the bath up to the normal operating level. Sodium cyanide (NaCN) levels are normally maintained as follows [8]:

High-speed steels and hot-working steels	15% minimum NaCN
Low-carbon and alloy steels	20% minimum NaCN

It is also necessary to maintain bath cleanliness by "dragging out" the bath with a scoop and disposing of the dragout in the appropriate manner, bearing in mind that the residue contains some cyanide residues as well as some carbonates. The bath contaminants originate not only from the salt residues but also from iron oxides on the inside of the salt pot, from oxide contaminants from the work surfaces, and from support fixtures or work suspension wire.

A. Nitriding in the Salt Bath

1. High-Speed and Hot-Working Tool Steels

As with gas or other methods of nitriding, tool steels are nitrided for longevity and performance. High-speed steel tools, such as drills, reamers, end mills, side and face cutters, and form tools, will benefit from the nitriding treatment. Cutting tools that are subjected to high vibration stresses are not considered suitable candidates for nitriding [8].

Very high surface hardnesses, up to 1000 VPN, are typically seen. However, embrittlement of the tool may occur if treatment times are prolonged. Cycle times are relatively short by normal nitriding standards, ranging from 5 to 60 min, but typically 10–20 min cycles are the norm. The life of a cutting tool can be extended appreciably, and usually in excess of 100%, by subjecting the tool to the nitriding treatment. Another practice has been to steam-treat the tool after nitriding it to further extend its life. The steam treatment will cause the nitrided case to diffuse further into the steel, thus reducing the surface hardness, and will very slightly oxidize the steel surface.

Aluminum extrusion dies and mandrels made from H.13 hot-working steel experience aggressive wear from the surface aluminum oxide of the extruded aluminum billet. In addition, they suffer from "metal pickup" from the billet. The mechanics of the wear pattern that occur are due to the aluminum billet oxidizing as a result of preheating as well as frictional forces due to the extrusion pressure of the press on both the billet and the die. Nitriding has successfully reduced the wear on the die from these areas. Surface hardness values up to 1100 VPN are typically observed. Cycle times at temperature can be up to, say, 90–120 min [8].

Press forging dies experience a longer life pattern after nitriding than those that are untreated. The press die is, like the extrusion die, typically manufactured from an H Series tool steel, usually of H.13 material. The case depth should not be large. Dies with a shallow case depth (usually up to 0.008 in.) perform considerably better than dies with a deep case (say, in excess of 0.010 in.). The reason for this is that as the case diffuses deeper into the steel, it begins to lose its flexibility and will very quickly begin to exhibit surface cracking. Cycle times are usually in the region of 2–4 h. Hammer forging dies are not normally good candidates for the nitriding process because of the case's inability to resist high impact values.

2. Safety in Operating Molten Salt Baths for Nitriding

The operation of salt baths of any description requires very careful handling in order to maintain a high degree of safety for both the operator and the work being processed [8]. The following is a list of simple, commonsense operating procedures and precautions:

1. Ensure that all operators observe good personal hygiene when handling any type of heat treatment salt.
2. Ensure that all operators who are involved with the use of heat treatment salts are fully aware that the salts are extremely poisonous and should be handled with the utmost care.
3. All operating personnel should wear appropriate safety clothing and protective equipment such as gloves, arm shields, eye protection and safety mask, fire-resistant apron, leggings, and safety shoes.
4. Do not in any way mix cyanides with nitrate salts; otherwise, there is a serious risk of an explosion or a fire.
5. Preheat all work thoroughly before introducing it into the molten bath. The reason for this is twofold: to remove all traces of moisture and to reduce thermal shock to the work.
6. Store and accurately label all storage drums and containers. Once again, do not mix cyanide sludge with nitrate sludge. Doing so poses a serious risk of explosion.
7. Ensure that adequate ventilation surrounds the top of the salt bath to remove fumes and to make the area free and clear of obstructions.

3. Maintenance of Nitriding Salt Baths

A strict maintenance schedule is necessary for the successful operation of a nitriding salt bath. Maintenance can be performed on a daily and weekly basis by the operator for the optimal performance of the bath.

Daily Schedule

1. Titrate the bath for cyanide/cyanate content. It is necessary that this operation be carried out at the commencement of each shift, the results recorded, and the necessary additions made.
2. Check that the temperature-measuring equipment (i.e., the control thermocouple, overtemperature thermocouple, control instrument, and overtemperature instrument) is functional.
3. If the bath is aerated, check that the air pumps and flowmeter are operating without restrictions.
4. Check the surface appearance of the part after treatment for surface color and surface pitting.

5. Desludge the bath at the commencement of each shift to remove free iron particles and oxides from the salt.

Weekly Schedule

1. Remove the salt pot each week, and examine the external surfaces of the pot for signs of stretching and for signs of external oxidation (in the form of scale).

2. Regenerate the bath each week by raising the temperature to approximately 600°C (1112°F) and increasing the aeration flow into the bath. This will cause the free irons and oxides to precipitate out of the salt and settle on the bottom of the bath, ready for desludging.

VIII. PRESSURE NITRIDING

The method of nitriding that makes use of pressures that are higher than those of atmospheric pressures was first used in the early to mid-1950s. The pressure nitriding process [8] is a relatively simple method of nitriding that requires the use of a sealed retort similar in construction to that of a gas nitriding retort, differing in that it can withstand higher than normal operating gas pressures. As with normal gas nitriding, part cleanliness is mandatory to the success of the procedure. After the cleaning of the part has been completed, the retort is sealed and ammonia gas is introduced into it to a predetermined operating pressure. The selection of the appropriate pressure is determined by a number of parameters, the main ones being (1) part geometry, (2) the surface area to be treated, (3) the retort volume, and (4) the process temperature.

The question of what the appropriate operating pressure is is one that will most probably be debated for many years to come. Intensive investigations were conducted by a leading German company into the use of very high operating pressures—as high as 100 bar (1470 psi). It was found that the higher pressures produced no real commercial value save that of penetrating very fine holes, such as those holes found in a diesel engine fuel injector nozzle. The company investigated the use of lower pressures (50 bar), again with no significant improvement. The method has not gained wide commercial acceptance as a method of nitriding.

IX. THE DILUTION METHOD OF NITRIDING

As with gas or salt bath nitriding, the formation of the compound zone is usually of prime concern, and many unique ideas have been explored. Machlet's original patent application reads "for the nitrogenization of irons and cast irons using ammonia diluted with hydrogen." This formed the basis of an operating technology developed in eastern Europe. The method takes into account the component ratio of ammonia gas, which is 3:1 (hydrogen to nitrogen). When hydrogen is added to the flow of ammonia, the nitrogen component of the ammonia is diluted. The compound zone and its quality at the surface of the steel are controlled by the carbon in the steel, which, if at a high enough level, will begin the nucleation of the epsilon phase in the compound zone. It is also influenced by nitrogen, which, at high enough concentrations, will encourage the nucleation of the gamma prime phase of the compound zone. It can be seen that very careful control of the gas chemistry is of great importance.

Care should be taken when using hydrogen as a diluent gas because of its high inflammability. Simple precautions should be taken when using hydrogen:

1. Use good effective retort seals.

2. Do not allow oxygen to mix with hydrogen because this will constitute a serious explosion or fire risk.

3. Ensure that when the cycle is complete, the retort has been given sufficient time to cool down before opening it to the atmosphere. Cool to below 150°C (300°F), then purge with nitrogen to remove all traces of hydrogen.

The above precautions also apply to conventional gas nitriding.

The advantages of using the dilution method of nitriding is that it offers a unique method of controlling the formation of the compound zone by reducing the amount of active and available nitrogen.

X. FURNACE EQUIPMENT FOR NITRIDING

The gas nitriding furnace is a relatively simple furnace construction that has been refined, mainly in the area of control rather than construction design, as previously shown in Figure 8.

A. The Furnace

For the process to be successful, it is mandatory that the furnace be constructed in such a manner as to have insignificant heat losses through the furnace walls and door and to maintain a uniform temperature for extended periods of time. Up to 100 h cycle times are not uncommon for gas nitriding. Good insulation materials will ensure that heat losses are kept low and will reduce the energy costs. Temperature should not vary by more than, say, 15°F.

The furnace should be equipped with a mechanical handling system for loading and unloading. This can be in the form of rollers in the furnace with a loading table outside the furnace for a horizontal furnace system or an overhead crane system for a top-loading furnace. Two bell-type nitriding furnaces are diagrammed in Figure 9.

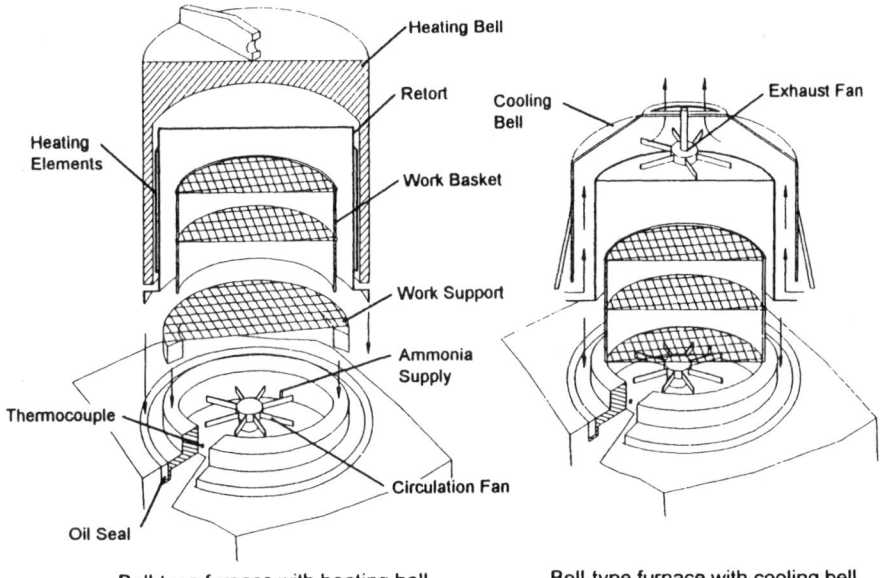

Bell-type furnace with heating bell   Bell-type furnace with cooling bell

Figure 9 A typical bell-type furnace construction. Note the internal recirculation fan. (Courtesy *ASM International Heat-Treatment Handbook* Vol. 4.)

B. The Retort

There are two aspects of the retort design that require careful consideration: the seal and material selection. Many different materials have been used and rejected. An all-glass retort was tried and proved to be the most effective from the point of view that it took no part in the gas reaction; however, it was completely impractical because of its inability to withstand possible mechanical damage. Brick has been tried but also found to be impractical because of its porosity and iron content.

The most successful retorts are metallic, manufactured from either Inconel or type 35-15 stainless steel. These retort materials start to deteriorate after extensive use and begin to dissociate ammonia. This can be remedied simply by heating the retort to a temperature of 1650°F and, in an oxidizing atmosphere, holding it there for several hours. After this procedure, all of the surface (both external and internal) should be sandblasted.

C. Gas Circulation

In order for effective nitriding to take place, it is necessary to have a good flow of ammonia gas throughout the retort during the complete nitriding cycle, and it is usually introduced at a slight positive pressure above atmospheric pressure. This is necessary to reduce the risk of gas stagnation due to pockets that may form as a result of shielding, which will cause inconsistent nitriding and, at the very least, a nonuniform case, or even, in some instances, no case at all.

To prevent gas stagnation and to ensure adequate gas circulation, it is necessary to fit a gas circulation fan into the retort. The material of construction for the fan is the same as or similar to that used for the furnace retort. The circulation fan wheel need only be a simple paddle wheel type of construction.

D. Temperature Control

As with any heat treatment process, good temperature control to maintain a uniform temperature is essential to the process, and careful consideration should be given to the size of the furnace heating chamber if the furnace is to be a new one. This is necessary in order to determine the number of control zones within the furnace. The temperature control can be either a single-zone or multizone system. With a single-zone control, a single thermocouple can be used in conjunction with an overtemperature control thermocouple. If the furnace is a multizone system, a master thermocouple is employed in conjunction with "slave thermocouples" and the necessary overtemperature thermocouples. The control system can be set up in such a manner that the master thermocouple will control not only within the "master" zone but also within each of the remaining zones.

When selecting the point of insertion of each of the thermocouples, be it for a single zone or multizone system, the thermocouple should be as close to the nitriding retort as is conveniently possible. In addition to this, provision for a retort thermocouple should be made. This is accomplished by fitting a stainless steel tube into the retort with the end of the tube inside the retort being sealed and gastight. This gives the operator the facility to measure the internal retort temperature and thus the work temperature. The tube length should be such as to protrude far enough into the retort to indicate a reasonable temperature average of the whole of the interior of the retort.

Temperature recording can be accomplished in a number of ways:

1. Conventional time–temperature control instruments transmitting a voltage signal to a data logging instrument.

2. Conventional time–temperature controllers transmitting a voltage signal through a microprocess controller to a data logging instrument. This type of system begins to take away some of the operator involvement, thus relieving the operator for other duties.

3. A newer development has been to make use of a programmable logic controller (PLC) and report all process activity signals through a liquid crystal display (LCD) screen.

4. The most recently developed control system makes use of the PLC in conjunction with a computer, such as a 486 PC. This allows many process control functions to be programmed and monitored as well as providing screens for, say, programmed furnace maintenance (see Figure 10).

E. Salt Bath Nitriding Furnaces

Salt bath nitriding furnaces [11] are also very simply constructed. Three modules make up the complete furnace system: (1) the salt container (salt pot), (2) the heating method, and (3) the control system.

1. The Salt Pot

In the past, the salt pot was constructed from low-carbon construction steels. This type of pot proved to be a great source of contamination to the nitriding salt because free iron oxides dislodged from the internal surfaces of the pot and entered into the nitriding salt. When work was introduced into the bath, the iron oxides migrated to the work surfaces and caused a non-uniform case to form on the work surface. To some extent, the iron oxides could be made to precipitate out of the salt by simply raising the temperature to approximately 570°C (1150°F) and holding at that temperature for approximately 1–1.5 h. The iron oxides would then pre-

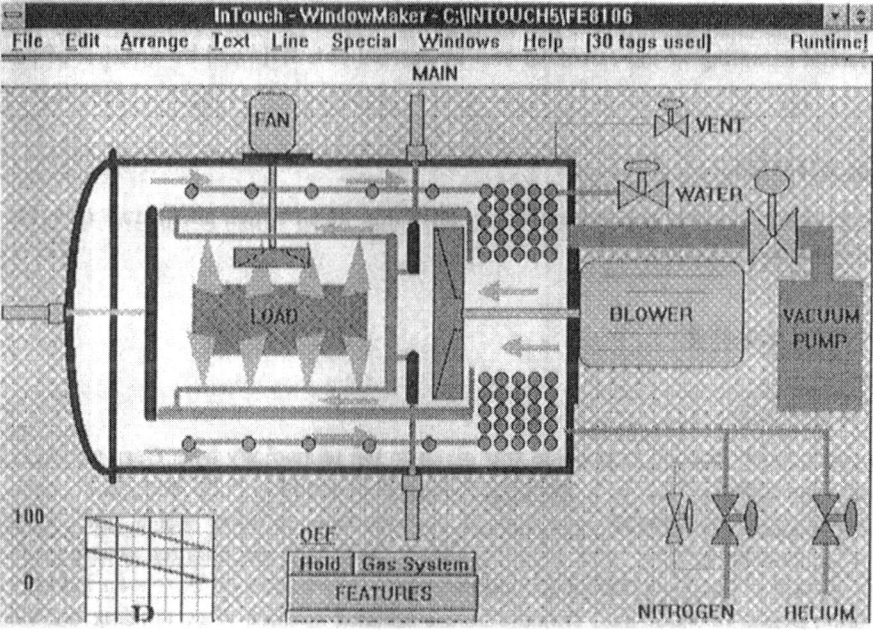

Figure 10 A typical PC screen configuration for a heat treatment furnace. (Courtesy of SECO/WARWICK Corp.)

cipitate out and settle on the bottom of the pot in the form of sludge, which could be scooped out and loaded into appropriate disposal containers.

As an alternative to the low-carbon steel, titanium was then used for the salt pot. Although the titanium pot was successful in terms of performance, it was not a cost-effective solution. From this design, the lined pot was developed. This alternative was a low-carbon pot with a thin titanium liner. This lined pot eliminated the iron oxide from the pot and at the same time proved to be cost-effective.

2. Heating Methods

The heating method can be (1) gas-fired (direct or indirect) or (2) electrical (external or immersed electrode). If the choice has been made for gas firing, great care should be exercised so that no direct flame impinges onto the external surfaces of the pot. This would lead to rapid oxidation of the outer surface of the pot, causing a serious reduction in the sectional thickness of the pot, leading in turn to "necking" and stretching and thus to premature failure of the pot.

If an electrical heating method has been selected, then a further choice needs to be made between immersed electrodes and external wall-mounted heating elements located within the heating chamber of the furnace. Making the choice for immersed electrodes means that, once started, the bath cannot be shut down without first removing the molten salt. As a result, the furnace is running 24 h a day, which means it must be designed as a high-production furnace. If an external heating element system is chosen, then the furnace is designed for batch-type operations and can be shut down after the work has been processed through the bath. Four types of salt bath systems are illustrated in Figure 11.

3. Control Systems

The control systems for nitriding salt baths are very much the same as for gas nitriding systems but with two exceptions. The first exception is that it is necessary to protect the thermocouple from attack by the salt. This is accomplished simply by placing a protective sheath over the thermocouple. The other exception is that none of the control systems can monitor the salt chemistry. Manual sampling of the salt must still be carried out, followed by titration of the salt sample.

XI. PLASMA NITRIDING

There are many names given to the process of ion nitriding, all of which adequately describe the process and the technology.

Ion nitriding
Glow discharge nitriding
Plasma nitriding
Plasma ion nitriding

It was in the early 1930s that the first name was given to the technology by a German physicist, Wehnhelt, who called it "the glow discharge" method of nitriding [12]. Wehnhelt's technology was based upon the familiar chemistry of gas nitriding. In other words, the "glow discharge" was to become a "tool" with which to nitride. Wehnhelt encountered severe problems with the control of the glow discharge, and Berghaus, a well-known European industrialist at that time, began to work with Wehnhelt on a suitable method of controlling the unstable glow. This partnership opened a new door for metallurgical processing, but their method was not used very extensively because it was considered too complex, too expensive, and too

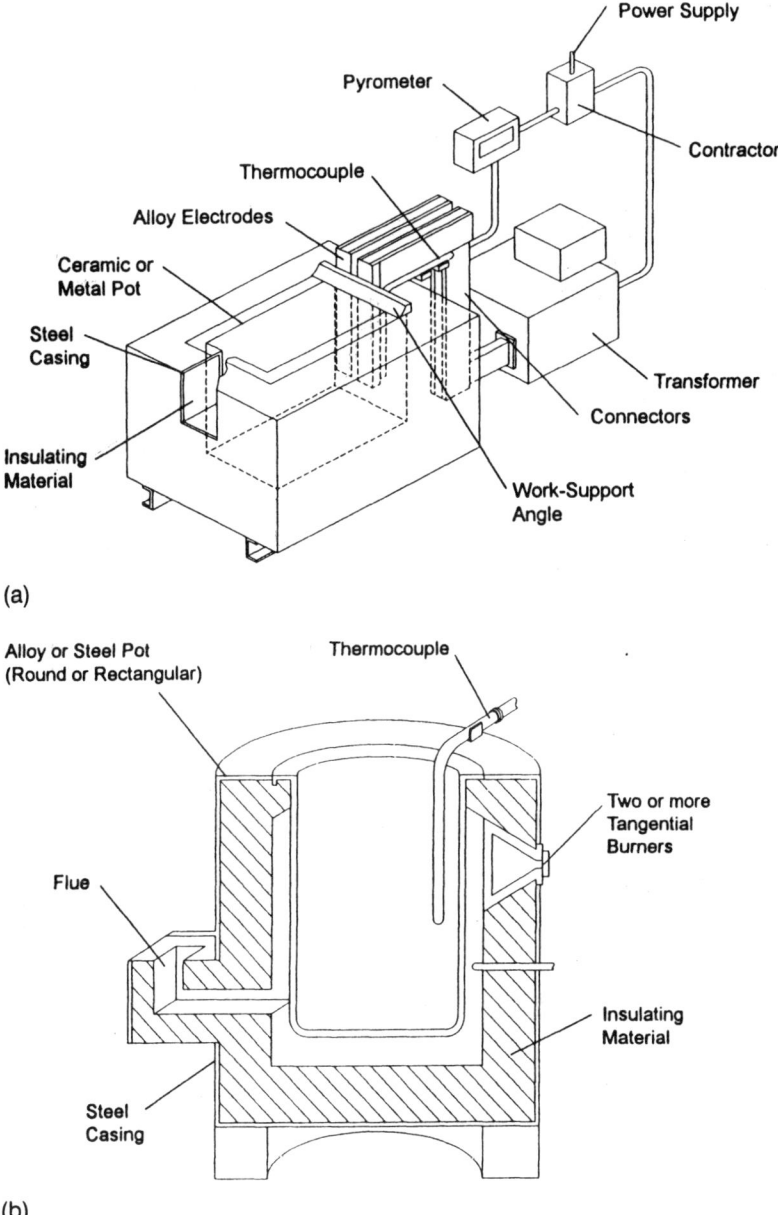

(a)

(b)

Figure 11 Salt bath furnaces for nitriding. (a) Immersed electrode for continuous operation. (b) Gas-fired batch-type furnace. (c) Electrically heated furnace with resistance heating elements. (d) Typical continuous salt bath nitriding line. (Courtesy *ASM International Heat-Treatment Handbook* Vol. 4.)

unreliable to guarantee consistent and repeatable results. The process was used to great effect during the World War II by German ordnance manufacturers.

In the late 1940s and early 1950s, two General Electric scientists in Massachusetts, Claude Jones and Derek Sturges, began an investigation into the technology of metallurgical processing. They successfully built a system with which they were able to not only control the process

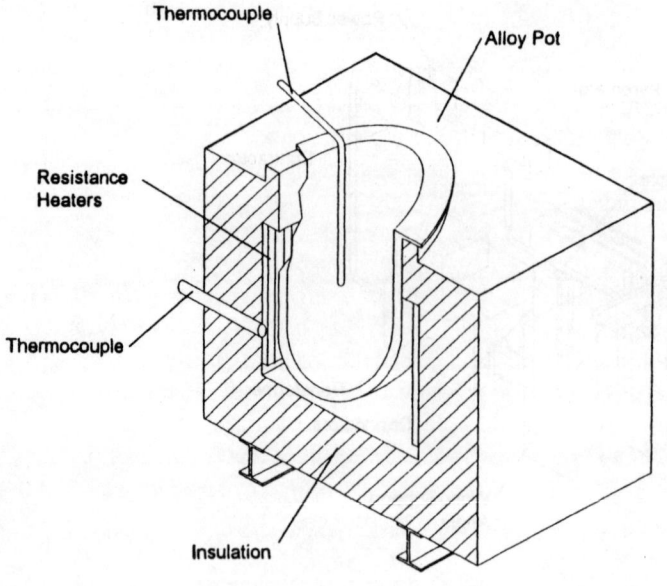

(c)

(d)

Figure 11 Continued

but also to obtain repeatable and consistent results. Many of GE's materials and parts were processed through their furnace [13].

A. Plasma Generation

When two electrodes are placed in a gas environment at partial atmospheric atmosphere and a voltage is applied to the electrodes, then the gas in the enclosed environment will begin to glow, emitting a light, the color of which will depend on the type of gas. An everyday example of this phenomenon is the fluorescent light.

To generate the plasma or glow discharge, reliance is placed on the fact that even at atmospheric temperature and pressure, gas molecules are always in a state of movement and are always colliding with each other. If we now place the gas in an enclosed vessel with two electrodes, seal that vessel in such a manner as to make it gastight, and apply a voltage across the two electrodes, then the gas molecules will become excited, liberate free electrons from their outer shell, and begin to move in a random manner, colliding with each other. If the gas is at atmospheric pressure, then when collision occurs, due to electrical excitation there will be a release of energy, albeit a very small amount of energy. The energy released will be insignificant because the probability of collision between the molecules is very high because the mean free path of the molecules is very small. An illustration of this can be seen in Figure 12. If, on the other hand, the internal pressure of the chamber is reduced to a high vacuum level, then the probability of molecular collision will be very low because the mean free path of the gas molecules will be very long. The direct result of this is that a large amount of energy is released but cannot be used because of the infrequency of collision. It follows then that somewhere between the two extremes of pressure there should be an ideal pressure band in which the glow phenomenon can exist. This pressure band has been found to be between 1 and 4 torr (134 and 535 P). Pressure has therefore been found to be one of the principal control elements of the glow discharge, others being voltage, the composition of the gas mixture, and surface area.

If, at nitriding temperature, a high operating pressure is used (one that is closer to atmospheric pressure), then the glow will be seen to have incomplete coverage of the surface be-

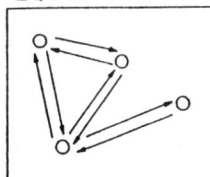

EQUAL VOLUME

HIGH VACUUM
(PARTIAL PRESSURE)

LOW COLLISION PROBABILITY
HIGH MEAN FREE PATH
HIGH ENERGY OUTPUT
1×10^{-6} TORR (0,001 MICRON)

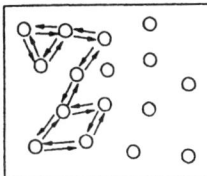

EQUAL VOLUME

LOW VACUUM
(PARTIAL PRESSURE)

HIGH COLLISION PROBABILITY
LOW MEAN FREE PATH
LOW ENERGY OUTPUT
100 TORR (10^5 MICRON)

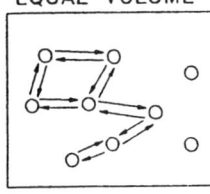

EQUAL VOLUME

MEDIUM VACUUM
(PARTIAL PRESSURE)

IDEAL COLLISION PROBABILITY
IDEAL MEAN FREE PATH
IDEAL ENERGY OUTPUT
1 TO 2 TORR (10^{-3} MICRON)

Figure 12   Probability of collision schematic.

ing treated. Conversely, if the pressure is low (that is, at a high vacuum), then the glow area will appear to be hazy and at a distance from the work surface being treated.

B. Glow Discharge Characteristics

In order to understand glow discharge characteristics, we refer to what is known as the Paschen curve [6, p. 422]. The curve reflects the relationship between the voltage on the vertical axis and the current density on the horizontal axis (see Figure 13). It is through observations of the relationship between the voltage and the current density that we can make use of the Paschen curve to begin to determine the appropriate process voltage necessary to ion nitride.

The Townsend Discharge Region. Within this region, if we apply a voltage, which we call the ignition voltage, electrons from the gas within the vacuum chamber can be accelerated toward the anode, which in this case is the vacuum vessel. Referring back to the "mean free path" principle, when collision occurs within the partial pressure environment, ionization will be caused with an appropriate release of energy. This is called ignition. If the voltage is increased slightly, the current density is increased. Conversely, if the voltage is decreased, then the current density will decrease, as will the energy output.

The Corona Region. If the voltage is increased sufficiently, more electrons can be released by further gas ionization in this region. This increase in released energy will cause further ionization, thus becoming self-maintained. It can almost be described as a perpetual chain reaction.

The Subnormal Glow Discharge Region. If the current-limiting resistance is reduced, then the current density will increase and continue to increase, causing a voltage drop between the cathode (the workpiece) and the anode (retort). Voltage stability cannot be maintained; hence the name transition region also used for this region.

Normal Glow Discharge Region. At this point, the glow will be seen to be covering the complete surface of the work being treated, with a uniform glow thickness, without variation and with a constant voltage drop.

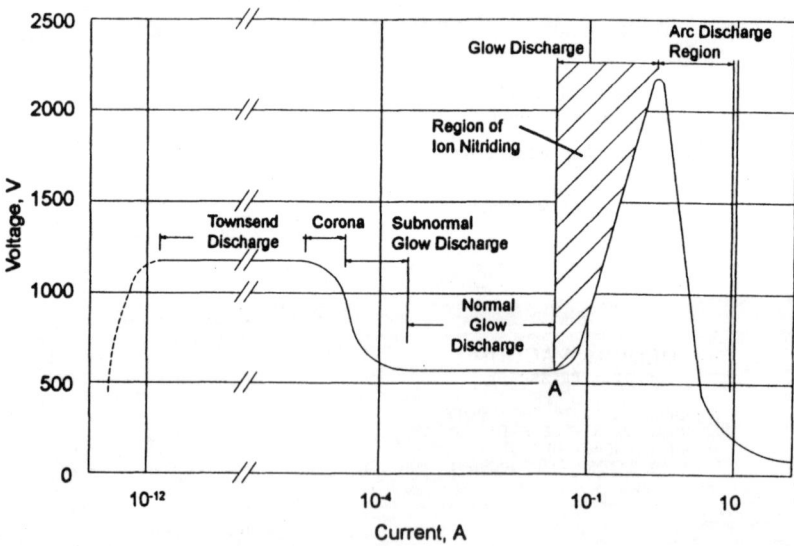

Figure 13 Paschen curve showing the relationship between voltage (V) and current (A).

Glow Discharge Region. Within this region, the total work surface will be completely covered with a uniform glow, following the geometrical form of the workpiece. It is within this region that the ideal conditions exist for plasma ion nitriding.

Arc Discharge Region. A noticeable increase in the voltage drop occurs as the current density increases, in addition to which the power density at the work surface increases. As this occurs, an increase in the surface temperature becomes noticeable, to the point that serious overheating can (and does) take place. The increase in work surface temperature can cause serious metallurgical problems to the point that melting can occur in localized areas, along with other problems such as sharp corners and thin wall sections, and can also cause surface pitting. The arc, when it occurs, can be likened to a lightning strike, being seen as a discharge between the work surface (cathode) and the retort (anode).

It can now be understood that very careful control of the power, current density, pressure, and gas composition is absolutely necessary for successful nitriding to take place.

C. Control Characteristics

If a constant voltage is applied in the partial pressure band where ionization takes place, then the electron collisions will not only generate a glow but will also generate energy in the form of heat. By using different gases, such as nitrogen, hydrogen, and methane (N_2, H_2, CH_4), and utilizing all of these phenomena, we now have the basis for different thermochemical process techniques—in this instance, the process of nitriding. If we now return to the basic formula of anhydrous ammonia and its decomposition when subjected to a heat source, we see that we have one nitrogen molecule and three hydrogen molecules from each two molecules of ammonia:

$$2NH^3 \rightleftharpoons N_2 + 3H_2$$

The act of decomposition (left to right above) liberates nitrogen and hydrogen as individual gases. Each gas exists momentarily in its atomic form (Equation 1 of Section III), from which a small proportion of nitrogen atoms are absorbed and dissolved into the steel. As the formula indicates, we are using ammonia as the soruce gas for nitrogen. The direct result of using ammonia with "fixed" gas ratios means "fixed" metallurgy.

If we now bring the two individual gases, nitrogen and hydrogen, together and vary their ratio it follows that we can manipulate the surface metallurgy of the steel we are treating. This was Machlet's concept in the early 1900s when he patented his "nitrogenization process" and used hydrogen as a diluent gas. With variable gas ratios, we can have variable surface metallurgy, which is a great advantage of the plasma nitriding process.

The gas nitriding process is controlled by controlling four process areas:

1. Time
2. Temperature
3. Gas dissociation
4. Surface area

The ion nitride process has many more process variables to control:

1. Time
2. Temperature (chamber)
3. Temperature (workpiece)
4. Gas flows (nitrogen, hydrogen, methane)

5. Surface area
6. Power voltage
7. Pressure
8. Current density
9. Amperage
10. Rate of temperature rise

If one is able to manage all of these parameters, then process results should be more repeatable and consistent. With the advent of the computer, this is both a possible and practical approach to process management.

D. Equipment Technology

There are two types of ion nitriding technology:

1. Cold-wall, continuous dc technology
2. Hot-wall, pulsed dc technology

1. Cold-Wall Technology

The cold-wall ion nitride furnace (Figure 14) is a system made up of many parts, which will now be described individually.

Retort.   The retort is usually of double-wall construction, consisting of an inner vacuum vessel surrounded by a water jacket. The vacuum vessel is usually fabricated from a stainless steel alloy, and the water jacket from a low-carbon material. Between the two is a space for a continuous flow of water for cooling the inner retort, which heats up because of workload heat losses. The two chambers are separated by the bottom shell flange, which mates with the top shell flange. The bottom flange is usually water-cooled for seal integrity. The top shell is also fitted with an observation port to view the workpiece and the glow seam.

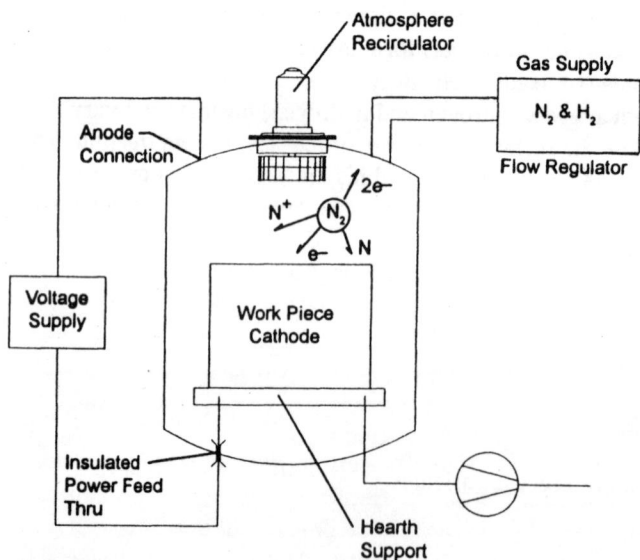

Figure 14   Schematic of a typical arrangement of a cold-wall continuous dc plasma nitriding system. (Courtesy of SECO/WARWICK Corporation.)

The power feedthrough, through which the power is fed to the cathode, is fitted in the bottom shell. The power feedthrough is suitably grounded (earthed) for safety. In addition, the bottom shell is a convenient location through which the process gas supply and thermocouple feedthrough can also be fed.

Vacuum Pumping System. The vacuum pumping system is usually a very simple pumping arrangement. The specific pumping arrangement is dependent on the volume of the chamber to be pumped down. It is either a simple mechanical vacuum pump or a combination of a mechanical pump and a booster pump.

Pump Sizing. Pump size can be determined according to [14]

$$D = \frac{V\,Fa}{t}$$

where D is the pump size in cubic feet per minute (cfm), V is the volume of the vacuum chamber, Fa is the pressure factor (the factor assumes a cold, clean, empty, and tight vacuum vessel), and t is the required pumpdown time in minutes to absolute pressure in torr (mm Hg).

Pumpdown time can thus be expressed as

$$t = \frac{V\,Fa}{D}$$

The type of pump required (a rotary vane mechanical pump or a combination of mechanical pump and booster) will be determined by the size of the anode retort.

Power Supply. To generate the plasma, a continuous dc power supply is necessary, which is constructed in such a manner that it will respond to signals from the control system. The dc power supply allows a continuous flow of voltage to the cathode, located in the process retort, which not only excites the gas electrons but also creates energy in the form of heat in the workpiece. The power is fed into the cathode through the wall of the retort via a power feedthrough, usually through the base of the retort. The design of the power feedthrough is such that it allows the uninterrupted passage of the power through to the cathode and insulates the cathode from the anode.

The power supply is set up in such a manner as to create a bias between the chamber wall and the workpiece. Some of the earlier generations used a straight dc power system to generate the glow discharge. These power supplies were at a disadvantage when operating in the lower portion of the glow region. This caused considerable problems of glow stability, particularly when nitriding complete geometries and blind holes. To process a more complex workpiece, it was necessary to use much higher process voltages. The danger of using the higher voltages (as can be seen from the Paschen curve) is the real and serious risk of arc discharge occurring. If the arc is not suppressed, there is a real risk of metallurgical damage to the workpiece.

Thermocouples. Because the process is temperature-related, it is necessary to measure the process temperature and adjust it if necessary. As in conventional heat treatment processes, it is common practice to measure the temperature as close to the workpiece as is practical. This will give what can be considered to be an almost accurate temperature. When plasma nitriding, and using the cold-wall dc method, the thermocouple is fitted into a specially designed ceramic insert that is, in turn, inserted into a steel blade of a sectional thickness representative of the material section being treated in order to give an appropriate accurate part temperature. The process and part temperature are usually held to within 10°F of the inserted thermocouple on the "dummy" block. The emf generated by the temperature millivolts is transmitted back to a temperature controller, and, if necessary, to a data logger to record the process parameters.

It is of no consequence if the temperature of the other parts of the process retort are indicating a retort temperature either below or in excess of the part temperature because it is only the part temperature that the process is concerned with. The question of temperature uniformity is of concern only at the port, not at the retort.

Gas Flow Control. The control of gas flow by flowmeters is now almost a thing of the past. Later methods of gas flow control used micrometer needle valves. This worked reasonably well but on a rather limited basis. More accurate methods of gas flow control are available today and are currently being used on modern nitriding equipment as well as with the more conventional methods of nitriding.

As gas flow and ratios can affect the current density at the work surface, it is necessary to monitor the gas flow very accurately. Gas flows can be anything up to, say, 100 L/h during ion nitriding. This is, of course, dependent on the surface area being treated. The ion nitriding process makes little demand on the gas consumption compared to conventional methods of nitriding. It can therefore be appreciated that the cost of process gas economics is very low because only the gas necessary for the process is used; there is no "sweep" gas usage as with conventional methods of nitriding. After completion of the process, cooling can be accomplished under vacuum or by gas cooling using nitrogen.

The quality of process gases is of critical importance to the success of the process. It is necessary to use metallurgical grade nitrogen up to 35 ppm maximum oxygen and metallurgical grade hydrogen to 5 ppm maximum oxygen.

2. Hot-Wall, Pulsed DC Current

Hot-wall, pulsed dc technology takes a completely different approach to the ion nitriding process from that of cold-wall technology in that it recognizes (from the Paschen curve) [6] that at ambient temperature the power voltage required to generate not only plasma but also heat makes plasma very dangerous to the integrity of the work. The high voltages necessary to generate heat require that the current density relative to the process voltage be almost at the arc discharge region. Most metallurgical damage to the work occurs at ambient temperature owing to the high voltages that are necessary for extended periods of time to generate the process energy to heat the work to the process temperature. A typical schematic layout is shown in Figure 15.

Because of the metallurgical sensitivity of the work to possible arc discharge, a very simple solution was found. The solution involved using a single-walled vacuum chamber and surrounding it with a conventional furnace with heating elements in its wall. This means that the heating of the work can now be separated from the generation of plasma, because plasma is not necessary to heat the part. The work and the retort can now be heated by the more conventional method of external heating and by convection inside the retort.

Power Generation. The method of generating power for the cathode work support hearth is different from that of the continuous dc system inasmuch as it produces a series of dc current impulses to the hearth at cathode potential. The power generation system is fitted with a variable-pulse system to allow the operator to manipulate the power impulses according to the geometric configuration of the workpiece. This allows better penetration of plasma into holes and reduces the risk of what is known as "hollow cathode." The power impulses can be adjusted to accommodate material sectional changes in the workpiece being treated. With a part that has a complex shape, and using continuous dc power, the thin section of that part will reach operating temperatures much faster than the thicker sections. As a result, thermal differences in temperature are introduced to the workpiece, thus causing the potential for stress

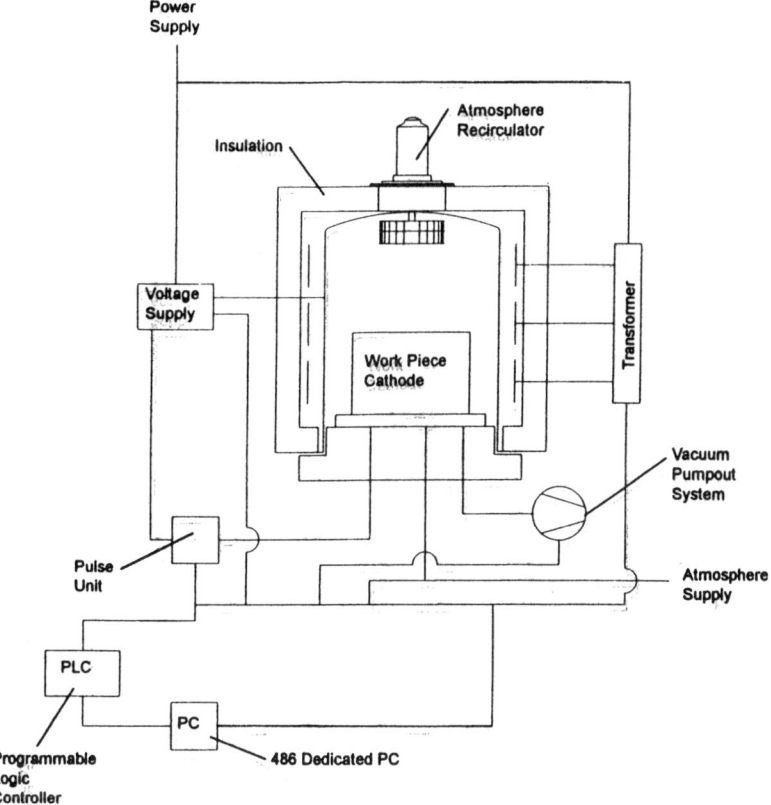

Figure 15   Schematic of a typical layout of a hot-wall plasma ion nitriding furnace. (Courtesy of SECO/WARWICK Corporation.)

risers between thick and thin sections. The pulsed power can be adjusted to accommodate the workpiece sectional changes so that when the power is off the thin section dissipates its heat while the thick section is absorbing heat. This can be controlled to the point that the temperature differentials can be kept to within approximately 10°F on either side of the set-point temperature. The frequency of the pulsed power should be variable enough to allow it to be adjusted to accommodate extreme changes in section thickness. In other words, it should have an adjustable duty cycle of power on to power off. Some systems have a power-on time as low as 20 μs, which is variable up to a power-on time of 1950 μm, which closely resembles a continuous dc power system (see Figure 16). Conversely, the power-off time would be from 20 μs up to 1950 μs. The power system incorporates a high-powered transistorized switching converter system that operates between 1000 and 10,000 Hz.

Temperature Control.   Temperature control of the hot-wall, pulsed dc method is approached in a different manner than that of cold-wall, continuous dc technology, in that the thermocouple is placed in a protective shield and made to be a cathode potential. This means that if the thermocouple were not enclosed within its shield, it, as well as the workpiece, would be nitrided. The thermocouple is electrically grounded (earthed) and electrically isolated for safety and operational integrity. The generated emf from the thermocouple is now transmitted to a thermocouple amplifier and out to the data logger or control screen for observation and control.

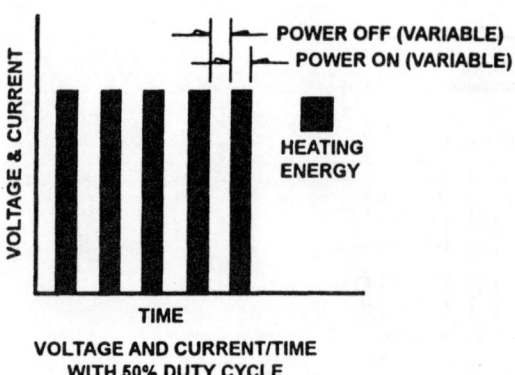

Figure 16   Variable-pulse duty cycle. (Courtesy of SECO/WARWICK Corp.)

External Heating Source.   The external preheating soruce can be located on the wall of the external furnace surrounding the process retort. Some systems operate with what are known as auxiliary heaters. There are a series of heaters within the process retort itself that are mounted facing the work area and are usually (but not always) backed by a heat shield. The problem with this method of auxiliary heating is that the voltage differential between cathode and anode is always influenced by the voltage placed on the auxiliary heating elements, which is not always constant. As a result, the current density at the workpiece surface will be directly influenced by the voltage applied to the auxiliary heating elements. As heat is generated at the workpiece surface, there is also a heat loss that migrates to the chamber wall. If the furnace is constructed as a double-walled vessel with water passing between the two vessels, then the heat migration to the wall will cause a rise in temperature of the water, depending, of course, on the volume of water and the flow speed of the water. Although the idea of supplementary heaters works, they are not always an easy system to work with.

The conventional hot-wall system is, quite literally, an external furnace surrounding a single-walled process retort. This means that there will be an air gap between the external wall of the retort and the inner wall of the furnace. The heating elements are attached to the insulated wall of the furnace and radiate heat onto its external wall. As previously stated, there will be a heat loss from the inside of the process retort to its wall. Any heat buildup on the wall and in the air gap is sensed by an external thermocouple, and an airblower is switched on that blows shop air over the external surfaces of the process retort.

E.   Process Control

The methods of control of the process are many and varied and can be both simple and complex. The degree of simplicity or complexity can be driven by the investment economics in relation to operational skills.

Low Capital Investment, High Operational Skills.   This method of control will take cognizance of a basic and simple pulsed power unit using a manual power output control system, with the temperature and pressure controls being a PID loop system that monitors both pressure and temperature. This type of control does require high operational skills, the operator must be able not only to recognize the process activities and fluctuations but also to correct them.

Moderate Capital Investment, Moderate Operation Skills. This method of control will recognize the use of an Eprom-developed chip to monitor, manage, and control the process parameters. The system operates on the basis of process-generated signals to a microprocess controller. The microprocess controller can be either a proprietary unit or a commercially developed unit. The process will usually manage itself with little or no operator involvement save that of loading, unloading, and programming. The control system can be programmed for security access.

High Capital Investment, Low Operational Skills. This method of control would require a programmable logic controller (PLC) operating in conjunction with the pulsed power pack and feeding the operational information to a programmable controller. This type of control system is extremely user-friendly and can be programmed for security to accommodate both poorly skilled operators and highly skilled operators capable of establishing, developing, and entering process programs into the computer.

XII. METALLURGICAL CONSIDERATIONS AND ADVANTAGES

Ion nitriding offers to the metallurgist many control and process advantages that have not been previously possible with conventional nitriding techniques. It is possible to control not only temperature (particularly at the work surface) but heating rate, gas composition, gas species, species activity at the steel surface, pressure within the retort, and process time.

In addition to improved process cycle times, the greatest advantage of the plasma nitriding process over conventional nitriding techniques is the ability to control the extent of the "white layer" or compound zone. In some instances it is advantageous to have the white layer form. It has been said that "to remove the compound zone, we can simply grind, or lap, it off." While this is a true statement, great care is needed in the grinding operation to ensure that the compound zone is completely removed and stress patterns are not set up through the grinding, which can lead to crack propagation.

A hydrogen atmosphere will heat the workload faster than nitrogen due to the high thermal heat transfer characteristics of hydrogen, but a mixture of 95% H_2 and 5% N_2 enables some nitriding activity to begin during the ramp to operating temperatures. Pressure is also a factor, but maximum heat input rates can generally be achieved at about 3 torr. White layer, or compound zone, development is much slower in ion nitriding, remaining almost constant at 0.0001–0.0002 in. up to 1020°F. At 750°F, time does not appear to influence the depth of the layer to any great extent.

It has been reported that ion nitriding has been accomplished without any white layer [15]. However, other sources indicate that the phenomenon is not very detrimental in service if the layer is less than 0.0005 in. in thickness. Indeed, there is evidence that the white layer is, in fact, beneficial with regard to both wear and fatigue properties.

The speed of nucleation and case development in ion nitriding using AISI 4340 compared with other forms of nitriding indications tends to show that ion nitriding is at least as fast as any other process and, in the vast majority of cases, very much faster.

Stainless steels and some of the more exotic steels have been successfully plasma ion nitrided when little or nothing could be accomplished with conventional methods. A very simple and general rule for plasma ion nitriding is: The lower the alloy content of the steel, the deeper the case and the lower the hardness values. The higher the alloy content of the steel, the shallower the case and the higher the hardness values.

XIII.   THE STRUCTURE OF THE ION NITRIDED CASE

As with the conventional nitriding process methods, ion nitriding is a diffusion-related process that relies on time and temperature. The basic difference between conventional methods and the plasma method lies not only in the time and temperature relationship but also the process gas composition and the composition of the steel as well as some of the process operating parameters that determine the final and total case formation.

In an alloy steel, the increased surface hardness gained by nitriding is attributable to the formation and fine dispersion of both coherent and semicoherent nitrides formed with the alloying elements of the parent material. Nitride-forming alloy elements such as vanadium, chromium, titanium, aluminum, molybdenum, and tungsten are also, to some extent, carbide formers (with the exception of aluminum) and will form carbides as well as nitrides during the nitriding process. They will also influence the rate of reaction and nucleation of the precipitates taking place at the steel surface due to the interaction with the nitrogen process gas. In addition, the amount of carbon present in the steel will also affect the rate of nitrogen diffusion due to the prior microstructure obtained as a direct result of austenitizing, quenching, and tempering prior to the nitriding operation. The transverse microhardness of the nitrided case will increase as the alloying content increases, and conversely, with the higher alloy concentrations, the steel will to some extent resist the diffusion of nitrogen.

In the ion nitriding process, it is accepted that the mixture of nitrogen and hydrogen gases can reduce, and in some cases eliminate, the formation of the compound zone by using low volumes of nitrogen relative to hydrogen. Conversely, the use of low nitrogen concentrations on a low-alloy steel, will encuorage formation of a compound zone but produce a smaller one. Hydrogen takes a catalytic role in terms of the formation of richer (ε) compounds (Fe_2N and Fe_3N). On the other hand, if the gas compositions are reversed to give high ratios of nitrogen to hydrogen (75:25) on, say, pure iron, or certainly on low-alloy steels, very thick compound zones can be produced with hardness values of 700 VPN or more. This involves the formation of an Fe_2N phase at the steel surface, which, through extended cycle times, will decompose into Fe_4 (γ') or $Fe_{2-3}N$ (ε). Work has been done with pure nitrogen atmospheres without the hydrogen catalyst, and case depths were seen to be less than with gaseous mixtures of both nitrogen and hydrogen, probably due to the lack of the catalytic effect of hydrogen in the process reaction [16]. The ion nitriding mechanism is illustrated in Figure 17.

It can be seen that the ability to vary the process gas nitrogen/hydrogen ratios affects both the formation of the compound zone and the diffusion zone of the total nitrided case. During the early 1970s, work investigations were instituted into the manipulation of the compound zone [10,17] by diluting the nitrogen compound ammonia with hydrogen, a process that is still practiced today. Research is still being carried out on the formation of the compound zone by both plasma nitriding and the dilution method.

Table 1 is a selection chart of various types of case compositions and duties. Four typical ion nitrided structures are shown in Figures 18–22.

Plasma ion nitriding offers the following benefits to both the engineer and the metallurgist:

1.   Environment.   The process is nontoxic, has no effluent problems, and the part usually does not require any other surface finishing technique.

2.   Operation costs.   The process is a cost-efficient method of heat treatment because it reduces operator supervision, floor space, consumables, and energy costs. The disadvantage has been, and still is, a higher initial capital investment than more conventional methods of heat treatment. However, this is offset by lower operating costs and improved metallurgy.

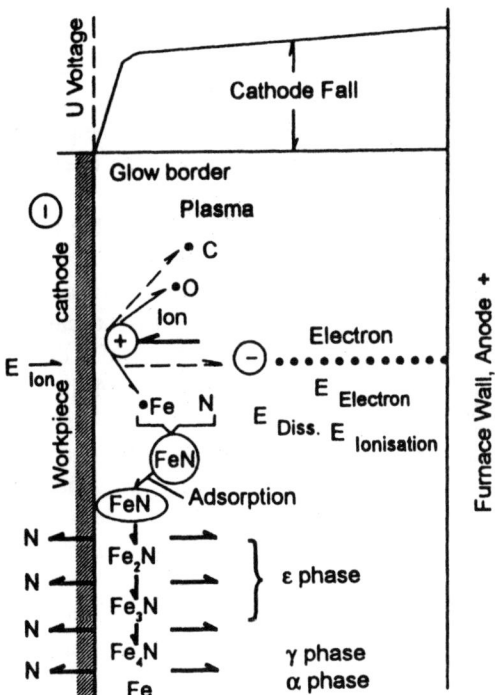

Figure 17 The ion nitriding mechanism. (Courtesy *ASM International Heat-Treatment Handbook* Vol. 4.)

3. Control. Close tolerance of the process, repeatability, and creative metallurgy can be accomplished to suit a particular application.

4. Post-treatment cleaning requirements. The workpieces do not require any cleaning, washing, or other surface cleaning after treatment, thus reducing supportive plant equipment requirements.

5. Metallurgical results. Metallurgical control of the process is much simpler than with conventional processes. The case formation can be either single-phase or dual-phase or diffusion zone only. However, if the compound zone is formed, the carbon composition of the process should be taken into account if an epsilon zone is required or if it is to be suppressed. Growth and distortion are less than with traditional nitriding treatments.

Finally, it should be recognized by a potential user that plasma nitriding is not the panacea of all processes. It has its place in the metallurgical processing arena, but it should be acknowledged that one would not plasma nitride a very simple wear plate when a salt bath nitriding process would suffice. Quite clearly, the user must review all aspects of the product under consideration for plasma nitriding, including, but not limited to, the value of the component, its geometrical complexity, the material of manufacture, the postmachining, the distortion factor requirement, and the metallurgy of the component.

XIV. STEELS FOR NITRIDING

A. Selecting a Steel for Nitriding

One of the most difficult tasks in nitriding is to select a steel in relation to the operating environment of the part that will not only ensure good nitriding results but will also be cost-effec-

Table 1 Case Formation Selection on Medium-Alloy Nitridable Steels

Surface deformation activity	Resistance of nitrided case to activity			
	Dual phase: ε and γ′	Mono phase		Diffusion zone only
		ε	γ′	
Abrasion and sliding: vertical and horizontal loading, sliding action	Poor	Excellent	Fair	Not recommended
Impact: repeated or cyclical impaction	Not recommended unless compound zone is removed	Poor	Good	Excellent

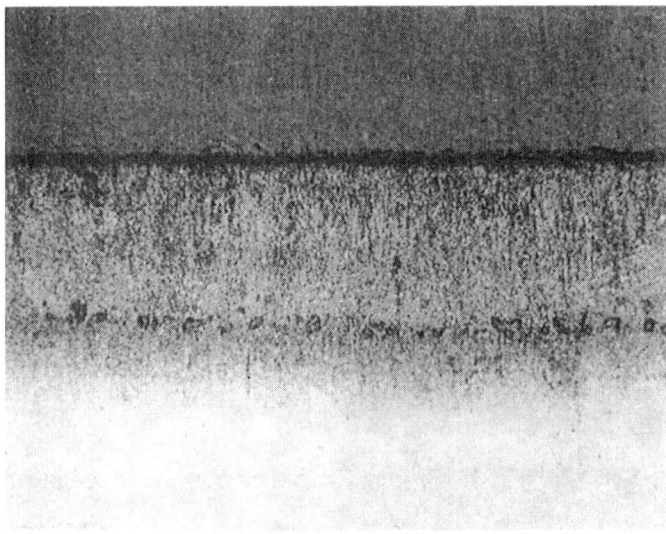

Figure 18   A typical case formation of AISI SS316 nitrided in a fluidized bed, 200 ×, 3% nital etch.

tive, easy to machine or fabricate, and functional. Below are some of the considerations that need to be addressed during the selection process.

1. What is the product to be manufactured, and how complex is the part geometry?
2. Under what type of operating conditions will the workpiece operate—i.e., compressive load, and to what extent; impact load, and to what extent; tensile load, and to what extent; and cyclic load, and to what extent?
3. Are there abrasiveness considerations to recognize?
4. Are there corrosiveness considerations to be taken into account?

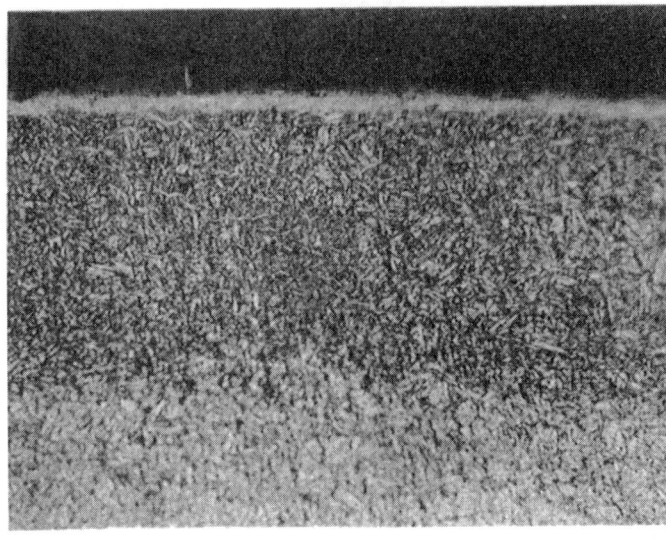

Figure 19   A typical case formation of AISI 4140, 400 ×, 3% nital etch.

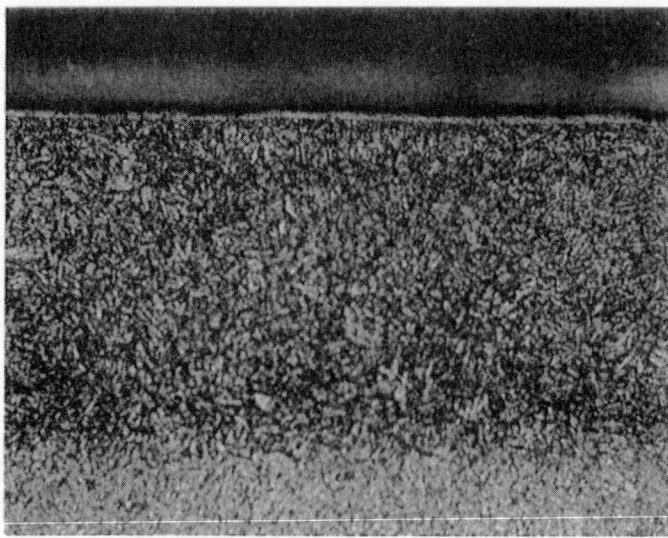

Figure 20   A typical case formation of H.13 hotwork die steel, 400 ×, 3% nital etch.

5.   Will the workpiece be thermally cycled (hot or cold)?
6.   Will there be adequate lubrication?
7.   Is further machining after nitriding a consideration?

Once all of these issues have been addressed, consideration can be given to the steel selection itself. Ideally, the classical Nitralloy steels originally developed by Fry at Krupp Steel contained combinations of aluminum, chromium, molybdenum, tungsten, vanadium, titanium, and manganese. The early investigatory work was on steels that contained aluminum, which were

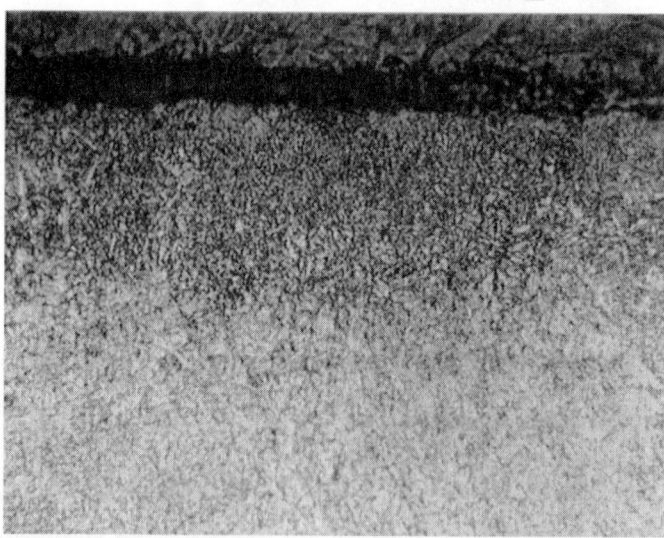

Figure 21   A typical case of AISI 422 stainless steel.

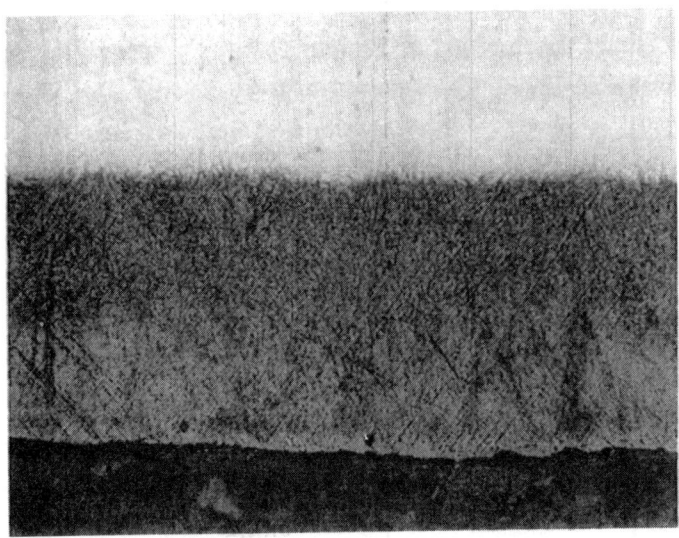

Figure 22 A typical case of 17.4 Ph stainless; note the absence of the compound zone.

found to exhibit greater surface hardness than the lower alloy steels. This led to the development of the Nitralloy grade of steels. Typical nitridable steels are listed in Table 2.

B. Stainless Steels

All stainless steels can be nitrided to some degree because of their high chromium content. It should be noted that nitriding will adversely affect the stainless steel's corrosion resistance although it will increase surface hardness and provide a lower coefficient of friction, thus improving abrasion resistance.

1. Austenitic and Ferritic Alloys

Austenitic stainless steels are usually the most difficult to nitride. These nonmagnetic alloys cannot be hardened by heat treatment; consequently, core material remains relatively soft and the nitrided surface is limited as to the loads it can support. This is equally true of the non-hardenable ferritic stainless steels. Alloys in this group that have been satisfactorily nitrided include Types 430 and 446. With proper prior treatment, these alloys are somewhat easier to nitride than the Series 300 alloys.

2. Hardenable Alloys

The hardenable martensitic alloys are capable of providing high core strength to support the nitrided case. Hardening, followed by tempering at a temperature that is at least 50°F higher than the nitriding temperature, should precede the nitriding operation. Precipitation-hardening alloys, such as 17.4 PH, 17.7 PH, 15.5 PH, and A-286, will also nitride successfully.

3. Prenitride Condition

Befor being gas nitrided, Series 300 steels and nonhardenable ferritic steels should be annealed and stress-relieved. The normal annealing treatments generally employed to obtain maximum corrosion resistance are usually adequate. Microstructure should be as nearly homogeneous

Table 2 Steels That Have Been Designed and Developed as Nitriding Steels[a]

Alloy Steels	C%	Cr%	Mo%	Si%	Mn%	Ni%	V%
SAE 4132	0.34	1	0.2	—	—	—	—
SAE 4137	0.35	1	0.2	0.25	0.8	—	—
SAE 4142	0.42	1	0.2	—	—	—	—
SAE 4140	0.40	1	0.2	0.25	0.85	—	—
SAE 9840	0.36	1	0.2	—	—	1	—
SAE 4150	0.5	1	0.2	—	—	—	—
28 Ni Cr Mo V 85	0.3	1.3	0.4	—	—	2	0.1
32 Ni Cr Mo 145	0.32	1	0.3	—	—	3.3	—
30 Cr Ni Mo 8	0.3	2	0.4	—	—	2	—
34 Cr Ni Mo 6	0.34	1.5	0.2	—	—	1.5	—
SAE 4337	0.38	0.8	0.4	—	—	1.5	—
SAE 4130	0.26	1	0.2	—	—	—	—

	C%	Si%	Mn%	P%	Cr%	Mo%	Ni%	V%	Al%
Nitralloy	0.20–0.30	0.10–0.35	0.40–0.65	0.05 max	2.90–3.50	0.40–0.70	0.40 max	—	—
Nitralloy M	0.30–0.50	0.10–0.35	0.40–0.80	0.05 max	2.50–3.50	0.70–1.20	0.40 max	0.10–0.30	—
Nitralloy 135	0.25–0.35	0.10–0.35	0.65 max	0.05 max	1.40–1.80	0.10–0.25	0.40 max	—	0.90–1.30
Nitralloy 135M	0.35–0.45	0.10–0.35	0.65 max	0.05 max	1.40–1.80	0.10–0.25	0.40 max	—	0.90–1.30

Cold tool steels	C	Si	Mn	Cr	Mo	Ni	V	W
F2	1.45	—	—	0.3	—	—	0.3	3.0

Special-purpose tool steels

	C	Si	Mn	Cr	Mo	Ni	V	W
H13	0.5	1.0	—	5	1.4	—	1.4	—
	1.2	—	—	0.2	—	—	0.1	1.0
	0.9	—	—	0.2	—	—	0.3	1.0

Dimensionally stable tool steels

	C	Si	Mn	Cr	Mo	Ni	V	W
D2	1.55	—	—	11.5	0.8	—	1.0	—
D3	2.0	—	—	12.0	—	—	—	—
A2	1.0	—	—	5.0	1.0	—	0.2	—
O1	0.95	—	10.5	—	—	0.1	0.5	—
O2	0.9	—	20.4	—	—	0.2	—	—
D6	2.1	—	—	11.5	—	—	0.2	0.7
D2	1.65	—	—	11.5	0.6	—	0.1	0.5
S1	0.59	—	—	1.1	—	—	0.2	1.9

Die block steels

	C	Cr	Mo	Ni	V
L6	0.55	5	0.5	1.7	0.1
L6	0.55	0.2	0.3	1.7	0.1

Hotwork tool steels

	C	Cr	Mo	V	W	Co
H12	0.36	5.2	1.4	1.4	0.4	1.3
H13	0.4	5	1.3	1.3	1	—
H11	0.4	5	1.3	1.3	0.6	—

(continued)

Table 2 Continued[a]

	C	Cr	Mo	V	W	Co	
H21	0.3	2.7	—	—	0.4	8.5	—
H19	0.4	4.3	0.4	—	2	4.3	4.3
H10	0.32	2.8	2.8	—	-0.5	0.3	—
High-speed steel tungsten base	C	Cr	Mo	V	W	Co	
T5	0.75	4	0.6	1.6	18	9.5	
T4	0.8	4	0.7	1.6	18	5	
T1	0.75	4	—	1	18	—	
T15	1.5	5	—	5	12.5	5	
M42	1.08	4	9.5	1.2	1.5	8	
M41	0.92	4	5	1.8	6.5	5	
M3	1.2	4	5	3	6.5	—	
M2	0.87	4	5	1.8	6.5	—	
M2	1.0	4	5	1.8	6.5	—	
M7	1.0	4	8.7	2	1.8	—	
M1	0.83	4	9	1.2	1.8	—	

[a]These are typical alloy steels that will gas or salt bath nitride.

as possible. Observance of these prior conditions will prevent flaking or blistering of the final nitrided case.

Martensitic steels, as previously noted, should be in the quenched and tempered condition.

4. Surface Preparation

Prior to nitriding, stainless steels require surface preparation that is not required of low-alloy steels. The film of chromium oxide that protects stainless alloys from oxidation and corrosion must be broken down. This may be accomplished by wet blasting, pickling, chemical reduction in a reducing atmosphere, or submersion in molten salts or by one of several proprietary processes. Parts must undergo surface treatment before they are placed in the nitriding furnace. Should there be any doubt as to the complete and uniform depassivation of the surface, further reduction of the surface oxide may be accomplished in the furnace by the introduction of a reducing hydrogen atmosphere.

Stainless parts must be perfectly clean and free of embedded foreign particles prior to nitriding. After depassivation, care should be exercised to avoid contaminating stainless surfaces with finger and hand markings. Sharp corners should also be avoided wherever possible.

5. Nitriding Cycles

In most cases, stainless steels are nitrided in single-stage cycles at temperatures of about 975–1025°F for periods of 20–48 h, depending on the depth of case required. Dissociation rates for the single-stage cycle range from 18 to 30%. Except for the depassivation of the metal surface, the nitriding of stainless steels is similar to the single-stage nitriding of low-alloy steels.

With nitrided stainless steels, the case almost always has lower corrosion resistance than the substrate material; nevertheless, it can be adequate for certain applications. Nitrided stainless steels are not resistant to mineral acids and are subject to rapid corrosion when exposed to halogen compounds.

XV. DISTORTION AND GROWTH DURING THE NITRIDING PROCESS

"Distortion" is a term that is familiar to those involved in all aspects of heat treatment. Many claims have been made about processes that will guarantee no distortion and no size change. These claims cannot be substantiated because distortion can be caused by machining stress, thermal stress from heating, thermal stress from rapid cooling, and volumetric size damage due to the surface chemistry of the steel being manipulated by the growth nucleation of nitrogen, carbon, sulfur, and oxygen. It is necessary then to have an understanding of distortion and growth.

The term *distortion* describes dimensional changes to the workpiece during heat treatment or quenching resulting from the inducement of stresses at some point during the manufacturing operation or heat treatment. The stresses are manifested during the heat treatment process or as a result of that process. There are two types of distortion:

1. Shape distortion, or more simply put, a change in the geometrical form, such as curving, twisting, or bending.
2. Size distortion, which is the change in the volume of the workpiece due to either growth or shrinkage.

A. Shape Distortion

When a piece of steel is worked, through rolling, forging, casting, filing, or machining, stresses are induced in the work that can be described as either applied stresses or residual stresses.

The only way induced, or residual, stresses can be relieved is through the application of heat. Stress can be relieved by the introduction of intermediate stress-relieving operations prior to the final heat treatment process. Stress can be created by quenching the part; therefore, any process that will eliminate the need for quenching will eliminate a major source of distortion problems.

B. Size Distortion

When a piece of steel is austenitized and cooled at various rates, a number of structures can be produced. Austenite has the smallest and martensite the largest specific volume, if we ignore carbides for the moment. In the heat treatment cycle, various changes occur [6, p. 761].

The effect of different hardening temperatures varies, depending on whether the steel is of high or low hardenability. Shallow-hardening steels undergo greater volume change when the hardening temperature is raised due to the increased grain size. In contrast, high-alloy steels show less volume change with an increase in hardening temperature due to the presence of refined austenite.

A very important factor that is often overlooked is the dependence of volume increase on penetration of hardness. When the section size increases, the degree of penetration is reduced; consequently the amount of austenite-to-martensite transformation increases, taking place only in the outer hardened zone. Therefore, the volume change of the whole piece is much less than that which would apply to the fully hardened condition. Volume increases in tools of different dimensions is completely dependent on hardenability.

Generally, when the tool is hardened, there is an overall increase in volume. The only exception to this would be if excess austenite had been retained after quench. Tempering must, of course, follow hardening, and during this operation, new changes occur in volume and consequently in size.

C. Distortion During Nitriding

Distortion due to residual stress can occur during nitriding if the workpiece has not been previously stress-relieved at a sufficiently high temperature. If stress relief has not been carried out, then the nitriding process will act as a stress-relieving operation, thus allowing stresses to manifest themselves in the form of twisting and bending. There will be a change in the surface of the workpiece due to the diffusion of nitrogen into the surface of the steel and phase changes.

The volume change is always due to growth. However, the growth is uniform in all directions. The volume change that takes place is dependent on the nitrided case penetration. The shallower the case, the less growth. The major growth will occur as a result of the formation of the compound zone. The greater the thickness of the compound zone, the greater the degree of growth. Growth of the part is usually related to the thickness of the compound zone and as a rule of thumb, is 15% of the compound zone thickness.

XVI. SELECTIVE NITRIDING

Nitriding can be prevented in a number of ways. The selection of the method depends on the method of nitriding that has been chosen [6, p. 401].

A. Gas Nitriding

One method of selective gas nitriding is to plate (or paint) the areas to be "stopped off" (the areas in which nitriding is to be prevented), by using bronze, nickel, silver, or copper. Due to the high cost of materials, copper is the most widely used. In addition, there are proprietary paints available for selective nitriding. In order for the painting method to be successful, it is extremely important that the manufacturer's instructions be followed.

Thickness and density of plated coatings are important in determining their effectiveness as stopoffs. Minimum thickness of bronze or copper plate should be 0.7 mil for ground surface finishes of 64 μin. or smoother; 1.0 mil for finishes between 64 and 125 μin.; and 1.5 mil for finishes of 125 μin. and rougher. Compared to copper and bronze, nickel is a more effective stop-off, permittnig a thinner coating.

B. Liquid Nitriding

Selective nitriding can be accomplished in liquid nitriding baths by stopping off nitrogen penetration with either copper plate or copper-base paint. Because cyanide-base salts can dissolve copper, formulations that operate with relatively low cyanide contents are necessary. One formulation that has met with considerable success operates at 8–10% sodium cyanide with approximately 45% barium chloride energizer; another is a low-cyanide, calcium type of bath.

C. Paint-On Methods

A simple method of selective nitriding, for both gas and liquid nitriding, involves the use of a copper-based paint. There are variations of this paint, with some using a cadmium additive and some a tin additive, and some use tin as the base material of the paint.

The secret of success of the paint method of stop-off lies in the preparation of the part to be stopped off and the preparation of the paint. The manufacturer's stop-off paint and part preparation instructions should be followed explicitly, and the part must be handled in a very delicate manner. If the part is subjected to mechanical damage, the paint may chip off, in which case the exposed area will nitride, which will leave hard spots on the surface after paint stripping. This can lead to tool damage if the part requires further machining such as turning or milling.

D. Plasma Ion Nitriding

Masking an area for plasma ion nitriding is extremely simple. Using the simple rule "What plasma can see, it will nitride; what plasma cannot see, it cannot nitride," all that is required is that the area to be masked be covered with a piece of shim stock steel attached simply to the component. Plating or painting is not recommended for plasma processes due to outgassing and stripping at the sputtering stage.

XVII. EVALUATION OF THE NITRIDED CASE

It is necessary to examine and evaluate the results of the nitrided case after the nitriding process has been carried out by the chosen method. Stress testing of the nitrided case is probably the most common method of examining the results. The selection of the hardness testing method is most critical. One cannot effectively use a hardness test and acquire a meaningful result using a 150 kg load hardness tester for a nitrided case. The load is too great for the shallow case and will penetrate it. It is usual to use a hardness testing machine that has a variable load applicator, usually a Vickers hardness testing machine, Knoop hardness tester, or a micro hardness tester for transverse hardness surveys of the case.

When presenting the workpiece to be tested to the penetrator, the surface to be tested should be clean and free from decarburization. It should be presented squarely at 90° to the central axis of the penetrator. Any deviation from this will result in a false reading. There are normally two areas to examine for hardness, the first being the surface and the second being a transverse section through the case using a micro hardness tester to establish the case hardness profile.

A. Hardness Tests

1. *The Vickers Hardness Test*

The Vickers hardness test is an extremely accurate method of testing and is used as the primary test. This test measures the resistance to penetration. The indenter is a square-based diamond pyramid, and the hardness value obtained when using this penetrator is frequently referred to as the diamond pyramid hardness (DPH) or VPN for Vickers pyramid numeral.

2. *Knoop Hardness Test*

The indenter used in the Knoop hardness test machine is known as the Knoop indenter. It is a pyramidal indenter developed by the U.S. National Bureau of Standards. The indentation produced is a long diamond-shaped impression. It is claimed that the advantage of this indenter lies in the fact that elastic recovery along the long axis of the indentation is very small, thus reducing variation from this source, which could be especially troublesome at very low loads. Hardness numbers are based on the long dimension of the indentation and are calculated as the ratio of the indenting load to the projected area of the indentation.

3. *Case Depth Measurement*

Case depth can be measured by one, or all, of the following three methods:

The first is hardness traverses through the case. This requires the sample to be cut, mounted, ground, and polished and then placed under the micro hardness tester with the case being in the horizontal plane and at 90° to the indenter. Very careful indentations are made through the case toward the core of the sample. Usually the effective case depth is from the surface of the case to a point of 500 VPN.

The second method involves step grinding the sample and checking the hardness at each step through the case to the core at 500 VPN for effective case depth.

The third method of determining case depth is to cut, mount, grind, polish, etch, and visually measure the case depth under the microscope or metallograph. This method can be a reasonably accurate means of establishing case depth.

The difference between "effective" case depth and "total" case depth is as follows:

Effective case depth: from the surface to 500 VPN

Total case depth: from the surface to a hardness of 50 VPN above that of core hardness

B. Microscopic Examination

The sample to be examined must first be cut in a manner that will not cause any localized overheating of either the case or the core. The area of the sample that is to be microscopically examined is then etched, with the type of etch employed depending on the purpose of the examination. Three common etches for microscopic examination of nitrided cases are the marbles etch, picral, and nital.

Marbles etch

---

4 g copper sulfate
20 mL hydrochloric acid
20 mL distilled water

Immerse sample or swab for 50–60 s. The etch is made more active by adding a few drops of sulfuric acid to the solution just before use.

Use: Shows true depth of nitride case, reveals phase formation, and is also used for etching stainless steels after nitriding.

Picral

---

100 mL ethyl or methyl alcohol (95% or absolute)
4–5 drops of zephiran chloride

Immerse the surface to be examined for from 5 to approximately 1 min.

Use: This can be used for all grades of carbon steel for the examination of tempered martensite and bainitic structures and will also detect carbides.

Nital

---

2 mL HNO_3 or 98 mL ethyl or methyl alcohol
Etch for 3 s up to 1 min.

Use: This etch is good for outlining in ferrite grain boundaries, cemenite networks, and general case depth measurement.

Note: Great care should be taken when handling and mixing acids. Special safety precautions should always be practiced when handling chemicals of any kind. **All chemicals are potentially dangerous**, and it is assumed that the person using any of the etchants is thoroughly familiar with all the chemicals involved and is aware of the proper procedures for handling and mixing these chemicals [20].

ACKNOWLEDGMENTS

My thanks to Mr. Paul Huber, Mrs. Georgia Minter, and Mrs. Robin Maloney of SECO/WARWICK Corporation, Meadville, Pennsylvania, Mr. Rodney Allwood; Mrs. Lynn Pye; The South African Institute of Foundrymen (Heat Treatment Specialist Division); and all of my colleagues in the industry.

REFERENCES

1.   D. Pye, *Industrial Heating Magazine*, September 1991.
2.   H. W. McQuaid and W. J. Ketcham, Some practical aspects of the nitriding process, in *Source Book on Nitriding*, ASM Int., Cleveland, 1977, p. 2.
3.   A. Fry, U.S. Patent 1,487,554 (Mar. 18, 1924).
4.   V. O. Homerberg and J. P. Walstead, A study of the nitriding process, Part 1, *Source Book on Nitriding*, ASM Int., Cleveland, 1977, p. 56.
5.   R. M. Falling, Control of the compound layer structure upon gaseous nitrocarburizing, AGA Innovation, Lidingo, Div. of Physical Metallurgy, Royal Institute of Technology, Stockholm.
6.   *ASM Handbook*, Vol. 4, *Heat Treating*, ASM Int., Materials Park, OH, 1991.
7.   C. F. Floe, A study of the nitriding process effect of ammonia on the case depth and structure (1943), in *Source Book on Nitriding*, ASM Int., Cleveland, 1977, p. 144.
8.   ICI Cassell, *Manual of Heat Treatment and Case Hardening,* 7th ed., ICI, U.K., 1964.
9.   U. S. Patents 2,596,981, 2,779,697, and 2,986,484.
10.   B. J. Lightfoot and D. H. Jack, Kinetics of nitriding with and without white layer formation, *Source Book on Nitriding*, ASM Int., Cleveland, 1977, p. 248.

11. D. Pye, *Nitriding for current and Next Generation Steels*, SECO/WARWICK Corporation, Meadville, PA, 1991.
12. D. Pye, Plasma nitriding, an alternative nitriding method for aluminum extrusion die, *Procedures E.T. (Extrusion Technology) '92*, p. 268.
13. C. K. Jones, D. J. Sturgess, and S. Martin, Glow discharge nitriding in production, *Metal Prog.*, December 1973.
14. Kinney Vacuum Pump Systems, *Design for Pumping*.
15. W. Kovacs and W. Russell, An introduction to ion nitriding: what is it, why is it used, where is it used, *Proceedings,* Conf. on Ion Nitriding, September 1986, ASM Int., Materials Park, OH, 1987, p. 12.
16. O. T. Inal, K. Ozbaysal, E. S. Metin, and N. Y. Pehlinvanturk, A review of plasma surface modification: process parameters and microstructrual characterization, *Ion Nitriding and Carburizing Proceedings*, Cincinnati, OH, September 1989, ASM Int., Cleveland, 1989, p. 58.
17. T. Bell, B. J. Birch, V. Korotchenko, and S. P. Evens, Controlled nitriding in ammonia hydrogen mixtures, *Source Book on Nitriding*, ASM Int. Cleveland, 1977, p. 259.
18. *Hardness Testing*, 3rd printing, ASM Int., Materials Park, OH, 1994, Chapter 4, p. 57.
19. Leco, *Metallurgical Preparation*, 1991.
20. R. L. Anderson, *Safety in the Metallography Laboratory*, Westinghouse Research Laboratory, Pittsburgh, PA.

11A

INDUCTION HEAT TREATMENT
Basic Principles, Computation, Coil Construction, and Design Considerations

Valery I. Rudnev, Raymond L. Cook, Don L. Loveless, and Micah R. Black
INDUCTOHEAT, Inc., Madison Heights, Michigan

I. INTRODUCTION

As modern technology has changed from the abacus to the computer, a similar change has occurred in the field of induction heat treatment. In modern industry the requirements for the induction heating process have become quite strict. Some of these requirements could not possibly have been satisfied 15 or even 10 years ago. In order to provide a successful design for modern induction heating it is now necessary to take into account many details of the process. Many years ago a basic knowledge of electromagnetic fields, a calculator, and engineering intuition were all that were available to create an induction heating system. Now they are not enough. The designer of induction heating systems today must have special software tools that allow effective simulation of the process. The designer must also be aware of new developments in theory and practice in the area of induction heat treatment.

Several useful books and lessons have been published on the basic principles of induction heating design [1–12]. This material is very useful for the first-time designer with limited experience in induction heating. A significant portion of the reference material describes various aspects of induction heating, such as mass heating for forging, extrusion, and rolling. Induction heat treatment has many features that make it a unique process from the standpoint of coil design as well as computational methods, process operation, and equipment maintenance.

Traditional methods of calculation for the induction heating process (e.g., Baker's method) were based on the equations for an infinitely long coil and workpiece. Unfortunately, this assumption is not always valid in induction heat treating, where induction coils typically have no more than a few turns and cannot be considered infinitely long.

During the last decade a considerable amount of experience has been accumulated on the computation of induction heating and heat treatment problems by using numerical techniques. Unfortunately, the descriptions of particular methods are contained in a variety of internal reports, scientific journals, or special literature concerned with numerical methods. They are usually presented in a form almost inaccessible to engineers who have limited experience in numerical analysis. Also, the textbooks on numerical analysis usually emphasize the mathematical methods. They do not get into the details of the physical aspects of the problem that are often crucial to the success of a simulation. It is not our aim to describe all of the available numerical methods. However, in order to make the right choice, one should have some knowledge of the possibilities as well as the limitations of the methods that are most suitable for the heat treatment process.

In most previous publications, induction heating was introduced as a so-called uncoupled process without emphasis on the interrelated features of electromagnetic and heat transfer phenomena. The studies often consider the two phenomena separately. Such an approach could lead to substantial errors in predicting the required power, frequency, and heat treatment pattern. The nature of induction heat treatment is a combination of the two phenomena, which are tightly interrelated because physical properties depend strongly on both magnetic field intensity and temperature; therefore, this effect should be taken into account in contemporary design. Besides that, today's heat treater should consider the nonconstant behavior of the thermal and electromagnetic properties of the workpiece such as specific heat, thermal conductivity, magnetic permeability, and electrical resistivity during the heating cycle.

In addition, some induction heating coils can work very effectively with certain types of power supplies and ineffectively with others. Therefore, the optimal design of a modern induction heat treatment system should consider the features of induction heat treatment not as a physical stand-alone process but as a combination of the inductor, load-matching station, and power supply (i.e. solid-state inverter).

The major goal of this chapter is to embark upon the next step in the study and design of the modern induction heat treatment process and equipment. We concentrate on the features of modern induction heat treatment design and introduce advanced methods used in the study and evaluation of different types of heat treatment processes. The study includes the systematization of new knowledge accumulated in recent years at INDUCTOHEAT Inc. and by our colleagues around the world. Some materials presented here are new and have never been published before. Others have existed only in the articles or internal reports of the INDUCTOHEAT Group.

Many useful and practical recommendations are presented in this chapter as well as a description of some subtle aspects of the electromagnetic and heat transfer phenomena that are imperative for modern heat treatment practitioners and engineers to know.

Upon completion of this chapter, readers will have an orientation in modern computational methods for electroheat problems encountered in induction heat treatment. They will be able to evaluate the important features of the induction process and have knowledge that will help to avoid many of the unpleasant surprises one might encounter in the design of induction heat treatment systems. A basic knowledge of electrical engineering and understanding of basic induction heating principals should be essential to grasping some of the modern computation introduced in this chapter. Current information will be presented on the use of magnetic field plots, temperature profiles as well as sketches, drawings and photographs of practical applications. An attempt will be made to bridge the gap between purely theoretical information and that which is of practical use to the heat treatment specialists.

A reader with great experience in the field of induction heating and heat treatment will also find this chapter useful because he or she will discover the reasons for intuitive engineering decisions made in the past. College students preparing for a career involving induction heating or metal heat treating will see how the theory and mathematical methods they are studying are used to solve problems encountered in everyday practice. This chapter consists of information that will be useful to a manager as well, because he or she will better understand the complexity of the process and the attention to detail required to obtain cost-effective, high quality induction heat treatment results.

This chapter is not intended to describe in detail the specific mathematical methods or deep theoretical aspects of electrodynamics, thermodynamics and optimal control involved in the process of designing modern induction heating equipment. For that the reader would need to be well versed in many advanced theoretical subjects. As we mentioned above, a basic knowledge of electrical engineering principles and common sense will be enough to understand the materials presented here. However, if someone finds this material too advanced and too difficult to understand or desires only a very basic or intuitive introduction to the subject, we would recommend the book *Basics of Induction Heating* by Chester Tudbury [1] which is available from INDUCTOHEAT, Inc.

Finally, it is a most comforting discovery for most engineers that someone has actually built a production system to accomplish the same task they are trying to perform. Plots of electromagnetic fields and photographs of a variety of production installations are provided to show not only that the task has been previously accomplished but also how it has been done.

For those who would like to acquire more information about certain subjects, we have provided several references. Readers are also welcome to contact the authors of this chapter directly at INDUCTOHEAT Inc. in Madison Heights, Michigan.

II. PRELUDE TO DISCUSSION OF INDUCTION HEAT TREATMENT

Induction heating is often one of the most effective heat treatment processes available for a variety of applications including

Surface hardening

Through hardening

Tempering and stress relief (low-temperature)

Annealing and normalizing (high-temperature)

Weld seam annealing

Sintering of powdered metals

In most of these applications, induction heating is used to selectively heat only the portion of the part that requires treatment. This usually means that the process can be accomplished in a relatively short time and with high efficiency because energy is applied to the part only where it is needed. The ability to heat treat in-line, as opposed to batch processing, with high productivity, less distortion, and a clean environment is an obvious benefit of induction heat treatment.

Induction heat treatment (IHT) is a segment of the much larger technical field of induction heating, which combines many other industrial processes using the phenomenon of heating by induction. Examples of these processes are

Heat treating

Melting

Levitation of molten metal

Through heating for forging, rolling, extrusion, etc.

Brazing, soldering and welding

Shrink fitting

Mold core meltout

Adhesive curing

Sealing

Induction plasma

This handbook is devoted to heat treatment. Therefore, this chapter focuses specifically on the features of modern heat treatment by induction.

Any electrically conductive material can be induction heated. The ability to couple energy into a specific part of a workpiece by induction and obtain the desired heat treatment depends on many factors that are explored in detail in this chapter. Some of the more important factors are

Thermal and electromagnetic properties of material

Power applied

Frequency selected

Rate of heating

Cooling process

Ability of the material to respond to heat treatment

Workpiece shape

Inductor and quench design

Process steps used to accomplish the heat treatment of a part (workpiece) by induction are usually considerably fewer than are required by alternative processes. As an example, contour induction hardening process steps are compared to carburizing in Figure 1 [13–17]. The induction contour hardening process is accomplished in a much shorter time with less energy and also results in far less part distortion and grain size growth than carburizing. Figure 2 shows the pattern obtained by induction contour hardening of gear teeth.

Induction heat treatment (IHT) is well suited to in-line processing and flexible manufacturing cells. In many cases, the IHT machine receives the part by conveyor from an automatic grinding machine and includes part washing, a go/no-go quality check of critical dimensions, induction hardening and tempering, eddy current check of the heat treatment, part marking, and data logging. The part is then transferred automatically by conveyor or robot to the next cell. A machine of this type is shown in Figure 3 processing automotive front wheel drive parts. Due to its unique features, IHT is very flexible and has been successfully used in a variety of heat treatment processes.

Let us briefly examine the most popular applications of heat treatment by induction.

A. Surface Hardening

Surface hardening of carbon steel and iron is the most common form of induction heat treatment because the heating can be localized to the areas where the metallurgical changes are desired. This process is a complex combination of electromagnetic, heat transfer, and metallurgical phenomena that occur when a workpiece is heated rapidly to a temperature above that which is required for a phase transformation to austenite and then rapidly quenched. As shown in Chapters 1–3, during the heating and quenching of the workpiece the metal actually goes through a change in crystalline structure. The goal in surface hardening is to provide a martensitic layer on specific areas of the workpiece to increase hardness and wear resistance while allowing the remainder of the part to be unaffected by the process [6,11,12,18–20]. Figure 4 shows a constant-velocity automotive front wheel drive component that has been cut and etched to show the pattern obtained by induction heat treating [93]. This component requires two areas of hardness with different strength, load, and wear requirements. The "stem" needs torsional strength as well as a hard outer surface, whereas the soft core must be ductile and therefore able to handle the mechanical shock from constant pulsing. The inner surface of the "bell" needs hardness for wear purposes, as ball bearings ride in the track or raceways. The thread of this component holds the wheel on. For heavy duty applications the thread is also hard-

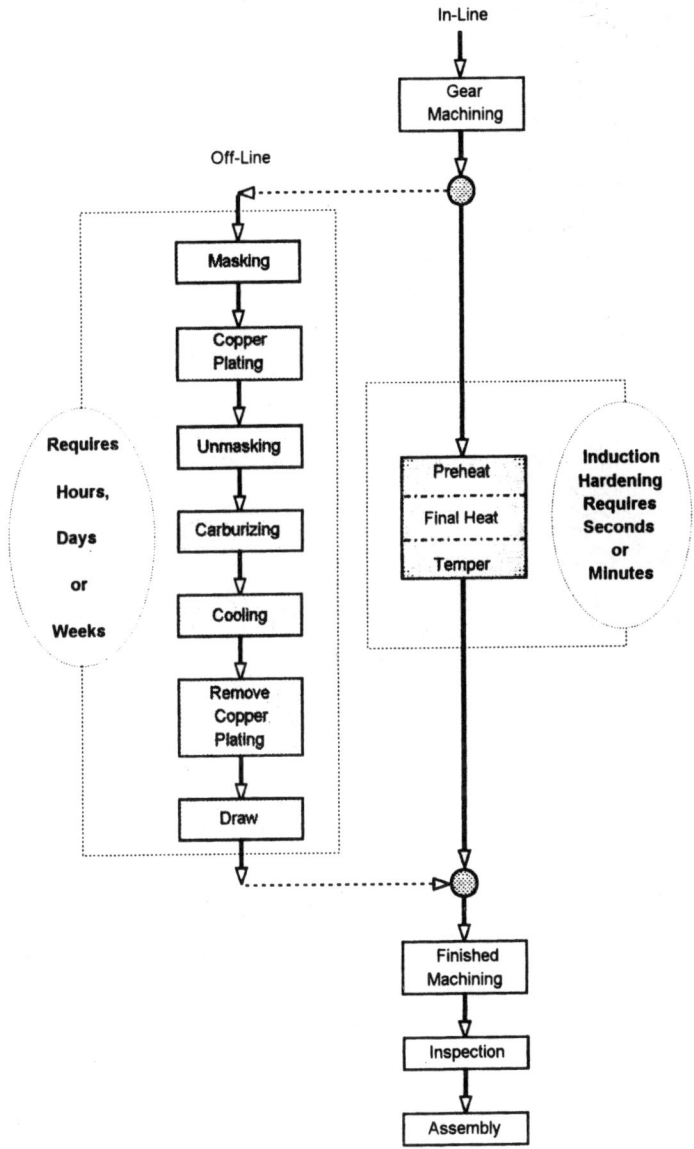

Figure 1 Steps required for carburizing compared with steps for induction hardening.

ened and then annealed to produce a very tough thread. The annealing is almost always done with induction, using a separate inductor.

Surface hardening to increase wear resistance requires a shallow-hardened case [6,18–24]. Case depth (hardness depth) is typically defined as the surface area where the microstructure is at least 50% martensite. Below the case depth the hardness begins to decrease drastically. This depth is measured in a cut section of the hardened workpiece using a hardness tester to determine the hardness at various distances from the surface. Distance below the part surface where the hardness drops to 10 Rc lower than the surface hardness is often called the effective case depth. Hardness distribution along the workpiece radius depends primarily on the

Figure 2 Pattern obtained by induction contour hardening of gear teeth.

temperature distribution, the microstructure of the metal, its carbon content, quenching conditions, and the hardenability of the steel.

Bearings, rocker arms, pump shafts, and skid plates are examples of parts that require a shallow-hardened case for wear resistance. Case depths for these parts that will enable them to handle light loads are usually in the range of 0.25–1 mm (0.010–0.040 in.). In the following sections of this chapter, control of the depth of induction heating by selection of frequency, power density, and workpiece/coil geometry is discussed in detail. Generally speaking, heating for shallow case depths requires high frequency, low energy and high power density. In some cases of shallow case depth surface hardening (less than 1 mm), it is possible to use self-quenching or mass quench. Because the heated surface layer is very fine and its mass is negligibly small compared to the mass of the cold core, it is possible to have rapid surface cooling due to heat being conducted toward the cold core. In this case, the mass of the cold core acts as a large heat sink. Therefore, self-quenching excludes the use of fluid quenchants except as a cool down to allow handling of parts.

Parts that require both wear resistance and moderate loading such as camshafts and crankshafts are usually induction hardened to case depths of 1–2.5 mm (0.040–0.1 in.). Deeper case depths strengthen the part dramatically because load stresses drop exponentially from the surface. The induction heating frequency that is required to obtain this case depth is usually between 8 and 50 kHz.

Parts that must withstand a heavy load require greater case depths; these include axle shafts, wheel špindles, and large heavy duty gears. The cross-sectional areas of the part and the magnitude of the load they must handle determine the appropriate case depth.

In many of these applications the induction heating pattern can encompass a significant part of the cross-sectional area. As a result, some distortion may appear. Where distortion is present, it may be necessary to provide additional stock and case depth to allow for a final

(a)

(b)

Figure 3 Induction heat treatment machine for automotive front wheel drive part.

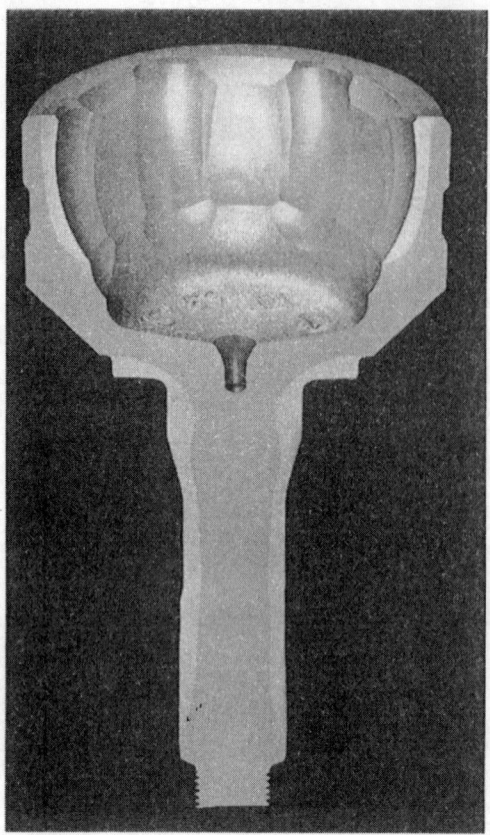

Figure 4 Constant-velocity automotive front wheel drive component.

grinding after hardening. These heavy workpiece applications usually require a case depth in the range from 3.5 mm to as much as 12 mm (0.125–0.47 in.). Here a higher energy at 500 Hz to 3 kHz is usually called for.

Induction surface hardening to these depths also causes residual compressive stresses induced by the transformation hardening of the surface. In the great majority of heat treating applications it is desirable to have compressive stresses at the surface. However, when it is necessary the effects of these stresses may be eliminated by a subsequent stress relief process for which induction is also quite a popular choice.

Induction surface hardening is characterized by high dimensional accuracy of the heat-treated parts. The ability to keep part distortion as low as possible after the heat treatment process is due to the fact that induction heating is a very fast process and concentrates the heat sources in a thin surface layer called the current penetration depth. An analysis of this feature is given in Section IV.

Surface hardening is accomplished by raising the required depth of material above the transformation temperature (A_{c3} from Fe–C on the phase diagram) to the point where it is transformed to austenite and then cooling the workpiece rapidly to produce martensite. This rapid cooling or quenching is covered in detail in Chapter 4. The heating time to complete this process normally ranges from 1 to 8 s per component. The end use of the part to be hardened, as discussed briefly above, wili determine the depth of hardening and may also indicate the specific type of material to be selected based on hardenability. The hardenability of metals is covered in detail in Chapter 3 of this handbook.

The general classification of steels surface-hardened by induction is as follows.

High-carbon steels are used primarily for tools such as drill bits and other cutting tools due to their ability to achieve high hardness. Commonly used steels are in the AISI 1050–1080 range.

Medium-carbon steels are often used in the automotive industry; they are used, for example, for front wheel drive components and drive shafts. These steels may be in the range of AISI 1035–1050.

Low-carbon steels are used where toughness rather than high hardness is required such as in clutch plates or pins for farm equipment. These may be AISI 1020–1035.

Alloy steels are used for such things as bearings and automotive transmissions. These components may have several functions and therefore require a combination of materials. More information about alloy steels is presented in Chapter 2.

When discussing surface hardening by induction it is necessary to mention the phenomenon that is usually called superhardening. Due to this phenomenon, for a given carbon steel the surface hardness of the part that has been heat treated by induction is typically 3–8 Rc higher than that for a furnace-hardened steel. Superhardening can be attributed to the finer grain size of steels that have been induction-hardened. This occurs because the steel is at the austenizing temperature for a very short time, which results in finer grain sizes [6,12]. Another factor that is involved in superhardening is the higher lattice strain from residual compressive stresses at the surface of the part when its internal regions and the core remain underheated or even cold [12]. This phenomenon is particularly true for steels with a carbon content of 0.35–0.6%.

B. Through-Hardening

Through hardening may be needed for parts requiring high strength such as snowplow blades, springs, chain links, truck bed frames, and certain fasteners such as nails or screws. In these cases the entire workpiece is raised above the transformation temperature and then quenched. Selection of the correct induction heating frequency is very important to achieve uniform temperature "surface-to-core" in the shortest time with the highest efficiency. The factors related to frequency selection are discussed in detail later in this chapter. Depending on the workpiece, the inductors may be in any of three stages: preheat, midheat, and final heat. The different stages, sometimes with different heating frequencies, allow the heat to soak into the workpiece. Induction heating for through hardening and for forging or extrusion are very similar; they all are accomplished by heating the entire workpiece uniformly to the desired temperature and at required process time.

It is necessary to mention that the cooling intensity of the surface layers during quenching of a surface-hardened workpiece is greater than the intensity of cooling of a through-hardened workpiece. This is so because in surface hardening, additional cooling of the surface layers takes place due to the cold core, which plays the role of a heat sink. This greater cooling rate results in the ability to obtain a deeper martensitic structure after induction surface hardening. Consequently, the surface compressive stresses and hardness of through-hardened parts is typically lower than the hardness of a surface-hardened workpiece.

C. Tempering

The tempering process takes place after steel is hardened, but is no less important in metal heat treatment. Tempering temperatures are usually below the lower transformation temperature. The main purposes of tempering are to increase the steel's toughness, yield strength and ductility, to relieve internal stresses, to improve homogenization, and to eliminate brittleness

[6,11,18,19,21,149-151]. The transformation to martensite through quenching creates a very hard and brittle structure. Untempered martensite is typically too brittle for commercial use and retains a lot of stresses. As discussed above, in surface hardening, only a thin surface layer of the workpiece is heated. The surface is raised to a relatively high temperature in a short period of time. A significant surface-to-core temperature difference and steel transformation phenomena results in the buildup of internal stresses. Reheating the steel for tempering after hardening and quenching leads to a decrease or relaxation of these internal stresses. In other words, because of tempering it is possible to improve the mechanical properties of the workpiece and to reduce the stresses caused by the previous heat treatment stage without losing too much of the achieved hardness.

Tempering temperatures are usually in the range of 120–600°C (248–1112°F). If the steel is heated to less than 120°C (248°F), there is no change in metal structure, and tempering will not take place. Low temperature tempering is typically done at temperatures of 120°C–300°C (248°F–572°F). The main purpose of low temperature tempering is stress reliving. Hardness reduction typically does not exceed 1-2 points Rc. If the tempering temperature is higher than 600°C (1112°F), essential changes in the structure of the steel may result that can lead to a significant loss in hardness. Therefore, tempering is always a reasonable compromise between maintaining the required hardness and obtaining a low-stress microstructure in the metal. Hardness reduction exceeds 15 points Rc and maximum hardness is typically in a range of 36-44 Rc.

There is a common misconception that tempering removes all internal stresses. It does not, but tempering significantly decreases stresses. Tempering makes the steel softer and reduces the chance of distortion and the possibility of cracking. It is important that the time from quench to temper be held to a minimum. If this "transient time" is long enough, the internal stresses may have enough time to allow shape distortion or cracking to take place. Therefore, a long transient time between quenching and tempering will decrease or eliminate the tempering benefits. The soaking time of the tempering is much longer than both the heating time of the hardening and quenching. It may take seconds to induction harden and minutes to induction temper.

If tempering has been done correctly, there will be only a slight loss in hardness. The benefits obtained, such as internal stress relief, the creation of the required ductility or toughness, shifting of the dangerous maximum of tensell stresses which is located under the hardened surface layer further down towards the core of the workpiece, and improvement in the machinability of the steel, will offset the slight reduction in hardness.

D. Normalizing

Normalizing can be done by heating the steel above the upper transformation temperature and then allowing it to cool to room temperature. Normalizing usually takes a very long time (much longer than tempering), and can take as much as several hours. Normalizing is done to return the metal structure to a state in which the next heat treatment application can take place. Some parts are spheroidized to make them easier to form. It is often necessary to normalize them before hardening.

E. Annealing

Annealing is very similar to normalizing except that the soak time during cooling is even longer than in the normalizing process. Annealing temperatures are usually 750–900°C (1350–1650°F). The purpose of annealing is much like that of tempering in that the hardness is decreased and the ductility is increased [12,19,25]. This can be done by applying higher temperatures and results in a more ductile structure. A good example is the annealing of the thread of a pinion. The entire pinion surface is hardened to approximately 58 Rc, then the thread is annealed to 35–40 Rc to produce a thread that is harder than the core material and is very ductile. This results in production of a very strong part without the possibility of thread breakage when a heavy torque is applied.

Normalizing and annealing reduce hardness and strength while providing following benefits:

1. Improved formability and machinability
2. Refined steel structure
3. Significant relief of internal stresses

F. Sintering

Sintering is the bonding of the molecular structure in powdered metal. It is done by heating the metal to a high temperature that is below its melting point to produce recrystallization. High-temperature sintering produces a more consistent structure than casting or forging and provides added hardenability with more consistent heat treatment results. Sintering powdered metal is like normalizing carbon steel and is recommended if a subsequent heat treatment is to be performed. The process of powder metallurgy is discussed in detail in Chapter 12.

There are many ways to heat treat metal parts including the use of gas-fired furnaces, fluidizing bed furnaces, salt baths, lasers, plasma systems, infra-red heaters, electric and fuel-fired furnaces, etc. Each method has its own advantages as well as disadvantages. There is obviously no universal method that is the best in all cases of metal heat treating.

In the past two decades heat treating by induction has become more and more popular. A major reason is the ability to create high heat intensity very quickly at very well defined locations on the part. This leads to low process cycle time (high productivity) with constant quality. Induction heating is also more energy efficient and environmentally friendly than most other heat sources used for heat treatment such as carburizing systems or gas-fired furnaces. It usually requires far less start-up and shutdown time and a lower labor cost for machine operators. In many cases, induction heat treatment will require a minimim floor space and produce less distortion in the workpiece.

Induction heat treatment is a complex combination of electromagnetic, heat transfer, and metallurgical phenomena involving many factors and components. The main components of an induction heat treatment system are an induction coil, power supply, load-matching station, quenching system, and the workpiece itself. The induction coil or inductor is usually designed for a specific application and is therefore found in widely varying shapes and sizes. Features involved in the design and operation of modern IHT processes are discussed later in this chapter.

III. THEORETICAL BACKGROUND AND MATHEMATICAL MODELING OF MODERN INDUCTION HEAT TREATMENT PROCESSES

Mathematical modeling is one of the major factors of successful induction heat treatment design. Theoretical models may vary from a simple hand-calculated formula to a very complicated numerical analysis, which can require several hours of computational work using modern supercomputers. The choice of a particular theoretical model depends on several factors, including the complexity of the engineering problems, required accuracy, time limitations, and cost.

Before an engineer starts to provide a mathematical simulation of any process he or she should have a sound understanding of the nature and physics of the process. Engineers should also be aware of the limitations of applied mathematical models, assumptions, and possible errors and should consider correctness and the sensitivity of the chosen model to some poorly defined parameters such as boundary conditions, material properties or initial temperature conditions. One model can work very well in certain applications and give absolutely unrealistic results in another. Underestimation of features of the process or overly simple assumptions can lead to an incorrect mathematical model (including chosen governing equations) that will not be able to provide the required accuracy. It is very important to remember that any computational analysis can at best produce only results that are derived from the governing equations. Therefore, the first and most important step in any mathematical simulation is to choose an appropriate theoretical model that will correctly describe the technological process or phenomena. Let's briefly look at the basic physical concepts related to the induction heat treatment process.

As mentioned above, an induction heat treatment is a complex combination of electromagnetic, heat transfer and metallurgical phenomena. The metallurgical aspects have been discussed in previous chapters. Therefore, in this chapter we will concentrate on the first two phenomena. Heat transfer and electromagnetics are tightly interrelated because the physical properties of heat-treated materials depend strongly on both magnetic field intensity and temperature. In conventional induction heating, all three modes of heat transfer—conduction, convection, and radiation—are present [26–36]. Heat is transferred by conduction from the high-temperature region of the workpiece toward the low-temperature region. The basic law that describes heat transfer by conduction is Fourier's law,

$$q_{cond} = -\lambda \; grad(T)$$

where q_{cond} is heat flux by conduction, λ is thermal conductivity, and T is temperature.

As one can see, according to Fourier's law the rate of heat transfer in a workpiece is proportional to the temperature gradient (temperature difference) and the thermal conductivity of the workpiece. In other words, a large temperature difference between surface and core (which, for example, typically takes place during surface hardening) and a high value of thermal conductivity of the metal result in intensive heat transfer from the hot surface of the workpiece toward the cold core. Conversely, the rate of heat transfer by conduction is inversely proportional to the distance between regions with different temperatures.

In contrast to conduction, heat transfer by convection is carried out by fluid, gas or air (i.e., from the surface of the heated workpiece to the ambient area). Convection heat transfer can be described by the well-known Newton's law. That law states that the heat transfer rate is directly proportional to the temperature difference between the workpiece surface and the ambient area,

$$q_{conv} = \alpha(T_s - T_a)$$

where q_{conv} is heat flux density by convection, W/m^2 or W/in^2; α is the convection surface heat transfer coefficient, W/(m$^2 \cdot {}^\circ$C) or W/(in.$^2 \cdot {}^\circ$F); T_s is surface temperature, $^\circ$C or $^\circ$F; and T_a is ambient temperature, $^\circ$C or $^\circ$F. The subscripts s and a denote surface and ambient, respectively.

The convection surface heat transfer coefficient is primarily a function of the thermal properties of the workpiece, the thermal properties of the surrounding fluids, gas or air, and their viscosity or the velocity of the heat-treated workpiece if the workpiece is moving at high speed (e.g., induction heat treating of a strip or spinning disk). It is particularly important to take this mode into account when designing low-temperature induction heating applications [i.e., tempering, stress relieving, paint curing or other processes with a maximun temperature <500°C (932°F)]. This heat transfer mode plays a particulary important role in the quenching process where the surface heat transfer coefficient describes the cooling process during quenching.

In the third mode of heat transfer, which is heat radiation, the heat may be transferred from the hot workpiece into a nonmaterial region (vacuum). The effect of heat transfer by radiation can be introduced as a phenomenon of the electromagnetic energy propagating due to a temperature difference. This phenomenon is governed by the Stefan–Boltzmann law of thermal radiation, which states that the heat transfer rate by radiation is proportional to a radiation loss coefficient C_s and the value of $T_s^4 - T_a^4$. Due to the fact that radiation losses are proportional to the fourth power of temperature, these losses are a significant part of the total heat losses in high temperature applications (for example, surface hardening, through hardening, induction heating before forging, rolling, etc.). The radiation heat loss coefficient includes emissivity, radiation shape factor, and surface conditions. For example, polished metal will radiate less heat to the surroundings than will nonpolished metal. The radiation heat loss coefficient can be determined approximately as $C_s = \sigma_s \varepsilon_1$, where ε_1 is the emissivity of the metal and σ_s is the Stefan–Boltzmann constant [$\sigma_s = 5.67 \times 10^{-8}$ W/(m²·K⁴)]. Values of ε_1 for metals that are typically used in heat treatment are given in Table 1.

The above-described determination of radiation heat loss is a valid assumption for mathematical modeling of a great majority of induction heating and heat treatment problems. However, there are a few applications where the radiation heat transfer phenomenon can be complicated and such a simple approach would not be valid. Complete details of all three modes of heat transfer can be found in several references [26–36]. In typical induction heating and heat treatment, heat transfer by convection and radiation reflects the value of heat loss. A high value of heat loss reduces the total efficiency of the induction heater.

In general, the transient (time-dependent) heat transfer process in a metal workpiece can be described by the Fourier equation

$$C\gamma \frac{\partial T}{\partial t} + \mathrm{div}(-\lambda \ \mathrm{grad} \ T) = Q \tag{1}$$

where T is temperature; γ is the density of the metal, C is the specific heat, λ is the thermal conductivity of the metal, and Q is the heat source density induced by eddy currents per unit time in a unit volume (so-called heat generation). This heat source density is obtained by solving the electromagnetic problem.

The values of the specific heat and thermal conductivity of some typical metals commonly used in heat treating are shown in Figure 5 [37,38]. As one may note, the specific heat and thermal conductivity are functions of temperature. Both specific heat and thermal conductivity have not only pure mathematical meaning but also a concrete engineering interpretation. The value of specific heat indicates the amount of energy that would have to be absorbed by the workpiece to achieve the required temperature change. A high value of specific heat corresponds to a higher required power.

Table 1 Emissivity of Metals

	Aluminum	Carbon steel	Copper	Brass and zinc
Polished	0.042–0.053	0.062	0.026–0.042	0.03–0.039
Commercial	0.082–0.40	0.71–0.8	0.24–0.65	0.21–0.50

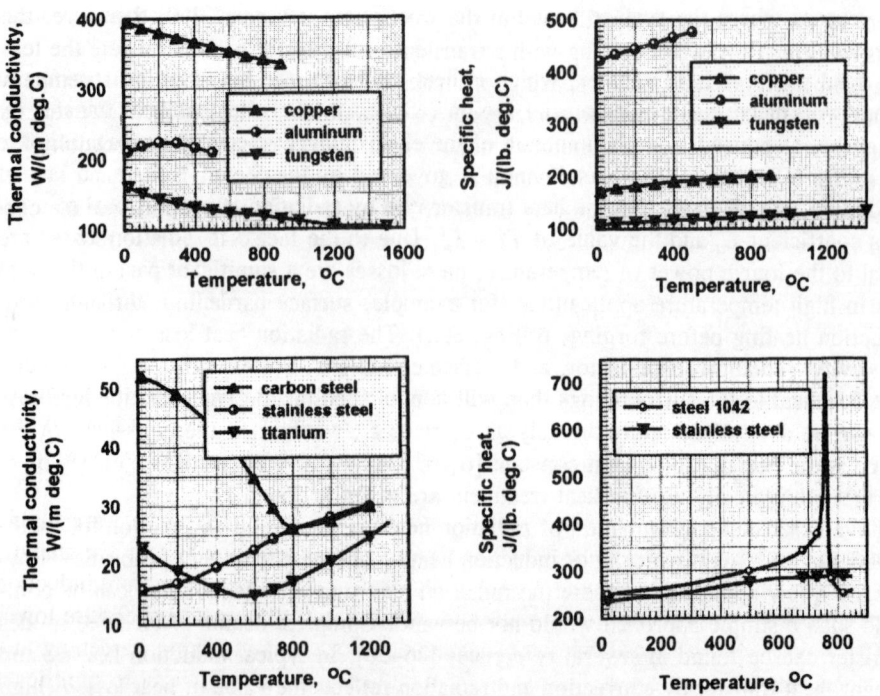

Figure 5 Thermal conductivity and specific heat of metals.

Thermal conductivity λ designates the rate at which heat travels across the workpiece. A material with a high λ value will conduct the heat faster than a material with a low λ. In choosing a material for an inductor's refractory, a lower value of λ is required and will correspond to higher efficiency and lower heat loss. Conversely, when the thermal conductivity of the metal is high, then it is easier to obtain a uniform temperature distribution along the workpiece thickness, which is important in through-hardening processes. Therefore, from the point of view of obtaining temperature uniformity, a higher thermal conductivity of metal is preferable. However, in surface hardening or selective hardening applications, a high value of λ is quite often a disadvantage because of its tendency to equalize the temperature distribution within the workpiece. As a result of temperature equalization the temperature rise will take place not only in the surface layer of the workpiece, which is to be hardened, but in internal areas as well, which are not. The temperature increase in the internal areas of the workpiece results in cooling of its surface and therefore requires more power to heat the surface layer to the desired final temperature. A large amount of heated mass in the workpiece can also cause a distortion of the workpiece geometry.

Our experience shows that in the great majority of tempering, stress relieving, annealing and normalizing applications, a rough approximation of thermal conductivity in simulations of the heat treatment process will not lead to significant errors in temperature distributions. However, in surface hardening and through hardening a rough approximation of λ can create unacceptable results. At the same time, regardless of application, a rough approximation of specific heat could create significant errors in obtaining the required coil power and temperature profile within the workpiece.

Equation (1), with suitable boundary and initial conditions, represents the three-dimensional temperature distribution at any time and at any point in the workpiece. The initial condition refers to the temperature profile within the workpiece at time $t = 0$; therefore, that condition is required only when dealing with a transient heat transfer problem where the temperature is a function not only of the space coordinates but of time also. The initial temperature distribution is usually uniform and corresponds to the ambient temperature. However, the initial temperature distribution is sometimes nonuniform due to the previous technological process (i.e., preheating, quenching, or continuous casting applications).

For most engineering induction heating problems, boundary conditions combine the convective and radiative losses. In this case the boundary condition can be expressed as

$$- \lambda \frac{\partial T}{\partial n} = \alpha(T_s - T_a) + C_s(T_s^4 - T_a^4) + Q_s \qquad (2)$$

where $\partial T/\partial n$ is the temperature gradient in a direction normal to the surface at the point under consideration, α is the convection surface heat transfer coefficient, C_s is the radiation heat loss coefficient, Q_s is the surface loss (i.e., during quenching), and n denotes the normal to the boundary surface.

As one may see from Equation (2), the heat losses at the workpiece surface are highly variable because of the nonconstant behavior of convection and the radiation losses. The analysis shows that convection losses are the major part of the heat loss in low-temperature induction heating applications [i.e., aluminum, lead, zinc, tin, magnesium, steel at a temperature lower than 350°C (or 662°F)]. In high-temperature applications (such as induction hardening and heating of steels, titanium, tungsten, nickel, etc.), radiation losses are much more significant than convection losses (Figure 6).

If the heated body is geometrically symmetrical along the axis of symmetry, the Neumann boundary condition can be formulated as

$$\frac{\partial T}{\partial n} = 0 \qquad (3)$$

The Neumann boundary condition implies that the temperature gradient in a direction normal to the axis of symmetry is zero. In other words, there is no heat exchange at the axis of sym-

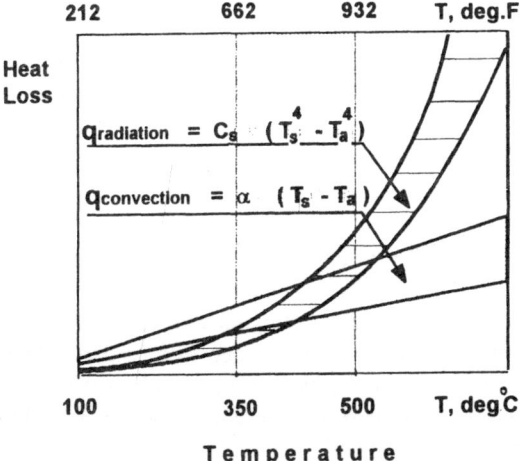

Figure 6   Convection and radiation heat losses in typical induction heating.

metry. This boundary condition can also be applied in the case of a perfectly insulated workpiece.

In the case of heating a cylindrical workpiece, Equation (1) can be rewritten as

$$C\,\gamma\,\frac{\partial T}{\partial t} = \frac{\partial}{\partial Z}\left(\lambda\,\frac{\partial T}{\partial Z}\right) + \frac{1}{R}\frac{\partial}{\partial R}\left(\lambda\,R\frac{\partial T}{\partial R}\right) + Q(Z,R) \tag{4}$$

The same Equation (1) can be shown in Cartesian coordinates (i.e., heat transfer in slab, strip, or plate) as

$$C\,\gamma\,\frac{\partial T}{\partial t} = \frac{\partial}{\partial X}\left(\lambda\,\frac{\partial T}{\partial X}\right) + \frac{\partial}{\partial Y}\left(\lambda\,\frac{\partial T}{\partial Y}\right) + \frac{\partial}{\partial Z}\left(\lambda\,\frac{\partial T}{\partial Z}\right) + Q(X,Y) \tag{5}$$

Equations (4) and (5) with boundary conditions (2) and (3) are the most popular equations for mathematical modeling of the heat transfer processes in conventional heat treatment.

The technique of calculating electromagnetic fields depends on the ability to solve Maxwell's equations. For general time-varying electromagnetic fields, Maxwell's equations in differential form can be written as [39–52]

$$\operatorname{curl}\mathbf{H} = \mathbf{J} + \frac{\partial D}{\partial t} \qquad \text{(from Ampere's law)} \tag{6}$$

$$\operatorname{curl}\mathbf{E} = -\frac{\partial \mathbf{B}}{\partial t} \qquad \text{(from Faraday's law)} \tag{7}$$

$$\operatorname{div}\mathbf{B} = 0 \qquad \text{(from Gauss' law)} \tag{8}$$

$$\operatorname{div}\mathbf{D} = \rho \qquad \text{(from Gauss' law, magenetic)} \tag{9}$$

where $\mathbf{E}$ is electric field intensity, $\mathbf{D}$ is electric flux density, $\mathbf{H}$ is magnetic field intensity, $\mathbf{B}$ is magnetic flux density, $\mathbf{J}$ is conduction current density, and ρ is electric charge density.

Maxwell's equations not only have a purely mathematical meaning, they have a concrete physical interpretation as well. For example, Equation (6) says that the curl of $\mathbf{H}$ always has two sources: conductive ($\mathbf{J}$) and displacement ($\partial D/\partial t$) currents. A magnetic field is produced whenever there are electric currents flowing in surrounding objects. From Equation (7) one can conclude that a time rate of change in magnetic flux density always produces the curling electric field. In other words, a changing magnetic field always produces induced current in the surrounding area, and therefore it produces an electric field in the area where such changes take place. The minus sign in Equation (7) determines the direction of that induced electric field. This fundamental result can be applied to any region in space.

Let us consider how Equations (6) and (7) can be used to explain the basic electromagnetic process that takes place in induction heat treatment. The application of alternating voltage to the induction coil will result in the appearance of an alternating current in the coil circuit. According to Equation (6), an alternating coil current will produce in its surrounding area an alternating (changing) magnetic field that will have the same frequency as the source current (coil current). That magnetic field's strength depends on the current flowing in the induction coil, the coil geometry, and the distance from the coil. The changing magnetic field induces eddy currents in the workpiece and in other objects that are located near that coil. By

Equation (7), induced currents have the same frequency as the source coil current; however, their direction is opposite that of the coil current. (This is determined by the minus sign in Equation 7.) According to Equation (6), alternating eddy currents induced in the workpiece produce their own magnetic fields, which have opposite directions to the direction of the main magnetic field of the coil. The total magnetic field of the induction coil is a result of the source magnetic field and induced magnetic fields.

As one would expect from an analysis of Equation (7), there can be an undesirable heating of tools or other electrically conductive structures that are located near the induction coil. From another perspective, Equation (7) can partially explain the undesirable existence of the magnetic field at the operator (the concern about the effects of electromagnetic field exposure on human beings). Both electric and magnetic fields induce weak electric currents in the operator's body.

In induction heating and heat treatment applications, an engineer should pay particular attention to such simple relations as (8) or (9). The popular saying "The best things come in the smallest packages" is particularly true here. The short notation of Equation (8) has real significance in induction heating and the heat treatment of metals. To say that the divergence of magnetic flux density is zero is equivalent to saying that **B** lines have no source points at which they originate or end, in other words **B** lines always form a continuous loop. A clear understanding of such a simple statement will allow one to explain and avoid many mistakes in dealing with induction heat treatment and the induction heating of workpieces with complicated geometry.

The above-described Maxwell's equations are in indefinite form because the number of equations is less than the number of unknowns. These equations become definite when the relations between the field quantities are specified. The following constitutive relations are additional and hold true for a linear, isotropic medium.

$$\mathbf{D} = \varepsilon\, \varepsilon_0\, \mathbf{E} \tag{10}$$

$$\mathbf{B} = \mu\, \mu_0\, \mathbf{H} \tag{11}$$

$$\mathbf{J} = \sigma\, \mathbf{E} \quad \text{(Ohm's law)} \tag{12}$$

where the parameters ε, μ, and σ denote respectively the relative permittivity, relative magnetic permeability, and electrical conductivity of the material; $\sigma = 1/\rho$, where ρ is electrical resistivity. The constant $\mu_0 = 4\pi \times 10^{-7}$ H/m [or Wb/(A·m)] is called the permeability of free space (the vacuum), and similarly the constant $\varepsilon_0 = 8.854 \times 10^{-12}$ F/m is called the permittivity of free space. Both relative magnetic permeability μ and relative permittivity ε are nondimensional parameters and have very similar meanings. Relative magnetic permeability indicates the ability of a material to conduct the magnetic flux better than vacuum or air. This parameter is very important in choosing materials for magnetic flux concentrators. Relative permittivity (or dielectric constant) indicates the ability of a material to conduct the electric field better than vacuum or air.

By taking Equations (10) and (12) into account, Equation (6) can be rewritten as

$$\text{curl } \mathbf{H} = \sigma\, \mathbf{E} + \frac{\partial(\varepsilon_0 \varepsilon \mathbf{E})}{\partial t} \tag{13}$$

For most practical applications of the induction heat treatment of metals, where the frequency of currents is less than 10 MHz, the induced conduction current density **J** is much greater than the displacement current density $\partial D/\partial t$, so the last term on the right-hand side of Equation (13) can be neglected. Therefore, Equation (13) can be rewritten as

$$\text{curl } \mathbf{H} = \sigma \mathbf{E} \tag{13a}$$

After some vector algebra and using Equations (6), (7), and (11), it is possible to show that

$$\text{curl}\left(\frac{1}{\sigma}\text{ curl } \mathbf{H}\right) = -\mu\mu_0\frac{\partial \mathbf{H}}{\partial t} \tag{14}$$

$$\text{curl}\left(\frac{1}{\mu}\text{ curl } \mathbf{E}\right) = -\sigma\mu_0\frac{\partial \mathbf{E}}{\partial t} \tag{15}$$

Since the magnetic flux density $\mathbf{B}$ satisfies a zero divergence condition (Equation 8), it can be expressed in terms of a magnetic vector potential $\mathbf{A}$ as

$$\mathbf{B} = \text{curl } \mathbf{A} \tag{16}$$

and then, from (7) and (16), it follows that

$$\text{curl } \mathbf{E} = -\text{ curl}\frac{\partial \mathbf{A}}{\partial t} \tag{17}$$

Therefore, after integration, one can obtain

$$\mathbf{E} = -\frac{\partial \mathbf{A}}{\partial t} - \text{grad } \varphi \tag{18}$$

where φ is the electric scalar potential. Equation (12) can be written as

$$\mathbf{J} = -\sigma\frac{\partial \mathbf{A}}{\partial t} + J_s \tag{19}$$

where $J_s = -\sigma \text{ grad}(\varphi)$ is the source (excitation) current density in the induction coil.

Taking the material properties as being piecewise continuous and neglecting the hysteresis and magnetic saturation, it can be shown that

$$\frac{1}{\mu\mu_0}(\text{curl curl } \mathbf{A}) = J_s - \sigma\frac{\partial \mathbf{A}}{\partial t} \tag{20}$$

It should be mentioned here that for the great majority of induction heat treating applications (such as through hardening) a heat effect due to hysteresis losses does not exceed 7% compared to the heat effect due to eddy current losses. Therefore, an assumption neglecting the hysteresis is valid. However, in some cases, such as induction tempering, paint curing, heating prior galvanizing and lacquer coating, hysteresis losses can be quite significant compared to eddy current losses. In these cases, hysteresis should be considered.

It can be shown that for the great majority of induction heat treatment applications it is possible to further simplify the mathematical model by assuming that the currents have a steady-state quality. Therefore, with this assumption we can conclude that the electromagnetic field quantities in Maxwell's equations are harmonically oscillating functions with a single frequency. Thus a time-harmonic electromagnetic field can be introduced. This field can be described by the following equations, which are derived after some vector algebra from Equations (14), (15), and (20), respectively.

$$\frac{1}{\sigma} \nabla^2 \mathbf{H} = \mathbf{j}\,\omega\,\mu\,\mu_0\,\mathbf{H} \tag{21}$$

$$\frac{1}{\mu} \nabla^2 \mathbf{E} = \mathbf{j}\,\omega\,\sigma\,\mu_0\,\mathbf{E} \tag{22}$$

$$\frac{1}{\mu\mu_0} \nabla^2 \mathbf{A} = -\mathbf{J}_s + \mathbf{j}\,\omega\,\sigma\,\mathbf{A} \tag{23}$$

where ∇^2 is the Laplacian, which has different forms in cartesian and cylindrical coordinates. In cartesian coordinates,

$$\nabla^2 \mathbf{A} = \frac{\partial^2 A}{\partial X^2} + \frac{\partial^2 A}{\partial Y^2} + \frac{\partial^2 A}{\partial Z^2} \tag{24}$$

In cylindrical coordinates (axisymmetric case),

$$\nabla^2 \mathbf{A} = \frac{1}{R}\frac{\partial}{\partial R}\left(R\frac{\partial A}{\partial R}\right) + \frac{\partial^2 A}{\partial Z^2} \tag{25}$$

In other words, it has been assumed that harmonics are absent in both the impressed and induced currents and fields. The governing equations, (21)–(23), for the time-harmonic field with the appropriate boundary condition can be solved with respect to $\mathbf{H}$, $\mathbf{E}$, or $\mathbf{A}$.

Equations (21)–(23) are valid for general three-dimensional fields and allow one to find all of the required parameters of the induction system such as current, power, coil impedance and heat source density induced by eddy currents. Although there is considerable practical interest in three-dimensional problems, a great majority of engineering problems in induction heating tend to be handled with a combination of two-dimensional assumptions. Several factors discourage three-dimensional field consideration:

1. Computing costs are much higher for three-dimensional cases (especially taking into account tightly interrelated features of electromagnetic and heat transfer phenomena in induction heating).
2. The user must have special knowledge in many theoretical subjects and should have specific experience working with three-dimensional software.
3. Representation of both results and geometric input data could create a well-known problem of working with 3-dimensional image.

For many induction heat treatment applications the quantities of the magnetic field (such as magnetic vector potential, electric field intensity, and magnetic field intensity) may be assumed to be entirely directed. For example, in the longitudinal cross section of the induction coil, both $\mathbf{A}$ and $\mathbf{E}$ vectors have only one component, which is entirely Z-directed. In the case of a transverse section, $\mathbf{H}$ and $\mathbf{B}$ vectors have only one component. This allows one to reduce the three-dimensional field to a combination of two-dimensional forms. For example, in the case of magnetic vector potential, Equation (23) can be expressed as shown below.

For a two-dimensional Cartesian system,

$$\frac{1}{\mu \mu_0} \left(\frac{\partial^2 A}{\partial X^2} + \frac{\partial^2 A}{\partial Y^2} \right) = -J_S + j \omega \sigma A \tag{26}$$

For the axisymmetric cylindrical system,

$$\frac{1}{\mu \mu_0} \left(\frac{\partial^2 A}{\partial R^2} + \frac{1}{R} \frac{\partial A}{\partial R} + \frac{\partial^2 A}{\partial Z^2} - \frac{A}{R^2} \right) = -J_S + j \omega \sigma A \tag{27}$$

The boundary of the region is selected such that the magnetic vector potential A is zero along the boundary (Dirichlet condition) or its gradient is negligibly small along the boundary compared to its value elsewhere in the region (Neumann condition $\partial A / \partial n = 0$). Therefore, Equations (4) and (27) with their initial and boundary conditions fully describe the electrothermal processes in a great majority of conventional cylindrical induction heat treatment systems.

By using analogous vector algebra manipulations it is possible to obtain governing equations similar to (26) and (27) that can be formulated with respect to E, B, or H. Therefore, any given electromagnetic problem in induction heat treatment may be worked in terms of either A, E, B, or H. Part of the art of mathematical modeling of electromagnetic fields derives from the right choice of field representation, which could be different for different applications.

Partial differential equations that are formulated with respect to A or E are very convenient for describing the electromagnetic field in a longitudinal cross section of the induction heating system. However, the electromagnetic field distribution in a transverse cross section of the workpiece can be more conveniently described by governing equations formulated with respect to B or H [49,50]. Field representations that are typically used by the induction heat treatment designer for describing electromagnetic processes in the conventional induction surface or through hardening of cylindrical workpieces are shown in Figure 7.

As mentioned above, contemporary design of induction heat treatment equipment requires that several features of the process be taken into account, which makes its analysis quite complicated.

One of the major difficulties in electromagnetic field and heat transfer computation is the nonlinearity of material properties. As mentioned earlier, this exists because the thermal conductivity, specific heat, and electrical conductivity of metals (i.e., steel) are functions of the temperature but their relative magnetic permeability is a function of two parameters: magnetic field intensity and temperature. Figure 8 shows electrical resistivities of typical metals as functions of temperature. Relative magnetic permeability as a function of temperature and magnetic field intensity is shown in Figure 9. In everyday engineering language, the induction heating specialists often call the relative magnetic permeability simply magnetic permeability. In this chapter also we will refer to it as magnetic permeability.

Even a cursory look at the behavior of the material properties in Figures 5, 8, and 9 reveals the danger in using methods (e.g., analytical or equivalent circuit coil design methods) that cannot take into account the total variation of properties. Owing to their nature, these methods can consider only average values, which can lead to unpredictable and often erroneous results, with possible financial loss to both the customer and the producer of the heat treatment equipment.

Another difficulty in the analysis and computation of the induction heat treatment process is related to the fact that the shapes of the workpiece and induction coil can be complicated. Therefore, to provide the required accuracy of evaluation it is necessary to reformulate the computational model for almost every problem.

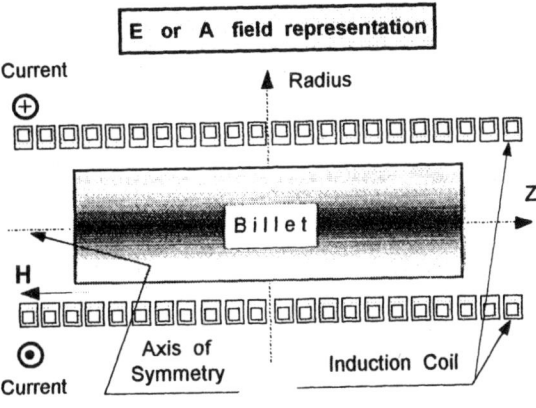

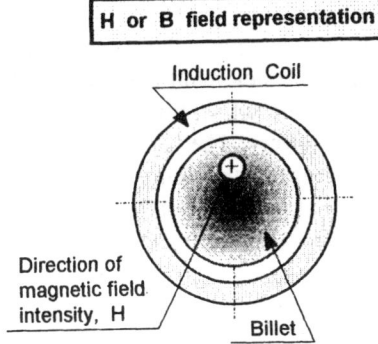

Figure 7   **E, A**, and **H, B** field representations in cylindrical system.

Besides the considerations mentioned above, the induction heater configuration can consist of several coils, flux concentrators, "flux robbers," shunts, and shells, and finally, there are tightly coupled (interrelated) effects of the electromagnetic and heat transfer phenomena. This can require the development of special computational algorithms that are able to deal with these coupled effects. The highly pronounced variation of physical properties makes the induction heat treatment problem nonlinear. There are several ways to couple electromagnetic and heat transfer problems. One of the most common approaches is called the indirect coupling method. This method calls for the iteration process. An iteration process consists of an

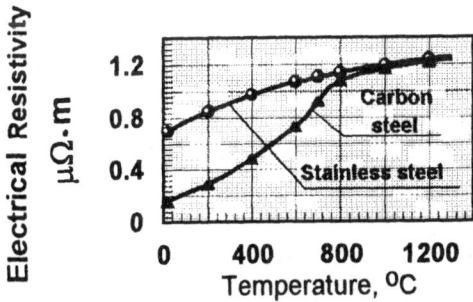

Figure 8   Electrical resistivity of steels.

Relative Magnetic Permeability

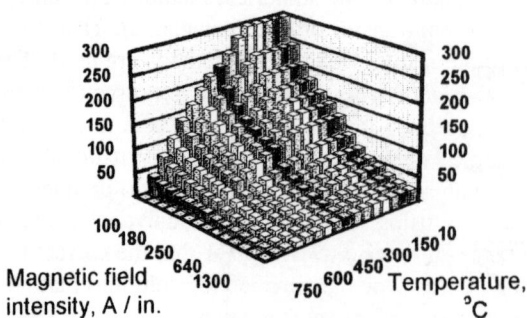

Figure 9 Magnetic permeability as a function of magnetic field intenstiy $\mathbf{H}$ = 100–1500 A/in. and temperature T = 10–750°C (T = 50–1382°F).

electromagnetic computation and then recalculation of heat sources in order to provide a heat transfer computation. This assumes temperature variations are not significant, which means the material properties remain approximately the same, and during certain times the heat transfer process continues to be solved without correcting the heat sources. The temperature distribution within the workpiece obtained from the time-stepped heat transfer computation is used to update the values of specific heat and thermal conductivity at each time step. As soon as the heat source variations become significant (due to the variations of electrical conductivity and magnetic permeability), the convergence condition will no longer be satisfied, and recalculation of the elcetro-magnetic field and heat sources will take place.

For most induction heating applications, an indirect coupling approach is valid and very effective. However, there are situations (i.e., intensive induction heating of carbon steel 1010 just above the Curie point, high power density induction contour hardening applications or low power density strip annealing) where this approach could lead to significant errors. In these cases, only the direct coupling method should be applied [51,53].

The analytical methods and equivalent circuit coil design methods popular in the 1960s and 1970s no longer satisfy the modern designer because of the inherent restrictions outlined above. The designer must be aware that in many applications, very erroneous and inadequate results can be obtained when such methods are used. The development of modern computers and the increasing complexity in most modern induction heating applications have significantly restricted the solution to the application of simple formulas, analytical and seminumerical methods. These methods can be useful only in obtaining approximate results in simple cases. Rather than use simple computatiion techniques with many restrictions, modern induction heat treatment specialists are currently turning to highly effective numerical methods such as finite-difference, finite-element, impedance boundary element, and integral equation methods that are widely and successfully used in the computation of electromagnetic and heat transfer problems in heat treatment. Each of these methods has certain advantages and has been used alone or in combination with others.

Because of the extraordinarily large amount of information that is available in special scientific literature, even an experienced engineer can be easily confused if too many of the nuances of computer modeling of induction heat treatment problems are introduced. Therefore, we decided to discuss the state-of-the-art of the subject of modern electroheat numerical computation while simplifying the materials so they are understood by induction heat treatment specialists who have limited experience in numerical modeling. Thus our goal is to provide the reader with a general orientation in advanced numerical simulation methods.

Before we discuss the features and applications of the most popular numerical methods it is necessary to point out one of their important qualities: All numerical methods give an approximate solution to the modeled problem (i.e., heat transfer or electromagnetic). Therefore, there is always a danger of obtaining inappropriate results when those methods are used. The fact that the solution is approximate, and is not *absolutely* accurate, should not discourage engineers from using numerical methods. On the contrary, it should stimulate them to carefully study the features of these methods and transform them into a powerful computational tool that will allow heat treatment analysts to control an accuracy of simulation and to produce information that cannot be measured or obtained by using analytical, semi-analytical or other kinds of methods, including physical experiments. It is wise to remember that the correct use of numerical methods will provide approximate, but acceptable, engineering solutions that will satisfy the requirements of modern technology from a practical standpoint.

Many mathematical modeling methods and programs exist or are under development. Work in this field is done in universities, research institutes, inside large companies such as INDUCTOHEAT Inc., and by specialized software companies such as MacNeal-Schwendler Corp., Integrated Engineering Software Inc., Infolytica Inc., Ansoft Corp., Structural Research and Analysis Inc., Cosmos Corp., Vector Fields Inc., and others.

For each particular problem or family of similar problems, certain numerical methods or software are preferred. It is obviously quite difficult if not impossible to find a single medicine that will be equally effective for both constipation and diarrhea. The medicine will most likely stop diarrhea or it will help constipation but not both. The same reasoning holds in searching for an efficient and universal computational tool. There is no cure-all or single universal computational method that fits all cases and is optimum for solving all induction heating problems. Our experience in the use of different numerical methods has shown that it is preferable to have a number of methods and programs rather than searching for one universal program for solving a wide variety of tasks. The right choice of computational method and software depends on the application and features of the specific problem or phenomenon. It is important for the induction heat treatment designer to know the advantages and limitations of a variety of computational techniques. This will allow the analyst to select an appropriate tool that will provide an accurate evaluation of the problem and allows the engineer to predetermine the temperature distribution within the workpiece in all stages of the process cycle, which includes heating, soaking, and quenching. This will result in obtaining the required heating conditions and desired metallurgical properties of the workpiece.

Because of space limitations, this chapter does not give an exhaustive list of the methods available for electromagnetic field and heat transfer calculations. There are many publications that describe the features and applications of mathematical modeling methods. An interested reader can study the description of the most popular computational techniques used for simulation of heat transfer and electromagnetic processes in References 5, 30, 47–92. Here we briefly analyze and describe some of the methods that are successfully used at INDUCTOHEAT. We hope that knowledge obtained will afford engineers and designers an understanding of how the numerical methods work and what kind of numerical method can be the most suitable for a particular induction heat treatment application.

A. Finite-Difference Method

The finite-difference method (FDM) was the earliest numerical technique [27,49,51–59] used for mathematical modeling of heat treatment processes. The finite-difference method has been used extensively for solving both heat transfer and electromagnetic problems. It is particularly easy to apply when the modeling area has cylindrical or rectangular shapes. The area of study is represented by the mesh of lines that are parallel to the coordinate axes (so-called

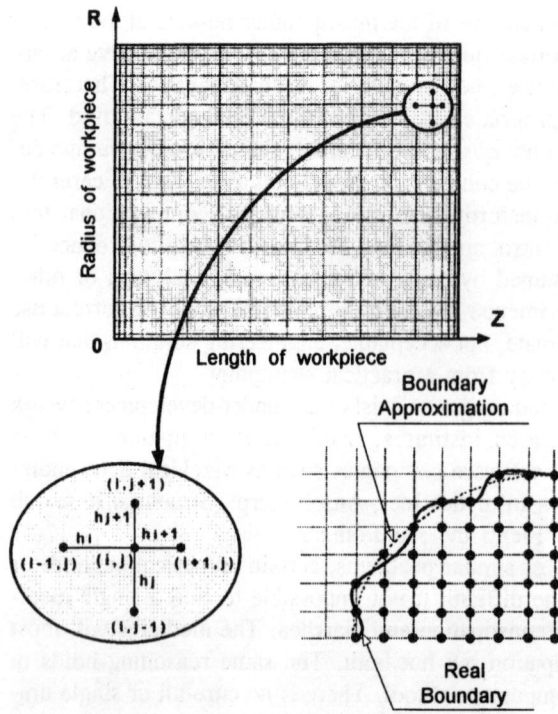

Figure 10 Finite-difference approximation and rectangular network of mesh.

orthogonal grid of mesh nodes). The orthogonal mesh discretizes the area of modeling (i.e., induction coil, workpiece, flux concentrator, etc.) into a finite number of nodes (Figure 10.)

Because of the orthogonal mesh, the discretization algorithm is quite simple. An approximate solution of the governing equation is found at the mesh points defined by intersections of the lines. The computation procedure consists of replacing each partial derivative of the governing equations (4), (5), (26), or (27) by a finite-difference "stencil" that couples the value of the unknown variable (i.e., temperature or magnetic vector potential) at a node of approximation with its value in the surrounding area. This method has universal application because of its generality and its relative simplicity to apply.

By Taylor's theorem for two variables, the value of a variable at a node on the mesh can be expressed in terms of its neighboring values and separation distance (space step) h as in the following expressions (stencils):

$$\frac{\partial T}{\partial X} \Rightarrow \frac{T_{i+1} - T_i}{h} + O(h) \qquad \text{(Forward difference)} \qquad (28)$$

$$\frac{\partial T}{\partial X} \Rightarrow \frac{T_i - T_{i-1}}{h} + O(h) \qquad \text{(Backward difference)} \qquad (29)$$

$$\frac{\partial T}{\partial X} \Rightarrow \frac{T_{i+1} - T_{i-1}}{2h} + O(h) \qquad \text{(Central difference)} \qquad (30)$$

$$\frac{\partial^2 T}{\partial X^2} \Rightarrow \frac{T_{i+1} - 2T_i + T_{i-1}}{h^2} + O(h^2) \tag{31}$$

Here the notation $O(h)$ is used to show that the error involved in the approximation is on the order of h. Similarly, $O(h^2)$ is for the approximation error on the order of h^2, which is more accurate than one on the order of h.

Substituting the finite-difference stencils into the electromagnetic and heat transfer partial differential equations gives the local approximation. By assembling all local approximations and taking into account the proper initial and boundary conditions, one can obtain a set of linear algebraic equations that can be solved with respect to unknown variables (i.e., T, $\mathbf{A}$, $\mathbf{E}$, $\mathbf{H}$, or $\mathbf{B}$) at each node of the mesh, either by iterative techniques or by direct matrix inversion methods. Since the matrices are sparsely occupied (nonzero only in the neighborhood of the diagonal), some simplification in the computational procedure is used.

In FDM, it is important that the boundaries of the mesh region coincide with the boundaries of the appropriate regions of the induction heat treatment system. Experience in using FDM in induction heat treatment computations has shown that noncoincidence of the boundaries has a strong negative effect on the accuracy of the calculation.

B. Finite-Element Method

As with FDM, the finite-element method (FEM) is a numerical technique for obtaining an approximate solution for different technical problems, including those encountered in induction heat treatment. FEM has had a particularly great deal of success in engineering practice and has become one of the most popular numerical tools for a variety of scientific and engineering tasks.

In the last two decades, several variations of finite-element models have been developed. Here are some of them:

- weighted residual method (weak form of the governing equations);
- different kinds of the Ritz method;
- different types of the Galerkin method;
- pseudo-variational methods;
- methods based on minimization of the energy functional, etc.

Those models are used in applications such as electric machines, motors, circuit breakers, transformers, magnetic recording systems, test equipment, electrical heating, and induction heat treatment. Even among the FEM only, there are situations where one type of FEM works just fine and in another case will not be able to do a good job at all.

As described above, the finite-difference method provides pointwise approximations; however, FEM provides an element-wise approximation to the governing equations. The finite-element method does not require direct calculation of the governing equation. Due to the general postulate of the variational principle, the solution of electro-magnetic field computation is typically obtained by minimizing the energy functional that corresponds to the governing equation (e.g., Equations 26 or 27) instead of solving that equation directly. The energy functional is minimized for the integral over the total area of modeling, which includes workpiece, coil, flux concentrators and surrounding area.

The well-known principle of minimum energy [55,60–64,66,68,71,75,80–82] requires that the vector potential distribution correspond to the minimum of the stored field energy per unit length. As a result of that assumption, it is necessary to solve the global set of resulting simultaneous algebraic equations with respect to the unknown, for example, magnetic vector potential at each node in the region of evaluation. The formulation of the energy functional, its minimization to obtain a set of finite-element equations, and the solution techniques (the solver) were created for both two-dimensional and axisymmetric heat treatment problems.

Many geometric arrangements and shapes of finite elements are possible. Their deformability allows them, in fact, to satisfy regions of any shape, i.e., any geometry of workpiece, inductors, flux concentrators, "flux robbers," etc. The most common two-dimensional finite element is the first-order triangle (Figure 11). In the axisymmetric cylindrical case, such a finite-element mesh may be represented as a set of rings. Each ring revolves around the axis of symmetry and has a triangular cross section (so-called triangular torus element).

The following are some general remarks concerning the finite-element discretization (mesh generation):

   The area of study is subdivided into nonoverlapping finite elements (so-called finite-element mesh, Figure 11). The sides of finite elements intersect at nodes. The number and location of these elements are matters of personal judgment, but to obtain reasonable accuracy of the numerical solution, the finite-element mesh has to be relatively fine (sizes of finite elements must be smaller) in the regions where the rate of change of the unknown (i.e. the magnetic vector potential) is high. Experience in using finite-element software at INDUCTOHEAT, Inc. has shown that special effort should

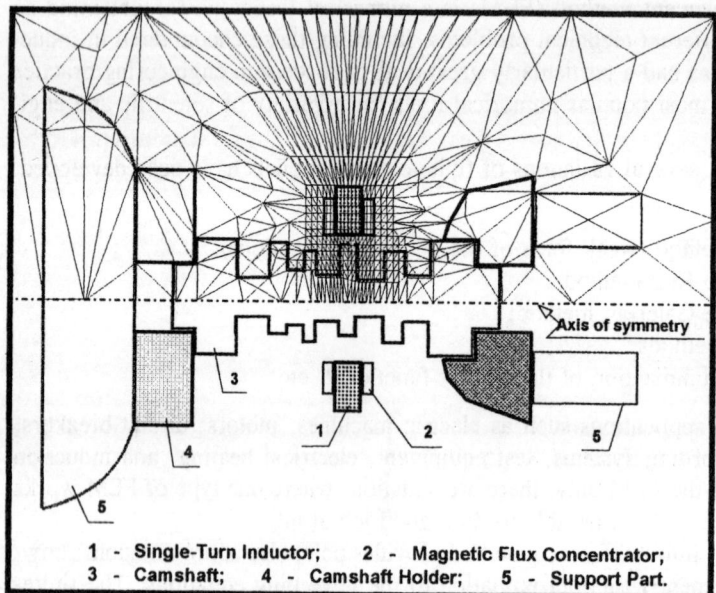

1 - Single-Turn Inductor;   2 - Magnetic Flux Concentrator;
3 - Camshaft;   4 - Camshaft Holder;   5 - Support Part.

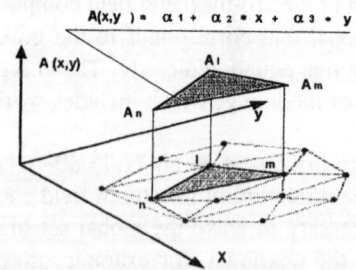

Figure 11 Finite-element discretization.

be made to obtain a fine mesh within three penetration depths in the workpiece. A higher frequency requires a finer mesh.

All the finite elements should have the same unit depth in the Z direction.

The current density, flux density, conductivity, magnetic permeability, and other material properties are postulated to be constant within each element. At the same time, they can be different from element to element.

The designer should take advantage of the symmetry involved in the system geometry, for example by disappearing normal derivative values of $\mathbf{A}$ along the symmetry.

In most typical cases, when it is necessary to obtain the electromagnetic field distribution and temperature profile along the length of a cylindrical workpiece, the finite-element method has been used in solving the governing equation with respect to magnetic vector potential for the electromagnetic problem, Equations (26) and (27), and with respect to the temperature for the heat transfer problem, Equations (4) and (5).

After solving the system of algebraic equations and obtaining the distribution of the magnetic vector potential in the region of modeling, it is possible to find all of the required output parameters of the electromagnetic field.

Induced current density in conductors:

$$\mathbf{J}_e = - j \, \omega \, \sigma \, \mathbf{A} \tag{32}$$

Total current density in the conductor:

$$\mathbf{J} = \mathbf{J}_s - j \, \omega \, \sigma \, \mathbf{A} \tag{33}$$

Magnetic flux density components $\mathbf{B}_x$ and $\mathbf{B}_y$ can be calculated from relationship (16) as follows [63,64]:

$$\frac{\partial \mathbf{A}}{\partial Y} = -\mathbf{B}_x; \qquad \frac{\partial \mathbf{A}}{\partial X} = \mathbf{B}_y$$

Therefore, flux density can be obtained as

$$\mathbf{B} = \left[\mathbf{B}_x^2 + \mathbf{B}_y^2 \right]^{1/2} \tag{34}$$

For the axisymmetric case of a cylindrical workpiece, the magnetic flux density components $\mathbf{B}_R$ and $\mathbf{B}_z$ can be calculated as

$$\mathbf{B}_R = -\frac{\partial \mathbf{A}}{\partial Z}; \qquad \mathbf{B}_z = \frac{\partial \mathbf{A}}{\partial R} + \frac{\mathbf{A}}{R}$$

Magnetic field intensity:

$$\mathbf{H} = \mathbf{B}/\mu\mu_o \tag{35}$$

Electric field intensity:

$$\mathbf{E} = - j \, \omega \, \mathbf{A} \tag{36}$$

Electromagnetic force density in current-carrying conductors and the workpiece can be computed from the cross product of the vector of total current density and the vector of magnetic flux density:

$$F_x = \mathbf{J} \times \mathbf{B}_y; \qquad F_y = - \mathbf{J} \times \mathbf{B}_x$$

From a vector potential solution it is possible to compute the other important quantities of the process such as stored energy, flux leakage, total power loss, and coil impedance.

Many worthwhile texts, conference proceedings, and articles have been written on the subject of finite-element modeling [55,60–83]. The large number of papers on the subject of FEM applications for electro-magnetic field computations makes it impossible to mention all of the contributions. At the same time, some of the proposed finite-element models are similar in form. However, we would like to mention here that the first general nonlinear variational formulation of magnetic field analysis using FEM was presented by Silvester and Chari [60,61]. Essential input into the development of FEM was provided by Lord, Udpa, Lavers and many others [63–66,69,70,72,81,83]. The software that was developed and is in use at INDUCTOHEAT Inc. is based on the finite-element concept proposed by Lord [63,64] and Udpa and coworkers [72,73]. This concept was adapted to induction heat treatment needs and successfully used for several years.

C. A Comparison

Superficially, the FDM and FEM appear to be different; however, they are closely related. As outlined above, FDM starts with a differential statement of the problem of interest, and requires that the partial derivative of the governing equation be replaced by a finite-difference stencil to provide pointwise approximation. FEM starts with a variational statement and provides an elementwise approximation. Both methods discretize a continuous function (e.g., magnetic vector potential or temperature) and result in a set of simultaneous algebraic equations to be solved with respect to its nodal values. Therefore, the two methods are, in fact, quite similar.

Finite-difference stencils overlap one another, and in the case of complicated workpiece geometry they could have nodes outside the boundary of the workpiece, coil or other components of the induction heating system. Finite elements do not overlap one another, do not have nodes outside the boundaries, and fit the complicated shape boundary perfectly. As shown above, in electro-magnetic field computation finite elements are usually introduced as a way to minimize a functional. In fact, FDM can also be described as a form of functional minimization (so called finite-difference energy method).

Therefore, FDM and FEM are different only in the choice of mesh generation and the way in which the global set of algebraic equations is obtained. They have approximately the same accuracy; however, required computer time and memory are often less when FDM is used. For example, the computer time needed to form global matrices is usually four to nine times greater with FEM than with FDM. As one would expect, a comparison of the efficiency of the two methods depends on the type of problem and program organization. Experience with both methods at INDUCTOHEAT shows that FDM is not well suited for an induction heat treatment system with complicated boundary shapes or in the case of a mixture of materials and forms (e.g. heat treating of camshafts, crankshafts, gears and other critical components). In this case, FEM has a distinct advantage over FDM.

As shown above, both finite-difference and finite-element methods require a network mesh of the area of modeling. That network includes induction coils, the workpiece, flux concentrators, etc. Unfortunately, to suit the condition of smoothness criteria and continuity of the governing differential equation it is also necessary to take into general consideration the airspace regions. Electromagnetic field distribution in the air, in most cases of coil designs, can be considered useless information, of interest only during the final design stage when evaluating electromagnetic field exposure of the induction heater. The need to always compute the electromagnetic field distribution in air can be considered a disadvantage of both the finite-difference and finite-element methods. Another difficulty which appears when using FDM or FEM for electromagnetic field computation is how to treat an exterior region that extends to infinity. This deals with an infinite nature of electromagnetic wave propagation. Several methods have been used taking into account this phenomena of the infinite exterior region. Some of those methods are "Ballooning" method, "Mapping" technique, and combination of finite elements and infinite elements. However, each of the above mention methods has certain shortcomings.

Therefore, one of the important improvements in using FDM or FEM in computation of induction heat treatment problems is the use of impedance boundary elements (IBE). The theory of impedance boundary elements is not easily explained. The mathematics required is more advanced than that needed for FDM or FEM. The interested reader will find several texts, conference proceedings, journal articles, and surveys [49,51,83–90] that describe various modifications of IBE. Here we will just mention that IBE will allow one to consider only conductive bodies in the computation. This significantly simplifies mesh generation, and decreases the computation time and required computer memory. Such improvements make this numerical software significantly more efficient.

D. Summary

We may summarize our introduction into numerical methods used for the simulation of induction heat treatment problems very simply. Each of the above-described methods has certain advantages. In many applications it is effective to use a combination of methods. The right choice of method depends on the specific application and features of the induction heat treating process.

The use of modern software does not guarantee correct computational results. It must be used in conjunction with experience in numerical techniques and engineering knowledge to achieve the required accuracy of mathematical modeling. This is especially so because even in modern commercial software, regardless of the amount of testing and verification, a computation program may never have all of its possible errors detected. The engineer must consequently be on guard against various kinds of errors. The more powerful the software, the greater the probability of errors. Common sense and engineering "gut feeling" are always the analyst's helpful assistants.

Computer modeling provides the ability to predict how different factors may impact the transitional and final heat treating conditions of the workpiece and what must be accomplished in the design of the induction heating system to improve the effectiveness of the process and guarantee the desired heat treating results. As an example, Figure 12 shows the results of the magnetic field computation in the selective hardening of a workpiece with a complicated shape (e.g., a section of a camshaft or crankshaft). The proper choice of the coil/magnetic flux concentrator geometry, frequency, and power density allows the heating of selected areas of the workpiece. Without using the flux concentrators, the electromagnetic field distribution will cause the eddy current to be induced not only in the desired areas of the workpiece but also significantly in the adjacent areas. Heat produced in these areas as a result of the induced current could cause undesirable metallurgical changes.

The correct choice of flux concentrators, their location, and properties allows the designer to avoid undesirable heating of adjacent areas of the workpiece. Furthermore, as a result of using flux concentrators, the heat source intensity in the adjacent areas decreases because the concentrator provides the preferred path for the magnetic field. We show this magnetic field distribution (Figure 12) as an example of the use of numerical techniques for modeling the heat treatment process and obtaining a desired heating pattern. Later we discuss in detail the

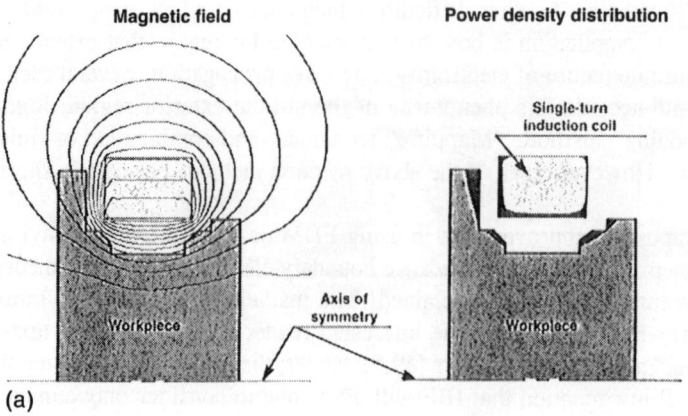

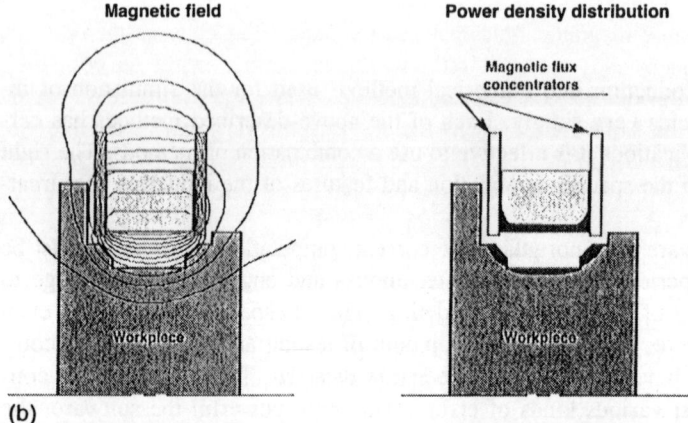

Figure 12   Results of numerical simulation of magnetic vector potential in selective hardening. Shown are field distributions (a) without and (b) with flux concentrator.

use of magnetic flux concentrators, their advantages and disadvantages. It will also be shown that in order to increase coil efficiency the flux concentrator should have a "C" shape. This will provide a preferred magnetic path in the sides and in the area external to the coil.

By combining the most advanced software with an outstanding computational and engineering background, modern induction heat treatment specialists possess a unique ability to analyze in a few hours complex induction heating problems that could take days or even weeks to solve using other methods including physical experiment. This leads to the saving of prototyping dollars and facilitates the building of reliable, competitive products in a short design cycle.

IV.   BASIC ELECTROMAGNETIC PHENOMENA IN INDUCTION HEATING

As shown earlier, an alternating voltage applied to an induction coil (e.g., solenoid coil) will result in an alternating current in the coil circuit. An alternating coil current will produce in

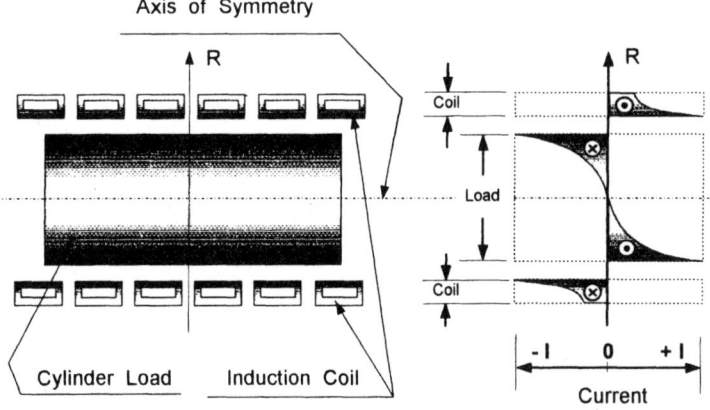

Figure 13   Current distribution in coil–workpiece induction heating system.

its surroundings a time-variable magnetic field that has the same frequency as the coil current. This magnetic field induces eddy currents in the workpiece located inside the coil. Eddy currents will also be induced in other electrically conductive objects that are located near the coil. Induced currents have the same frequency as the coil current; however, their direction is opposite to the direction of the coil current. These currents produce heat by the Joule effect (I^2R). A conventional induction heating system that consists of a cylindrical load surrounded by a multiturn induction coil is diagrammed in Figure 13.

Due to several electromagnetic phenomena, the current distribution within an inductor and workpiece is not uniform. This heat source nonuniformity causes a nonuniform temperature profile in the workpiece. Nonuniform current distribution can be caused by to several electromagnetic phenomena, including (1) the skin effect, (2) the proximity effect, and (3) the ring effect. These effects play an important role in understanding the induction heating phenomena [1–12,40,50–53,92,118–121].

A.   Skin Effect

As one may know from the basics of electricity, when a direct current flows through a conductor that stands alone (bus bar or cable), the current distribution within the conductor's cross section is uniform. However, when an alternating current flows through the same conductor, the current distribution is not uniform. The maximum value of the current density will always be located on the surface of the conductor; current density will decrease from the surface of the conductor toward its center. This phenomenon of nonuniform current distribution within the conductor cross section is called the skin effect. The skin effect always occurs when there is an alternating current. Therefore, the skin effect will also be found in the workpiece located inside the induction coil (Figure 13). This is one of the major factors that cause the concentration of eddy current in the surface layer (skin) of the workpiece. Due to the circumferential nature of the eddy current induced in the workpiece, there is no current flow in its center.

Because of the skin effect, approximately 86% of the power will be concentrated in a surface layer of the conductor. This layer is called the reference (or penetration) depth (δ). The degree of skin effect depends on the frequency and material properties (electrical resistivity and relative magnetic permeability) of the conductor. There will be a pronounced skin effect when a high frequency is applied or when the radius of the workpiece is relatively large

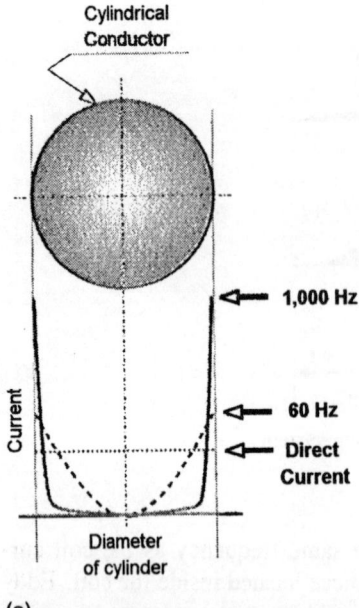

(a)

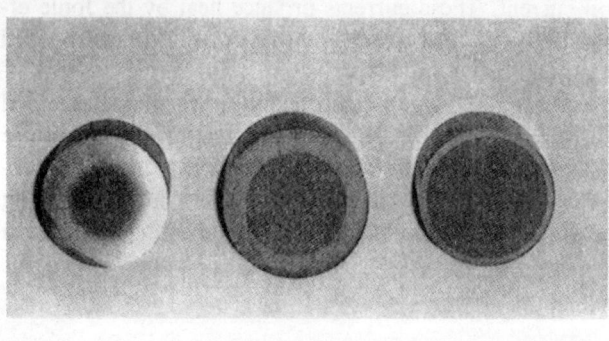

(b)

Figure 14 Skin effect. (a) Current distribution as function of frequency; (b) variation in depth of skin effect results in different hardness patterns.

(Figure 14). The distribution of the current density along the workpiece thickness (radius) can be calculated by the equation

$$I_R = I_o e^{-y/\delta} \tag{37}$$

where I_R is current density at distance R from the surface, A/m^2; I_o is current density at the workpiece surface, A/m^2; y is the distance from the surface toward the core, m; and δ is penetration depth, m.

Penetration depth is described as

$$\delta = 503(\rho/\mu F)^{1/2} \quad [\text{m}] \tag{38}$$

where
ρ is the electrical resistivity of the metal, $\Omega{\cdot}m$;
μ is the relative magnetic permeability; and
F is frequency, Hz (cycles per second), or

$$\delta = 3160(\rho/\mu F)^{1/2} \quad [\text{in.}] \tag{39}$$

where electrical resistivity is in $\Omega{\cdot}in$.

Mathematically speaking, the penetration depth δ in Equation (37) is the distance from the surface of the conductor toward its core at which the current decreases exponentially to "1/exp" its value at the surface. The power density at this distance will decrease to "$1/\exp^2$" its value at the surface. Figure 15 shows the percentage reduction of current density and power density from the surface toward the core. As one can see from Figure 15, at one penetration depth from the surface ($y = \delta$), the current will equal 37% of its surface value. However, the power density will equal 14% of its surface value. From this we can conclude that 63% of the current and 86% of the power in the workpiece within a surface layer of thickness δ, will be concentrated.

Analysis of Equations (38) and (39) shows that the penetration depth has different values for different materials and is a function of frequency. The magnetic permeability μ of a nonmagnetic workpiece is equivalent to that of air and is assigned a value of 1. The electrical resistivity of metals ρ is a function of temperature (Figure 8). During the heating cycle ρ can increase to 4–5 times its initial value. Therefore, even for nonmagnetic metals, during the heating cycle the penetration depth can increase significantly. Table 2 shows some penetration depths of metals that are most commonly used with induction heating.

In contrast to nonmagnetic metals, the permeability of magnetic steels commonly used in heat treating can vary from 1 to more than 500, depending on the magnetic field intensity $\mathbf{H}$ and temperature. The magnetization curve describes the nonlinear relationship between magnetic flux density $\mathbf{B}$ and magnetic field intensity $\mathbf{H}$. The nonlinear variation of $\mu = \mathbf{B}/(\mathbf{H}\mu_0)$ for a typical carbon steel is shown in Figure 16. Permeability is given by the ratio $\mathbf{B}/\mathbf{H}$. The maximum permeability occurs at the "knee" of the curve. The magnetic field intensity $\mathbf{H}_{cr}$ that corresponds to the maximum permeability is called a critical value of H. When $\mathbf{H} > \mathbf{H}_{cr}$, the magnetic permeability will decrease with increasing $\mathbf{H}$. If $\mathbf{H} \rightarrow \infty$, then $\mu \rightarrow 1$. In con-

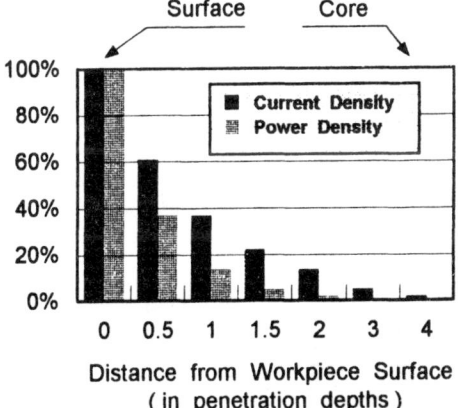

Figure 15   Current density and power density distribution due to the skin effect.

Table 2 Penetration Depth of Non-Magnetic Metals (mm)

Metal	T		ρ		Frequency (kHz)										
	°C	°F	μΩ·m	μΩ·in.	0.06	0.5	1	2.5	4	8	10	30	70	200	500
Aluminum	20	68	0.027	1.06	10.7	3.70	2.61	1.65	1.30	0.92	0.83	0.48	0.31	0.18	0.12
	250	482	0.053	2.09	15.0	5.18	3.66	2.32	1.83	1.29	1.16	0.67	0.44	0.26	0.16
	500	932	0.087	3.43	19.2	6.64	4.69	2.97	2.35	1.66	1.48	0.86	0.56	0.33	0.21
Copper	20	68	0.018	0.71	8.81	3.05	2.16	1.36	1.08	0.76	0.68	0.39	0.26	0.15	0.10
	500	932	0.050	1.97	14.5	5.03	3.56	2.25	1.78	1.26	1.12	0.65	0.43	0.25	0.16
	900	1652	0.085	3.35	19.3	6.67	4.72	2.98	2.36	1.67	1.49	0.86	0.56	0.33	0.21
Brass	20	68	0.065	2.56	16.6	5.74	4.06	2.56	2.03	1.43	1.28	0.74	0.48	0.29	0.18
	400	752	0.114	4.49	21.9	7.60	5.37	3.40	2.69	1.90	1.70	0.98	0.64	0.38	0.24
	900	1632	0.203	7.99	29.3	10.1	7.17	4.53	3.58	2.53	2.27	1.31	0.86	0.51	0.32
Stainless steel	20	68	0.690	27.2	53.9	18.7	13.2	8.36	6.61	4.67	4.18	2.41	1.58	0.93	0.59
	800	1472	1.150	45.3	69.6	24.1	17.1	10.8	8.53	6.03	5.39	3.11	2.04	1.21	0.76
	1200	2192	1.240	48.8	72.3	25.1	17.7	11.2	8.86	6.26	5.60	3.23	2.12	1.25	0.79
Silver	20	68	0.017	0.67	8.34	2.89	2.04	1.29	1.02	0.72	0.65	0.37	0.24	0.14	0.09
	300	572	0.038	1.50	12.7	4.39	3.10	1.96	1.55	1.10	0.98	0.57	0.37	0.22	0.14
	800	1472	0.070	2.76	17.2	5.95	4.21	2.66	2.10	1.49	1.33	0.77	0.50	0.30	0.19
Tungsten	20	68	0.050	1.97	14.5	5.03	3.56	2.25	1.78	1.26	1.12	0.65	0.43	0.25	0.16
	1500	2732	0.550	21.7	48.2	16.7	11.8	7.46	5.90	4.17	3.73	2.15	1.41	0.83	0.53
	2800	5072	1.040	40.9	66.2	22.9	16.2	10.3	8.11	5.74	5.13	2.96	1.94	1.15	0.73
Titanium	20	68	0.500	19.7	45.9	15.9	11.3	7.11	5.62	3.98	3.56	2.05	1.34	0.80	0.50
	600	1112	1.400	55.1	76.8	26.6	18.8	11.9	9.41	6.65	5.95	3.44	2.25	1.33	0.84
	1200	2192	1.800	70.9	87.1	30.2	21.3	13.5	10.7	7.54	6.75	3.90	2.55	1.51	0.95

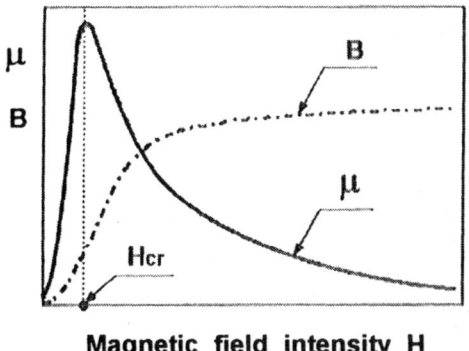

Magnetic field intensity H

Figure 16 Magnetic field density (**B**) and relative magnetic permeability (μ).

ventional induction heat treating, the magnetic field intensity H_{surf} at the workpiece surface is much greater than H_{cr}. However, there are some heat treatment applications where this is not true and this phenomenon will play an important role.

Because of the items discussed above, the same kind of carbon steel at the same temperature and frequency can have different penetration depths due to differences in the intensity of the magnetic field. Similar to the current distribution, the magnetic field intensity is at its maximum value at the surface of the homogeneous workpiece and falls off exponentially toward the core (Figure 17). As a result, the magnetic permeability varies within the magnetic body. At the surface, μ_{surf} corresponds to the surface magnetic field intensity H_{surf}. In quick calculations H_{surf} can be considered as the field intensity in the air gap between the coil and the workpiece. With increasing distance from the surface, μ increases and after reaching its maximum value at $H = H_{cr}$ begins to fall off (Figure 17).

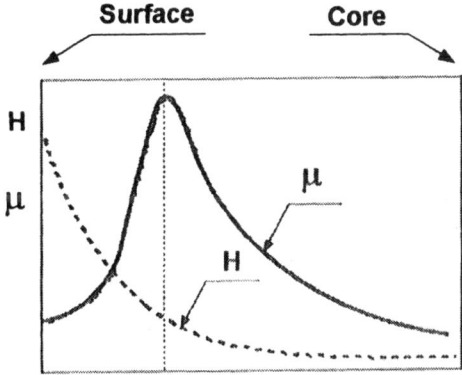

Radius of carbon steel cylinder

Figure 17 Distribution of magnetic field intensity (**H**) and relative magnetic permeability (μ) along the radius of a carbon steel homogeneous cylinder.

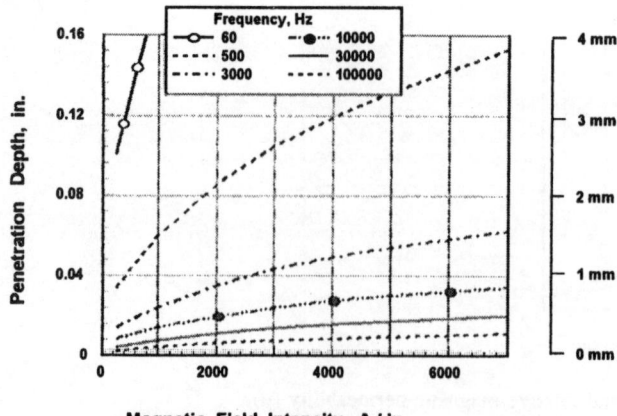

Magnetic Field Intensity, A / in.

Figure 18 Current penetration depth in carbon steel (1045) at ambient temperature 21°C (70°F).

While discussing the behavior of μ within the ferromagnetic workpiece, it is necessary to mention that a definition of penetration depth of current into a magnetic body in its classical forms, Equations (38) and (39), does not have a fully determined meaning because of the nonconstant distribution of μ within the workpiece. In engineering practice, the value of relative magnetic permeability at the surface of the workpiece is typically used to give a determination of those equations in definite form. Here we will also use the value of μ_{surf} to determine the penetration depth in the magnetic workpiece. Figure 18 and Table 3 show the value of the penetration depth in carbon steel (1045) at an ambient temperature (21°C or 70°F) as a function of frequency and magnetic field intensity **H** at the workpiece surface.

From another perspective, penetration depth is a function of temperature as well. At the beginning of the heating cycle, the current penetration into the carbon steel workpiece will increase slightly (Figure 19) because of the increase in electrical resistivity of the metal with temperature. With a further rise of temperature (at approximately 550°C or 1022°F), μ starts to decrease more and more. Near a critical temperature T_c known as the Curie temperature or Curie point, permeability drastically drops to unity because the metal becomes nonmagnetic.

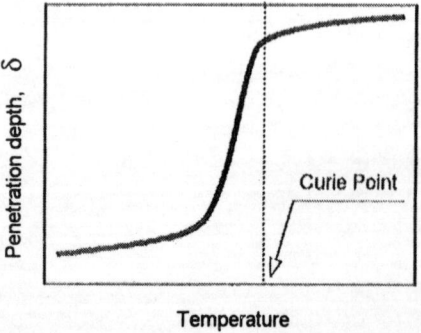

Figure 19 Typical variation of current penetration depth during induction heating of carbon steel workpiece.

Table 3 Penetration Depth of Carbon Steel 1040 at Ambient Temperature of 21°C (70°F)

Magnetic field intensity		Frequency (Hz)											
		60		500		3,000		10,000		30,000		100,000	
		Penetration depth											
A/mm	A/in.	mm	in.	mm	in.	mm	in.	mm	in.	mm	in.	mm	in.
10	250	2.5	0.100	0.88	0.034	0.36	0.014	0.2	0.008	0.11	0.004	0.06	0.002
40	1000	4.7	0.185	1.63	0.064	0.67	0.026	0.36	0.014	0.21	0.008	0.12	0.005
80	2000	6.3	0.249	2.2	0.086	0.9	0.035	0.49	0.019	0.28	0.011	0.16	0.006
120	3050	7.76	0.306	2.69	0.106	1.1	0.043	0.6	0.024	0.35	0.014	0.19	0.007
160	4050	8.76	0.345	3.03	0.119	1.24	0.049	0.68	0.027	0.39	0.015	0.21	0.008
200	5100	9.63	0.379	3.33	0.131	1.36	0.054	0.75	0.029	0.43	0.017	0.24	0.009
280	7100	11.2	0.442	3.89	0.153	1.59	0.062	0.87	0.034	0.50	0.02	0.27	0.011

As a result, the penetration depth will increase significantly (Table 4). After heating above the Curie temperature, the penetration depth will continue to increase due to the increase in electrical resistivity of the metal (Figure 19). However, the rate of growth will not be as significant as it was during the transition through the Curie temperature.

Typically, the variation of δ during the induction heating of a carbon steel workpiece drastically changes the degree of skin effect. It is especially important to take this phenomenon into account when designing for induction through hardening of carbon steel where the core-to-surface temperature difference is primarily a result of the skin effect.

B. Electromagnetic Proximity Effect

When we discussed the skin effect in conductors or cables, we assumed that a conductor stands alone and that there were no other current-carrying conductors in the surrounding area. In most practical applications this is not the case. Most often there are other conductors in close proximity. These conductors have their own magnetic fields, which interact with nearby fields, and as a result the current and power density distributions will be distorted.

An analysis of the effect on current distribution in a conductor when another conductor is placed nearby is given below. Figure 20a shows the skin effect and magnetic field distribution in a conductor (i.e., cylindrical bar) that stands alone. When another conductor is placed near the first one, the currents in both conductors will redistribute. If the currents flowing in the bars have opposite directions, then both currents (Figure 20b) will be concentrated in the

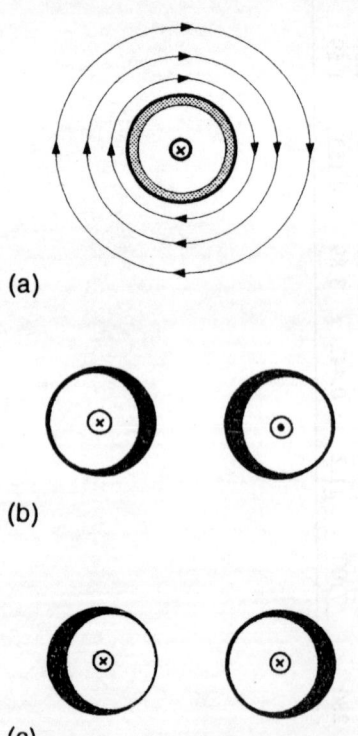

(a)

(b)

(c)

Figure 20 Proximity effect in cylindrical conductors (a) Single cable; (b) cables with opposite currents; (c) cables with similar currents.

Table 4 Penetration Depth of Carbon Steel 1040 at Ambient Temperature of 621°C (1150°F)

Magnetic field intensity		60		500		Frequency (Hz) 3,000		10,000		30,000		100,000	
						Penetration depth							
A/mm	A/in.	mm	in.	mm	in.	mm	in.	mm	in.	mm	in.	mm	in.
10	250	8.6	0.337	2.97	0.117	1.21	0.048	0.66	0.026	0.38	0.015	0.21	0.008
40	1000	15.5	0.611	5.38	0.212	2.20	0.086	1.20	0.047	0.69	0.027	0.38	0.015
80	2000	20.9	0.824	7.25	0.285	2.96	0.117	1.62	0.064	0.94	0.037	0.51	0.02
120	3050	24.5	0.966	8.50	0.335	3.47	0.137	1.9	0.075	1.1	0.043	0.60	0.024
160	4050	27.4	1.08	9.48	0.373	3.87	0.152	2.12	0.083	1.22	0.048	0.67	0.026
200	5100	29.8	1.17	10.3	0.406	4.2	0.166	2.31	0.091	1.33	0.052	0.73	0.029
280	7100	33.5	1.32	11.6	0.457	4.74	0.187	2.60	0.102	1.50	0.059	0.82	0.032

areas facing each other (internal areas). However, if the currents have the same direction, then these currents will be concentrated on opposite sides of the conductors (Figure 20c). The same will be true with bus bars (Figure 21).

When the currents flow in opposite directions, a strong magnetic field forms in the area between the bus bars (Figure 21). This occurs because in this area the magnetic field lines that are produced by each bus bar have the same direction. Therefore, the resulting magnetic field between the bus bars will be very strong. However, because the current is concentrated in the internal areas, the external magnetic field will be weak. The external magnetic fields will have opposite directions and will tend to cancel each other. This phenomena is used in co-axial cables. The opposite is true if the currents have the same direction, for then the magnetic field lines will have opposite directions in the area between bus bars, and therefore they will cancel each other in that area. Because of this cancellation, a weak magnetic field will exist between the bus bars. However, the external magnetic field will be quite strong because the magnetic lines produced by the two conductors will have the same direction in the external area.

If the distance between bars increases, then the strength of the proximity effect will decrease. Proximity effect in the case of nonsymmetrical systems is shown in Figure 22.

The phenomenon of proximity effect can be directly applied in induction heating. Induction systems consist of two conductors [141]. One of these conductors is an inductor that carries the source current (Figure 23), and the other is the workpiece that is located near the inductor. Eddy currents are induced in the workpiece by an external alternating magnetic field

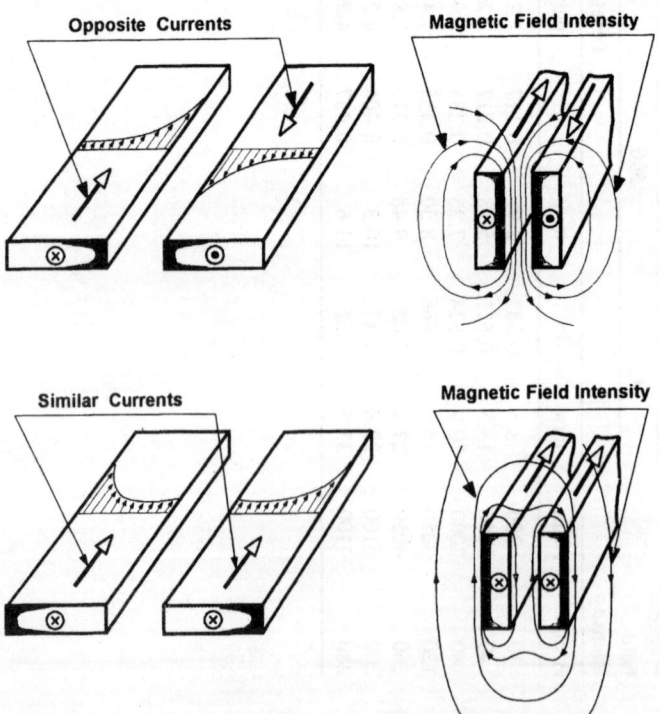

Figure 21   Current distribution in bus bars due to proximity effect.

Opposite Currents   **Similar Currents**

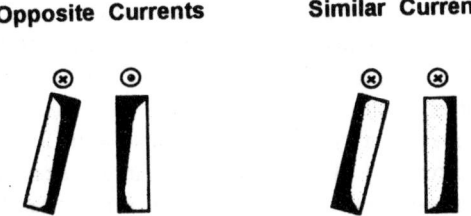

Figure 22   Proximity effect in nonsymmetrical systems.

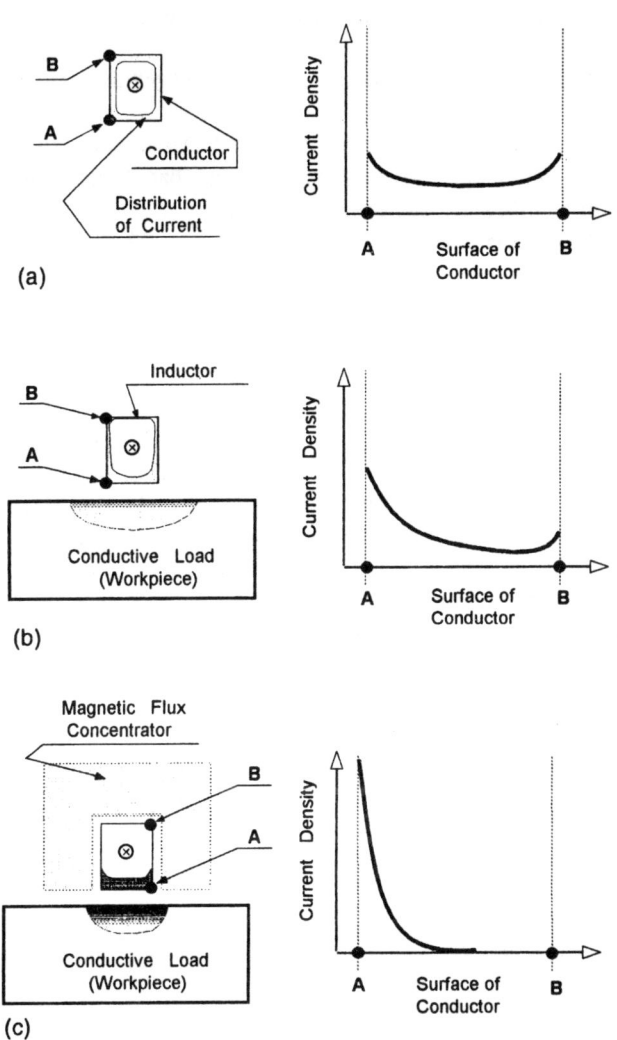

Figure 23   (a) Current distribution in a straight conductor. (b) Current redistribution due to the proximity effect. (c) The slot effect.

of the source current (Figure 23b). As shown above, due to Faraday's law (Equation 7), eddy currents induced within the workpieces have a direction opposite that of the source current of the inductor. Therefore, due to the proximity effect, the coil current and workpiece eddy currents will concentrate in the areas facing each other (Figure 23a). This is the second factor that causes a current redistribution in an induction heating system as shown in Figure 13.

Figure 24 shows how the electromagnetic proximity effect produces different heating patterns. A carbon steel cylinder is located nonsymmetrically inside a single-turn inductor. If the cylinder is static (does not spin), then two different patterns will develop in its cross section. The appearance of these patterns is caused by a difference in the eddy current distribution in the cylinder. As shown in Figure 24, the eddy currents have a higher density in the workpiece area where the coil–workpiece air gap is small ("good coupling"). Therefore, there will be intense heating due to the Joule effect. As a result, the heat pattern will be relatively narrow and deep. A lower frequency will result in a wider heat pattern and vice versa.

In the area with the larger air gap ("poor coupling") the temperature rise will not be as significant as in the case of good coupling. Also, the heat pattern will be much wider and more shallow.

In the case of an unequal coil–workpiece air gap, an almost identical heat pattern can be obtained by rotating the workpiece.

An understanding of the physics of the electromagnetic proximity effect and skin effect is important not only in induction heating but also in power supply and bus bar design. The proper design of a bus bar network will significantly decrease its impedance.

There is another electromagnetic effect that is related to the proximity effect. This is called the "slot" effect.

C. Electromagnetic Slot Effect

When we discussed the proximity effect, we first introduced the current distribution in a stand-alone conductor (Figure 23a) and then observed the current redistribution when a conductive load (workpiece) was located near this conductor (Figure 23b). As shown in Figure 23b, a significant part of the conductor's current will flow near the surface of the conductor that faces the load. The remainder of the current will be concentrated in the sides of the conductor [141].

Continuing our study, let us locate an external magnetic flux concentrator (e.g., "C"-shaped laminations) around this conductor as shown in Figure 23c. As a result, practically all of the conductor's current will be concentrated on the surface facing the workpiece. The magnetic concentrator will squeeze the current to the "open surface" of the conductor, in other

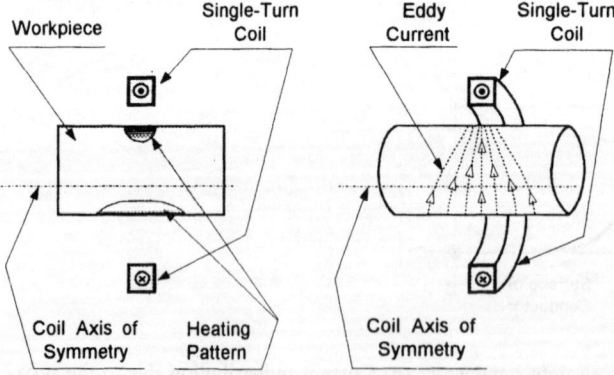

Figure 24 Proximity effect in nonsymmetrical single-turn inductors.

words to the open area of the slot. It is necessary to mention here that the slot effect will also take place without the workpiece. In this case, the current will be slightly redistributed in the conductor, but most of it will still be concentrated in the "open surface" area. This effect always occurs when there is a conductor located within the magnetic slot. The actual current distribution in the conductor depends on the frequency, magnetic field intensity, geometry, and electromagnetic properties of the conductor and the concentrator.

Slot and proximity effects play a particularly important role in the proper design of coils for selective induction hardening including channel, hairpin, odd-shaped, spiral-helical, and pancake types of inductors.

The slot effect is widely used not only in connection with induction heating but also in the design of other industrial machines such as motor generators and AC and DC machines.

D. Electromagnetic Ring Effect

Up to now we have discussed current density distribution in straight conductors. One such conductor, a rectangular bus bar, and its current distribution are shown in Figure 25. If that bar is bent to shape it into a ring, then its current will be redistributed also. Magnetic flux lines will be concentrated inside the ring, and therefore the density of the magnetic field will be higher inside the ring. Outside the ring, the magnetic flux lines will be disseminated. As a result, most of the current will flow within the thin inside surface layer of the ring [4]. As one can see, this ring effect is somewhat similar to the proximity effect. Figure 26 also shows the appearance of the electromagnetic ring effect in cylinders. As one can see, this effect leads to a concentration of current on the inside surface of the induction coil. The ring effect takes place not only in single-turn inductors but also in multiturn coils. Therefore, it is the third electromagnetic effect that is responsible for the current distribution in an induction system shown in Figure 13.

The appearance of the ring effect can have a positive or negative effect on the process. For example, in conventional induction heating of cylinders, when the workpiece is located inside the induction coil this effect plays a positive role because in combination with the skin and proximity effects it will lead to a concentration of the coil current on the inside diameter of the coil. As a result, there will be close coil–workpiece coupling, which leads to good coil efficiency.

The ring effect plays a negative role in the induction heating of internal surfaces (so called I.D.-Heating), where the induction coil is located inside the workpiece. In this case, this effect leads to a coil current concentration on the inside diameter of the coil. This makes the coil–workpiece coupling poor and therefore decreases coil efficiency. However, despite the ring effect, the proximity effect here tends to move the coil current to an outside surface of

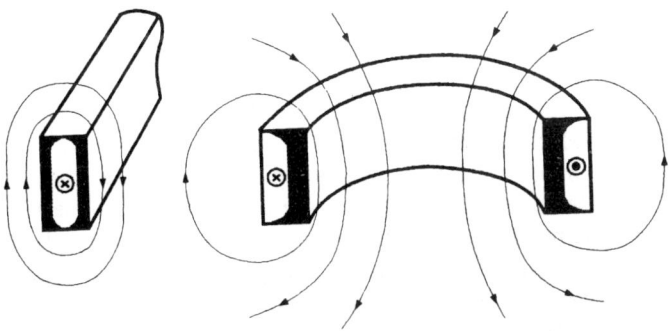

Figure 25   Ring effect in rectangular conductors.

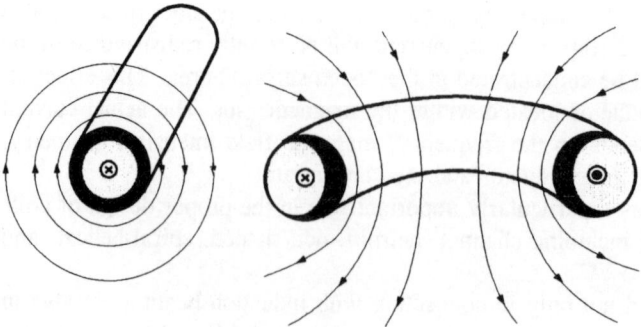

Figure 26 Ring effect in round conductors.

the coil. Therefore, the coil current distribution in such applications is a result of two counteracting phenomena: the proximity and ring effects. It should be mentioned here, that in order "to help" the proximity effect dominate the ring effect, in a great majorigy of I.D. Heating applications, a magnetic flux concentrator is located inside of the coil. This allows a slot effect to appear and "assist" the proximity effect to increase the coil efficiency and overcome the ring effect. The ring effect plays an important role in power supply design also. Because of this effect, the current is concentrated in areas where bus bars are bent, which leads to undesirable overheating of certain areas of the bus bar. To avoid local overheating it is necessary to take this effect into account when designing the cooling circuit for the bus bar.

E. Heating and Cooling During Induction Heat Treatment

Up to this point we have been describing the electromagnetic process that takes place in induction heating. This must now be combined with the generation of heat that is produced by the Joule effect in the workpiece (I^2R). To obtain a complete picture of the induction heating process, we must also examine the heat transfer process. To simplify this, let's discuss the features of induction heating of a carbon steel cylinder located inside a solenoidal coil. The process of induction heating of a magnetic cylinder from an ambient temperature or a temperature below the Curie point to a temperature above it has several features that should be taken into account when designing a modern induction system.

Figure 27 shows the dynamics of the change of surface and core temperatures with constant current applied to the coil. As shown above, if the temperature of carbon steel is changed, some of the properties that affect the electroheating process also change. The most significant of these are magnetic permeability, electrical resistivity, and specific heat (Figures 5, 8, and 9). At the first stage of the heating cycle the entire workpiece is magnetic, magnetic permeability is quite large, current penetration depth is very small (according to Equations 38 and 39), and therefore the skin effect is pronounced. At the same time, because of the relatively low temperature, the heat losses from the cylinder surface at this stage are relatively low. Because of the pronounced skin effect, the induced power appears in the fine surface layer of the workpiece. This leads to a rapid increase in temperature at the surface with no change at the core. Figure 28a shows the temperature and heat source distribution along the radius of the workpiece at this stage. The maximum temperature is located at the surface. This stage is characterized by intensive heating and the existence of a large temperature differential within the workpiece. As one can see from Figure 28a, the temperature profile does not match the heat source profile because of thermal conductivity λ, which spreads the heat from the surface toward the core.

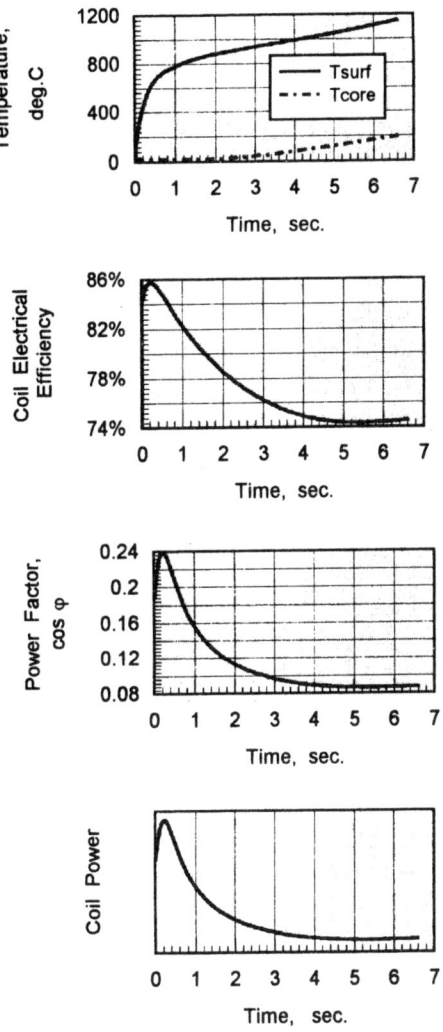

Figure 27   Change of coil parameters with time during surface hardening of carbon steel cylinder. (Coil current is constant.)

During this stage, the electrical efficiency (Figure 27b) increases due to an increase in electrical resistivity ρ of the metal with temperature (Figure 8). At the same time, the magnetic permeability μ remains high, and a slight reduction of μ does not affect the rise in electrical efficiency. After a short time, electrical efficiency reaches its maximum value and then starts to decrease. The surface will reach the Curie temperature first, and after that the heat intensity at the surface will significantly decrease. This will take place primarily for two reasons:

1. Specific heat has its maximum value (a peak) near the Curie point (Figure 5). The value of the specific heat denotes the amount of energy that must be absorbed by the metal to achieve the required heat. Therefore, that peak leads to a decrease in heat intensity in the surface.

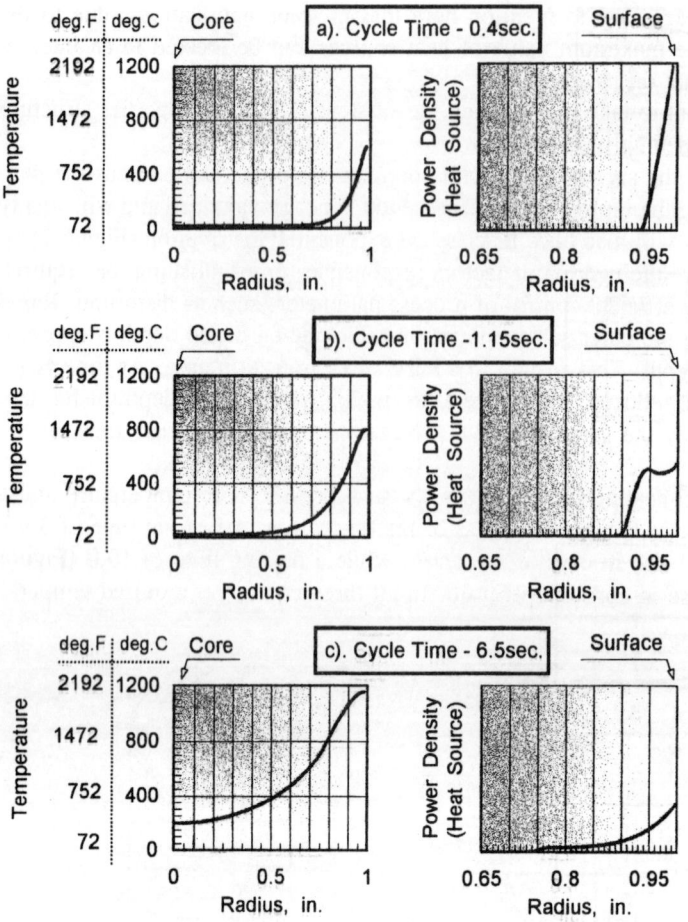

Figure 28 Temperature and power density (heat source) profiles at different stages of induction heating of 2 in. cylinder (frequency is 25 kHz).

2. Steel in the surface area loses its magnetic properties and μ drops to 1. As a result, the surface power density and heat sources will decrease also.

Figure 28b shows the temperature profile and heat source distribution along the radius of the cylinder after the surface temperature passes the Curie point (second stage). At this stage, the electrical resistivity of the carbon steel increases approximately two- to threefold compared to its value in the initial stage. A decrease of μ and an increase of ρ cause a six- to tenfold increase in penetration depth from its value in the initial stage [93]. Most of the power is now induced in the surface and the internal layers of the workpiece. This stage can be characterized as the biproperties stage of steel. The workpiece surface becomes nonmagnetic; however, the internal layers of the bar remain magnetic. This stage takes place until the thickness of the nonmagnetic layer is less than the penetration depth in hot steel. Heat sources have a unique waveshape (waveform) that is different from the classical exponential distribution. Figure 28b shows that at the cylinder surface there is a maximum of the heat sources. Then the heat sources decrease toward the core. However, at a distance of 1.4 mm from the surface, the heat sources start to increase again. This takes place because of the remaining magnetic properties of the steel at this distance. It is necessary to mention here that in some applications, due to the biproperties phenomenon, the maximum value of heat sources can be located in an internal layer of the workpiece and not on its surface.

It is particularly important to take into account the existence of the biproperties structure in designing the contour hardening processes.

Finally, the thickness of the surface layer with nonmagnetic properties exceeds the penetration depth in hot steel and the biproperties phenomenon is not pronounced and will finally disappear. The power density will then have its classical exponential distribution (Figure 28c).

Time and temperature are the two major factors responsible for establishing the required metal structure and providing efficient control of process parameters such as distortion. Rapid heating will tend to heat only the outer surface layer; in effect the hardened depth is approximately equal to the heated depth. This results in a very small transition area. As time is increased and power density is reduced, the temperature within the required depth of hardening will remain approximately the same; however, the temperature of internal areas of the workpiece will increase.

As seen in Figure 29a, a heating time of 1.5 s shows no rise in core temperature and a slight increase in temperature in the internal areas of the workpiece. A heating time of 3.0 s (Figure 29b) leads to a small rise in core temperature, while a heating time of 10.0 (Figure 29c) leads to a significant rise in core temperature. In all three cases, the required tempera-

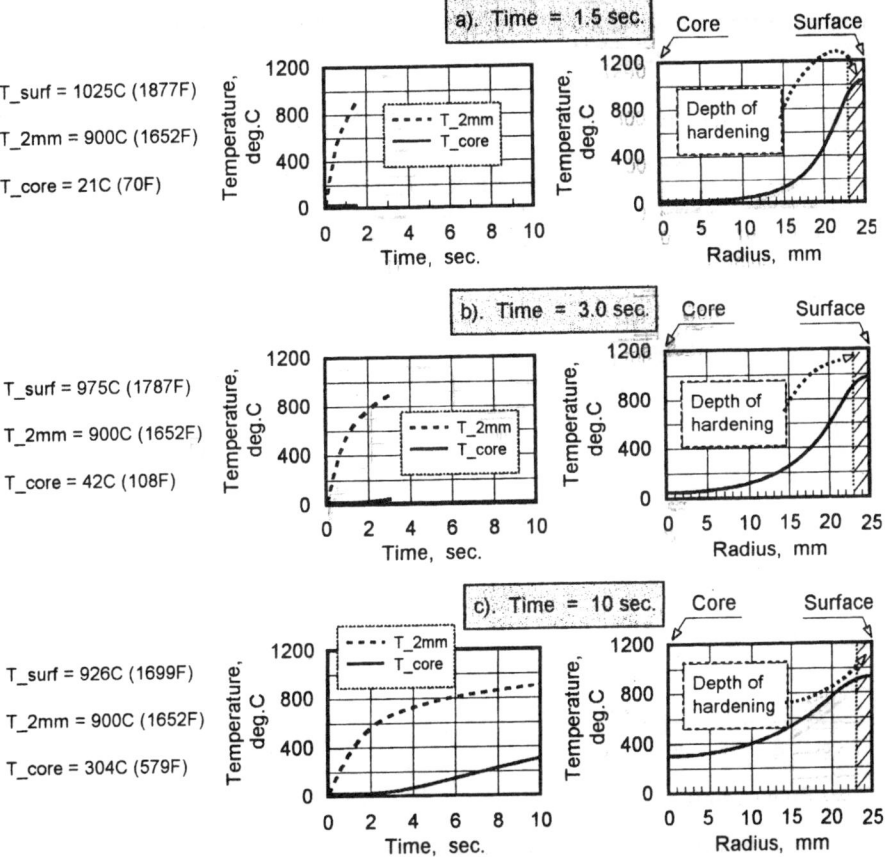

Figure 29 Temperature profiles during different stages of induction hardening (10 kHz). T_22 mm-internal temperature (2 mm below surface).

ture at the hardened depth (2 mm from the surface of the workpiece) is approximately the same. Below the hardened depth is where the difference is seen as the temperature of internal layers begins to increase. Physics indicates that the more heat induced in the workpiece, the greater the mass heated and the greater the expansion, thus leading to more distortion.

From the point of view of decreasing part distortion, it is desirable to have the heating time as short as possible. However, there are some limitations. First, the material must reach the minimum required temperature at the depth to be hardened. If the frequency or surface power density are too high and the skin effect is pronounced when the temperature at this depth reaches the required value, then the surface could overheat or even melt. Second, as a result of the short cycle time, large temperature gradients will be present and thermal stresses can reach their critical value. Chapter 5 of this handbook consists of a detailed analysis of the cracking, distortion, and residual stresses that appear during heat treatment of metals.

As one can see, the choice of frequency is not as easy a task as it would seem to be at first glance and requires a detailed evaluation of the entire process. Practical recommendations in choosing the frequency and other design parameters are presented in the next section.

Up to now, we have discussed the heating process during induction heat treatment. However, as shown in Chapter 4, quenching is also one of the most important components of the heat treatment process. As an example, Figure 30 shows the dynamics of the induction heating of a carbon steel cylinder and its cooling during quenching [93]. As discussed later in this chapter, with induction heat treatment of carbon steel, the quenching must typically begin immediately after the required heating temperature is reached. As shown in Figure 30, after 4.1 s of induction heating the surface layer reached its required final temperature (approximately 1050°C or 1922°F). The core temperature did not rise significantly because of several factors such as a pronounced skin effect, the quite intensive heating process, and short heating time. Because of these factors the heat soak from the surface of the workpiece toward its core was not sufficient.

After the heating stage is complete, the quenching process begins. In the first stage of quenching the high temperature of the workpiece surface layer begins to lessen. Figure 30 shows that after 2 s of quenching the surface temperature will be reduced by as much as 450°C.

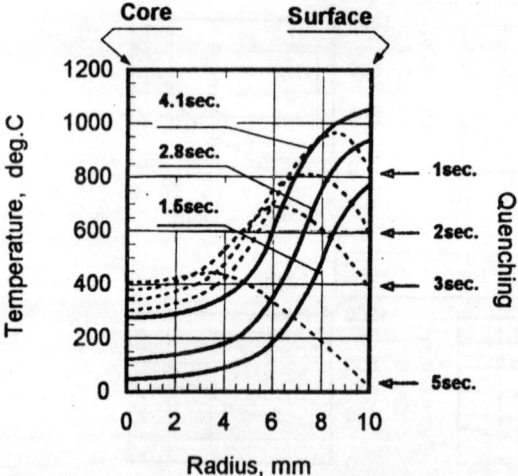

Figure 30 Dynamics of induction heating (——) of a carbon steel cylinder (20 mm O.D.) and its cooling during quenching (- - -) ($F = 10$ kHz).

This results in a workpiece surface temperature of 600°C (1112°F) At this point, the maximum temperature will be located at a distance of 3 mm inward from the surface. It is necessary to note that at this stage two heat transfer phenomena will take place. First, the surface layers will cool as discussed above. The second phenomenon is the heating (soaking) effect of the core. After 5 s of quenching the surface temperature will decrease almost to the temperature of the quenchant. At the same time, the core is still quite warm. Core temperature is >400°C (752°F). In some cases, heat treaters do not cool the part completely. After the part is unloaded from the induction coil, it is kept for some time on the shop floor at the ambient temperature. During that time, the heat of the warm core soaks toward the surface. In time, the temperature distribution within the part will equalize. In this case, the remaining heat is used for a slight temper-back, which gives the part a nonbrittle structure.

The rate of cooling during the quenching process depends on several factors, including type of quenchant, temperatures, flow rate, geometry, and material properties of the metal. The design features of quenching are discussed in the following section.

V. COIL CONSTRUCTION AND DESIGN CONSIDERATIONS

The term inductor, coil, or work coil refers to a current-carrying conductor in close proximity to a part or workpiece to be heated. The alternating current in this inductor creates a time-varying magnetic field that links the inductor and workpiece.

As shown earlier, by Faraday's law (Equation 7), this varying magnetic field causes a voltage to be induced or magnetically coupled from the inductor to the workpiece. The magnitude of this induced voltage is determined by the rate of change of the magnetic flux in the workpiece as well as the number of turns in the induction coil ($e = -N \, d\Phi/dt$). The presence of a voltage or potential difference in the workpiece leads to the flow of current and subsequent I^2R loss or power dissipated in the workpiece. This power loss depends on the workpiece resistivity and manifests itself in the heating of the workpiece. This is the unique process referred to as induction heating.

The heat-treating coil is usually mounted to a transformer for impedance matching. Impedance (load) matching in induction heat treatment are discussed in Chapter 11B. In this section we discuss the basics of inductor design and some of the thought processes it involves. The major intent of this section is to discuss some principles of coil design and to share experience and knowledge in building induction heat treatment equipment. We also discuss cases in which workpieces or parts can only be heat treated by induction and others in which induction can be considered an option along with other kinds of heat treatment. Practical advice and recommendations are presented here as well.

Inductors or coils are typically made of copper because of its high electrical conductivity and therefore low power loss. Other important reasons for using copper in coil manufacture are that copper is relatively inexpensive and has good thermal and mechanical properties. Oxygen-free high-conductivity (OFHC) copper is commonly used for high-power or high-frequency applications. Coils made of OFHC copper are expected to have a longer life than coils constructed of commercial grade copper. Inductors are constructed of tubing or solid machined blocks and are almost always water-cooled. The tubing may be as small as 1/8 in. in diameter and is usually no larger than a rectangular size of 0.5 × 1.5 in. Tubing can be wrapped or brazed together to produce wider turns. It must be large enough to permit adequate flow of water for cooling. In cases where water passages will not allow adequate flow, a high-pressure booster pump is required. With machined coils, water passages must be smooth and free of burrs.

The formula that expresses adequate flow in the coil is

$$\text{gpm} = \frac{PK_1 K_2}{K_3 \, \Delta T} \tag{40}$$

where gpm = gallons per minute, P is total coil power, kW; K_1 is a tubing coefficient (for the great majority of high-frequency induction heat treating applications $K_1 = 0.5$); $K_2 = 3415$ is a conversion constant that is derived from Btu/kWh; K_3 is a conversion constant that represents the heat capacity of water (typically $K_3 = 500$); and ΔT is the permissible temperature rise in the cooling water, normally 40°F or less, to allow for proper cooling of the copper.

ΔT can be measured by installing an in-line thermometer as close to the water outlet as possible. However, it is necessary to keep in mind that some coil designs in heat treatment applications are dictated by the required heat treatment pattern. Therefore, there could be situations where adequate water flow is not obtained, thus the coil life will be limited.

In past years, induction heat treatment equipment was cooled by using process water from cooling towers or from a local source such as city water. Water quality would vary widely from facility to facility. It was found that electrolysis can occur in power equipment with high DC voltage potentials in the water paths, which can result in corroded water pipes and plugged circuits. This problem is compounded greatly by the presence of any ferrous metals in the water. Thus there was a need for a closed-loop water system, in which water quality can be controlled very closely. The entire loop can be constructed of nonferrous metals. Once the closed-loop system is charged with nonconductive, acid-free water it will stay relatively nonconductive with very little maintenance. Sometimes slightly alkaline water can be used.

In applications that require a water-recirculating system, that system can be bolted directly to the power supply or it can be a stand-alone unit. It may also cool more than one induction system, including coils and solid-state power supplies. Usually closed-loop water-recirculating systems have a plate heat exchanger, which allows heat to be removed quickly and efficiently. Typically, the heat exchanger is made with carefully selected, well-engineered components that resist corrosion and ion contribution, ensuring a long life and trouble-free operation. Plumbing requires a special consideration. Ferrous metals or aluminium should not be used for plumbing. Copper, brass, PVC or certain types of stainless steel are good choices for pipes, tubes, hoses, etc. Because ionized water or any magnetic particles in the system can cause electrolysis or arcing, deionized water is used to protect the electronic devices within the power supply. However, in newly designed power supplies (i.e., IGBT and FET styles) the water does not come in contact with high AC voltage potentials, thus eliminating the electrolysis problem. The water system, if used, should be sized for the entire induction system (i.e., based on the total heat dissipation required). Two of the most popular compact closed-loop water-cooling and -recirculating systems are shown in Figure 31 along with some data on three models of each [95,96].

Coil-cooling design is a very important part of induction heater design. Because of electromagnetic phenomena discussed earlier (skin effect, proximity effect, and ring effect), the current distribution in the conductors can be significantly nonuniform. Therefore, in some applications the location of cooling is critical. It is obvious that the coil cooling should be as close as possible to the heating face. The heating face wall thickness should increase as the frequency decreases. This fact is directly related to the penetration (reference) depth of current in the copper and holds for both solid machined and tubing coils. For quick estimation of the penetration depth in the copper, δ_1 can be calculated with the formulas

$$\delta_1 = \frac{70}{\sqrt{F}} \quad \text{[mm]} \qquad \text{or} \qquad \delta_1 = \frac{2.75}{\sqrt{F}} \quad \text{[in.]} \tag{32}$$

where F is frequency, Hz.

(a)

STAND ALONE WATER SYSTEM

MODEL NO.	PLANT WATER FLOW (GPM)	OUTPUT WATER FLOW (GPM)	HEAT DISSIPATION (BTU/HR)	FLOOR PLAN DIMENSIONS
46P0210	37	100	200,000	29" x 53"
46P0410	73	100	400,000	29" x 53"
46P0815	146	150	800,000	29" x 53"

(b)

Figure 31A INDUCTOHEAT's stand alone closed-loop water cooling and recirculating systems.

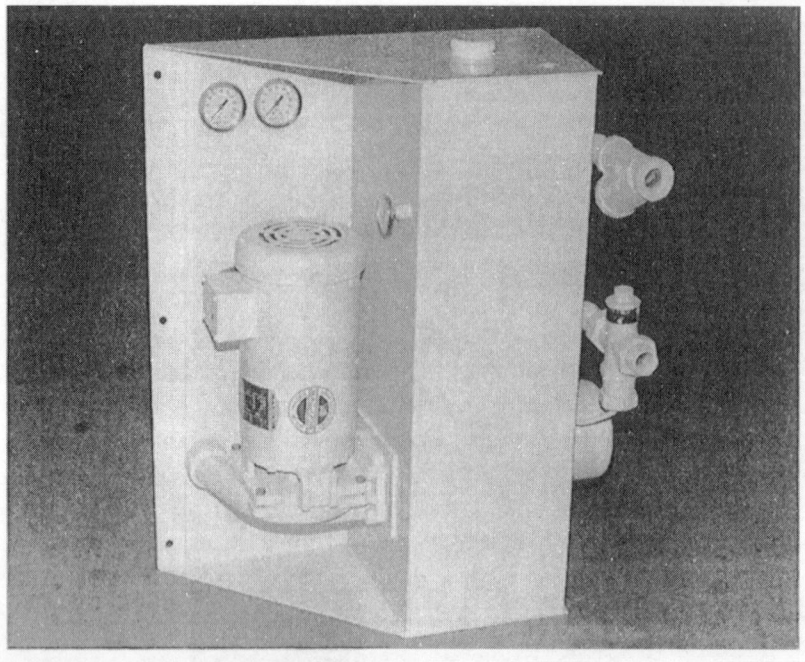

(a)

BOLT ON STLYE

MODEL DESC.	MINIMUM FLOW (GPM)	HEAT DISSIPATION (BTU/HR)	FLOOR PLAN DIMENSIONS
UNICOOL	10	90,000	12" x 23.5"
UNICOOL	25	170,000	12" x 23.5"
UNICOOL	40	265,000	12" x 23.5"

(b)

Figure 31B INDUCTOHEAT's compact closed-loop water cooling and recirculating systems.

The most effective thickness of the conductive part of the tubing wall (d_1) can be calculated as $d_1 \cong 1.6\delta_1$. A tubing wall of the induction coil smaller than $1.6\ \delta_1$ results in a reduction in coil efficiency.

In some cases the tubing wall may be thicker than the calculated penetration depths. This is because it may not be mechanically practical to use a tubing wall thickness of 0.01 in. (e.g., in the case of high or radio frequency). Some guidelines for wall thickness are shown in the following table (where RF = radio frequency, af = audio frequency, and lf = low frequency).

Copper wall thickness (in.)	Frequency (kHz)
0.032–0.048	450–50 (RF)
0.065–0.090	25–8.3 (af)
0.125–0.156	10–3 (af)
0.156–0.250	3–1 (lf)

As shown above, the penetration depth is the actual depth at which the current flows in the copper surface. The rest of the coil serves other mechanical purposes such as cooling, quench pocket design, and support against mechanical flexing caused by electromagnetic forces. At higher frequencies, coil currents are typically lower. As frequency is lowered, more attention must be paid to coil support and brazed joints. There is also more vibration at lower frequencies, especially at the turns near both ends of the multiturn solenoidal coil or split-return inductor. Nonmagnetic metal studs held together with an insulator can be added for support. Brazed joints and copper may work harder and develop fatigue from the on–off cycling of power. The clearance in silver solder joints should be held to a minimum. The silver should flow from capillary action in joints that are critical. The areas that do not participate in the actual heating can be less precise. These include cooling tubes, covers for water pockets, studs for coil support, and any other items that are not expected to carry current. Commonly used silver solder contains 35–45% silver. Experience in using silver solder with this silver content has shown that it flows well and has relatively low electrical resistance.

A.   Foreword to Inductor Design

Before the inductor design can begin, some design parameters must be established. The first would be the style of inductor. Several things must be analyzed such as current path, workpiece geometry, cycle time, power, and frequency. Electric current, like water in a stream, takes the path of least impedance. This is true in the workpiece as well as the inductor. The workpiece must provide a complete path for current (Figure 32). Care must be taken to avoid current paths that cancel each other (Figure 33, left), because the current cancellation can lead to a significant decrease in electrical efficiency.

Typically, frequencies used in surface-hardening applications result in a pronounced skin effect, which prevents eddy currents induced within the workpiece from cancelling each other. Therefore, in surface hardening, current cancellation typically does not take place.

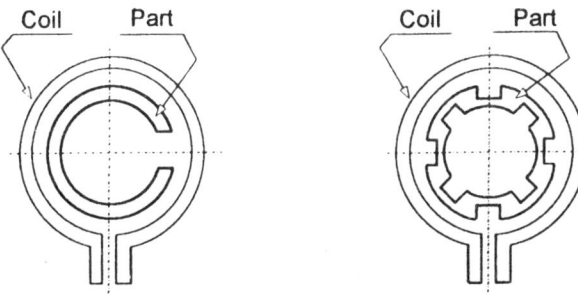

Figure 32   C-shaped and slotted parts.

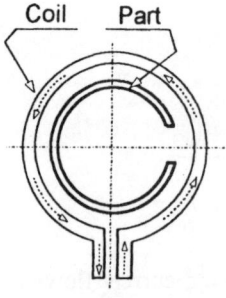 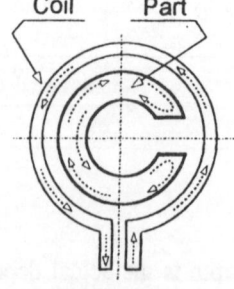

Workpiece (part) is too
thin compared to current
penetration depth (there
is current cancellation)

Workpiece (part) is thick
enough compared to current
penetration depth. There is
a distinguished current path
(no cancellation of current)

Figure 33 Current cancellation in induction heating of C-shaped parts.

In heat treatment applications such as through hardening, tempering, normalizing, and annealing, applied frequencies are much lower than in surface-hardening applications, and care should be taken to avoid current cancellation. Table 5 lists the recommended frequencies for induction heating of steel cylinders above the Curie temperature.

If the workpiece has an irregular shape (e.g., C-shaped tubes, odd-shaped parts, or slotted cylinders, Figure 32), the eddy current will flow on the inside area of the part in order to provide an uninterrupted current loop and current cancellation can take place. The basic rule of thumb that will allow one to avoid current cancellation is that the current penetration depth should be no more than one-third of the thickness of the current-conducting path.

In some applications, it is very effective to use a dual-frequency design. This requires the use of a low frequency during the stage when workpiece surface layers retain their magnetic properties (below the Curie point). In the second stage, when the workpiece becomes nonmagnetic and penetration depth is increased up to 2–5 times, it is more efficient to use a higher frequency.

Workpiece geometry and hardness patterns are two of the major factors that can determine the inductor shape and style. For instance, some patterns may be obtained only with a single-shot inductor. Along the same line of thinking, the desired cycle time may be obtained only with a particular style of inductor. The coil geometry is affected by the frequency also.

In induction surface hardening the required frequency is primarily dictated by the hardened depth and hardness pattern and is directly related to the reference depth. As a basic rule of thumb, the required frequency can be found from the condition

$$\left(\frac{4}{X_{hd}}\right)^2 < \text{frequency} < \left(\frac{16}{X_{hd}}\right)^2$$

where X_{hd} is required hardened depth, mm.

Table 5 Frequency for Deep Heating Steel Cylinders Above Curie

Frequency, kHz	0.06	0.5	1	3	10	30
O.D., in.	>12	3.5–7.5	2.5–5	1.5–3.5	1–2	0.5–1.5

Table 6 Heating Frequency and Power Density to Obtain Various Hardening
Depths in Carbon Steel

Frequency (kHz)	Hardened depth (in.)	Power density (kW/in.^{2})	
		Low	High
450	0.015–0.045	7	12
	0.045–0.090	3	8
10	0.060–0.090	8	15
	0.090–0.160	5	13
3	0.090–0.120	10	17
	0.160–0.200	5	14
1	0.200–0.280	5	12
	0.280–0.350	5	12
Contour gear hardening			
450–200[a]	0.015–0.045	15	25

[a]A low power density preheat at 3 or 10 kHz is recommended for contour gear hardening.

Required power can be approximated by using 5–15 kW per square inch of workpiece
surface area (Table 6). High power density will produce a shallow pattern; conversely, low
power density will produce a deep pattern. When higher power density is used, the life of
the inductor is limited.

The data in Tables 5 and 6 are very useful for quick ballpark estimates of inductor parameters. The entire machine concept and material handling are based on the coil style selected for the particular application. For this reason, it is important to establish the above features before the first line is drawn on paper.

B. Typical Procedure for Designing Inductor-to-Workpiece Coupling Gaps

The first step in designing an inductor is to draw an outline of the workpiece to be heated.
This must be drawn accurately and must show the minimum and maximum case depths. Special attention should be given to features such as holes, fillets, sharp corners, snap ring grooves,
and keyways. These areas have a tendency to overheat or underheat due to heat conduction
and electromagnetic edge effects. The features of induction heating a part with such elements
are discussed later.

The inductor can then be drawn to conform to the workpiece, depending on the required
heat pattern. If the inductor is machined, the copper wall may be made thicker to allow for
some shaping of the inductor in the test or development stage. Once the coil is developed,
the wall thickness should be sized according to the reference depths. The coupling gaps should
be held to a minimum. However, they are sometimes dictated by material handling, workpiece
tolerances, and thermal expansion of the workpiece during heating. Frequency is also a consideration in establishing coil-to-workpiece coupling gaps. With higher frequency the coupling
gap is more critical and more sensitive. Any change in the coupling gap has much less effect
at lower frequency than in higher frequency applications. This is because the flux lines stay
closer to the coil at higher frequencies. In cases where high frequencies (10–450 kHz) are
used, it is extremely important that the relationship between the workpiece and the inductor
remain the same from cycle to cycle.

Figure 34 shows a comparison of the heat patterns with proper and improper gaps. In
the case of a proper coupling gap the result will be a uniform heating pattern. Changing the
coupling gap by as little as 1 mm can result in nonuniform heating. Overheating will occur in

(a)

(b)

Figure 34 Heat patterns with a proper coupling gap compared to an improper one.

the area with the larger diameter (bottom part of the workpiece). At the same time, the workpiece area with the small diameter (top part of the workpiece) will be underheated.

Choosing a proper coupling gap and keeping the necessary tool tolerances will ensure consistent results. Otherwise, severe changes can occur in the heat treatment patterns. Coupling gaps may be as small as 1/16 in. (1.5 mm) to as large as 1 in. (25.4 mm), depending on frequency and type of heating. Table 7 gives the recommended coupling gaps for three frequency ranges. However, as mentioned above, those gaps may vary for different applications.

Inductors intended to heat fillet areas such as those found on axle shafts should have a larger gap on the shaft diameter and be coupled closely to the perpendicular flange surface. Scan inductors in general should have slightly more clearance because of the possibility of workpiece warping. This possibility increases as the part length increases because more heat is put into the part and more material is being transformed into martensite. In the case of through heating, the coupling usually has to be larger than recommended in Table 7. There

Table 7 Typical Coil-to-Workpiece Gap with Various Heating
Frequencies

	Frequency (kHz)		
	1–3	10–25	50–450
Coupling gap			
mm	6–3	3–2	1.5
in.	0.23–0.12	0.12–0.08	0.06

are several reasons for this. First of all, a low frequency is typically used for induction through hardening. Usually this type of inductor has a refractory cast liner to protect the coil from the high-temperature workpiece. The coil may also have a water-cooled stainless steel liner. These liners are required to protect the coil from wear and mechanical damage from moving workpieces. Because of these features, through-hardening, annealing or tempering inductors could have a relatively large coil coupling (i.e., more than 0.5 in.). These types of coils are typically multi-turn, relatively long, and use lower power density than surface hardening coils.

C. Mounting Styles

The electrical connection is sometimes referred to as the mounting foot. There are several basic styles for this connection. A standard keyed foot has been developed to ensure good electrical contact. It is usually fastened with four 3/8-16 stainless steel bolts. The bolts should be tightened to approximately 35–40 ft lb. Overtightening of these bolts will pull the threads from the soft copper, and undertightening will not produce a good electrical contact. Stainless steel threaded inserts should always be used where the electrical contact is made. When fast or frequent coil change-over is required, a quick-change type of mounting foot may be recommended. This can be in the form of a toggle clamp (Figure 35) [97], dovetail, or pneumatic. It can also be keyed for accurate location. One of the main advantages of this design is that no tools are required. The major shortcoming is that there are power and frequency limitations. For example, the toggle clamp and dovetail type should be limited to 300 kW with a frequency range of 3–450 kHz. All electrical contact areas must be clean and free of nicks or burrs, and they therefore require additional care and cleaning.

D. Spray Quench

The spray quench design can be just as important as the coil design. The quench must be designed for rapid heat removal to develop the desired hardness and metal structure. Nonintensive quenching results in soft pearlite-type and bainite-type metal structures. Uneven quenching makes the distortion problem more pronounced. The intensity of quenching depends upon the flow rate, the angle at which the quenchant strikes the workpiece surface, temperature, purity and type of quenchant. There are three main considerations for quench design: heat treatment pattern requirements, style of inductor, and workpiece geometry. Spray quenching works best if the workpiece is rotated during the quenching operation. This will ensure uniformity in quenching. The point of impingement will have a faster rate of cooling than an area that is quenched from a flood of quenching fluid. Rotating the workpiece simulates a constant impingement rather than many small impingements. Small orifices are required to agitate quench, to prevent steam pockets from developing.

Various fluids, such as aqueous polymer solutions, salt water, oils, and even straight water, are used in quenching. Features of quenchants were discussed in detail in Chapter 4. Here we will just mention that straight water is seldom the best choice for quenching. Typically,

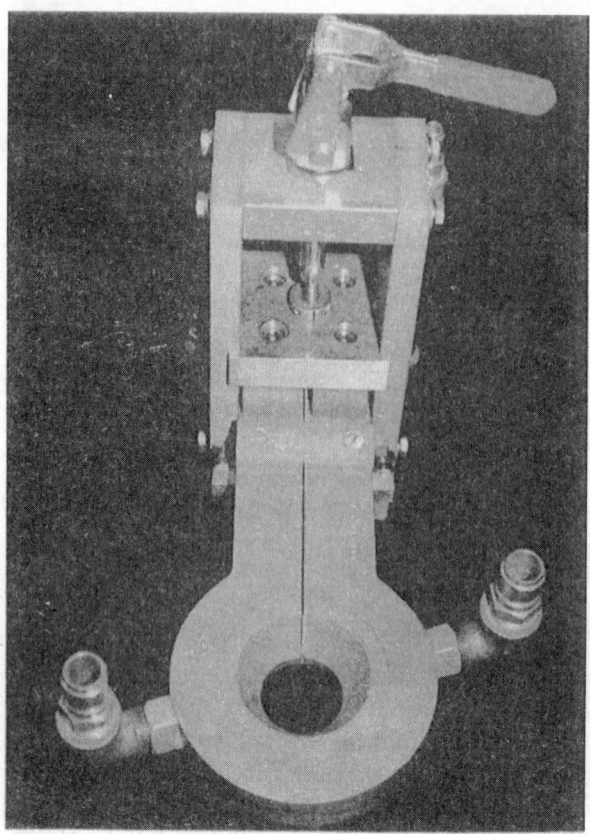

Figure 35 Quick-change coil design.

steels with very low hardenability require the use of straight water. However, the use of water results in rust, and the workpiece distortion and cracking also become more pronounced.
Polymer fluids are usually used as quenchants in induction heat treatment. Using a small percentage of polymer can make quenching even faster than straight water. The polymer helps
to eliminate a vapor barrier that can occur with straight water.

The quench pattern is generally conformed to the part, just as the inductor is. When using a single-shot inductor it is preferable to quench from two sides of the workpiece. The
quench holes should be placed facing the part at 3/16–1/4 in. intervals and have a staggered
pattern. The orifice size is related to the shaft or workpiece diameter.

Shaft diameter (in.)	Orifice size (in.)
0.25–0.50	0.046–0.063
0.50–1.50	0.063–0.094
>1.50	0.125–0.156

For inductors where quench holes are drilled through the heating face, the holes should
not occupy more than 10% of the heating face area. To control the direction of the stream,
the wall thickness must be at least 1.5 times the orifice diameter. The inlet/outlet ratio should

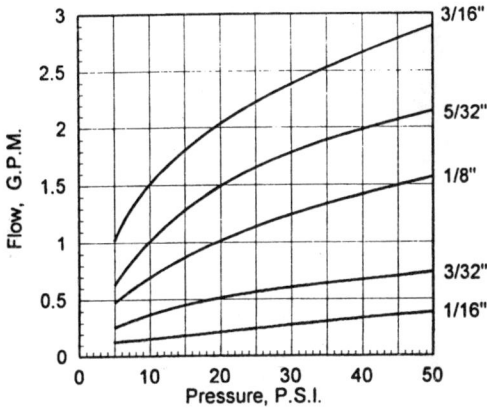

Figure 36 Flow of a single-quench orifice. Example: At 15 psi, with seventy-two 1/8 in. holes, gpm = 0.87 × 72 = 62.64.

be at least 1:1. Baffles or deflectors should be added in the area of the inlet. These will help to obtain a more uniform pressure distribution. The baffle hole area should also have at least a 1:1 ratio. The total quench flow may need to be estimated for pump sizing and flowmeter sizing. This can be approximated by finding the flow of one quench hole from the chart shown in Figure 36. The flow found from this chart should be multiplied by the number of holes in the inductor.

The temperature of the quenchant can be critical. A great majority of induction hardening applications require quench temperatures of 24–35°C (75–95°F). Therefore, with start-up of equipment, some warm-up time can be required. Otherwise, start-up with cold quenchant can cause cracking and part distortion. On the other hand, the use of hot quenchant results in a reduction in hardness.

E.   Inductor Styles—Scan Inductors

There are many types of inductors. Three of the most common are scanning, static, and single-shot. Scanning is sometimes called progressive heating because either the workpiece moves through the inductor or the inductor moves and the workpiece is stationary [98–102]. With a static inductor, neither the workpiece nor the inductor move, except when the workpiece is moved in and out of the inductor. Thus during heating the workpiece remains in a static position. Single-shot inductors are also said to be static in the same regard [103]. The main difference lies in the inductor design and construction. Single-shot inductors are discussed in the next subsection.

The main advantage of the scanning type of inductor is its flexibility in running various lengths of parts. This type of inductor provides a repeatable, easily automated process that can quickly adapt to new heat treatment tasks and be easily integrated into the work cell.

Scan inductors are usually single-turn (Figure 37) or have two or three turns. The number of turns is determined by the coil calculation. From another perspective, the number of coil turns plays an important role in the tuning of the heat station and coil with the power source. This problem is discussed in detail in Chapter 11B.

The more turns there are, the faster the scan rate will be. Single-turn inductors are used where a sharp pattern runout is required, as is often the case, adjacent to a snap ring groove. Single-turn inductors can be machined from a solid cooper bar, thus making them very rigid and durable.

Figure 37   Scan inductor.

Quench holes can be integrated into a single-turn scan inductor. This type of inductor is sometimes called an MIQ inductor (machined integral quench; see Figure 38). The quench spray should hit the part approximately 3/4 in. from the heating face and be angled down to prevent the quench from washing back into the inductor. This dimension will vary with different types of steel and with different scan rates [98–102]. An additional quench follower must be added to inductors designed to run parts with varying diameters and/or grooves. This ensures a good quench in the fillet and radius areas that otherwise might be shadowed from the primary quench. A quench follower may also help eliminate the "barber pole" effect. A narrow heating face is required for a sharp cutoff in pattern because of the small flux field and higher power density it provides. A wider heating face or more coil turns can be used where a faster scan is desired. The main disadvantage to this is that it will produce a gradual pattern runout and may not meet some pattern specifications.

Scan inductors that are intended to heat a fillet area as well as the shaft and the perpendicular flange must be focused into the fillet area. They should also have flux intensifiers (flux

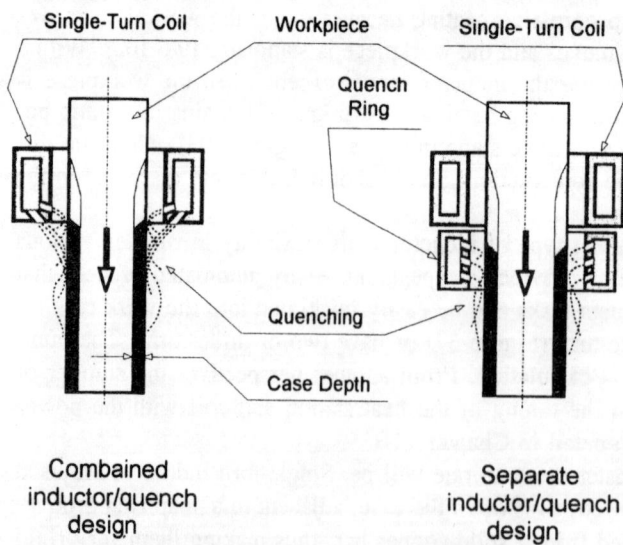

Figure 38   Combined and separate inductor/quench design.

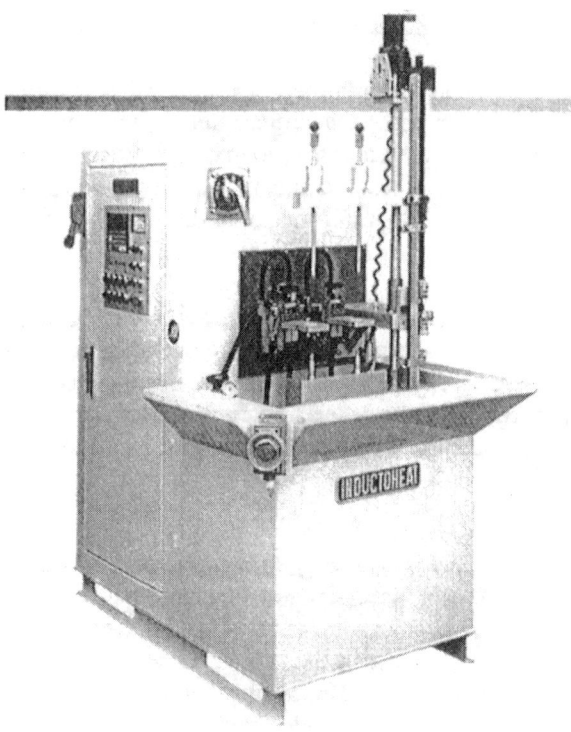

Figure 39 Statiscan unitized vertical dual spindles scanning system.

concentrators) to aid in focusing into the fillet. These are critical applications that require careful design to make the inductor work because the current will try to take the shortest path and stay in the shaft area. Therefore, all efforts must be made to focus it into the fillet. Typically, lower frequencies work better in this application. Scan inductors have a good chance of working without coil modifications. This is due to the flexibility that results from being able to vary power, scan rate, and delay of quench to achieve the desired hardness pattern. They are also able to run various lengths. A disadvantage of this inductor type is that it is generally limited to the shaft type of parts. Scanning is recommended for shafts where power is limited and/or shaft length is over 12 in.

Applications may call for horizontal or vertical scanning systems. As an example, Figure 39 shows a Statiscan vertical induction scanner with dual spindles. This system has been used for hardening and tempering a wide variety of parts such as shafts, rasp bars, axles, and hubs. Parts are located and held between centers while they are scanned during the cycle. The part is rotated during the cycle by using a variable-speed motor. This particular model is provided with several power ratings: from 50 kW to 250kW at frequencies from 3kHz to 200 kHz. A user-friendly keyboard on the Statiscan provides quick setup, change-over, and diagnostics capability. To simplify setup, upper tooling centers can be adjusted without tools. A selection of standard quick-change inductor mountings are also helpful in minimizing change-over time.

F.   Inductor Styles—Single-Shot

Single-shot inductors [103] are made of tubing or machined from a solid, but unlike scan inductors they produce an axial rather than radial current path. Typically, they have two horseshoe-shaped loops that join the two legs. This type of inductor requires the most care in manufacturing because of the high power densities it must accommodate. Because of this, the brazed joints must accurately mate. Tongue-in-groove joints are preferred. The flux fields can usually be shaped to produce the exact hardness pattern desired; however, altering that shape is not always easy. Sometimes a completely new inductor head must be made.

Single-shot inductors are designed for a specific heat treatment pattern on a specific workpiece. The heat cycle for single-shot inductors is usually much shorter than for other inductor styles. This leads to higher production rates. At high rates of production, a single-shot inductor typically has a shorter life than scan coils.

Electrically speaking, it is always important to keep in mind that the inductor is the weak link in an induction system and should be likened to a fuse. The inductor will fail if power is increased to the point at which water cannot adequately cool it. Additional cooling passages may be needed with single-shot coils, and a high-pressure booster pump is almost always required. Single-shot inductors should be used where the workpiece has varying diameters, radii, and fillets.

1. The Striping Phenomenon

When discussing specifics of induction hardening, it is necessary to mention an effect that, because of its complexity, is not usually discussed in induction heat treatment publications: the "striping" phenomenon. Striping typically occurs during intensive induction hardening of carbon steels where high power densities are used. Because of this effect, the workpiece area under the coil may unexpectedly start to heat nonuniformly. The striping can be seen even in the case of a single-turn coil with a conventional cylindrical load (Figure 40). Shortly after the heating cycle begins, alternating hot bright areas (bright stripes) and cold areas (dark stripes) become visible. These bright and dark stripes encircle the cylinder and thus have the shape of rings.

The barber-pole effect and the striping phenomenon have never been obtained by mathematical modeling. They have been viewed only in practical applications or during laboratory experiments on the induction heating of magnetic steels. In some applications striping suddenly occurs and then disappears. There is no single explanation of this phenomenon. The only attempt to explain it was made by Lozinskii [2] in the early 1940s. It was a very simple description based on the knowledge available to induction heat treaters at that time. It should be mentioned that Lozinskii's hypothesis concerning striping has a certain logic and can be accepted as an introduction to this effect. However, from our point of view, that explanation oversimplifies the mechanism of the striping phenomenon. Since that time, there have been

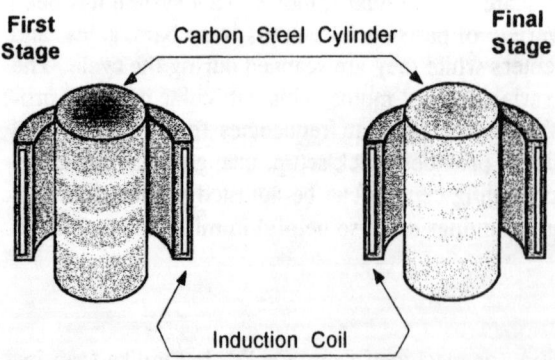

Figure 40   Striping phenomenon in induction heating of carbon steel cylinder.

no further attempts to explain this effect. Therefore, here we will briefly introduce Lozinskii's hypothesis along with our own point of view, which is based on modern experience and new theoretical knowledge accumulated at INDUCTOHEAT during the development of various induction heat treatment processes.

Assume that a magnetic cylinder is located inside a single-turn inductor (Figure 40). As a result of the electromagnetic field produced by the induction coil, eddy currents will flow within the workpiece. Due to the skin effect, these eddy currents will appear primarily in the surface layer of the workpiece located inside the coil and cause the surface temperature of the workpiece to increase.

Realistically speaking, any workpiece has certain nonuniformities, microscopic defects, impurities, and nonhomogeneities. This includes structural/mechanical nonuniformities, metallurgical nonhomogeneities, etc. As a result, different surface regions of the workpiece will be heated slightly differently. Some will reach the Curie temperature first and lose their magnetic properties. The relative magnetic permeability of these areas will dramatically drop to unity ($\mu = 1$). This leads to a significant increase in the penetration depth in those areas. The resistance of these nonmagnetic regions will drastically decrease compared to neighboring surface areas that retain their magnetic properties. As a result, the density of induced currents in the low-resistance regions will increase. This leads to an increase in power density and an increase in heat sources in these areas. At the same time, there will be a redistribution of eddy currents in the workpiece surface. Eddy currents induced in areas that retain their magnetic properties (dark rings) will have a tendency to rush to complete their loops through the low-resistance paths (bright rings). This current redistribution leads to a further heat source reduction in the magnetic areas with low temperature (dark rings) and appears as additional heat sources in the nonmagnetic areas with high temperature (bright rings). Therefore, positive feedback will occur. As a result, one can view in the workpiece a mixture of ring-shaped stripes. Hot bright stripes will alternate with the relatively cold dark stripes. Experience shows that usually the thickness of the bright and dark stripes depends primarily on the frequency and power density and equals 1–3 current penetration depths in hot steel.

Besides the current redistribution, the striping phenomenon is a result of several other electromagnetic and heat transfer effects, including the electromagnetic edge effect of joining materials with different properties (i.e., magnetic and nonmagnetic metals). The electromagnetic edge effect of joining materials with different properties (EEJ effect) occurs when two different metals are located in a common magnetic field. To simplify the study of this effect, let us consider the electromagnetic process in a conventional solenoidal induction coil with two workpieces, for example, two cylindrical billets (Figure 41). Assume that the billets have

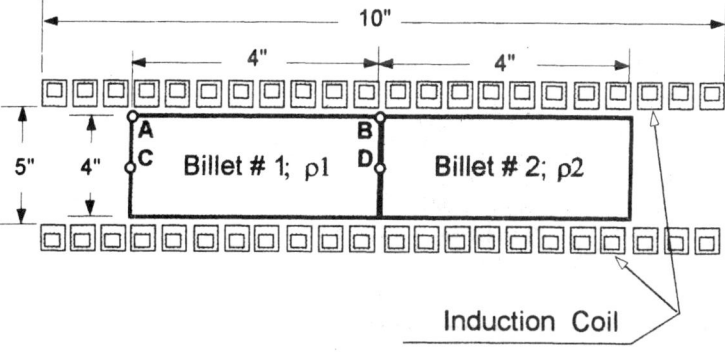

Figure 41   Sketch of the induction heating system.

different material properties (e.g., different electrical resistivity ρ or magnetic permeability μ). When two billets with different material properties are joined together and are located inside an induction coil, the electromagnetic field in the joint area (so-called transition area) becomes distorted [10,50,51]. For example, if one billet has been heated above the Curie point (has become nonmagnetic) and the other still maintains its magnetic properties, then the distribution of the electromagnetic field will be as shown in Figure 42. If the billets are long enough, then the magnetic field intensity at their central areas will be approximately the same and correspond to the coil current. At the same time, the power densities at the surface of the magnetic and nonmagnetic billets will be rather different (Figure 42).

At the left tail end of the nonmagnetic billet (billet 1) and at the right tail end of the magnetic billet (billet 2), there will be a nonuniform power density distribution due to the end effects of the nonmagnetic and magnetic workpieces. At the transition area of the billets, the field distribution is quite complicated. At the right end of the nonmagnetic cylinder (billet 1), the magnetic field intensity and power density sharply increase. At the left end of the magnetic cylinder (billet 2), those parameters sharply decrease. This phenomenon is called the electromagnetic edge effect of joining materials with different properties (EEJ effect). Obviously, this effect plays an important role in the appearance of the striping phenomenon and leads to a significant redistribution of the electromagnetic field in the area of the dark rings (which retain their magnetic properties) and in the bright high-temperature rings (which have become nonmagnetic). As one might expect, this electromagnetic effect will cause the heat source distribution to differ from the classical form that is traditionally assumed in the study and design of induction heat treatment processes.

When discussing the EEJ effect, it is necessary to mention that it also occurs when both workpieces are nonmagnetic but have different electrical resistivities (ρ). Figure 43 shows the power density distribution within billet 1 for the induction system shown in Figure 41. Both billets are nonmagnetic, but they have different electrical resistivities (ρ_1 and ρ_2). When the electrical resistivity of billet 1 (ρ_1) is three times that of billet 2, (Figure 43a), then power density is reduced in the joint area of billet 1. When $\rho_1 = 0.33 \rho_2$, (Figure 43b), there is an increase in power density (heat source).

The effect of joining materials with different properties should be taken into account when designing other induction heat treatment and induction heating systems also. For example, it may have a significant influence not only on the striping phenomenon but also on the final temperature distribution, which should be considered when designing the billet heater, espe-

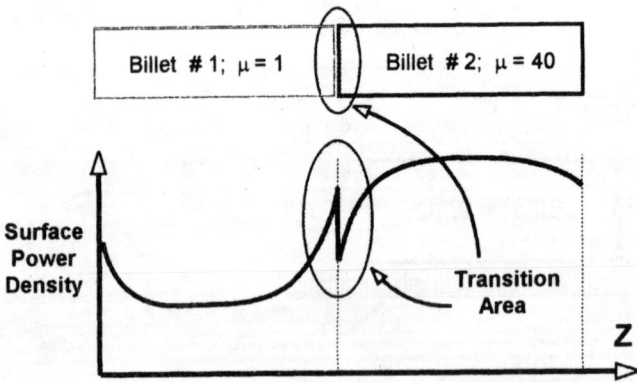

Figure 42 Electromagnetic field distortion at joint of magnetic–nonmagnetic steels.

Power Density Ratio, P / P₀

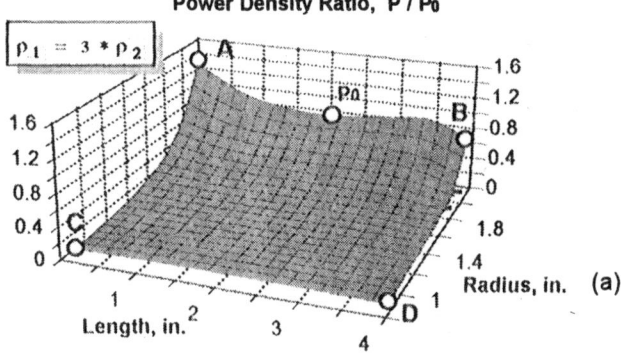

Power Density Ratio, P / P₀

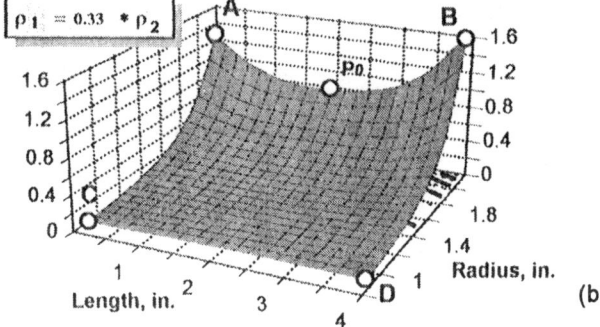

Figure 43   Power density distribution along the length of billet 1. (Frequency $\doteq$ 60 Hz; $\rho_1 = 1.1\ \mu\Omega{\cdot}$in.)

cially when the induction equipment will operate below or just above the Curie point [50,51,108].

Experience shows that striping can appear in several different ways. However, in the great majority of cases, very narrow bright stripes (rings) will appear at the beginning of the heating cycle (Figure 40). With time, the narrow stripes will widen. At this stage, the maximum temperatures will move from the center of each ring toward the edges of each bright hot ring. During the heating process, the stripes sometimes move back and forth along the workpiece surface area under the coil. When the length of the heating cycle is increased, typically the striping effect will not be pronounced and temperatures will equalize in the workpiece surface.

The appearance of the stripes depends on a complex function of the frequency, magnetic field intensity, and thermal, electrical, and magnetic properties of the steel. It can occur only when high power density is applied. If the power density is relatively low, then the temperature will equalize between the neighboring bright (high-temperature) and dark (low-temperature) rings because of the thermal conductivity of the steel.

2.   Gear Hardening

Gears are induction heated by either encircling the part with a coil or, in larger gears, heating them "tooth-by-tooth" (Figure 44). Heating tooth-by-tooth lends itself to the ability to achieve a soft core in the center of the tooth. The inductor can be designed to heat only the root and ramp of the tooth, leaving the tip soft and ductile. These are both desirable due to

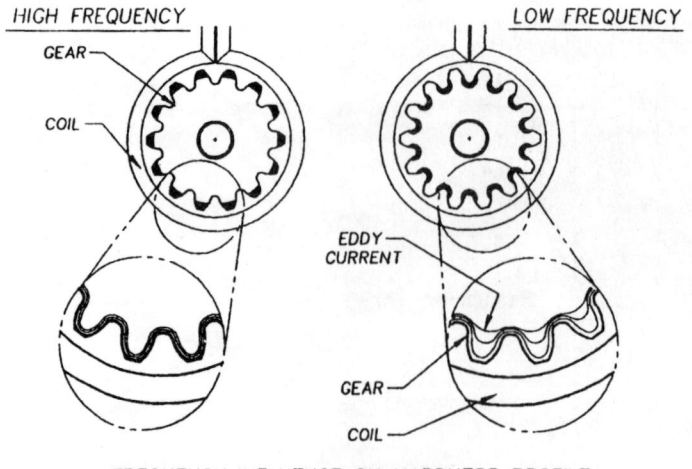

(a)   FREQUENCY INFLUENCE ON HARDNESS PROFILE
        WITH AN EN−CIRCLING INDUCTION COIL.

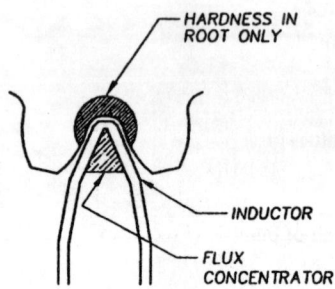

(b)   TOOTH BY TOOTH HARDENING OF GEAR

Figure 44   (a) Encircling and (b) tooth-by-tooth gear hardening.

the loads applied to the teeth. However, this process is very time-consuming and requires
constant inductor maintenance. Inductors that encircle the part can heat much faster and have
a more robust design. In the past, the disadvantage of this has been through hardening of the
tooth. For this reason, contour hardening was developed. In this process, the gear is preheated
within an induction coil (Figure 45,) [14], to a temperature determined by the process. This
is accomplished by using a medium frequency (3 to 10 kHz) and/or high frequency (30 to
450 kHz) depending on the type and size of the part. Figure 46 shows a "dual pulse" contour
hardening system designed to preheat, harden and temper in the same coil using one high frequency power supply. Figure 47 illustrates the process cycle with moderate power preheat,
short high power final hardening heat and quench followed by low power heat for temper.
Preheating ensures a reasonable heated depth at the roots of the gear, enabling the attainment
of the desired metallurgical results and decreasing distortion in some materials.

Preheating also reduces the amount of energy required in the final heat and allows the
high-powered RF current applied during the final heat stage to follow the contour of the gear
teeth (Figure 48) [14,109–114]. For the final heating stage, the frequency is selected to allow
the current to penetrate only to an exact repeatable depth. Quenching is then done to bring

Figure 45 Final-stage contour gear-hardening machine.

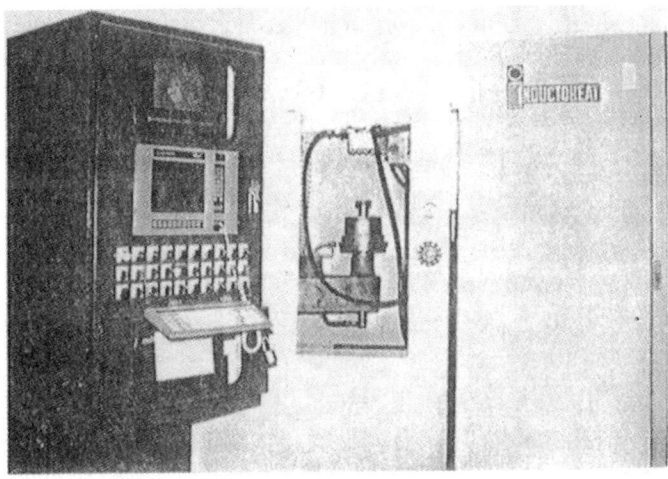

Figure 46 Dual-pulse contour hardening system.

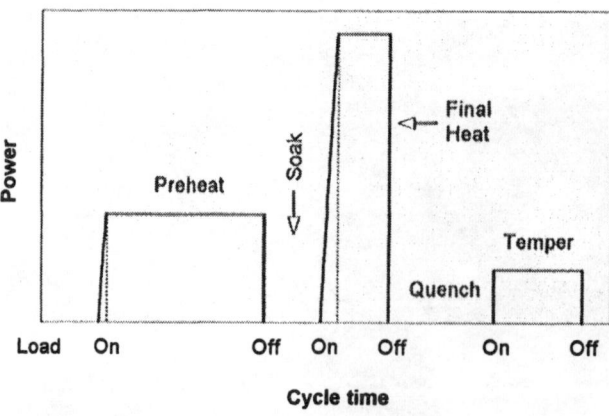

Figure 47   Dual-pulse induction hardening.

the part to ambient temperature. The part can then be induction tempered. Tempering is typically accomplished in-line with the hardening machine, usually in part of the same machine. A medium or low frequency is typically applied for gear tempering. The choice of frequency depends on the metal and the size and shape of the part.

Development, like coil design, is largely based on induction principles, results of mathematical evaluation, and experience with previous jobs. The development not only establishes cycle times and power levels, it also establishes coil contouring. When single-shot or static coils are used, it is sometimes necessary to contour the coil. As the coil is contoured, so is the flux field (Figure 49). The portion of the workpiece that is closest to the inductor will usually be in a stronger magnetic field, will have higher power density, and will therefore be heated more rapidly. Altering the power and the duration of heating also alters the flux field. A high power generally gives a long shallow pattern; conversely, a low power level will pro-

Figure 48   Contour-hardened parts.

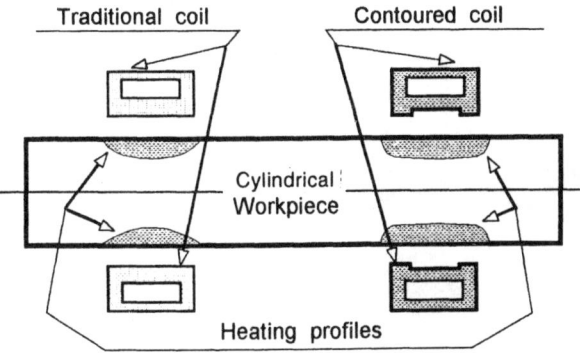

Figure 49 Temperature profiles with traditional and contoured single-turn induction coils.

duce a deep pattern. If extra length is needed on a pattern, keep the total energy (kilowatt seconds) the same and increase the power and decrease the time.

It is important to note what happens after every parameter change. Understanding the results of what did not work is just as important as understanding what did work. The evaluation of why certain results occur will speed the process along. It is suggested that a journal be kept of any development. This should have the parameter changes, a brief description of the results, and some hardness readings. Understanding the principles of induction and coil design and using modern computational techniques are imperative for a successful process development.

G.   Inductor Styles—Specialty Inductors

Some special types of inductors include I.D. coils, pancakes, split returns, clamshells, and hairpin inductors. Generally speaking, internal heating inductors are not very efficient, mainly because the current travels on the inside diameter of the inductor. This phenomenon automatically leads to the fact that the real coil coupling is larger than the actual air gap between the internal surface of the workpiece and the coil O.D. Physics of this phenomena has been previously explained in our discussion of electromagnetic ring effect.

Internal diameter (I.D.) coils almost always require the use of flux concentrators that are located inside the coil. If the frequency is less than 15 kHz, efficient heating can be provided for a workpiece I.D. greater than 1 in. (25.4 mm). Slightly smaller diameters may work with radio frequencies (Figure 50).

Pancake inductors have been used for selective surface heating of flats, discs or plates. This type of inductor has the appearance of an electric stove burner. In fact, induction cooktop stoves are commercially available, all using the pancake coil (Figure 51).

Split-return inductors offer a unique distribution of current. The center leg of the inductor carries twice as much current as the side legs. This type of inductor produces a narrow band of heating.

Clamshell inductors are generally used for crankshafts, where part geometry will not allow close coupling or for unusually shaped cams (i.e. cams for racing engines). This type of inductor has a hinge on one side and opens and closes when the part is in position.

Hairpin inductors are usually in the form of bent tubing and conform to a workpiece to heat a selected area. This type is usually used with radio frequency. Hairpin inductors are often shaped during manufacturing using the workpiece as a pattern.

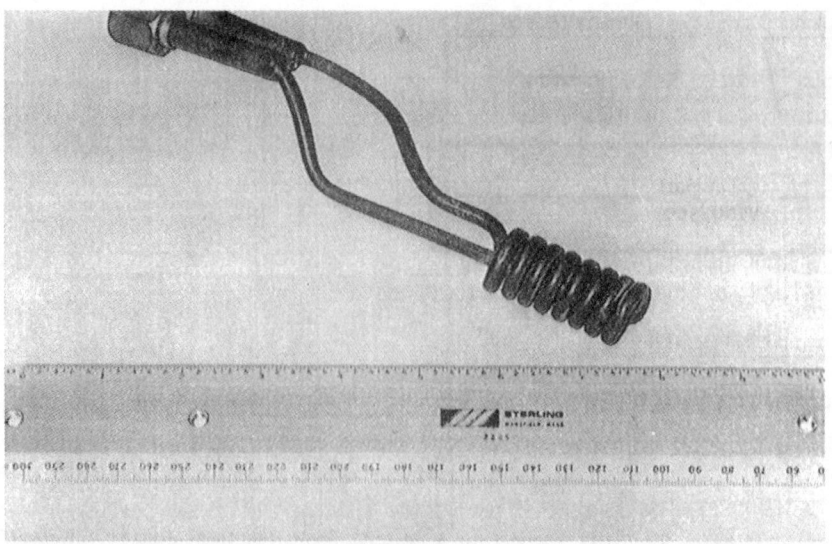

Figure 50 Inner diameter type of inductor.

H.   Temper Inductor Styles

The decision to induction temper should be carefully considered. It may not be optimal in all cases. Suprisingly, in many cases the parts can be successfully induction tempered. The main purposes of tempering have been discussed in the introduction of this chapter.

Before we discuss the features of induction tempering, let us examine how stresses appear during induction hardening. Detailed analyses of the nature of residual stresses are presented in References 6,12,19, and 146–151. Because of certain aspects of induction hardening, the mechanism of formation of residual stresses here is slightly different than in other

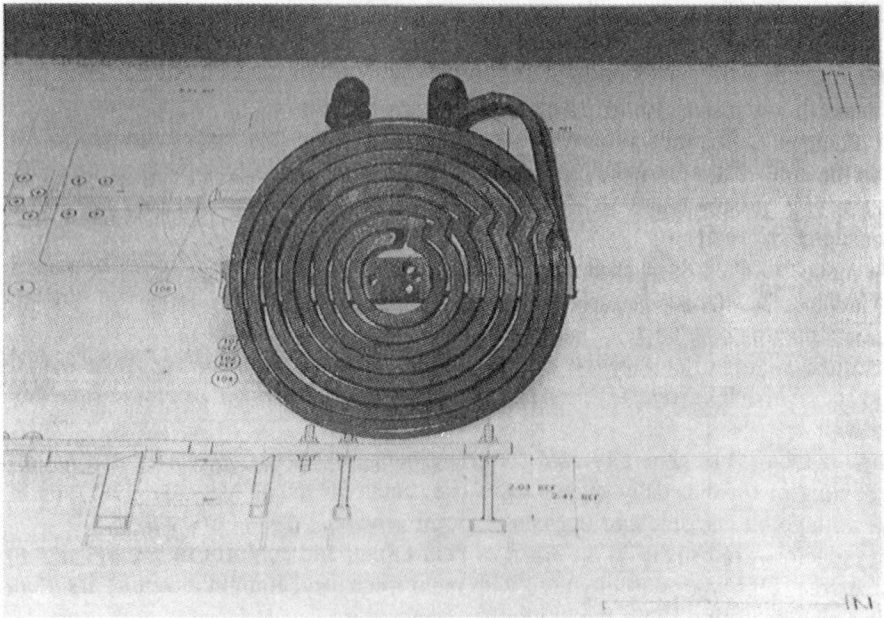

Figure 51 Pancake inductor.

heat treatment processes such as carburizing. Substantial work in developing a commercial induction tempering system and study of the mechanism of residual stresses has been done by HWG (Germany) [149–151], VNIITVCh (Russia) [12] and Colorado School of Mines. Here we give only a short description of the features of induction tempering primarily based on materials obtained from HWG and many years of experience developing induction tempering systems at INDUCTOHEAT, Inc.

Generally speaking, there are two different types of stresses: thermal stresses and transformation stresses. Thermal stresses are caused by different magnitudes of temperature and temperature gradients. Transformation stresses occur due to microstructural changes taking place as a result of the formation of austenite or martensite. The total stress is a combination of both components. At different stages of heat treating the impact of both components into total stresses is different.

Figure 52 is a sketch of an induction-heated part (i.e., a steel cylinder). As a result of heating, the section of the part located under the coil will try to expand. At the first stage of the heating cycle, the temperature of the workpiece is relatively low (less than 500°C or 932°F). At this stage, carbon steels have a nonplastic condition and cannot easily expand. As a result of that, stresses build up within the workpiece. The temperature rise will result in the appearance of high compressive stresses at the surface (Figure 53). In the temperature range 580–750°C (1076–1382°F) the steels become more elastic and start to expand. As a result, the stresses start to decrease, and since the temperature exceeds 850°C (1562°F) the steel attains

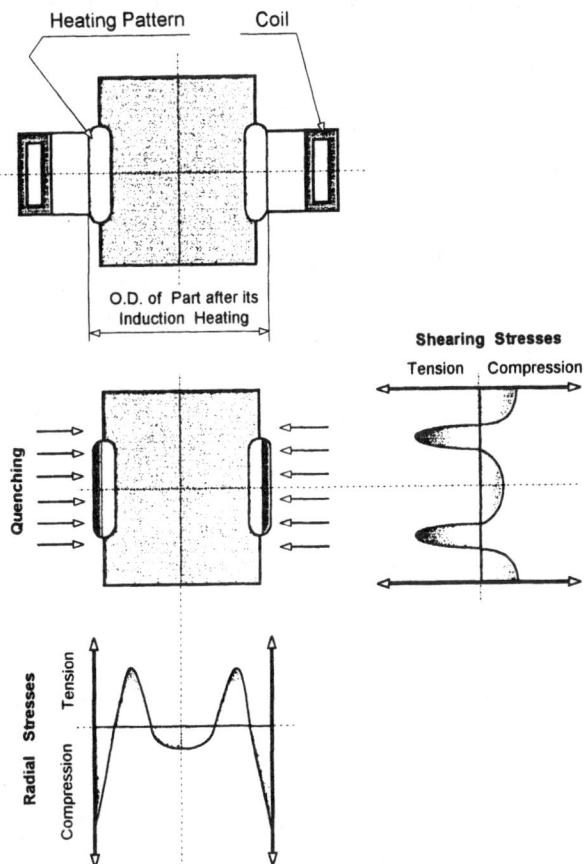

Figure 52 Formation of residual stresses after induction hardening.

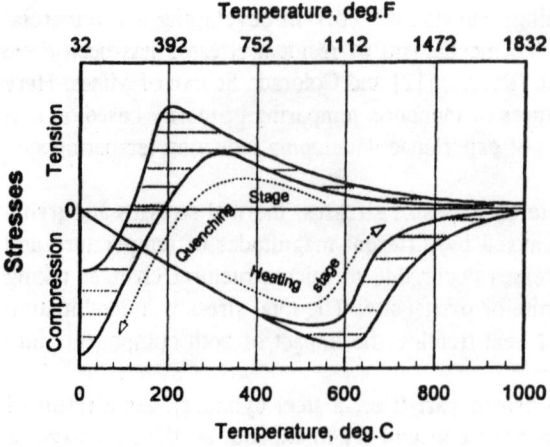

Figure 53   Stresses at the surface of carbon steel cylinder during heating–quenching cycle.

a plastic condition, the diameter of the heated area becomes slightly greater than its initial diameter, and stresses at the surface significantly decrease.

After the quenching fluid is sprayed onto the heated surface of the part, an intensive surface cooling will take place. Quite quickly, the surface layer loses its elastic properties and tension stresses appear at the surface of the workpiece. There is a pronounced maximum of tensile stresses at the surface of the workpiece. This maximum typically takes place at a surface temperature from 380°C (716°F) to 200°C (392°F). In time, the austenite transforms to martensite or pearlite depending on the cooling conditions. This results in the appearance of compression stresses at the surface (Figure 53).

Figure 52 shows the appearance of the thin surface layer where the austenite-to-martensite transformation takes place. Two mutually opposite stresses act in this layer. Because of the drastic temperature decrease in that surface layer, it tries to reduce its volume during the austenite-to-martensite transformation. As the process of quenching continues and martensite forms in the internal layers below the surface of the cylinder, stresses occur due to the fact that the outer layers are already hardened and in a nonplastic condition. Finally, when the part is cooled, a combination of compression and tension stresses exists in the workpiece (Figure 52).

The compression stresses in the surface may show higher hardness values than are normally achieved with the given steel and also afford some protection against cracks caused by microscopic scratches. The overall stress condition, however, increases brittleness and notch sensitivity, which reduces part reliability. Therefore, it is necessary to relieve stress on the part.

A conventional way of relieving stress is to run the parts through a tempering furnace, which is typically located in a separate production area and therefore requires extra space, labor, and time for parts transportation. In addition, tempering in the furnaces is a time-consuming process that may take up to 2–3 h. To overcome these disadvantages, in-line induction tempering was developed.

Basically, there are two ways to perform induction tempering: self-tempering (or tempering by residual heat) and induction tempering.

1. Self-Tempering

The principles of self-tempering after induction hardening are illustrated in Figure 54. As described in Section IV, during the initial stage of induction heating of the steel cylinder, an intensive heating of the surface layers takes place (Figure 54a). On completion of the heating cycle, the temperature profile will be as shown in Figure 54b. In this case, the temperature

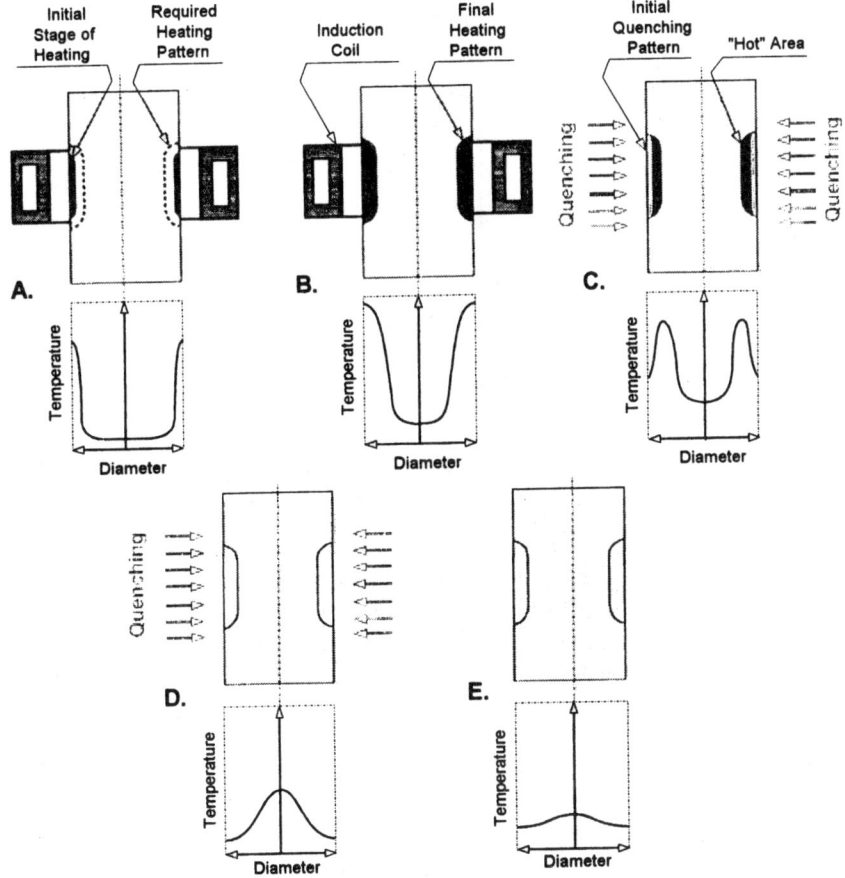

Figure 54   Induction heating, quenching, and self-tempering of a carbon steel cylinder.

within the surface layer that is required for hardening should be within the limits of the hardening temperatures for the given steel. At the same time, the core temperature of the cylinder will be a little different from its initial temperature. After the heating stage is complete, the quenching begins. A hardened layer appears on the surface as a result of the intense heat reduction. The maximum temperature of the workpiece will be in the internal layer below the surface. At the same time, because of the thermal conductivity fo the steel there will be an increase in the core temperature (the heat soaks from the surface toward the core). When the surface layer has been hardened and the temperature at the surface is reduced below the lower martensitic temperature, a considerable amount of heat is still retained inside the cylinder (Figure 54d). If at this moment the supply of quenching fluid is cut off, the part will begin to be heated through due to the accumulation of internal heat. After a certain time, the surface temperature will be increased to a value higher than it had when the quench was cut off. With proper selection of the quenching condition, the heat that is retained inside the part can be used for carrying out the tempering. Typically, self-tempering temperatures do not exceed 260–290°C (500–554°F). At the same time, in the case of surface hardening of selected areas of the workpiece, the self-tempering temperatures usually do not exceed 210–240°C (410–464°F). Figure 55 shows a typical temperature–time diagram of this process for carbon steel shaft hardening (shaft O.D. is 1 in.).

To ensure that the process is always performed correctly, several precautions must be taken. The energy introduced to the part as well as the heating time have to be monitored to

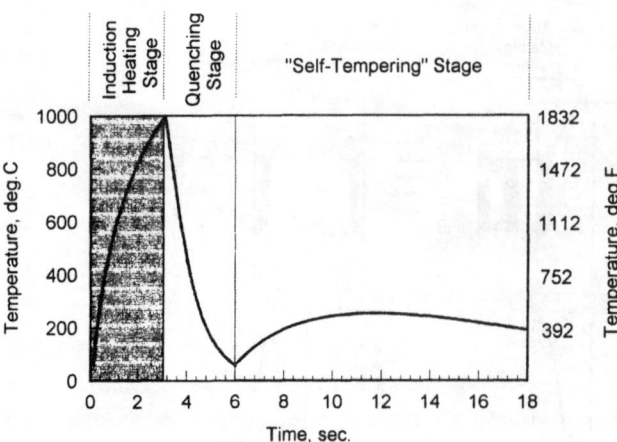

Figure 55 Surface temperature change during three stages of induction surface hardening: heating, quenching, and self-tempering.

ensure a constant amount of residual heat. The quench flow, quench time, and quench temperature should be monitored and held within close tolerances to ensure that the same surface temperature after reheating is always achieved. Moreover, in many cases an infrared pyrometer is used to monitor the surface tempering temperature. More details of process monitoring are discussed in Section III of Chapter 11B.

Typically, self-tempering is used when the part has a large mass and single-shot induction hardening is applied. In the case of scan hardening, it is more complicated to use this type of tempering because the core temperatures are different over the scanning length. The amount of heat stored as well as the heat sink underneath each section of the hardened case must be the same; otherwise, the temperatures achieved after heating balance will be different and the tempering result will be unacceptable.

Sometimes a quench–soak cycle is used. This includes the heating stage, first quenching stage, first self-tempering, second quenching stage, and final self-tempering. Such a method allows one to obtain unique properties of the workpiece.

2. Induction Tempering Method

For those parts that cannot be self-tempered, the induction tempering method can be applied. Typically, it is not recommended to use the same inductor for hardening and tempering. There are three reasons for this. The first reason has to do with the fact that in order to obtain the required hardness pattern of the workpiece, because of the workpiece shape, it is necessary to introduce more energy within certain areas. Second, the power densities during hardening are much higher than with tempering. With tempering it is necessary to heat the surface at a much slower rate to achieve a low temperature gradient from the surface to case depth for otherwise the surface could exceed the required tempering temperature, which would result in an unacceptable soft surface. The third reason has to do with the fact that quite often it is preferable to use a lower frequency for tempering because the tempering temperatures are below the Curie point. As a result, the heated part retains its magnetic properties and the skin effect is pronounced. Therefore, in induction tempering, in order to increase the current penetration depth in the part, it is more effective to use a lower frequency compared to induction hardening. The temper inductor should heat the entire workpiece, not just the selected hardened area. A loosely coupled multiturn inductor can be used for this purpose. Because of the physics of the tempering process, the total tempering cycle is usually much longer than the hardening cycle.

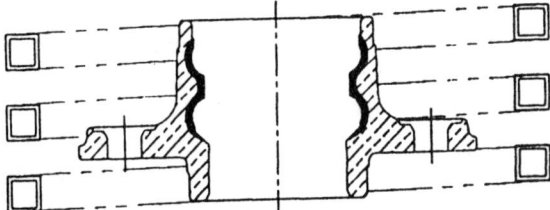

Figure 56   Induction tempering of a part that has been hardened on the inside surface.

In the case of induction tempering of complex shaped parts, such as gears or other critical components, the choice of frequency, power density and coil geometry is dictated by a
need to apply enough energy into the certain areas of the part. For example, in gear tempering applications, it is necessary to induce enough energy into the root area of the tooth without overheating its tip. The root of the gear is the most critical area because the maximum
concentration of residual stress is located there. As a result, cracks and distortion occure primarily in the root area. Therefore, this area needs to be stress relieved in the first order.
However, there are three factors which make this task quite complicated. First of all, the root
area does not typically have a good coupling with the induction coil compared to the gear tip.
Because of that, it is more difficult to induce energy there. Secondary, the tempering temperatures are below the Curie point, therefore, the gear is magnetic and the skin effect is pronounced. Use of high frequency will result in power surplus in the tip of the tooth compared
to its root. The third factor deals with a fact that there is a significant heat sink located under
the gear root (under the base circle). In order to overcome the above mentioned difficulties,
induction heat treatment manufactures have developed several new design concepts which have
resulted in the development of advanced induction gear tempering machines.

Figures 56 and 57 show two of the most typical examples of induction tempering. Figure 56 shows a part that has been hardened on the inside surface. This is the most effective
use of induction tempering. The tempering coil is located around the part, so the temperature
can slowly increase from the outside surface toward the hardened layer on the inside surface.
By using the proper energy control, this is a very effective method to carry out the tempering
without the danger of overheating the hardened surface. The same idea can be applied for a
hollow workpiece that has been hardened on the outside surface. In this case, it is very effective to locate the tempering coil inside the hollow workpiece.

Figure 57 shows a hub. This kind of workpiece requires a specially developed induction
coil that will allow the predetermined amount of energy to be induced into each section. This

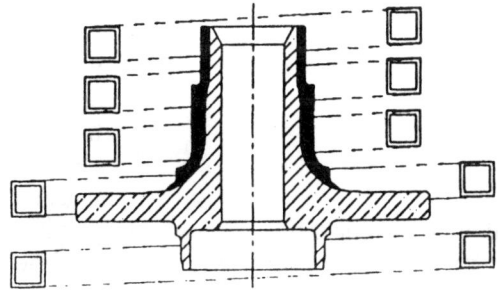

Figure 57   Induction tempering of a part that has been hardened on the outside surface.

allows one to bring each hardened section to the required tempering temperature. It will not be possible to design an inductor that heats only the hardened case like the induction hardening inductor does because of the wide variation of mass distribution in the hardened area, which results in a different heat sink effect. The inductor for such a workpiece should also induce the heat into areas that are not heat treated such as the flange. These areas will then act as a heat buffer and may be at a temperature slightly above the tempering temperature. This will allow for inconvenient areas to also be heated. With these types of inductors it is possible to induction temper workpieces with complicated shapes and mass distributions such as CV joints, which are hardened inside the bell and on the outside of the shaft.

Typically, the tempering inductor is loosely coupled to the workpiece. This allows the tempering coils the ability to heat lightly even the edges, grooves, and other critical regions.

Often a heat–soak cycle is used. This allows the heat to soak all the way through the workpiece. The number of heat cycles can be determined by taking hardness readings in the sectioned workpiece.

As a rule of thumb, the heating time for tempering is at least twice the hardening cycle time (heating and quenching). This means that an induction heat treatment machine can have one station for hardening (heat and quench) and two stations for tempering. In this case, the indexing time between the two tempering stations acts as soak time. As mentioned above, the frequency selected for induction tempering is quite often lower than the frequency for hardening, thus resulting in a deeper penetration of the electromagnetic field in the part.

A maximum of tension stresses is typically located below the hardened area, somewhere within the transition zone. This requires expanding the heated area not only to the case depth (hardened depth) but into the transition zone as well.

A cooling cycle may follow completion of the tempering cycle. A cooling station may be located just after tempering. Another practice is to have a cooling station separate from the tempering machine. In this case, the cooling station is located on an exit conveyor. This reduces the number of stations involved in the machine design.

The correct process parameters for induction tempering should be found by hardness measurements. Table 8 shows residual stresses in steel 1045 after surface hardening and tempering [151]. As the transformation from the tetragonal martensite into tempered martensite is a function of time and temperature, one can see from the data that for the same drop in hardness it is necessary to have a higher temperature when using induction tempering. However, at the same time it will have better reduction in residual stresses.

Induction tempering can be commercially successful primarily because it offers the following advantages:

Table 8 Effect of induction tempering on 1045 steel [151]

Heat treatment after hardening	Surface hardness (Rc)	Maximum residual stresses (kg/mm*mm)			
		Shearing		Axial	
		Compress.	Tensile	Compress.	Tensile
Ordinary tempering					
At 100°C (212°F)	60	70	45	48	28
At 200°C (392°F)	55	48	35	30	23.5
Induction tempering					
At 200°C (392°F)	60	63	38	43	23
At 300°C (572°F)	55	40	26	25	28

Compress. = compressive

1. Less tempering time than furnace tempering
2. Time and labor savings
3. Low floor space requirements
4. Savings in investment cost
5. Can be incorporated in-line or in a work cell
6. Environmentally friendly

Fatigue and failure testing for induction tempering should be compared to that in furnace tempering of individual workpieces. It is important to remember that the entire workpiece, not just the hardened area, usually needs to achieve the desired temperature for proper tempering. The surface temperature alone is not a valid indication of a proper temper.

I. End and Edge Effects, Longitudinal and Transverse Holes, Key Ways

1. End and Edge Effects

Coil–workpiece geometry has a significant influence on the heat treatment pattern. Nonuniformity of the heating pattern at the coil or workpiece ends is related to the distortion of the electromagnetic field in those areas. This distortion is called the electromagnetic end effect. Electromagnetic end effect can result in either overheating or underheating of the workpiece end. Basically, it is a function of frequency, coil–workpiece geometry, and material properties. This effect will be discussed in the following subsection. Here we briefly discusss its appearance in the case of the single-turn coil.

Figure 58 shows the distribution of the coil current and induced eddy current for different coil locations relative to the end of the part. When low frequency is applied, the eddy current distribution and heat pattern will be different.

Generally speaking, electromagnetic end effects are considered one of the most complicated problems in induction heat treatment. Electromagnetic end effects can be studied by using advanced computational techniques such as finite-element analysis or impedance boundary elements.

As mentioned earlier, induction heat treatment is a very flexible process and has been used successfully in a variety of applications. However, in applications where it is required to heat treat parts (camshaft, crankshaft, axle, transmission shafts, etc.) that contain longitudinal and transverse holes, sharp corners, and key ways (Figure 59), it presents certain difficulties. Existence of these features can result in an undesirable appearance of hot and cold spots, cracks, and shape distortion. In these cases it is necessary to make a careful evaluation of the eddy current and temperature fields in order to achieve the required heat uniformities and meet process specifications.

2. Longitudinal Holes

The existence of longitudinal holes (Figure 60) or longitudinally oriented hollow areas within the part can cause a redistribution of the eddy current flow that can result in overheating of certain regions ("hot spots"). The left-hand part of Figure 61 shows a segment of a cylindrical part and the normal current flow within it. If a longitudinal hole is located within the current penetration depth, then it blocks the normal eddy current path and leads to a current redistribution (Figure 61, right). Because of that redistribution, areas between the part's surface and the hole can become overheated or even melt.

Actually, there are two factors that cause the overheating effect. The first is an increase in the power density in that areas due to the redistribution of induced current. The second is a lack of adjacent mass in that area. As a result, less heat will soak from the surface of the

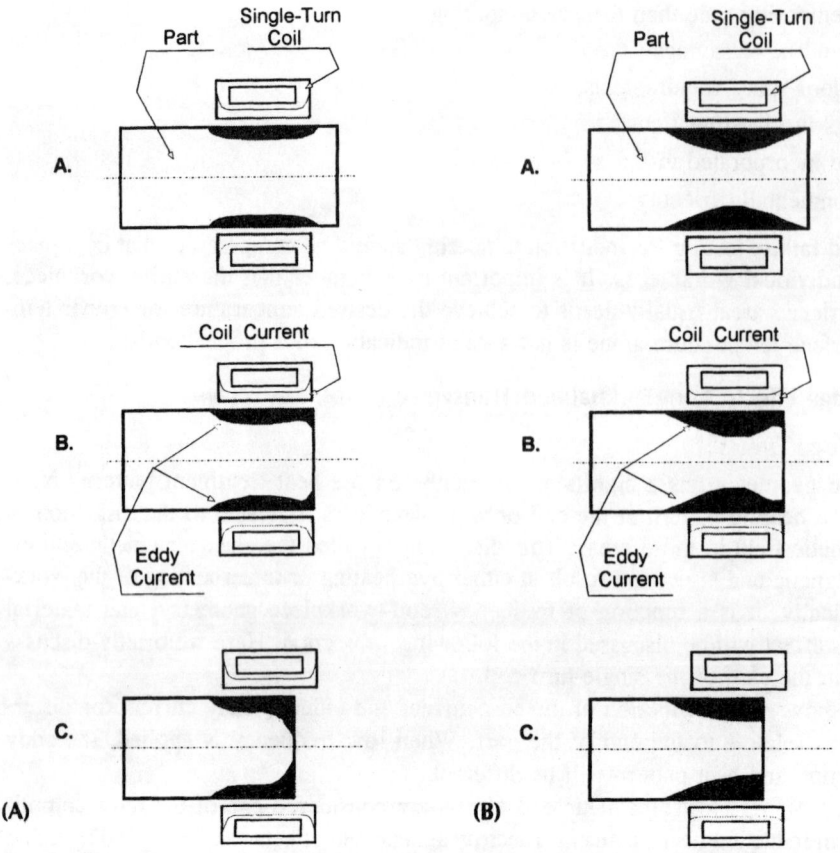

Figure 58 Electromagnetic end effect in induction heat treatment. (A) High frequency applied, (B) Low frequency applied.

part toward its core. The decrease in this heat soaking is due to the fact that the thermal conductivity of the air is much smaller than the thermal conductivity of the metal. As one can see, these two factors may coincide and result in the overheating of certain areas of the part. In different applications the influence of the two may be different. If a longitudinal hole is located near the part surface within one penetration depth, then the first factor typically prevails and is primarily responsible for the heat surplus in that area (Figure 62, hole A). Intense heating with higher power density makes this overheating more pronounced. When the hole is located within one or two current penetration depths (hole B), then both factors have approximately the same influence on the heat surplus. If the longitudinal hole is located within two or three penetration depths and the heat cycle time is relatively long (8–12 s or more), the second factor makes a major contribution to the overheating of the area. When the hole is located within three to five current penetration depths or more under the surface of the part (Figure 62, hole D), the heat surplus due to the existence of a longitudinal hole is minor and the probability of overheating is very small.

3. Transverse Holes, Key Ways, Various Oriented Hollow Areas.

Transverse holes can also cause a redistribution of eddy current flow (Figure 60). Unlike the case of longitudinal holes, eddy current redistribution due to transverse holes can result in both

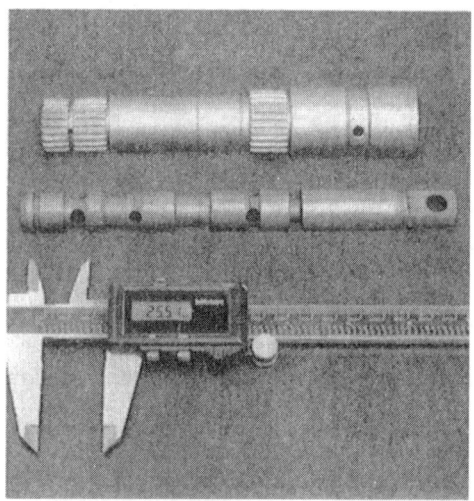
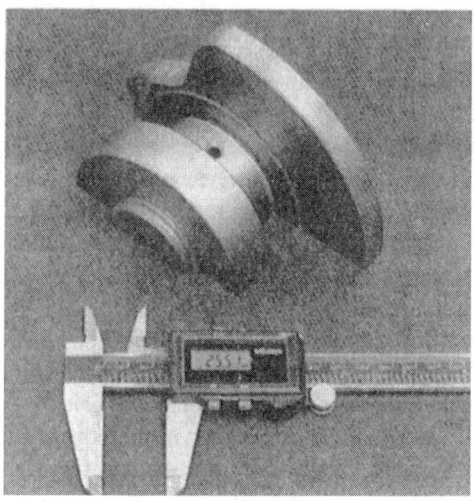

Figure 59   Induction heat treatment parts with holes, keyways and sharp corners.

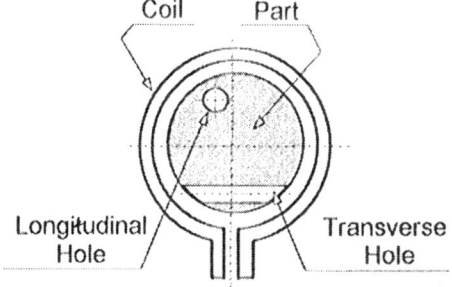

Figure 60   Longitudinal and transverse holes in induction heat treatment.

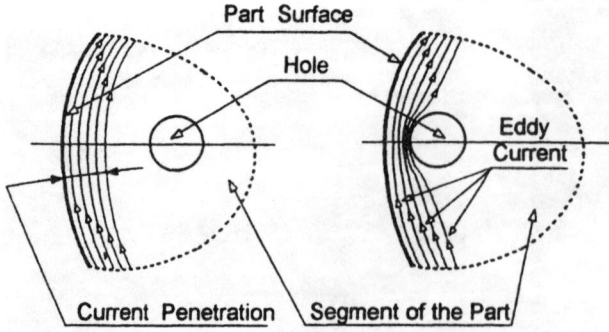

Figure 61 Eddy current redistribution due to presence of longitudinal hole.

an underheating and an overheating of the hole edges (Figure 63a). Because of the current concentration, overheating can occur at the hole edges which are parallel to the eddy current flow [12,107]. On the other hand, the hole edges that are perpendicular to the eddy current flow will be underheated. With an increase of both the inside diameter of the transverse hole and the frequency, the nonuniformity of the temperature distribution along the perimeter of the hole will be more pronounced. These heat nonuniformities can cause cracks and distortion in the vicinity of the hole edges.

It is possible to obtain a relatively uniform temperature distribution along the hole perimeter by putting a plug in the hole. If the plug is made of the same metal as the part, the heat nonuniformities will be negligible and the temperature distribution can be considered uniform (Figure 63b). Despite their use in avoiding over- and underheating at the hole edges, steel plugs present problems because it is often difficult to insert them and remove them after heat treatment because they may become welded in the hole.

An alternative to the use of steel plugs is the use of copper plugs. In this case, a nonuniform current distribution still occurs, which can lead to local overheating and underheating. However, the overheating/underheating phenomenon is opposite to that seen with steel plugs. Figure 63c shows that the eddy current will gather in the copper plug from the neighboring carbon steel regions. This takes place because the electrical resistivity of the copper is much less (approximately a factor of 10) than the resistivity of any steel and it is much easier for

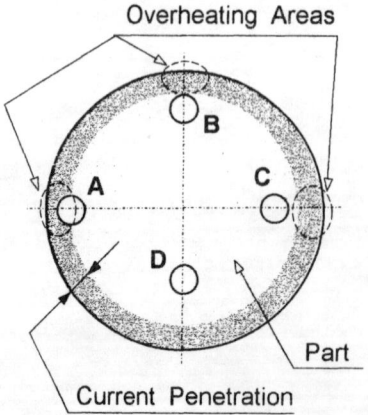

Figure 62 Overheating areas due to presence of longitudinal holes.

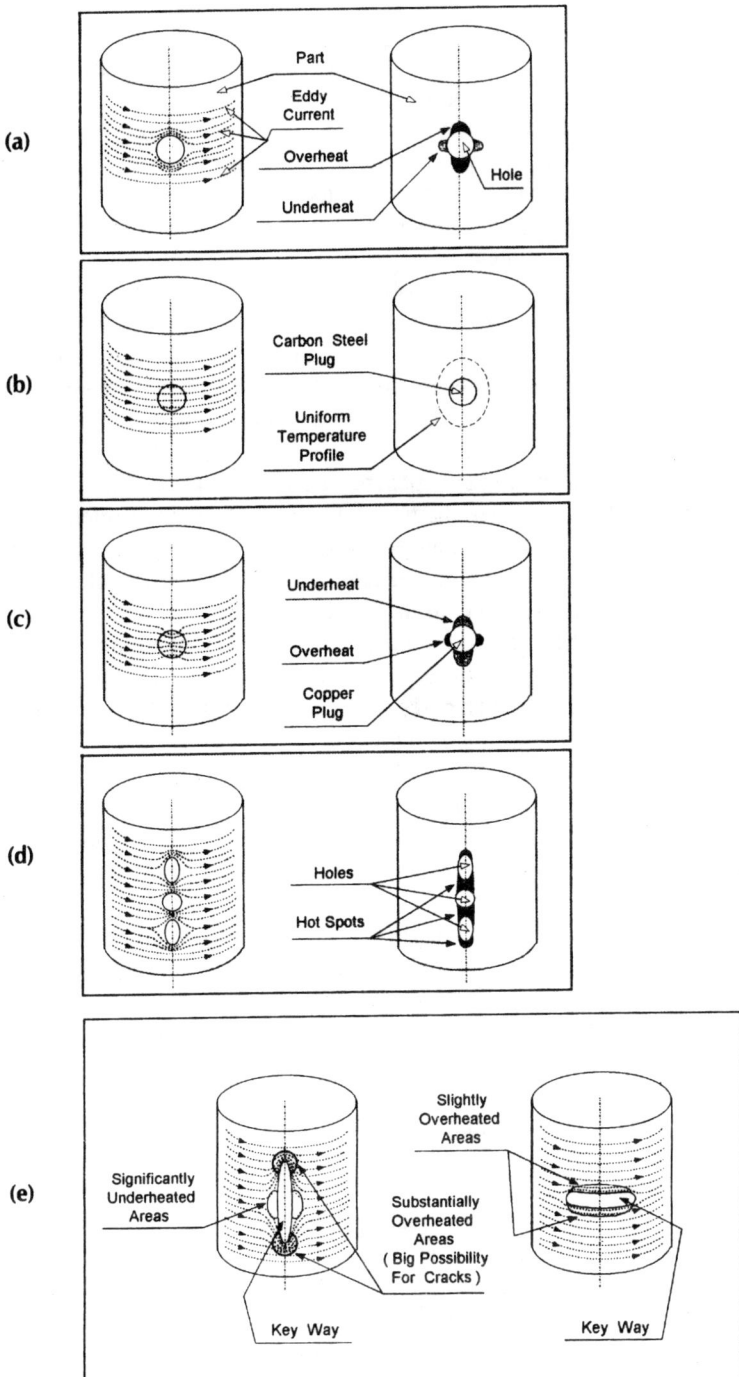

Figure 63 Eddy current distribution and heat nonuniformities due to presence of transverse holes. (a) Transverse hole, no plug; (b) carbon steel part and carbon steel plug; (c) carbon steel part and copper plug; (d) multi-holed part, no plugs; (e) keyways in part.

the eddy currents to flow through the low-resistance copper than through the high-resistance steel. Therefore, when copper plugs are used, the hole edges that are parallel to the eddy current flow are underheated and the hole edges that are perpendicular to the current flow are overheated. An actual distortion of the hardness patterns due to the use of copper plugs is shown in Reference 107. It is necessary to mention that when using copper plugs (Figure 63c), overheating of the hole edges is much less pronounced compared to the case when plugs are not used (Figure 63a). At the same time, the copper plugs eliminate the appearance of cracks in the hole areas during induction heating and quenching.

In some applications, instead of metal plugs, water-soaked wooden plugs have been used. Because wood is not electrically conductive, the wooden plugs do not change the eddy current distribution in the hole edge areas. Therefore, there will still be regions with high and low current densities as shown in Figure 63a. However, the use of water-soaked wooden plugs allows one to decrease the overheating of the hole edges up to a certain point due to the heat transfer from the high-temperature regions into the water-soaked wood by thermal conductivity. There is a benefit in using any plug that limits exposure of the hole to quenching.

Special care should be taken in the case when the heat-treated part contains angled holes or when there is a combination of closely located transverse holes. The temperature field in angled hole areas combines the features of nonuniform temperature distribution along the perimeter of the hole and an overheating effect due to the increase of current density and the lack of mass on one side of the hole. The last two effects are typical for longitudinal holes.

Certain difficulties can appear when the part consists of several closely located holes (Figure 63d). Applying induction heat treatment in such cases could cause a sequence of cold spots (poorly heated areas) and very hot spots (almost melted areas). In such cases, quite often, an alternative heat treatment processes can be recommended.

Key ways can be considered extreme cases of longitudinal holes. Size, shape, and orientation of key ways have a substantial effect on the ability to obtain the required temperature profiles within the key way area and to avoid undesirable hot and cold spots in these areas (Figure 63e).

Inserting plugs into holes and taking them out are very delicate and time-consuming processes. A popular saying among heat treaters who deal with parts containing holes is that if there is a possibility of avoiding nonuniform heating of the hole edges without using plugs (e.g., by using a special heating regime or coil design), then the extra development effort required is justified. Therefore plugs should be used only as a last resort.

When discussing heat nonuniformities due to electromagnetic end and edge effects, longitudinal and transverse holes, and temperature field nonuniformities due to keyways, sharp edges, and corners, it is necessary to mention that smooth chamfers on edges and rounding of sharp corners can be a great help in decreasing the possibility of overheating or cracking.

Experience at INDUCTOHEAT, Inc. during heat treating of different parts with the above mentioned features shows that, surprisingly, the proper choice of design parameters (applied frequency, power density, coil geometry, etc.) allows the heat treater to obtain the required heat treating pattern even in cases that seem unsuitable for heat treating by induction. For example, in some cases, even such complicated shaped parts as a ball bearing cage, which consists of a number of closely located large holes, can be surface hardened, or contour hardened, by induction instead of using time and space consuming carburizing processes.

J.   Induction Bar End Heater

The majority of induction heating applications discussed so far in this chapter have dealt with complex shapes. However, a significant portion of heat treatment applications deal with classical cylindrical workpieces such as bars and tubes. Although many of the cylindrical work-

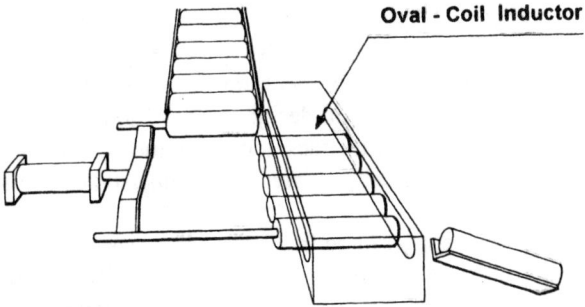

Figure 64   Oval coil bar end heater.

pieces being manufactured today lend themselves to automatic processes in which a bar or tube can be fed into a roll former or other type of forming system, certain parts require the forming of only the end of the bar. Some examples of these types of parts are "sucker rods" for oil country goods or various structural linkages in which an eye or a thread may be added to one or both ends of the bar [104,106,115–117].

Induction bar end heating is generally accomplished by placing the end of the bar into a multiturn coil and heating it for a specified amount of time. Multiple bar ends can be heated in a channel type of coil, a single multiturn oval coil, or a two-, three-, or four-coil arrangement that is configured out of individual conventional solenoidal coils.

In the case of the oval coil design shown in Figure 64, bars are loaded into a magazine from which they are pushed, one at a time, into the upper end of an inclined oval coil. They roll down as previous bars are removed. By the time they reach the bottom position they have reached the desired temperature and are pushed out for delivery to the forming machine. In the case of a channel coil (Figure 65), typically the magazine-loaded bars are removed by carriers on a belt, and the end of the bar to be heated passes through the channel coil. Upon leaving the coil, the end of bar is at the required temperature, and the bar moves to the forging operation.

The choice of a particular type of induction bar end heater arrangement depends on the customer's requirements. Besides the variety of coil arrangements, the loading/unloading operation can be fully automated or semiautomated, the choice depending on the required production rate and the cost of equipment. The least expensive is the semiautomated design where the operator simply removes the hottest bar and replaces it with a cold bar before it is moved to the forming hammer, press, or upsetter.

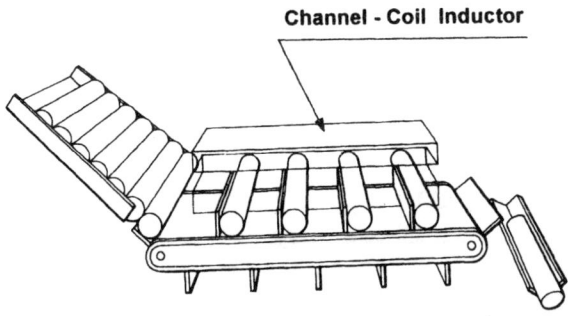

Figure 65   Channel coil bar end heater.

Although there are a variety of bar end heating coil designs, the basic principles and rules of thumb in obtaining the required temperature profile within the bar end are quite similar. Consequently, we will discuss the features of the design and operation of the induction bar end heater by analyzing a conventional multiturn solenoidal coil design (Figure 66) [104,115]. However, the reader should again bear in mind that the principles and recommendations that will be discussed for the conventional solenoidal heater can be extended to other types of induction bar end heater designs.

In progressive induction heating applications (e.g., in-line inductors), it is easier to predict the expected temperature distribution and obtain the required temperature profile in the bar than in the case of bar end heating applications. Basically, this is so because each part of the bar in a progressive induction heater [50,104,105,115,116] experiences the same magnetic flux with respect to time. On a bar end heater it is a much different story, and more variables are involved in obtaining the required temperature profiles within the bar. The large number of design parameters makes the design process quite complicated. Let's have a closer look at the features of the induction bar end heating process.

Typically, in the induction bar end heater the bar may be only partially inserted in the heating coil (Figure 66). This shows the difficulty of using analytical and equivalent circuit coil design methods [1–4,6–11] to calculate this process, because those methods are based on the assumptions of an infinitely long coil and symmetrically located workpiece within the coil. Unfortunately, such assumptions are not valid for the induction bar end heater, which cannot be considered an infinitely long system. Nor, in most cases, is the bar located symmetrically within the coil. As a result of such limitations, those computation techniques cannot be used to accurately predict the power density and temperature distribution within the bar end.

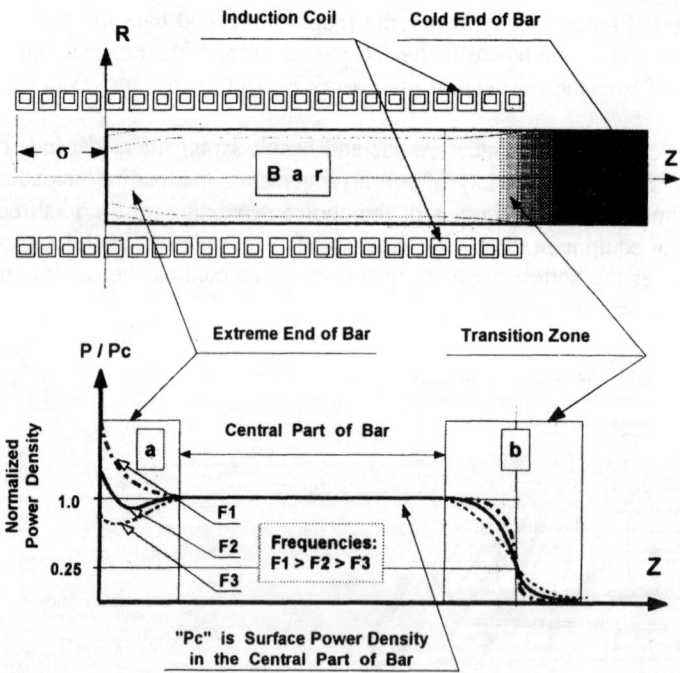

Figure 66   Sketch of induction heater and power distribution along the length of bar.

From another perspective, a difference in the effective coil/workpiece resistance and re-actance may affect the tuned frequency of the load and the available power. It also causes a much different situation for analysis of the workpiece thermal condition and temperature distribution because the extreme end of the workpiece (Figure 66) offers no path for the conduction of heat, while its cold end provides a ready heat sink and easy conduction path.

For the designer, all of the above-mentioned variables usually cause unpleasant surprises and a cut the try method of coil design rather than a true ability to predict what will actually happen.

The required temperature distribution within the bar end depends on the frequency, the coil and bar geometry (including the part-to-coil air gap and coil overhang), the material properties of the bar, emissivity, refractory, power density, and cycle time. Obviously, only modern numerical methods can be applied as a computational technique that allows accurate calculations of results, saves prototyping time and dollars, and facilitates the manufacture of reliable, competitive products with short design cycles.

In the case of the induction bar end heater, the temperature distribution within the bar is affected, among other factors, by a distortion of the electromagnetic field in the extreme end of the bar and induction coil end zone (Figure 66). This field distortion and corresponding distribution of induced current and power are referred to as electromagnetic end effects. These end effects may be illustrated by the curves in the lower diagram of Figure 66, zone a [5,49–53,104,115,118]. Basically, the electromagnetic end effect in the extreme end of the cylindrical bar is defined by three variables, R/δ, the skin effect; σ, the coil overhang; and the ratio R_i/R, where R is the radius of the bar and R_i is the inside radius of the coil; δ is the current penetration depth which can be defined by Equation (38) or (39).

An incorrect combination of these factors can lead to underheating or overheating the extreme end of the bar. Studies show that typically the electromagnetic end effect area extends toward the central region of the bar (Figure 66) no further than 1.5 times the bar diameter. Higher frequency and large coil overhang will lead to a power surplus in the extreme end of the bar. As a result, significant overheating may take place in that area. A low frequency and small coil overhang will cause a power deficit at the extreme end of bar, which will therefore be underheated.

It should be pointed out here that a uniform power distribution along the extreme end of the bar will not correspond to its uniform temperature profile because of the additional heat losses (radiation and convection losses) at the bar end area compared to its central part. By the proper choice of design parameters, it is possible to obtain a situation where the additional heat losses at the end of the bar are compensated for by the additional power (power surplus) due to the electromagnetic end effect. This will allow the designer to obtain a reasonably uniform temperature distribution within the bar at the end as well as at the center.

The magnetic field distribution in the bar near the right end of the coil in Figure 66 (zone b) depends primarily on the radii ratio R_i/R and the skin effect, with the latter being most prominent. Due to the physics of the electromagnetic end effect in that area, there is always a power deficit under the coil tail end at any frequency. It is possible to show that in the case of a long multiturn induction coil with a long homogeneous bar (zone b), the density of the induced eddy current in the bar area under the coil tail end is only half that in the central part. Therefore, the power density in that area is only one-fourth of the power in its central part. Typically, the length of zone b is equal to 0.5–3.5 times the coil radius. High frequency and pronounced skin effect lead to a shorter end effect zone. An external magnetic flux concentrator will also give a shorter length of zone b.

It should be emphasized that in order to compensate for the power deficit within the bar area located near the cold end of the bar (zone b), the extreme end of the coil should be extended further toward the bar cold end. In other words, the coil length should be longer than the required uniformly heated area of the bar.

Another important feature that defines the coil length is the fact that in zone b, which is often called the transition zone, there is a significant temperature gradient along the length of the bar. As a result, heat is conducted from the high-temperature region of the bar toward its cold area, the "heat sink" phenomenon. This in turn results in a cooling effect at the other end of the required heated area (zone b). The proper choice of coil geometry (primarily coil length), power density, and frequency will allow one to compensate for this cooling effect and obtain the required uniform temperature profile for the bar.

Typically, there is a limitation on the permissible length of the transition zone. Therefore, the choice of coil length is always a reasonable compromise between obtaining the required heating profile and the minimum length of the transition zone. A short cycle time, high power density, small coil–workpiece air gap, and high frequency help to make this zone shorter.

It is important to understand that the choice of frequency is defined not only by the required temperature profile in the bar, but also by the requirement for high electrical efficiency.

Figure 67 shows a sketch of an induction bar end heater for a carbon steel bar (AISI-SAE 1035, bar O.D. is 1 in.) that has been heated at a frequency of 9000 Hz. Frequently, the need for a bar end heater involves equipment that can be used for a wide range of part sizes. This particular coil arrangement can provide satisfactory results in induction end heating for bar diameters as small as 3/8 in. The dynamics of an induction carbon steel bar end heating process (bar O.D. 1 in.) are shown in Figure 68.

The analysis shows that in the first heating stage the whole bar is magnetic (cycle time < 3 s); therefore, the value of relative magnetic permeability is high. As a result, the skin effect is pronounced ($R/\delta > 20$). At the same time, because of the relatively low temperature, the heat losses from the bar surface are relatively low. Because of the pronounced skin effect, the induced power appears in the fine surface layer of the bar. No power is induced into its inner layers. The core of the bar is heated from the surface only by thermal conductivity. Consequently, the surface of the bar is heated much faster than the core. This stage is characterized by a high temperature gradient along the radius of the bar which can cause cracks in steels with 0.5% or more carbon.

With time (cycle time < 5 s), as a result of surface temperature rise, the surface of the bar starts to lose its magnetic properties and the relative magnetic permeability in the surface

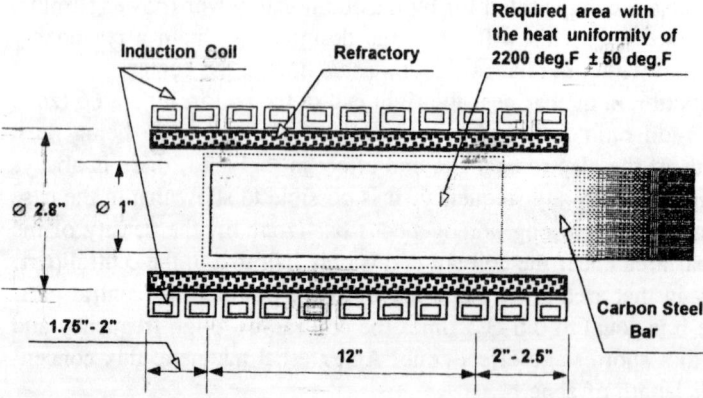

Figure 67   Geometry of induction bar end heater (Frequency, 9 kHz; heating cycle, 36; 9 coil turns; length of billet, 18 in.)

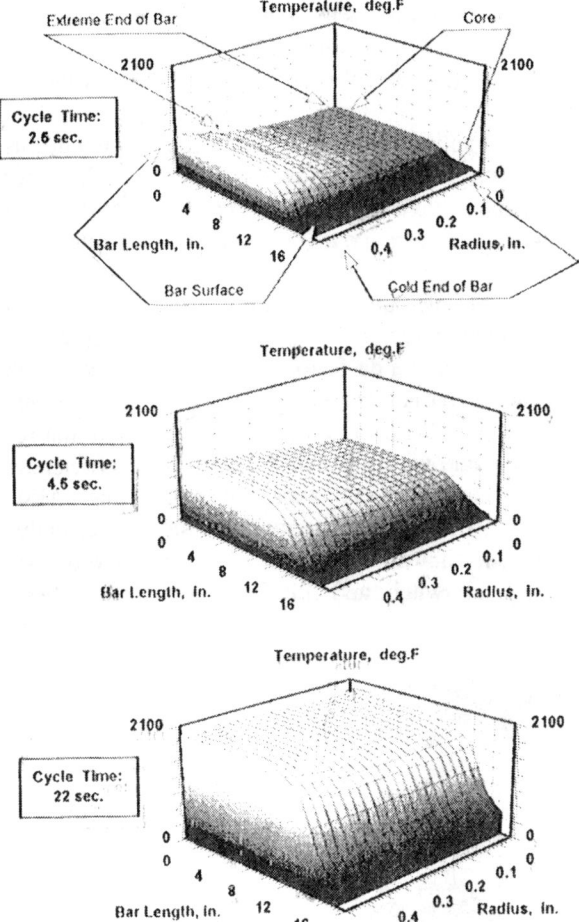

Figure 68 Dynamics of induction heating of carbon steel bar end.

layer drops to 1. Furthermore, because of the temperature rise, the electrical resistivity of the carbon steel increases to approximately 1.5–2.5 times its initial stage value. The decrease of relative magnetic permeability and increase of electrical resistivity cause an increase in the penetration depth. As a result, the essential portion of power is now induced in the internal layers of the bar. This stage can be characterized as the biproperties stage of the bar [4,51,93,104,152]. In this stage the bar surface becomes nonmagnetic; however, the internal area of the bar and its core remain magnetic. The surface-to-core temperature difference here is not as significant as it was in the first stage.

Finally (cycle time $\cong$ 15 s), the total bar area under the coil, including the core of the bar, becomes nonmagnetic. The penetration depth becomes quite large. Skin effect becomes less pronounced ($R/\delta < 2.5$). As a result, a significant amount of energy is induced within the internal area of the bar, and the core starts to heat much more intensely than in the previous two stages. Furthermore, the heat losses from the bar surface become relatively high (primarily due to radiation losses). Because of the large penetration depth, increased heat losses from the surface of the bar, and thermal conductivity within the bar, the temperature profile within the bar thickness will equalize.

The correct choice of induction coil geometry, power density, and frequency allows the induction heating designer to obtain the required temperature uniformity, high efficiency, short transition zone, small equipment space requirements, less scale, and high production rate for a variety of stock sizes.

One of the most successful developments of an induction bar end heater is shown in Figure 69 [106,115,116]. This portable induction system (Unipower UPF) is produced by INDUCTOHEAT and can be used in forging applications as well as for bright annealing stainless steel tubing, stress-relieving pipe and wire, and other induction heating applications. Machines with output frequencies of 1–30 kHz providing output power of 25–400 kW are available. Unipower UPF systems have a built-in control, output transformer (depending on the application), and heat station for stand-alone operation. Unipower and Uniforge types of heaters (both produced by INDUCTOHEAT Inc.) are actually a de facto standard for use in a large variety of induction heating applications including bar end heating. They have a broad power-matching capability to deliver full power throughout the heating cycle. Precise power settings are ensured within ±2% accuracy with ±10% line variance. All the controls necessary for independent operation are built into machines or located on an optional remote operator's panel. Controls include a single meter with a selector switch for monitoring kilowatts, volts, or frequency; light-emitting diodes to indicate heat on, kilowatts, capacitor volts, output volts, and output current limits; a manual/timed/auto selector switch; and built-in digital timer. Optional

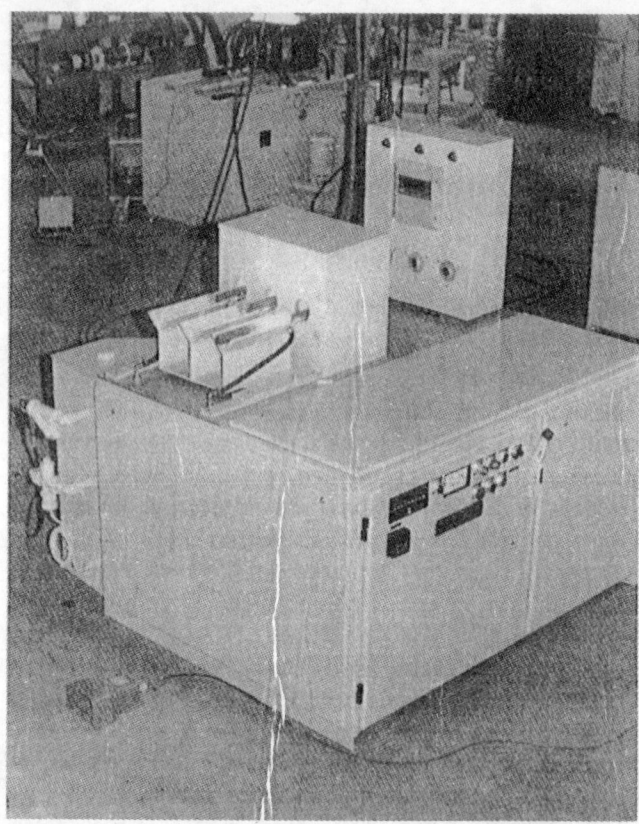

Figure 69 Portable induction bar end heating system Unipower UPF.

equipment includes diagnostic indicator lights for door open, water temperature, water pressure, capacitor pressure, and side-mounted water-recirculating system. Additional heat levels and special coil assemblies can be added. Input power factors of such systems are 0.94 under all operating conditions.

Utilizing the Unipower and Uniforge design concept, the power supply, matching components, and heating coil are all grouped together in a single package that can easily be moved from place to place. The secret of the success of both machines is that their design combines outstanding practical experience, modern theoretical knowledge, and a full understanding of the intricacies of the process. Unipower and Uniforge machines offer the ultimate induction heating technology in the most economical package.

K. Induction Strip Heating

As shown above, the induction phenomenon applies to heat treatment of a variety of metalworking processes. Induction strip heating is another area in which induction heating has been successfully used. This includes full and partial annealing, galvanizing and galvannealing, preheating prior to furnace heating, tempering, paint curing, lacquer coating, and drying. A particular application requires determination of the optimal coil design. There are three basic induction strip heating coil designs [119–121]:

1. Longitudinal flux inductor
2. Transverse flux inductor
3. Traveling wave inductor

Generally speaking, these designs are distinguished by the orientation of the main magnetic flux. Each has certain advantages, and all have been used either alone or in combination with others.

Due to several electromagnetic phenomena (end and edge effects), the distribution of heat sources (power density) induced in a rectangular workpiece (i.e., slab, plate, strip) is not uniform [51,53,73,92,104,119–121]. As a result, there is a problem obtaining the required temperature profile. However, unlike the case of the induction heating of slabs or thick plates [92,104,119–121], the heat source nonuniform distribution along the strip thickness will not typically cause a surface-to-core nonuniform temperature profile. This is because the thermal conductivity of the metal is able to quickly equalize the temperature nonuniformities within the strip thickness.

From another perspective, unlike the induction heating of round bars, strip heating will lead to temperature nonuniformity along the strip surface (width). The temperature distribution along the strip width is primarily affected by a distortion of the electromagnetic field in its edge areas. This is another instance of the electromagnetic edge effect which can cause a major problem in developing effective induction strip heating equipment.

1. Longitudinal Flux Inductor

A longitudinal flux inductor can be described as a solenoidal heater (Figure 70) similar to that used in the induction heating of cylinders. The strip is surrounded by the induction coil. An alternating current flows through the coil turns and produces a longitudinally oriented time-variable magnetic field that induces eddy currents to circulate within the strip thickness. These currents produce heat by the Joule effect.

Traditionally, longitudinal flux induction heaters have high efficiency and uniform temperature distribution within the strip. There will be high coil efficiency when the ratio of strip

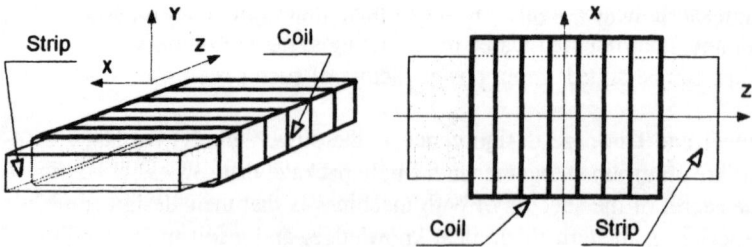

Figure 70 Longitudinal flux induction heater.

thickness d_{st} to penetration depth δ is 2.5 or more. The optimal value of frequency that corresponds to the maximum coil efficiency can be determined as follows:

Nonmagnetic strip:

$$\frac{d_{st}}{\delta} \cong 3\text{--}3.3$$

Magnetic strip:

$$\frac{d_{st}}{\delta} \cong 2.7\text{--}2.9$$

The use of a frequency higher than the optimal will change the coil efficiency only slightly. At the same time, the use of very high frequencies tends to decrease the total electrical efficiency due to higher power losses in the coil and bus bars. If the chosen frequency is lower than the optimal value, the efficiency will dramatically decrease. This is due to cancellation of the induced currents circulating in the opposite sides of the strip cross section.

This type of induction heating does not normally demand a very tight air gap, and it does not require big adjustments of the heater for strips with different widths and thicknesses. Longitudinal flux inductors are effectively used for low-temperature induction heating of thin magnetic strips when the final temperature of the strip is below the Curie point. However, in the case of nonmagnetic thin strip heating (i.e., aluminum, stainless steel, etc.) or for heating of magnetic strips above the Curie point, this type of inductor will require sufficient power at high frequencies of up to several hundred kilohertz or even megahertz and sufficient power. A new family of solid-state power supplies developed by INDUCTOHEAT can produce 1.5 MW at 400 kHz. In some applications this frequency is not high enough. The use of tube generators in the megawatt range of power can lead to well-known problems, such as concern for safety, low reliability, and low efficiency yet they are the only choice until higher frequency solid-state power supplies are developed.

The process of induction heating of magnetic strips from an ambient temperature, or a temperature below the Curie temperature, to a temperature above the Curie point has several features that could call for a special design of the induction system. At the first stage, the whole strip is magnetic and the heating process is very efficient and intensive. After the strip is heated above the Curie point, the heat intensity will significantly decrease and the heating process will be inefficient. To avoid this, one could use a dual inductor. In the second stage when the load becomes nonmagnetic it is more efficient to use another type of inductor, such as the transverse flux inductor.

As mentioned earlier, one of the major difficulties in obtaining uniform temperature distribution within the strip is caused by the various geometries of the products (i.e., thickness or width) that are heated in the same coil. However, this problem is not as pronounced with the longitudinal flux inductor as with the other types of induction coils (i.e., transverse flux or traveling wave inductors).

2. Transverse Flux Inductor

The transverse flux inductor (Figure 71) is one of the oldest induction heating techniques, having been developed for use in the induction aluminum alloy strip heating industry in the early 1940s. The principles of the process, simulation procedures, and experience of industrial utilization of transverse flux induction heaters were reported by Baker and Lamourdedieu as early as 1950 [122–125]. This process was established as a way to overcome the problems of induction heating of thin nonmagnetic strips in longitudinal flux inductors.

In the conventional transverse flux coil design, the strip passes through induction coil pairs that are located on both sides of the strip, and they create a common magnetic flux. This magnetic flux passes perpendicularly through the strip width. Unlike the longitudinal flux induction heater, in the transverse flux induction heater the induced eddy currents are circulated in the plane of the strip but not within the strip thickness. This allows induction heating of the thin strip to be carried out with high power densities using low and sometimes even line frequency. The transverse flux inductor often uses external magnetic flux concentrators (i.e., laminations or other high-permeability, low-reluctance materials). This type of inductor does not strictly require a small coil–strip air gap. However, electrically efficient heating can be provided only when strip thickness is 1.5–2 times the penetration depth or less. Without satisfying this condition, the transverse flux effect will disappear and typical proximity heating will take place.

The most difficult problem in using transverse flux inductors is in obtaining temperature uniformity along the width of the strip. The eddy current paths in the strip match the shape of the induction coils. When current reaches the edge area it must flow along the strip edge. As a result, the strip edges become overheated. Because of this natural phenomenon, the current's concentration will be higher in the strip edge area than in the central area of the strip. This undesirable effect can be corrected by the proper choice of inductor design parameters, such as pole step (pole pitch), coil opening, and frequency. Frequency that allows high electrical efficiency in the use of transverse flux induction heating of a nonmagnetic strip can be determined as

$$F = 2 \times 10^6 \frac{\rho h}{\tau^2 t}$$

where ρ is electrical resistivity $\Omega \cdot$m; h is coil opening, m; τ is the pole step of the coil (pole pitch), m; t is the strip thickness, m. This formula is valid when

$$h/\tau = 0.2-0.55$$

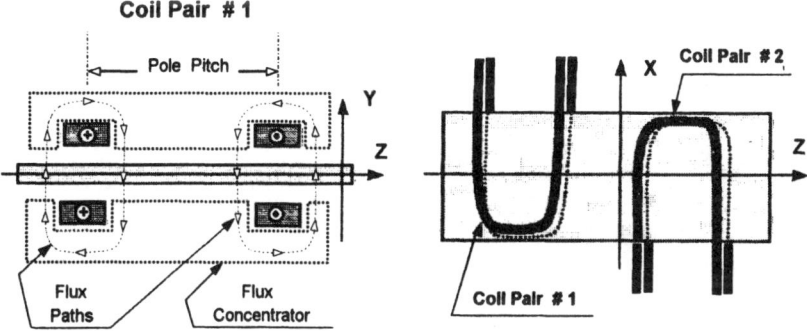

Figure 71 Conventional transverse flux induction heater.

One of the transverse flux inductor design parameters that has a significant influence on current and therefore on the heat source distribution in the plane of the strip is the shape of the induction coil. Over the last three decades, different three-dimensionally distributed transverse-flux coils have appeared quite regularly [126–140]. In practice, most coil pairs have certain limitations in providing precise and effective induction heating when the strip width and thickness vary widely.

3. Traveling Wave Inductor

The traveling wave inductor is not as commonly used for strip heating applications as any of the heaters already discussed. The main reasons are probably the complexity of the process and the fact that there has not been enough experimental and research work done in developing the traveling wave induction heating process. The fundamental concept of that process is quite simple [10,119] and is similar to that of the conventional three-phase electric machine where the strip takes the place of the rotor and the induction coil can be considered a stator (Figure 72). The inductor turns are located quite close to each other and carry different phase currents. The inverse connections of the middle phases and the external magnetic flux concentrator have been used to eliminate current cancellation in neighboring coils. The coils are located in the slots of the flux concentrator. The three-phase current flows through the coil turns, producing an electromagnetic field and heat sources in the strip. One of the main advantages of this system is its low levels of vibration and industrial noise compared to other types of inductors. This is primarily important when induction heating nonmagnetic (aluminum, copper, etc.) thin strips. However, there are several difficulties that prevent common use of this type of heater.

First, the travelling wave heater strictly requires a small air gap between the coil and load. This can lead to difficult mechanical design problems. Second, even with the use of magnetic flux concentrators, because of the closeness of multiphase turns, there is a certain magnetic tie between the turns with different phases, and therefore magnetic fields still cancel in adjacent sections. This leads to an overall decrease in electrical efficiency of the inductor compared to the transverse flux or longitudinal induction heater. Finally, there is still the difficulty of producing a uniform temperature profile for strips of various widths.

No single type of coil will provide acceptable heating results in all strip heating applications. Therefore, the selection of the best type of coil and its specific design must rely on a detailed analysis of the application.

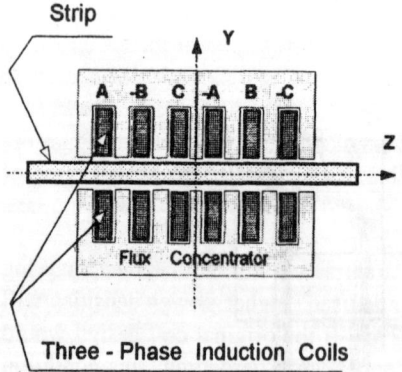

Figure 72   Traveling wave inductor.

L. Magnetic Flux Concentrators (Flux Intensifiers)

Magnetic flux concentrators (flux intensifiers) have become an acknowledged standard in induction heat treatment design. Modern, high-permeability, low-power-loss materials are now routinely used in a manner similar to that of magnetic flux diverters (cores) in power transformers or motors. A traditional function of flux concentrators in induction heat treatment has been to improve the magnetic coupling efficiency (by loss reduction) and to obtain effective selective heating in workpiece areas that are difficult to heat. Successful development of powdered metal concentrators based on Fe, Ni, Co, and other elements have dramatically increased the popularity of magnetic flux concentrators.

Let us examine what happens when a magnetic flux concentrator is applied. Without a concentrator, the magnetic flux would spread around the coil or current-carrying conductor and link with the electrically conductive surroundings (i.e., auxiliary equipment, metal support, tools, etc.). The flux concentrator forms a magnetic path to channel the main magnetic flux of the inductor in a well-determined area outside the coil. Figure 23a shows the current distribution in an isolated conductor. The current redistribution within this conductor after locating a conductive load (i.e., workpiece) near the conductor can be observed in Figure 23b. Due to the proximity effect, a significant part of the conductor's current will flow near the surface of the conductor that faces the load. The remainder of the current will be concentrated in the sides of the conductor.

When an external magnetic flux concentrator is placed around the conductor (Figure 23c), practically all of the conductor's current will be concentrated on the surface facing the workpiece. The magnetic concentrator will squeeze the current to the "open surface" of the conductor—in other words, to the open area of the slot (slot effect).

Current concentration within the coil surface facing the workpiece results in good coil-workpiece coupling and therefore improves the coil efficiency.

The actual current distribution in the conductor (Figure 23c) depends on the frequency, magnetic field intensity, geometry, and material properties of the conductor, the workpiece, and the concentrator.

There is a common misconception that the use of flux concentrators automatically leads to an increase in efficiency. Flux concentrators improve the efficiency of the process partly by reducing the coupling distance between the conductive part of the coil and the workpiece but also by reducing the stray losses (by reducing the reluctance of the air path). However, because the flux concentrator is an electrically conductive body and conducts high-density magnetic flux, there is some power loss generated as heat within it due to the Joule effect. This phenomenon could cause a reduction of electrical efficiency and the need to design a special water-cooling system to remove the heat from the concentrator. The first two factors will counteract the third, and the change in electrical efficiency will be a result of all three factors. In some applications, efficiency can actually be reduced. However, the appropriate use of a magnetic flux concentrator will typically allow an increase in process efficiency to be achieved. This can also result from the flux concentrator's ability to localize the magnetic field in a specific area. Because of this, the major portion of the magnetic field will not propagate behind the concentrator and the heated area will be localized. As a result, the heated mass of metal will be smaller, and therefore less power will be required to accomplish the desired heat treatment.

In most cases an application of magnetic flux concentrators does not require reengineering of the induction system. However, when a concentrator is used, higher power densities will be generated on the inside diameter of the coil (Figure 23c). If the original coil design would be susceptible to stress cracking over time due to the copper work hardening, this condition will occur sooner. Therefore, consideration must be given to the coil I.D. wall thickness and the location of quench holes, which are frequently located near the coil surface edges. The propagation of stress fractures can almost always be minimized or eliminated by a well thought out coil design. From another side, impedance of the straight coil can be much different compared to the coil with a flux concentrator. Therefore, after the flux concentrator has been installed in the straight coil it is necessary to check that the load properly matches to the power supply. Problem of load matching will be discussed in Chapter 11B.

Using a magnetic flux concentrator leads to an increase in power density not only in the induction coil but also in the workpiece areas that must be heated selectively. Because of this, the slot effect plays a particularly important role in the proper design of coils for selective induction hardening, including channel, hairpin, odd-shaped, spiral-helical, and pancake inductors. Special care should be taken when applying flux concentrators to a multi-turn coil. With this type of coil the voltage across the coil turns can be significant, and a short current path can develop through the concentrator. In this case, the reliability of the electrical insulation of concentrators plays an essential role in the coil design.

One of the major problems in the induction hardening of steel up to the austenite temperature range is that of undesirable heating of adjacent areas that have previously been hardened (the so-called temper-back or annealing effect of adjacent parts). This is particularly important in the induction hardening of crankshafts, camshafts, gears, and other critical components [152]. The complexity of this problem arises from the fact that, due to electromagnetic field propagation, the eddy currents are induced not only in the workpiece that is located under the inductor but in adjacent areas as well. A sketch of the induction carbon steel camshaft hardening system is shown in Figure 73. Figure 74 shows the magnetic vector potential field distribution around a two-turn cylindrical induction coil. Without a concentrator, the magnetic flux would spread around the coil and link with electrically conductive surroundings, which include neighboring areas of the part (i.e., cam lobes, journals) and possibly areas of the machine or fixture. As a result of induced eddy currents, heat will be produced. This heat can cause undesirable metallurgical changes in these areas. At different stages of the heating cycle, the pronunciation of the heating rate of the adjacent areas can vary.

At the first stage of the heating cycle, the entire workpiece is magnetic, the inductor has good efficiency, and intensive heating of the areas located under the coil takes place. Because of better coupling, any surface areas of the workpiece located under the coil (Figure 73, region A or B) will have much more intense heating than any other areas in the coil surroundings (for example, region C).

After a short time, the surface reaches the Curie temperature, the magnetic permeability μ drops to 1, the surface layer becomes nonmagnetic, and its heating intensity drastically

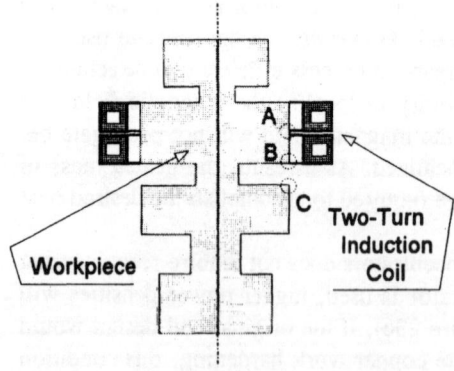

Figure 73   Sketch of induction surface-hardening system.

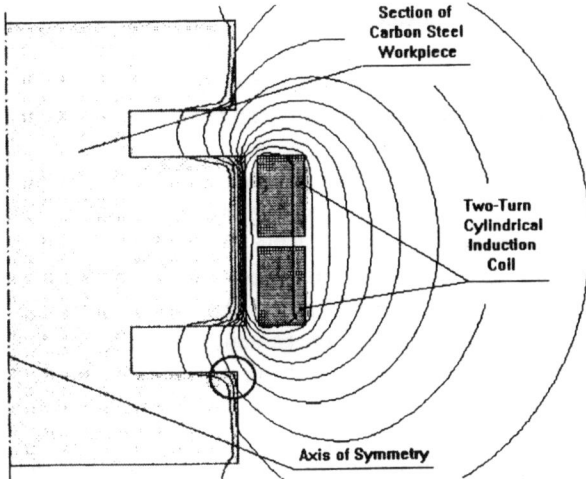

Figure 74 Electromagnetic field distribution in two-turn cylindrical coil without magnetic flux concentrator.

decreases. At this stage, the coil will not have as good a coupling factor as it had during the first stage, when the whole workpiece was magnetic. Although the surface of the part has lost its magnetic properties, the adjacent areas retain theirs. Consequently, the coupling factor of these areas will not decrease, and a greater portion of the electromagnetic field will link with the adjacent areas.

In addition, in order to have a short cycle time and to keep the heat intensity of the surface area located under the coil within the same range as during the first stage of heating, the system can automatically increase the coil power after the surface temperature passes the Curie point. This also will result in additional heating of magnetic parts located in the coil surroundings, which leads to the temper-back of these areas.

After a magnetic flux concentrator is located around the inductor (Figure 75), a small portion of the electromagnetic field of the coil will link with adjacent areas. For example, in the induction surface hardening of carbon steel camshafts, appropriate use of magnetic flux

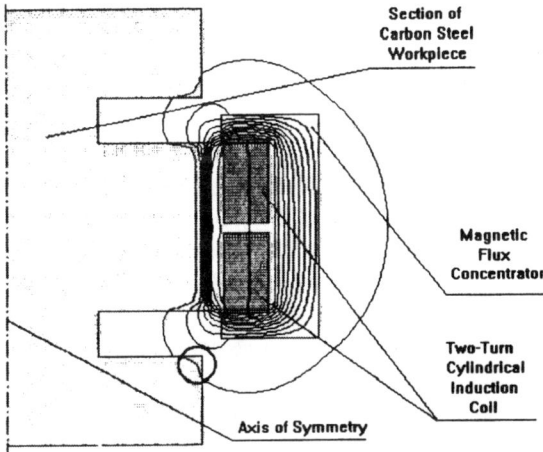

Figure 75 Electromagnetic field distribution in two-turn cylindrical coil with magnetic flux concentrator.

concentrators allows a 4–12-fold reduction in the power density induced in adjacent cams compared to using a bare coil.

Therefore, magnetic flux concentrators allow decoupling of the induction coil and the adjacent electrically conductive areas. This eliminates undesirable heating of these areas and as a result the temper-back (annealing) effect.

As one can see from Figures 12, 74, and 75, because of the magnetic field redistribution, a bare induction coil cannot provide the required heating pattern in the workpiece. The areas with high power density are observed in adjacent areas of the part but not in the required area. With the addition of a flux concentrator at the coil edges, practically all of the coil's current will be concentrated on the surface facing the workpiece. The magnetic flux concentrator will squeeze the current to the "open internal face" of the coil. Current concentration within the coil surface facing the workpiece results in good "coil-workpiece" coupling, with a consequent improvement in process efficiency. At the same time, the correct choice of flux concentrators and their location, geometry, material properties, and frequency allows the designer to decrease the heat intensity in the adjacent areas of the workpiece and therefore avoid their undesirable heating. In contrast to locating concentrators only at the end areas of the coil (Figure 12), the application of a "C"-shaped flux concentrator which warps around the coil will lead to some coil efficiency improvement. However, this will of course increase the cost of the design.

In some heat treatment applications, several coils are involved. Because of the relatively small distances between coils, strong magnetic ties can form between coils. This can lead to some negative electromagnetic effects (i.e., power transfer between coils). From another perspective, the stray flux might cause an undesirable temperature profile in the workpiece. In these applications, magnetic concentrators can be used as electromagnetic shields, which will allow undesirable coil interactions and their negative results to be eliminated. In general, the effectiveness of magnetic flux shields depends on various parameters such as frequency, magnetic field intensity, material properties, and the geometry of the induction system. Care should be taken at the corners of flux concentrators because of their tendency to overheat due to electromagnetic end effects.

Manufacturers of induction heat treatment equipment have found the development and use of flux control technology increasingly important in reducing the size and improving the quality of induction heat treatment systems. Before beginning a project, detailed mathematical modeling and lab tests should be conducted to determine the cost effectiveness of using flux concentrators in a particular application. Different applications may call for different flux concentrator materials (i.e., Laminations, Alphaflux, Fluxtrol, Ferrotron, Alphaform, etc.) Interested readers will find an analysis of the use of various materials for flux concentrators in References 141–152. Here we give only a short description of the features of using flux concentrators based on materials obtained from Fluxtrol Manufacturing Inc., IHS, HWG, Alpha 1, and many years of experience using flux concentrators at INDUCTOHEAT, Inc.

The choice of concentrator material depends on several factors where the higher the value, the better the situation, including [152] relative magnetic permeability, electrical resistivity, thermal conductivity, Curie point, saturation flux density and ductility. Other important factors rely on lower values for a better situation including hysteresis losses and eddy current losses.

Additional factors that should be considered include the ability to be cooled and to withstand high temperatures; resistance to chemical attack by quenchants; machinability; formability; ease of installation and removal and cost, which is dependent on the concentrator material, frequency, power density and geometry of the heat treating system.

In heat treating, the materials typically used as magnetic flux concentrators are "soft magnetic" in nature, meaning that they are magnetic only when an external magnetic field is applied. In an electromagnetic field these materials can change their magnetization rapidly without much friction. They are charactertized by a tall and narrow hysteresis loop of small area.

Magnetic materials that are soft magnetic usually have a uniform structure, low anisotropy, and magnetic domains that are randomly arranged. Random arrangement corresponds to a minimum energy configuration when the magnetic effects of the domains cancel each other. Therefore, the overall result is zero magnetization. The magnetic domains can be easily rearranged by applying an external magnetic field. The direction of domain rearrangement corresponds to the direction of the applied field. In this case, the magnetic materials behave as a temporary magnet [40,44–46].

As mentioned above, magnetic materials used for flux concentrators should have both a high slope of magnetic permeability and a high saturation flux density. Besides the magnetic permeability there are other important material properties of the concentrator material such as electrical resistivity and thermal conductivity. Magnetic materials with high electrical resistivity reduce the eddy current losses of the flux concentrator, thereby reducing its temperature increase. High thermal conductivity flux concentrators usually have a longer life because they aren't sbuject to local overheating. Local overheating can be caused by heat radiation from the heated workpiece or the high-density flux in certain areas of the magnetic concentrator.

One of the most important magnetic properties of concentrator materials is hysteresis loss. This quality is derived from the magnetization curve [40]. A typical magnetization curve, representing the magnetization process of a magnetic material, shows

1. A cycle of magnetization in one direction
2. A reversal of the applied magnetic field, which results in demagnetization of previously magnetized material and its magnetization in the opposite direction
3. Another reversal process resulting in magnetization in the original direction

Hysteresis loss is characterized by the conversion of electromagnetic energy into thermal energy due to the rearrangement of the magnetic domains during the hysteresis cycle. This loss should be as small as possible because it signifies a temperature rise in the flux concentrator, which can cause a loss of its magnetism, and therefore a reduction in coil efficiency.

Hysteresis loss is proportional to the area of the hysteresis loop and the frequency. Materials used for flux concentrators should have a coercive force as small as possible. A perfect flux concentrator with maximum efficiency would have no magnetization remaining after the external magnetic field has decreased to zero. A wide opening in the magnetization curve and a high frequency correspond to a high value of hysteresis loss. The flux concentrator properties can be determined from the manufacturer's data sheet or can be measured with the appropriate test equipment.

Materials most commonly used in induction heat treatment for flux concentrators and flux diverters are of the following types:

1. Laminations
2. Electrolytic iron-based materials
3. Carbonyl iron-based materials
4. Pure ferrites and ferrite-based materials
5. Soft formable materials

Laminations have been adapted for use in heat treatment from the motor and transformer industry. Grain-oriented magnetic alloys used in laminations are nickel-iron alloys and cold-rolled and hot-rolled silicon-iron alloys. Packets of laminated steel punchings are used effectively from line frequency to 30 kHz. Hovever, there are cases where laminations have been successfully used at higher frequencies (i.e., 100 kHz plus). Laminations must be electrically isolated from each other and used at the proper frequency. Laminations are insulated with mineral and organic coatings. The thickness of the individual laminations should be held to a minimum to keep eddy-current losses in the concentrators low. Generally, laminations are 0.06–0.8 mm thick. Thin laminations are used for higher frequencies. Laminations with a thickness greater than 0.5 mm are typically used for frequencies below 3 kHz. Compared to most available magnetic flux concentrator materials, laminations have the highest relative magnetic permeability and saturation flux density. This is considered an important advantage. When laminations are applied, there are some problems that can occur. Laminations are particularly sensitive to aggressive environments such as quenchants, that leads to rust and degradation problems. Degradation of the magnetic properties of laminations are caused by an increase of coercive force and hysteresis loss. If the laminations are not firmly clamped, the punchings could start vibrating, resulting in mechanical damage, noise, and subsequent failure of the coil or process. Care should be taken with the corners of laminations because of their tendency to overheat due to electromagnetic end effects.

One of the main advantages of using laminations is that they are relatively inexpensive and can withstand high temperatures better than other materials. Lamination packets can also be used to support the induction coil while remaining insulated from it.

Electrolytic iron-based materials [143] were developed in the 1980s and early 1990s specifically for induction heat treatment applications. They can be machined by conventional methods, come in different sizes, and are available in two types of alloys with permeability up to 56. Some alloys are rated for higher frequencies (50–450 kHz) and others for lower frequencies (50 Hz to 50 kHz). They do not significantly degrade over time and can be easily removed and replaced for coil repairs.

Carbonyl iron-based materials came to the induction industry from products developed for the radio industry in the 1960s. They are easily machined but are available in only a few sizes having a low permeability ($\mu = 15$). These products were developed for the higher frequencies (200–450 kHz) and are easy to machine because of their high plastic content [143]. They cannot be soft soldered and must be acid etched. They have temperature characteristics similar to those of the electrolytic products but usually produce case depths about 10–20% less than the electrolytic products and laminations.

Other materials are pure ferrites or ferrite-based. Ferrites are dense ceramic structures made by mixing iron oxide (Fe_2O_3) with oxides or carbonates of one or more metals such as nickel, zinc, or magnesium. They are pressed, then fired in a kiln at high temperature and machined to suit the coil geometry. In relatively weak magnetic fields ferrits have very high magnetic pemeability ($\mu = 2000$ plus). Ferrites are quite brittle materials and this is their main drawback. Other disadvantages of ferrites deal with the low saturation flux density (typically 3000 gauss–4000 gauss), low Curie point (approximately 220°C or 450°F) poor machinability and the inability to withstand thermal shocks. Because of their high resistance, ferrite-based magnetic concentrators are particularly attractive for use in high power density magnetic fields or with high frequencies (50 kHz and higher).

Some concentrator materials are provided in a soft formable state (e.g., Alphaform [145]) that can be easily molded to a desired shape for development purposes and later machined, if desired, to exact tolerances. Alphaform is an advanced composite of insulated iron micro-particles, space-age polymers, and a thermally sensitive catalyst.

In some applications, magnetic flux concentrators can be made from a single material. Others may be constructed of several materials. For example, in a split-return coil, laminations can be located in the middle area of the coil and iron- or graphite-based materials can be placed at the coil ends. Such designs are cost-effective and electrically efficient because they take into account the field distortion due to the electromagnetic end effect that results in additional losses within laminations at the coil ends.

As stated earlier, the use of magnetic flux concentrators delivers substantial profits to the modern heat treater, such as:

Reducing the operating power levels required to obtain the desired heating of the work-pieces

Improving the electrical efficiency of the process and decreasing the amount of energy used

Making it possible to selectively heat specific areas of the workpiece

Obtaining a superior heat pattern (i.e., more consistent hardness of the workpiece or more uniform heat patterns) and improving the physical and metallurgical properties

Minimizing geometric distortion of the workpiece

Preventing undesirable heating and annealing of adjacent parts

Reducing the number of rejected parts, rework, and scrap

Improving equipment life

Reducing cycle time

Eliminating the negative biological effects of electromagnetic field exposure on humans

Adding a flux concentrator to an existing inductor will result in an additional expense. A common saying among people who deal with induction is that if a good part can be produced without a flux concentrator, there is no reason to add to the coil cost.

One of the major concerns in using magnetic flux concentrators has to do with the reliability of their installation. Typically, flux concentrators are soldered or screwed or sometimes simply glued to the induction coil. With time, due to different factors, there can be unexpected shifting or movement of the concentrator to an improper position. Usually, magnetic concentrators are located in the areas with high magnetic flux density. Because of this, concentrators are affected by electromagnetic forces. Over time, as a result of those forces, the concentrator can become loose.

In addition to electromagnetic forces, this looseness can take place due to unstable temperature conditions. During the operation cycle, the flux concentrator could be heated to 250°C (482°F); then during quenching it could be cooled to the ambient temperaure. In typical surface-hardening applications, such heating–cooling cycles can be repeated up to 200–600 times per hour. After a series of such heating-cooling cycles an expansion–reduction of the volume of the flux concentrator takes place. Unstable temperature conditions can also result in the flux concentrator moving or relocating itself in an improper position. As a result, the heating and hardening patterns will change.

Unexpected changes in the hardening pattern can cause very serious damage. For example, in the automotive industry this can result in the car being recalled for replacement of the defective part. To prevent such a situation, the flux concentrators should be examined from time to time and repaired if necessary. In some cases, special monitoring systems can be installed to indicate changes in concentrator operation; however, such monitoring systems substantially increase the cost of the equipment.

In order to choose the right concentrator material, a heat treater can be confused by a wide variety of offerings available in today's market. Some manufacturers of magnetic flux concentrators may claim that their products are better suited to a certain induction heat treating application than another.

The following examples illustrate some of the practical benefits and features of using flux concentrators in various heat treatment applications [142].

Application 1. Single-Shot Hardening of Long Shafts

Purpose: To increase the torsional strength of the shaft by case hardening to a depth of one-fourth of the shaft diameter (one-half the radius).

Single-shot applications can increase productivity by heating the entire hardened area at one time. This application will require a larger power supply. When the power input is too low, the heat is conducted too deep; when power is too high, the surface overheats. By using a flux concentrator, the power requirement is minimized and better control is established over process parameters such as cycle time and pattern termination at the ends of the heated zone. Lower power also increases the life of the inductor. The shaft shown in Figure 76 (a dummy for test purposes only) was heated with less than 200 kW. Quench blocks have been removed to permit viewing of the heated area.

Application 2. Channel Coil for Continuous Processing

Purpose: To anneal parts that are moving through a long channel on a chain conveyor.

In multiple-station processes where conveyor speed is constant through all stations, the induction processing operation can be tuned to the constraints of the system by adjusting either the power input or the length of the inductors. Applications such as shell case and spark plug annealing are aided by flux concentrators on the channel coils (Figure 77), which increase the amount of energy that actually reaches the workpieces. Thus both the power requirement and coil length can be reduced.

Application 3. Single-Shot Hardening of the Drive Stem

Purpose: To increase strength of torque-transmitting component; to harden into fillet area and up to, but not into, the snap-ring groove.

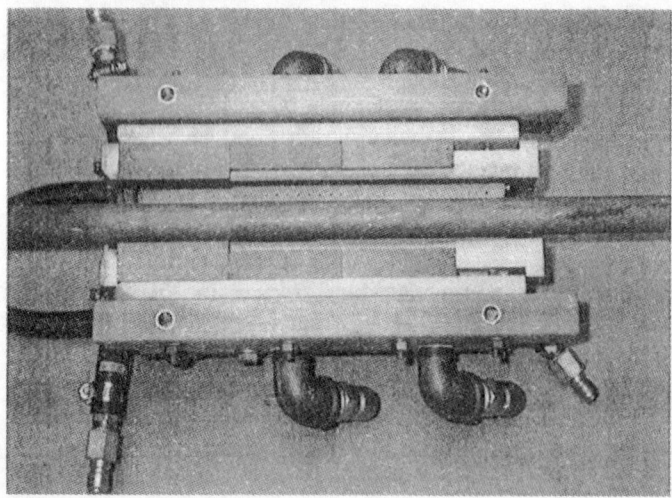

Figure 76   Single-shot induction hardening of long shafts.

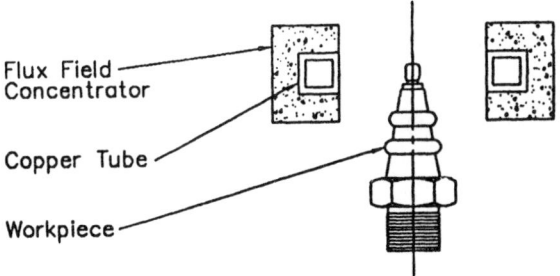

Flux Field
Concentrator

Copper Tube

Workpiece

Figure 77   The use of flux concentrators in channel-type induction coils, multistation setups.

Traditional scanning methods for induction hardening have created several problems, including (1) overheating of the corner of the large diameter near the fillet; (2) formation of a shallow and/or weak pattern in the fillet; and (3) a long, tapered pattern terminating below the snap-ring groove.

The single-shot inductor enhanced heating characteristics but still did not consistently harden to specified depths over the entire part. Hot bands were observed at each end of the stem, and underheating occurred on the corner of the large diameter, in the fillet, and in the shank. With the addition of flux concentrators, the shank, large diameter, and fillet between them are heated at the same rate to produce the required hardness pattern. With the flux concentrators placed correctly at the top of the inductor, the pattern termination near the snap-ring groove is sharper and more controllable. The improved heating pattern was clearly visible in the parts as they were heated. Figure 78a shows the cut samples, the effects of improper heating (through heating at the top of the stem), and the improved pattern produced by the inductor after adding flux concentrators.

As inductor designers gain experience and familiarity with families of workpiece configurations, inductors can be designed with flux concentrator materials installed during initial fabrication (Figure 78b). Minor changes are all that may be required for final pattern development. Typical heat time for short stems like this, at 10 kHz, is 6 s at 100 kW.

Application 4. Hardening a Valve Seat

Purpose: To improve surface wear to avoid marking by the valve.

Using a magnetic concentrator in this case turned an almost impossible development task into a fairly reasonable one. There were two obvious complexities. First, the diameters adjacent to the seat are the same size as the seat's outside diameter. The heat pattern produced by a simple inductor small enough to pass through these bores will not reach the outside diameter on the surface of the valve seat. Second, the tube inside the seat must not be hardened, yet the field intensity in a round inductor is greatest within the bore of the coil.

Magnetic flux concentrator material applied in the center and top of the inductor protects the center tube. It causes the heated pattern to flatten out across the seat and "pushes" the pattern across the entire surface. The pattern shown was accomplished with a 400 kHz unit at 6.0 kW in 4.0 s. Figure 79 shows the placement of the powdered iron concentrator.

Application 5. Surface Hardening of Rocker Arm Tip

Purpose: To improve surface wear on nodular iron rocket arm.

Previous methods of heating rocker arm tips usually involved a circular inductor that heated the entire end of the rocker arm. The pattern was not optimal, and the time required to con-

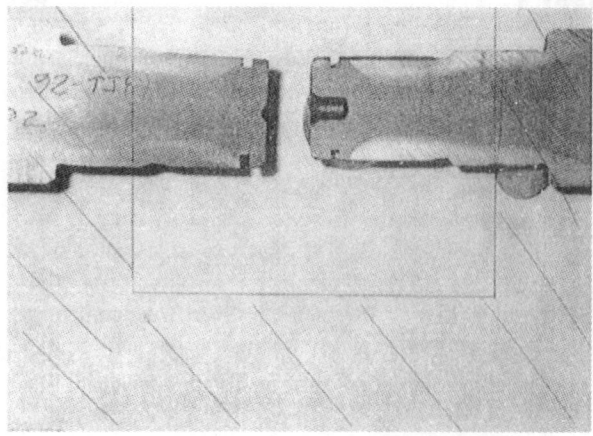

(a)

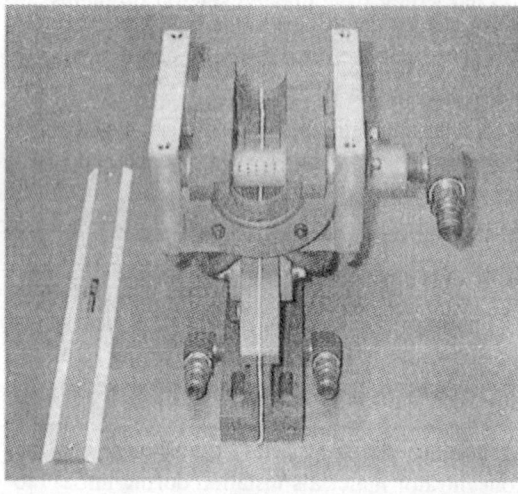

(b)

Figure 78 Single-shot hardening of the drive stem.

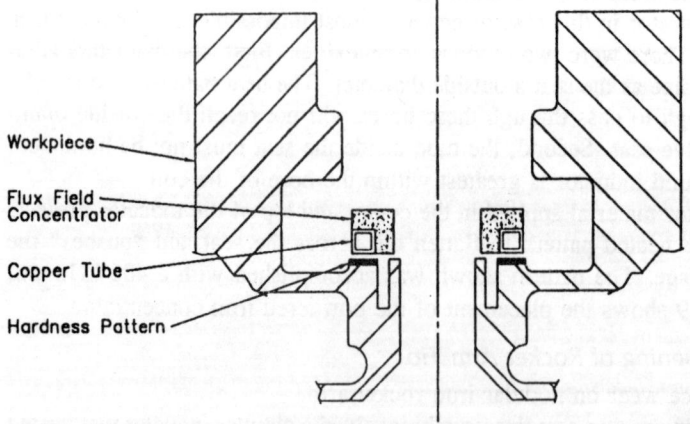

Figure 79 Placement of powdered iron–based concentrator for induction heat treatment of complex shaped part.

duct heat to the center of the wear surface was excessive. A split-return type of inductor improved results, and the use of flux concentrators allows the wear surface to reach the hardening temperature before the remainder of the workpiece is appreciably heated. Heating time for this application was 1.0 s at 27 kW at a frequency of 25 kHz.

From the examples described above, one can conclude that when choosing a magnetic flux concentrator for a particular application, the selection factors discussed here should be carefully considered. At the same time, major attention should be given to the location of the concentrator, its shape, and applied frequency. When many factors are involved in obtaining the required heating pattern, the computational ability of the induction heat treatment manufacturer becomes an ultimate advantage over companies that rely upon intuition and the experience of past mistakes.

VI. SUMMARY

To briefly summarize the chapter, induction heat treatment may be viewed as a very specialized area of induction heating. Well-known basic principles of heat transfer and electromagnetics that apply to induction heat treatment have been discussed, although no attempt has been made to provide an exhaustive analysis because of space limitations.

Modern technology requires a deeper study of the processes involved in induction heat treatment. The use of the most advanced computational methods and software allows engineers to obtain more detailed knowledge of the process. This knowledge helps them to generate innovative ideas, discover new processes, and better understand the unique industrial technologies.

As mentioned in the introduction, one of the goals of this chapter has been to emphasize the interrelated features of electromagnetic and heat transfer phenomena. However, the metallurgical aspect is the third but no less important part of the heat treating. We specifically did not discuss the metallurgical aspect here because it is discussed in previous chapters.

The variety of induction heat treatment applications necessitates the use of many different combinations of power supplies, load matching, process control, and monitoring equipment. Some induction coils can be very effective with certain inverters and ineffective with others. Therefore, the optimal design of a modern induction heat treatment system must take into consideration not only the features of the physical stand-alone process but a combination of the inductor, load-matching station, and inverter. The operational characteristics of the power supply must properly match the coil requirements to obtain the desired results. These subjects are discussed in Chapter 11B.

Should further details be required or questions arise regarding particular applications, the authors welcome any reader inquiries or suggestions at INDUCTOHEAT, Inc.

REFERENCES

1. C. A. Tudbury, *Basics of Induction Heating*, Rider, New York, 1960.
2. M. G. Lozinskii, *Industrial Applications of Induction Heating*, Pergamon, London, 1969.
3. E. J. Davies, *Induction Heating Handbook*, McGraw-Hill, New York, 1979.
4. A. E. Slukhotskii and S. E. Ryskin, *Inductors for Induction Heating*, Energy Publ., St. Petersburg, Russia, 1974 (in Russian).
5. A. E. Slukhotskii, V. S. Nemkov, et al., *Induction Heating Equipment*, St. Petersburg, Russia, 1981 (in Russian).
6. S. L. Semiatin and D. E. Stutz, *Induction Heat Treatment of Steel*, ASM International, Metals Park, OH, 1986.
7. ASM Int., *Induction Heating*, Course 60, ASM International, Metals Park, OH, 1986.
8. M. Orfeuil, *Electric Process Heating*, Battelle Press, 1987.

9. S. Zinn, *Elements of Induction Heating: Design, Control, and Applications*, ASM Int., Metals Park, OH, 1988.
10. E. J. Davies, *Conduction and Induction Heating*, Peter Peregrinus, London, UK, 1990.
11. A. D. Demichev, *Surface Induction Hardening*, St. Petersburg, Russia, 1990 (in Russian).
12. G. F. Golovin and M. M. Zamjatin, *High-Frequency Induction Heat Treating*, Mashinostroenie, St. Petersburg, Russia, 1990 (in Russian).
13. D. F. Mellon, Contour gear hardening using induction heating with RF and thermographic control, *Industrial Heating*, July 1988.
14. INDUCTOHEAT Bulletin, *Contour Gear Hardening Eliminates Carburizing Problems*, November 1988.
15. D. F. Mellon, Induction contour gear hardening for efficient in-line heating, *Precision Metals*, June 1988, pp. 49–53.
16. INDUCTOHEAT Bulletin, *Gear Hardening Integrated with Workpiece Inspection*, August 1991.
17. INDUCTOHEAT Bulletin, *Induction Heating Update: Cellular Manufacturing*, August 1992.
18. INDUCTOHEAT Bulletin, *Hardening and Tempering of Automotive Spindles*, 1991.
19. ASM, *Metals Handbook*, 9th ed., Vol. 4, *Heat Treating*, ASM, Cleveland, OH, 1991.
20. INDUCTOHEAT Bulletin, *A New Wave of Induction Hardening Automation*, 1992.
21. INDUCTOHEAT Bulletin, *Hardening and Tempering Steel Chain*, 1991.
22. INDUCTOHEAT Bulletin, *Shrink Rear Axle Drive Gear*, 1991.
23. INDUCTOHEAT Bulletin, *Three-Stage Hardening of Transmission Parts Shifts into Gear*, 1992.
24. INDUCTOHEAT Bulletin, *Harden and Temper ID and OD of Cams*, 1992.
25. INDUCTOHEAT Bulletin, *Annealing: Marine Engine Propeller Draft Shafts*, 1992.
26. B. Gebhart, *Heat Transfer*, McGraw-Hill, New York, 1970.
27. S. Patankar, *Numerical Heat Transfer and Fluid Flow*, Hemisphere, New York, 1980.
28. F. P. Incropera and D. P. Dewitt, *Fundamentals of Heat Transfer*, Wiley, New York, 1981.
29. R. F. Myers, *Conduction Heat Transfer*, McGraw-Hill, New York, 1972.
30. J. A. Adams and D. F. Rogers, *Computer Aided Analysis in Heat Transfer*, McGraw-Hill, New York, 1973.
31. W. M. Rohsenow and J. P. Hartnett, *Handbook of Heat Transfer*, McGraw-Hill, New York, 1973.
32. R. Siegel and J. R. Howell, *Thermal Radiation Heat Transfer*, 2nd ed., McGraw-Hill, New York, 1980.
33. J. A. Wiebelt, *Engineering Radiation Heat Transfer*, Holt, Rinehart and Winston, New York, 1966.
34. E. M. Sparrow and R. D. Cess, *Radiation Heat Transfer*, Wadsworth, Englewood Cliffs, NJ, 1966.
35. N. M. Beljaev and A. A. Rjadno, *Methods of the Heat Transfer*, Part 1-2, Vysshaja Schola, Moscow, 1982 (in Russian).
36. J. R. Howell, *A Catalog of Radiation Configuration Factors*, McGraw-Hill, New York, 1982.
37. ASM, *Metals Handbook*, Vol. 2, *Properties and Selection: Nonferrous Alloys and Pure Metals*, ASM, Cleveland, 1979.
38. ASM, *High Temperature Property Data: Ferrous Alloys*, ASM, Metals Park, OH, 1988.
39. L. V. Bewley, *Flux Linkages and Electromagnetic Induction*, Dover, New York, 1964.
40. *Encyclopedia of Physics*, 2nd ed., VCH, 1991.
41. R. W. Chabay and B. A. Sherwood, *Electric and Magnetic Interactions*, Wiley, New York, 1995.
42. P. Hammond, *Electromagnetism for Engineers*, Pergamon, New York, 1978.
43. M. A. Plonus, *Applied Electromagnetics*, McGraw-Hill, New York, 1978.
44. I. E. Tamm, *Fundamentals of the Theory of Electricity*, Moscow, Russia, 1981 (in Russian).
45. W. H. Hayt, Jr. *Engineering Electromagnetics*, McGraw-Hill, New York, 1981.
46. H. G. Booker, *Energy in Electromagnetism*, Peter Peregrinus, London, UK, 1982.
47. N. N. Rao, *Elements of Engineering Electromagnetics*, Prentice-Hall, Englewood Cliffs, NJ, 1987.
48. R. Ehrlich, J. Tuszynski, L. Roelofs, and R. Storner, *Electricity and Magnetism Simulations: The Consortium and Upper-Level Physics Software*, Wiley, New York, 1995.
49. V. S. Nemkov, B. S. Polevodov, and S. G. Gurevich, *Mathematical Modeling of High-Frequency Heating Equipment*, 2nd ed., St. Petersburg, Russia, 1991 (in Russian).
50. V. I. Rudnev and K. L. Schweigert, Designing induction equipment for modern forge shops, *Forging*, Winter 1994, pp. 56–58.
51. V. S. Nemkov and V. B. Demidovich, *Theory of Induction Heating*, Energoatomizdat, St. Petersburg, Russia, 1988 (in Russian).
52. D. J. Griffiths, *Electrodynamics*, Prentice-Hall, Englewood Cliffs, NJ, 1989.
53. V. I . Rudnev, Mathematical simulation and optimal control of induction heating of large-dimensional cylinders and slabs, Ph.D. Thesis, Dept. Electrical Technology, St. Petersburg Electrical Eng. Univ., Russia, 1986 (in Russian).
54. G. D. Smith, *Numerical Solution of Partial Differential Equations: Finite Difference Methods*, 3rd ed., Oxford Univ. Press, Oxford, UK, 1985.
55. K. J. Binns, P. J. Lawrenson, and C. W. Trowbridge, *The Analytical and Numerical Solution of Electric and Magnetic Fields*, Wiley, New York, 1992.

56. A. A. Samarskii, *Theory of Finite Difference Schemes*, Moscow, Russia, 1977 (in Russian).
57. S. V. Patankar and B. R. Baliga, A new finite-difference scheme for parabolic differential equations, *Num. Heat Transfer 1*: 27–30 (1978).
58. K. S. Demirchian and V. L. Chechurin, *Computational Methods for Electromagnetic Field Simulations*, Moscow, 1986 (in Russian).
59. J. C. Strikwerda, *Finite Difference Schemes and Partial Differential Equations*, Wadsworth & Brooks, Belmont, CA, 1989.
60. M. V. K. Chari, Finite element analysis of nonlinear magnetic fields in electric machines, Ph.D. Dissertation, McGill Univ., Montreal, PQ, Canada, 1970.
61. M. V. K. Chari and P. P. Silvester, Finite element analysis of magnetically saturated DC machines, *IEEE Trans. PAS 90*: 2362 (1971).
62. J. Donea, S. Giuliani, and A. Philippe, Finite elements in solution of electromagnetic induction problems, *Int. J. Num. Methods Eng. 8*: 359–367 (1974).
63. W. Lord, Application of numerical field modeling to electromagnetic methods of nondestructive testing, *IEEE Trans. Magn. 19*(6): 2437–2442 (1983).
64. W. Lord, Development of a finite element model for eddy current NDT phenomena, Elec. Eng. Dept., Colorado State Univ., 1983.
65. C. Marchand and A. Foggia, 2D finite element program for magnetic induction heating, *IEEE Trans. Magn. 19*(6): 2647–2649 (1983).
66. P. P. Silvester and R. L. Ferrari, *Finite Elements for Electrical Engineers*, Cambridge Univ. Press, New York, 1983.
67. W. Muller, C. Kramer, and J. Krueger, Calculation of 2- or 3-dimensional linear or nonlinear fields by the CAD-program profi, *IEEE Trans. Magn. 19*(6): 2670–2673 (1983).
68. R. K. Livesley, *Finite Elements: An Introduction for Engineers,* Cambridge Univ. Press, New York, 1983.
69. N. Ida, 3-D finite element modelling of electromagnetic NDT phenomena, Ph.D. Thesis, Colorado State Univ., 1983.
70. J. D. Lavers, Numerical solution methods for electroheat problems, *IEEE Trans. Magn. 19*(6): 2566–2572 (1983).
71. D. A. Lowther and P. P. Silvester, *Computer Aided Design in Magnetics,* Springer, Berlin, 1986.
72. W. Lord, Y. S. Sun, S. S. Udpa, and S. Nath, A finite element study of the remote field eddy current phenomenon, *IEEE Trans. Magn. 24*(1): 435–438 (1988).
73. A. Muhlbauer, S. S. Udpa, V. I. Rudnev, and A. F. Sutjagin, Software for modeling induction heating equipment by using finite elements method, Proc. 10th All-Union Conf. High-Frequency Application, St. Petersburg, Russia, 1991, Part 1, pp. 36–37 (in Russian).
74. E. J. W. ter Maten and J. B. M. Melissen, Simulation of inductive heating, *IEEE Trans. Magn. 28*: 1287–1290 (1992).
75. J. M. Jin, *The Finite Element Method in Electromagnetics,* Wiley, New York, 1993.
76. S. Mandayam, L. Udpa, S. S. Udpa, and Y. S. Sun, A fast iterative finite element model for electrodynamic and magnetostrictive vibration absorbers, Proc. 9th Conf. Comput. Electromagnetic Fields, COMPUMAG, Miami, 1993, pp. 8–10.
77. T. Dreher and G. Meunier, A 3D line current model of voltage driven coils, Proc. 9th Conf. Comput. Electromagnetic Fields, COMPUMAG, Miami, 1993, pp. 50–52.
78. M. Hano, An improved finite element analysis of eddy current problems connected to voltage source, *IEEE Trans. Magn. 29*(2): (1993).
79. E. Thimpson, The finite element method, Class Notes CE-665, Colorado State Univ., Progress in 1978–1990.
80. O. C. Zienkiewicz and R. L. Taylor, *The Finite Element Method*, 4th ed., Vol. 1, *Basic Formulation and Linear Problems*, McGraw-Hill, New York, 1989.
81. L. J. Segerlind, *Applied Finite Element Analysis*, Wiley, New York, 1976.
82. C. S. Desai, *Elementary Finite Element Method*, Prentice-Hall, Englewood Cliffs, NJ, 1979.
83. S. J. Salon and J. M. Schneider, A hybrid finite element–boundary integral formulation of the eddy current problem, *IEEE Trans. Magn. 18*(2): 461–466 (1982).
84. T. H. Fawzi, K. F. Ali, and P. E. Burke, Boundary integral equations analysis of induction devices with rotational symmetry, *IEEE Trans. Magn. 19*(1): 36–44 (1983).
85. S. Cristina and A. Di Napoli, Combination of finite and boundary elements for magnetic field analysis, *IEEE Trans. Magn. 19*(6): 2337–2339 (1983).
86. C. A. Brebbia, *The Boundary Element Method for Engineers*, Pentech Press, London, 1978.
87. K. R. Shao and K. D. Zhou, Boundary element solution to transient eddy current problems, *Eng. Anal. 1*: 182–187 (1984).
88. M. H. Lean, Electromagnetic field solution with the boundary element method, Ph.D. Thesis, Univ. Manitoba, Winnipeg, MB, Canada, 1981.

89. Y. B. Yildir, A. boundary element method for the solution of Laplace's equation in three-dimensional space, Ph.D. Thesis, Univ. Manitoba, Winnipeg, Canada, 1985.

90. T. Inuki and S. Wakao, Novel boundary element analysis for 3-D eddy current problems, *IEEE Trans. Magn.* 29(2): 1520–1523 (1993).

91. K. F. Wang, S. Chandrasekar, and H. T. Y. Yang, Finite element simulation of induction heat treatment and quenching of steels, *Trans. NAMRI/SME XX*: 83–90 (1992).

92. V. I. Rudnev, Characteristics of transverse electromagnetic edge effect in induction heating of magnetic and nonmagnetic slabs, in *The Study of Electrothermal Processes*, Cheboscary, Russia, 1985, pp. 30–34 (in Russian).

93. V. I. Rudnev, D. L. Loveless, M. R. Black, P. J. Miller, Progress in study of induction surface hardening of carbon steels, gray irons and ductile (nodular) irons, *Industrial Heating*, March, 1996.

94. G. Roen, Metallurgical considerations in induction hardening, Proceedings of 6th International induction heating seminar, Nashville, September, 1996.

95. INDUCTOHEAT Bulletin, *UNICOOL, Closed Loop Water and Recirculating System*, 1992.

96. INDUCTOHEAT Bulletin, *Water Recirculating System, Plate Type Closed Loop*, 1991.

97. INDUCTOHEAT Bulletin, *Quick Change Coil Adapters*, 1993.

98. INDUCTOHEAT Bulletin, *UNISCAN-I*, 1991.

99. INDUCTOHEAT Bulletin, *UNISCAN-II*, 1992.

100. INDUCTOHEAT Bulletin, *UNISCAN-IV*, 1992.

101. INDUCTOHEAT Bulletin, *STATISCAN VSS-20*, 1991.

102. INDUCTOHEAT Bulletin, *Horizontal Scanning Machine*, 1991.

103. INDUCTOHEAT Bulletin, *Single-Shot Hardening*, 1991.

104. V. I. Rudnev, *The Art of Computation of the Induction Heating Process*, INDUCTOHEAT Booklet, INDUCTOHEAT, Inc., Madison Heights, MI, 1994.

105. V. I. Rudnev, New induction heat technology in Russia, in *Heat Treating: Equipment and Processes*, Conf. Proc., ASM International, 1994, pp. 209–213.

106. INDUCTOHEAT Bulletin, *UNIPOWER UPF, Self Contained Induction Heating System*, 1990.

107. V. I. Rudnev, R. L. Cook, and J. LaMonte, Induction heat treating: Keyways and holes, *Metal Heat Treating*, March-April, 1996.

108. V. I. Rudnev, D. L. Loveless, and R. L. Cook, Striping phenomena, *Industrial Heating*, November, 1995.

109. Gear Research Institute, Review of Literature on Induction Hardening of Gears, Project A-1051 (C553), 1994.

110. American Gear Manufacturers Association, *Gear Materials and Heat Treatment Manual*, ANSI/AGMA 2004-B29, Section 5.2, Flame and Induction Hardening, AGMA, 1989.

111. AGMA, *Design Guide for Vehicle Spur and Helical Gears*, ANSI/AGMA 6002-B93, AGMA, Section 3.

112. K. Namiki, T. Urita, I. Machida, and T. Takagi, The application of hardenability assured cold forging medium carbon steels to CVJ outer race, SAE Tech. Paper 930965, SAE Int.

113. J. M. Storm and M. R. Chaplin, Dual frequency induction gear hardening, *Gear Technol.* 10(2): 22–25 (1993).

114. Y. Matsubara, M. Kumagawa, and Y. Watanabe, Induction hardening of gears by dual frequency induction heating, *J. Jpn. Soc. Heat Treatment* 29(2): 92–98 (1989).

115. V. I. Rudnev and R. L. Cook, Bar end heating, *Forging*, Winter 1995, pp. 27–30.

116. INDUCTOHEAT Bulletin: *Bar Heating*, INDUCTOHEAT, Inc., 1994.

117. V. I. Rudnev and W. B. Albert, Continuous aluminum bar re-heating prior to reducing mill, *33 Metal Producing*, January 1995, p. 50.

118. V. S. Nemkov, V. B. Demidovich, V. I. Rudnev, and O. Fishman, Electromagnetic end and edge effects in induction heating, UIE Congress, Montreal, Canada, 1991.

119. V. I. Rudnev and D. L. Loveless, Induction slab, plate and bar edge heating for continuous casting lines, *33 Metal Producing*, October 1994, pp. 32–34, 43–44.

120. V. I. Rudnev and D. L. Loveless, Longitudinal flux induction heating of slabs, bars and strips is no lnoger "black magic," Part 1, *Industrial Heating*, January 1995, pp. 29–34.

121. V. I. Rudnev and D. L. Loveless, Longitudinal flux induction heating of slabs, bars and strips is no longer "black magic," Part 2, *Industrial Heating*, February 1995, pp. 46–50.

122. R. M. Baker, Transverse flux induction heating. *Elec. Eng.*, October 1950, pp. 922–924.

123. R. M. Baker, Transverse flux induction heating, *Trans. AIEE* 69(2): 711–719 (1950).

124. M. Lamourdedieu, Continuous heat treatment of aluminum alloy strip, *Metal Prog.*, October 1951, pp. 88–92.

125. M. Lamourdedieu, Continuous heat treatment of aluminum alloys of the Duralumin type, *80*: 335–338 (1951/1952).

126. R. C. Gibson and R. H. Johnson, High efficiency induction heating as a production tool for heat treatment of continuous strip metal, *Sheet Metal Ind.*, December 1982, pp. 889–892.

127. R. Waggott, D. J. Walker, R. C. Gibson, and R. H. Johnson, Transverse flux induction heating of aluminum alloy strip, *Metals Technol.*, *9*: 493–498 (1982).

128. J. Blacklung, Induction heating in rolling mills, new ideas and applications, *Rev. Metall., Cah. Inf. Tech.* *84*: 67–71 (1987).

129. N. V. Ross, R. H. Kaltenhauser, and G. A. Walzer, Transverse flux induction heating of steel strip, Electroheat Congr., Toronto, Canada, 1992, pp. 110–119.

130. T. Yamagishi, Y. Kitajima, K. Nagahama, H. Ishii, and H. Ikeda, TFX induction heating CAL for rolled aluminum sheet, Jpn. Light Metal Assoc., Tokyo, April 1988.

131. R. C. J. Ireson, Induction heating with transverse flux in strip metal process lines, *IEE Power Eng. J.,* Vol. 3, March 1989.

132. R. C. Gibson, W. B. R. Moore, and R. A. Walker, TFX—an induction heating process for the ultra rapid heat treatment of metal strip, ASM Heat Treatment and Surface Engineering Conf., Amsterdam, May 1991.

133. R. C. J. Ireson, Developments in transverse flux (TFX) induction heating of metal strip, *Metallurgia 58*:68 (February 1991).

134. S. Wilden, Inductive high-capacity annealing furnaces for the continuous treatment of strips, *Metallurgia,* Vol. 43, August 1989.

135. I. Oku, M. Inokuma, and K. Awa, Application of induction heating to continuous treatment for aluminum alloy strip, *J. Jpn. Inst. Light Metals*, Vol. 40, August 1990.

136. S. B. Lasday, Work processing for continuous annealing of sheet by transverse flux induction heating at steel plant, *Induction Heating*, October 1991, pp. 43–45.

137. K. Standford, Transverse flux induction heating, *Eng. Digest*, September 1987, pp. 23–25.

138. P. W. Ainscow, Inductive heating for production of galvannealed steel strip, *Steel Times*, April 1992, pp. 173–175.

139. E. Balle, J. Calas, and A. B. Wilhelm, Lacquer coating line for tin mill products, *Iron Steel Eng.,* May 1991, pp. 36–38.

140. W. Kolakowski, Economical production of hot strip with the compact strip production process, Improved Technologies for the Rational Use of Energy in the Iron and Steel Industry, NEC Birmingham, UK, 1992.

141. V. I. Rudnev and R. L. Cook, Magnetic flux concentrators: myths, realities, and profits, *Metal Heat Treating*, March/April 1995, pp. 31–34.

142. J. S. LaMonte and M. R. Black, How flux concentrators improve inductor efficiency, *Heat Treating*, June 1989.

143. R. S. Ruffini, Production and concentration of magnetic flux for eddy current heating applications, Report presented to 1993 Int. Fed. Heat Treatment and Surface Engineering, Beijing, China, 1993.

144. R. S. Ruffini, Induction heating with magnetic flux technology, *Modern Appl. News*, 1991.

145. Alpha 1 Induction Service Center, General Presentation, *Industrial Heating*, pp. 54–55.

146. *Handbook of Experimental Stress Analysis*, Wiley, New York, 1963.

147. R. D. Cook, *Finite Element Modelling for Stress Analysis*, Wiley, New York, 1995.

148. A. J. Fletcher, *Thermal Stress and Strain Generation in Heat Treating*, Elsevier Science, London, 1989.

149. K. Weiss, In-line tempering on induction heat treating equipment, Proc. 1st Int. Induction Heating Seminar, São Paulo, Brazil, 1995.

150. General Presentation of HWG, Germany, 1993.

151. K. Weiss, In-line tempering on induction heat treating equipment relieves stresses advantageously, *Industrial Heating*, December, 1995.

152. V. I. Rudnev, R. L. Cook, D. L. Loveless, Keeping your temper with magnetic flux concentrators, *Modern Application News*, November, 1995.

11B

INDUCTION HEAT TREATMENT
Modern Power Supplies, Load Matching, Process Control, and Monitoring

Don L. Loveless, Raymond L. Cook, and Valery I. Rudnev
INDUCTOHEAT, Inc., Madison Heights, Michigan

I. POWER SUPPLIES FOR MODERN INDUCTION HEAT TREATMENT

Induction heating power supplies are frequency changers that convert the available utility line frequency power to the desired single-phase power at the frequency required by the induction heating process. They are often referred to as converters, inverters, or oscillators, but they are generally a combination of these. The converter portion of the power supply converts the line frequency alternating current input to direct current, and the inverter or oscillator portion changes the direct current to single-phase alternating current of the required heating frequency.

Many different power supply types and models are available to meet the heating requirements of a nearly endless variety of induction heating applications [1–10]. The specific application will dictate the frequency, power level (Figure 1), and other inductor parameters such as coil voltage, current, and power factor (cos φ) or Q factor.

For purposes of efficient communication, it is best at this point to provide a brief glossary of terms commonly used by induction heat treatment specialists that may be encountered in this section.

Buttons/Taps/Studs   The individual terminal connections on the load-matching capacitors above the porcelain insulators. Connection of a button involves installing a conducting copper strap or washer between an existing copper bus and the capacitor terminal, thus making an electrical connection between a section of the capacitor and the bus.

Cans   Load-matching capacitors.

Efficiency   Ratio of the actual power output of the power supply or coil to the required input power.

KVAR   A rating term for capacitors that specifies the amount of reactive volt-amperes that the capacitor can supply to the circuit when run at a specified voltage and frequency.

873

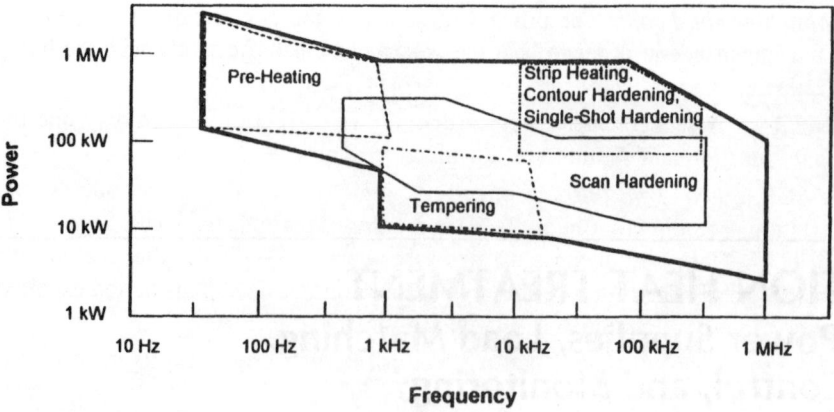

Figure 1   Typical induction heat treatment applications.

Power Factor   The cosine of the phase angle between the voltage and current of a given circuit. This is most often of interest to the user of induction heating as the ratio of kilowatts to the actual volt-ampere product required to produce that power. A higher power factor (cos φ) generally means a lower volt-ampere product required from the selected frequency converter.

Reactance and Q Factor   Inductive reactance is the opposition to the flow of alternating current by a pure inductance. A mechanical analogy would liken the inertia of a flywheel to the inductive reactance of a coil while the friction of the shaft and bearing of the flywheel is like the resistance of the coil. Using the same analogy, a high-Q coil is like a large flywheel with little shaft friction, and a low-Q coil is like a small flywheel with much friction. Q is the ratio of inductive reactance to the resistance of the coil.

Resonance   For a series circuit, resonance is the frequency at which the vector sum of capacitive and inductive reactance is equal to zero, the power factor (cos φ) is equal to 1 and the impedance is at its minimum value. For a parallel circuit containing resistance in either or both of the circuit paths, resonance may be defined as the frequency at which capacitive and inductive reactance is equal to zero, cos φ = 1, or the impedance is at its maximum value. For a low-Q circuit these points may be slightly different from one another, but at higher Q values they are essentially the same. For practical purposes, in induction heating it is possible to consider them equivalent.

Over the Hump   Refers to a condition where a swept frequency power supply is operating beyond the resonant frequency of the oscillating circuit.

Tank Circuit   The components in the resonant portion of the induction heating circuit, usually consisting of capacitors, transformers, and the induction coil. These components may be connected in series or parallel, depending on the type of power supply used.

Trombone Bus   A bus that is designed to slide in and out of a specified receiving bus. As the trombone bus is moved in and out of the receiving bus, the circuit inductance will vary to allow adjustment of the running frequency, voltage, and power.

Tuning or Matching the Load   The process of adjusting the circuit capacitance, inductance, and impedance in order to heat the workpiece most effectively with the hardware available.

Diode or Rectifier   A semiconductor device that allows current to flow in only one direction. Its function and symbol are illustrated in Figure 2.

Thyristor or SCR (Silicon-controlled rectifier)   A semiconductor device that allows current to flow in one direction but only after a firing signal or positive voltage is applied to its gate terminal. It turns off or blocks current flow only after the current supplied to it has stopped for a certain minimum period of time. This period is called the turn-off time. Its function and symbol are illustrated in Figure 3.

Transistor   A semiconductor device that will control current flow from its input terminal to its output terminal by application of a control signal to its gate or base terminal. The transistor can be switched on or off at any time. Its function and symbol are illustrated in Figure 4.

Frequency is a very important parameter in induction heat treatment because it is the primary control over the depth of current penetration and therefore of the depth and shape of the resulting heat pattern. Frequency is also important in the design of induction heating power supplies because the power components must be rated for operation at the specified frequency. The power circuit must ensure that these components are operated with adequate margin to yield high reliability at this frequency. The inverter circuits that convert direct current to alternating current use solid-state switching devices such as thyristors (SCRs) and transistors. For high power and lower frequencies, large thyristors are commonly used. For frequencies above 25 kHz or low power, transistors are used because of their ability to be turned on and off very fast with low switching losses.

Vacuum tube oscillators have been used extensively for many years at frequencies above 300 kHz. However, the tube oscillators have a low conversion efficiency of typically 50–60% compared to 83–95% for an inverter using transistors. Power vacuum tubes have a useful life of 2000–4000 h and are therefore a costly maintenance item. The high voltage (over 10,000

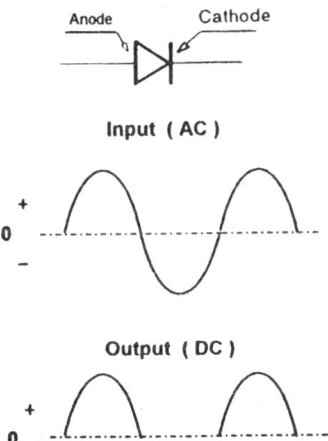

Figure 2   Symbol and waveshapes of diode.

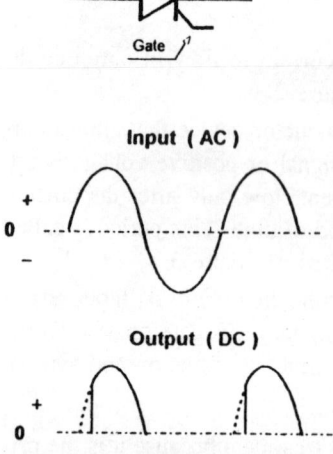

Figure 3 Symbol and waveshapes of thyristor.

required for tube operation is more dangerous than the 1000 V or less present in typical transistorized inverters. These negative features of tube oscillators have brought about a dramatic move toward the use of transistorized power supplies in most heat treatment applications that require a frequency of less than 1 MHz.

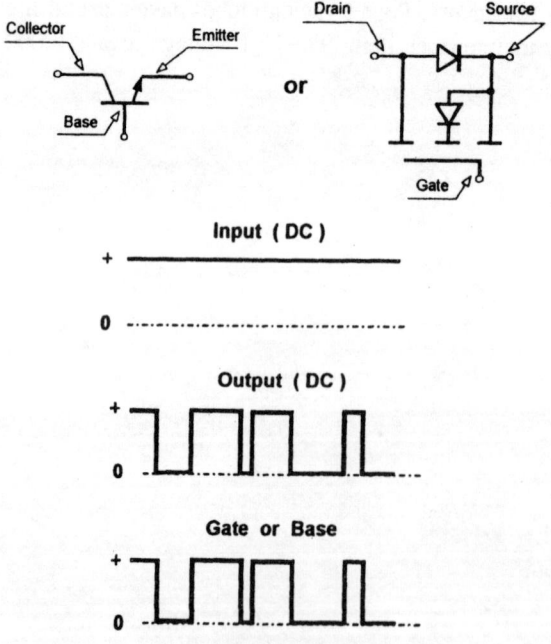

Figure 4 Symbol and waveshapes of transistor.

Figure 5 shows in graphical form the various power and frequency combinations that are covered by thyristor, transistor, and vacuum tube power supplies. There are obviously large areas of overlap where more than one type of power supply can be used.

The power required for a given application depends on the volume and kind of metal to be heated, the rate of heating, and the efficiency of the heating process. Small areas heated to a shallow depth may require as little as 1 or 2 kW, whereas heating wide, fast-moving steel strips to temperatures above their Curie point may require many megawatts of induction heating power. It is therefore necessary to define the process and its power requirements by using the numerical techniques described in the computation section of this chapter or by careful extrapolation from similar applications.

The part/coil geometry and the electrical properties of the material to be heated determine the specific coil voltage, current, and power factor. Defining these parameters is necessary to ensure that the output of the power supply is capable of matching the requirements of the coil. Most induction heating power supply systems have the ability to match a reasonable range of heating coil parameters.

Physical constraints imposed by the environment in which the induction heat treatment is to be done can also play an important part in the selection or application of the power supply. Each type of power supply, described in detail later in this section, has specific advantages that may directly affect its suitability in the overall heat treatment system.

Floor space, machine design, and plant layout are important factors in many induction heat treatment installations. For example, in highly automated machines with a number of hardening and tempering stations, the very compact unitized construction of a transistor-based power supply with self-contained load-matching transformer and capacitors is a definite advantage. On the other hand, for installations requiring a long distance between the power supply and the work coil, the heat station or load-matching portion should be separated from the rest of the power supply and located at the work coil.

A. Types of Induction Heating Power Supplies

An understanding of the various types of power supply circuits used for induction heating is necessary to select the best type for a given application or to assess the suitability of an available power supply to the application. A very basic block diagram that applies to nearly all induction heating power supplies is shown in Figure 6. The input is generally three-phase 50

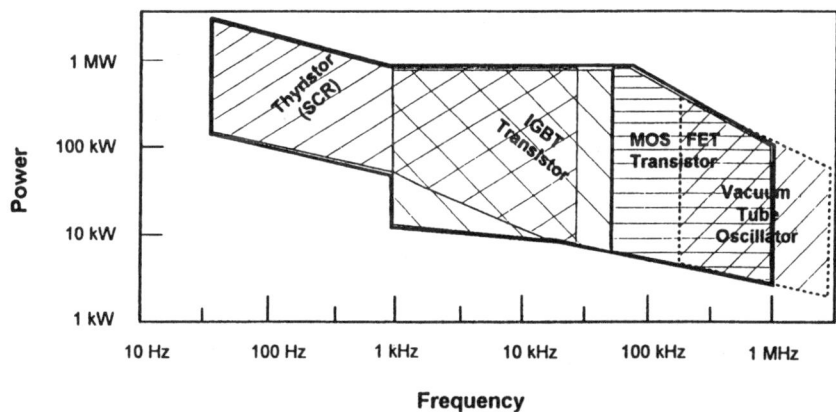

Figure 5 Modern inverter types for induction heat treatment.

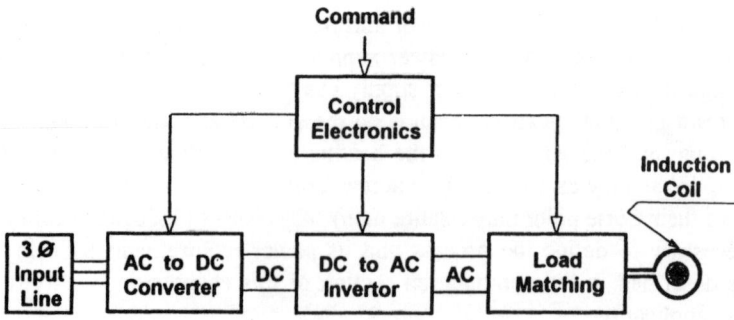

Figure 6 Induction heat treatment power supply, basic block diagram.

or 60 Hz at a voltage that is between 220 and 575 V. The first block represents the AC to DC converter or rectifier. This section may provide a fixed DC voltage, a variable DC voltage, or a variable direct current. The second block represents the inverter or oscillator section, which switches the direct current to produce a single-phase AC output. The third block represents the load-matching components, which adapt the output of the inverter to the level required by the induction coil. The control section compares the output of the system to the command signal and adjusts the DC output of the converter, the phase or frequency of the inverter, or both to provide the desired heating.

1.   Full-Bridge Inverter

The most common inverter configuration is the full bridge as shown in Figure 7. Often referred to as an H bridge, it has four legs that each contain a switch. The output is located in the center of the H so that when switches S1 and S2 are closed, current flows from the DC supply through the output circuit from left to right. When switches S1 and S2 are opened and switches S3 and S4 are closed, current flows in the opposite direction, from right to left. As this process is repeated, an alternating current is generated at a frequency determined by the rate at which the switches are opened and closed.

2.   Half-Bridge Inverter

The half-bridge inverter, as its name implies, requires only two switches and two filter capacitors to provide a neutral connection for one side of the output circuit as shown in Figure 8. The other side of the output circuit is then switched between positive DC supply by S1 and the negative supply by S2, thus generating an AC voltage across the output. This configuration is used in place of the full bridge where lower output voltage or output power is desired.

Many books and technical papers have been written about the detailed design and theory of operation of the various types of induction heating power supplies [1–10,24]. Inclusion of

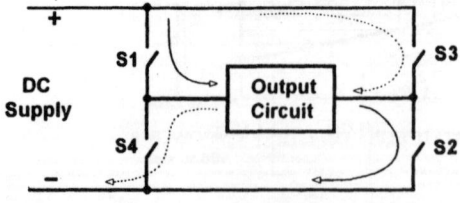

Figure 7 Basic full-bridge inverter.

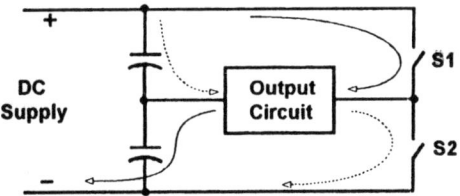

Figure 8 Basic half-bridge inverter.

such detail here would likely be of little help to the heat treater. Therefore, the following paragraphs will only categorize the commonly used power circuit and control combinations. This will give some insight into the advantages and disadvantages of modern induction heating power supplies and their applications and features.

Figure 9 shows the principal design features of the inverter configurations most commonly used in induction heating power supplies. The two major types are the voltage-fed and the current-fed. The chart further subdivides each of these by the DC source (fixed or variable), the mode of inverter control, and the load circuit connection (series or parallel).

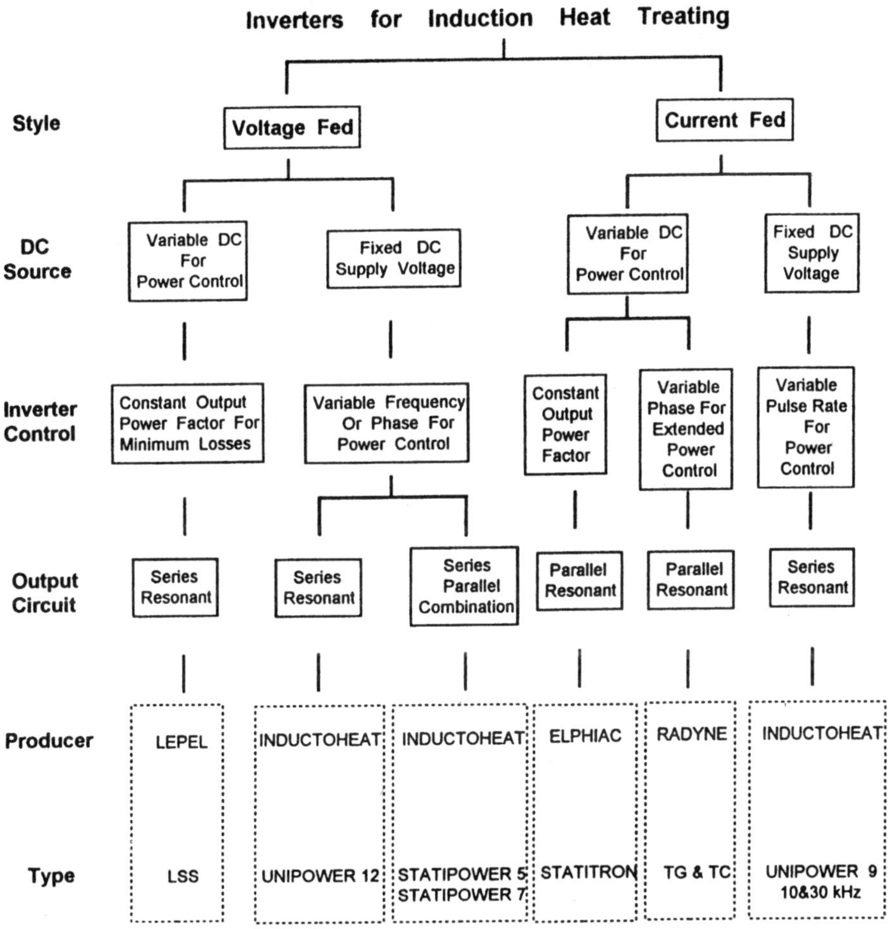

Inverters for Induction Heat Treating

	Voltage Fed			Current Fed		
Style						
DC Source	Variable DC For Power Control	Fixed DC Supply Voltage		Variable DC For Power Control		Fixed DC Supply Voltage
Inverter Control	Constant Output Power Factor For Minimum Losses	Variable Frequency Or Phase For Power Control		Constant Output Power Factor	Variable Phase For Extended Power Control	Variable Pulse Rate For Power Control
Output Circuit	Series Resonant	Series Resonant	Series Parallel Combination	Parallel Resonant	Parallel Resonant	Series Resonant
Producer	LEPEL	INDUCTOHEAT	INDUCTOHEAT	ELPHIAC	RADYNE	INDUCTOHEAT
Type	LSS	UNIPOWER 12	STATIPOWER 5 STATIPOWER 7	STATITRON	TG & TC	UNIPOWER 9 10&30 kHz

Figure 9 Induction heat treatment inverters.

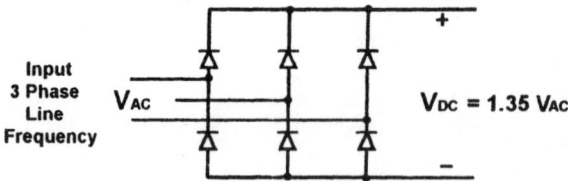

Figure 10   Uncontrolled rectifier.

3.   Direct Current Section

All of the power supplies outlined in the Figure 9 chart have a converter section that converts the line frequency alternating current to direct current [2,3]. Nearly all induction heating power supplies use one of three basic converters. The simplest is the uncontrolled rectifier shown in Figure 10. The output voltage of this converter is a fixed value relative to the input line-to-line voltage, and no control of the output is provided by the converter section. The uncontrolled rectifier must therefore be used with an inverter section capable of regulating the power supply output.

The phase-controlled rectifier, shown in Figure 11, has thyristors that can be switched on in a manner that provides control of the DC output relative to the input line voltage. This relatively simple converter can be used to regulate the output power of the inverter by controlling the DC supply voltage. The circuit has two disadvantages. First, the input line power factor is reduced to values that are not acceptable to modern power quality specifications when the DC output voltage of the converter is less than maximum [11,12]. Second, the control response time is necessarily slower because it is not able to respond faster than the frequency of the input line it is acting upon. There are, however, schemes that require additional power components for alleviating both of these disadvantages [2,11].

The third converter type has an uncontrolled rectifier followed by a switch mode regulator as shown in Figure 12. The switch mode regulator shown in the diagram is one of the simplest forms and is called a buck regulator [2]. The level of DC voltage or current at the output is regulated by rapidly switching the pass transistor on and off. The greater the on-time/off-time ratio, the higher the output voltage or current. This converter can therefore regulate the output power of the inverter by controlling the supply of direct current. The input line power factor is maximum at all power levels, and the response time can be very fast due to the relatively high switching rate of the buck regulator. Therefore, this converter overcomes the disadvantages of the simple controlled rectifier while being more complex, more costly, and slightly less efficient [2].

All of the converters just discussed draw non–sine wave current from the input AC line. This means that there are harmonics or multiples of the line frequency present in the current waveshape. The harmonic distortion of the current waveshape can adversely affect supply trans-

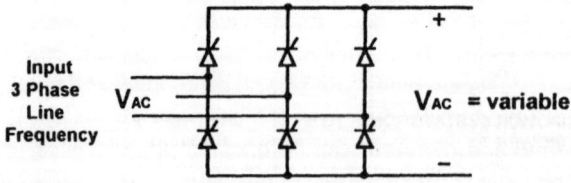

Figure 11   Phase-controlled rectifier.

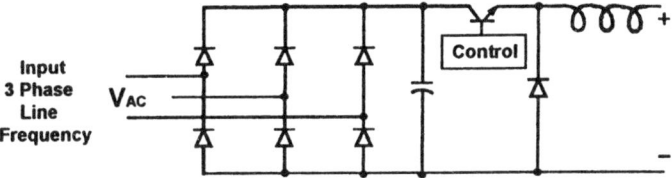

Figure 12 Uncontrolled rectifier with switch mode regulator.

formers and other electronic equipment connected to the same line. In most heat treatment situations where the power supply rating is less than 600 kW and the plant power distribution system provides a low source impedance or "stiff line," a six-pulse rectifier as described above is acceptable. For higher power systems or where utility requirements require reduced harmonic content, a 12-pulse rectifier, which requires a six-phase input and 12 rectifiers, can be used. The chart below compares the typical line current harmonics as a percent of the fundamental for these rectifier configurations.

Harmonic	6-pulse	12-pulse
5th	17.5%	2.6%
7th	11.1%	1.6%
11th	4.5%	4.5%
13th	2.9%	2.9%
17th	1.5%	1.5%

As shown in the chart, the 5th and 7th harmonics are nearly eliminated in the 12-pulse case, resulting in a dramatic reduction of the total harmonic distortion of the line current. Use of higher pulse configurations such as 18 or 24 would lead to a further reduction but at considerable expense.

4. Inverter Section

Voltage-Fed Inverters with Simple Series Load Voltage-fed inverters are distinguished by the use of a filter capacitor at the input of the inverter and a series-connected output circuit as shown in the simplified power circuit schematic of Figure 13. The voltage-fed inverter is used in induction heating to generate frequencies from 90 Hz to as high as 1 MHz. Thyristors, which are also called silicon-controlled rectifiers (SCRs), can be used to switch the current at frequencies below 10 kHz. Below 50 kHz, insulated gate bipolar transistors (IGBTs) are commonly used. Above 50 kHz power, MOSFET transistors are chosen for their very fast switching speeds.

The voltage-fed inverter can be switched below resonance as illustrated by the bridge output voltage (Figure 14, trace 1) and the output current waveshape (trace 2). This must be the case when thyristor switches are used because diode conduction must follow thyristor conduction for sufficient time to allow the thyristor to turn off. This minimum turn-off time requirement limits the practical use of thyristors to frequencies below 10 kHz. The INDUCTOHEAT Statipower 6 is an example of this type of inverter [4].

Transistors do not require turn-off time and therefore can be operated at resonance as illustrated by the output current waveshape (Figure 14, trace 3). In this case, there is little or no diode conduction, and the transistor is switched while the current is at zero, thus minimiz-

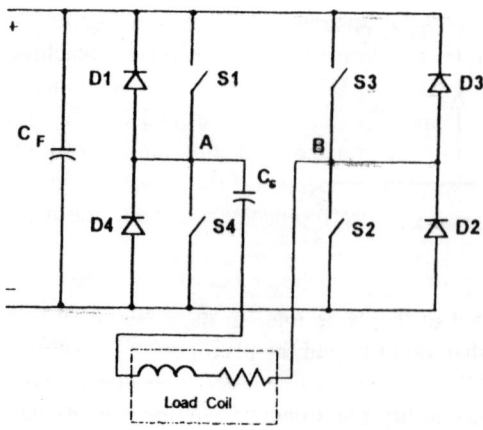

Figure 13   Voltage-fed series-connected output.

ing switching losses and maximizing inverter power rating and efficiency. Operation at resonance means that the output power factor is unity and maximum power is being transferred from the DC source to the load. To regulate power in this case the DC supply voltage must be controlled. The LSS family of induction heating power supplies (produced by Lepel Corp.) is an example of this type; they are operated at resonance with power controlled by variable direct current supplied by a switch mode regulator [5].

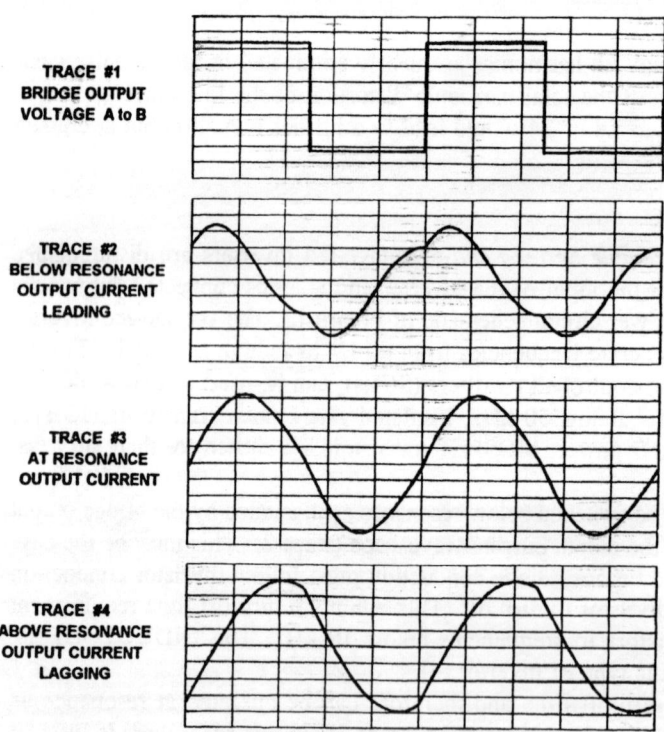

TRACE #1
BRIDGE OUTPUT
VOLTAGE A to B

TRACE #2
BELOW RESONANCE
OUTPUT CURRENT
LEADING

TRACE #3
AT RESONANCE
OUTPUT CURRENT

TRACE #4
ABOVE RESONANCE
OUTPUT CURRENT
LAGGING

Figure 14   Voltage-fed inverter waveshapes with series-connected output.

Transistors can also be switched above resonance as illustrated by trace 4 in Figure 14. In this case, the conducting switches (S3 and S4) are turned off prior to the current reaching zero. This forces the current to flow in the diodes (D1 and D2) that are across the nonconducting switches (S1 and S2). These switches (S1 and S2) can then be turned on and will conduct as soon as the load current changes direction. This mode of operation minimizes transistor and diode switching losses while allowing the inverter to operate off resonance to regulate power. Regulation of output power by control of the inverter frequency relative to the natural resonant frequency of the load will be discussed in Section II.

The voltage-fed inverter supplies a square wave voltage at the output of the bridge, and the impedance of the load determines the current drawn through the bridge to the series load circuit. In nearly all heat treatment applications, an output transformer is required to step up the current available from the inverter to the higher level required by the induction heating coil. The secondary of this transformer is connected directly to the heating coil when the heating frequency is 30 kHz or less and the coil voltage is less than 250 V. In higher frequency applications where the coil voltage is necessarily greater, the series resonant capacitor is usually placed in the secondary circuit of the transformer and in series with the heating coil.

Figure 15 shows Lepel's model of a solid-state air-cooled portable RF power supply. This low-power (300 w), high-frequency (180–350 kHz) unit is specifically designed for such heat treatment applications as preheating, soft soldering, brazing, shrink fitting, annealing, through hardening, and epoxy curing of small parts. While capable of functioning as a low-power, in-place unit, the generator's light weight enables it to be hand carried to wherever it is needed. A built-in electronic digital timer allows automatic timing of the heat cycle to a preset interval and system shutoff.

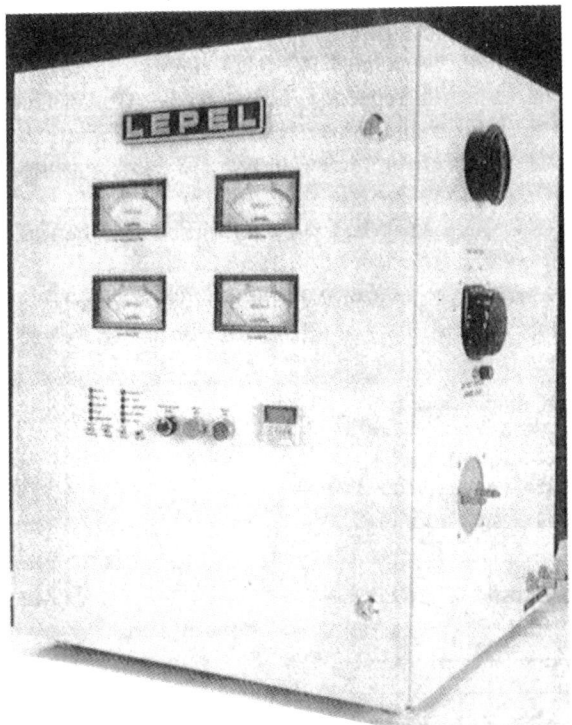

Figure 15 Air-cooled portable radio-frequency solid-state induction heating power supply.

The salient features of the voltage-fed inverter with a simple series resonant induction heating load are compared to those of the current-fed bridge inverter and summarized in Figure 16.

Voltage-Fed Inverter with Series Connection to a Parallel Load   A popular variation of the voltage-fed inverter for induction heating has an internal series-connected inductor and capacitor that couple power to a parallel resonant output or "tank" circuit as shown in Figure 17. The values of the internal series inductor and capacitor are selected to be resonant above the operating or "firing" frequency of the inverter with an impedance at this firing frequency that will allow sufficient current to flow from the bridge to permit full-power operation. A very important feature of this style of inverter is that the internal series circuit isolates the bridge from the load. This protects the inverter from load faults caused by shorting or arcing and from badly tuned loads, making it one of the most robust thyristor-based induction power supplies available for heat treatment.

A second feature of this series-parallel configuration is realized when the internal series circuit is tuned to the third harmonic of the firing frequency. The power supply is then capable of developing full power into the parallel tank circuit tuned to either the fundamental firing frequency or the third harmonic. For example, the INDUCTOHEAT Statipower 5 family of induction heat treating power supplies are produced in three dual-frequency models, 1 and 3 kHz, 3.2 and 9.6 kHz, and 8.3 and 25 kHz [6] with a power range of 10–1500 kW (Figure 18). Because load current is not used for commutation, this system can be operated with the output shorted for easy troubleshooting. Solid-state accuracy ensures output power regulation of $\pm 1\%$ with a line variance of up to $\pm 10\%$. Reliability is further enhanced by placement of 95% of all circuitry on one control board that is accessible without entering the high-voltage section of the power supply.

The voltage-fed inverter with series connection to a parallel load commonly uses thyristors for power switching in the bridge and has an unregulated DC input supply. Regulation of output power is accomplished by varying the firing frequency relative to the parallel load resonant frequency. The waveshapes present in this style of inverter are shown in Figure 19. Trace 1 shows the voltage waveshape at the output of the bridge. Trace 2 shows the bridge current to the load; trace 3 is the load current and the current when the load is tuned to the fundamental or firing frequency. The corresponding waveshapes for operation with the load tuned to the third harmonic of the firing frequency are shown in Figure 19, trace 4.

Current-Fed Inverters   Current-fed inverters are distinguished by the use of a variable-voltage DC source followed by a large inductor at the input of the inverter bridge and a par-

Bridge Inverter Features	
Voltage Fed	Current Fed
DC Filter Capacitor Square Wave Voltage Sine Wave Current Series Resonant Output Load Current = Output I. Voltage x "Q" Best For Low "Q" Loads	DC Inductor Sine Wave Voltage Square Wave Current Parallel Resonant Output Load Voltage = Output V. Current x "Q" Best For High "Q" Loads

Figure 16   Bridge inverter features.

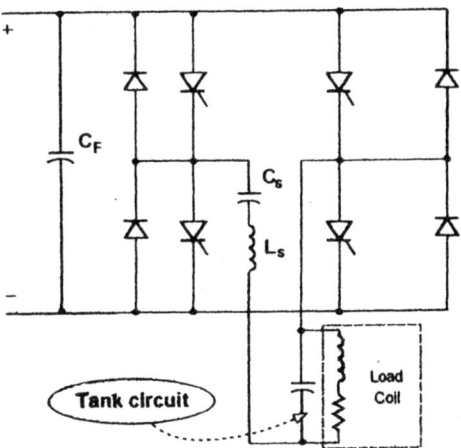

Figure 17 Voltage-fed inverter with series connection to parallel load.

Figure 18 Solid-state induction heating power supply Statipower BSP 5&7.

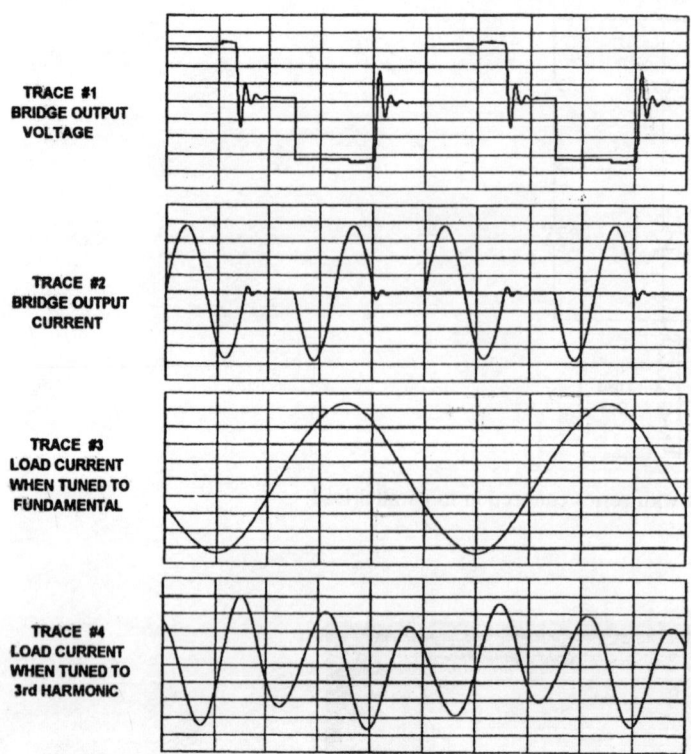

TRACE #1
BRIDGE OUTPUT
VOLTAGE

TRACE #2
BRIDGE OUTPUT
CURRENT

TRACE #3
LOAD CURRENT
WHEN TUNED TO
FUNDAMENTAL

TRACE #4
LOAD CURRENT
WHEN TUNED TO
3rd HARMONIC

Figure 19   Waveshapes of voltage-fed inverter with series connection to parallel load.

allel resonant load circuit at the output as shown in the simplified power circuit schematic of Figure 20. Current-fed inverters are available in models that cover the entire 90 Hz to 1 MHz range of frequencies used for induction heat treatment. Thyristors are commonly used below 10 kHz, whereas transistors are chosen for the higher frequencies.

When the power switching is done with thyristors, the current-fed inverter must be operated above the resonant frequency of the parallel resonant load. As illustrated by the waveshapes of Figure 21, the voltage across the output of the bridge is a sine wave (trace 1) and the current (trace 2) is a square wave. It is interesting to note that this is just the reverse of the voltage-fed inverter, where the voltage is a square wave and the current is a sine wave. The DC bus voltage across the bridge after the large inductor L_{dc} (trace 3) resembles a full wave rectified sine wave. The bus voltage is forced negative from the time the bridge is

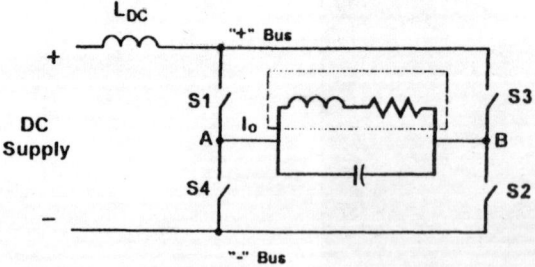

Figure 20   Current-fed full-bridge inverter.

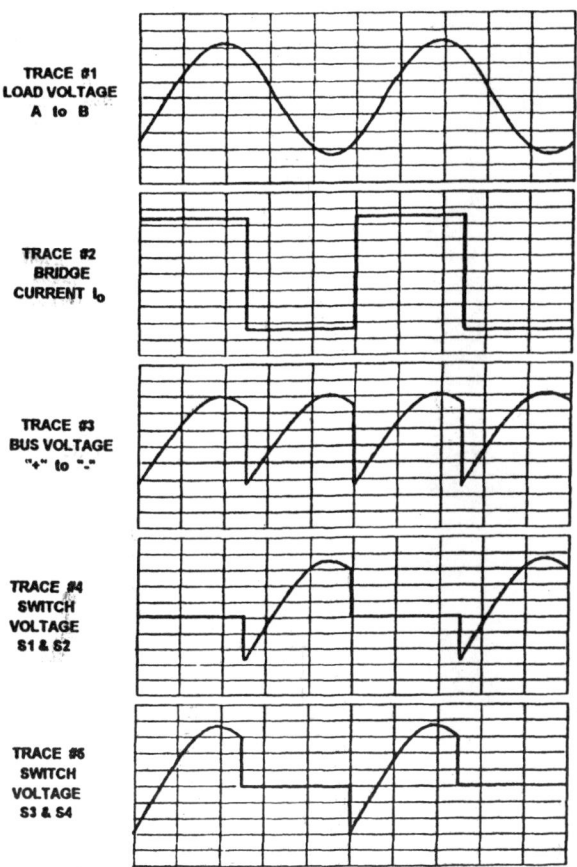

Figure 21   Current-fed inverter wareshapes above resonance.

switched until the load voltage reaches zero. This time must be sufficiently long to provide turn-off time to thyristors that are no longer conducting. The voltage across the thyristor switches are shown in traces 4 and 5 of Figure 21 with the negative portion of the waveshape noted as turnoff time. The TG and TC family of induction heat treatment power supplies (produced by Radyne Limited, UK) are of this design (Figure 22) and have been in use since 1970 [7].

For operation of the current-fed inverter at frequencies above 10 kHz, transistors are used in the inverter bridge because they can be switched very fast and do not require turn-off time. In this case the inverter can be operated at the resonant frequency of the parallel resonant tank circuit as shown in Figure 23. One diagonal of the bridge containing transistors T1 and T2 is turned on as transistors T3 and T4 of the other diagonal are turned off. This switching or commutation is done at a time when the voltage across the load, inverter bus, and transistors is zero. The inverter waveshapes obtained in this mode of operation are shown in Figure 24. Switching at zero voltage minimizes the switching losses in the transistors and therefore allows for higher frequency operation. When the inverter frequency is locked to the natural resonant frequency of the load, the output power must be regulated by controlling the input current to the inverter. This is accomplished by using one of the variable-voltage DC supplies described earlier. The Statitron 3 (produced by Inducto Elphiac, Belgium) uses MOSFET

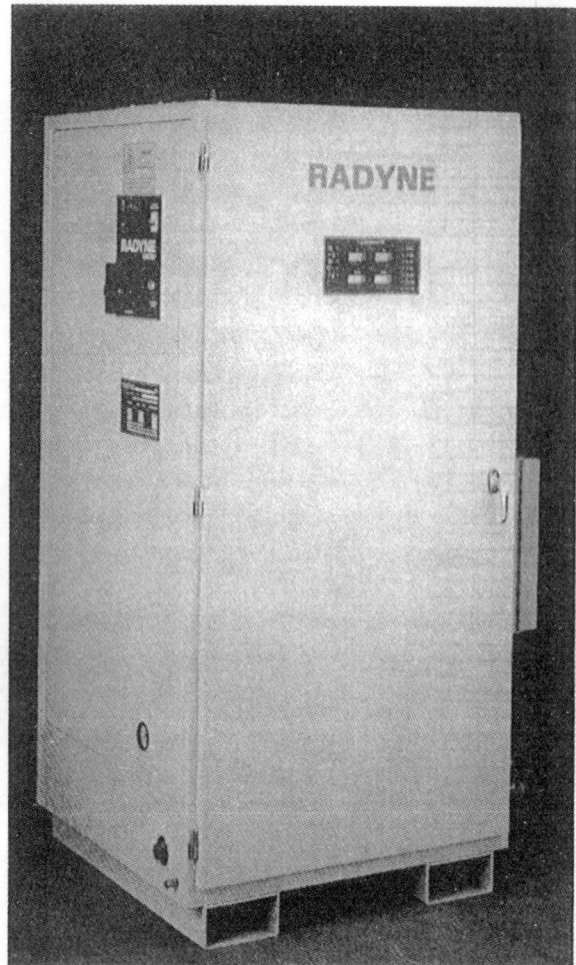

Figure 22   Radyne TG-type solid-state power supply.

transistors in a current-fed inverter configuration for heat treating at frequencies from 15 to 800 kHz with power levels up to 1 MW [8].

Another inverter configuration that has been used extensively for heat treating at 10 and 30 kHz uses only one thyristor and is referred to as a chopper or quarter-bridge. Figure 25

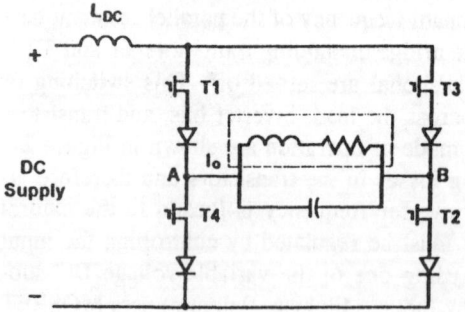

Figure 23   Current-fed full-bridge transistor inverter.

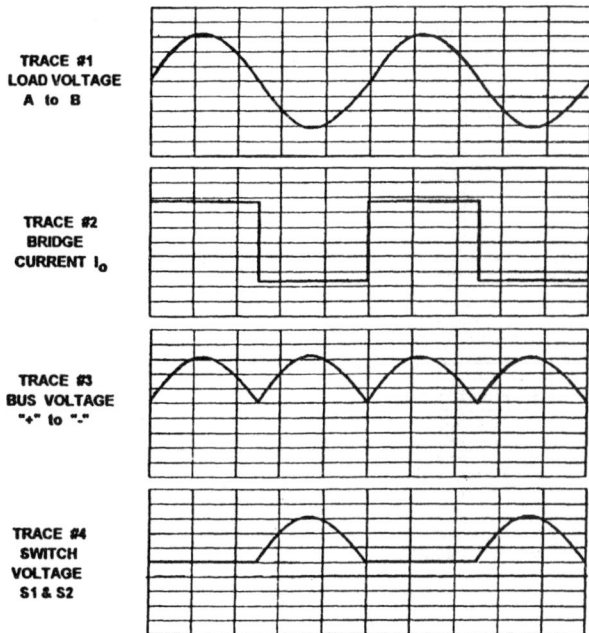

Figure 24   Current-fed inverter waveshapes at resonance.

shows a simplified circuit diagram. It is classified as a current-fed inverter because it has a large inductor in series with the DC supply to the inverter. Unlike the conventional full-bridge current-fed inverter, the chopper has a series-connected output circuit. When the thyristor is switched on, current flows both from the DC source through the large inductor and from the series load-tuning capacitor, discharging it through the load coil. The resulting load current pulse (Figure 26, trace 2) is nearly sinusoidal, with the first half-cycle of current passing through the thyristor and the second half-cycle through the diode. During this part of the period, current is rising in the input inductor. When current stops flowing in the diode, the energy stored in the input inductor causes direct current to flow in the output circuit, recharging the series load-tuning capacitor. The frequency of the output sine wave is determined by the series capacitor and the load coil inductance. It is this frequency that determines the penetration depth of the induction heating current. The firing rate of the inverter regulates the output power, and therefore a simple fixed voltage DC source may be used. The INDUCTOHEAT Unipower 9 and Uniscan induction scan hardening machine both make use of this simple inverter [2,9,13–15].

Operational considerations that impact on the suitability of each type of power supply include initial cost, operating cost or overall efficiency, reliability, maintainability, flexibility, cooling water availability, and the power supply's impact on utility power quality.

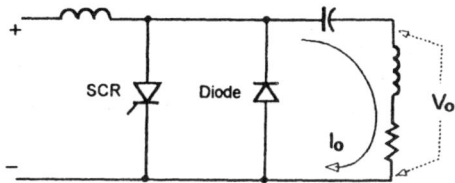

Figure 25   Current-fed chopper or quarter-bridge.

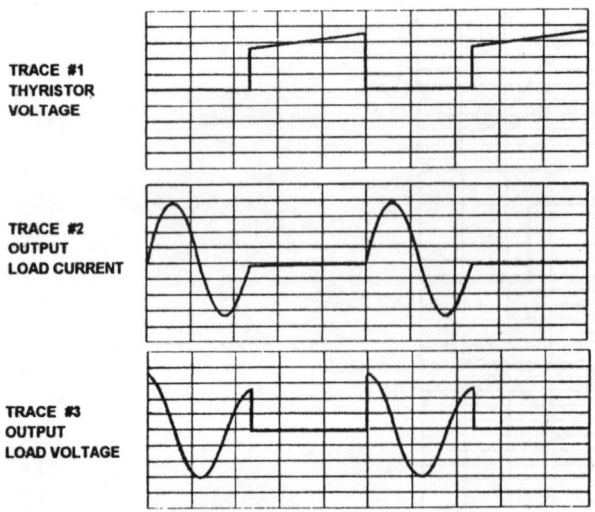

TRACE #1
THYRISTOR
VOLTAGE

TRACE #2
OUTPUT
LOAD CURRENT

TRACE #3
OUTPUT
LOAD VOLTAGE

Figure 26   Current-fed chopper or quarter-bridge waveshapes.

Initial cost is important but should be a deciding factor only when all of the inverter types considered meet the other operational requirements. In general, the chopper or quarter-bridge power supply has the lowest purchase price. For power levels below 250 kW, the voltage-fed inverter with series resonant load is the next choice based on cost. The current-fed inverter has a low cost per kilowatt when high power at low frequency is required. The most expensive is usually the voltage-fed inverter with a series connection to a parallel load. It has more power components per kilowatt than any other type of inverter in its frequency range but is the most robust and flexible for induction heating applications.

Operating cost, which is usually determined by the power conversion efficiency, is also a consideration. Modern semiconductor-based induction heat treatment power supplies, however, all have reasonably high conversion efficiency compared to their motor generator and vacuum tube predecessors. Most have a conversion efficiency of 80–93% when running at rated output power. The conversion efficiency referred to here is that of the power supply from the input power connection to the output terminals and therefore does not include, in some cases, the output-matching transformer and load-tuning capacitors. Measurement and specification of power conversion efficiency can be accomplished in many ways with differing results. At one extreme, only the losses in the inverter portion are used in the calculation of efficiency. At the other extreme, all the losses from line to load are used by taking the ratio of the output power delivered to a calorimeter load to the input line power to the system. This method includes the losses in the inductor coil, which can be relatively high, resulting in a much lower stated efficiency. It is therefore essential to know specifically what portions of the system are included in the specified efficiency to make direct comparisons of power supply efficiency.

Reliability and maintainability and a power supply's tolerance to input and load perturbations is more a function of power component design margin and control circuit design than the general type of power supply circuit used. Without carrying out a detailed analysis of a power supply it is very difficult to assess its reliability. Without this analysis, the best guide to equipment reliability is an assessment of the manufacturer's reputation, how long it has been in the business of producing induction heating power supplies, and the amount of its equipment in field use. Maintainability is affected by many features of power supply design, including the level of self-diagnostics provided, accessibility of components for inspection and measurement, and ease of component and subassembly removal and replacement. When power components, subassemblies, and control boards are interchangeable without adjustment or modification, troubleshooting and repair can be quickly and effectively accomplished by electrical maintenance personnel with only minimal training. Self-diagnostic systems can be very helpful in locating failures in a power supply. However, the inclusion of diagnostic circuitry, which can also fail, has a negative impact on reliability, and therefore a balance between the level of fault diagnostics and power supply reliability is necessary. A very reliable power supply design should require only very basic fault indicators, while more failure-prone designs should be equipped with more extensive diagnostics to speed the repair process even though an incremental decrease in reliability will result.

Flexibility or the ability of a power supply to operate under varying load conditions or in different applications is an important factor in some situations. If the heat treatment machine is a general-purpose one such as a scan hardening machine used in a job shop, the ability to match a wide range of coils at more than one frequency is attractive if not essential. In this case, a dual-frequency power supply with a versatile load-matching system, including both transformer tap switches and dual-frequency capacitor banks, is recommended. The relatively new transistorized power supplies with external transformer tap switching are also attractive where their small size, light weight, and minimal cooling water requirements allow them to be portable and used by multiple machines. The Unipower 12 shown in Figure 27 is an example of such a multiple-application power supply [10].

II. LOAD MATCHING

A. Prelude to Discussion of Load Matching

A very important facet of induction heat treatment that is often overlooked in the initial design stages is the ability to successfully deliver to the workpiece the maximum available power from a given power supply at the minimum cost. Circumstances do not always allow for optimal design of a complete induction heat treatment system in which the power supply design is based on the application including the specific induction coil parameters. Quite often, the induction coil is designed to achieve the desired induction heat treatment pattern without regard for the power supply that will be used. When this is the case, a flexible interface is required to match the output characteristics of the power supply to the input characteristics of the induction coil and workpiece combination [11]. If this match is not provided, the power supply will not be able to develop its rated power if the coil requires more voltage or current than the supply can deliver.

There are many factors involved, any of which can cause complications in arriving at the stated goal. Variable ratio transformers, capacitors, and sometimes inductors are connected between the output of the power supply and the induction coil. The adjustment of these components is commonly referred to as "load matching" or "load tuning."

B. Four Steps in Understanding Load Matching for Solid-State Power Supplies

1. Step One

The most common example of matching a power source and load would be a simple lighting circuit application where a 6V light bulb is available for use on a 120 V_{ac} power line (Figure 28). Obviously there is a need for some type of interface hardware to prevent the 120 V_{ac}

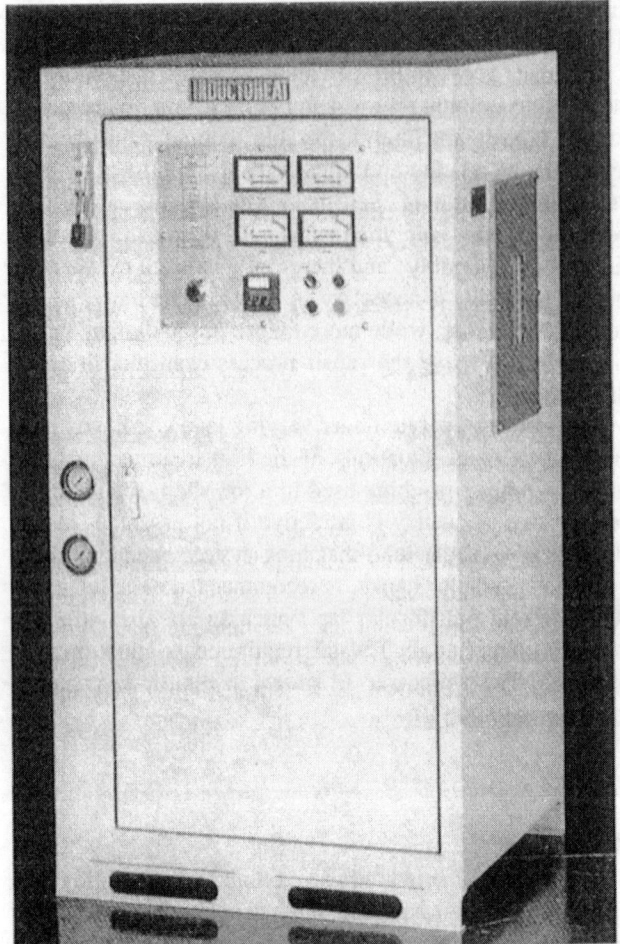

Figure 27 Multiple application solid-state power supply Unipower-12.

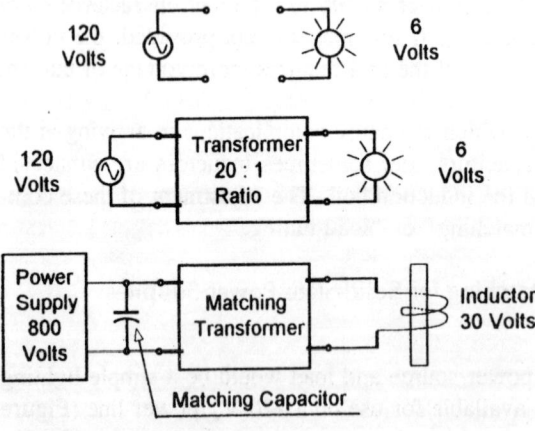

Figure 28 Load tuning—impedance matching.

from destroying the light bulb. This would commonly be accomplished by inserting a transformer between the light bulb and the power line.

Induction heating circuits have not only a resistive element but also considerable inductance. As a part of the electric circuit, any inductor can be introduced as a combination of resistance and reactance (inductance). Both the resistance and the reactance of the inductor are nonlinear functions of several parameters such as coil–workpiece geometry, material properties, and frequency. Furthermore, the electrical resistivity and magnetic permeability of metals are nonlinear functions of the temperature (see Figures 8 and 9 of Chapter 11A). As shown in Chapter 11A, electrical resistivity and magnetic permeability vary during the heating cycle. In addition, modern metalworking processes require that workpieces of different sizes be heated in the same inductor. Combinations of production mix and variation of material properties result in changing coil resistance and reactance, which affects the tuning and performance of the power supply. Generally speaking, a change in coil resistance and reactance results in a change of the phase angle between the coil voltage and coil current of a given circuit. Such a change can be characterized by the coil power factor, which refers to the cosine of the phase angle (cos φ). Power factors of different types of inductors are affected differently by the various factors. At the same time, for different frequencies, the power factor can be significantly different (i.e., from cos $\varphi = 0.02$ up to cos $\varphi = 0.6$, which makes the Q factor range from $Q=50$ down to $Q=1.7$.

In addition to these factors, the process itself usually requires that the part be heated at some frequency other than the line frequency. In conventional heat treatment, the applied frequency typically ranges from 200 Hz to 400 kHz. Since a relatively large current is required to successfully heat a workpiece, it is necessary to build power sources with extremely high output current capability or to use a simple resonant circuit to minimize the actual current or voltage requirement of the frequency converter. A simple example may help at this point.

Example   Given a work coil that requires 100 kW, 40 V 10,000 A at 10 kHz and a power source that is rated at 100 kW, 440 V, and 350 A, are the two incompatible?

By using an isolation transformer we might select a ratio of 440/40 or 11:1 to match the work coil to the power source. This would leave us with a current requirement of 10000/11 or 909 A, which is too high for the given power source.

By the addition of a specific capacitance to the load circuit it is possible to lower the current requirement and still accomplish the heating task. The addition of sufficient capacitance to tune the circuit to unity power factor (cos $\varphi = 1$) would result in a required current from the power source of 100 kW/440 V or 227 A, well within the limitations of our selected power source. This relaxes the requirements not only on the power source but also on interconnecting cables, contactors, and transformers operating in the area of the improved power factor.

As shown in the previous section, there are two basic types of resonant frequency converters that use parallel and series resonant circuits. Figure 29 shows the characteristics of a series and parallel resonant circuits. Looking first at the parallel circuit, it is easy to see that if the capacitor value is equal to zero, then a given voltage applied to the circuit at a fixed frequency will result in a specific amount of power dependent on the circuit impedance. When sufficient capacitance is added to the circuit to tune the load circuit near resonance, the circuit impedance rises dramatically and the amount of current drawn from the power source falls off dramatically. The circuit voltage required to achieve a specific power level is the same as it was in the initial case of zero capacitance, but now the higher current required by the load is being supplied by the capacitors rather than the power source.

In a parallel-tuned load circuit we have a Q rise in current in the tank circuit compared with the input line from the power source (Figure 30). This analogy can be repeated for the case of the series circuit to realize that with the calculated change in circuit impedance the

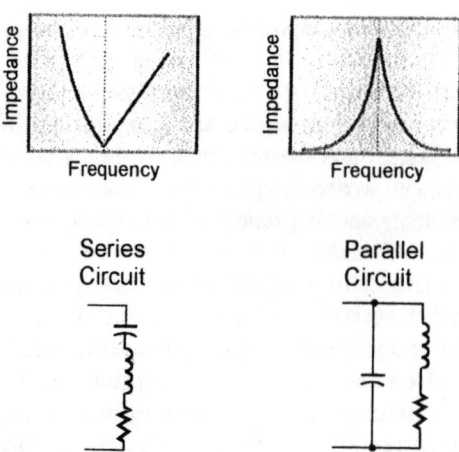

Figure 29 Resonance at series and parallel circuits.

circuit current will be much higher for a given input voltage when the circuit is tuned near the resonant frequency because the impedance is approaching zero. The load coil current required for a given power is the same for the given load circuit regardless of whether the connection is series or parallel, but because the overall impedance has fallen and the required current is fixed, the required driving voltage is approximately a factor of Q lower than the coil voltage. Hence we have a Q rise in current in the parallel circuit and a Q rise in voltage with the series-connected circuit (Figure 29). It is therefore imperative to have an understanding of what type of circuit connection exists in order to understand the effect that tuning changes will have on the power source and workstation components.

2. *Step Two*

Turning now to look at the output power characteristic for a given load circuit versus the circuit operating frequency (Figure 31), it is easy to see that if we begin at a low frequency and gradually increase the operating frequency to the point of resonance, we will have an increase in output power. Beyond the point of resonance, an increase in the operating frequency will result in a decrease in the output power. This characteristic is often used to accomplish the required regulation mode for the power source. The goal in tuning the workstation is to de-

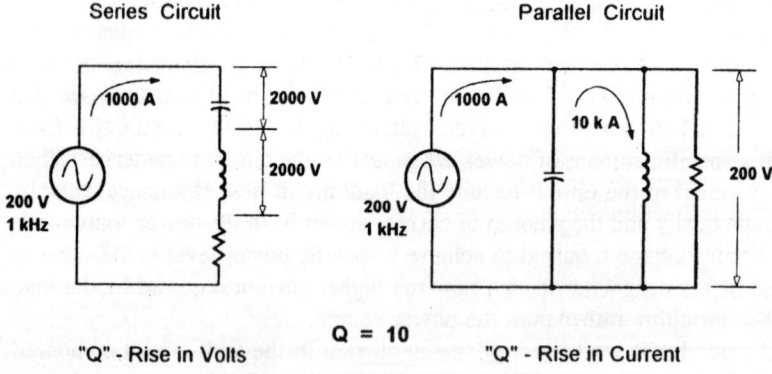

Figure 30 Series and parallel circuits.

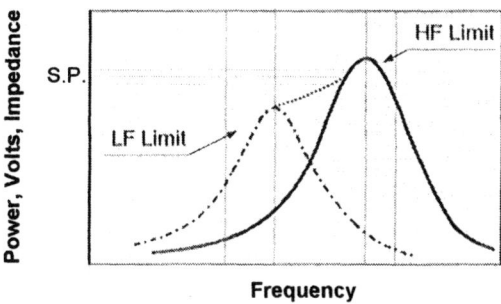

Figure 31   Load resonant.

liver the required power to heat the workpiece without exceeding any of the power source parameters. Figure 32 is a schematic for a typical workstation with capacitors, an autotransformer, and an isolation transformer.

A useful way to approach the tuning of a workstation is first to determine the ratings of the power source and record the available workstation hardware. The next step would be to make an estimate of the required coil voltage for the desired output power. This may be done with previous data or rule-of-thumb extrapolations. Then select an isolation transformer ratio to yield the proper voltage match. The next step would be to set up the load in the work coil as it will be heated and use a load frequency analyzer or signal generator to determine the resonant frequency of the load circuit [12]. Figure 33 shows the load frequency analyzer, which is a solid-state portable instrument that quickly determines the resonant frequency of the induction heater without heating the workpiece. This analyzer eliminates hours of setup time and prevents the waste of production parts. Instead of guesswork, the load frequency analyzer can easily and precisely determine the resonant frequency for any induction heating or heat treatment system.

After obtaining a resonant frequency it is necessary to add or subtract capacitance from the circuit to match the tuned frequency of the workstation to the rated frequency of the power source. At this time it is useful to run a cycle and record data for use in extrapolating to the desired setup. Figure 34 gives a simple MathCad sheet for use in extrapolations when tuning the workstation. Variables are presented for power, voltage, coil length, coil turns, frequency, and coil diameter. The variables P1, V1, L1, N1, F1 and D1 are the initial values recorded. The remaining variables are the desired values except for one variable that is chosen as the one to solve for. The equation can be solved manually for any of the given variables. An example at this point might prove helpful.

From previous data:

P1 = 100 kW;   V1 = 30V;   F = 9.6 kHz

N1,L1,D1 = 1

Desired: What voltage is required for the same power at 1.2 kHz?

P2 = 100 kW;   V2 =?   F = 1.2 kHz

N2, L2, D2 = 1 (same as above)

Solve for V2.
Answer: 15 V (approximately).

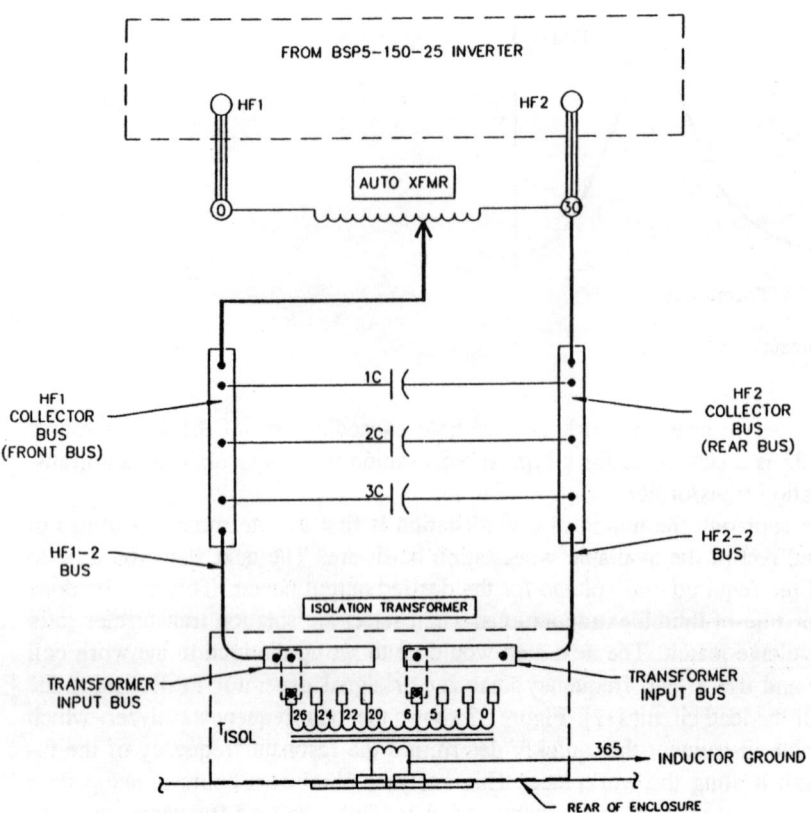

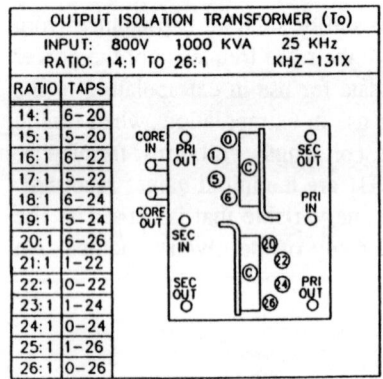

OUTPUT ISOLATION TRANSFORMER (To)		
INPUT: 800V 1000 KVA 25 KHz		
RATIO: 14:1 TO 26:1 KHZ-131X		

RATIO	TAPS
14:1	6-20
15:1	5-20
16:1	6-22
17:1	5-22
18:1	6-24
19:1	5-24
20:1	6-26
21:1	1-22
22:1	0-22
23:1	1-24
24:1	0-24
25:1	1-26
26:1	0-26

HIGH FREQUENCY CAPACITORS										
					KVAR PER TAP					
NO.	INDUCTOHEAT PART NO.	FREQ.	KVAR	RATED VOLTAGE	1	2	3	4	5	6
1-2C	10310-026	25KHZ	450	800	45	45	90	135	135	-

CAPACITOR LOCATIONS

Figure 32 Typical load-tuning component interconnection.

Note: For most load-matching calculations, an answer within 10% is close enough to accomplish the job. Although not pleasing scientifically, it is very practical and will save considerable time probing for exact answers.

It may be profitable at this point to consider the effect of changes in various components or component values in the workstation. The induction coil may be varied in shape and size with respect to the workpiece being heated. The effect of changing a simple cylindrical coil and workpiece is shown in Figure 35. The value of circuit Q is directly affected by the part-

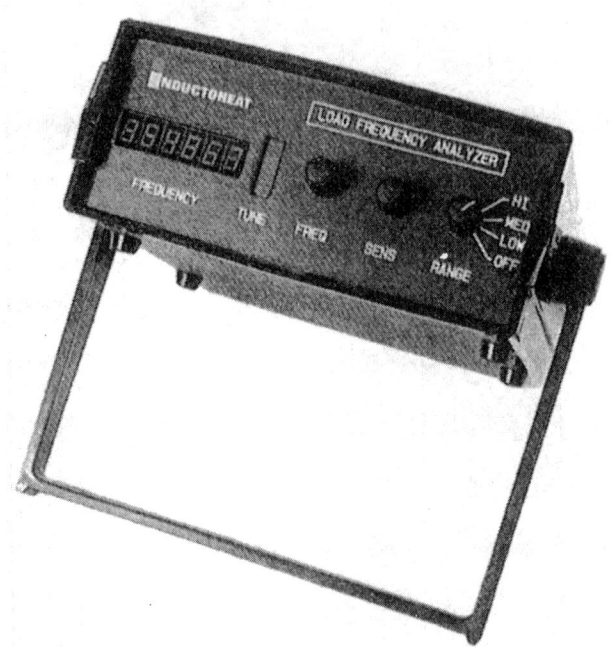

Figure 33 Load frequency analyzer.

to-coil coupling and will cause a need for increased capacitive volt-amperes to balance the inductive portion of the circuit. The larger the gap between the part and the coil, the greater the required value of heat station capacitance (Figure 35).

The heat station isolation transformer can also be varied by physically changing tap connections to change the turns ratio. A change in this transformer will produce a significant change

---

SOLVER ROUTINE FOR THE EXTRAPOLATION EQUATION: RLC 4/5/94

To use this program:
1) input all values that are known
2) input 1's for all remaining variables

$P1 := 100$ $P2 := 100$

3) Insert the name of the variable that
   you are solving for in the Find(x)
   statement below.

$V1 := 30$ $V2 := 1$

4) The Answer is displayed below

$F1 := 9600$ $F2 := 1200$

NOTE: TO INPUT VARIABLES, POINT WITH YOUR
MOUSE TO THE NUMBER YOU WANT TO
CHANGE AND CLICK THE LEFT BUTTON.
THEN DELETE OR ENTER NUMBERS AS
REQUIRED. CLICK ON THE NEXT POSITION
OR PRESS RETURN.

$L1 := 1$ $L2 := 1$

$N1 := 1$ $N2 := 1$

$D1 := 1$ $D2 := 1$

Given

STANDARD KNOWN VOLTAGES
FOR SINGLE TURN INDUCTOR

$$\frac{P1}{P2} = \left(\frac{V1}{V2}\right)^2 \cdot \left(\frac{F2}{F1}\right)^{.674} \cdot \frac{L1}{L2} \cdot \left(\frac{N2}{N1}\right)^2 \cdot \frac{D2}{D1}$$

VOLTS	FREQ	KW
30	9.6KHz	100
20	3KHz	100
15	1.2KHz	100

Answer := Find(V2)

Answer = 14.886

Figure 34 Solver routine.

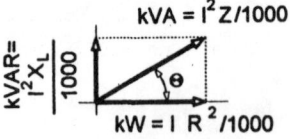

$$Q = \frac{kVAR}{kW} = TAN\,\Theta = \frac{X_L}{R}$$

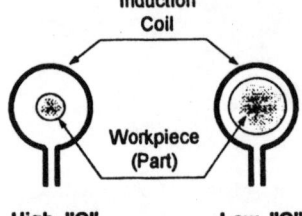

Induction
Coil

Workpiece
(Part)

High "Q"   Low "Q"

The higher the "Q", the more
matching capacitors required

Figure 35   Definition of the parameter Q.

in the circuit inductance reflected to the primary side of the circuit and subsequently the amount of capacitance required to balance the load properly.

The amount of capacitance connected in the circuit can also be changed. Increasing the circuit capacitance will lower the resonant frequency of the circuit, and a decrease in capacitance will result in an increase in the load resonant frequency. There is sometimes confusion in recording the actual amount of capacitance in a given load circuit due to the use of the terms KVAR and microfarads to describe the quantity of capacitance. The term KVAR can be expressed by the formula

$$KVAR = 2\pi FCV^2/1000$$

and is obviously a function of the frequency F, microfarad value C, and operating voltage V of the capacitor. For a given capacitor, the KVAR rating is at the name plate frequency and voltage. If the capacitor is used at another voltage or frequency, the actual required microfarad value (C) will change according to the formula above.

One might ask what the real benefit of the KVAR term is. It can facilitate calculations when transformers are installed between the coil and the capacitor bank. As shown in Figure 36, if we have a load that has a Q value of 8, the required value of the load or tank capacitor is approximately the Q value multiplied by the required kilowatts. For a 150 kW unit this load would require 1200 KVAR of heat station capacitance at the rated voltage and frequency. This is true regardless of the position of the capacitor in the circuit. If the microfarad value were used here, then the effective microhenry value of the coil would have to be calculated through each of the transformer connections and then the capacitor microfarad value would be calculated.

One of the major causes of misapplication of this information is that most users assume that the KVAR value of the capacitor is constant, but, in fact, it changes in value as the operating frequency and voltage change. It is then necessary to specify the KVAR at the operating voltage and select the capacitors so that they provide the required KVAR value at the desired

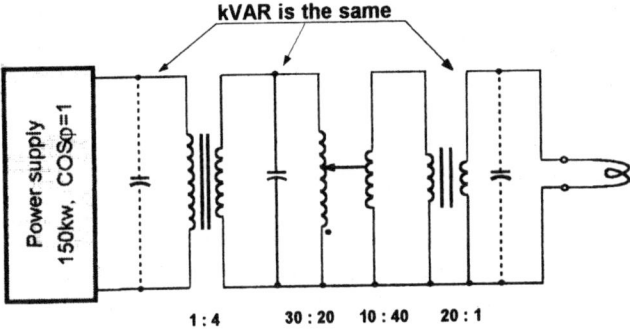

Figure 36 Schematic of induction system.

voltage and operating frequency. For example, using the information above, if the circuit requires 1200 KVAR at 600 V and 10 kHz, then the nearest standard capacitor that would be acceptable would probably be a 2133 KVAR, 800 V, 10 kHz capacitor.

Manufacturers for many years have standardized on 220, 400, 440, and 800 V operating voltages to reduce the variety of capacitors produced and to match the existing standard output ratings of motor generators.

Another component that can be changed in the matching process is the autotransformer. A change in the tapping of this transformer will, for all practical purposes, affect the output voltage of the circuit but not the operating frequency. This is stated with the qualifier "for all practical purposes" because with solid-state power supplies the leakage reactance of the autotransformer may sometimes be significant. Care should be taken in choosing the placement of the autotransformer because it can greatly affect the required KVA rating of the transformer. As shown in Figure 37a, if the transformer is installed between the power supply and the capacitor bank it is operating at the power factor of the power source with relatively low current. If this same transformer is installed between the capacitor bank and the isolation transformer or work coil, (Figure 37b) the current is much higher and the required KVA rating of the transformer is much higher, roughly Q times the kilowatt operating point of the power supply.

3. *Step Three*

Historically, for the motor generator set, tuning was approached by trying to add enough capacitance to read unity power factor or zero phase angle on the panel meters. With solid-state

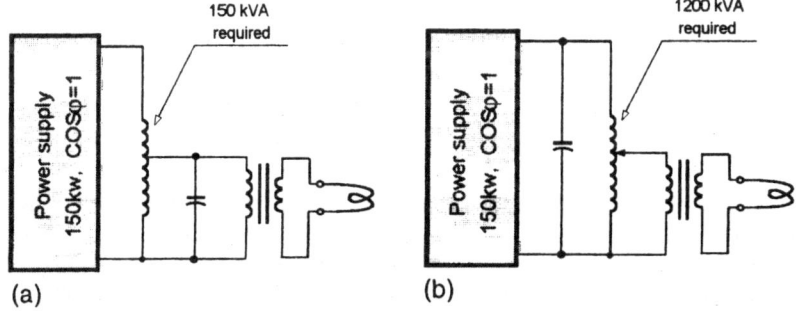

Figure 37 Features of the reactive power compensation due to different autotransformer location.

power supplies, the power source often operates at less than unity power factor, and any inductance added in the transmission lines becomes more of a factor. Often a reactance located in the power supply must be considered part of the tuned circuit. To complicate matters further, each type of power source has a variety of limiting conditions that could prevent delivery of maximum power to the workpiece. It is advisable, before purchasing a power supply, to check with the manufacturer as to how much reserve capacity is available in the power supply. More than one user has been cut short by buying a 150 kW power supply only to find out that the maximum power it will deliver into the load circuit is 90 kW. Figure 31 shows a typical tuning curve for a swept frequency power supply operating into a parallel tank circuit. This type of power supply most often begins at a lower frequency, called the low frequency limit (LF in Figure 31) and begins sweeping up in frequency until the preset power level is attained or a limit is reached. Typical limits would include a high-frequency (HF) limit, a phase or low impedance limit, output voltage limit, output current limit, maximum power limit, etc.

One complication that can arise as a result of mistuning the load is that the frequency may be increased beyond the resonant frequency of the tuned circuit. This results in confusing the control circuit, which is normally in a mode of increasing the frequency to increase the output power. Since the power will decrease for an increase in frequency beyond the load resonant frequency, the power supply frequency will continue to increase until it reaches the high-frequency limit. This condition is referred to as "going over the hump." The remedy is generally to reduce the value of the heat station capacitor or vary the inductance to produce an increase in the load resonant frequency.

It should be noted that on power supplies with an output series capacitor this "over the hump" condition will result when the series capacitance is too small relative to the kilowatt rating of the power supply. By adjusting the tank circuit capacitor and transformer or coil turns, it is possible to shift the curve in Figure 31 to the left or right to move away from expected limiting conditions. It should be noted that the power delivered is that shown on the curve and that if the curve is shifted too far to the left, a higher power than that desired may be delivered with no apparent control by the power potentiometer. Most solid-state converters will not run at a zero power output even though the potentiometer may be set to zero.

A current-fed power supply operating into a parallel tank circuit operates at a fixed frequency by phase locking itself to the tank circuit resonant frequency. One might think that this would eliminate the need for tuning. Unfortunately this is not the case. The problem of matching impedances still exists. On many current-fed units the maximum allowable current is only slightly higher than that which would be calculated for unity power factor at full voltage. This means that unless the impedance of the tank circuit is exactly right to deliver full current at full voltage the power supply will deliver less than full power to the load. This sometimes requires the insertion of a special tuning bus to adjust the impedance for the correct value. If the current as a percent of maximum is higher than the percent of voltage, more inductance is required in the circuit. If the voltage is higher than the current, then inductance must be removed from the circuit. Another solution provided by the control of some current-fed inverters is to operate the inverter above the resonant frequency of the tank circuit. This reduces the load impedance to better match the output of the inverter.

Effort spent on load matching can be reduced by using a power supply that has more rated capacity than required or one that can demonstrate the capacity to run at 120% of its power rating. This will ensure ease of tuning when the applications calls for 100% power or less.

Another general guide in load matching is to aim approximately for a voltage match as outlined in the following table.

% Voltage	% Power
10	1
20	4
30	9
40	16
50	25
60	36
70	49
80	64
90	81
100	100

Since the power varies as the square of the output voltage, striving for these values will give a setting that will allow easy adjustment to higher or lower values without continual limiting conditions.

4. The Final Step

A final caveat in load matching has to do specifically with the transmission lines from the power source to the load-matching (or heat) station and those from the load-matching capacitors or output transformer to the heating coil. Large inductances in these areas can cause considerable problems because much of the voltage generated by the power supply is dropped across the high-inductance elements of the circuit and not across the load itself. This can result in a considerable reduction in allowable output power and possibly in the inability to complete the desired heating task. This inductance is particularly critical in the higher kVA portion of the circuit (between the matching capacitors and the coil), especially at higher frequencies and higher currents. A good practice is to minimize the transmission line inductance within the required cost and size constraints.

C. Medium- and High-Frequency Transformers for Modern Heat Treatment

A transformer is an important part of the induction heating machine. Different types of transformers are used in inverters and heat stations. The total efficiency of the power supply is primarily affected by the transformer's efficiency. Years ago, when motor generators were widely used, the design of isolation transformers was a straightforward process. Some basic information, such as frequency, kilowatts, kilowatt-amperes, and input/output voltages, was all that was required. Today, with many different types of solid-state inverters and heat stations, the task of designing efficient transformers becomes more complex. The successful design of contemporary transformers should involve such features as the current/voltage waveforms, which can be square, sinusoidal, or sawtooth and often contain harmonics.

In 1955, the Jackson Transformer Company first started to manufacture transformers for induction heating applications. Since that time, thousands of different transformers have been developed to match the variety of applications. Here we briefly introduce some basic types of medium- and high-frequency transformers for modern heat treatment based on the materials provided by the Jackson Transformer Company [16,17].

The transformer's main purpose is to change one voltage to another, making it possible to operate a great variety of loads at suitable voltages. In a transformer the turns of the primary and secondary coils are coupled closely together so that their respective turns ratios determine very closely the output voltages and volt-ampere characteristics. The coils are usually wound on a laminated core of magnetic material, and the transformer is then known as an iron-core transformer. Sometimes, as in many radio-frequency transformers, there is no magnetic core; the transformer then may be described as an air-core transformer.

Transformer manufacturers provide the induction heating industry with a wide range of transformers and other magnetic products from line frequency to 800 kHz, from a few volt-amperes to over a megawatt, and with water- and air-cooled designs. Products include isolation, auto, current, potential, and RF transformers, along with AC-DC reactors and integrated magnetic devices. As a general rule, most of the magnetic devices are water-cooled. This is because of size limitations, cost factors, power requirements, and frequency ranges.

1. AC-DC Reactors

Alternating current reactor designs take into consideration any DC component, from a few hertz to several hundred kilohertz, water-cooled or dry, open construction or encapsulated. The legs of the inductors have distributed gaps to minimize flux leakage and to reduce noise. The legs are normally encapsulated to minimize vibrations. They are available from a few microhenries to several millihenries, and from a few amperes to several thousand amperes.

Direct current reactors are designed to handle an AC ripple component that may be present. They are available in both dry and water-cooled designs. Typically, they have a shell-type construction, whereby the gaps are distributed in the center leg to reduce fringing. The center leg is encapsulated to minimize noise and vibrations. They are available from a few microhenries to several millihenries and from a few amperes to several thousand amperes.

2. Variable Impedance Transformer

The variable impedance transformer (VIT) is an integrated magnetic device in which the primary windings and the magnetic amplifier windings are placed on common cores. Its purpose is to provide stepless power to electric furnaces that have silicon carbide elements, vacuum furnaces for deposition of metal, plating power supplies, and load banks. The VIT is a current-control device that requires minute signals to control a large amount of power. The VIT can operate with large unbalanced loads or with an open-circuit phase. The VIT can withstand short circuits for prolonged periods of time without incurring component failure. VITs are available in single-, two-, or three-phase designs that provide from a few hertz to 200 kHz, from a few volts to 2000 V up to 500 kV.A.

3. Heat Station Transformers

Jackson heat station transformers such as the 52V1, 51V1, and 531V1 have been used in the heating, hardening, and annealing industry. 52V1 transformers are normally used where the voltage needs to be stepped down anywhere from 5:1 to 22:1 or from 5:2 to 22:2 or other ratio combinations depending on customer requirements. The input voltages are anywhere from 220 to 1200 V or thereabouts, and frequencies from 500 H to 10 kHz. The kVA can range from 50 to over 10,000 kV.A.

The construction of the windings can be either open or epoxy-encapsulated. The output connections (secondary terminals) are generally referred to as fishtails. The input side of the transformer, which is referred to as the primary winding, is tapped to cover the required turns ratio. The windings use rectangular copper tubing of the thin wall because of the skin effect of the current at medium frequency. A typical profile of the tubing used would be $0.25 \times 1 \times 0.048$ wall. The primary and secondary windings are of an interleaved design to take advantage of the shape of the tubing and to reduce the resistance and impedance of the transformer. One of the unique features of this design is that the losses in the primary and secondary windings are equal. For a typical 22:1 ratio transformer there are 22 primary turns in a series and there are 22 secondary turns connected in parallel in a one-turn construction. Therefore, the secondary resistance reflected to the primary is 22^2 times the secondary resistance. The total resistance is the primary resistance plus the reflected secondary resistance.

The construction of the core uses thin electrical steel (0.006 or 0.007 in. thick) of EE- or E1-type laminations. The core is water-cooled by means of copper cooling plates sandwiched in between the steel laminations. It has been concluded after many tests that the flux generated by the ampere turns in the magnetic circuit flows along the inside legs of the laminations just as current in a circuit takes the least resistive path. Therefore, the width of the outside leg of a shell-type transformer operated at medium frequencies does not need to be one-half the tongue (center leg), as is required for low-frequency designs, because the outside legs are narrower than the center leg and core losses of the outside legs are higher than those of the center leg. The core loss of the transformer varies as the square of the input voltage, inversely as the square of the input turns, and approximately as the fourth root of the frequency.

4. Toroidal Transformers

Typically toroidal transformers are totally encapsulated and are used in through hardening, tempering, forging, and annealing. Normally the output voltages are higher in the heat station transformers, and in many instances the output voltage is equal to or much higher than the applied voltage. Input voltages can be from 100 V to 2000 V or higher. The output voltages can range from 50 to several thousand volts. Taps are provided within the voltage range. The frequencies can be anywhere from 200 Hz to 10 kHz. The kVA can range from 50 to 3000 kV.A or higher. They are more efficient than laminated transformers and have virtually no air gaps. A disadvantage of being encapsulated is that they are not easily repaired and in most cases must be replaced.

Toroidal autotransformers usually are smaller in size and have lower exciting current, better regulation, and higher efficiency than an isolation transformer. The reason for this is that in an isolation transformer all of the KVA is transferred to the secondary, whereas in an autotransformer only a portion of the total KVA is transformed, the rest flowing directly from the primary to the secondary without transformation. The windings in an autotransformer are wound around the same core and are used to step up or step down the input voltage. The core of toroidal transformers consists of thin steels wound in a cylindrical or toroidal form. Water-cooled copper heat sinks are used on the flat surface of the cores to carry away the heat generated by the core. Without water cooling, the physical size of the core would increase drastically. The windings are hand-wound over the core, using round copper tubing, its size being determined by the design current.

5. Integrated Magnetic Transformers

Jackson Transformer Company has developed and patented a method of combining a transformer and inductor in a single package whereby the inductor and the primary of the transformer have a common core. This product is referred to as a Transinductor and can be designed to provide a fixed inductance in the primary or secondary or both. Variable ratios can be provided on the transformer portion. By combining the two components the size of the product is reduced, the overall efficiency is increased, and the leakage flux of the magnetic device is minimized.

6. Rectangular (C-Core) Transformers

The construction of a rectangular transformer uses a C core and interleaved windings. Normally, the unit is epoxy-encapsulated. The design of the rectangular core transformer is usually at low to medium frequencies with input voltages from a few hundred to a few thousand volts, output voltages from a few hundred to a few thousand volts, and input power up to several thousand kilovolt-amperes. Specific requirements for this type of transformer are low leakage inductance and high efficiency.

7. Narrow-Profile Transformers

Narrow-profile transformers are designed to deliver high power at medium frequencies within narrow physical constraints. A typical example is the induction heating bearing surfaces of an engine crankshaft. A series of narrow-profile transformers can be placed in side by side for simultaneous induction heating of a number of different bearing surfaces of an engine crankshaft. The construction of this style of transformer uses the interleaved winding design and ferrite cores and is epoxy-encapsulated. This allows the transformer to achieve its narrow-profile, high-efficiency, low-leakage inductance and be completely protected from its harsh environment and physical abuse.

8. Ferrite-Core Transformers

Ferrite-core transformers are similar to heat station transformers in that they have an interleaved winding construction. One of the differences is that in place of the steel lamination used in the core, ferrite material is used. Ferrites offer advantages over steel laminations in that they have low eddy currents and high permeability over a wide frequency range. Having a homogenous ceramic structure and inherent low core loss, the ferrites become very attractive at frequencies above 10 kHz for transformer applications. In some cases, even though the ferrite core loss is low, they may still need to be water-cooled because of the frequencies at which they are used. In applications when the output power from the power supply is fairly low and the frequency may be under 10 kHz, ferrites are more advantageous than steel because of the lower loss of the ferrite.

9. Radio-Frequency Transformers

The radio-frequency (RF) transformer is normally referred to as a current transformer and is designed without any core material. The critical element in the design is to obtain the highest current transfer ratio from primary to secondary. Generally, the primary winding is encapsulated silicon rubber, which is a moisture-resistant material. This is required because of the high dielectric strength needed between the primary and secondary and also to protect the windings from the environment. Great care must be taken in the construction and selection of material. A clean room environment is highly recommended.

10. Maintenance, Sizing, and Specification of Transformers

As a general rule, when transformers are water-cooled, most failures occur because of a breakdown of the insulation between the windings. Normally, this is due to lack of water, poor quality water, too high a water inlet temperature, or operation of the transformer outside its designed rating. Sometimes insulation breaks down because of the harsh environment to which the transformer is subjected. Another failure that commonly occurs is the melting of the output connection (fishtail), which is caused by improper tightening or poor maintenance of the inductor (e.g., dirty and oxidized surfaces on the mating inductor or fishtail). Sometimes the core fails. Again, this may be due to lack of water, poor quality water, too many input volts per turn (voltage per turn exceeds core loss temperature limitation), and improper use of frequency. A well designed water system will pay for itself with reductions in component failures and downtime. Proper maintenance of the inductor/transformer connections will also help greatly.

To properly size or specify a transformer the following information is generally required: input voltage to the transformer, power source wattage; the frequency range of the power source and the frequency at which the transformer will operate, the turns ratio or the output voltage required at full load (or no load), the input KVA at the minimum and maximum turns ratios, and expected efficiency (based on the kilowatt rating of the power source) or loss of the transformer. It is also helpful to know any unique characteristics of the power source, type of waveform, and if any direct current will be present. The more information the designer has available, the more assurance the customer has of getting the proper, most efficient transformer.

III. PROCESS CONTROL AND MONITORING

A. Prelude to Discussion of Process Control and Monitoring

One of the most important features of a modern induction heat treatment machine is that it has an effective control and adequate monitoring system. The monitoring system provides an operator with information about what is actually happening during the process and whether the heat treatment of the workpiece has been successful [18,19].

In earlier years, controls as simple as dual set-point meters were used to determine whether a given parameter was running between two preset points at the time the circuit was interrogated. With the advent of the PLC, a much larger number of points could be monitored in real time during the heat treatment cycle. In the early 1980s, HWG, in Germany (this company is now a part of the INDUCTOHEAT Group), developed the coil signature system, which was sold on many commercial machines and has been effectively applied for repeatable heat treatment processes.

The general idea of the "signature" concept is rather simple and can be described as follows. The monitoring system observes one of the unregulated variables related to the process and stores the most important parameters during the machine cycle. These values are compared to set points stored in the information bank within the PLC ("ideal signature"), and an output indication is given on the CRT readout. In normal performance all subsequent signatures of cycles are compared to the ideal one and must remain between the upper and lower limits (Figure 38). If any signature goes outside the area of limits, then the operator can see exactly during which part of the heat treatment cycle the signature was not repeated and the process exceeded the set limits. The operator knows immediately what the problem is and what should be done to adjust the machine to get the cycle signature back into the correct setup. It is not necessary for the operator to know in detail the electromagnetic, heat transfer, or quenching features of the process. He or she merely needs to know how to adjust the machine to get the signatures back to the ideal shape.

As microprocessor technology continued to grow, this signature technology was employed by manufacturers of strain-measuring equipment and eddy current testers. Modern signature

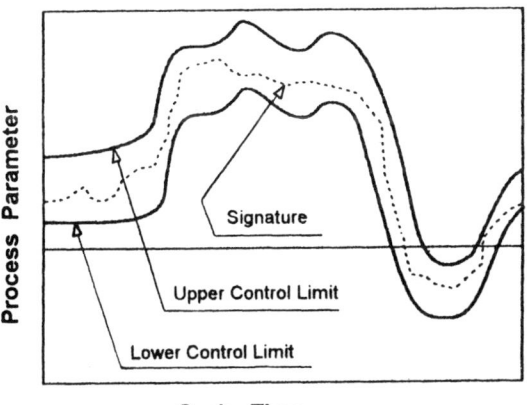

Figure 38 Control limits and a sample of a signature.

systems will allow the measurement of many variables at once with corresponding real-time graphing of the function within the preset set points as well as any required SPC analysis.

The question always remains as to which are the most significant variables to monitor to ensure that the process has been successfully repeated and what the correlation is between the variables and the heat treatment process. A list of variables for induction hardening would include the following:

Workpiece chemistry	Workpiece geometry
Coil geometry	Coil material
Workpiece-to-coil location	Frequency
Power, voltage, current	Cycle time
Workpiece temperature	Quench flow
Quench data (media, temperature, purity, concentration)	

One manufacturer who was purchasing a number of machines took the refreshing approach of doing a design of experiments (DOE) for his particular process. The DOE determined that there were relatively few parameters that significantly affected the process. The decision was made to control the significant parameters rather than focusing on the larger number of insignificant items usually pursued.

During workpiece induction heating there is a large variation in material properties, including electrical resistivity and the relative magnetic permeability of ferromagnetic metals. These changes result in a significant redistribution of heat sources within the workpiece during heat treatment. Variation in electrical resistivity and magnetic permeability during the heating cycle leads to changes in the reference depth of current in the workpiece. These changes can be observed by monitoring the coil voltage, current, and phase angle (cos φ) at the induction coil (Figures 39 and 40).

Although at first glance the changes may appear to be very dramatic, there are some factors that may mask the expected effect. For a mass heating system running at relatively low field intensity (i.e., induction heating before forging, rolling, or extrusion), the relative magnetic permeability of the workpiece surface may change from 200 to 1 during the transition from cold to hot. For a surface-hardening application running at 10 kW/in.^{2} or more, the change of μ will typically be only from 8 to 1. Since the reference depth varies as the inverse square root of the permeability change, the actual change in inductance from cold to hot load may be relatively small. In many cases of induction hardening, the actual change in inductance and impedance is relatively small and is greatly reduced for stronger magnetic fields and higher power densities (see Figure 16 of Chapter 11A).

Cold Workpiece   **Hot Workpiece**

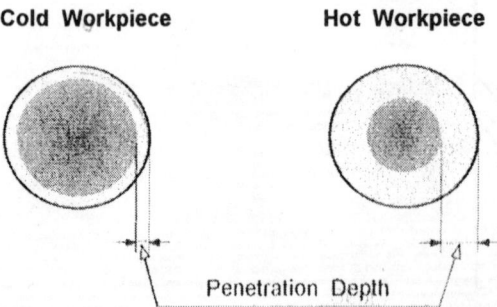

Penetration Depth

Figure 39   Depth of current penetration in cold and hot steel.

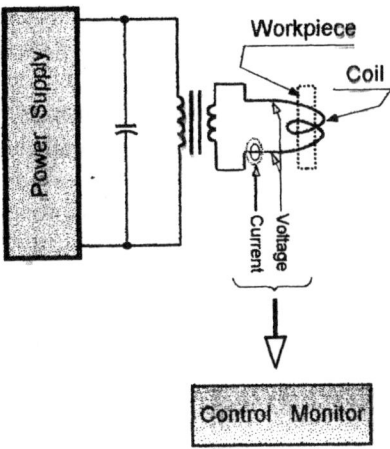

Figure 40 Monitoring of induction heat treatment system.

Different types of monitoring systems are available in the market. The choice of a particular monitoring system is a matter of operational features of the process, cycle time, technological requirements, and cost. In some applications a relatively simple energy monitor will be sufficient. Other applications, however, may require advanced signature monitoring devices.

B. Energy Monitor

The simple energy monitor is shown in Figure 41 [18,20]. This monitor measures and displays the actual energy delivered to the induction coil in kilowatt-seconds. It is a relatively inexpensive device. Once a workpiece heating pattern is developed and the correct power and

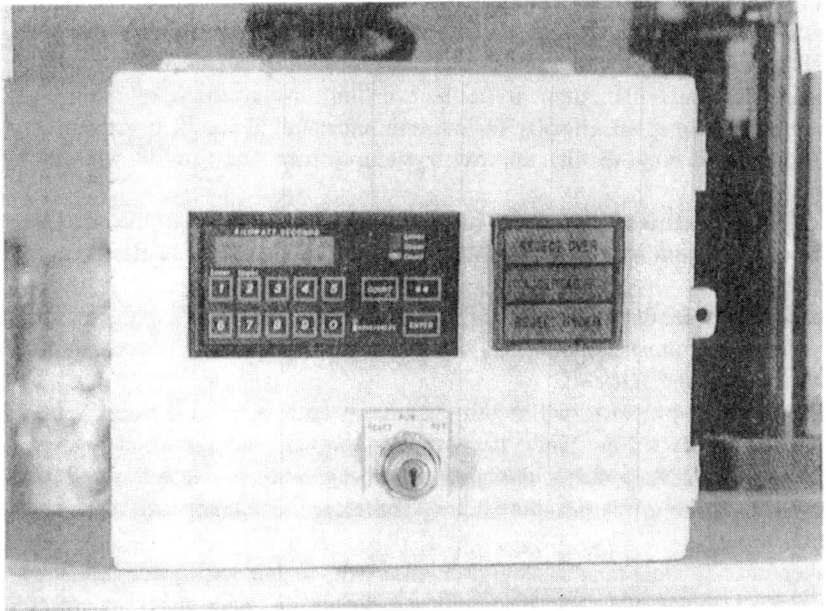

Figure 41 Energy monitor.

heating time ar established, this information is preset into the monitor. The acceptable lower and upper kilowatt-second limits are then entered by the user. If insufficient or excessive energy is applied to the load, the display will show "REJECT/UNDER" or "REJECT/OVER," respectively. Auxiliary contacts can be used to reject the part in automated lines or sound an alarm in manual operations. If the count falls within the preset range, the "ACCEPTABLE" indicator will be displayed. The energy monitor can be used as an induction process controller to turn off the power automatically when the desired amount of energy is delivered to the load. The energy monitor circuitry accurately measures and displays the output of the power supply. While most earlier RF monitoring was on the input, fiber optics make it possible to monitor the high voltages and frequencies on the output safely.

C. Advanced Monitoring

In many cases, energy or coil monitors can monitor the heating cycle effectively. However, in some applications, these monitors give only a partial picture because the quenching phase is as critical to the proper hardening of the part as the heating phase. It would be desirable to have more advanced monitoring equipment available. This equipment could monitor several parameters simultaneously in real time and indicate which parameters may cause an improperly hardened part.

The Stativision concept has been used extensively in developing advanced monitoring devices [21,22]. Monitors that use this concept have a real-time interface that monitors the energy into a workstation and checks it against the ideal system set points. The signature of load parameters is polled, varying from 30 ms to 2 s, and 75 readings are taken. The ideal signature, which is based on previous successful lab developments and the history of operations, is stored on-line. During each successive machine cycle a new signature is compared to the stored ideal signature. The difference in the two is displayed along with the actual running signature. If the two don't match, a fault is logged, and a signal is sent to the machine control.

Operator screens also display process data. The main screen has several graphs representing major process parameters such as power, frequency, and current. The fault screen displays a list of system problems, including faults as well as limits. From this screen, the operator or maintenance personnel can move the cursor to the fault or limit displayed. A key stroke will bring up a help screen developed specifically for the item indicated. These help screens provide user-friendly diagnostic routines that improve system-up time and provide the tuning needed for a particular load.

It is wise to remember that each parameter that is monitored adds cost to the machine. For a great majority of induction hardening applications, four signatures are sufficient to define the process [19].

Scanning induction surface hardening typically requires the following signatures: scan speed, quench pressure, rotation speed, and load power. As an example, scan speed and load power signatures are shown in Figure 42.

Single-shot heat treatment processes will require parameters such as rotation speed, quench pressure, load power, and quench flow. Since the part is not moving during the heating cycle, the scan speed is irrelevant to the process, although monitoring these parameters might indicate a tendency toward a failure of the machine index. Therefore, for maintenance scan speed could be monitored.

Other parameters that it would be advisable to monitor with a signature system are shown in Figure 43. Workpiece temperature can be monitored during the cycle using an infrared monitor. In some applications, in addition to the parameters shown in Figure 42, signatures

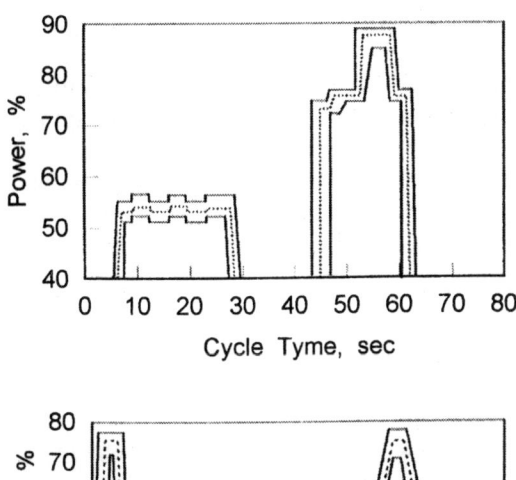

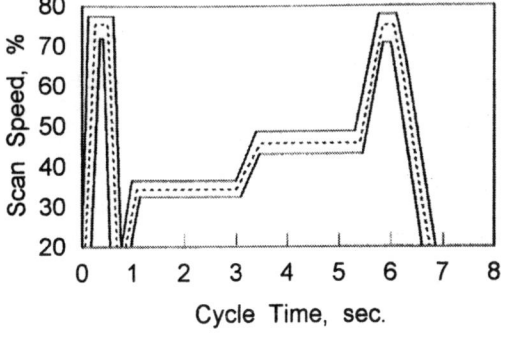

Figure 42 Samples of signatures for scanning induction hardening.

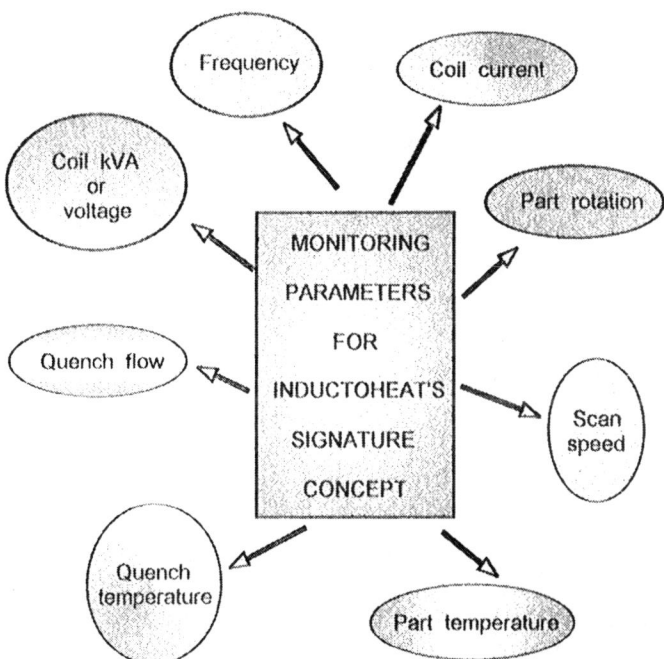

Figure 43 INDUCTOHEAT's monitoring parameters for signature concept.

can involve other parameters that are essential to particular processes and change during each cycle.

An advanced commercial system such as INDUCTOHEAT's *QA* Ultra 8000 (Figure 44) is capable of producing four or eight simultaneous process parameter signatures. This system verifies all machine settings to provide confidence in the quality of processing for the part. It can detect high/low power, high/low rotation speed, excessive/slack quench, high/low heat time, etc. The features and benefits of this and other similar monitoring systems are:

Machine repeatability on every component

Real-time parameter sample monitoring

Cost savings through reduced destructive testing

Immediate test results

Easy setup and operation with user-friendly software

Help in troubleshooting, allowing identification of the fault condition and the point in the cycle where the problem occurred

Figure 44   Quality assurance monitor (signature monitor) *QA* Ultra 8000.

Trend analysis

Data collection for SPC software.

Advanced monitoring systems provide precise verification of the real process parameters. As with most processes, it is advantageous to use a number of techniques to manage the process. The combination of a reliable power supply, an efficient induction coil, and a full process monitoring system ensures a proper hardening profile.

IV. SUMMARY

The unique viewpoint presented in this chapter is that of the integration of modern power supply selection, load matching, process control and monitoring, and contemporary inductor design techniques. Current information has been presented on the use of different types of solid-state power supplies with their basic block diagrams, waveshapes, and resonant characteristics as well as photographs of practical applications. An attempt has been made to bridge the gap between induction coil design information and power supply, load-matching, and process control and monitoring. This information is truly useful to the modern heat treatment engineer.

Should further details be required or questions arise regarding different types of induction heat treatment power supplies in particular applications, the authors welcome any reader inquiries or suggestions at INDUCTOHEAT, Inc.

REFERENCES

1. C. A. Tudbury, *Basics of Induction Heating*, Vol. 2., Rider, New York, 1960.
2. D. L. Loveless, R. L. Cook, and V. I. Rudnev, Considering nature and parameters of power supplies for efficient induction heat treating, *Ind. Heating*, June 1995, pp. 33–37.
3. D. L. Loveless, Solid state power supplies for induction heating prior to rolling, forging and extrusion, *33 Metal Producing*, August 1995.
4. INDUCTOHEAT Bulletin, *Statipower 6*, 1991.
5. General presentation of activity of Lepel Corp. 1990.
6. INDUCTOHEAT Bulletin, *Statipower 5*, 1991.
7. General presentation of activity of Radyne Ltd., UK, 1992.
8. General presentation of activity of Elphiac, Belgium, 1990.
9. INDUCTOHEAT Bulletin, *Unipower 9*, 1991.
10. INDUCTOHEAT Bulletin, *Unipower 12*, 1993.
11. R. L. Cook, D. L. Loveless, and V. I. Rudnev, Load matching in modern induction heat treating, *Ind. Heating*, September 1995.
12. INDUCTOHEAT Bulletin, *Load Frequency Analyzer*, 1991.
13. INDUCTOHEAT Bulletin, *Uniscan-I*, 1991.
14. INDUCTOHEAT Bulletin, *Uniscan-II*, 1992.
15. INDUCTOHEAT Bulletin, *Uniscan-IV*, 1992.
16. W. E. Terlop and S. Cassagrande, Special transformer technology for medium and high frequency applications, Proc. 1st Int. Induction Heating Seminar, São Paulo, Brazil, 1995.
17. General presentation of Jackson Transformer Company, 1993.
18. R. L. Cook. R. J. Myers, and V. I. Rudnev, Process monitoring for more effective induction hardening control, *Mod. Appl. News*, August 1995.
19. R. J. Myers, Induction control with smoke and mirrors, Proc. Int. Heat Treating Conf.: Equipment and Processers, ASM Int. 1994, pp. 295–297.
20. INDUCTOHEAT Bulletin, *Energy Monitor: Quality Control Energy Monitor*, 1993.
21. INDUCTOHEAT Bulletin, *QA Ultra 8000*, 1992.
22. INDUCTOHEAT Bulletin, *Stativision*, 1993.
23. P. J. Miller, Automation is key to efficiency for many truck, auto makers, *Heat treat*, May 1982.
24. D. Oxbrough, Current fed inverters, Proc. 6th International Induction Heating Seminar, Nashville, September, 1996.

12

HEAT TREATMENT OF POWDER METALLURGY STEEL COMPONENTS

Joseph M. Capus
Consultant in Powder Metallurgy, Beaconsfield, Quebec, Canada

Howard Ferguson
Metal Powder Products Company, Coldwater, Michigan

I. INTRODUCTION

In this chapter we explain the basics of powder metallurgy (P/M) and the how and why of its application to the manufacture of steel components. The place of heat treatment in the manufacture of P/M parts is discussed, followed by a review of the heat treatment processes that are suitable for P/M steels with details of recommended procedures where these differ from those applicable to wrought steel. Finally, a summary is given of typical mechanical properties for the main types of heat-treated P/M steels and the standards published by MPIF (the Metal Powder Industries Federation), ASTM, and ISO. Much of this chapter is based on a previous review of the heat treatment of P/M steels by one of the present authors [1].

Powder metallurgy is a process in which solid components are fabricated by the consolidation of metallic materials in powder form. It is also frequently used to include the manufacture of the metal or alloy powders themselves. As currently practiced, the technology is relatively modern, most of the applications having been developed since World War II. The process is applied overwhelmingly to ferrous materials; the principal uses are found in the manufacture of self-lubricating bearings, gears, sprockets, cams, and numerous other small parts in the construction of automobiles, where the primary driving force has been cost saving.

The most common methods for the consolidation of P/M steels are known as conventional press and sinter, powder forging (P/F), and metal injection molding (MIM). Other powder consolidation methods include powder rolling and isostatic pressing, but these are less likely to be encountered in the manufacture of steel components that would be subjected to heat treatment.

II. HOW P/M PARTS ARE MADE

The vast majority of P/M steel components are manufactured by compacting the powder in a shaped die at room temperature, followed by sintering at a temperature above 1100°C

(2010°F). In this process, ferrous-based powders are first blended together to form a uniform mixture having the desired alloy content. Carbon is usually added as finely powdered graphite. A dry, powdered lubricant, such as a wax or metallic stearate, is usually added in quantities of 0.5-1% to provide lubrication of the tools during compaction.

The powder mixture is conveyed to a compacting press, where it is fed into a die cavity formed by the body of the die and the lower punch (see schematic diagram of the die compaction step in Figure 1). After removing excess powder, the die cavity is closed by the upper punch. Exertion of pressure between the upper and lower punches forces the powder particles together, eventually resulting in partial cold welding and interlocking of the particles. Sufficient pressure must be applied to densify the powder until the degree of cohesion between the particles enables the compacted powder shape to be handled without breaking after ejection from the die. For ferrous materials, compacting pressures are typically in the range of 30-50 tons per square inch (tsi), resulting in pressed densities of 6.5-7.2 g/cm³ (83-92% of the pore-free density). The compaction cycle is completed by release of the pressure and subsequent ejection of the powder compact from the die by the lower punch. Finally the lower punch is moved back to its starting position to provide the cavity ready to receive the next fill of powder. This description is for the making of simple cylindrical compacts. The tooling and press movements for more complicated components, e.g., those having several levels (of thickness), would naturally be more complex; however, the principles remain the same. After the removal from the press, the "green" compacts are inspected and then transferred to a sintering furnace (Figure 2), where they are subjected to a multiple-stage heating and cooling cycle, under a protective atmosphere, to bond the powder particles together without melting the main constituent.

The sintering process generally includes an intermediate heating step to remove the lubricant, which is volatilized and burned. During the subsequent "high heat" part of the sintering cycle, the metal powder particles fuse together, carbon from the admixed graphite becomes dissolved in the iron, and other alloying elements such as copper, nickel, and molybdenum diffuse into the solid iron matrix, with or without actually melting. The third phase of the sintering cycle is the cooling step, in which the components pass through a slow-cooling zone, then a water-cooled jacket, still under protective atmosphere so that they can

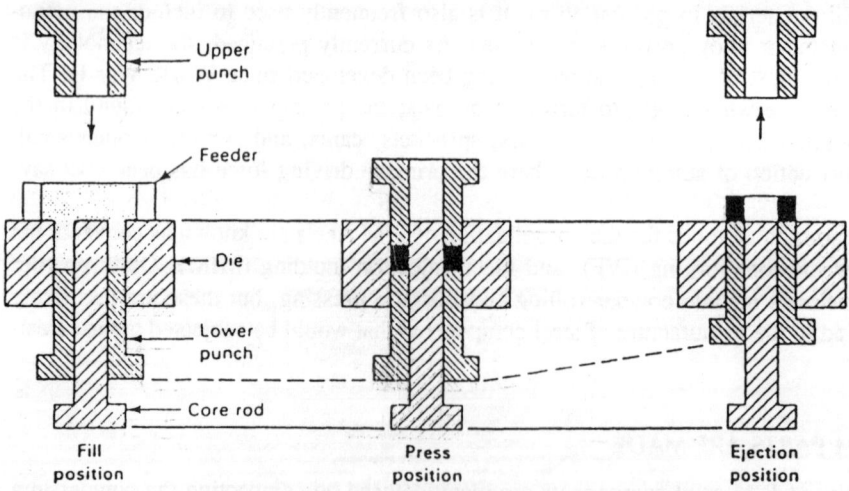

Figure 1 Compaction of metal powder in a die: schematic. (From Ref. 2.)

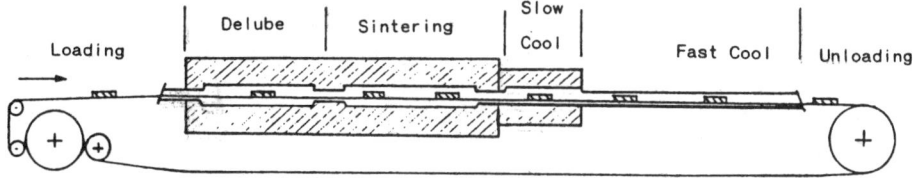

Figure 2 Sintering of P/M parts in a mesh belt furnace: schematic.

be brought back to a temperature where they can exit the furnace without the risk of oxidation. Cooling is sometimes intentionally accelerated (sinter-hardening) to improve mechanical properties. Following the sintering process, P/M steel components are usually heterogeneous in both composition and microstructure and also contain pores. Porosity is diminished during sintering through solid-state diffusion; porosity does not usually close up completely, and in fact additional pores can arise from the melting out of lower melting point metals such as copper, the disappearance of graphite particles, and the burning off of lubricant particles.

It may be wondered why constituents such as carbon and metallic alloying ingredients are added in elemental form to the powder mix instead of being alloyed in the original powder particles before they were compacted. Although there are some historical reasons [e.g., the manufacture of (unalloyed) sponge iron powder from iron oxide by solid-state reduction], the main reason is that in solution carbon and most alloying elements harden the iron powder particles so that the powder would become significantly less compressible. As a result of this lack of compressibility, desired compact densities would not be achieved with conventional compacting presses, and the manufacture of many existing components from powder would become impractical. Much effort has been expended during the past few decades to develop commercial grades of iron powder with improved compressibility so that higher densities and hence improved mechanical properties can be achieved in sintered parts.

Other reasons for admixing some of the constituents of P/M steels include adjustment of shrinkage (e.g., copper produces a "growth" effect that compensates for shrinkage during sintering, allowing better control of dimensions in the sintered part) and improvement of machinability (by including admixed powdered constituents that would not necessarily be present in the desired dispersion if added as prealloyed ingredients in the base powder).

Sintered components may be subjected to a variety of secondary operations, including re-pressing, resintering, coining or sizing, infiltration, impregnation, machining, and heat treatment as well as finishing operations such as deburring, burnishing, steam treatment, and plating. These operations are generally employed as needed to improve dimensional tolerances, mechanical properties, or surface appearance. Machining is frequently employed where one or more design features (e.g., undercuts, screw threads) of the component can not feasibly be molded in uniaxial compaction.

Re-pressing and resintering of P/M components allows further densification of the material and can result in significantly improved mechanical properties. (The improvement is particularly important for heat-treated properties.) Depending on composition, the double-press-double-sinter process enables the densities of P/M steels to be increased to about 7.4 g/cm^3. The limitation is usually the pressure required to close up the porosity in the harder matrix of sintered steel. An alternative route for the densification of pressed and sintered P/M steels is to partially or completely fill the pores with a lower melting point alloy. These metallic infiltrants are usually copper-based alloys prepared in powder form for compaction and subsequent melting into the pores of the P/M steel component by making a second pass through a sintering furnace. Infiltration and sintering of the base material are sometimes accomplished in the same pass. Infiltration is used, for example, in the production of ferrous structural parts requiring densities in excess of 7.4 g/cm^3 and mechanical properties superior to those obtained by compacting, sintering, and coining.

Where the properties of fully densified P/M steels are required, hot forming of sintered preforms is usually the preferred method for eliminating the porosity. Powder metallurgy hot forming, also referred to as powder forging (P/F) or sinter forging, is usually intended to produce materials with structures and properties that are able to replace conventional wrought steel. In recent years the method has found increasing applications, for example in automobile engine connecting rods and other highly stressed components. In outline, the P/F process consists of first producing a carefully designed sintered preform, which is then reheated to a forging temperature and transferred to a hot-forging press and forged in a closed die to densify it. The preform is usually closely similar in shape to the finished component, since the main purpose is to eliminate porosity and achieve mechanical properties close to or equal to those of wrong material. For this reason, heat treatment of hot-forged P/M steels is no different from that of their wrought counterparts. The only distinction would be where the chemical compositions of P/M steels vary slightly from those of corresponding wrought alloy designations, thus affecting hardenability. The influence of compositional differences on hardenability, and, of course, the resulting properties, would have to be taken into account in setting heat treatment parameters.

Another relatively recent development in powder metallurgy is the application of injection molding technology to the consolidation of metal powders. In metal injection molding (MIM), ultrafine metal or alloy powders are mixed with an organic binder to form a plastic mix with the consistency of toothpaste. The mix, or feedstock, is injected into a split mold in a machine similar to a plastics injection molding machine. The molded piece is then treated to remove the organic binder by solvent and/or thermal processes and finally subjected to a high-temperature sintering operation where it sinters to near full density. The fine particle size of the powder promotes the sintering reaction, which enables a high-density part to be fabricated even though the injection molding feedstock may contain only 50% by volume of metal powder particles. The attraction of MIM is that it permits the fabrication of components with intricate shapes that cannot be formed by conventional P/M compaction methods. MIM parts have been made in a wide range of alloy systems, including iron-nickel, low-alloy, stainless, and tool steels. Applications are concentrated in small complex components that are expensive to manufacture by investment casting or machining. Such parts as the tilt ring in the IBM Selectric typewriter, computer and other business machine parts, fire-arm components, and orthodontic devices and other medical applications are cited as well-known examples of this technology. In general, because of the problem of binder removal, the size of the parts is limited to pieces up to 0.5 in. in thickness and weighing no more than a few ounces, typically 1 oz or less. The market for MIM has so far been mostly in specialized high-value-added components. Although the application of MIM technology is growing at a faster rate than conventional P/M, it so far represents only a minute fraction (less than 1%) of the total North American P/M market.

Since MIM steel powder parts are produced at high density levels, with little or no interconnected porosity, heat treatment practices are similar to those for small wrought or cast steel components.

III. TYPES OF STEEL PARTS MADE FROM POWDERS

The major benefit of the P/M process is to be able to manufacture net-shape (i.e., ready-to-use) components at an economical cost. As a result, the applications tend to be concentrated in automobile and applicable areas where large numbers of small parts are used in assembly. The typical North American automobile of the 1990s contains several dozen ferrous structural P/M parts in the engine, transmission, braking system, and chassis. (These components are in addition to P/M self-lubricating bearings.) Other common application areas include agricultural equipment, off-road machinery, lawn and garden equipment, business machines, domestic hardware and appliances, power and hand tools, sporting goods, and military hardware.

The compacting process generally dictates the design options that are available for pressed and sintered P/M parts. Thus small gears, pinions, sprockets, cams, rings, pulleys, cylinders, and levers (Figure 3) are the types of P/M parts most frequently met in practice. Examples of these and more complex parts are scattered throughout the literature. A more detailed review of the shapes of components that are suitable for manufacture by powder metallurgy is given in the *Powder Metallurgy Design Manual* [4] and in Reference 5.

IV. COMPOSITIONS OF P/M STEELS

As with conventional (wrought or cast) low-alloy steels, there are many compositions for ferrous P/M parts, some of which are proprietary. The most common compositions for ferrous structural parts are iron-copper-carbon, iron-nickel-carbon, and iron-carbon. P/M parts made in these compositions are generally produced from blends of elemental powders of iron, copper, graphite, nickel, etc. Other types of low-alloy steel compositions such as nickel-molybdenum (modified 4600 type), nickel-copper-molybdenum, and chromium-molybdenum are more likely to be manufactured from partially or fully alloyed powders to which graphite and/or other elemental powders may be admixed for special properties. At suitable densities, all of

Figure 3 Examples of steel parts manufactured by powder metallurgy. (From Ref. 3. Courtesy of Metal Powder Industries Federation, Princeton, NJ.)

these P/M steels are amenable to heat treatment by quenching and tempering and/or to surface-hardening processes.

Compositions of the main types of ferrous P/M structural materials used in North American
have been standardized in designations and specifications published by the MPIF under its
Standard 35 [6]. Composition limits for a selection of the pressed and sintered iron and steel
materials are listed in Table 1. Comparable standards have been published by ASTM and SAE.
The materials in Table 1 have been chosen to include those compositions for which heat-treated
properties are listed in the standard. For additional information on ferrous powder metallurgy
materials, readers are referred to the article by Pease in ASM *Metals Handbook* [7].

V. HEAT TREATMENT OF P/M STEELS

A. General

For conventional wrought or cast steels, response to heat treatment is governed primarily by
chemical composition and grain size. With P/M steels, the presence of porosity is an additional factor to be taken into account, particularly if the pores are large enough to be interconnected. Interconnected (or "open") porosity is a network of connecting pores remaining
in a P/M part after compacting and sintering that can let a fluid or gas penetrate or pass through
the part. As we shall see, porosity can influence the practice and results of heat treatment in
several ways. On the other hand, for powder-forged or MIM parts, as already explained,

Table 1 Compositions of P/M Structural Steels

Description	MPIF designation	Composition limits (%)[a]			
		C	Ni	Cu	Fe
P/M iron and	F-0000	0–0.3			
carbon steels	F-0005	0.3–0.6			
	F-0008	0.6–0.9			
Iron-copper and	F-0200	0–0.3		1.5–3.9	93.8–98.5
copper steels	F-0205	0.3–0.6		1.5–3.9	93.5–98.2
	F-0208	0.6–0.9		1.5–3.9	93.2–97.9
Iron-nickel and	FN-0200	0–0.3	1.0–3.0	0–2.5	92.2–99.0
nickel steels	FN-0205	0.3–0.6	1.0–3.0	0–2.5	91.9–98.7
	FN-0208	0.6–0.9	1.0–3.0	0–2.5	91.6–98.4
	FN-0405	0.3–0.6	3.0–5.5	0–2.0	89.9–96.7
	FN-0408	0.6–0.9	3.0–5.5	0–2.0	89.6–96.4
Copper-infiltrated	FX-1005	0.3–0.6		8.0–14.9	82.5–91.7
steels	FX-1008	0.6–0.9		8.0–14.9	82.2–91.4
	FX-2005	0.3–0.6		15.0–25.0	72.4–84.7
	FX-2008	0.6–0.9		15.0–25.0	72.1–84.4
				Mo	
Low-alloy	FL-4205	0.4–0.7	0.35–0.55	0.50–0.85	95.9–98.75
steels	FL-4405	0.4–0.7		0.70–1.00	96.3–98.9
	FL-4605	0.4–0.7	1.70–2.00	0.40–0.80	94.5–97.5
	FLN-4205	0.4–0.7	1.35–2.50[b]	0.50–0.85	93.95–97.75

[a]Other elements: Total by difference equals 2.0% max., which may include other minor elements added for specific
purposes.
[b]At least 1% of the nickel is admixed as elemental powder.
Source: Ref. 6.

porosity is either minimal or entirely absent and should not be a factor in heat treatment. If heat treatment procedures are adopted that take into account the porosity factor of pressed and sintered parts, then satisfactory results should be obtained.

B. Influence of Porosity on Heat Treatment of P/M Steels

Porosity can influence the heat treatment process through its effect on

Density
Permeability (to gases and liquids)
Thermal conductivity
Electrical resistivity (in induction hardening)

These factors have an immediate bearing on the hardenability of P/M steels.

Maximum hardness in a quenched steel is achieved with a fully martensitic structure. This microstructure can only be produced, however, if the diffusion-dependent transformation of austenite to other phases can be suppressed by sufficiently rapid cooling. Hardenability for a wrought steel can be defined as the capacity of the metal to transform from austenite to martensite at a given depth when cooled rapidly. The core hardenability depends directly on the thermal conductivity of the steel to dissipate the heat content.

Thermal conductivity in P/M parts has been shown to be strongly influenced by porosity, which acts as an insulator in slowing heat transfer:

$$\text{Conductivity} = \text{theoretical conductivity of metal} \times \left(1 - \frac{2 \times \% \text{ porosity}}{100}\right)$$

As a result, hardenability of a P/M steel declines linearly as porosity increases. This is illustrated by the experimental data in Figure 4, where the results of a series of Jominy tests on sintered carbon steel test pieces are compared with those for wrought steel [8]. The P/M specimens were prepared from atomized iron mixed with 0.95 graphite, pressed and sintered at various densities from 6.0 to 7.1 g/cm^3. These corresponded to porosity levels ranging from 9 to 24%. The test specimens, along with a test bar machined from wrought C1080 steel having

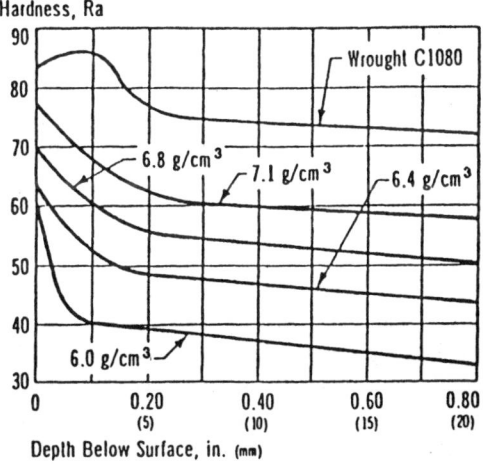

Figure 4   Effect of density on the hardenability of P/M steels. (From Ref. 8.)

a composition similar to that of the P/M steel, were austenitized at 870°C (1600°F) for 30 min in a protective neutral atmosphere. They were then end-quenched in a water column according to the Jominy test described in ASTM A255. Hardness readings were taken every 2.5 mm (0.10 in.) From the quenched end and plotted as shown in Figure 4. This figure graphically illustrates two important features of P/M heat treatment: Not only is hardenability directly affected by porosity through its influence on thermal conductivity, but there is also an important secondary effect related to density because porosity also reduces the apparent hardness of P/M materials. There is a third effect at an even more practical level: The presence of porosity makes the use of liquid salt bath hardening unsuitable because of the problem of removing the salt from the parts afterward. Even with careful washing, it is extremely difficult to avoid subsequent corrosion of the parts due to traces of salt left behind in the surface pores. As a result, P/M parts are normally heat treated in gaseous atmospheres or in a vacuum.

C.   Influence of Porosity on Case Depth

At porosity levels above about 8%, the pores are interconnected by small channels. This allows gases to penetrate quite rapidly. When case hardening P/M parts with gases containing a high carbon potential, this interconnected porosity allows diffusion of carbon into the internal pore surfaces as well as to the external surfaces of the part. This can result in carburization to a significant depth and consequent loss of any sharply defined case-hardening effect. The same applies to other gaseous surface-hardening procedures such as nitriding or carbonitriding.

Figure 5 shows microstructures of P/M test specimens that were carbonitrided. The 25 mm (1 in.) diameter slugs were pressed and sintered to increasing density levels from atomized iron with 0.20% graphite added. After sintering, the chemistry of the P/M slugs approximated that of AISI/SAE C1018 carbon steel. A test specimen was also machined from C1018 bar stock, and all specimens were then carbonitrided at 870°C (1600°F) for 60 min and oil quenched. As shown, increasing density was accompanied by a proportional decrease in carbon penetration, up to approximately 7.0 g/cm^3. At this point, the interconnected pores began to close off, preventing the gases from penetrating into the internal volume of the part. By plotting case depth against density (Figure 6), it became evident that for optimum control of case-hardening processes, P/M parts should be specified at densities of 7.0 g/cm^3 or higher.

These test specimens were then microhardness tested. From the graph shown in Figure 7 it is apparent that problems can occur when case hardening low-density parts. Because of the fast penetration of carburizing gases, a controlled carbon potential cannot be sustained at the surface of the part. This may allow soft spots to occur on quenching.

The primary purpose of case hardening low-carbon steels is to provide a hard, wear-resistant surface while maintaining a soft ductile core. This combination imparts optimum wear resistance and toughness properties to these materials. Carbon penetration should be controlled to avoid embrittlement and excessive dimensional distortion on quenching.

D.   Effect of Alloy Content on P/M Hardenability

It is known that increasing the carbon content of steel raises the tensile strength and hardness levels in proportion to the added carbon. In wrought steel, this ratio is maintained to about 1.2% C. In sintered P/M steels, the maximum tensile strength is reached at the eutectoid composition of 0.8% C. When carbon content is increased above this level, carbide networks begin to form at grain boundaries and along porosity channels, which causes embrittlement and loss of strength. As alloying elements such as nickel, molybdenum, chromium, and copper are added, the optimum carbon level content is lowered. In sintered steels the most common alloying elements are copper and nickel.

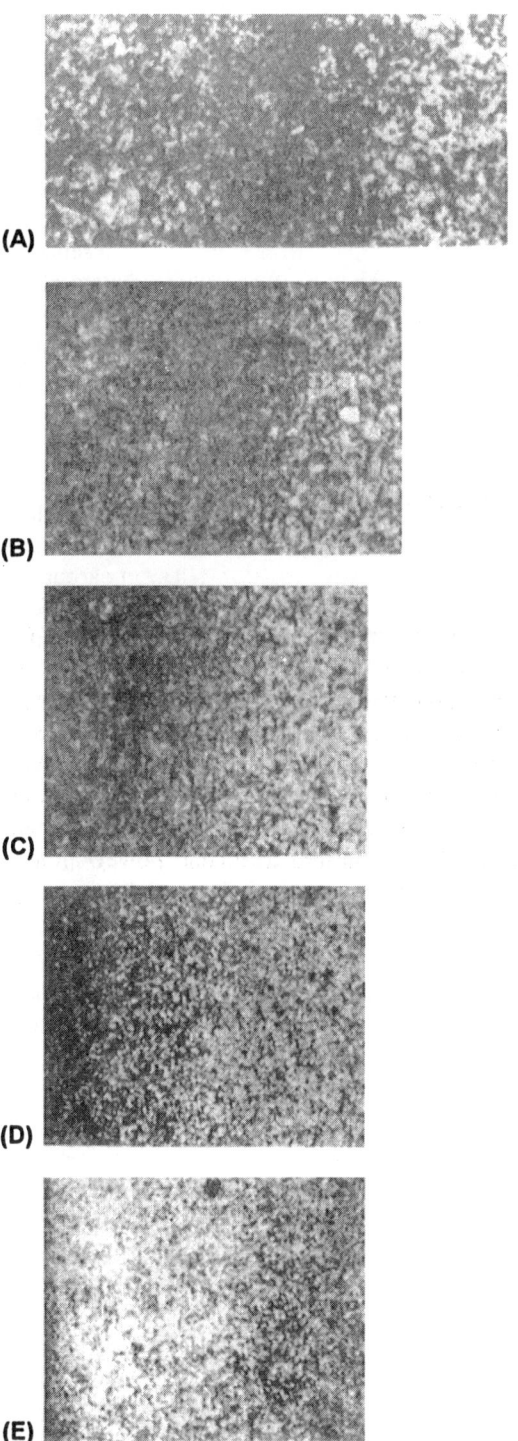

(A)

(B)

(C)

(D)

(E)

Figure 5 Cross-section micrographs illustrating the relationship between carbon penetration and density for carbonitrided P/M steels versus similar carbonitrided C1018 steel. (From Ref. 8.)

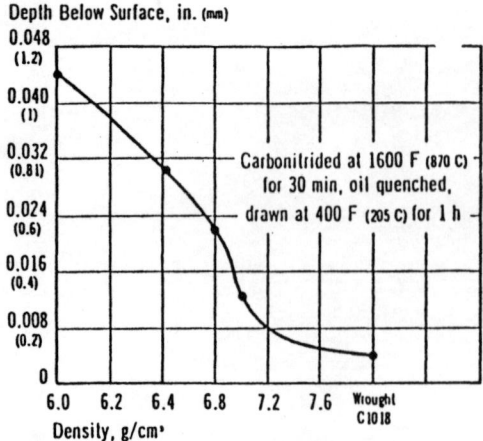

Figure 6 Effect of density on case depth for carbonitrided P/M steel. (From Ref. 8.)

1. *Copper Content*

Additions of copper increase both hardness and tensile strength in the sintered condition. On heat treatment they increase depth of hardness but reduce toughness and elongation. The relationship between strength, copper content, and combined carbon is illustrated in Figure 8. As copper content increases, rupture strength rises to an optimum level and then declines.

In the plain iron-carbon compositions, peak strength occurs near the euctectoid composition in the as-sintered condition and at approximately 0.65% C in the heat-treated condition. When copper is added to the iron-carbon compositions, there is a significant increase in strength with increasing copper content to approximately 5% Cu. On heat treatment, however, optimum strength is found at a lower carbon level as the copper content is increased.

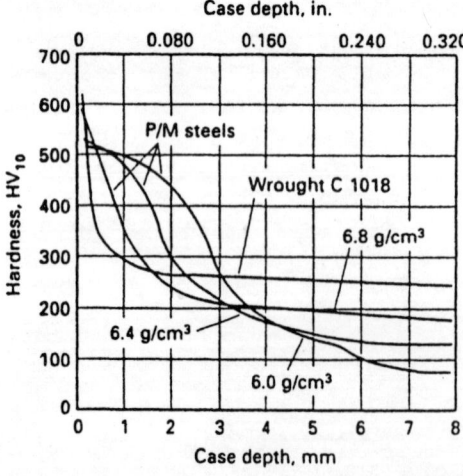

Figure 7 Plots of hardness versus case depth as a function of density for P/M steels carbonitrided and then oil-quenched. Comparison with carbonitrided C1018 steel. (From Ref. 1.)

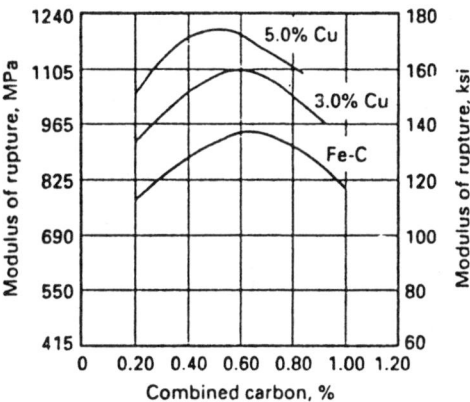

Figure 8 Relationship between carbon content, copper content, and strength for heat-treated P/M steels. (From Ref. 1.)

2. Nickel Content

Nickel increases tensile properties of as-sintered P/M steels approximately one-half as much as copper but provides significantly higher strengths after heat treatment. This is because of the heterogeneous nature of nickel alloy steels made from blended elemental powders. Copper melts at 1083°C (1980°F) and forms a liquid phase on sintering at 1120°C (2050°F), creating a more homogeneous iron-copper alloy. Nickel has a higher melting point than the sintering temperature, and it alloys by solid-state diffusion. This produces a duplex microstructure consisting of partially alloyed iron surrounding nickel-rich islands. On quenching, the matrix transforms to martensite, but the nickel-rich areas remain austenitic. This results in a strengthening of the matrix with some toughness and ductility retained.

3. Nickel–Copper Content

The effect of increasing nickel and copper contents on the hardenability of P/M steels is illustrated in Figure 9. Jominy bars were pressed to a density of 6.7 g/cm³ with increasing levels of copper and nickel while carbon content was maintained at 0.5% C. After sintering, all bars were austenitized at 850°C (1560°F) for 2 h before quenching. A dramatic increase in surface hardness but with a relatively small increase in hardenability was seen when 2.5% copper was added to the iron-carbon alloy. Adding nickel to the iron–copper–carbon alloy produced a slight gain in surface hardness but a significant improvement in hardenability. Many P/M components that require optimum heat-treated properties contain both nickel and copper.

E. Effect of Type of P/M Alloy

Reference has already been made to one of the factors restricting the choice of P/M steel compositions. This was the fact that prealloyed steel powders are less compressible than blended mixtures of elemental powders. This affects the achievement of the higher densities that are desirable for successful heat treatment. Several alternative routes have been devised to circumvent this restriction.

One of the earliest developments that allowed higher densities in low-alloy steel powders was the production of "partially alloyed" powders in which particles of alloying elements such

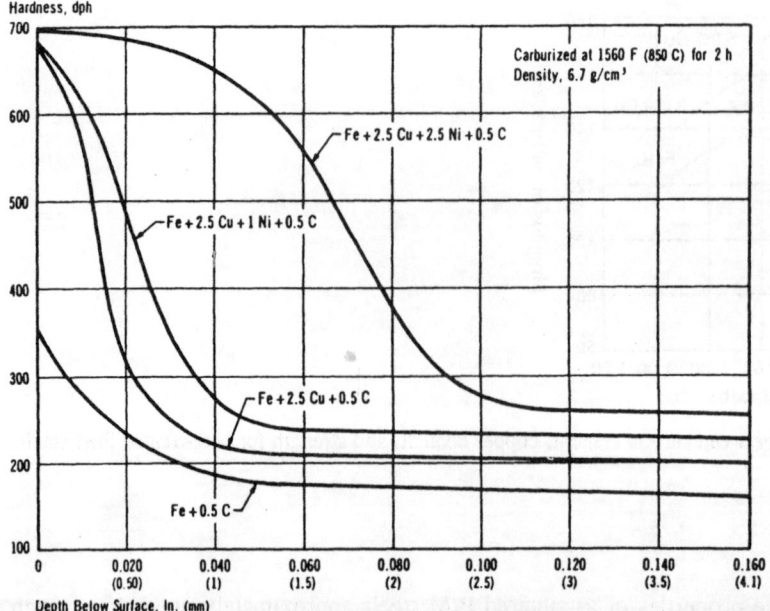

Figure 9 Effect of copper and nickel additions on the hardenability of P/M steels. (From Ref. 8.)

as nickel and copper were diffusion-bonded to the iron powder particles during the powder production process. Such powders, also known as diffusion-alloyed powders, show better compressibilities than comparable prealloyed powders, although they are less compressible than blended elemental powders. The diffusion-alloyed powders have two other advantages. The alloy elements, being already partially diffused into the matrix, can be more easily dissolved and hence more homogeneously distributed during sintering, than with premixed powders. The other advantage is that the tendency of fine alloy particles to segregate during powder handling is eliminated; this enhances the effectiveness of a given alloy addition.

In a later development, the use of "bonded" powders, in which elemental alloy particles as well as graphite particles are bonded to the iron powder particles with an organic polymer, enables the segregation to be largely suppressed as well as allowing finer alloy powders to be used, thus enhancing the alloying efficiency by accelerating diffusion during sintering. Bonded powders have compressibilities that are close to those of conventional blended elemental premixes.

An alternative route to achieving higher densities with both premixed and prealloyed powders is that of double pressing and sintering. The difficulty here is that normally, after the sintering operation, the P/M steel is too hard to compress, primarily because of the dissolved carbon. By lowering the initial sintering temperature until the solution of graphite is minimized, the "presintered" compacts can be more easily densified by re-pressing. After a second sintering treatment at or above the normal temperature, a satisfactory alloyed steel can be produced with densities up to 95% of theoretical.

F. High-Temperature Sintering

Optimization of alloy efficiency and promotion of densification can also be achieved by the use of so-called high-temperature sintering. Sintering of P/M steel compacts is mostly done in mesh belt furnaces with a maximum operating temperature of about 1150°C (2100°F). Furnaces are now being employed that eliminate the temperature restriction of the alloy mesh belt.

These include pusher-type and walking beam furnaces as well as batch and continuous vacuum furnaces, providing temperature capabilities up to 1380°C (2515°F).

By sintering at higher temperatures, greater uniformity in alloying is realized, along with improved pore rounding and coalescence, hence densification. This can provide added hardenability and more consistent heat treatment response for the same alloy content. Mechanical properties of 4600 type prealloyed P/M steel, sintered at 1260°C (2300°F) and subsequently heat treated, are compared in Table 2 with those for a double-pressed and sintered alloy for which the sintering temperature was the standard 1120°C (2050°F). Heat-treated mechanical properties (UTS, elongation, fatigue strength, and toughness) were substantially improved by the use of higher temperature sintering, and the hardness value was increased by 20%.

The use of high-temperature sintering opens up an additional opportunity in the selection of P/M alloy compositions that has been closed to traditional practice. As indicated earlier, the choice of alloying agents in P/M steels has been more or less restricted to carbon and those metallic elements with easily reduced oxides such as nickel, copper, and molybdenum. This is because oxides of these elements that may be formed during the powder atomization process or in subsequent processing are easily reduced in the normal sintering cycle at 1120°C in an endothermic gas or dissociated ammonia atmosphere. Oxides of elements such as chromium and manganese, on the other hand, are not reduced in the conventional sintering cycle and result in impaired response to heat treatment (less alloy in solution) and inferior mechanical properties.

On the other hand, oxides of chromium and manganese can be converted back to the metallic state if heated at higher temperatures in a sufficiently reducing atmosphere. As a consequence, high-temperature sintering, either in atmosphere or in vacuum, allows the use of the less expensive alloying elements such as chromium and manganese in 4100 type alloys.

Other recent developments include the use of higher compressibility prealloyed iron-molybdenum powders in various combinations and very fine ferro-alloy powders in bonded premixes.

VI. HEAT TREATMENT PROCESSES APPLIED TO P/M STEELS

A. Neutral Hardening

Many P/M parts heat treated today are neutral-hardened. Neutral hardening was used to develop the properties data shown in MPIF Standard 35 (see Table 6). These alloys contain 0.5–0.8% C and have densities exceeding 88% of theoretical. They are hardened primarily for wear resistance and improved core strength. These parts are normally hardened in belt-type or batch-type sealed quench furnaces. Because they are net-shape parts, they must not touch or impinge on one another during the heating cycle, for that would tend to distort the parts and cause soft spots. The most widely used atmosphere for these furnaces is endothermic gas, with both methane and air additions to control carbon potential. Because these gases can rapidly penetrate the porosity, carbon diffusion is quite rapid, and short time cycles at relatively low temperatures are therefore used.

A general rule for selection of hardening cycles when maximum wear resistance and core strength are desired is to determine the heat-treating cycle on the basis of the P/M part density (see Table 3). Low-density parts are susceptible to slack quenching and require a fast transfer to the quench medium. The ideal furnace for these parts would be a mesh belt or shaker hearth furnace where the parts can fall freely into the quench. In these furnaces, it is recommended that a high temperature and a slightly carburizing atmosphere be used to counterbalance loss of control at the quench end of the furnace.

Medium- to high-density parts are better suited to batch-type sealed quench furnaces where temperature uniformity and carbon content of the atmosphere are better controlled. Here lower

Table 2 Effect of High-Temperature Sintering on the Heat-Treated Mechanical Properties of Prealloyed 4600 Type P/M Steel

P/M alloy	Sintering temp.		Ultimate tensile strength		Yield strength		Elongation (%)	Hardness (HRC)	Fatigue strength		Impact resistance (unnotched Charpy test)	
	°C	°F	MPa	ksi	MPa	ksi			MPa	ksi	J	ft·lbf
Prealloyed 4600 type with 0.5% C, pressed at 40 tsi (550 MPa), re-pressed at 30 tsi (410 MPa)	1120	2050	895	130	—	—	1.0	40	345	50	11–14	8–10
	1260	2300	1170	170	1035	150	1.5	48	425	62	16–19	12–14

Source: Ref. 1.

Table 3 Recommended Heat Treatment Cycles for Optimum Wear Resistance and Core Strength According to P/M Steel Density

Density (g/cm³)	Quenching					Tempering	
	Temperature		Time at temperature (min)	Transfer time (s)	Transfer medium		
	°C	°F				°C	°F
6.4–6.8	870–890	1600–1635	30–45	<8	Fast oil	—	—
6.8–7.2	850–870	1560–1600	45–60	<12	Fast oil	150–180	300–355
>7.2	820–850	1510–1560	60–75	<25	Medium to fast oil	170–220	340–430

Source: Ref. 1.

temperatures can be used that minimize part distortion, and atmosphere carbon content can be neutral to the part. The one drawback to batch-type furnaces is the longer transfer time to the quench. Many of the new batch furnaces on the market have automated internal handling equipment to transfer the hot load to the quench. These are not suitable for many P/M parts because of the excessive time delay in reaching the quench.

For quenching, a fast oil is recommended (defined as having a General Motors' Quench-ometer rating of 10 s or less). Load size is also a critical factor in attaining uniform hardness on quenching. Most sealed quench furnaces are designed to quench out 0.5 kg (1 lb) of parts per gallon of quench oil. For P/M parts it is recommended that 11–15 L (3–4 gal) of quench oil be used per 0.5 kg (1 lb) of parts. Overloading is probably the most common problem in attaining uniform heat-treated properties.

B. Carburizing

Powder metallurgy parts with large cross-sectional thickness are carburized to obtain a maximum fatigue and impact properties. The material usually specified for carburizing contains hardenability agents such as nickel, molybdenum, and copper, with relatively low carbon content. To develop optimum dynamic properties at porosity levels between 10 and 15%, a combined carbon level of 0.30–0.35% is recommended. As porosity is reduced below 10%, combined carbon can be reduced to 0.15–0.25%. Because improved dynamic properties are also associated with high densities, it is recommended that combined carbon be adjusted to a level best suited to re-pressing after sintering. Parts with porosity levels much higher than 15% are not recommended for carburizing because of the penetration of the pores by the carburizing gas. Liquid carburizing is not suitable for P/M parts because of the problem of washing the parts free of salt after the treatment.

In wrought steel, carburization is normally characterized by a surface hardness range and an effective case depth. Microhardness measurements can accurately show the hardness profile in a cross section of wrought steel but can be erratic when used on P/M steels, where subsurface porosity can influence the hardness readings. It is recommended that at least three readings be taken at each level below the surface and averaged to determine effective case depth. (A detailed review of this procedure can be found in reference 13).

P/M steels are usually gas carburized at temperatures between 900 and 930°C (1650 and 1705°F). Time cycles are normally short because of the rapid diffusion of carbon through the interconnected porosity. Therefore, atmospheric carbon potentials need to be somewhat higher than those required for wrought steels of similar composition.

C. Carbonitriding

The most common case-hardening treatment used on P/M parts is carbonitriding. Carbonitriding is a modified form of gas carburizing rather than a form of nitriding. The principal process modification consists of introducing up to 10% ammonia into the gas carburizing atmosphere, which results in the addition of nitrogen to the carburized case as it is being produced. Nascent nitrogen forms at the workpiece surface by dissociation of the ammonia in the furnace atmosphere. Nitrogen diffuses into the steel surface simultaneously with the carbon, retarding the critical cooling rate on quenching and leading to a more consistent martensite transformation. It also produces a more consistent surface hardness profile, which improves wear resistance and toughness of the P/M steel. Process temperatures for carbonitriding are lower, 800–850°C (1470–1560°F), providing better control of distortion than carburizing. Care must be taken when adding ammonia, however, as excessive nitrogen diffusion into the internal pore surfaces can cause embrittlement.

Carbonitriding is a shallow case-hardening treatment. Case depths greater than 0.5 mm (0.02 in.) are seldom specified. For this reason cycle times are relatively short, usually on the order of 30–60 min. As in neutral hardening, carbon control is a critical aspect of the treatment. Normally, carbon potentials of 1.0–1.2% are specified to maintain the carbon profile in the part.

D. Tempering

When part density exceeds 90%, tempering is usually necessary after hardening, because significantly high stresses are developed on quenching. As porosity is increased, this stress level is reduced to a level at which tempering is not required. However, if a substantial amount of retained austenite is present, tempering may be advisable. This would also be the case if the part has thin cross sections, sharp corners, or undercuts that can act as stress raisers.

Recommended tempering temperatures for P/M parts range from 105 to 200°C (220–390°F). Above this temperature, entrained quench oil can ignite, creating a hazardous condition in the furnace. Tempering above 200°C will result in improved toughness and fatigue properties of the heat-treated P/M steel, but furnaces need special adaptations to handle the high volume of smoke created by the ignition of the quench oil.

E. Induction Hardening

Spur gears, bevel gears, splined hubs, and cams are ideal components for the use of P/M production techniques. These parts usually require hard, wear-resistant surfaces in some areas, with retention of the ductility of the sintered matrix in the remainder of the part. Induction hardening is commonly specified for these applications. The process can be placed in an automated machining line to reduce handling and can be a cost-effective hardening treatment when high volumes of parts are being produced. Because the inductance of P/M materials is typically less than that of a wrought material of similar composition due to porosity, a higher power setting is normally required to reach a given depth of hardening. Furthermore, because the heat is rapidly dissipated, a rapid transfer to the quench is mandatory.

As with wrought steels, the response to hardening by induction is dependent on combined carbon content, alloy content, and surface decarburization. This latter variable can be a major concern with P/M parts. With today's conventional belt-type sintering furnaces using an endogas atmosphere, decarburization can occur as the parts leave the hot zone and cool slowly through the 1100 to 800°C (2010 to 1470°F) temperature range.

In most instances, induction-heated P/M parts are quenched in a water-based solution containing some type of rust preventative to forestall internal corrosion. In those applications where induction hardening is considered, densities above 90% should be specified. With a decrease in density, the resistivity of the steel increases and permeability decreases. For this reason, integral quench coils using a high-velocity spray quench are generally used to attain maximum surface hardness in the P/M part.

F. Nitrocarburizing

Nitrocarburizing is rapidly growing in popularity as a treatment for P/M parts. Gaseous ferritic nitrocarburizing is a diffusion heat treatment that involves the addition of nitrogen and carbon to the surfaces of steel parts. To this extent, the process resembles carbonitriding. It differs from carbonitriding mainly in that the temperatures used are completely within the ferritic phase region, typically 570–600°C (1060–1110°F). At these temperatures no austenitic transformation occurs, thereby significantly reducing the dimensional changes and distortion. Nitrogen is diffused into the surfaces of the steel in sufficient concentration to form a thin layer of epsilon-iron nitride on the surface of the part, producing a thin, hard, wear-resistant case.

The nitrocarburizing process uses conventional integral quench atmosphere furnaces. The atmosphere usually consists of a 50:50 mixture of endothermic gas and anhydrous ammonia. Control of the nitrided layer thickness, as with other treatments, is dependent on density. If the nitrided layer is allowed to form on the internal pore surfaces to any significant extent, a volume expansion can occur. For this reason, density of the P/M part should be above 90% of the pore-free density. This nitrided layer, when properly applied, can reduce the coefficient of friction at the surface of the part and provide better wear resistance than conventional quench-hardening treatments. The process is best used in applications where sliding wear and fretting are involved.

Because the hard nitrided layer is relatively thin, the process should not be applied where high indentation or impact loading is involved. The epsilon-nitride layer that is formed can attain a file hardness in excess of HRC 60, depending on alloy content of the steel. Indentation hardness testing is not recommended when evaluating this process. Since no transformation occurs, the P/M parts can be air cooled without loss of surface hardness. Also no oil absorption occurs, which leaves the porosity open for impregnation if desired. Nitrocarburizing also improves strength and reduces notch sensitivity in P/M parts. Figure 10 shows the fatigue improvement of two low-carbon P/M steels after nitrocarburizing.

G.   Sinter-Hardening

An alternative route that has been found to be practical and commercially successful for smaller P/M parts such as gears is to choose material compositions with sufficient alloy content to be air-hardenable. This allows parts to be hardened in the sintering furnace while cooling from the hot zone. P/M parts made from 4600 type nickel-molybdenum low-alloy steel powders with blended additions of 2% copper and 0.8% carbon have been found to develop satisfactory properties by sintering at 1120° (2050°F) followed by tempering in air at 177°C (350°F) [9]. These parts transform largely, but not completely, to martensite during cooling from the sintering temperature. This incomplete transformation has been found to be beneficial for mechanical properties, the optimum tensile strength being achieved with 70–80% tempered martensite (the balance of the microstructure being bainite or fine pearlite and the porosity).

There are several advantages to the sinter-hardening approach that are likely to make it increasingly attractive in the future as applications grow and specification requirements for P/M parts are raised. Thus, there are substantial economic advantages in eliminating the separate heat treatment steps (e.g., quenching and tempering) that offset the cost of additional al-

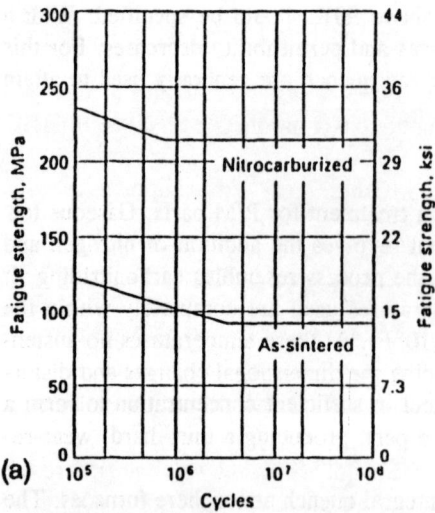

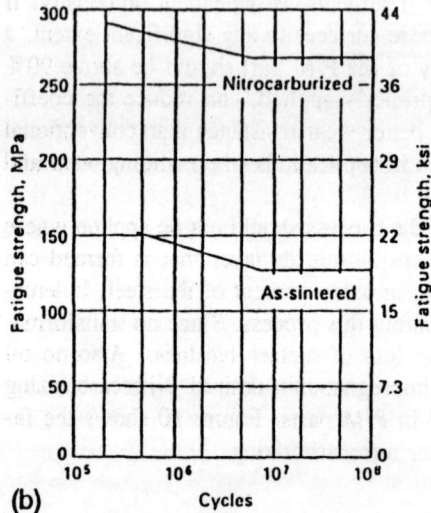

Figure 10 Effect of nitrocarburization on the notched fatigue strength of two low-carbon P/M steels. (a) F-0000 carbon steel; (b) FC-0205 copper-carbon steel. P/M density: 7.1 g/cm³. (From Ref. 1.)

loying elements. The elimination of oil quenching makes tempering in air much easier and cleaner. The reduction in distortion due to quenching improves the dimensional uniformity and hence yield and quality of parts. It is still possible to do some limited machining on the sinter-hardened parts, although these will necessarily be designed as net-shape components.

H. Steam Treating

Many P/M parts have traditionally been steam treated for improved wear resistance, corrosion resistance, and sealing capacity. In this process, P/M parts are heated in a specific manner under a steam atmosphere at temperatures between 510 and 595°C (950–1100°F) to form a layer of black iron oxide (magnetite, or ferrous-ferric oxide, $FeO \cdot Fe_2O_3$) in the surface porosity, according to the chemical reaction

$$3Fe + 4H_2O \text{ (steam)} \rightarrow Fe_3O_4 + 4H_2\text{(gas)}$$

Steam treating cannot truly be described as a heat treatment because no structural changes occur in the matrix. In the process, magnetite (Fe_3O_4) is formed at the interconnecting surface porosity, filling the porosity with a second phase. Magnetite has a hardness equivalent to HRC 50.

The process itself is straightforward, the primary variables being temperature, time, and steam pressure. Caution must be exercised to prevent the formation of hydroxides and lower oxides such as ferrous oxide (FeO) and ferric oxide (Fe_2O_3), which is red rust.

The recommended procedure for steam treating is as follows.

1. Preclean parts to remove any oil or lubricants that may have been absorbed into the porosity from prior machining, sizing, or finishing operations.

2. Load clean sintered parts in loosely packed baskets, and place fixture into a furnace preheated to 315°C (600°F).

3. Heat parts in air until the center of the load has stabilized as the set temperature.

4. Introduce superheated steam at a line pressure of 35–105 kPa (5–15 psi), and allow furnace to purge for at least 15 min.

5. Increase furnace temperature to desired steam treatment temperature, and hold for no longer than 4 h at heat.

6. On completion of treatment, reduce furnace temperature to 315°C (600°F). When parts reach this temperature, the steam can be shut off and the parts unloaded.

Caution should be used when opening the furnace door after the steam cycle. As indicated above, hydrogen is produced during this process and can ignite. It is recommended that a nitrogen purge be applied prior to unloading. This process, when correctly applied, can impart improved surface properties, and, depending on steel composition, increased compressive yield strength.

In steam-treated P/M steels, the ductility is significantly reduced due to the internal stresses created by the formation of the iron oxide. Care must be taken when steam treating high-carbon P/M steels because these internal stresses can initiate microcracking and cause severe loss of ductility. Cases have been reported in which steam-treated parts accidentally dropped on the floor shattered in a brittle fashion. The best recommendation for preventing such an incident is to specify a maximum carbon content of 0.5% for P/M parts that are to be steam treated.

The increase in density and apparent hardness produced by steam treating is indicated in Table 4 and Figure 11 [10]. With its porosity filled with a hard second phase, the P/M steel offers better support to the indentation hardness tester. Figure 12 illustrates that the transverse rupture strength is increased significantly by steam treatment for low-carbon P/M steels but only modestly for high-carbon (0.8%) P/M steels.

VII. GUIDELINES FOR HEAT TREATING OF P/M PARTS

Whether the heat treatment of P/M parts is done in-house or sent to an outside commercial heat treater, some guidelines need to be followed to ensure that parts get properly treated. In the past, heat treatment of P/M parts was primarily specified to improve wear resistance. With the advent of high-compressibility powders and high-temperature sintering, heat treatments are now added to provide improved dynamic properties as well. The following is a list of recommendations for heat treating porous P/M parts:

1. Always degrease parts prior to heat treating. In many manufacturing plants, P/M parts are dipped in a rust preventive after sintering. Also, many P/M parts are machined

Table 4 Effect of Steam Treatment on the Density and Hardness of P/M Steels

Material	Density (g/cm³)		Apparent hardness	
	Sintered	Steam-treated	Sintered	Steam-treated
F-0000-N	5.8	6.2	7 HRF	75 HRB
F-0000-P	6.2	6.4	32 HRF	61 HRB
F-0000-R	6.5	6.6	45 HRF	51 HRB
F-0008-M	5.8	6.1	44 HRB	100 HRB
F-0008-P	6.2	6.4	58 HRB	98 HRB
F-0008-R	6.5	6.6	60 HRB	97 HRB
FC-0700-N	5.7	6.0	14 HRB	73 HRB
FC-0700-P	6.35	6.5	49 HRB	78 HRB
FC-0700-R	6.6	6.6	58 HRB	77 HRB
FC-0708-N	5.7	6.0	52 HRB	97 HRB
FC-0708-P	6.3	6.4	72 HRB	94 HRB
FC-0708-R	6.6	6.6	79 HRB	93 HRB

Source: Ref. 10.

prior to heat treatment and can retain some of the lubricant. These oils can contaminate the heat-treating atmosphere and cause discoloration.

2. Do not use dense loads. Overlapping parts can cause soft spots and distortion. For best results, parts should be single-spaced in layers.

3. Use highly agitated quench oil with a quench severity rating (H) of 0.7–1.5.

4. When tempering, hold at 205°C (400°F) for 2 h before proceeding to higher temperatures. This is to minimize the evolution of smoke and to prevent ignition.

If it is planned to send parts to an outside commercial heat treater, a technical review of the part requirements and prior processing history is recommended. Many commercial shops have had bad experiences with P/M because they did not fully understand the implications of

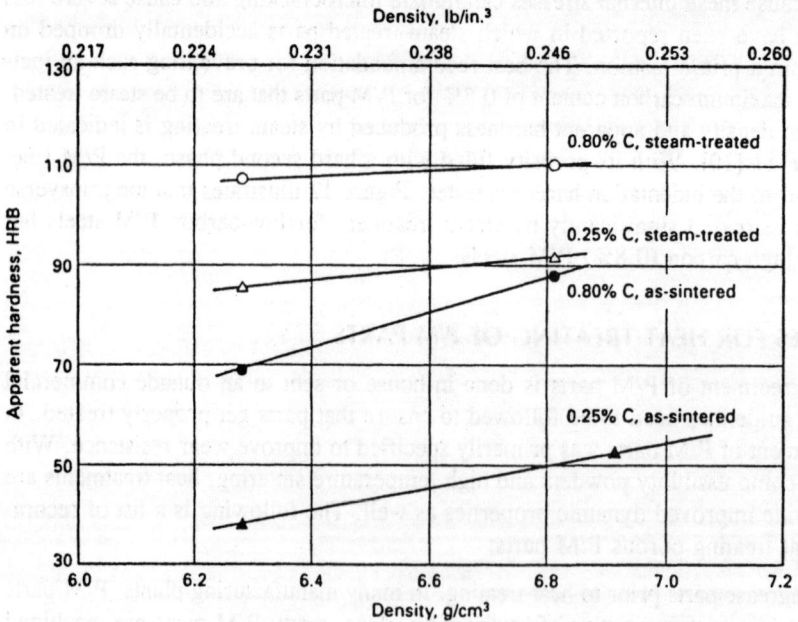

Figure 11 Effect of steam treating on apparent hardness of P/M carbon steels. (From Ref. 1.)

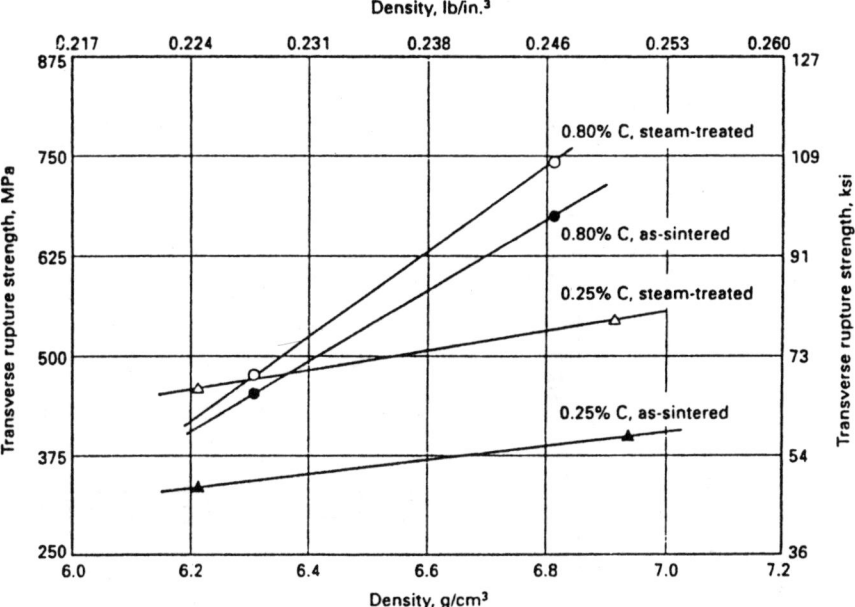

Figure 12 Effect of steam treating on transverse rupture strength of sintered P/M carbon steels. (From Ref. 1.)

porosity and tried to treat the parts as they would wrought steel. It is important that you establish a good communication network with your heat treater so that you can alert him to any structural changes in your part caused by changes in powder lot, tooling, or sintering cycle.

A good procedure is to send sample lots to the heat treater for him to establish process capability. After he has met the print requirements, he should be requested to run the parts in the same furnace under the exact same conditions as were established. As a parts fabricator, it would be wise to give as much consideration to the heat-treating process as would be given to the primary processes. Reworking because of improper heat treatment not only adds more cost but can also significantly affect the physical properties of the P/M part.

VIII. TYPICAL MECHANICAL PROPERTIES OF HEAT-TREATED P/M STEELS

A. General

In this section we provide a summary of the most recently published data on the key properties of selected P/M steels after heat treatment. Also included is a listing of industry standards and specifications where additional information on the properties of P/M steels can be found. Since, at the time of writing, this topic is still under intense development, it is advisable to consult the latest editions of standards documents and other sources of information such as materials suppliers and consultants.

B. Mechanical Properties Data

Table 5 shows minimum and typical values of heat-treated mechanical properties for the P/M carbon and alloy steels listed in Table 1. This information is taken from MPIF Standard 35 [6]. Reference should be made to Table 1 for an explanation of the MPIF material codes (composition designations).

Table 6 lists compositions and mechanical properties of selected heat-treated iron-copper, iron-nickel, and nickel-molybdenum P/M steels provided by Quebec Metal Powders Limited and based on its ATOMET steel powder grades [1]. Table 7 lists compositions and mechanical

Table 5 Mechanical Properties of Heat-Treated P/M Carbon and Low-Alloy Steels

Material designation code	Minimum Values[a] Minimum ultimate strength[c] (MPa)	Typical Values[b]										
		Tensile properties[d]		Elastic constants		Unnotched Charpy impact energy (J)	Transverse rupture strength (MPa)	Compressive yield strength (0.1%) (MPa)	Rockwell hardness		Fatigue limit (estimated as 38% of tensile strength) (Mpa)	Density (g/cm^3)
		Ultimate strength (MPa)	Elongation (in 25.4 mm) (%)	Young's modulus (GPa)	Poisson's ratio				Macro (apparent)	Micro (converted)		
F-0005												
-50HT	340	410	<0.5	115	0.25	4	720	300	20HRC	58HRC	156	6.6
-60HT	410	480	<0.5	130	0.27	5	830	360	22	58	182	6.8
-70HT	480	550	<0.5	140	0.27	5	970	420	25	58	209	7.0
F-0008												
-55HT	380	450	<0.5	115	0.25	4	690	290	22HRC	60HRC	171	6.3
-65HT	450	520	<0.5	115	0.25	5	790	400	28	60	198	6.0
-75HT	520	590	<0.5	135	0.27	6	900	520	32	60	224	6.9
-85HT	590	660	<0.5	150	0.27	7	1000	590	35	60	248	7.1
FC-0205												
-60HT	410	480	<0.5	110	0.25	3	660	390	19HRC	58HRC	182	6.2
-70HT	480	550	<0.5	105	0.25	5	760	490	25	58	209	6.5
-80HT	550	620	<0.5	130	0.27	6	830	590	31	58	236	6.8
-90HT	620	690	<0.5	140	0.27	7	930	660	36	58	262	7.0
FC-0208												
-50HT	340	450	<0.5	105	0.25	3	660	400	20HRC	60HRC	171	6.1
-65HT	450	520	<0.5	120	0.27	5	760	500	27	60	198	6.4
-80HT	550	620	<0.5	130	0.27	6	900	630	35	60	236	6.8
-95HT	660	720	<0.5	150	0.27	7	1030	720	43	60	274	7.1
FN-0205												
-80HT	550	620	<0.5	115	0.25	5	830	530	23HRC	55HRC	236	6.6
-105HT	720	830	<0.5	135	0.27	6	1110	620	29	55	315	6.9
-130HT	900	1000	<0.5	150	0.27	8	1310	680	33	55	380	7.1
-155HT	1070	1100	<0.5	155	0.28	9	1480	710	36	55	418	7.2
-180HT	1240	1280	<0.5	170	0.28	13	1720	770	40	55	486	7.4

Material	Designation											
FN-0208	-80HT	550	620	<0.5	120	0.25	5	830	680	26HRC	57HRC	236	6.7
	-105HT	720	830	<0.5	135	0.27	6	1030	850	31	57	315	6.9
	-130HT	900	1000	<0.5	140	0.27	7	1280	940	35	57	380	7.0
	-155HT	1070	1170	<0.5	155	0.28	9	1520	1120	39	57	445	7.2
	-180HT	1240	1340	<0.5	170	0.28	11	1720	1300	42	57	509	7.4
FN-0405	-80HT	550	590	<0.5	105	0.25	5	790	460	19HRC	55HRC	224	6.5
	-105HT	720	760	<0.5	130	0.27	7	1000	610	25	55	289	6.8
	-130HT	900	930	<0.5	140	0.27	9	1380	710	31	55	353	7.0
	-155HT	1070	1100	<0.5	160	0.28	13	1690	850	37	55	418	7.3
	-180HT	1240	1280	<0.5	170	0.28	18	1930	910	40	55	486	7.4
FL-4205	-80HT	550	620	<0.5	115	0.25	5	930	550	28HRC	60HRC	236	6.60
	-100HT	690	760	<0.5	130	0.26	5	1100	760	32	60	289	6.80
	-120HT	830	900	<0.5	140	0.26	5	1280	970	36	60	342	7.00
	-140HT	970	1030	<0.5	155	0.27	6	1480	1170	39	60	390	7.20
FL-4405	-100HT	690	760	<1.0	120	0.25	7	1100	–e	24HRC	60HRC	289	6.70
	-125HT	860	930	<1.0	135	0.27	9	1380	–e	29	60	353	6.90
	-150HT	1030	1100	<1.0	150	0.27	12	1590	–e	34	60	418	7.10
	-175HT	1210	1280	<1.0	160	0.28	15	1930	–e	38	60	486	7.30
FL-4605	-80HT	550	590	<0.5	110	0.24	5	900	630	24HRC	60HRC	224	6.55
	-100HT	690	760	<0.5	125	0.25	6	1140	790	29	60	289	6.75
	-120HT	830	900	<0.5	140	0.26	8	1340	960	34	60	342	6.95
	-140HT	970	1070	<0.5	155	0.27	9	1590	1170	39	60	407	7.20
FLN-4205	-80HT	550	620	<1.0	115	0.25	7	900	–e	24HRC	60HRC	236	6.60
	-105HT	720	790	<1.0	130	0.27	9	1170	–e	30	60	300	6.80
	-140HT	970	1030	<1.0	145	0.27	12	1590	–e	36	60	391	7.05
	-175HT	1210	1280	1.0	160	0.28	19	2000	–e	42	60	486	7.30

Notes:

[a] Suffix numbers represent minimum strength values in 10^3 psi.

[b] Mechanical property data derived from laboratory-prepared test specimens sintered under commercial manufacturing conditions.

[c] Tempering temperatures: 260°C for P/N nickel steels, 177°C for all others.

[d] Yield and ultimate tensile strength are approximately the same for heat-treated materials.

[e] Additional data in preparation will appear in subsequent editions of this standard.

Source: Ref. 6. The above table is based on MPIF Standard 35, "Materials Standards for P/M Structural Parts," 1994 Edition, published by the Metal Powder Industries Federation, Princeton, New Jersey 08540-6692.

Table 6 Composition and Mechanical Properties of Selected P/M Heat-Treated Copper, Nickel, and Nickel-Molybdenum Steels

P/M alloy steel	MPIF material	code[a]	Composition, wt%					Density		UTS		Transverse rupture strength		Apparent hardness,	Impact strength	
			Fe	C	Cu	Ni	Mo	g/cm^3	lb/in.^{3}	MPa	ksi	MPa	ksi	HRC	J	ft·lbf
Mixed elemental powders based on ATOMET 1001 steel powder[b]																
Iron-copper	FC-0205-HT		97.5	0.5	2	—	—	6.8	0.246	786	114	1170	170	27	—	—
								7.0	0.253	869	126	1345	195	30	—	—
			96.5	0.5	3			6.8	0.246	765	111	1235	179	27	—	—
								7.0	0.253	883	128	1370	199	29	—	—
	FC-0208-HT		97.3	0.7	2			6.8	0.246	862	125	1360	197	35	—	—
								7.0	0.253	1030	149	1595	231	40	—	—
			96.3	0.7	3			6.8	0.246	848	123	1435	208	34	—	—
								7.0	0.253	979	142	1745	253	38	—	—
Iron-nickel	FN-0205-HT		97.4	0.6	—	2	—	6.8	0.246	792	115	1235	179	36	—	6
								7.0	0.253	993	144	1545	224	41	—	8
								7.2	0.260	1165	169	1795	260	44	—	10
Based on 4201 and 4601 prealloyed low-alloy steel powders																
Nickel-molybdenum	FL-4205-HT		98.45	0.5	—	0.45	0.60	6.8	0.246	765	111	1480	215	34	—	—
								7.0	0.253	889	129	1780	258	38	—	—
								7.1	0.256	979	142	1930	280	40	—	—
	FL-4605-HT		97.15	0.5	—	1.8	0.55	6.8	0.246	876	127	1505	218	33	—	—
								7.0	0.253	1035	150	1795	260	39	—	—
								7.1	0.256	1150	167	1950	283	42	—	—

[a]MPIF. Metal Powder Industries Federation.
[b]All mixes contained 0.5% zinc stearate [Zn(C$_{18}$H$_{35}$O$_2$)$_2$]; sintered in endogas at 1125°C (2050°F) for 30 min. Heat treatment: austenitized at 815°C (1500°F) for 15 min. quenched in oil at 65°C (150°F), tempered at 175°C (350°F) for 60 min.
Source: Ref. 1. Data from Quebec Metal Powders Ltd.

Table 7 Mechanical Properties of High-Strength P/M Alloy Steels Based on Prealloyed Powders

Composition additives (wt %)	Compacting pressure (tsi)	Density (g/cm³)	Apparent hardness (HRC)	Yield strength (0.2%) (ksi)	UTS (ksi)	Elongation (%)	Impact strength (ft-lb)	TRS (ksi)	Fatigue limit (ksi)
Part I									
4600 type (1.8% Ni, 0.55% Mo) prealloyed powder base (*Source:* Ref. 11)									
0.5% C[a]	40	6.96	36	131.4	143.5	1.1	9	204.6	
2% Ni, 0.5% C	40	7.00	36.5	120.6	146.9	1.3	12	234.1	
0.85% Mo prealloyed powder base (*Source:* Ref. 11)									
2% Ni, 0.5% C	40	7.07	40	145.7	177.9	1.5	11	231.1	
4% Ni, 0.5% C	40	7.15	39.5	131.6	172.3	1.6	13	252.8	
Part II									
Single and double pressed 0.85% Mo prealloyed powder base (*Source:* Ref. 12)									
0.6% C, single press/sinter	45	7.15	32	–	108.8	0.5	9		40.0
0.6% C, double press/sinter	45/45	7.51	36	152.0	182.7	0.9	22		63.3

[a]Carbon added as graphite.

properties of heat-treated P/M alloy steels based on prealloyed powders containing molybdenum or nickel plus molybdenum. Part I of the table illustrates the variations in density and mechanical properties resulting from various combinations of nickel (1.8–4.0%) and molybdenum (0–0.85%) at constant carbon levels (0.5% admixed graphite) and compacting pressure (40 tsi) [11]. In Part II of the table [12], mechanical properties of heat-treated P/M molybdenum steels containing 0.85% prealloyed molybdenum and 0.6% admixed graphite are compared for single pressing and for double pressing and sintering, illustrating the substantial improvement in properties obtainable by increasing the density to 7.5 g/cm^3.

Table 8, taken from Reference 13, lists typical mechanical properties quoted by Hoeganaes Corp. for fully densified powder-forged nickel-molybdenum low-alloy P/M steels containing 0.24–0.70% carbon. As indicated in the table, tempering temperatures varied between 350 and 825°F.

C. Standardization

Because of the small size or complexity of most P/M parts, it is not often feasible to make test pieces for measuring tensile and other mechanical properties by cutting them out of the finished parts. Standardized test pieces have therefore been developed that can be pressed and sintered and subsequently heat treated alongside production parts (see [14]). Details of the test pieces and the test methods are given in the following standards:

Test	MPIF	ASTM	ISO
Tensile strength	10	E8	2740
Transverse rupture	41	B528	3325
Impact strength	40	—	—

For heat-treated materials, round tensile test pieces should be machined from specially molded sintered bars, because heat-treated unmachined specimens may yield erroneously low values.

As already indicated, the MPIF has published a succession of standards listing properties data for the most common P/M steels. The same specifications have been adopted by ASTM (Standard B 783-91). The equivalent SAE specification is J471d, which lists ferrous P/M materials. International standard specifications for P/M steels for structural parts are included

Table 8 Typical Mechanical Properties for Powder-Forged Nickel-Molybdenum Low-Alloy P/M Steels

Property[a]	FL-4200		FL-4600	
Carbon (%)	0.28	0.70	0.24	0.60
Tempering temp. (°F)	350	650	350	825
Ultimate tensile strength (ksi)	152.7	261.9	227.0	211.0
Yield strength 0.2% offset (ksi)	130.0	226.5	207.0	170.0
Elongation (% in 1 in.)	10.6	5.0	13.0	10.0
Reduction in area (%)	42.8	11.8	42.0	32.0
Room temp. Charpy V-notch impact energy (ft-lb)	16	5	12	10
Core hardness (HRC)	35	52	49	48

[a]Properties taken from Hoeganaes brochures.
Source: Ref. 13.

in ISO 5755, Part 2. There are corresponding national standards in several European countries; these, however, may be replaced in the near future by the adoption of "European" standards, probably based on the ISO standard.

REFERENCES

1. H. A. Ferguson, Heat treating of powder metallurgy steels, in *Metals Handbook, 10th Ed.*, Vol. 4, *Heat Treating*, ASM, Materials Park, OH, 1991, p. 229.
2. R. Burns, Production presses and tooling, in *Metals Handbook, 9th Ed.*, Vol. 7, *Powder Metallurgy*, ASM, Materials Park, OH, 1984, p. 329.
3. J. M. Capus, Metal powder manufacturers adapt to changing needs, *ASTM Standard. News 18*(3):54 (1990).
4. Anon, *Powder Metallurgy Design Manual*, MPIF, Princeton, NJ, 1989, p.65.
5. S. Mocarski, D. W. Hall, J. Khanuja, and S.-K. Suh, Parts for automotive applications. Part III, *Int. J. Powder Met. 25*: 103 (1989).
6. *MPIF Standard* 35, Materials Standards for P/M Structural Parts, 1994 Ed., Metal Powder Industries Federation, Princeton, NJ, 1994.
7. L. F. Pease III, Ferrous powder metallurgy materials, *Metals Handbook,* 10th ed., Vol. 1, ASM, Materials Park, OH, 1990, p. 801.
8. H. A. Ferguson, Heat treatment of P/M parts, *Met. Prog. 107*(6):81 (1975); *108*(1):66 (1975).
9. A. de Rege, D. Pantano, and L. F. Pease III, Air hardening of low alloy powder metallurgy materials, *Advances in Powder Metallurgy* (1989), Vol. 1, MPIF, Princeton, NJ, 1989, p. 223.
10. L. F. Pease III, J. P. Collette, and D. A. Pease, Mechanical properties of steam-blackened P/M materials in *Modern Developments in P/M*, Vol. 21, MPIF, Princeton, NJ, 1988, p. 275.
11. R. J. Causton, J. A. Hamill, Jr., and S. O. Shah, Properties of heat treated P/M alloy steels, *Advances in Powder Metallurgy and Particulate Materials—1993*, Vol. 4, MPIF, Princeton, NJ, 1993, p. 61.
12. J. J. Fulmer and R. J. Causton, Tensile, Impact and fatigue performance of a new water atomized low-alloy powder—Ancorsteel 85 HP, *Advances in Powder Metallurgy 1990*, Vol. 2, MPIF, Princeton, NJ, p. 459.
13. R. A. Ketterer, Production aspects of powder metal forging, *Advances in Powder Metallurgy 1989*, Vol. 1, MPIF, Princeton, NJ, p. 321.
14. *Standard Test Methods for Metal Powders and Powder Metallurgy Products, 1993 Edition*, MPIF, Princeton, NJ, 1993.

13

METALLURGICAL PROPERTY TESTING

Xiwen Xie
Beijing University of Aeronautics & Astronautics, Beijing,
People's Republic of China

I. INTRODUCTION

Various testing methods may be selected to verify the quality of heat-treated parts or specimens. Hardness testing is perhaps the most commonly used testing method for heat-treated parts; however, additional tests (such as impact and tension tests) are required for some critical parts. In some cases, the microstructure of heat-treated parts should be examined under a metallurgical microscope because any property change during heat treatment is closely related to microstructural change. Sometimes microstructural measurements are needed in order to obtain a more informative and quantitative result. These may be made manually with an image analyzer. This chapter provides an overview of the various metallurgical testing methods, how they are conducted, and the physical principles involved in testing. Helpful tables and testing strategies are also provided.

II. METALLOGRAPHIC TECHNIQUE FOR STEELS

The true microstructure of a steel specimen can be observed under a microscope only when the necessary preparation procedures are properly performed. In the past, the quality of a prepared metallographic specimen depended on the experience and skillfulness of the operator. This metallographic technique was an art, not a science. Besides, traditional methods of specimen preparation involve many lengthy steps creating an excessive expenditure of consumables.

There has recently been tremendous progress in understanding the physical nature of specimen preparation. The principles and techniques of specimen preparation are now scientifically established. Based on extensive work, Nelson [1] developed a new concept of specimen preparation producing superior polished surfaces on all materials, with the least number of steps and in the shortest possible time. This is based on the following premises:

1. Each stage of preparation is important. Any mistake at one stage is difficult to correct during subsequent steps, as each stage relies on the quality of the previous stage.
2. It is very important to attain a proper balance between the material removal rate and the depth of the deformed layer that remains at the end of each step.

3. The polishing parameters (surface, abrasive type and size, and fluid media) should be selected according to the physical properties of the sample material.

4. Other parameters such as pressure, rotational direction and speed, and time should also be carefully controlled at each step.

Generally, specimen preparation can be divided into several stages: sampling, sectioning, mounting, plane grinding, fine grinding, polishing, and etching. The first three stages may be regarded as preliminary stages of specimen preparation. Plane grinding, fine grinding, and polishing are the main stages of specimen preparation through which a smooth, scratch-free deformed surface layer of minimal depth is obtained. Etching is the final stage of specimen preparation in which the true microstructure is revealed by using suitable reagents.

A. Preliminary Stages of Specimen Preparation

1. Sampling

Although random sampling sounds more reasonable, it is usually impractical. In many cases, the top priority is convenience in sectioning. For some important heat-treated parts, e.g., the turbine axis of a jet engine, test specimen locations are usually specified at critical sites. For the examination of wrought materials, it is important that the orientation of the prepared surface be chosen in accordance with the appropriate specification. Information about metal flow patterns and inclusion deformability can be easily seen on longitudinal sections and is not obtainable on transverse sections. The microstructure often appears more uniform and the grain structure more equiaxed on transverse sections. Sometimes separate test specimens are loaded into the furnace together with the parts to be treated. In this case, the test specimens should be of the same heat and located in the same temperature zone as the parts to be treated. This is because steels of the same grade but different heats differ slightly from each other in chemical composition; although still within allowable limits, their responses to heat treatment may not be the same.

2. Sectioning

The surface condition of a metallographic specimen after sectioning can be considered the starting point for specimen preparation. The quality of this initial surface condition is often overlooked. If sectioning is done on a handsaw or dry-cutting machine, a very rough surface with a deep deformed layer and, in many cases, excessive surface damage due to overheating will be produced. This in turn will lead to extended preparation time, higher consumption of consumables, and incorrect microstructure. Therefore, proper sectioning is critical in specimen preparation. Two basic requirements must be fulfilled: flatness of the resultant surface and minimal depth of the damaged or deformed layer. The sectioning operation must be fast, easy, and inexpensive. Various types of cutoff machines are available for sectioning; Figure 1 shows a typical one. The cutoff wheel usually consists of a suitable abrasive and bonding material; a Bakelite bonded aluminum oxide (alumina) cutoff wheel is suitable for cutting ferrous metals. During sectioning, effective cooling of the specimen must be maintained.

To minimize the contact area between the abrasive cutoff wheel and the workpiece, a so-called Exicut technique (see Figure 2) is adopted for the cutoff machine shown in Figure 1. In this design, the abrasive cutoff wheel moves forward and backward through the workpiece while rotating. The result is simpler and faster cutting and a significant reduction in the consumption of abrasive cutoff wheels. The combined rotating and oscillating movement of the wheel also improves the cooling by allowing easier access of the coolant to the surface being cut. Figure 3 shows the Exicut technique in operation.

Figure 1 The Struers Exotom cutoff machine. (Courtesy of Struers, Inc.)

Another type of cutoff machine is the precision cutoff machine or precision saw. This machine is suitable for sectioning medium-sized (up to 50 mm in diameter) to very small samples. The diameter of the wafering blades and abrasive cutoff wheels varies from 100 to 180 mm. This cutoff machine provides a much thinner deformed layer of the surface being cut, e.g., 25 μm for carbon steel at a low cutting speed of 150 rpm as compared with 120 μm for the same material using a conventional abrasive wheel [2]. The wheel speed may be increased to 5000 rpm. A fivefold increase in cutting speed reduces cutting time by one-fifth without increasing the deformation depth by more than 10%.

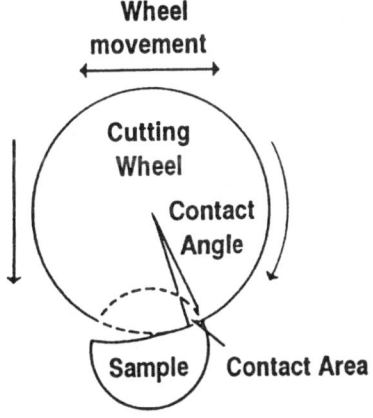

Figure 2 The Exicut technique (schematic). (Courtesy of Struers, Inc.)

Figure 3   The Exicut technique in actual operation. (Courtesy of Struers, Inc.)

3. *Mounting*

Mounting of metallographic specimens is necessary in several cases, for

1. Small parts or specimens such as thin sheets and coil springs that are difficult to handle during grinding and polishing.
2. When edge retention is needed (e.g., for case-hardened or decarburized parts); otherwise edge rounding will result.
3. Specimen holders of certain semiautomatic and automatic grinding/polishing machines that require the use of mounted specimens.

Mounted specimens are especially suitable for storage.

There are two plastic mounting techniques: hot mounting and cold mounting. Hot mounting requires heat and pressure during molding, whereas cold mounting is conducted at room temperature in open atmosphere. Hot mounting or compression molding is the fastest and most efficient mounting method. It provides the best quality and allows the use of less expensive mounting resins, among which Bakelite is most popular. A hydropneumatic mounting press is shown in Figure 4. During hot mounting, the applied heating time, molding temperature, and cooling time, which are different for different resins, can be accurately adjusted.

Cold mounting is used when the specimen cannot withstand any heat or pressure during mounting, e.g., as-quenched martensite, which would transform to tempered martensite. Materials for cold mounting consist of fluid resin (e.g., epoxies) and hardener, which are mixed thoroughly before pouring. The specimen is placed in a mold, and the mixed mounting material is poured over the specimen. After curing, the mount is usually removed from the mold. Maximum adhesion between embedding and mold is ensured by using a plastic mold with ridges on the inside. Such a mold is prepared together with the mount (Figure 5). It is important that the cold-mounting medium adhere well to the specimen and be free of hardening shrinkage; otherwise cracking will occur [3]. Also, the hardness of the mounting medium should be matched with that of the specimen, as a harder specimen will result in edge rounding. A

Figure 4 Hydroneumatic mounting press for mounts up to 50 mm in diameter. (Courtesy of Struers, Inc.)

series of sintered aluminum oxide pellets dispersed in the resin will match the hardness of a wide range of materials [3].

If a specimen with quenching cracks is to be examined, vacuum impregnation is a necessary step to remove the air from the cracks during mounting, thus allowing the epoxy to enter so that complete bonding is effected. Otherwise the crack may be enlarged and crack edges may become rounded during later preparation steps [3]. Polishing abrasives, solvent, and etchants may also be entrapped in open cracks, causing staining problems.

Figure 5 A cold embedding mount (Monoform) for fast cold mounting. (Courtesy of Struers, Inc.)

B.   Main Stages of Specimen Preparation

1.   *Plane or Coarse Grinding*

The aim of plane grinding is to make the surface plane and to remove scratches and damaged layers produced during cutting. Scratches and disturbed layers produced in this stage should be easily removed during subsequent steps. Therefore, it is important to select the most effective type of abrasive and the finest grit to flatten and smooth the specimen in the shortest time.

Traditionally, plane grinding of steel specimens is performed with 120 or 180 grit SiC abrasive paper. However, the shortcoming of conventional abrasive paper is its limited service life, especially when it is used with semiautomatic or automatic machines. For example, sometimes even a new sheet of SiC abrasive paper cannot complete a single step of plane grinding for six specimens clamped in a specimen holder.

When plane grinding is carried out in a multipurpose grinding/polishing machine, a ZrO_2 grinding paper may be used, replacing conventional SiC paper. The abrasion rate of ZrO_2 paper is much faster than that of ordinary SiC paper of the same grit, and its service life is considerably longer.

An Al_2O_3 grinding stone is most effective for plane grinding of steel specimens. Figure 6a shows a high-capacity automatic grinding machine. The wheel speed is 1500 rpm, the specimen holder runs at 140 rpm, and the pressure on the specimens can be adjusted between 0 and 700 N. Figure 6b shows a benchtop grinding/polishing machine that is microprocessor-controlled. Data for eight individual steps can be programmed and stored, and a combination of these steps allows the preparation of several different materials without reprogramming. Most steps are completed in 2–4 min, and the total preparation time ranges from 6 to 12 min.

2.   *Fine Grinding*

The fine grinding stage is the most important of the entire preparation sequence. Any previously deformed layers from plane grinding not removed here and any excessive damage created here are likely to remain to the end of the preparation sequence, which may cause misinterpretation of the microstructure. Therefore, the purpose of fine grinding is to remove all the deformed layer from plane grinding without introducing any more than necessary [4].

Traditionally, fine grinding of steel specimens employs a series of three or four SiC papers with the grit sequence 220–320–400–600 (the corresponding abrasive sizes are approximately 65, 46, 30, and 18 μm, respectively). Wet grinding with water as coolant should be used to minimize frictional heat and metal entrapment between abrasive particles and to maximize the rate of metal removal and the service life of abrasive papers. Wet grinding is usually carried out at 250–300 rpm. As mentioned earlier, the service life of SiC papers is rather limited, and the time consumed in replacing exhausted papers accounts for a significant portion of the total processing time when semiautomatic grinding machines are used. In spite of this, this method is still widely used with simple grinding machines.

Fine grinding can be effected in one step by using a rigid grinding disk of a synthetic material to which diamonds (usually 6 or 9 μm) are applied in the form of a spray [5–7]. The individual diamond particles will adhere to the disk, providing a relatively high rate of material removal. Consequently, the time for this one-step operation will be much shorter than that for the four-step operation of SiC paper grinding. Excellent planarity and edge retention will also be obtained at one-half the cost per specimen.

3.   *Polishing*

The purpose of polishing is to remove all the scratches and deformed layer from the previous grinding step. The resulting surface should be scratch-free, relief-free, and planar. The edge

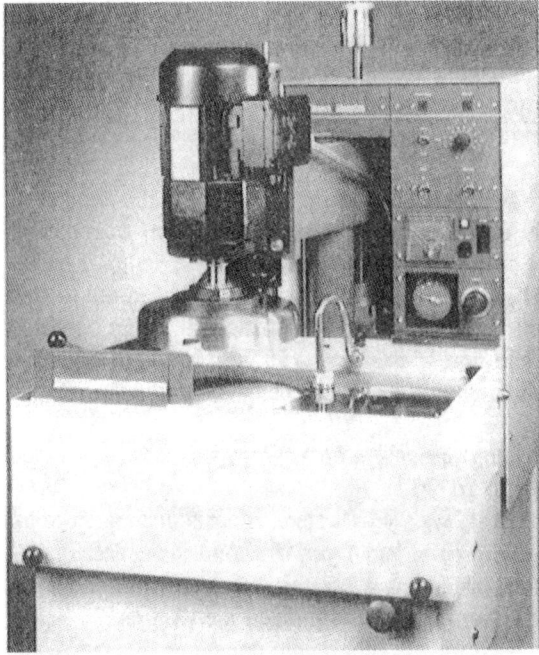

(a)

(b)

Figure 6 Examples of grinding machines. (a) Struers Abraplan; (b) Struers Abramin. (Courtesy of Struers, Inc.)

retention must be acceptable. For routine examination, very fine and well-dispersed scratches that will not hinder observation of the true microstructure are permissible.

The most important parameters during polishing are time and pressure. Polishing should be completed within the shortest possible time to avoid relief, and the pressure should be minimal to obtain the highest surface perfection. A careful balance between time and pressure can be achieved by trial and error.

For specimens of plain carbon steel and low-alloy steel, polishing with 3 μm diamond spray on a synthetic long-napped cloth is generally sufficient to completely remove all traces of retained deformed layer. The specimen is then ready for routine examination. Diamond polishing is usually carried out at 125–300 rpm. Higher speeds cause the polishing medium to flow off the disk too early, and at lower speeds the rate of removal is insufficient. Many operators prefer to polish manually at 250–300 rpm and automatically at 150 rpm to avoid excessive frictional heat. The grinding/polishing machine shown in Figure 6b is also suitable for polishing.

If final polishing with colloidal silica polishing suspension on a synthetic short-napped cloth is added, a very good polish can be obtained in 10–30 s.

For specimens of high strength and toughness, e.g., Ni-Cr steel, more polishing steps are needed. First, there is a three-step polishing with 6, 3, and 1 μm diamond in succession followed by colloidal silica polishing suspension, producing a scratch-free and brilliant surface finish. The use of diamonds as a polishing medium has revolutionized metallographic preparation; diamonds are the abrasive most often used for metallographic polishing. There are two different types of synthetic diamonds: monocrystalline and polycrystalline. Polycrystalline diamonds are far superior because individual polycrystalline diamond grains are produced from a number of very small crystals that are sintered together to form a single grain. Therefore, each grain has many minute cutting edges, resulting in a surface with much finer scratches than can be achieved with monocrystalline diamonds of the same grain size, although the rate of material removal is the same for both types. The use of one-step polishing for routine examination of steels, as mentioned earlier, was made possible only after the introduction of polycrystalline diamonds [8]. Figure 7 shows a scanning electron micrograph of 15 μm polycrystalline diamonds. Note the very even grain size.

Colloidal silica is an extremely useful final polishing suspension that is suitable for nearly all materials [9]. Although it has been used for polishing single crystals of silicon for many years, it was not popular with metallographers until recently [10]. With colloidal silica, final polishing of almost any material will result in a surface without any scratches or deformation, and problems related to edge rounding and poor planarity will not occur due to the short duration, usually from a few seconds to about 3 min of this polishing step.

Colloidal silica contains very fine amorphous, rather than crystalline, particles that remain in suspension for such a long time that shaking or stirring before use is never needed. Figure 8 shows a transmission electron micrograph of amorphous silica particles, which are nearly spherical in shape, with a mean diameter of 27.2 nm.

Schwuttke and Oster [11] suggested that polishing with colloidal silica involves a combination of chemical and mechanical action. Owing to the development of additives to minimize evaporation, silica crystallization, and freezing problems, colloidal silica is much easier to use. Although the polishing rate is not high, colloidal silica is indispensable for those metallographers who seek the highest quality micrographs [9].

C. Etching

When a properly prepared as-polished specimen is examined under a metallurgical microscope, it appears essentially featureless. Only features with a 10% or greater difference in reflectivity,

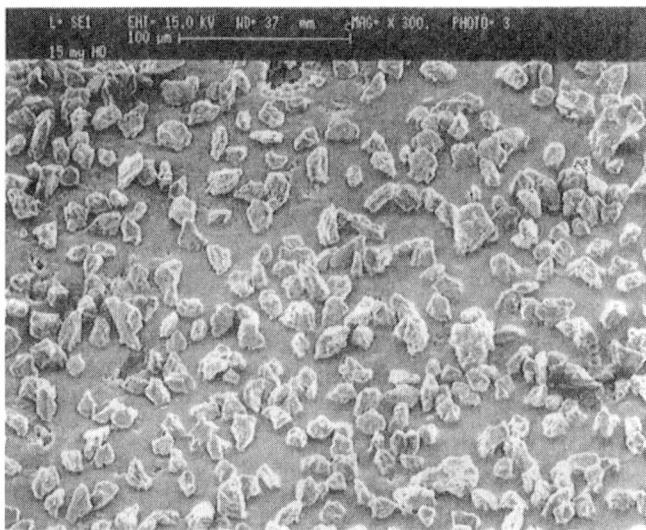

Figure 7   SEM micrograph of 15 μm polycrystalline diamonds. (Courtesy of Struers, Inc.)

such as nonmetallic inclusions, cracks, pores, and scratches, can be detected without etching. The microstructure is usually visible only after suitable etching, and therefore the purpose of etching is to produce a structure with sufficient contrast to delineate as much detail as possible. Before a specimen is ready for etching, its surface must be flat and free of deformed layers, scratches, pullout, and edge rounding.

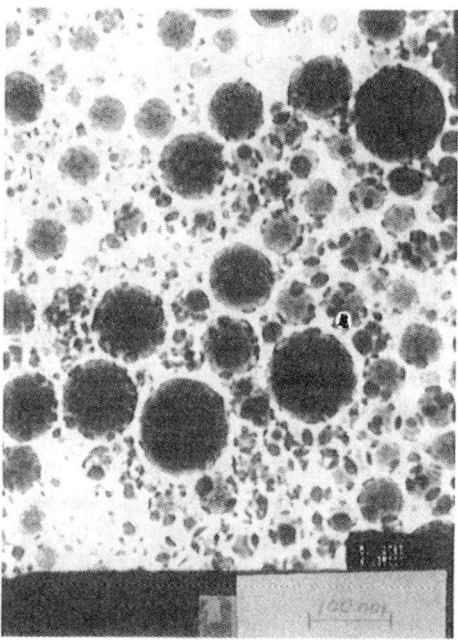

Figure 8   A TEM micrograph of amorphous silica particles in Syton HT-40. (Courtesy of G. F. Vander Voort, Carpenter Technology Corp.)

1. Chemical Etching

Chemical etching is the oldest and most commonly used etching method in which the specimen is usually immersed in the etchant. Microstructural contrast is produced through selective dissolution among various constituent phases due to differences in electrochemical potential. When etching occurs, local cells are formed where the more positive constituent phase acts as anode and reacts more strongly with the etchant than the more negative constituent. Potential differences arise chiefly from differences in composition; however, differences in the physical state may also cause potential differences. For example, grain boundaries are etched more easily than bulk grains; also, deformed metal reacts more strongly with the etchant than underformed metal.

Most etching is performed at room temperature. As etching proceeds, the surface becomes less reflective or duller, and the appropriateness of etching can be judged by the appearance of the surface being etched. Owing to the empirical nature of chemical etching, it is difficult to reproduce a distinctive etching contrast.

After etching has been done, the specimen should be rinsed first in distilled water and then in alcohol, which accelerates drying and avoids the formation of water spots. The specimen is then dried in a stream of warm air. For mounted specimens, the use of ultrasonic cleaner is recommended to more effectively clear away the etchant remaining in the pores, cracks, or interfaces between specimen and mount. Specimens can then be stored in desiccators.

The most widely used general-purpose etchants for steels are nital and picral. The former consists of a 0.5–10% (usually 2–3%) solution of HNO_3 in ethanol or methanol, and the latter consists of a 4% solution of picric acid in ethanol, sometimes with about 0.25–1% zephiran chloride (wetting agent) added to improve structure delineation, etching rate, and uniformity. Nital is more often used, but picral is superior for certain structures. Nital attacks both ferrite grain boundaries and ferrite/cementite phase boundaries, but picral attacks the latter only. Accordingly, for mild steels, nital should be used, whereas for spheroidized carbide structures, the use of picral is preferable because the sensitivity of nital to orientation often produces faint etching at the ferrite/carbide interface within certain grains.

Table 1 lists etchants for steels selected from Reference 3. More information can be obtained in References 3 and 12 and ASTM Standard E407.

2. Tint Etching

Tint etching produces microstructural contrast through selective coloration rather than selective dissolution as in chemical etching, due to deposition of a thin film of oxide, sulfide, complex molybdate, etc., on the polished surface of the specimen. The deposited film generally has a thickness of 40–500 nm. Coloration is produced by interference, which is visible under bright-field illumination. The color contrast can be enhanced further with the use of polarized light. Film thickness controls film color, which varies with the orientation of the substrate phase and etching time. Tint etching is performed at room temperature by immersing the specimen, never by swabbing. Often the surface is not etched beneath the interference film; therefore, a light pre-etch with nital or picral to reveal phase boundaries is recommended.

Beraha and Shpigler [13] developed a series of tint etchants for steels and many other metals and alloys. The tint etching system can be classified as anodic, cathodic, or complex depending on the nature of film precipitation. Most tint etchants are anodic systems that color the anodic phases only. They are usually acidic solutions using either water or alcohol as the solvent, and sodium metabisulfite ($Na_2S_2O_5$), potassium metabisulfite ($K_2S_2O_5$), and sodium thiosulfate ($Na_2S_2O_3 \cdot 5H_2O$) as common ingredients. Tint etchant based on sodium molybdate ($Na_2MoO_4 \cdot 2H_2O$) generally colors the cathodic phase, such as cementite in steels [14]. Some tint etchants for steels are listed in Table 1.

Table 1 Selected Etchants for Steels

Etchant composition	Comments
	General-purpose etchants
1. 1–10 mL HNO$_3$, 90–99 mL methanol or ethanol	Nital. Most common etchant for Fe, carbon and alloy steels, cast iron. Reveals alpha grain boundaries and constituents. The 2% solution is most common; 5–10% used for high-alloy steels (do not store). Use by immersion of sample for up to about 60 s.
2. 4 g picric acid, 100 mL ethanol	Picral. Recommended for structures consisting of ferrite and carbide. Does not reveal ferrite grain boundaries. Addition of about 0.5–1% zephiran chloride improves etch rate and uniformity.
3. 100 mL water, 0.5 g picric acid	Etch for producing contrast between as-quenched martensite and ferrite and other transformation products. Use at 71–77°C for 15–20 s. Saturated solution also used in the same manner.
4. 10 parts 4% nital, 1 part 4% picral	Combinations of nital and picral. used by immersion of sample. Used to differentiate austenite, martensite, and tempered martensite and to determine the depth of nitrided layers.
5. 100 mL ethanol, 5 mL HCl, 1 g picric acid	Vilella's reagent. Good for ferrite-carbide structures. Produces grain contrast for estimating prior austenite grain size. Results best on martensite tempered at 300–500°C. Occasionally reveals prior austenite grain boundaries in high-alloy steels. Outlines constituents in stainless steels.
6. 10 g picric acid, 100 mL alcohol	Superpicral. Above saturation limit. Need to heat to get picric acid into solution. Immerse sample for up to 1 min or more. Used with high-alloy steels, tool steels, and stainless steels.
7. Solution a: 5 mL H$_2$SO$_4$, 8 g oxalic acid,100 mL water Solution b: H$_2$O$_2$ (30%)	Marshall's reagent. Mix equal parts of stock solution a with solution b. Use fresh, 1–3 s. Etch has short life. A 2 s pre-etch with nital is useful if no reaction occurs with Marshall's reagent. A 20 s post-etch with nital increases etch attack. Uniform ferrite grain boundary etch. Colors cementite, attacks inclusions. Reveals prior austenite grain boundaries in martensitic low-carbon steels. Hold sample vertical in solution to reduce pitting.
8. 100 mL alcohol, 5 g FeCl$_3$	For tempered steels.

Table 1 *Continued*

Etchant composition	Comments
9. 1 g FeCl₃, 2 ml HCl, 0.3 mL zephiran chloride, 100 mL alcohol	Etch for bainitic steels, 1–5 min.
10. 5 g FeCl₃, 5 drops HCl, 100 mL water	General-purpose etchant. Sometimes reveals prior austenite grain boundaries in martensitic medium-carbon steels.
11. 20 mL HCl, 65 mL ethanol, 15 mL water, 1 g CuCl₂	For high-speed steels and martensitic stainless steels,
12. Solution a: 98 mL ethanol, 2 mL HNO₃ Solution b: 20 g sodium metabisulfite, 100 mL water	For austenitic Mn steels. Etch in solution a for 5 s, rinse, dry. Etch in solution b until surface is darkened. Produces excellent grain contrast, reveals depth of surface decarburization.
13. 3 parts glycerin, 3 parts HCl, 1 part HNO₃	"Glyceregia." For high-alloy steels, austenitic Mn steels, stainless steels. Use fresh; do not store. For slower action, use 2 parts HCl.
	Dual phase steel etchants
14. 2 g ammonium persulfate, 2 mL HF, 50 mL acetic acid, 150 mL water	Martensite is darkened; retained austenite is lighter than ferrite.
15. Solution a: 4 g picric acid, 100 mL methanol Solution b: 8 g CrO₃, 40 g NaOH, 72 mL water	Etch first with picral solution a, then use solution b boiling. Martensite is stained black, "old" ferrite gray, "new" ferrite white.
16. Solution a: 1 g sodium metasulfite, 100 mL water Solution b: 4 g picric acid, 100 mL ethanol	Mix equal parts of solutions a and b. Etch for 7–12 s. Surface colored blue-orange. Bainite is black, ferrite tan, martensite white.
17. Solution a: 2 mL HNO₃, 98 mL ethanol Solution b: 10 g sodium metabisulfite, 100 mL water	Pre-etch 1–2 s in nital solution a, then for 20 s in solution b. Martensite is dark, ferrite off-white, austenite white.
18. 50 mL sat. aq. sodium thiosulfate, 1 g potassium metabisulfite.	Klemm's I. Immerse sample 60–90s until surface is violet. Light pre-etch with nital is helpful. Ferrite is light or dark blue, martensite brown and black, retained austenite white.
	Prior-autenite grain size etchants
19. 100 mL ethanol, 1 g picric acid, 5 mL HCl	Vilella's etch. Works best on martensite tempered at 300–500°C. Immerse sample at room temperature. Sometimes produces grain contrast (results can be improved with several polish–etch cycles). For high-alloy steels, grain boundary attack is sometimes obtained. HCl sometimes added to 4% picral.

20. Sat. aq. picric acid plus small amount of a wetting agent

Bechet and Beaujard's etch. Most successful etchant, good for martensitic and bainitic steels. Many wetting agents have been used; sodium tridecylbenzenesulfonate is one of most successful. Use at 20–100°C. Swab or immerse sample for 2–60 min. Etch in ultrasonic cleaner. Additions of 0.5 g $CuCl_2$ per 100 mL solution or about 1% HCl have been used for higher alloy steels to produce etching. Room-temperature etching most common. Lightly back-polish to remove surface smut.

21. 34 g sodium bisulfite, 100 mL water

Used to reveal grain boundaries in fine-grained, heavily deformed steels. Immerse sample 1–2 s to produce a thin cinnamon yellow film,

22. 50 mL HCl, 25 mL HNO_3, 1 g $CuCl_2$, 150 mL

Used for 18% Ni maraging steels.

23. 100 mL sat. sq. picric acid, 2 g Calsoft-90, 2 g ammonium persulfate, 6 drops H_2O_2

Used to reveal grain boundaries in partly or fully recrystallized HSLA steels water-quenched after rolling. Reveals extent of recrystallization and size of grains.

24. 10 mL HCl, 3 mL HNO_3, 80–100 mL alcohol

For high-speed steels. Also used for as-quenched high-carbon steels. Examine under polarized light. Sensitive tint emphasizes grain contrast effect.

25. 80 mL water, 28 mL 10% aq. oxalic acid, 4 mL H_2O_2 (30%)

For quench and tempered low-carbon steels.

26. 400 mL water, 5 mL HCl, 10 g $FeCl_3$, 10 mL zephiran chloride

For martensitic stainless steels and high-Cr alloy steels.

27. 100 mL HCl, 120–140 mL ethanol, 8 g $FeCl_3$, 7 g $CuCl_2$

For tool steels. Etch 10–120 s. Remove gray surface deposit with cotton moistened with 4% alcoholic HCl.

Stainless steel etchants

28. 10 mL HNO_3, 20 mL HCl, 10 ml H_2O_2 (30%), 20 mL glycerin

Vilella's mixed-acid etch. Immerse sample at 20°C. Applicable to most stainless steels. Can vary amount of HCl. Use under a hood; do not store.

29. 1 part HNO_3, 1 part HCl, 1 part water

General-purpose etch for most stainless steels. Stir solution during etching (20°C) for uniform, stain-free results. Outlines constituents, reveals grain structure. Can be stored.

30. 5 g $CuCl_2$, 100 mL HCl, 100 mL ethanol

Kalling's No. 2 etch for duplex stainless steels. Ferrite attacked most readily, carbides unattacked, austenite slightly attacked. Use at 20°C by immersion of sample.

31. 1.5 g $CuCl_2$, 33 mL ethanol, 33 mL water, 33 mL HCl

Kalling's No. 1 etch for martensitic stainless steel. Martensite dark, ferrite colored, austenite not attacked. Use at 20°C by immersion of sample.

32. 40 mL HCl, 5 g $CuCl_2$, 30 mL water, 25 mL ethanol

Fry's reagent for martensitic and precipitation-hardenable grades. Use at 20°C by immersion of sample.

33. 4 g $CuSO_4$, 20 mL HCl, 20 mL water

Marble's reagent used for most stainless steels. Use at 20°C for 3–10 s. Reveals grain structure, attacks sigma.

34. HCl saturated with $FeCl_3$

For ferritic and austenitic stainless steels. Activate with small amount of HNO_3 (optional). Reveals grain structure, sometimes colors delta ferrite.

Table 1 *Continued*

Etchant composition	Comments
35. 25 mL HCl, 5–50 mL 10% aq. CrO_3	For austenitic stainless steel. Control speed of attack by amount of 10% CrO3. Good, rapid, even etch.
36. 20 g $CuSO_4$, 50 mL H_2SO_4, 100 mL HCl, 100 mL water	Modified Marble's reagent for austenitic stainless steels. Swab sample.
37. 15 mL HCl, 5 mL HNO_3	Aqua regia. For austenitic grades. Use fresh. Use at 20°C for 5 s. Attacks sigma, outlines carbides. After 20 s, sigma completely dissolved. Reveals grain boundaries. Do not store etchants.
	Tint etchants
38. 50 mL sat. sq. sodium thiosulfate, 1 g potassium metabisulfite	Klemm's I reagent. Use at 20°C for 40–100 s. Reveals P segregation (white), colors ferrite (blue or red), cementite and austenite unaffected, colors martensite brown., Light nitral pre-etch helpful. Can reveal overheating.
39. 50 mL sat. aq. sodium thiosulfate, 5 g potassium metabisulfite	Klemm's II reagent. Use at 20°C for 30–90 s. P-rich regions darkened when etched >15 s. For austenitic Mn steels, γ yellow to brown, α-martensite dark brown, ε-martensite white. Tint etch for lath or plate martensite. use at 20°C for 2 min.
40. 1 g $Na_2S_2O_3$, 100 mL water	Darkens untempered martensite, carbides and phosphides unaffected. Lightly pre-etch sample with nitral, immerse in etchant solution about 1–15 s.
41. 10 g sodium metabisulfite, 100 mL water	
42. Solution a: 80 g sodium thiosulfate, 60 g ammonium nitrate, 1000 mL water, Solution b: 1 part HNO_3 2.5 parts H_3PO_4	Beraha's tint etch for ferrite. Before use, heat 100 mL of solution a to 70–75°C and add 0.4–0.5 mL of solution b with vigorous stirring. Reagent becomes turbid and is useful for 15 min. Pre-etch with nitral, then immerse sample in reagent, moving it slowly until the surface turns dark blue, usually 1–3 min. Wash and remove any loosely adhering sulfur particles. Ferrite colored dark red or blue, cementite and phosphide sharply outlined and clear, sulfides brightened.
Beraha's	
43. 20 g anhydrous sodium pyrophosphate, 13 mL H_3PO_4, 3 g sodium molybdate, 6 g sodium nitrate, 1000 mL water (Final pH 3.5–4)	Beraha's stain etch for pearlitic steels. Solution is stable for 48 h. (For stock solution, leave out the sodium nitrate.) Use light nitral pre-etch. For grain contrast in low-carbon steels, etch 5–20 s; for high-magnification work, etch <30 s. Carburized and nitrided layers unaffected.
44. 100 mL ethanol, 1–2 mL HCl, 0.5 mL selenic acid	Beraha's tint etch for steel, tool steel, and martensitic or PH stainless steels. Ferrite and austenite colored bright; nitrides and carbides are colored. Pre-etch with nitral before using. Cementite colored red, blue, or green, and ferrite colored yellow or brown.

45. Solution a: 100 mL water, 3 g potassium metabisulfite, 1 g sulfamic acid
Solution b: 100 mL water, 6 g potassium metabisulfite, 2 g sulfamic acid
Solution c: 1000 mL water, 3 g potassium metabisulfite, 2 g sulfamic acid
Solution d: 100 mL water, 3 g potassium metabisulfite, 2 g sulfamic acid, 0.5–1 g ammonium bifluoride

Beraha's tint etchants for steels, tool steels, Mn steels, and ferritic and martensitic stainless steels. Use all at 20°C. Reagents active for 2–4 h. Discard when yellow. Cd or Zn coatings inhibit staining. Use solution a 15 s to 4 min for carbon and alloy steels. Solution b is similar to solution a but faster acting. Use solution c for carbon, alloy, and Mn steels; 5–30 s. To detect retained austenite (brown or blue), use 45–90 s. Use solution d for stainless steels, Mn steel, and some tool steels; 30 s to 3 min. Use plastic container and forceps with solution d.

46. 3 g potassium metabisulfite, 1–2 g sulfamic acid, 0.5–1 g ammonium bifluoride, 100 mL water

Beraha's tint etch for ferritic and martensitic stainless steels, Mn. steels, and tool steels. Immerse sample at 20°C for 30–180 s.

47. Solution a: 1000 mL water, 200 mL HCl
Solution b: 0.5–1.0 g potassium metabisulfite per 100 mL of solution a

Beraha's tint etch for austenitic grades, maraging steels, and PH grades. Immerse sample at 20°C for 30–120 s with agitation. Colors austenite, carbides unaffected.

48. 1000 mL water, 200 mL HCl, 24 g ammonium bifluoride

Beraha's tint etch for stainless steels. Before use, add 600–800 mg potassium metabisulfite to 100 mL of this stock solution (100–200 mg for martensitic grades). After mixing, reagent is active for about 2 h. Use plastic tongs and beaker. Immerse sample for 20–90 s at 20°C; shake gently while etching. Longer times intensify colors. Grain and twin boundaries clearly revealed. Second-phase particles bright and uncolored.

49. 75–98 mL water, 25–2 mL HF (tot. vol. 100 mL) Heat to boiling point and add molybdic acid until saturation, then cool.

New molybdic etchant (Weck and Leistner [14]). After cooling, the etchant is ready for use. Serves as sole etchant for all nonaustenitic steels. Colors martensitic needles; retained austenite remains white. Martensitic matrix of hardened high-speed steel mottled, establishing good contrast to the white carbide. In dual-phase steels, martensitic areas appear dark brown, ferritic grains only slightly tinted.

Source: Ref 3, pp. 632–655.

When preparing tint etchants, the formula and the order of mixing should be followed strictly, and cautions in handling hazardous reagents should be carefully obeyed. Specimens undergoing a tint etch should be prepared very carefully, since even very light scratches undetectable after polishing will be visible after tint etching. For better results, automatic polishing or vibratory polishing is preferable [15].

D. Electrolytic Polishing

As a specimen preparation method, electrolytic polishing for quality control and research work has been employed in many laboratories. In this method, the specimen surface is leveled by anodic solution in an electrolytic cell consisting of a tank for the electrolyte, a direct current source, an anode (specimen), and a cathode.

Although the detailed micromechanism of electropolishing is not clearly understood, the operating conditions of a specific specimen–electrolyte combination can be evaluated by plotting the current density versus the applied voltage curve as shown schematically in Figure 9. Etching occurs at the initial segment (A–B) of the curve, polishing occurs at the plateau (C–D) of the curve, and gas evolution and pitting occurs at the final part (D–E) of the curve. Optimum polishing effect usually occurs at C.

A desirable electrolyte should be a somewhat viscous solution, a good solvent during electrolysis, and simple to mix, stable, and safe. It should contain one more large ions, e.g., PO_4^{3-}, Cl_4^{4-}, or SO_4^{2-}, or large organic molecules. It should be operable at room temperature and insensitive to temperature changes and should not attack the specimen with the current off.

The advantages of electrolytic polishing are as follows:

1. It is the fastest and most reproducible specimen preparation method once the polishing parameters have been established and can be precisely controlled.
2. No scratches or deformed or damaged layer is left after completion of the operation.
3. In some cases, once electropolishing has been done, the cell with the same electrolyte can be immediately switched to the etching condition by simply reducing the applied voltage to about 10% of that required for polishing and continuing electrolysis for a few seconds.

However, one should be aware of the following limitations of electropolishing in order to make full use of its advantages:

1. Only micro scratches can be removed. The specimen must undergo fine grinding or even prepolishing prior to electropolishing.

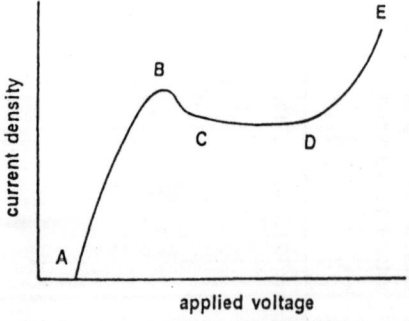

Figure 9 Typical plot of current density versus applied voltage of an electropolishing solution (schematic).

2. Specimens with multiphase or heterogeneous structures such as nonmetallic inclusions in steels or carbides in stainless steel are not well suited for electropolishing because of differences in electrochemical potentials among constituent phases.

3. Some of the electrolytes are toxic, highly corrosive, or otherwise hazardous and can even be explosive. Therefore, precautions for the safe use of the electrolytes should be strictly observed.

4. Edge retention cannot be maintained during electropolishing unless the side of the specimen is protected by a nonconductive epoxy resin.

5. The size of the area that can be polished successfully is limited because of a potential change from the center to the edge of the area to be polished. For example, the polished area is limited to 0.5–2 cm^2 for stainless steel and some high-alloy steels, but up to 5 cm^2 of some carbon steels can be polished.

The mechanical and electrolytic polishing methods each have merits and specific applications. They are complementary and not mutually exclusive. One should select the methods that will give the best result for each specific specimen.

Table 2 provides a listing of electrolytic polishing solutions for steels selected from Reference 3.

E. Vibratory Polishing

The vibratory polishing method was first introduced by Krill [16] and Long and Gray [17] in the 1950s. A vibratory polisher is a torsional vibrational system in which the helical motion of the polishing bowl is produced approximately at resonance by electromagnetic force from a solenoid pulsed with a half-wave rectified current. Weighted or unweighted specimens are placed in the polishing bowl. The vibrations cause the specimens to move around the periphery of the bowl and about its own axis, producing a polishing action. The vibrational amplitude of the polishing bowl must be adjusted to ensure smooth movement of the specimen around the bowl without bouncing. The speed of the specimen around the bowl is usually 18–20 rpm, which is much slower than that of a conventional polishing machine.

Due to the inherently low polishing rate, vibratory polishing often lasts for many hours. On the other hand, as many as eighteen 40 mm diameter specimens, or specimens of other diameters comprising an equivalent area, may be prepared simultaneously in a single polishing bowl. Once the specimens are loaded, little operator attention is needed, and any specimen can be removed and inspected while others are still being polished. Therefore, the vibratory polisher behaves as a semiautomatic polishing machine.

The polishing quality obtained on the vibratory polisher is very good when measured in terms of a scratch-free surface, retention of inclusions, specimen flatness, and minimal depth of disturbed layer [18]. Figure 10 shows a photomicrograph of an annealed 0.45% carbon steel, vibratory polished and etched with 2% nital.

Recently Vander Voort [9] showed that a brief (less than 30 min) final vibratory polish with colloidal silica may be used for carbon and low-alloy steels. The same procedure with longer duration for AISI type 52100 bearing steel (1% C–1.5% Cr) is the only effective way to eliminate minor staining problems around inclusions. Another example is boronated stainless steels that undergo image analysis measurements. A final vibratory polish with colloidal silica is also adapted. It seems that vibratory polishing is a useful complement to conventional mechanical polishing. Figure 11 illustrates a more powerful drive mechanism that produces nearly 100% horizontal motion (i.e., motion that causes polishing) with virtually no vertical motion. The vibration frequency is doubled, compared with the previous model, increasing the polishing rate.

Table 2 Electrolytic Polishing Solutions for Steels

Electrolyte composition	Current density (A/cm²)	Voltage (V, dc)	Temp. (°C)	Time (min)	Comments
1. 185 mL H₃PO₄, 765 mL acetic anhydride	0.04-0.06	50 (ext.)	<30	4-5	Wide applicability. Age solution 24 h before use. Use Fe or Al cathode, 20 times as large as sample. For gamma stainless steel, use 0.1 A/cm².
2. 54 mL perchloric acid, 146 mL water, 800 mL ethanol plus 3% ether	4	110 (ext.)	<35	15 s	Good for many metals. Use Fe or austenitic stainless steel cathode. Pump electrolyte. Cool.
3. 420 mL H₃PO₄, 470 mL glycerol, 150 mL water	1.5-20		100	8-15	For stainless steel.
4. 25 g CrO₃, 133 mL acetic acid, 7 mL water	0.09-0.22	20	17-19	6	Wide applicability. Dissolve CrO₃ in solution using water bath at 60-70°C. Samples mounted in Bakelite can be safely electropolished. Grind to 600 grit. Cool bath during use. Will attack inclusions and cracks. Etch stainless steel samples at 0.025 A/cm² for 5-20 min. Store solution in airtight bottle.
5. 1000 mL acetic acid, 50mL perchloric acid, 5-15 mL water, (optional)	0.01	45 (ext.)	25		Wide applicability. Best polishing without water. Can produce preferential attack in two-phase alloys.
6. 650 mL H₃PO₄, 150 mL H₂SO₄, 150 mL water, 50 g CrO₃	66-100		40-60	3-7	For carbon steels up to 1.1% C.
7. 62 mL perchloric acid, 700 mL ethanol, 100 mL butyl Cellosolve, 137 mL water	1.2			20 s	Wide applicability, including high- and low-carbon steels and high-speed steels. Add perchloric acid carefully to ethanol and water. Add butyl Cellosolve immediately after electropolishing.
8. 40% H₂SO₄, 46% H₃PO₄, 4% dextrose, 10% water	23-70		28-40	5-10	For carbon steels. Etch at 12-16 A/dm² for 10 min.

Source: Ref. 3.

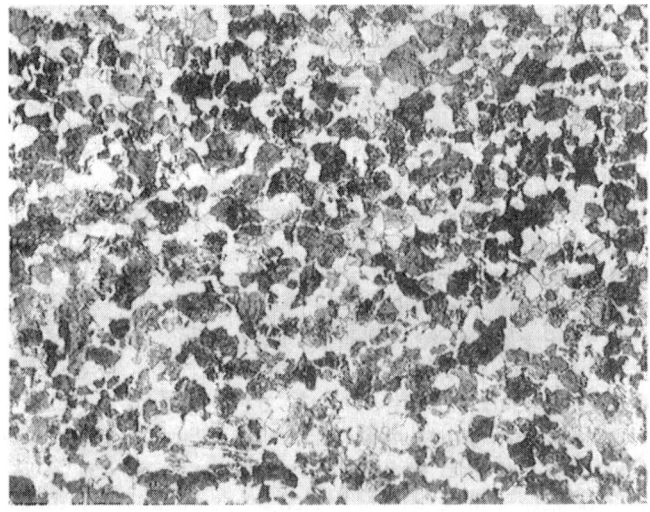

Figure 10 Photomicrograph of an annealed 0.45% C carbon steel, vibratory polished and etched with 2% nital, 200×. (Reduced 15% in reproduction)

III. THE METALLURGICAL MICROSCOPE

It is well known that the behavior of a steel depends strongly on the volume fraction, size, and distribution of its constituent phases. During heat treatment, various changes among constituent phases or microstructural constituents may occur. Therefore, examination of the microstructure of a steel is a very important and effective measure in both quality inspection of

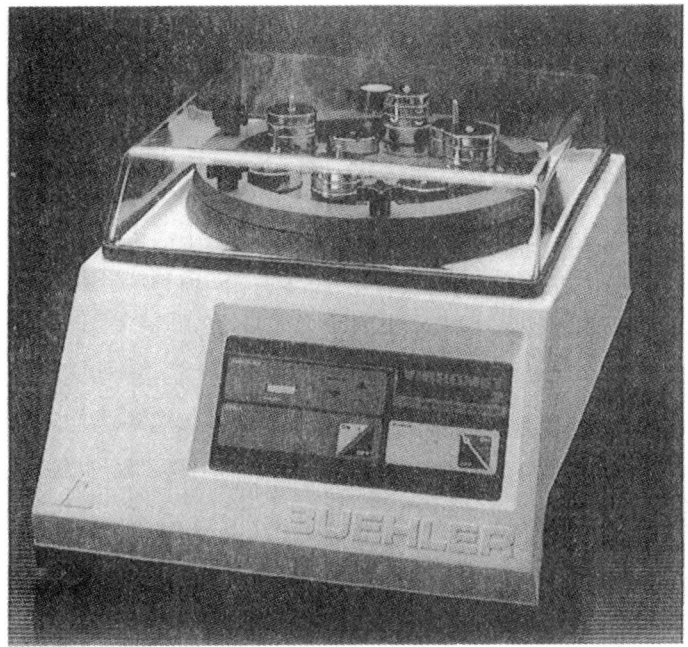

Figure 11 VIBROMET I vibratory polisher. (Courtesy of Buehler Ltd.)

heat-treated parts and related research work. Owing to its ease of operation, its larger field of view and therefore its yield of more general information, and its relatively lower cost, the optical metallurgical microscope is still one of the most commonly used instruments in routine inspection and research work.

A. Image Formation of Metallurgical Microscope

The principle of image formation of a metallurgical microscope during visual observation is essentially the same as that of an ordinary microscope, as shown in Figure 12. Magnification is accomplished by an objective with very short focus and an ocular or eyepiece with somewhat longer focus. To minimize various aberrations, both objectives and eyepieces are highly corrected compound lenses, with the former far more complicated. For simplicity, both objective and eyepiece are treated as simple thin lenses as shown in Figure 12. The object (*ab*) is located just outside the front focal point of the objective (F_{ob}) so that a real magnified image (*a'b'*) can be formed. This intermediate or primary image is located within the focal point of the eyepiece (F_{oc}), which functions as a magnifier and forms a further enlarged but virtual image (*a''b''*). This image becomes the object for the eye itself, which forms the final real image on the retina (*a'''b'''*).

The overall magnification (*M*) of a microscope is the product of the linear magnification (m_1) of the objective and the angular magnification (m_2) of the eyepiece:

$$M = m_1 \times m_2 \tag{1}$$

From geometrical optics,

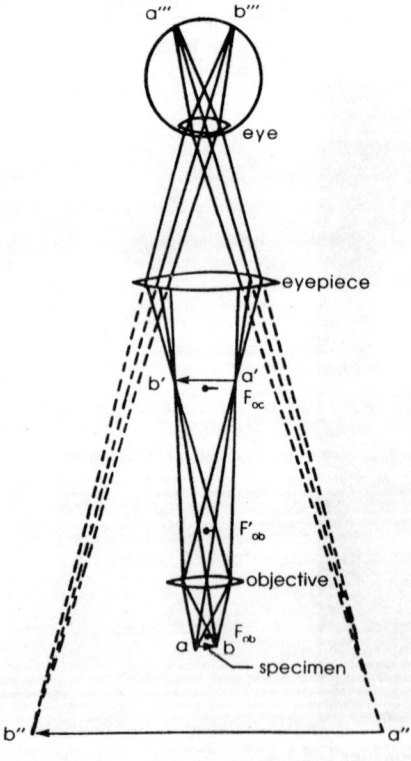

Figure 12 Image formation of a microscope during visual observation.

$$m_1 = -L/f_1 \quad \text{and} \quad m_2 = 250/f_2 \tag{2}$$

where f_1 and f_2 are focal lengths of the objective and eyepiece, respectively; L is the so-called optical tube length of the microscope, which is measured from the back focal point of the objective to the position where the intermediate image stands; and 250 is the distance of most distinct vision, in mm. f_1, f_2, and L are all expressed in mm. The negative sign of the magnification m_1 signifies an inverted image.

The overall magnification is therefore

$$M = m_1 \times m_2 = -\frac{L \times 250}{f_1 \times f_2} \tag{3}$$

The optical tube length is inconvenient in practical use. Therefore, the mechanical tube length, which is defined as the distance from the screw flange of the objective to the end of the eyepiece tube, is commonly used instead. The mechanical tube length is normally 160–250 mm. Since even a slight deviation from the correct tube length will result in a notable reduction of image quality, it is not advisable to use an objective designed for a different tube length from another manufacturer.

Actually, the principle of image formation of a microscope is not as simple as described above. Owing to the fineness of the structure to be resolved by an objective, diffraction and interference must be considered in the theory of image formation first put forward by E. Abbé over a century ago, which shows that the object of the image is the result of interference of direct or undiffracted and diffracted beams [19].

B. Objectives

The image quality of a microscope is determined, to a great extent, by the quality of the objective, which forms the primary image of the specimen. Therefore, the objective is the most important part of a microscope.

1. Numerical Aperture

The numerical aperture (NA) of an objective is a measure of its light-collecting ability and can be expressed by the equation

$$NA = n \sin \alpha \tag{4}$$

where n is the refractive index of the medium between the specimen and the front lens of the objective and α is half the aperture angle of the most oblique rays entering the front lens of the objective (see Figure 13).

Equation (4) shows that the larger the angle α, the greater the light-collecting ability of the front lens of the objective. The magnitude of angle α depends on the size of the front lens and the working distance of the objective, i.e., the distance between the specimen and the front surface of the objective when the image is in sharp focus. Since the angle α cannot exceed 90°, the numerical aperture of a dry objective cannot exceed 1, and normally it is within 0.95. The numerical aperture of oil immersion objectives can be as high as 1.4.

2. Resolution

The resolution or resolving power of an objective is defined as the nearest resolvable distance between two parallel lines or separated points in the object. The smaller the resolvable distance, the higher the resolution. The resolution can be expressed by the equation

$$d = \frac{\lambda}{2NA} \tag{5}$$

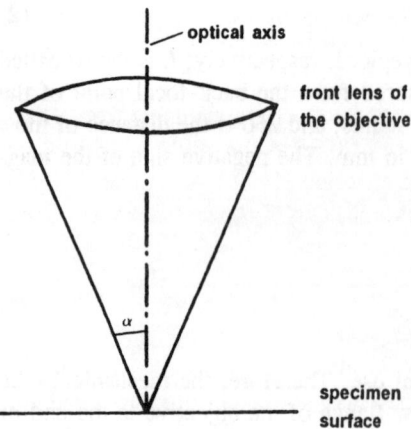

Figure 13 The aperture angle of an objective (schematic).

where d is the resolution, λ is the wavelength of the light used, and NA is the numerical aperture of the objective as described earlier.

From Equation (5) it can be seen that with incident light of known wavelength, the resolution of an objective is totally determined by its numerical aperture. The larger the numerical aperture, the higher the resolution. As an example, an objective with the highest attainable numerical aperture of 1.4, using yellow-green light of a wavelength of 550 nm, the resolution is about 200 nm or 0.2 μm.

The role of the eyepiece is to magnify the intermediate image formed by the objective. Accordingly, any unresolvable structural details cannot be resolved through the magnification of the eyepiece. Therefore, the resolving power of a microscope is governed chiefly by the resolution of the objective.

To fully utilize the resolution of the objective so that the already resolved structural detail can be viewed clearly by the human eye, suitable magnification of the microscope should be incorporated. According to Abbé's theory, the limit of resolving angle of the human eye under the best illuminating conditions is about 1 min of arc, but more realistic upper and lower limits are 4 and 2 min of arc, which correspond to a separation of about 0.30 and 0.15 mm, respectively, at a distance of 250 mm, i.e., the distance of most distinct vision. Using $\lambda =$ 550 nm, the approximate range of magnification M necessary for the microscope can be obtained:

$$550 \, \text{NA} < M < 1100 \, \text{NA} \tag{6}$$

Equation (6) is also called the useful range of magnification. Below the lower limit, the human eye cannot clearly see the structural detail already resolved by the objective; above the upper limit, no more structural detail can be observed, and blurring of the image may even occur. This latter kind of magnification, often called *empty magnification*, should be prevented during observation.

However, Vander Voort [3, p. 282] argued that Equation (6) is derived using an ideal contrasting image with an ideal eye and should not be adhered to dogmatically. In fact, resolution depends on both resolving power and image contrast. Besides, illumination, magnification, lens quality, and observation conditions also affect image contrast and therefore resolution. It is suggested that the lowest useful magnification for maximum resolution should be four times greater than under optimum condition, i.e., 2200 NA, and the use of photographs with magnifications up to 4000× or more is fully justified.

Owing to the improved resolution, oil immersion objectives are useful in metallography. However, such lenses should be cleaned with a special solvent after every use, which is inconvenient. High-power dry objectives (such as 100/0.95 and 150/0.95) are used more frequently. Recently, dry objectives with high magnification (200×) are available from Nikon Corporation of Japan. With the same numerical aperture of 0.95 as the 100× and 150× dry objectives, an overall magnification well above the old criterion (1100 NA) can be easily attained with this 200× objective during visual observation. This further justifies the above viewpoint.

3. Common Types of Objectives

Plano achromatic and apochromatic objectives are the most commonly used objectives in metallurgical microscopes. The prominent characteristic of a plano objective is its significantly enlarged flatness of the image field, which is due to extensive correction of image field curvature. During visual observation with plano objectives, fewer fields of view are needed for a given area than during observation with conventional objectives. As a result, eyestrain is notably lessened. Besides, it is self-evident that plano objectives are particularly useful for photomicrography. However, the accuracy of the sliding mechanism is the object stage and the quality of specimen preparation should be higher when plano objectives are used.

Plano achromatic objectives, or planachromats, are corrected spherically for red and blue colors. Therefore, there are still residual chromatic aberrations, but they can be minimized by the use of yellow-green filters.

Plano apochromatic objectives, or planapochromats, are corrected chromatically in the range of red, green, and blue colors, i.e., over almost the entire range of visible light. The correction of spherical aberration can reach the range of green and blue colors.

Plano objectives have more lens elements and are therefore more expensive.

C. Eyepieces

As stated earlier, the function of the eyepiece is to magnify the intermediate image formed by the objective so that the resolved structural detail of the specimen can be viewed clearly by the human eye. Wide-field compensating eyepieces have been developed to accommodate the plano objectives. With a significantly larger field of view number (diameter of field diaphragm), such eyepieces can fully utilize the enlarged flat area of the intermediate image. Since plano objectives are undercorrected with respect to lateral chromatic aberration, compensating eyepieces are overcorrected.

Ordinary eyepieces do not suit eyeglass wearers, especially those with astigmatic eyes, because the designed eye clearance (the distance between the eye lens of the eyepiece and the eye) of 10 mm is inadequate for eyeglass viewing. As a result, with eyeglasses the image size would be severely limited due to increased eye clearance. If the eyeglass wearer views without eyeglasses, the image quality will be affected by astigmatism. High-eyepoint eyepieces are specially designed for eyeglass wearers. with such eyepieces, the entire field of view can be obtained at a distance of about 20 mm from the eye lens, which is sufficient for viewing with eyeglasses.

D. Parfocality

Microscopes are often provided with a parfocal lens system so that there is no need to refocus after changing objectives and/or eyepieces except for minor adjustment of the fine focusing knob. The optical-mechanical dimensions of both the objectives and eyepieces should meet the following requirements of parfocality:

1. The distance between the specimen and the intermediate image should be identical for all objectives; for a given mechanical tube length, the distance between the specimen and the screw flange of the objective, also called the parfocal length of the objective, should be held constant. A length of 45 mm is generally taken as the parfocal length.
2. The distance between the intermediate image and the end of the eyepiece tube should be the same for all objectives.
3. The focal plane of all eyepieces must always coincide with the intermediate image plane.

Obviously, parfocality is not an intrinsic property of an objective or eyepiece, but rather a measure adopted in the design of modern microscopes to ease manipulation and to prevent possible damage to the objective lens or the specimen surface during operation.

E. Illumination System

1. Köhler's Principles of Illumination

Due to the opaqueness of a metal specimen, it is necessary that an incident light illumination system, also called a vertical illuminator, be used in a metallurgical microscope (see the left-hand portion of Figure 14). It primarily consists of a light source with a collector lens, several auxiliary lenses that provide the desired path of the beam, an aperture diaphragm for controlling the illumination aperture, a field diaphragm for controlling the illuminated area of the object, a plane glass reflector that inclines toward the direction of the beam with an angle of 45° and reflects part of the beam toward the objective, and an objective that acts as a con-

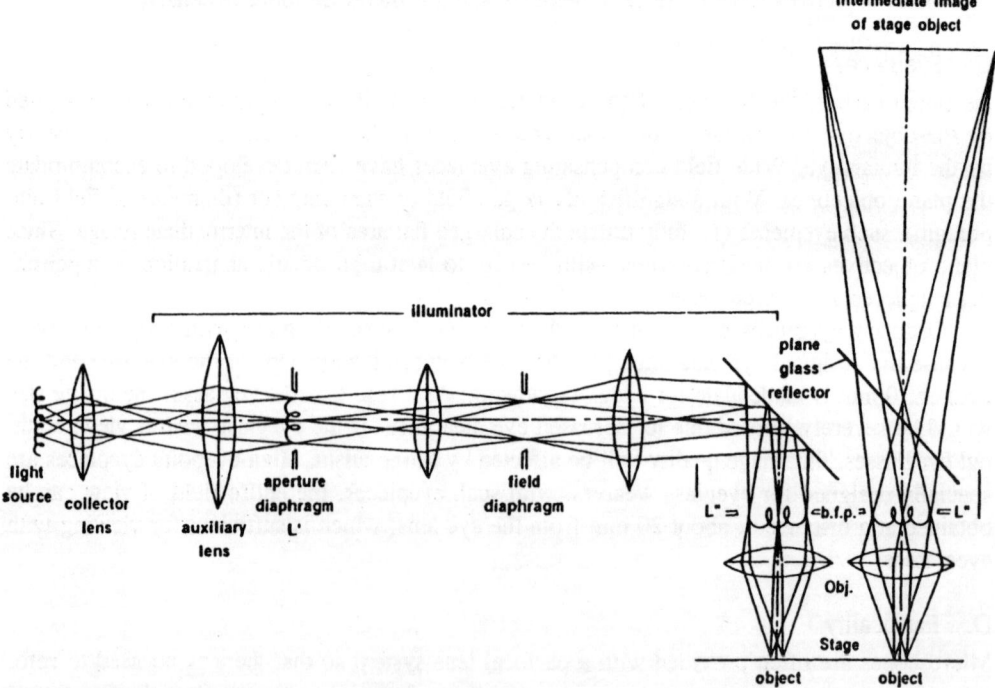

Figure 14 Ray path in an illumination system according to Koehler's principles. (Courtesy of W. Heffer & Sons, Ltd.)

denser in the illumination system.

Illumination according to Köhler's principles, also shown schematically in Figure 14, is used in most metallurgical microscopes. To obtain the best image that a given microscope is capable of providing, the requirements of Köhler illumination should be fulfilled. The left-hand portion of Figure 14 shows the illuminating beam path, and the right-hand portion shows the part of the imaging beam path that ends at the intermediate image plane. The light source is centered and imaged at the aperture diaphragm, which is then imaged together with the light source image at the back focal plane (b.f.p.) of the objective. By adjusting the setting of the aperture diaphragm, the illumination can be matched with the aperture of each objective, bringing the resolution of the objective into full play. However, glare and marginal lens aberrations still exist at this setting. Therefore, one should strive for the ideal compromise between optimum resolution and the best possible contrast; the aperture diaphragm should be as open as possible to achieve the best resolution and as closed as necessary to achieve the contrast essential for observation of the resolved detail [20].

The aperture diaphragm setting can easily be adjusted by viewing down the eyepiece tube with the eyepiece removed. If there is a built-in Berrand lens in the microscope, after swinging it into the light path and properly focusing it, a magnified image of the back focal plane can be seen through the eyepiece, and the aperture diaphragm setting can be adjusted more easily. Although it is usually recommended that the setting of the aperture diaphragm should be such that its image covers about two-thirds to three-fourths of the rear surface of the objective, this should not be regarded as a hard and fast rule.

It should be kept in mind that the aperture diaphragm can never be used to adjust the brightness of the image, which can be done by using neutral density filters or by adjusting the lamp voltage. Figure 15 shows the effect of aperture diaphragm setting on the image quality of pearlite. In Figure 15a, the aperture diaphragm is in its optimum setting, producing the best contrast and good resolution. In Figure 15b, the aperture diaphragm is fully open. The contrast is inferior because of excessive glare, and although the resolution is expected to be the best, it is almost the same as in Figure 15a. In Figure 15c, the aperture diaphragm is closed to its minimum position. Structure with interference fringes can be seen in many places. The impairment of resolution is most clearly illustrated in a pearlite nodule located at the lower right of the figure.

The field diaphragm is imaged at the object surface, which is then imaged at the intermediate image plane. The field diaphragm should be so adjusted that the image of its edges is located just outside the field of view, thus cutting off unnecessary light and minimizing image glare. With the field diaphragm, the object field being observed is evenly illuminated up to its edge and can be matched exactly with the field of view.

2. Brightfield Illumination

In the above-described method of illumination, the light beam strikes the specimen surface vertically. If the specimen surface has a mirror polish, it will appear bright, because nearly all the reflected light can enter the objective again. After etching, depressions or grooves with sloping surfaces are produced along grain or phase boundaries, and the surface of certain phases becomes somewhat roughened. Part of the reflected light beam from these regions cannot enter the objective. Dark constituents of various gray levels against a bright background are now seen. This kind of illumination is called brightfield illumination and is the most commonly used illumination method.

3. Darkfield Illumination

Figure 16 shows schematically the optical path in darkfield illumination. The objective shown in this figure is of infinite optics, which will be described in Section III. F. With darkfield

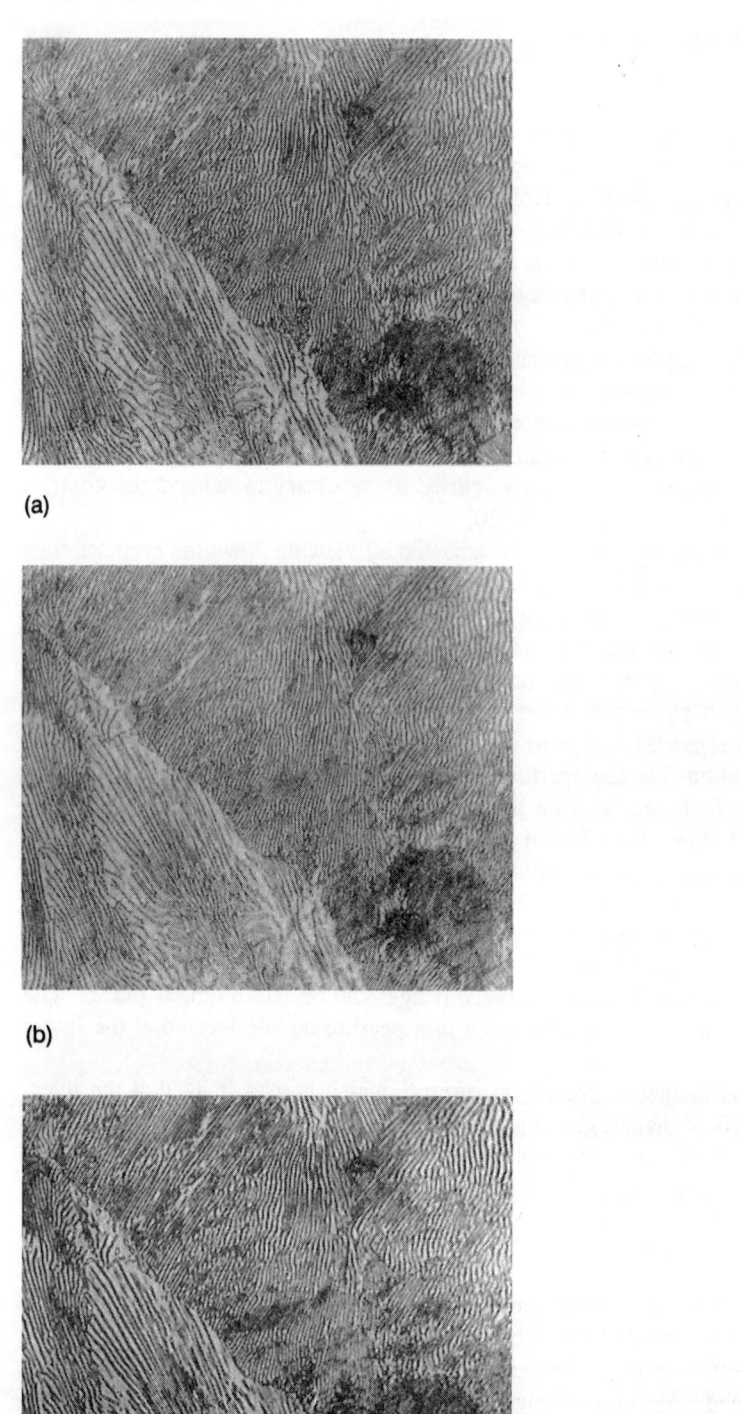

(a)

(b)

(c)

Figure 15 Effect of aperture diaphragm setting on the image quality of pearlitic structure. (600×) (a) Optimum setting; (b) maximum setting; (c) minimum setting. (Reduced 20% in reproduction)

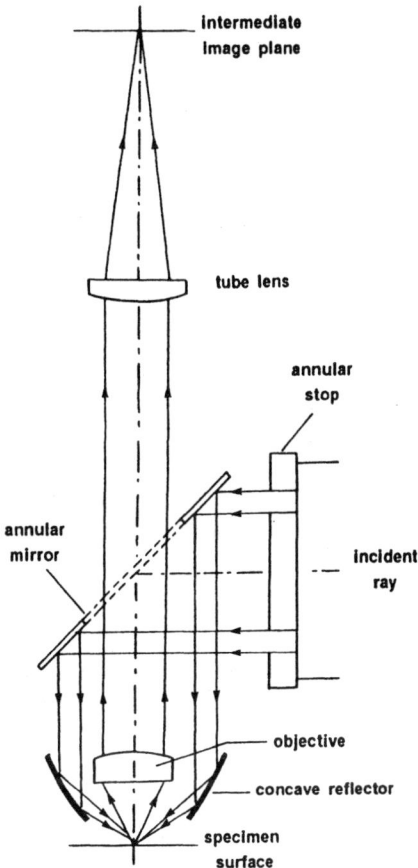

intermediate
image plane

tube lens

annular
stop

annular
mirror

incident
ray

objective

concave reflector

specimen
surface

Figure 16 Optical path of darkfield illumination (schematic).

illumination, a ring of light, which is produced after passing an annular stop with the aperture fully open, falls on the annular mirror that is located around the plane glass reflector. Thus, the light beam does not pass through the objective but traverses outside the objective and is directed rather obliquely onto the specimen surface by a concave reflector. If the specimen surface has a mirror polish, it will appear dark because the reflected light does not enter the objective. Structural constituents with sloping surfaces, such as grain and phase boundaries, will appear light because the reflected light may enter the objective, at least partly. The image contrast is completely reversed with darkfield illumination as compared with brightfield illumination. Figure 17 shows the prior-austenite grain structure of a high-strength low-alloy steel under darkfield illumination.

The advantages of darkfield illumination are as follows:

1. Higher resolution. Under darkfield illumination, transparent particles within the specimen that are undetectable under brightfield illumination become self-luminous and are detectable even at 0.006 μm (6 nm).

2. Higher image contrast. The objective no longer acts as a condenser in the illumination system, reducing glare.

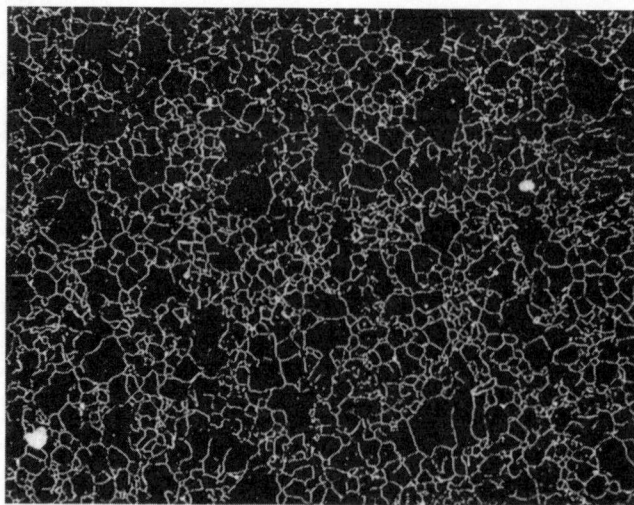

Figure 17 Prior austenite grain structure of a high-strength low-alloy steel under darkfield illumination (200×). (Reduced 20% in reproduction)

3. Transparent inclusions can be identified accurately according to their inherent colors; e.g., silica (SiO_2) in steels appears dark gray in brightfield illumination but bright yellow in darkfield illumination.

4. As already shown in Figure 17, grain boundaries are more easily delineated in darkfield than in brightfield illumination and can be used for the rating of grain size when compared with standard charts.

However, there are disadvantages in darkfield illumination. Image brightness is much lower, making exposure during photomicrography significantly longer and necessitating high-intensity light sources and automatic exposure devices. A very well prepared surface is necessary for darkfield observation because even the slightest scratches, undetectable in brightfield, will be evident.

F. The Infinity-Corrected Optical System

There is a tendency to use objectives designed for infinite image distance instead of conventional objectives computed for finite image distance, and the color correction of such objectives is also made for infinite image distance. Such optical systems are called infinity-corrected optical systems or simply infinite optics [21]. When the incident light beam is reflected from the specimen surface and passes through an infinite objective, it does not converge and remains a parallel light ray until it passes through the tube lens; then an intermediate image is formed, as shown in Figure 18.

The main advantage of infinite optics is that modules of various contrasting modes—incident light beam splitter for bright- and darkfield, polarizing beam splitter, Wollaston prisms for interference contrast, analyzers, additional filters, and so on—can be engaged in the extended space of the infinite light path between the flange of the objective and the tube lens. Since there is no additional lens system to disturb the part of the light path that is used for both imaging and incident light illumination, the quality of the optical image will not be impaired. Moreover, when infinite optics are used, the tube length coefficient remains at unity,

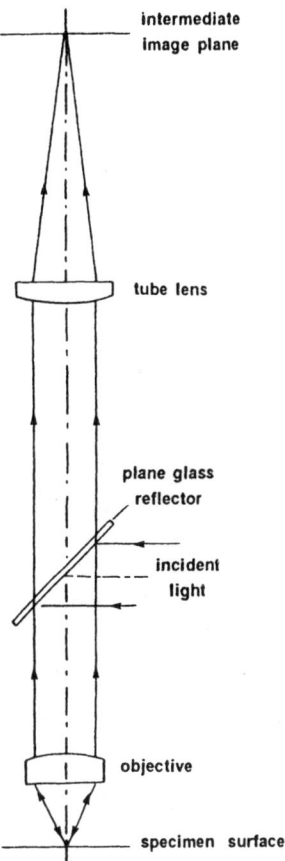

intermediate
image plane

tube lens

plane glass
reflector

incident
light

objective

specimen surface

Figure 18 Formation of intermediate image in a microscope with infinite objective (schematic).

and a fixed lens relay system is not necessary no matter how far the objective is from the eyepiece.

G. Polarized Light Illumination

Polarized light has been used in metallography to identify inclusions. This method has been replaced by electronic microprobe analysis and energy-dispersive spectroscopy. Polarized light is also very useful for the examination of noncubic anisotropic metals such as beryllium, titanium, zinc, and zirconium. The use of polarized light for the examination of heat-treated specimens is rather limited; therefore, only a brief introduction of polarized light and its characteristics will be given.

Natural lights and most artificial lights are transverse waves vibrating in all directions at right angles to the direction of propagation (Figure 19a). Light passing through a polarizer vibrates in only a certain plane along the direction of propagation (Figure 19b) and is called plane-polarized or linearly polarized light.

Although both Nicol prisms and polarizing filters can be used as polarizers in microscopes, the latter are most commonly used because of their lower cost. In a polarizing filter, polyvinyl alcohol film is stretched to line up the complex molecules and then impregnated with iodine. This film is usually mounted between two thin plates of optical glass.

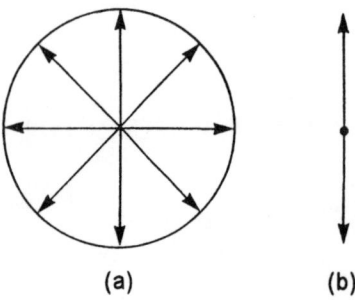

Figure 19 Vibration of (a) ordinary light and (b) plane-polarized wave.

Polarized light illumination is attained by placing the polarizer and analyzer, which are made of the same material, into the optical path of the microscope under brightfield illumination. The polarizer is situated ahead of the plane glass reflector and is usually nonrotatable, while the analyzer is placed after the objective and is usually rotatable. When the polarizing direction of the analyzer is at right angles to the polarizer, no light will be observed if a polished isotropic metal specimen, e.g., a metal with cubic structure, is used. This condition is known as cross-polarized illumination. However, if a polished anisotropic metal specimen is used instead, an image of microstructures such as grain structure, twinning, and the distribution of phases can be delineated.

If the polished surface of an optically isotropic metal specimen has a thin oxide film deposited on it by anodizing the specimen in a standard electropolishing cell, the optical anisotropy and surface irregularities of the film may exhibit birefringence, so that the same effect can be produced as with an optically anisotropic metal. Figure 20 shows the microstructure of a sand-cast Al – 1 wt % Ni aluminum alloy, electropolished and anodized. Figure 20a was obtained under brighfield illumination. The specimen has a typical dendritic structure with intermetallic phase $NiAl_3$ located at the dendritic arm interface, but the extent of individual grains can hardly be differentiated. Figure 20b shows the same field under polarized light illumination. In this case grains with different orientations can easily be recognized according to their gray levels. Figure 20c shows grain boundaries of the same field traced out according to Figure 20b; the four grains are denoted by numbers. It can be seen that a dendritic cell of grain 2 is isolated within grain 3, while another dendritic cell of grain 3 is mostly within grain 2 except for a small segment of its boundary neighboring grain 1.

H. Phase Contrast Illumination

Although the phase contrast method was developed first and primarily for transmitted light microscopy, it has found application in incident light microscopy. When both phases in a metallographic specimen have similar etching and reflection behaviors, the resulting microstructure shows a lack of intensity contrast. Such a specimen is called a phase object as compared with an amplitude object, in which sufficient contrast is obtained owing to a difference in amplitude or intensity of the reflected light from different phases. For example, in a plain carbon steel containing 1.2% C after spheroidizing, both cementite particles and ferrite matrix appear bright after etching with nital under brightfield illumination, as shown in Figure 21a. However, cementite is much harder than ferrite and is less attacked by the etchant; therefore, cementite particles protrude slightly over the surrounding ferritic matrix after polishing even without etching. As a result, there is a path difference between lights reflected from cementite and ferrite: the light reflected from cementite is 2δ ahead of the light reflected from

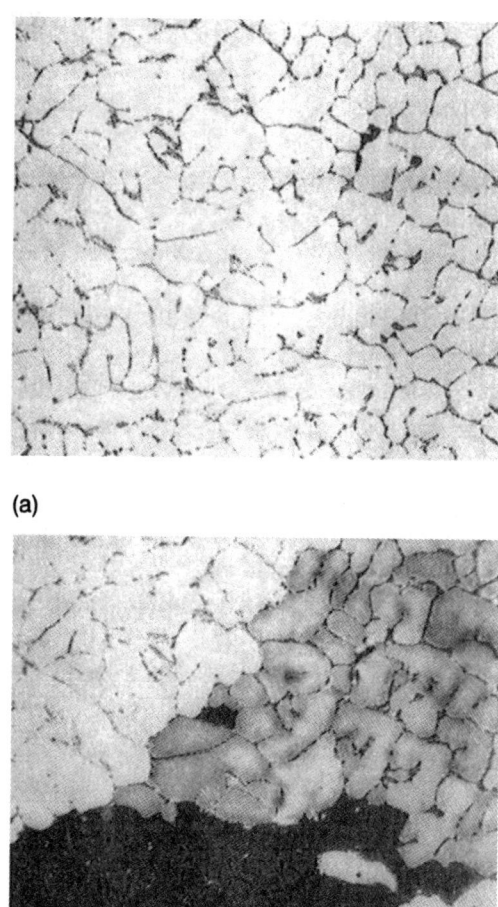

(a)

(b)

(c)

Figure 20 Microstructure of a sand cast Al–1 wt % Ni aluminum alloy, electropolished and anodized. (a) Brightfield illumination; (b) polarized light illumination; (c) grain boundaries traced out according to (b). (165×) (Reduced 20% in reproduction)

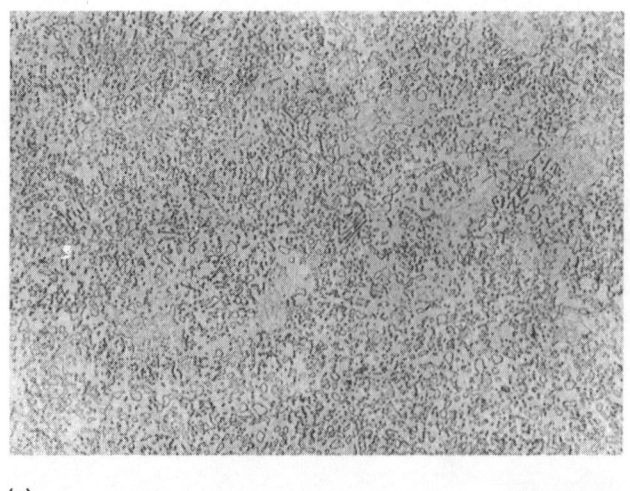

(a)

(b)

Figure 21 Photomicrograph of a plain carbon steel containing 1.2% C after spheroidizing (500×). (a) Brightfield illumination; (b) phase contrast illumination. (Reduced 30% in reproduction)

its surrounding ferritic matrix, where δ is the protruded height of cementite over ferrite. This path difference corresponds to a phase angle of $\phi = 2\delta \times 2\pi/\lambda = 4\pi\delta/\lambda$, where λ is the wavelength of the incident light. The phase angle between these light paths can be represented in a reference circle as shown in Figure 22, where, for simplicity, the amplitudes of the two light rays are assumed equal.

Lights with only a difference in phase angle cannot be differentiated by the human eye directly. However, under phase contrast illumination, it is possible to transform such a phase difference into an amplitude or intensity difference in the final image, so that the image contrast can be enhanced significantly and thus can be detected readily by the human eye.

The accessories used for phase contrast illumination consist of an annular diaphragm and a phase plate with a phase ring on it. The former is located in place of the aperture diaphragm, making the incident ray a hollow cone of parallel light, while the latter is located at the rear

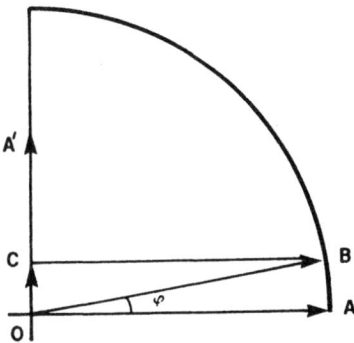

Figure 22 Schematic representation of the lights reflected from ferrite and cementite phases before and after leaving the phase plate. *OA*, direct light, i.e., light reflected from the ferritic matrix. *OB*, light reflected from cementite. *OC*, diffracted light. *OA'*, direct light of reduced amplitude and shifted phase after leaving the phase ring (positive phase contrast).

focal plane of the objective, where the phase ring just covers the image of the annular diaphragm. Figure 23 schematically illustrates the appearance of the annular diaphragm and phase plate.

From Figure 22 it can be easily seen that light with a small shift in phase *OB* can be regarded as the sum of the direct light *OA*, i.e., light reflected from the matrix, and a new diffracted light *OC*, which is ahead of phase by approximately $\pi/2$.

The role of the annular diaphragm is to separate the direct light from the diffracted light by allowing the former to pass only through the phase ring while the latter passes through the whole phase plate, including the phase ring. The phase ring comprises only a rather small fraction (about 15%) of the area of the phase plate, and hence the influence of this small part of the diffracted light can be neglected.

The phase ring is made thinner than its surrounding phase plate so that the direct light is advanced in phase by $\pi/2$ with respect to the diffracted light. Metal film is deposited on the phase ring so that the intensity of the direct light is reduced significantly, e.g., by approximately 80%, and is made comparable to that of the diffracted light.

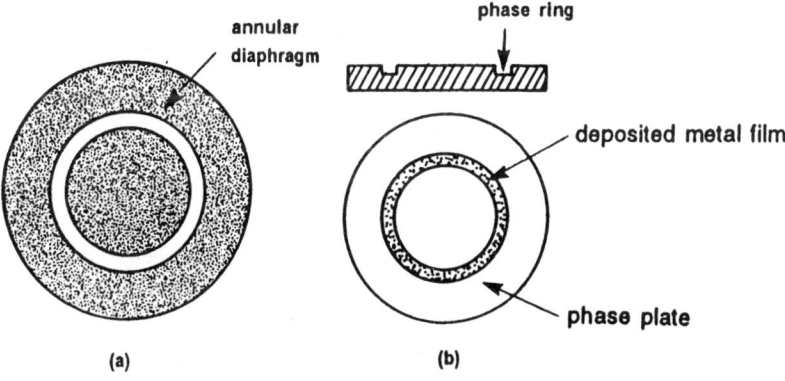

Figure 23 Appearance of (a) annular diaphragm and (b) phase plate (schematic).

After passing through the phase ring, the direct light of reduced intensity and shifted phase, represented by the vector OA' in Figure 22, is approximately in phase with the diffracted light OC. As a result of constructive interference, the image intensity of cementite is approximately proportional to $(OA' + OC)^2$. Therefore, the resulting contrast (K) between cementite and ferrite under phase contrast illumination can be represented by the expression

$$K = 1 - (OA')^2/(OA' + OC)^2 \tag{7}$$

which has been enhanced greatly compared with $K \approx 0$ under brightfield illumination.

Figure 21b shows the photomicrograph of the same specimen and same field of view as in Figure 21a but under phase contrast illumination. The bright particles are the protruded cementite phase, and the dark background is the ferritic matrix.

To ascertain the correctness of austenitizing temperature and soaking time during heat treatment of high-speed steels, it is often necessary to determine the austenitic grain size and the degree of carbide dissolution. The latter can best be evaluated under phase contrast illumination. Figure 24 shows photomicrographs of a Fe-10W-4Mo-4Cr-3V-Al high-speed steel after quenching from 1240°C. Figure 24a was obtained under brightfield illumination, and Figure 24b under phase contrast illumination. The two photographs are of the same field of view. Under phase contrast illumination, the bright carbide particles ar clearly delineated against the dark matrix composed of martensite and retained austenite.

When the height difference between two phases on the specimen surface is around 500 nm, good contrast can be obtained under phase contrast illumination. Usually a light etch of the specimen is needed to attain this condition. However, for best results, a higher quality of specimen preparation is required under phase contrast illumination, because even a very light scratch that is undetectable under brightfield illumination will be clearly revealed.

The phase contrast method in metallurgical microscopes has been gradually replaced by the differential interference contrast technique, which reveals height differences among phases as a variation in color or gray level over a much wider range.

I. Differential Interference Contrast Illumination

The principles of differential interference contrast (DIC) illumination were set forth by G. Nomarski about 40 years ago. This method uses a modified Wollaston prism located ahead of the objective in conjunction with polarized light at extinction; i.e., the polarizer and the analyzer are in the 90° crossed position as shown in Figure 25.

A Wollaston prism [22] is made of a birefringent, uniaxial material such as quartz and consists of two wedge-shaped prisms cemented together (Figure 26). The optic axes of the two prisms are at right angles to each other. When a plane-polarized light wave with its vibration plane inclined by 45° to the plane of drawing reaches the Wollaston prism perpendicularly, it will be split into two plane-polarized waves in the lower prism. Each of their vibration planes makes an angle of 45° with the incident wave. At the cemented interface of the Wollaston prism, these two component waves are deflected in two directions at a relatively small angle (less than 0.5 min) and encounter different indices of refraction so that their wavefronts travel at different speeds. It is impossible to locate the interference plane of the Wollaston prism in the image-side focal plane of the objective, at least with high-power objectives. Therefore, a modified Wollaston prism with its interference plane outside the compound prism is used instead and can be located relatively farther from the objective.

The beam from the light source is plane-polarized after passing through the polarizer; then it is deflected to the Wollaston prism by the plane glass reflector of the vertical illuminator as shown in Figure 25. After emerging from the objective the two component waves become parallel to each other and with a slight lateral separation. The light rays reflected from the

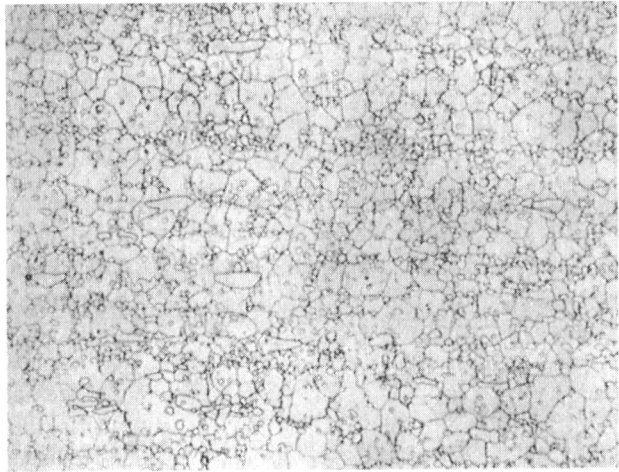

(a)

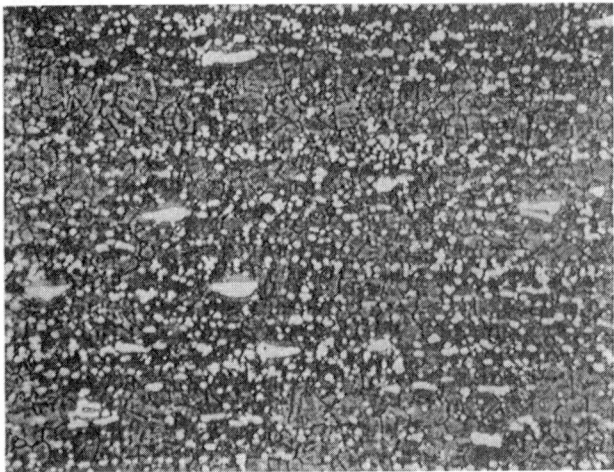

(b)

Figure 24   Photomicrograph of a Fe-10W-4Mo-4Cr-3V-Al high-speed steel after quenching from 1240°C (400×). (a) Brightfield illumination; (b) phase contrast illumination. (Reduced 15% in reproduction)

specimen are then recombined by means of the objective and the Wollaston prism, pass through the plane glass reflector, and, with the aid of the analyzer, are brought to travel in one plane again, making them capable of interference. The interference contrast image of the specimen can then be formed in the intermediate image plane (not shown) and can be viewed through the eyepiece.

The path difference can be altered by moving the Wollaston prism laterally, i.e., perpendicular to the optical axis. Using white light to illuminate a structureless object, e.g., an optically flat polished surface, the image field (also called the background image) appears dark when the Wollaston prism is at a position with zero path difference. However, once there is a path difference, a colored background image appears; every path difference is associated with a characteristic interference color.

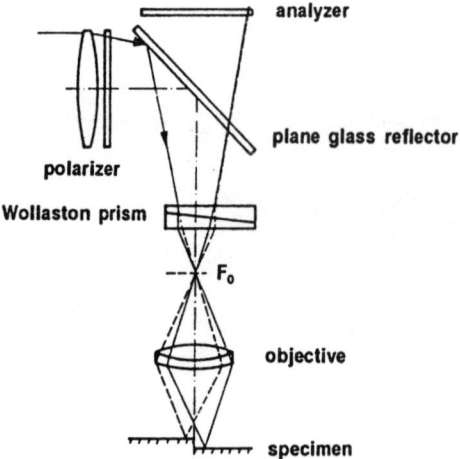

Figure 25 Beam path in reflected light DIC. (Courtesy of Carl Zeiss.)

With regard to a phase object, the path difference depends on the geometrical profile of the object, the phase retardation resulting from reflection of the waves from the opaque object, and the relative lateral position of the Wollaston prism.

Figure 27 shows schematically the profile of the object, the wavefronts of the DIC image before and after reaching the analyzer, and the resulting intensity distribution. Let phase B be the matrix of the object; then the color of the interference background image depends on the path difference (δ_0). Assuming that the difference in phase retardation resulting from reflection of the waves from the two phases is negligibly small, then the DIC image color of phase A is nearly the same as that of the background (phase B). However, the path differences at phase boundaries vary with δ_0; therefore, by adjusting the path difference δ_0, it is possible to observe images similar to darkfield (when $\delta_0 = 0$), relief-like, and/or color images. Thus, the best result can be obtained by selecting the type of image suitable for the object to be observed, and the contrast of the object is increased by interference colors.

Figure 28 shows a comparison of photomicrographs of a commercial pure iron specimen under (a) brightfield illumination and (b) DIC illumination. The specimen has undergone im-

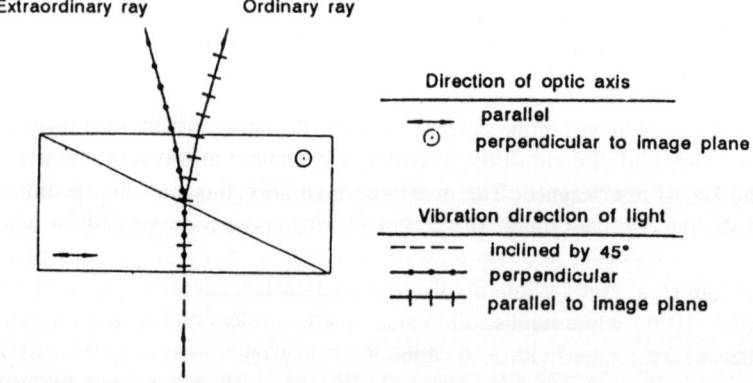

Figure 26 Beam splitting of plane-polarized light at the interface of the Wollaston prism. (Courtesy of Carl Zeiss.)

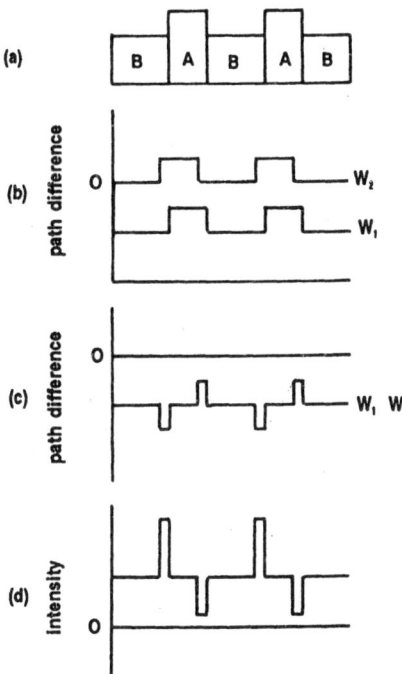

Figure 27 Schematic representation of the formation of DIC image. (a) Profile of the specimen; (b) and (c) wavefronts of the DIC image before and after reaching the analyzer; (d) resulting intensity distribution.

pact loading at low temperature, and the deformation twins or Neumann bands are more clearly delineated under DIC illumination.

Under phase contrast illumination, the relief of the specimen surface for producing an image with optimal contrast varies within relatively narrow limits because the phase shift and absorption of direct light are fixed by the phase ring of the phase plate. In contrast, by moving the Wollaston prism laterally, DIC images with excellent contrast can be obtained over a far greater range of path differences in the object. Since the DIC method is obviously superior to the phase contrast method in incident light microscopy, modern metallurgical microscopes are equipped predominantly with DIC illumination, and phase contrast microscopy can be found only in some older models.

J. Photomicrography and Photomacrography

Photomicrography is the reproduction of microstructures observed under a microscope. Photomacrography is the reproduction of macro-etched sections or broken parts such as castings, forgings, and heat-treated parts at a magnification from less than 1 to 20 or so with the aid of a camera or stereomicroscope. The aim of photomicrography and photomacrography for heat treaters is to faithfully reproduce what has been observed in heat-treated specimens or parts for reporting or publication.

1. Photomicrography

Equipment for Photomicrography Every modern metallurgical microscope is equipped with either a camera built into the microscope stand as shown in Figure 29 or exits for cameras as shown in Figure 30. With even the simplest bench-type microscope, a camera attach-

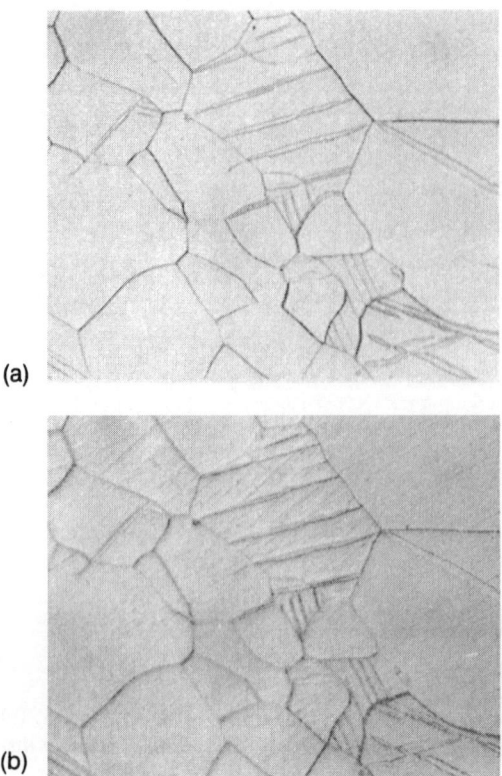

(a)

(b)

Figure 28 Photomicrographs of a commercial pure iron specimen that had undergone impact loading at low temperature (600×). (a) Brightfield illumination; (b) DIC illumination. (Reduced 15% in reproduction)

Figure 29 The Reichert MEF4A widefield metallograph. (Courtesy of Reichert Division of Leica Aktiengesellschaft.)

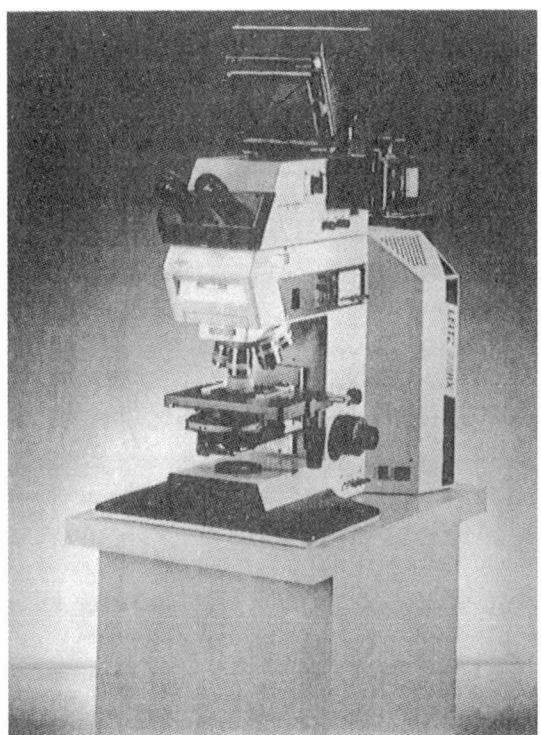

Figure 30 Phototube Leitz DM RD mounted to universal stand DM RXE. (Courtesy of Leica Mikroskopie und Systeme GmbH.)

ment can be fitted onto the binocular phototube, forming a rigid link with the microscope as shown in Figure 31. The field of view on the film plane of the camera is usually the same as that of the eyepiece, and images at both places are in sharp focus simultaneously so that no special focusing eyepiece is necessary. However, the intermediate image must be located just outside the focal plane of the photo-eyepiece so that a real final image can be formed on the film plane. Two types of cameras are used in photomicrography: large format (4 × 5 in. or 9 × 12 cm) and 35mm.

Films Used in Photomicrography One variety of Polaroid instant film that produces a negative and a positive print simultaneously is perhaps the most extensively used large format film because of its availability, efficiency, and convenience (no darkroom work is necessary). The negative can be kept for future use when more copies are needed. If only one instant positive print is produced per exposure, the photographic process must be repeated for each copy needed before the specimen is removed from the microscope.

Increasingly, 35mm film is being used because of lower processing cost and time, but enlargement is necessary for obtaining a print. For best quality of a positive print, large format sheet film is recommended although it is more expensive and requires more darkroom time.

Black-and-white films are more commonly used than color films in photomicrography due to their lower cost. The two major types of black-and-white film used are panchromatic and orthochromatic. Panchromatic films are sensitive to the entire visible spectrum and should be

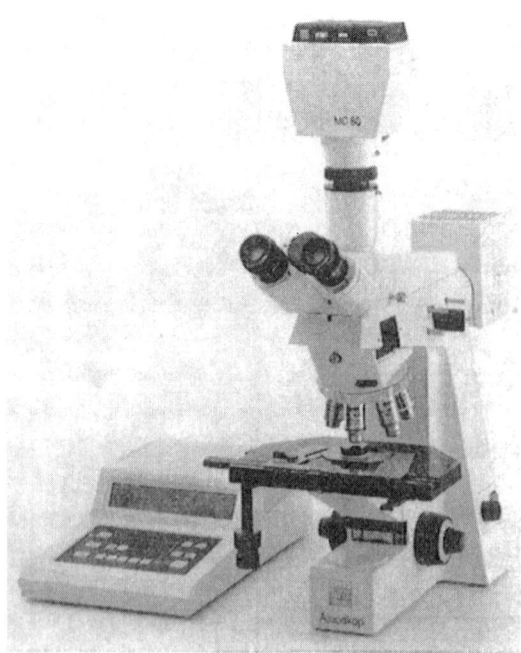

Figure 31 The MC 80 microscope camera incorporated with Zeiss Axiovert 100A. (Courtesy of Carl Zeiss.)

developed in total darkness, whereas orthochromatic films are sensitive to all colors except orange and red and can be developed under dark red safelight. Since a yellow-green filter is often used with plano-achromatic or achromatic objectives to minimize lens aberrations, or-thochromatic films are useful for photomicrography.

The "speed" of a film is an indication of its sensitivity to light and is expressed as a number intended to be used to determine the exposure needed to produce an image of satisfactory quality. The International Organization for Standardization (ISO) adopted both the American Standards Association (ASA), formerly the American National Standards Institute (ANSI), and Deutsche Industrie Norm (DIN) systems for rating film speed. For example, ISO 100/21° corresponds to an ASA rating of 100 and a DIN rating of 21°.

In the arithmetic expression used by the ASA, the film speed value varies inversely with the exposure required to produce the specified density. For example, films with values of 25, 50, 100, and 200 ASA each require half the exposure of the preceding film; in other words, a speed of 100 ASA is double that of 50 ASA.

In the DIN system's logarithmic expression, the film speed numbers increase by 3 as sensitivity is doubled. For example, values of 18°, 21°, and 24° DIN represent materials that each require half the exposure of the preceding material. In other words, a speed of 24° DIN is double that of 21° DIN.

ASA and DIN numbers can be easily interconnected from

$$DIN = 10 \log_{10} ASA + 1 \tag{8}$$

$$ASA = 10^{(DIN - 1)/10} \tag{9}$$

It is desirable that the size of silver grains in the photographic emulsion after develop-ment be as small as possible so that it will not mask the details of the image. High-speed films

(ISO 200/24°–ISO 400/27°) are always accompanied by coarser grains. Therefore, fine-grained slow films (ISO 32/16°–ISO 80/20°) are preferred for photomicrography, where very fine detail is required and slow shutter speeds are possible. However, if the negative is not going to be enlarged substantially, good prints with sufficient resolution can be obtained from medium-/or normal speed films (ISO 100/21°–ISO 160/23°), which are more easily available on the market. Increasingly, color films in photomicrography are being used because more information can be obtained in a color photomicrograph. It is known that the human eye can distinguish more than 1000 different colors but can recognize no more than 20–30 tones of gray. Color photomicrographs are useful for the identification of inclusions and carbides in steels and for the identification of phases in some stainless steels [15].

There are two types of color films: color reversal and color negative films. Positive color slides are obtained from a color reversal film after processing and are used primarily for illustrating lectures or demonstrations. True color reproduction of the original microstructure can be obtained only at a specific color temperature, which is a scale for rating the color quality of illumination expressed in kelvins (K). Therefore, it is important to choose a type of film to suit the light for which it is "balanced." For a xenon light source, daylight film balanced for a color temperature of about 5600 K should be used, and for a halogen light source, artificial light film balanced for a color temperature of about 3200 K should be used. If daylight film is used in halogen light, a suitable filter known as a conversion filter (e.g., Kodak 80B) should be added to alter the color temperature of the incident light and render it suitable for daylight film.

Color negative film produces a negative with complementary colors. For example, a blue object is recorded as yellow, but the complementary relationship is partly obscured by the presence of color-masking dyes often used in color negatives. For the same reason as for color reversal film, daylight or artificial light film should be chosen according to the light source used in photomicrography; otherwise, a suitable conversion filter should be used. True color reproduction in the prints of photomicrographs from negatives is not as easy as for ordinary prints because the photolab technicians are not familiar with the subject matter and as a result, proper correction cannot often be made in printing. It would be best if metallographers could do the processing themselves, but this depends much on the volume of work and available funds.

It should be kept in mind that when color film is used in photomicrography, the light source should be operated at its rated voltage so as to attain the correct color temperature. Therefore, lamp voltage must not be used as a means of adjusting the image intensity. In such a case, neutral density filters, which reduce the intensity of incident light without altering its color temperature, with various transmissivities should be used instead.

Photographic Exposure and the Reciprocity Effect   Correct exposure is an important step for obtaining a negative with all necessary details and good contrast. Photographic exposure (H) is defined as the quantity of light per unit area received by the photographic emulsion of the film and is commonly expressed as the product of illumination (E) and time (t):

$$H = E \times t \tag{10}$$

The correct exposure is determined by a light-metering system. The most frequently used system in photomicrography is integrated metering, which measures the light reaching the film, giving an average reading over the entire format area. Sometimes it is desirable to obtain an extremely accurate exposure in a very localized area of the image, which may be darker or brighter than the average reading. In such a case, spot metering can be used in which the angle of acceptance is so narrow that the size of the measured area can be as small as 1% of the total image area. For ordinary photography, however, a center-weighted metering system is used in most single-lens reflex (SLR) cameras. That is, light from all parts of the image field is measured by the silicon photocell, but the influence of the central zone is greatest.

Experiments show that equal exposures do not produce photographic images of equal densities and equal contrast if the time of exposure is extremely long or extremely short (and the corresponding light level is too weak or too strong). This is known as the reciprocity effect or Schwartzschild effect. This effect must be taken into account when taking photomicrographs under darkfield, polarized light, and phase contrast illumination where a longer or even a much longer exposure is needed. Fortunately, modern automatic photographic systems are capable of correcting for the reciprocity effect automatically.

2. Photomacrography

Some metallurgical microscopes are provided with special accessories for photomacrography at magnifications of around $1-10\times$. In some models, the magnification is continuously adjustable by altering the zoom setting and the object distance, and separate macro illumination is required.

Photomacrography with 35mm SLR cameras at an image ratio of about 1:10 (1/10 life size) to 1:1 (life size) or larger is rather easy and convenient. Figure 32 shows the cutoff view of a 35mm SLR camera [23]. Light entering the lens (A) falls on a movable reflex mirror (B), which reflects it upward onto a ground glass screen (C), then enters the pentaprism (D) and reflects twice, so that an upright and correctly oriented image is produced for viewing through the eyepiece (E). The through-the-lens (TTL) light-metering system has a light-sensitive cell (G) that measures the amount of light entering through the lens. When the shutter release button is pressed, the reflex mirror swings upward, allowing the incident light to fall on the film frame (F) and expose the film once the shutter (H) is released.

The focal length of the lens and its maximum aperture, e.g., 50mm 1:2, are shown on the front of the lens. The focal length of a standard lens of a 35mm camera is 50 mm, and the angle of view is 46°, roughly the same as that of the human eye. The shorter the focal length, the wider the angle of view, and vice versa.

The brightness of the image on the film is controlled by aperture size expressed by f number, which is defined as the quotient of the lens focal length and the effective aperture (diameter of the light beam entering the lens and just filling the opening in the diaphragm of a camera lens as shown in Figure 33). For example, f8 or f/8 means that the diameter of the effective aperture is 1/8 of the lens focal length. The smaller the f number, the larger the aperture.

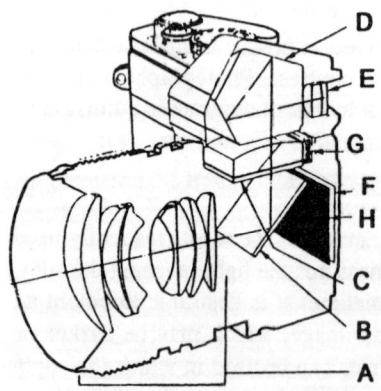

Figure 32 Cutoff view of a 35mm SLR camera. (Courtesy of Fountain Press Ltd.)

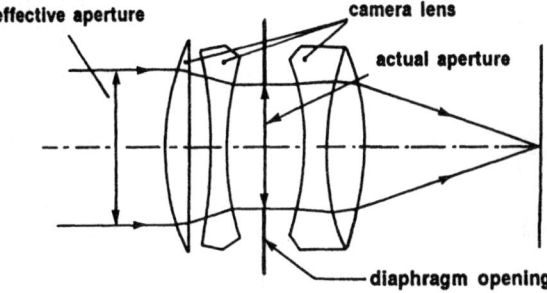

effective aperture

camera lens

actual aperture

diaphragm opening

Figure 33 Effective aperture and actual aperture of a camera (schematic).

The maximum aperture of a lens corresponds to its smallest f number. For example, if the maximum aperture of a 50mm lens is 25 mm, the corresponding f number is f2 or f/2, which is usually marked as 1:2.

The f numbers for any lens are engraved on the lens mount in a series as follows: f 1.4, 2, 2.8, 4, 5.6, 8, 11, 16, 22. The series is so arranged that as the f number is increased by one stop, the amount of light allowed to enter is halved. In other words, the image illuminance is inversely proportional to the square of the f number, and the required exposure time is directly proportional to the square of the f number.

Lenses of the same f number are said to have the same speed. That is, they produce images of essentially equal luminance.

Aperture size can also be used to control depth of field, which is defined as the range of object distances within which objects are imaged with acceptable sharpness [24]. Depth of field increases as focal length decreases and as the f number increases.

Several equipment options have been devised for taking close-up and photomacrographs with SLR cameras. The basic options are close-up lens, extension tubes or bellows, reversing ring, and macro lens [25,26].

Close-Up Lens The close-up lens is by far the cheapest accessory for photomacrography. It works rather like a magnifying glass and can be screwed onto the front rim of the camera lens. It focuses the lens at a closer distance than normal so that a larger image can be formed on the film plane. The close-up lens is inferior optically to the expansive camera lens, so image quality may be slightly compromised. Normally this is hardly noticeable, especially when a large f number, e.g., f8–f16, is used.

The strength of a close-up lens is expressed in diopters, which is defined as 1 m divided by the focal length of the lens. Its usual range is 0.4–5 diopters. The larger the number, the closer the lens can focus.

Extension Tubes or Bellows When a close-up lens is not used in photomacrography, something is needed to extend the distance between lens and camera body. For example, in order to get a life-size reproduction (1:1) of a small object with a standard 50mm lens, an extension of 50 mm between lens and film plane is needed. An extension tube is a tube or set of tubes that can be fitted between lens and camera body as shown in Figure 34. So-called automatic extension tubes are so designed that the action of the fully automatic diaphragm works normally. Since extension tubes are not optical devices, no deterioration in lens performance will result.

An extension bellows is basically a continuously variable extension tube mounted on a special focusing rail and allowing a full range of extension between 30 and 200 mm as shown in Figure 35.

Figure 34 One or more extension tubes can be fitted between the camera and lens. (Courtesy of AMPHOTO.)

Reversing Ring The reversing ring is screwed into the filter ring of the lens and allows the lens to be mounted backward on the camera. Since the normal lens is not highly corrected in close-up positions, turning the lens backward changes the optical corrections in the lens and results in better definition of the close-up pictures; hence, the lens behaves as though it were really a macro lens (see next paragraph). However, when the lens is mounted backward,

Figure 35 Camera fitted with extension bellows and focusing rail. (Courtesy of AMPHOTO.)

the fully automatic diaphragm ceases working, and consequently the aperture must be adjusted manually.

Macro Lens A macro lens is so designed and highly corrected as to focus an object closer than normal lenses without special attachments. For example, a 50mm macro lens can be extended to such a length that an object located about 10 cm from the lens can be focused. Most macro lenses can produce at least an image ratio of 1:2, and some can even produce an image ratio of about 1:1. Macro lenses with focal lengths of 50 and 100 mm are most commonly used. A macro lens usually has a maximum aperture of about f/3.5 or f/4, which is about three to four times slower than an f/2 lens. When used with extension tubes or bellows, a macro lens still performs better than normal lenses.

IV. QUANTITATIVE METALLOGRAPHY

Metallography has long been a means of obtaining qualitative or semiquantitative descriptions of microstructures for quality control in production and research. However, the rapid progress of science and technology over recent decades has placed more stringent demands on the mechanical properties of heat-treated parts and requires a more accurate quantitative description of microstructures. Quantitative metallography deals with the quantitative relationships between measurements made on the two-dimensional plane of polish and the magnitudes of the microstructural features in the three-dimensional metal or alloy [27]. It is an important branch of stereology applied specifically to metals and alloys.

A. Some Stereology Basics

According to Elias [28], stereology is a system of methods to obtain information about three-dimensional structure from two-dimensional images, whether sections or projections. Although easily understood and accepted by most practical stereologists and microscopists, this definition is limited because it considers only the procedures from a "two-dimensional image" to a "three-dimensional structure." As indicated by Miles [29] in 1972, all basic stereological equations are special cases of equations in a generalized n-dimensional space. This implies that any n-dimensional structure can be studied by sectioning it with any s-dimensional section, and the principles involved are always the same. It follows that the increasing attention of mathematicians to the problems in stereology is indispensable to the establishment of a solid foundation for this interdisciplinary field of study.

The basic symbols used in stereology are P, L, A, S, V, and N, which represent point, line, plane surface, curved surface, volume, and number of features, respectively. Compound symbols are used to represent stereological parameters, in which the capital letter refers to the microstructural quantity and the subscript pertains to the test quantity. For example, the symbol P_P refers to point fraction, which is equivalent to P/P_T, with P the number of points (in areal features) and P_T the number of test points used during measurement; the symbol S_V refers to surface area per unit test volume. Table 3 is a list of basic symbols used in stereology and their definitions quoted from Reference 30.

Equations (11)–(13) are statistically exact expressions that relate points, lines, surfaces, and volumes in three-dimensional structures to measurements made on two-dimensional sections [27].

$$V_V = A_A = L_L = P_P \tag{11}$$

$$S_V = (4/\pi)L_A = 2P_L \tag{12}$$

$$L_V = 2P_A \tag{13}$$

Table 3 List of Basic Stereological Symbols and Their Definitions

Symbol	Dimensions[a]	Definition
P		Number of point elements or test points.
P_P		Point fraction. Number of points (in real features) per test point.
P_L	mm^{-1}	Number of point intersections per unit length of test line.
P_A	mm^{-2}	Number of points per unit test area.
P_V	mm^{-3}	Number of points per unit test volume.
L	mm	Length of lineal elements, or test line length.
L_L	mm/mm	Lineal fraction. Length of lineal intercepts per unit length of test line.
L_A	mm/mm^2	Length of lineal elements per unit test area.
L_V	mm/mm^3	Length of lineal elements per unit test volume.
A	mm^2	Planar area of intercepted features, or test area.
A_A	mm^2/mm^2	Area fraction. Area of intercepted features per unit test area.
S_V	mm^2/mm^3	Surface area per unit test volume.
V	mm^3	Volume of three-dimensional features, or test volume.
V_V	mm^3/mm^3	Volume fraction. Volume of features per unit test volume.
N		Number of features (as opposed to points).
N_L	mm^{-1}	Number of interceptions of features per unit length of test line.
N_A	mm^{-2}	Number of interceptions of features per unit test area.
N_V	mm^{-3}	Number of features per unit test volume.
$\overline{L}$	mm	Average lineal intercept, L_L/N_L.
$\overline{A}$	mm^2	Average areal intercept, A_A/N_A.
$\overline{S}$	mm^2	Average surface area, S_V/N_V.
$\overline{V}$	mm^3	Average volume, V_V/N_V.

[a]Arbitrarily shown in millimeters
Source: Ref. 29.

The derivation of the above equations is beyond the scope of this book and can be found in standard textbooks on stereology [30–32]. Stereological parameters with subscripts other than V are measurable and are called measured quantities; stereological parameters with subscript V cannot be measured and are called calculated quantities. However, the measured parameters in Equations (11) and (12), i.e., A_A, L_L, and L_A, can also be calculated.

The exactness of Equations (11)–(13) signifies that in the derivation of these equations, no simplified assumptions on the size, shape, distribution, spacing, etc., of microstructural constituents are required. The microstructure under investigation should be representative of the bulk specimen, since microstructures of different fields of view in the same specimen can in no way be identical. Measurements of stereological parameters should be made randomly or with statistical uniformity.

B. Volume Fraction

Volume fraction is one of the most important and most commonly used stereological measurement. From Equation (11) it can be seen that the volume fraction of a phase or microstructural constituent in a microstructure can be obtained by measuring area, lineal, or point fraction, but for manual operation the use of point counting is most effective and has sufficient accuracy.

Usually the intersections of a test grid are used as test points. Figure 36 illustrates a 16-point (4 × 4) test grid. The grid can be inserted in the eyepiece of the microscope or superimposed on a photomicrograph. The number of points (P_α) that falls within a certain phase

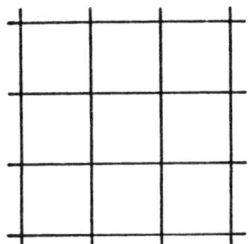

Figure 36 A 16-point (4 × 4) test grid.

(say α) is counted, the point fraction (P_P) of phase α is P_α divided by the number of grid points P_T. Points that fall on phase boundaries are counted as 1/2 [33]. This process is repeated several times so as to obtain required accuracy (see Section IV. C). Figure 37 shows an example of point counting, with a 4 × 4 test grid superimposed on a granular microstructure (schematic) with phase α randomly dispersed in a matrix. There are two test points that fall within phase α and five points located on phase boundaries. Therefore, the point fraction of phase α in this field of view is $(2 + 5 \times \frac{1}{2})/16 = 0.281$ or 28.1%.

Low-density test grids, such as 3 × 3, 4 × 4, or 5 × 5, are usually employed for point counting. Test grids with more than 100 test points are seldom used because of operator eye fatigue. Generally, higher density test grids should be used for microstructures with lower volume fraction. The magnification should be so chosen that no more than one grid point falls on a given particle of interest [3, p. 427].

Point counting can be used to estimate the carbon content of fully annealed hypoeutectoid and eutectoid plain carbon steels. For instance, in low-carbon steels, pearlite is the minor constituent compared with proeutectoid ferrite. The volume fraction of pearlite can be easily calculated by point counting of pearlite in the microstructure. Let $(P_P)_P$ be the measured point fraction of pearlite. Then, according to the lever rule, the carbon content x in weight percent of the steel can be obtained from the expression

$$x = 0.77(P_P)_P \tag{14}$$

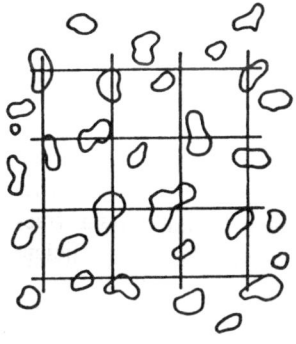

Figure 37 An example of point counting.

where 0.77 is the carbon content in weight percent of eutectoid steel. In deriving Equation (14), the very small amount of carbon in ferrite (0.0218 wt % C) is neglected.

Similarly, the carbon content of a spheroidized steel can be calculated. Let $(P_P)_{cem}$ be the measured point fraction of cementite particles. Then, according to the lever rule, the carbon content in weight percent of the steel can be obtained from the expression

$$x = 6.69 \ (P_P)_{cem} \tag{15}$$

where 6.69 is the carbon content of cementite. The carbon content in ferrite is also neglected in deriving Equation (15).

Volume fraction is such an important stereological measurement that the first standard on quantitative metallography developed by the American Society for Testing and Materials (ASTM) deals with volume fraction. ASTM E562 (Standard Test Method for Determining Volume Fraction by Systematic Manual Point Count) was first adopted in 1976.

C. Statistical Analysis[1]

As stated earlier, there is no possibility of having two identical microstructures in a specimen. Therefore, no stereological measurements can be exact, and stereology is deeply statistical in nature [32, Vol. 1, p. 2]. Thus, without proper statistical analysis, stereological measurements and calculations cannot be regarded as complete. We shall not go into details on statistics; interested readers may refer to Reference 34. Following the example of the determination of volume fraction by point counting given below, the basic procedures of statistical analysis will be shown.

Suppose the point count for phase α in a specimen were performed on 20 fields of view using a 4 × 4 test grid like that shown in Figure 37. The results of 20 counts of P are as follows (the numbers in parentheses are number of occurrences): 3(1), 3.5(3), 4(5), 4.5(5), 5(3), 5.5(3). The mean volume fraction of phase α can be obtained by averaging the 20 counts:

$$V_V = P_P = \Sigma P_\alpha / \Sigma P_T = 87.5/(20 \times 16) = 0.273 \quad \text{or} \quad 27.3\%$$

1. Standard Deviation

The standard deviation s of the data is expressed as

$$s = \left[\frac{\Sigma(x_i - \bar{x})^2}{n - 1} \right]^{1/2} \tag{16}$$

where $\bar{x}$ is the mean value of the individual counts x_i and n is the number of counts. In this example, $\bar{x}$ and x_i correspond to P_α and $(P_\alpha)_i$, respectively, and $n = 20$.

Standard deviation is an absolute measure of data dispersion. It can be easily obtained by using a pocket scientific calculator in the statistical calculation (STAT) mode. In this example, $s = 0.7232$. The standard deviation of point fraction of phase α is $s/P_T = 0.7232/16 = 0.0452$ or 4.52%.

2. Coefficient of Variation (CV)

When the mean values of various measurements differ substantially, the comparison of standard deviation among those measurements seems difficult. Therefore, it is more useful to use a new parameter, i.e., the coefficient of variation (CV), obtained by normalizing the standard deviation and expressed in percent:

$$CV = (s/\bar{x}) \times 100 \tag{17}$$

---

[1]This section is derived from Reference 3, pp. 428–432.

In this example, CV = (0.0452/0.273)(100) = 16.6%.

3. *95% Confidence Limit*

For the nominal 95% confidence limit (95% CL) of the mean,

$$95\% \text{ CL} = 2s_{\bar{x}} \tag{18}$$

where $s_{\bar{x}}$ is called the standard deviation of the mean,

$$s_{\bar{x}} = s/(n - 1)^{1/2} \tag{19}$$

Therefore,

$$95\% \text{ CL} = 2s/(n - 1)^{1/2} \tag{20}$$

This means that if the test data exhibit a normal distribution about the mean and counting is repeated more times, then 95% of the mean value of the counts will be within 95% CL about the mean. In this example,

$$95\% = 2(4.52)/\sqrt{19} = 2.07\%$$

The final volume fraction determination of this example is

27.3% ± 2.07%   (95% CL)

4. *Percent Relative Accuracy*

The percent relative accuracy (%RA) of the test is obtained by using the equation

$$\%\text{RA} = (95\%\text{CL}/\bar{x}) \times 100 \tag{21}$$

Percent relative accuracy is also called 95% relative confidence limit or 95% RCL. In this example, we have

$$\%\text{RA} = (2.07/27.3) \times 100 = 7.58\%$$

If higher accuracy (i.e., lower %RA) is needed, then the required number of tests can be computed. Substituting n for $n - 1$ in Equation (20), we have $95\%\text{CL} = 2s/n^{1/2}$; then by substitution in Equation (21), a simple formula proposed by DeHoff [35] can be obtained:

$$n = [(200/\%\text{RA}) \times (s/\bar{x})]^2 \tag{22}$$

As long as n is not too small, the difference between $(n - 1)^{1/2}$ and $n^{1/2}$ is negligible and the above approximation is allowable.

In our example, if a %RA of 5% is desired, the required number of tests can be found from Equation (22):

$$n = [(200/5) \times (4.52/27.3)]^2 = 43.86 \approx 44$$

That is, 24 more tests are required to attain a %RA of 5% instead of 7.58% for 20 tests.

D. Surface Area per Unit Test Volume and Length of Lineal Traces per Unit Test Area[2]

Equation (12) is the combination of two important equations requiring a P_L measurement.

$$S_V = 2P_L \tag{23}$$

$$L_A = \pi P_L/2 \tag{24}$$

---

[2]This section is based on material in Reference 27.

When surface area per unit test volume (S_V) is of interest, Equation (23) is used. Figure 38 [36] shows a plot of Brinell hardness of a 0.8% C pearlite steel versus S_V. After austenitizing, specimens of the same steel were held isothermally at different temperatures below A_1 in the pearlitic transformation region. As a result, pearlitic structures with different lamellar spacing can be obtained. It follows from Figure 38 that the Brinell hardness of pearlite is proportional to S_V, the lamellar interface area between ferrite and cementite per unit test volume.

The stereological parameter S_V can also be used for analyzing grain boundary precipitates or transformation products, because grain boundaries are preferential nucleation sites at a certain stage of precipitation of phase transformation in the solid.

When lineal traces in the plane of polish are of interest, Equation (24) is useful. In corrosion studies, for example, grain boundaries are more easily attacked by the corroding medium owing to their higher free energy with respect to their neighboring grains. The magnitude of L_A of the surface exposed to the corroding medium is important.

E. Grain Size

During heat treatment of steels, grain sizes may undergo various changes due to phase transformation or recrystallization. It is well known that mechanical properties and behaviors of steels depend strongly on their grain sizes, and in most cases fine grain is preferred because of higher strength, ductility, and toughness. Therefore, grain size measurement is indispensable for quality control of heat-treated parts.

1. Grain Size Characterization

Up to now, there has been no statistically exact relationship for the determination of three-dimensional grain size from its two-dimensional sections. However, several parameters have been used to characterize grain size. Those used in ASTM E112 are listed below:

ASTM grain size number G

Number of grains per unit area N_A

Average area $\overline{A}(\overline{A} = 1/N_A)$

Average diameter $\overline{D}$ (also called nominal diameter, $\overline{D} = \sqrt{\overline{A}}$)

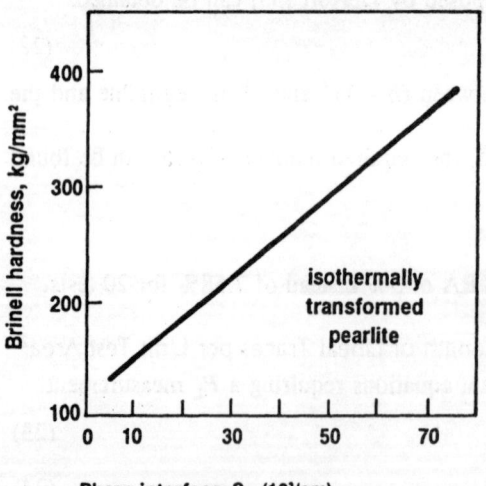

Figure 38 Brinell hardness of a 0.8% C pearlite steel as a function of S_v.

Intercept count N_L (number of interceptions of grains per unit length of test line)

Average intercept length $\bar{L}_3$ ($\bar{L}_3 = 1/N_L$)

Feret's diameter d_f ($d_f = \bar{A}/\bar{L}_3$)

Calculated number of grains per unit test volume N_V ($N_V = 0.5659\, N_L^3$)

To facilitate testing in production, test standards on grain size measurement were first developed by ASTM in the early 1930s. The adoption of ASTM E112 (Methods for Estimating the Average Grain Size of Metals) in 1961 merged all previous test standards into a single one. Since then it has been revised 10 times. It is one of the most extensively quoted ASTM standards [37]. Many other countries have their standards for determining grain size; all are based on ASTM E112 with only minor differences.

2. Comparison Method for Rating Grain Size

The simplest way to estimate grain size is to compare the grain structure under a microscope or a photomicrograph with the standard charts at magnification of 100× provided by ASTM and select the one that tallies most with that of the specimen or interpolate between two standard charts. Estimations should be made on three or more representative areas of the specimen.

To minimize estimation error, the appearance of the standard charts should be reasonably close to that of the specimen. For this reason, four categories of standard charts are available from ASTM. Figure 39 shows two standard charts having the same ASTM grain size number ($G = 3$) but from different periods. Chart (a) is from ASTM E19-46; (b) is from ASTM E112-63 and is still in use. The grain size in (a) looks more uniform but too idealized. Grain size distribution in (b) is more realistic and stereologically correct, because when a single-phase grain structure is sectioned randomly the sectioning plane can cut through the grains anywhere between maximum section and extremely small corners but with different probabilities. Therefore, it is natural that a range of grain sizes will be observed even in a specimen composed of extremely uniform grains.

Sometimes mixed grain sizes, i.e., grains with two distinctly different sizes or the so-called bimodel or duplex grain structures, do exist, particularly in hot-worked metals. Averaging from measurements of duplex grain structures is meaningless, because such grain size does not exist

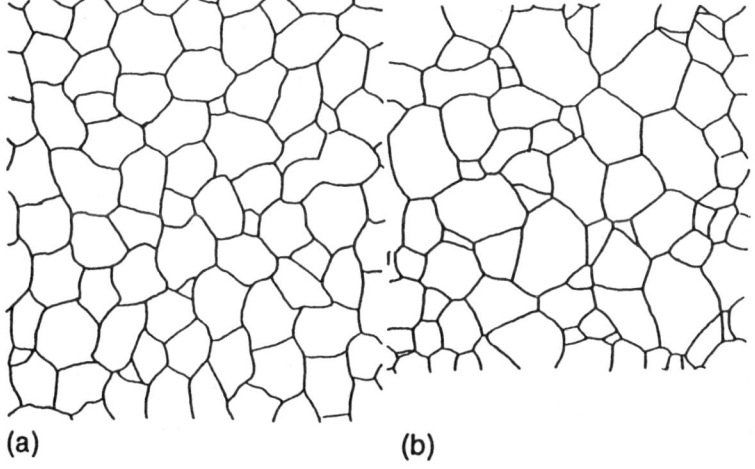

(a)   (b)

Figure 39 Standard charts of untwinned grains (flat etch) at 100× ($G = 3$) (a) from ASTM E19-46 and (b) from ASTM E112-63. (Reduced 45% in reproduction)

in the specimen. In such cases, two ranges of sizes should be reported, e.g., 40% of ASTM Nos. 8–10 and 60% of ASTM Nos. 3–4. Interested readers may refer to ASTM E1181 (Standard Test Methods for Characterizing Duplex Grain Sizes).

3. *Basic Equations for Rating the ASTM Grain Size Number*

Although the first standard chart for rating grain size appeared in 1930, the basic equation used to define the ASTM grain size number was not introduced until the adoption of ASTM E91 in 1951 in the form

$$N_A = 2^{G-1} \tag{25}$$

where N_A is the number of grains per square inch at $100\times$ and G is the ASTM grain size number. Equation (25) can also be expressed as

$$G = 1.0000 + \frac{\log N_A}{\log 2} = 1.0000 + 3.3219 \log N_A \tag{26}$$

When the metric system is used, Equation (26) can be expressed as

$$G = -2.9542 + 3.3219 \log N_A \tag{27}$$

where N_A is the number of grains per square millimeter at $1\times$.

ISO 643 defined grain size with the two equations

$$m = 8 \times 2^G \tag{28}$$

or

$$G = -3.0000 + 3.3219 \log m \tag{29}$$

where m is the number of grains per square millimeter at $1\times$ and G is the ISO grain size number. It follows that $G = 1$ when $m = 16$.

Comparing Equations (27) and (29), we have

$$G(\text{ASTM}) - G(\text{ISO}) = 0.0458 \tag{30}$$

That is, the ISO grain size number is smaller than the ASTM grain size number by 0.0458. This difference is insignificant and hence can be neglected in practice.

When grain structure of a specimen at magnification other than $100\times$ is compared with the standard charts, the following correction should be made:

$$G = G' + Q \tag{31}$$

where G is the actual grain size number after correction, G' is the apparent grain size number obtained by comparing the grain structure as viewed at magnification M with the standard chart at the basic magnification M_b ($100\times$), and Q is the correction factor that should be added to G',

$$Q = 2 \log_2 (M/M_b) = 6.6439 \log(M/M_b) \tag{32}$$

In actual practice, the alternative magnifications are usually selected as simple multiples of the basic magnification, and consequently the correction factors are also simple integers. Thus, for magnifications of $400\times$ and $50\times$, the corresponding values of Q are 4 and –2, respectively.

Table 4 gives micro-grain size relationship from ASTM E112, with the symbols characterizing grain size replaced by those adopted for stereology; i.e., d_n, $\bar{l}$, n/l, $\bar{a}$, n/v, n/a in the original table from ASTM E112 are replaced by $\bar{D}$, $\bar{L}_3$, N_L, A, N_V, N_A respectively.

Table 4 Micro-Grain Size Relationships Computed for Uniform Randomly Oriented Equiaxed Grains

grain size number G	"Diameter" of average grain section		Average Intercept distance, $\bar{L}_3 m$(mm)	Intercept count, N_L(mm^{-1})	Area of grain section, $\bar{A}$(mm^2)	Calculated number of grains per mm^3, N_V (mm^{-3})	Average number of grains N_A	
	Nominal $\bar{D}$(mm)	Feret's d_f(mm)					At 1× (mm^{-2})	At 100× (in^{-2})
00	0.51	0.570	0.453	2.210	0.258	6.11	3.83	0.250
0	0.36	0.403	0.320	3.125	0.129	17.3	7.75	0.500
0.5	0.30	0.339	0.269	3.716	0.0912	29.0	11.0	0.707
1.0	0.25	0.285	0.226	4.42	0.0645	48.8	15.50	1.00
1.5	0.21	0.240	0.190	5.26	0.0456	82	21.9	1.414
2.0	0.18	0.202	0.160	6.25	0.0323	138	31.0	2.000
2.5	0.15	0.170	0.135	7.43	0.0228	232	43.8	2.828
3.0	0.125	0.143	0.113	8.84	0.0161	391	62.0	4.000
3.5	0.105	0.120	0.095	10.51	0.0114	657	87.7	5.657
	μm	μm	μm		mm^2 × 10^{-3}			
4.0	90	101	80.0	12.5	8.07	1105	124	8.000
4.5	75	85	67.3	14.9	5.70	1859	175	11.31
5.0	65	71	56.6	17.7	4.03	3126	248	16.00
5.5	55	60	47.6	21.0	2.85	5258	351	22.63
6.0	45	50	40.0	25.0	2.02	8842	496	32.00
6.5	38	42	33.6	29.7	1.43	14871	701	45.25
7.0	32	36	28.3	35.4	1.008	25010	992	64.00

(continued)

Table 4 Continued

grain size number G	"Diameter" of average grain section		Average Intercept distance, $\bar{L}_3 m$(mm)	Intercept count, N_L(mm^{-1})	Area of grain section, $\bar{A}$(mm^2)	Calculated number of grains per mm^3, N_V (mm^{-3})	Average number of grains N_A	
	Nominal $\bar{D}$(mm)	Feret's d_f(mm)					At 1× (mm^{-2})	At 100× (in^{-2})
7.5	27	30	23.8	42.0	0.713	41061	1403	90.51
	μm	μm	μm		mm^2 × 10^{-6}	×10^6	×10^3	
8.0	22	25	20.0	50.0	504	0.0707	1.98	128.0
8.5	19	21	16.8	59.5	356	0.1190	2.81	181.0
9.0	16	18	14.1	70.7	252	0.200	3.97	256.0
9.5	13	15	11.9	84.1	178	0.336	5.61	362.0
10.0	11	13	10.0	100	126	0.566	7.94	512.0
10.5	9.4	10.6	8.41	119	89.1	0.952	11.22	724.1
11.0	8	8.9	7.07	141	63.0	1.600	15.87	1024
11.5	6.7	7.5	5.95	168	44.6	2.692	22.45	1448
12.0	5.6	6.3	5.00	200	31.5	4.527	31.7	2048
12.5	4.7	5.3	4.20	238	22.3	7.61	44.9	2896
13.0	4.0	4.5	3.54	283	15.8	12.80	63.5	4096
13.5	3.3	3.7	2.97	336	11.1	21.54	89.8	5793
14.0	2.8	3.2	2.50	400	7.88	36.2	127	8192

Source: ASTM E112.

For macro-grain size scale, Equation (26) is also used, but N_A represents number of grains per square inch at $1\times$. Macro-size number is designated as, for example, M-6. For macro size numbers higher than M-14, micro-grain size number should be used. Micro-size numbers may be converted to macro-size numbers by adding 13.288 size numbers. Macro-size numbers are useful for castings where the grains are much coarser.

4. Jefferies Planimetric Method

The planimetric method for more accurate rating of grain size was introduced by Jefferies but has been replaced by the more convenient intercept method. In the planimetric method, the number of grains within a circle or rectangle with an area of 5000 mm^2 is counted. The ASTM grain size number can then be computed from Equation (27), in which N_A is obtained from the expression given below:

$$N_A = M^2(n_1 + 0.5n_2)/5000 \tag{33}$$

where M is the magnification of the microstructure or of the photomicrograph, n_1 is the number of grains included completely within the measured area, and n_2 is the number of grains intercepted by the circumference of the test area. A minimum of 50 grains per given area is required. For higher orders of accuracy, areas containing 500–1000 or more grains may be used.

5. Heyn Intercept Method

In the intercept method, the number of grains intercepted by test lines of known length is counted, and the average intercept length of grains can be obtained from the expression.

$$\bar{L}_3 = \frac{1}{N_L} = \frac{1}{P_L} = \frac{L_T}{NM} \tag{34}$$

where $\bar{L}_3$ is the average intercept length of grains in mm at $1\times$; N_L is the number of grains intercepted by unit length of test line in mm^{-1}; P_L is the number of intersections of grain boundaries cut by unit length of test line in mm^{-1} (for closed test lines, $P_L = N_L$); L_T is the total length of test lines in mm; N is the number of intersections of grain boundaries cut by test lines of length L_T; and M is the magnification of the microstructure or photomicrograph.

Since the ASTM grain size number was initially computed from N_A, and no exact relationship exists between $\bar{L}_3$ and N_A (or $\bar{A}$), a formal relation between $\bar{L}_3$ and average grain area $\bar{A}$ was assumed:

$$\bar{L}_3 = (\pi\bar{A}/4)^{1/2} \tag{35}$$

Hence,

$$N_A = 1/\bar{A} = \pi/4\bar{L}_3 \tag{36}$$

Substituting into Equation (27),

$$G = -3.3027 - 6.6439 \log \bar{L}_3 \tag{37}$$

Equation (37) is precise only for spheres and practically exact only for uniform equiaxed grains. To eliminate the problem of variable conversion factors, the relationship between ASTM grain size number and average intercept length was redefined by ASTM E112 in 1980:

$$G = 2 \log_2 (\bar{L}_0/\bar{L}_3) \tag{38}$$

where $\bar{L}_0$ is the average intercept length for $G = 0$. For the macro size scale, $\bar{L}_0 = 32$ mm at $1\times$. Substituting into Equation (38) we have

$$G = 10.0000 - 6.6439 \log \bar{L}_3 \qquad (39)$$

For the micro-size scale, $\bar{L}_0 = 32$ mm at $100\times$ or 0.32 mm at $1\times$. Then, substituting into Equation (38) we have

$$G = 10.000 - 6.6439 \log \bar{L}_3 \qquad (\bar{L}_3 \text{ in mm at } 100\times) \qquad (40)$$

or

$$G = -3.2877 - 6.6439 \log \bar{L}_3 \qquad (\bar{L}_3 \text{ in mm at } 1\times) \qquad (41)$$

Comparing Equation (41) with Equation (37), it can be seen that for the same value of $\bar{L}_3$, the ASTM grain size number G according to the redefined Equation (38) is only 0.015 smaller than that calculated from Equation (37). This difference is again insignificant.

The test pattern most appropriate for manual operation was developed by Abrams [38]. It consists of three concentric and equally spaced circles having a total circumference of 500 mm, as shown in Figure 40. The magnification is so selected that the test pattern will make approximately 100 intersections with the grain boundaries of the specimen. The number of intersections per field should be no less than 70 and no more than 150, the lower limit being set for obtaining enough accuracy and the upper limit for more efficient operation.

When the three-circle test pattern is used, the measured value of $\bar{L}_3$ is greater than that using the straight-line test pattern. This is because a straight line connecting two neighboring intercepts is always shorter than a circular arc. For example, when $N = 4$, 6, and 10, the measured values of $\bar{L}_3$ are 10, 4.7, and 1.7% larger, respectively. However, as N increases, the difference decreases significantly. For example, when $N = 18$, $\bar{L}_3$ is only 0.5% larger.

When the test pattern meets a triple point, i.e., the junction of three grains, a score of 1.5 intersections should be recorded. A tangential intersection with a grain boundary should be scored as 1 intersection.

6. Statistical Analysis of the Grain Size Measurement Data

Usually five randomly selected and widely spaced fields should be measured for every specimen with the three-circle test pattern, and the count of intercepts should be recorded separately for each of the five tests.

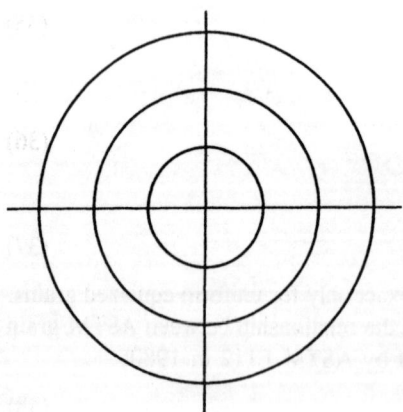

Figure 40   The three-circle test pattern for intercept count. (Reduced 40% in reproduction)

Let the measured count of intercepts be denoted by $N_1, N_2, \ldots, N_i, \ldots, N_n$. Then the standard deviation (s) and coefficient of variation (CV) of the counts and the 95% confidence limit (95% CL) of the mean $\bar{N}$ can be calculated from Equations (16)–(20), where x_i, x, and $s_{\bar{x}}$ are replaced by N_i, N, and $s_{\bar{N}}$, respectively. Since grain size is usually characterized by average intercept length $\bar{L}_3$ and/or ASTM grain size number G, the following procedures of statistical analysis should also be carried out.

Calculation of the 95% Relative Confidence Limit on $\bar{L}_3$ Since the limiting values of N are $\bar{N} \pm 2s_{\bar{N}}$ (95% CL), substituting into Equation (34) the limiting values of $\bar{L}_3$ (denoted by $\bar{L}_3^{\pm}$) can be obtained:

$$\bar{L}_3^{\pm} = \frac{L_T}{M\bar{N}(1 \mp 2s_{\bar{N}}/\bar{N})} = \frac{\bar{L}_3}{1 \mp 2s_{\bar{N}}/\bar{N}} \tag{42}$$

Averaging the difference on the limiting values of $\bar{L}_3$ and dividing by $\bar{L}_3$, the 95% RCL on $\bar{L}_3$ is obtained:

$$\frac{\Delta\bar{L}_3}{2\bar{L}_3} = \frac{\bar{L}_3^+ - \bar{L}_3^-}{2\bar{L}_3} = \frac{2s_{\bar{N}}/\bar{N}}{1 - (2s_{\bar{N}}/\bar{N})^2} \tag{43}$$

From Equations (17) and (19), we have

$$\frac{s_{\bar{N}}}{\bar{N}} = \frac{CV}{(n-1)^{1/2}} \tag{44}$$

Substituting into Equation (43), we can get 95% RCL on $\bar{L}_3$ in terms of the CV of the count and number of tests.

$$95\% \text{ RCL on } \bar{L}_3 = \frac{2(n-1)^{1/2}(CV)}{(n-1) - 4(CV)^2} \tag{45}$$

Usually, after determining the value of CV of the count, the 95% RCL on $\bar{L}_3$ can be obtained directly from the chart in ASTM E112 at a given value of CV and certain numbers of tests ($n = 5, 10, 15, 20, 26, 37, 65, 101$).

Calculation of the 95% Confidence Limit on $G(\Delta G)$ Substituting the limiting values of $\bar{N}$ (95% CL) into Equation (41), the limiting values of G (denoted by $G^{\pm}$) can be obtained:

$$G^{\pm} = -3.2877 - 6.6439 \log\left[\frac{L_T}{M(\bar{N} \pm 2s_{\bar{N}})}\right] = G + 6.6439 \log\left(\frac{1 \pm 2s_{\bar{N}}/\bar{N}}{\bar{N}}\right) \tag{46}$$

Hence, the 95% CL on G can be obtained:

$$\Delta G = (G^+ - G^-)/2$$
$$= 3.3219 \ \log[(\bar{N} + 2s_{\bar{N}})/(\bar{N} - 2s_{\bar{N}})]$$
$$= 3.3219 \ \log\left[\frac{(n-1)^{1/2} + 2(CV)}{(n-1)^{1/2} - 2CV}\right] \tag{47}$$

Equation (44) is also used in obtaining the final form of Equation (47).

Usually, after determining the value of CV of the count, the 95% CL on G can be obtained directly from the chart in ASTM E112 at a given value of CV and certain numbers of tests ($n = 5, 10, 15, 20, 26, 37, 65, 101$).

Calculation of Necessary Number of Tests According to Required 95% CL on G   If the calculated value of 95% CL on G is larger than required, it is possible to calculate the number of tests n from Equation (47) at a given value of CV and required 95% CL on G.

From Equation (47), the number of tests n can be expressed as a function of ΔG (95% CL) and CV:

$$n = 4\left(\frac{Z+1}{Z-1}\right)^2 CV^2 + 1 \tag{48}$$

where $Z = 10^{\Delta G/3.3219}$.

For example, if the required accuracy of grain size scale is $0.5G$, then ΔG (95% CL) should be 0.25, and substituting into Equation (48) yields

$$n = (23.14\ CV)^2 + 1 \tag{49}$$

From the chart in ASTM E112, one can also find the approximate necessary number of tests at a given CV value and the required ΔG (95% CL), because in the chart of ΔG versus CV, the curves of only eight different numbers of tests are given.

The following example illustrates procedures for determining confidence limit from grain size measurement. The data are taken from ASTM E112-84.

The five intercept counts with a three-circle test pattern at a magnification of 200 are $N = 92, 78, 109, 74$, and 117. It follows that

$$\Sigma N_i = 470 \quad \text{and} \quad \bar{N} = \Sigma N_i/n = 94.0$$

1. From Equation (34), the average intercept length $\bar{L}_3 = 0.0266$ mm or 26.6 μm, and the corresponding ASTM grain size number can be obtained from Equation (41), i.e., $G \approx 7.177 - 7.2$.
2. Using Equation (16) or a pocket calculator, the standard deviation of the count $s = 18.80$, and from Equation (17) the coefficient of variation (CV) = 0.20 or 20%.
3. From Equation (20), the 95% confidence limit of the mean is $2s/(n-1)^{1/2} = s = 18.80$. Therefore, the limiting values of $\bar{N}$ (95% CL) are 94.0 ± 18.80.
4. From Equation (45), the 95% relative confidence limit on $\bar{L}_3$ is found to be 0.2083 or 20.83%. Therefore, the limiting values of $\bar{L}_3$ (95% CL) are 0.0266 (1 ± 0.2083) = 0.0266 ± 0.0055 mm.
5. From Equation (47), the 95% confidence limit on G is found to be 0.5850. Hence, the limiting values of G (95% CL) are 7.2 ± 0.6.
6. If the required accuracy of grain size scale is 0.5 G, then ΔG (95% CL) should be 0.25. Hence, we can use Equation (49) for CV = 0.20, and the necessary number of tests is found to be $n = 22.4 \approx 23$; i.e., 18 more tests are needed.
7. Suppose 21 more tests are made, i.e., $n = 26$, and the mean of the counts $\bar{N} = 101$ instead of 94. The corresponding average intercept length $\bar{L}_3 = 0.0248$ mm, and ASTM grain size number $G = 7.385 \approx 7.4$; the new CV is known to be 0.19. From Equation (45), the 95% RCL on $\bar{L}_3$ is found to be 0.0764. Hence, $\Delta \bar{L}_3 =$

Table 5 Test Results for $n = 5$ and 26

Number of tests n	Average intercept length $\bar{L}_3 \pm \Delta\bar{L}_3$ (95% CL) (mm)	ASTM grain size number	
		$G \pm \Delta G$ (95% CL)	Rounded to the nearest 0.5 G
5	0.027 ± 0.006	7.2 ± 0.6	7.0
26	0.025 ± 0.002	7.4 ± 0.2	7.5

$0.0764 \times 0.0248 = 0.0019 \approx 0.002$ mm. The limiting values of $\bar{L}_3$ are 0.025 ± 0.002 mm. From Equation (47), the 95% CL on G is found to be 0.22. Hence, the limited values of G are 7.4 ± 0.2.

8. If the test result is rounded to the nearest $0.5G$, then, for $n = 5$, $G = 7.0$, and for $n = 26$, $G = 7.5$.

Table 5 lists the test and calculated results for $n = 5$ and 26.

It should be mentioned that all of the results calculated above can be obtained or approximately obtained from the charts in ASTM E112 once the number of tests n and the coefficient of variation (CV) are known.

F. Automatic Image Analysis[3]

As described in earlier sections, quantitative metallography using manual methods has the advantage of convenience at much lower cost because no sophisticated instrument other than a microscope is needed. Besides, the feature recognition ability of the human eye is far superior to that of an image analyzer. However, manual operation is much slower, and it is also tedious, causing operator fatigue.

With the advent of the automatic image analyzer and its continuous improvement and perfection over recent decades, this computer-aided alternative to manual image analysis has occupied a firm and increasingly important place in research, quality assurance, and process control. Figure 41 shows a modern image processing and analysis system. On the left is a microscope with an attached video camera on the trinocular tube; in the middle is a powerful industry standard PC, and on the right the monitor, keyboard, and mouse.

Automatic image analysis consists of the four main steps: image acquisition, image processing, detection of image details, and quantitative measurements.

1. Image Acquisition

The image to be analyzed usually comes from an optical microscope. To obtain the best image contrast from different types of specimens, various methods of illumination, e.g., brightfield and darkfield, polarized light, DIC, and phase contrast illumination, should be provided.

The video camera attached to the optical microscope converts the optical image into a digitized image, which has a discrete number of pixels (picture elements), each containing an intensity value, usually scaled from 0 (black) to 255 (white). Camera resolution is determined by the number of pixels, with 512×512 being the most common, although 1024×1024 is sometimes available for high-resolution cameras that can be used for analyzing large and small features at the same time. It is self-evident that the resolution of the microscope is also a vital importance in determining the overall resolution of the imaging system. Digitized images can be saved on disk for later processing and analysis or just for archival purposes.

---

[3]This section is based on References 39 and 40.

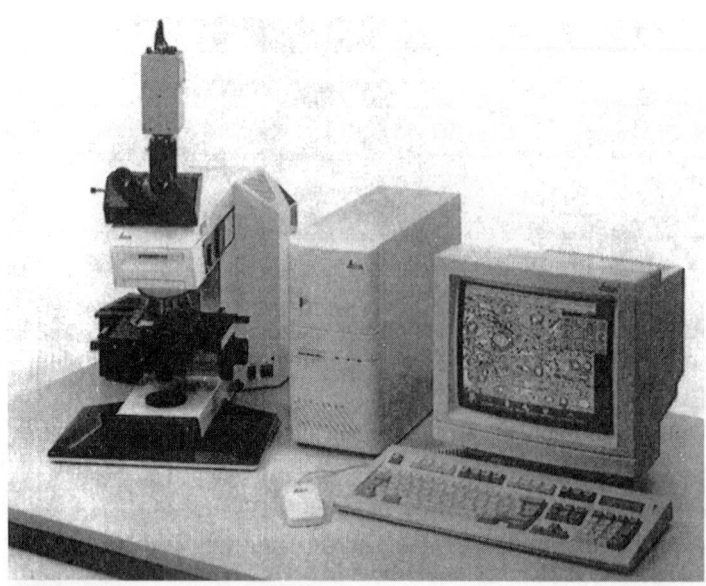

Figure 41 The Quantimet 500+ Image Processing and Analysis System. (Courtesy of Leica Cambridge Ltd.)

2. Image Processing

Image processing, or image enhancement, is employed to enhance the image details and thus improve the image contrast. To extract maximum information from the image, it is necessary that an image analyzer be equipped with a rich library of image processing capabilities, so that the image can be modified in a controlled manner. Any image processing operation can be shown on the screen and the results displayed in a window. The one operation that is applied to almost every image is the removal of noise and nonuniformity of illumination.

Although the image is changed to some extent as a result of image processing operations, no additional information can be gained that was not present in the original image.

The following is an example of image processing procedures used for grain sizing. Although it is quite easy to determine the ASTM grain size number of an etched single-phase specimen using the manual intercept method as described in earlier sections, automatic grain sizing does encounter difficulties so that some of the following image processing procedures are needed prior to automatic image analysis [40].

1. *Grain boundary detection* Pixels are identified as grain boundary or grain based on local image characteristics. If the grains are lighter or darker than the boundaries as in the case of brightfield or darkfield illumination, respectively, thresholding defines the boundary. If contrast etch is used to delineate grains, then edge-enhancement algorithms must be used prior to thresholding.

2. *Inclusion removal* Inclusions may be mistaken for incomplete segments of grain boundaries. Therefore, inclusions smaller than a specified size should be eliminated.

3. *Grain boundary completion or reconstruction* This is the most important procedure in automatic grain sizing. The absence of even a single grain boundary pixel may result in erroneous counting. For example, two neighboring grains may be regarded as one grain if a pixel of their common grain boundary is absent. Figure 42 shows the microstructure of a cold-rolled ferritic steel (a) before and (b) after image processing. The measured average grain areas in (a) and (b) are 35.21 and 25.01 mm, respec-

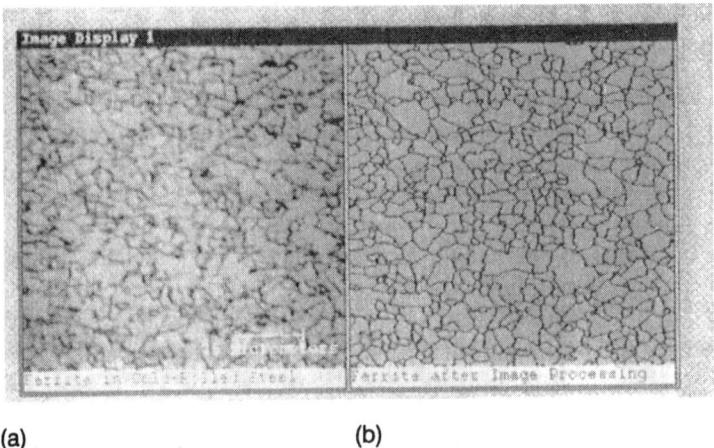

(a)   (b)

Figure 42 Microstructure of a cold-rolled ferritic steel (a) before and (b) after image processing. (Courtesy of Gamma-Tech, Inc.)

tively, which correspond to ASTM grain size numbers of 11.8 and 12.3, respectively. Using the intercept method, the manually measured ASTM grain size number of the same field is 12.4, which is very close to that of automatic image analysis after image processing.

4. *Grain boundary thinning*   This is an operation for minimizing the area fraction of grain boundaries. It is well known that grain boundaries are only a few atomic diameters in width, but the etched grain boundaries observed under the microscope are much thicker. During manual grain sizing the intersections of test lines with grain boundaries are considered points; hence, the thickness of grain boundaries is not a problem. However, in a digital image, the area occupied by grain boundaries cannot be ignored, particularly at small grain sizes. The goal of grain boundary thinning is to minimize the boundaries to a width of 2 pixels. For example, in a grain structure of $G = 6$, the area fraction of the grain boundaries can be reduced from 19% to 8% by a boundary-thinning routine. According to Vander Voort and Friel [41], under these conditions each grain will still be undersized slightly, causing an insignificant increase of about 0.2 in ASTM grain size number.

5. *Detection and removal of twin boundaries*   Annealing twins are often found in certain face-centered cubic metals and alloys, e.g., austenitic stainless steels. During automatic grain sizing, the twin boundaries should not be mistaken for grain boundaries. They should be detected and removed before measurements are made. In austenitic stainless steels most of the twin boundaries are straight lines. Using an artificial intelligence approach [41], this type of twin boundary can be detected according to the straightness of the boundary and its length, and hence can be removed. However, such straight twin boundaries become curved after cold deformation and cannot be identified. Likewise, lenticular deformation twins such as Neumann bands in ferritic grains (Figure 28) cannot be detected.

3. Detection of Image Details

Image details or features to be measured are selected by applying thresholds to highlight the features of interest. In the case of a multiphase specimen, several thresholds are set, one for each phase. Two phases can be differentiated only if they have different gray levels; therefore, the etching step during specimen preparation is critical.

As stated earlier, the human eye can discriminate only about 20 levels of intensity at the same time, but it can distinguish hundreds of different shades of color. Therefore, pseudocolor can be used in an image to show details that would be lost in the gray scale image. Also, a pseudocolor display of the image gives more visual comfort; even detection on the basis of gray level is satisfactory. However, true color facilities should be used when color is used as the basis for the identification and detection of features.

4. Quantitative Measurements

Usually an image analyzer is capable of providing a generous range of measurement parameters that covers a wide spectrum of applications. In field measurements, each individual feature is measured separately, and the results are summed for each field of view. In feature measurements, multiple features are measured, and a cumulative statistical distribution of results for all features in the measured fields is presented. Results can be printed or filed to disk in the most suitable format.

To meet the needs of the increasing use of automatic image analysis, three ASTM metallographic test standards were adopted in 1986–1990 [37].

> ASTM E1112-86: Standard Practice for Obtaining JK Inclusion Rating Using Automatic Image Analysis
>
> ASTM E1245-89: Standard Practice for Determining Inclusion Content of Steel and Other Metals by Automatic Image Analysis
>
> ASTM E1382-91: Standard Test Methods for Determining the Average Grain Size Using Semiautomatic and Automatic Image Analysis.

REFERENCES

1. J. A. Nelson, A new direction for metallography, *Pract. Metallogr. 26*:225 (1989).
2. P. Wellner, Investigation on the effect of the cutting operation on the surface deformation of different materials, *Pract. Metallogr. 17*:525 (1980).
3. G. F. Vander Voort, *Metallography—Principles and Practice*, McGraw-Hill, New York, 1984.
4. T. Palmer, Development of methods for automatic preparation, Structure 24, *Struers, J. Materialogr. 2*:3 (1991).
5. R. Hoeg and I. Liebl, *Metal Prog. 120*(3):66 (1981).
6. R. Hoeg and I. Liebl, *Pract. Metallogr. 19*:391 (1982).
7. S. D. Glancy et al., A study in the dynamics and wear of rigid disc system, Structure 25, *Struers J. Materialogr. 1*:3 (1992).
8. C. Ekstrøm, Diamonds don't last forever—if a metallographer gets hold of them!, Structure 21, *Struers Metallogr. News 1*:14 (1990).
9. G. F. Vander Voort, Polishing with colloidal silica, Structure 26, *Struers J. Materialogr. 2*:3 (1992).
10. J. A. Nelson, *New Abrasives for Metallography* (Microstruct. Sci. Vol. 11), Elsevier, New York, 1983, pp. 251–261.
11. G. H. Schwuttke and A. Oster, Damage removal on silicon surfaces: A comparison of polishing techniques. Tech. Rep. 7, Part 1 of Damage Profiles in Silicon and Their Impact on Device Reliability, IBM Corp., Ref. No. TR 22, 1989, January 1976. (From Ref. 9.)
12. J. H. Richardson, *Optical Microscopy for the Materials Sciences*, Marcel Dekker, New York, 1971, pp. 432–449.
13. E. Beraha and B. Shpigler, *Color Metallography*, American Society for Metals, Materials Park, OH, 1977.
14. E. Weck and E. Leistner, Molybdic etching reagent—almost universal, Structure 11, *Struers Metallogr. News*, August 1985, pp. 11–16.
15. G. F. Vander Voort, Tint etching, *Metal Prog. 127*(4):31–41 (1985).
16. F. M. Krill, *Metal Prog. 70*(1):81 (1956).
17. E. L. Long, Jr. and R. J. Gray, *Metal Prog. 74*(4):145 (1958).
18. P. Rothstein and F. R. Turner, Metallographic specimen preparation by vibratory polishing, in *Symposium on Methods of Metallographic Specimen Preparation*, ASTM STP 285, ASTM, Philadelphia, 1960, pp. 90–102.
19. H. Modin and S. Modin, *Metallurgical Microscopy*, Butterworths, London, 1973, pp. 61–66.

20. O. Goldberg, Koehler illumination, *Microscope* 28:15 (1st qtr 1980).
21. P. Euteneuner, A. Muller-Rentz, and K.-H. Schade, DELTA—The new system of microscope optics from Leica, *Sci. Tech. Inf. 10*(4): 114–122 (1992).
22. W. Lang, *Nomarski Differential Interference Contrast Microscopy* (Reprint), Collection of four articles from Zeiss Information, S 41-210.2-5-e, 1979.
23. E. Voogel and P. Keyzer, *200 SLR Tips*, Fountain Press, Windsor, 1984, p.1.
24. L. Stroebel and H. N. Todd, *Dictionary of Contemporary Photography*, Morgan & Morgan, Dobbs Ferry, NY, 1974, p. 49.
25. M. J. Langford (Consulting Editor), *The Camera Book*, Ziff-Davis, New York, 1980, pp. 152–173.
26. L. Ericksenn and E. Sincebough, *Adventures of Closeup Photography*, AMPHOTO, New York, 1983, p. 10.
27. E. E. Underwood, Quantitative metallography, in *Metals Handbook*, 9th ed., Vol. 9, *Metallography and Microstructures*, ASM, Materials Park, OH, 1985, pp. 123–134.
28. H. Elias, Introduction: problems of stereology, in *Stereology* (Proc. 2nd Int. Congr. Stereology, Chicago, Apr. 8–13, 1967), H. Elias, Ed., Springer-Verlag, Berlin, 1967, pp. 1–11.
29. R. E. Miles, Multidimensional perspectives on stereology, *J. Microsc. 95*(2):181–195 (1972).
30. E. E. Underwood, *Quantitative Stereology*, Addison-Wesley, Reading, MA, 1970.
31. R. T. DeHoff and F. N. Rhines (Eds.), *Quantitative Microscopy*, McGraw-Hill, New York, 1968.
32. E. R. Weibel, *Stereological Methods*, Vol. 1, *Practical Methods for Biological Morphometry*, Vol. 2; *Theoretical Foundations*, Academic, New York, 1978, 1980.
33. J. E. Hilliard, Measurement of volume in volume, in *Quantitative Microscopy*, R.T. Hoff and F. N. Rhines, Eds., McGraw-Hill, New York, 1968, p. 45.
34. R. T. DeHoff, The statistical background to quantitative metallography, in *Quantitative Microscopy*, R.T. Hoff and F. N. Rhines, Eds., McGraw-Hill, New York, 1968, pp. 11–44.
35. R. T. DeHoff, Quantitative metallography, *Techniques of Metals Research*, Vol. II, Part 1, Interscience, New York, 1968, pp. 221–253. Quoted from Reference 3, p. 503.
36. F. N. Rhines, Geometry of microstructures. Part I, *Metal Prog. 112*(3):60 (1977).
37. G. F. Vander Voort, Progress in metallographic standards, *Adv. Mater. Process. 138*:30 (November 1990).
38. H. Abrams, Grain size measurements by the intercept method, *Metallography 4*:59 (1971).
39. K. A. Leithner, Basics of quantitative image analysis, *Adv. Mater. Process 144*(5):18–23 (1993).
40. J. J. Friel and E. B. Prestridge, Grain sizing by image analysis, *Adv. Mater. Process. 139*(2):33–37 (1991).
41. G. F. Vander Voort and J. J. Friel, Image analysis measurements of duplex grain structure, *Mater. Charact. 29*:293–312 (1992).

14

MECHANICAL PROPERTY TESTING METHODS

D. Scott MacKenzie
Consultant, Villa Ridge, Missouri

I. INTRODUCTION

Mechanical testing is necessary to ensure that components will not fail in service. Testing is performed so that the designer can predict the performance of a part or a component in the field. In this chapter, a brief description of design theories and selected testing methods that are available to the designer to predict component serviceability under realistic conditions are provided. Often, mechanical testing is done for quality control. Since mechanical testing is a dynamic field, new tests are continually being devised, and older tests are being revised, and reinterpreted, for cost or performance reasons. Therefore, a review of the literature should be conducted to conserve materials, achieve reliability, and minimize environmental impact.

The prediction of the behavior of a stressed material is dependent on the applied loading and the relative magnitudes of the principal stresses and strains. The material can fail in either a ductile or brittle fashion. The predictive theory used is dependent on the type of loading and the expected response.

When a component is loaded, it responds either elastically or plastically depending on the material, amount of stress, strain rate, and geometry.

In elastic deformation, there is a temporary change in the distances between atoms or crystallographic planes. In plastic deformation, there is a permanent change in the relative position of the atoms. This is caused by the displacement of atoms within the crystal lattice.

If the loads are sufficiently high to cause failure or fracture, the material behaves in either a brittle or ductile fashion. Gensamer [1] summarized the terms used to characterize fracture. These terms and the described behavior are shown in Table 1.

A. Types of Loading

The method of loading affects the manner in which a body reacts under stress. A body can experience four types of loading:

1. Static load, short time. The load is gradually applied until failure occurs. The loads on the body are in equilibrium. A tensile test is typical of this type of loading.
2. Static load, long duration. The load is applied gradually until it reaches a maximum and is then maintained. Creep testing is done in this manner.

Table 1   Terms Used to Describe Fracture Surfaces

Crystallographic mode:	Shear, cleavage
Appearance:	Fibrous, granular
Strain to fracture:	Ductile, brittle

3. Dynamic loading.   The load is applied rapidly, and the loads do not have time to reach equilibrium within the body. The part momentum is considered as the part is accelerated. The stresses are considered static but are applied in a short time impulse. Charpy impact testing is typical of this type of loading.

4. Repeated loading.   Loads are applied to the body repetitively, with the load being either completely or partially removed after each cycle. There are also two main types of loading: high stress applied for a few cycles or low stress applied for many cycles.

Loadings of various types may be applied at the same time or sequentially. The order is unimportant unless it is necessary to determine the source of failure. The local stress distribution and the relative magnitudes of the stresses and strains present are also important.

B. Stress and Strain

1. Plane Stress

If a small differential body is stressed (Figure 1), there are three main stresses normal to the faces of the cube (σ_x, σ_y, and σ_z) and six shear stresses along each face (τ_{xy}, $\tau_{xz}\tau_{yz}$, τ_{yx}, τ_{zx}, and τ_{zy}). If σ_z is much smaller than σ_x and σ_y, then the stress state is a condition called *plane stress*, since the normal stresses of significant magnitude lie in only one plane.

Assume that a triaxial stress state occurs at a uniaxially loaded notched plate (Figure 2). Because no load is applied to the surface or at the root of the notch, the stress in the z direction is zero. Stress in the z direction rises to a maximum at the center of the plate. If the plate is thin, then the increase in this stress is small relative to stress in the other directions. A plane stress condition exists (Figure 1b). This is very common in thin materials such as sheets.

While the stress in the z direction is small compared to that in the x and y directions, there is a large maximum shear stress that lies at a 45° angle to the load direction. Plastic deformation occurs by shear along this inclined plane, creating a shear-lip or slant fracture.

2. Plane Strain

If the strain ε in one direction is much smaller than in the other two directions, then a condition of plane strain exists (Figure 1c). If the notched plane is thick, then the strain in the z direction is constrained, resulting in a plane strain condition. Because strain in the z direction is limited, the plastic zone at the notch tip is small. Therefore the axial stresses at the notch tip are large, and the shear stresses are small. Since the axial stresses are much larger than the shear stresses, the axial stresses control the fracture. A flat fracture results that is normal to the axial stresses.

II. PLASTICITY AND DUCTILE FRACTURE

When a material behaves in a ductile or plastic manner, gross deformation occurs. This may manifest as a failure to obey Hooke's law throughout the load history or as plasticity in an initially elastic region, as is typical in a common tensile test. It occurs by portions of grains sliding over one another. Energy is absorbed by the plastic deformation. Ductile fractures are

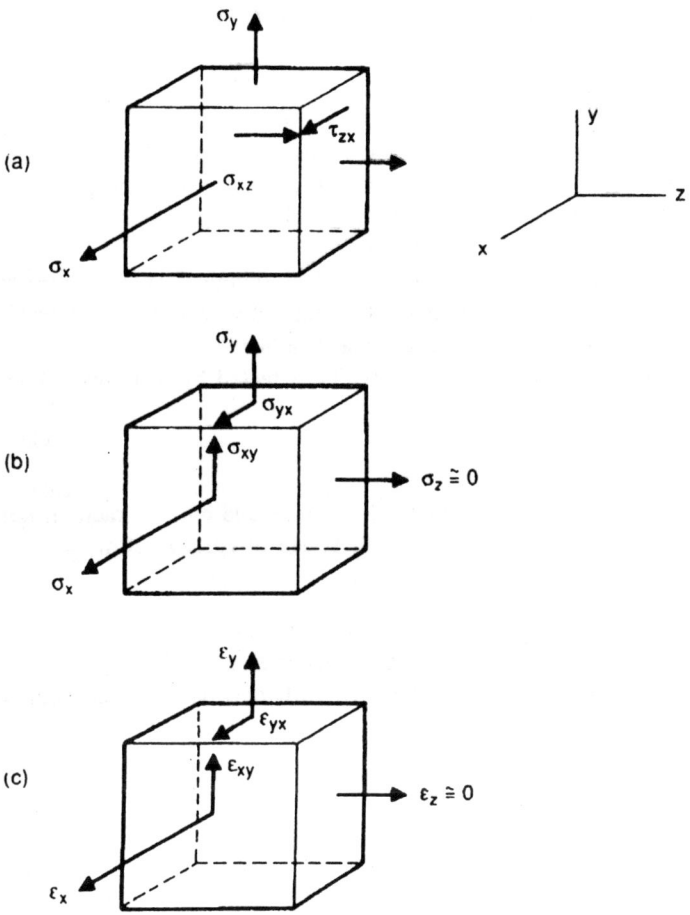

Figure 1 (a) Principal stresses on a small differential element. (b) Plane stress. (c) Plane strain.

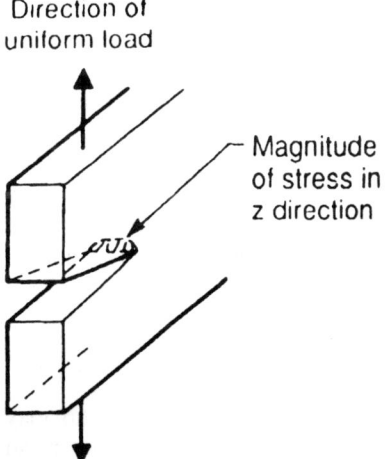

Figure 2 Stress distribution in a notched thin plate, showing stress distribution at notch.

characterized by yielding along the edges of the fracture. The final fracture is usually at 45°
to the original plane of fracture.

A. Ductile Fracture Appearance

On a macroscopic scale, a ductile fracture is accompanied by a relatively large amount of plastic
deformation before the part fails. After failure, the cross section is reduced or distorted. Shear
lips are observed at the latter part of the fracture and indicate the final failure of the part. The
fracture surface is dull, with a fibrous appearance.

   Microscopically, a ductile fracture surface is characterized by dimples. These are voids
that join internally under load. The formation of dimples is associated with slip along crystallographic planes, with decohesion at second-phase particles such as inclusions or precipitates.
This is shown schematically in Figure 3. Dimples can only be detected by using electronic
microscopy (see Figure 4).

B. Theories of Plasticity and Ductile Fracture

Many theories have been put forth in an attempt to explain plasticity and ductile fracture, but
only a few have survived rigorous examination. Three of these are discussed below.

1. *Maximum Normal Stress Theory*

According to the maximum normal stress theory, the failure of a part occurs when the largest
principal stress equals the yield strength. Consider a small material element that is loaded as
in Figure 5. If $\sigma_1 > \sigma_2 > \sigma_3$, then failure occurs when $\sigma_1 = \sigma_y$, where σ_y is the yield strength

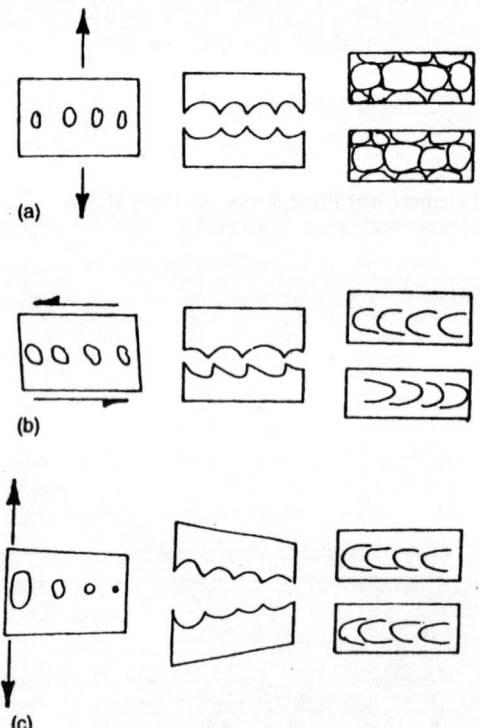

Figure 3 Schematic representation of the creation of dimples in a loaded member by (a) simple tension, (b) shear loading, and (c) tearing.

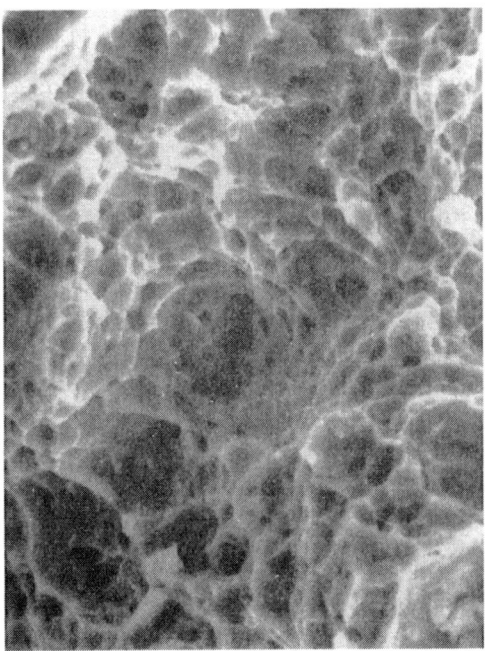

Figure 4 Scanning electron photomicrograph of dimples in a high strength steel (5000×).

of the material. This is illustrated schematically in Figure 6. This theory is applicable for simple stress states such as tension. It is not applicable to torsion. In torsion, this theory implies that failure occurs when $\tau_{max} = \sigma_y$. This does not fit experimental data.

2. Maximum Shear Stress Theory

The maximum shear stress theory is easy to use and errs on the conservative side. It has been used as the basis for many design codes and adequately predicts yielding. This theory says

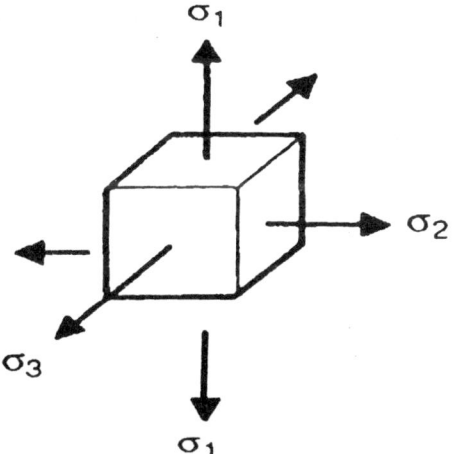

Figure 5 Principal stresses on a small differential element.

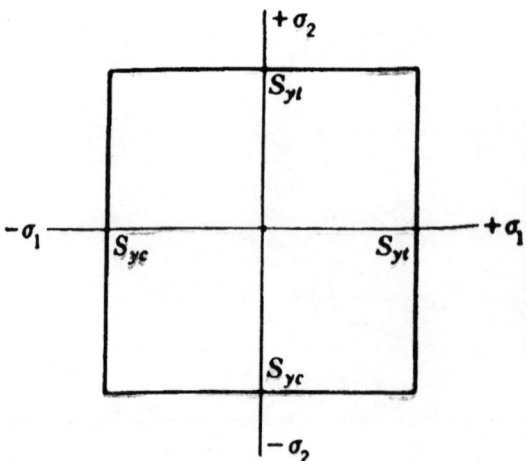

Figure 6 The maximum normal stress theory of failure based on the yield strength of a material.

that yielding occurs whenever the maximum shear stress in a small element is equal to the maximum shear stress in a tensile specimen at the onset of yielding.

If the Mohr's circle for a tensile test is considered (Figure 7), then yielding occurs when

$$\tau_{max} = \sigma_y/2 \tag{1}$$

For a three-dimensional mechanical element, the shear stresses are

$$\tau_{12} = \frac{\sigma_1 - \sigma_2}{2}, \qquad \tau_{23} = \frac{\sigma_2 - \sigma_3}{2}, \qquad \text{and} \qquad \tau_{13} = \frac{\sigma_1 - \sigma_3}{2} \tag{2}$$

The largest of the above stresses is designated τ_{max}; whenever $\tau_{max} = \sigma_y$, then yielding will occur. This is illustrated in Figure 8. The maximum normal stress theory and the maximum shear stress theory are identical when the principal stresses have the same sign.

3. Distortion Energy Theory (von Mices–Hencky Theory)

The von Mices–Hencky theory is the best theory to use for ductile materials because it defines accurately the beginning of yielding. This theory was originally proposed by von Mices

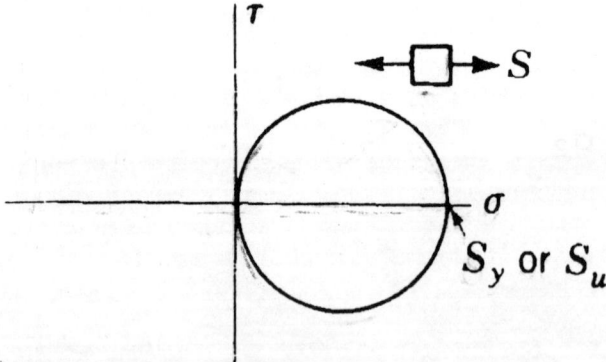

Figure 7 Mohrs circle for a tensile test.

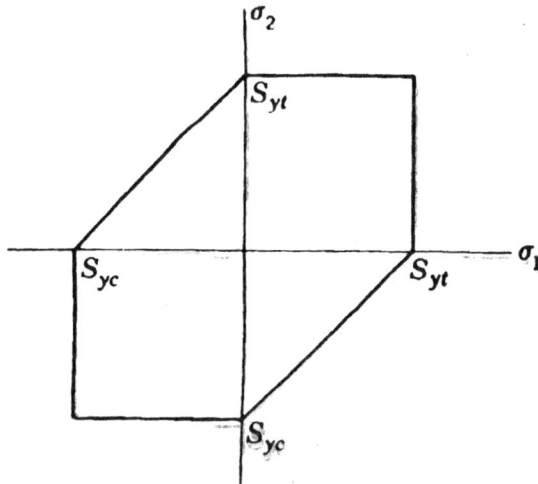

Figure 8 The maximum shear stress theory of failure.

[2] in 1913. In essence, the von Mices yield criterion states that all the principal (tensile and shear) must be considered to create the von Mices stress, σ_0:

$$\sigma_0 = \frac{1}{\sqrt{2}}[(\sigma_1 - \sigma_2)^2 + (\sigma_2 - \sigma_3)^2 + (\sigma_3 - \sigma_1)^2 + 6(\tau_{12}^2 + \tau_{23}^2 + \tau_{13}^2)^{1/2} \qquad (3)$$

Yielding is predicted whenever the von Mices stress $\sigma_0 = \sigma_y$. If the stress state is biaxial ($\sigma_3 = 0$) and the member is in torsion ($\sigma_2 = -\sigma_1$ and $\tau = \sigma_1$), then Equation (3) shows that

$$\tau_{max} = 0.577\sigma_y \qquad (4)$$

This definition of yielding is not dependent on any normal stress or shear stress but depends on all components of the stress. Because the terms are squared, it is independent of the direction of the stress. It is also not necessary to know the largest principal stress to learn if yielding will occur.

4.   A Comparison

The three yielding criteria are shown schematically in Figure 9. Comparing Equation (4) from the distortion energy theory and the maximum shear stress theory Equation (1), the distortion energy theory predicts that significantly higher stresses must be experienced before yielding occurs. The maximum normal stress theory predicts results equivalent to those predicted by the maximum shear stress theory whenever the directions of the principal stresses are the same but fails to accurately predict yielding when the signs of the stresses are opposite. The maximum shear stress theory always gives conservative results. Of the three theories, the distortion energy theory predicts yielding with the greatest accuracy in all four quadrants. The maximum normal stress theory should not be used in the design of a component.

III.   ELASTICITY AND BRITTLE FRACTURE

Since the early 1940s there has been tremendous growth in the number of large welded structures. Many of these structures have failed catastrophically in service—most notably the "Lib-

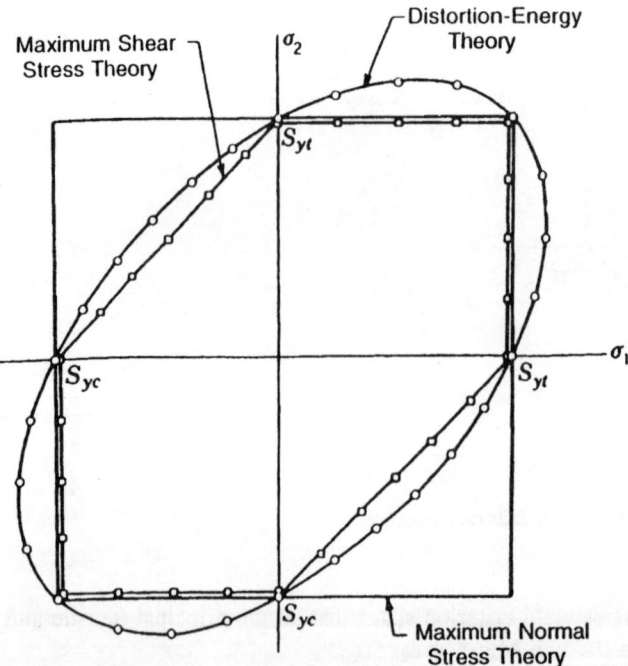

Maximum Shear
Stress Theory

Distortion-Energy
Theory

σ_2

S_{yt}

S_{yc}

σ_1

S_{yt}

S_{yc}

Maximum Normal
Stress Theory

Figure 9 Comparison of the distortion energy theory and the maximum shear stress theory.

erty ships" [3] used to transport war material during World War II. Analysis of the fracture surfaces of the failures [4] indicated that they initiated at a notch and propagated with no plastic deformation. These notches were of three types:

1. Design features. Structural members that were rigidly joined at angles less than 90° and then welded.
2. Fabrication details. Procedures used during the manufacture of the part caused the formation of notches. Welding arc strikes, gouges, and fitting procedures created physical notches. Weld procedures and heat treatment caused metallurgical or microstructural notches to occur from abrupt changes in microstructure or the production of microstructures that were brittle. Features such as porosity from welding or casting also caused brittle fracture initiation.
3. Material flaws. These flaws resulted from melt practice at the mill and appeared as large inclusions, internal oxidation, porosity, or segregation.

In brittle fractures, limited energy is absorbed by the fracture. Energy is absorbed through regions of small plastic deformation. Individual grains separate by cleavage along specific crystallographic planes.

A. Brittle Fracture Appearance

Visually, brittle fractures are characterized by little or no plastic deformation or distortion of the shape of the part. The fracture is usually flat and perpendicular to the stress axis. The fracture surface is shiny, with a grainy appearance. Failure occurs rapidly, often with a loud report. Because the brittle cleavage is crystallographic in nature, the fracture appearance is faceted (Figure 10). Often other features are present, such as river patterns [5]. These are shown in Figure 11.

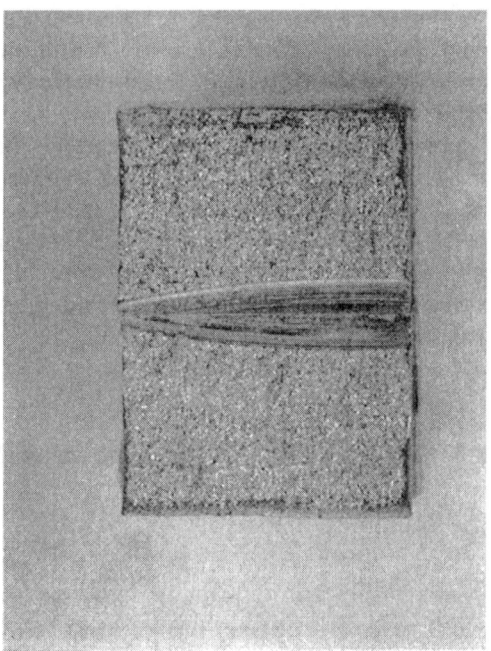

Figure 10 Fracture surface of AISI 1020 steel broken by impact at −73°C.

B. Theories of Elasticity and Brittle Fracture

When calculating stresses on a body, it is assumed that the body is elastic and homogeneous and that it conforms to Hooke's law,

$$\varepsilon = \sigma/E \tag{5}$$

where ε is the strain (in./in.); σ is the stress (lb/in.^{2}); and E is Young's modulus. However, a material is not completely elastic or completely homogeneous. It is a combination of a collection of fibers or microstructures, either randomly aligned or with an organized orientation.

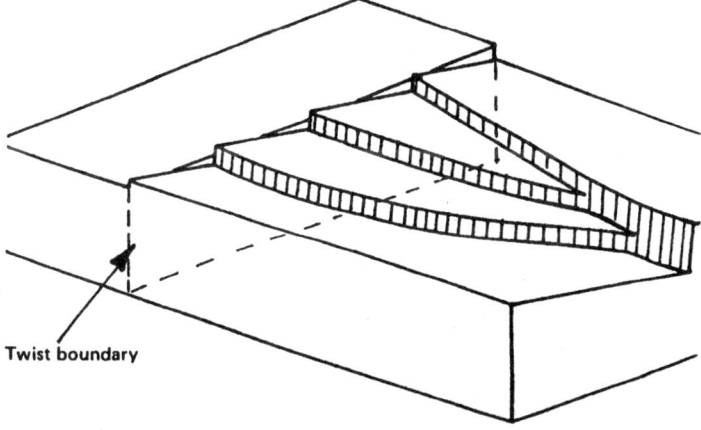

Twist boundary

Figure 11 Schematic of river patterns formed in brittle materials. (After McCall and French [5].)

This orientation can provide a directionality to the properties of a material. A material is isotropic (no directionality of properties) if the component is much larger than the constituent parts. If the organization of the constituent parts is nonrandom, the material is anisotropic and will show directionality in properties such as tensile, impact, or electrical conductivity.

If a material is perfectly elastic, there is no permanent deformation, and fracture occurs at the maximum stress, σ_{UTS}. Characteristics of brittle materials include the following. (1) They do not under go plastic deformation, i.e., $\sigma_y = \sigma_{UTS}$; (2) they follow Hooke's law until fracture; (3) their compressive strength is much greater than their tensile strength; and (4) their torsional strength is equal to their tensile strength. Many of the theories proposed to explain brittle fracture make the assumption that the material is perfectly elastic. A few of the more important theories explaining brittle fracture are described below.

1. Coulomb–Mohr Theory

The Coulomb–Mohr theory, proposed by Coffin [6], is based on tensile properties and compressive strength. It says that failure will occur for any stress state for an elastic material whenever

$$\frac{\sigma_1}{\sigma_{UTS}} + \frac{\sigma_3}{\sigma_{UCS}} \geq 1 \tag{6}$$

This theory was confirmed by Grassi and Cornet [7], who stressed gray iron cast iron tubes biaxially until failure. Note that the maximum normal stress theory produces similar results when both of the principal stresses, σ_1 and σ_2, are positive. This theory was modified by Burton [8], who produced similar results. The results are not as conservative as the Coulomb–Mohrs theory in that fracture was more closely predicted. These theories are shown schematically in Figure 12.

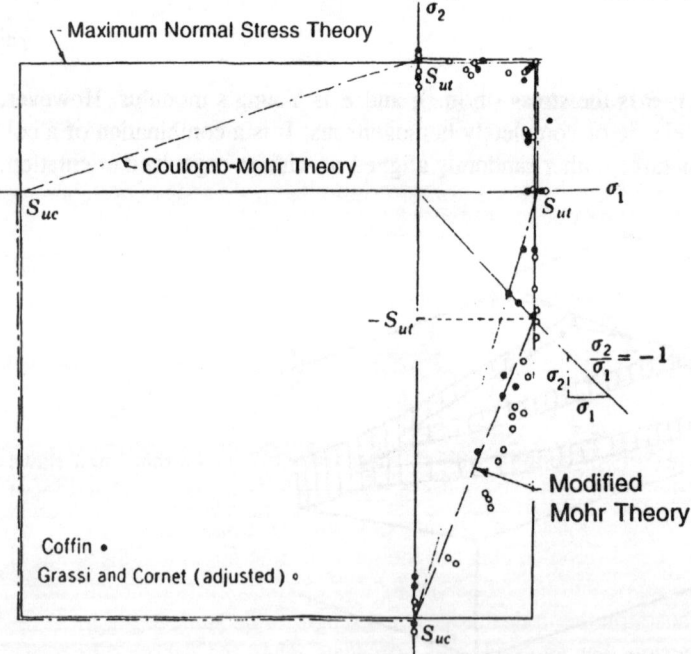

Figure 12   Coulomb–Mohr theory and maximum normal stress theory.

2. Griffith Microcrack Theory

The theoretical strength of a material is based on the cohesive force between atoms. If the cohesive strength is a sine curve, the following is obtained:

$$\sigma = \sigma_{max} \sin \frac{2\pi x}{\lambda} \tag{7}$$

where σ_{max} is the maximum theoretical cohesive strength, x is the atomic displacement due to the applied force, and λ is the wavelength of the lattice spacing. Since the changes in the atomic displacement are small, $\sin x \approx x$, resulting in a modification of Equation (7):

$$\sigma = \sigma_{max} \frac{2\pi x}{\lambda} \tag{8}$$

Since only brittle materials are being considered,

$$\sigma = E\varepsilon = Ex/\lambda \tag{9}$$

Combining Equations (9) and (8) gives

$$\sigma_{max} = \frac{\lambda}{2\pi} \left(\frac{E}{a_0} \right) \tag{10}$$

When the part fractures, two surfaces are created. Each of these surfaces has a surface energy γ_s, so that the work done per unit area in creating the fracture surface is the area under the stress–displacement curve:

$$U_0 = \int_0^{1/2} \sigma_{max} \sin \frac{2\pi x}{\lambda} \, dx = \lambda \frac{\sigma_{max}}{\pi} \tag{11}$$

Since energy is required to create two fracture surfaces,

$$\lambda = 2\pi\gamma_s/\sigma_{max} \tag{12}$$

The maximum theoretical stress from atomic cohesive forces is

$$\sigma_{max} = (E\gamma_s/a_0)^{1/2} \tag{13}$$

Using typical values for the variables above and expressing results in terms of the elastic modulus E, estimates of σ_{max} vary between $E/4$ and $E/15$. Engineering steels rarely exceed fracture stresses above 300,000 psi, or $E/100$. Common construction steels typically have a fracture stress of approximately $E/1000$. The only materials that approach the theoretical values are defect-free metallic whiskers or ceramic fibers. Therefore, this shows that small flaws or cracks are responsible for the tremendous decrease in the theoretical strength. These small flaws or cracks reduce the fracture strength due to stress concentrations [9].

Griffith [10] proposed that the difference between the theoretical strength and the strength realized in practice was due to a population of fine cracks that produce stress concentrations. These stress concentrations cause the theoretical cohesive strength to be achieved in local regions. Griffith crated the criterion that "a crack will propagate when the decrease in strain energy is at least equal to the energy required to create a new crack surface" [11]. This statement is used to establish when a flaw of a specific size exposed to a tensile stress will propagate in a brittle fashion.

Using the crack model in Figure 13, the flaw or crack is assumed to have an elliptical cross section of length $2c$. This shape is typical of many types of flaws. The thickness of the plate is very small compared to the width and length of the plate (conditions of plane stress predominate). Using the stress concentration of this elliptical crack (determined by Inglis [12]), the maximum stress at the tip of the crack is

$$\sigma_{max} = \sigma\left[1 + 2\left(\frac{c}{\rho_t}\right)^{1/2}\right] \approx 2\sigma\left(\frac{c}{\rho_t}\right)^{1/2} \tag{14}$$

where ρ_t is the radius of the crack tip. The reduction of the elastic strain energy is equivalent to

$$U_E = -\pi c^2 \sigma^2 / E \tag{15}$$

where σ is the applied tensile stress normal to the crack. Since energy is released by propagation of the crack, the term to the right of the equal sign is negative. The theoretical stress between the atomic planes needs to be exceeded at only one point, and the applied stress will be significantly lower than the theoretical stress. If the crack has a length of $2c$, and it is elliptical in shape and is relieved of stress in a roughly circular area of radius c, the increase in surface energy is made up for by the decrease in strain energy. The condition when elastic strain energy equals the increase in surface energy due to propagation of the crack is provided by

$$\frac{d\Delta U}{dc} = 0 = \frac{d}{dc}\left(4c\gamma_s - \frac{\pi c^2 \sigma^2}{E}\right) \tag{16a}$$

where

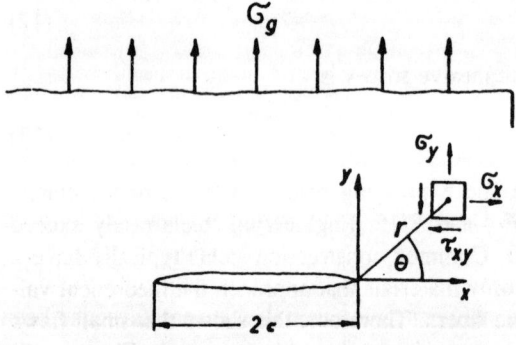

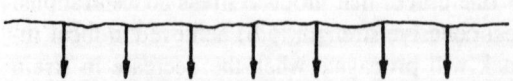

Figure 13 Griffith crack model used to determine the stress required to propagate a crack.

$$4\gamma_s - \frac{\pi c^2 \sigma^2}{E} = 0 \tag{16b}$$

This leads to the Griffith formula,

$$\sigma = \left(\frac{2E\gamma_s}{\pi c}\right)^{1/2} \tag{17}$$

which is the stress required to propagate a crack of size c in a brittle material. If the material is thick in relation to the crack, and plane strain conditions predominate, then the Griffith equation is

$$\sigma = \left(\frac{2E\gamma_s}{(1-v)^2 \pi c}\right)^{1/2} \tag{18}$$

where v is Poisson's Ratio.

IV. FRACTURE MECHANICS

Metals that fail in a brittle manner experience plastic deformation before failure [13–15]. There are three types of loading common in most engineering structures. They are shown in Figure 14. Mode I is the most common and is discussed here. Because of plastic deformation prior to fracture in metals (even when failing in a brittle manner), the Griffith microcrack theory does not apply to metals. One method of making the Griffith criterion of brittle fracture compatible with the plastic deformation evident in metals was suggested by Orowan [16]. He suggested the inclusion of a surface energy term due to the plasticity at the crack tip, γ_p. This results in a modification of the Griffith equation:

$$\sigma = \left(\frac{2E(\gamma_s + \gamma_p)}{\pi c}\right)^{1/2} \approx \left(\frac{2E\gamma_p}{c}\right)^{1/2} \tag{19}$$

The elastic surface energy term, γ_s, is neglected because $\gamma_p \gg \gamma_s$.

Irwin [17] proposed that stress at the crack tip were a function of the applied stress and the crack size. He developed the relationship

$$K = \sigma\sqrt{\pi c} \tag{20}$$

where K is the stress intensity factor. K is completely defined by the crack geometry, applied stress, and specimen geometry. The value of the stress intensity factor when unstable crack growth occurs is the critical stress intensity factor, K_{Ic} (for mode I), where the value of K_{Ic} is a material property. While the above is for an elliptical flaw, other flaw shapes have also been calculated [18,19].

This assumes that plane strain conditions have been realized. If plane stress conditions are present, then the stress is relaxed by the increased plastic zone at the crack tip. Further, the state of stress is no longer triaxial and is diminished. The cases of plane stress and plane strain for modes I, II, and III are more fully described by Hertzberg [20] and Rolfe and Barsom [21].

Toughness is a measure of the energy required to resist fracture in a material. Often this property is more important than the actual tensile properties, particularly if the part is to be

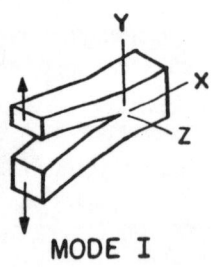

MODE I

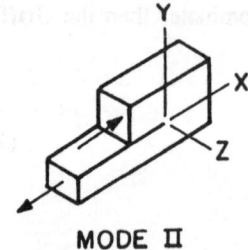

MODE II

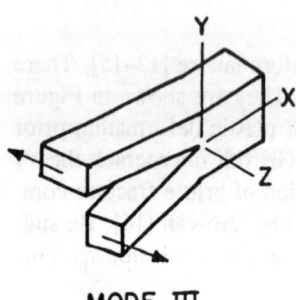

MODE III

Figure 14 Types of loading typically experienced by engineering materials.

used in a dynamic environment. The term "impact strength" is used to denote the toughness of the material. This term is actually a misnomer; it should really be "impact energy." However, the term impact strength is so established that it makes little sense to change it. Toughness is strongly dependent on the rate of loading, temperature, and the presence of stress concentrations. Several standardized tests have been developed since World War II to measure the resistance to brittle fracture, notably the Charpy V notch test [22], the dynamic tear test [23], and the plane strain fracture toughness test [24]. Essentially, these tests are attempts to quantify the behavior of the material in service and how the material is expected to fail in service. The first two are discussed more fully later.

Under impact loading, there is a limited time for uniform plastic flow to occur. Locally, the deformation may exceed the fracture stress required at the grain boundaries, geometric irregularities, or other discontinuities. Once the crack has initiated, the crack itself becomes a stress riser, propagating until it is blunted or complete failure has occurred. Fracture toughness of steels is dependent on a variety of variables that affect the mechanical properties such as test, temperature, chemistry and melt practice, strain rate, section size, notch acuity. As these variables are changed, the transition temperature between ductile and brittle behavior may change. An excellent review of the effect of processing variables on high strength steels was conducted by Thurston [25]. Over 100 papers were reviewed, and 77 references are provided.

A.   Effect of Test Temperature

In body-centered cubic metals, during impact loading, a transition from ductile to brittle fracture occurs that is dependent on temperature. The temperature at which it occurs is called the transition temperature. Other variables such as geometry, grain size, and alloying elements affect the ductile to brittle transition temperature, but only within a given alloy. The toughness versus temperature curve (Figure 15) has three basic regions: the upper shelf, the lower shelf, and the transition region. The upper shelf is characterized by primarily ductile fracture. High impact energies are associated with this regime. The lower shelf is a region where fracture is brittle; low impact energies are found in this region. The third region is the transition region, which displays a reduction in impact energies required to fracture the specimen. It is in this region that the fracture changes from ductile to brittle. Because this transition often occurs over a wide temperature range, various criteria have been developed to define the transition temperature.

One criterion of the ductile to brittle transition temperature that has common acceptance is the ductility transition temperature [26]. This standard was developed because of spectacular failures of World War II Liberty ships mentioned earlier. This standard established the acceptance criteria of Charpy V notch energy, C_v, greater than 15 ft·lb. It was learned that this value would not result in a brittle fracture at a test temperature of –40°F. However, it is not applicable for many materials that have lower shelf impact energies greater than 15 ft·lb when tested at –40°F. The transition temperature is often much lower. For example, AISI 4340 steel quenched and tempered to HRC 35 exhibits a transition temperature of approximately –150°F and at 40°F has a C_v of 60 ft·lb.

Another definition of the ductile to brittle transition temperature, and the one used in this chapter, is the fracture appearance transition temperature (FATT). For a given test, the ductile to brittle transition temperature is proportional to the plastic zone at the crack tip. As the

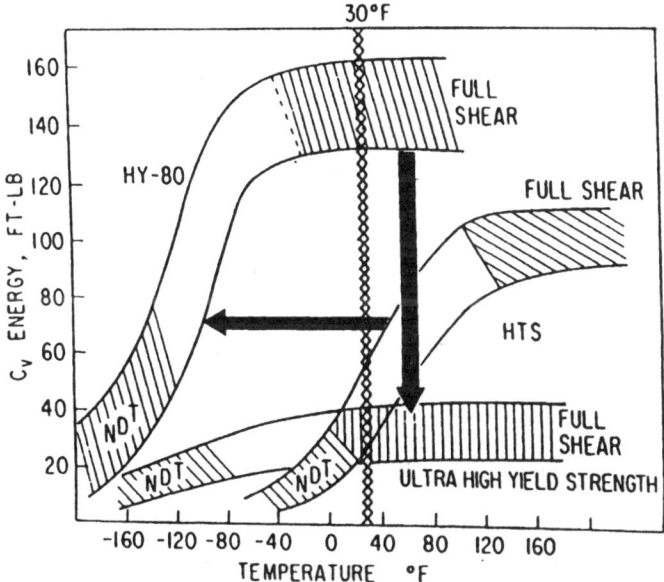

Figure 15 Typical Charpy impact curves for high strength steels.

test temperature is decreased, increasing amounts of cleavage occur. When the temperature is increased, increasing amounts of dimpled rupture occur. At the temperature at which there are equal amounts of cleavage and dimpled rupture, the temperature is known as the 50% fracture appearance transition temperature, T_{50}. This temperature is determined by plotting the amount of brittle or ductile fracture as a function of temperature. The amount of brittle or ductile fracture is usually decided by looking at the fracture surface either with the naked eye or with a low power (5×) magnifying glass. Ductile or shear fractures appear dull, and brittle fractures usually appear bright and shiny. A schematic example of the fracture surface of an impact test is shown in Figure 16.

The percentage of shear fracture in an impact test can be determined by any of several methods. The first method involves measuring the length and width of the cleavage portions of the fracture surface. The amount of ductile fracture is determined from a series of tables (assuming a standardized test geometry), depending on the type of measurement used. The amount of shear can also be measured by magnifying the fracture surface and comparing it to a calibrated overlay, or by photographing the fracture surface and measuring the ductile fracture by means of a planimeter. Another method compares the fracture surface of the specimen with a comparative fracture appearance feature chart. Because of the subjective nature

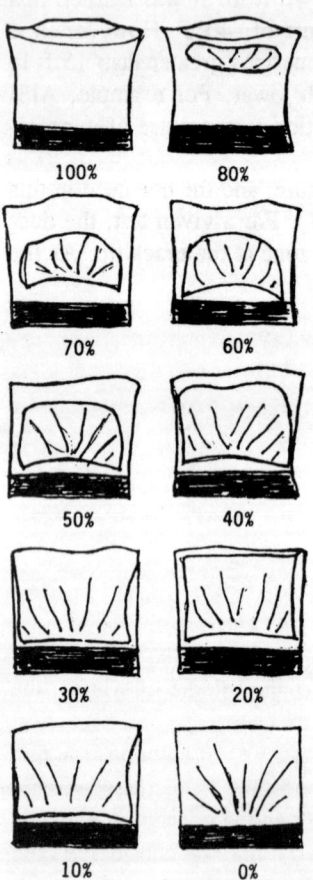

Figure 16   Schematic of Charpy impact fracture surfaces showing change in percentage of brittle fracture.

of the measurements, the transition temperature, defined by the above methods, is not recommended for acceptance testing, material specifications, or procurement documents.

B.   Effect of Chemistry, Melt Practice, and Grain Size

Changes in the ductile to brittle transition temperature of up to 100°F can be made by simple changes in the chemistry of the steel [27,28]. The greatest chemical changes are due to the addition of carbon (Figure 17) and manganese. It has been found that the transition temperature will increase by 25°F for each additional 0.1% carbon added to the steel. Cotrell et al. [29] tested a very controlled chemistry 1% Cr-Mo steel and found that the Charpy impact strength sharply decreased as the carbon content was raised from 0.30 to 0.44% C. These findings are similar to tests performed on a 43XX steel [30]. However, for an H-11 steel, the notch strength ratio increased carbon contents up to 0.43% contrary to the results of Cotrell. (Espey). Increases in the manganese content will decrease the transition temperature by 10°F for each 0.1% increase of manganese. However, these limits must be kept in mind because of hardenability constraints and the presence of retained austenite in the steel.

Other alloying additions also tend to change the ductile to brittle transition temperature in steels. Nickel decreases the transition temperature, whereas chromium has little effect. Silicon raises the transition temperature, as does phosphorus. Phosphorus in excessive amounts will also tend to form grain boundary precipitates, further increasing the transition temperature. Additions of columbium and vanadium in small amounts to the alloy decrease the transition temperature by forming carbide precipitates. These precipitates play a role in dispersion strengthening and act to pin the grain boundaries, preventing grain growth.

Sulfur greatly affects the fracture toughness of steel. Wei [31] measured the fracture toughness in a 4745 steel under controlled melting conditions at several different sulfur levels and strengths. He found that the fracture toughness decreased smoothly as the sulfur level was increased from 0.008% to 0.049%. This is shown schematically in Figure 18. Interestingly, in this study, silicon was found not to increase the fracture toughness.

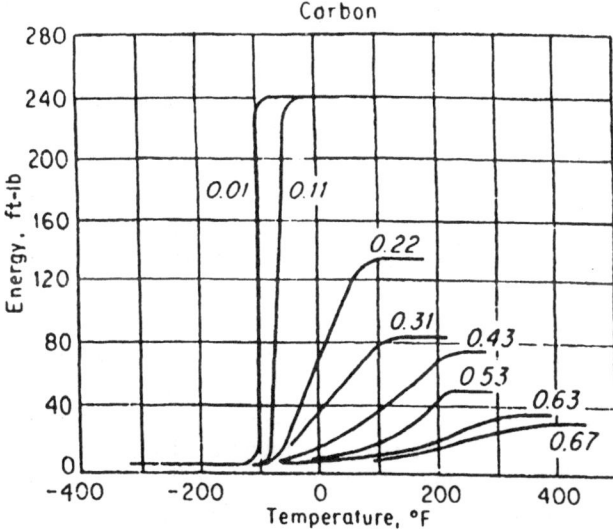

Figure 17   Effect of cabon content in steel on the impact energy and transition temperature. (After Dieter [26].)

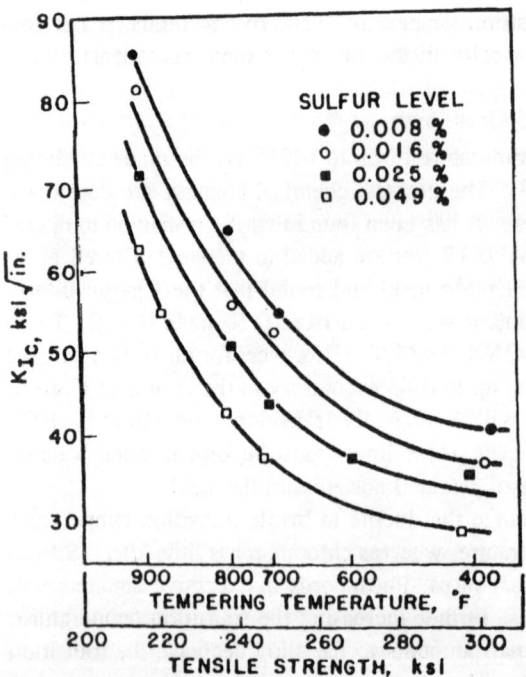

Figure 18 Charpy impact curves, showing the effect of sulfur content on an AISI 4730 steel. (After Dieter [114].)

The oxygen content of the steel also plays an important part in determining the ductile to brittle transition temperature. Rimmed steel with a high content of oxygen shows a high transition temperature. Silicon-deoxidized (semikilled) steels have lower transition temperatures, and the oxygen is tied up as inclusions. Fully killed steels with aluminum and silicon have even lower transition temperatures.

Vacuum melting produces alloys with excellent fracture toughness. This is accomplished by reducing the gas content and the number of nonmetallic inclusions. The effect varies with composition and the heat-treated tensile strength of the alloy. It is not always beneficial. Gilbert and Brown [32] reported an improvement of 100% in the transverse fracture strength, with the longitudinal fracture strength increasing by 50% over that of conventional air-melted 4340 steels heat treated to 280 ksi ultimate tensile strength. They also found that the property directionality was almost completely removed. Cotrell [33] investigated consumable electrode vacuum melting on 3%Cr-Mo-V steel (yield strength greater than 250 ksi). The fracture strength was increased by almost 20 ksi. Different results were found with HP9-4-30 maraging steel (0.30% C, 9% Ni, and 4% Co). Vacuum remelting reduced directionality but showed no increase in the notch strength over that of a conventional silicon and aluminum deoxidized heat. However, a vacuum carbon deoxidation showed significant improvements in notch strength. Tests on vacuum-remelted 4340 steel showed in excess of 100% improvement over conventional air melting when the steel was heat treated to the 280 ksi ultimate tensile strength range [34]. The fracture toughness of vacuum-remelted H-11 steel was found to increase by 10% over air-melted H-11 steel heat treated to 250 ksi ultimate strength. No effect was found when it was heat treated to 300 ksi ultimate strength.

Grain size of the steel plays an important role in the temperature of the ductile to brittle transition. Decreasing the grain size significantly decreases the transition temperature [35]. Change in processing that create or promote fine ferrite grain size, such as normalizing, with subsequent tempering will decrease the transition temperature. The rolling temperatures and other processing variables all play important roles in the finished grain size of the product and therefore the grain size. Careful manipulation of the process will yield a low transition temperature.

Decarburization has a negative influence on fatigue properties and fracture toughness in high strength steels. Warke and Elsea [36] conducted an extensive review and found decarburization to be advantageous in some materials (due to reducing the yield strength, while significantly reducing the fatigue strength). Banding and segregation effects have been found [37] to have serious detrimental effects, with cracks forming and propagating along segregation bands.

Finally, significant variation in the fracture toughness in connection with chemistry has been found [38] between heat lots and between vendors of the same alloy. Charpy V notch testing was found insensitive to these subtle changes. K_{Ic} testing was incorporated into the procurement specifications.

C. Microstructure

The propagation of a crack is impeded by changes in the orientation of the cleavage plane at the grain boundary. This change in orientation is accomplished by [39] creating cleavage steps, localized deformation, or tearing near grain boundaries. These changes in orientation cause additional energy to be expended by the crack tip, slowing the propagation of the crack.

D. Effect of Strain Rate

An increase in the strain rate increases the transition temperature in low strength steels. This is because the yield point in low strength steels is dependent on the strain rate. In addition, the transition between ductile and brittle becomes sharper as the strain rate is increased. In high strength steels (above 200 ksi), the yield strength is not as dependent on the strain rate, primarily due to the presence of additional alloying elements.

There has not been a significant amount of literature regarding the effects of strain rate on the toughness of high strength steels. A martensitic grade stainless, 422M, with a tensile strength of 250 ksi was tested [40] using center-cracked specimens over a range of temperatures, and it was found that the fracture strength transition temperature was not affected. However, when the material was tested at room temperature it showed a slight decrease in strength. Others [41] found decreases of up to 43% in a similar steel heat treated to a higher strength (290 ksi ultimate tensile strength).

E. Effect of Section Size

The measurements of transition temperature obtained in different types of tests do not coincide. This is partially because there is no single definition of the transition temperature. For example, for the Charpy V notch test, the ductile to brittle transition can be defined as the temperature at which some specific energy is absorbed or the fracture appearance transition temperature (FATT) where 50% ductile and brittle fracture appears. A ductile to brittle transition temperature occurs because the yield stress σ_y lowers as temperature is increased. The size of the plastic zone surrounding the crack tip also increases as the testing temperature increases. Because of the larger plastic zone surrounding the crack tip, there is a corresponding loss of constraint.

For a given test, the ductile to brittle transition temperature T_{DB} is proportional to the thickness B and the size of the plastic zone. In ferritic or martensitic steels, the transition temperature is an increasing function of temperature. To obtain a correlation between the results of the various fracture tests and the size of the plastic zone, Francois [42] defined the parameter β, where

$$B = \beta(K_{IC}/\sigma_y)^2 \tag{21}$$

For a given thickness B and a chosen value of β, a particular temperature on the K_{Ic} transition curve can be defined as the transition temperature. If β is taken to be equal to 2.5 (the ASTM limit of validity of K_{Ic} specimen), then there will be a certain temperature where plane strain conditions predominate and the test is valid. The transition temperature is then defined in terms of both thickness B and the chosen value of β. The K_{Ic} transition curve can be approximated by various empirical formulas. Ikeda and Kihara [43] developed a good correlation of experimental data,

$$K_{IC} = Ae^{T/T_0} \tag{22}$$

Eliminating the fracture toughness between Equations (22) and (21) leads to a formulation for the transition temperature:

$$T_{DB} = \frac{T_0}{2} \ln\left[\frac{B}{\beta}\left(\frac{\sigma_y}{A}\right)^2\right] \tag{23}$$

This approach has its limitations: It does not make sense that the transition temperature, T_{DB}, disappears when the thickness B becomes smaller or that the transition temperature increases to infinity when B becomes very big. Based on these assertions, the validity of this equation is limited to thicknesses between 5 and 500 mm.

Using the 50% FATT as the basis for the measurement of the ductile to brittle transition temperature, Francois found that the dynamic tear test has a $\beta = 0.6$. For the Charpy V notch test (discussed later), a lower value of $\beta = 0.4$ was determined. This is expected because of the smaller size of the Charpy V notch specimen. Francois showed a reasonable correlation between the observed transition temperatures and those calculated from Equation (23). However, the scatter is large, yielding a variation of $\pm 20°C$.

The transition temperatures of different tests are related to the logarithm of the thickness ratio by the expression

$$T_{DB_1} - T_{DB_2} = \frac{T_0}{2} \ln\left(\frac{\beta_1}{\beta_2}\right) \tag{24}$$

The transition temperature for a particular test and section thickness can be obtained if the transition temperature has already been learned for two different thicknesses, B_1 and B_2:

$$\frac{T_{DB} - T_{DB_1}}{T_{DB_2} - T_{DB_1}} \frac{\ln(\beta/\beta_1)}{\ln(\beta_2/\beta_1)} = 1 \tag{25}$$

For a given steel, the transition temperature of one test can be used to determine the transition temperature of a different test (e.g., Charpy V notch and dynamic tear test) by using

$$T_{DB_1} - T_{DB_2} = \frac{T_0}{2} \ln\left[\frac{B_1\beta_2}{B_2\beta_1}\right] \tag{26}$$

If the transition temperature from K_{Ic} testing is unknown, the transition temperature for plane strain fracture could be estimated from knowledge of the transition temperature from other tests, using

$$\frac{T_{K_{Ic}} - T_{t_1}}{T_{t_2} - T_{t_1}} = \frac{\ln(B\beta_1/B\beta)}{\ln(B_2\beta_1/B_1\beta_2)} \tag{27}$$

In this case, $\beta_1 = 0.4$ (Charpy) and $\beta_2 = 0.6$ (dynamic tear). This method is very useful in determining K_{Ic} values and critical flaw sizes that, because of size limitations, might preclude K_{Ic} testing (such as a field failure).

Other approaches have been taken to correlate simpler impact tests to K_{Ic}. Rolfe and Barsom [21] and Hertzberg [20] show that for steel exhibiting 100% brittle fracture, the fracture toughness K_{Ic} can be estimated from the yield strength at the testing temperature:

$$K_{IC_{est}} = 0.45\sigma_{ys} \tag{28}$$

where K_{Ic} is in the customary units of $\sqrt{in.}$ and σ_{ys} is the yield strength at the testing temperature. If the yield strength is not available at the testing temperature, room temperature properties can be used with conservative results. For materials showing 100% ductile failure, the following expression can be used:

$$\left(\frac{K_{IC}}{\sigma_{ys}}\right)^2 = 5\left(\frac{CVN}{\sigma_{ys}}\right) - 0.05 \tag{29}$$

where CVN is the Charpy V notch impact energy (ft-lbs).

This correlation is useful for learning the effects of different variables such as thickness, microstructure, and transition temperatures from different kinds of tests or for determining the K_{Ic} transition curve from different tests. It should not be used for procurement specifications.

V. FATIGUE

Parts are subject to varying stresses during service. These stresses are often in the form of repeated or cyclic loading. After enough applications of load or stress, the components fail at stresses significantly less than their yield strength. *Fatigue* is a measure of the decrease in resistance to repeated stresses.

Fatigue failures are brittle appearing, with no gross deformation. The fracture surface is usually normal to the main principal tensile stress. Fatigue failures are recognized by the appearance of a smooth rubbed type of surface, generally in a semicircular pattern. The progress of the fracture (and crack propagation) is generally suggested by "beach marks." The initiation site of fatigue failures is generally at some sort of stress concentration site or stress riser. Typical fracture appearance is shown schematically in Figure 19.

Three factors are necessary for fatigue to occur. First the stress must be high enough that a crack is initiated. Second, the variation in the stress application must be large enough that the crack can propagate. Third, the number of stress applications must be sufficiently large that the crack can propagate a significant distance. The fatigue life of a component is affected by a number of variables, including stress concentration, corrosion, temperature, microstructure, residual stresses, and combined stresses.

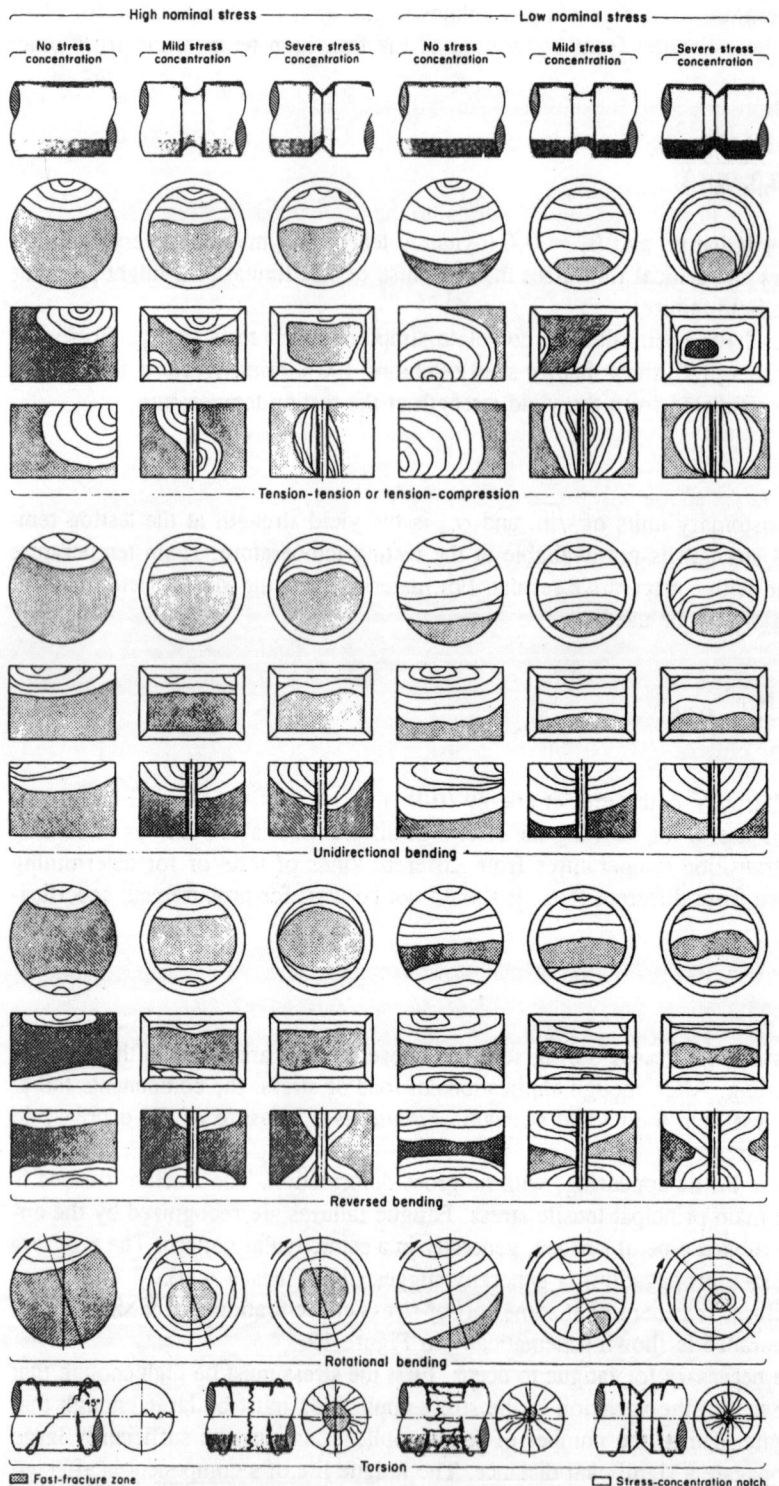

Figure 19 Schematic of fatigue failures commonly found on stressed members.

A. Fatigue Mechanisms

The structural features of fatigue failures are generally divided into four distinct areas [44]:

1. Crack initiation, the early development of fatigue damage.
2. Slip band crack growth, the early stages of crack propagation. This is often called stage I crack growth.
3. Stable crack growth, which is usually normal to the applied tensile stress. This is called stage II crack growth.
4. Unstable crack growth with final failure from overload. This is called stage III crack growth.

Fatigue usually occurs at a free surface, with the initial features of stage I growth, fatigue cracks, being initiated at slip band extrusions and intrusions [45,46]. Cottrell and Hull [47] proposed a mechanism for the formation of these extrusions and intrusions (shown schematically in Figure 20) that depends on the presence of slip, with slip systems at 45° angles to each other operating sequentially on loading and unloading. Wood [48] suggested that the formation of the intrusions and extrusions was the result of fine slip and buildup of notches (Figure 21). The notch created on a microscopic scale would be the initiation site of stable fatigue crack growth.

In stage II, stable fatigue crack growth, striations (Figure 22) often striations show the successive position of the crack front at each cycle of stress. Fatigue striations are usually detected using electron microscopy and are visual evidence that fatigue occurred. The absence of fatigue striations does not preclude the occurrence of fatigue, however.

Striations are formed by a plastic blunting process [49]. At the end of the stage I crack tip, there exist sharp notches due to the presence of slip. These sharp notches cause stress to be concentrated at the crack tip. The application of a tensile load opens the crack along slip planes by plastic shearing, eventually blunting the crack tip. When the load is released, the slip direction reverses, and the crack tip is compressed and sharpened. This provides a sharp notch at the new crack tip where propagation can occur. This is shown schematically in Figure 23.

An alternative hypothesis on striation formation was presented by Forsyth and Ryder [50]. In their model, the triaxial stress state at the crack tip forms a dimple ahead of the crack front. The material between the crack tip and the dimple contracts and eventually ruptures, forming a fatigue striation. This is shown schematically in Figure 24.

In mild steel, striations are observed, but not in as well defined or as spectacular a manner as in aluminum. This was first assumed to be due to the crystal lattice structure, since

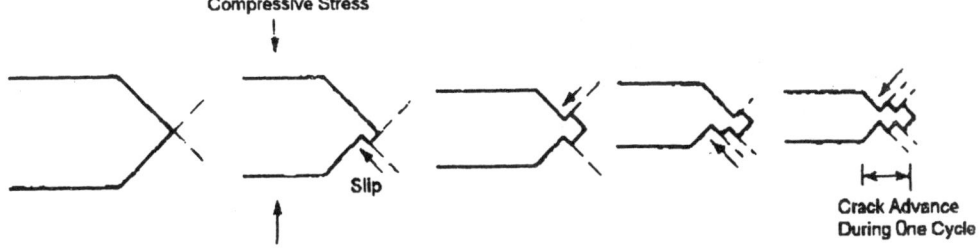

Figure 20 Schematic representation of the mechanism of fatigue intrusions and extrusions. (After Cottrell and Hull [47].)

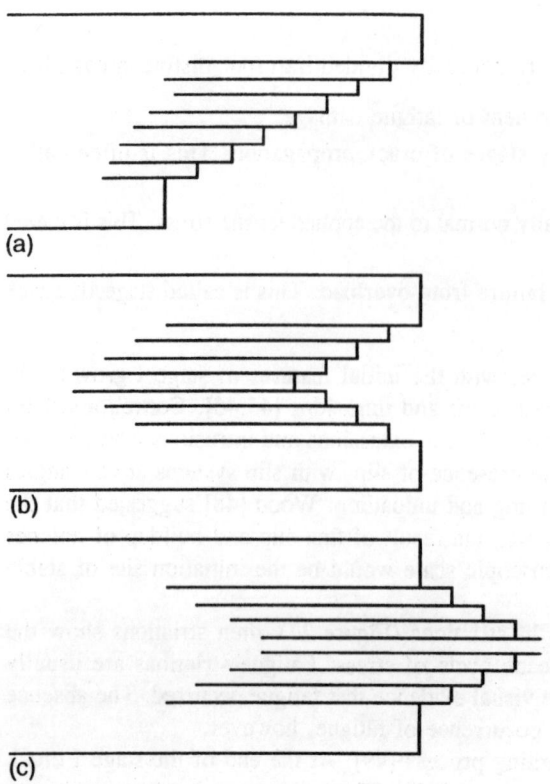

(a)

(b)

(c)

Figure 21 Mechanism of intrusions and extrusions. (After Wood [48].)

Figure 22 Typical fatigue striations, these in 7075 aluminum. Striations in steels, particularly high-strength steels, are difficult to observe.

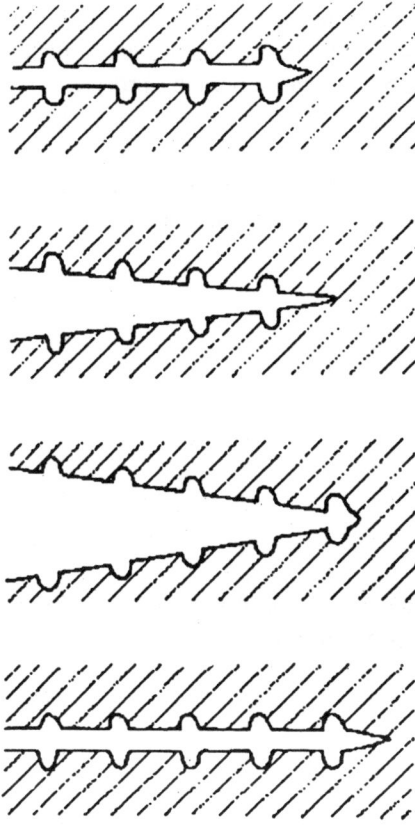

Figure 23 Mechanism for striation formation.

face-centered cubic austenitic steels show well-defined striations, and mild steels (base-structured cubic) do not [51]. Other alloys, such as titanium alloys [52], with a hexagonal close-packed (hcp) crystal structure show very defined striations. However, β-titanium alloys (bcc) [53] show strongly defined striations. Therefore, attributing defined striations to crystal lattice alone was discounted as a viable theory.

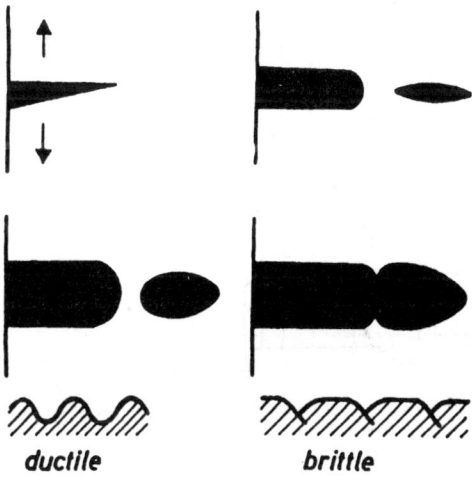

Figure 24 Striation formation from ductile dimple formation ahead of crack front.

Deformation and available slip systems were presumed to be more significant [54]. However, this does not follow, because mild carbon steels are more ductile than austenitic steels. It is now generally accepted that fatigue striations form by the plastic blunting process.

B. Design for Fatigue

Fatigue is caused by a series of loading and unloading, in a variety of waveforms. The load application can be sinusoidal, triangular (sawtooth waveform), or spectral (random) loading. The range of stress application σ_r is suggested by

$$\sigma_r = \sigma_{max} - \sigma_{min} \tag{30}$$

and the alternating stress, σ_a, is represented by

$$\sigma_a = \sigma_r/2 \tag{31}$$

The mean stress σ_m is calculated as

$$\sigma_m = \frac{\sigma_{max} + \sigma_{min}}{2} \tag{32}$$

When representing fatigue data, two quantities, the stress ratio R and the stress amplitude A, completely describe the stress applied. The stress ratio is given by

$$R = \sigma_{max}/\sigma_{min} \tag{33}$$

and the amplitude is

$$A = \sigma_a/\sigma_m \tag{34}$$

Engineering fatigue data are usually displayed graphically in an S-N curve (Figure 25), where the applied stress σ (or s') is plotted against logarithm of the number of cycles. As the stress is increased, the number of cycles until failure decreases. For steels, the S-N curve becomes horizontal at some low stress level. This is called the *fatigue limit*. Aluminum alloys do not show a fatigue limit, and failure occurs at some extended number of cycles at low applied stress.

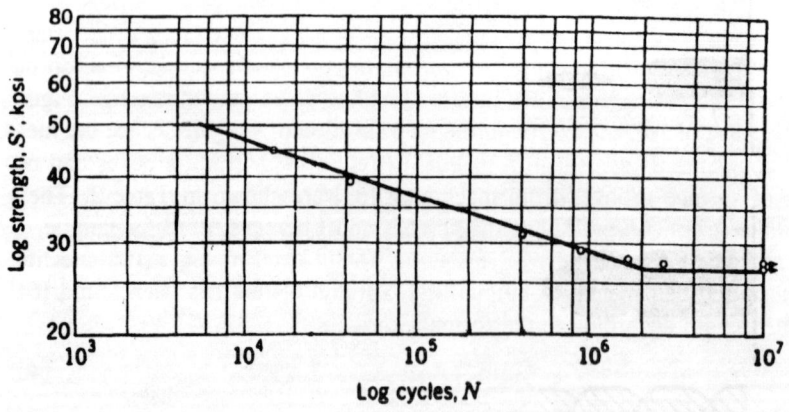

Figure 25 Typical presentation of fatigue data in an S-N diagram. Data shown are for an annealed AISI 1040 steel.

For steels, the fatigue limit σ_e depends on the ultimate tensile strength (UTS) [55] in a rotating, reversed fatigue test:

$$\sigma_e = \frac{\sigma_{UTS}}{2} \tag{35}$$

In other loading types, the endurance limit for alternating axial loading (completely reversed) is [56]

$$\sigma_{e_a} = 0.35\sigma_{UTS} \tag{36}$$

and for reversed torsional testing, the torsional endurance limit is

$$\tau_e = 0.3\sigma_{UTS} \tag{37}$$

The fatigue limit and fatigue tests show considerable variation [57–59]. Because of this variability, the fatigue life must be examined in terms of probability. This requires testing many samples so that the mean and standard deviation can be determined. In one study [60], 200 steel specimens were tested, and it was found that fatigue life followed a gaussian distribution when it was examined using a log scale. The fatigue endurance limit was examined by Ransom [61]. In this study, 10 S–N curves were developed using 10 specimens for each, for a total of 100 fatigue specimens. The specimens were prepared identically and were from the same bar of material. Ransom found that there was a 20% scatter around the fatigue limit.

When designing for cyclic loading, it is necessary to allow for the factors that influence fatigue. One such method was proposed by Shigley [62]. In this method, correction terms for each of the important deleterious effects are multiplied together to obtain a maximum design stress, σ_e', for the machine element. This maximum design stress is obtained from data representing a smooth polished rotating beam specimen at the endurance limit σ_e as

$$\sigma_e' = k_a k_b k_c k_d k_e K \ k_n \sigma_e \tag{38}$$

The correction terms k_i are related to the stress concentration factor by

$$k_i = 1/K_i \tag{39}$$

where K_i is the stress concentration.

1. Surface Treatments

Since fatigue failures usually begin at the surface, the surface condition is very important. Surface roughness is a key factor influencing fatigue. Highly polished specimens exhibit the longest fatigue life, with increasingly rougher surfaces yielding decreased fatigue life. Figure 26 shows the correction factor for surface roughness as a function of strength. Since the measurement of surface roughness is statistical in nature, extremes of machine grooves can occur. Typical depths of surface grooves from machining [63] are shown in Table 2. These extremes, as well as the stress concentrations they create, must be considered for fatigue.

The surface treatment of a steel must be accounted for in determining its fatigue life. Electroplating is considered detrimental to fatigue life. Chrome plating has been found [64] to reduce the fatigue strength according to the formula

$$k = 1 + Y \tag{40}$$

where Y is the correction factor

$$Y = 0.3667 - 9.193 \times 10^{-3}\sigma_e \tag{41}$$

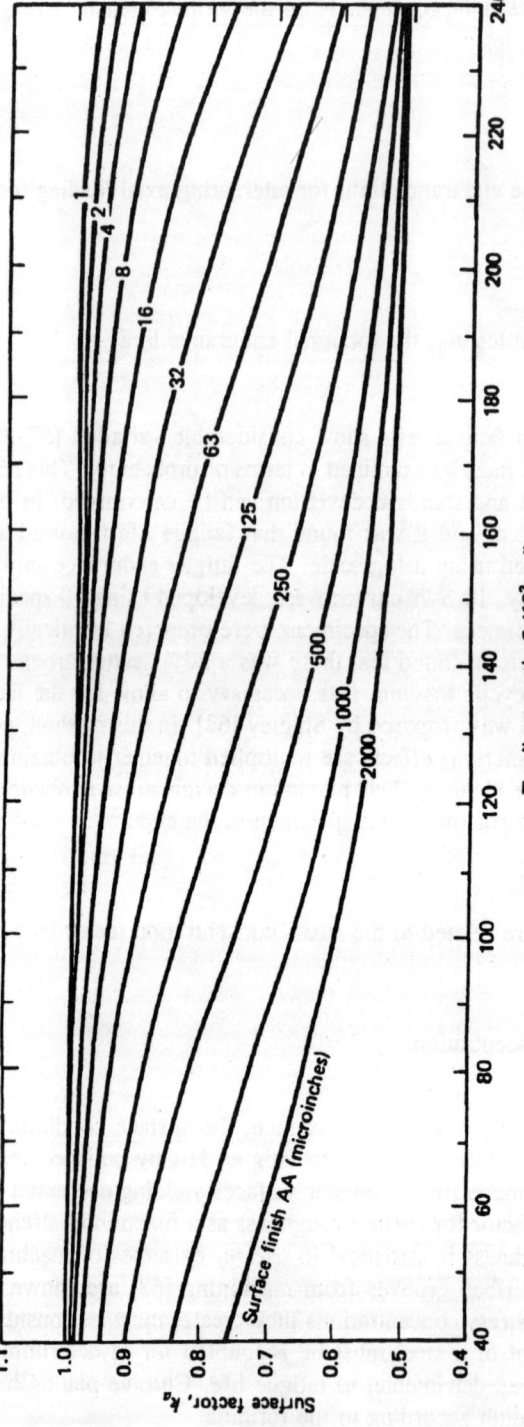

Figure 26 Fatigue stress concentration due to surface finish and surface roughness. (After Faupel [69].)

Table 2 Machine Grooves Commonly Found on Machined Surfaces

Method	Surface roughness (rms)	Groove depth (in. $\times 10^{-3}$)
Polished	8	0.04
Fine grind	10	0.08
Rough grind	70	0.2–0.4
Fine turn	70–90	0.4–0.8
Rough turn	90–500	0.8–2

Source: Ref. 69.

If the part is nickel plated the correction factor $Y = -0.01$ for 1008 steel and $Y = -0.33$ for 1063 steel [65]. If shot peening is carried out after nickel or chromium plating, then the fatigue strength is often increased beyond that of the base metal [66]. Cadmium plating is not considered to affect the fatigue life of steels because it is very soft and ductile. The plating conditions can greatly affect the fatigue properties of a steel. A comprehensive review of the effect of plating on fatigue life is given by Hammond and Williams [67].

Surface treatments of steel are not limited to electroplating. Other surface treatments such as carburizing and nitriding are used to impart surface hardness and improved wear characteristics [68]. Improvements in fatigue life are also obtained. These effects are due to the increased hardness of the surface (with increased endurance limit) and the creation of a favorable residual stress at the surface. Flame hardening and induction hardening have similar effects. Using Equation (40) as a basis for determining the surface improvements, values of the correction factor Y for surface treatments have been determined [69]; these are shown in Table 3. Because of the compressive residual stress fields at the surface, surface-hardened steels generally do not initiate fatigue at the surface but at the case/core interface.

2. Residual Stresses

As indicated above, residual stresses at the surface can improve or decrease the fatigue life of a steel. Residual stresses occur when a part is plastically deformed nonuniformly over its cross section. If a part is plastically deformed at only the surface (by rolling or peening), a compressive surface stress results. This compressive stress constrains the interior of the part, placing it under a tensile stress. These stresses are superimposed on the applied stress field, reducing the total stress. Unfortunately, a uniform compressive stress field is difficult to produce on a complex geometry and difficult to measure. It is measured using strain gauges on X-ray diffraction or by noting part displacement when material is removed.

Typically shot peening is used to impart a compressive residual stress at the surface. Again, using Equation (40), the correction factor improvement in the fatigue life from shot peening

Table 3 Values of Y from Surface Hardening

	Flame and induction hardening	Carburizing	Nitriding
Layer thickness	0.125–0.500 (induction) ≈ 0.125 (flame)	0.03–0.1	0.004–0.02
Steel	0.66–0.80	0.62–0.85	
Alloy steel	0.06–0.64	0.02–0.36	0.30–1.00

Source: Ref. 69.

Table 4 Value of Y from Shot
Peening for Steels

Surface	Y
Polished	0.04–0.22
Machined	0.25
Rolled	0.25–0.5
Forged	1–2

Source: Ref. 69.

is shown in Table 4. Surface rolling creates a thicker compressive residual stress layer (0.040–0.5 in.) and therefore creates a larger improvement in the fatigue life. For steel shafts, $Y \approx 0.2$–0.8, while for polished steel parts, $Y \approx 0.06$–0.5 (Faupel). This is further discussed by Lipson and Juvinall [70]. It has been shown [71] that grinding produces tensile residual stresses at the surface.

3. Size Effect

In fatigue, the larger the part tested, the lower the fatigue limit. Fatigue starts at the surface, and as part diameter increases the surface area also increases, providing more sites for fatigue initiation. Also because of the larger diameter, the stress gradient across the plastic zone at the crack tip decreases, making it difficult to model larger structures on small specimens. A correction for fatigue in larger parts is provided by Juvinall [72]. For parts with diameters less than or equal to 0.030 in., the correction factor $k = 1$. If the part diameter is greater than 0.3 in. but less than 2 in., then $k = 0.85$. For parts larger than 2 in. in diameter, the decrement in fatigue strength is [73]

$$k_b = 1 - \frac{d - 0.03}{15} \qquad (42)$$

It has also been found [74] that the thicker the test piece, the faster the crack propagation rate. It is likely that the propagation rates for thicker pieces are due to increased plane strain conditions, with a small plastic zone at the crack tip. Since for a small plastic zone there is a greater stress gradient, a faster crack propagation rate might be expected. Also, in thicker panels there is a higher state of triaxial stress, which would also tend to increase crack growth rates.

4. Stress Concentrations

The fatigue life of a component is seriously affected by the presence of stress concentrations. Often, it is found that fatigue initiates at the site of a stress concentration. Stress concentrations are determined from geometry effects and are either calculated from elasticity theory [75] or with numerical methods or determined experimentally using stress analysis techniques. A very comprehensive set of stress concentration factors has been collected by Peterson [76] and Young [77]. Besides the stress concentration, the notch sensitivity of the material is necessary to calculate the effect of the stress concentration. The notch sensitivity of a material is related to the stress concentration by

$$q = \frac{K_f - 1}{K_t - 1} \qquad (43)$$

where q is the notch sensitivity, K_t is the theoretical stress concentration, and K_f is the stress concentration for fatigue. This is related to the correction factor k for the fatigue endurance limit by

$$K_f = 1 + q(K_t - 1) \tag{44}$$

or

$$k = 1/K_f \tag{45}$$

The notch sensitivity, q, of a material has been determined for many materials [78], and some data are plotted in Figure 27. The notch sensitivity varies with the type of notch and notch sensitivity. The notch sensitivity increases with increasing tensile strength.

An alternative approach was taken by Neuber [79], who suggested that the stress concentration for fatigue be expressed as

$$K_f = 1 + \frac{K_t - 1}{1 + \left(\dfrac{\pi}{\pi - \omega}\right)\left(\dfrac{a}{r}\right)^{1/2}} \tag{46}$$

where r is the root radius of the notch, ω is the angle of the notch, and a is the elementary block size. The elementary block size idea has no real basis but is a convenient empirical tool.

The above is one approach to determining the fatigue life of a component. Fracture mechanics has also been used to study crack propagation [80,81]. The crack propagation rate in a material follows an equation of the form

$$\frac{da}{dN} - C\sigma_a^m a^N \tag{47}$$

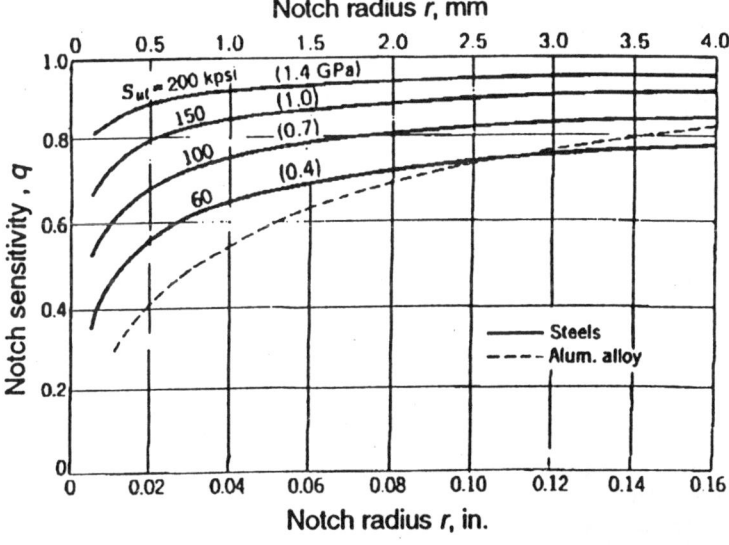

Figure 27 Notch sensitivity of steels as a function of strength. (After Dieter [114].)

where C is a constant based on the geometry of the crack and test, σ_a is the alternating stress, a is the crack length, m is a material constant found during testing, and N is the cycles to failure. The growth of a crack can also be expressed as a function of the stress intensity factor K from fracture mechanics,

$$\frac{da}{dN} - C(\Delta K)^n \tag{48}$$

where C and n are material constants found during testing. For martensitic steels, it has been found [82] that

$$\frac{da}{dN} = 0.66 \times 10^{-8} \Delta K^{2.25} \tag{49}$$

and for ferritic-pearlitic steels

$$\frac{da}{dN} = 03.6 \times 10^{-10} \Delta K^{3.0} \tag{50}$$

Equation (48) is illustrated in Figure 28. This has been found to be an extremely useful concept in high strength steels.

While the above provides information regarding how a crack propagates in a solid, it can also be used to predict fatigue life. Assume a solid, with some elliptical flaw. The critical size of the flaw is given by Brock [83] as

$$a_{cr} = \frac{K^2 Q}{1.21 \pi \sigma^2} \tag{51}$$

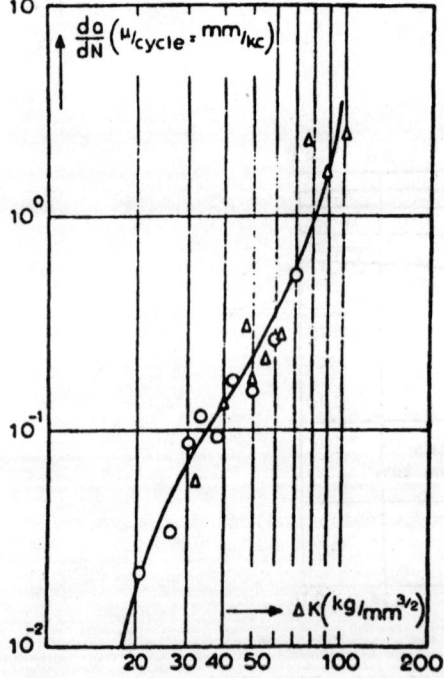

Figure 28 Schematic crack growth, da/dN, curve.

where a_{cr} is the critical flaw depth, K is the critical stress intensity, and σ is the applied stress. The growth rate of a crack is expressed as Equation (48) and in its simplest form as

$$\frac{da}{dN} = CK_1^n \qquad (52)$$

where da/dN is the growth of a flaw per cycle, a is the depth of the flaw, N is the number of cycles sustained, and K_1 is the stress range,

$$K_1 = \sigma(1.21\pi a/Q)^{1/2} \qquad (53)$$

The stress range is obtained using specially designed specimens related to a specific geometry [84–86]. The material constants c and n are obtained through fatigue crack growth testing. The value of the flaw shape factor is obtained from the references cited above as well as others [87–89]. In this instance the stress range is for an elliptical flaw in the inside diameter of a tube. By combining Equations (52) and (53) and rearranging, the following equation is obtained:

$$dN = \frac{da}{c[\sigma(1.21\pi a/Q)^{1/2}]^n} \qquad (54)$$

The cycle life is obtained by integrating Equation (54) between the limits of the initial flaw depth a_i and the critical flaw depth a_{cr}:

$$N = \int_{a_i}^{a_{cr}} da \left\{ c \left[\sigma \left(\frac{1.21\pi a}{q} \right)^{1/2} \right]^n \right\}^{-1} \qquad (55)$$

The initial flaw depth a_i could be either the detectable flaw size from nondestructive testing or an assumed flaw size.

There are very important differences between the two fatigue prediction methods described above. Each has found a home according to the industry served. For the most part, the aerospace industry, because of its use of ultrahigh strength steels, relies on the fracture mechanism approach, while most other industries tend to use the notch analysis approach. A brief comparison of the two methods is presented in Table 5 [90]. A more complete review of the differences is found in Reference 91.

VI. CREEP AND STRESS RUPTURE

The effects of elevated temperature on mechanical properties and material behavior are commonplace in everyday living. Examples include pipes bursting in the middle of winter, the expansion of a bridge in the middle of summer, and the sagging of a fireplace grate. Each of these examples is an indication that properties change with temperature. In addition, the discussion above indicated that steels become more brittle as the temperature is decreased. There are many other effects of temperature that have been cited [92]. Even the concept of elevated temperature is relative [93]. What is considered hot for one material may be considered cold for another; for instance, gallium has a melting point of 30°C and Tungsten a melting point of about 3400°C.

Table 5 Comparison of Notch Analysis and Fracture Mechanics in Fatigue Analysis

	Fracture mechanics	Notch analysis
Fundamentals	Elastic energy Equivalent stress Concentration	Stress concentration
Applied theory	Linear elastic theory	Linear elastic theory, elastic-plastic behavior
Principles	$\sigma \sqrt{a}$ = constant Correction for plastic zone at crack tip	Size effect

Source: Ref. 90.

Creep is the continuous deformation of a material as a function of time and temperature. This topic is treated very throughly by Finnie and Heller [94]. The creep of a material is shown in Figure 29. It can be seen from the figure that creep in a material occurs in three stages:

1. Stage I, where a rapid creep rate is seen at the onset of load application, then gradually decreasing
2. Stage II, where creep remains at a steady-state rate
3. Stage III, where the creep rate shows an increasing rate until failure occurs

The behavior, and the creep rate, are sensitive to the temperature to which the material is exposed, the surrounding atmosphere, and the prior strain history. Andrade and Chalmers [95] were pioneers in the study of creep and proposed that creep followed the equation

$$\varepsilon = \varepsilon_0 (1 + \beta t^{1/3}) e^{Kt} \tag{56}$$

where β and K are material constants that can be evaluated by several different methods [96]. A better fit for the creep of materials was proposed by Garofalo [97]. He indicated that

$$\varepsilon = \varepsilon_0 + \varepsilon_t (1 - e^{-n}) + \frac{d\varepsilon}{dt} t \tag{57}$$

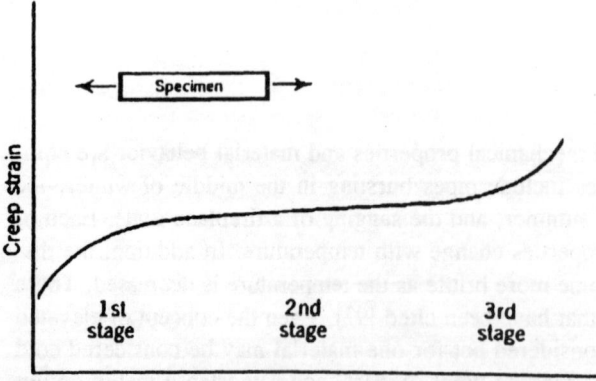

Figure 29 Creep of a material showing the stages of creep. (After Dieter [114].)

where $d\varepsilon/dt$ is the steady-state creep rate, ε_0 is the strain on loading, r is the ratio of the transient creep rate to the transient creep strain, and ε_t is the transient creep strain.

Very early it was recognized that fractures at elevated temperatures occurred intergranularly [98]. In stage III creep, intergranular wedge cracks and cavities form. Wedge-shaped cracks and creep cavities usually initiate at or near grain boundary triple points and propagate along grain boundaries normal to the applied tensile stress. Creep cavities form at higher temperatures and lower working stresses. These structural features are shown in Figure 30.

Creep testing is usually performed for 1000–10,000 h with strains of up to 0.5%. Stress rupture testing, or testing to failure, uses much higher loads and temperatures, and the test is usually terminated after 1000 h. In stress rupture testing, the time to failure is measured at a constant stress and constant temperature. This test has gained acceptance for elevated temperature testing of turbine blade materials in jet engines.

Using a tensile machine and high temperature furnace (Figure 31), the strain is measured in creep testing by special extensometers suited for elevated temperatures. In stress rupture testing, simple apparatus such as dial calipers are used, since only the overall strain at constant time and temperature are needed.

VII. TENSION TESTING

Tensile testing is the most generally useful of all mechanical tests. Both the strength and a measure of ductility are obtained. In this test, a specimen is loaded axially and the load is increased continuously until failure occurs. The load and elongation are continuously plotted during testing and then converted to engineering stress σ,

$$\sigma = P/A_0 \tag{58}$$

and engineering strain e,

$$e = \delta/L_0 \tag{59}$$

where P is the load applied, A_0 is the original cross-sectional area δ is the measured elongation, and L_0 is the original gauge length. The resulting data are plotted as engineering stress and strain, schematically shown in Figure 32. For a rigorous mathematical treatment of the tensile test and the changes that occur, the reader is referred to Nadai [99].

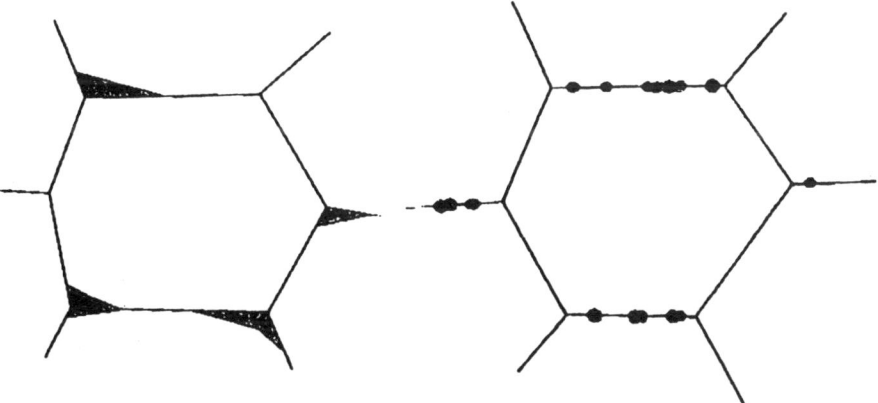

Figure 30 Creep cavities and wedges forming at grain boundaries.

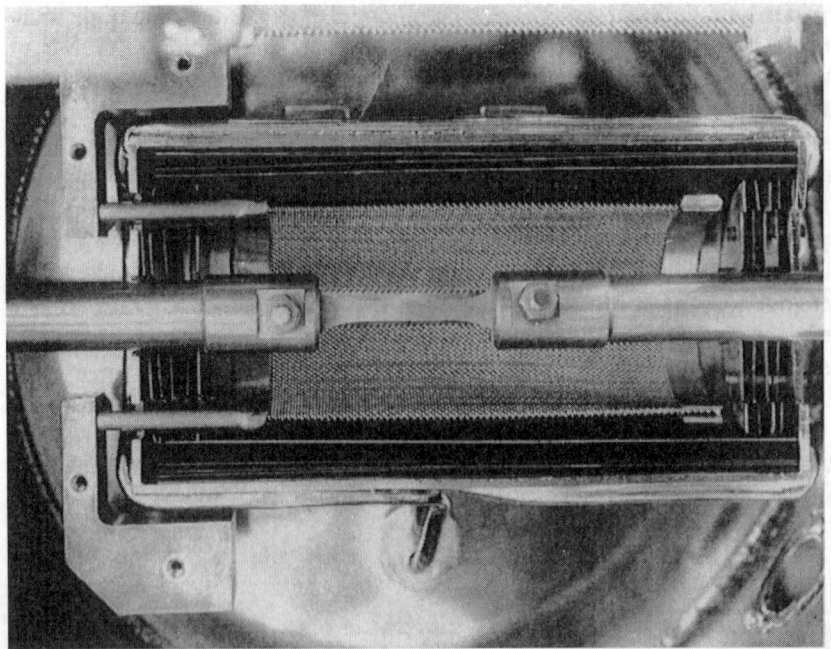

Figure 31 Typical creep testing apparatus.

At the beginning of the test, there is a linear region where Hooke's law is followed. The slope of this linear region is the elastic modulus E (for steels, the elastic modulus is about 30 $\times 10^6$ psi). The yield stress σ_{ys}, which is the limit of elastic behavior, is defined as the point at which a small amount of permanent deformation occurs. This deformation is defined in the United States as a strain of 0.002 in./in. or 0.2% strain [100]. Loading past this point causes plastic deformation. As the plastic deformation increases, strain hardening occurs, making the

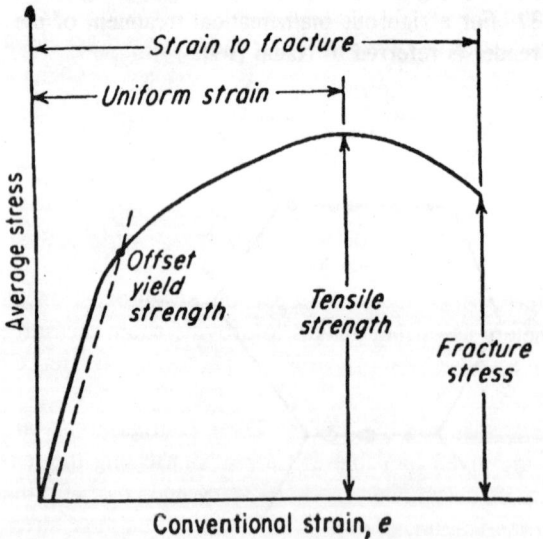

Figure 32 Schematic of tensile test.

material stronger. Eventually, the load reaches a maximum value and failure occurs. The ultimate tensile strength σ_{UTS} is determined by taking the maximum load experienced and dividing it by the original area:

$$\sigma_{UTS} = \frac{P_{max}}{A_0} \tag{60}$$

In ductile materials, the strain may increase after the maximum load is reached and the applied load is decreased. At this point is the onset of necking.

The ultimate strength is the most quoted property but is the least useful of all the properties determined from the tensile test. In ductile materials, ultimate strength is a measure of the maximum load that the material can experience in uniaxial loading. However, in brittle materials, the ultimate tensile strength is valid design information The ultimate tensile strength used to be the basis for many designs and design codes (with a margin of safety), but now design codes rely on the yield strength instead. Because of its reproducibility, the ultimate tensile strength is often used for procurement, specifications, and quality documents.

The yield strength σ_{ys} is the stress required to obtain a small (0.2%) permanent strain. In other words, if the specimen were unloaded at the yield stress, it would be 0.2% longer than originally. The yield strength is now the basis for many design codes and is used extensively in the determination of many other properties such as fracture toughness and fatigue strength.

Ductility in the tensile specimen is measured and reported two different ways: In terms of elongation,

$$\%E_f = \frac{L_f - L_0}{L_0} \times 100 \tag{61}$$

and in terms of the reduction in area,

$$\%RA = \frac{A_0 - A_f}{A_0} \times 100 \tag{62}$$

where L_f is the final gauge length after fracture, L_0 is the original gauge length, A_0 is the original cross-sectional area, and A_f is the final cross-sectional area after fracture. Since the final strain will be concentrated in the necked region, the elongation is dependent on the gauge length L_0. For that reason, in reporting elongation, the gauge length is always provided. The change in elongation as a function of gauge length is provided in Figure 33.

Testing machines used for tensile testing are generally simple. Either screw-type (Figure 34) or hydraulic (Figure 35) machines are used. The load is measured by a load cell composed of strain gauges or an LVDT (linear velocity displacement transducer). This is recorded by a chart recorder as a plot of load vs. strain.

The strain is measured by extensometers, which measure in situ the strain experienced by the tensile specimen. Modern extensometers use strain gauges to provide strain measurements. This is an advantage because it sends an electric signal proportional to the experienced strain to the chart recorder.

The shapes of tensile specimens are standardized by ASTM. Their configurations are shown in Figure 36, and their dimensions are given in Table 6. Care in fabricating the tensile specimen is necessary to preclude faulty or inaccurate results. Heating and cold working of the specimens must be minimized during fabrication, or inaccurate yield and ultimate stress values may result. The specimens must be straight and flat; otherwise a distortion of the elas-

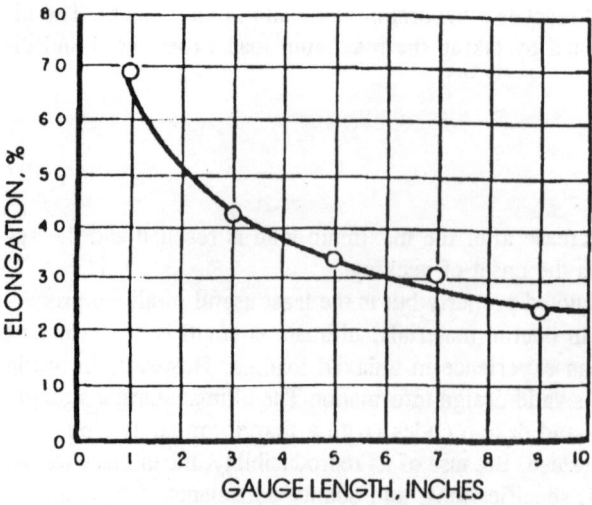

Figure 33　Correction factor for percent elongation vs. gauge length in steels.

tic region and an inaccurate yield strength may result. During machining, the tensile specimens must be symmetrical about the load axis. If not, bending may occur, leading to erroneous results because of the combined stresses of tension and bending.

The tensile specimen is held during testing by either hydraulic or mechanical grips. Both
methods grip the specimen by wedges with serrated surfaces. In mechanical grips, the load
applied to the specimen forces the wedges tighter against the specimen by inclined surfaces
inside the grip. With hydraulic wedges, hydraulic pressure is used to force the wedges against
the specimen and hold it in place. For high strength steels or steels that have hard or highly

Figure 34　Screw-type tensile machine.

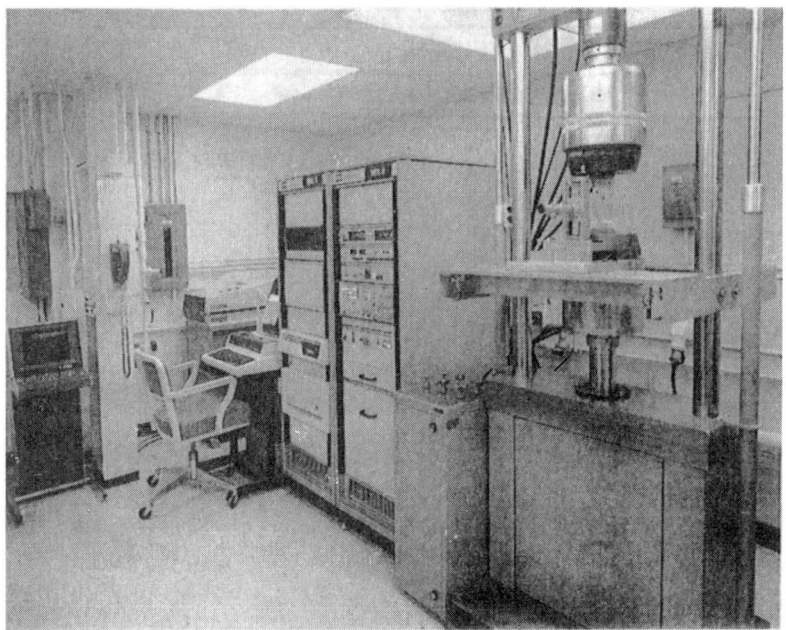

Figure 35 Hydraulic tensile machine.

polished surfaces, hydraulic grips are preferred to prevent the specimen from slipping in the grip.

The tensile properties of annealed and normalized steels are dependent on the flow characteristics of ferrite. The strength of ferrite is a function of the alloying elements in solid solution in it and the ferrite grain size [101]. The percent carbon in the steel has a strong effect because it influences the amount of cementite (Fe_3C) present as pearlite or spherodite. As a general rule, the strength increases and the ductility decreases as the percent carbon is increased.

Normalized steels tend to show higher strengths than annealed steels, since a finer pearlite spacing results from the faster cooling rate in normalizing. Expressions for the ultimate tensile strength, yield strength, and reduction in area have been published [102] and are reproduced here:

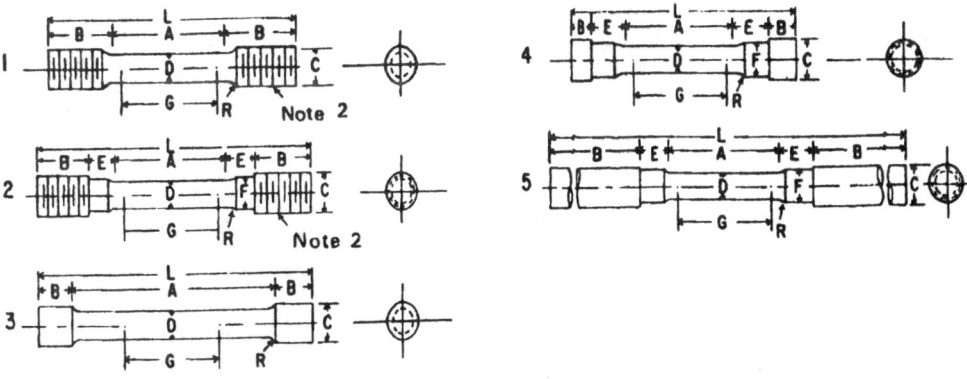

Figure 36 Typical tensile specimens per ASTM E8.

Table 6   Dimensions of Full-Size Tensile Specimens per ASTM E8

	Specimen dimensions (mm)				
	1	2	3	4	5
G, Gauge length	50	50	50	50	50
D, Diameter	12.5	12.5	12.5	12.5	12.5
R, Fillet radius	10	10	2	10	10
A, Length of reduced section	60	60	100	60	60
L, Overall length	125	140	140	120	240
B, Length of end section	35	25	20	13	75
C, Diameter of end section	20	20	18	22	20
E, Length of shoulder	—	16	—	20	16
F, Diameter of shoulder	—	16	—	16	15

$$\sigma_{UTS}[ksi] = 42.8 + 4(\%Mn) + 12(\%Si) + 5.64V_p + 0.224d^{-1/2} \tag{63}$$

$$\sigma_{ys}[ksi] = 15 + 4.73(\%Mn) + 12.2(\%Si) + 0.55d^{-1/2} \tag{64}$$

$$\%RA - 78.5 + 5.39(\%Mn) - 0.53V_p - 8.399d \tag{65}$$

where alloying elements are expressed as weight percent, V_p is the volume percent pearlite, and d is the grain intercept in inches.

Quenched and tempered steels, heat treated to tempered martensite, offer the best balance of strength and ductility. For a given composition, the properties are altered by changing the tempering temperature. Composition plays a minor role, with the function of alloying elements being to increase the hardenability of the steel or to increase the depth to which a fully hardened structure can be obtained. Relationships for the strength of steel as a function of quench rate, tempering temperature, and composition are published elsewhere [103,104].

VIII. HARDNESS TESTING

Hardness testing is probably the most common type of mechanical test performed in the United States, perhaps the most common worldwide. The term "hardness" is poorly defined and is relative to the measuring device. There are three basic types of hardness tests: the scratch test, the indenter, and the dynamic rebound test. The scratch test is familiar to mineralogists, who use the Mohs scale.

The Mohs scale is the relative hardness of 10 minerals arranged in order. Talc is the softest (Mohs 1), and diamond is the hardest (Mohs 10). Most hard metals fall in the range of Mohs 4–8. There is inadequate differentiation along the scale to be of much use to a metallurgist. In dynamic tests, an indenter is dropped on the material, and hardness is defined as the energy of impact. The Mohs scale is commonly used for rubbers and polymers. One exception is the Shore sceleroscope which is used for metals. The indenter type of hardness test is the most widely accepted for metals.

In the indenter type of hardness test, an indenter is pressed into the material and released, and either the diameter or the depth of the impression is measured. The load and the impression measurement determine the hardness. Since a hardness impression is made, there is a plastic zone around the hardness indentation that is surrounded by undisturbed elastic material. This elastic zone hinders plastic flow. Since the plastic region is constrained by the elastic region, the compressive strength of the material in the area of the hardness impression is higher than the value of simple compression. This is a classic problem in plasticity and should be able to be explained by slip line theory. The load required to indent a specific distance δ by a punch is given by Hill [105] as

$$P = \frac{4a\sigma_{ys}}{\sqrt{3}}\left[1 + \frac{\pi}{2} - \left(\frac{\delta}{2R}\right)^{1/2}\right]$$ (66)

where R is the radius of the punch, P is the applied load, and a is the resulting radius after punching. Nadai [99] determined the pressure to indent, using slip line theory, and found

$$p = \frac{2\sigma_{ys}}{\sqrt{3}}\left[\frac{1}{2} + \frac{\pi}{4} + \alpha + \frac{\cot\alpha}{2}\right]$$ (67)

where α is the included angle of the conical indenter.

In hardness testing, the impression is asymmetric, so that the slip line theory is not applicable [106,107], but an elastic/plastic boundary problem, best explained from the Hertz theory of contact stresses [108]. This model accounts for the material displaced by the indenter by the decrease (by compression) in volume of the elastic underlying material. No upward flow around the indenter is predicted, which agrees with observation. This explanation is the basis for all indention hardness tests used for metallic materials.

A. Brinnell Test

The Brinnell test was first proposed in 1900 and has since become widely accepted throughout the world. It is accomplished by indenting the surface with a 10 mm steel ball with a 3000 k load. For soft metals a 500 k load is usually used because otherwise the impression is too deep. For hard metals, a tungsten carbide ball is used to prevent distortion of the indenter. The diameter of the round indentation is measured with a low power microscope after loading. At least two measurements of the diameter are made, and the results are averaged. To ensure that accurate measurements are made, the surface must be free of dirt and scale. The hardness expressed as a Brinnell hardness number (BHN) is determined from the equation

$$\text{BHN [kg/mm}^2] = \frac{P}{\pi D/2}[D - (D^2 - d^2)^{1/2}]$$ (68)

where P is the applied load (kg), D is the diameter of the indenter (10 mm), and d is the measured diameter of the impression. The Brinnell number could also be calculated by measuring the depth of the impression, t:

$$\text{BHN} = P/\pi Dt$$ (69)

or

$$\text{BHN} = \frac{P}{(\pi/2)D^2(1 - \cos\phi)}$$ (70)

since $d = D \sin\phi$, where ϕ is the included angle of the chord of the impression.

Because of the large size of the impression made in the Brinnell test, it averages out any local inhomogeneities. It also precludes the testing of small objects or objects in which the Brinnell impression can be a site of crack initiation. Therefore, it is commonly used for castings, forgings, or raw stock. The size of the piece tested for Brinnell hardness should be at least 10 times the depth of the impression. Because of the plastic zone surrounding the impression and the elastic constraint, additional Brinnell hardnesses should not be measured any closer to the impression than 4 times the impression diameter. The distance from an edge when taking a hardness reading should also be at least 4 times the impression diameter.

B. Vickers Hardness or Diamond Pyramid Hardness

In the Vickers hardness test, a square-based pyramid shaped diamond penetrator is used. The included angle of the pyramid is 136° between opposite faces. The test is taken by indenting, under load, with the penetrator, and after release of the load, measuring the width of the diagonals. The Vickers hardness number (VPH) or diamond pyramid hardness is calculated as

$$DPH = \frac{2P\sin(\theta/2)}{L^2} \qquad (71)$$

where P is the load (kg), L is the average length of the diagonals, and θ is the angle between opposite faces (136°).

The Vickers hardness test has gained wide acceptance in the world for research because the loads can be varied between 1 and 120 kg. It is a continuous scale of hardness that is internally consistent; i.e., hardness numbers determined from one test with one load can be compared to those determined in another test using a different load (except for very light loads) because the impressions are geometrically similar. Other tests require changing loads and indenters, and the hardnesses obtained cannot be strictly compared with others.

This test has not been widely accepted in the United States because it is slow and requires careful surface preparation. In addition, the opportunity exists for operator error when measuring the diagonals.

Sometimes, impressions are obtained that are not perfectly square. These fall into two categories: pincushion-shaped impressions and barrel-shaped impressions (Figure 37). The pincushion-shaped impression is caused by the tested metal sinking in around the flat faces of the pyramid. This often occurs when testing very soft or annealed metals, with the result that inaccurate low hardness numbers are obtained by overestimating the diagonal lengths. Barrell-shaped impressions are usually found when testing highly cold worked materials and are caused by the piling up of material around the indenter. This produces low diagonal values, with the hardnesses erring on the high side.

C. Rockwell Hardness Test

The Rockwell hardness test is very widely used in the United States because of its speed and its freedom from errors by operating technicians. The impression is small, making it possible to test a wide variety of parts.

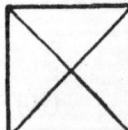  

Figure 37   Vickers indenter defects. (After Kehl [109].)

In this test, a minor load of 10 kg is applied to the part to seat the indenter and part; then the major load is applied. After 30 s the depth of the impression is measured and exhibited on a rotary dial or, in newer machines, on a digital readout. Some machines are equipped with printer ports or RS-232C interfaces for communicating with a computer.

The scale used in Rockwell testers is based on 100 divisions, with each division equal to a depth of 0.00008 in. The scale is reversed, so that a small impression results in a high hardness number. The number from the Rockwell test is an arbitrary number that is only consistent within the same scale. This is unlike the Brinnell and Vickers hardness tests, which provide numbers that are based on mass per unit area (kg/mm^2).

More than one indenter can be used with the Rockwell test. The most common is the Brale indenter, which is a 120° diamond cone and is used for the Rockwell C test. Other indenters used are 1/16 and 1/8 in. diameter steel balls. Major loads can be varied from 60 to 150 kg. Table 7 lists the application, indenter, and major load used for the three types of Rockwell hardness tests.

If a carburized surface is being measured, the hardness reading may be lower than expected because of a greater depth of impression. The material supporting the test piece may also influence the hardness reading. As a rule of thumb, it is wise to test only test pieces that are at least 10 times as thick as the depth of the impression [109]. If the impression of the hardness indenter shows on the other side of the tested piece, the test is obviously invalid.

Because the scale on a Rockwell hardness machine is arbitrary, it is necessary to ensure that one machine records the same hardness as another machine. This is accomplished by the use of standardized test blocks calibrated by the manufacturer. Generally, three test blocks of values throughout the range are adequate to maintain the hardness machine in calibration.

D. Rockwell Superficial Hardness Test

The superficial tester operates in an identical fashion to the Rockwell hardness tester. In fact, they look very similar. Two types of indenters are used: a 1/16 in. diameter steel ball (used for surface hardnesses of brasses, aluminum sheet, etc.) and the Brale penetrator. This Brale penetrator is similar to the one used in the Rockwell hardness tests except that its spherical end is shaped to higher tolerances. This penetrator is designated the N-Brale indenter.

The minor load used to seat the indenter and specimen is 3 kg. The major loads used are 15, 30, and 45 kg. The major load is applied for 30 s, and the depth of the penetration is measured. The scale of the superficial Rockwell test is arbitrary like that of the Rockwell test, except that each division represents 0.001 in. in depth. The scale is reversed, so deeper impressions mean lower hardness numbers. Since the loads used on the superficial Rockwell tester are different from those used for the Rockwell tests, different scales were established defined by the load and indenter as shown in Table 8.

Since the impression left by the superficial test is not very deep, it is important that the test surface be smoother than is necessary for the standard Rockwell test. The superficial hard-

Table 7 Description of Standard Rockwell Hardness Scales

	Rockwell A	Rockwell B	Rockwell C
Indenter	Brale	1/16 in. diameter ball	Brale
Major load (kg)	60	100	150
Application	Very hard materials; tungsten carbide, hard thin materials, etc.	Rolled steel sheet, brass, aluminum, annealed steels, etc.	Fully hardened steels, quenched and tempered steels, etc.

Table 8 Indenters and Major Loads for the Rockwell Superficial Hardness
Test Scales

Scale	Penetrator	Major load (kg)
15-N	N Brale	15
30-N	N Brale	30
45-N	N Brale	45
15-T	1/16 in. diameter steel ball	15
30-T	1/16 in. diameter steel ball	30
45-T	1/16 in. diameter steel ball	45

ness test is more accurate when the part has a good surface finish. The test is sensitive to dirt
and hard particles under the test piece or the indenter, giving unusually high readings. Soft
particles underlying the test piece would provide lower readings.

E. Tukon Microhardness Test

The Tukon or Knoop microhardness test was developed in 1939 [110] by the National Bureau of Standards as a method of measuring the hardness of very small constituent phases,
segregation effects, and hard and brittle materials. With proper selection of the load, the depth
of the impression will not exceed 1 μm.

The Knoop penetrator is a pyramidal diamond, cut to have an included transverse angle
of 130° and an included longitudinal transverse angle of 170° 30′. The resulting impression
is rhombic, with the long diagonal approximately seven times the length of the transverse
diagonal, as shown in Figure 38.

The test is conducted by placing the prepared metallographic specimen on a microscope
stage, and the desired location for the impression is determined with the aid of a metallographic
microscope. Once located, the indenter stage is located over the desired area, and the tester
is actuated. The specimen is moved upward automatically by means of an elevating screw until
it makes contact with the indenter. The preload, is applied for approximately 20 s, then the
selected major load is applied gradually, reaching maximum load in approximately 20 s. The
load is removed, and the specimen is lowered. The operator moves the microscope to view
the impression, and the impression's long diagonal is measured.

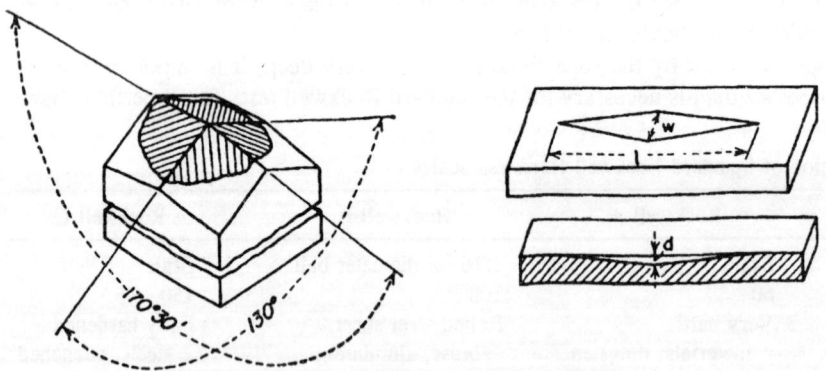

Figure 38 Schematic of Knoop indenter. (After Kehl [109].)

The longitudinal diagonal of the Knoop impression is measured with a filar eyepiece, which is part of the microhardness tester. This eyepiece has a fixed micrometer scale with a movable vertical hairline. The micrometer screw is divided into 100 divisions, with each division corresponding to a lateral movement of 0.01 mm. Movement is maintained in one direction to eliminate any errors due to backlash in the micrometer gear train.

Using the filar eyepiece, the long diagonal is measured, and the Knoop hardness number (KHN) is calculated:

$$\text{KHN} = \frac{P}{0.7028l^2} \tag{72}$$

where p is the applied load and l is the length of the long diagonal (mm). Since the impression left by the microhardness tester is very small, it is necessary that the surface to be measured be prepared metallographically and be free of any surface scratches or other detrimental defects. The surface may be either etched or unetched. It is very important that the specimen be perpendicular to the penetrator. If the impression is lopsided, the measurements of the long diagonal will be inaccurate because of asymmetric elastic recovery. The specimen must be removed and releveled and oriented properly so that accurate measurements can be taken.

IX. TOUGHNESS TESTING

A. Charpy V Notch Test

The Charpy V notch test is a test for measuring impact strength in which a small notched bar is loaded dynamically in three-point bending. The specimen has a square cross section of 10 mm and a length of 55 mm. The bar contains a sharp notch with an included angle of 45° and a depth of 2 mm. The notch radius is small, 0.25 mm. The use of subsize specimens is permitted provided that they conform to ASTM E23-88.

European specifications for Charpy-type impact testing include the ISO-U and ISO-V specifications. The ISO-U specification calls for a U-shaped notch, which shows a gradual degradation in the impact energy absorbed as the testing temperature is lowered. The ISO-V test is identical to the U.S. standard ASTM E23-88. There have been studies [111] that indicate that in the upper shelf area the absorbed energy is directly proportional to the width of the specimen. In other words, a specimen 5 mm across with a 10 mm depth will show 50% of the impact strength exhibited by the full-size specimen [112]. This is applicable to both transverse and longitudinal loading. There is very little difference in the lower shelf region.

Details of the testing procedure are covered in ASTM E23-88 [113]. In this test, the test specimen is removed from its cooling (or heating) bath and placed on the specimen fixture (Figure 39). The pendulum is released, and the specimen is broken within 5 s after removal from the bath. The calibrated dial of the impact machine is read, and the broken specimen is retrieved. If high strength, low energy specimens are tested at low temperatures, the specimens have a tendency to leave the machine perpendicular to the swing of the pendulum. This may cause errors in reading (as well as pose hazards to the operator) from the specimens hitting the pendulum. Because of conservation of energy, the specimens may leave the machine at speeds in excess of 50 ft/s. If the specimen hits the pendulum with sufficient energy, the pendulum will slow down and the machine will record a higher impact energy absorbed than truly occurred. This has been cited in ASTM E23-88 as the cause of much of the scatter in Charpy V notch testing in the 10–25 ft·lb range.

Being hit by the pendulum forces the specimen to bend and fracture. The strain rate of loading is high, approximately 10^3 s^{-1}. Because of the high strain rate, a considerable plastic

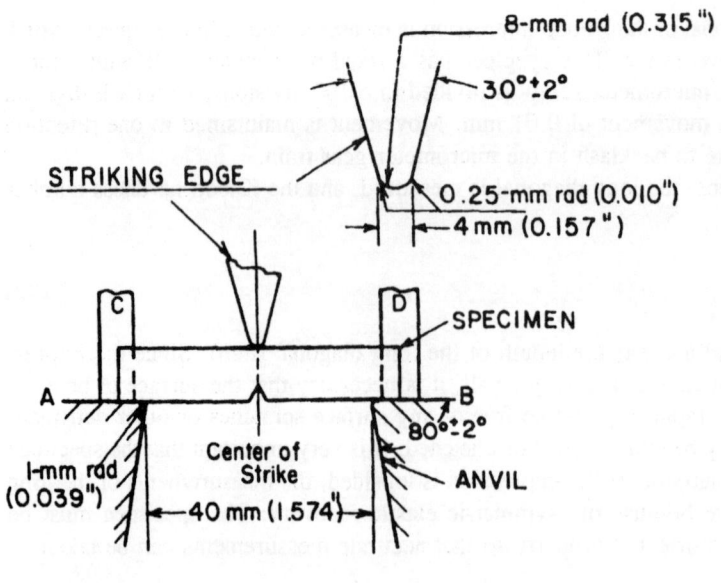

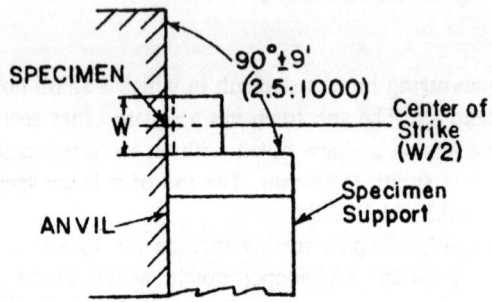

Figure 39 Fixture used for holding Charpy impact specimens.

constraint exists at the notch. This plastic constraint yields a triaxial stress state at the notch tip. Dieter [114] indicates that the maximum plastic stress concentration at the notch tip is given by

$$K_\sigma = 1 + \frac{\pi}{2} + \frac{\omega}{2} \tag{73}$$

where ω is the included angle of the notch. Because of the high strain rate, section thickness, and notch radius, a large amount of plane strain loading and triaxial stresses exist. There have been numerous correlations of the Charpy V notch and plane strain fracture toughness K_{Ic} testing [115–116].

B. Izod Test

The Izod notched bar test was developed at the beginning of this century for measuring notch-toughness. It was assumed originally that the impact energy of a material was proportional to its notch toughness and that there was advantage to testing at temperatures other than room temperature. Because the Izod specimen is held in a large vise (see schematic, Figure 40), there is a large heat sink present, making testing at temperatures other than room temperature problematic. Now the test is used primarily for the plastics industry. It is rarely used for the testing of metallic materials.

C. Dynamic Tear Test

The dynamic tear test was developed to measure the transition temperature of the steels used in the pipe industry. It is an ASTM standard [117] and is essentially a large scale Charpy V notch test. ASTM E604 has established the width of the specimen to 0.625 in., but thicknesses as great as 12 in. have been tested. Typical specimen dimensions are shown in Figure 41.

The test involves the dynamic tear specimen being impact loaded in three-point bending, and the energy absorbed is measured. Other measurements performed include the percentage of brittle or ductile failure to determine the FATT temperature. The testing machine can either be as pendulum type like a Charpy V notch machine or a drop-weight type with the capacity to break the specimen in a single blow. For most steels, with a 0.625 in. (16 mm) thick specimen, the capacity necessary to conduct the dynamic tear test is approximately 2700 J (2000 ft·lb$_f$).

The notch on the specimen is prepared by machining a notch 0.475 in. deep with a radius of 0.0625 in. The angular root section of the notch is usually made by a precisely ground saw or an electric discharge machine. The notch if further sharpened by pressing a hardened high-speed tool steel (HRC 60 min) knife blade into the machined notch to a depth of 0.130 in. Notches have also been made with an electron beam welder. The high energy density of the beam creates a very sharp, brittle, and well-defined heat-affected zone. The notch is also embrittled by the use of alloying elements.

The dynamic tear test is conducted at a variety of temperatures, and the percentage of brittle failure is plotted as a function of temperature. This determines the transition tempera-

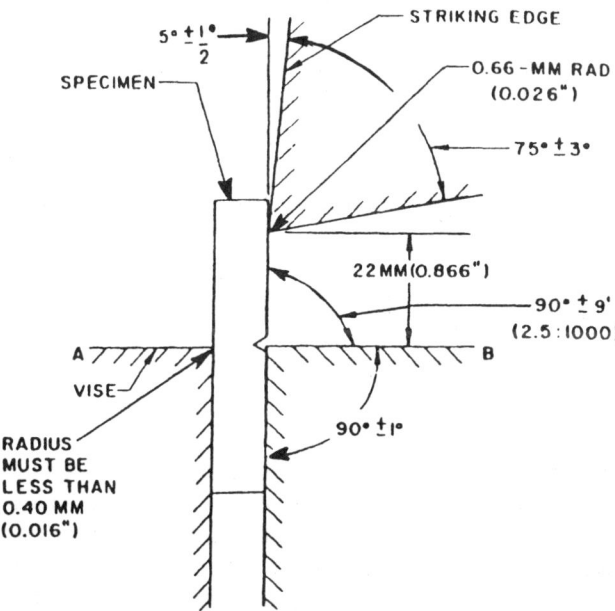

Figure 40 Izod device.

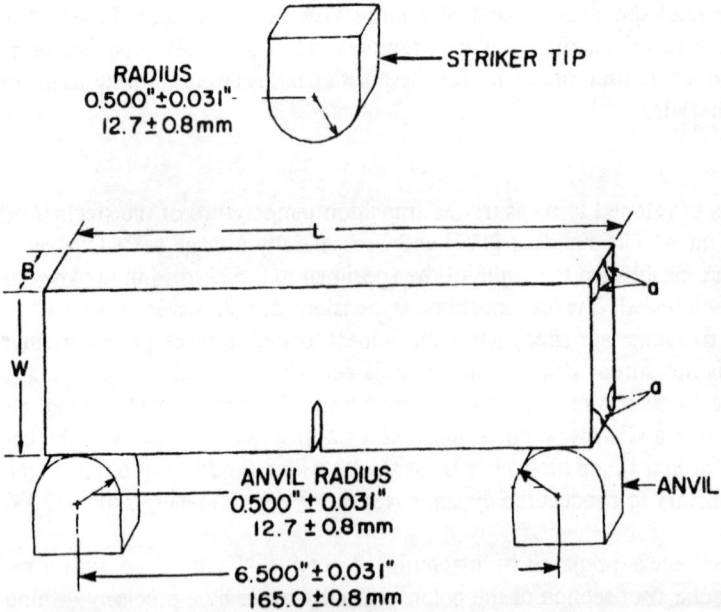

RADIUS
0.500"±0.031"
12.7±0.8 mm

◄— STRIKER TIP

ANVIL RADIUS
0.500"±0.031"
12.7±0.8 mm

◄—ANVIL

6.500"±0.031"
165.0±0.8 mm

Figure 41 Dynamic tear test specimens.

Dimensions and Tolerance for Specimen Blank

Parameter	Units	Dimension	Tolerance
Lenge L	in.	7.125	0.125
	mm	181.0	3.2
Width W	in.	1.60	0.10
	mm	38.0	2.5
Thickness B	in.	0.625	0.033
	mm	15.8	0.8
Angularity α	deg	90	2

ture (FATT) of the material being tested. The energy absorbed during fracture is also plotted, yielding similar information.

The strain rate of the dynamic tear test is similar to that of the Charpy V notch test at approximately 10^{-3} s^{-1}. Because of the greater section thickness, a higher state of triaxial stress exists in the specimen during testing than in the Charpy V notch test. A plane strain condition will be reached earlier than in the Charpy test as the testing temperature is decreased.

D.   Fracture Toughness (K_{Ic}) Testing

As indicated above, the use of fracture mechanics is important in determining the maximum flaw size that a material can withstand before failing catastrophically. As has been noted, cracking in a thick plate is worse than in a thin plate. This is because of plane strain conditions. At the crack tip, the plastic zone is small, with a high stress gradient across the plastic zone. A schematic of the plastic zone is shown in Figure 42. In addition, very high triaxial stresses are present. Because of this, the fracture appearance changes with specimen thickness. This is shown in Figure 43.

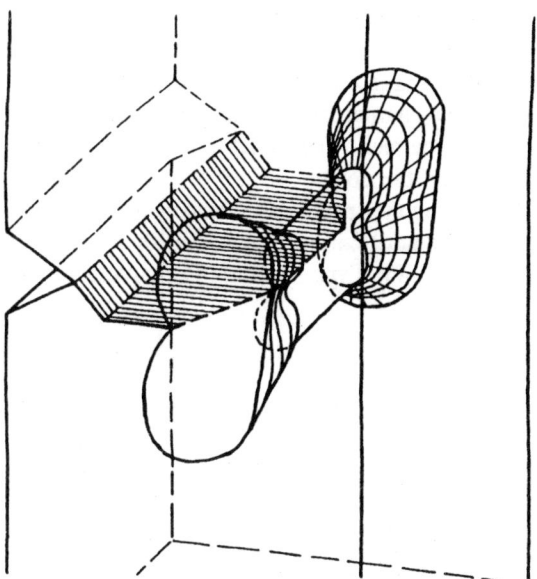

Figure 42 Plastic zone at crack tip.

In thin plates, the fracture is characterized by a mixed mode ductile and brittle fracture, with the presence of shear lips. Under plane strain conditions, when the plate is thick enough the fracture is flat, and the fracture stress is a constant with increasing thickness. The minimum thickness for plane strain conditions to occur is given by

$$B = 2.5(K_{IC}/\sigma_{ys})^2 \tag{74}$$

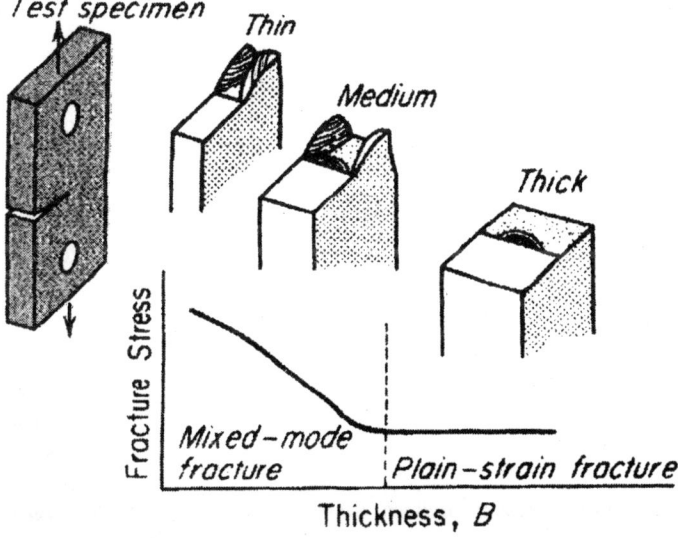

Figure 43 Thickness of specimen and resulting fracture surface.

Different configurations are used to determine the plane strain fracture toughness, K_{Ic}. These are shown in Figure 44. Other specimens include the center-cracked plate [118]. The center-cracked plate is also used in determining the fatigue crack growth rate.

The notch is machined in the specimen and made sharper by fatiguing at low cycle, high strain until the crack is about the width of the test specimen. The initial crack length is measured by including the length of the fatigue crack and the notch.

Testing of the specimen is accomplished by loading the specimen in tension, with the load and crack opening displacement continuously recorded until failure. In general, there are three types of load responses to the testing, shown in Figure 45. Type I loading is characteristic of

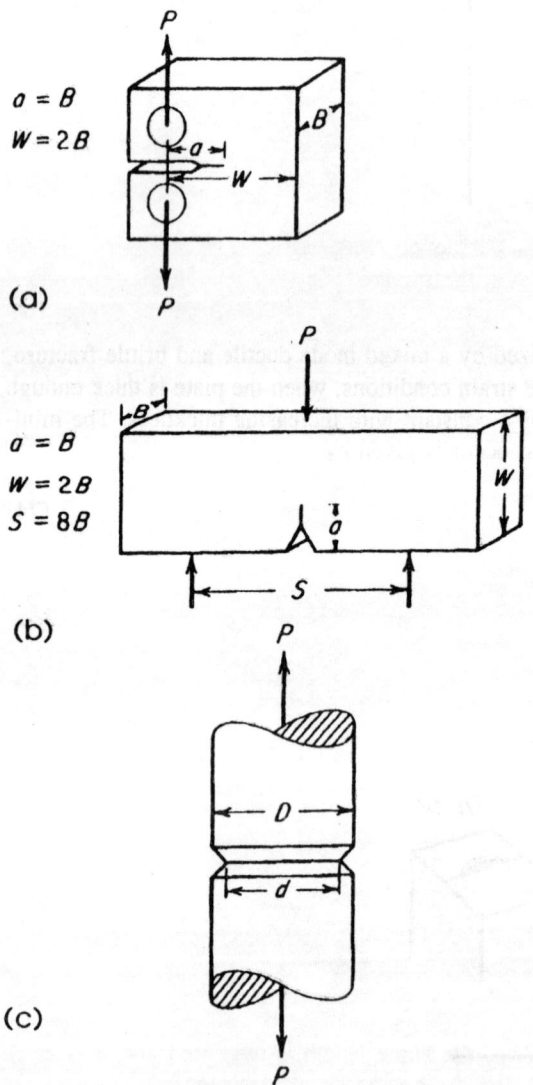

$a = B$
$W = 2B$

(a)

$a = B$
$W = 2B$
$S = 8B$

(b)

(c)

Figure 44 Typical K_{Ic} specimens. (a) Compact tension specimen; (b) bend specimen; (c) notched round specimen. (After Dieter [114].)

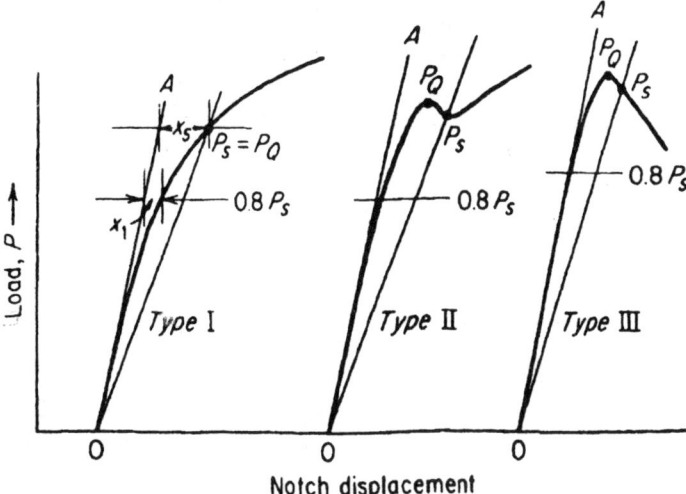

Figure 45 Typical K_{Ic} load responses to determine fracture toughness.

ductile materials, with no onset of unstable brittle fracture. For this type of response, the K_{Ic} value is difficult to obtain, and very thick specimens must be used before plane strain conditions occur. Type II response shows a sharp drop in the load with some load recovery. In this mode, the crack becomes unstable and propagates partially through the material until the crack front is blunted by plastic tearing. For this response, P_Q is considered the maximum load for K_{Ic} determination. In type III loading, a maximum load is reached, with brittle crack propagation occurring rapidly. In this response, plane strain conditions exist.

A conditional value of fracture toughness, K_Q is calculated using (for the compact specimen)

$$ K_Q = \frac{P_Q S}{Bb} \left[2.9 \left(\frac{a}{b} \right)^{1/2} - 4.6 \left(\frac{a}{b} \right)^{3/2} + 21.8 \left(\frac{a}{b} \right)^{5/2} - 37.6 \left(\frac{a}{b} \right)^{7/2} + 38.7 \left(\frac{a}{b} \right)^{9/2} \right] \tag{75} $$

For the three-point bending specimen,

$$ K_Q = \frac{P_Q}{Bb^{1/2}} \left[2.9.6 \left(\frac{a}{b} \right)^{1/2} - 185.5 \left(\frac{a}{b} \right)^{3/2} + 655.7 \left(\frac{a}{b} \right)^{5/2} - 1017 \left(\frac{a}{b} \right)^{7/2} + 638.9 \left(\frac{a}{b} \right)^{9/2} \right] \tag{76} $$

For the notched round specimen,

$$ K_Q = \frac{P}{D^{3/2}} \left(1.72 \frac{D}{d} - 1.27 \right) \tag{77} $$

To determine the fracture toughness, K_{Ic}, the crack length is measured and B is calculated:

$$ B = 2.5(K_Q/\sigma_{ys})^{1/2} \tag{78} $$

Table 9 Typical Fracture Toughness K_{Ic} of Some Steels

Steel	Yield strength σ_{ys} (ksi)	Fracture toughness K_{Ic} (ksi $\sqrt{in.}$)
SAE 4340	230	60
Cr-Mo-V	101	100
Ni-Mo-V	89	80
304 Stainless	30	100
HP9420 (9%Ni-4%Co-0.20% C)	180	170

If both B and a are less than the width b of the specimen, then $K_Q = K_{Ic}$. If not, then a thicker specimen is required, and K_Q is used to determine the new thickness using Equation (78). Typical K_{Ic} values for steels are shown in Table 9.

REFERENCES

1. M. Gensamer, *Fatigue and Fracture in Metals*, Wiley, New York, 1952.
2. R. von Mices, Der kritischer aussendruck Zylinderscher Rohre, *Z. Ver. Deut. Ing.* 58:750 (1914).
3. M. L. Williams, Analysis of brittle behavior in ship plates, in Symposium on the Effect of Temperature on the Brittle Behavior of Metals with Particular Reference to Low Temperatures, ASTM STP 158, 1954, pp. 11–44.
4. M. E. Shank, A critical survey of brittle fracture in carbon steel structures other than ships, in Symposium on the Effect of Temperature on the Brittle Behavior of Metals with Particular Reference to Low Temperatures, ASTM STP 158, 1954, pp. 45–110.
5. J. McCall and P. French, *Metallography in Failure Analysis*, Plenum, New York, 1978, p. 6.
6. L. F. Coffin, Flow and fracture of a brittle material, *J. Appl. Mech.* 17:233 (1950).
7. R. C. Grassi and I. Cornet, Fracture of gray cast iron tubes under biaxial stresses, *J. Appl. Mech.* 16:178 (1949).
8. P. Burton, A modification of the Coulomb–Mohs theory of fracture, *J. Appl. Mech., Ser. E* 28:259 (1961).
9. E. Orowan, *Welding J.* 34:157 (1955).
10. A. Griffith, *Phil. Trans. Roy. Soc. Lond.* 221A:163 (1920).
11. G. Dieter, *Mechanical Metallurgy*, McGraw-Hill, New York, 1976, p. 253.
12. C. E. Inglis, *Trans. Inst. Nav. Archit.* 55:219 (1913).
13. E. Klier, *Trans. Am. Soc. Met.* 43:935 (1951).
14. L. C. Chang, *J. Mech. Phys. Solids* 3:212 (1955).
15. D. K. Felbeck and E. Orowan, *Welding J.* 34:570s (1955).
16. E. Orowan, Ed., *Fatigue and Fracture of Metals*, Symposium at MIT, Wiley, New York, 1952.
17. G. R. Irwin, Fracture, *Encyclopedia of Physics*, Vol. VI, Springer, Heidelberg, 1958, p. 561.
18. P. C. Paris and G. C. Sih, *Fracture Toughness Testing*, ASTM STP 381 (J. E. Shawley and W. F. Brown, Eds.), American Society for Testing and Materials, Philadelphia, 1965.
19. J. W. Faupel and F. E. Fisher, *Engineering Design*, Wiley, New York, 1981.
20. R. W. Hertzberg, *Deformation and Fracture Mechanics of Engineering Materials*, Wiley, New York, 1976.
21. S. T. Rolfe and J. M. Barsom, *Fracture and Fatigue Control in Structures*, Prentice-Hall, Englewood Cliffs, NJ, 1977.
22. Anon., Standard Test Methods for Notched Bar Impact Testing of Metallic Methods, E23-88, American Society for Testing and Materials, Philadelphia, PA.
23. Anon., Standard Test Method for Dynamic Tear Testing of Metallic Materials, E604-83, American Society for Testing and Materials, Philadelphia, PA.
24. Anon., Standard Test Method for Plane Strain Fracture Toughness, K_{Ic}, of Metallic Materials, ASTM E399-74, American Society for Testing and Materials, Philadelphia, PA.
25. R. C. A. Thurston, The notch toughness of ultrahigh-strength steels in relation to design considerations, *Problems in the Load-Carrying Application of High Strength Steels*, DMIC Report 210, Oct. 26, 1964.
26. G. E. Dieter, *Mechanical Metallurgy*, McGraw-Hill, New York, 1976, p. 496.

27. J. A. Rinebolt and W. J. Harris, Jr., *Trans. ASM 43*:1175 (1951).
28. A. S. Tetelman and A. J. McEvily, *Fracture of Structural Materials*, Wiley, New York, 1967.
29. C. L. Cotrell, P. F. Langstone, and J. H. Rendall, *Iron Steel Inst. 20*(1):1032–1037 (1963).
30. E. P. Klier, ASTM STP 287, 1961, p. 196.
31. R. P. Wei, ASTM Preprint, ASTM Annual Meeting, June 1964.
32. L. L. Gilbert and J. A. Brown, *High Strength Steels for the Missile Industry*, ASM 3-39, 1961.
33. C. L. Cottrell, *High Strength Steels*, Iron Steel Inst. Spec. Rep. 76, 1962, pp. 1–6.
34. J. L. Shawin, G. B. Espey, L. J. Repko, and W. F. Brown, *Proc. ASTM 60*:761–777 (1960).
35. W. S. Owen, P. H. Whitman, M. Cohen, and R. L. Averbach, *Welding J. 36*:503s (1957).
36. W. R. Warke and A. R, Elsea, DMIC 154, Battelle Memorial Institute, Columbus, OH, 18 June 1962.
37. G. E. Pellissier, 3rd Annual Maraging Steel Project Review, RTD-TDR-63-4048, November 1963, p. 407.
38. W. F. Payne, ASTM Annual Meeting, 1964.
39. R. E. Smallmann, *Modern Physical Metallurgy*, Butterworth-Heinemann, Oxford, UK, 1985, p. 483.
40. J. E. Shawley and C. D. Beacham, NRL Rep. 5127, April 9, 1958.
41. C. W. Marshall, DMIC 147, Battelle Memorial Institute, Columbus, OH, 6 February 1961.
42. D. Francois, Relation between various fracture transition temperatures and the K_{Ic} fracture toughness transition curve, *Eng. Frac. Mech. 23*(2):455–465 (1986).
43. K. Ikeda and H. Kihara, Proceedings of the Second International Conference on Fracture (P. L. Pratt et al., Eds.), Chapman and Hall, London, 1969, pp. 851–867.
44. W. J. Plumbridge and D. A. Ryder, *Metall. Rev. 14*:136 (1969).
45. P. J. Forsyth and C. A. Stubbington, *J. Inst. Met. 83*:395 (1955).
46. W. A. Wood, *Some Basic Studies of Fatigue in Metals*, Wiley, New York, 1959.
47. A. H. Cottrel and D. Hull, *Proc. Roy. Soc. Lon. 242A*:211 (1953).
48. W. A. Wood, *Bull. Inst. Met. 3*:5 (1955).
49. C. Laird, Fatigue crack propagation, in ASTM STP 415, American Society of Testing and Materials, Philadelphia, 1967, p. 136.
50. P. J. Forsyth and D. A. Ryder, Some results of the examination of aluminum specimen fracture surfaces, *Metallurgica 63*:117 (1961).
51. G. Jacoby, Observations of crack propagation on the fracture surface, in *Current Aeronautical Fatigue Problems* (J. Schijve, Ed.), Pergamon, New York, 1965, p. 78.
52. W. R. Warke and J. M. McCall, *Fractography Using the Electron Microscope*, ASM Tech. Rep. W3-2-65, American Society of Metals, Metals Park, OH, 1965.
53. G. Jacoby, Fractographic methods, *Exp. Mech.*, 1965, p. 65.
54. P. J. Forsyth, A two stage process of fatigue crack growth, Symp. Crack Propagation, Cranfield, UK, 1961, Vol. II, p. 76.
55. H. J. Grover, S. A. Gordon, and L. R. Jackson, *Fatigue of Metals and Structures*, NAVWEPS Rep. 00-25-534, Bureau of Naval Weapons, Dept. of the Navy, Washington, DC, 1960.
56. K. E. Thelning, *Steel and Its Heat Treatment*, Butterworths, London, UK, 1984.
57. J. T. Ransom and R. F. Mehl, *Trans. AIME 185*:364 (1949).
58. P. H. Armitage, *Metall. Rev. 6*:353 (1964).
59. R. E. Little and E. H. Jebe, *Statistical Design of Fatigue Experiments*, Wiley, New York, 1975.
60. H. Muller-Stock, *Mitt. Kohle Eisenforsch GmbH 8*:83 (1938).
61. J. T. Ransom, ASTM STP 121, 1952, p. 59.
62. J. E. Shigley, *Mechanical Engineering Design*, McGraw-Hill, New York, 1977.
63. P. G. Forest, *Fatigue in Metals*, Addison-Wesley, Reading, PA, 1962.
64. N. E. Frost, K. J. Marsh, and L. P. Pook, *Metal Fatigue*, Oxford Univ. Press, London, UK, 1974.
65. L. Sors, *Fatigue Design of Machine Components*, Pergamon, Oxford, UK, 1971.
66. V. M. Fairies, *Design of Machine Elements*, Macmillan, New York, 1965.
67. R. A. Hammond and C. Williams, *Metall. Rev. 5*:165 (1960).
68. Anon., *Fatigue Durability of Carburized Steel*, American Society for Metals, Metals Park, OH, 1957.
69. J. H. Faupel and F. E. Fisher, *Engineering Design*, Wiley-Interscience, New York, 1981.
70. C. Lipson and R. C. Juvinall, *Stress and Strength*, Macmillan, New York, 1963.
71. L. P. Tarasov, W. S. Hyler, and H. R. Letner, *ASTM Proc. 57*:601 (1957).
72. R. C. Juvinall, *Stress, Strain and Strength*, McGraw-Hill, New York, 1967.
73. G. Castleberry, *Mach. Des. 50*:108 (Feb. 23, 1978).
74. D. Broek and J. Schijve, *The Influence of Sheet Thickness in the Fatigue Crack Propagation in* 2024-T3 *Alcad Sheet Material*, NLR Tech. Rep. M2129, Amsterdam, 1963.
75. I. S. Sokolnikoff, *Mathematical Theory of Elasticity*, McGraw-Hill, New York, 1956.
76. R. E. Peterson, *Stress Concentration Design Factors*, Wiley, New York, 1974.
77. W. C. Young, *Roark's Formulas for Stress and Strain*, McGraw-Hill, New York, 1989.

78. R. E. Peterson, in *Metal Fatigue* (G. Sines and J.L. Waisman, Eds.), McGraw-Hill, New York, 1959, p. 138.
79. H. Neuber, *Theory of Notch Stresses*, J. W. Edwards, Ann Arbor, MI 1946.
80. D. Walton and E. G. Ellison, *Inst. Metal Rev. 17*:100 (1972).
81. T. J. Crooker and E. A. Lange, *Inst. Metal Rev. 17*:94 (1972).
82. R. I. Stephens, Linear elastic fracture mechanics and its application to fatigue, SAE Paper 740220, Automotive Engineering Congress, Detroit, MI, 1974.
83. D. Broek, *Elementary Engineering Fracture Mechanics*, Noordhoff, Leyden, The Netherlands, 1974.
84. W. F. Brown and J. E. Shawley, *Plane Strain Fracture Toughness Testing of High Strength Metallic Materials*, ASTM STP 410, ASTM, Philadelphia, PA, 1966.
85. A. S. Kobayashi, Ed., *Experimental Techniques in Fracture Mechanics*, Vols. I and II, Soc. Experimental Stress Analysis, Westport, CT, 1975.
86. J. E. Campbell, W. E. Berry, and C. E. Feddergen, Eds., *Damage Tolerance Handbook—A Compilation of Fracture and Crack Growth Data for High Strength Alloys*, MCIC-HB-01, MCIC, Battelle Memorial Institute, Columbus, OH, January 1972.
87. T. P. Rich and D. J. Cartwright, Eds., *Case Studies in Fracture Mechanics*, AMMRC MS 77-5, Army Materials and Mechanics Research Center, Watertown, MA, June 1977.
88. H. Tada, P. C. Paris, and G. R. Irwin, *The Stress Analysis of Cracks Handbook*, Del Research Corp., Hellertown, PA, 1973.
89. H. Liebowitz, Ed., *Fracture Mechanics of Aircraft Structures*, AGARD-AG-176, NTIS, Springfield, VA, 1974.
90. G. Jacoby, *Application of Microfractography to the Study of Crack Propagation Under Fatigue Stress*, AGARD Rep. 541, NATO, 1966.
91. P. Kuhn, A comparison of fracture mechanics and notch analysis, presented to the ASTM Special Committee on Fracture Testing, January 1965.
92. J. E. Dorn, Ed., *Mechanical Behavior of Materials at Elevated Temperatures*, McGraw-Hill, New York, 1961.
93. R. W. Guard, *Prod. Eng. 27*(10):160–174 (1956).
94. I. Finnie and W. R. Heller, *Creep of Engineering Materials*, McGraw-Hill, New York, 1959.
95. E. N. da C. Andrade and B. Chalmers, *Proc. Roy. Soc Lond. 138A*:348 (1932).
96. J. B. Conway, *Trans. Metall. Soc. AIME 223*:2018 (1965).
97. F. Garofalo, *Properties of Crystalline Solids*, ASTM STP 283, ASTM, Philadelphia, 1965.
98. W. Rosenhahn and D. Ewen, *J. Inst. Met. 10*:119 (1913).
99. A. Nadai, *Theory of Flow and Fracture of Solids*, Vol. I, McGraw-Hill, New York, 1950.
100. Anon., *Standard Methods of Tension Testing Metallic Materials*, ASTM E8, American Society for Testing and Materials, Philadelphia, PA, 1979.
101. C. E. Lacy and M. Gensamer, *Trans. ASM 32*:88 (1944).
102. E. C. Bain and H. W. Paxton, *Alloying Elements in Steel*, ASM, Metals Park, OH, 1966.
103. P. Maynier, B. Jungmann, and J. Dollet, Creusot–Loire system for the prediction of the mechanical properties of low alloy steel products, in H*ardenability Concepts with Applications to Steel* (D. V. Doane and J. S. Kirkaldy, Eds.), Met. Soc. AIME, 1978, p. 518.
104. D. Venugopalan and J. S. Kirkardy, New relations for predicting the mechanical properties of quenched and tempered low alloy steels, in H*ardenability Concepts with Applications to Steel* (D. V. Doane and J. S. Kirkaldy, Eds.), Met. Soc. AIME, 1978, p. 249.
105. R. Hill, *The Mathematical Theory of Plasticity*, Clarendon Press, Oxford, UK, 1950.
106. M. C. Shaw and G. J. DeSalvo, *J. Eng. Ind. 92*:469 (1970).
107. M. C. Shaw and G. J. DeSalvo, *Met. Eng. Q. 121*(5):1 (1972).
108. S. Timoshenko and J. N. Goodier, *Theory of Elasticity*, 2nd ed., McGraw-Hill, New York, 1972, p. 372.
109. G. L. Kehl, *The Principles of Metallographic Laboratory Practice*, McGraw-Hill, New York, 1949.
110. F. C. Knoop, C. G. Peters, and W. B. Emerson, A sensitive pyramidal-diamond tool for indentation measurements, *J. Res. Natl. Bur. Stand. 23*:49 (1959).
111. G. Robiller, Influence of the width of the specimen on the results of the notched bar impact bending test, *Stahl Eisen 100*(19):1132–1138 (1955).
112. M. Lai and W. Ferguson, Effect of specimen thickness on fracture toughness, *Eng. Fracture Mech. 23*(4):649 (1986).
113. Anon., Standard Test for Notched Bar Impact Testing of Metallic Materials, ASTM E23-88, American Society for Testing and Materials, Philadelphia, PA, 1988.
114. G. E. Dieter, *Mechanical Metallurgy*, McGraw-Hill, New York, 1976, p. 496.
115. R. H. Sailors and H. T. Corten, Relationship between material facture toughness using fracture mechanics and transition temperature test, in *Fracture Toughness*, Proc. 1971 Natl. Symp. Fracture Mech., Part

II, ASTM STP 514, American Society for Testing and Materials, 1972, pp. 164–191.
116. D. Francois and A. Krasowsky, Relation between various fracture transition temperatures and the K_{Ic} fracture toughness transition curve, *Eng. Fracture Mech. 23*(2):455 (1986).
117. Anon., Standard Test Method for Dynamic Tear Testing of Metallic Materials, E604-83, American Society for Testing and Materials, Philadelphia, PA, 1983.
118. K. M. Kraft, *Techniques of Metal Research*, Vol. V, Wiley, New York, 1971, Chap. 7.

15

STEEL NOMENCLATURE

Anil Kumar Sinha
AKS Associates, Fort Wayne, Indiana

I. INTRODUCTION

Steels are the alloys of carbon and iron with between 0.05 and 2.85% carbon content; some manganese and sometimes other alloying elements are added to obtain the desired properties [1–3]. Steels are the most complex and widely used engineering materials because of (1) the abundance of iron in the earth's crust, (2) the high melting temperature of iron (1534°C), (3) a range of mechanical properties such as from moderate (200–300 MPa or 30–40 ksi) yield strength with excellent ductility to in excess of 1400 MPa or 200 ksi yield strength with fracture toughness up to 100 MPa m^{-2} or 100 ksi in.$^{-2}$, and (4) associated microstructures produced by solid-state phase transformations by varying the cooling rate from the austenitic condition [1].

This chapter describes the effects of alloying elements on the properties and/or characteristics of steels, reviews the various systems used to classify steels, and provides extensive tabular data relating to the designation of steels (both domestic and international) [2].

II. EFFECTS OF ALLOYING ELEMENTS

Steels contain alloying elements and impurities that must be associated with austenite, ferrite, and cementite. The combined effect of alloying elements and heat treatment produce an enormous variety of microstructures and properties. Given the limited scope of this chapter, it would be difficult to include a detailed survey of the effects of alloying elements on the iron-carbon equilibrium diagram. This complicated subject, which lies in the domain of ferrous physical metallurgy, has been reviewed extensively in Chapter 2 of this handbook and elsewhere in the literature [1, 4–8].

In this section, the effects of various elements on steelmaking (deoxidation) practices and steel characteristics will be briefly outlined. It should be noted that the effects of a single alloying element on either practice or characteristics are modified by the influence of other elements. These interrelationships must be considered when evaluating a change in the composition of a steel. To simplify the matter, the effects of various alloying elements listed below are summarized separately.

Carbon The amount of carbon required in the finished steel limits the type of steel that can be made. As the C content of rimmed steels increases, surface quality deteriorates. Killed steels in the approximate range of 0.15–0.30% C may have poorer surface quality and require special processing to attain surface quality comparable to steels with higher or lower carbon contents. Carbon has a moderate tendency to segregate, and segregation of carbon is often more significant than that of any other alloying elements. Carbon is the main hardening element in all steels except the austenitic PH stainless steels and maraging steels. Tensile strength in as-rolled conditions increases as the C content increases up to about 0.85%. Ductility and weldability decrease with increasing C content [2].

Manganese Manganese is present in virtually all steels in amounts of 0.3% or more [8]. Manganese is essentially a deoxidizer and desulfurizer [9]. It has a lesser tendency for macrosegregation than any of the common elements. Steels above 0.60% Mn cannot be readily rimmed.

Manganese is beneficial to surface quality in all carbon ranges (with the exception of extremely low carbon rimmed steels) and is especially beneficial in resulfurized and free-cutting steels due to a reduction in the risk of red shortness. Mn increases the strength and hardness and, also favorably affects forgeability and weldability. It is a solid-solution strengthener in steel and is very effective in increasing hardenability. However, large quantities (>2%) result in an increased tendency toward cracking and distortion during quenching [1,10].

Phosphorus Phosphorus segregates, but to a lesser extent than C and S. A small amount of P dissolves in the ferrite and slightly increases the steel's strength and hardness. A large quantity decreases ductility and notch impact toughness in the as-rolled condition, rendering the steel cold-short (i.e., having the tendency to crack during cold working) [1]. The latter deleterious effects are greater in quenched and tempered higher carbon steels. Higher P content is often specified in low-carbon free-machining steels to improve machinability. In low-alloy structural steels containing ~0.1% C, P increases strength and atmospheric corrosion resistance (rust-resistant steels). In austenitic Cr-Ni steels, the addition of phosphorus can cause precipitation effects and an increase in yield point [10].

Sulfur Increased amounts of S have a detrimental effect on transverse ductility, notch impact toughness, weldability, and surface quality (particularly in the lower carbon and lower manganese steels) but has a slight effect on longitudinal mechanical properties. It can cause reduction in hot-working properties (i.e., increased red/hot shortness) due to the low-melting sulfide eutectics surrounding the grains in reticular fashion [10, 11]. Higher S grades (>0.05%) are more susceptible to quench cracking than the low-S grades. There are several reasons for this.

1. Sulfur, mainly present in the form of sulfide inclusions, has a greater segregation tendency than any other common elements. Obviously, greater segregation or frequency of such inclusions can be expected in the resulfurized grades.
2. The surface of the hot-rolled high-sulfur steel has a greater tendency to form seams, which act as stress raisers during quenching.
3. They are usually coarse grained for better machinability, which increases brittleness and therefore promotes quench cracking [12].

Hence, only a low S content (<0.05%) is maintained in most carbon steels, where good weldability, fabrication properties, and minimum quench cracking tendency during hardening are desired. However, S in the range of 0.08–0.35% is added intentionally to steels (called free-machining grades) for increased machinability [13].

Sulfur improves the fatigue life of bearing steels [14], because (1) MnS inclusions produce compressive stresses in the surrounding matrix, (2) the thermal coefficient of these inclusions is higher than that of steel, and (3) they coat or cover other inclusions (notably silicate and alumina), thereby reducing the tensile stresses [5, 15].

Silicon Silicon is one of the principal deoxidizers used in steelmaking; therefore, Si content also determines the type of steel produced. Killed carbon steels may contain any amount of Si up to 0.60% maximum. Semikilled steels may contain moderate amounts of Si, although there is a definite maximum amount that can be tolerated in these steels. Rimmed and capped steels contain no significant amounts of Si. For example, in rimmed steel, the Si content is generally less than 0.10%.

Silicon is somewhat less effective than Mn in increasing as-rolled strength and hardness. Silicon has only a slight tendency to segregate. It has a detrimental effect on surface quality in low-carbon steels, and this condition is pronounced in low-carbon resulfurized grades.

When Si content in steel is below 0.30%, it dissolves completely in ferrite, increasing its strength without greatly decreasing ductility. Beyond 0.40%, a marked decrease in ductility is noticed in plain carbon steels [1]. However, it increases wear resistance in Si-Mn heat-treated steels, elastic limit in spring steels, and scale resistance in heat-resistant steels. Silicon is also used in electrical quality steel sheet [10].

Cobalt Cobalt inhibits grain growth at high temperatures and significantly improves the retention of temper and high-temperatures strength, resulting in an increase in tool life. The use of Co is generally restricted to high-speed steels, hot-forming tool steels, and creep-resistant and high-temperature materials [8, 10].

Copper Copper addition has a moderate tendency to segregate. Above 0.30%, Cu can cause precipitation hardening. It increases hardenability. If present in appreciable amounts, it is detrimental to hot-working operations. Cu is detrimental to surface quality and exaggerates the surface defects inherent in resulfurized steels. However, Cu improves the atmospheric corrosion resistance (when in excess of 0.20%) and the tensile properties in alloy and low-alloy steels and reportedly helps the adhesion of paint [2, 9]. In acid-resistant high-alloy steels, a Cu content above 1% results in improved resistance to HCl and H_2SO_4 acids [10].

Lead Lead is sometimes added (in the range of 0.2–0.5%) to carbon and alloy steels through mechanical dispersion during teeming to improve machinability.

Boron Boron, in very small amounts (0.0005–0.0035%), has a startling effect on the hardenability of steel. It also improves the hardenability of other alloying elements, and in the United States at least, it is being used as a very economical substitute for some of the more expensive elements. The beneficial effects of boron are only apparent with lower and medium carbon steels, there being no real increase in hardenability above 0.6% C [9]. The weldability of boron-alloyed steels is another principal reason for their use. However, large amounts of boron result in brittle, unworkable steels.

Chromium Chromium is a strong carbide former. It increases hardenability and corrosion and oxidation resistance, improves high-temperature strength and high-pressure hydrogenation properties, and enhances abrasion resistance in high-carbon grades. Chromium carbides are hard and wear-resistant and increase the edge-holding quality. Complex chromium-iron carbides slowly go into solution in austenite; therefore, a longer time at temperature is necessary to allow solution to take place before quenching is accomplished [2, 9]. Chromium is, in fact, the most efficient of the common hardening elements and is frequently used with a toughening element such as Ni to produce superior mechanical properties.

Nickel Nickel is a ferrite strengthener and toughener. As a result of the open austenite phase field, Ni contents > 7% produce austenitic structure to chemically resistant steels down to well below room temperature [10]. They also improve the impact strength at low temperature when other materials become embrittled [14].

The good ductility, toughness, and flexible heat treatment of low-carbon nickel steels make them good case-hardening materials [14]. In combination with Cr, Ni produces alloy steels with greater hardenability, impact strength, and fatigue resistance than are possible with carbon steels.

Molybdenum Molybdenum is a pronounced carbide former. The addition of Mo produces fine-grained steels, increases the depth of hardness, and improves the fatigue strength. It is added in constructional steels, usually in the range of 0.10–0.60%. Mo can induce secondary hardening during the tempering of quenched steels and improves the creep strength of low-alloy steels at elevated temperatures. Alloy steels containing 0.15–0.30% Mo and/or V display a delayed temper embrittlement (TE) due to the slow precipitation of alloy carbides of increasing stability. Molybdenum increases corrosion resistance and is thus used a great deal with high-alloy Cr steels and with austenitic Cr-Ni steels. High Mo contents reduce the steel's susceptibility to pitting [10].

Tungsten Tungsten is a very important carbide former. In steel, W forms very hard, abrasion-resistant carbides. It promotes hot strength and red hardness and thus cutting ability. It improves toughness and prevents grain growth. This combination of properties makes it very useful in high-speed cutting tools [8]. However, W impairs scaling resistance.

Vanadium Vanadium is an excellent deoxidizer, carbide former, and grain refiner, but it is very expensive and scarce [13]. It dissolves to some extent in ferrite, imparting strength and toughness. Vanadium increases fatigue strength on the one hand but improves notch sensitivity on the other; it has no appreciable effect on the steel's corrosion resistance. Vanadium also forms nitrides and is present in most nitriding steels.

Vanadium increases wear resistance, edge-holding quality, and high-temperature strength. It is therefore used mainly as an additional alloying element in high-speed, hot-forming, and creep-resistant steels. It promotes the weldability of heat-treatable steels.

Vanadium steels exhibit a much finer structure than steels of a similar composition without V. Vanadium provides other important alloying effects such as increased hardenability, secondary hardening effect on tempering, and increased elevated temperature hardness. The presence of V retards the rate of temper embrittlement in molybdenum-bearing steels by a factor of 10; the mechanism has not yet been established.

Niobium and Tantalum Niobium and tantalum are ferrite formers and therefore reduce the austenite phase. Small amounts of Nb increase the yield strength and, to a lesser extent, the tensile strength of carbon steel. A 0.02% Nb addition can increase the yield strength of medium-carbon steel by 70–100 MPa (10–15 ksi). This increased strength may be accompanied by considerably reduced notch toughness unless special measures are employed to refine grain size during hot rolling. Grain refinement during hot rolling involves special thermo-mechanical treatment techniques such as controlled rolling practices, low finishing temperature for final reduction passes, and accelerated cooling after the completion of rolling.

Aluminum Aluminum is widely used as a deoxidizer and a grain refiner [4]. It has the drawback of a tendency to promote graphitization and is therefore undesirable in steels to be used for high-temperature applications. As Al forms very hard nitrides with nitrogen, it is usually an alloying element in nitriding steels. It increases scaling resistance and is therefore often added to alloy ferritic heat-resistant steels. Of all the alloying elements, Al is the most effective in controlling grain growth prior to quenching.

Titanium The effects of titanium are similar to those of V and Nb, but Ti is only beneficial in fully killed (aluminum-deoxidized) steels due to its strong deoxidizing effects. It is used widely in stainless steels as a carbide former for stabilization against intergranular corrosion. It increases creep rupture strength through formation of special nitrides and tends significantly to segregation and banding [10].

Ti, Zr, and V are effective grain growth inhibitors; however, for structural steels that require heat treatment (quenching and tempering), these three elements may have adverse effects on hardenability, because their carbides are quite stable and difficult to dissolve in austenite prior to quenching.

Zirconium   Zirconium is added to killed high-strength low-alloy steels to obtain improvements in inclusion characteristics, particularly sulfide inclusions, where modifications in inclusion shape improve ductility in transverse bending. Zr increases the life of heat-conducting materials and produces a contracted gamma-phase field [10].

Tin   Tin in relatively small amounts is harmful to steels for deep drawing, but for most uses the effects of tin in the amounts usually present are negligible [16]. It tends toward increased segregation and limits the gamma-phase field [10].

III. CLASSIFICATION OF STEELS

Steels can be classified by several different systems depending on [1–3]

1. The compositions, such as carbon, low-alloy, alloy, or stainless steels
2. The manufacturing methods, such as basic and acid open hearth or electric furnace methods
3. The finishing methods, such as hot rolling or cold rolling
4. The product shape, such as bar, plate, strip, tubing, or structural shape
5. The application, such as structural, spring, and high tensile steels
6. The oxidation practice employed, such as killed, semikilled, capped, and rimmed steels
7. The microstructure, such as ferritic, pearlitic, and martensitic (Figure 1)
8. The required strength level, as specified in ASTM standards
9. Heat treatment, such as annealing, quenching and tempering, and thermomechanical processing
10. Quality descriptors/classifications, such as forging quality and commercial quality

Among the above classification systems, chemical composition is the most widely used basis for designation and is given due emphasis in this chapter. Classification systems based on oxidation practice and quality descriptors are also briefly discussed.

A. Types of Steels Based on Deoxidation Practice

Steels, when cast into ingots, can be classified into four types according to the deoxidation practice used or, alternatively, by the amount of gas evolved during solidification. These types are called killed, semikilled, capped, and rimmed steels. If practically no gas is evolved, the steel is termed "killed" because it lies quietly in the molds. Increasing extents of gas evolution results in semikilled, capped, or rimmed steels [2, 3].

1. Killed Steel

Killed steel is a type of steel from which there is only slight or practically no evolution of gas during solidification of the steel ingot after pouring. The top of the ingot solidifies much faster than semikilled or rimmed steel. Killed steel is characterized by a homogeneous struc-

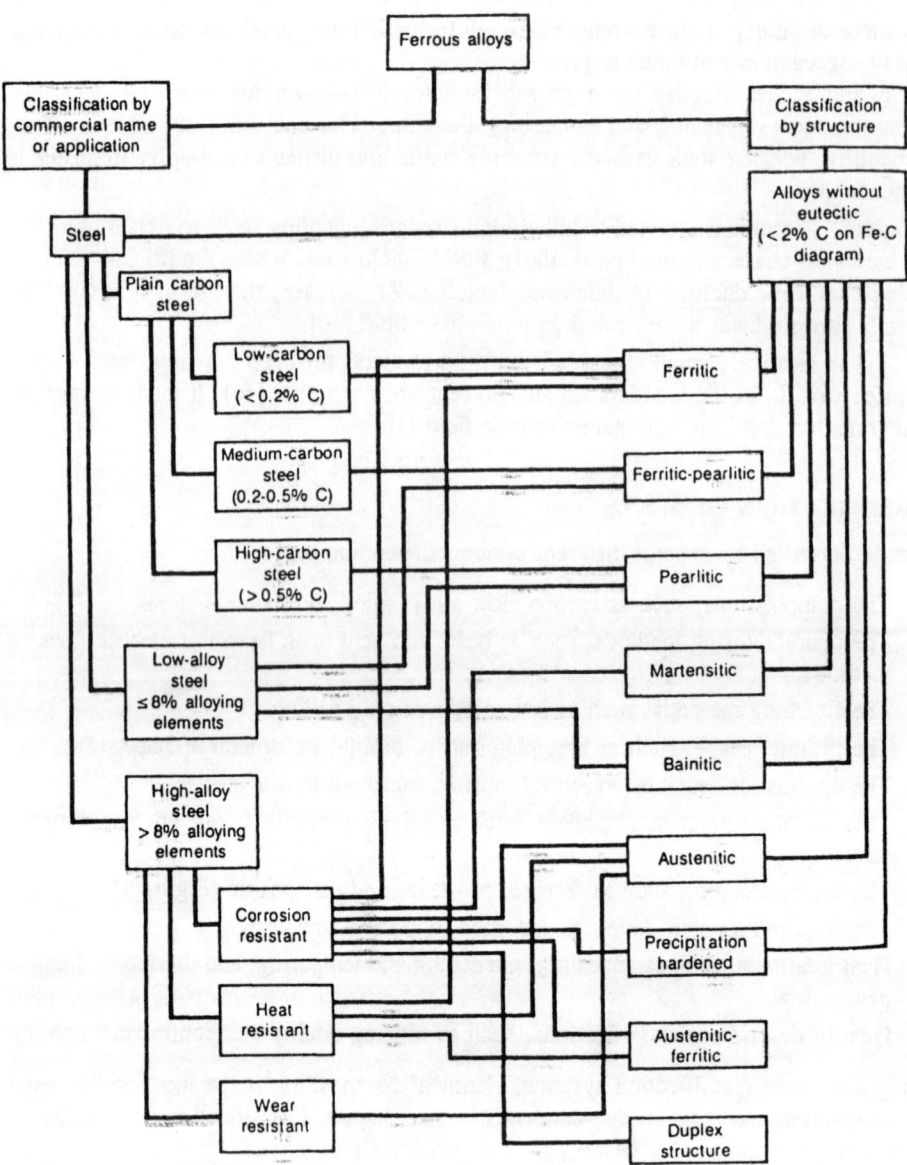

Figure 1 Classification of steels. (Courtesy of D.M. Stefanescu, University of Alabama, Tuscaloosa, AL.)

ture (free of blowholes and segregation), even distribution of chemical composition and properties, and formation of pipe in the upper central portion of the ingot, which is later cut off and discarded. All steels having more than 0.3% carbon are killed. Alloy steels, forging steels, and carburizing grades of steels are usually killed, where the essential quality criterion is soundness [16–18].

Killed steel is produced by various steel melting practices involving the use of certain deoxidizing elements that act with varying intensities. The most common deoxidizing elements are Al and ferroalloys of Mn and Si; however, calcium silicide and other special strong deoxidizers such as V, Ti, and Zr are sometimes used. Deoxidation practices in the manufacture of killed steels are generally left to the discretion of the producer.

2. Semikilled Steel

Gas evolution is not completely suppressed by deoxidizing additions in semikilled steel, because the steel is partially deoxidized. There is a greater degree of gas evolution than in killed steel, but less than in capped or rimmed steel. An ingot skin of considerable thickness is formed prior to the beginning of gas evolution. A correctly deoxidized semikilled steel ingot does not have a pipe but does have well-scattered large blowholes in the top-center half of the ingot; however, the blowholes weld shut during rolling of the ingot. Semikilled steels generally have a carbon content in the range of 0.15–0.30%. They find a wide range of uses in structural shapes, skelp, and pipe applications.

The main features of semikilled steels are (1) variable degrees of uniformity in composition, which are intermediate between those of killed and rimmed steels and less segregation than rimmed steel, and (2) a pronounced tendency for positive chemical segregation at the top center of the ingot (Figure 2).

3. Rimmed Steel

Rimmed steel is characterizied by a greater degree of gas evolution during solidification in the mold and a marked difference in chemical composition across the section and from top to bottom of the ingot (Figure 2). This results in the formation of an outer ingot skin or rim of relatively pure iron (hence the name rimmed steel) and an inner liquid (core) portion of the ingot with higher concentration of alloying elements, especially C, N, S, and P, having lower melting temperature. The high-purity zone in the surface is preserved during rolling [18]. Most low-carbon steels (<0.25% C) are rimmed steels. Rimmed ingots are best suited for the manufacture of flat-rolled products (plates, sheets) as well as wires, rods, and other products in the cold-rolled or subcritically annealed condition, where good surface or ductility is required [18].

The technology of producing rimmed steels limits the maximum content of C and Mn, and these maximum contents vary among producers. Rimmed steels are less expensive to make than killed or semikilled steels (because they are tapped without deoxidizer being added in the furnace and with only a small addition of deoxidizer made in the ladle) and have better ingot surfaces. Rimmed steels do not retain any significant amount of highly oxidizable elements such as Al, Si, or Ti.

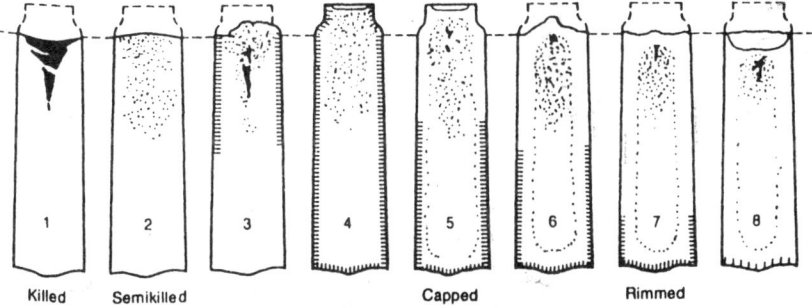

Figure 2 Eight typical conditions of commercial steel ingots, cast in identical bottle-top molds, in relation to the degree of suppression of gas evolution. The dotted line denotes the height to which the steel originally was poured in each ingot mold. Based on the carbon, and more significantly, the oxygen content of the steel, the ingot structures range from that of a completely killed ingot (No. 1) to that of a violently rimmed ingot (No. 8). (From Ref. 16.)

4. Capped Steel

Capped steel is a type of steel with characteristics similar to those of a rimmed steel but to a degree intermediate between that of rimmed and semikilled steels. Less deoxidizer is used to produce a capped ingot than to produce a semikilled ingot [19]. This induces a controlled rimming action when the ingot is cast. The gas entrapped during solidification is in excess of that required to counteract normal shrinkage, resulting in a tendency for the steel to rise in the mold.

Capping is a variation of rimmed steel practice. The capping operation confines the time of gas evolution and prevents the formation of an excessive number of gas voids within the ingot. The capped ingot practice is usually applied to steels with carbon contents greater than 0.15% that are used for sheet, strip, tin plate, skelp, wire, and bars.

Mechanically capped steel is poured into bottle-top molds using a heavy cast iron cap to seal the top of the ingot and to stop the rimming action [19].

Chemically capped steel is cast in open-top molds. The capping is accomplished by the addition of aluminum or ferrosilicon to the top of the ingot, causing the steel at the top surface to solidify rapidly. The top portion of the ingot is cropped and discarded.

B. Quality Descriptors/Classifications

Quality descriptors are names applied to various steel products to indicate that a particular product possesses certain characteristics that make it especially well suited for specific applications or fabrication processes. The quality designations/descriptors for various carbon steel products and alloy steel plates are listed in Table 1. Forging quality and cold-extrusion quality descriptors for carbon steels are self-explanatory. However, others are not explicit; for example, merchant quality hot-rolled carbon steel bars are made for noncritical applications requiring modest strength and mild bending or forming but not requiring forging or heat-treating operations. The quality classification for one particular steel commodity is not necessarily extended to subsequent products made from the same commodity; for example, standard quality cold-finished bars are produced from special quality hot-rolled carbon steel bars. Alloy steel plate qualities are described by structural, drawing, cold-working, pressure vessel, and aircraft qualities [17].

The various physical and mechanical characteristics indicated by a quality descriptor result from the combined effects of several factors such as (1) the degree of internal soundness; (2) the relative uniformity of chemical composition; (3) the number, size, and distribution of nonmetallic inclusions; (4) the relative freedom from harmful surface imperfections; (5) the extensive testing during manufacture; (6) the size of the discard cropped from the ingot; and (7) hardenability requirements. Control of these factors during manufacture is essential to achieve mill products with the desired characteristics. The degree of control over these and other related factors is another segment of information conveyed by the quality descriptor.

Some, but not all, of the basic quality descriptors may be modified by one or more additional requirements as may be appropriate, namely macroetch test, special discard, restricted chemical composition, maximum incidental (residual) alloying elements, austenitic grain size, and special hardenability. These limitations could be applied to forging quality alloy steel bars but not to merchant quality bars.

Understanding the various quality descriptors is difficult because most of the prerequisites for qualifying a steel for a specific descriptor are subjective. Only limitations on chemical composition ranges, residual alloying elements, nonmetallic inclusion count, austenitic grain size, and special hardenability are quantifiable. The subjective evaluation of the other attributes depends on the experience and skill of the individuals who make the evaluation. Although the

Table 1 Quality Descriptions[a] of Carbon and Alloy Steels

Carbon steels			Alloy steels

Semifinished for forging
Forging quality
  Special hardenability
  Special internal soundness
  Nonmetallic inclusion
    requirement
  Special surface
 	
Carbon steel structural sections
Structural quality
 	
Carbon steel plates
Regular quality
Structural quality
Cold-drawing quality
Cold-pressing quality
Cold-flanging quality
Forging quality
Pressure vessel quality
 	
Hot-rolled carbon steel bars
Merchant quality
Special quality
  Special hardenability
  Special internal soundness
  Nonmetallic inclusion
    requirement
  Special surface
Scrapless nut quality
Axle shaft quality
Cold extrusion quality
Cold-heading and cold-forging
  quality
 	
Cold-finished carbon steel bars
Standard quality
  Special hardenability
  Special internal soundness
  Nonmetallic inclusion
    requirement
  Special surface
Cold-heading and cold-forging
  quality
Cold extrusion quality
	Hot-rolled sheets
	Commercial quality
	Drawing quality
	Drawing quality special killed
	Structural quality
	
	Cold-rolled sheets
	Commercial quality
	Drawing quality
	Drawing quality special killed
	Structural quality
	
	Porcelain enameling sheets
	Commercial quality
	Drawing quality
	Drawing quality special killed
	
	Long terne sheets
	Commercial quality
	Drawing quality
	Drawing quality special killed
	Structural quality
	
	Galvanized sheets
	Commercial quality
	Drawing quality
	Drawing quality special killed
	Lock-forming quality
	
	Electrolytic zinc coated sheets
	Commercial quality
	Drawing quality
	Drawing quality special killed
	Structural quality
	
	Hot-rolled strip
	Commercial quality
	Drawing quality
	Drawing quality special killed
	Structural quality
	
	Cold-rolled strip
	Specific quality descriptions are not
	  provided in cold-rolled strip
	  because this product is largely
	  produced for specific end use
		Tin mill products
		Specific quality descriptions are not
		  applicable to tin mill products
		
		Carbon steel wire
		Industrial quality wire
		Cold extrusion wires
		Heading, forging, and roll-threading
		  wires
		Mechanical spring wires
		Upholstery spring construction wires
		Welding wire
		
		Carbon steel flat wire
		Stitching wire
		Stapling wire
		
		Carbon steel pipe
		
		Structural tubing
		
		Line pipe
		
		Oil country tubular goods
		
		Steel specialty tubular products
		Pressure tubing
		Mechanical tubing
		Aircraft tubing
		
		Hot-rolled carbon steel wire rods
		Industrial quality
		Rods for manufacture of wire intended
		  for electric welded chain
		Rods for heading, forging, and
		  roll-threading wire
		Rods for lock washer wire
		Rods for scrapless nut wire
		Rods for upholstery spring wire
		Rods for welding wire
			Alloy steel plates
			Drawing quality
			Pressure vessel quality
			Structural quality
			Aircraft physical quality
			
			Hot-rolled alloy steel bars
			Regular quality
			Aircraft quality or steel subject to
			  magnetic particle inspection
			Axle shaft quality
			Bearing quality
			Cold-heading quality
			Special cold-heading quality
			Rifle barrel quality, gun quality, shell
			  or A.P. shot quality
			
			Alloy steel wire
			Aircraft quality
			Bearing quality
			Special surface quality
			
			Cold-finished alloy steel bars
			Regular quality
			Aircraft quality or steel subject to
			  magnetic particle inspection
			Axle shaft quality
			Bearing shaft quality
			Cold-heading quality
			Special cold-heading quality
			Rifle barrel quality, gun quality, shell
			  or A.P. shot quality
			
			Line pipe
			
			Oil country tubular goods
			
			Steel specialty tubular goods
			Pressure tubing
			Mechanical tubing
			Stainless and head-resisting pipe,
			  pressure tubing, and mechanical
			  tubing
			Aircraft tubing
			Pipe

[a]In the case of certain qualities, P and S are usually finished to lower limits than the specified maximum.
Source: Ref. 3.

use of these subjective quality descriptors might appear impractical and imprecise, steel products made to meet the requirements of a specific quality descriptor can be relied upon to have those characteristics necessary for that product to be used in the suggested application or fabrication operation [2].

C. Classification of Steel Based on Chemical Composition

1. Carbon and Carbon-Manganese Steels

In addition to carbon, plain carbon steels contain the following other elements: Mn up to 1.65%, S up to 0.05%, P up to 0.04%, Si up to 0.60%, and Cu up to 0.60%. The effects of each of these elements in plain-carbon steels have been summarized in Section II.

Carbon steels can be classified according to various deoxidation practices (see Section III.A). Deoxidation practice and steelmaking process will have an effect on the characteristics and properties of the steel (see Section II). However, variations in C content have the greatest effect on mechanical properties, with increasing C additions leading to increased hardness and strength. As such, carbon steels are generally grouped according to their C content. In general, carbon steels contain up to 2% total alloying elements and can be subdivided into low-carbon steels, medium-carbon steels, high-carbon steels, and ultrahigh-carbon steels; each of these designations is discussed below.

As a group, carbon steels constitute the most frequently used steel. Table 2 lists various grades of standard carbon and low-alloy steels with SAE-AISI designations. Table 3 shows some representative standard carbon steel compositions with SAE-AISI and corresponding UNS designations [2, 3, 20].

Low-carbon steels contain up to 0.25% C. The largest category of this class is flat-rolled products (sheet or strip), usually in the cold-rolled or subcritically annealed condition and usually with final temper rolling treatment. The carbon content for high-formability and high-drawability steels is very low (< 0.10% C) with up to 0.40% Mn. These lower carbon steels are used in automobile body panels, tin plates, appliances, and wire products.

The low-carbon steels (0.10–0.25% C) in this group have increased strength and hardness and reduced cold formability compared to the lowest carbon group. They are designated as carburizing or case-hardening steels [4]. Selection of these grades for carburizing applications depends on the nature of the part, the properties required, and the processing practices preferred. An increase in carbon content of the base steel results in greater core hardness with a given quench. However, an increase in Mn increases the hardenability of both the core and the case.

A typical application for carburized plain carbon steel is for parts with hard wear-resistant surface but without any need for increased mechanical properties in the core, e.g., small shafts, plungers, or highly loaded gearing [3]. Rolled structural steels in the form of plates and sections contain ~ 0.25% C, with up to 1.5% Mn and Al, if improved toughness is required. When used for stampings, forgings, seamless tubes, and boiler plate, Al addition should be avoided. An important type of this category is the low-carbon free-cutting steels containing up to 0.15% C and up to 1.2% Mn, a minimum of Si and up to 0.35% S with or without 0.30% Pb. These steels are suited to automotive mass production manufacturing methods [1].

Medium-carbon steels containing 0.30–0.55% C and 0.60–1.65% Mn are used where higher mechanical properties are desired. They are usually hardened and strengthened by heat treatment or by cold work. Lower carbon and manganese steels in this group find wide applications for certain types of cold-formed parts that need annealing, normalizing, or quenching and tempering treatment prior to use. The higher carbon grades are often cold drawn to specified mechanical properties for use without heat treatment for some applications.

All of these steels can be used for forgings, the selection being dependent on the section size and the mechanical properties needed after heat treatment [3]. These grades, generally produced as killed steels, are used for a wide range of applications that include automobile parts for body, engines, suspensions, steering, engine torque converter, and transmission [21]. Some Pb and/or S additions make them free-cutting grades, whereas Al addition produces grain refinement and improved toughness. In general, steels containing 0.4–0.60% C are used as rails, railway wheels, tires, and axles.

High-carbon steels containing 0.55–1.00% C and 0.30–0.90% Mn have more restricted applications than the medium-carbon steels because of higher production cost and poor formability (or ductility) and weldability. High-carbon steels find applications in the spring industry (as light and thicker flat springs, laminated springs, and heavier coiled springs), farm implement industry (as plow beams, plowshares, scraper blades, discs, mowers, knives, and harrow teeth), and high-strength wires where improved wear characteristics and higher strength than those attainable with lower carbon grades are needed.

Ultrahigh-carbon (UHC) steels are experimental plain carbon steels with 1.0–2.1% C (15–32 vol % cementite) [22–24]. Optimum superplastic elongation has been found at about 1.6% C content [4]. These steels have the capability of emerging as important technological materials because they exhibit superplasticity. The superplastic behavior of these materials is at-

Table 2 SAE-AISI Designation System for Carbon and Low-Alloy Steels

Numerals and digits	Type of steel and nominal alloy content, %
Carbon steels	
10xx(a)	Plain carbon (Mn 1.00 max)
11xx	Resulfurized
12xx	Resulfurized and rephosphonized
15xx	Plain carbon (max Mn range: 1.00–1.65)
Manganese steels	
13xx	Mn 1.75
Nickel steels	
23xx	Ni 3.50
25xx	Ni 5.00
Nickel-chromium steels	
31xx	Ni 1.25; Cr 0.65 and 0.80
32xx	Ni 1.75; Cr 1.07
33xx	Ni 3.50; Cr 1.50 and 1.57
34xx	Ni 3.00; Cr 0.77
Molybdenum steels	
40xx	Mo 0.20 and 0.25
44xx	Mo 0.40 and 0.52
Chromium-molybdenum steels	
41xx	Cr 0.50, 0.80, and 0.95; Mo 0.12, 0.20, 0.25, and 0.30
Nickel-chromium-molybdenum steels	
43xx	Ni 1.82; Cr 0.50 and 0.80; Mo 0.25
43BVxx	Ni 1.82; Cr 0.50; Mo 0.12 and 0.25; V 0.03 min
47xx	Ni 1.05; Cr 0.45; Mo 0.20 and 0.35
81xx	Ni 0.30; Cr 0.40; Mo 0.12
86xx	Ni 0.55; Cr 0.50; Mo 0.20
87xx	Ni 0.55; Cr 0.50; Mo 0.25
88xx	Ni 0.55; Cr 0.50; Mo 0.35
93xx	Ni 3.25; Cr 1.20; Mo 0.12
94xx	Ni 0.45; Cr 0.40; Mo 0.12
97xx	Ni 0.55; Cr 0.20; Mo 0.20
98xx	Ni 1.00; Cr 0.80; Mo 0.25
Nickel-molybdenum steels	
46xx	Ni 0.85 and 1.82; Mo 0.20 and 0.25
48xx	Ni 3.50; Mo 0.25
Chromium steels	
50xx	Cr 0.27, 0.40, 0.50, and 0.65
51xx	Cr 0.80, 0.87, 0.92, 0.95, 1.00, and 1.05
Chromium (bearing) steels	
50xxx	Cr 0.50 ⎫
51xxx	Cr 1.02 ⎬ C 1.00 min
52xxx	Cr 1.45 ⎭
Chromium-vanadium steels	
61xx	Cr 0.60, 0.80, and 0.95; V 0.10 and 0.15 min
Tungsten-chromium steel	
72xx	W 1.75; Cr 0.75
Silicon-manganese steels	
92xx	Si 1.40 and 2.00; Mn 0.65, 0.82, and 0.85; Cr 0 and 0.65
High-strength low-alloy steels	
9xx	Various SAE grades
Boron steels	
xxBxx	B denotes boron steel
Leaded steels	
xxLxx	L denotes leaded steel

(a) The xx in the last two digits of these designations indicates that the carbon content (in hundredths of a percent) is to be inserted.

Source: Courtesy of ASM International, Materials Park, OH.

Table 3 Standard Carbon Steel Compositions with SAE-AISI and Corresponding UNS Designations.
Applicable to semifinished products for forging, hot-rolled and cold-finished bars, wire rods, and seamless tubing.

Plain Carbon Steel (Nonresulfurized, 1.0% Mn max)[a]

UNS number	SAE-AISI number	Cast or heat chemical ranges and limits, %(a)			
		C	Mn	P max	S max
G10060	1006	0.08 max	0.45 max	0.040	0.050
G10080	1008	0.10 max	0.50 max	0.040	0.050
G10090	1009	0.15 max	0.60 max	0.040	0.050
G10100	1010	0.08–0.13	0.30–0.60	0.040	0.050
G10120	1012	0.10–0.15	0.30–0.60	0.040	0.050
G10150	1015	0.12–0.18	0.30–0.60	0.040	0.050
G10160	1016	0.12–0.18	0.60–0.90	0.040	0.050
G10170	1017	0.14–0.20	0.30–0.60	0.040	0.050
G10180	1018	0.14–0.20	0.60–0.90	0.040	0.050
G10190	1019	0.14–0.20	0.70–1.00	0.040	0.050
G10200	1020	0.17–0.23	0.30–0.60	0.040	0.050
G10210	1021	0.17–0.23	0.60–0.90	0.040	0.050
G10220	1022	0.17–0.23	0.70–1.00	0.040	0.050
G10230	1023	0.19–0.25	0.30–0.60	0.040	0.050
G10250	1025	0.22–0.28	0.30–0.60	0.040	0.050
G10260	1026	0.22–0.28	0.60–0.90	0.040	0.050
G10300	1030	0.27–0.34	0.60–0.90	0.040	0.050
G10330	1033	0.29–0.36	0.70–1.00	0.040	0.050
G10350	1035	0.31–0.38	0.60–0.90	0.040	0.050
G10370	1037	0.31–0.38	0.70–1.00	0.040	0.050
G10380	1038	0.34–0.42	0.60–0.90	0.040	0.050
G10390	1039	0.36–0.44	0.70–1.00	0.040	0.050
G10400	1040	0.36–0.44	0.60–0.90	0.040	0.050
G10420	1042	0.39–0.47	0.60–0.90	0.040	0.050
G10430	1043	0.39–0.47	0.70–1.00	0.040	0.050
G10450	1045	0.42–0.50	0.60–0.90	0.040	0.050
G10460	1046	0.42–0.50	0.70–1.00	0.040	0.050
G10490	1049	0.45–0.53	0.60–0.90	0.040	0.050
G10500	1050	0.47–0.55	0.60–0.90	0.040	0.050
G10550	1055	0.52–0.60	0.60–0.90	0.040	0.050
G10600	1060	0.55–0.66	0.60–0.90	0.040	0.050
G10640	1064	0.59–0.70	0.50–0.80	0.040	0.050
G10650	1065	0.59–0.70	0.60–0.90	0.040	0.050
G10700	1070	0.65–0.76	0.60–0.90	0.040	0.050
G10740	1074	0.69–0.80	0.50–0.80	0.040	0.050
G10750	1075	0.69–0.80	0.40–0.70	0.040	0.050
G10780	1078	0.72–0.86	0.30–0.60	0.040	0.050
G10800	1080	0.74–0.88	0.60–0.90	0.040	0.050
G10840	1084	0.80–0.94	0.60–0.90	0.040	0.050
G10850	1085	0.80–0.94	0.70–1.00	0.040	0.050
G10860	1086	0.80–0.94	0.30–0.50	0.040	0.050
G10900	1090	0.84–0.98	0.60–0.90	0.040	0.050
G10950	1095	0.90–1.04	0.30–0.50	0.040	0.050

Table 3 Continued

Free-Cutting (Resulfurized) Carbon Steel Compositions[a]

UNS number	SAE-AISI number	Cast or heat chemical ranges and limits, %			
		C	Mn	P max	S
G11080	1108.............	0.08–0.13	0.50–0.80	0.040	0.08–0.13
G11100	1110.............	0.08–0.13	0.30–0.60	0.040	0.08–0.13
G11170	1117.............	0.14–0.20	1.00–1.30	0.040	0.08–0.13
G11180	1118.............	0.14–0.20	1.30–1.60	0.040	0.08–0.13
G11370	1137.............	0.32–0.39	1.35–1.65	0.040	0.08–0.13
G11390	1139.............	0.35–0.43	1.35–1.65	0.040	0.13–0.20
G11400	1140.............	0.37–0.44	0.70–1.00	0.040	0.08–0.13
G11410	1141.............	0.37–0.45	1.35–1.65	0.040	0.08–0.13
G11440	1144.............	0.40–0.48	1.35–1.65	0.040	0.24–0.33
G11460	1146.............	0.42–0.49	0.70–1.00	0.040	0.08–0.13
G11510	1151.............	0.48–0.55	0.70–1.00	0.040	0.08–0.13

Standard Resulfurized and Rephosphorized Carbon Steels[a]

UNS number	SAE-AISI number	Cast or heat chemical ranges and limits, %(a)				
		C max	Mn	P	S	Pb
G12110	1211............	0.13	0.60–0.90	0.07–0.12	0.10–0.15	. . .
G12120	1212............	0.13	0.70–1.00	0.07–0.12	0.16–0.23	. . .
G12130	1213............	0.13	0.70–1.00	0.07–0.12	0.24–0.33	. . .
G12150	1215............	0.09	0.75–1.05	0.04–0.09	0.26–0.35	. . .
G12144	12L14[b].........	0.15	0.85–1.15	0.04–0.09	0.26–0.35	0.15–0.35

Standard Nonresulfurized Carbon Steels (Over 1.0% Manganese)

UNS number	SAE-AISI number	Cast or heat chemical ranges and limits, %			
		C	Mn	P max	S max
G15130	1513.............	0.10–0.16	1.10–1.40	0.040	0.050
G15220	1522.............	0.18–0.24	1.10–1.40	0.040	0.050
G15240	1524.............	0.19–0.25	1.35–1.65	0.040	0.050
G15260	1526.............	0.22–0.29	1.10–1.40	0.040	0.050
G15270	1527.............	0.22–0.29	1.20–1.50	0.040	0.050
G15360	1536.............	0.30–0.37	1.20–1.50	0.040	0.050
G15410	1541.............	0.36–0.44	1.35–1.65	0.040	0.050
G15480	1548.............	0.44–0.52	1.10–1.40	0.040	0.050
G15510	1551.............	0.45–0.56	0.85–1.15	0.040	0.050
G15520	1552.............	0.47–0.55	1.20–1.50	0.040	0.050
G15610	1561.............	0.55–0.65	0.75–1.05	0.040	0.050
G15660	1566.............	0.60–0.71	0.85–1.15	0.040	0.050

[a]It is not common practice to produce the 12xx series of steels to specified limits for silicon because of its adverse effect on machinability.

[b]Contains 0.15–0.35% lead; other steels listed here can be produced with similar amounts of lead.

Source: Ref. 20.

tributed to the structure consisting of uniform distribution of very fine, spherical, discontinuous particles (0.1–1.5 μm diameter) in a very fine-grained ferrite matrix (0.5–2.0 μm diameter) that can be readily achieved by any of four thermomechanical treatment routes described elsewhere [1].

2. Low-Alloy Steels

Alloy steels may be defined as those steels that owe their improved properties to the presence of one or more special elements or to the presence of larger proportions of elements such as Mn and Si than are ordinarily present in carbon steels [16]. Alloy steels contain Mn, Si, or Cu, in quantities greater than the maximum limits (e.g., 1.65% Mn, 0.60% Si, and 0.60% Cu) of carbon steels, or they contain specified ranges or minimums of one or more other alloying elements. The alloying elements increase the mechanical and fabrication properties. Broadly, alloy steels can be divided into (1) low-alloy steels containing 2–8 wt % total noncarbon alloy addition and (2) alloy steels with more than 8 wt % total noncarbon alloy addition. In some cases, an overlap exists between the low-alloy and alloy steels. Table 4 lists the low-alloy steel compositions with SAE-AISI and corresponding UNS designations.

Low-alloy steels constitute a group of steels that exhibit superior mechanical properties compared to plain carbon steels as the result of addition of such alloying elements as Ni, Cr, and Mo. For many low-alloy steels, the main function of the alloying elements is to increase hardenability in order to optimize mechanical properties and toughness after heat treatment. In some instances, however, alloying elements are used to reduce environmental degradation under certain specified conditions.

Low-alloy steels can be classified according to (1) chemical composition such as nickel steels, nickel-chromium steels, molybdenum steels, chromium-molybdenum steels, and so forth, based on the principal alloying element(s) present and as described in Table 2; (2) heat treatment such as quenched and tempered, normalized and tempered, annealed, and so on; and (3) weldability.

Because of the large variety of chemical compositions possible and the fact that some steels are employed in more than one heat-treated condition, some overlap exists among the low-alloy steel classifications. However, these grades can be addressed into four major groups such as (1) low-carbon quenched and tempered (QT) steels, (2) medium-carbon ultrahigh-strength steels, (3) bearing steels, and (4) heat-resistant Cr-Mo steels (see Table 5).

Low-carbon quenched and tempered steels (also called low-carbon martensitic steels) are characterized by a relatively high yield strength with minimum yield strength of 690 MPa (100 ksi) and good notch toughness, ductility, corrosion resistance, or weldability. The chemical composition of typical low-carbon QT (or martensitic) steels are listed in Table 5. These steels are not included in SAE-AISI classification. However, they are covered by ASTM designations, and a few steels, such as HY-80 and HX-100, are included by military specifications. The steels listed are primarily available in the form of plate, sheet, bar, structural shape, or forged products. They are extensively used for a wide variety of applications such as pressure vessels, earth moving and mining equipment and as major members of large steel structures. They are also used for cold-headed and cold-forged parts such as fasteners or pins and heat treated to the desired properties [16].

Medium-carbon ultrahigh-strength steels are structural steels with very high strength. These steels exhibit a minimum yield strength of 1380 MPa (200 ksi). Table 5 lists typical compositions such as AISI/SAE 4130, higher-strength 4140, deeper hardening higher-strength 4340, 300M [a modification of 4340 steel with increased (1.6%) Si content to prevent embrittlement when the steel is tempered at the low temperatures required for very high strength], and Ladish D-6a and Ladish D-6ac steels (another modification of 4340 with grain refiner V and higher C, Cr, and Mo contents, developed for aircraft and missile structural applications). Other less

Table 4 Low-Alloy Steel Compositions Applicable to Billets, Blooms, Slabs, and Hot-Rolled and Cold-Finished Bars. (Slightly wider ranges of compositions apply to plates.)

UNS number	SAE number	Corresponding AISI number	Ladle chemical composition limits, %(a)								
			C	Mn	P	S	Si	Ni	Cr	Mo	V
G13300	1330	1330	0.28–0.33	1.60–1.90	0.035	0.040	0.15–0.35	…	…	…	…
G13350	1335	1335	0.33–0.38	1.60–1.90	0.035	0.040	0.15–0.35	…	…	…	…
G13400	1340	1340	0.38–0.43	1.60–1.90	0.035	0.040	0.15–0.35	…	…	…	…
G13450	1345	1345	0.43–0.48	1.60–1.90	0.035	0.040	0.15–0.35	…	…	…	…
G40230	4023	4023	0.20–0.25	0.70–0.90	0.035	0.040	0.15–0.35	…	…	0.20–0.30	…
G40240	4024	4024	0.20–0.25	0.70–0.90	0.035	0.035–0.050	0.15–0.35	…	…	0.20–0.30	…
G40270	4027	4027	0.25–0.30	0.70–0.90	0.035	0.040	0.15–0.35	…	…	0.20–0.30	…
G40280	4028	4028	0.25–0.30	0.70–0.90	0.035	0.035–0.050	0.15–0.35	…	…	0.20–0.30	…
G40320	4032	…	0.30–0.35	0.70–0.90	0.035	0.040	0.15–0.35	…	…	0.20–0.30	…
G40370	4037	4037	0.35–0.40	0.70–0.90	0.035	0.040	0.15–0.35	…	…	0.20–0.30	…
G40420	4042	…	0.40–0.45	0.70–0.90	0.035	0.040	0.15–0.35	…	…	0.20–0.30	…
G40470	4047	4047	0.45–0.50	0.70–0.90	0.035	0.040	0.15–0.35	…	…	0.20–0.30	…
G41180	4118	4118	0.18–0.23	0.70–0.90	0.035	0.040	0.15–0.35	…	0.40–0.60	0.08–0.15	…
G41300	4130	4130	0.28–0.33	0.40–0.60	0.035	0.040	0.15–0.35	…	0.80–1.10	0.15–0.25	…
G41350	4135	…	0.33–0.38	0.70–0.90	0.035	0.040	0.15–0.35	…	0.80–1.10	0.15–0.25	…
G41370	4137	4137	0.35–0.40	0.70–0.90	0.035	0.040	0.15–0.35	…	0.80–1.10	0.15–0.25	…
G41400	4140	4140	0.38–0.43	0.75–1.00	0.035	0.040	0.15–0.35	…	0.80–1.10	0.15–0.25	…
G41420	4142	4142	0.40–0.45	0.75–1.00	0.035	0.040	0.15–0.35	…	0.80–1.10	0.15–0.25	…
G41450	4145	4145	0.41–0.48	0.75–1.00	0.035	0.040	0.15–0.35	…	0.80–1.10	0.15–0.25	…
G41470	4147	4147	0.45–0.50	0.75–1.00	0.035	0.040	0.15–0.35	…	0.80–1.10	0.15–0.25	…
G41500	4150	4150	0.48–0.53	0.75–1.00	0.035	0.040	0.15–0.35	…	0.80–1.10	0.15–0.25	…
G41610	4161	4161	0.56–0.64	0.75–1.00	0.035	0.040	0.15–0.35	…	0.70–0.90	0.25–0.35	…
G43200	4320	4320	0.17–0.22	0.45–0.65	0.035	0.040	0.15–0.35	1.65–2.00	0.40–0.60	0.20–0.30	…
G43400	4340	4340	0.38–0.43	0.60–0.80	0.035	0.040	0.15–0.35	1.65–2.00	0.70–0.90	0.20–0.30	…
G43406	E4340(b)	E4340	0.38–0.43	0.65–0.85	0.025	0.025	0.15–0.35	1.65–2.00	0.70–0.90	0.20–0.30	…
G44220	4422	…	0.20–0.25	0.70–0.90	0.035	0.040	0.15–0.35	…	…	0.35–0.45	…
G44270	4427	…	0.24–0.29	0.70–0.90	0.035	0.040	0.15–0.35	…	…	0.35–0.45	…
G46150	4615	4615	0.13–0.18	0.45–0.65	0.035	0.040	0.15–0.25	1.65–2.00	…	0.20–0.30	…
G46170	4617	…	0.15–0.20	0.45–0.65	0.035	0.040	0.15–0.35	1.65–2.00	…	0.20–0.30	…
G46200	4620	4620	0.17–0.22	0.45–0.65	0.035	0.040	0.15–0.35	1.65–2.00	…	0.20–0.30	…
G46260	4626	4626	0.24–0.29	0.45–0.65	0.035	0.04 max	0.15–0.35	0.70–1.00	…	0.15–0.25	…

(continued)

Table 4 Continued

UNS number	SAE number	Corresponding AISI number	Ladle chemical composition limits, %(a)								
			C	Mn	P	S	Si	Ni	Cr	Mo	V
G47180	4718	4718	0.16–0.21	0.70–0.90	...	...	...	0.90–1.20	0.35–0.55	0.30–0.40	...
G47200	4720	4720	0.17–0.22	0.50–0.70	0.035	0.040	0.15–0.35	0.90–1.20	0.35–0.55	0.15–0.25	...
G48150	4815	4815	0.13–0.18	0.40–0.60	0.035	0.040	0.15–0.35	3.25–3.75	...	0.20–0.30	...
G48170	4817	4817	0.15–0.20	0.40–0.60	0.035	0.040	0.15–0.35	3.25–3.75	...	0.20–0.30	...
G48200	4820	4820	0.18–0.23	0.50–0.70	0.035	0.040	0.15–0.35	3.25–3.75	...	0.20–0.30	...
G50401	50B40(c)	...	0.38–0.43	0.75–1.00	0.035	0.040	0.15–0.35	...	0.40–0.60	...	...
G50441	50B44(c)	50B44	0.43–0.48	0.75–1.00	0.035	0.040	0.15–0.35	...	0.40–0.60	...	...
G50460	5046	...	0.43–0.48	0.75–1.00	0.035	0.040	0.15–0.35	...	0.20–0.35	...	...
G50461	50B46(c)	50B46	0.44–0.49	0.75–1.00	0.035	0.040	0.15–0.35	...	0.20–0.35	...	...
G50501	50B50(c)	50B50	0.48–0.53	0.75–1.00	0.035	0.040	0.15–0.35	...	0.40–0.60	...	...
G50600	5060	...	0.56–0.64	0.75–1.00	0.035	0.040	0.15–0.35	...	0.40–0.60	...	...
G50601	50B60(c)	50B60	0.56–0.64	0.75–1.00	0.035	0.040	0.15–0.35	...	0.40–0.60	...	...
G51150	5115	...	0.13–0.18	0.70–0.90	0.035	0.040	0.15–0.35	...	0.70–0.90	...	...
G51170	5117	5117	0.15–0.20	0.70–0.90	0.040	0.040	0.15–0.35	...	0.70–0.90	...	...
G51200	5120	5120	0.17–0.22	0.70–0.90	0.035	0.040	0.15–0.35	...	0.70–0.90	...	...
G51300	5130	5130	0.28–0.33	0.70–0.90	0.035	0.040	0.15–0.35	...	0.80–1.10	...	...
G51320	5132	5132	0.30–0.35	0.60–0.80	0.035	0.040	0.15–0.35	...	0.75–1.00	...	...
G51350	5135	5135	0.33–0.38	0.60–0.80	0.035	0.040	0.15–0.35	...	0.80–1.05	...	...
G51400	5140	5140	0.38–0.43	0.70–0.90	0.035	0.040	0.15–0.35	...	0.70–0.90	...	...
G51470	5147	5147	0.46–0.51	0.70–0.95	0.035	0.040	0.15–0.35	...	0.85–1.15	...	...
G51500	5150	5150	0.48–0.53	0.70–0.90	0.035	0.040	0.15–0.35	...	0.70–0.90	...	...
G51550	5155	5155	0.51–0.59	0.70–0.90	0.035	0.040	0.15–0.35	...	0.70–0.90	...	...
G51600	5160	5160	0.56–0.64	0.75–1.00	0.035	0.040	0.15–0.35	...	0.70–0.90	...	...
G51601	51B60(c)	51B60	0.56–0.64	0.75–1.00	0.035	0.040	0.15–0.35	...	0.70–0.90	...	...
G50986	50100(b)	...	0.98–1.10	0.25–0.45	0.025	0.025	0.15–0.35	...	0.40–0.60	...	...
G51986	51100(b)	E51100	0.98–1.10	0.25–0.45	0.025	0.025	0.15–0.35	...	0.90–1.15	...	...
G52986	52100(b)	E52100	0.98–1.10	0.25–0.45	0.025	0.025	0.15–0.35	...	1.30–1.60	...	...
G61180	6118	6118	0.16–0.21	0.50–0.70	0.035	0.040	0.15–0.35	...	0.50–0.70	...	0.10–0.15
G61500	6150	6150	0.48–0.53	0.70–0.90	0.035	0.040	0.15–0.35	...	0.80–1.10	...	0.15 min
G81150	8115	8115	0.13–0.18	0.70–0.90	0.035	0.040	0.15–0.35	0.20–0.40	0.30–0.50	0.08–0.15	...
G81451	81B45(c)	81B45	0.43–0.48	0.75–1.00	0.035	0.040	0.15–0.35	0.20–0.40	0.35–0.55	0.08–0.15	...

UNS No.	SAE/AISI	C	Mn	P	S	Si	Ni	Cr	Mo	
G86150	8615	0.13–0.18	0.70–0.90	0.035	0.040	0.15–0.35	0.40–0.70	0.40–0.60	0.15–0.25	...
G86170	8617	0.15–0.20	0.70–0.90	0.035	0.040	0.15–0.35	0.40–0.70	0.40–0.60	0.15–0.25	...
G86200	8620	0.18–0.23	0.70–0.90	0.035	0.040	0.15–0.35	0.40–0.70	0.40–0.60	0.15–0.25	...
G86220	8622	0.20–0.25	0.70–0.90	0.035	0.040	0.15–0.35	0.40–0.70	0.40–0.60	0.15–0.25	...
G86250	8625	0.23–0.28	0.70–0.90	0.035	0.040	0.15–0.35	0.40–0.70	0.40–0.60	0.15–0.25	...
G86270	8627	0.25–0.30	0.70–0.90	0.035	0.040	0.15–0.35	0.40–0.70	0.40–0.60	0.15–0.25	...
G86300	8630	0.28–0.33	0.70–0.90	0.035	0.040	0.15–0.35	0.40–0.70	0.40–0.60	0.15–0.25	...
G86370	8637	0.35–0.40	0.75–1.00	0.035	0.040	0.15–0.35	0.40–0.70	0.40–0.60	0.15–0.25	...
G86400	8640	0.38–0.43	0.75–1.00	0.035	0.040	0.15–0.35	0.40–0.70	0.40–0.60	0.15–0.25	...
G86420	8642	0.40–0.45	0.75–1.00	0.035	0.040	0.15–0.35	0.40–0.70	0.40–0.60	0.15–0.25	...
G86450	8645	0.43–0.48	0.75–1.00	0.035	0.040	0.15–0.35	0.40–0.70	0.40–0.60	0.15–0.25	...
G86451	86B45(c)	0.43–0.48	0.75–1.00	0.035	0.040	0.15–0.35	0.40–0.70	0.40–0.60	0.15–0.25	...
G86500	8650	0.48–0.53	0.75–1.00	0.035	0.040	0.15–0.35	0.40–0.70	0.40–0.60	0.15–0.25	...
G86550	8655	0.51–0.59	0.75–1.00	0.035	0.040	0.15–0.35	0.40–0.70	0.40–0.60	0.15–0.25	...
G86600	8660	0.56–0.64	0.75–1.00	0.035	0.040	0.15–0.35	0.40–0.70	0.40–0.60	0.15–0.25	...
G87200	8720	0.18–0.23	0.70–0.90	0.035	0.040	0.15–0.35	0.40–0.70	0.40–0.60	0.20–0.30	...
G87400	8740	0.38–0.43	0.75–1.00	0.035	0.040	0.15–0.35	0.40–0.70	0.40–0.60	0.20–0.30	...
G88220	8822	0.20–0.25	0.75–1.00	0.035	0.040	0.15–0.35	0.40–0.70	0.40–0.60	0.30–0.40	...
G92540	9254	0.51–0.59	0.60–0.80	0.035	0.040	1.20–1.60	...	0.60–0.80	...	...
G92600	9260	0.56–0.64	0.75–1.00	0.035	0.040	1.80–2.20	...	...	...	...
G93106	9310(b)	0.08–0.13	0.45–0.65	0.025	0.025	0.15–0.35	3.00–3.50	1.00–1.40	0.08–0.15	...
G94151	94B15(c)	0.13–0.18	0.75–1.00	0.035	0.040	0.15–0.35	0.30–0.60	0.30–0.50	0.08–0.15	...
G94171	94B17(c)	0.15–0.20	0.75–1.00	0.035	0.040	0.15–0.35	0.30–0.60	0.30–0.50	0.08–0.15	...
G94301	94B30(c)	0.28–0.33	0.75–1.00	0.035	0.040	0.15–0.35	0.30–0.60	0.30–0.50	0.08–0.15	...

(a) Small quantities of certain elements that are not specified or required may be found in alloy steels. These elements are to be considered as incidental and are acceptable to the following maximum amount: copper to 0.35%, nickel to 0.25%, chromium to 0.20%, and molybdenum to 0.06%. (b) Electric furnace steel. (c) Boron content is 0.0005–0.003%.

Source: Ref. 20.

Table 5 Chemical Compositions for Typical Low-Alloy Steels

Steel	C	Si	Mn	P	S	Ni	Cr	Mo	Other
Low-carbon quenched and tempered steels									
A 514/A 517 grade A	0.15–0.21	0.40–0.80	0.80–1.10	0.035	0.04	...	0.50–0.80	0.18–0.28	0.05–0.15 Zr(b) 0.0025 B
A 514/A 517 grade F	0.10–0.20	0.15–0.35	0.60–1.00	0.035	0.04	0.70–1.00	0.40–0.65	0.40–0.60	0.03–0.08 V 0.15–0.50 Cu 0.0005–0.005 B
A 514/A 517 grade R	0.15–0.20	0.20–0.35	0.85–1.15	0.035	0.04	0.90–1.10	0.35–0.65	0.15–0.25	0.03–0.08 V
A 533 type A	0.25	0.15–0.40	1.15–1.50	0.035	0.04	...	...	0.45–0.60	...
A 533 type C	0.25	0.15–0.40	1.15–1.50	0.035	0.04	0.70–1.00	...	0.45–0.60	...
HY-80	0.12–0.18	0.15–0.35	0.10–0.40	0.025	0.025	2.00–3.25	1.00–1.80	0.20–0.60	0.25 Cu 0.03 V 0.02 Ti
HY-100	0.12–0.20	0.15–0.35	0.10–0.40	0.025	0.025	2.25–3.50	1.00–1.80	0.20–0.60	0.25 Cu 0.03 V 0.02 Ti
Medium-carbon ultrahigh-strength steels									
4130	0.28–0.33	0.20–0.35	0.40–0.60	0.035	...	...	0.80–1.10	0.15–0.25	...
4340	0.38–0.43	0.20–0.35	0.60–0.80	0.035	...	1.65–2.00	0.70–0.90	0.20–0.30	...
300M	0.40–0.46	1.45–1.80	0.65–0.90	...	...	1.65–2.00	0.70–0.95	0.30–0.45	0.05 V min
D-6a	0.42–0.48	0.15–0.30	0.60–0.90	...	...	0.40–0.70	0.90–1.20	0.90–1.10	0.05–0.10 V
Carburizing bearing steels									
4118	0.18–0.23	0.15–0.30	0.70–0.90	0.035	0.040	...	0.40–0.60	0.08–0.18	...
5120	0.17–0.22	0.15–0.30	0.70–0.90	0.035	0.040	...	0.70–0.90	...	...
3310	0.08–0.13	0.20–0.35	0.45–0.60	0.025	0.025	3.25–3.75	1.40–1.75	...	...
Through-hardened bearing steels									
52100	0.98–1.10	0.15–0.30	0.25–0.45	0.025	0.025	...	1.30–1.60	...	...
A 485 grade 1	0.90–1.05	0.45–0.75	0.95–1.25	0.025	0.025	0.25	0.90–1.20	0.10	0.35 Cu
A 485 grade 3	0.95–1.10	0.15–0.35	0.65–0.90	0.025	0.025	0.25	1.10–1.50	0.20–0.30	0.35 Cu

(a) Single values represent the maximum allowable. (b) Zirconium may be replaced by cerium. When cerium is added, the cerium/sulfur ratio should be approximately 1.5/1, based on heat analysis.

Source: Ref. 2.

prominent steels that may be included in this family are AISI/SAE 6150 steel (a tough shock-resistant, shallow-hardening Cr-V steel with high fatigue and impact resistance in the heat-treated condition) and 8640 steel (an oil-hardening steel exhibiting properties similar to those of 4340 steel) [25]. Product forms include billet, bar, rod, forging, plate, sheet, tubing, and welding wire.

These steels are used for gears, aircraft landing gear, airframe parts, pressure vessels, bolts, springs, screws, axles, studs, fasteners, machinery parts, connecting rods, crankshafts, piston rods, oil well drilling bits, high-pressure tubing, flanges, wrenches, sprockets, etc. [25].

Bearing steels used for ball and roller bearing applications comprise low-carbon (0.10–0.20% C) case-hardened steels and high-carbon ($\sim 1\%$ C) through-hardened or surface-induction hardened steels (Table 5). Many of these steels are covered by SAE-AISI designations.

Chromium-molybdenum heat-resistant steels contain 0.5–9% Cr, 0.5–1.0% Mo, and usually < 0.20% C. They are ordinarily supplied in the normalized and tempered, quenched and tempered, or annealed condition. Cr-Mo steels are extensively used in oil refineries, oil and gas industries, chemical industries, electric power generating stations, and fossil fuel and nuclear power plants for piping, heat exchangers, superheater tubes, and pressure vessels. Various product shapes and corresponding ASTM specifications for these steels are provided in Table 6. Nominal chemical compositions are given in Table 7.

3. High-Strength Low-Alloy Steels

A general description of high-strength low-alloy (HSLA) steel is one that contains (1) low carbon (0.03–0.25%) content to obtain good toughness, formability, and weldability; (2) one or more of the strong carbide-forming microalloying elements (MAEs) (e.g., V, Ti, or Nb); (3) a group of solid solution strengthening elements (e.g., Mn up to 2.0% and Si); and (4)

Table 6 ASTM Specifications for Chromium-Molybdenum Steel Product Forms

Type	Forgings	Tubes	Pipe	Castings	Plate
½Cr-½Mo	A 182-F2	· · ·	A 335-P2 A 369-FP2 A 426-CP2	· · ·	A 387-Gr 2
1Cr-½Mo	A 182-F12 A 336-F12	· · ·	A 335-P12 A 369-FP12 A 426-CP12	· · ·	A 387-Gr 12
1¼Cr-½Mo	A 182-F11 A 336-F11/F11A A 541-C11C	A 199-T11 A 200-T11 A 213-T11	A 335-P11 A 369-FP11 A 426-CP11	A 217-WC6 A 356-Gr6 A 389-C23	A 387-Gr 11
2¼Cr-1Mo	A 182-F22/F22a A 336-F22/F22A A 541-C22C/22D	A 199-T22 A 200-T22 A 213-T22	A 335-P22 A 369-FP22 A 426-CP22	A 217-WC9 A 356-Gr10	A 387-Gr22 A 542
3Cr-1Mo	A 182-F21 A 336-F21/F21A	A 199-T21 A 200-T21 A 213-T21	A 335-P21 A 369-FP21 A 426-CP21	· · ·	A 387-Gr 21
3Cr-1MoV	A 182-F21b	· · ·	· · ·	· · ·	· · ·
5Cr-½Mo	A 182-F5/F5a A 336-F5/F5A A 473-501/502	A 199-T5 A 200-T5 A 213-T5	A 335-P5 A 369-FP5 A 426-CP5	A 217-C5	A 387-Gr 5
5Cr-½MoSi	· · ·	A 213-T5b	A 335-P5b A 426-CP5b	· · ·	· · ·
5Cr-½MoTi	· · ·	A 213-T5c	A 335-P5c	· · ·	· · ·
7Cr-½Mo	A 182-F7 A 473-501A	A 199-T7 A 200-T7 A 213-T7	A 335-P7 A 369-FP7 A 426-CP7	· · ·	A 387-Gr7
9Cr-1Mo	A 182-F9 A 336-F9 A 473-501B	A 199-T9 A 200-T9 A 213-T9	A 335-P9 A 369-FP9 A 426-CP9	A 217-C12	A 387-Gr9

Source: Ref. 2.

Table 7 Nominal Chemical Compostions for Heat-Resistant Chromium-Molybdenum Steels

Type	UNS designation	Composition, %(a)						
		C	Mn	S	P	Si	Cr	Mo
½Cr-½Mo	K12122	0.10–0.20	0.30–0.80	0.040	0.040	0.10–0.60	0.50–0.80	0.45–0.65
1Cr-½Mo	K11562	0.15	0.30–0.60	0.045	0.045	0.50	0.80–1.25	0.45–0.65
1¼Cr-½Mo	K11597	0.15	0.30–0.60	0.030	0.030	0.50–1.00	1.00–1.50	0.45–0.65
1¼Cr-½Mo	K11592	0.10–0.20	0.30–0.80	0.040	0.040	0.50–1.00	1.00–1.50	0.45–0.65
2¼Cr-1Mo	K21590	0.15	0.30–0.60	0.040	0.040	0.50	2.00–2.50	0.87–1.13
3Cr-1Mo	K31545	0.15	0.30–0.60	0.030	0.030	0.50	2.65–3.35	0.80–1.06
3Cr-1MoV(b)	K31830	0.18	0.30–0.60	0.020	0.020	0.10	2.75–3.25	0.90–1.10
5Cr-½Mo	K41545	0.15	0.30–0.60	0.030	0.030	0.50	4.00–6.00	0.45–0.65
7Cr-½Mo	K61595	0.15	0.30–0.60	0.030	0.030	0.50–1.00	6.00–8.00	0.45–0.65
9Cr-1Mo	K90941	0.15	0.30–0.60	0.030	0.030	0.50–1.00	8.00–10.00	0.90–1.10
9Cr-1MoV(c)	⋯	0.08–0.12	0.30–0.60	0.010	0.020	0.20–0.50	8.00–9.00	0.85–1.05

(a) Single values are maximums. (b) Also contains 0.02–0.030% V, 0.001–0.003% B, and 0.015–0.035% Ti. (c) Also contains 0.40% Ni, 0.18–0.25% V, 0.06–0.10% Nb, 0.03–0.07% N, and 0.04% Al

Source: Ref. 2.

one or more of the additional microalloying elements (e.g., Ca, Zr) and the rare earth elements, particularly Ce and La, for sulfide inclusion shape control [1, 26, 27]. In many other HSLA steels, small amounts of Ni, Cr, Cu, and particularly Mo are also present, which increase atmospheric corrosion resistance and hardenability. It is interesting to note that a very fine ferrite grain structure in the final product produced by a combination of controlled rolling procedure with an optimum utilization of microalloying additions, in HSLA steels, is an important factor in simultaneously increasing strength and toughness and decreasing the ductile-to-brittle transition temperature (to as low as $-70°C$). Carbides (NbC, TiC), nitrides (AlN), and carbonitrides [e.g., Nb(CN), V(CN), Ti(CN), (Nb,V)CN, (Nb,Ti)CN] are the dispersed second-phase particles that act as grain size refiners in HSLA steels.

HSLA steels are successfully used as strip, plate, bar, structural section, and forged bar products. They find applications in several diverse fields such as in oil and gas pipelines; in the automotive, agricultural, and pressure vessel industries; in off-shore structures and platforms; and in the construction of crane, bridges, buildings, shipbuildings, railroad, tank cars, and power transmission and TV towers [26].

Classification of HSLA Steels   Several special terms are used to describe various types of HSLA steels [27, 28]:

1. *Weathering steels*: Steels with ~0.1% C, 0.15–0.30% Cu, 0.5–1.0% Mn, 0.05–0.15% P, and 0.15–0.90% Si and exhibiting superior atmospheric corrosion resistance. Typical applications include railroad cars, bridges, and unpainted buildings.

2. *Control-rolled steels*: Steels designated to develop a highly deformed austenite structure by hot rolling (according to a predetermined rolling schedule) that will transform to a very fine equiaxed ferrite structure on cooling.

3. *Pearlite-reduced steels*: Steels strengthened by very fine-grained ferrite and precipitation hardening but with low carbon content and therefore exhibiting little or no pearlite in the microstructure.

4. *Microalloyed steels*: Conventional HSLA steels as defined above. They exhibit discontinuous yielding behavior.

5. *Acicular ferrite steels*: Very low carbon (typically 0.03–0.06%) steels with enough hardenability (by Mn, Mo, Nb, and B additions) to transform on cooling to a very fine, high-strength acicular ferrite (also called low-carbon bainite) structure rather than the usual polygonal ferrite structure. In addition to high strength and good toughness, these steels have continuous yielding behavior.

6. *Dual-phase steels*: Steels comprising essentially fine dispersion of hard, strong martensite (with ~0.3% C) but sometimes also retained austenite or even bainite in a soft and fine-grained ferrite matrix. They are characterized by continuous yielding, (i.e., no yield point elongation), low yield stress (the YS/UTS ratio being around 0.50), high UTS, superior formability, and rapid initial work-hardening rate. Additionally, they possess greater resistance to onset of necking (i.e., plastic instability) in the uniaxial sheet material forming process to provide large uniform strain [29–32].

Table 8 lists HSLA steels according to chemical composition and minimum mechanical properties requirements.

4. Tool Steels

A tool steel is any steel used to shape other metals by cutting, forming, machining, shearing, battering, or die casting or to shape and cut wood, paper, rock, or concrete. Hence tool steels

Table 8 Composition Ranges and Limits for SAE
HSLA Steels

SAE designation(b)	Heat composition limits, %(a)		
	C max	Mn max	P max
942X..................	0.21	1.35	0.04
945A..................	0.15	1.00	0.04
945C..................	0.23	1.40	0.04
945X..................	0.22	1.35	0.04
950A..................	0.15	1.30	0.04
950B..................	0.22	1.30	0.04
950C..................	0.25	1.60	0.04
950D..................	0.15	1.00	0.15
950X..................	0.23	1.35	0.04
955X..................	0.25	1.35	0.04
960X..................	0.26	1.45	0.04
965X..................	0.26	1.45	0.04
970X..................	0.26	1.65	0.04
980X..................	0.26	1.65	0.04

(a) Maximum contents of sulfur and silicon for all grades: 0.050%
S, 0.90% Si. (b) Second and third digits of designation indicate
minimum yield strength in ksi. Suffix X indicates that the steel
contains niobium, vanadium, nitrogen, or other alloying elements.
A second suffix K indicates that the steel is produced fully killed
using fine-grain practice; otherwise, the steel is produced semikilled.

Source: Ref. 20.

are designed to have high hardness and durability under severe service conditions. They comprise a wide range from plain carbon steels with up to 1.2% carbon without appreciable amounts of alloying elements to the highly alloyed steels in which alloying additions reach 50%. Although some carbon tool steels and low-alloy tool steels have a wide range of carbon content, most of the higher alloy tool steels have a comparatively narrow carbon range. A mixed classification system is used to classify tool steels based on the use, composition, special mechanical properties, or method of heat treatment.

According to AISI specification, there are nine main groups of wrought tool steels. Table 9 lists the compositions of these tool steels with corresponding designated symbols [33], which are discussed herein.

High-speed steels are used for applications requiring long life at relatively high operating temperatures such as for heavy cuts or high-speed machining. High-speed steels are the most important alloy tool steels because of their very high hardness and good wear resistance in the heat-treated condition and their ability to retain high hardness at the elevated temperatures often encountered during the operation of the tool at high cutting speeds. This red- or hot-hardness property is an important feature of a high-speed steel [34, 35].

High-speed steels are grouped into molybdenum type M and tungsten type T. Type M tool steels contain Mo, W, Cr, V, Mo, and C as major alloying elements, while type T tool steels contain W, Cr, V, Mo, Co, and C as the main alloying elements. In the United States, type M steels account for 95% of the high-speed steels produced. There is also a subgroup consisting of intermediate high-speed steels in the M group. The most popular grades among molybdenum types are M1, M2, M4, M7, M10, and M42, while those among tungsten types are T1 and T15.

The main advantage of type M steels is their lower initial cost (approximately 40% cheaper than that of similar type T steels), but they are more susceptible to decarburizing, thereby necessitating better temperature control than type T steels. By using salt baths and sometimes

Table 9 Composition Limits of Principal Types of Tool Steels

Designation AISI	UNS	C	Mn	Si	Composition(s), % Cr	Ni	Mo	W	V	Co
Molybdenum high-speed steels										
M1	T11301	0.78–0.88	0.15–0.40	0.20–0.50	3.50–4.00	0.30 max	8.20–9.20	1.40–2.10	1.00–1.35	...
M2	T11302	0.78–0.88; 0.95–1.05	0.15–0.40	0.20–0.45	3.75–4.50	0.30 max	4.50–5.50	5.50–6.75	1.75–2.20	...
M3, class 1	T11313	1.00–1.10	0.15–0.40	0.20–0.45	3.75–4.50	0.30 max	4.75–6.50	5.00–6.75	2.25–2.75	...
M3, class 2	T11323	1.15–1.25	0.15–0.40	0.20–0.45	3.75–4.50	0.30 max	4.75–6.50	5.00–6.75	2.75–3.75	...
M4	T11304	1.25–1.40	0.15–0.40	0.20–0.45	3.75–4.75	0.30 max	4.25–5.50	5.25–6.50	3.75–4.50	...
M7	T11307	0.97–1.05	0.15–0.40	0.20–0.55	3.50–4.00	0.30 max	8.20–9.20	1.40–2.10	1.75–2.25	...
M10	T11310	0.84–0.94; 0.95–1.05	0.10–0.40	0.20–0.45	3.75–4.50	0.30 max	7.75–8.50	...	1.80–2.20	...
M30	T11330	0.75–0.85	0.15–0.40	0.20–0.45	3.50–4.25	0.30 max	7.75–9.00	1.30–2.30	1.00–1.40	4.50–5.50
M33	T11333	0.85–0.92	0.15–0.40	0.15–0.50	3.50–4.00	0.30 max	9.00–10.00	1.30–2.10	1.00–1.35	7.75–8.75
M34	T11334	0.85–0.92	0.15–0.40	0.20–0.45	3.50–4.00	0.30 max	7.75–9.20	1.40–2.10	1.90–2.30	7.75–8.75
M35	T11335	0.82–0.88	0.15–0.40	0.20–0.45	3.75–4.50	0.30 max	4.50–5.50	5.50–6.75	1.75–2.20	4.50–5.50
M36	T11336	0.80–0.90	0.15–0.40	0.20–0.45	3.75–4.50	0.30 max	4.50–5.50	5.50–6.50	1.75–2.25	7.75–8.75
M41	T11341	1.05–1.15	0.20–0.60	0.15–0.50	3.75–4.50	0.30 max	3.25–4.25	6.25–7.00	1.75–2.25	4.75–5.75
M42	T11342	1.05–1.15	0.15–0.40	0.15–0.65	3.50–4.25	0.30 max	9.00–10.00	1.15–1.85	0.95–1.35	7.75–8.75
M43	T11343	1.15–1.25	0.20–0.40	0.15–0.65	3.50–4.25	0.30 max	7.50–8.50	2.25–3.00	1.50–1.75	7.75–8.75
M44	T11344	1.10–1.20	0.20–0.40	0.30–0.55	4.00–4.75	0.30 max	6.00–7.00	5.00–5.75	1.85–2.20	11.00–12.25
M46	T11346	1.22–1.30	0.20–0.40	0.40–0.65	3.70–4.20	0.30 max	8.00–8.50	1.90–2.20	3.00–3.30	7.80–8.80
M47	T11347	1.05–1.15	0.15–0.40	0.20–0.45	3.50–4.00	0.30 max	9.25–10.00	1.30–1.80	1.15–1.35	4.75–5.25
M48	T11348	1.42–1.52	0.15–0.40	0.15–0.40	3.50–4.00	0.30 max	4.75–5.50	9.50–10.50	2.75–3.25	8.00–10.00
M62	T11362	1.25–1.35	0.15–0.40	0.15–0.40	3.50–4.00	0.30 max	10.00–11.00	5.75–6.50	1.80–2.10	...
Tungsten high-speed steels										
T1	T12001	0.65–0.80	0.10–0.40	0.20–0.40	3.75–4.50	0.30 max	...	17.25–18.75	0.90–1.30	...
T2	T12002	0.80–0.90	0.20–0.40	0.20–0.40	3.75–4.50	0.30 max	1.00 max	17.50–19.00	1.80–2.40	...
T4	T12004	0.70–0.80	0.10–0.40	0.20–0.40	3.75–4.50	0.30 max	0.40–1.00	17.50–19.00	0.80–1.20	4.25–5.75
T5	T12005	0.75–0.85	0.20–0.40	0.20–0.40	3.75–5.00	0.30 max	0.50–1.25	17.50–19.00	1.80–2.40	7.00–9.50
T6	T12006	0.75–0.85	0.20–0.40	0.20–0.40	4.00–4.75	0.30 max	0.40–1.00	18.50–21.00	1.50–2.10	11.00–13.00
T8	T12008	0.75–0.85	0.20–0.40	0.20–0.40	3.75–4.50	0.30 max	0.40–1.00	13.25–14.75	1.80–2.40	4.25–5.75
T15	T12015	1.50–1.60	0.15–0.40	0.15–0.40	3.75–5.00	0.30 max	1.00 max	11.75–13.00	4.50–5.25	4.75–5.25
Intermediate high-speed steels										
M50	T11350	0.78–0.88	0.15–0.45	0.20–0.60	3.75–4.50	0.30 max	3.90–4.75	...	0.80–1.25	...
M52	T11352	0.85–0.95	0.15–0.45	0.20–0.60	3.50–4.30	0.30 max	4.00–4.90	0.75–1.50	1.65–2.25	...

(continued)

Table 9 Continued

AISI	UNS	C	Mn	Si	Cr	Ni	Mo	W	V	Co
Chromium hot-work steels										
H10	T20810	0.35–0.45	0.25–0.70	0.80–1.20	3.00–3.75	0.30 max	2.00–3.00	...	0.25–0.75	...
H11	T20811	0.33–0.43	0.20–0.50	0.80–1.20	4.75–5.50	0.30 max	1.10–1.60	...	0.30–0.60	...
H12	T20812	0.30–0.40	0.20–0.50	0.80–1.20	4.75–5.50	0.30 max	1.25–1.75	1.00–1.70	0.50 max	...
H13	T20813	0.32–0.45	0.20–0.50	0.80–1.20	4.75–5.50	0.30 max	1.10–1.75	...	0.80–1.20	...
H14	T20814	0.35–0.45	0.20–0.50	0.80–1.20	4.75–5.50	0.30 max	...	4.00–5.25	...	...
H19	T20819	0.32–0.45	0.20–0.50	0.20–0.50	4.00–4.75	0.30 max	0.30–0.55	3.75–4.50	1.75–2.20	4.00–4.50
Tungsten hot-work steels										
H21	T20821	0.26–0.36	0.15–0.40	0.15–0.50	3.00–3.75	0.30 max	...	8.50–10.00	0.30–0.60	...
H22	T20822	0.30–0.40	0.15–0.40	0.15–0.40	1.75–3.75	0.30 max	...	10.00–11.75	0.25–0.50	...
H23	T20823	0.25–0.35	0.15–0.40	0.15–0.60	11.00–12.75	0.30 max	...	11.00–12.75	0.75–1.25	...
H24	T20824	0.42–0.53	0.15–0.40	0.15–0.40	2.50–3.50	0.30 max	...	14.00–16.00	0.40–0.60	...
H25	T20825	0.22–0.32	0.15–0.40	0.15–0.40	3.75–4.50	0.30 max	...	14.00–16.00	0.40–0.60	...
H26	T20826	0.45–0.55(b)	0.15–0.40	0.15–0.40	3.75–4.50	0.30 max	...	17.25–19.00	0.75–1.25	...
Molybdenum hot-work steels										
H42	T20842	0.55–0.70(b)	0.15–0.40	...	3.75–4.50	0.30 max	4.50–5.50	5.50–6.75	1.75–2.20	...
Air-hardening, medium-alloy, cold-work steels										
A2	T30102	0.95–1.05	1.00 max	0.50 max	4.75–5.50	0.30 max	0.90–1.40	...	0.15–0.50	...
A3	T30103	1.20–1.30	0.40–0.60	0.50 max	4.75–5.50	0.30 max	0.90–1.40	...	0.80–1.40	...
A4	T30104	0.95–1.05	1.80–2.20	0.50 max	0.90–2.20	0.30 max	0.90–1.40	...	...	...
A6	T30106	0.65–0.75	1.80–2.50	0.50 max	0.90–1.20	0.30 max	0.90–1.40	...	...	...
A7	T30107	2.00–2.85	0.80 max	0.50 max	5.00–5.75	0.30 max	0.90–1.40	0.50–1.50	3.90–5.15	...
A8	T30108	0.50–0.60	0.50 max	0.75–1.10	4.75–5.50	0.30 max	1.15–1.65	1.00–1.50	...	...
A9	T30109	0.45–0.55	0.50 max	0.95–1.15	4.75–5.50	0.30 max	1.30–1.80	...	0.80–1.40	...
A10	T30110	1.25–1.50(c)	1.60–2.10	1.00–1.50	...	1.55–2.05	1.25–1.75	...	...	...
High-carbon, high-chromium, cold-work steels										
D2	T30402	1.40–1.60	0.60 max	0.60 max	11.00–13.00	0.30 max	0.70–1.20	...	1.10 max	...
D3	T30403	2.00–2.35	0.60 max	0.60 max	11.00–13.50	0.30 max	...	1.00 max	1.00 max	...
D4	T30404	2.05–2.40	0.60 max	0.60 max	11.00–13.00	0.30 max	0.70–1.20	...	1.00 max	...
D5	T30405	1.40–1.60	0.60 max	0.60 max	11.00–13.00	0.30 max	0.70–1.20	...	1.00 max	2.50–3.50
D7	T30407	2.15–2.50	0.60 max	0.60 max	11.50–13.50	0.30 max	0.70–1.20	...	3.80–4.40	...
Oil-hardening cold-work steels										
O1	T31501	0.85–1.00	1.00–1.40	0.50 max	0.40–0.60	0.30 max	...	0.40–0.60	0.30 max	...
O2	T31502	0.85–0.95	1.40–1.80	0.50 max	0.50 max	0.30 max	0.30 max	...	0.30 max	...
O6	T31506	1.25–1.55(c)	0.30–1.10	0.55–1.50	0.30 max	0.30 max	0.20–0.30	...	...	...
O7	T31507	1.10–1.30	1.00 max	0.60 max	0.35–0.85	0.30 max	0.30 max	1.00–2.00	0.40 max	...

Shock-resisting steels

S1 T41901	0.40–0.55	0.10–0.40	1.00–1.80	0.30 max	0.50 max	1.50–3.00	0.15–0.30
S2 T41902	0.40–0.55	0.30–0.50	0.90–1.20	0.30 max	0.30–0.60	…	0.50 max
S5 T41905	0.50–0.65	0.60–1.00	1.75–2.25	…	0.20–1.35	…	0.35 max
S6 T41906	0.40–0.50	1.20–1.50	2.00–2.50	…	0.30–0.50	…	0.20–0.40
S7 T41907	0.45–0.55	0.20–0.90	0.20–1.00	3.00–3.50	1.30–1.80	…	0.20–0.30(d)

Low-alloy special-purpose tool steels

L2 T61202	0.45–1.00(b)	0.10–0.90	0.50 max	…	0.25 max	…	0.10–0.30
L6 T61206	0.65–0.75	0.25–0.80	0.50 max	1.25–2.00	0.50 max	…	0.20–0.30(d)

Low-carbon mold steels

P2 T51602	0.10 max	0.10–0.40	0.10–0.40	0.75–1.25	0.15–0.40	…	…
P3 T51603	0.10 max	0.20–0.60	0.40 max	0.40–0.75	…	…	…
P4 T51604	0.12 max	0.20–0.60	0.10–0.40	4.00–5.25	0.40–1.00	…	…
P5 T51605	0.10 max	0.20–0.60	0.40 max	2.00–2.50	…	…	…
P6 T51606	0.05–0.15	0.35–0.70	0.10–0.40	3.25–3.75	…	…	…
P20 T51620	0.28–0.40	0.60–1.00	0.20–0.80	1.40–2.00	0.30–0.55	…	…
P21 T51621	0.18–0.22	0.20–0.40	0.20–0.40	0.50 max	…	0.15–0.25	1.05–1.25Al

Water-hardening tool steels

W1 T72301	0.70–1.50(e)	0.10–0.40	0.10–0.40	0.15 max	0.10 max	0.15 max	0.10 max
W2 T72302	0.85–1.50(e)	0.10–0.40	0.10–0.40	0.15 max	0.10 max	0.15 max	0.15–0.35
W5 T72305	1.05–1.15	0.10–0.40	0.10–0.40	0.40–0.60	0.10 max	0.15 max	0.10 max

(a) All steels except group W contain 0.25 max Cu, 0.03 max P, and 0.03 max S; group W steels contain 0.20 max Cu, 0.025 max P, and 0.025 max S. Where specified, sulfur may be increased to 0.06 to 0.15% to improve machinability of group A, D, H, M, and T steels. (b) Available in several carbon ranges. (c) Contains free graphite in the microstructure. (d) Optional. (e) Specified carbon ranges are designated by suffix numbers.

Source: Ref. 33.

Table 10 Compositions of Standard Stainless Steels

Type	UNS designation	Composition, %(a)							
		C	Mn	Si	Cr	Ni	P	S	Other
Austenitic types									
201	S20100	0.15	5.5–7.5	1.00	16.0–18.0	3.5–5.5	0.06	0.03	0.25 N
202	S20200	0.15	7.5–10.0	1.00	17.0–19.0	4.0–6.0	0.06	0.03	0.25 N
205	S20500	0.12–0.25	14.0–15.5	1.00	16.5–18.0	1.0–1.75	0.06	0.03	0.32–0.40 N
301	S30100	0.15	2.00	1.00	16.0–18.0	6.0–8.0	0.045	0.03	…
302	S30200	0.15	2.00	1.00	17.0–19.0	8.0–10.0	0.045	0.03	…
302B	S30215	0.15	2.00	2.0–3.0	17.0–19.0	8.0–10.0	0.045	0.03	…
303	S30300	0.15	2.00	1.00	17.0–19.0	8.0–10.0	0.20	0.15 min	0.6 Mo(b)
303Se	S30323	0.15	2.00	1.00	17.0–19.0	8.0–10.0	0.20	0.06	0.15 min Se
304	S30400	0.08	2.00	1.00	18.0–20.0	8.0–10.5	0.045	0.03	…
304H	S30409	0.04–0.10	2.00	1.00	18.0–20.0	8.0–10.5	0.045	0.03	…
304L	S30403	0.03	2.00	1.00	18.0–20.0	8.0–12.0	0.045	0.03	…
304LN	S30453	0.03	2.00	1.00	18.0–20.0	8.0–12.0	0.045	0.03	0.10–0.16 N
302Cu	S30430	0.08	2.00	1.00	17.0–19.0	8.0–10.0	0.045	0.03	3.0–4.0 Cu
304N	S30451	0.08	2.00	1.00	18.0–20.0	8.0–10.5	0.045	0.03	0.10–0.16 N
305	S30500	0.12	2.00	1.00	17.0–19.0	10.5–13.0	0.045	0.03	…
308	S30800	0.08	2.00	1.00	19.0–21.0	10.0–12.0	0.045	0.03	…
309	S30900	0.20	2.00	1.00	22.0–24.0	12.0–15.0	0.045	0.03	…
309S	S30908	0.08	2.00	1.00	22.0–24.0	12.0–15.0	0.045	0.03	…
310	S31000	0.25	2.00	1.50	24.0–26.0	19.0–22.0	0.045	0.03	…
310S	S31008	0.08	2.00	1.50	24.0–26.0	19.0–22.0	0.045	0.03	…
314	S31400	0.25	2.00	1.5–3.0	23.0–26.0	19.0–22.0	0.045	0.03	…
316	S31600	0.08	2.00	1.00	16.0–18.0	10.0–14.0	0.045	0.03	2.0–3.0 Mo
316F	S31620	0.08	2.00	1.00	16.0–18.0	10.0–14.0	0.20	0.10 min	1.75–2.5 Mo
316H	S31609	0.04–0.10	2.00	1.00	16.0–18.0	10.0–14.0	0.045	0.03	2.0–3.0 Mo
316L	S31603	0.03	2.00	1.00	16.0–18.0	10.0–14.0	0.045	0.03	2.0–3.0 Mo
316LN	S31653	0.03	2.00	1.00	16.0–18.0	10.0–14.0	0.045	0.03	2.0–3.0 Mo: 0.10–0.16 N
316N	S31651	0.08	2.00	1.00	16.0–18.0	10.0–14.0	0.045	0.03	2.0–3.0 Mo: 0.10–0.16 N
317	S31700	0.08	2.00	1.00	18.0–20.0	11.0–15.0	0.045	0.03	3.0–4.0 Mo
317L	S31703	0.03	2.00	1.00	18.0–20.0	11.0–15.0	0.045	0.03	3.0–4.0 Mo
321	S32100	0.08	2.00	1.00	17.0–19.0	9.0–12.0	0.045	0.03	5 × %C min Ti
321H	S32109	0.04–0.10	2.00	1.00	17.0–19.0	9.0–12.0	0.045	0.03	5 × %C min Ti
330	N08330	0.08	2.00	0.75–1.5	17.0–20.0	34.0–37.0	0.04	0.03	…
347	S34700	0.08	2.00	1.00	17.0–19.0	9.0–13.0	0.045	0.03	10 × %C min Nb
347H	S34709	0.04–0.10	2.00	1.00	17.0–19.0	9.0–13.0	0.045	0.03	8 × %C min – 1.0 max Nb
348	S34800	0.08	2.00	1.00	17.0–19.0	9.0–13.0	0.045	0.03	0.2 Co: 10 × %C min Nb: 0.10 Ta
348H	S34809	0.04–0.10	2.00	1.00	17.0–19.0	9.0–13.0	0.045	0.03	0.2 Co: 8 × %C min – 1.0 max Nb: 0.10 Ta
384	S38400	0.08	2.00	1.00	15.0–17.0	17.0–19.0	0.045	0.03	…

Type	UNS No.	C	Mn	Si	Cr	Ni	P	S	Other elements
Ferritic types									
405	S40500	0.08	1.00	1.00	11.5–14.5	...	0.04	0.03	0.10–0.30 Al
409	S40900	0.08	1.00	1.00	10.5–11.75	0.50	0.045	0.045	6 × %C min – 0.75 max Ti
429	S42900	0.12	1.00	1.00	14.0–16.0	...	0.04	0.03	...
430	S43000	0.12	1.00	1.00	16.0–18.0	...	0.04	0.03	...
430F	S43020	0.12	1.25	1.00	16.0–18.0	...	0.06	0.15 min	0.6 Mo(b)
430FSe	S43023	0.12	1.25	1.00	16.0–18.0	...	0.06	0.06	0.15 min Se
434	S43400	0.12	1.00	1.00	16.0–18.0	...	0.04	0.03	0.75–1.25 Mo
436	S43600	0.12	1.00	1.00	16.0–18.0	...	0.04	0.03	0.75–1.25 Mo: 5 × %C min – 0.70 max Nb
439	S43035	0.07	1.00	1.00	17.0–19.0	0.50	0.04	0.03	0.15 Al; 12 × %C min – 1.10 Ti
442	S44200	0.20	1.00	1.00	18.0–23.0	...	0.04	0.03	...
444	S44400	0.025	1.00	1.00	17.5–19.5	1.00	0.04	0.03	1.75–2.50 Mo; 0.025 N: 0.2 + 4 (%C + %N) min – 0.8 max (Ti + Nb)
446	S44600	0.20	1.50	1.00	23.0–27.0	...	0.04	0.03	0.25 N
Duplex (ferritic-austenitic) type									
329	S32900	0.20	1.00	0.75	23.0–28.0	2.50–5.00	0.040	0.030	1.00–2.00 Mo
Martensitic types									
403	S40300	0.15	1.00	0.50	11.5–13.0	...	0.04	0.03	...
410	S41000	0.15	1.00	1.00	11.5–13.5	...	0.04	0.03	...
414	S41400	0.15	1.00	1.00	11.5–13.5	1.25–2.50	0.04	0.03	...
416	S41600	0.15	1.25	1.00	12.0–14.0	...	0.06	0.15 min	0.6 Mo(b)
416Se	S41623	0.15	1.25	1.00	12.0–14.0	...	0.06	0.06	0.15 min Se
420	S42000	0.15 min	1.00	1.00	12.0–14.0	...	0.04	0.03	...
420F	S42020	0.15 min	1.25	1.00	12.0–14.0	...	0.06	0.15 min	0.6 Mo(b)
422	S42200	0.20–0.25	1.00	0.75	11.5–13.5	0.5–1.0	0.04	0.03	0.75–1.25 Mo; 0.75–1.25 W: 0.15–0.3 V
431	S43100	0.20	1.00	1.00	15.0–17.0	1.25–2.50	0.04	0.03	...
440A	S44002	0.60–0.75	1.00	1.00	16.0–18.0	...	0.04	0.03	0.75 Mo
440B	S44003	0.75–0.95	1.00	1.00	16.0–18.0	...	0.04	0.03	0.75 Mo
440C	S44004	0.95–1.20	1.00	1.00	16.0–18.0	...	0.04	0.03	0.75 Mo
Precipitation-hardening types									
PH 13-8 Mo	S13800	0.05	0.20	0.10	12.25–13.25	7.5–8.5	0.01	0.008	2.0–2.5 Mo; 0.90–1.35 Al: 0.01 N
15-5 PH	S15500	0.07	1.00	1.00	14.0–15.5	3.5–5.5	0.04	0.03	2.5–4.5 Cu: 0.15–0.45 Nb
17-4 PH	S17400	0.07	1.00	1.00	15.5–17.5	3.0–5.0	0.04	0.03	3.0–5.0 Cu: 0.15–0.45 Nb
17-7 PH	S17700	0.09	1.00	1.00	16.0–18.0	6.5–7.75	0.04	0.04	0.75–1.5 Al

(a) Single values are maximum values unless otherwise indicated. (b) Optional

Source: Ref. 36.

surface coatings, decarburization can be controlled. The mechanical properties of type M and type T steels are similar except that type M steels have slightly greater toughness than type T steels at the same hardness level [1].

Hot-work tool steels (AISI H series) fall into three major groups: (1) chromium-base, types H1–H19; (2) tungsten-base, types H20–H39; and (3) molybdenum-base, types H40–H59. The distinction is based on the principal alloying additions; however, all classes have medium carbon content and Cr content varying from 1.75 to 12.75%. Among these steels, H11, H12, and H13 are produced in large quantities. These steels possess good red hardness and retain high hardness (~50 Rc) after prolonged exposures at 500–550°C. They are used extensively for hot-work applications, which include parts for aluminum and magnesium die casting and extrusion, plastic injection molding, and compression and transfer molds [34].

Cold-work tool steels comprise three categories: (1) air-hardening, medium-alloy tool steels (AISI A series), (2) high-carbon, high-chromium tool steels (AISI D series), and (3) oil-hardening tool steels (AISI O series). AISA A series tool steels have high hardenability and harden readily on air cooling. In the air-hardened and tempered condition, they are suitable for applications where improved toughness and reasonably good abrasion resistance are required such as for forming, blanking, and drawing dies. The most popular grade is A2. AISI D series tool steels possess excellent wear resistance and nondeforming properties, thereby making them very useful as cold-work die steels. They find applications in blanking and cold-forming dies, drawing and lamination dies, thread rolling dies, shear and slitter blades, forming rolls, and so forth. Among these steels, D2 is by far the most popular grade [34]. AISI O series tool steels are used for blanking, coining, drawing, and forming dies and punches, shear blades, gauges, and chuck jaws after oil quenching and tempering [34]. Among these grades, O1 is the most widely used.

Shock-resisting tool steels (AISI S series) are used where repetitive impact stresses are encountered such as in hammers, chipping and cold chisels, rivet sets, punches, driver bits, stamps, and shear blades in quenched and tempered conditions. In these steels, high toughness is the major concern and hardness the secondary concern. Among these grades, S5 and S7 are perhaps the most widely used.

Low-alloy special-purpose tool steels (AISI L series) are similar in composition to the W-type tool steels, except that the addition of Cr and other elements render greater hardenability and wear-resistance properties. Type L6 and the low-carbon version of L2 are commonly used for a large number of machine parts.

Mold steels (AISI P series) are mostly used in low-temperature die casting dies and in molds for the injection or compression molding of plastics [33].

Water-hardening tool steels (AISI W series) Among the three compositions listed, W1 is the most widely used as cutting tools, punches, dies, files, reamers, taps, drills, razors, woodworking tools, and surgical instruments in the quenched and tempered condition.

5. Stainless Steels

Stainless steels may be defined as complex alloy steels containing a minimum of 10.5% Cr with or without other elements to produce austenitic, ferritic, duplex (ferritic-austenitic), martensitic, and precipitation-hardening grades. AISI uses a three-digit code for stainless steels. Tables 10 and 11 list the compositions of standard and nonstandard stainless steels, respectively, with corresponding designated symbols, which are discussed below [36].

Austenitic stainless steels constitute about 65–70% of the total U.S. stainless steel production and have occupied a dominant position because of their higher corrosion and oxidation resistance and fabricability, enhanced mechanical properties such as strength and toughness at both elevated and ambient temperatures, excellent cryogenic properties, aesthetic appeal,

Table 11 Compositions of Nonstandard Stainless Steels

Designation(s)	UNS designation	Composition, %(b)							
		C	Mn	Si	Cr	Ni	P	S	Other
Austenitic stainless steels									
Gall-Tough	S20161	0.15	4.00–6.00	3.00–4.00	15.00–18.00	4.00–6.00	0.040	0.040	0.08–0.20 N
203 EZ (XM-1)	S20300	0.08	5.0–6.5	1.00	16.0–18.0	5.0–6.5	0.040	0.18–0.35	0.5 Mo: 1.75–2.25 Cu
Nitronic 50 (XM-19)	S20910	0.06	4.0–6.0	1.00	20.5–23.5	11.5–13.5	0.040	0.030	1.5–3.0 Mo: 0.2–0.4 N: 0.1–0.3 Nb: 0.1–0.3 V
Tenelon (XM-31)	S21400	0.12	14.5–16.0	0.3–1.0	17.0–18.5	0.75	0.045	0.030	0.35 N
Cryogenic Tenelon (XM-14)	S21460	0.12	14.0–16.0	1.00	17.0–19.0	5.0–6.0	0.060	0.030	0.35–0.50 N
Esshete 1250	S21500	0.15	5.5–7.0	1.20	14.0–16.0	9.0–11.0	0.040	0.030	0.003–0.009 B: 0.75–1.25 Nb: 0.15–0.40 V
Type 216 (XM-17)	S21600	0.08	7.5–9.0	1.00	17.5–22.0	5.0–7.0	0.045	0.030	2.0–3.0 Mo: 0.25–0.50 N
Type 216 L (XM-18)	S21603	0.03	7.5–9.0	1.00	17.5–22.0	7.5–9.0	0.045	0.030	2.0–3.0 Mo: 0.25–0.50 N
Nitronic 60	S21800	0.10	7.0–9.0	3.5–4.5	16.0–18.0	8.0–9.0	0.040	0.030	0.08–0.18 N
Nitronic 40 (XM-10)	S21900	0.08	8.0–10.0	1.00	19.0–21.5	5.5–7.5	0.060	0.030	0.15–0.40 N
21-6-9 LC	S21904	0.04	8.00–10.00	1.00	19.00–21.50	5.50–7.50	0.060	0.030	0.15–0.40 N
Nitronic 33 (18-3-Mn)	S24000	0.08	11.50–14.50	1.00	17.00–19.00	2.50–3.75	0.060	0.030	0.20–0.40 N
Nitronic 32 (18-2-Mn)	S24100	0.15	11.00–14.00	1.00	16.50–19.50	0.50–2.50	0.060	0.030	0.20–0.45 N
18-18 Plus	S28200	0.15	17.0–19.0	1.00	17.5–19.5	...	0.045	0.030	0.5–1.5 Mo: 0.5–1.5 Cu: 0.4–0.6 N
303 Plus X (XM-5)	S30310	0.15	2.5–4.5	1.00	17.0–19.0	7.0–10.0	0.20	0.25 min	0.6 Mo
MVMA(c)	...	0.05	0.60	1.30	18.5	9.50	...	...	0.15 N: 0.04 Ce
304B1(d)	S30415	0.08	2.00	0.75	18.00–20.00	12.00–15.00	0.045	0.030	0.10 N: 1.00–1.25 B
304 HN (XM-21)	S30424	0.04–0.10	2.00	1.00	18.0–20.0	8.0–10.5	0.045	0.030	0.16–0.30 N
Cronifer 1815 LCSi	S30452	0.018	2.00	3.7–4.3	17.0–18.5	14.0–15.5	0.020	0.020	0.2 Mo
RA 85 H(c)	S30600	0.20	0.80	3.50	18.5	14.50	...	...	1.0 Al
253 MA	S30615	0.05–0.10	0.80	1.4–2.0	20.0–22.0	10.0–12.0	0.040	0.030	0.14–0.20 N: 0.03–0.08 Ce: 1.0 Al
Type 309 S Cb	S30815	0.08	2.00	1.00	22.0–24.0	12.0–15.0	0.045	0.030	10 × %C min to 1.10 max Nb
Type 310 Cb	S30940	0.08	2.00	1.50	24.0–26.0	19.0–22.0	0.045	0.030	10 × %C min to 1.10 max Nb + Ta
254 SMO	S31254	0.020	1.00	0.80	19.50–20.50	17.50–18.50	0.030	0.010	6.00–6.50 Mo: 0.50–1.00 Cu: 0.180–0.220 N
Type 316 Ti	S31635	0.08	2.00	1.00	16.0–18.0	10.0–14.0	0.045	0.030	5 × %(C + N) min to 0.70 max Ti: 2.0–3.0 Mo: 0.10 N
Type 316 Cb	S31640	0.08	2.00	1.00	16.0–18.0	10.0–14.0	0.045	0.030	10 × %C min to 1.10 max Nb + Ta: 2.0–3.0 Mo: 0.10 N
Type 316 HQ		0.030	2.00	1.00	16.00–18.25	10.00–14.00	0.030	0.015	3.00–4.00 Cu: 2.00–3.00 Mo
Type 317 LM	S31725	0.03	2.00	1.00	18.0–20.0	13.5–17.5	0.045	0.030	4.0–5.0 Mo: 0.10 N
17-14-4 LN	S31726	0.03	2.00	0.75	17.0–20.0	13.5–17.5	0.045	0.030	4.0–5.0 Mo: 0.10–0.20 N
Type 317 LN	S31753	0.03	2.00	1.00	18.0–20.0	11.0–15.0	0.030	0.030	0.10–0.22 N
Type 370	S37000	0.03–0.05	1.65–2.35	0.5–1.0	12.5–14.5	14.5–16.5	0.040	0.010	1.5–2.5 Mo: 0.1–0.4 Ti: 0.005 N: 0.05 Co
18-18-2 (XM-15)	S38100	0.08	2.00	1.5–2.5	17.0–19.0	17.5–18.5	0.030	0.030	
19-9 DL	S63198	0.28–0.35	0.75–1.50	0.03–0.8	18.0–21.0	8.0–11.0	0.040	0.030	1.0–1.75 Mo: 0.1–0.35 Ti: 1.0–1.75 W: 0.25–0.60 Nb

(continued)

Table 11 Continued

Designation(s)	UNS designation	Composition, %(b)							Other
		C	Mn	Si	Cr	Ni	P	S	
20Cb-3	N08020	0.07	2.00	1.00	19.0–21.0	32.0–38.0	0.045	0.035	2.0–3.0 Mo: 3.0–4.0 Cu: 8 × %C min to 1.00 max Nb
20Mo-4	N08024	0.03	1.00	0.50	22.5–25.0	35.0–40.0	0.035	0.035	3.50–5.00 Mo: 0.50–1.50 Cu: 0.15–0.35 Nb
20Mo-6	N08026	0.03	1.00	0.50	22.00–26.00	33.00–37.20	0.03	0.03	5.00–6.70 Mo: 2.00–4.00 Cu
Sanicro 28	N08028	0.02	2.00	1.00	26.0–28.0	29.5–32.5	0.020	0.015	3.0–4.0 Mo: 0.6–1.4 Cu
AL-6X	N08366	0.035	2.00	1.00	20.0–22.0	23.5–25.5	0.030	0.030	6.0–7.0 Mo
AL-6XN	N08367	0.030	2.00	1.00	20.0–22.0	23.50–25.50	0.040	0.030	6.00–7.00 Mo: 0.18–0.25 N
JS-700	N08700	0.04	2.00	1.00	19.0–23.0	24.0–26.0	0.040	0.030	4.3–5.0 Mo: 8 × %C min to 0.5 max Nb: 0.5 Cu: 0.005 Pb: 0.035 S
Type 332	N08800	0.01	1.50	1.00	19.0–23.0	30.0–35.0	0.045	0.015	0.15–0.60 Ti: 0.15–0.60 Al
904L	N08904	0.02	2.00	1.00	19.0–23.0	23.0–28.0	0.045	0.035	4.0–5.0 Mo: 1.0–2.0 Cu
Cronifer 1925 hMo	N08925	0.02	1.00	0.50	24.0–26.0	19.0–21.0	0.045	0.030	6.0–7.0 Mo: 0.8–1.5 Cu: 0.10–0.20 N
Cronifer 2328	…	0.04	0.75	0.75	22.0–24.0	26.0–28.0	0.030	0.015	2.5–3.5 Cu: 0.4–0.7 Ti: 2.5–3.0 Mo
Ferritic stainless steels									
18-2 FM (XM-34)	S18200	0.08	1.25–2.50	1.00	17.5–19.5	…	0.040	0.15 min	1.5–2.5 Mo
Type 430 Ti	S43036	0.10	1.00	1.00	16.0–19.5	0.75	0.040	0.030	5 × %C min to 0.75 max Ti
Type 441	S44100	0.03	1.00	1.00	17.5–19.5	1.00	0.040	0.040	0.3 + 9 × (%C) min to 0.90 max Nb: 0.1–0.5 Ti: 0.03 N
E-Brite 26-1	S44627	0.01	0.40	0.40	25.0–27.0	0.50	0.020	0.020	0.75–1.5 Mo: 0.05–0.2 Nb: 0.015 N: 0.2 Cu
MONIT (25-4-4)	S44635	0.025	1.00	0.75	24.5–26.0	3.5–4.5	0.040	0.030	3.5–4.5 Mo: 0.2 + 4 (%C + %N) min to 0.8 max (Ti + Nb): 0.035 N
Sea-Cure (SC-1)	S44660	0.025	1.00	1.00	25.0–27.0	1.5–3.5	0.040	0.030	2.5–3.5 Mo: 0.2 + 4 (%C + %N) min to 0.8 max (Ti + Nb): 0.035 N
AL 29-4C	S44735	0.030	1.00	1.00	28.0–30.0	1.00	0.040	0.030	3.60–4.20 Mo: 0.20–1.00 Ti + Nb and 6 (%C + %N) min Ti + Nb: 0.045 N
AL 29-4-2	S44800	0.01	0.30	0.20	28.0–30.0	2.0–2.5	0.025	0.020	3.5–4.2 Mo: 0.15 Cu: 0.02 N: 0.025 max (%C + %N)
18 SR(c)	…	0.04	0.30	1.00	18.0	…	…	…	2.0 Al: 0.4 Ti
12 SR(c)	…	0.02	…	0.50	12.0	…	…	…	1.2 Al: 0.3 Ti
406	…	0.06	1.00	0.50	12.0–14.0	0.50	0.040	0.030	2.75–4.25 Al: 0.6 Ti
408 Cb	…	0.03	0.2–0.5	0.2–0.5	11.75–12.25	0.45	0.030	0.020	0.75–1.25 Al: 0.65–0.75 Nb: 0.3–0.5 Ti: 0.03 N

Ferritic stainless steels (continued)

Alloy	UNS	C	Mn	Si	Cr	Ni	P	S	Other
ALFA IV	...	0.03	0.50	0.60	19.0–21.0	0.45	0.035	0.005	4.75–5.25 Al: 0.005–0.035 Ce: 0.03 N
Sealmet 1	...	0.08	0.5–0.8	0.3–0.6	28.0–29.0	0.40	0.030	0.015	0.04 N

Duplex stainless steels

Alloy	UNS	C	Mn	Si	Cr	Ni	P	S	Other
44LN	S31200	0.030	2.00	1.00	24.0–26.0	5.50–6.50	0.045	0.030	1.20–2.00 Mo: 0.14–0.20 N
DP-3	S31260	0.030	1.00	0.75	24.0–26.0	5.50–7.50	0.030	0.030	2.50–3.50 Mo: 0.20–0.80 Cu: 0.10–0.30 N: 0.10–0.50 W
3RE60	S31500	0.030	1.20–2.00	1.40–2.00	18.00–19.00	4.25–5.25	0.030	0.030	2.50–3.00 Mo
2205	S31803	0.030	2.00	1.0	21.0–23.0	4.50–6.50	0.030	0.020	2.50–3.50 Mo: 0.08–0.20 N
2304	S32304	0.030	2.50	1.0	21.5–24.5	3.0–5.5	0.040	0.040	0.05–0.60 Mo: 0.05–0.60 Cu: 0.05–0.20 N
Uranus 50	S32404	0.04	2.00	1.0	20.5–22.5	5.5–8.5	0.030	0.010	2.0–3.0 Mo: 1.0–2.0 Cu: 0.20 N
Ferralium 255	S32550	0.04	1.50	1.00	24.0–27.0	4.50–6.50	0.04	0.03	2.00–4.00 Mo: 1.50–2.50 Cu: 0.10–0.25 N
7-Mo PLUS	S32950	0.03	2.00	0.60	26.0–29.0	3.50–5.20	0.035	0.010	1.00–2.50 Mo: 0.15–0.35 N

Martensitic stainless steels

Alloy	UNS	C	Mn	Si	Cr	Ni	P	S	Other
Type 410S	S41008	0.08	1.00	1.00	11.5–13.5	0.60	0.040	0.030	...
Type 410 Cb (XM-30)	S41040	0.15	1.00	1.00	11.5–13.5	...	0.040	0.030	0.05–0.20 Nb
E4	S41050	0.04	1.00	1.00	10.5–12.5	...	0.045	0.030	0.10 N
CA6NM	S41500	0.05	0.5–1.0	0.60	11.5–14.0	0.60–1.1	0.030	0.030	0.5–1.0 Mo
416 Plus X (XM-6)	S41610	0.15	1.5–2.5	1.00	12.0–14.0	3.5–5.5	0.060	0.15 min	0.6 Mo
Type 418 (Greek Ascolloy)	S41800	0.15–0.20	0.50	0.50	12.0–14.0	1.8–2.2	0.040	0.030	2.5–3.5 W
TrimRite	S42010	0.15–0.30	1.00	1.00	13.5–15.0	0.25–1.00	0.040	0.030	0.40–1.00 Mo
Type 420 F Se	S42023	0.3–0.4	1.25	1.00	12.0–14.0	...	0.060	0.060	0.15 min Se: 0.6 Zr: 0.6 Cu
Lapelloy	S42300	0.27–0.32	0.95–1.35	0.50	11.0–12.0	0.50	0.025	0.025	2.5–3.0 Mo: 0.2–0.3 V
Type 440 F	S44020	0.95–1.20	1.25	1.00	16.0–18.0	0.75	0.040	0.040	0.08 N
Type 440 F Se	S44023	0.95–1.20	1.25	1.00	16.0–18.0	0.75	0.040	0.10–0.35	0.15 min Se: 0.60 Mo

Precipitation-hardening stainless steels

Alloy	UNS	C	Mn	Si	Cr	Ni	P	S	Other
PH 14-4 Mo	S14800	0.05	1.00	1.00	13.75–15.0	7.75–8.75	0.015	0.010	2.0–3.0 Mo: 0.75–1.50 Al
PH 15-7 Mo (Type 632)	S15700	0.09	1.00	1.00	14.0–16.0	6.5–7.75	0.040	0.030	2.0–3.0 Mo: 0.75–1.5 Al
AM-350 (Type 633)	S35000	0.07–0.11	0.5–1.25	0.50	16.0–17.0	4.0–5.0	0.040	0.030	2.5–3.25 Mo: 0.07–0.13 N
AM-355 (Type 634)	S35500	0.10–0.15	0.5–1.25	0.50	15.0–16.0	4.0–5.0	0.040	0.030	2.5–3.25 Mo: 0.07–0.13 N
Custom 450 (XM-25)	S45000	0.05	1.00	1.00	14.0–16.0	5.0–7.0	0.030	0.030	1.25–1.75 Cu: 0.5–1.0 Mo: 8 × %C min Nb
Custom 455 (XM-16)	S45500	0.05	0.50	0.50	11.0–12.5	7.5–9.5	0.040	0.030	1.5–2.5 Cu: 0.8–1.4 Ti: 0.1–0.5 Nb: 0.5 Mo

(a) XM designations in this column are ASTM designations for the listed alloy. (b) Single values are maximum values unless otherwise indicated. (c) Nominal compositions. (d) UNS designation has not been specified. This designation appears in ASTM A 887 and merely indicates the form to be used.

Source: Ref. 36.

and varying specific combination of properties that can be obtained by different compositions within the group [37].

In general, austenitic stainless steels are Fe-Cr-Ni-C and Fe-Cr-Ni-Mn-C alloys containing 16–26% Cr, 0.75–19.0% Mn, 1–37% Ni, and 0.03–0.35% C. The 2xx series (Cr-Mn-Ni) steels contain nitrogen, 5.5–15.5% Mn, and up to 6% Ni. The 3xx (Cr-Ni) types contain higher amounts of Ni and up to 2% Mn. Mo, Cu, Si, Al, Ti, and Nb may be added to confer certain properties such as halide pitting resistance or oxidation resistance. Sulfur or selenium may be added to certain grades to enhance machinability.

Broadly, austenitic stainless steels can be classified into 10 groups [1,38]. These classifications are not straightforward because of the overlapping effects.

Ferritic stainless steels contain essentially 10.5–30% Cr with additions of Mn and Si and occasionally Ni, Al, Ti, or Mo to confer particular characteristics. As they remain ferritic at room and elevated temperatures, they cannot be hardened by heat treatment. Their yield strengths in the annealed condition are usually in the range 275–415 MPa (40–60 ksi). They are used because of their good ductility and cold-formability, good resistance to general liquid corrosion and high-temperature oxidation, resistance to pitting and stress corrosion cracking, and generally lower cost than the austenitic grades [5]. As in the austenitic grades, S and Se may be added to improve machinability.

The standard ferritic stainless steels are types 405, 409, 429, 430, 430F, 430F-Se, 434, 436, 439, 444, and 446 (Table 10).

Duplex stainless steels contain 18–29% Cr, 2.5–8.5% Ni, and 1–4% Mo [1]. They possess a mixed structure of ferrite and austenite. The volume fractions of ferrite and austenite vary between 0.3 and 0.7 in a duplex structure. The ratio of the ferrite and austenite phase (also called phase balance) determines the properties of duplex stainless steels. Compared to austenitic grades, they can offer improved strength (yield strength is about 2–3 times greater), better corrosion resistance, and greater resistance to stress corrosion cracking (SCC). Compared to ferritic grades, they can provide improved toughness, formability, and weldability. Thus, the duplex steels combine some of the merits and disadvantages of the austenitic and ferritic steels; however, for certain applications, they are considered to be the optimum selection.

Types AISI 329 and Carpenter 7-Mo and 7-Mo-plus (UNS S32950) are the more popular duplex steels (Tables 10 and 11). They find applications as welded pipe products for handling wet and dry CO_2 and sour gas and oil products in the petrochemical industry, as welded tubing for heat exchangers, for handling chloride-containing coolants, and for handling hot brines and various organic chemical in the chemical, electrical power, and other industries [1].

Martensitic stainless steels contain 11.5–18% Cr and 0.08–1.2% C. They can be hardened and tempered to yield strength in the range of 550–1900 MPa (80–275 ksi). The Cr content provides these steels with such high hardenability that they can be air hardened even in large sections. If they are to be heat treated for maximum strength, the amount of δ-ferrite should be minimized [5].

The standard martensitic grades are types 403, 410, 414, 416, 416 Se, 420, 422, 431, 440 A, 440B, and 440C (Table 10). They are used in manifold stud bolts, heat control shafts, steam valves, Bourdon tubes, gun mounts, water pump parts, carburetor parts, wire cutter blades, garden shears, cutlery, paint spray nozzles, glass and plastic molds, bomb shackle parts, drive screws, aircraft bolting, cable terminals, diesel engine pump parts, instrument parts, crankshaft counterweight pins, valve trim, ball bearings, and races.

Precipitation-hardening (PH) stainless steels are high-strength alloys with appreciable ductility and good corrosion resistance that are developed by a simple heat treatment comprising martensite formation and low-temperature aging (or tempering) treatment; the latter heat treatment step may be applied after fabrication. PH stainless steels may be divided into three broad groups: (1) martensitic type, (2) semiaustenitic type, and (3) austenitic type (Tables 10 and 11). A majority of these steels are classified by a three-digit number in the AISI 600 series or by a five-digit UNS designation. However, most of them are better known by the trade name or their manufacturer. All steels are available in sheet, strip, plate, bar, and wire.

Martensitic PH stainless steels (also called single-treatment alloys) are most widely used and include 17-4PH (AISI 630 or UNS S17400), stainless W (AISI 635 or UNS S17600), 15-5PH (UNS S15500), PH 13-8Mo (UNS S13800), and Custom 450 (UNS S45000). These steels have a predominantly austenitic structure at the solution-annealing temperature, but they undergo an austenite-to-martensite transformation during cooling to room temperature. These steels can be readily welded [36].

Semiaustenitic PH stainless steels (also called double-treatment alloys) were developed for increased formability prior to the hardening treatment. Important alloys are 17-7PH (UNS S17700) and PH 15-7Mo (UNS S15700). These alloys are completely austenite in the as-quenched condition after solution annealing (which displays good toughness and ductility in the cold-forming operations), and eventually martensite can be formed by simple thermal or thermomechanical treatment.

Austenitic PH stainless steels possess their austenite structures at all temperatures. The most important steels in this class include A-286 (AISI 600 or UNS S66286), 17-10P, and 14-17Cu-Mo alloys. Of these grades, A286 is the most extensively used in the aerospace applications.

6. Maraging Steels

Maraging steels are a specific class of low-carbon ultrahigh-strength steels that derive their strength not from carbon but from precipitation of intermetallic compounds [37–39]. The commonly available maraging steels contain 17–19% Ni, 8–12% Co, 3–5% Mo, and 0.2–1.6% Ti. Since these steels develop very high strength by martensitic transformation and subsequent age-hardening, they are termed maraging steels [40].

There are four types of maraging steels, namely, 200, 250, 300, and 350; the number refers to the ultimate tensile strength in ksi (kpsi). The tensile strength is based on the Ti content, which varies between 0.2 and 1.85%. Table 12 lists the compositions of these grades [41]. In these grades, C content is maintained at a very low level (< 0.03%); Si and Mn total is lower (<0.2%); and P and S contents are also very small (< 0.005% and < 0.008%, respectively) [1].

Maraging steels have found applications where lightweight structures with ultrahigh strength and high toughness are essential and cost is not a major concern. Maraging steels have been extensively used in two general types of applications:

1. Aerospace and aircraft industry for critical components such as missile cases, load cells, helicopter flexible drive shafts, jet engine drive shafts, and landing gear.
2. Tool manufacturing industries for stub shafts, flexible drive shafts, splined shafts, springs, plastic molds, hot forging dies, aluminum and zinc die casting dies, cold-heading dies and cases, diesel fuel pump pins, router bits, clutch disks, gears in the machine tools, carbide die holders, autofrettage equipment, etc.

IV. DESIGNATIONS FOR STEELS

A designation is the specific identification of each grade, type, or class of steel by a number, letter, symbol, name, or suitable combination thereof unique to a certain steel. It is used in a

Table 12   Nominal Compositions of Commercial Maraging Steels

Grade	Composition, %(a)					
	Ni	Mo	Co	Ti	Al	Nb
Standard grades						
18Ni(200)	18	3.3	8.5	0.2	0.1	⋯
18Ni(250)	18	5.0	8.5	0.4	0.1	⋯
18Ni(300)	18	5.0	9.0	0.7	0.1	⋯
18Ni(350)	18	4.2(b)	12.5	1.6	0.1	⋯
18Ni(Cast)	17	4.6	10.0	0.3	0.1	⋯
12-5-3(180)(c)	12	3	⋯	0.2	0.3	⋯
Cobalt-free and low-cobalt bearing grades						
Cobalt-free 18Ni(200)	18.5	3.0	⋯	0.7	0.1	⋯
Cobalt-free 18Ni(250)	18.5	3.0	⋯	1.4	0.1	⋯
Low-cobalt 18Ni(250)	18.5	2.6	2.0	1.2	0.1	0.1
Cobalt-free 18Ni(300)	18.5	4.0	⋯	1.85	0.1	⋯

(a) All grades contain no more than 0.03% C. (b) Some producers use a combination of 4.8% Mo and 1.4% Ti, nominal. (c) Contains 5% Cr

Source: Ref. 41.

specific document as well as in a particular country. In the steel industries, these terms have very specific uses: "grade" is used to describe chemical composition; "type" is used to denote deoxidation practice; and "class" is used to indicate some other attributes such as tensile strength level or surface quality [3].

In ASTM specifications, however, these terms are used somewhat interchangeably. For example, in ASTM A 434, "grade" identifies chemical composition and "class" indicates tensile properties. In ASTM A 515, "grade" describes strength level; the maximum carbon content allowed by the specification is dependent on both the plate thickness and the strength level. In ASTM A 533, "type" indicates chemical analysis, while "class" denotes strength level. In ASTM A 302, "grade" identifies requirements for both chemical composition and tensile properties. ASTM A 514 and A 517 are specifications for high-strength quenched and tempered alloy steel plate for structural and pressure vessel applications, respectively; each has a number of grades identifying chemical composition capable of developing the required mechanical properties. However, all grades of both designations have the same composition limits.

By far the most widely used basis for classification and/or designation of steels is the chemical composition. The most commonly used system of designating carbon and alloy steels in the United States is that of the American Iron & Steel Institute and the Society of Automotive Engineers (AISI and SAE) numerical designation. The Unified Numbering System (UNS) is also being increasingly employed. Other designations used in the specialized fields include Aerospace Materials Specification (AMS) and American Petroleum Institute (API). These designation systems are discussed below.

A.   SAE-AISI Designations

As stated above, the SAE-AISI system is the most widely used designation for carbon and alloy steels. The SAE-AISI system is applied to semifinished forgings, hot-rolled and cold-finished bars, wire rod, seamless tubular goods, structural shapes, plates, sheet, strip, and welded tubing. Table 2 lists the SAE-AISI system of numerical designations for both carbon and low alloy steels.

Carbon and Alloy Steels   With few exceptions, the SAE-AISI system uses a four-digit number to designate carbon and alloy steels, specified to chemical composition ranges. Certain types of alloy steels are designated by five digits (numerals). Table 2 shows an abbreviated listing of four-digit designations of the SAE-AISI carbon and alloy steels. The first digit, 1, of this designation indicates a carbon steel; i.e., carbon steels comprise 1xxx groups in the SAE-AISI system and are subdivided into four series due to the variance in certain fundamental properties among them. Thus, the plain carbon steels comprise 10xx series (containing 1.00% Mn maximum); resulfurized carbon steels comprise the 11xx series; resulfurized and rephosphorized carbon steels comprise the 12xx series; and nonresulfurized high-manganese (up to 1.65%) carbon steels comprise the 15xx series. Both the 11xx and 12xx groups of steels are produced for applications requiring good machinability.

Carbon and alloy steel designations showing the letter B inserted between the second and third digits indicate that the steel has 0.0005–0.003% boron. Likewise, the letter L inserted between the second and third digits indicates that the steel has 0.15–0.35% lead for enhanced machinability. Sometimes the prefix M is used for merchant quality steels and the suffix H is used to comply with specific hardenability requirements. In alloy steels, the prefix letter E is used to designate steels that are produced by the basic electric furnace process.

The major alloying element in an alloy steel is indicated by the first two digits of the designation (Table 2). Thus, a first digit of 2 denotes a nickel steel; 3, a nickel-chromium steel; 4, a molybdenum, chromium-molybdenum, nickel-molybdenum, or nickel-chromium-molybdenum steel; 5, a chromium steel; 6, a chromium-vanadium steel; 7, a tungsten-chromium steel; 8, a nickel-chromium-molybdenum steel; and 9, a silicon-manganese steel or a nickel-chromium-molybdenum steel. In the case of a simple alloy steel, the second digit represents the approximate percentage of the predominant alloying element. For example, 2520 grade indicates a nickel steel of approximately 5% Ni (and 0.2% carbon).

The last two digits of four-numeral designations and the last three digits of five-numeral designations indicate the approximate carbon content of the allowable carbon range in hundredths of a percent. For example, 1020 steel indicates a plain carbon steel with an approximate mean of 0.20% carbon varying within acceptable carbon limits of 0.18 and 0.23%. Similarly, 4340 steels are Ni-Cr-Mo steels and contain an approximate mean of 0.40% carbon varying within an allowable carbon range of 0.38–0.43%, and 51100 steel is a chromium steel with an approximate mean of 1.00% carbon varying within an acceptable carbon range of 0.98–1.10% [2, 20, 42].

Potential standard steels are listed in SAE J1081 and Table 13. They are experimental steels to which no regular AISI-SAE designations have been assigned. The numbers consist of the prefix PS followed by a sequential number starting with 1. Some were developed to minimize the amount of nickel and others to enhance a particular attribute of a standard grade of alloy steel [20].

HSLA Steels Several grades of HSLA steels have been described in the SAE Recommended Practice J410. Their chemical composition and minimum mechanical properties requirements are provided in Table 8 [20].

Formerly Listed SAE Steels A number of grades of carbon and alloy steels have been excluded from the list of standard SAE steels because of their inadequate applications. A detailed list of formerly used SAE carbon and alloy steels are given in SAE J1249, and producers should be contacted concerning their availability.

B. UNS Designations

The Unified Numbering System (UNS) has been developed by the American Society for Testing and Materials (ASTM E 527), the Society of Automotive Engineers (SAE J1086), and several other technical societies, trade associations, and U.S. government agencies [19]. A UNS

Table 13 SAE Potential Standard Steel Compositions

SAE PS number(a)	Ladle chemical composition limits, wt%								
	C	Mn	P max	S max	Si	Ni	Cr	Mo	B
PS 10	0.19–0.24	0.95–1.25	0.035	0.040	0.15–0.35	0.20–0.40	0.25–0.40	0.05–0.10	...
PS 15	0.18–0.23	0.90–1.20	0.035	0.040	0.15–0.35	...	0.40–0.60	0.13–0.20	...
PS 16	0.20–0.25	0.90–1.20	0.035	0.040	0.15–0.35	...	0.40–0.60	0.13–0.20	...
PS 17	0.23–0.28	0.90–1.20	0.035	0.040	0.15–0.35	...	0.40–0.60	0.13–0.20	...
PS 18	0.25–0.30	0.90–1.20	0.035	0.040	0.15–0.35	...	0.40–0.60	0.13–0.20	...
PS 19	0.18–0.23	0.90–1.20	0.035	0.040	0.15–0.35	...	0.40–0.60	0.08–0.15	0.0005–0.003
PS 20	0.13–0.18	0.90–1.20	0.035	0.040	0.15–0.35	...	0.40–0.60	0.13–0.20	...
PS 21	0.15–0.20	0.90–1.20	0.035	0.040	0.15–0.35	...	0.40–0.60	0.13–0.20	...
PS 24	0.18–0.23	0.75–1.00	0.035	0.040	0.15–0.35	...	0.40–0.60	0.20–0.30	...
PS 30	0.13–0.18	0.70–0.90	0.035	0.040	0.15–0.35	0.70–1.00	0.45–0.65	0.45–0.60	...
PS 31	0.15–0.20	0.70–0.90	0.035	0.040	0.15–0.35	0.70–1.00	0.45–0.65	0.45–0.60	...
PS 32	0.18–0.23	0.70–0.90	0.035	0.040	0.15–0.35	0.70–1.00	0.45–0.65	0.45–0.60	...
PS 33(b)	0.17–0.24	0.85–1.25	0.035	0.040	0.15–0.35	0.20 min	0.20 min	0.05 min	...
PS 34	0.28–0.33	0.90–1.20	0.035	0.040	0.15–0.35	...	0.40–0.60	0.13–0.20	...
PS 36	0.38–0.43	0.90–1.20	0.035	0.040	0.15–0.35	...	0.45–0.65	0.13–0.20	...
PS 38	0.43–0.48	0.90–1.20	0.035	0.040	0.15–0.35	...	0.45–0.65	0.13–0.20	...
PS 39	0.48–0.53	0.90–1.20	0.035	0.040	0.15–0.35	...	0.45–0.65	0.13–0.20	...
PS 40	0.51–0.59	0.90–1.20	0.035	0.040	0.15–0.35	...	0.45–0.65	0.13–0.20	...
PS 54	0.19–0.25	0.70–1.05	0.035	0.040	0.15–0.35	...	0.40–0.70	0.05 min	...
PS 55	0.15–0.20	0.70–1.00	0.035	0.040	0.15–0.35	1.65–2.00	0.45–0.65	0.65–0.80	...
PS 56	0.080–0.13	0.70–1.00	0.035	0.040	0.15–0.35	1.65–2.00	0.45–0.65	0.65–0.80	...
PS 57	0.08 max	1.25 max	0.040	0.15–0.35	1.00 max	...	17.00–19.00	1.75–2.25	...
PS 58	0.16–0.21	1.00–1.30	0.035	0.040	0.15–0.35	...	0.45–0.65	0.13–0.20	...
PS 59	0.18–0.23	1.00–1.30	0.035	0.040	0.15–0.35	...	0.70–0.90	0.13–0.20	...
PS 61	0.23–0.28	1.00–1.30	0.035	0.040	0.15–0.35	...	0.70–0.90	0.13–0.20	...
PS 63	0.31–0.38	0.75–1.10	0.035	0.040	0.15–0.35	...	0.45–0.65	0.13–0.20	...
PS 64	0.16–0.21	1.00–1.30	0.035	0.040	0.15–0.35	...	0.70–0.90	0.13–0.20	...
PS 65	0.21–0.26	1.00–1.30	0.035	0.040	0.15–0.35	...	0.70–0.90	0.13–0.20	0.0005–0.003
PS 66(c)	0.16–0.21	0.40–0.70	0.035	0.040	0.15–0.35	1.65–2.00	0.45–0.75	0.08–0.15	...
PS 67	0.42–0.49	0.80–1.20	0.035	0.040	0.15–0.35	...	0.85–1.20	0.25–0.35	...

(a) Some PS steels may be supplied to a hardenability requirement. (b) Supplied to a hardenability requirement of 15 HRC points within the range of 23–43 HRC at J4 (4/16 in. distance from quenched end). (c) PS 66 has a vanadium content of 0.10–0.15%. subject to agreement between producer and user.

Source: Ref. 20.

number, which is a designation of chemical composition and not a specification, is assigned to each chemical composition of the standard carbon and alloy steel grades for which controlling limits have been established by the SAE-AISI [16, 20, 43].

The UNS designation consists of a single-letter prefix followed by five numerals (digits). The letters denote the broad class of alloys; the numerals define specific alloys within that class. The prefix letter G signifies standard grades of carbon and alloy steels; the prefix letter H indicates standard grades that meet certain hardenability requirements limits (SAE-AISI H steels); the prefix T includes tool steels, wrought and cast; the prefix letter S relates to heat- and corrosion-resistant steels (including stainless steel), valve steels, and iron-base superalloys; the prefix letter J is used for cast steels (except tool steels); the prefix letter K identifies miscellaneous steels and ferrous alloys; and the prefix W denotes welding filler metals (for example, W00001–W59999 series represent a wide variety of steel compositions) [43]. The first four digits of the UNS number usually correspond to the standard SAE-AISI designations, while the last digit (except zero) of the five-numeral series denotes some additional composition requirements, such as boron, lead, or nonstandard chemical ranges. Tables 3 and 4 list the UNS numbers corresponding to SAE-AISI numbers for various standard carbon and alloy steels, respectively, with composition ranges.

V. SPECIFICATIONS FOR STEELS

A specification is typically an acronym or abbreviation for a standards organization plus a specific written statement of both technical and commercial requirements that a product must satisfy. It is a document that restrains or controls procurement that is issued by that standards organization. All material specifications contain general and specific information [44]. Any reasonably adequate specification will furnish the information about the items stated below [2, 3].

Scope may include product classification, required size range, condition, and any comments on product processing considered helpful to either the supplier or user. An informative title and a statement of the required form may be employed instead of a scope item.

Chemical composition may be described, or it may be denoted by a well-known designation based on chemical composition. The SAE-AISI designations are normally used.

A quality statement covers any appropriate quality descriptor and whatever additional prerequisites might be necessary. It may also include the type of steel and the steelmaking processes allowed.

Quantitative requirements recognize permissible composition ranges and all physical and mechanical properties necessary to characterize the material. Testing methods employed to check these properties should also be included or reference made to standard test methods. This section should only address those properties that are vital for the intended application.

Additional requirements can cover surface preparation, special tolerances, and edge finish on flat-rolled products as well as special packaging, identification, and loading instructions.

Engineering societies, trade associations, and institutes whose members make, specify, or purchase steel products publish standard specifications; many of them are well recognized and highly respected. Some of the notable specification-writing groups and/or standard organizations in the United States are listed below. It is clear from these names that a particular specification-writing group is limited to its own specialized field.

Organization	Acronym
Association of American Railroads	AAR
American Bureau of Shipbuilding	ABS
Aerospace Materials Specification (of SAE)	AMS
American National Standards Institute	ANSI
American Petroleum Institute	API
American Railway Engineering Association	AREA
American Society of Mechanical Engineers	ASME
American Society for Testing and Materials	ASTM
American Welding Society	AWS
Society of Automotive Engineers	SAE

A. ASTM (ASME) Specifications

The most widely used standard specifications for steel products in the United States are those published by ASTM, many of which are complete specifications, usually adequate for procurement purposes. These specifications frequently apply to specific products, which are usually oriented toward the performance of the fabricated end product. They begin with the prefix ASTM, followed by letter A, identifying a ferrous material, then a number indicating the actual specification, which may be followed by letters or numbers subdividing the material by analysis. The AISI code is sometimes used for this purpose. Finally the year of origin is mentioned. A letter T after this denotes a tentative specification. Generally, each specification includes a steel in a specific form or for a special purpose rather than by analysis.

ASTM specifications represent a consensus drawn from producers, specifiers, fabricators, and users of steel mill products. In many cases, the dimensions, tolerances, limits, and restrictions in the ASTM specifications are the same as the corresponding items of the standard practices in the AISI steel products manuals. Many of the ASTM specifications have been adopted by the American Society of Mechanical Engineers (ASME) with slight or no modification. ASME uses the prefix S with the ASTM specifications; for example, ASME SA 213 and ASTM A 213 are the same.

Steel products can be distinguished by the ASTM specification number to which they are produced. Sometimes, citing the ASTM specification is not sufficient to completely identify a steel product. For example, A 434 is a specification used for heat-treated (hardened and tempered) alloy steel bars. To fully identify steel bars indicated by this specification, the grade/AISI-SAE designation and class (the required strength level) must also be quoted. The ASTM specification A 434 also covers, by reference, two standards for test methods (A 370 for mechanical testing and E 112 for grain size determination) and A 29 for specifying general requirements for bar products.

SAE-AISI designations for the chemical compositions of carbon and alloy steels are sometimes included in the ASTM specifications for bars, wires, and billets for forging. Some ASTM specifications for sheet products incorporate SAE-AISI designations for chemical composition. ASTM specifications for plates and structural shapes normally specify the limits and ranges of chemical composition directly without the SAE-AISI designations. Table 14 incorporates a list of some ASTM specifications that include SAE-AISI designations for compositions of different steel grades.

Table 14 ASTM Specifications That Cover SAE-AISI Designations

A 29	Carbon and alloy steel bars, hot rolled and cold finished		A 510	Carbon steel wire rods and coarse round wire
A 108	Standard quality cold-finished carbon steel bars		A 534	Carburizing steels for antifriction bearings
A 295	High carbon–chromium ball and roller bearing steel		A 535	Special quality ball and roller bearing steel
A 304	Alloy steel bars having hardenability requirements		A 544	Scrapless nut quality carbon steel wire
A 322	Hot-rolled alloy steel bars		A 545	Cold-heading quality carbon steel wire for machine screws
A 331	Cold-finished alloy steel bars		A 546	Cold-heading quality medium high carbon steel wire for hexagon-head bolts
A 434	Hot-rolled or cold-finished quenched and tempered alloy steel bars		A 547	Cold-heading quality alloy steel wire for hexagon-head bolts
A 505	Hot-rolled and cold-rolled alloy steel sheet and strip		A 548	Cold-heading quality carbon steel wire for tapping or sheet metal screws
A 506	Regular quality hot-rolled and cold-rolled alloy steel sheet and strip		A 549	Cold-heading quality carbon steel wire for wood screws
A 507	Drawing quality hot-rolled and cold-rolled alloy steel sheet and strip		A 575	Merchant quality hot-rolled carbon steel bars
			A 576	Special quality hot-rolled carbon steel bars
A 646	Premium quality alloy steel blooms and billets for aircraft and aerospace forgings			
A 659	Commercial quality hot-rolled carbon steel sheet and strip			
A 682	Cold-rolled spring quality carbon steel strip, generic			
A 684	Untempered cold-rolled high-carbon steel strip			
A 689	Carbon and alloy steel bars for springs			
A 711	Carbon and alloy steel blooms, billets, and slabs for forging			
A 713	High-carbon spring steel wire for heat-treated components			
A 752	Alloy steel wire rods and coarse round wire			
A 827	Carbon steel plates for forging and similar applications			
A 829	Structural quality alloy steel plates			
A 830	Structural quality carbon steel plates			

Source: Ref. 3.

B. AMS Specifications

Aerospace Materials Specifications (AMS), published by SAE, are procurement documents, not design specifications. The majority of the AMS specifications pertain to materials intended for aerospace applications. These specifications generally include mechanical property requirements and limits significantly more severe than those for materials or steel grades with identical compositions but meant for nonaerospace applications. Their compliance will ensure procurement of a specific form and condition or a specific material (or steel grade) or process. Tables 15 and 16 show the AMS designations of carbon and alloy steels, respectively, indicating the chemical composition, title of specification (covering specific form, chemical composition, process, and condition), and equivalent UNS number, nearest proprietary or AISI-SAE grade, and similar military (MIL) or federal (FED) specifications [45].

Table 15 AMS Number, Title of Specification, and Equivalent UNS Number, Proprietary/AISI-SAE Alloy, and Similar Specification for Wrought Carbon Steels

AMS No.	Title of Specification	UNS No.	Alloy	Similar Specification
5100II	Bars, Screw Stock, Free Machining, Cold Drawn	G12120	1212	
		G12150	1215	
5020C	Bars, Forgings and Tubing, 1.5Mn 0.25Pb (0.32-0.39C), Free Cutting	G11374	111.37	
5022L	Bars, Forgings, and Tubings, 0.14-0.20C, Free Cutting	G11170	1117	
5024F	Bars, Forgings and Tubing, 1.5Mn (0.32-0.39C), Free Cutting	G11370	1137	
5027C	Wire, Welding, 1.05Cr 0.55Ni 1.0Mo 0.07V (0.26-0.32C), Vacuum melted, Environment Controlled Packaging	K24728	D6AC	
5028B	Wire, Welding, 1.05Cr 0.55Ni 1.0Mo 0.07V (0.34-0.40C), Vacuum melted, Environment Controlled Packaging	K23725	D6AC	
5029B	Wire, Welding, 0.78Cr 1.8Ni 0.35Mo 0.20V (0.33-0.38C), Vacuum melted, Environment Controlled Packaging	K23577		
5030F	Wire, Welding, 0.06 Carbon Maximum	K00606		
5031C	Welding Electrodes, Covered, Steel, 0.07-0.15C	W06013	s6013	FED-QQ-E-450, Type 6013
5032E	Wire, 0.18-0.23C, Annealed	G10200	1020	FED-QQ-W-461
5036G	Sheet and Strip, Aluminum Coated, Low Carbon			MIL-S-4174, Type 1, Grade B
5040J	Sheet and Strip, 0.15 Carbon Maximum, Deep Drawing Grade	G10100	1010	
5042J	Sheet and Strip, 0.15 Carbon Maximum, Forming Grade	G10100	1010	
5044G	Sheet and Strip, 0.15 Carbon Maximum, Half Hard Temper	G10100	1010	
5045F	Sheet and Strip, 0.25 Carbon maximum, Hard Temper	G10200	1020	
5046A	Sheet, Strip, and Plate, Annealed	G10200	1020	MIL-S-7952
		G10250	1025	
5047D	Sheet and Strip, 0.08-0.13C, Al Killed, Deep Forming Grade	G10100	1010	
5050J	Tubing, Seamless, 0.15 Carbon Maximum, Annealed	G10100	1010	
5053G	Tubing, Welded, 0.13 Carbon Maximum, Annealed	G10100	1010	
5060F	Bars, Forgings and Tubing, 0.13-0.18C	G10150	1015	
5061D	Bars and Wire, Low Carbon	K00802		
5062E	Bars, Forgings, Tubing, Sheet, Strip, and Plate, Low Carbon	K02508		
5069E	Bars, Forgings, and Tubing, 0.15-0.20C	G10180	1018	
5070G	Bars and Forgings, 0.18-0.23C	G10220	1022	
5075E	Tubing, Seamless, 0.22-0.28C, Cold Drawn & Stress Relieved	G10250	1025	MIL-T-5066
5077E	Tubing, Welded, 0.22-0.28C, Normalized or Stress Relieved	G10250	1025	MIL-T-5066
5080II	Bars, Forgings, and Tubing, 0.31-0.38C	G10350	1035	
5082E	Tubing, Seamless, 0.31-0.38C, Stress Relieved	G10350	1035	
5085D	Sheet, Strip, and Plate, 0.47-0.55C, Annealed	G10500	1050	
5110F	Wire, Carbon, Spring Temper, Cold Drawn, 0.75-0.88C	G10800	1080	
5112J	Wire, Spring Quality Music Wire, 0.70-1.00C, Cold Drawn	G10900	1090	
5115G	Wire, Valve Spring Quality, 0.60-0.75C, Hardened & Tempered	G10700	1070	
5120J	Strip, 0.68-0.80C	G10740	1074	
5121G	Sheet and Strip, 0.90-1.40C	G10950	1095	MIL-S-7947
5122G	Strip, 0.90-1.04C, Hard Temper	G10950	1095	MIL-S-7947, Hard Temper
5132G	Bars, 0.90-1.30C	G10950	1095	

Source: Ref. 45.

Table 16 AMS Number, Title of Specification, and Equivalent UNS Number, Nearest Proprietary or AISI-SAE Grade, and Similar Specification for Wrought Alloy Steels

AMS No.	Title of Specification	UNS No.	Alloy	Similar Specification
6250H	Bars, Forgings, and Tubing, 1.5Cr 3.5Ni (0.07-0.13C)	K44910	3310	MIL-S-7393, Composition 1
6255A	Bars, Forgings, and Tubing, 1.1Si 1.45Cr 1.0Mo 0.08Al (0.16-0.22C), Premium Air Quality, Double Vacuum Melted	K21940	CBS 600	
6256A	Bars, Forgings, and Tubing, 1.0Cr 3.0Ni 4.5Mo 0.08Al 0.38V (0.10-0.16C), Premium Air Quality, Double Vacuum Melted	K71350	CBS 1000M	
6257	Bars, Forgings, and Tubing, 1.6Si 0.82Cr 1.8Ni 0.40Mo 0.08V (0.40-0.44C), Consumable Electrode Vacuum Remelted, Normalized and Tempered			
6260L	Bars, Forgings, and Tubing, Carburizing Grade, 1.2Cr 3.2Ni 0.12Mo (0.07-0.13C)	G93106	9310	
6263H	Bars, Forgings, and Tubing, Carburizing Grade, 1.2Cr 3.2Ni 0.12Mo (0.11-0.17C)	G93150	9315	
6264G	Bars, Forgings, and Tubing, Carburizing Grade, 3.2Ni 1.2Cr 0.12Mo (0.14-0.20C)	K44414	9317	
6265H	Bars, Forgings, and Tubing, 1.2Cr 3.25Ni (0.07-0.13C), Vacuum Consumable Electrode Remelted	G93106	9310	
6266G	Bars, Forgings, and Tubing, 0.50Cr 1.82Ni 0.25Mo 0.003B 0.06V (0.08-0.13C)	K21028	43BV12	
6267D	Bars, Forgings, and Tubing, 1.2Cr 3.25Ni 0.12Mo (0.07-0.13C), Electroslag Remelted or Vacuum Remelted, Consumable Electrode	G93106	9310	
6270L	Bars, Forgings, and Tubing, 0.5Cr 0.55Ni 0.20Mo (0.11-0.17C)	G86150	8615	
6272H	Bars, Forgings, and Tubing, 0.50Cr 0.55Ni 0.20Mo (0.15-0.20C)	G86170	8617	
6274L	Bars, Forgings, and Tubing, 0.50Cr 0.55Ni 0.20Mo (0.18-0.23C)	G86200	8620	
6275F	Bars, Forgings, and Tubing, 0.40Cr 0.45Ni 0.12Mo 0.002B (0.15-0.20C)	G94171	94B17	
6276F	Bars, Forgings, and Tubing, 0.50Cr 0.55Ni 0.20Mo (0.18-0.23C), Consumable Electrode Vacuum Melted	G86200	8620	
6277D	Bars, Forgings, and Tubing, 0.50Cr 0.55Ni 0.20Mo (0.18-0.23C), Vacuum Arc or Electroslag Remelted	G86200	8620	
6278A	Bars, Forgings, and Tubing, 4.1Cr 3.4Ni 4.2Mo 1.2V (0.11-0.15C), Premium Aircraft Quality for Bearing Applications, Double Vacuum Melted			
6280H	Bars, Forgings, and Rings, 0.50Cr 0.55Ni 0.20Mo (0.28-0.33C)	G86300	8630	MIL-S-6050
6281G	Tubing, Mechanical, 0.50Cr 0.55Ni 0.20Mo (0.28-0.33C)	G86300	8630	
6282G	Tubing, Mechanical, 0.50Cr 0.55Ni 0.25Mo (0.33-0.38C)	G87350	8735	
6290F	Bars and Forgings, Carburizing Grade, 1.8Ni 0.25Mo (0.11-0.17C)	G46150	4615	MIL-S-7493, Composition 4615
6292F	Bars and Forgings, Carburizing Grade, 1.8Ni 0.25Mo (0.14-0.20C)	G46170	4617	MIL-S-7493 Composition 4617
6294F	Bars and Forgings, Carburizing Grade, 1.8Ni 0.25Mo (0.17-0.22C)	G46200	4620	
6299C	Bars, Forgings, and Tubing, 0.50Cr 1.8Ni 0.25Mo (0.17-0.23C)	H43200	4320H	
6300C	Bars and Forgings, 0.25Mo (0.35-0.40C)	G40370	4037	
6302E	Bars, Forgings, and Tubing, Low Alloy, Heat-Resistant, 0.65Si, 1.25Cr 0.50Mo 0.25V (0.28-0.33C)	K23015	17-22A(S)	
6303E	Bars and Forgings, Low Alloy, Heat-Resistant, 0.65Si 1.25Cr, 0.50Mo 0.85V (0.25-0.30C)	K22770	17-22A(V)	
6304G	Bars, Forgings, and Tubing, Low Alloy, Heat-Resistant, 0.95Cr 0.55Mo 0.30V (0.40-0.50C)	K14675	17-22A	MIL-S-24502
MAM 6304	Bars, Forgings, and Tubing, Low Alloy, Heat Resistant, 0.95Cr 0.55Mo 0.30V (0.40-0.50C)	K14675	17-22A	MIL-S-24502
6305B	Bars, Forgings, and Tubing, Low Alloy, Heat Resistant, 0.95Cr 0.55Mo 0.30V (0.40-0.50C), Vacuum Arc Remelted	K14675	17-22A	

(continued)

C. Military and Federal Specifications

Military (MIL) specifications and standards are produced and adopted by the U.S. Department of Defense. MIL specifications are used to define materials, products, and services. MIL standards provide procedures for design, manufacturing, and testing instead of giving only a particular material description. MIL specifications begin with the prefix MIL, followed by a

Table 16 Continued

AMS No.	Title of Specification	UNS No.	Alloy	Similar Specification
6308A	Bars and Forgings, 0.90Si 1.0Cr 2.0Ni 3.2Mo 2.0Cu 0.10V (0.07-0.13C), Vacuum Arc or Electroslag Remelted	K71040	Pyrowear, Alloy 53	
6312E	Bars, Forgings, and Tubing, 1.8Ni 0.25Mo (0.38-0.43C)	K22440	4640	
6317F	Bars and Forgings, 1.8Ni 0.25Mo (0.38-0.43C), Heat Treated, 125 ksi (862 MPa) Tensile Strength	K22400	4640	
6320J	Bars, Forgings, and Rings, 0.50Cr 0.55Ni 0.25Mo (0.33-0.38C)	G87350	8735	
6321D	Bars, Forgings, and Tubing, 0.42Cr 0.30Ni 0.12Mo 0.003B (0.38-0.43C)	K03810	81B40	
6322K	Bars, Forgings, and Rings, 0.50Cr 0.55Ni 0.25Mo (0.38-0.43C)	G87400	8740	MIL-S-6049
6323H	Tubing, Mechanical, 0.50Cr 0.55Ni 0.25Mo (0.38-0.43C)	G87400	8740	
6324E	Bars, Forgings, and Tubing, 0.65Cr 0.70Ni 0.25Mo (0.38-0.43C)	K11640	8740 Mod	
6325F	Bars, Forgings, and Rings, 0.50Cr 0.55Ni 0.25Mo (0.38-0.43C), Heat Treated, 105 ksi (724 MPa) Tensile Strength	G8740	8740	MIL-S-6049
6327G	Bars and Forgings, 0.50Cr 0.55Ni 0.25Mo (0.38-0.43C), Heat Treated, 125 ksi (862 MPa) Tensile Strength	G8740	8740	MIL-S-6049
6328H	Bars, Forgings, and Tubing, 0.50Cr 0.55Ni 0.25Mo (0.48-0.53C)	K13550	8750	
6330E	Bars, Forgings, and Tubing, 0.65Cr 1.25Ni (0.33-0.38C)	K22033		
6331	Wire, Welding, 0.50Cr 0.55Ni 0.20Mo (0.33-0.38C), Vacuum Melted, Environment Controlled Packaging	G87350	8735	
6342H	Bars, Forgings, and Tubing, 0.80Cr 1.0Ni 0.25Mo (0.38-0.43C)	G98400	9840	
6348A	Bars, 0.95Cr 0.20Mo (0.28-0.33C), Normalized	G41300	4130	MIL-S-6758
6349B	Bars, 0.95Cr 0.20Mo (0.38-0.43C), Normalized	G41400	4140	MIL-S-5626
6350H	Sheet, Strip, and Plate, 0.95Cr 0.20Mo (0.28-0.33C)	G41300	4130	MIL-S-18729
6351E	Sheet, Strip, and Plate, 0.95Cr 0.20Mo (0.28-0.33C), Spheroidized	G41300	4130	
6352F	Sheet, Strip, and Plate, 0.95Cr 0.20Mo (0.33-0.38C)	G41350	4135	
6354D	Sheet, Stip, and Plate, 0.75Si 0.62Cr 0.20Mo 0.10Zr (0.10-0.17C)	K11914	NAX 9115-AC	
6356D	Sheet, Strip, and Plate, 0.95Cr 0.20Mo (0.30-0.35C)	G41320	4132	
6357G	Sheet, Strip and Plate, 0.50Cr 0.55Ni 0.25Mo (0.33-0.038C)	G87350	8735	
6358F	Sheet, Strip, and Plate, 0.50Cr 0.55Ni 0.25Mo (0.38-0.43C)	G87400	8740	
6359F	Sheet, Strip, and Plate, 0.80Cr 1.8Ni 0.25Mo (0.38-0.43C)	G43400	4340	
6360J	Tubing, Seamless, 0.95Cr 0.20Mo (0.28-0.33C), Normalized or Stress Relieved	G41300	4130	MIL-T-6736 Condition N
6361C	Tubing, Seamless Round, 0.95Cr 0.20Mo (0.28-0.33C), 125 ksi (860 MPa) Tensile Strength	G41300	4130	MIL-T-6736
6362D	Tubing, Seamless, 0.95Cr 0.20Mo (0.28-0.33C), 150 ksi (1034 MPa) Tensile Strength	G41300	4130	MIL-T-6736 Condition HT-
6365H	Tubing, Seamless, 0.95Cr 0.20Mo (0.33-0.38C), Normalized or Stress Relieved	G41350	4135	MIL-T-6735
6370K	Bars, Forgings, and Rings, 0.95Cr 0.20Mo (0.28-0.33C)	G41300	4130	MIL-S-6758
6371H	Tubing, Mechanical, 0.95Cr 0.20Mo (0.28-0.33C)	G41300	4130	MIL-T-6736
6372H	Tubing, Mechanical, 0.95Cr 0.20Mo (0.33-0.38C)	G41350	4135	
6373C	Tubing, Welded, 0.95Cr 0.20Mo (0.28-0.33C)	G41300	4130	MIL-T-6736
6374A	Tubing, Seam-Free, Round, 0.95Cr 0.20Mo (0.28-0.33C), 95 ksi (655 MPa) Tensile Strength	G41300	4130	
6375	Wire, Welding, 0.50Cr 0.55Ni 0.20Mo (0.18-0.23C), Vacuum Melted, Environment Controlled Packaging	G86200	8620	
6378E	Bars, 1.0Cr 0.20Mo 0.045Se (0.39-0.48C), Die Drawn, 130 ksi (896 MPa) Yield Strength, Free Machining	K11542	4142H Mod	
6379A	Bars, Die Drawn, 0.95Cr 0.20Mo 0.05Te (0.40-0.53C), Tempered, 165 ksi (1140 MPa) Yield Strength	K11546	4140 Mod	
6381E	Tubing, Mechanical, 0.95Cr 0.20Mo (0.38-0.43C)	G41400	4140	
6382K	Bars, Forgings, and Rings, 0.95Cr 0.20Mo (0.38-0.43C)	G41400	4140	MIL-S-5626
6385E	Sheet, Strip, and Plate, Low Alloy, Heat Resistant, 1.25Cr 0.50Mo 0.65Si 0.25V (0.27-0.33C)	K23015	17-22A/S	
6386B	Sheet and Plate, Heat Treated, 90 ksi and 100 ksi Yield Strength	K11856	....	
6390C	Tubing, Mechanical, 0.95Cr 0.20Mo (0.38-0.43C)	G41400	4140	
6395D	Sheet, Strip, and Plate, 0.95Cr 0.20Mo (0.38-0.43C)	G41400	4140	

code letter that represents the first letter of the title for the item, followed by a hyphen and then the serial numbers or digits. Some examples of MIL specifications for steels with corresponding AMS numbers, UNS numbers, and nearest proprietary or AISI-SAE grades are listed in Tables 15 and 16.

Table 16 Continued

AMS No.	Title of Specification	UNS No.	Alloy	Similar Specification
6396B	Sheet, Strip, and Plate, 0.80Cr 1.8Ni 0.25Mo (0.49-0.55C), Annealed	K22950		
6406C	Sheet, Strip, and Plate, 2.1Cr 0.58Mo 1.6Si 0.05V (0.41-0.46C), Annealed	K34378	X200	
6407E	Bars, Forgings, and Tubing, 1.2Cr 2.0Ni 0.45Mo (0.27-0.33C)	K33020	HS-220	
6408	Bars and Forgings, Tool, Hotwork, 5.2Cr 1.5Mo 1.0V (0.35-0.45C), Electroslag Remelted (ESR) or Consumable Electrode Vacuum Arc Remelted (VAR), Annealed	T20813	H-13	
6409	Bars, Forgings, and Tubing, 0.80Cr 1.8Ni 0.25Mo (0.38-0.43C), Special Aircraft Quality Cleanliness, Normalized and Tempered	G43400	4340	MIL-S-5000
6411D	Bars, Forgings, and Tubing, 0.88Cr 1.8Ni 0.42Mo 0.08V (0.28-0.33C), Consumable Electrode Remelted	K23080	4340 Mod	
6412J	Bars and Forgings, 0.80Cr 1.8Ni 0.25Mo (0.35-0.40C)	G43370	4337	
6413H	Tubing, Mechanical, 0.80Cr 1.8Ni 0.25Mo (0.35-0.40C)	G43370	4337	
6414F	Bars, Forgings, and Tubing, 0.80Cr 1.8Ni 0.25Mo (0.38-0.43C), Vacuum Consumable Electrode Remelted	G43400	4340	
6415M MAM	Bars, Forgings, and Tubing, 0.80Cr 1.8Ni 0.25Mo (0.38-0.43C)	G43400	4340	MIL-S-5000
6415	Bars, Forgings, and Tubing, 0.80Cr 1.8Ni 0.25Mo (0.38-0.43C)	G43400	4340	MIL-S-5000
6417D	Bars, Forgings, and Tubing, 0.82Cr 1.8Ni 0.40Mo 1.6Si 0.08V (0.38-0.43C), Consumable Electrode Remelted	K44220	300M	
6418G	Bars, Forgings, Tubing, and Rings, 0.30Cr 1.8Ni 0.40Mo 1.3Mn 1.5Si (0.23-0.28C)	K32550	Hy-Tuf	MIL-S-7108
6419C	Bars, Forgings, and Tubing, 0.82Cr 1.8Ni 0.40Mo 0.08V 1.6Si (0.40-0.45C), Consumable Electrode Vacuum Remelted	K44220	300M	MIL-S-8844
6421C	Bars, Forgings, and Tubing, 0.80Cr 0.85Ni 0.20Mo 0.003B (0.35-0.40C)	....	98B37 Mod	
6422F	Bars, Forgings, and Tubing, 0.80Cr 0.85Ni 0.20Mo 0.003B 0.04V (0.38-0.43C)	K11940	98BV40 Mod	
6423D	Bars, Forgings, and Tubing, 0.92Cr 0.75Ni 0.52Mo 0.003B 0.04V (0.40-0.46C)	K24336	98BV40 Mod	
6424B	Bars, Forgings, and Tubing, 0.80Cr 1.8Ni 0.25Mo (0.49-0.55C)	K22950	....	
6425	Bars, Forgings, and Tubing, 0.30Cr 1.8Ni 0.40Mo 1.4Mn 1.5Si (0.23-0.28C), Consumable Vacuum Electrode Remelted	K32550	Hy-Tuf	
6426D	Bars, Forging, and Tubing, 1.0Cr 0.58Mo 0.75Si (0.80-0.90C), Consumable Electrode Melted	K18597	52CB	
6427H	Bars, Forgings, and Tubing, 0.88Cr 1.8Ni 0.42Mo 0.08V (0.28-0.33C)	K23080	4330 Mod	
6428D	Bars, Forgings, and Tubing, 0.80Cr 1.8Ni 0.35Mo 0.20V (0.32-0.38C)	K23477	4335 Mod	
6429D	Bars, Forgings, Tubing, and Rings, 0.78Cr 1.8Ni 0.35Mo 0.20V (0.33-0.38C), Consumable Electrode Vacuum Melted	K33517	4335 Mod	
6430D	Bars, Forgings, Tubing, and Rings, 0.78Cr 1.8Ni 0.35Mo 0.20V 0.75Mn (0.32-0.38C)	K33517	4335 Mod	
6431J	Bars, Forgings, and Tubing, 1.05Cr 0.55Ni 1.0Mo 0.11V (0.45-0.50C), Consumable Electrode Vacuum Melted	K24728	D6	MIL-S-8949
6432A	Bars, Forgings, and Tubing, 1.05Cr 0.55Ni 1.0Mo 0.12V (0.43-0.49C)	K24728	D6A	
6433D	Sheet, Strip, and Plate, 0.80Cr 1.8Ni 0.35Mo 0.20V 0.75Mn (0.33-0.38C)	K33517	4335 Mod	
6434D	Sheet, Strip, and Plate, 0.78Cr 1.8Ni 0.35Mo 0.20V (0.33-0.38C)	K33517	4335 Mod	
6435C	Sheet, Strip, and Plate, 0.78Cr 1.8Ni 0.35Mo 0.20V (0.33-0.38C), Vacuum Consumable Electrode Melted, Annealed	K33517	4335 Mod	
6436B	Sheet, Strip, and Plate, Low-Alloy, Heat Resistant, 0.65Si 1.25Cr 0.50Mo 0.85V (0.25-0.30C), Annealed	K22770	17-22A(V)	

(continued)

Federal (QQ) specifications are identical to the military, except that they are provided by the General Services Administration (GSA) and are used by federal agencies as well as by military establishments when there are no separate MIL specifications available. Federal specifications begin with the prefix FED-QQ, followed by the letter and code numbers. Examples of federal specifications for steels with equivalent UNS numbers in parentheses are FED-QQ-

Table 16   Continued

AMS No.	Title of Specification	UNS No.	Alloy	Similar Specification
6437D	Sheet, Strip, and Plate, 5.0Cr 1.3Mo 0.50V (0.38-0.43C)	T20811	H-11	
6438D	Sheet, Strip, and Plate, 1.05Cr 0.55Ni 1.0Mo 0.12V (0.45-0.50C), Consumable Electrode Vacuum Melted	K24728	D6	
6439B	Sheet, Strip, and Palte, 1.05Cr 0.55Ni 1.0Mo 0.12V (0.42-0.48C), Consumable Electrode Vacuum Melted, Annealed	K24728	D6AC	MIL-S-8949
6440J	Bars, Forgings, and Tubing, 1.45Cr (0.98-1.10C), for Bearing Applications	G52986	52100	
6442E	Bars, Forgings, 0.50Cr (0.98-1.10C), for Bearing Applications	G50986	50100	MIL-S-7420
6443E	Bars, Forgings, and Tubing, 1.0Cr (0.98-1.10C), Consumable Electrode Vacuum Melted	G51986	51100	
6444H	Bars, Forgings, and Tubing, 1.45Cr (0.98-1.10C), Premium Aircraft-Quality, Consumable Electrode VAcuum Melted	G52986	52100	
6445E	Bars, Forgings, and Tubing, 1.05Cr 1.1Mn (0.92-1.02C), Consumable Electrode Vacuum Melted	K22097	51100 Mod	
6446C	Bars, Forgings, and Tubing, 1.0Cr (0.98-1.10C), Electroslag Remelted	G51986	51100	
6447D	Bars, Forgings, and Tubing, 1.4Cr (0.98-1.10C), Electroslag Remelted	G52986	52100	
6448F	Bars, Forgings, and Tubing, 0.95Cr 0.22V (0.48-0.53C)	G61500	6150	MIL-S-8503
6449C	Bars, Forgings, and Tubing, 1.0Cr (0.98-1.10C), for Bearing Applications	G51986	51100	MIL-S-7420
6450F	Wire, Spring, 0.95Cr 0.22V (0.48-0.53C), Annealed and Cold Drawn	G61500	6150	
6451A	Wire, Spring, 1.4Si 0.65Cr (0.51-0.59C), Oil Tempered	G92540	9254	
6452A	Wire, Welding, 0.95Cr 0.20Mo (0.38-0.43C), Vacuum Melted, Environment Controlled Packaging	G43406	E4340	MIL-R-5632, Type II
6453	Wire, Welding, 0.30 Cr 1.8Ni 0.40Mo (0.23-0.28C), Vacuum Melted, Environment Controlled Packaging	K 32550	Hy Tuf	
6454B	Sheet, Strip, and Plate, 1.8Ni 0.8Cr 0.25Mo (0.38-0.43C), Consumable Electrode Melted	G43400	4340	
6455G	Sheet, Strip, and Plate, 0.95Cr 0.22V (0.48-0.53C)	G61500	6150	MIL-S-18731
6456A	Wire, Welding, 0.8Cr 1.8Ni 0.25Mo (0.35-0.40C), Vacuum Melted, Environment Controlled Packaging		4340 Mod	MIL-R-5632, Type III
6457A	Wire, Welding, 0.95Cr 0.20Mo (0.28-0.33C), Vacuum Melted, Environment Controlled Packaging	K13147	4130	MIL-R-5632, Type I
6458F	Wire, Welding, 1.25Cr 0.50Mo 0.30V 0.65Si (0.28-0.33C), Vacuum Melted, Environment Controlled Packaging	K23015	17-22A(S)	
6459B	Wire Welding, 1.0Cr 1.0Mo 0.12V (0.18-0.23C), Vacuum Induction Melted	K22720		
6460D	Wire, Welding, 0.62Cr 0.20Mo 0.75Si 0.10Zr (0.10-0.17C)	K11365	NAX-915-AC	
6461G	Wire, Welding, 0.95Cr 0.20V (0.28-0.33C), Vacuum Melted, Environment Controlled Packaging	K13148	6130	
6462F	Wire, Welding, 0.95Cr 0.20V (0.28-0.33C)	K13149	6130	
6463B	Wire, Welding, 18.5Ni 8.5Co 5.2Mo 0.72Ti 0.10Al, Vacuum Environment Controlled Packaging	K93130	Mar 300	
6464E	Electrodes, Welding, Covered, 1.05Mo 0.20V (0.06-0.12C)	W10013	10013 (AWS)	MIL-E-6843, Class E-10013
6465B	Wire, Welding, 2.0Cr 10Ni 8.0Co 1.0Mo 0.02Al 0.06V (0.10-0.14C), Vacuum Melted, Environment Controlled Pakaging	K91971	HY-180	
6466D	Wire, Welding, Corrosion Resistant, 5.2Cr 0.55Mo	S50280	Type 502	
6467C	Electrode, Welding, Covered, 5Cr 0.55Mo	W50210	Type 502	
6468B	Wire, Welding, 1.0Cr 3.8Co 0.45Mo 0.08V(0.14-0.17C), Vacuum Melted, Environment Controlled Packaging	K91461	HP 9-4-20	
6469A	Wire, Welding, 1.75Mn 0.80Cr 2.8Ni 0.85Mo (0.09-0.12C), Vacuum Melted, Environment Controlled Packaging	....	....	
6470J	Bars, Forgings, and Tubing, Nitriding Grade, 1.6Cr 0.35Mo 1.1Al (0.38-0.43C)	K24065	135 Mod	MIL-S-6709
6471D	Bars, Forgings, and Tubing, Nitriding Grade, 1.6Cr 0.35Mo 1.2Al (0.38-0.43C), Consumable Electrode Vacuum Melted	K24065	135 Mod	

S-700 (C1030) (G10300); FED-QQ-S-700 (C1085) (G10850); FED-QQ-S-763 (309) (S30900); and FED-QQ-S-766 (316L) (S31603) [43]. (See Table 16 also.)

D.   API Specifications

The American Petroleum Institute (API) fosters the development of standards, codes, and safe practices within the petroleum industries. The API standards appear with the prefix API be-

Table 16 Continued

AMS No.	Title of Specification	UNS No.	Alloy	Similar Specification
6472C	Bars and Forgings, Nitriding Grade, 1.6Cr 0.35Mo 1.1Al (0.38-0.43C), Hardened and Tempered, 112 ksi (772 MPa) Tensile Strength	K24065	135 Mod	MIL-S-6709
6473	Wire, Welding, 0.88Cr 1.8Ni 1.6Co 0.42Mo 0.08V (0.28-0.33C), Vacuum Melted, Environment Controlled Packaging			
6475F	Bars, Forgings, and Tubing, Nitriding Grade, 1.1Cr 3.5Ni 0.25Mo 1.25Al (0.21-0.26C)	K52355		
6476	Bars, Forgings, and Tubing, 0.50Cr 0.12Mo (0.89-1.01C), for Bearing Applications			
6477	Bars, Forgings, and Tubing, 0.80Cr (0.90-1.03C), for Bearing Applications			
6485G	Bars and Forgings, 5.0Cr 1.3Mo 0.50V (0.38-0.43C)	T20811	H-11	FED-QQ-T-570 Class H-11
6487G	Bars and Forgings, 5.0Cr 1.3Mo 0.50V (0.38-0.43C), Consumable Electrode Vacuum Melted	T20811	H-11	
6488E	Bars and Forgings, 5.0Cr 1.3Mo 0.5V (0.38-0.43C)	T20811	H-11	
6490D	Bars, Forgings, and Tubing, 4.0Cr 4.2Mo 1.0V (0.77-0.85C), Premium Aircraft Quality for Bearing Applications, Consumable Electrode Vacuum Melted	T11350	M-50	
6491A	Bars, Forgings, and Tubing, 4.1Cr 4.2Mo 1.0V (0.80-0.85C), Premium Aircraft Quality for Bearing Applications, Double Vacuum Melted	T11350	M-50	
6501A	Wire, Welding, Maraging Steel, 18Ni 8.0Co 4.9Mo 0.40Ti 0.10Al, Vacuum Induction Melted, Environment Controlled Packaging	K92890	Maraging 250	
6512C	Bars, Forgings, Tubing, and Rings, 18Ni 7.8Co 4.9Mo 0.40Ti 0.10Al, Consumable Electrode Melted, Annealed	K92890	Maraging 250	MIL-S-46850 TypeIII, Grade
6514C	Bars, Forgings, Tubing, and Rings, Maraging, 18.5Ni 9Co 4.9Mo 0.65Ti 0.10Al, Consumable Electrode Melted, Annealed	K93120	Maraging 300	MIL-S-46850 Type 300 MIL-S-13881, Type II, Class I
6518A	Sheet, Strip, and Plate, Maraging, 19Ni 3.0Mo 1.4Ti 0.10Al, Double Vacuum Melted, Solution Heat Treated			
6519A	Bars, Forgings, Tubing, and Rings, Maraging, 19Ni 3.0Mo 1.4Ti 0.10Al, Double Vacuum Melted, Annealed			
6520B	Sheet, Strip, and Plate, Maraging 250, 18Ni 7.8Co 4.9Mo 0.40Ti 0.10Al, Consumable Electrode Melted, Solution Heat Treated	K92890	Maraging 250	
6521A	Sheet, Strip, and Plate, 18.5NI 9.0Co 4.9Mo 0.65Ti 0.10Al, Consumable Electrode Melted, Solution Heat Treated	K93120	Maraging 300	MIL-S-46850 Grade 300
6522A	Plate, 2.0Cr 10Ni 14Co 1.0Mo (0.13-0.17C), Vacuum Melted, Normalized and Overaged	K92571	AF-1410	
6523C	Sheet, Strip, and Plate, 0.75Cr 9.0Ni 4.5Co 1.0Mo 0.09V (0.17-0.23C), Vacuum Consumable Electrode Melted, Annealed	K91472	HP 9-4-20	
6524C	Wire, Welding, 1.0Cr 7.5Ni 4.5Co 1.0Mo 0.09V (0.29-0.34C), Consumable Electrode Vacuum Melted	K91313	HP 9-4-30	
6525A	Bars, Forgings, Tubing, and Rings, 0.75Cr 9.0Ni 4.5Co 1.0Mo 0.09V (0.17-0.23C), Consumable Electrode Vacuum Melted	K91472	HP 9-4-20	
6526C	Bars, Forgings, Tubing, and Rings, 1.0Cr 7.5Ni 4.5Co 1.0Mo 0.09V (0.29-0.34C), Consumable Electrode Vacuum Melted, Annealed	K91283	HP 9-4-30	
6527B	Bars and Forgings, 2.0Cr 10Ni 14Co 1.0Mo (0.15-0.19C), Vacuum Melted, Normalized and Overaged	K92571	AF 1410	
6528	Bars, 0.95Cr 0.20Mo (0.28-0.33C), Special Aircraft Quality Cleanliness, Normalized	G41300	4130	
6529	Bars, 0.95Cr 0.20Mo (0.38-0.43C), Special Aircraft Quality Cleanliness, Normalized	G41400	4140	
6530H	Tubing, Seamless, 0.50Ni 0.55Cr 0.20Mo (0.28-0.33C)	G86300	8630	
6532	Bars and Forgings, 3.1Cr 11.5Ni 13.5Co 1.2Mo (0.21-0.25C), Vacuum Melted, Annealed	K92580	Aermet 100	

(continued)

fore the specification. For example, API Spec 5D covers all grades of seamless drill pipe (for use in drilling and producing operations), process of manufacture, chemical composition and mechanical property requirements, testing and inspection methods, and requirements for dimensions, weights, and lengths [46]. API Spec 5L covers all grades of seamless and welded steel line pipe and requirements for dimensions, weights, lengths, strengths, threaded ends, plain ends, belled ends, and thread protectors, and testing and inspection methods. This speci-

Table 16 Continued

AMS No.	Title of Specification	UNS No.	Alloy	Similar Specification
6533	Wire, Welding, 2.0Cr 10Ni 14Co 1.9Mo (0.13-0.17C), Vacuum Melted, Environment Controlled Packaging	K92571	AF 1410	
6535G	Tubing, Seamless, 0.50Cr 0.55Ni 0.20Mo (0.28-0.33C)	G86300	8630	
6543B	Bars and Forgings, 2.0Cr 10Ni 8.0Co 1.0Mo (0.10-0.14C), Double Vacuum Melted, Solution Heat Treated	K92571	AF 1410	
6544B	Plate, Maraging, 2.0Cr 10Ni 8.0Co 1.0Mo (0.10-0.14C), Double Vacuum Melted, Heat Treated	K91970		
6546D	Sheet, Strip, and Plate, 0.48Cr 0.80Ni 4.0Co 0.48Mo 0.09V (0.24-0.30C), Consumable Electrode Melted, Annealed	K91122	HP 9-4-25	
6550II	Tubing, Welded, 0.55Cr, 0.50Ni 0.20Mo (0.28-0.33C)	G86300	8630	MIL-T-6734

Source: Ref. 45

fication includes A25, A, B, X42, X46, X52, X56, X60, X65, X70, and X80 grades and grades intermediate to grade X42 and higher. It provides the standards for pipe suitable for use in conveying gas, water, and oil in both the oil and natural gas industries [47]. API Spec 5LC covers seamless, centrifugal cast, and welded corrosion-resistant alloy line pipe (austenitic stainless steels, martensitic stainless steels, duplex stainless steels, and nickel-base alloys), dimensions, weights, strengths, process of manufacture, chemical and mechanical properties requirements, and testing and inspection methods [48]. API Spec 5LD covers seamless, centrifugal cast, and welded clad steel line pipe and lined steel pipe with increased corrosion-resistant properties. The clad and lined steel line pipes are composed of a base metal outside and a CRA layer inside the pipe; the base material conforms to API Spec 5L, except as modified in the API Spec 5LC document. This specification provides standards for pipe with improved corrosion resistance suitable for use in conveying gas, water, and oil in both the oil and natural gas industries [49].

E. ANSI Specifications

An American National Standards Institute (ANSI) standard begins with the prefix ANSI, followed by an alphanumeric code with an uppercase letter, subsequently followed by one to three digits and additional digits that are separated by decimal points. ANSI standards can also have a standard developer's acronym in the title. Examples are ANSI H35.2, ANSI A156.2, ANSI B18.2.3.6M, ANSI/ASME NQ2-1989, ANSI/API Spec 5CT-1992, ANSI/API Spec 5D-1992, ANSI/API Spec 5L-1992, and ANSI/API Spec 5LC-1991 [44, 46–49].

F. AWS Specifications

American Welding Society (AWS) standards are used to support welding design, testing, quality assurance, and other related joining functions. These standards begin with the prefix AWS followed by the letter and numerals with decimal point. Examples of AWS specifications with corresponding AISI/SAE or proprietary grade and UNS number in parentheses are AWS A5.1 (E6010, W06010); AWS A5.2 (RG65, WK00065); and AWS A5.5 (E9018-D3, W19118).

VI. INTERNATIONAL SPECIFICATIONS AND DESIGNATIONS

Since steelmaking technology is available worldwide, familiarity with international specifications and designations for steels is necessary. Table 17 cross-references SAE steels with those of a selected group of international specifications and designations, which are described in the

Table 17 Cross-Reference to Steels

United States (SAE)	Fed. R. of Germany (DIN)	Japan (JIS)	United Kingdom (BS)	France (AFNOR NF)	Italy (UNI)	Sweden (SS₁₄)
Carbon steels						
1005	1.0288, D5-2 1.0303, QSt32-2 1.0312, D5-1 1.0314, D6-2 1.0393, ED3 1.0394, ED4 1.1012, RFe120	· · ·	970 015A03	· · ·	5598 3CD5	1160
1006	1.0311, D7-1 1.0313, D8-2 1.0317, RSD4 1.0321, St23 1.0334, StW23 1.0335, StW24 1.0354, St14Cu3 1.0391, EK2 1.0392, EK4 1.1009, Ck7	· · ·	970 030A04 970 040A04 970 050A04	A35-564 XC6FF	5598 3CD6 5771 C8	1147 1225
1008	1.0010, D9 1.0318, St28 1.0320, St22 1.0322, USD8 1.0326, RSt28 1.0330, St2, St12 1.0333, St3, St13 1.0331, RoSt2 1.0332, StW22 1.0336, USt4, USt14 1.0337, RoSt4 1.0344, St12Cu3 1.0347, RRSt13 1.0357, USt28 1.0359, RRSt23 1.0375, Feinstblech T57, T61, T65, T70 1.0385, Weissblech T57, T61, T65, T70 1.0744, 6P10 1.0746, 6P20 1.1116, USD6	G3445 STKM11A (11A)	1449 3CR 1449 3CS 1449 3HR 1449 3HS 1717 ERW101 3606 261	A35-551 XC10 XC6 XC6FF	5598 3CD8	1142 1146
1010	1.0204, UQSt36 1.0301, C10 1.0328, USD10 1.0349, RSD9 1.1121, Ck10 1.1122, Cq10	G4051 S10C G4051 S9Ck	1449 40F30, 43F35, 46F40, 50F45, 60F55, 68F62, 75F70 (available in HR, HS, CS conditions) 1449 4HR, 4HS, 4CR, 4CS 970 040A10 (En2A, En2A/1, En2B) 970 045A10, 045M10 (En32A) 970 050A10 970 060A10 980 CEW1	A33-101 AF34 CC10 C10	5331 C10 6403 C10 7065 C10 7846 C10 5598 1CD10 5598 3CD12 5771 C12 7356 CB10FF, CB10FU	1232 1265 1311
1012	1.0439, RSD13	G4051 S12C	1449 12HS, 12CS 1501 141-360 970 040A12 (En2A, En2A/1, En2B) 970 050A12 970 060A12	A33-101 AF37 A-35 551 XC12 C12	· · ·	1332 1431
1013	1.0036, USt37-2 1.0037, St37-2 1.0038, RSt37-2 1.0055, USt34-1 1.0057, RSt34-1 1.0116, St37-3 1.0218, RSt41-2 1.0219, St41-3 1.0307, StE210.7 1.0309, St35.4 1.0315, St37.8 1.0319, RRStE210.7 1.0356, TTSt35 1.0417 1.0457, StE240.7	· · ·	3059 360 3061 360 3603 360	A35-551 XC12 CC12	5869 Fe360-1KG, Fe360-2KW 6403 Fe35-2 7070 Fe34CFN 7091 Fe34	1233 1234 1330

(continued)

Table 17 Continued

United States (SAE)	Fed. R. of Germany (DIN)	Japan (JIS)	United Kingdom (BS)	France (AFNOR NF)	Italy (UNI)	Sweden (SS_{14})
Carbon steels (continued)						
1015	1.0401, C15 1.1132, CQ15 1.1135, Ck16Al 1.1140, Cm15 1.1141, Ck15 1.1144 1.1148, Ck15Al	G4051 F15Ck G4051 S15C	970 040A15 970 050A15 970 060A15 970 080A15, 080M15 970 173H16	XC15	5331 C16 7065 C16 7356 CB15 7846 C15	1370
1016	1.0419, RSt44.2 1.0467, 15Mn3 1.0468, 15Mn3Al 1.1142, GS-Ck16	. . .	3059 440 3606 440 970 080A15, 080M15 970 170H15 970 173H16	. . .	. . .	1370 2101
1017	. . .	G4051 S17C	1449 17HS, 17CS 970 040A17 970 050A17 970 060A17	A35-551 XC18 A35-552 XC18 A35-566 XC18 A35-553 XC18S A35-554 XC18S	. . .	1312
1018	1.0453, C16.8	. . .	970 080A17	A33-101 AF42 C20	. . .	. . .
1019	. . .	. . .	. . .	. . .	. . .	. . .
1020	1.0402, C22 1.0414, D20-2 1.0427, C22.3 1.0460, C22.8 1.1149, Cm22 1.1151, Ck22	G4051 S20C G4051 S20CK	970 040A20 970 050A20 (En2C, En2D) 970 060A20	A35-551 XC18 A35-552 XC18 A35-566 XC18 A35-553 C20 A35-553 XC18S A35-554 XC18S CC20	5598 1CD20 5598 3CD20 6922 C21 7356 CB20FF	1450
1021	. . .	. . .	970 070M20 970 080A20	A35-551 21B3 A35-552 21B3 A35-553 21B3 A35-557 21B3 A35-566 21B3	5332 C20 7065 C20	. . .
1022	1.0432, C21 1.0469, 21Mn4 1.0482, 19Mn5 1.1133, 20Mn5, GS-20Mn5 1.1134, Ck19	. . .	3111 Type 9 970 120M19 970 170H20	A35-551 20MB5 A35-552 20 MB5 A35-553 20MB5 A35-556 20MB5 A35-557 20MB5 A35 566 20MB5 A35-566 20M5	5771 20Mn4	. . .
1023	1.1150, Ck22.8 1.1152, Cq22	G4051 S22C	1449 2HS, 22CS 970 040A22 (En2C, En2D) 970 050A22 970 060A22 970 080A22	. . .	5332 C20 7065 C20	. . .
1025	1.0406, C25 1.0415, D25-2, D26-2 1.1158, Ck25	G4051 S25C	. . .	A35-552 XC25 A35-566 XC25	5598 1CD25 5598 3CD25	. . .
1026	1.1155, GS-Ck25 1.1156, GS-Ck24	. . .	970 070M26 970 080A25 970 080A27	. . .	7845 C25 7847 C25	. . .
1029	1.0562, 28Mn4	G3445 STKM15A (15A), STKM15C (15C) G4051 S28C	970 060A27 970 080A27 (En5A)	A33-101 AF50 CC28 C30	. . .	. . .
1030	1.0528, C30 1.0530, D30-2 1.1178, Ck30 1.1179, Cm30 1.1811, G-31Mn4	G4051 S30C	1449 30HS, 30CS 970 060A30 970 080A30 (En5B) 970 080M30 (En5)	A35-552 XC32 A35-553 XC32	5332 C30 6403 C30 7065 C30 7845 C30 7874 C30 5598 3CD30 6783 Fe50-3 7065 C31	. . .
1035	1.0501, C35 1.0516, D35-2 1.1172, Cq35 1.1173, Ck34 1.1180, Cm35 1.1181, Ck35	G4051 S35C	1717 CDS105/106 970 060A35 970 080A32 (En5C) 970 080A35 (En8A) 980 CFS6	A33-101 AF55 A35-553 C35 A35-553 XC38 A35-554 XC38 XC35 XC38TS C35	5333 C33 5598 1CD35 5598 3CD35 7065 C35 7065 C36 7847 C36 7356 CB35	1550 1572
1037	1.0520, 31Mn4 1.0561, 36Mn4	G4051 S35C	3111 type 10 970 080M36 970 170H36	. . .	. . .	. . .
1038	No international equivalents					

Table 17 Continued

United States (SAE)	Fed. R. of Germany (DIN)	Japan (JIS)	United Kingdom (BS)	France (AFNOR NF)	Italy (UNI)	Sweden (SS₁₄)

Carbon steels (continued)

United States (SAE)	Fed. R. of Germany (DIN)	Japan (JIS)	United Kingdom (BS)	France (AFNOR NF)	Italy (UNI)	Sweden (SS₁₄)
1039	1.1190, Ck42Al	. . .	970 060A40 970 080A40 (En8C) 970 080M40 (En8) 970 170H41	40M5 A35-552 XC38H2 A35-553 38MB5 A35-556 38MB5 A35-557 38MB5 A35-557 XC38H2 XC42, XC42TS	. . .	. . .
1040	1.0511, C40 1.0541, D40-2 1.1186, Ck40 1.1189, Cm40	G4051, S40C	1287 1449 40HS, 40CS 3146 Class 1 Grade C 3146 Class 8 970 060A40 970 080A40 (En8C) 970 080M40 (En8)	A33-101 AF60 C40	5598 1CD40 5598 3CD40 6783 Fe60-3 6923 C40 7065 C40 7065 C41	. . .
1042	1.0517, D45-2	G4051 S43C	970 060A42 970 080A42 (En8D)	A35-552 XC42H1 A35-553 C40 CC45 XC42, XC42TS	. . .	. . .
1043	1.0558, GS-60.3	G4051 S43C	970 060A42 970 080A42 (En8D) 970 080M46	A35-552 XC42H2	7847 C43	. . .
1044	1.0517, D45-2	. . .	. . .	. . .	. . .	. . .
1045	1.0503, C45 1.1184, Ck46 1.1191, Ck45, GS-Ck45 1.1192, Cq45 1.1194, Cq45 1.1201, Cm45 1.1193, Cf45	G4051 S45C G5111 SCC5	970 060A47 970 080A47 970 080M46	A33-101 AF65 A35-552 XC48H1 A35-553 XC45 A35-554 XC48 XC48TS C45	3545 C45 5332 C45 7065 C45 7845 C45 7874 C45 5598 1CD45 5598 3CD45 7065 C46 7847 C46	1672
1046	1.0503, C45 1.0519, 45MnAl 1.1159, GS-46Mn4	. . .	3100 AW2 970 080M46	45M4TS A35-552 XC48H1 A35-552 XC48H2 XC48TS	. . .	. . .
1049	. . .	G3445 STKM17A (17A) G3445 STKM17C (17C)	970 060A47 970 080A47	A35-552 XC48H1 A35-554 XC48 XC48TS	6403 C48 7847 C48	. . .
1050	1.0540, C50 1.1202, D53-3 1.1206, Ck50 1.1210, Ck53 1.1213, Cf53 1.1219, Cf54 1.1241, Cm50	G4051 S50C G4051 S53C	1549 50HS 1549 50CS 970 060A52 970 080A52 (En43C) 970 080M50 (En43A)	A35-553 XC50	5332 C50 7065 C50 7065 C51 7845 C50 7874 C50 5598 1CD50 5598 3CD50 6783 Fe70-3 7847 C53	1674
1053	1.1210 Ck53 1.1213 Cf53 1.1219 Cf54	G4051 S53C	970 080A52 (En43C)	52M4TS A35-553 XC54	7847 C53	1674
1055	1.0518, D55-2 1.0535, C35 1.1202, D53-3 1.1203, Ck55 1.1209, Cm55 1.1210, Ck53 1.1213, Cf53 1.1219, Cf54 1.1220, D55-3 1.1820, C55W	G4051 S53C G4051 S55C	3100 AW3 970 060A57 970 070M55 970 080A52 (En43C) 970 080A57	A33-101 AF70 A35-552 XC55H1 A35-552 XC55H2 A35-553 XC54 XC55 C55	5598 3CD55 7065 C55 7845 C55 7874 C55 7065 C56 7847 C53	. . .
1059	1.0609, D58-2 1.0610, D60-2 1.0611, D63-2 1.1212, D58-3 1.1222, D63-3 1.1228, D60-3	. . .	970 060A62	A35-553 XC60	. . .	. . .
1060	1.0601, C60 1.0642, 60Mn3 1.1221, Ck60 1.1223, Cm60 1.1740, C60W	G4051 S58C	1449 60HS 1449 60CS 970 060A57 970 080A57	A35-553 XC60	3545 C60 7064 C60 7065 C60 7845 C60 7874 C60 5598 3CD60 7065 C61	1678

(continued)

Table 17 Continued

United States (SAE)	Fed. R. of Germany (DIN)	Japan (JIS)	United Kingdom (BS)	France (AFNOR NF)	Italy (UNI)	Sweden (SS$_{14}$)
Carbon steels (continued)						
1064	1.0611, D63-2 1.0612, D65-2 1.0613, D68-2 1.1222, D63-3 1.1236, D65-3	· · ·	970 060A62 970 080A62 (En43D)	· · ·	5598 3CD65	· · ·
1065	1.0627, C68 1.0640, 64Mn3 1.1230, Federstahldraht FD 1.1233 1.1240, 65Mn4 1.1250, Federstahldraht VD 1.1260, 66Mn4	· · ·	970 060A67 970 080A67 (En43E)	XC65	· · ·	· · ·
1069	1.0615, D70-2 1.0617, D73-2 1.0627, C68 1.1232, D68-3 1.1237 1.1249, Cf70 1.1251, D70-3 1.1520, C70W1 1.1620, C70W2	· · ·		A35-553 XC68 XC70		· · ·
1070	1.0603, C67 1.0643, 70Mn3 1.1231, Ck67	· · ·	1449 70HS, 70CS 970 060A72 970 070A72 (En42) 970 080A72	XC70	3545 C70	1770
1074	1.0605, C75 1.0645, 76Mn3 1.0655, C74 1.1242, D73-3	· · ·	970 070A72 (En42) 970 080A72	A35-553 XC75 XC70	3545 C75 7064 C75	1774
1075	1.0614, D75-2 1.0617, D73-2 1.0620, D78-2 1.1242, D73-3 1.1252, D78-3 1.1253, D75-3	· · ·	· · ·	A35-553 XC75 XC70	3545 C75 7064 C75 5598 3CD70 5598 3CD75	· · ·
1078	1.0620, D78-2 1.0622, D80-2 1.0626, D83-2 1.1252, D78-3 1.1253, D75-3 1.1255, D80-3 1.1262, D83-3 1.1525, C80W1	G4801 SUP3	970 060A78	XC80	5598 3CD80	· · ·
1080	1.1259 80Mn4 1.1265 D85-2	· · ·	1449 80HS, 80CS 970 060A78 970 060A83 970 070A78 970 080A78 970 080A83	XC80	5598 3CD80 5598 3CD85	· · ·
1084	1.1830, C85W	· · ·	970 060A86 970 080A86	XC85	· · ·	· · ·
1085	1.0647, 85Mn3 1.1273, 90Mn4 1.1819, 90Mn4	· · ·	970 080A83	· · ·	· · ·	· · ·
1086	1.0616, C85, D85-2 1.0626, D83-2 1.0628, D88-2 1.1262, D83-3 1.1265, D85-3 1.1269, Ck85 1.1272, D88-3	· · ·	970 050A86	A35-553 XC90	5598 3CD85 5598 3CD90	· · ·
1090	1.1273, 90Mn4 1.1819, 90Mn4 1.1282, D95S3	· · ·	1449 95HS 1449 95CS 970 060A96	· · ·	3545 C90 7064 C90 5598 3CD95	
1095	1.0618, D95-2 1.1274, Ck101 1.1275, Ck100 1.1282, D95S3 1.1291, MK97 1.1545, C105W1 1.1645, C105W2	G4801 SUP4	1449 95HS 1449 95CS 970 060A99	A35-553 XC100	3545 C100 7064 C100	1870

Table 17 Continued

United States (SAE)	Fed. R. of Germany (DIN)	Japan (JIS)	United Kingdom (BS)	France (AFNOR NF)	Italy (UNI)	Sweden (SS₁₄)
Carbon-manganese steels						
1513.........	1.0424, Schiffbaustahl CS:DS 1.0479, 13Mn6 1.0496, 12Mn6 1.0513, Schiffbaustahl A32 1.0514, Schiffbaustahl B32 1.0515, Schiffbaustahl E32 1.0549 1.0579 1.0583, Schiffbaustahl A36 1.0584, Schiffbaustahl D36 1.0589, Schiffbaustahl E36 1.0599 1.8941, QStE260N 1.8945, QStE340N 1.8950, QStE380N	· · ·	1449 40/30 HR 1449 40/30 HS 1449 40/30 CS 1453 A2 2772 150M12 970 125A15 970 130M15 970 130M15 (En201)	12M5 A33-101 AF50-S A35-501 E35-4 A35-501 E36-2 A35-501 E36-3	· · ·	· · ·
1522.........	1.0471, 21MnSi5 1.0529, StE350-Z2 1.1120, GS-20Mn5 1.1138, GS-21Mn5 1.1169, 20Mn6 1.8970, StE385.7 1.8972, StE415.7 1.8978 1.8979	G4106 SMn21	1503 221-460 1503 223-409 1503 224-490 3146 CLA2 980 CFS7	A35-551 20MB5 A35-552 20M5 A35-556 20M5 A35-552 20MB5 A35-553 20MB5 A35-556 20MB5 A35-557 20MB5 A35-566 20MB5	4010 FeG52 6930 20Mn6 7660 Fe510	2165 2168
1524.........	1.0499, 21Mn6Al 1.1133, 20Mn5, GS-20Mn5 1.1160, 22Mn6	G4106 SMn21 G5111 SCMn1	1456 Grade A 970 150M19 (En14A, En14B) 970 175H23 980 CDS9, CDS10	· · ·	· · ·	· · ·
1526.........	· · ·	· · ·	970 120M28	A35-566 25MS5		2130
1527.........	1.0412, 27MnSi5 1.1161, 26Mn5 1.1165, 30Mn5 1.1165, GS-30Mn5 1.1170, 28Mn6	G5111 SCMn2	1453 A3 1456 Grade B1, Grade B2 3100 A5 3100 A6 970 150M28 (En14A, En14B)	· · ·	4010 FeG60 7874 C28Mn	· · ·
1536.........	1.0561, 36Mn4 1.1165, 30Mn5 1.1165, GS-30Mn5 1.1166, 34Mn5 1.1167, 36Mn5, GS-36Mn5 1.1813, G-35Mn5	G4052 SMn1H G4052 SMn433H G4106 SMn1 G4106 SMn433 G5111 SCMn2 G5111 SCMn3	1045 3100 A5, A6 970 120M36 (En15B) 970 150M36 (En15)	A35-552 32M5 A35-552 38MB5 A35-553 38MB5 A35-556 38MB5 A35-557 38MB5	4010 FeG60	
1541.........	1.0563, E 1.0564, N-80 1.1127, 36Mn6 1.1168, GS-40Mn5	G4106 SMn2, SMn438 G4052 SMn2H, SMn438H G4106 SMn3, SMn443 G4052 SMn3H, SMn443H G5111 SCMn5	970 135M44 970 150M40	40M5 45M5 A35-552 40M6	· · ·	2120 2128
1548.........	1.1128, 46Mn5 1.1159, GS-46Mn4	· · ·	· · ·	· · ·	· · ·	· · ·
1551.........	1.0542, StSch80	· · ·	· · ·	24M4TS	· · ·	· · ·
1552.........	1.0624, StSch90B 1.1226, 52Mn5	· · ·	· · ·	55M5	· · ·	· · ·
1561.........	1.0908, 60SiMn5	· · ·	· · ·	· · ·	· · ·	· · ·
1566.........	1.1233 1.1240, 65Mn4 1.1260, 66Mn7	· · ·	· · ·	· · ·	· · ·	· · ·
Resulfurized carbon steels						
1108.........	1.0700, U7S10 1.0702, U10S10	G4804 SUM12	· · ·	A35-562 10F1	· · ·	· · ·
1110.........	1.0703, R10S10	G4804 SUM11	· · ·	· · ·	· · ·	· · ·
1117.........	· · ·	G4804 SUM31	970 210A15 970 210M17 (En32M) 970 214A15 970 214M15 (En202)	· · ·	· · ·	· · ·
1118.........	· · ·	· · ·	970 214M15 (En201)	· · ·	· · ·	· · ·
1137.........	· · ·	G4804 SUM41	970 212M36 (En8M) 970 216M36 (En15AM) 970 225M36	35MF4 A35-562 35MF6	4838 CF35SMn10	· · ·
1139.........	1.0726, 35S20	· · ·	970 212A37 (En8BM) 970 212M36(En8M) 970 216M36 (En15AM) 970 225M36	35MF4 A35-562 35MF6	· · ·	1957

(continued)

Table 17 Continued

United States (SAE)	Fed. R. of Germany (DIN)	Japan (JIS)	United Kingdom (BS)	France (AFNOR NF)	Italy (UNI)	Sweden (SS_{14})
Resulfurized carbon steels (continued)						
1140	No international equivalents					
1141	...	G4804 SUM42	970 212A42 (En8DM) 970 216A42	A35-562 45MF4	...	...
1144	1.0727, 45S20	G4804 SUM43	970 212A42 (En8DM) 970 212M44 970 216M44 970 225M44 970 226M44	A35-562 45MF6	4838 CF44SMn28	1973
1146	1.0727, 45S20	...	970 212M44	45MF4	...	...
1151	1.0728, 60S20 1.0729, 70S20	...		...	...	1973
Resulfurized/rephosphorized carbon steels						
1211	No international equivalents					
1212	1.0711, 9S20 1.0721, 10S20 1.1011, RFe160K	G4804 SUM21	...	10F2 12MF4 S200	4838 10S20 4838 10S22 4838 CF9S22	...
1213	1.0715, 9SMn28 1.0736, 9SMn36 1.0740, 9SMn40	G4804 SUM22	970 220M07 (En1A) 970 230M07 970 240M07 (En1A)	A35-561 S250 S250	4838 CF9SMn28 4838 CF9SMn32	1912
1215	1.0736, 9SMn36	G4804 SUM23	970 240M07 (En1B)	A35-561 S300	4838 CF9SMn32 4838 CF9SMn36	...
12L14	No international equivalents					
Alloy steels						
1330	No international equivalents					
1335	1.5069, 36Mn7	...	...	...	...	...
1340	1.5223, 42MnV7	...	...	...	...	...
1345	1.0625, StSch90C 1.0912, 46Mn7 1.0913, 50Mn7 1.0915, 50MnV7 1.5085, 51Mn7 1.5225, 51MnV7	...	...	...	...	...
4023	1.5416, 20Mo3	...	...	...	...	...
4024	1.5416, 20Mo3	...	...	...	...	...
4027	1.5419, 22Mo4	...	...	...	...	...
4028	...	...	970 605M30	...	...	...
4032	1.5411	G5111 SCMnM3	970 605A32 970 605H32 970 605M30 970 605M36 (En16)	...	...	...
4037	1.2382, 43MnSiMo4 1.5412, GS-40MnMo4 3 1.5432, 42MnMo7	...	3111 Type 2/1 3111 Type 2/2 970 605A37 970 605H37	...	...	...
4042	1.2382, 43MnSiMo4 1.5432, 42MnMo7	...	...	...	...	...
4047	No international equivalents					
4118	1.7211, 23CrMoB4 1.7264, 20CrMo5	G4052 SCM15H G4105 SCM21H G4052 SCM418H G4105 SCM418H	970 708H20 970 708M20	...	7846 18CrMo4	...
4130	...	G4105 SCM1 G4105 SCM432 G4105 SCM2 G4105 SCM430 G4106 SCM2	1717 CDS110 970 708A30	A35-552 30CD4 A35-556 30CD4 A35-557 30CD4	30CrMo4 6929 35CrMo4F 7356 34CrMo4KB 7845 30CrMo4 7874 30CrMo4	2233
4135	1.2330, 35CrMo4 1.7220, 34CrMo4 1.7220, GS-34CrMo4 1.7226, 34CrMoS4 1.7231, 33CrMo4	G4054 SCM3H G4054 SCM435H G4105 SCM1 G4105 SCM432 G4105 SCM3 G4105 SCM435	970 708A37 970 708H37	35CD4 A35-552 35CD4 A35-553 35CD4 A35-556 35CD4 A35-557 34CD4	5332 35CrMo4 6929 35CrMo4F 7356 34CrMo4KB 7845 35CrMo4 7874 35CrMo4	2234
4137	1.7225, GS-42CrMo4	G4052 SCM4H G4052 SCM440H G4105 SCM4	3100 type 5 970 708A37 970 708H37 970 709A37	40CD4 42CD4 A35-552 38CD4 A35-557 38CD4	5332 40CrMo4 5333 38CrMo4 7356 38CrMo4KB	...

Table 17 Continued

United States (SAE)	Fed. R. of Germany (DIN)	Japan (JIS)	United Kingdom (BS)	France (AFNOR NF)	Italy (UNI)	Sweden (SS₁₄)

United States (SAE)	Fed. R. of Germany (DIN)	Japan (JIS)	United Kingdom (BS)	France (AFNOR NF)	Italy (UNI)	Sweden (SS$_{14}$)
Alloy steels (continued)						
4140	1.3563, 43CrMo4 1.7223, 41CrMo4 1.7225, 42CrMo4 1.7225, GS-42CrMo4 1.7227, 42CrMoS4	G4052 SCM4H G4052 SCM440H G4103 SNCM4 G4105 SCM4 G4105 SCM440	3100 Type 5 4670 711M40 970 708A40 970 708A42 (En19C) 970 708H42 970 708M40 970 709A40 970 709M40	40CD4 A35-552 42CD4, 42CDTS A35-553 42CD4, 42CDTS A35-556 42CD4, 42CDTS A35-557 42CD4, 42CDTS	3160 G40CrMo4 5332 40CrMo4 7845 42CrMo4 7847 41CrMo4 7874 42CrMo4	2244
4142	1.3563, 43CrMo4 1.7223, 41CrMo4	· · ·	970 708A42 (En19C) 970 708H42 970 709A42	40CD4 A35-552 42CD4, 42CDTS A35-553 42CD4, 42CDTS A35-556 42CD4, 42CDTS A35-557 42CD4, 42CDTS	7845 42CrMo4 7874 42CrMo4	2244
4145	1.2332, 47CrMo4	G4052 SCM5H G4052 SCM445H G4105 SCM5, SCM445	970 708H45	A35-552 45SCD6 A35-553 45SCD6	· · ·	· · ·
4147	1.2332, 47CrMo4 1.3565, 48CrMo4 1.7228, 50CrMo4 1.7228, GS-50CrMo4 1.7230, 50CrMoPb4 1.7238, 49CrMo4	G4052 SCM5H G4052 SCM445H G4105 SCM5, SCM445	970 708A47	A35-552 45SCD6 A35-553 45SCD6 A35-571 50SCD6	· · ·	· · ·
4150	1.3565, 48CrMo4 1.7228, 50CrMo4 1.7228, GS-50CrMo4 1.7230, 50CrMoPb4 1.7238, 49CrMo4	· · ·	· · ·	A35-571 50SCD6	· · ·	· · ·
4161	1.7229, 61CrMo4 1.7266, GS-58CrMnMo4 4 3	G4801 SUP13	3100 BW4 3146 CLA12 Grade C	· · ·	· · ·	· · ·
4320	· · ·	G4103 SNCM23 G4103 SNCM420 G4103 SNCM420H	· · ·	20NCD7 A35-565 18NCD4 A35-565 20NCD7	3097 20NiCrMo7 5331 18NiCrMo7 7846 18NiCrMo7	2523 2523-02
4340	1.6565, 40NiCrMo6	G4103 SNCM8 G4103 SNCM439 G4108 SNB23-1-5 G4108 SNB24-1-5	4670 818M40 970 2S.119	· · ·	5332 40NiCrMo7 6926 40NiCrMo7 7845 40NiCrMo7 7874 40NiCrMo7 7356 40NiCrMo7KB	· · ·
E4340	1.6562, 40NiCrMo7 3	· · ·	970 2S.119	· · ·	· · ·	· · ·
4422	1.5419, 22Mo	· · ·	· · ·	23D5	3608 G20Mo5	
4427	No international equivalent					
4615	· · ·	· · ·	· · ·	15ND8	· · ·	· · ·
4617	· · ·	· · ·	970 665A17 970 665H17 970 665M17 (En34)	· · ·	· · ·	· · ·
4620	· · ·	· · ·	970 665A19 970 665H20 970 665M20	2ND8	· · ·	· · ·
4626	· · ·	· · ·	970 665A24 (En35B)	· · ·	· · ·	· · ·
4718	No international equivalent					
4720	· · ·	· · ·	· · ·	18NCD4	· · ·	· · ·
4815	No international equivalent					
4817	No international equivalent					
4820	No international equivalent					
50B40	1.7003, 38Cr2 1.7023, 28CrS2	G4052 SMnC3H G4052 SMnC443H G4106 SMnC3 G4106 SMnC443 G5111 SCMnCr4	· · ·	A35-552 38C2 A35-556 38C2 A35-557 38C2 A35-552 42C2 A35-556 42C2 A35-557 42C2	7356 41Cr2KB	· · ·
50B44	· · ·	· · ·	· · ·	45C2	7847 45Cr2	· · ·
5046	1.3561, 44Cr2	· · ·	· · ·	· · ·	· · ·	· · ·
50B46	No international equivalent					
50B50	1.7138, 52MnCrB3	· · ·	· · ·	55C2	· · ·	· · ·
5060	1.2101, 62SiMnCr4	· · ·	970 526M60 (En11)	61SC7 A35-552 60SC7	· · ·	· · ·

(continued)

Table 17 Continued

United States (SAE)	Fed. R. of Germany (DIN)	Japan (JIS)	United Kingdom (BS)	France (AFNOR NF)	Italy (UNI)	Sweden (SS₁₄)

United States (SAE)	Fed. R. of Germany (DIN)	Japan (JIS)	United Kingdom (BS)	France (AFNOR NF)	Italy (UNI)	Sweden (SS14)
Alloy steels (continued)						
5115	1.7131, 16MnCr5, GS-16MnCr5 1.7139, 16MnCrS5 1.7142, 16MnCrPb5 1.7160, 16MnCrB5	G4052 SCr21H G4052 SCr415H G4104 SCr21 G4104 SCr415	970 527A17 970 527H17 970 527M17	16MC5 A35-551 16MC5	7846 16MnCr5	2127
5117	1.3521, 17MnCr5 1.7016, 17Cr3 1.7131, 16MnCr5, GS-16MnCr5 1.7139, 16MnCrS5 1.7142, 16MnCrPb5 1.7168, 18MnCrB5	...	...	18Cr4 A35-551 16MC5	...	...
5120	1.2162, 21MnCr5 1.3523, 19MnCr5 1.7027, 20Cr4 1.7028, 20Cr5 4 1.7121, 20CrMnS3 3 1.7146, 20MnCrPb5 1.7147, GS-20MnCr5 1.7149, 20MnCrS5	G4052 SCr22H G4052 SCr420H G4052 SMn21H G4052 SMn421H G4104 SCr22 G4104 SCr420	...	A35-551 20MC5 A35-552 20MC5	7846 20MnCr5	...
5130	1.8401, 30MnCrTi4	G4052 SCr2H G4052 SCr430H G4104 SCr2 G4104 SCr430	970 530A30 (En18A) 970 530H30	28C4	...	...
5132	1.7033, 34Cr4 1.7037, 34CrS4	G4104 SCr3 G4104 SCr435	970 530A32 (En18B) 970 530A36 (En18C) 970 530H32	A35-552 32C4 A35-553 32C4 A35-556 32C4 A35-557 32C4	7356 34Cr4KB 7874 34Cr4	...
5135	1.7034, 37Cr4 1.7038, 37CrS4 1.7043, 38Cr4	G4052 SCr3H G4052 SCr435H	3111 Type 3 970 530A36 (En18C) 970 530H36	38C4 A35-552 38Cr4 A35-553 38Cr4 A35-556 38Cr4 A35-557 38Cr4	5332 35CrMn5 6403 35CrMn5 5333 36CrMn4 7847 36CrMn4 7356 38Cr4KB 7845 36CrMn5 7874 36CrMn5 7847 38Cr4	...
5140	1.7035, 41Cr4 1.7039, 41CrS4 1.7045, 42Cr4	G4052 SCr4H G4052 SCr440H G4104 SCr4 G4104 SCr440	3111 Type 3 970 2S.117 970 530A40 (En18D) 970 530H40 970 530M40	A35-552 42C4 A35-557 42C4 A35-556 42C4	5332 40Cr4 7356 41Cr4KB 7845 41Cr4 7874 41Cr4	2245
5147	1.7145, GS-50CrMn4 4	...	3100 BW2, BW3 3146 CLA 12 Grade A 3146 CLA 12 Grade B	50C4	...	...
5150	1.7145, GS-50CrMn4 4 1.8404, 60MnCrTi4	...	3100 BW2 3100 BW3 3146 CLA 12 Grade A 3146 CLA 12 Grade B	...	...	2230
5155	1.7176, 55Cr3	G4801 SUP11 G4801 SUP9	...	A35-571 55C3	...	...
5160	1.2125, 65MnCr4	G4801 SUP9A	970 527A60 (En48) 970 527H60	...	...	...
51B60	No international equivalent					
E50100	1.2018, 95Cr1 1.3501, 100Cr2	...	...	A35-565 100C2	...	...
E51100	1.2057, 105Cr4 1.2109, 125CrSi5 1.2127, 105MnCr4 1.3503, 105Cr4	...	...	...	3160 G90Cr4	...
E52100	1.2059, 120Cr5 1.2060, 105Cr5 1.2067, 100Cr6 1.3505, 100Cr6 1.3503, 105Cr4 1.3514, 101Cr6 1.3520, 100CrMn6	...	970 534A99 (En31) 970 535A99 (En31)	...	100C6 3097 100Cr6	2258
6118	No international equivalent					
6150	1.8159, GS-50CrV4	G4801 SUP10	970 735A50 (En47) 970 S.204	A35-552 50CV4 A35-553 50CV4 A35-571 50CV4	3545 50CrV4 7065 50CrV4 7845 50CrV4 7874 50CrV4	2230

Table 17 Continued

United States (SAE)	Fed. R. of Germany (DIN)	Japan (JIS)	United Kingdom (BS)	France (AFNOR NF)	Italy (UNI)	Sweden (SS₁₄)

Alloy steels (continued)

United States (SAE)	Fed. R. of Germany (DIN)	Japan (JIS)	United Kingdom (BS)	France (AFNOR NF)	Italy (UNI)	Sweden (SS₁₄)
8115............	No international equivalents					
81B45..........	No international equivalents					
8615............	...	...	...	15NCD2 15NCD4	3097 16NiCrMo2 5331 16NiCrMo2 7846 16NiCrMo2	...
8617............	...	...	970 805A17 970 805H17 970 805M17 (En 361)	18NCD4 18NCD6	...	...
8620............	1.6522, 20NiCrMo2 1.6523, 21NiCrMo2 1.6526, 21NiCrMoS2 1.6543, 21NiCrMo2 2	G4052 SNCM21H G4052 SNCM220H G4103 SNCM21 G4103 SNCM220	2772 806M20 970 805A20 970 805H20 970 805M20 (En362)	18NCD4 20NCD2 A35-551 19NCDB2 A35-552 19NCDB2 A35-551 20NCD2 A35-553 20NCD2 A35-565 20NCD2 A35-566 20NCD2	5331 20NiCrMo2 6403 20NiCrMo2 7846 20NiCrMo2	2506-03 2506-08
8622............	1.6541, 23MnNiCrMo5 2	...	2772 806M22 970 805A22 970 805H22 970 805M22	23NCDB4 A35-556 23MNCD5 A35-556 23NCDB2 A35-566 22NCD2	...	...
8625............	...	...	970 805H25 970 805M25	25NCD4 A35-556 25MNCD6 A35-566 25MNDC6	...	...
8627............	No international equivalents					
8630............	1.6545, 30NiCrMo2 2	...	...	30NCD2	7356 30NiCrMo2KB	...
8637............	...	...	970 945M38 (En100)	40NCD3	5332 38NiCrMo4 7356 38NiCrMo4KB 7845 39NiCrMo3 7874 39NiCrMo3	...
8640............	1.6546, 40NiCrMo2 2	...	3111 Type 7, 2S.147 970 945A40 (En 100C)	40NCD2 40NCD2TS 40NCD3TS 40NCD3	5333 40NiCrMo4 7356 40NiCrMo2KB 7845 40NiCrMo2 7874 40NiCrMo2 7847 40NiCrMo3	...
8642............	No international equivalents					
8645............	No international equivalents					
86B45..........	No international equivalents					
8650............	No international equivalents					
8655............	No international equivalents					
8660............	...	...	970 805A60 970 805H60	...	...	...
8720............	No international equivalents					
8740............	1.6546, 40NiCrMo2 2	...	3111 Type 7, 2S.147	40NCD2 40NCD2TS 40NCD3TS	7356 40NiCrMo2KB 7845 40NiCrMo2 7874 40NiCrMo2	...
8822............	No international equivalents					
9254............	No international equivalents					
9260............	...	G4801 SUP7	970 250A58 (En45A) 970 250A61 (En45A)	60S7 61S7	...	...
E9310..........	1.6657, 14NiCrMo13 4	...	970 832H13 970 832M13 (En36C) S.157	16NCD13	6932 15NiCrMo13 9335 10NiCrMo13	...
94B15..........	No international equivalents					
94B17..........	No international equivalents					
94B30..........	No international equivalents					

Source: Ref. 2.

following paragraphs. More elaborate information on cross-referencing are available in References 10, 44, and 50.

DIN standards are developed by Deutsches Institut für Normung in Germany. All German steel standards/specifications are preceded by the letters DIN and followed by an alphanumeric or numeric code. An uppercase letter sometimes precedes this code. German designations are reported in one of two methods. The one method uses the descriptive code with chemical symbols and numbers in the designation. The second, called the Werkstoff number, uses numbers only, with a decimal point after the first digit. There are four figures after the decimal point, the first two of which are used to identify the alloy, the last two the quantity. Most steels are covered by the significant figure 1, but some have no significant figure before the decimal point. Examples of both methods are provided in Table 17, which cross-references SAE and DIN designations and indicates chemical composition for DIN steels. However, standards for heat-resistant steels are prefixed with the letters SEW (Stahl-Eisen-Werkstoff Blatter, steel-iron material sheets). As examples of DIN designations in both methods with equivalent UNS numbers in parentheses, DIN 40NiCrMo6 or DIN 1.6565 (G43400) is a Ni-Cr-Mo steel that contains 0.35–0.45% C, 0.9–1.4% Cr, 0.5–0.7% Mn, 0.2–0.3% Mo, 1.4–1.7% Ni, 0.035% P, 0.035% S, and 0.15–0.35% Si; DIN 17200 1.1149 or DIN 17200 Cm22 (G10200) is a nonresulfurized carbon steel containing 0.17–0.24% C, 0.3–0.6% Mn, 0.035% max P, 0.02–0.035% S, and 0.4% max Si.

JIS standards are developed by the Japanese Industrial Standards Committee (JISC) in Tokyo, Japan. The JIS specifications begin with the prefix JIS, followed by a letter G for carbon and low-alloy steels. This is followed by a space and a series of numbers and letters indicating the particular steel. JIS designations are provided in Table 17. As examples of JIS designations with equivalent UNS/AISI numbers in parentheses, JIS G3445 STKM11A (G10080) is a low-carbon tube steel containing 0.12% C, 0.35% Si, 0.60% Mn, 0.04% P, and 0.04% S; JIS G3445 STK 17A (G10490) is a medium-carbon nonresulfurized steel containing 0.45–0.55% C, 0.40–1.0% Mn, 0.04% P, 0.04% S, and 0.04% Si; JIS G4403 SKH2 (AISI T1 grade) is a tungsten high-speed tool steel containing 0.73–0.83% C, 3.8–4.5% Cr, 0.4% Mn, 0.4% Si, 0.8–1.2% V, and 17–19% W; and JIS G4403 SKH59 (AISI M42 grade) is a molybdenum ultrahard high-speed tool steel containing 1–1.15% C, 7.5–8.5% Co, 3.5–4.5% Cr, 0.4% max Mn, 9–10% Mo, 0.5% max Si, 0.9–1.4% V, 1.2–1.9% W, 0.25% max Ni, 0.03% max P, 0.03% max S, and 0.25% Cu.

BS standards are developed by the British Standards Institute (BSI) in London, England. The letters BS precede the standard numerical code, and, like JIS standards, each British designation covers a product form and an alloy code. Table 17 lists steels identified by British standards. Some example of BS designations with equivalent AISI designations in parentheses are given: BS 970 708A30 (4130) is a Cr-Mo low-alloy steel containing 0.28–0.33% C, 0.9–1.2% Cr, 0.4–0.6% Mn, 0.15–0.25% Mo, 0.035% P, 0.04% S, and 0.1–0.35% Si; and BS 970 304S15 (304) is a wrought austenitic stainless steel (sheet, strip, plate) containing 0.06% C, 17.5–19% Cr, 0.5–2.0% Mn, 8–11% Ni, 0.05% P, 0.03% S, and 0.2–1.0% Si.

AFNOR standards are developed by the Association Française de Normalisation in Paris, France. The AFNOR standards, which are given in Table 17, usually begin with the letters NF, followed by an alphanumeric code constituting an uppercase letter followed by a series of digits, which are subsequently followed by an alphanumeric sequence. For example, resulfurized (free-cutting) steel is listed in AFNOR NF A35-562 standard or specification, and 35MF6 designation (equivalent to SAE 1137) represents the steel bar containing 0.33–0.39% C, 1.30–1.70% Mn, 0.10–0.40% Si, 0.040% P, and 0.09–0.13% S; whereas 45MF4 designation (equivalent to SAE 1146) contains 0.42–0.49% C, 0.8–1.1% Mn, 0.1–0.4% Si, 0.04% max P, 0.09–0.13% S. Similarly, AFNOR NF A35-573 Z6CN 18.09 is a wrought (SAE 304) stainless steel (sheet, strip, plate) and contains 0.07% C, 17–19% Cr, 2% Mn, 8–10% Ni, 0.04% P, 0.03% S, and 1% Si.

UNI standards are developed by the Ente Nazionale Italiano di Unificazione in Milan, Italy. Italian standards are preceded by the prefix UNI followed by a four-digit product form code subsequently followed by alphanumeric alloy identification as given in Table 17. For example, UNI 5598 3CD5 (equivalent to SAE 1005 grade) is a low-carbon steel used for wire rod containing 0.06% C, 0.25–0.50% Mn, 0.035% P, 0.035% S, and 0.007% N; and UNI 5332 40Cr4 (equivalent to SAE 5140 steel) is a chromium low-alloy steel containing 0.37–0.44% C, 0.9–1.2% Cr, 0.5–0.8% Mn, 0.035% P, 0.035% S, and 0.4% Si.

Swedish standards (SS) are developed by the Swedish Standards Institution (SIS) in Stockholm. Swedish standards are preceded by the letters SS followed by the number 14 for carbon and alloy steels, subsequently followed by a four-digit numerical sequence similar to the German Werkstoff number. Swedish designations are indicated in Table 17. Thus, SS14 2242 is a hot-work tool steel (equivalent to AISI H13 tool steel) containing 0.35–0.42% C, 5–5.5% Cr, 0.3–0.6% Mn, 1.2–1.6% Mo, 0.8–1.2% Si, and 0.85–1.15% V; SS14 2260 is a medium-alloy air-hardening cold-work tool steel (equivalent to AISI A2 tool steel) containing 0.95–1.05% C, 5–5.5% Cr, 0.45–0.75% Mn, 1–1.2% Mo, 0.15–0.3% Si, 0.15–0.25% V, 0.03% max P, and 0.02% max S; and SS14 2310 is a high-carbon, high-chromium cold-work tool steel (equivalent to AISI D2 tool steel) containing 1.45–1.65% C, 11–13% Cr, 0.3–0.6% Mn, 0.7–0.9% Mo, 0.03% max P, 0.02% max S, 0.2–0.4% Si, and 0.7–0.9% V.

REFERENCES

1. A. K. Sinha, *Ferrous Physical Metallurgy*, Butterworths, 1989.
2. Anon., Classification and Designation of Carbon and Low-Alloy Steels, *ASM Handbook,* 10th ed., Vol. 1, ASM International, Materials Park, OH, 1990, pp. 140–194.
3. Anon., Carbon and Alloy Steels, SAE J411, *1993 SAE Handbook*, Vol. 1, Materials, Society of Automotive Engineers, Warrendale, PA, pp. 2.01–2.04.
4. G. Krauss, *Steels—Heat Treatment and Processing Principles*, ASM International, Materials Park, OH, 1990.
5. W. C. Leslie, *The Physical Metallurgy of Steels*, McGraw-Hill, New York, 1981.
6. E. C. Bain and H. W. Paxton, *Alloying Elements in Steel*, American Society for Metals, Cleveland, OH, 1966.
7. R. W. K. Honeycombe, *Steels—Microstructure and Properties*, Edward Arnold, London, 1982.
8. H. E. Boyer, in *Fundamentals of Ferrous Metallurgy*, Course 11, Lesson 12, Materials Engineering Institute, ASM International, Materials Park, OH, 1981.
9. R. B. Ross, *Metallic Materials Specification Handbook*, 4th ed., Chapman and Hall, London, 1992.
10. C. W. Wegst, *Stahlschlüssel (Key to Steel)*, Verlag Stahlschlüssel Wegst GmbH, 1992.
11. W. J. McG. Tegart and A. Gittins, in *Sulfide Inclusions in Steel*, J. T. Deabradillo and E. Snape, Eds., American Society for Metals, Cleveland, OH, 1975, p. 198.
12. R. Kern, *Heat Treating 17*:(4), 38–42 (1985).
13. C. W. Kovach, in *Sulfide Inclusions in Steel*, J. T. Deabradillo and E. Snape, Eds., American Society for Metals, Cleveland, OH, 1975, p. 459.
14. C. M. Lyne and A. Kazak, *Trans. ASM 61*:10 (1968).
15. D. Brovoksbank and K. W. Andrews, *JISI 206*:595 (1968).
16. W. D. Lankford and H. E. McGannon, Eds., *The Making, Shaping, and Treating of Steel*, 10th ed., U.S. Steel, 1985.
17. *Steel Products Manual: Plates; Rolled Floor Plates: Carbon, High Strength Low Alloy, and Alloy Steel*, American Iron and Steel Institute, Washington DC, August 1985.
18. E. Hornbogen, in *Physical Metallurgy*, R. W. Cahn and P. Haasen, Eds., Elsevier, New York, 1983, pp. 1075–1138.
19. J. D. Smith, *Fundamentals of Ferrous Metallurgy*, Course 11, Lessons 2 and 5, Materials Engineering Institute, ASM International, Materials Park, OH, 1979.
20. Numbering System, Chemical Composition, *1993 SAE Handbook,* Vol. 1, *Materials*, Society of Automotive Engineers, Warrendale, PA, pp. 1.01–1.189.

21. W. F. Smith, *Structure and Properties of Engineering Alloys*, McGraw-Hill, New York, 1981.
22. O. D. Sherby, B. Walser, C. M. Young, and E. M. Cady, *Scr. Metall. 9:*569 (1975).
23. E. S. Kayali, H. Sunada, T. Oyama, J. Wadsworth, and O. D. Sherby, *J. Mater. Sci. 14:*2688–2692 (1979).
24. D. W. Kum, T. Oyama, O. D. Sherby, O. A. Ruano, and J. Wadsworth, Superplastic forming, Conference Proceedings, 1984, American Society for Metals, Cleveland, OH, 1985, pp. 32–42.
25. T. A. Philip, in *ASM Handbook*, 10th ed., Vol. 1, ASM International, Materials Park, OH, 1990, pp. 430–448.
26. A. K. Sinha, in *Production of Iron/Steel and High Quality Product Mix* (Conf. Proc. 1991), ASM International, Materials Park, OH, 1992, pp. 195–206.
27. L. F. Porter and P. E. Repas, *J. Met. 34*(4):14–21 (1982).
28. L. F. Porter, in *Encyclopaedia of Materials Science and Engineering*, Pergamon, Oxford, 1986, pp. 2157–2162.
29. A. H. Nakagawa and G. Thomas, *Metall. Trans. 16A:*831–840 (1985).
30. D. Z. Yang, E. L. Brown, D. K. Matlock, and G. Krauss, *Metall. Trans. 16A:*1523–1526 (1985).
31. R. C. Davies, *Metall. Trans. 10A:*113–118 (1979)
32. A. R. Marder, *Metall. Trans. 13A:*85–92 (1982)
33. A. M. Bayer and L. R. Walton in *ASM Handbook*, Vol. 1, ASM International, Materials Park, OH, 1990, pp. 757–779.
34. M. G. H. Wells, in *Encyclopaedia of Materials Science and Engineering*, Pergamon, Oxford, 1986, pp. 5115–5120.
35. R. Higgins, *Engineering Metallurgy*, 5th ed., Krieger, Malabar, FL, 1983.
36. S. D. Washko and G. Aggen, in *ASM Handbook*, 10th ed., Vol. 1, ASM International, Materials Park, OH, 1990, pp. 841–907.
37. R. F. Decker, J. T. Each, and A. J. Goldman, *Trans. ASM 55:*58 (1962).
38. S. Floreen, *Met. Rev. 13:*115–128 (1968).
39. S. Floreen, in *Encyclopaedia of Materials Science and Engineering*, Pergamon, Oxford, 1986, pp. 5171–5177.
40. T. Morrison, *Metall. Mater. Technol. 8:*80–85 (1976).
41. K. Rohrbach and M. Schmidt, in *ASM Handbook*, 10th ed., Vol. 1, ASM Int., Materials Park, OH, 1990, pp. 793–800.
42. *Heat Treater's Guide—Standard Practices and Procedures for Steel*, ASM, Cleveland, OH, 1982.
43. *Metals and Alloys in the Unified Numbering System*, 6th ed. Society of Automotive Engineers, Warrendale, PA, 1993.
44. *Worldwide Guide to Equivalent Irons and Steels*, ASM Int., Materials Park, OH, 1992.
45. *1994 SAE AMS Index*, Society of Automotive Engineers, Warrendale, PA.
46. *Specification for Drill Pipe*, API Specification 5D, 3rd ed., Aug. 1, 1992, American Petroleum Institute, Washington, D.C.
47. *Specification for Line Pipe*, API Specification 5L, 40th ed., Nov. 1, 1992, American Petroleum Institute, Washington, D.C.
48. *Specification for CRA Line Pipe*, API Specification 5LC, 2nd ed., Aug. 1, 1991, American Petroleum Institute, Washington, D.C.
49. *1994 Publications, Programs & Services*, American Petroleum Institute, Washington, D.C.
50. D. L. Potts and J. G. Gensure, *International Metallic Materials Cross-Reference*, Genium, 1989.

16

ENVIRONMENTAL AND SAFETY REGULATIONS AFFECTING HEAT TREATERS

D. Randy Junkins
Junkins Engineering, Inc., Morgantown, Pennsylvania

I. RCRA HAZARDOUS WASTE REGULATIONS

A. Scope

In 1976, Congress enacted the Resource Conservation and Recovery Act (RCRA) to protect human health and the environment from improper waste management practices. The passage of RCRA mandated the adoption of regulations by the U.S. Environmental Protection Agency (EPA) governing the following aspects of hazardous waste management:

1. Identification and listing of hazardous waste
2. Requirements for generators of hazardous waste
3. Requirements for transporters of hazardous waste
4. Standards for hazardous waste treatment, storage, and disposal (TSD) facilities
5. Permitting of TSD facilities

The intent of the federal government was to develop a broad-based, comprehensive regulatory program that would properly track the movement of hazardous waste from the point of generation to its ultimate disposal site. In addition, standards for acceptable methods of handling, treatment, and disposal of hazardous waste were developed. The individual states were then encouraged to develop their own programs that would be equivalent to (or more stringent than) that of the federal government. Federal funds were allocated to the states in the form of grants to be used for the development of such programs. *It is therefore imperative that companies become familiar with their state's hazardous waste regulations as well as the federal RCRA provisions.*

The federal hazardous waste management system is a "cradle-to-grave" system. This means that the generation, transportation, treatment, and disposal of hazardous wastes are regulated by federal and state laws. It is important to note that companies are responsible "from cradle to grave" for the hazardous wastes they generate. This makes it important for companies to responsibly manage not only the generation and storage of their hazardous wastes but also their transportation and disposal in order to minimize associated liabilities.

B. Hazardous Waste Determination

A waste is considered hazardous if

1. It is listed as a hazardous waste in the RCRA regulations (Subpart D or 40 CFR Part 261), *or*
2. It exhibits one of the characteristics (i.e., ignitability, corrosivity, reactivity, TCLP toxicity) of a hazardous waste as specified in the regulations (Subpart C of 40 CFR Part 261).

Therefore, if a material is not listed as a hazardous waste, it should be tested to confirm that it does not exhibit one of the characteristics of a hazardous waste.

Companies are responsible for determining if any of the wastes they generate are hazardous. *Note*: If a nonhazardous material becomes contaminated with a hazardous substance, then the entire mixture becomes hazardous. This is a very important consideration relative to the proper management of waste materials generated in the heat treatment industry.

Typical hazardous wastes generated by the heat treating industry include

Spent sodium cyanide salt

Spent nitrate/nitrite salt

Vapor degreasing solvents

Solvent still bottoms

Brazing fluxes

Mineral spirits

Paints

C. Large Quantity Generator Requirements

Companies that generate more than 2200 lb (1000 kg) of hazardous waste or 2.2 lb (1 kg) of acutely hazardous waste per month are classified as large quantity generators (LQGs). LQGs may store a maximum of 13,200 lb (6000 kg) of hazardous waste on-site for up to 90 days without obtaining a storage permit. The federal hazardous waste regulations require large quantity generators of hazardous waste to comply with the following rules:

1. Determine if their waste is hazardous.
2. Obtain an EPA Identification Number from the EPA using Notification of Hazardous Waste Activity Form 8700-12.
3. Comply with RCRA preparedness and prevention requirements (i.e., maintain spill cleanup and safety equipment, post emergency phone numbers, and establish emergency communication procedures).
4. Prepare a written contingency plan describing emergency response procedures.
5. Prepare a written personnel training program and conduct employee training concerning the proper handling of hazardous wastes.
6. Obtain a treatment, storage, and disposal (TSD) facility permit if hazardous waste is stored on the generator's property for more than 90 days.
7. Use appropriate containers that have been properly marked and labeled for shipment.
8. Use the National Uniform Manifest for tracking hazardous waste.
9. Ensure, through the manifest system, that the hazardous waste arrives at the authorized hazardous waste TSD facility designated by the generator.

10. Submit an exceptance report to the EPA for any shipments of hazardous waste that do not reach the TSD facility designated on the hazardous waste manifest.

11. Submit a biennial report by March 1 of each even-numbered year to EPA that indicates types and quantities of hazardous waste generated and the associated disposal sites.

12. Contract transporters that have proper EPA identification numbers.

13. Prepare a manifest for off-site treatment, storage, and disposal of hazardous waste.

14. Label, mark, package, and placard the waste according to EPA and Department of Transportation (DOT) regulations.

15. Keep records (i.e., manifests) of all hazardous waste shipments for a period of 3 years.

16. Store hazardous wastes on-site according to RCRA regulations.

Recycling and/or reclamation of hazardous waste is permitted, even encouraged, by EPA. The recycling process itself is exempt from regulation as long as the waste to be recycled has not been stored on-site for more than 90 days.

D. Small Quantity Generator Requirements

In issuing hazardous waste regulations under RCRA, EPA first focused on those large generators who produce the greatest portion of the hazardous waste generated in the United States. Regulations the EPA published on May 19, 1980, exempted "small quantity generators" from most hazardous waste management requirements. However, on November 8, 1984, amendments to RCRA that mandated several new requirements for small quantity generators were signed into law.

One of the new RCRA provisions directed the EPA to promulgate regulations for the generators of small quantities of hazardous waste. Previously, the EPA regulated only those establishments generating more than 2200 lb of hazardous waste per month. Under the 1984 law, establishments that generate between 220 and 2200 lb in a calendar month are considered small quantity generators and also have to comply with certain requirements that address the storage, transportation, and disposal of hazardous waste. Any generator of hazardous waste who produces between 220 and 2200 lb of hazardous waste and less than 2.2 lb of acutely hazardous waste in a month is considered a small quantity generator (SQG) and is exempted from certain requirements such as the preparation of written personnel training plans, written contingency plans, and the submission of biennial reports and exception reports.

Small quantity generators may store a maximum of 13,200 lb of hazardous waste on-site up to 180 days (270 days, if the waste must be shipped over 200 mi to a disposal site) without obtaining a storage permit. A company that accumulates more than 13,200 lb of hazardous waste at any time or exceeds the 180 day (of 270 day) storage time limit becomes subject to full RCRA regulation, including permitting as a hazardous waste storage facility.

SQGs must comply with the following rules:

1. File a Notification of Hazardous Waste Activity Report with EPA and obtain an EPA identification number.

2. Identify all hazardous waste generated on-site in quantities between 220 lb (100 kg) and 2200 lb (1000 kg) per month.

3. Classify, package, mark, label, and placard waste material as required by the DOT and EPA.

4. Prepare a Uniform Hazardous Waste Manifest for all hazardous waste shipments and maintain manifests on file for 3 years.

5. Use EPA registered transporters and EPA-permitted TSD facilities.
6. Document the types and quantities of hazardous wastes generated on-site.
7. Store wastes in proper containers and follow proper waste storage procedures.
8. Designate a company Emergency Coordinator.
9. Post emergency telephone numbers and the location of emergency response equipment near company telephones.
10. Familiarize employees with proper waste handling and emergency response procedures.

Companies in the SQG category can recycle, reclaim, or treat their hazardous waste without obtaining a special EPA permit as long as the recycling/reclamation is conducted in tanks or containers that comply with RCRA standards and does not extend past the 180/270 day accumulation period.

E. Conditionally Exempt Small Quantity Generators

Companies that generate less than 220 lb (100 kg) of hazardous waste and 2.2 lb (1 kg) of acutely hazardous waste per calendar month are classified as Conditionally Exempt Small Quantity Generators and are generally exempted from full RCRA regulation. Companies in this hazardous waste generation category are allowed to accumulate up to 2200 lb (1000 kg) of hazardous waste on-site without being subject to EPA regulation. If more than 2200 lb (1000 kg) of hazardous waste is accumulated at any time, the company becomes subject to full EPA hazardous waste regulation. No specific time limit is applied to this accumulation limit. Companies in this category are allowed to mix up to 220 lb (100 kg) per calendar month of hazardous waste with other, nonhazardous waste unless the resulting mixture exhibits one of the characteristics (i.e., ignitability, corrosivity, reactivity, or TCLP toxicity) of a hazardous waste. If the company's hazardous waste is mixed with used oil and the mixture is to be burned for energy recovery, the resulting mixture is subject to special regulations for used fuel oil.

It is very important to note that, unlike OSHA's hazardous communication standard, EPA's hazardous waste regulations do *not* preempt state regulations. Many states have enacted their own hazardous waste regulations, and some are more strict than the RCRA requirements. Therefore, it is imperative that companies become familiar with their state's hazardous waste regulations as well as the federal EPA regulations, because companies must comply with the regulations that are the most stringent.

F. Waste Oil Regulations

The Environmental Protection Agency defines used oil as follows:
"Used oil" is petroleum-derived or synthetic oil including, but not limited to, oil that is used as a

Lubricant (engine, turbine or gear)

Hydraulic fluid (including transmission fluid)

Metalworking fluid (including cutting, grinding, machining, rolling, stamping, quenching, and coating oils); or

Insulating fluid or coolant that is contaminated through use of subsequent management.

The EPA has adopted the following definition for recycled oil:

Recycled oil means used oil that is either burned for energy recovery, used to produce a fuel, reclaimed (including used oil that is reprocessed or re-refined), or otherwise recycled, or that is collected, accumulated, stored, transported, or treated prior to recycling.

Currently, EPA does not consider used oil a hazardous material unless it is contaminated with a listed hazardous waste or it exhibits one of the four characteristics of a hazardous waste. EPA has determined that used oil being recycled should not be listed as a hazardous waste under RCRA. It is noted, however, that some states do consider used oil a hazardous waste and regulate it as such. Even in those states where waste oil is not classified as hazardous, its storage, handling, transportation, and disposal are strictly regulated. EPA has also promulgated a Used Oil Management Standard that regulates waste oil that is recycled.

II. EPA'S EMERGENCY PLANNING AND COMMUNITY RIGHT-TO-KNOW ACT

A. Provisions of EPCRA

The Emergency Planning and Community Right-To-Know Act (EPCRA) became law on October 17, 1986, as Title III of the Superfund Amendments and Reauthorization Act (SARA). It establishes new requirements for facilities at which certain substances are present in specified amounts called threshold planning quantities. These established quantities take into account the amount of the substance that, if released at a facility, would likely pose a hazardous substance emergency. The specified substances and their associated threshold planning quantities (TPQs) are listed in 40 CFR Part 355 Appendixes A and B. The provisions of the law address the following:

1. Emergency planning notification
2. Emergency chemical release notification
3. Community right-to-know reporting
4. Toxic chemical release reporting

B. Emergency Planning Notification

Although EPCRA does not preempt state and local laws, it does require states to establish a state emergency response commission, which in turn must designate local emergency planning districts and committees. If a facility has on its premises, at any time, a quantity of an extremely hazardous substance (EHS) greater than the TPQ stipulated by the EPA, the facility is subject to the emergency planning and notification provisions of the law. These provisions require the owner or operator to do the following:

1. Notify the state emergency response commission that the facility is subject to the provisions of the law.
2. Designate a representative of the company who will act as a facility emergency coordinator and participate in the local emergency planning process.
3. Promptly inform the local emergency planning committee of any relevant changes occurring at the facility.
4. Upon request from the local emergency planning committee, promptly provide the community any information it needs to develop and implement a district emergency plan.

EPA's Extremely Hazardous Substance list includes approximately 400 chemicals that can potentially cause serious and irreversible health effects from accidental releases. *Note*: Ammonia and sodium cyanide are listed as extremely hazardous substances and have TPQs of 500 lb and 100 lb, respectively.

C.   Emergency Chemical Release Notification

Facilities that use extremely hazardous substances must immediately notify the local emergency planning committee and the state emergency response commission if there is a spill or other release of a listed extremely hazardous substance that exceeds that chemical's "reportable quantity" and that will result in exposure to persons outside the facility boundaries. Reportable quantities will vary from substance to substance. In addition to the listed extremely hazardous substances, this emergency notification requirement also applies to releases of substances subject to the emergency notification requirements under the Superfund law (CERCLA; see Section III). Chemicals subject to CERCLA reporting are contained in a separate listing of approximately 700 substances, some of which may be found in the average heat-treating facility. Included on the CERCLA list are such commonly used substances as ammonia, trichloroethylene, and 1,1,1-trichloroethane.

As with the reportable quantities for the extremely hazardous substances, reportable quantities for the CERCLA substances will vary from substance to substance. Reportable quantities for the products mentioned above are

Ammonia	100 lb
Trichloroethylene	100 lb
1,1,1-Trichloroethane	1000 lb

In the event of a release of a CERCLA chemical that is in excess of the reportable quantity and that results in a release to the environment, you must immediately notify the local emergency planning committee and the state emergency response commission, and you must also notify the National Response Center at (800) 424-8802. The initial notification can be by telephone but must include

The identity of the spilled or released chemical

An indication of whether the substance is extremely hazardous

An estimate of the quantity released

The time and duration of the release

The medium into which the release occurred

Any known or anticipated acute or chronic health risks associated with the emergency

Proper precautions that should be taken

The name and telephone number of the contact person at your plant

The law also requires a follow-up written report to be sent to the state and local emergency planning committees that updates the information concerning the actual response actions taken.

D.   Community Right-to-Know Reporting

The community right-to-know reporting provisions of SARA Title III covers those chemicals designated hazardous according to OSHA's Hazard Communication Standard and therefore impacts all heat treaters. There are two separate community right-to-know reporting requirements that apply to users of hazardous chemicals. The first involves submitting MSDSs or a listing of the OSHA hazardous chemicals stored in your plant. You must inform local and state agencies of any OSHA hazardous chemicals present in your plant in amounts equal to or greater than 10,000 lb. For extremely hazardous substances, the reporting amount is the material's TPQ or 500 lb, whichever is less. Copies of the MSDSs in your possession or a list of these chemicals must be provided to

1.   The local emergency planning committee

2. The state emergency response commission
3. The local fire department

According to the regulations, you may choose to submit a list of MSDS chemicals in lieu of copies of the actual MSDSs. However, that list must include the chemical name or common name of each substance and any hazardous component as provided on the MSDS. Even if a list is submitted, you must still provide a copy of any MSDS for which a request is made by the local planning committee. However, you should be aware that many states have passed strict laws requiring submission of lists in response to this requirement and are either directly prohibiting MSDS copies or levying fines against companies who submit MSDS copies. Make sure you are familiar with your individual state's submission requirements.

In addition to initially providing a chemical listing or copies of MSDSs on file for hazardous substances in your plant in amounts equal to or greater than 10,000 lb, any time a new chemical is introduced into your plant in amounts equal to or greater than the threshold amount and for which you receive an MSDS, you must provide a copy of that MSDS or a revised chemical list to the three agencies mentioned above within 3 months.

The second community right-to-know reporting requirement that may apply to your facility involves submission of an annual Emergency and Hazardous Chemical Inventory Report (Tier II report) to the local emergency planning committee, the state emergency response commission, and the local fire department. Hazardous chemicals covered by this annual inventory report are the same as those for which MSDSs are required to be submitted. The first annual inventory report was due March 1, 1988, and covered chemicals present in your plant during the preceding year in amounts equal to or greater than 10,000 lb. The same 10,000 lb inventory reporting threshold applies to the annual inventory due on March 1 each year. Again, for extremely hazardous substances, the reporting threshold is the material's TPQ or 500 lb, whichever is less.

Companies must provide the following Tier II information for each substance covered:

1. The chemical name or the common name indicated on the MSDS
2. An estimate (in ranges) of the maximum amount of the specific chemical present at any time during the preceding year
3. A brief description of the manner of storage of the chemical
4. An indication of whether the owner elects to withhold location information from disclosure to the public

Note: Some states have developed their own state form that must be filed, and therefore companies must investigate their state's specific reporting requirements.

E. Toxic Chemical Release Reporting (Form R)

The fourth and final reporting requirement imposed on companies as a result of SARA Title III requires certain facilities to complete a Toxic Chemical Release Inventory Report (Form R) for specified chemicals. This reporting requirement applies to companies (1) within SIC codes 20–39 (which includes the heat treating industry); (2) that have 10 or more employees; and (3) that manufactured, processed, or otherwise used a listed toxic chemical in excess of the specified threshold quantity. Chemicals subject to this reporting requirement are contained in yet another EPA list that includes over 300 chemicals. Included on this list are commonly used materials such as sulfuric acid, hydrochloric acid, nitric acid, ammonia, methanol, trichloroethylene, and 1,1,1-trichloroethane.

Different threshold quantities are used for determining whether or not your usage of listed toxic chemicals requires reporting under Section 313 of SARA Title III. Two thresholds exist, depending on whether you manufacture or process any of the listed chemicals or whether you otherwise use any of the listed chemicals.

The Environmental Protection Agency defines "manufacture" as producing, preparing, importing, or compounding one of more of the listed toxic chemicals. This definition covers the classic activities of a true chemical manufacturer. "Process" means to make mixtures, repackage, or use a listed toxic chemical as a feedstock, raw material, or starting material for making another chemical.

If you manufacture or process any of the listed toxic chemicals, the threshold quantity is

75,000 lb during calendar year 1987

50,000 lb during calendar year 1988

25,000 lb during calendar year 1989 and subsequent years.

If you otherwise use a listed toxic chemical, the threshold quantity is

10,000 lb during calendar year 1987 and subsequent years

These reporting thresholds apply to toxic chemicals known by the owner/operator to be used in amounts above the thresholds. Beginning in 1989, Section 313 required suppliers of mixtures and name brand products to notify customers of the presence of Section 313 listed chemicals in their products.

Under certain circumstances, some or all of the reporting requirements under Section 313 may not apply. The following are the major exemptions:

1. De minimis amounts of listed toxic chemicals found in mixtures and brand name products are exempt from the quantity threshold determinations. EPA defines this de minimis level as concentrations of less than 1% of the mixture or less than 0.01% if the chemical is defined by OSHA as carcinogenic.

2. You are not required to count toxic chemicals present in "articles" at your facility. "Article" is defined as a manufactured item that is formed to a specific shape or design during manufacture, that has end use functions dependent in whole or in part upon its shape or design during end use, and that does not release a toxic chemical under normal conditions of processing or use of that item at the facility.

3. In considering whether a reporting threshold has been exceeded, you are not required to count toxic chemicals that are used at your facility for any of the following purposes:

    a. As a structural component of the facility

    b. In routine janitorial or facility grounds maintenance

    c. In food, drugs, cosmetics, or other items for personal use

    d. In motor vehicle maintenance

    e. In process water and noncontact cooling water, or in air used either as compressed air or as part of combustion

The EPA has determined that toxic chemicals that are recycled on-site should not be counted more than once. Consequently, you would count only the amount of chemical added to a recycle/reuse system during the reporting year, not the quantity that might have been brought to the facility during the year. (In other words, storage does not constitute use.)

If your facility is covered by Section 313, you must complete and return the annual Toxic Chemical Release Report, EPA Form R, no later than July 1 of each year, covering the preceding calendar year. Copies must be sent to U.S. EPA and to your state's Emergency Response Commission. You are not required to measure or monitor releases for purposes of Section 313 reporting. You are allowed to use readily available data to report the quantities of chemicals that you use and the amounts released into the environment. If no data exists or are available, the law allows you to report reasonable estimates.

You must report the following information for each listed chemical manufactured, imported, processed, or used at your facility in yearly amounts that exceed the threshold:

1. The name and location of your facility.
2. Whether you manufacture, import, or process the chemical or use it in any other way.
3. The maximum quantity of the chemical on site at any time during the year.
4. The total quantity of the chemical released during the year, including both accidental spills and routine emissions. Separate estimates must be provided for releases to air, water, and land (e.g., deep well injection, permitted landfill).
5. Off-site locations to which you shipped wastes containing the chemical and the quantities of that chemical sent to those locations.
6. Treatment or disposal methods used for wastes containing the chemical and estimates of their efficiency for each chemical (efficiency of treatment methods used on site).

F. Penalties

Stiff civil and criminal penalties are authorized for failure to comply with certain provisions of the act. These include civil penalties of up to $25,000 per day for failure to participate in the planning process and failure to give initial emergency notification, and criminal penalties of up to $25,000 or up to 2 years imprisonment, or both, for knowingly and willfully failing to provide emergency notice in the event a chemical release occurs.

G. Key Dates

Key compliance dates pertaining to this law are listed below.

Date	Required action
May 17, 1987	Notice to state emergency response commission by owner or operator that his facility uses one or more extremely hazardous substances in quantities that exceed the threshold planning quantity.
September 17, 1987 (or within 30 days of establishing a local emergency planning committee)	Notice to the local committee of the name of a facility emergency coordinator (EC).
October 17, 1987	Submission of MSDSs or an alternative hazardous list to the local emergency planning committee, state emergency response commission, and local fire department.
March 1, 1988 (and annually thereafter)	Submission of emergency and hazardous chemical inventory forms (i.e., Tier I/Tier II) to the local committee, state commission, and local fire department.
July 1, 1988 (and annually thereafter)	Submission of Toxic Chemical Release Inventory forms (i.e., Form R) to EPA and state governor's designee.

Companies that should have but didn't file Form Rs beginning in 1988 must file past due reports.

The SARA Title III regulations require companies that use and/or store certain substances above designated threshold planning quantities to comply with various notification and reporting requirements. Agencies that must receive these notifications and reports include state emergency response commissions (SERCs), local emergency planning committees (LEPCs), fire departments, and EPA. Specific requirements include

1. Submitting notifications if any extremely hazardous substances (EHS) are present on site in quantities that exceed chemical specific threshold planning quantities.
2. Submitting notifications if chemical releases occur that exceed reportable quantities (RQs).
3. Filing a Tier II Form by March 1 each year if any hazardous chemicals are stored on-site above threshold planning quantities or 10,000 lb, whichever is less.
4. Filing a Form R by July 1 each year if certain hazardous chemicals are used in quantities that exceed 25,000 lb if "processed or manufactured" or 10,000 lb if "otherwise used."

Penalties for failure to meet the above requirements are severe, and many companies, large and small, have been fined for noncompliance. The heat treating industry uses numerous materials regulated by SARA Title III, and therefore companies must be cognizant of their responsibilities under this regulation.

III.   EPA'S UNDERGROUND STORAGE TANK REGULATIONS

A.   Scope

In 1988, the EPA initiated the Underground Storage Tank (UST) program to regulate the underground storage of petroleum products and hazardous substances regulated under the Comprehensive Environmental Response Compensation and Liability Act (CERCLA; also known as the Superfund Act). These environmental regulations address the underground storage of hazardous raw materials but does not apply to the underground storage of hazardous waste. The regulations define an underground storage tank as any tank with at least 10% of its volume buried belowground, including any piping attached to the tank. Thus, aboveground tanks with extensive underground piping may be regulated under this law. The UST program prohibits the installation of unprotected tanks, initiates a tank notification program, sets federal technical standards for all tanks, and provides for federal inspection and enforcement programs. The final UST regulations are published in 40 CFR Part 280.

B.   Exemptions

The following types of underground storage tanks are exempt from the federal EPA regulations:

1. Farm and residential tanks holding less than 1000 gal of motor fuel used for non-commercial purposes
2. Tanks storing heating oil burned on the premises where it is stored
3. Tanks on or above the floor of underground areas such as basements or tunnels
4. Septic tanks and systems for collecting stormwater and wastewater
5. Flow-through process tanks (i.e., quench tanks)

6. Tanks holding 110 gal or less
7. Emergency spills and overfill tanks

The regulations also exclude storage areas that may be considered tanks, such as surface impoundments and pits.

The definition of a *flow-through process tank* includes tanks that form an integral part of a production process through which there is an intermittent flow of materials during the operation of the process. A production process is defined as any process at manufacturing, commercial, or industrial facilities where a tangible good or service is produced or performed. EPA included an exemption for process tanks because the intent of the agency is to regulate storage tanks, not tanks used in a production process (such as oil quenching).

C. Notification

Owners of underground storage tanks are required to notify their designated sate agency if the facility presently has a UST already in the ground, if the facility is going to install UST, or if the facility owns a UST that was taken out of service after January 1, 1974, but remains in the ground.

D. Requirements for New Petroleum USTs

New UST systems are those installed after December 1988. The regulations create five minimum requirements for all new petroleum USTs:

1. All tanks and associated piping must be properly designed and constructed for underground use and made of fiberglass-reinforced plastic, steel, or a composite of the two.
2. The tank and associated piping must be protected from corrosion. Steel tanks and piping must be "cathodically" protected and coated with a corrosion-resistant coating. Other tanks and piping must be made totally of fiberglass-reinforced plastic or of a composite of steel and a fiberglass-reinforced plastic.
3. The UST must be equipped with devices that prevent spills and overfills associated with product transfers to the UST. Also, correct tank-filling practices must be followed.
4. The owner or operator must certify that the UST is installed properly by a licensed tank installer.
5. The UST must have a leak detection method that provides monitoring for leaks at least every 30 days.

A leak detection method must be able to detect a leak in any part of the UST, including the piping.

E. Requirements for Existing Petroleum USTs

Existing UST systems are those installed before December 1988. One goal of the UST regulations was, over a 10 year period, to improve petroleum USTs already in the ground so that they will meet the more demanding requirements for new USTs as described above. By December 1998, all existing USTs will need to show three required improvements:

1. They must meet the same requirements for corrosion protection that apply to new USTs.
2. They must meet the new UST requirements for having a leak detection method.
3. They must be equipped with devices that prevent spills and overfills.

The leak detection requirements apply before this 10 year period ends and are being phased in depending on the age of the UST. This phased-in approach will ensure that older USTs, which are more likely to leak, have leak detection first.

F.   USTs Storing Hazardous Chemicals

A number of chemicals are designated as "hazardous" in Section 101(14) of the Comprehensive Environmental Response, Compensation, and Liability Act of 1980, better known as CERCLA or Superfund. With one exception, the UST regulations apply to the same hazardous chemicals as those identified by CERCLA. The exception concerns substances that are hazardous wastes, which are already regulated under Subtitle C of the Solid Waste Disposal Act. Therefore, these hazardous wastes are not covered by the UST regulations.

In addition to the requirements for new petroleum USTs previously described, new chemical USTs and piping must have "secondary containment" and "interstitial monitoring." The UST itself makes up the first or "primary" containment. With only primary containment, a leak can escape into the environment. By enclosing a UST within a second wall, leaks can be contained in a relatively small and controllable area. There are several ways to construct secondary containment:

1.   Placing one tank inside another tank (making them double-walled tank systems)
2.   Placing the tank inside a concrete vault
3.   Lining the excavation area surrounding the tank with natural or synthetic liners that cannot be penetrated by the chemical

A chemical UST must have a leak detection system that can indicate the presence of a leak in the confined area between the primary and secondary walls. Several devices are available to monitor this confined "interstitial" area. The regulations describe these various methods and the requirements for their proper use.

You can apply for an exception, called a variance, from the requirements for secondary containment and interstitial monitoring. This variance will be granted only if the chemical UST meets the following requirements:

1.   A leak detection method must be available for the stored chemical. The leak detection method must meet the same general requirements for leak detection discussed earlier.
2.   Single-walled USTs must be protected from corrosion.
3.   All pressurized piping not provided with interstitial or continuous monitoring must have an emergency cutoff pressure monitor.

The goal of the UST regulations is to improve chemical USTs already in the ground so that they will meet the requirements for new chemical USTs, as described above. By December 22, 1998, all existing chemical USTs will be required to show these improvements:

1.   They must meet the same requirements for secondary containment and interstitial monitoring that apply to new chemical USTs and piping. As with new chemical USTs, a variance can be granted for single-walled USTs protected from corrosion and equipped with an approved leak detection method.
2.   They must be equipped with devices that prevent spills and overfills.

Certain leak detection requirements apply before this 10 year period ends. These requirements are the same as those described earlier for petroleum USTs already in the ground.

However, the leak detection methods noted for petroleum USTs can be applied to a chemical UST only if those methods are compatible with the stored chemical. If an effective leak detection method is not in place by the scheduled deadlines, a UST must be closed or upgraded.

G. Spill Cleanup Procedures

If a UST leak is suspected, the following actions must be implemented:

1. Report the discovery immediately to the state UST regulatory authority.
2. Conduct tightness testing of the entire UST system.
3. Investigate the UST site to determine if any environmental damage has occurred and, if so, the nature and extent of the damage.
4. If the above actions confirm that a leak has occurred, report this to your state UST agency and any other appropriate regulatory authority as directed by your state UST agency and follow the spill response steps described below.

Under the UST regulations, response to a petroleum leak or spill should occur in two stages: immediate and long-term. The following immediate steps are required for all *petroleum* leaks or spills:

1. Take immediate action to stop and contain the leak or spill, and make sure it poses no hazard to human health and safety by removing explosive vapors and fire hazards.
2. Tell the regulatory authority within 24 h that there is a leak or spill. The only exception is for aboveground petroleum spills and overfills of less than 25 gal. These small petroleum spills do not have to be reported if they are immediately contained and cleaned up.
3. Remove any visibly contaminated soil, making sure that the material is handled properly so no hazards are posed.
4. Determine how far the leaked petroleum has moved, and begin to recover this material.
5. Report your spill investigation and clean up actions to date to the regulatory authority no later than 20 days after a leak is confirmed.
6. Conduct a complete site investigation to determine the extent of the leak and how much it has damaged nearby soil and groundwater. Within 45 days of confirming a leak, submit a written plan for removing leaked petroleum if it has been detected in the groundwater.

Some leaks will require long-term attention to correct the problem. In long-term cases, sites must be cleaned up to meet environmental requirements established at each site by the regulating authority. Long-term actions that may be necessary include the following.

1. Develop a corrective action plan, an officially approved plan for cleaning up your site that shows how you will meet cleanup requirements established at your site by the regulating authority.
2. Make sure you meet the cleanup requirements approved by the regulatory authority for your site.

The immediate and long-term steps described above also apply for chemical USTs except for the following important difference. You must report all aboveground chemical leaks or spills immediately to the National Response Center at (800) 424-8802 and within 24 h to the state regulatory agency unless the spills are smaller than the "reportable quantities" identified under CERCLA and they are immediately contained and cleaned up.

H. Tank Repair

A tank can be repaired if standard industrial codes for making repairs are strictly followed. Damaged metal piping cannot be repaired and must be replaced. FRP piping, however, can be repaired in accordance with manufacturer's instructions of national codes of practice. The regulations set minimum standards for tank repair; for example, an open seam more than 3 in. long cannot be repaired. These minimum standards follow codes used by the tank industry. The tank must also pass tests proving that it was repaired correctly. Within 30 days of completing a UST system (tank or piping) repair, it must be shown that the repairs have worked by conducting either an internal inspection, a tightness testing, or a monthly leak detection monitoring. Within 6 months of repair, tanks with cathodic protection must be tested to show that the cathodic protection is working properly.

I. Tank Closure

You may decide to close your UST temporarily or permanently. Any tank not used for 3–12 months must meet four requirements for temporary closure:

1. You must maintain corrosion protection systems at the tank.
2. You must continue to operate a leak detection method to identify leaks. If a leak is found, you will have to respond just as you would for a leak from an active UST, as described previously.
3. You must cap all lines, except the vent line, attached to the UST.
4. You must notify your state UST agency of your actions.

If your tank is not protected from corrosion and remains closed for more than 12 months or you decide to close it permanently, then you must follow the requirements for permanent closure:

1. Notify the regulatory authority 30 days prior to the closure.
2. Collect soil and/or groundwater samples to determine if leaks from your tank have occurred and damaged the surrounding environment. If so, follow the corrective action steps previously described.
3. You can either remove the UST from the ground or leave it in the ground. Before taking the UST out of the ground, you must empty and clean it by removing all liquids, vapors, and sludges. If you leave the UST in the ground, then you must cap attached piping and remove all liquids and gases from the tank and fill it with a harmless, chemically inactive solid such as sand.

J. Financial Responsibility

Owners or operators of petroleum USTs must be able to demonstrate, at any time, their ability to pay for damages that could be caused if their tanks leaked. These payments would have to cover the costs of cleaning up a site and compensating other people for bodily injury and property damage.

The amount of financial responsibility you must demonstrate depends on the type of business you operate, the amount of tank throughput, and the number of USTs owned. If your tanks are not used in petroleum production, refining, or marketing; your facility has a monthly throughput of less than 10,000 gal,; and you own or operate fewer than 100 USTs, then you must show that you have $500,000 or "per occurrence" coverage and $1 million of annual aggregate coverage. "Per occurrence" refers to the amount of money that must be available to pay the costs of one occurrence. The annual aggregate amount is the total amount of financial responsibility you must demonstrate to cover all leaks that might occur during one year. If your facility has a monthly throughput of more than 10,000 gal, then you must show that you have $1 million of "per occurrence" coverage as well as the annual aggregate coverage. You can use one or a combination of several mechanisms to demonstrate your financial responsibility—financial tests, insurance, guarantees, indemnity contracts, trust funds, risk retention groups, surety bonds, letters of credit, and state-assured mechanisms.

To date, EPA has not promulgated financial responsibility requirements for chemical UST owners and operators.

K. Recordkeeping/Reporting

Underground storage tank (UST) recordkeeping requirements include the following.

Leak detection method Maintain records of any manufacturers' claims of how well their leak detection devices will perform for 5 yr. Keep records of monitoring results for at least 12 months. Tank tightness test results must be kept until the tank is tested again.

Corrosion protection system If a protection system was applied at a factory, the last service check conducted under the tank warranty must be kept. If the system was applied "in the field," then reporting periods depend on the type of protection installed. For the "impressed current" type, the results of the last three inspections and two annual service readings must be kept; for the "sacrificial anode" type, the last two annual service readings must be kept.

Tank repair After a repaired tank is put back into service, records showing that the tank was properly repaired and subsequently passed repair confirmation tests must be kept for as long as the tank is in use.

Closing a UST A record of the test results for either temporary or permanent closure must be maintained for 1 yr and 3 yr, respectively. (These test results show what impact a UST has had on the surrounding area.)

Generally, you should follow this rule of thumb for maintaining records: When in doubt, keep it!

The UST reporting requirements stipulate the following:

Submit a UST Notification for new and existing tanks.

Report suspected releases.

Report confirmed releases.

Report the leak follow-up actions taken to correct any damage caused.

Notify the state regulatory agency 30 days prior to permanently closing a UST.

L. State and Local UST Regulations

State and local agencies monitor and enforce UST regulations. Companies must therefore be cognizant of their state and local UST regulations because they may be more stringent than the federal requirements. Each state has a UST agency that can provide information concerning state and local regulations.

M. Summary

The EPA's UST regulations define an underground storage tank as any tank with at least 10% of its volume, including attached piping, buried below ground. USTs that contain petroleum products and CERCLA hazardous chemicals (which include several chemicals used in the heat-treating industry) are regulated. The UST regulations require

1. Notification to state UST agency
2. Leak detection
3. Spill and overfill prevention
4. Corrosion protection
5. Proper installation
6. Recordkeeping and reporting
7. Proper closure procedures
8. Demonstration of financial responsibility

Owners of USTs must also be familiar with state and local UST regulations, which can be more stringent that federal regulations.

IV.   WASTEWATER TREATMENT/PRETREATMENT REGULATIONS

A.   Wastewater Discharge to Streams

The discharge of wastewater into streams and sewers is regulated by the Federal Clean Water Act, state water quality regulations, and municipal sewer use rules and regulations. Before discharging wastewater directly into a stream or some other body of water, a company must obtain a National Pollutant Discharge Elimination System (NPDES) permit. This includes the discharge of cooling tower and boiler blowdowns to storm sewers that drain to surface waters.

The allowable concentrations of substances in wastewater that can be discharged under an NPDES permit is dependent on several variables, including the water quality classification of the receiving stream. If the receiving stream is water quality limited (such as a small trout stream), then the allowable concentrations could be very low. Regulations governing the direct discharge of wastewaters to rivers and streams or any surface water are very site-specific and are considered by state and federal agencies on a case-by-case basis.

B.   Wastewater Discharge to Sewers

Due to the stringent discharge requirements generally imposed by an NPDES permit, the long permitting process required to obtain permission to discharge wastewaters directly to a receiving stream, high treatment costs, and strict monitoring requirements, it is generally advantageous for companies to discharge their wastewaters to a municipal sewer system if that is possible. The federal government requires municipalities to adopt sewer use ordinances that regulate the discharge of industrial wastewaters into their sewer systems. Generally, companies must obtain an industrial discharge permit from the local sewer authority before discharging wastewaters into a municipal sewer system. Depending on the characteristics of the wastewater, monitoring may be required (for flow, metals, pH, suspended solids, oil and grease, biochemical oxygen demand, etc.). This information is required for sewer use billing purposes and to ensure compliance with the sewer use ordinance.

Types of materials that are typically prohibited from being discharged to a municipal sewer system include explosives, acids, caustics, corrosive substances, solids or viscous wastes, oils and greases, noxious or malodorous substances, radioactive wastes, hot wastes, toxic wastes, clean waters (cooling water), dilution water (i.e., cannot dilute discharges to sewers), excessive flows, and any material that would adversely affect the performance of the municipal wastewater treatment plant.

Any wastewaters that do not meet the requirements of a sewer use ordinance must be pretreated before being discharged to the sewer. Surcharges may also be added to a sewer bill if wastewater characteristics exceed designated limits.

C. EPA Effluent Guidelines

The EPA has promulgated regulations that address the discharge of industrial wastewater into the waters of the United States and into publicly owned treatment works (POTWs). These wastewater treatment standards have been prepared for various categories of manufacturing processes and dictate the allowable discharge limits for pollutants typically found in wastewaters generated by these processes. EPA's effluent guidelines also include categorical pretreatment standards for wastewaters generated from new and existing sources that discharge to POTWs. These pretreatment standards require certain industries to pretreat their wastewaters prior to discharge to a municipal wastewater treatment plant.

The EPA has issued proposed categorical pretreatment standards that cover the heat treatment industry. These regulations are titled "Metal Products and Machinery (MP&M) Effluent Guidelines" and are scheduled to be finalized in late 1996. The MP&M effluent guidelines establish stringent discharge limits for aluminum, cadmium, chromium, copper, iron, nickel, zinc, cyanide, oil and grease, suspended solids and pH.

D. Spill Prevention Control and Countermeasure Plans

The Clean Water Act stipulates that all non-transportation-related facilities that could reasonably be expected to discharge oil into navigable waters of the United States or the adjoining shorelines must prepare and implement a Spill Prevention Control and Countermeasure (SPCC) plan. Exemptions include

1. Facilities having a total aboveground storage capacity of 1320 gal or less of oil, provided no single container has a capacity in excess of 660 gal.
2. Facilities having a total underground storage capacity of 42,000 gal or less of oil.
3. Facilities that, due to their location, could not reasonably be expected to discharge oil into or upon navigable waters of the United States or adjoining shorelines.

It is noted that, generally speaking, every body of water or continuous stream should be considered navigable. EPA's definition includes the following:

Waters that are navigable in fact

Waters declared navigable by a federal agency or court

Tributaries of navigable waters

Intrastate lakes, rivers, and streams from which fish or shellfish are taken and sold in interstate commerce or that are used by interstate travelers for recreational or other purposes

A written SPCC plan must be prepared that describes how oil spills will be contained and managed and must be reviewed and certified by a registered professional engineer.

V. CLEAN AIR ACT

A. The Clean Air Act Amendments of 1990

In order to control and improve the overall quality of air within the United States, Congress enacted the Clean Air Act (CAA), which provides EPA with the authority to regulate and monitor major air pollution sources. This act establishes national ambient air quality standards to protect the public health and welfare. Since the intial enactment of the law, there have been several amendments to the act, most recently, the CAA Amendments of 1990. These latest amendments establish many new provisions for reducing air pollution in the United States, and greatly expand the number of industries (including small businesses) regulated by the act.

The CAA is designed to curb three major threats to the nation's environment and public health: acid rain, urban air pollution (smog), and toxic air emissions. More specifically, the goal of EPA's CAA is to reduce emissions of sulfur dioxide, nitrogen oxides, carbon monoxide, particulates, lead, volatile organic compounds (VOCs), and 189 other substances designated as hazardous (i.e., toxic) air pollutants.

B. CAA Provisions

The provisions of the CAA are stipulated in various titles of the law, as summarized below.

1. Title I. Attainment and Maintenance of National Ambient Air Quality Standards (NAAQS)

Ozone (smog), nitrogen oxides, sulfur dioxide, particulates, carbon monoxide, and lead (also called criteria pollutants) are regulated under Title I. Volatile organic compounds (VOCs) and nitrogen oxides (NO_x) are precursors to the formation of ozone and are therefore regulated under the CAA. The CAA establishes NAAQS and criteria for designating air quality attainment and nonattainment areas for these pollutants. Nonattainment area classifications are ranked according to the severity of the area's air pollution problem. These classifications are marginal, moderate, serious, severe, and extreme. Nonattainment area classifications dictate the requirements which facilities located in these ares must satisfy in order to comply with the CAA. The emissions levels that determine whether a facility is a major air pollution source also vary according to nonattainment area classification.

A source that emits or has the potential to emit 100 tons per year (tpy) of any Title I air pollutant is designated a major source. EPA defines "potential to emit" as the emissions levels that could be reached is a source operated continuously at maximum capacity 24 h/day for 365 days/yr. Sources located in nonattainment areas are designated major sources if their potential to emit exceeds the following thresholds:

Nonattainment area classifications	Major Source Threshold*, tpy			
	VOC**	NO_x	CO	PM
Marginal	100	100	–	–
Moderate	100	100	–	–
Serious	50	50	50	70
Severe	25	25	–	–
Extreme	10	10	–	–

*VOC and NO_x thresholds apply to ozone nonattainment areas.
**Sources located in an Ozone Transport Region have a VOC major source threshold = 50 tpy.

It is noted that some states have enacted more stringent thresholds under their state air regulations.

2. Title II. Mobile Sources

The provisions of this title relate to motor vehicles and do not affect stationary sources such as heat treat facilities. To curb air pollution from mobile sources, the CAA requires certain nonattainment areas to implement exhaust-testing programs, switch to alternative fuels, and establish car pooling ordinances.

3. Title III. Air Toxics

Toxic air pollutants (also called hazardous air pollutants—HAPS) are substances deemed particularly hazardous to human health and the environment. The CAA includes a list of 189 toxic air pollutants, and establishes a comprehensive plan for significantly reducing emissions of these materials. Under Title III, EPA must identify categories of sources that emit these pollutants, and establish discharge limits called National Emissions Standards for Hazardous Air Pollutants (NESHAP). An NESHAP has been enacted for cold and vapor degreasers that use solvents included on the air toxics list. Acceptable air pollution control technologies also will be stipulated under Title III. Materials used by heat treaters that are classified as air toxics include methanol, 1,1,1-trichloroethane, trichloroethylene, perchloroethylene, methylene chloride, and cyanide compounds. A major source of toxic air pollutants is defined as a stationary source that emits or has the potential to emit 10 tpy or more of a listed air toxic or 25 tpy or more of a combination of listed air toxics. Most heat treaters do not emit air toxics in significant amounts.

4. Title IV. Acid Rain

This provision of the CAA applies primarily to the utilities industry.

5. Title V. Permits

The CAA introduces an operating permits program modeled after EPA's national pollutant discharge elimination system (NPDES) program for wastewater discharge. Its purpose is to ensure compliance with all applicable requirements of the CAA and to enhance EPA's enforcement of the act. Facilities covered under the CAA (i.e., major sources) are required to obtain Title V operating permits for air pollution sources.

6. Title VI. Stratospheric Ozone Protection

The provisions under this title require the phase-out in production of substances that deplete the ozone layer, and the labeling of ozone-depleting materials. Some heat treaters may be affected by these provisions, since the degreaser 1,1,1-trichloroethane is a regulated substance. Other degreasers such as trichloroethylene, perchloroethylene and methylene chloride are not regulated under Title IV of the CAA.

7. Title VII. Enforcement

Under the CAA of 1990, environmental law enforcement has greater authority. The EPA is empowered to issue administrative penalty orders up to $25,000 per day per violation (with a maximum of $200,000) and in-field citations up to $5,000 for lesser infractions. Civil judicial penalties are enhanced, and criminal penalties for knowing/willful violations are upgraded from misdemeanors to felonies. In addition, air pollution sources must certify their compliance with the law. The EPA also has authority to issue administrative subpoenas for emissions data and compliance orders to correct violations.

C. Clean Air Act Requirements

The CAA requires facility owners and operators to perform the following:

1. Determine if their facility is located in an attainment or nonattainment area.
2. Assess all air emissions from the facility.
3. Determine if it is a major source of air pollution.
4. Determine if it emits any toxic air contaminants into the atmosphere.
5. Obtain air permits for their air pollution emission sources.
6. Install air pollution control equipment (where applicable) to reduce the amount of air contaminants emitted from their facility.

The CAA Amendments of 1990 have generated new regulations that significantly affect new and existing air pollution sources. Companies can anticipate many additional regulations to be enacted under the CAA in the coming years.

D. Accidental Release Prevention Program

The CAA of 1990 amendments also include an accident prevention provision. Under this provision, regulated companies have to assess the hazards or accident risks of certain hazardous chemicals used on-site, then develop risk management plans to prevent the accidental release of these materials. Three states—California, Delaware, and New Jersey—already have similar state laws in effect. The list of regulated substances include ammonia and propane. The requirements of this regulation are almost identical to OSHA's Process Safety Management regulation discussed in Section XI.

VI. EPA'S STORMWATER DISCHARGE PERMIT REGULATIONS

A. Scope

In 1977, Congress enacted the Clean Water Act to protect the surface waters of the United States. The act provided the EPA with the authority to regulate wastewater discharges into lakes, streams, or other surface water bodies. All such discharges were required to be permitted under the National Pollutant Discharge Elimination System (NPDES). The EPA considers stormwater discharges from industrial facilities to be a major contributor of pollution in the United States and has therefore issued regulations to control and reduce contaminated stormwater discharges.

The EPA's stormwater discharge permit regulations establish NPDES permit application requirements for stormwater discharges associated with industrial activity and discharges from municipal stormwater sewer systems. The stormwater discharge permit application regulations require companies in Standard Industrial Classification (SIC) Codes 1011–1499, 2011–4581 (except 3111), and 5171 to apply for an NPDES permit for their stormwater discharges. This includes the heat treatment industry (SIC Code 3398).

B. Definition of Stormwater

Stormwater discharge associated with industrial activity is defined as the discharge from any conveyance that is used for collecting and conveying stormwater and is directly related to manufacturing, processing, or materials storage at an industrial facility. The definition includes, but is not limited to, stormwater discharges from

Industrial plant yards
Immediate access roads and rail lines used or traveled by carriers of raw materials, manufactured products, waste materials, or by-products used or created by a facility
Material-handling sites
Refuse sites
Sites used for the application or disposal of process wastewaters
Sites used for the storage and maintenance of material-handling equipment
Sites used for residual treatment, storage, or disposal
Shipping and receiving areas
Manufacturing buildings
Storage areas (including tank farms) for raw materials and intermediate and finished products
Areas where industrial activity has taken place in the past and significant materials remain that are exposed to stormwater

The definition pertains to the areas listed above only if they are exposed to stormwater.

C. Stormwater Discharge Classification

The regulations separate stormwater discharges into five categories:

1. Direct discharge to surface waters
2. Discharge to a municipal separate stormwater sewer system (i.e., one that carries only stormwater)
3. Discharge to a privately owned separate stormwater system
4. Discharge to a publicly owned treatment works (POTW)
5. Discharge to a municipal combined sewer overflow system (i.e., one that carries stormwater and sanitary flows)

Facilities that discharge stormwater directly to surface waters (i.e., stream, lake, or river), to a municipal separate stormwater system, or to a privately owned separate stormwater system are required to obtain NPDES permits for their stormwater discharges. It is noted that companies that discharge stormwater to a municipal separate stormwater sewer system must also notify the local municipality about their stormwater discharge. This notification process identifies the facilities that use the municipal collection system and provides information on contaminant loadings.

D. Stormwater Outfalls

A stormwater outfall (i.e., discharge point) is the gathering point for stormwater runoff just before it leaves the site. This definition includes discharges from pipes, ditches, or storm drains. If a facility has multiple collection points (i.e., storm drains) on-site that collect and combine the stormwater discharges in a single pipe prior to leaving the site, and sampling can be conducted at the end of the pipe, then the facility needs to permit only the one (i.e., end-of-pipe) outfall. However, if the multiple collection points discharge individually from the site or discharge into an underground collection system that is inaccessible for end-of-pipe sampling, a separate permit would be required for each storm drain.

For the purpose of sampling, the regulations allow certain outfalls from similar drainage areas with the same discharge characteristics to be grouped together as "like discharges." An example would be drainage from a truck docking area with five stormwater drains. Only one of the five drains would require sampling, provided the characteristics of the stormwater discharges from all the drains were similar in nature. That is, you would be permitting the five drains but sampling only one. Areas such as an employee parking lot, where there are no industrial activities taking place, are not required to be included in the permit.

It is very important for facilities to carefully analyze their site to accurately assess the number of outfalls and "like discharges" from their property. This procedure can greatly reduce the complexity and cost of obtaining stormwater permits.

E. Types of Stormwater Permit Applications

Facilities may apply for stormwater permits in one of two ways:

1. Submit a Notice of Intent (NOI) form to be covered under a general stormwater permit
2. Submit an individual application to be covered under an individual stormwater permit

The general permit requires the least costly and easiest form of application, the notice of intent (NOI). The NOI requires only general site information and a site map. No sampling is required during the application process. However, the resultant general permit requirements may be more stringent than those stipulated by an individual permit because the latter is site- and industry-specific.

F. Types of Stormwater Permits

There are two types of NPDES permits for stormwater discharges that federal or state authorities will issue to industrial facilities. A general permit provides coverage to many facilities in a particular state, region, or industrial category. An individual permit is site-specific and is issued on a case-by-case basis to individual facilities. The two types of permits are discussed below.

1. Individual Permits

Individual permits are issued to individual facilities for stormwater discharges associated with industrial activity at specified locations. The permit is tailored to encompass the unique discharge characteristics of the facility and/or the special requirements of the receiving waters. A facility may be required to obtain or continue to operate under an individual permit

1. When denied coverage under a general permit
2. When an individual permit for wastewater and/or stormwater discharge has already been issued to the facility
3. When the permitting authority has determined that the facility is contributing to the violation of a water quality standard or is a significant contributor of pollutants to waters of the United States

Facilities with existing NPDES permits for stormwater and nonstormwater discharges must maintain compliance with the terms of those permits. However, a facility may apply for coverage under a new stormwater discharge permit 180 days prior to the expiration of the existing permit.

An individual permit to discharge stormwater associated with industrial activity will likely specify some form of compliance monitoring and sampling parameters specific for the individual site based on raw materials, materials handling practices, and processes used at the facility. Each site will also be required to prepare and implement a Stormwater Pollution Prevention Plan. There may be additional requirements for these plans to include provisions of other emergency plans such as spill prevention, control, and countermeasure (SPCC) plans or contingency plans.

2. General Permits

The EPA has developed two types of general permits. The "baseline" general permit is designed to cover most industrial facilities throughout each NPDES delegated state (i.e., states with authority to issue NPDES permits) and EPA region. Under the baseline general permit, all companies are subject to the same permit conditions, with some additional requirements for certain industries. Also, some NPDES states have added state-specific requirements.

The "multisector" general permit is designed to cover 29 industrial sectors as a result of the group application process. This general permit stipulates sector-specific (rather than industry-specific) permit requirements for the industries grouped in the individual sectors. The heat treatment industry is included in the Primary Metals sector. This multisector general permit has been finalized, and is being used by EPA in the 12 non-NPDES delegated states and territories where EPA is the permitting authority and by those NPDES delegated state that chooses to use it as a model for a state-issued permit.

Both of EPA's general permits require some form of compliance monitoring and the development and implementation of Stormwater Pollution Prevention Plans (SWP^{3}s). The SWP^{3}s in both cases must identify potential sources of stormwater pollution and describe engineering and management practices used to reduce and control those sources. The specific requirements contained in each general permit, however, are quite different.

G. Summary

The EPA's stormwater regulations establish permitting requirements for stormwater discharges from heat treatment facilities. Heat treaters that discharge stormwater runoff to surface waters or separate stormwater sewer systems must obtain a discharge permit from either their state regulatory agency or the EPA. Companies may either file a Notice of Intent (NOI) form to be covered under a general stormwater permit or submit an application to be covered under an individual stormwater permit.

Individual permits are site-specific and are issued on a case-by-case basis to individual facilities. The permit is tailored to encompass the unique discharge characteristics of the facility and/or the special requirements of the receiving stream. An individual permit to discharge stormwater will likely specify some form of compliance monitoring and sampling parameters specific for the individual site based on raw materials, materials handling practices, and processes used at the facility. Each site will also be required to prepare and implement a stormwater pollution prevention plan.

The EPA has developed two types of general permits. The *baseline general permit* is designed to provide broad coverage and does not specify industry-specific requirements. Under the baseline general permit, all companies are subject to the same permit conditions regardless of the materials and processes they use. The *multisector general permit* is designed to cover 29 industrial sectors and stipulates sector-specific permit requirements for the industries grouped in the individual sectors. The heat treatment industry is included in the primary metals sector. The multisector general permit has been finalized and is being used by EPA and some states as a model for issuing general permits to industrial sites.

Both of EPA's general permits require some form of compliance monitoring and the development and implementation of stormwater pollution prevention plans (SWP³s). The SWP³s in both cases must identify potential sources of stormwater pollution and describe engineering and management practices used to reduce and control those sources.

Heat treaters' stormwater permitting requirements will vary from state to state because some states have the authority to issue NPDES permits and others do not. NPDES states have the option to issue individual stormwater discharge permits, adopt one of EPA's model general permits, or draft their own general permit using EPA's general permits as models. Non-NPDES states will use EPA's multisector general permit for heat treaters because the EPA is the permitting authority in these states and heat treaters are covered under the primary metals sector. Heat treaters should contact the stormwater permitting authority in their state to determine their permitting options and specific compliance requirement.

VII. HAZARD COMMUNICATION STANDARD

A. Scope

On November 25, 1983, the Occupational Safety and Health Administration (OSHA) of the U.S. Department of Labor promulgated a new Hazard Communication Standard (HCS) (29 CFR 1910.1200). The purpose of this standard, commonly known as OSHA's Right-to-Know Law, is to "ensure that the hazards of all chemicals produced or imported by chemical manufacturers or importers are evaluated and that information concerning their hazards is transmitted to affected employers and employees." *Note*: The HCS covers any permanent, temporary, and office employees who could potentially be exposed to hazardous substances in the workplace. The standard preempts any state law pertaining to the same subject and applies to chemical manufacturers, importers, and distributors as well as to all employers (manufacturers and nonmanufacturers), including heat treaters.

The goal of the Hazard Communication Standard is to remove, as much as possible, the mystery surrounding the risks that are due to chemical hazards in the workplace. The HCS thus requires that chemical manufacturers, importers, and distributors and all employers institute risk management and safety programs. These programs are to inform employees of the hazards they work with and how they can minimize both the probability and severity of potential harm. Although the standard is aimed at enhancing safety, OSHA did not attempt to define an unacceptable level of risk. Rather, the intent is to provide employees with enough information to make their own safety judgments.

The standard applies to the following groups:

Chemical manufacturers and importers, who must determine the hazards of their products and prepare material safety data sheets and labeling in accordance with the HCS requirements.

Distributors of hazardous substances, who must transmit to employers pertinent information about hazardous chemicals.

Manufacturing and nonmanufacturing employers (including heat treaters), who must establish a hazard communication program to provide employees with information about chemical hazards to which they are exposed.

B. HCS Requirements

The standard requires (1) development of a written hazard communication program, (2) preparation of a hazardous substance list, (3) availability of material safety data sheets (MSDSs), (4) labeling of containers, and (5) training of employees. All these requirements are aimed at the general goal of providing employees with reliable information about the various materials hazards they may encounter on the job. The five required components of a hazard communication program per OSHA's HCS are described below.

1. Written Hazard Communication Program

The HCS requires employers to develop a written hazard communication program (HCP) for their workplace. The HCP must detail how the standard's stipulations will be met and must address labeling, MSDS files, employee training, hazards associated with nonroutine tasks, and procedures for informing nonemployees who come on-site about any chemical hazards present in the workplace. The entire written program must be available for inspection upon request by employees or their designated representatives or OSHA.

2. Hazardous Substance List

Employers must prepare a list of hazardous substances used in the workplace. This list must be made available to the employees. The hazardous substance list should include each chemical's scientific and/or trade name and any other name(s) commonly used in the workplace so employees will recognize it when reviewing the list. It is recommended that the hazardous substance list also indicate each chemical's MSDS number (assigned by the company). The list can then serve as both a listing of hazardous chemicals used in the workplace and an index sheet for the company's MSDS file. This helps maintain an organized file of MSDSs, which is extremely important during emergency situations when chemical hazard information must be retrieved quickly.

3. Material Safety Data Sheets

Manufacturers and importers of hazardous chemicals must obtain or prepare a Material Safety Data Sheet (MSDS) for each hazardous chemical they produce or import. An MSDS must contain certain product information specified by OSHA, and chemical manufacturers/importers must provide MSDS copies with the first shipment to companies that purchase their chemicals. If an MSDS is not provided, the shipment can be refused. Chemical manufacturers and importers are responsible for identifying any hazards presented by their chemicals. Users of hazardous chemicals, such as heat treaters, can rely on the hazards information presented on an MSDS supplied to them; they are not responsible for researching a chemical's hazards. Employers who use hazardous chemicals must obtain and ensure that MSDSs are readily available at all times to employees exposed to chemical hazards by maintaining a file of MSDSs on-site. It is recommended that a file of MSDS originals, those provided by the chemical suppliers, be kept in the office and that copies be strategically located in the plant to allow quick and easy access by employees. For instance, copies could be maintained in chemical storage and production areas.

The HCS stipulates that MSDSs must be in English and do not have to be provided in a foreign language. However, other OSHA regulations require companies to provide a safe workplace for their employees (and any other personnel on-site). Therefore, it is recommended that other safety provisions such as labelling and training be provided in a foreign language if non-English-fluent personnel are employed. Chemical suppliers are responsible for updating their MSDSs to ensure that the information presented is current and for sending revised MSDSs to their customers.

4. Labeling Provisions

Manufacturers, importers, distributors, and employers must ensure that all hazardous chemicals are labeled or otherwise identified to inform employees of exposure hazards.

Labeling information must include, as a minimum, the identity of the substance, hazard warning, and the name and address of a responsible party from whom additional information can be obtained, if needed. Drums and containers of hazardous substances delivered to the workplace should already be properly labeled. Employers should be primarily concerned with labeling any tanks, vats, sumps, or other containers of hazardous substances in the workplace. Employees must also be made aware of any piping systems in the workplace that carry hazardous substances. All containers, regardless of size, that contain a hazardous substance should be labeled.

The HCS requires labels to be in English. If a significant number of employees do not read or speak English well, it may be desirable to add a second label with the required information translated into the appropriate language. Some local regulations may even require this. Remember, OSHA requires employers to provide a safe workplace for all employees.

OSHA allows warning labels to be posted adjacent to a tank as long as the label is clearly visible. This may be desirable if placing a label directly on a tank is impractical or the label would become unreadable over a short period of time. Containers of raw materials designated as hazardous substances such as quench oil and containers of hazardous wastes such as spent solvents should also be labeled.

5. Employee Training

Employers must inform all employees who could potentially be exposed to hazardous chemicals about their rights under the HCS and where chemical hazards are present in the workplace. They must also inform employees about the company's written hazard communication program (HCP), hazardous substance list, and MSDS files—specifically that these documents are available for review and where they are located in the facility. Employers must further train employees concerning methods and observations used to detect releases of hazardous chemicals in the workplace, the physical and health hazards of any chemicals used by the company, safe chemical handling practices and protective actions available to employees, nonroutine tasks such as spill cleanup, and an explanation of methods by which warning notices (labels, MSDSs, signs, etc.) appear and are used in the workplace. Hazards must also be communicated to any outside contractor's employees who may be working on company grounds. It is noted that new (and temporary) employees must be trained before they begin working at the facility if they could be exposed to any hazards in the workplace. Office personnel should also be given training if they could potentially be exposed to any hazardous substances.

The federal HCS requires initial training at time of employment, before potential exposure to any hazardous chemicals and follow-up training if changes in the workplace (i.e., new chemicals used, worker transfers to different job in plant, etc.) occur that could cause employees to be exposed to a new hazard. The HCS does not require annual training, but it is recommended as part of the company's safety program.

Training sessions must be conducted in English and a foreign language if some workers who could potentially be exposed to hazardous chemicals in the workplace do not speak English. All training sessions should be documented concerning subjects covered, times, dates, instructors, and names of employees attending. It is recommended that employees sign an attendance sheet. Training information should also be maintained in each employee's personnel file.

C. Federal HCS Preemption of State Right-to-Know Laws

An important distinction regarding the Federal Hazard Communication Standard is that it is called a "standard" rather than a "rule" or a "regulation". As such, it has precedence, within the boundaries of the workplace, over any state-enacted right-to-know (RTK) laws. Therefore, companies must comply with the federal HCS rather than state laws in matters regarding employee safety on company grounds. An exception is those states that have OSHA-approved state programs. In these states, employers must comply with the state right-to-know law. States that currently have OSHA-approved job safety and health programs include Alaska, Arizona, California, Connecticut (state and municipal employees only), Hawaii, Indiana, Iowa, Kentucky, Maryland, Michigan, Minnesota, Nevada, New Mexico, New York (state and municipal employees only), North Carolina, Oregon, South Carolina, Tennessee, Utah, Vermont, Virginia, Washington, and Wyoming. Puerto Rico and the U.S. Virgin Islands also have these programs.

It is noted that although the federal HCS preempts state RTK laws in the workplace, the community aspects of state RTK laws are still valid and enforceable. The United States Court of Appeals for the Third circuit ruled that the "public rights" aspect of state laws, which is not covered by the OSHA Hazard Communication Standard, are not preempted by the federal HCS. As a result, employers may be required to comply with certain requirements of state RTK acts, such as the preparation of certain survey forms, to ensure public access to this information.

VIII. LOCKOUT/TAGOUT

A. Scope

On January 2, 1990, the Occupational Safety and Health Administration (OSHA) of the U.S. Department of Labor enacted OSHA Standard 1910.147, Control of Hazardous Energy Sources (Lockout/Tagout). The purpose of this standard is to prevent the unexpected energization, startup, or release of stored energy during equipment servicing and maintenance activities. The standard establishes requirements for the control of hazardous energy and applies to all employees engaged in normal equipment/machinery servicing and maintenance activities if

The employees must remove or bypass a guard or other safety device; or

The employee must place any part of his or her body into an area on a machine or piece of equipment where work is actually performed upon the material being processed (point of operation), or where an associated danger zone exists during a machine operating cycle.

Workplace servicing and maintenance activities that require the use of energy control devices include

Lubricating, cleaning, or unjamming equipment/machinery

Constructing, installing, or setting up equipment/machinery

Equipment/machinery inspections

Equipment modifications and repairs

Machine adjustments and tool changes

The lockout/tagout standard does not apply to normal production operations such as minor tool changes and adjustments if the procedures are routine, repetitive, and integral to the use of the equipment. Also excluded from the standard are any maintenance procedures performed on cord- or plug-connected electric equipment where the employee has exclusive control of the disconnected plug and hot tap operations involving pressurized transmission/distribution systems for substances such as gas, steam, water, or petroleum products. However, the employer must demonstrate that continuity of service is essential, shutdown of the system is impractical, and special equipment that provides effective employee protection is used.

B. Hazardous Energy Control Program Requirements

Standard 1910.147 requires employers to develop a Hazardous Energy Control Program (HECP) that establishes specific lockout/tagout procedures for isolating hazardous energy sources and otherwise disabling machinery and equipment to prevent unexpected energization, start-up, or release of stored energy during service or maintenance activities. Provisions of the standard require employers to

1. Develop a written energy control program.
2. Ensure that energy sources to new or overhauled equipment can be locked out.
3. Institute a tagout program when lockout is impracticable.
4. Prepare written lockout/tagout procedures for equipment and machinery.
5. Institute procedures for applying and removing lockout/tagout devices.
6. Provide and use standardized locks and tags.
7. Conduct annual audits of the company lockout/tagout program.
8. Conduct employee training.
9. Prepare specific safety procedures that address equipment servicing, outside contractors working at the site, multiple lockouts, and shift or personnel changes.

Facilities are required to institute a lockout system for all machinery and equipment unless the energy-isolating device is not capable of being locked out, in which case a tagout system must be utilized.

C. Hazardous Energy Sources

A hazardous energy source is defined as any source of electrical, mechanical, hydraulic, pneumatic, chemical, thermal, or other energy that could unexpectedly energize or start up a machine or piece of equipment during maintenance activities and cause serious injury to employees. Each source of hazardous energy associated with a machine or piece of equipment must be physically isolated (if applicable) prior to beginning any maintenance procedures. Controls that physically prevent the transmission or release of energy such as an electric circuit breaker, a manually operated switch, a line valve, a block, or a wedge may be used as hazardous energy–isolating devices (lockout/tagout devices).

D. Lockout/Tagout Devices

Owners and operators must provide all locks, tags, chains, wedges, key blocks, adapter pins, self-locking fasteners, and other hardware necessary for isolating, securing, or blocking machinery or equipment from hazardous energy sources. Lockout/tagout devices may not be used for any purpose except equipment lockout/tagout activities, and no other devices except those provided by the owner/operator may be used for controlling hazardous energy sources. All lockout/tagout devices must comply with the following criteria.

Durable Lockout/tagout devices must be capable of withstanding the environmental conditions (heat, cold, humidity, corrosiveness) to which the devices are exposed for the maximum period of time that exposure is expected. Tagout devices shall be printed and constructed so that exposure to wet, damp, or corrosive environments will not cause the tag to deteriorate or become illegible.

Standardized Lockout/tagout devices shall be standardized by either color, shape, or size throughout the facility. Print and format must be standardized for tagout devices.

Substantial Lockout devices shall be substantial enough to prevent removal without use of excessive force such as bolt cutters or other metal-cutting tools. Tagout devices, including their means of attachment, must be substantial enough to prevent inadvertent or accidental removal. Materials used to attach tagout devices must be nonreusable, attachable by hand, self-locking, and nonreleasable and must be at least equivalent to a one-piece, all-environment-tolerant nylon cable tie.

Identifiable Lockout/tagout devices must indicate the identity of the employee who applied them. Tagout devices must warn against hazardous conditions if the machine is energized and must include a legend such as "DO NOT START," "DO NOT OPEN," "DO NOT CLOSE," "DO NOT ENERGIZE," or "DO NOT OPERATE."

Examples of common lockout/tagout devices include keyed locks, tags, cord locks, multiple lock hasps, valve wheel locks, ball valve locks, chains, blocks, pins, and wedges. It is noted again that only materials specifically designated as lockout/tagout devices may be used for controlling hazardous energy sources.

IX. CONFINED SPACE ENTRY

A. Scope

On January 14, 1993, the Occupational Safety and Health Administration (OSHA) published its Final Rule concerning requirements for Confined Space Entry (CSE). The CSE standard applies to all of general industry, including heat treaters. This regulation took effect April 15, 1993, and requires employers to prepare a written program and implement a confined space entry permit system to protect employees from hazards associated with entry into and work within confined spaces. The regulatory requirements were developed from several CSE standards including those from numerous state agencies, the American National Standards Institute (ANSI), and the National Institute for Occupational Safety and Health (NIOSH). These standards define the practices and procedures necessary to protect employees from the potential hazards inherent in confined space entry.

B. Definition of a Confined Space

The OSHA standard defines a "confined space" as an enclosed area that is large enough that an employee can completely enter the space to perform an assigned task but has limited or restricted means for entrance and exit and is not designed for continuous employee occupancy. A "permit-required confined space" ("permit space") is defined as a confined space that exhibits one or more of the following characteristics:

1. Contains or has the potential to contain a hazardous atmosphere (toxic, flammable, asphyxiating).
2. Contains a material that may engulf the entrant.
3. Is configured such that an entrant could be trapped or asphyxiated by inwardly converging walls or has a downward sloping floor that tapers to a smaller cross-sectional area.
4. Contains any other recognized serious safety or health hazards.

If one or more permit spaces are present at a facility, the employer must develop a confined space entry permit system to restrict and regulate entry into those spaces.

A confined space that presents or has the potential to present an atmospheric hazard is considered a permit-required confined space. An atmospheric hazard is defined by OSHA as an atmospheric concentration of any substance that may expose employees to the risk of death, incapacitation, impairment of ability to self-rescue, injury, or acute illness. Atmospheric hazards include flammable gas, vapor, or mist in excess of 10% of its lower flammable limit (LFL), airborne combustible dust in concentrations exceeding its LFL, oxygen concentrations below 19.5% or above 23.5%, concentrations of substances in excess of their published acceptable doses or permissible exposure limits (PELs), and any other atmospheric condition that is immediately dangerous to life or health (IDLH).

C. Types of Confined Spaces

Confined spaces can be generally categorized as areas with open tops that restrict the natural movement of air or enclosed spaces with a limited number of openings for entry or exit. In both cases, the space may contain mechanical equipment with moving parts that compounds the hazards present within the confined space. Furnaces, quench tanks, degreasers, and the access pits around these vessels may be classified as confined spaces. Gases that are heavier than air (propane, carbon dioxide, argon) can collect at the bottom of these spaces. Even open-topped water tanks that appear harmless may develop toxic atmospheres such as hydrogen sulfide from the vaporization of contaminated water. Gases heavier than air are a primary concern where entry into a confined space is being planned. Other hazards may develop due to the work being performed within a confined space. For instance, welding can generate toxic gases or use up the available oxygen and change the status of a confined space from safe to hazardous. Confined spaces such as sewers, manholes, storage tanks, baghouses, and furnaces usually have limited access. The problems arising in these areas are similar to the hazards that occur in open-topped, confined spaces; however, the limited access increases the risk of injury. Gases that are heavier than air, such as argon and propane, may lie in a confined space for hours or even days after the area has been opened. Because some gases are odorless, the hazard may be overlooked, with fatal results. Gases that are lighter than air, such as natural gas, may also be trapped within an enclosed type of confined space, especially one with access from the bottom or side.

Under certain conditions, excavations, trenches, or natural depressions may be considered confined spaces because they trap vapor and restrict the flow of oxygen. Activities associated with storage tanks that may involve confined space entry include tank lining, trench digging for access to piping associated with tanks, excavating or repairing beneath tanks, and entering manways to access instrumentation.

Hazards specific to a confined space are dictated by

1. The material stored or used in the confined space
2. The activity carried out, such as heating using gases that are toxic or flammable or displace oxygen
3. The presence of moving parts
4. The internal configuration, which may hamper entry or exiting

The most hazardous form of confined space is one that combines limited access and mechanical devices. All the hazards associated with open-top and limited access confined spaces may be present together with the additional hazard of moving parts. Vessels that contain power-driven equipment, unless properly isolated, may be inadvertently activated after entry. Such equipment may also contain physical hazards that further complicate the work environment and the entry and exit process.

Employers are required to review their facilities to determine if the workplace contains *permit-required confined spaces*, and if so the employer must establish as CSE permit program.

D. Confined Space Entry Requirements

The hub of OSHA's rule is a required entry permit system that applies to "permit-required confined spaces" (permit spaces). Employers must evaluate their workplace to identify all confined spaces that meet OSHA's definition for permit space. If one or more permit spaces are on-site, an employer must comply with remaining provisions of the confined space rule, including preparation of a written confined space entry program, prior to entry of any permit space by an employee. The rule requires employers to develop a permit system and written program that include atmospheric testing before allowing workers to enter potentially hazardous spaces.

Employers that do not allow employees to enter permit spaces must still take action to prevent unauthorized entry—for example, by introducing barriers supplemented by signs or informing employees of the existence and location of permit spaces. Also, employers must inform contractors hired to perform entry operations about permit space hazards.

Requirements of OHSA's Confined Space Entry Rule include the following.

1. Identification of all permit confined spaces in the workplace
2. Preparation of a written confined space entry program that describes company confined space entry procedures
3. Use of a confined space entry permit (i.e., checklist)
4. Use of proper safety and atmospheric testing equipment
5. Employee training
6. Atmospheric testing
7. Isolation of confined spaces prior to entry
8. Notification of outside contractors and emergency rescue services about permit confined spaces present in the workplace

E. Alternative Entry Procedures

OSHA's CSE regulation provides for alternative entry procedures if the only hazard presented by the permit space is an actual or potentially hazardous atmosphere that can be controlled by continuous forced air ventilation and the work to be performed introduces no additional hazards. For these procedures to apply, an employer must first develop supportive monitoring and inspection data and demonstrate that forced air ventilation alone is sufficient to maintain the permit space for entry. OSHA suggests using as a safe entry guideline 50% of the permissible exposure level (PEL) for flammable or toxic substances in determining whether an atmosphere is safe after ventilation.

The alternative entry procedures do not require compliance with many of the permit space provisions including the attendant requirement. However, an employer still must train entry personnel, test the permit space atmosphere prior to entry, use forced air ventilation until all entrants have left the space, periodically test the atmosphere within the space to ensure that ventilation is preventing atmospheric hazards, and certify in writing (available to entrants) prior to entry that required measures have been taken.

X. IN-PLANT AIR QUALITY

A. Scope

To protect employees from short-term and long-term exposure to hazardous materials, the Occupational Safety and Health Administration (OSHA) promulgated Permissible Exposure Limits (PELs) for many substances found in the workplace. These PELs were initially adopted from the 1968 threshold limit values (TLVs) established by the American Conference of Governmental Industrial Hygienists (ACGIH).

Subsequently, however, industrial experience, new developments in technology, and scientific data all indicated that the threshold limits initially adopted were inadequate and obsolete. Also, newly developed toxic materials frequently used by many industries were not regulated by the adopted limits, and many technical, professional, industrial, and governmental organizations recommended lowering the allowable exposure limits. In addition, these agencies identified many new or existing substances for which allowable exposure limits were needed. In some cases, large industrial facilities supplemented the OSHA PELs with their own internal guidelines.

For these reasons, in 1989 OSHA amended the Air Contaminants Standard (29 CFR 1910.1000). The amended standard lowered the PELs for 212 substances, established new PELs for 164 hazardous substances, and retained certain PELs as they currently existed. However, the courts ruled that EPA had followed improper procedures in establishing the amended PELs for in-plant air contaminants and allowable PELs reverted back to the preamendment levels. Many states with OSHA-approved programs continue to enforce the 1989 levels.

B. In-Plant Air Quality Requirements

The in-plant air quality standards require employers to ensure that employees are not exposed to chemical concentrations above the "time-weighted average" (TWA), the "short-term exposure limit" (STEL), or the "ceiling limit" for a given substance. The definitions for these terms are provided below.

Time-weighted average (TWA): An employee's allowable average airborne contaminant exposure limit during an 8 h workshift of a 40 h work week. The TWA represents the average contaminant concentration that cannot be exceeded during an 8 h workshift.

Short-term exposure limit (STEL): An employee's 15 min time-weighted average exposure limit that shall not be exceeded at any time during a workday. The STEL represents the average contaminant concentration that cannot be exceeded during any 15 min interval over an 8 h workday.

Ceiling limit: An employee's maximum exposure limit that may not be exceeded at any time during the workday.

Employers are required to provide all engineering controls, equipment, work practices, and personal protective equipment necessary to comply with the permissible exposure limits.

XI. PROCESS SAFETY MANAGEMENT STANDARD

A. Scope

On February 24, 1992, OSHA published the final version of its process safety management standard covering companies using chemicals designated by OSHA as highly hazardous. The standard is entitled "Process Safety Management of Highly Hazardous Chemicals" and seeks to make accident prevention as important to plant managers and workers as maximizing production.

This standard affects facilities handling any of the 137 highly hazardous chemicals (HHCs) listed in the standard above their respective threshold quantities (TQs). *Anhydrous ammonia is on the list, with a TQ of 10,000 lb.* In addition, facilities handling 10,000 lb or more of flammable liquids or gases also must comply with the process safety management standard in certain circumstances. Natural gas or propane used to heat a furnace containing ammonia would be regulated by the standard if more than 10,000 lb of the gas were on site. Compliance requirements are triggered by the presence of a listed chemical at or above its threshold quantity in a single process at one point in time. The rule's definition of "process" reads: "any activity involving a highly hazardous chemical including any use, storage, manufacturing, handling, or the on-site movement of such chemicals." Any group of interconnected or separate vessels, which are located such that a highly hazardous chemical could be involved in a release, is considered a single process.

Storage or use of a listed chemical at various sites within a facility may not trigger compliance if the volume at each site is below the chemical's threshold quantity. OSHA reasoned that the potential hazard of a catastrophic release exists when the highly hazardous chemical is concentrated in a process, not scattered throughout an entire facility. For example, if a dike around a liquid storage vessel fully contains a released material and prevents it from interacting with another vessel outside the dike and neither vessel by itself contains a TQ of a listed chemical, the vessels would not be considered part of the same process and would not be covered by the process safety management standard.

Processes using a highly hazardous chemical listed in the standard but not covered by the rule (i.e., exempted) include

1. Processes using hydrocarbon fuels used solely for workplace consumption as a fuel (i.e., comfort heating, vehicle refueling) and not part of a process containing another listed chemical
2. Storage of flammable liquids in atmospheric tanks below their normal boiling points without chilling or refrigeration
3. Operations and processes at retail facilities
4. Processes at oil or gas well drilling or servicing operations
5. Operations and processes occurring at normally unoccupied remote facilities where not employees are permanently stationed

The standard does not include exemptions for small businesses. If a process (including just storing or handling the chemical on-site) exceeds a specified threshold, then the standard applies to that activity. Each plant process that exceeds the threshold falls under the standard's provisions.

B. Process Safety Management Requirements

The OSHA process safety management standard is designed to guide facilities in the safe management of processes that use highly hazardous chemicals. The process safety management standard is a performance standard, which means that it sets performance requirements for management programs rather than prescribing specific methods that must be used to control the hazardous chemicals. The requirements are intended to prevent or reduce the risk of major industrial incidents that might expose employees to the hazards of catastrophic release of extremely toxic, flammable, reactive, or explosive chemicals.

Regulated facilities must study each process that uses a highly hazardous chemical above its TQ, analyze associated hazards, develop written operating procedures, and provide periodic training to employees working with the process. Once any process hazards are identified, appropriate risk reduction measures must be implemented. Compliance with the process safety management standard also requires broad employee participation. The major provisions of OSHA's process safety management standard include

Employee participation
Pre-start-up safety review
Process safety information
Mechanical integrity
Process hazard analysis
Hot work permits
Risk reduction
Management of change
Operating procedures
Incident investigation
Employee training
Emergency planning and response
Contractors safety
Compliance audits

These provisions of the standard are described in the paragraphs below.

1. *Employee Participation*

To mitigate chemical releases of these highly hazardous chemicals and the hazards they present, employers need to develop the necessary expertise, experience, judgment, and proactive initiative within their workforce to properly implement and maintain an effective process safety management program. Employee participation is an important element of OSHA's standard. Employers must determine appropriate methods of participation and write an Employee Participation Plan specifying how employees and their representatives will be given access to required information and documentation including process hazard analyses. Employers must consult with employees and their representatives on the conduct and development of process hazard analyses, process safety information, and other elements of the standard.

2. *Process Safety Information*

Companies must compile information concerning the hazards and characteristics of chemicals, technology, and equipment used in processes covered by the standard. Employers may use material safety data sheets (MSDSs) to comply with the requirements for information on the hazards of chemicals involved in a process to the extent that they contain the information specified in the standard.

Information pertaining to hazardous chemical hazards must cover toxicity, permissible exposure limits (PELs), physical data, reactivity, corrosivity, thermal and chemical stability, and the hazardous effects of inadvertent mixing of different materials. Information relative to process technology must include

1. A block flow diagram or simplified process flow diagram
2. Process chemistry
3. Maximum intended inventory
4. Safe upper and lower process operating limits, such as temperature, pressure, flow, and composition
5. Evaluation of the consequences of deviations including those affecting worker health and safety

If such technical information does not exist, it may be developed in conjunction with the process hazard analysis.

Information must also be developed to ensure that equipment being used in regulated processes is appropriate and meets applicable standards and codes.

3. *Process Hazard Analysis*

A process hazard analysis is an organized and systematic effort to identify and analyze the significance of potential hazards associated with the processing or handling of highly hazardous chemicals. Such an analysis provides information to assist employers and employees in making decisions for improving safety and reducing the consequences of unplanned releases of hazardous chemicals. A hazard analysis is directed toward analyzing potential causes and consequences of fires, explosions, release of toxic or flammable chemicals, and major spills of hazardous chemicals. The hazard analysis focuses on equipment, instrumentation, utilities, human actions (routine and nonroutine), and external factors that might impact the process. These considerations assist in determining the hazards and potential failure points or failure modes in a process.

Companies must analyze the significance of potential hazards associated with processing and handling hazardous chemicals on-site. The analyses must be performed for each process, documented, updated every 5 yr, and kept on file for the life of the process. As a general rule, hazard analyses cannot be initiated until corresponding safety information has been developed.

A process hazard analysis must

1. Identify process hazards
2. Identify previous incidents with potential for catastrophic consequences
3. Discuss engineering and administrative controls applicable to hazards and their interrelationships
4. Discuss consequences of failures in engineering and administrative controls
5. Describe the facility and its location
6. Discuss such human factors as the potential for error
7. Provide a qualitative evaluation of the range of possible safety and health effects on employees in the event of control failures

When process hazard analyses identify potential worker safety problems, steps to correct or resolve them must be implements. *Note*: The process safety management standard allows a trade association that has members with similar regulated processes to develop and use a generic process hazard analysis procedure that can be used by each member.

4. Risk Reduction Measures

Employers must develop a system to ensure that risk reduction measures recommended as a result of the hazard analysis are addressed in a timely manner and that all actions taken to remediate a potential problem area are communicated to those affected.

5. Operating Procedures

Owners must develop and implement written operating procedures that provide clear instructions for safely conducting activities involved in each process consistent with the process safety information. These written procedures must address steps for each operating phase, operating limits, safety and health considerations, and safety systems and their functions.

6. Employee Training

The rule establishes a variety of training requirements covering process employees, contractors, management of changes, emergency response, and process maintenance. All employees, including maintenance and contractor employees, involved with highly hazardous chemicals need to fully understand the safety and health hazards presented by those chemicals and processes they work with for the protection of themselves, their fellow employees, and the residents of the surrounding community.

7. Contractors

Employers must inform each contract employer of potential hazards related to the contractor's work and applicable provisions of the plant safety rules. The contractor is then responsible for training his employees in work practices necessary to perform their work safely. It also is the contractor's responsibility to verify that each employee understands the training and documents that training was conducted.

8. Pre-Start-Up Safety Review

Companies must perform a pre-start-up review for new facilities and those that have been modified to determine if any process safety information has changed. The review is designed to make sure that certain important considerations have been addressed before any highly hazardous chemical is introduced into a process.

9. Mechanical Integrity

The process safety management standard requires companies to test and maintain the integrity of critical process equipment to ensure that listed chemicals are contained within the process and not released in an uncontrolled manner. Employees who maintain the ongoing integrity of process equipment also must be trained. Such training must include a process overview and cover associated hazards and safe operating procedures.

Specifically, the mechanical integrity provisions of the standard require written procedures, training for process maintenance employees, and inspection and testing for process equipment including pressure vessels and storage tanks, piping systems, relief and vent systems, emergency shutdown systems, pumps, and controls such as monitoring devices, sensors, alarms, and interlocks. Correction of equipment deficiencies and assurance that new equipment, maintenance materials, and spare parts are suitable for the process and that new equipment is properly installed are also required. Records must be kept on all inspections and tests.

10. Hot Work Permits

Companies must issue a permit for all hot work operations conducted on or near a covered process. Hot work includes activities such as electric or gas welding, cutting, or brazing. The purpose of this provision is to ensure that employers are aware of hot work being performed and that appropriate safety precautions are taken.

11. Management of Change

Companies must develop written procedures to manage changes to process chemicals, technology, equipment, procedures, and other changes to facilities that affect the covered process. Employees must be informed about the changes and the issues that must be addressed before a change is implemented. Written procedures do not apply to replacements in kind, which are replacements satisfying a design specification.

12. Incident Investigation

Companies must investigate every incident that results in or easily could have resulted in a catastrophic release of highly hazardous chemicals in the workplace. Investigations must be initiated within 48 h after the incident occurs. The standard stipulates that an investigative team must be used, including at least one person knowledgeable in the process (a contractor employee, if appropriate). A written incident report must also be developed. Employers must address and document the response to report findings and recommendations and review findings with affected employees and contractor employees. Reports must be retained for 5 yr.

13. Emergency Planning and Response

Employers must prepare a written emergency action plan and train employees in emergency response procedures. The written plan must address what actions employees are to take when there is an unwanted release of highly hazardous chemicals. Employers must review the plan with each employee when it is developed, whenever an employee's responsibility or designated actions under the plan changes, and whenever the plan is changed.

Employers must, at a minimum, have an emergency action plan that will facilitate the prompt evacuation of employees due to an unwanted release of a highly hazardous chemical.

14. Compliance Audits

Processes covered by the standard must be audited every 3 yr to verify compliance with the standard's various requirements. Employers must retain the two most recent audit reports on-site. The audit must be conducted by at least one person knowledgeable in the process, and the employer must respond to the audit findings. Employers must select a trained individual or assemble a trained team of people to audit their process safety management program. A small plant may need only one knowledgeable person to conduct an audit.

Following an audit, employers must determine and document appropriate responses to each finding. They also must document that all deficiencies have been corrected.

C. Compliance Deadlines

Most of the major provisions of OSHA's process safety management standard were to be completed by August 1992. Process safety information and process hazard analyses were to be completed in 25% increments over a 4 yr period; with the first compliance deadline occurring in May 1994. The first compliance audit was to be completed by May 1995 and repeated every 3 yr thereafter.

D. Minimizing Obligations

Companies can take certain self-help steps to minimize their compliance obligations. These steps include inventory reduction, process modification, and chemical substitution. By reducing inventories of any highly hazardous chemicals on-site to below their respective threshold quantities, a company can avoid compliance requirements. In addition, a company may disperse its inventory of listed chemicals to several locations on-site and avoid compliance requirements as long as no single location's inventory exceeds listed thresholds.

A facility can also make minor physical modifications to redefine the boundaries of a process and thereby avoid triggering compliance requirements. For example, a process linked to adjacent liquid storage tanks each containing half the threshold quantity of a liquid chemical would trigger compliance requirements. However, by constructing a dike around each tank, the facility effectively could cut the process chemical inventory in half and avoid triggering compliance requirements. To qualify as an acceptable barrier, OSHA requires the area within the dike to be capable of fully containing released storage tank material and preventing it from interacting with another vessel or tank outside the dike.

Companies also may reduce compliance requirements by using nonlisted substitutes for any listed chemicals covered by the standard.

XII. RESPIRATORY PROTECTION

A. Scope

To control occupational hazards associated with breathing air contaminated with harmful dusts, fogs, fumes, gases, mists, smokes, sprays, or vapors, OSHA requires employers to provide their employees with the proper respiratory equipment. Respiratory protective equipment must be provided, used, and maintained in a sanitary and reliable condition in all areas where hazards capable of causing injury or impairment of the body through inhalation of air contaminants exist.

The regulations define a respirator as a piece of equipment that removes contaminants from the air or provides purified air for breathing in order to protect the respiratory system. Respirators used either to remove airborne contaminants or to provide clean air in hazardous atmospheres are required in all areas where hazardous chemical gases, fumes, particulates, or vapors are present. Respirators may also be required for confined space entry (CSE) tasks. If a company provides respirators as part of its CSE program, for emergency response situations or for any other reason, then a Respiratory Protection Program must also be implemented even if in-plant air quality meets all PEL allowable limits.

B. Respiratory Protection Program

The OSHA regulations require employers to develop a respiratory protection program to protect their employees from any respiratory hazards (i.e., when contaminant exposure levels exceed allowable limits) present in the workplace. The use of respiratory protection equipment for confined space entry activities would also trigger the requirement for a respiratory protection program. The regulations require companies that provide respiratory protection equipment to perform the following tasks.

1. Develop a written program governing the operating procedures, use, storage, maintenance, cleaning, and selection of respirators.
2. Monitor work area conditions and evaluate the degree of employee exposure and stress. Worker stress and exposure are to be minimized.
3. Evaluate written operating procedures to determine the effectiveness of the program. Changes should be made as necessary to improve the program.
4. Provide the proper respiratory equipment selected according to the specific job and specific hazard encountered at the worksite.
5. Provide periodic medical check-ups (e.g., annually) and employee training in the proper use of respirators. Proper fitting and instruction in the care and maintenance of the respirator must also be provided.

6. Inspect respirators regularly for broken parts and general cleanliness. Non-emergency use respirators are to be inspected before and after each use. Emergency use respirators must be inspected monthly and after each use, and/or inspection of their maintenance records must be updated.

7. Develop procedures for the proper storage of air respirators. Respirators must be cleaned, disinfected, and put in storage after each use.

OSHA regulations require employers to conduct workplace hazard analyses to determine if employees should be provided personal protective equipment (such as a respirator).

XIII. EMERGENCY ACTION/RESPONSE PLANS

A. Scope

To protect employees during emergency situations such as fires and accidental chemical releases, OSHA requires employers to develop and implement emergency action/response plans. The emergency action/response plans must address each type of emergency situation that could potentially occur at the facility, including

Fires/explosions
Hazardous chemical releases
Accidents (vehicle or equipment)
Equipment failure
Floods, hurricanes, tornados, earthquakes
Others

In the heat treatment plant, potential emergency situations include chemical leaks, release of chemical vapors, oil fires, chemical spills, and gas explosions, among others. All emergency action/response plans must be in writing, and each employee must be informed of the proper emergency action/response implementation procedures.

B. Emergency Action Plans

Employers are required to develop an emergency action plan that covers the designated actions that employers and employees must implement to ensure employee safety from fire and other emergency situations. The emergency action plan must be in writing except at facilities with 10 or fewer employees, in which case the procedures can be communicated orally. The emergency action plan must contain, at a minimum, the following information:

Emergency evacuation procedures and escape route assignments
Procedures to be followed by employees who remain in the facility to operate critical plant operations during emergency situations
Procedures to account for all employees after emergency evacuation has been completed
Rescue and medical duties for those employees required to perform such tasks
A means of reporting fires and other emergencies
The names and job titles of persons or departments who are to be contacted for further information

Employers must also provide alarm systems to alert employees about emergency situations and training in company emergency action procedures.

C. Fire Prevention Plans

Employers are required to develop a fire prevention plan that covers the appropriate actions that are to be conducted at the facility to limit the possibility of fire. As with the emergency action plan, the fire prevention plan must be in writing unless the facility employs 10 or fewer employees, in which case the plan can be communicated orally. Key aspects that must be incorporated into the first prevention plan include

1. A list of the major workplace fire hazards and their proper handling and storage procedures
2. A list of potential ignition sources (i.e., welding, equipment, burners, etc.) and their control procedures
3. The type of fire protection equipment or systems used to control fires within the facility
4. Names or job titles of those personnel responsible for maintaining equipment and systems used to prevent or control ignitions and fires
5. Names or job titles of those personnel responsible for the control of fuel source hazards

Housekeeping procedures must also be established to control and prevent the accumulation of flammable and combustible waste materials and residues so that these materials do not contribute to a fire emergency. Heat-producing equipment capable of causing accidental ignition of combustible materials should be equipped with monitoring devices and/or controls to prevent such accidents from occurring.

D. Chemical Release Response Plans

Employers are required to develop and implement an emergency response plan if the potential exists for the release of hazardous substances in the workplace. The emergency response program must address each type of emergency situation that may potentially occur at the facility and contain, at a minimum, the following information:

Pre-emergency planning and coordination with outside parties. The emergency response plan must be compatible and integrated with the disaster, fire, and/or emergency response plans of local, state, and federal agencies.

Personnel roles during emergencies, lines of authority and communications, and procedures for reporting incidents to local, state, and federal agencies.

Emergency recognition and prevention procedures, including a means to notify employees during an emergency situation.

Emergency evacuation procedures including emergency exit routes, safe distances, and regrouping areas.

Site security systems and a means to control access to the site.

Emergency medical treatment and first air procedures.

Personal protective equipment (PPE) and emergency response equipment available on-site.

Procedures for decontaminating the hazard zone, personal protective equipment, and other equipment used to control and clean up the emergency area.

Site topography map and facility layout drawing showing chemical/gas storage areas, emergency exits, emergency response equipment, prevailing wind direction, and other information pertinent to emergency response procedures.

Based on the information available at the time of an emergency, the employer must evaluate the incident and the facility's response capabilities and proceed with the appropriate steps to implement the site emergency response plan. If the facility is not capable of controlling the incident, outside emergency response agencies must be notified immediately.

The emergency response plan must be reviewed annually and updated with new or changing site conditions or information. The emergency response plan must be available on-site for review by emergency response agencies and facility employees.

Emergency response personnel must receive training prior to responding to an emergency situation. The training must include a review of the company's emergency response plan, the facility's standard operating procedures, personal protective equipment necessary for emergency response, and procedures for handling emergency incidents. In addition, the emergency response plan must be rehearsed regularly (e.g., semiannually) as part of the company's training program.

XIV. BLOODBORNE PATHOGENS STANDARD

A. Scope

On December 6, 1991, the Occupational Safety and Health Administration (OSHA) promulgated its Occupation Exposure to Bloodborne Pathogens Standard (BPS). The objective of this standard is to eliminate or minimize occupational exposure to hepatitis B virus (HBV), human immunodeficiency virus (HIV), and other bloodborne pathogens, which are microorganisms carrying bloodborne disease.

The BPS, which became effective March 6, 1992, was initially aimed at healthcare facilities where blood is routinely encountered. However, its scope was written so that it applies to virtually all industrial facilities. It is estimated that over half a million business establishments will be required to achieve compliance with this standard.

Although healthcare facilities are the most obvious workplaces affected, the BPS was written to apply to all situations in which an employee has potential occupational exposure to blood or other potentially infectious materials. It is not the type of facility nor the particular occupation that triggers the applicability of the standard, but the employee's occupational activity that involves a risk of exposure to bloodborne pathogens.

"Occupational exposure," as defined in the standard, means reasonably anticipated exposure that may result from the performance of an employee's duties. OSHA's position is that for exposure to be reasonably anticipated and result from performance of an employee's duties, the individual's job description must make some mention of specific acts that would bring that person into contact, or potential contact, with contaminated or infectious materials.

In addition, OSHA has stated that Good Samaritan acts, such as helping a coworker with a minor cut or injury, are exempt. Coverage under OSHA's Bloodborne Pathogens Standard is therefore based on whether any employee has been specifically trained and designated by an employer through a written job description s a first air provider. For example, if an employee is trained in first aid and designated by the employer as responsible for providing medical assistance as part of his or her job duties, that employee is covered by the standard.

However, a key criterion to determining coverage is that according to OSHA Regulation 1910.151:

> In the absence of an infirmary, clinic or hospital in near proximity to the workplace that is used for the treatment of all injured employees, a person or persons shall be adequately trained to render first aid

This means that a company is required to have on-site someone trained in first aid if its facility is not in close proximity (i.e., within 3–5 min) to a medical facility. Therefore, some heat treaters may be required to have such a trained person on-site, who would consequently be covered by the Bloodborne Pathogens Standard.

Even though OSHA's position is that Good Samaritan acts are not covered by the standard if any employee happens to contract a serious disease while providing minor first aid during an emergency response activity, an employer could be in a poor liability position if the infected employee decided to pursue legal action against the company.

In view of today's very liberal court decisions and associated astronomical awards, employers must do all they can to protect themselves. It is advisable that all heat treaters, even if they are within close proximity to a clinic or hospital, provide personal protection equipment that will allow personnel to take universal precautions to protect themselves in the event an accident requiring even minor first aid occurs at the site. As a minimum, a company should provide medical gloves with all first aid kits at its facility.

B. BPS Requirements

The following is a list of the major requirements of OSHA's Bloodborne Pathogens Standard (BPS).

1. Implement universal precautions.   Treat all human blood and body fluids as if known to be infectious for HIV, HBV, or other bloodborne pathogens.

2. Prepare written exposure control plan.   Identify employees covered by the standard, specific measures taken to minimize risk of exposure, and procedures to follow if an exposure incident occurs.

3. Provide employee training.   Provide annual training to affected employees concerning the provisions of the company's exposure control plan and appropriate safety information required to protect themselves from exposure.

4. Keep adequate records.   Establish and maintain medical and training records for each affected employee, and make them available to OSHA upon request.

5. Institute engineering and work practice controls.   Institute controls that isolate or remove the bloodborne pathogens hazard from the workplace, and reduce the likelihood of exposure by altering the manner in which a task if performed.

6. Supply personal protective equipment.   Provide appropriate personal protective equipment such as gloves, gowns, eye protection, face shields, and mouthpieces that prevent employee exposure.

7. Practice good housekeeping.   Ensure that the workplace is maintained in a clean and sanitary condition, decontamination procedures are followed when appropriate, and contaminated wastes are disposed of properly.

8. Follow a hepatitis B vaccination program.   Make the hepatitis B vaccine and vaccination series available to affected employees as well as the postexposure evaluation and follow-up report.

9. Use required labels and signs.   Affix *Biohazard* warning labels to containers of potentially infectious material and regulated wastes.

All of these requirements are aimed at training employees to recognize the risks of bloodborne diseases and minimizing the risks to employees who may encounter these pathogens on the job. *Note*: Only those heat treaters who designate one or more employees as first aid responders as part of their job description must comply with OSHA's Bloodborne Pathogen Standard. The proximity of the heat treatment plant to a medical facility dictates whether the regulatory requirement to make such an assignment is applicable.

XV. GENERAL SAFETY REGULATIONS

A. Scope

In 1970, Congress enacted the Occupational Safety and Health Act. This act requires all employers to provide each employee with a workplace free from recognized hazards that cause or are likely to cause death or serious physical harm. The purpose of the act is to encourage employers to develop and implement specific safety programs appropriate to their operation and to comply with all OSHA safety standards.

B. Safety Regulations

The general safety provisions of OSHA cover the posting of notices; recordkeeping; emergency reporting; walking and working surfaces; exits; machine guards; portable hand and power tools; flammable and combustible liquids; fire extinguishers; materials handling and storage; forklifts; welding, cutting, and brazing; noise exposure; personal protective equipment; electrical wiring and equipment installation; compressed gas cylinders; cranes; and chains. States with "approved state plans" may have additional regulations that apply to a heat treatment facility.

XVI. ENVIRONMENTAL AND SAFETY REGULATIONS SUMMARY

A listing of the major EPA and OSHA regulations that affect heat treaters is presented in Table 1. Summaries of these regulations' requirements concerning written programs, training and equipment is presented in Table 2.

Table 1   Environmental/Safety Regulations Affecting the Heat Treating Industry

- EPA's Resource Conservation and Recovery Act (RCRA)—Hazardous Waste
- EPA's SARA Title III—Hazardous Chemicals
- EPA's Underground Storage Tank (UST) Program
- EPA's Aboveground Storage Tank (AST) Program
- EPA's Clean Water Act (CWA)—Wastewater
- EPA's Clean Air Act (CAA)—Air Pollution
- EPA's Clean Air Act—Accident Prevention Program (APP)
- EPA's Stormwater Discharge Regulations (SW)
- EPA's Spill Prevention Control and Countermeasure Plan (SPCC)—Oil Storage
- EPA's Used Oil Management Regulations (UOM)
- EPA's Vapor Degreaser Regulation
- OSHA's Hazard Communication Standard (HCS, Right-To-Know)
- OSHA's In-Plant Air Quality Standards
- OSHA's Lockout/Tagout Regulations (LO/TO)
- OSHA's Confined Space Entry Regulations (CSE)
- OSHA's HAZWOPER Regulations
- OSHA's Lab Safety Plan (LS)
- OSHA's Respirator Maintenance Program (RMP)
- OSHA's Process Safety Management Regulations (PSM)
- OSHA's Bloodborne Pathogens Standard (BBP)
- OSHA's Emergency Response Plan (ERP)
- OSHA's Personal Protection Equipment (PPE) Standard
- OSHA's General Safety Regulations

Table 2 Major EPA Regulations Compliance Requirements Summary

EPA Regulation	Requirement*
1) Hazardous Waste (RCRA)	Contingency plan
	Training
	Safety equipment
	Labeling
2) SARA Title III ·	Hazardous substance list
	Tier II
	Form R
3) Clean Water Act	Permits
	Discharge reports
4) Underground Storage Tanks	Tank registration
	Leak detection
	Spill/overflow control
	Recordkeeping
	Financial responsibility
5) Clean Air Act (CAA)	Emissions estimates
	Permits
	Emissions controls
6) CAA Accident Prevention	Written program
	Hazard analysis procedures
	Training program
7) Stormwater (SW)	Sampling procedures
	Permitting
	SW pollution prevention plan
8) Spill Prevention/Control (SPCC)	Written plan
	Training
9) Used Oil Management	Preparation/prevention procedures
	Contingency plan
	Labeling
	Recordkeeping
	Biennial report
10) Vapor Degreasers	Register degreaser with EPA/State
	Upgrade degreaser with required modifications
	Comply with emissions limitations

OSHA Regulation	Requirement*
1) Hazard Communication Standard	Written program
	MSDS file
	Hazardous substance list
	Labeling
	Training program
2) Lockout/Tagout	Written program
	LO/TO equipment
	Training program
3) Confined Space Entry	Written program
	Entry [ermit
	Training program
	CSE equipment
	CSE signs

(*continued*)

Table 2 Continued

OSHA Regulation	Requirement*
4) In-Plant Air	Sampling
	Safety equipment
	Engineering controls
5) Process Safety Management	Written program
	Hazard analysis procedures
	Training program
6) Respiratory Protection	Written procedures
	Respiratory protection equipment
7) Emergency Response Plan	Written procedures
	Training program
	Safety equipment
8) Bloodborne Pathogens	Written program
	Training program
	Safety equipment
9) HAZWOPER	Written program
	Training
	Safety equipment
	Labeling
10) Lab Safety	Chemical hygiene plan
	Training
11) Personal Protection Equipment	Conduct in-plant hazard analysis
	Identify PPE needs
	Conduct employee training

*Note: Partial list, requirements concerning written programs, training and equipment shown.

Appendix 1

COMMON CONVERSION CONSTANTS

Quantity	Traditional units	SI equivalent
1 atmosphere	1 atm = 760 mm Hg	101,325 N/m^2
Avogadro's constant	N_A = 6.0225 × 10^{23}	6.0225 × 10^{23} mol^{-1}
1 angstrom	1 Å = 10^{-8} cm	1 × 10^{-10} m
1 bar	1 bar = 1 dyn/cm^2	10^5 N/m^2
Boltzmann's constant	k = 1.380 × 10^{-16} erg/°C	1.380 × 10^{-23} J/K
1 calorie	1 cal = 2.61 × 10^{19} eV	4.184 J
1 dyne	1 dyn = 2.25 × 10^{-6} lb	1 × 10^{-5} N
1 dyne/cm^2	1.45 × 10^{-5} lb/in.^{2}	10^1 N/m^2
1 day	86,400 s	86.4 ks
1 degree (angle)	1° = 0.017 rad	17 mrad
1 erg	6.24 × 10^{11} eV	10^{-7} J
	2.39 × 10^{-8} cal	
1 erg/cm^2	6.24 × 10^{11} eV/cm^2	10^{-3} Jm^{-2}
Gas constant	R 8.3143 × 10^7 erg-atom^{-1}	8.3143 J Mol^{-1} K^{-1}
	1.987 cal (deg^{-1} g-atom^{-1})	
Electronic charge	e$^-$ 4.8 × 10^{-10} esu	1.6021 × 10^{-19} C
1 electron volt	eV 3.83 × 10^{-20} cal,	1.6021 × 10^{-19} J
	1.6021 × 10^{-12} erg	
Faraday constant	F = $N_A e$	9.648 × 10^4 C/mol
1 inch	1 in. = 2.54 cm	2.54 × 10^{-2} m
1 kilocalorie	1 kcal = 4.186 × 10^{10} erg	
1 kilogram	1 kg = 2.21 lb	1 kg
1 kilogram/cm^2	1 kg/cm^2 = 14.22 lb/in.^{2}	10^4 kg/m^2
1 liter	1 L = 1.057 qt	1 dm^3
1 micron	μm 10^4 Ångstroms	10^{-6} M
	10^{-4} cm	
1 minute (angle)	min 2.91 × 10^{-4} radians	min=2.91 × 10^{-4} Rad
Planck's constant	h 6.6256 × 10^{-27} ergs	6.6256 × 10^{-34} Js
1 pound	lb 453.59 g	0.45359 kg
Mass of electron	m^e = 9.1091 × 10^{-28} g	9.1091 ×10^{-31} kg
1 pound (force)	1 lb$_f$	4.4482 N
1 psi	1 lb$_f$/in.^{2}	6.895 × 10^3 N/m^2
1 radian	1 rad = 57.296°	1 rad
1 ton (force)	1 ton$_f$	9.96402 kN
1 tsi	1 ton$_f$/in.^{2} = 1.5749 kg/mm^2	15.4443 MN/m^2
		15.443 MPa
1 tonne (metric ton)	1 t = 2200 lb = 1.1 tons	10^3 kg
1 torr	1 torr = 1 mm Hg	133.322 N/m^2
Velocity of light	c = 2.997925 × 10^{10} cm/s	2.997925 × 10^8 m/s

Appendix 2

TEMPERATURE CONVERSION TABLE

°C	°F/°C[a]	°F	°C	°F/°C[a]	°F
−273	−459.4		−67.8	−90	−130
−268	−450		−62.2	−80	−112
−262	−440		−56.7	−70	−94
−257	−430		−51.1	−60	−76
−251	−420		−45.6	−50	−58
−246	−410		−40.0	−40	−40
−240	−400		−34.4	−30	−22
−234	−390		−28.9	−20	−4
−229	−380		−27.8	−18	−0.4
−223	−370		−26.7	−16	1.8
−218	−360		−25.6	−14	5.4
−212	−350		−24.4	−12	9
−207	−340		−23.3	−10	14
−201	−330		−17.8	0	32
−196	−320		−17.2	1	33.8
−190	−310		−16.7	2	35.6
−184	−300		−16.1	3	37.4
−179	−290		−15.6	4	39.2
−173	−280		−15.0	5	41.0
−169	−273	−459.4	−14.4	6	42.8
−168	−270	−454	−13.9	7	44.6
−162	−260	−436	−13.3	8	46.4
−157	−250	−418	−12.8	9	48.2
−151	−240	−400	−12.2	10	50.0
−146	−230	−382	−11.7	11	51.8
−140	−220	−364	−11.1	12	53.6
−134	−210	−346	−10.6	13	55.4
−129	−200	−328	−10.0	14	57.2
−123	−190	−310	−9.4	15	59.0
−118	−180	−292	−8.89	16	60.8
−112	−170	−274	−8.33	17	62.6
−107	−160	−256	−7.78	18	64.4
−101	−150	−238	−7.22	19	66.2
−95.6	−140	−220	−6.67	20	68.0
−90.0	−130	−202	−6.11	21	69.8
−84.4	−120	−184	−5.56	22	71.6
−78.9	−110	−166	−5.00	23	73.4
−73.3	−100	−148	−4.44	24	75.2

°C	°F/°C[a]	°F	°C	°F/°C[a]	°F
−3.89	25	77.0	25.0	77	170.6
−3.33	26	78.8	25.6	78	172.4
−2.78	27	80.6	26.1	79	174.2
−2.22	28	82.4	26.7	80	176.0
−1.67	29	84.2	27.2	81	177.8
−1.11	30	86.0	27.8	82	179.6
−0.56	31	87.8	28.3	83	181.4
0.	32	89.6	28.9	84	183.2
0.56	38	91.4	29.4	85	185.0
1.11	34	93.2	30.0	86	186.8
1.67	35	95.0	30.6	87	188.6
2.22	36	96.8	31.1	88	190.4
2.78	37	98.6	31.7	89	192.2
3.33	38	100.4	32.2	90	194.0
3.89	39	102.2	32.8	91	195.8
4.44	40	104.0	33.3	92	197.6
5.00	41	105.8	33.9	93	199.4
5.56	42	107.6	34.4	94	201.1
6.11	43	109.4	35.0	95	203.0
6.67	44	111.2	35.6	96	204.8
7.22	45	113.0	36.1	97	206.6
7.78	46	114.8	36.7	98	208.4
8.33	47	116.6	37.2	99	210.2
8.89	48	118.4	37.8	100	212.0
9.44	49	120.2	38	100	212
10.0	50	120.0	43	110	230
10.6	51	123.8	49	120	248
11.1	52	125.6	54	130	266
11.7	53	127.4	60	140	284
12.2	54	129.2	66	150	302
12.8	55	131.0	71	160	320
13.3	56	132.8	82	180	356
13.9	57	134.6	88	190	374
14.4	58	136.4	93	200	392
15.0	59	138.2	99	210	410
15.6	60	140.0	100	212	413
16.1	61	141.8	104	220	428
16.7	62	143.6	110	230	446
17.2	63	145.4	116	240	464
17.8	64	147.2	121	250	482
18.3	65	149.0	127	260	500
18.9	66	150.8	132	270	518
19.4	67	152.6	138	280	536
20.0	68	154.4	143	290	554
20.6	69	156.2	149	300	572
21.1	70	158.0	154	310	590
21.7	71	159.8	160	320	608
22.2	72	161.6	166	330	626
22.8	73	163.4	171	340	644
23.3	74	165.2	177	350	662
23.9	75	167.0	182	360	680
24.4	76	168.8	188	370	698

(continued)

°C	°F/°Cª	°F	°C	°F/°Cª	°F
193	380	716	488	910	1670
199	390	734	493	920	1688
204	400	752	499	930	1706
210	410	770	504	940	1724
216	420	788	510	950	1742
208	430	806	516	960	1760
227	440	824	521	970	1778
232	450	842	527	980	1796
238	460	860	532	990	1814
243	470	878	538	1000	1832
249	480	896	543	1010	1850
254	490	914	549	1020	1868
260	500	932	554	1030	1886
266	510	950	560	1040	1904
271	520	968	566	1050	1922
282	540	1004	571	1060	1940
288	550	1022	577	1070	1958
293	560	1040	582	1080	1976
299	570	1058	588	1090	1994
304	580	1076	593	1100	2012
310	590	1094	599	1110	2030
316	600	1112	604	1120	2048
321	610	1130	610	1130	2066
327	620	1148	616	1140	2084
332	630	1166	621	1150	2102
338	640	1184	627	1160	2120
343	650	1202	632	1170	2138
349	660	1220	638	1180	2156
354	670	1238	643	1190	2174
360	680	1256	649	1200	2192
366	690	1274	654	1210	2210
371	700	1292	660	1220	2228
377	710	1310	666	1230	2246
382	720	1328	671	1240	2264
388	730	1346	677	1250	2282
393	740	1364	682	1260	2300
399	750	1382	688	1270	2318
404	760	1400	693	1280	2336
410	770	1418	699	1290	2354
416	780	1436	704	1300	2372
421	790	1454	710	1310	2390
427	800	1472	716	1320	2408
432	810	1490	721	1330	2426
438	820	1508	727	1340	2444
443	830	1526	732	1350	2462
449	840	1544	738	1360	2480
454	850	1562	743	1370	2498
460	860	1580	749	1380	2516
446	870	1598	754	1390	2534
471	880	1616	760	1400	2552
477	890	1634	766	1410	2570
482	900	1652	771	1420	2588

°C	°F/°C[a]	°F	°C	°F/°C[a]	°F
777	1480	2606	1066	1950	3542
782	1440	2624	1071	1960	3560
788	1450	2642	1077	1970	3578
793	1460	2660	1082	1980	3596
799	1470	2678	1088	1990	3614
804	1480	2696	1093	2000	3632
810	1490	2714	1099	2010	3650
816	1500	2732	1104	2020	3668
821	1510	2750	1110	2030	3686
827	1520	2768	1116	2040	3704
832	1530	2786	1121	2050	3722
838	1540	2804	1127	2060	3740
843	1550	2822	1132	2070	3758
849	1560	2840	1138	2080	3776
854	1570	2858	1143	2090	3794
860	1580	2876	1149	2100	3812
866	1590	2894	1154	2110	3830
871	1600	2912	1160	2120	3848
877	1610	2930	1166	2130	3866
882	1620	2948	1171	2140	3884
888	1630	2966	1177	2150	3902
893	1640	2984	1182	2160	3920
899	1650	3002	1188	2170	3938
904	1660	3020	1193	2180	3956
910	1670	3038	1199	2190	3974
916	1680	3056	1204	2200	3992
921	1690	3074	1210	2210	4010
927	1700	3092	1216	2220	4028
932	1710	3110	1221	2230	4046
938	1720	3128	1227	2240	4064
949	1730	3146	1232	2250	4082
949	1740	3164	1238	2260	4100
954	1750	3182	1243	2270	4118
960	1760	3200	1249	2280	4136
966	1770	3218	1254	2290	4154
971	1780	3236	1260	2300	4172
977	1790	3254	1266	2310	4190
982	1800	3272	1271	2320	4208
988	1810	3290	1277	2330	4226
993	1820	3308	1282	2340	4244
999	1830	3326	1288	2350	4262
1004	1840	3340	1293	2360	4280
1010	1850	3362	1299	2370	4298
1016	1860	3380	1304	2380	4316
1021	1870	3398	1310	2390	4334
1027	1880	3416	1316	2400	4352
1032	1890	3434	1321	2410	4370
1038	1900	3452	1327	2420	4388
1043	1910	3470	1332	2430	4406
1049	1920	3488	1338	2440	4424
1054	1930	3506	1343	2450	4442
1060	1940	3524	1349	2460	4460

(continued)

°C	°F/°C[a]	°F	°C	°F/°C[a]	°F
1354	2470	4478	1504	2740	4964
1360	2480	4496	1510	2750	4972
1366	2490	4514	1516	2760	5000
1371	2500	4532	1521	2770	5018
1377	2510	4550	1527	2780	5036
1382	2520	4568	1532	2790	5054
1388	2530	4586	1538	2800	5072
1393	2540	4604	1543	2810	5090
1399	2550	4622	1549	2820	5108
1404	2560	4640	1554	2830	5126
1410	2570	4658	1560	2840	5144
1416	2580	4676	1566	2850	5162
1421	2590	4694	1571	2860	5180
1427	2600	4712	1577	2870	5198
1432	2610	4730	1582	2880	5216
1438	2620	4748	1588	2890	5234
1443	2630	4766	1593	2900	5252
1449	2640	4784	1599	2910	5270
1454	2650	4802	1604	2920	5288
1460	2660	4820	1610	2930	5306
1466	2670	4838	1616	2940	5324
1471	2680	4856	1621	2950	5342
1477	2690	4874	1627	2960	5360
1482	2700	4892	1632	2970	5378
1488	2710	4910	1638	2980	5396
1493	2720	4928	1643	2990	5414
1499	2730	4946	1649	3000	5432

[a]Identify the desired value to be converted in the °F/°C column and then look to either the °C column or the °F column for the conversion. For example, if the –200 in the °F/°C column is being read as –200°F, then look to the °C column to find the value converted to –129°C, but if the –200 in the °F/°C column is being read as –200°C, then look to the °F column to find the value converted to –328°F.

Appendix 3

VOLUME CONVERSION TABLE

Cubic feet	a	Cubic meters	Cubic feet	a	Cubic meters
35.3	1	0.0283	1342	38	1.08
70.6	2	0.0566	1377	39	1.10
105.9	3	0.0850	1413	40	1.13
141.3	4	0.113	1448	41	1.16
176.6	5	0.142	1483	42	1.19
211.9	6	0.170	1519	43	1.22
247.2	7	0.198	1554	44	1.25
282.5	8	0.227	1589	45	1.27
317.8	9	0.255	1625	46	1.30
353.2	10	0.283	1660	47	1.33
388.5	11	0.312	1695	48	1.36
423.8	12	0.340	1730	49	1.39
459.1	13	0.368	1766	50	1.42
494.4	14	0.396	1801	51	1.44
529.7	15	0.425	1836	52	1.47
565.1	16	0.453	1872	53	1.50
600.4	17	4.481	1907	54	1.53
635.7	18	0.510	1942	55	1.56
671.0	19	0.538	1978	56	1.59
706.3	20	0.566	2013	57	1.61
741.6	21	0.595	2048	58	1.64
777.0	22	0.623	2084	59	1.67
812.3	23	0.651	2119	60	1.70
847.6	24	0.680	2154	61	1.73
882.9	25	0.708	2190	62	1.76
918.2	26	0.736	2225	63	1.78
953.5	27	0.765	2260	64	1.81
988.9	28	0.793	2296	65	1.84
1024	29	0.821	2331	66	1.87
1059	30	0.850	2366	67	1.90
1095	31	0.878	2402	68	1.93
1130	32	0.906	2437	69	1.95
1165	33	0.935	2472	70	1.98
1201	34	0.963	2507	71	2.01
1236	35	0.991	2543	72	2.04
1271	36	1.02	2578	73	20.7
1307	37	1.05	2613	74	2.10

Identify the desired value to be converted in the m^3/ft^3 column and then look to either the ft^3 column or the m^3 for the conversion. For example, if the 1 in the m^3/ft^3 column is being read as 1 m^3, then look to the ft^3 column to find the value converted to 35.3 ft^3, but if the 1 in the m^3/ft^3 column is being read as 1 ft^3, then look to the m^3 column to find the value converted to 0.0283 m^3.

Appendix 4

HARDNESS CONVERSION TABLES: HARDENED STEEL AND HARD ALLOYS

	Rockwell Scale						Dph 10 kg	KHN* 500 g and over	BHN 3000 kg	Tensile strength 10^3 psi (approx.)
C	A	D	15 m	30 m	45 m	g				
80	92.0	86.5	96.5	92.0	87.0	–	1865	–	–	–
79	91.5	85.5	–	91.5	86.5	–	1787	–	–	–
78	91.0	84.5	96.0	91.0	85.5	–	1710	–	–	–
77	90.5	84.0	–	90.5	84.5	–	1633	–	–	–
76	90.0	83.0	95.5	90.0	83.5	–	1556	–	–	–
75	89.5	82.5	–	89.0	82.5	–	1478	–	–	–
74	89.0	81.5	95.0	88.5	81.5	–	1400	–	–	–
73	88.5	81.0	–	88.0	80.5		1323	–	–	–
72	88.0	80.0	94.5	87.0	79.5	–	1245	–	–	–
71	87.0	79.5	–	86.5	78.5	–	1160	–	–	–
70	86.5	78.5	94.0	86.0	77.5	–	1076	972	–	–
69	86.0	78.0	93.5	85.0	76.5	–	1004	946		
68	85.5	77.0	–	84.5	75.5	–	942	920	–	–
67	85.0	76.0	83.0	83.5	74.5	–	894	895	–	–
66	84.5	75.5	92.5	83.0	73.0	–	854	870	–	
65	84.0	74.5	92.0	82.0	72.0	–	820	846	–	–
64	83.5	74.0	–	81.0	71.0	–	789	822	–	–
63	83.0	73.0	91.5	80.0	70.0	–	763	799	–	–
62	82.5	72.5	91.0	79.0	69.0	–	739	776	–	–
61	81.5	71.5	90.5	78.5	67.5	–	716	754	–	–
60	81.0	71.0	90.0	77.5	66.5	–	695	732	614	–
59	80.5	70.0	89.5	76.5	65.5	–	675	710	600	–
58	80.0	69.0	–	75.5	64.0	–	655	690	587	–
57	79.5	68.5	89.0	75.0	63.0	–	636	670	573	–
56	79.0	67.5	88.5	74.0	62.0	–	617	650	560	–
55	78.5	67.0	88.0	73.0	61.0	–	598	630	547	301
54	78.0	66.0	87.5	72.0	59.5	–	580	612	534	291
53	77.5	65.5	87.0	71.0	58.5	–	562	594	522	282
52	77.0	64.5	86.5	70.5	57.5	–	545	576	509	273
51	76.5	64.0	86.0	69.5	56.0	–	528	558	496	264
50	76.0	63.0	85.5	68.5	55.0	–	513	542	484	255

(*continued*)

			Rockwell Scale				Dph	KHN* 500 g and	BHN	Tensile strength 10³ psi
C	A	D	15 m	30 m	45 m	g	10 kg	over	3000 kg	(approx.)
49	75.5	62.0	85.0	67.5	54.0	–	498	526	472	246
48	74.5	61.5	84.5	66.5	52.5	–	485	510	460	237
47	74.0	60.5	84.0	66.0	51.5	–	471	495	448	229
46	73.5	60.0	83.5	65.0	50.0	–	458	480	237	221
45	73.0	59.0	83.0	64.0	49.0	–	446	466	426	214
44	72.5	58.5	82.5	63.0	48.0	–	435	452	415	207
43	72.0	57.5	82.0	62.0	46.5	–	424	438	404	200
42	71.5	57.0	81.5	61.5	45.5	–	413	426	393	194
41	71.0	56.0	81.0	60.5	44.5	–	403	414	382	188
40	70.5	55.5	80.5	59.5	43.0	–	393	402	372	182
39	70.0	54.5	80.0	58.5	42.0	–	383	391	362	177
38	69.5	54.0	79.5	57.5	41.0	–	373	380	352	171
37	69.0	53.0	79.0	56.5	39.5	–	363	370	342	166
36	68.5	52.5	78.5	56.0	38.5	–	353	360	332	162
35	68.0	51.5	78.0	55.0	37.0	–	343	351	322	157
34	67.5	50.5	77.0	54.0	36.0	–	334	342	313	153
33	67.0	50.0	76.5	53.0	35.0	–	325	334	305	148
32	66.5	49.0	76.0	52.0	33.5	–	317	326	297	144
31	66.0	48.5	75.5	51.5	32.5	–	309	318	290	140
30	65.5	47.5	75.0	50.5	31.5	–	301	311	283	136
29	65.0	47.0	74.5	49.5	30.0	91.0	293	304	276	132
28	64.5	46.0	74.0	48.5	29.0	90.0	285	297	270	129
27	64.0	45.5	73.5	47.5	28.0	89.0	278	290	265	126
26	63.5	44.5	72.5	47.0	26.5	88.0	271	284	260	123
25	63.0	44.0	72.0	46.0	25.5	87.0	264	278	255	120
24	62.5	43.0	71.5	45.0	24.0	86.0	257	272	250	117
23	62.0	42.5	71.0	44.0	23.0	84.5	251	266	245	115
22	61.5	41.5	70.5	43.0	22.0	83.5	246	261	240	112
21	61.0	41.0	70.0	42.5	20.5	82.5	241	256	235	110
20	60.5	40.0	69.5	41.5	19.5	81.0	236	251	230	108

*Knoop Hardness Conversion - The values of the Knoop hardness number are approximate only, since they were determined on a limited number of tests and samples. These values are only for loads of 500 g or heavier.

NOTE: Although conversion tables dealing with hardness can only be approximate and never mathematically exact, it is of considerable value to be able to compare different hardness scales in a generaly way.

Source: "Hardness Testing Handbook," by Vincent E. Lysaght and Anthony DeVellis, American Chain & Cable Co.

Appendix 5

RECOMMENDED MIL 6875 SPECIFICATION STEEL HEAT TREATMENT CONDITIONS

Table 1 Heat Treatment Procedure for Class A (Carbon and Low-Alloy) Steel

SAE, AISI or producer's designation	Heating and cooling requirements				Approximate tempering temperature °F — Tensile strength range – Ksi[5]									
	Normalizing temperature range[2]	Annealing temperature range[3,4]	Austenitizing temperature range	Approved Quenchant	90–125	125–150	150–170	160–180	180–200	200–220	220–240	240–260	260–280	280–300
1025	1600/1700	1575/1650	1575/1650	Water or polymer	700	—	—	—	—	—	—	—	—	—
1035	1600/1700	1575/1650	1525/1575	Oil, water, polymer	900	—	—	—	—	—	—	—	—	—
1045	1600/1700	1550/1600	1475/1550	Oil, water, polymer	1100	—	—	—	—	—	—	—	—	—
1095	1500/1600	1450/1525	1450/1525	Oil, water, polymer	—	—	1000	850	750	[6]	—	—	—	—
3140	1600/1700	1450/1525	1475/1525	Oil, polymer	1250	1100	1000	825	700-	—	—	—	—	—
4037	1600/1700	1525/1575	1525/1575	Oil, water, polymer	1200	1100	925	—	—	—	—	—	—	—
4130	1600/1700	1500/1600	1550/1600	Oil, water, polymer	1250	1050	925	850	725	—	—	—	—	—
4135	1600/1700	1525/1575	1550/1600	Oil, polymer	—	1125	1025	900	800	725	—	—	—	—
4140	1600/1700	1525/1575	1550/1600	Oil, polymer	1300	1175	1075	950	850	—	—	—	—	—
4150	1525/1650	1500/1550	1500/1550	Oil	—	1200	1100	975	800	—	—	—	—	—
4330V	1600/1700	1525/1600	1550/1650	Oil	—	—	—	—	1000	800	—	—	—	—
4335V	1600/1700	1525/1600	1550/1650	Oil	—	—	—	—	1000	800	—	—	—	—
4340	1600/1700	1525/1575	1500/1550	Oil	—	1200	1100	1050	925	850	—	—	—	—
4640	1600/1700	1525/1575	1500/1550	Oil	1200	1100	1000	900	750	—	—	—	—	—
6150	1600/1700	1525/1575	1550/1625	Oil	—	—	—	—	725	[8]	—	—	—	—
8630	1600/1700	1525/1575	1525/1600	Oil, water, polymer	1250	1050	925	850	—	—	—	—	—	—
8735	1600/1700	1525/1575	1525/1600	Oil, polymer	—	1125	1025	800	785	—	—	—	—	—
8740	1600/1700	1500/1575	1525/1575	Oil	—	1175	1075	950	850	725	—	—	—	—
Hy-Tuf[7]	1700/1750	1375/1425	1575/1625	Oil	—	—	—	—	—	—	550	—	—	—
300m[7]	1675/1725	1525/1575	1575/1625	Oil	—	—	—	—	—	—	—	—	[10]	575
H-11[9]	-	1550/1650	1825/1875	Air, oil, polymer	—	—	—	—	1150	1100	1025	—	—	—
98BV40	1550/1650	1525/1575	1540/1560	Oil	—	—	—	—	—	—	—	—	600	500
D6AC	1700/1750	1525/1575	1675/1725	Oil	—	—	—	—	1150	1100	1025	—	—	—
9ni-4Co-.20c	1600/1700	1250/1150[11]	1525/1575	Oil, water, polymer[12]	—	—	—	—	1050	1000	—	—	—	—
9Ni-4Co-.30C	1625/1675	1250/1150[11]	1475/1525	Oil[12]	—	—	—	—	—	—	—	—	—	—
52100	1600/1700[13]	1400/1450[14]	1500/1575[15]	Oil	—	—	—	—	—	—	—	—	—	—
AF1410	1625/1675[17]	-	1500/1550	Oil[12]	—	—	—	—	—	—	950	—	—	[16]

[1]Steel alloys listed are the more frequent ones used. Alloys not listed should be heat treated as recommended by their manufacturers.

[2]For the purpose of this specification, normalizing describes a metallurgical process rather than a set of properties. All steels are air quenched from temperature range.

[3]Furnace cool to 1000°F or below, except furnace cool 4330V, 4335V to 800°F, 4640 to 750°F, 4340 to 800°F and 300M to 600°F. Rate of furnace cool for alloy steels, except 4130, 8630, 4037 and 8735 should be 50°F per hour or slower.

[4]Recommended subcritical anneal temperature is 1250°F.

[5]Absence of values indicates the respective steel is not recommended for this tensile strength range.

[6]In general - for spring temper, temper at 700°F - 800°F for RC 40-45.

[7]a. 4340, 260 - 280 tempering must be between 425°F and 500°F. b. 300M and Hy Tuf - tempering is mandatory.

[8]In general - for spring temper, temper at 725°F - 900°F fir Rc 43-47.

[9]Multiple cyclic annealing may be permitted to prevent grain growth.

[10]Final tempering shall be at or above 1000°F. No tempering temperature shall be less than that of previous temper nor more than 25°F higher than the previous temper.

[11]Duplex anneal - hold 4 hours ± 0.25 hrs. at 1250°F + 25°F, air cool to room temperature, then reheat to 1150 + 25°F and hold for 8 hrs ± 0.25 hrs and air cool to room temperature.

[12]Cool to -100°F for 1 hour minimum within 2 hours after quenching and before tempering.

[13]Normalizing is not recommended practice for 52100 steel.

[14]The following annealing treatment for 52100 steel should be used:
Heat to 1430°F, hold for 20 minutes, and cool at controlled rates as follows:
1430° to 1370°F at a rate not to exceed 20°F per hour;
1370° to 1320°F at a rate not to exceed 10°F per hour;
1320° to 1250°F at a rate not to exceed 20°F per hour.

[15]Size stability may be enhanced by refrigeration. When required, cool to -100°F for 1 hour within 2 hours after quenching and before tempering.

[16]For antifriction bearings, temper to Rc 58 to 65 at 300°F - 450°F.

[17]Overage to facilitate machining by normalizing plus 1250°F + 25°F for not less than 6 hrs. and air cool.

Table 2   Heat-Treatment Procedure for Class B (Martensitic Corrosion-Resistant) Steel

SAE, AISI or producer's designation	Annealing °F		Transformation hardening cycle °F		Recommended subcritical anneal	Approximate tempering temperature °F for tensile strength - KSI				
	Temperature	Furnace cool to approx. temp. shown or below	Austenitizing temp.	Quenchant[2]		100 (minimum)	120 (minimum)	Avoid tempering or holding within range from[1]	180 (minimum)	200 (minimum)
403	1500 to 1600	Furnace cool 25 to 50° per hour to 1100	1750 to 1850	Oil Air Polymer	1200 to 1450, air cool	1300	1100	700 to 1100	500	
410	1500 to 1600	Furnace cool 25 to 50° per hour to 1100	1750 to 1850	Oil Air Polymer	1200 to 1450, air cool	1300	1100	700 to 1100	500	
416	1500 to 1650	Furnace cool 25 to 50° per hour to 1100	1750 to 1850	Oil Air Polymer	1200 to 1450, air cool	1300	1075	700 to 1075	500	
420	1550 to 1650 for 6 hours	Furnace cool 25 to 50° per hour to 100 followed by water quenching	1750 to 1850	Oil Air Polymer	1350 to 1450, air cool	1300[3]	1075	700 to 1075		600
440C	1550 to 1600 for 6 hours, or 1650 for 2 hrs, + 1300 for 4 hours	Furnace cool 25 to 50° per hour to 1100	1900 to 1950	Oil Air Polymer	1250 to 1350, air cool	Temper at 325 for Rockwell C 58 minimum, 375 for Rockwell C 57 minimum, 450 for Rockwell C 55 minimum				

[1] When approved by the cognizant engineering organization, parts may be tempered in 1000–1050°F range when 135–145 Ksi tensile strength is required, providing the parts are not subject to substantial impact loading or stress-corrosion conditions. Tempering these alloys in the range listed results in decreased impact strength and also reduced corrosion resistance. However, tempering in this range is sometimes necessary to obtain the strength and ductility required. When approved by the purchaser, material may be tempered in this range.

[2] Controlled atmosphere quench is optional for small parts. The quench for 440C shall be followed by refrigeration to −100°F or lower for 2 hours. Double temper to remove retained austenite.

[3] Temper 420 steel: 300°F for Rc 52 minimum; 400°F for Rc 50 minimum; 600°F for Rc 48 minimum.

Table 3 Annealing Procedure for Class C (Austenitic Corrosion–Resistant) Steel

SAE, AISI or producer's designation	Annealing treatment	
	Heating °F	Cooling[1]
201 and 202[2]	1850 to 2050	Water quench
301, 302 and 303[2]	1850 to 2050	Water quench
304 304L and 308[2]	1850 to 2050	Water quench
309[2]	1900 to 2050	Water quench
310 316 and 316L[2]	1900 to 2050	Water quench
321[3]	1750 to 2050	Air or water quench
347[3] and 348	1800 to 2050	Air or water quench

[1]Other means of cooling permitted provided it is substantiated by tests that the rate is rapid enough to prevent carbide precipitation.
[2]Stress relieving of unstabilized grades, except 304L and 316L between 875 + 25°F and 1500°F is prohibited. Stress relieving of stabilized grades should be at 1650°F for 1 hours.
[3]When stress relieving after welding is specified, hold for ½ hour minimum at temperature specified in Table 3 or holding for 2 hours at 1650°F ± 25°F.

Table 4 Heat Treatment Procedure for Class D (Precipitation-Hardening and Maraging) of Steel

Steel	1	Temp. °F	Hold[2]	Quench[3]	1	Max Temp °F	Hold min.[4]	Quench[3]	Temp. F°	Minimum min	Min. Time hr
		Solution Treatment				**Austenite Conditioning**				**Transformation treatment**	
Wrought 17-4 PH[3] 15.5PH[20]	A	1900	7,8	Air polymer Oil[9]							4 1
Cast 17-4 PH[2] 15-5 PH	HC A	2100 1900	8	Air[9] Air[9]							1½
PH 13-8 Mo	A	1700	8	Air or Oil[10]							4
17-7 PH[11,7]	A	1950		Water, polymer, or air[12]	T R	1400 1750	90 10	Air[13] Air[13]	+ 60 -100	30 480	1½ 1
PH 15-7Mo	A	1950	7 8	Water	T R R	1400 1750 1750	90 10 10	Air[13] Air[13] Air[13]	+ 60 -100 -100	30 480 480	1½ 1 1
PH 14-8 Mo	A	1825	8	Water	SR	1700	60	Air[13]	-100	480	1
AM-350	A	1925	7	Water oil or polymer	SCT	1710	14	Water	-100[15]	180	3
AM-355	A	1900	16	Water, Oil or polymer	T SCT SCT	1710 1800 1710	17 6	Air Air Water	1400 -100 -100[15]	180[10] 180 180	3 2 3
A-286	A	1800	8	Air-blast, oil, water				or oil			16
Custom 450 455[18] 200[19] 250[19,20] 300[19,20]	A A A A	1700 1525 1500 1500 1500	60 30 8 8 8	Water, polymer or oil Air Air Air							4 4 3 3 3

[1]For the purpose of this specification, normalizing describes a metallurgical process rather than a set of properties. All steels are air quenched from temperature range.

[2]Recommended subcritical anneal temperature is 1250°F.

[3]Size stability may be enhanced by refrigeration. When required, cool to –100°F for 1 hour within 2 hours after quenching and before tempering.

[4]Stress relieving of unstabilized grades, except 304L and 316L between 875 + 25°F and 1500°F is prohibited. Stress relieving of stabilized grades should be at 1650°F for 1 hours.

[5]Furnace cool to 1000°F or below, except furnace cool 4330V, 4335V to 800°F, 4640 to 750°Fm 4340 to 800°F and 300M to 600°F. Rate of furnace cool for alloy steels, except 4130, 8630, 4037 and 8735 should be 50°F per hour or slower.

[6]Normalizing is not recommended practice for 52100 steel.

[7]Cool to –100°F for 1 hour minimum within 2 hours after quenching and before tempering.

[8]For antifriction bearings, temper to Rc 58 to 65 at 300°F–450°F.

[9]The following annealing treatment for 52100 steel should be used:

Heat to 1430°F, hold for 20 minutes, and cool at controlled rates as follows:

1430° to 1370°F at a rate not to exceed 20°F per hour;

1370° to 1320°F at a rate not to exceed 10°F per hour;

1320° to 1250°F at a rate not to exceed 20°F per hour;

Heat Treatment Procedure for Class D (Precipitation-Hardening and Maraging) Steel

Aging temperature (°F)															
For minimum tensile strength - Ksi															
130	135	140	145	150	155	165	170	175	180	185	190	200	205	210	220
	1150	1100	1075	1050	1025		925								
									900						
1100				1000			925		900						
							935								
1150			1100					1050		1025			1000		950
									1050			950			
			1100	1080			1060		1050						
											1050		1000		
															950
															950
														1050	950
						1000	950			850					
						850					(for castings only)				
									850						
							1000				(For wrought products only)				
1325 (Air Cool)															
1100		1050	1025				950		900				1000		950
													900		

[10]Absence of values indicates the respective steel is not recommended for this tensile strength range.

[11]When stress relieving after welding is specified, hold for ½ hour minimum at temperature specified in Table 1C or holding for 2 hours at 1650°F ± 25°F.

[12]Multiple cyclic annealing may be permitted to prevent grain growth.

[13]Temper 420 steel: 300°F for Rc 52 minimum; 400°F for Rc 50 minimum; 600°F for Rc 48 minimum.

[14]In general - for spring temper, temper at 700°F - 800°F for Rc 40–45.

[15]Controlled atmosphere quench is optional for small parts. The quench for 440C shall be followed by refrigeration to –100°F or lower for 2 hours. Double temper to remove retained austenite.

[16]Steel alloys listed are the more frequent ones used. Alloys not listed should be heat treated as recommended by their manufacturers.

[17]In general - for spring temper, temper at 725°F - 900°F for Rc 43–47.

[18]a. 4340, 260–280 tempering must be between 425°F and 500°F. b. 300M and Hy Tuf - tempering temperature is mandatory.

[19]Final tempering shall be at or above 1000°F. No tempering temperature shall be less than that of previous temper nor more than 25°F higher than the previous temper.

[20]Other means of cooling permitted provided it is substantiated by tests that the rate is rapid enough to prevent carbide precipitation.

Appendix 6

COLORS OF HARDENING AND TEMPERING HEATS

Table 1 Correlation of Hot Steel Color with Temperature

Temperature		Hot Steel Color
°F	°C	
752	400	Red; visible in the dark
885	474	Red; visible in twilight
975	525	Red; visible in daylight
1077	581	Red; visible in sunlight
1292	700	Dull red
1472	800	Turning to cherry red
1652	900	Cherry proper
1832	1000	Bright cherry red
2012	1100	Orange red
2192	1200	Orange yellow
2372	1300	White
2552	1400	Brilliant white
2732	1500	Dazzling white
2912	1600	Bluish white

Table 2 Correlation of Tempering Temperature with Heated Steel Color

Temperature held for 1 h		Color of Oxide	Temperature held for 8 min.	
°F	°C		°F	°C
370	188	Faint yellow	460	238
390	199	Light straw	510	265
410	210	Dark straw	560	293
430	221	Brown	610	321
450	232	Purple	640	337
490	254	Dark blue	660	349
510	265	Light blue	710	376

Appendix 7

WEIGHT TABLES FOR STEEL BARS

Round and Square Bars, Specific Gravity 7.85

Size[a] (mm)	Weight (kg/m) Round bars	Square bars	Size[a] (mm)	Weight (kg/m) Round bars	Square bars	Size[a] (mm)	Weight (kg/m) Round bars	Square bars
5	0.15	0.20	47	13.6	17.3	89	48.8	62.2
6	0.22	0.28	48	14.2	18.1	90	49.9	63.6
7	0.30	0.38	49	14.8	18.9	91	51.1	65.0
8	0.39	0.50	50	15.4	19.6	92	52.2	66.4
9	0.50	0.64	51	16.0	20.4	93	53.3	67.9
10	0.62	0.79	52	16.7	21.2	94	54.5	69.4
11	0.75	0.95	53	17.3	22.1	95	55.6	70.9
12	0.89	1.13	54	18.0	22.9	96	56.8	72.4
13	1.04	1.33	55	18.7	23.8	97	58.0	73.9
14	1.21	1.54	56	19.3	24.6	98	59.2	75.4
15	1.39	1.77	57	20.0	25.5	99	60.4	76.9
16	1.58	2.01	58	20.7	26.4	100	61.7	78.5
17	1.78	2.27	59	21.5	27.3	105	68.9	86.6
18	2.00	2.54	60	22.2	28.3	110	74.6	95.0
19	2.23	2.83	61	22.9	29.2	115	81.5	104
20	2.47	3.14	62	23.7	30.2	120	88.8	113
21	2.72	3.46	63	24.5	31.2	125	96.3	123
22	2.98	3.80	64	25.3	32.2	130	104	133
23	3.26	4.15	65	26.1	33.2	135	112	143
24	3.55	4.52	66	26.9	34.2	140	121	154
25	3.85	4.91	67	27.7	35.2	145	130	165
26	4.17	5.31	68	28.5	36.3	150	139	177
27	4.49	5.72	69	29.4	37.4	155	148	189
28	4.83	6.15	70	30.2	38.5	160	158	201
29	5.19	6.60	71	31.1	39.6	165	168	214
30	5.55	7.07	72	32.0	40.7	170	178	227
31	5.92	7.54	73	32.8	41.8	175	189	240
32	6.31	8.04	74	33.8	43.0	180	200	254
33	6.71	8.55	75	34.7	44.2	185	211	269
34	7.13	9.07	76	35.6	45.3	190	223	283
35	7.55	9.62	77	36.6	46.5	195	234	299
36	7.99	10.2	78	37.5	47.8	200	247	314

(*continued*)

Size[a] (mm)	Weight (kg/m) Round bars	Weight (kg/m) Square bars	Size[a] (mm)	Weight (kg/m) Round bars	Weight (kg/m) Square bars	Size[a] (mm)	Weight (kg/m) Round bars	Weight (kg/m) Square bars
37	8.44	10.8	79	38.5	49.0	210	272	346
38	8.90	11.3	80	39.5	50.2	220	298	380
39	9.38	11.9	81	40.5	51.5	230	326	415
40	9.86	12.6	82	41.5	52.8	240	355	452
41	10.4	13.2	83	42.5	54.1	250	386	491
42	10.9	13.9	84	43.5	55.4	260	417	531
43	11.4	14.5	85	44.5	56.7	270	449	572
44	11.9	15.2	86	45.6	58.1	280	483	615
45	12.5	15.9	87	46.7	59.4	290	518	660
46	13.1	16.6	88	47.7	60.8	300	555	707

Molybdenum high speed steels weigh approximately 3% more. High speed steels of 18-4 type weigh approximately 10% more.

[a]Size is diameter of circular cross section, side of square cross section.

Hexagonal and Octagonal Bars, Specific Gravity 7.85

Size[a] (mm)	Weight (kg/m)	
	Hexagonal	Octagonal
5	0.17	0.16
6	0.25	0.23
7	0.33	0.32
8	0.44	0.42
9	0.55	0.53
10	0.68	0.65
11	0.82	0.79
12	0.98	0.94
13	1.15	1.10
14	1.33	1.27
15	1.53	1.46
16	1.74	1.66
17	1.97	1.88
18	2.20	2.11
19	2.45	2.35
20	2.72	2.60
21	3.00	2.87
22	3.29	3.15
23	3.60	3.44
24	3.92	3.74
25	4.25	4.06
26	4.60	4.40
27	4.96	4.74
28	5.33	5.09
29	5.72	5.47
30	6.12	5.85
31	6.53	6.25
32	6.96	6.66
33	7.41	7.08
34	7.86	7.52
35	8.33	7.97
37	9.31	8.90
39	10.3	9.89
41	11.4	10.9
43	12.6	12.0
45	13.8	13.2
47	15.0	14.4
50	17.0	16.3
53	19.1	18.3
56	21.3	20.4

[a]Dimensions are the length between opposite sides of the bar.

INDEX